GEOCHIMICA ET COSMOCHIMICA ACTA

Supplement 3

PROCEEDINGS

OF THE

THIRD LUNAR SCIENCE CONFERENCE

Houston, Texas, January 10–13, 1972

GEOCHIMICA ET COSMOCHIMICA ACTA

Journal of The Geochemical Society and The Meteoritical Society

SUPPLEMENT 3

PROCEEDINGS

OF THE

THIRD LUNAR SCIENCE CONFERENCE

Houston, Texas, January 10–13, 1972

Sponsored by
The Lunar Science Institute

VOLUME 1

MINERALOGY AND PETROLOGY

Edited by
ELBERT A. KING, JR.
University of Houston, Houston, Texas

THE MIT PRESS

Cambridge, Massachusetts, and London, England

Set in Monophoto Times Roman.
Printed and bound by the Colonial Press Inc.
in the United States of America.

Library of Congress Cataloging in Publication Data

Lunar Science Conference, 3d, Houston, Tex., 1972. Proceedings.

   (Geochimica et cosmochimica acta. Supplement 3)
   "Sponsored by the Lunar Science Institute."
   Includes bibliographies.
   CONTENTS: v. 1. Mineralogy and petrology, edited by Elbert A. King,
Jr.—v. 2. Chemical and isotope analyses, organic chemistry, edited by
Dieter Heymann.—v. 3. Physical properties, edited by David R. Criswell.
   1. Lunar geology—Congresses. 2. Moon—Congresses. I. King, Elbert A.,
ed. II. Heymann, Dieter, ed. III. Criswell, David R., ed. IV. Lunar Science
Institute. V. Series.

QB592.L85 1972   559.9′1   72–5496

ISBN 0–262–12060–7 (vol. 1)
        0–262–12062–3 (vol. 2)
        0–262–12063–1 (vol. 3)
        0–262–12064–X (3-vol. set)

Supplement 3
GEOCHIMICA ET COSMOCHIMICA ACTA
Journal of The Geochemical Society and The Meteoritical Society

BOARD OF ASSOCIATE EDITORS

Astronaut David R. Scott preparing to take samples near the Lunar Roving Vehicle. Hadley Rille is at right-center. Hadley Delta is the mountain in the background that rises approximately 3,500 meters above the plain. The direction of view is almost due south. (NASA photograph AS15–82–11121 by Astronaut James B. Irwin).

Preparations to take the double core tube sample by Astronaut Alan B. Shepard Jr. on the top of the Fra Mauro formation at the Apollo 14 landing site. Other sampling equipment, sample bags, and tools are attached to the Modularized Equipment Transporter. (NASA photograph AS14–68–9405 by Astronaut Edgar D. Mitchell).

Preface

In the early summer of 1969, few scientists would have guessed that the Apollo program would be so successful in exploring the lunar surface and returning samples for detailed studies. This volume contains the first substantial results of mineralogical and petrological investigations of the exceedingly heterogeneous Apollo 14 samples from the Fra Mauro landing site. In addition, initial results of the Apollo 15 samples are included, as well as continuing studies of Apollo 11, Apollo 12, and Luna 16 samples. The preliminary data on the Apollo 14 samples and landing site were published in *Science* **173**, p. 681–693, and similar data for the Apollo 15 samples also appeared in *Science* **173**, p. 363–375.

Although the three volumes in this series constitute the *Proceedings of the Third Lunar Science Conference*, not all of the papers presented at the conference are included. Several papers were not accepted for publication by the editor, and a few authors chose to take more time with their work before publication. Their results will be published elsewhere. Authors (and editors) were under considerable schedule pressure to produce final manuscripts, which had to arrive in press no later than May 1, 1972. During the final editing and preparation of these papers, the Apollo 16 Mission returned *another* cargo of samples from the moon, approximately 100 kilos from the lunar highlands.

The arrangement of papers in this volume is approximately as follows: general geology, surface and orbital observations; investigations of rocks with igneous texture; pyroxenes; plagioclases; breccias; glasses; fines samples; and miscellaneous studies. A subject index and an author index are provided for easy location of information; also included are complete sample inventories for Apollo 11, Apollo 12, Apollo 14, and Apollo 15. Sample cross-reference indices are provided in Volumes 2 and 3. Jeffrey L. Warner (NASA, MSC) deserves most of the credit for providing indices and sample inventories.

On behalf of the Lunar Science Institute, sponsor of these *Proceedings*, I express thanks to NASA for its cooperation and enthusiasm, and to the authors who, in spite of their world-wide locations, mostly managed to submit manuscripts within the required times. Special thanks are due to the Associate Editors and numerous Technical Reviewers for insuring the high quality of the papers that are contained herein, and to Pergamon Press, Ltd., which has allowed these *Proceedings* to be designated as Supplement 3 to *Geochimica et Cosmochimica Acta* with The MIT Press as the publisher. Mrs. Ann Geisendorff, Olene Edwards, Mildred Armstrong, and Elizabeth Howe provided excellent secretarial support.

The University of Houston
Houston, Texas
May 1972   Elbert A. King, Jr.

Contents

Contents

Contents

Contents

GEOCHIMICA ET COSMOCHIMICA ACTA

SUPPLEMENT 3

PROCEEDINGS

OF THE

THIRD LUNAR SCIENCE CONFERENCE

Houston, Texas, January 10–13, 1972

Proceedings of the Third Lunar Science Conference
(Supplement 3, *Geochimica et Cosmochimica Acta*)
Vol. 1, pp. 1–14
The M.I.T. Press, 1972

Geology of Hadley Rille*

KEITH A. HOWARD

U.S. Geological Survey, Menlo Park, California 94025

JAMES W. HEAD

Bellcomm, Inc., Washington, D.C. 20024

GORDON A. SWANN

U.S. Geological Survey, Flagstaff, Arizona 86001

Abstract—Apollo 15 data support the concept that Hadley Rille is a giant collapsed lava tube that originated at its south end. The south half of the rille is sinuous and the bends are not structurally controlled, whereas the north half of the rille follows pre-mare structural troughs. Most of the rille has a V-shaped profile apparently formed by recession of the rims and coalescence of talus from both sides. The rille is scalloped and discontinuous locally, and the deepest parts consistently are the widest. This is a different relationship from that shown by river channels, but can instead be explained by collapse, with most extensive foundering at the deepest points.

The upper 60 m of the walls of the rille at the Apollo 15 site expose several layers of mare basalt. Most are massive units averaging at least 10 m thick; others are layered. Numerous talus blocks derived from the massive outcrops are 10–30 m across. The lunar flows are thus both thick and little jointed. The outcrop ledge and blocky talus derived from it are absent where the rille abuts Apennine massifs, showing that the rille cuts through the mare basalt and against massif material.

Several features suggest that the mare lavas subsided differentially as much as 100 m after partly congealing. This could have resulted from drainage into and along a lava conduit that became Hadley Rille.

INTRODUCTION

GEOLOGICAL DESCRIPTIONS, samples, photographs, and derivative topographic maps from Apollo 15 allow a new characterization of the sinuous Hadley Rille. These data provide constraints on hypotheses of origin for sinuous rilles, and further give a view of lunar strata exposed in cross section in the rille walls. Before the Apollo 15 mission Greeley (1971) proposed that the rille is a lava tube and channel, and this interpretation was also favored by Carr and El-Baz (1971) and Howard (1971). Data from the mission support and refine this interpretation. A revision to Greeley's model is required by new topographic data, which place at least part of the rille in a low rather than high part of the mare. In this report, we describe the form of the rille and the materials exposed in the walls, and discuss how these features bear on a lava-tube explanation. Finally we briefly propose a model relating the rille to drainage of the mare lavas into Palus Putredinus.

SETTING

Hadley Rille (Fig. 1) lies at the base of the Apennine Mountains, which form the southeast boundary of the large multiringed Imbrium Basin. The mountains have

---

* Publication authorized by the Director, U.S. Geological Survey.

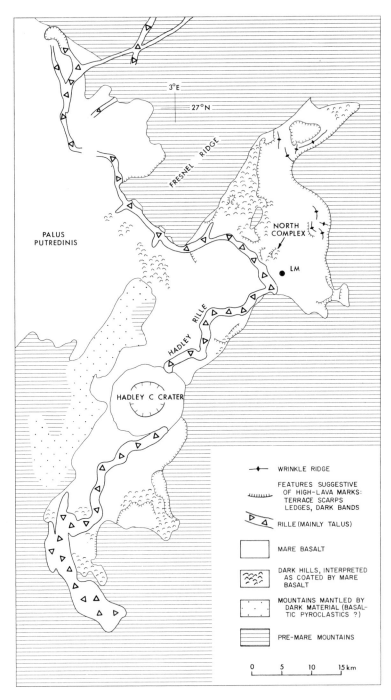

Fig. 1. Geologic sketch map of the Hadley Rille area (in part after Carr and El-Baz, 1971). Apollo 15 Lunar Module is designated LM.

prominent fault patterns trending northeast and northwest, respectively concentric and radial to the Imbrium Basin (Hackman, 1966; Carr and El-Baz, 1971). For much of its length, Hadley Rille occupies a mare-filled graben, trending northeast between two high mountain massifs (Fig. 1). Most of the rille is in mare material, but locally the rille cuts against the pre-mare mountains (Fig. 2). To the northwest the mare plain and the rille extend through a gap in the mountains to join the main part of Palus Putredinis.

The south end of Hadley Rille adjoins an elongate cleft-like depression that cuts into highlands. This cleft was compared to the source craters of terrestrial lava channels by Greeley (1971). Part of the cleft appears to be controlled by fractures

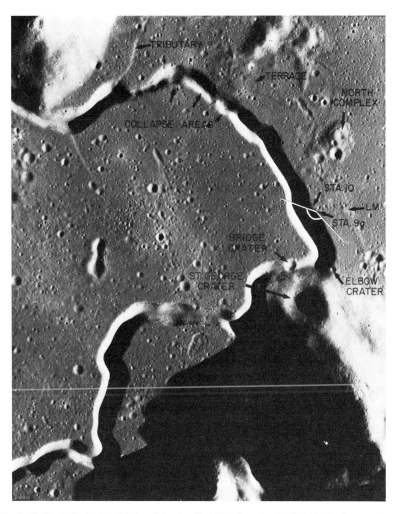

Fig. 2. Hadley Rille in the vicinity of the Apollo 15 landing site (LM). NASA photograph by the Apollo 15 panoramic camera. Area shown is about 20 by 24 km. Arc symbol shows area of panorama in Fig. 6.

related to the Imbrium Basin (Carr and El-Baz, 1971). Dark mare materials may have been derived, in part at least, from the cleft, as suggested by particularly dark areas on either side of its north end (Howard and Head, 1972).

The ridge that extends northwest of the cleft is darker than the mare and apparently is mantled by material that predates the mare (Carr and El-Baz, 1971; Howard and Head, 1972). Several low dark hills (such as North complex) near the landing site have the same albedo as surrounding maria. These hills may be coated with mare lava (see Table 1, nos. 4–7, 9).

RILLE MORPHOLOGY

Study of the azimuthal frequency of individual segments of the sinuous rille shows that the southern half of the rille, between the cleft and the Apollo 15 landing site, zigs and zags mainly to the north and east-northeast (Howard and Head, 1972).

Table 1. Evidence for subsidence of mare lavas.

	Feature	Interpretation	Reference
1.	Horizontal ledges near base of mountains east of LM	High-lava marks	Howard, 1971
2.	Dark band 80–90 m high at base of Mount Hadley	High-lava mark	Astronauts' transcript; Swann et al., 1972
3.	Grabens at the mare border east of LM	Pulling away of lava from its enclosing basin	Howard, 1971
4.	Low albedo (same as mare) for North complex and several other hills as high as 100 m	Coated by mare lava	Howard, 1971; Swann, et al., 1972
5.	Faint outline of a 600-m crater on North Complex	Buried by coating of mare	Schaber and Head, 1971; Howard, 1971
6.	Subhorizontal ledges on the dark hills	High-lava marks	Howard, 1971
7.	Some blocks on walls of Pluton crater (North complex) have albedos typical of mare basalts; others have higher albedos typical of pre-mare breccias	Pluton crater pierced a cap of mare basalt draped over an older hill	H. E. Holt, in Swann et al., 1972
8.	Mare surface slopes away from Hadley Delta mountain	Lavas subsided after congealing at their margin	Howard and Head, 1972
9.	From one dark hill, mare surface descends in a fan toward the rille	Lava caps the hill and is draped toward the rille because of subsidence	Howard and Head, 1972
10.	Low winding scarps locally terrace the mare	Subsidence of a lava lake	See Holcomb, 1971
11.	South of Hadley C crater, mare surface descends toward the rille from mountains on either side	Mare subsided owing to drainage into the rille	Unpub. photogrammetry by Wu et al., 1972
12.	Near LM, mare level is 30–40 m higher on NE side of rille than on SW side	Lava on low SW side drained more extensively into rille	See Fig. 3; Howard and Head, 1972
13.	Several small trenches intersect the rille near Fresnel ridge	Tributary lava conduits formed by drainage into rille	Howard, 1971

These directions are at substantial angles to the trend of the graben that the rille follows, and are oblique also to Imbrium radial structures. Individual segments of this part of the rille, therefore, are not structurally controlled but instead resemble bends in a sinuous flow channel.

In the northern half of the rille northwest of the Apollo 15 site, individual segments tend to follow the main course of the rille, which is northwest and roughly parallel to Imbrium radial structures (Howard and Head, 1972). This part of the rille therefore is less sinuous and appears more closely related to highland structures, suggesting the lava channel followed old fault valleys.

At bends in the rille the outside has less curvature than the inside (Fig. 2). This geometry cannot result from simple distention by fracturing, because the two sides do not match. It can, however, be caused by drainage and erosion in the rille, through carrying material along its course and by back wasting of the slopes.

The south half of Hadley Rille is characterized by a rounded V-shaped profile, which appears to be a consequence of recession of the walls by mass wasting so that the opposing talus aprons coalesce (Fig. 3). Thick rock ledges crop out in the upper part of the walls. Between the LM site and Fresnel Ridge to the northwest, the rille is discontinuous and consists of a series of coalescing bowl-shaped depressions interrupted by septa, comparable to partly collapsed lava tubes (Fig. 2) (Greeley, 1971; Howard, 1971). The immediate mare surface slopes down toward the bowl-shaped depressions that are locally bordered by slumps. These features indicate that the mare rocks have slumped toward the rille. Beyond Fresnel Ridge to the northwest, the rille is U-shaped in profile. Apparently the rille here is shallow enough that talus from the two sides does not coalesce.

Throughout the area shown by Fig. 2, the rille consistently is deepest where it is widest (Fig. 4). This relation is opposite to that shown by river channels, in which depth and width vary inversely so that the cross-sectional area remains approximately constant. The rille geometry could result if Hadley Rille formed by collapse of a buried tube, with more extensive foundering at the deepest points. Talus deposits suggest that before collapse and mass wasting the original conduit or trench was narrower, and therefore deeper, than now. Where the rille is shallow and flat-bottomed, northwest of Fresnel Ridge, it may never have been roofed.

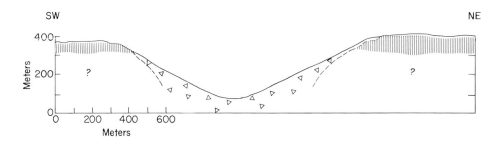

Fig. 3. Cross section of the rille at Apollo 15 site (at station 10). Vertical ruling is mare basalt; triangles are talus deposits. Scale in meters. Topography by R. M. Batson, in Swann *et al.*, 1972.

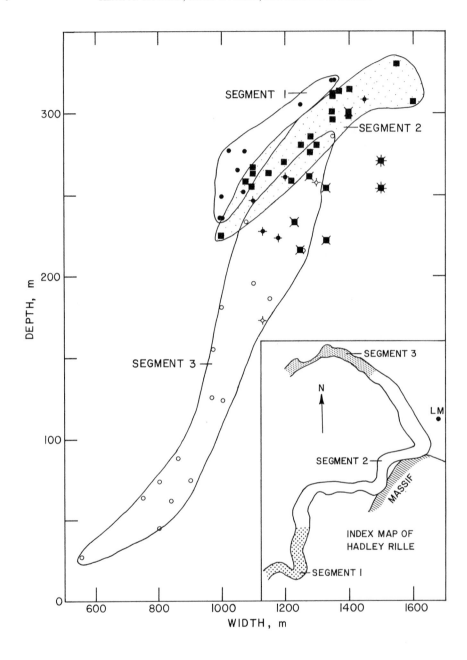

Fig. 4. Width and depth of Hadley Rille in the area of Fig. 2. Dots indicate localities along segment 1 (see inset); squares indicate segment 2; circles indicate segment 3. Trends for the three segments are outlined. The rille has anomalously low depth-to-width ratios at localities that are within 500 m of massif (× through symbol), and at sharp bends in the rille (+ through symbol). Derived from topography by Wu and others (1972).

The depth-to-width ratios increase to the south and are lowest in the discontinuous northwestern segment (Fig. 4). This indicates that the slopes of the rille walls steepen toward the south, where photographs show that outcrops are thickest and most continuous. Places where the rille abuts the Apennine massif have lower depth-to-width ratios than normal for the segment of the rille in which they occur (Fig. 4). This is explained by a partial filling of mass-wasted fine debris from the massif (Swann and others, 1972; Howard and Head, 1972). Sharp bends in the rille also have low depth-to-width ratios.

MATERIALS IN THE RILLE NEAR THE APOLLO 15 SITE

Near the Apollo 15 site (Fig. 3), regolith approximately 5 m thick is exposed at the top of the rille. Below that bedrock extends to approximately 55 m, and talus covers the remaining ~ 300 m to the bottom, with blocks as much as 30 m in diameter. Below a semicontinuous scarp formed by outcrops near the top there generally is a narrow, rather flat bench of coarse talus. Lower in the walls there is commonly one, or more, subtle rounded benches and inflections in the talus, which possibly represent sub-talus stratigraphy or structure.

Outcrops and blocky talus are present in the rille wall only where the rille is in mare material, and not where the rille abuts Apennine massifs. This demonstrates that hard mare rock accounts for most or all outcrops and talus blocks. This rock is basalt as shown by samples from an exposure at the lip of the rille, from regolith, and from the rims of two craters that excavated to depth of 50 to 100 m. Apennine massif materials (breccias) on the rille wall below St. George Crater are distinguished from mare materials by the general lack of blocks. The lack of blocks indicates a large degree of disintegration and a relatively thick regolith.

The regolith, as exposed in the rille, normally is approximately 5 m thick and has an irregular base. A similar thickness is estimated from the shapes of craters in the mare. The regolith is exposed in cross section at the rille rim due to net erosion into the rille by small impacts (Swann *et al.*, 1972). At the bottom of the rille (Fig. 5) is a filling of fine-grained material, probably consisting of regolith that has been ejected into the rille or comminuted from talus by impacts.

Talus blocks and fines of mare basalt spalled off the bedrock line the walls of the rille. Recent instability of the blocks is shown by several tracks made by rolling boulders. Talus is especially blocky where penetrated by fresh craters.

The talus deposits generally are more poorly sorted and contain a larger component of fine-grained debris than terrestrial talus slopes, especially in the lower parts of the rille walls (Fig. 5). Some patches of talus in the rille have accumulated so recently, however, that fine-grained debris does not fill the interstices (Figs. 6, 7, 8). Where aligned with fractured outcrops, some of these block fields appear jumbled but not far out of place, similar to frost-heaved blocks that cover outcrops on some terrestrial mountain peaks. In other patches, blocks have accumulated on gentle benches or inflections in the slope.

Several blocks in the talus are more than 10 m across (Fig. 5). These are about the same size as, or a little larger than, the thickness of the unbroken outcrops.

Fig. 5. Bottom of the rille, looking north from station 2 near Elbow crater (telephoto picture). Material in the bottom of the rille has bimodel grain size; it consists of a filling of fine regolith, and large talus blocks that have rolled to the bottom. The largest block is 15 m across.

Unbroken blocks of basalt this large are uncommon on earth, indicating that the Hadley basalt flows are both thicker and much less jointed than many terrestrial flows.

Outcrops in the rille were photographed through the 500 mm telephoto lens at three places. A two-layered outcrop lies on the north wall just east of Bridge Crater (Fig. 2). Rocks also crop out discontinuously along the top of the northeast wall of the rille between Elbow Crater and station 9a. In side view, looking along the slope, these outcrops appear as long subhorizontal slabs; they have a very slight northeasterly dip, outward from the rille, parallel to the mare surface here (Fig. 3) An exposure sampled at station 9a is continuous with this line of outcrops.

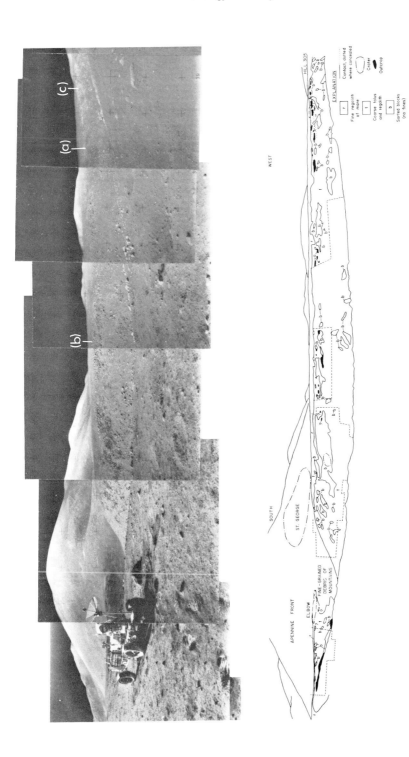

Fig. 6. Top: panorama of the rille from station 9a (location shown in Fig. 2). Letters indicate locations of telephoto pictures as follows: (a) Fig. 7a, (b) Fig. 7b, (c) Fig. 8. Bottom: sketch of same area showing distribution of outcrops. Dashed lines indicate areas photographed with telephoto lens. Note that coarse talus is absent under Elbow Crater and where rille abuts Apennine massif. NASA photographs.

Fig. 7. Outcrops of massive rock at top of rille wall across from station 9a (localities shown in Fig. 6). Top: Outcrop, 15–20 m thick, has horizontal planar parting, and vertical joints. Bottom: Two massive units, separated by a covered interval; vertical distance from mare surface (at top) to base of exposures is approximately 60 m. NASA telephoto pictures.

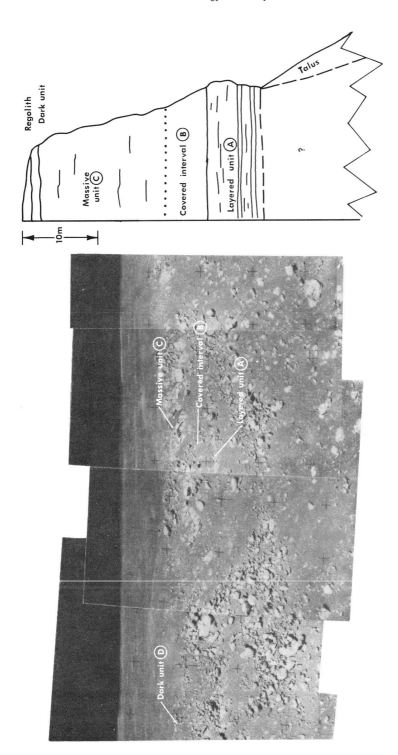

Fig. 8. Partial panorama of upper rille wall (locality shown in Fig. 6). On the right is an idealized stratigraphic section. Unit A is well layered; unit C is massive and breaks along joints that dip to the right; unit D is dark, hackly weathering, and crudely layered. NASA telephoto pictures.

The third area of outcrops (Fig. 6), and the best documented, is across the rille from stations 9a and 10 where much of a 2 km stretch was photographed from both stations, allowing the photographs to be viewed stereoscopically. Although most of the outcrops expose massive rock 10–20 m thick (Fig. 7a), local discontinuous exposures are less massive. In one area, three stratigraphic intervals are recognized (Fig. 8). In the middle is a single massive flow unit 15–20 m thick, below it is a thinly layered sequence, and on top is a poorly exposed dark unit (possibly the top of the massive unit?). Locally, two superposed massive units can be distinguished (Fig. 7b). The upper unit shown in Fig. 7b breaks cleanly along joints; the lower unit breaks on irregular rounded faces and is pitted. Massive units such as these provide most of the outcrops as well as a high proportion of the talus blocks in the rille.

Discontinuous horizontal parting planes or fractures are present within some of the massive outcrops. Layering within the basal sequence in Fig. 8 is expressed by horizontal niches and ribs. The niches may be analogous to eroded-back vesicular zones seen by the astronauts in blocks at the edge of the rille.

Layering of a different type occurs in one irregular slabby outcrop near the mare surface. An open crack underlies the thin slab and parallels its curving top surface, very much like shelly pahoehoe where the fluid lava drained from beneath the cooled crust. The astronauts described pahoehoe-like surfaces on some blocks on the mare, and a photographed talus block has crescentic ribs that resemble pahoehoe festoons. These observations bear on the origin of the rille, for essentially all terrestrial lava tubes are formed in fluid pahoehoe lavas. A lava tube of the dimensions of Hadley Rille implies a very fluid lava, comparable in viscosity to the Apollo 11 basalts (Murase and McBirney, 1970).

Jointing in the outcrops is inclined locally (Fig. 8) but crude vertical joints also are common (Fig. 7). The vertical joints may be cooling joints, as in many terrestrial lava flows. Rarely the outcrops break in irregular column-like blocks; the top of one nearly horizontal exposure is characterized by interlocking polygons about 3 m across that resemble columnar jointing.

STRATIGRAPHY EXPOSED IN THE RILLE WALL

To summarize the stratigraphy shown by outcrops in the southwest wall of the rille near the Apollo 15 site: several units or layers can be recognized, most of them thick and massive but some show internal horizontal layering. Whether individual layers are correlative across the rille to the sampling locality is uncertain. It seems likely, however, that most of the collected samples of mare basalt originated in one or more thick massive units such as recognized on the far southwest side of the rille. Photographs are insufficient to distinguish whether the multiple units or layers are separate flows or a series of flow units from a single eruption. Conceivably, some of the lower layers are younger than higher layers, as in some lava tubes and lava lakes where freezing progressed from the surface downward. Multiple flow units are common in large terrestrial lava tubes, and possibly the layers exposed in Hadley Rille are analogous. Additional stratigraphic complexities may be present if, as in some terrestrial lava tubes, the rille has been used and reused by lava from more than one eruption.

The base of the persistent outcrop ledge on either side of the rille may represent a stratigraphic discontinuity, or alternatively, the top of rather uniform talus slopes. Perhaps seismic studies could decipher whether the rille is cut entirely in hard mare basalts or extends into a different substratum.

At Elbow Crater the outcrop ledge thins until the mare basalt pinches out against the Apennine massif, as indicated by a change in the character of debris in the rille wall (Fig. 6). This suggests that the massif material dips shallowly northward under the mare fill. The full thickness of the basalt apparently was not penetrated by Elbow Crater, however, because all samples from its rim and ejecta blanket are either basalt, or breccia and fines derived primarily from basalt.

Regionally, mare outcrops are progressively thicker and more continuous as the rille is followed to the south toward the cleft-shaped depression. Northwest beyond Fresnel Ridge the depth of the rille decreases and slopes evidently are not steep enough to expose bedrock.

The thickness of soft material above the top of the major outcrop ledge also increases toward the south, from approximately 5 m at the landing site to 15–25 m south of Hadley C. This thickened interval above the ledge apparently includes, in addition to about 5 m of normal regolith, thin flow units and/or pyroclastic materials that thin northward toward the Apollo 15 landing site.

Where the rille abuts massifs, it cuts through the mare rock, implying lateral erosion. A possible terrestrial analog is a lava tube described by Greeley and Hyde (in press) that locally undercut its banks into older rocks, but erosion such as this is rare in terrestrial lavas. However, the lower viscosity of lunar basalts compared with terrestrial basalts (Murase and McBirney, 1970) and the large cross section of Hadley Rille as compared with terrestrial lava channels indicate that the Reynolds number for mare lavas in the rille would be orders of magnitude higher than that of terrestrial lavas. Thus, the lunar lavas would be in the regime of turbulent flow and could have substantial erosive capabilities.

SUBSIDENCE OF MARE LAVAS

Several lines of evidence suggest that the mare lavas subsided differentially after partly congealing (Table 1). Figure 1 pictures many of the features attributed to subsidence. Much of this subsidence may have resulted from lava drainage via Hadley Rille. The Hadley mare is terraced locally as in partly drained lava lakes (no. 10, Table 1). Holcomb (1971) found evidence that two similarly terraced mare areas had drained via sinuous rilles. South of Hadley C Crater the surface of the mare slopes down toward the rille from the mountains on either side (no. 11, Table 1), in contrast to Greeley's (1971) pre-mission data. This configuration is unlike most terrestrial feeder lava channels and tubes, which are leveed or lie atop broad low ridges built by overflows from the channel. Instead it supports the concept that Hadley Rille drained the mare lava.

When the Hadley lavas were first erupted, presumably from the cleft-shaped depression, they probably were trapped in the graben valley at the base of the Apennine Mountains. To reach the main part of Palus Putredinis the lavas eventually had to surmount the low gap south of Fresnel ridge. Overflow through this gap could then have caused accelerated flow, channeling, downcutting, and partial drainage of the ponded lava. Hadley Rille may have resulted from collapse of the drainage conduit.

This brief explanation raises numerous questions. How completely was the lava channel roofed to form a tube, and when did the roof form? Did the channel work its way headward to the source cleft, or did it originate at the source? How can a large, partly drained lava lake be reconciled with the exposed layered lavas? Does the outward slope of the northeast rim of the rille at the Apollo 15 site (Fig. 3) represent a local levee? Why does the mare surface drop 30–40 m in level across the rille here (Table 1, no. 12)? Several subsided areas of the mare seem unrelated to the rille (Fig. 1); how were they drained? These questions will have to be answered to assess fully the hypothesis that Hadley Rille represents a collapsed tube that drained the mare lavas.

Acknowledgments—We thank C. A. Hodges and H. G. Wilshire for helpful comments. Supported by NASA contracts T-65253-G; W-13,130; and NASW-417.

REFERENCES

Carr M. H. and El-Baz F. (1971) Geologic map of the Apennine-Hadley region of the moon: Apollo 15 pre-mission map. U.S. Geol. Survey Misc. Geol. Inv. Map I-723, sheet 1.

Greeley R. (1971) Lunar Hadley Rille: Considerations of its origin. *Science* **172**, 722–725.

Greeley R. and Hyde J. H. (1971) Lava tubes of the Cave Basalt, Mount St. Helens, Washington. NASA Tech. Memo. TM X-62,022, 33pp.; *Bull. Geol. Soc. Amer.* (in press).

Hackman R. J. (1966) Geologic map of the Montes Apenninus region of the moon. U.S. Geol. Survey Misc. Geol. Inv. Map I-463.

Holcomb R. (1971) Terraced depressions in lunar maria. *J. Geophys. Res.* **76**, 5703–5711.

Howard K. A. (1971) Geologic map of part of the Apennine-Hadley region of the moon: Apollo 15 pre-mission map. U.S. Geol. Survey Misc. Geol. Inv. Map I-723, sheet 2.

Howard K. A. Head J. W. and Swann G. A. (1972) Geology of Hadley Rille: Preliminary report. U.S. Geol. Survey open-file rept., 57 pp.

Howard K. A. and Head J. W. (1972) Regional geology of Hadley Rille. In *Apollo 15 Preliminary Science Report*, NASA Spec. Paper 289, p. 25.53–25.58.

Murase T. and McBirney A. R. (1970) Viscosity of lunar lavas. *Science* **167**, 1491–1493.

Schaber G. G. and Head J. W. (1971) Surface operational map of the Apennine-Hadley landing site, Apollo 15. U.S. Geol. Survey open-file map.

Swann G. A. *et al.* (1972) Preliminary geologic investigation of the Apollo 15 landing site. In *Apollo 15 Preliminary Science Report*, NASA Spec. Paper 289, p. 5.1–5.112.

Wu S. S. C. Schafer F. J. Jordan R. Nakata G. M. and Derick J. L. (1972) Photogrammetry of Apollo 15 photography. In *Apollo 15 Preliminary Science Report*, NASA Spec. Paper 289, p. 25.36–25.48.

Proceedings of the Third Lunar Science Conference
(Supplement 3, *Geochimica et Cosmochimica Acta*)
Vol. 1, pp. 15–25
The M.I.T. Press, 1972

Lineaments of the Apennine Front—Apollo 15 landing site

E. W. WOLFE AND N. G. BAILEY

U.S. Geological Survey, Flagstaff, Arizona 86001

Abstract—The well-developed systems of lineaments that suggest regional fracturing or layering of the bedrock in the mountains at the Apollo 15 site may be illusions created by oblique lighting rather than expressions of true geologic structure. Supporting evidence includes: (1) occurrence of similar dominant lineament trends on many of the slopes even though the slope orientations differ widely; (2) presence of a thick mantling regolith that should conceal bedrock structures; (3) approximate symmetry of some of the lineament sets about the sun line; (4) lack of clear-cut topographic expression for many of the lineaments; and (5) general similarity of the lunar lineaments to lineament patterns produced by low oblique lighting on model surfaces mantled by fine sifted powder. Locally, as at Silver Spur, some lineament trends are related to distinct topographic features and are more confidently interpreted to reflect underlying geologic structures.

INTRODUCTION

PROMINENT LINEAMENT SYSTEMS that resemble the surface expressions of systematically fractured or layered rocks were observed and photographed by the Apollo 15 crew on the flanks of Hadley Delta and Mt. Hadley (Fig. 1). The lineament sets are visible both in photographs taken from the lunar surface (Figs. 2 and 3) and in orbital photographs taken with the high resolution panoramic camera.

If they truly represent the surface traces of layers or of regional fracture sets, the lineaments have great significance. However, an alternative hypothesis that some may be illusions created by low-angle illumination of randomly irregular surfaces is suggested by both lunar surface data and model experiments.

LUNAR SURFACE DATA

Orbital and lunar surface photographs of the Apennine Front show that lineament azimuths on several different mountain slopes form two dominant trends that are insensitive to slope attitude and may be sensitive to sun position. For the most part, the lineaments lack clear topographic expression and occur in areas where a thick regolith would be expected to conceal geologic structures.

A preliminary statistical analysis of lineament trends was made by measuring the orientations of approximately 1,300 lineaments in five separate areas (Fig. 1) on orbital panoramic camera photographs, in which the lineaments appear as short, discontinuous lines. The results are summarized in azimuth frequency diagrams (Fig. 4). Panoramic camera photographs from three different orbits were used to provide a variety of sun elevations, 18°, 38.5°, and 43°, with sun azimuths, respectively, of 99°, 112.6°, and 117°.

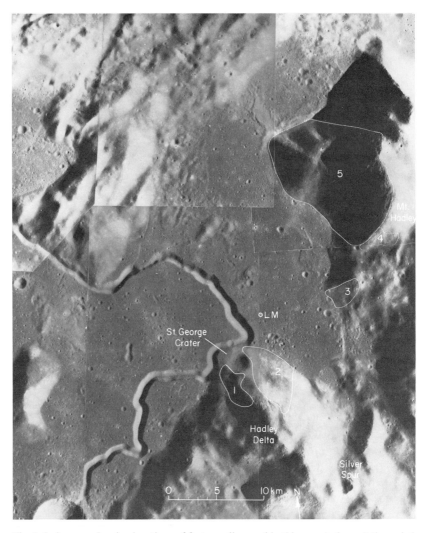

Fig. 1. Index map showing locations of features discussed in this report. Areas 1 through 5 are the sites for which azimuth frequency diagrams (Fig. 4) were prepared. Uncontrolled mosaic compiled from Apollo 15 orbital panoramic camera photographs.

Areas 1 through 4 (Fig. 1) have fairly uniform slopes with different orientations (NASA Lunar Topographic Map, Rima Hadley, Sheets A and B, 1971). The lineament patterns are dominated by major north and northeast trends and the obtuse angles of their intersections are roughly bisected by the sun azimuth (Fig. 4). At the two higher sun elevations (38.5° and 43°) the dominant lineament orientations are similar. Only in area 2 were lineaments measured under both low (18°) and high (43°) sun angles. At low sun the obtuse angle facing the light source is about 20° wider than it is at high sun (Figs. 4b and 4c). As sun elevation increased, the northeast set

Fig. 2a. View south from LM toward Hadley Delta (approximately 3.5 km high) showing the northeast-trending lineament set (sloping gently left) and the north-trending lineament set (sloping more steeply to the right). Rectangle outlines area of Fig. 2b on the rim of St. George Crater. (NASA photograph AS15-85-11374.)

apparently shifted clockwise about 20°, a movement approximately equivalent in magnitude and direction to the shift in sun azimuth from low sun to high sun.

The lineaments of areas 1 through 4 have no constant relation to slope such as gravitational effects might produce. Stereonet rotation of each slope to horizontal produces only minor changes in the lineament trends, but in some cases it produces major shifts in apparent sun azimuth. This suggests that the relation between sun azimuth and lineament azimuths is not affected by changes in slope attitude. Apparently, if the lineaments are lighting illusions, they are sensitive only to the lighting direction and not to the orientation of the slope on which they occur.

Figure 2a, a 60 mm photograph of area 2 taken from the lunar surface, shows the two lineament sets. The northeast set slopes gently left in the photograph, and the north set, which is particularly evident on the rim and flanks of St. George Crater, slopes steeply to the right. Figure 2b, a 500 mm photograph of the St. George Crater rim taken from the lunar surface, shows more detail of the lineament patterns. Parallelism of the long edges of crater rim shadows with the north-trending (upper left to lower right) lineament set suggests that this set may be a lighting illusion. This effect is particularly enhanced by foreshortening of the view on the more distant (south) wall of St. George Crater. The large thickness of mature regolith, implied by the scarcity of blocks even near the largest and freshest crater on the northeast rim

Fig. 2b. View south from LM to rim of St. George Crater showing northeast-trending lineament set (sloping gently left) and the north-trending lineament set (sloping steeply right).
The bright crater on the east rim of St. George Crater is about 50 m in diameter. (NASA
500 mm photograph AS15-84-11236.)

of St. George Crater (Fig. 2b) as well as in the areas that the astronauts visited on the
lower part of Hadley Delta, suggests that neither of the lineament sets reflects layering
of the bedrock.

Area 5, which includes the west face of Mount Hadley, is partly shown in Figs.
3a and 3b. Lineaments measured in a panoramic-camera photograph taken with

Fig. 3a. View to northeast of Mount Hadley (approximately 4.5 km high). Prominent
lineaments sloping steeply to left form the northeast-trending set. Rectangle shows location
of area in Fig. 3b. (NASA photograph AS15-90-12208.)

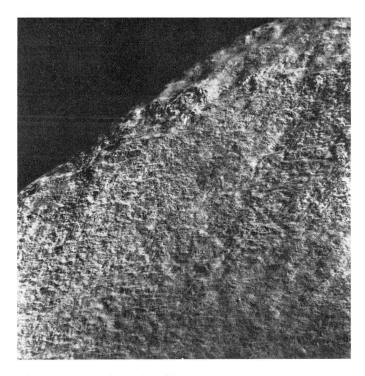

Fig. 3b. View to northeast of a portion of Mount Hadley. Shows northeast-trending lineaments that dip steeply to left and the cross-cutting nearly horizontal regolith-covered
benches. Vertical lineaments (north-northeast-trending) obscure in this view. Approximate
width of photographed area is 2.4 km. (NASA 500 mm photograph AS15-84-11321.)

intermediate (38°) sun elevation are grouped around three major trends (Fig. 4f)
that correlate with the lineaments photographed from the lunar surface (Figs. 3a and
3b). The predominant set trends N55°E and corresponds with the well-defined set
that dips steeply to the left in Figs. 3a and 3b. Although no single lineament can be
traced across the entire mountain face, the parallelism and crisp definition of some
of the lineaments that can be traced for distances of tens or hundreds of meters
create the impression that they are the surface traces of a system of well-defined,
uniform fractures or layers. The impression is strengthened in a few places where the
same lineament apparently emerges both upslope and downslope from beneath the
cover of one of the many subhorizontal, smooth, regolith-covered benches on the
mountain face (Fig. 3b). In other places faint traces of the lineaments extend across
these benches.

    A second set of lineaments, trending N25°E, appears as a vertical set in Figs. 3a
and 3b. It is less prominent than the northeast set in both orbital and lunar surface
photographs.

    The third set, trending northwest along the sun line, is approximately parallel to
the abundant subhorizontal benches prominent in Fig. 3b. The benches are visible in
the orbital photograph as discontinuous, narrow, somewhat sinuous bands that are

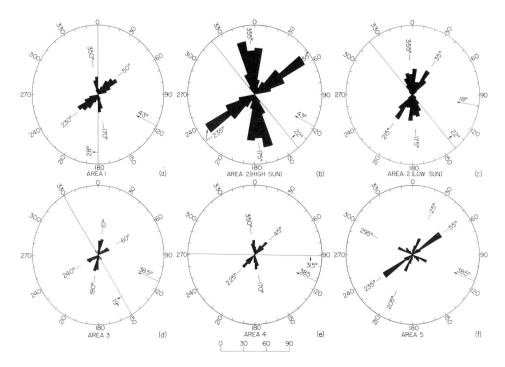

Fig. 4. Lineament azimuth frequency diagrams for the areas shown in Fig. 1. Line through center of plot and attached arrow show generalized strike and dip of slope as determined from NASA Lunar Topographic Map, Rima Hadley, Sheets A and B, 1971. Short arrow at edge of plot in southeast quadrant indicates sun azimuth with elevation in degrees. Length of bar, measured from the origin, represents the number of lineaments as shown by the scale. Lineaments plotted in 10 degree intervals. Azimuths plotted in both directions from the origin. (a and b) Lineament frequencies determined from 3X enlargement of NASA orbital panoramic camera photograph, 9809. (a: 188 lineaments measured; b: 459 lineaments measured.) (c) Lineament frequencies determined from contact print of NASA orbital panoramic photograph 9375 (252 lineaments measured). (d and e) Lineament frequencies determined from 3X enlargement of NASA orbital panoramic camera photograph 9425 (d: 108 lineaments measured; e: 107 lineaments measured). (f) Lineament frequencies determined from contact print of NASA orbital panoramic camera photograph 9425 (166 lineaments measured).

roughly parallel to the slope contours. They are distinct from the lineaments, which are straighter and more regular, and may be benches formed where the slope profile has been locally flattened by cratering or downslope mass movement. The northwest-trending lineaments are subordinate on the southwest mountain face, but on the northwest face they are prominent.

In areas 1 through 5, the lineaments do not clearly correspond to systematic topographic features such as closely spaced parallel troughs and ridges. At Silver Spur (Fig. 1), however, some prominent topographic lineaments occur. Figure 5, a 500 mm photograph of Silver Spur taken from the LM, shows massive ledges that apparently dip gently to the left and are crossed by finer more nearly horizontal

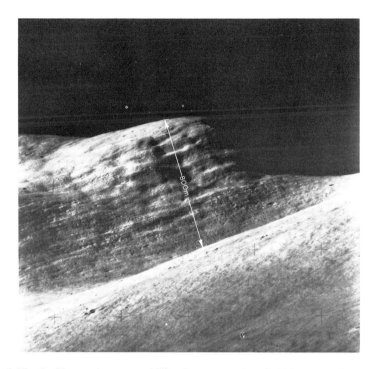

Fig. 5. View looking southeast toward Silver Spur, approximately 20 km away. Shows detail
of massive ledges that coincide with the north-northeast lineament system (sloping gently
left) and the possible fractures of the northwest lineament system (dark deeply shadowed
depressions sloping steeply right). Slope of Hadley Delta in the foreground. (NASA 500 mm
photograph AS15-84-11250.)

lineaments that give the impression of cross bedding. The finer lineaments are probably caused by the foreshortened view across an undulating cratered surface, an
effect similar to that on the south wall of St. George Crater (Fig. 2b).

The orbital photograph of Silver Spur (Fig. 6) shows two major sets of topographic
lineaments that intersect in a diamond-shaped pattern with boundaries that trend
about N35°W and N25°E. These topographic lineaments may reflect steeply dipping
geologic structure such as high-angle fractures or faults. The northwest lineament set
coincides with the steeply right-sloping shadowed bands of Fig. 5, and the north-
northeast set coincides with the prominent topographic ledges that appear to dip
gently to the left.

The preceding observations of Apennine Front lineaments lead to the following
generalizations and inferences:

(1) Sets of lineaments with similar dominant north and northeast trends occur on
widely separated slopes that have significantly different attitudes (areas 1 through 4).
Regardless of slope, these dominant trends are roughly bisected by the sun azimuth.
In the single area in which lineaments were measured under two widely different
illumination angles, a significant angular change between the dominant lineament
directions occurred.

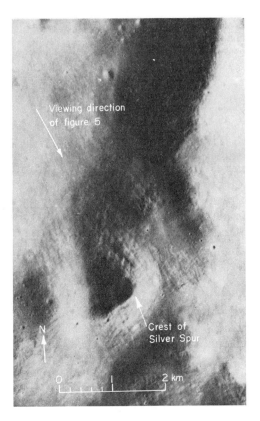

Fig. 6. Approximately vertical view of Silver Spur from orbit showing the diamond-shaped topographic pattern related to intersection of two sets of lineaments. (NASA orbital panoramic camera photograph 9430.)

(2) The major northeast trend of areas 1 through 4 may be duplicated by the prominent northeast-trending lineament set on Mt. Hadley (area 5), but neither of the other two Mt. Hadley sets corresponds in azimuth with the north-trending set of areas 1 through 4.

(3) In areas 1 through 5, the lineaments cannot be identified as the expressions of systematic topographic features such as parallel ridges and troughs. At Silver Spur, however, prominent northwest- and north-northeast-trending topographic lineaments may be related to steeply dipping geologic structures.

(4) Lunar surface photographs show that at least some lineaments occur in areas of relatively fine-grained, mature regolith that should conceal systematic layering in the underlying bedrock.

(5) Widespread occurrence of lineament segments with orientations that are similar over a variety of slope attitudes suggests that the lineaments are either illusions or are the surface expressions of steeply dipping repetitive geologic structures that form a regionally extensive pattern.

(6) Probability of extensive mantling by mature regolith suggests that if the lineaments are not illusions, they must represent the traces of bedrock structures propagated through the regolith to the surface as proposed by Schaber and Swann (1971).

MODEL EXPERIMENTS

An independent line of evidence that some of the Apennine Front lineaments may be illusions related to lighting angle is derived from model experiments in which lineaments similar to those of the Apennine Front have been produced. Howard and Larsen (1972) experimented with low oblique illumination on surfaces mantled by fine dry powders. They showed that:

(1) Low oblique lighting produces conjugate sets of discontinuous lineaments. The major lineament trends intersect at angles of 30° to 110° that are bisected by the azimuth of the light source. Increasing the light elevation tends to broaden the angle between the lineament sets. (This effect is the opposite of that observed in area 2 on the Apennine Front—see Figs. 4b and 4c.)

(2) The size of the individual topographic irregularities must be near the limit of resolution in the view of the model for the illusion of lineaments to be produced. For powder models, fortuitous alignment of topographic irregularities produces fine, discontinuous, randomly oriented topographic lineaments. The oblique light enhances those segments that coincide with the optimum acute angle to the light, thus producing the illusion of lineaments with preferred orientations.

To further investigate this phenomenon, the authors constructed and photographed models similar in shape to Mt. Hadley. Figure 7 shows such a model photographed with camera and illumination angles similar to those of Fig. 3. Approximately 15 cm high, the model was produced by sifting a thick mantle of dry cement

Fig. 7. Cement powder model approximately 15 cm high. Illumination and camera angles approximately similar to those in the view of Mt. Hadley in Fig. 3.

powder onto a preformed surface. As in the views of Mt. Hadley (Fig. 3), the model shows one lineament set dipping steeply to the left and a second set that is nearly horizontal. In a plan view of this model, there are two lineament sets that intersect at an acute angle facing the light source and continue with little deviation in azimuth across surfaces with widely divergent orientations.

The model experiments show that conjugate sets of lineaments resembling those of the Apennine Front can be produced on surfaces having abundant, random topographic irregularities near the limit of resolution. Geometric relations between conjugate sets and the light source on the models are not identical to the relations shown on the moon. For example, increasing light elevation in the models causes spreading of the lineament intersection angles that face the light, whereas in area 2 on the Apennine Front the intersection angles facing the sun decrease as the sun elevation increases (Figs. 4b and 4c). In general these angles tend to be broader on the Apennine Front (Fig. 4) than in the models of Howard and Larsen. These differences may perhaps be accounted for by the fact that the fine scale irregularities of the model surfaces are predominantly positive, grains and clots of powder, whereas those of the moon's surface are dominated by craters, which are topographically negative.

CONCLUSIONS

The lineaments on the mountain slopes at the Apollo 15 site may be caused by the selective enhancement of fortuitous alignments of small topographic irregularities by the obliquely incident sunlight. This possibility is consistent with: (1) occurrence of similar lineament trends on many slopes with widely different orientations, (2) approximate symmetry of those trends around the sun line, (3) shift in measured lineament trends with illumination change in area 2, (4) the recognition of somewhat similar lineament patterns in small-scale models, (5) absence of distinct topographic features coincident with the lineaments, and (6) evidence that the slopes are mantled by a thick regolith that would be expected to conceal geologic structures of the underlying bedrock.

If either of the prominent north or northeast trends does, however, represent bedrock structure, that structure is more likely to be regional fracturing propagated through the regolith than to be a surface expression of layering. The extensive areal distribution of the pattern and its azimuthal constancy regardless of slope orientation would suggest that the pattern, if structural in origin, represents a regional set of nearly vertical conjugate fractures.

Some locally prominent lineament patterns observed on the Apennine Front more certainly represent the surface expressions of true geologic structure. Most notable is Silver Spur, where the lineaments coincide with topographic scarps suggestive of a steeply dipping conjugate fracture pattern.

Acknowledgments—Prepared under NASA contract T-65253-G. Publication authorized by the Director, U.S. Geological Survey.

REFERENCES

Howard K. A. and Larsen B. R. (1972) Lineaments that are artifacts of lighting. In *Apollo 15 Preliminary Science Report*, National Aeronautics and Space Administration, NASA SP-289, p. 25-58–25-62.

NASA Lunar Topographic Map, Rima Hadley, Sheets A and B, 1971.
Schaber G. G. and Swann G. A. (1971) Surface lineaments at the Apollo 11 and 12 landing sites. *Proc. Second Lunar Sci. Conf., Geochim. Cosmochim. Acta*, Suppl. 2, Vol. 1, pp. 27–38. M.I.T. Press.

Proceedings of the Third Lunar Science Conference
(Supplement 3, *Geochimica et Cosmochimica Acta*)
Vol. 1, pp. 27–38
The M.I.T. Press, 1972

Geology of the Apollo 14 landing site*

R. L. SUTTON, M. H. HAIT, AND G. A. SWANN

U.S. Geological Survey
601 East Cedar Avenue, Flagstaff, Arizona 86001

Abstract—Apollo 14 landed in the Fra Mauro region of the moon, within about 1,100 m of a 90 m high ridge of the Fra Mauro formation, interpreted as being ejecta from the Imbrium Basin. The primary geologic objective of the mission was to sample ejecta from Cone Crater, which is 340 m in diameter and penetrates at least 60 m into the ridge. Data from the mission strongly support the Imbrian ejecta origin for the Fra Mauro formation.

Returned samples and photographs show that ejecta from Cone Crater is composed of composite breccias that include multiple clasts of still older, pre-Imbrian, breccias. Cone Crater ejecta displays a wide range of thermal metamorphism effects. Samples from the valley where the Lunar Module landed, which was not on a recognizable ray of ejecta from Cone Crater, are predominantly fines and poorly consolidated breccias formed by the disintegration of Fra Mauro rocks and probably are not volcanic rocks as had been postulated before the mission.

INTRODUCTION

APOLLO 14 ASTRONAUTS Alan B. Shepard and Edgar D. Mitchell landed on the moon on February 2, 1971, in an area referred to as the Fra Mauro region. The landing site (Lat. 3° 40'S, Long. 17° 27'W) is located approximately 500 km south of the mountainous block-faulted southern (and outer) rim of the multiringed Imbrium Basin, and about 1,100 km from the center of the Imbrium Basin (Fig. 1). The region is underlain by the Fra Mauro formation (Eggleton, 1964; Wilhelms, 1970; Wilhelms and McCauley, 1971) represented by a surface topography of hummocks, ridges, and valleys that trend roughly north-south radial to the Imbrium Basin (Fig. 2) and are believed to have been formed by deposits of ejecta from the Imbrium Basin. Post-Imbrian mare basalts have flooded large parts of the Imbrium Basin as well as its ejecta, leaving unburied only the higher, more mountainous features such as the uplifted circumferential ring structures and the Fra Mauro highlands.

LANDING SITE

The Lunar Module (LM) landed in a broad valley 1,100 m west of Cone Crater, which is located on a Fra Mauro ridge (Fig. 3). Cone Crater, 340 m in diameter, is young and sharp rimmed. Blocks of ejected material up to 15 m across occur near the rim. Ejecta are presumed to have been excavated from depths of as much as 80 m, well below the regolith. It is also presumed, using terrestrial crater studies for comparison (Gault *et al.*, 1968; Roberts, 1968), that ejecta closest to the rim of Cone Crater were derived from the greatest depths, and that a mixture of ejecta from different depths occurs in blocky rays that extend outward from the crater. Such a ray of

*Publication authorized by the Director, U.S. Geological Survey

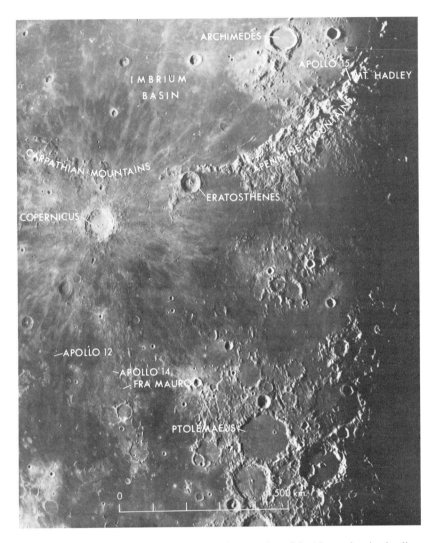

Fig. 1. The southern Imbrium Basin and Fra Mauro region of the Moon, showing landing sites of Apollo missions 12, 14, and 15. Apollo 12 landed in Oceanus Procellarum on a bright ray from Copernicus. Apollo 14 landed north of crater Fra Mauro in a ridge and valley highland area believed to be ejecta from the Imbrium Basin. Apollo 15 landed in a mare embayment at the base of the Apennine Mountains. Photograph by F. G. Pease, Sept. 15, 1919; Mt. Wilson Observatory, 100-inch telescope. Number 261, Catalog of Photographs and slides from Hale Observatories, Pasadena, California.

blocky ejecta extends westward from Cone Crater past the landing site, to the north, The "North Boulder Field" is a bouldery patch on this ray, where it was sampled at station H (Figs. 3, 4). Off the ray, the surface at the landing site is relatively smooth and fine grained (see Fig. 3), with very little Cone Crater ejecta on it. Larger craters in the valley, such as Doublet and Triplet, have penetrated through the regolith into

Fig. 2. Hummocks, ridges and valleys of the Fra Mauro highlands, showing the Apollo 14 landing site. NASA photograph AS12-52-7597.

material that we consider to be also the Fra Mauro formation. Their ejecta may be represented by samples from Stations G and G1, as well as from the comprehensive sample collected near ALSEP and Doublet Crater.

A regolith of mixed rock and glassy debris covers most of the surface at the landing site, the exceptions being fresh crater walls and crater rims. The regolith was derived primarily by repeated meteorite bombardment and fragmentation of local Fra Mauro materials (Carr and Meyer, 1972). This surficial debris layer probably is thinner on topographic ridges and slopes than in the valleys, with average thicknesses of 5–12 m in the region of the Apollo 14 landing site (Offield, 1970). A regolith thickness of 8.5 m at the ALSEP site is indicated by seismic studies of Watkins and Kovach (1972).

Fra Mauro Formation

The Fra Mauro formation is interpreted as material excavated by the enormous impact event that produced the Imbrium Basin (Gilbert, 1893; Eggleton, 1964). The ejecta from this 500 km diameter crater were transported radially away from the

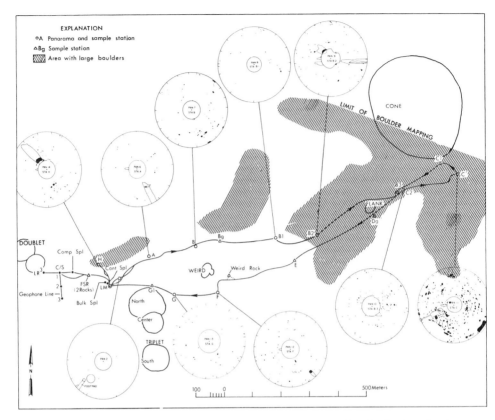

Fig. 3. Map of the Apollo 14 landing site and traverse routes, showing distribution of rock fragments at photographic panorama stations (circular insets) and rays of blocky ejecta from Cone Crater (line pattern). Circular insets represent 10 m radius areas around photographic panorama stations in which all rock fragments larger than 10 cm were mapped from the panorama photographs. Boulder rays were mapped from Lunar Orbiter photographs. The patchy distribution probably represents the limit of resolution in photographs rather than absence of rayed ejecta. EVA 1 traverse was west from the LM to deploy ALSEP and collect samples. EVA 2 traverse was east from the LM to within 20 m of the rim crest of Cone Crater. The White Rocks are located at Station C1.

basin and deposited during a complex episode involving base surge, debris flow, and the ballistic transport of materials (Eggleton, 1970; Offield, 1970). The Imbrian ejecta blanket thins with increasing distance from the basin. In the area of the Apollo 14 landing site, which is approximately 1.6 crater diameters from the edge of the excavated Crater Imbrium, the Fra Mauro formation probably is 100–200 m thick (Offield, 1970), being thinner beneath the valleys than beneath the ridges. The ridge on which Cone Crater is located rises about 90 m above the valley where the LM landed. Cone Crater excavated into Fra Mauro material to a depth of 60–80 m beneath the ridge, not deep enough to excavate materials down to valley level.

Before the mission, the relatively smooth surfaced valleys in the Fra Mauro region were thought possibly to be floored by volcanic rocks, or alternatively, to represent a

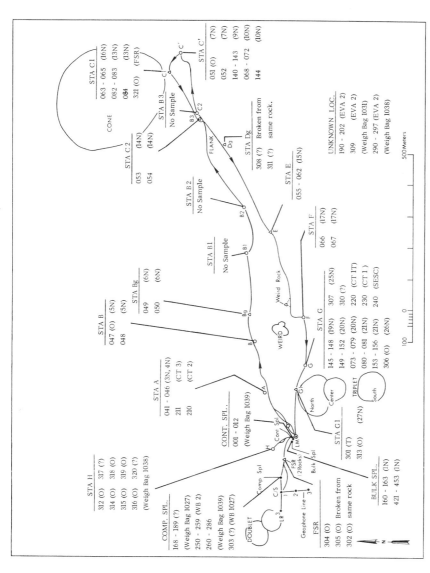

Fig. 4. Map of the Apollo 14 traverses showing stations and sample locations (Sutton *et al.*, 1971). Only the last three digits of sample numbers are shown. Numbers with letters in parentheses (17N) denote documented bags used during sampling. Core tubes are indicated by the symbol (CT2). Samples for which the lunar orientation is known are indicated by (O); tentative orientation is shown by (T).

fine-grained facies of the Fra Mauro formation that flowed into and was ponded in topographically low areas (Eggleton, 1970). Watkins and Kovach (1972) suggest that beneath the regolith in the valley is a layer of Fra Mauro material that is 16 to 76 m thick.

Samples

The primary geologic objective of Apollo 14 was to collect samples from the Fra Mauro formation. A relatively fresh, deep crater (Cone Crater) was chosen as a target for sampling, because it produced a natural drill hole through the regolith into the Fra Mauro formation to a depth of perhaps 80 m.

Apollo 14 returned approximately 43 kg of lunar samples, including 35 rocks that weigh more than 50 g each, and about 30 smaller rocks weighing between 10 and 50 g, as well as rock chips and fines collected with a scoop and with drive tubes (LSPET, 1971). Essentially all of the pre-mission sampling objectives were fulfilled. Ejecta from Cone Crater, believed to be ejecta from the Imbrium Basin, was sampled at different places during both the first and second Extravehicular Activity (EVA) traverses, from northwest of the LM to within 20 m of the rim crest of Cone Crater (Fig. 4). In addition, samples of regolith and poorly consolidated breccias were collected in the valley where the LM landed.

Returned rock samples were predominantly complex breccias (LSPET, 1971), collected from the patchy ray of Cone Crater ejecta in the valley as well as from Cone Crater ridge itself. The regolith fines were formed predominantly by the fragmentation of complex breccias of the Fra Mauro formation (Carr and Meyer, 1972), and poorly consolidated breccias collected in the valley probably were formed by the sub-sequent shock-lithification of regolith fines (Quaide, 1972). Breccia samples collected at Stations A, B, Bg, E, and near the LM in the bulk and contingency samples were mostly of the "regolith breccia" type.

Apollo 14 samples were classified according to their clast and matrix composi-tions, in addition to their relative coherence (Jackson and Wilshire, 1972; Warner, 1971, 1972). Coherence increases with increasing crystallization or annealing of the breccia matrix by thermal metamorphism. Thermal metamorphism is presumed to have occurred after the deposition of the Fra Mauro formation in much the same way as devitrification and annealing would occur in a thick, hot terrestrial ash flow tuff (Warner, 1971, 1972; Wilshire and Jackson, 1972). The presence of older metaclastic fragments as clasts in the Fra Mauro formation indicates that pre-Imbrian breccias, such as ejecta from the Serenitatis Basin thrown into the Imbrium impact site, had thermal histories similar to those of the Fra Mauro formation.

The Fra Mauro samples are composed almost entirely of complex breccias that contain compound clasts of still older breccias (LSPET, 1971; Chao et al., 1972; Dence et al., 1971; Warner, 1972; Wilshire and Jackson, 1972). Figure 7 shows sample 14321, the largest rock returned by Apollo 14. It is a dark-clast-dominant breccia that includes a compound clast at least 12 cm long, as large as many individual samples. In thin sections, as many as four stages of breccias-within-breccias have been described (Duncan et al., 1972; Wilshire and Jackson, 1972).

The samples returned by Apollo 14 do not support the hypothesis that the valley in which the LM landed is floored by volcanic rocks. Volcanic rock (basalt) is rare in samples of rocks and fines collected from Cone Crater ejecta. McKay *et al.* (1972) show that there is a relative abundance of basalt grains in the trench and core samples from Station G, near Triplet crater. This also is the place where the large basalt sample 14310 was collected. Because Station G is approximately one crater diameter from Triplet Crater (Copernican in age), the station may be located on the continuous ejecta blanket from Triplet Crater. It seems probable that the increased amount of basalt in Station G samples reflects either (1) an increased amount of basalt in Fra Mauro breccias that underlie the valley below the depth of excavation of Cone Crater or, (2) a basalt flow beneath the regolith of the valley, which was excavated by the 100 m diameter Triplet Crater. The first case seems more likely, because the basalt samples from Triplet Crater are plagioclase rich and are similar to those found with breccia near the rim of Cone Crater.

The evidence is not conclusive that there are, or are not, volcanic rocks associated with the valley. Certainly, breccias are dominant at the surface. Valleys are, however, collection troughs for especially thick accumulations of regolith derived from cratering events throughout the region, because ridges and slopes tend to shed loose debris.

BOULDERS—THE WHITE ROCKS

The size of boulders in the ejecta from Cone Crater increases toward the rim of the crater, from about 2 m maximum diameter at Station H, north of the LM, to about 15 m maximum diameter near the rim. Light-colored rocks unique to the crater rim were recognized and sampled by the crew at Station C1, within 20 m of the rim crest (Fig. 5; The "White Rocks" Hait, 1972; Swann *et al.*, 1971a, b). Like all other large boulders photographed on the EVA 2 traverse, they appeared to be layered and fractured breccias that include fragments of older rocks that are both lighter and darker in color than their surrounding groundmass. The White Rocks include primarily dark clasts. Because white rocks are rare and were seen only on the crater rim, we believe that they are among those excavated from the greatest depth in Cone Crater.

The White Rocks are some of the largest boulders of any so far to be directly photographed and sampled (sample no. 14082) during an Apollo mission. The photographs show three major features that can be seen, at different scales, from boulders to hand specimens, to thin sections: (1) clastic texture, (2) stratification, (3) jointing or fracturing. We consider these to be characteristics of the Fra Mauro formation.

CLASTIC TEXTURE

The clastic texture reflects the mixed assemblage of fragments of older rocks (mostly metaclastic) that are included as clasts in a fine-grained matrix. Some clasts are darker than the matrix and others are lighter. In the White Rocks there are both

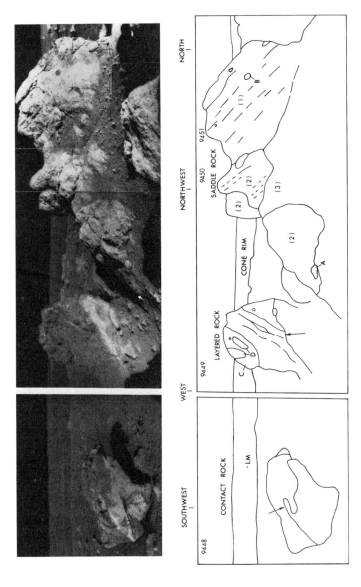

Fig. 5. "White Rock" group at station C1 (see Figs. 3, 4) on rim of Cone Crater. Contact Rock is about 3 m long; Layered Rock is about 3 m wide, and Saddle Rock is about 5 m long. The clasts and layers are prominent features. Most of the clasts are 10 cm or smaller, a few about 25 cm (A and B) and one (C) atop Layered Rock is a 1½ m long compound clast that contains at least two dark clasts about ½ m long at D. The dark and light layers that show in the rocks range in thickness from about 10–15 cm (dark layers in Layered Rock) to about 2 m (unit 1 in Saddle Rock). The structures expressed by these layers are a tight fold in Contact Rock, a truncated fold in Layered Rock, and in Saddle Rock, a possible fault where units 1 and 2 are in contact with unit 3. The White Rocks also contain evidence of jointing and a pervasive fracturing that probably was generated during or before the Cone Crater event. Parts of NASA photographs AS14-68-9448 to AS14-68-9451.

Fig. 6. Laminated and contorted ejecta of white sandstone, shale, and limestone in a quarry at Meteor Crater, Arizona. Note the apparent similarity of bedding and structure to the "White Rocks" at Cone Crater. (Photograph by J. F. McCauley, U.S. Geological Survey.)

dark and light clasts that occur in sizes from the limit of resolution to approximately 1.5 m. The largest clasts appear to be fragments of older breccias. Most clasts are less than 10 cm in diameter.

STRATIFICATION

Layering is clearly visible in the White Rocks and some of the other large boulders near Cone Crater. Jointing or fracturing may also resemble layering, but color contacts in the White Rocks help to identify one from the other. Lunar weathering, mostly erosion by micrometeorite bombardment, has etched the boulders along lines of weakness such as fractures and bedding planes, so that they look much like weathered terrestrial boulders. Layering on a microscopic scale is reported by Wilshire and Jackson (sample no. 14082, oral communication). Contorted layering within the White Rocks resembles irregular layering in ejecta near Meteor Crater, Arizona (Fig. 6), which are believed to have been deposited during the outward flowage of ejecta, as if in a fluid suspension, producing the observed structures (McCauley and Masursky, 1969).

Evidence for gross stratification within the Fra Mauro formation is circum-stantial, based on the areal distribution of samples, their compositions, and the probability that inverted stratigraphy exists within the ejecta from Cone Crater. Results of sample classification by Chao *et al.* (1972), Warner (1971, 1972), Jackson and Wilshire (1972), Wilshire and Jackson (1972) lead to the conclusion that there

Fig. 7. Sample 14321, collected near Station C1 on the rim of Cone Crater, was the largest rock returned by Apollo 14 (weight 8.9 kg). It is a dark-clast-dominant multiple breccia (Jackson and Wilshire, 1972). Note the 12 cm long dark clast that includes both darker and lighter fragments of still older rocks. Four stages of brecciation have been described in this sample by Duncan *et al.* (1972). NASA photograph S-71-28416.

are at least two major subdivisions of breccia types within the Fra Mauro formation to the depth excavated by Cone Crater. Breccias with predominantly dark clasts are thought to underlie breccias with predominantly light clasts in the Fra Mauro strata (Wilshire and Jackson, 1972).

Jointing or Fracturing

The Fra Mauro breccias, including the White Rocks, are jointed or fractured. It is not clear whether the jointing was a primary feature of the Fra Mauro formation imposed by cooling, or by lunar tectonism, or whether it resulted from the shock of the Cone Crater event. Many of the Apollo 14 samples and larger rocks that appear in the returned photographs have shapes indicating that they were broken along fractures; and some have internal fractures that are filled, or partially filled, with glass (Fig. 8, sample no. 14306, Wosinski *et al.*, 1972). The glass probably resulted from shock-melting that accompanied the Cone Crater impact.

Fig. 8. Sample 14306 (weight, 584.5 g) is a dark-clast-dominant coherent breccia (Jackson and Wilshire, 1972) whose shape is strongly influenced by bounding fractures. One fracture, at the bottom of the photograph, is filled with black glass, probably shock melted and injected during the Cone Crater event. NASA photograph S-71-34147.

Acknowledgments—We wish to thank H. G. Wilshire for helpful discussions pertaining to the Apollo 14 samples, and also R. M. Batson, G. E. Ulrich, and W. R. Muehlberger for reviewing the manuscript. Work performed under NASA contract T-65253-G.

REFERENCES

Carr M. H. and Meyer C. E. (1972) Petrologic and chemical characterization of soils from the Apollo 14 landing site (abstract). In *Lunar Science—III* (editor C. Watkins), pp. 116–118, Lunar Science Institute Contr. No. 88.

Chao E. C. T. Minkin J. A. and Boreman J. A. (1972) The petrology of some Apollo 14 breccias (abstract). In *Lunar Science—III* (editor C. Watkins), pp. 131–132, Lunar Science Institute Contr. No. 88.

Dence M. R. Plant A. G. and Traill R. J. (1972) Impact-generated shock and thermal metamorphism in Fra Mauro lunar samples (abstract). In *Lunar Science—III* (editor C. Watkins), pp. 174–176, Lunar Science Institute Contr. No. 88.

Duncan A. R. Lindstrom M. M. Lindstrom D. J. McKay S. M. Stoesser J. W. Goles G. G. and Fruchter J. S. (1972) Comments on the genesis of breccia 14321 (abstract). In *Lunar Science—III* (editor C. Watkins). pp. 192–194, Lunar Science Institute Contr. No. 88.

Eggleton R. E. (1964) Preliminary geology of the Riphaeus quadrangle of the moon and definition of the Fra Mauro formation. In *Astrogeol. Studies Ann. Prog. Rept.,* Aug 25, 1962 to July 1, 1963, Pt. A, U.S. Geol. Survey open-file report, pp. 46–63.

Eggleton R. E. (1970) Geologic map of the Fra Mauro region of the moon—Apollo 14 pre-mission map. U.S. Geol. Survey Misc. Geol. Inv. Map I-708, scale 1:250,000.

Gault D. E. Quaide W. L. and Oberbeck V. L. (1968) Impact cratering mechanics and structures. In *Shock Metamorphism of Natural Materials* (editors B. M. French and N. M. Short), pp. 87–99, Mono Book Corp., Baltimore, Md.

Gilbert G. L. (1893) The moon's face: A study of the origin of its features. *Phil. Soc. Washington Bull.* **12**, 241–292.

Hait M. H. (1972) The White Rock group and other boulders of the Apollo 14 site: A partial record of Fra Mauro history (abstract). In *Lunar Science—III* (editor C. Watkins), p. 353, Lunar Science Institute Contr. No. 88.

Jackson E. D. and Wilshire H. G. (1972) Classification of the samples returned from the Apollo 14 landing site (abstract). In *Lunar Science—III* (editor C. Watkins), pp. 418–420, Lunar Science Institute Contr. No. 88.

LSPET (1971) (Lunar Sample Preliminary Examination Team) Preliminary examination of lunar samples from Apollo 14. *Science* **173**, 681–693.

McCauley J. F. and Masursky H. (1969) The bedded white sands at Meteor Crater, Arizona (abstract). *Meteoritics* **4**, No. 3, 196–197.

McKay D. S. Clanton U. S. Heiken G. H. Morrison D. A. Taylor R. M. and Ladle G. (1972) Characterisation of Apollo 14 soils (abstract). In *Lunar Science—III* (editor C. Watkins), pp. 447–449, Lunar Science Institute Contr. No. 88.

Offield T. W. (1970) Geologic map of part of the Fra Mauro region of the moon—Apollo 14 pre-mission map. U.S. Geol. Survey Misc. Geol. Inv. Map I-708, scale 1:25,000.

Quaide W. L. (1972) Mineralogy and origin of Fra Mauro fines and breccias (abstract). In *Lunar Science—III* (editor C. Watkins), pp. 627–629, Lunar Science Institute Contr. No. 88.

Roberts W. A. (1968) Shock crater ejecta characteristics. In *Shock Metamorphism of Natural Materials* (editors B. M. French and N. M. Short), pp. 101–114, Mono Book Corp., Baltimore, Md.

Sutton R. L. Batson R. M. Larson K. B. Schafer J. P. Eggleton R. E. and Swann G. A. (1971) Documentation of the Apollo 14 samples. U.S. Geol. Survey open-file report, 37 pp.

Swann G. A. Trask N. J. Hait M. H. and Sutton R. L. (1971a) Geologic setting of the Apollo 14 samples. Science **173**, 716–719.

Swann G. A. Bailey N. G. Batson R. M. Eggleton R. E. Hait M. H. Holt H. E. Larson K. B. McEwen M. C. Mitchell E. D. Schaber G. G. Schafer J. P. Shepard A. B. Sutton R. L. Trask N. J. Ulrich G. E. Wilshire H. G. and Wolfe E. W. (1971b) Preliminary geologic investigation of the Apollo 14 landing site. In *Apollo 14 Preliminary Science Report*, NASA SP-272, pp. 39–85.

Warner J. L. (1971) Progressive metamorphism of Apollo 14 breccias (abstract). Geol. Soc. America 1971 Ann. Mtg., Washington, D.C., *Abstracts with Programs* **3**, No. 7, p. 744.

Warner J. L. (1972) Apollo 14 breccias: Metamorphic origin and classification (abstract). In *Lunar Science —III* (editor C. Watkins), pp. 782–784, Lunar Science Institute Contr. No. 88.

Watkins J. S. and Kovach R. L. (1972) Apollo 14 active seismic experiment. *Science* **175**, 1244–1245.

Wilhelms D. E. (1970) Summary of lunar stratigraphy—telescopic observation. U.S. Geol. Survey Prof. Paper 599-F, pp. F1–F47.

Wilhelms D. E. and McCauley J. F. (1971) Geologic map of the near side of the moon. U.S. Geol. Survey Misc. Geol. Inv. Map I-703. scale 1:5,000,000.

Wilshire H. G. and Jackson E. D. (1972) Petrology of the Fra Mauro formation at the Apollo 14 landing site (abstract). In *Lunar Science—III* (editor C. Watkins), pp. 803–805, Lunar Science Institute Contr. No. 88.

Wosinski J. F. Williams J. P. Korda E. J. Kane W. T. Carrier G. B. and Schreurs J. W. H. (1972) Inclusions and interface relationships between glass and breccia in lunar sample No. 14306,50 (abstract). In *Lunar Science—III* (editor C. Watkins), pp. 811–813, Lunar Science Institute Contr. No. 88.

Proceedings of the Third Lunar Science Conference
(Supplement 3, *Geochimica et Cosmochimica Acta*)
Vol. 1, pp. 39–61
The M.I.T. Press, 1972

New geological findings in Apollo 15 lunar orbital photography

Farouk El-Baz

Bell Telephone Laboratories, Washington, D.C. 20024

Abstract—The panoramic and metric camera systems, which were flown for the first time on Apollo 15, obtained a total of 4,140 photographs with a resolution of 1–3 m and 25–30 m respectively. Data derived from the metric camera system will allow mapping the overflown 12% of the lunar surface at 1:250,000 scale with 50 m contours; those of the panoramic camera will produce large scale maps (up to 1:10,000) with 5–10 m contours. The Hasselblad camera was also utilized to obtain oblique views with color film. The combined data represent the most thorough photographic coverage of any Apollo mission to date. Photogeologic interpretation of these data and correlation with other remotely sensed data, both from earth and from lunar orbit, will allow extrapolation of knowledge gained by surface exploration to large segments of the moon.

Several new features of probable volcanic origin were detected: (1) dark-haloed cones in the Apollo 17 Taurus-Littrow landing site. These "cinder cones" may have brought to the surface pyroclastic fragments from deep within the moon in the post-mare period of lunar surface history; (2) a D-shaped structure with light-colored units and blister-like smooth domes in its floor. The structure is believed to be a young collapsed caldera; (3) what appears to be a lava lake on the lunar farside with "lava marks" and evidence of lava drainage into prominent fissures; (4) unusually large (up to 40 km in diameter) domes near Rima Schroedinger I on the southern farside; and (5) numerous lava flows in western Mare Imbrium, some of which cross mare ridges.

Other features on which new data were obtained include: (1) two lineated units that are interpreted as landslides or rock avalanches: one on the northwestern rim of Tsiolkovsky (approximately 80 km long), and a smaller one (approximately 5 km long) in the Taurus-Littrow site; (2) unusual light-colored swirls in Mare Inginii, Mare Marginis, and Oceanus Procellarum. These sinuous markings may have been produced by alteration of the materials at the antipodal areas of impact points; and (3) man-caused changes in albedo, observed for the first time from lunar orbit, which include: brightening of the surface area beneath the LM, probably due to compaction during descent; and darkening of the areas around Rover and astronaut foot tracks, probably due to destruction of the (less than 1 mm) photometric layer.

INTRODUCTION

Two new photographic systems were carried for the first time on Apollo mission 15: a 24-inch panoramic camera and a 3-inch mapping camera system. The new cameras were housed in the scientific instrument module (SIM) bay and operated from the Command Module. The Command Module pilot (CMP) also operated the Hasselblad camera to take oblique views of specific targets using color film that accentuates subtle color-tone and textural variations of lunar surface units. The combined results of these systems constitute the most thorough photographic coverage of any Apollo mission to date.

The metric camera system is composed of (1) a terrain camera that took stereo photographs of the lunar surface at 25–30 m resolution; (2) a stellar camera that simultaneously took photographs of a star field for orientation; (3) a laser altimeter that made simultaneous measurements of the spacecraft altitude with 2 m precision; and (4) a clock that recorded the time of terrain camera exposures. This combination of data allows mapping the overflown 12% of the lunar surface (Fig. 1) at 1:250,000 scale with 15 m heighting accuracy, which will allow drawing 50 m contours.

The 24-inch panoramic camera took stereo photographs along all the ground-tracks at 1–3 m resolution. A single frame covered an area 22 km × 340 km. The central portion of the frames can be rectified to produce high quality photobases at 1:10,000 scale with 3 m accuracy, which will allow drawing 10 m contours. The basic data reduction scheme for Apollo 15 photography is summarized by Doyle (1972).

Because of the high inclination of Apollo 15 orbits, the returned photographs contain an unmatched wealth of data pertaining to numerous lunar surface features and processes. Because of space limitations, only a selection of the new findings will be given in this paper. Additional examples are given in El-Baz (1972a).

GEOLOGIC FEATURES OF APOLLO 17 LANDING SITE

On February 16, 1972 the National Aeronautics and Space Administration announced that Taurus-Littrow is to be the landing site for Apollo mission 17. The announcement came after thorough analyses and investigations of the scientific and operational aspects of the last Apollo lunar landing. The Taurus-Littrow area was considered by the Apollo site selection board because of the new findings from Apollo 15: Among the important considerations were the visual sightings from lunar orbit of cinder cones in this area and the variety of geologic features detected in the pano-ramic camera photography.

The Taurus-Littrow site is located on the southeastern rim of Mare Serenitatis south of the crater Littrow (20° 10′N, 30° 48′E). The Taurus Mountains, which bound eastern Mare Serenitatis, are clusters of massive peaks interconnected with old crater rims and light-colored, plains-forming highland materials. The relation-ship between the Taurus Mountains and the Serenitatis Basin is much like that between the Apennine Mountains and the Imbrium Basin. The Imbrium Basin, however, is younger than the Serenitatis Basin (Wilhelms and McCauley, 1971) and the Apennines are, in general, less degraded than the Taurus Mountains.

The geomorphological characteristics of Mare Serenitatis and the surrounding topography indicate that Mare Serenitatis is an impact basin that was filled with volcanic lava flows. The mare materials appear to have been deposited in two major episodes: an early event that left a light-colored mare surface in the middle portion of the basin and a later occurrence that deposited a dark-colored mare unit in an outer annulus or concentric zone (Fig 2). A concentric ridge system corresponds closely with this division of the mare units of Serenitatis. The ridges may have been formed by relatively viscous, extrusive volcanic materials along buried concentric fractures. This hypothesis is supported by the fact that the outer zone of the basin displays numerous concentric fractures and faults in the form of arcuate rilles, for example, the Sulpicius Gallus Rilles on the west rim, the Menelaus and Plinius Rilles to the south, and the Littrow and Chacornac Rilles to the east.

As shown in Fig. 2, a still darker unit occurs within the southwestern corner of the Taurus Mountains (the southeastern corner of the Serenitatis rim). In fact, this

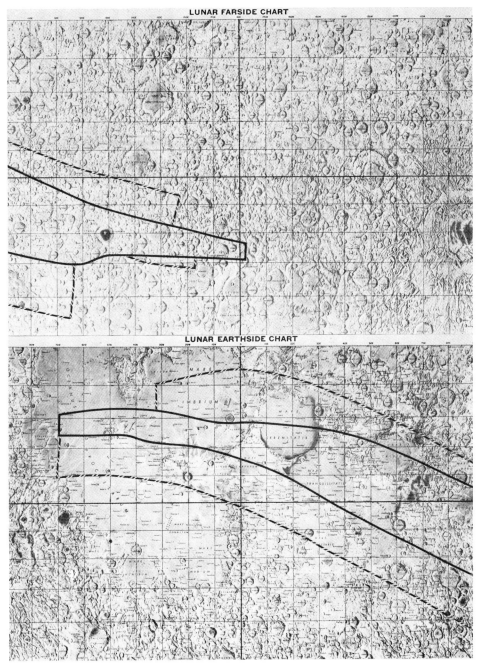

Fig. 1. Photographic coverage of the service module camera systems on Apollo 15. Area enclosed by black lines (about 12% of the lunar surface) is that covered vertically by the mapping camera. This coverage also corresponds with that of the rectifiable portions of the panoramic camera photography. Enclosed by the dashed lines are areas covered obliquely by the mapping camera.

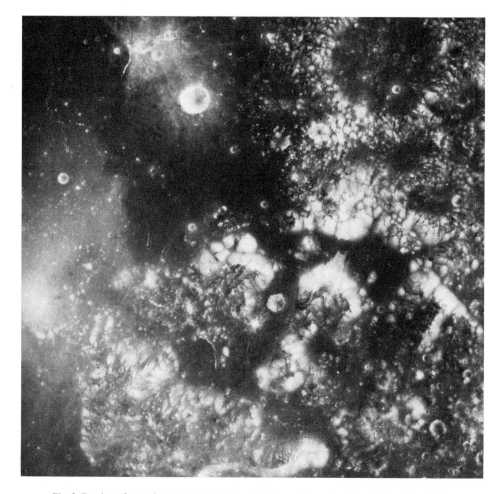

Fig. 2. Portion of mapping camera photograph of the south-eastern rim of Mare Serenitatis.
The crater Littrow is in the upper right corner of the photograph; Mt. Argaeus is at the
bottom. Note the dark-colored unit that fills the flat areas between the highland units; a
detail is shown in Fig. 3 (metric camera frame 1113).

unit constitutes the darkest surface material on the lunar near side (Pohn and Wildey,
1970) and, probably, on the whole Moon. The unit is centrally located between the
crater Littrow to the northeast, the crater Vitruvius to the southeast, and Mt. Argaeus
to the southwest. It appears to mantle the highland materials as well as the Serenitatis
mare materials. The unit was mapped as the Sulpicius Gallus formation by Carr
(1966) and is considered volcanic in origin.

Earth-based multispectral data indicate that the dark color of this unit is probably
related to compositional characteristics as discussed by McCord (1969). Radar data
also indicate that the unit is relatively smooth, with a small number of blocks. The
apparent young age of the dark unit is indicated by the relatively low density of
craters on its surface (R. Greeley, personal communication).

The regional geologic setting of the area is discussed by Carr (1972). The local setting and the relationship of the dark deposit to surrounding geologic units is discussed by El-Baz (1972b). During the Apollo 15 mission the dark deposit and its probable source were the objects of visual observations, which indicate that the unit probably consists of volcanic ash or pyroclastic deposits that came to the surface through a multitude of cinder cones.

As shown in Fig. 2, the dark material appears to embay the highland massif units. A panoramic camera view of part of the unit is shown in Fig. 3, in which the dark material is visible both in the lowlands between the mountains and, rarely, on top of massif units that display nearly level or depressed surfaces. Thus, the dark material probably was deposited from above; that is, by settling after ejection from beneath the surface.

In addition, remnants of the dark deposit can be distinguished on some of the massif slopes. This observation also tends to support deposition from above, followed by mass wasting down steep slopes. The dark material, therefore, would be mixed with the light-colored, fine-grained highland material as evidenced by the fact that gradational contacts occur between the dark lowland fill and the mountain fronts.

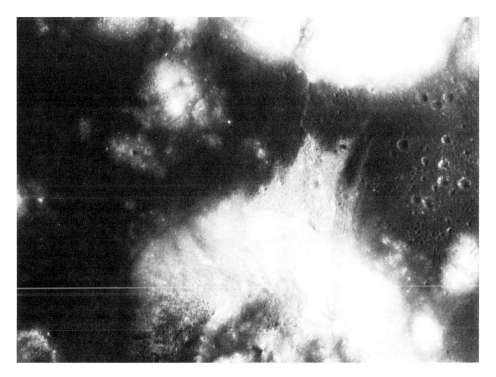

Fig. 3. Apollo 15 panoramic camera view of the Apollo 17 Taurus-Littrow landing site. The two massif units to the north and south constitute the southwestern corner of the Taurus Mountains. The bright area between the two units is interpreted as a landslide that originated at the southern massif unit (see also Fig. 4). (Central portion of panoramic camera frame 9559).

Based on the Apollo 15 photographs (Figs. 2 and 3), a geologic sketch map shown as Fig. 4 was made of the area. From this data the geologic history of the region may be summarized as follows:

1. The Serenitatis Basin was formed by a giant impact, which formed a 700 km-wide depression and at least two major systems of fractures, one concentric with the basin and the other radial to it. Massif units of the Taurus Mountains appear to have been uplifts along the concentric fault system, and they display fractures parallel to both.

2. A major episode of basin filling by Imbrian basaltic flows occurred. This episode probably ended about 0.5 billion years later (estimated by extrapolation of data acquired at previous Apollo landing sites). Another episode of filling followed, ending perhaps another 0.5 billion years later (Eratosthenian in age), during which the darker annulus surrounding the older mare material was formed (estimated on the basis of age dates of Apollo 12 and 15 basalts as opposed to Apollo 11 basalts, as well as relative ages in Wilhelms and McCauley, 1971). One of the latest manifestations of mare material extrusion was the formation of the wrinkle ridges now visible on the mare surface.

3. After the completion of basin filling by mare materials, volcanic eruptions began in the southwestern corner of the Taurus Mountains. Cinder cones, located mainly in the lowlands between the massif units, deposited the dark blanket that mantles the highland materials, the mare materials, and the grabens and wrinkle ridges in the southeastern corner of Mare Serenitatis.

4. From the time of formation, scarps of massif units of the Taurus Mountains were subjected to mass wasting. (Blocks on the foothills and their tracks on the slopes are shown in Fig. 3). One of the manifestations of this process is a unit, with north-south lineaments, which is interpreted as a landslide that originated from the southern massif unit and which has a distinct trough at its base (Fig. 4).

5. A fault scarp was formed by the relative upward movement of the materials west of the fault line and/or by the downward movement of the materials east of it. The fault is among the youngest tectonic features in the area; it bisects the oldest material (Taurus Mountains massif units) as well as the younger formation (the dark, ashlike mantling material). However, within the latter the scarp looks like a wrinkle ridge and the fault may have been draped by a thin ash deposit.

New Volcanic Features and Landforms

In addition to the cinder cones in the Taurus-Littrow region, Apollo 15 orbital photography exhibits numerous volcanic features that were observed on the moon for the first time. Detailed study of these structures will further our understanding of lunar endogenetic processes. Following is a brief description of four of these features.

D-shaped structure southwest of the Haemus Mountains

The hilly upland units between the circum-Serenitatis Haemus Mountains and the circum-Imbrium Apennine Mountains is characterized by isolated mare-like

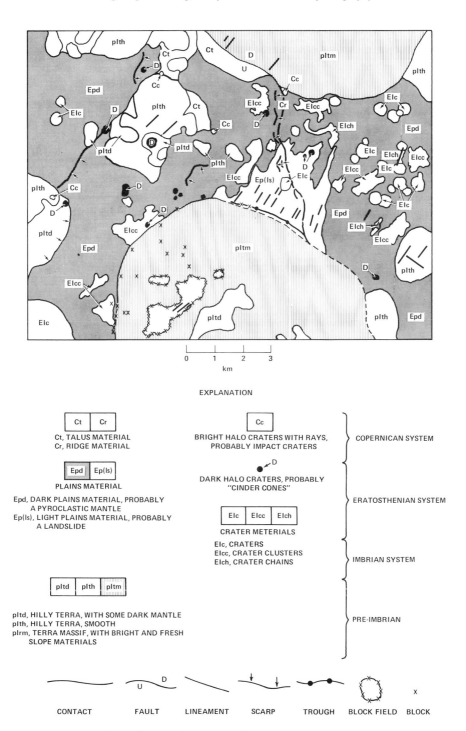

Fig. 4. Geologic sketch map of the area shown in Fig. 3.

patches. These dark patches are relatively younger mare flows that are superposed
on the highland materials (Fig. 5). Within one of these patches the Apollo 15 panoramic camera photographs reveal a D-shaped structure with unusual characteristics
(Whitaker, 1972a).

The structure is located at 18° 40′N, 5° 20′E and is about 3 km along the straight
part of the rim. It is a depression with a raised rim and an outer topographic rise that
extends to a maximum of about 4 km. The rim deposit is somewhat darker than the
surrounding mare material, but exhibits similar textural characteristics. As shown
in Fig. 6, the floor of the depression displays three different units: (1) a slightly
undulating, hilly and domical, light-gray unit that occupies the central and northeastern portions of the floor, (2) a very bright, almost white unit that makes an
annulus occupying the outer part of the depression, and (3) a unit made of about
50 disconnected, slightly sloping positive structures, which produce a blister-like
appearance. They are reminiscent of volcanic domes and a few display distinct
summit craters in the middle.

These characteristics indicate that the structure is of volcanic origin. It is my
opinion that we are dealing here with a caldera with several stages of extrusion and
intrusion in the central part. The blister-like domes appear to constitute the latest
events. The unusual low density of craters within the structure suggests that it may
be among the youngest lunar formations of volcanic origin.

Fig. 5. Portion of Lunar Orbiter IV photograph H-102 showing the area north of Mare
Vaporum on the foothills of the Haemus and Apennine Mountains. Several pools of mare
units overlie the low highland hills. The "bimat" defects in the center of the photograph
partly mask the structure shown in Fig. 6 (arrow).

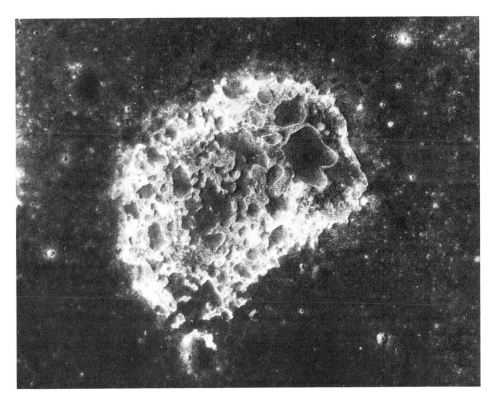

Fig. 6. D-shaped structure that is interpreted as a caldera. The straight segment of the rim is approximately 3 km long. Note the light-colored annulus in the floor, which is also exposed in the wall of the crater on the lower left corner. The blister-like domical hills have summit vents and appear to be completely void of impact craters, suggesting a very young age. Apollo 15 panoramic camera frames 0176 and 0181 make a stereo pair of the structure.

The most ambiguous feature in the structure is the light-colored unit. It is suggested by Whitaker (1972a) that this highly reflective material probably is sublimates. It is evident from Fig. 5 that the exposure on the bright crater wall near the southwest corner of the structure represents a difference of materials within a unit that was dug up by that crater.

It must be stated that the general area between the Haemus and Apennine Mountains exhibits an anomalous Al/Si ratio. As reported by Adler *et al.* (1972), the average ratio over mare areas is 0.67 and over highland areas 1.13. The region in question shows a median ratio of about 0.85. This apparently is due to the mare patches within the highland units. The area also is the seat of the A-7 zone of moonquake activity as reported by Latham *et al.* (1972). Movements along fractures within this area of intersection between the Imbrium and Serenitatis fracture systems is not unexpected. The apparent young age of the D-shaped caldera within the region also indicates a possibility of volcanic eruptions and/or readjustments to relatively recent volcanic deposits.

Lava lake southwest of the crater Perelman

At 103°E, 25°S, southwest of the crater Perelman, occurs an isolated patch of dark mare-like material. The area is elongate, measuring 120 km by 40 km, and the patch is surrounded by cratered plains (Fig. 7). The high resolution panoramic photography displays numerous probable volcanic craters, domes, and viscous flows, as shown in Fig. 8.

The northern part of the region is enclosed in a breached crater and displays a relatively smooth surface with numerous rimless depressions. The dark material appears on at least three different levels (Fig. 8). The highest level topographically is

Fig. 7. Dark mare-like material that floods an area southwest of the farside crater Perelman, upper right corner. The image area is approximately 160 km on each side. A detail is shown in Fig. 8 (metric camera photograph 2628).

Fig. 8. Apollo 15 panoramic camera view of portion of the caldera-like structure showing a "lava mark" on the western wall (arrow) and a level of subsidence that is marked by an annulus of topographically higher floor material near the rims (arrowheads). Drainage channels may be seen near the fissure, which trends in a northwesterly direction (panoramic camera frame 9965).

on the wall of the crater and shows best in the partly shadowed western wall. It is represented by a thin layer that produces a sharp scarp, and is best described as a "lava mark" where the infilling lava must have filled the crater to that high level and later receded leaving a mark, much like the "water mark" described by the Apollo 15 crew on the slope of Mt. Hadley.

The second level is represented by an annulus about one-tenth of the way inward. It produced an inward facing slope toward the third and lowest level, which makes up the rest of the area. This second level also is indicative of lava recession following extrusion. Flow channels leading into the fissure, which bounds the area to the south and trends in a northwesterly direction, suggest drainage.

These characteristics suggest that the features are the result of drainage of a lava lake. Textures and evidence of flow indicate that the original lava probably resembles that of the basaltic maria. Drainage of the lava following extrusion is a common occurrence on earth. This usually happens through the fractures that served as channelways for lava extrusion. It is, therefore, reasonable to conclude that the fissures and rille-like structures served as conveyors of the lava as well as drainage channels following extrusion.

Gigantic domes near Rima Schroedinger I

Schroedinger is a farside basin, 300 km in diameter, near the moon's south pole and from its rim emanate two large straight depressions or rilles. The western rille displays sharp boundaries and is as much as 20 km wide. Photographs of this rille, which show the domes described below, were obtained following trans-earth injection (TEI) using the Hasselblad camera with the 250 mm lens and color film.

As shown in Fig. 9, the photographs reveal in an area centered around 105°E, 60°S, a cluster of about six unusually large domes. The larger domes are 40 km in diameter. They display somewhat undulating surfaces and are reminiscent of mare domes in western Oceanus Procellarum on the moon's nearside. These features also appear to be made of extrusive volcanic deposits.

Fig. 9. Post-trans-earth injection view showing probable volcanic domes near Rima Schroedinger I (arrow). Note the dark mare-like material that fills the level areas in and around craters (Apollo 15 Hasselblad camera frame 13090).

The domes are located in an area of cratered highlands. However, relatively dark plains-forming materials fill most craters as well as flat areas between them. One unusual feature is a wrinkle ridge, which starts at the area where the domes are clustered. This occurrence of a wrinkle ridge is in the highlands and suggests that ridges of mare-like material may represent extrusive volcanism along pre-existing fractures, whether in the maria or the highlands.

Flow front in western Mare Imbrium

Apollo 15 photographs of the southwestern portion of Mare Imbrium are an excellent illustration of the utility of oblique metric camera photography (Fig. 10). In this area a system of lava flows stretches diagonally within Mare Imbrium near the crater Euler (23°N, 29°W). The system has been known from earth-based telescopic observations and photography (Strom, 1965).

As discussed by Whitaker (1972b), examination of the Apollo 15 photographs leaves little doubt that the source of the material comprising the major flows is situated to the west of the crater Euler and south of the limits of the photographic

Fig. 10. Oblique view (40° north) of the flow edges in western Mare Imbrium. The largest flows that extend from the left to the right are deformed by a prominent mare ridge (Apollo 15 metric camera frame 1556).

coverage. Some lava-flow channels located west-northwest of Euler are similar to features found on Mauna Loa and other terrestrial volcanic fields. The lavas, in the present case, have flowed in a general northeasterly direction, indicating the direction of slope that existed in Mare Imbrium at the epoch of the eruption. The point of termination of the flow system, near the crater Le Verrier, is close to the center of Mare Imbrium. This implies that the older filling of lavas never completely obliterated the shape of the pre-lava basin and that the later flows simply ran downhill to the lowest area at the center.

Several flows appear to cross ridges of substantial elevation without deviation, ponding, or change in thickness. This would have been unlikely if the ridges had preceded the flows; thus most, if not all, of the mare ridges included were formed after the solidification of the flows.

It must be stated that the area, where the flows appear to originate, corresponds with the highest γ-ray counts on Apollo 15 (Metzger et al., 1972). Although it was suggested that the anomaly may be due to ray materials in the area, this author proposes a subsurface origin at the source of the flows, e.g., higher concentration of radioactive materials in the last stages of magmatic extrusion, which resulted in the formation of the complex system of broad wrinkle ridges. It must be stated that the anomaly is removed from the fairly thick Lambert ejecta by approximately 300 km. Also, rays from larger craters (Copernicus, Kepler, and Aristarchus) are neither more abundant nor any different than in other overflown regions of both Mare Imbrium and Oceanus Procellarum.

Strom (1965) and Whitaker (1972b) have shown that these Imbrium flow boundaries correspond with distinct color differences. The younger units are bluer and the older units are redder. This distinction also corresponds with the moon-wide geological classification of mare units into older (redder) units of Imbrian age, and younger (bluer) flows of Eratosthenian age.

LANDSLIDE OF THE CRATER TSIOLKOVSKY

In the previous discussions, examples were given of photogeologic interpretation based on the examination of returned photographs. The purpose of this brief section is to illustrate the utilization of metric camera photography to make topographic measurements and provide terrain profiles.

The case in question is that of the lineated flow unit on the northwestern rim of the crater Tsiolkovsky (Fig. 11). This 80 km-long unit is interpreted as a landslide that originated near the rim crest of the crater and flowed in a westerly direction on the floor of the crater Fermi.

Preliminary terrain profiles of this unit are shown in Fig. 12. These were generated by partial levelling of a model on a stereo plotter (APC-type) and plotting elevation points of the surface along a traverse. The profiles show that the unit slopes gently with a double lip or terminus at one point. The flow front is also somewhat raised, which is not uncommon in landslides or debris flows on earth.

The landslide exhibits a larger number of fresh, small craters than on the older Fermi crater floor. Many of these are rimless and appear to have been the result of

Fig. 11. The landslide on the northwestern rim of the farisde crater Tsiolkovsky. The lineations are parallel to the direction of flow. The two lines AA′ and BB′ indicate the traces of the terrain profiles in Fig. 12 (Apollo 15 metric camera photograph 1034).

drainage of the fine-grained material into void spaces initially sealed over by larger blocks in the landslide.

The smaller (5 km) landslide in the Taurus Littrow region (described above) is not identical to that of Tsiolkovsky. The Littrow landslide appears more hummocky and shows more small-scale undulations. Both landslides, however, are blocky. This has significant implications in the case of the Apollo 17 landing site where the landslide may provide blocks of varying sizes from the whole of the stratigraphic section from which it was derived. Thus, its sampling would provide the best method of sampling the mountain scarp from which it originated.

UNUSUAL AND MAN-MADE ALBEDO VARIATIONS

The photometric properties of lunar surface materials are not yet fully understood. Gross variations in albedo can be correlated with material (chemical and/or physical) differences. In some cases, however, the surface brightening and darkening are caused by unknown factors. This is due primarily to the lack of knowledge of the characteristics of the lunar photometric layer—the upper few hundred microns of the soil. An understanding of this layer is fundamental to a better comprehension of the albedo of lunar surface materials and its variations. It is also important in interpreting results of many remote sensing techniques, which depend on the properties of the uppermost layer of the surface. Apollo mission 16 will provide us, for the first time, with a sample of this layer for detailed analysis and study. This section will be devoted to descriptions of unusual light-colored markings in three lunar surface regions and to man-caused albedo variations in the Apollo 15 landing site.

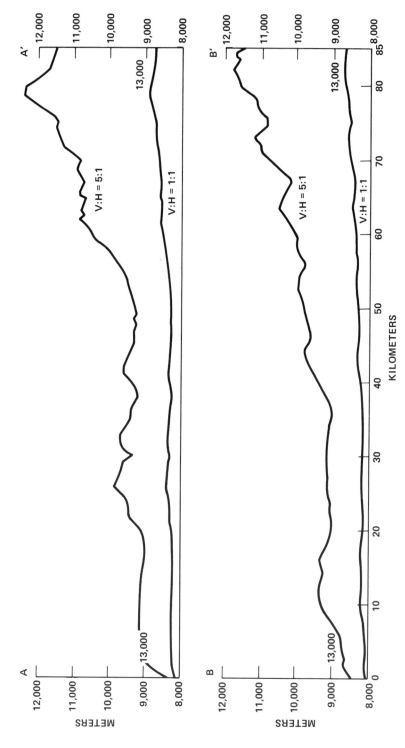

Fig. 12. Two profiles of the Tsiolkovsky landslide with the two different ratios of the vertical to horizontal scales (1 and 5); Courtesy of S. C. Wu, U.S. Geological Survey.

Unusual markings in the lunar maria

Rays and ejecta blankets of impact craters make up the majority of very bright and light-colored units on the Moon. An excellent example is given in Fig. 13, which shows a 5.5 km crater and surrounding ejecta. The crater is located on light-colored highlands near Gagarin on the lunar farside (145°E, 17°S). The ejecta, which is very blocky and hummocky, extends to about 1.5 crater diameters from the rim crest. Beyond this zone the rays extend in a radial pattern. Alternating light-and-dark streaks are common; the dark streaks are due to surface roughness and small-scale undulations.

The aforementioned brightness is at least easily correlated with ejecta, even if not fully understood. In three regions on the moon, there exists surface brightness that is not related to observable features. These are swarms of light-colored markings that display sinuous patterns in Mare Inginii, Mare Marginis, and Oceanus Procellarum. The swirls of bright materials in Mare Inginii on the lunar farside are as much as 120 km long, and occur in the southwest quadrant of the mare basin (Fig. 14). It must be noted that they appear old because secondary crater chains from the crater O'Day, to the northwest, cut through the bright material. Because the latter is exposed on the walls of the crater chains, the brightness seems to extend downward.

Fig. 13. Apollo 15 panoramic camera view of a sharp, bright-haloed crater near the northwestern rim of the farside crater Gagarin. Note the ejecta blanket and the radial ray system (left), and the blocky, streaked ejecta (right).

Fig. 14. Left, portion of Lunar Orbiter II photograph M-75 showing Mare Inginii with the
light-colored swirls in the southwest quadrant. Secondary crater chains of the crater O'Day
on the western rim of the basin cut across the swirls. Right, oblique view (looking southward)
of Mare Inginii showing the lack of topographic relief of the light-colored sinuous markings,
upper right (Apollo 15 Hasselblad camera frame 11724).

A similar group of light-colored swirls occur in the northern part of Mare Marginis; however, individual swirls are smaller in this case (Fig. 15). They are concentrated in two areas: west of the crater Goddard, in the mare materials; and
northeast of Ibn Yunus, in the highland materials. A somewhat atypical occurrence
is that of the Reiner-γ structure and associated bright markings in western Oceanus
Procellarum. This is atypical because only one swirl and a long bright line form the
structure.

In all three cases the swirls are best seen at small phase angles; however, because
of their very high albedo they are detectable at low sun (Fig. 14). From both photography and visual observations from orbit, these markings appear to have no topographic expressions associated with them. Also as pointed out previously (El-Baz,
1971), the following relationships are significant: (1) the swirls in Mare Inginii lie

Fig. 15. Swirls of light-colored markings in the area north of Mare Marginis on the eastern
limb of the moon (Apollo 15 Hasselblad camera frame 13094).

diametrically opposed to the center of Mare Imbrium, (2) the bright swarms in Mare Marginis lie on the opposite side of the moon from Mare Orientale, and (3) the Reiner-γ markings in Oceanus Procellarum lie opposite to a point within one crater diameter of the crater Tsiolkovsky.

Because of these characteristics and relationships it was proposed by El-Baz (1971) that these swirls probably are produced as a result of disturbances caused by major impacts; the disturbances may have been caused or triggered seismically at the antipodal areas of impact points. It is further speculated that converging seismic waves may have facilitated the release of trapped gases in the lunar interior; the gases may have penetrated the materials through fractures and chemically altered them. "Alteration" along fractures, although on a much smaller scale, appears feasible based on the occurrence of such cases in the Apollo 15 basalts (see Swann *et al.*, 1972, p. 82).

Man-caused albedo variations on the lunar surface

High resolution panoramic photographs of the Apollo 15 landing site show, for the first time, the LM structure on the moon as evidenced by reflected light and by a shadow. This was most convincingly demonstrated by before and after photography of the landing site (Hinners and El-Baz, 1972). The post-landing photographs (Fig. 16) show a bright halo, approximately 150 m in diameter, roughly centered on the

Fig. 16. Panoramic camera view of the Apollo 15 landing site taken two revolutions after landing. The circle encloses the landing site with a prominent bright halo, which may have been produced as a result of compaction of the surface material during the Lunar Module descent.

LM. The halo is attributed to an increase in mare surface brightness caused by the landing. The symmetry of the halo precludes reflected sunlight as a cause; reflected light should be observed largely only to the east. A relatively bright area southeast of the LM also is visible. This area, in contrast to the halo, is also relatively bright in the pre-landing photographs, and this brightness is attributed to the eastward-sloping wall of an old, subdued crater. However, some enhancement of the brightness as a result of the LM landing is not ruled out.

That a surface alteration occurred during landing is not surprising, because, in each Apollo mission, the descent engine exhaust plume has caused significant lunar surface erosion at LM altitudes below approximately 30 to 50 m. However, most surface disturbances seen at both the Surveyor and Apollo sites resulted in a darkening of the surface. This darkening appears to be caused by the destruction, by removal or covering, of a very thin (less than 1 mm) high albedo skin layer by darker sub-surface material. Such a disturbance can be seen in Fig. 17, which shows a panoramic camera photograph taken during the first extravehicular activity. It shows a dark path leading from the LM to the Apollo lunar surface experiments package (ALSEP) deployment site. This darkening is caused by a coating of subsurface soil kicked up by the lunar roving vehicle (Rover) wheels and by the astronaut walking next to the Rover en route to the ALSEP site and by the Rover wheels alone on the return trip.

Considering the above, one would predict that exhaust-induced erosion would destroy the thin, high albedo skin, leaving a dark halo surrounding the landing point, in direct contrast to what is observed. As detailed by Hinners and El-Baz (1972), the probable answer to the problem lies in a consideration of the lunar photometric function: a decrease in porosity will result in a photometric brightening of the surface.

It is suggested, therefore, that the bright halo surrounding the LM is a photometric effect caused by the compaction of the lunar soil under the influence of the dynamic pressure of the descent engine exhaust gases. The darkening of the lunar surface (seen

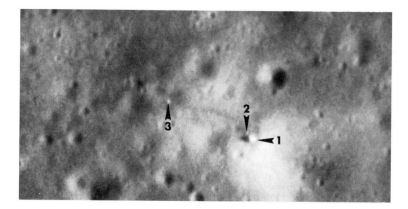

Fig. 17. Fifty times enlargement of panoramic camera frame 9430 of Apollo 15 landing site showing the LM with its prominent shadow (1) and the Rover which is parked near the northwest corner of the LM structure (2) and the ALSEP station (3). The tracks of the Rover to and from ALSEP are displayed by subtle surface darkening.

from orbital photographs, Fig. 17; and photographs taken on the surface, Fig. 18) along the Rover traverses is due to the destruction of the uppermost photometric layer.

From the soil standpoint, there is evidence of a coarser grain size in the Apollo 15 LM area than in adjacent undisturbed areas (McKay *et al.*, 1972). One would expect, therefore, that LM exhaust would both erode the uppermost photometric layer as well as compact the layers beneath it. The Apollo 16 sampling of this photometric layer, discussed above, should shed light on these relationships.

Fig. 18. Surface disturbances in the Apollo 15 landing site; Mt. Hadley is in the background. Astronaut footprints are bright, probably due to compaction; the areas where the dust had spread over around the footprints are dark. A similar situation is portrayed by the Rover tracks in the foreground (Apollo 15 Hasselblad camera frame 11793).

CONCLUSIONS

Several new features of probable volcanic origin were detected in the Apollo 15 photographs: (1) cinder cones in the Taurus-Littrow region that may have brought to the surface pyroclastic fragments from deep within the moon in the post-mare period of lunar surface history; (2) a D-shaped structure which appears to be a collapsed caldera with light-colored floor units with smooth and relatively younger domes. The feature may be among the youngest lunar formations of volcanic origin; (3) a lava lake on the lunar farside near the crater Perelman with lava marks and evidence of subsidence due to drainage into prominent fissures that display drainage channels; (4) domes near Rima Schroedinger I, which range in size to as much as 40 km in diameter; and (5) numerous lava flows in western Mare Imbrium, some of which cross mare ridges.

Other features include two lineated units that are interpreted as landslides. The larger (approximately 80 km) originated at the northwestern rim of the farside crater Tsiolkovsky; and the smaller (approximately 5 km) apparently originated on the steep slope of a 2 km high massif unit of the Taurus Mountains in the Apollo 17 landing site area. The landslide in the Taurus-Littrow region will provide an excellent opportunity to sample the massif materials during the Apollo 17 mission.

Unusual, light-colored markings in three mare regions (Mare Inginii on the farside, Mare Marginis on the eastern limb, and western Oceanus Procellarum) are interpreted as indications of chemical differences. These swirls may have been caused by alteration of the material by escaping gases at the antipodal areas of large and relatively young impact sites. Other albedo markings seen for the first time from orbit include: brightening of the surface area beneath the LM probably due to compaction; and darkening of the areas around the Rover tracks, probably due to destruction of the thin (less than 1 mm) photometric layer of the lunar surface.

Geologic interpretation of orbital photography is best done by correlation with remotely sensed data. Examples are given of successful correlation with data obtained by the geochemical sensors from orbit, e.g., the x-ray and γ-ray data. Photographic results may also be correlated with earth-based remote sensing; examples were given of good correlation between photogeologic interpretations and earth-based multi-spectral and radar studies. The ultimate value of photogeologic interpretation of the returned data is the extrapolation of knowledge gained by direct sampling to larger areas of the moon.

REFERENCES

Adler I. Trombka J. Gerard J. Lowman P. Yin L. and Blodgett H. (1972) Preliminary results from the S-161 x-ray fluorescence experiment (abstract). In *Lunar Science—III* (editor C. Watkins), pp. 4-6, Lunar Science Institute Contr. No. 88.

Carr M. H. (1966) Geologic map of the Mare Serenitatis region of the moon. U.S. Geol. Surv., Misc. Geol. Inv., Map I-489.

Carr M. H. (1972) Sketch map of the region around candidate Littrow Apollo landing sites. In *Apollo 15 Preliminary Science Report*, NASA SP-289, pp. 25-63–25-65.

Doyle F. J. (1972) Photogrammetric analysis of Apollo 15 records. In *Apollo 15 Preliminary Science Report*, NASA SP-289, pp. 25-27–25-36.

El Baz F. (1971) Light colored swirls in the lunar maria. *Trans. Amer. Geophys. Union* **52**, No. 11, 856.

El-Baz F. (1972a) Geologic conclusions from Apollo 15 photography (abstract). In *Lunar Science—III* (editor C. Watkins), pp. 214–216, Lunar Science Institute Contr. No. 88.

El-Baz F. (1972b) The cinder field of the Taurus Mountains. In *Apollo 15 Preliminary Science Report*, NASA SP-289, pp. 25-66–25-71.

Hinners N. W. and El-Baz F. (1972) Surface disturbances at the Apollo 15 landing site. In *Apollo 15 Preliminary Science Report*, NASA SP-289, pp. 25-50–25-53.

Latham G. Ewing M. Press F. Sutton G. Dorman J. Nakamura Y. Toksoz N. Lammlein D. and Duennebier F. (1972) Moon quakes and lunar tectonism—results from the Apollo passive seismic experiment (abstract). In *Lunar Science—III* (editor C. Watkins), pp. 478–479, Lunar Science Institute Contr. No. 88.

McKay D. S. Clanton U. S. Heiken G. H. Morrison D. A. and Taylor R. M. (1972) Vapor phase crystallization in Apollo 14 breccias and size analysis of Apollo 14 soils (abstract). In *Lunar Science—III* (editor C. Watkins), pp. 529–531, Lunar Science Institute Contr. No. 88.

Metzger A. E. Trombka J. I. Peterson L. E. Reedy R. C. and Arnold J. R. (1972) A first look at the lunar orbital gamma ray data (abstract). In *Lunar Science—III* (editor C. Watkins), pp. 540–541, Lunar Science Institute Contr. No. 88.

Pohn H. A. and Wildey R. L. (1970) Photoelectric-photographic study of the normal albedo of the moon. U.S. Geol. Surv., Prof. Pap. 599-E.

Strom R. G. (1965) Map, in Interpretation of Ranger VII Records. Calif. Inst. of Tech. JPL-TR-32-700, p. 32.

Swann G. A. Bailey N. G. Batson R. M. Freeman V. L. Hait M. H. Head J. W. Holt E. A. Howard K. A. Irwin J. B. Larson K. B. Muehlberger W. R. Reed V. S. Rennilson J. J. Schaber G. G. Scott D. R. Silver L. T. Sutton R. L. Ulrich G. H. Wilshire H. G. and Wolfe E. W. (1972) Preliminary geologic investigation of the Apollo 15 landing site. In *Apollo 15 Preliminary Science Report*, NASA SP-289, pp. 5-1–5-112.

Whitaker E. A. (1972a) An unusual mare feature. In *Apollo 15 Preliminary Science Report*, NASA SP-289, pp. 25-84–25-85.

Whitaker E. A. (1972b) Mare Imbrium lava flows and their relationship to color boundaries. In *Apollo 15 Preliminary Science Report*, NASA SP-289, pp. 25-83–25-84.

Wilhelms D. E. and McCauley J. F. (1971) Geologic map of the near side of the moon. U.S. Geol. Surv., Misc. Geol. Inv., Map I-703.

Proceedings of the Third Lunar Science Conference
(Supplement 3, *Geochimica et Cosmochimica Acta*)
Vol. 1, pp. 63–83
The M.I.T. Press, 1972

Significant results from Apollo 14 lunar orbital photography

Farouk El-Baz

Bell Telephone Laboratories, Washington, D.C. 20024

S. A. Roosa

NASA, Manned Spacecraft Center, Houston, Texas 77058

Abstract—Apollo 14 obtained 950 photographs from lunar orbit using the Hasselblad and Hycon cameras. The photographs reveal a number of new geologic features as well as previously unrecognized details of the morphology, structure, and stratigraphy of lunar surface units. The primary result is the verification of the extensive role of volcanism in the formation and modification of the lunar highlands, especially on the farside.

Terra volcanism appears to be manifest in the formation of (1) constructional units of hilly and furrowed materials of regional extent as in the Kant Plateau in the central nearside highlands and northwest of the crater Pasteur near the eastern limb of the moon; (2) somewhat viscous lava flows and pools associated with fracture systems and/or what appear to be volcanic craters; (3) craters, crater chains, and irregular depressions, particularly on the lunar farside.

The first photographs of a flow channel, a leveed sinuous rille that apparently originated by lava flowage on the surface, were obtained by Apollo 14. Another first is a high resolution photograph of the interior of what appears to be the youngest lunar crater yet photographed in the 20–40 km size range. Also, photographic sequences were made at zero phase conditions and at low sun elevation angles, near the terminator. The latter conditions are most suitable for studying small-scale topographic variations of lunar surface units.

Introduction

On Apollo 14, the 70 mm Hasselblad camera (with 80 mm, 250 mm, and 500 mm lenses) was utilized to obtain a total of 758 photographs from lunar orbit. In addition to this equipment, the lunar topographic camera (LTC) was carried for the first time to obtain high resolution stereo strip photography of the lunar surface, particularly of the Apollo 16 Descartes landing site. The LTC is a Hycon KA-74 camera which has a 450 mm (18-inch) focal length f/4 lens. It has an automatic rocking mount that compensates for forward motion of the spacecraft over the surface. The Hycon camera was operated during the fourth orbit when the Command/Service Module (CSM) was approximately 18 km above the lunar surface. A malfunction was noted by the Command Module pilot (CMP) during acquisition of the strip; it resulted in overexposure of the last half of the film magazine. The correctly exposed 192 frames will be discussed in this paper.

The 950 orbital photographs returned by Apollo 14 reveal new geologic features as well as previously unrecognized details of the morphology, structure, and stratigraphy of the overflown lunar surface units. The most significant results of Apollo 14 photography are the data that pertain to the role of volcanism in the formation and modification of the lunar highlands, particularly on the farside. This role has been for a long time a matter of controversy. The relatively younger volcanic features in the maria generally are well preserved, whereas in the highlands most landforms are somewhat degraded. However, in a few cases, geomorphology and spatial relationships identify certain highlands features as being of volcanic origin.

Terra volcanism has been described in several areas on the nearside of the moon. Two regional volcanic units, west of Mare Nectaris and of Mare Humorum, are described by Wilhelms and McCauley (1971) and Trask and McCauley (1972). In addition to these units, plains-forming materials and deposits that are associated with some craters and rilles mantle the lunar highlands. These also are attributed to volcanism as summarized in Mutch (1970) and Wilhelms (1970). Landforms of probable volcanic origin also are quite prevalent on the lunar farside. Some were revealed for the first time in Lunar Orbiter photography (Kosofsky and El-Baz, 1969). Others were depicted for the first time by Apollo 8 orbital photography; for example, flow scarps and cones (El-Baz and Wilshire, 1969).

PHOTOGRAPHIC RESULTS

To test some of the theories regarding farside upland volcanism, photographic strips were planned for the Apollo mission 14. The authors previously have published summaries of the significant results (Roosa and El-Baz, 1971; El-Baz and Roosa, 1971; El-Baz, 1971; and El-Baz and Head, 1971). We intend to present in detail some of the data with special emphasis on probable volcanic features. The latter will be discussed in the order in which they appear relative to the ground tracks. As illustrated in Fig. 1, these are:

1. Two conjugate craters west of the farside crater Chaplygin (146°E, 5°S).
2. Volcanic flows within a farside crater called "the bright one" (123°E, 5°S).
3. A probable lava lake and flow scarps associated with the crater King (121°E, 6°N).
4. Furrowed and grooved terra west of the crater Pasteur (100°E, 8°S).
5. Furrowed and hilly terra in the Apollo 16 landing site area (16°E, 9°S).
6. The Davy crater chain (6°W, 11°S).
7. A sinuous flow channel near the crater Lansberg (25°W, 3°S).
8. Flow scarps and rimless depressions near the crater Kunowsky (32°W, 3°N).

Conjugate craters west of Chaplygin

Two unnamed conjugate craters occur approximately 40 km from the north-western rim of the farside crater Chaplygin and display unique characteristics. The crater to the north is approximately 35 km in diameter, and the one to the south is 25 km from rim to rim (Fig. 2). Both craters display fairly smooth rims including those portions at the juncture between the two craters. The craters are shallow and the floors are contiguous. The floor material is distinctly different from that in neighboring craters. It is somewhat darker than the surrounding material and displays fractures that appear to be endogenetic in origin. The fractures in the crater to the north are concentric with the rim, whereas those in the southern crater form a network that gives a "turtleback" appearance.

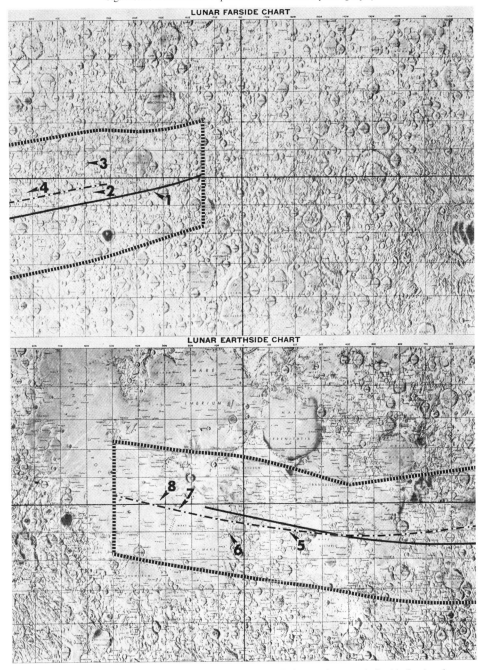

Fig. 1. Lunar orbital groundtracks of the Apollo 14 mission on the moon's farside (top) and nearside (bottom). The solid and dashed lines represent the tracks of the sunlit portions of the first and last (34th) orbits respectively. The outer envelope represents the spacecraft horizons, maximum area of visibility from the Command Module windows. The features discussed in this paper are numbered: (1) conjugate craters west of Chaplygin, (2) "the bright one," (3) the crater King, (4) terra units northwest of Pasteur, (5) the Kant Plateau and the Apollo 16 Descartes landing site, (6) Davy crater chain, (7) sinuous channel near Lansberg, and (8) flow scarps near Kunowsky in Oceanus Procellarum.

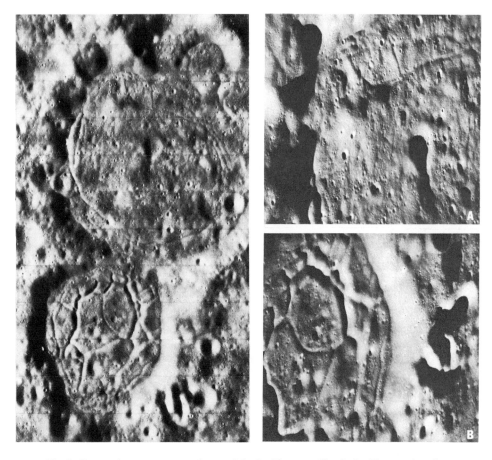

Fig. 2. Two conjugate craters northwest of the farside crater Chaplygin. The sun elevation angle is about 18°, the larger crater is about 35 km in diameter (Lunar Orbiter I photograph H-115).

Fig. 3. Two Apollo 14 high resolution frames of portions of the craters shown in Fig. 2. (A) The northwestern quadrant of the larger crater showing a central rimless depression (frame 9959). (B) The eastern half of the smaller crater displaying V-shaped, smooth cracks which give a "turtleback" appearance; sun elevation angle is about 6° (frame 9954).

The Apollo 14 high resolution photographs of these two conjugate craters were taken at low sun elevation angles (about 6°), resulting in the enhancement of contrast and a clear display of the small-scale features. As shown in Fig. 3, the floor of the larger crater displays an elongate depression, approximately 5 km in length, and a number of rimless and low-rimmed craters. The floor of the smaller crater is dissected by a network of fractures or rilles. These display V-shaped cross sections and fairly smooth walls.

The aforementioned characteristics lead us to the conclusion that the floor material is made of volcanic flows, with fractures resulting from shrinkage produced during the cooling of lava.

Flows within "the bright one"

One of the craters that dominates the scene on the lunar farside at high sun illumination angles is only 35 km in diameter. Its bright halo is spread over approximately 150 km (Fig. 4). For this reason, and lacking a formal name, it has been informally called "the bright one." Its rays as well as those of a neighboring, somewhat smaller crater approximately 50 km to the northeast, join the rays from the crater Giordano Bruno to the north. This bright area of the lunar farside (at high sun) erroneously was interpreted as the Soviet Mountains, a mountain range not related to a circular basin, but Apollo 8 photographs showed that the bright region is due mainly to crater rays (Whitaker, 1969).

The crater displays an irregular shape because of slumping of wall materials especially the western segment and the southeastern corner (Fig. 5). The rim crest is very sharp and the many, wavy crenulations suggest structurally controlled slumping. As shown in Fig. 6, the interior of the crater is very hummocky. The lack of a flat floor is striking and the excessive terracing is unique. Prominent flow scarps can be seen, and all flow fronts are not lobate: one major front is unusually straight. Small fractures, apparently caused by drag, can be seen in the main flow in the upper right portion of the frame. These are also traced for clarity in Fig. 6. The unusually large quantity of blocks on the terraces and the virtual lack of craters suggest that this is an extremely young crater. These characteristics, when coupled with the gross morphology, provide evidence that "the bright one" is the youngest lunar crater of the 20–40 km size range ever to be photographed on the moon.

Lava lake and flow scarps of the crater King

The crater King (formerly identified by the International Astronomical Union no. 211) is approximately 75 km in diameter and displays a generally round, partly crenulated rim.

The crater is situated in as yet undivided highland materials in the general area previously known as the Soviet Mountains. The crater exhibits a raised, wavy, and sculptured rim and terraced interior walls. The morphology suggests, although not unequivocally, an impact origin. The picture is complicated, however, by (1) the lack of ejecta and extensive fields of secondary craters; and (2) the presence of probable volcanic features within and in the immediate vicinity of the crater.

The crater is a few kilometers deep, and the depth of the floor in relation to the rim crest varies with the amount of fill. The crater wall displays as many as six terraces, and the highest terrace is steeper than most. This feature is common to craters of similar size. The floor of the crater displays a prominent central peak that is forked. It forms a unique Y-shape (Fig. 7), with the right arm trending nearly due north. The central peak displays a ropy appearance, especially along the southern part. This appears to be due to a fracture system that bisects the massive peak. Two bright tabular units appear to the north of the crater as if they are extensions of the two arms of the Y-shaped peaks. The nature of these units is not fully understood and must await further examination and photography on Apollo mission 16.

Fig. 4. View of the 150 km bright halo of the crater informally named "the bright one"; the dark-floored crater to the south is Tsiolkovsky (Apollo 8 frame 2506).

Fig. 5. Apollo 14 photograph of "the bright one," which is a 35 km in diameter crater with numerous terraces and evidence of wall slumping (frame 9671). A detail of the flow unit in the eastern portion of the floor is shown in Fig. 6.

Fig. 6. High resolution (500 mm lens) photograph of the hummocky interior and flow unit of the crater shown in Fig. 5 (left, Apollo 14 frame 9975). As shown in the drawing to the right, the direction of flow appears to have been mainly to the southwest (arrows); small feeder channels (arrowheads) seem to have added material to the main flow unit from topographically higher wall and floor masses (dotted). The fractures within the flow unit (thin lines and dashes) indicate a liquid flow (probably lava), that terminates at a sharp and unusually straight scarp (black).

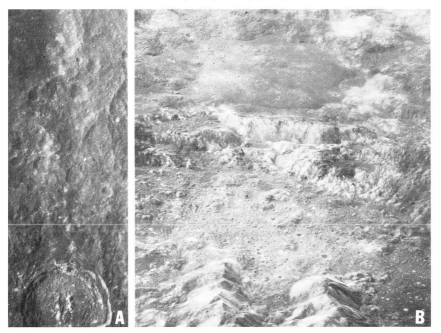

Fig. 7. Photographs of the unusual farside crater King. (A) View of the Y-shaped central peak and the light-colored tabular bodies that extend to the north of the crater (Apollo 13 frame 8664). (B) Oblique view showing the smooth lava pool on the north rim of the crater and four different color-tones of the wall materials (Apollo 10 frame 4351).

On the basis of Apollo 10 photography, four different types of materials had already been noted in the walls of the crater (El-Baz, 1969). These were distinguished by color, texture, and morphology (Table 1). Also, some wall-like tabular bodies of low albedo materials that cross the walls, rim, and floor of the crater were described as igneous intrusions. The interpretation of these features as probable dikes was used as supportive indications of the heterogeneity of lunar materials, as well as the plausibility of intrusive igneous activity on the moon (El-Baz, 1969).

The Apollo 14 high resolution photographs constitute a useful complement to existing imagery of the crater and its environs. The high obliquity of the photographs (approximately 60°) enhances details that were unnoticeable previously. A mosaic of the oblique photographic strip is shown in Fig. 8. This view provides additional details of the dark pool on the north rim of the crater. The flat, relatively smooth appearance of this material and the collapsed sinuous depressions within the unit indicate that the material originated as a lava lake. Some flow scarps are noticeable within and around the borders of this unit.

An additional flow scarp can be seen for the first time to the southeast of the crater. The nature of the lobate flow front is made obvious by the obliquity of the view. This particular scarp, although it may have been a landslide, is more reminiscent of lava flow fronts. The numerous probable volcanic features suggest that the crater may be endogenetic in origin. Apollo 16 should provide high resolution photographs of the crater and its surroundings.

Furrowed and grooved terra west of Pasteur

Lunar terra volcanism is manifest in the form of craters, plains-forming units, domical and hilly materials, and furrowed and grooved units. The latter are represented by elongate vents, crater chains, and irregular structures with discernible rim deposits. Several small elongate grooves and depressions were noted within the hilly terra northwest of the crater Pasteur. These grooves and furrows range in size between

Table 1. Geomorphological units in the northern wall of the crater King.

	Unit	Characteristics
A.	Western segment	Low albedo material which appears to mantle four terraces. The mantling material displays characteristics similar to the dark material which forms the pool north of the crater. This unit exhibits indications of massive downslope movement.
B.	Middle segment	Very light colored, almost white material which is segmented into domical hills that are separated by shallow furrows.
C.	Central part of middle segment	Very dark, almost black material which occurs within the uppermost terrace and atop the hills of the light-colored material.
D.	Eastern segment	Medium albedo unit with a larger number of craters and numerous small-scale lineaments. This finely textured unit is more representative of the rest of the crater wall than other units.

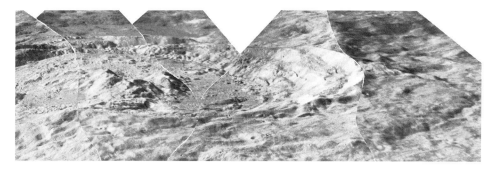

Fig. 8. Highly oblique view (about 60°) of the crater King showing a probable lava lake on the north rim and a large flow unit that encircles the crater, especially prominent to the right side (mosaic of Apollo 14 frames 9967 to 9971).

5 and 30 km as revealed by Lunar Orbiter photography (hundreds of meters resolution). The exact morphology of these grooves and furrows was, however, not discernible and their origin(s) not known.

It was planned on Apollo 14 to obtain photographs of the area using the Hasselblad camera with the 500 mm lens. These high resolution stereo photographs reveal that the area is strewn with elongate features of probable volcanic origin. The unit in which the grooves and probable volcanic vents are located is somewhat hilly and displays characteristics reminiscent of the Kant Plateau materials mapped by Milton (1968). The Apollo 14 Hasselblad/80 mm stereo strip photography also shows similar units in the highlands farther east. Most of the grooves display a raised rim, and many have the appearance of conjugate crater chains (Fig. 9).

Terrestrial analogs to these features occur north of Flagstaff, Arizona (for example Janus Crater, Sprul Crater, and many fissure cones east of SP flow). Although

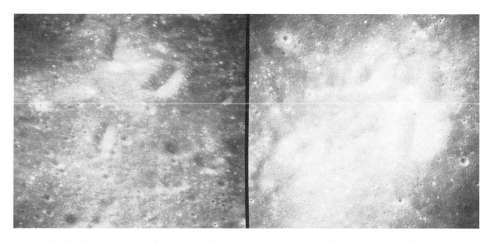

Fig. 9. Elongate vents, furrows, and grooves in hilly terra units northwest of the crater Pasteur (Apollo 14 frame 9993, left, and 9997, right).

the nature of the lunar features is not known, the morphology strongly suggests a volcanic origin.

The larger of the furrow-like craters and most smaller grooves are oriented in a north-south direction. It previously was suggested (El-Baz, 1971) that the furrows are structurally controlled, and that a tectonic belt with north-south fractures exists in this area of the moon.

Recognition of these probable volcanic features in the farside highlands is significant. It indicates that terra volcanism is a moon-wide process and its products are not restricted to the two areas west of Nectaris and Humorum, but may constitute large areas of the farside highlands. Probable volcanic units will be sampled on Apollo mission 16 and this will allow extrapolation of surface data to a not well-understood part of the moon—the farside.

Volcanic highlands of the Kant Plateau

One of the prime objectives of the Apollo 14 mission was to obtain photographs of the Apollo 16 Descartes landing site. Photography of the Kant Plateau materials was to be obtained utilizing the Hycon camera. As discussed above, the camera malfunctioned early during the mission and its coverage stopped about 70 km east of the landing site. The Hasselblad camera with the 500 mm lens was utilized in obtaining the required stereo photographs by rotating the spacecraft. Data obtained by both photographic systems will be briefly summarized and their geologic significance will be pointed out.

The Hycon camera's high resolution stereo photography (about 60% sidelap) covers a strip approximately 5 km wide on the lunar surface from 28.2° E, 11.3° S to 18.7° E, 8.3° S. The contrast is somewhat low because the sun elevation angle varied between 45° at the beginning of the strip to 35° at the end of the properly exposed frames.

The highland terrain covered by the photographs, at about 2 m resolution, includes two major units: Theophilus Crater and Kant Plateau materials. Theophilus materials are represented by the walls, floor, and the western ejecta blanket of the crater. The ejecta blanket may in turn be divided into a hummocky unit near the rim crest and a smoother facies farther away. The second major unit is that of the Kant Plateau materials, which include plains-forming units and subdued terra ridges. Figure 10 shows examples of these major units.

Perhaps the most significant contribution of the Hycon strip photography is that it provides necessary information for detailed crater studies. It provides rim-to-rim coverage of the crater Theophilus, which is about 100 km in diameter. This coverage will allow detailed studies of all units related to this large crater that is probably of impact origin and is Copernican in age.

Photogrammetric reduction by MSC of the Apollo 14 data confirms the expected topographic rise of the Kant Plateau. However, the actual measured height of 6 km above the nearest mare surface (Fig. 11) was surprising. The area is not associated with a gravity anomaly. It probably represents a segment of the lunar crust that is thicker than that beneath the rest of the nearside highlands. Constructional volcanic

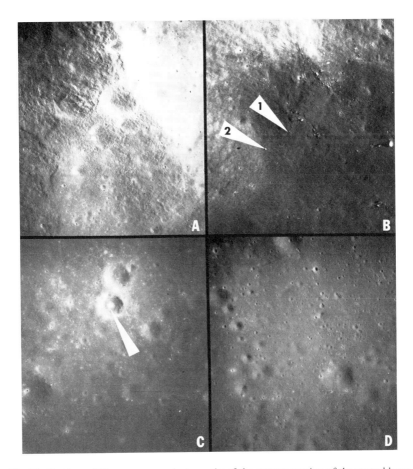

Fig. 10. Portions of Hycon camera photographs of the eastern portion of the central lunar highlands; the width of the photographs is about 3 km. (A) Mass wasting along the eastern rim of the crater Theophilus (frame 10437). (B) Large blocks on the northeastern wall of Theophilus; note track (point 1) of boulder (point 2) that rolled down slope (frame 10448, after Scott *et al.*, 1971). (C) Prominent, bright central peak in small crater on the floor of Theophilus (frame 10454, after Scott *et al.*, 1971). (D) Subdued craters in the gently undulating plains near Kant B (frame 10620).

rocks further increase the topographic expression of this ridge in the Kant Plateau area. The bright material on the north rim of the crater Descartes is anomalous to both radar and IR. It is a very rugged unit with practically no level surfaces.

The Apollo 16 landing site is targeted on the western border of the Kant Plateau approximately 300 km west-northwest of the crater Theophilus.

There are four geomorphologic units in the Descartes area. The characteristics of these units are well-displayed in the Apollo 14 photography of the site (Fig. 12). Relative to the proposed landing point these units are to the north (sculptured terra), south (furrowed terra), east (hilly terra), and west (cratered plains), as shown in the geological sketch map (Fig. 13).

Farouk El-Baz and S. A. Roosa

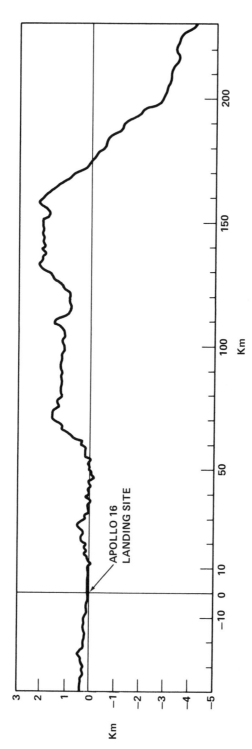

Fig. 11. Terrain profile of the Kant Plateau based on photogrammetric data reduction of the Apollo 14/80 mm stereo strip photography. The scale is in kilometers and the reference point is the Apollo 16 target landing point. Note that the Kant Plateau is as much as 6 km higher than the ejecta deposits of Theophilus to the east (data compiled by Mapping Sciences Branch, NASA-MSC).

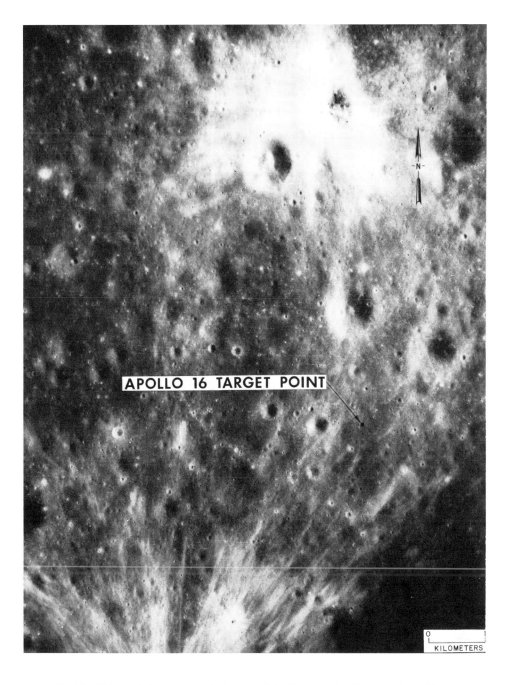

Fig. 12. High resolution, near vertical, view of the Descartes landing site taken with the Hasselblad camera and 500 mm lens. The sun elevation angle is about 56°. Note the bright halo and somewhat subdued rays of the 850 m North Ray Crater. Rays from South Ray Crater can be seen at the lower edge of the photograph (frame 9521).

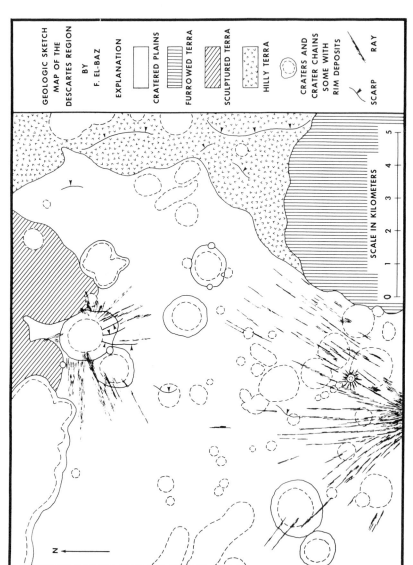

Fig. 13. Geologic sketch map of the Descartes landing site region. The "cratered plains" unit is a relatively smooth basin fill material which displays characteristics similar to the Cayley formation; the "furrowed terra" appear to be viscous flows that are continuous with the unit that mantles the crater Descartes to the south and displays north-south lineaments; the "sculptured terra" is a unit controlled by northwest trending structures that may be part of the Imbrium sculpture system; and the "hilly terra" is a unit similar to the "cratered plains" but somewhat darker (probably also younger) and more undulating. It also has numerous northwest trending furrows and crater chains and displays several westward facing scarps.

"Sculptured terra" are represented by a large hill, which is named "Smoky Mountain" by the Apollo 16 crew, as well as smaller spurs that are interpreted in two ways: (1) because the continuation of the unit to the north apparently is controlled by Imbrium sculpture, it is probably made of constructional volcanics along the Imbrium fracture system; (2) because the unit may have been part of an old crater rim, it probably is mantled only thinly by the young volcanic units. The two interpretations naturally reflect on the significance of the samples to be brought back by Apollo 16. In the vicinity, and to the west of this unit is North Ray Crater, one of the two relatively young, bright-haloed craters that are the prominent landmarks of the site.

The "furrowed terra" are made of grooved and hilly topography with prominent N-S structural control. This unit is interpreted as constructional volcanic highlands, and physically is connected to the bright material that obliterates and mantles the crater Descartes to the south. The part of the unit that is closest to the landing site is named "Stone Mountain" by the Apollo 16 crew.

The "hilly terra" is a unit characterized by low albedo, numerous westward facing scarps (Fig. 13), and a lower density of craters than in the surrounding plains. It probably is a volcanic fill of relatively young age. The "cratered plains" is a similar unit but more typical of the upland basin fill, or Cayley formation, a widespread unit in the central highlands. The distribution of crater sizes and ages within this unit will allow good sampling of the stratigraphy of the area. Although the units in this region are described somewhat differently by Elston *et al.* (1972) and Muehlberger *et al.* (1972), the basic geologic structure and stratigraphic concepts of the two descriptions are similar.

The most important objective of the Apollo 16 mission is to sample and investigate both the plains material, which represents 7% of the lunar nearside, and the furrowed and sculptured materials, which represent 4% of the lunar nearside highlands. The samples will allow study of the composition, age and history, and evolution of a large segment of the lunar highlands. Apollo 14 photographs of the area were fundamental to selection of this important site and planning of the surface mission.

Craters of Rima Davy I

Rima Davy I, northwest of the crater Alphonsus, is not a rille but rather a chain of some 30 small craters along a line, much like beads on a string. Most of the craters are about 1 km in diameter with a few larger ones, extending a total length of approximately 65 km. Nearly half of the craters are in the floor of Davy Y, a large old, filled crater; the rest extend across the highland ridge that marks the western edge of the central lunar highlands. The Davy crater chain is believed to be volcanic in origin. Low crater rims and characteristically smooth rim deposits suggest that the craters were formed by explosive gaseous eruptions, analogs of some terrestrial crater chains.

Prior to the flight of Apollo 14, the Davy crater chain was considered a candidate Apollo landing site (Beattie and El-Baz, 1970). The objective would have been to sample the materials around the craters and from the nearby highlands, including fragments from the subsurface that should reveal the composition of the lunar materials at great depths. Photographs (see, e.g., Fig. 14) and observations of the crater chain were made from lunar orbit. No particular distribution pattern of rim

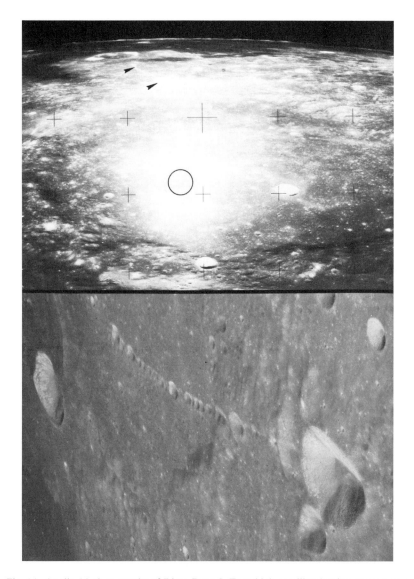

Fig. 14. Apollo 14 photographs of Rima Davy I. Top: high sun illumination at near zero phase conditions shows that the craters in the chain (arrowheads) are very bright indicating surface roughness. The circle surrounds the zero phase point in the floor of the crater Ptolemaeus (frame 10255). Bottom: high resolution (500 mm lens) photograph of the crater chain that ends to the right with the irregular Davy G Crater. The crater chain crosses both highland materials and the basin fill of the old crater Davy Y (frame 10103).

deposits was found. The raised rims of the craters appear to be the same—no distinguishable difference between larger and smaller craters on the one hand, and between sharper and more subdued craters on the other hand. Also, no extensive ejecta blankets or dark halos were observed around the craters.

Sinuous flow channel near Lansberg

Three Apollo 14 color photographs (10120 to 10122) reveal a sinuous channel that has not been recognized in previous photography. This flow channel is a gently meandering rille that displays distinct continuous levees extending along the entire length although they are more obvious along the northern part (Fig. 15). As discussed by Scott *et al.* (1971), only one sinuous rille with a distinct, but only partly leveed, channel in the Marius Hills region previously has been described.

The channel is about 15 km long and 700 m wide. It begins and ends in mare material that embays the southeast rim of the crater Lansberg. Individual levees are only about 100 m across, as discussed by Scott *et al.* (1971). This contrasts sharply with the relatively broad ridges commonly associated with terrestrial collapsed lava tubes. The levees of this channel resemble more closely those associated with some stream channels, lava channels, and certain types of highly fluid debris and mud flows (Fig. 15).

The morphological characteristics of this rille indicate that it must have originated as a result of surface fluid flow in open channels with the fluid medium probably being lava. The significance of this new feature is bifold: (1) it indicates that sinuous rilles are not all the product of the same process—some, with levees or overbank deposits, are clearly the result of surface fluid flow; and (2) it suggests that other leveed rilles may exist on the lunar surface but have not been detected in existing photographs due to inadequate resolution or lighting problems.

Mare flows near the crater Kunowsky

It has been shown by Kuiper *et al.* (1967) that low sun illumination accentuates topographic variations on the lunar surface. Some of the most revealing telescopic photographs were taken at low sun elevation angles. However, historically it has been

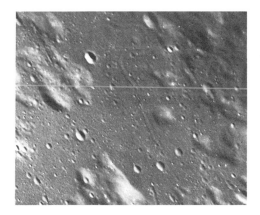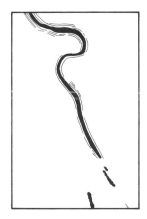

Fig. 15. Sinuous rille with leveed channels in the mare materials southeast of the crater Lansberg. As shown in the drawing to the right the multiple levees (thin lines) indicate surface flow of lava to form the channel (black). (Left, Apollo 14 frame 10120.)

difficult to obtain photographs from lunar orbit any nearer to the terminator than approximately 10° without severe underexposure.

Apollo 14 obtained the first near-terminator photographs by utilizing the Hasselblad camera with the 80 mm lens and a very high speed, black-and-white film (Kodak 2485). The technical details are treated by Head and Lloyd (1971), and the results are extremely satisfactory.

The Apollo 14 CMP took photographs of an area in the south-central portion of Oceanus Procellarum in the vicinity of the crater Kunowsky (Fig. 16). The photographs reveal details that are not discernible in higher sun photography of about the same resolution. As illustrated in Fig. 17, numerous flow fronts and irregular ridges are clearly visible. At this low sun elevation angle very small changes in topography are brought out, and small scale undulations are emphasized.

In addition to the flow fronts and scarps and the gentle swales and ridges, one can also distinguish the detailed characteristics of small craters in the area. Most small craters in the area display a raised, somewhat hummocky rim—suggesting impact origin. One crater, however, is distinctly rimless, and probably is a collapse depression.

The Apollo 14 near-terminator photographs proved the utility of this type of data for detailed photogeologic interpretations of lunar surface features. The success of the experiment prompted planning of several near-terminator sequences on Apollo 15, 16, and 17.

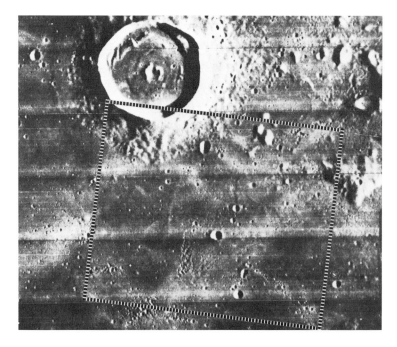

Fig. 16. Mare material south of the crater Kunowsky in Oceanus Procellarum (Lunar Orbiter photograph H-133). Dashed lines mark area of Fig. 17.

Fig. 17. Apollo 14/80 mm lens, near-terminator view of the mare southeast of Kunowsky. The very low sun elevation angle (0.5°) accentuates small variations in topographic relief as indicated by the numerous flow fronts and scarps that are not detectable in Fig. 17 (there is no significant difference in the resolution of both photographs). Distinctions between craters with raised rims and rimless depressions could be made easily in this photograph (frame 10377).

SUMMARY AND CONCLUSIONS

The 950 photographs taken from lunar orbit on Apollo 14 reveal a plethora of new geologic features as well as previously unrecognized details pertaining to the morphology, structure, and stratigraphy of lunar surface formations. Approximately one-fifth of the photographs were taken at an average of 2 m resolution with the Hycon KA-74 camera (LTC), prior to its malfunction during the flight. The remainder were taken with the Hasselblad camera with the 80 mm, 250 mm, and 500 mm lenses at resolutions that ranged between 10 and 60 m.

Obliquity of a great number of views enhances their use in photogeologic interpretations. The photographs strengthen and in some cases refine previous interpretations of the origin and stratigraphic relationships of geologic map units on the moon's nearside. The data also will be valuable in mapping presently undivided surface units on the lunar farside.

The Apollo 14 photography provided especially significant data on numerous volcanic features: some were observed for the first time; others that were photographed more suitably than before. The more important features discussed in this paper include (1) what appear to be cooling cracks that form a "turtleback"-like fracture pattern on the floor of two conjugate craters west of Chaplygin on the lunar farside. The craters themselves are thought to be volcanic in origin, and probably represent the largest structures of this type on the moon; (2) flow scarps and what appear to be drag fractures in lava flows within a bright-haloed, 35 km crater on the farside named "the bright one." The crater itself is very sharp and probably is the youngest lunar crater photographed in the 20–40 km size range. (3) A dark, flat, and smooth deposit on the north rim of the farside crater King, which is interpreted as a lava lake, also a distinct flow front surrounding the same crater. Although the origin of the crater is not yet well understood, a number of volcanic features suggest an endogenetic origin. (4) Numerous grooves, elongate vents, and crater chains in the terra units northwest of the crater Pasteur, near the eastern limb of the moon. These features apparently are controlled by north-south trending structures. (5) Four distinct units of probable volcanic origin on the Apollo 16 Descartes landing site; (6) rims and rim deposits of the volcanic craters that form Rima Davy I on the western edge of the central lunar highlands; (7) a sinuous rille in the mare material southeast of the crater Lansberg. The rille exhibits distinct levees and is interpreted as a channel produced by surface flowage of lava; and (8) flow scarps and a rimless depression in the mare material southeast of the crater Kunowsky, which was photographed at near-terminator conditions.

In summary, the primary result of Apollo 14 orbital photography is the recognition of the extensive role of volcanism in the formation and modification of lunar highlands, especially on the farside. This is particularly significant in view of the fact that the Apollo 16 mission to Descartes will sample highlands volcanic units. Surface knowledge of this segment of the central nearside will increase our confidence in extrapolating the data to larger segments of the lunar terra.

REFERENCES

Beattie D. A. and El-Baz F. (1970) Apollo landing sites. *Military Engineer* **62**, No. 410, 370–376.

El-Baz F. (1969) Lunar igneous intrusions. *Science* **167**, 49–50.

El-Baz F. (1971) Volcanic features in the farside highlands. In *Apollo 14 Preliminary Science Report*, NASA SP-272, 267–274.

El-Baz F. and Head J. W. (1971) Hycon photography of the central highlands. In *Apollo 14 Preliminary Science Report*, NASA SP-272, 283–290.

El-Baz F. and Roosa S. A. (1972) Apollo 14 in lunar orbit (abstract). In *Lunar Science—III* (editor C. Watkins), pp. 217–218, Lunar Science Institute Contr. No. 88.

El-Baz F. and Wilshire H. G. (1969) Possible volcanic features—Landforms. In *Analysis of Apollo 8 Photography and Visual Observations*, NASA SP-201, 32–33.

Elston D. P. Boudette E. L. Schafer J. P. Muehlberger W. R. Hodges C. A. and Milton D. J. (1972) Geology of the region of the Descartes (Apollo 16) site (abstract). In *Lunar Science—III* (editor C. Watkins), pp. 227–229, Lunar Science Institute Contr. No. 88.

Head J. W. and Lloyd D. D. (1971) Near-terminator photography. In *Apollo 14 Preliminary Science Report*, NASA SP-272, 297–300.

Kosofsky L. J. and El-Baz F. (1969) *The Moon as Viewed by Lunar Orbiter*. NASA SP-200, 152 pp.

Kuiper G. P. Whitaker E. A. Strom R. G. Fountain J. W. and Larsen S. M. (1967) Consolidated lunar atlas. Lunar and Planetary Laboratory, Univ. of Ariz., Contr. No. 4.

Milton D. J. (1968) Geologic map of the Theophilus quadrangle of the moon. U.S. Geol. Surv., Misc. Geol. Inv., Map I-546.

Muehlberger W. R. Boudette E. L. Elston D. P. Hodges C. A. and Milton D. J. (1972) Geologic setting of Apollo 16 landing site: Descartes (abstract). In *Lunar Science — III* (editor C. Watkins), pp. 227–229, Lunar Science Institute Contr. No. 88.

Mutch T. A. (1970) *Geology of the Moon: A Stratigraphic View*. Princeton Univ. Press, 324 pp.

Roosa S. A. and El-Baz F. (1971) Significant results of orbital photography and visual observations on Apollo 14 (abstract). Geol. Soc. Amer. Ann. Meet., Abstracts with Program, pp. 687–688.

Scott D. H. West M. N. Lucchitta B. K. and McCauley J. F. (1971) Preliminary geologic results from orbital photography. In *Apollo 14 Preliminary Science Report*, NASA SP-272, 274–283.

Trask N. J. and McCauley J. F. (1972) Differentiation and volcanism on the lunar highlands in relation to the Apollo 16 site. *Earth Planet. Sci. Lett.* **13**, 201–206.

Whitaker E. A. (1969) Comparison with Luna III photographs. In *Analysis of Apollo 8 Photography and Visual Observations*, NASA SP-201, 9–10.

Wilhelms D. E. (1970) Summary of lunar stratigraphy; telescopic observations. *Contr. to Astrogel. U.S. Geol. Surv. Prof. Pap. 599-F*, 47 pp.

Wilhelms D. E. and McCauley J. F. (1971) Geologic map of the near side of the moon. U.S. Geol. Surv., Misc. Geol. Inv., Map I-703.

Proceedings of the Third Lunar Science Conference
(Supplement 3, *Geochimica et Cosmochimica Acta*)
Vol. 1, pp. 85–104
The M.I.T. Press, 1972

Astronaut observations from lunar orbit
and their geologic significance

Farouk El-Baz

Bell Telephone Laboratories, Washington, D.C. 20024

A. M. Worden and V. D. Brand

NASA, Manned Spacecraft Center, Houston, Texas 77058

Abstract—To supplement orbital photography and other remotely sensed data, visual observations were made of 15 lunar surface targets during Apollo mission 15. The 30 m resolving power of the eye and its special sensitivities to subtle differences in texture and color-tone, when coupled with the interpretative powers of the brain provide a system of unmatched quality for lunar exploration. The extraordinary success of performing the task proves the outstanding capabilities of man and his use in spaceflight.

Among the significant results are (1) characterization of the floor material of Tsiolkovsky as no darker than the average (Eratosthenian) mare material, and interpretation of the lineated unit on the crater rim as a rock avalanche; (2) identification of layers on the wall of the crater Picard, which is probably volcanic in origin, (3) explanation of the ray-excluded zone of the crater Proclus as the result of structurally controlled ray shadowing; (4) observation of cinder cones in the Littrow area with dark haloes that probably are composed of pyroclastic deposits; and (5) recognition that the termini of numerous sinuous rilles in Oceanus Procellarum are flooded with younger mare materials that may have covered older terminal deposits.

Introduction

Visual observations by an astronaut from lunar orbit can be used in conjunction with photography and other remotely sensed data. The unique capabilities of the trained human observer are well known to all of us from everyday experience. It is not surprising that a highly trained observer can enhance significantly the scientific returns from the orbital flight portion of a lunar mission. Special characteristics of the human eye and interpretive powers of the brain give man an observing capability that cannot be matched by camera systems in certain respects.

The unaided eye has a resolving power of about one-sixtieth of a degree. This corresponds to recognition of an object 30 m in diameter from an orbital altitude of 110 km. The eye also detects patterns or lines formed by objects that are smaller or beyond the limit of visual resolution. (For example, there have been visual sightings of railroad beds and narrow roads from earth orbit.)

The eye is very sensitive to subtle differences in brightness, color tone, texture, and changes in topographic expression. Given sufficient time, the eye adapts to extreme lighting conditions. Visibility is possible in areas that would appear as hard shadows or bright "washout" regions in photographs. The eye also can see the lunar surface in earthshine.

The human eye has the capability to scan broadly over a wide field of view or to concentrate on small local areas as required. Visual targets can be studied continuously as the viewing angles and sun elevation angles are changing. In this way the observer can obtain impressions of a target that could never be obtained from one or a few photographs.

The trained astronaut can integrate all of the above mentioned information to make on the scene comparisons and interpretations. By making successive observations he can refine or verify his first impressions. He can also gain insight to be used for subsequent picture taking and photo interpretation.

PURPOSE

The objective of visual observations from lunar orbit was fulfilled for the first time, and with extraordinary· success, on Apollo mission 15. The purpose of the observations was to complement photographic and other remotely sensed data from lunar orbit. Geologic descriptions of the regional or local settings of particular lunar surface features and processes were obtained. These descriptions provided solutions to geologic problems that were difficult to solve by any other means. As explained above, they constitute a significant complement to instrument-gathered data and can be accomplished only by man.

Of the many interesting features overflown by Apollo 15, the first author selected 15 targets for detailed study (Table 1). The targets were discussed in detail with the second author (the Command Module Pilot, CMP) and the third author (the back-up CMP). Segments of time in the flight plan were allocated to the task, and photographs of the features and a list of the questions to be answered were carried onboard the spacecraft Endeavour. An example of these photographs and questions will be presented later in this report.

Observations were made from the Command Module windows without disturbing the operations of the Scientific Instrument Module. Data were acquired on all visual observation targets, including two not scheduled in the flight plan (Table 1). On farside passes, observations were recorded on the onboard tape recorder; and

Table 1. Apollo 15 visual observation targets.

| Target No. | Target designation | Spacecraft Longitude | |
		Start	Stop
1	Northwest of Tsiolkovsky	135°E	125°E
2	Crater Picard	60°E	55°E
3	Crater Proclus	53°E	45°E
4	Cauchy Rilles	51°E	40°E
5	Littrow Area	35°E	27°E
6	Crater Dawes	31°E	26°E
7	Sulpicius Gallus Rilles	18°E	9°E
8	Hadley-Apennines	10°E	2°E
9	Imbrium flows	15°W	24°W
10	Harbinger Mountains	35°W	44°W
11	Aristarchus Plateau (in earthshine)	46°W	52°W
11	Aristarchus Plateau	44°W	54°W
12 and 13	Post-transearth-injection views	not applicable	
14	Mare Ingenii (unscheduled)	170°E	157°E
15	Ibn Yunus area (unscheduled)	96°E	80°E

on nearside passes, observations were recorded by real-time voice communications with the Mission Control Center (NASA, 1971a). The third author communicated questions to the CMP during the mission, in real-time. After the mission, a debriefing was held during which the features studied and their geologic significance were discussed (NASA, 1971b). Excerpts from both the real-time comments and the debriefing statements (edited for clarity) will be given in the following discussion.

SIGNIFICANT RESULTS

We have recently summarized results of Apollo 15 observations (Worden and El-Baz, 1971; and El-Baz *et al.*, 1971). An attempt will be made here to present in detail some results and to point out their geological significance. Because of the limitations of space, only five visual observation targets will be discussed in the order of location relative to the mission groundtracks, from east to west (Fig. 1).

Dark floor and rim flow of Tsiolkovsky

Tsiolkovsky is a 200 km diameter crater that dominates a large area on the lunar farside. It is a relatively fresh crater, probably of Eratosthenian age, and its ejecta blanket is bright at high sun illumination angles. The crater displays a relatively flat, smooth floor, which appears in all photographs to be extremely dark. On the floor is a large central peak consisting of segments arranged in the shape of the letter "W."

The continuous ejecta blanket of Tsiolkovsky can be traced about one crater diameter out from the rim crest; secondary craters and crater chains originating from Tsiolkovsky are superposed on older surface units and formations for greater distances. On the northwestern rim of the crater is a lineated unit that displays characteristics of a flow of some sort.

Most of the aforementioned features of the crater Tsiolkovsky and its environs were the subject of visual observation target number 1 on Apollo 15 (Fig. 2). Excerpts from real-time CMP descriptions arranged in the sequence in which they appear in the air-to-ground transcript follow. The lunar revolution (LR) and ground-elapsed time (GET) in days, hours, minutes, and seconds will be provided for each entry (NASA, 1971a).

Tsiolkovsky is large enough that the CMP had the impression that the central peak was higher than the rim. The CMP also noted indications of layering on the central peaks.

"The central peak of Tsiolkovsky is very large; spur peaks on the south and east sides, being blocky on the north side. What appears to be some layering is visible on the south and west exposed scarps of the peak, dipping to the north at about 30°." (LR34/05: 23: 36: 31 GET).

The lineated segment of the northwestern rim of Tsiolkovsky, which has an area of approximately 80 km², was interpreted during the flight as a landslide or rock avalanche.

"I look at it every time I go by, and there is no question in my mind at all that it is a rock avalanche. It does have some interesting qualities about it. And it is

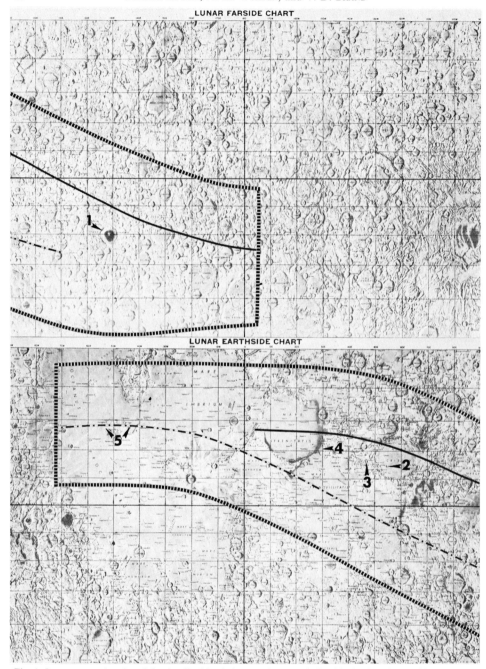

Fig. 1. Lunar groundtracks of the Apollo 15 mission on the moon's farside (top) and nearside (bottom). The solid and dashed-dotted lines represent the groundtracks of the first and last (75th) orbits respectively. The outer envelope marks the spacecraft horizon—the maximum area visible from the Command Module windows. Numbered are the features discussed in this paper: (1) the farside crater Tsiolkovsky, (2) the crater Picard in Mare Crisium, (3) the crater Proclus, (4) the Littrow region on the southeastern rim of Mare Serenitatis, and (5) the Harbinger Mountains-Aristarchus Plateau region.

somewhat hard for me to decipher right now, but it seems like the density of crater impacts in that slide is greater than in the surrounding terrain, even though the slide had to be emplaced on top of the surrounding terrain. May be it is just that the craters are fresher looking in that particular material, but there is no question about the lineaments being parallel to the direction of travel of the flow; all the characteristics of a rock avalanche." (LR33/05: 22: 26: 28 GET).

"On the west side, the rim is a very large, steep scarp; it continues from the basin floor to the rim crest in one large chunk. That scarp appears to define the limits of two fault zones that go through the rim of the crater Tsiolkovsky. But they are very distinct in the wall itself. The fault zones coincide with or occur in the same location as the southernmost edge of what appears to be a rock glacier. The latter has all the flow banding and the loping toes, which we consider characteristic of a rockslide. However, one feature about that slide that I mentioned before is that it has what looks like fairly fresh craters; in other words, a higher density of craters than on the surrounding floor of Fermi, although the Fermi floor looks much older; it is much smoother and more like the Cayley Formation." (LR34/05: 23: 37: 30 GET).

The mention during the flight of the higher frequency of craters on the rock avalanche or landslide on the northwestern rim of the crater Tsiolkovsky was the one item related to this visual observation target that could best be checked by studying the returned photographs. A photograph of the terminus of the landslide where the lineaments are radial to Tsiolkovsky and parallel to the flow direction is shown in Fig. 3. It clearly shows that a greater number of more sharply defined and smaller craters occur on the relatively younger flow unit than on the relatively older floor fill of the Fermi Basin to the west.

The discrepancy in crater density is restricted only to small craters. The excessive population of these craters on the younger of the two units may be due to one or both of the following reasons:

1. The presence of drainage craters, which may have developed by the seismic shaking of the surface and by drainage of the material in the void spaces initially sealed over by flow layers and large blocks.

2. The absence of a thick regolith on the rock avalanche because of the relative youth (small impact craters would tend to appear fresher and to "live longer" on such a unit) as compared to the floor of the crater Fermi that exhibits the characteristics of a relatively thick regolith.

During the post-mission visual observation debriefing one important aspect of Tsiolkovsky was stressed; namely, the apparent color of the crater floor. "The floor is a gun-metal-gray color. It is about the color of a metal cabinet. Everything was that same gray color; so the only apparent aspects were the differences in texture. In fact, the floor of Tsiolkovsky is lighter than some other mare areas." (NASA, 1971b, pp. 1–2).

Orbital photography of the crater Tsiolkovsky, obtained both by the unmanned Lunar Orbiters and by Apollo crews, show the floor of Tsiolkovsky to be extremely dark. Therefore, this mare unit was thought to be among the youngest on the moon. However, the visual impressions indicate that the mare material on the floor of

V-1A TSIOLKOVSKY REGION (128.5°E, 20°S)

Describe pertinent details relative to:
1. Structures and possible layering on the central peaks of Tsiolkovsky.
2. Nature of light-colored floor material and relationship to surrounding units.
3. Variations in texture and structure along segments of the wall of Tsiolkovsky.
4. Rim deposits due south of the crater and possible volcanic fill of the crater Waterman.
5. Origin and inter-relationship of crater pair due north of Tsiolkovsky.

Fig. 2. The crater Tsiolkovsky and its environs; an example of the materials that were carried onboard the spacecraft on Apollo 15 to support the task of visual observations from lunar orbit.

Fig. 3. Detail of the terminal portion of the lineated landslide on the northwestern rim of Tsiolkovsky. Note the relatively high density of small, fresh-appearing craters on the landslide, right, as compared to that on the older floor materials of the crater Fermi, left (Apollo 15 frame 12818).

Tsiolkovsky is no darker than other mare regions of Eratosthenian age. A post-mission crater density comparison between the floor of Tsiolkovsky and Eratosthenian mare units in Oceanus Procellarum indicates a similarity in age.

The reason for the apparently spurious darkness of the floor material in Tsiolkovsky is therefore that photographic exposures depend on average scene brightness. Because Tsiolkovsky is surrounded by very bright highlands, photographs taken of the crater and its environs will expose mostly for the surrounding highlands and, therefore, underexpose the mare fill in the floor. The human eye has an advantage, however. It responds to the scene as a whole and makes the relative brightness levels of each and every unit distinctly separable.

Layering in the crater Picard

Picard is a 30 km crater located in the western part of Mare Crisium. The interest in the crater itself and the surrounding mare material (Fig. 4) springs from the variations in texture and albedo, and perhaps color-tone, as indicated by existing

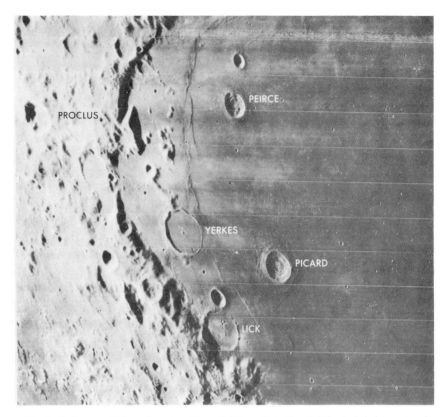

Fig. 4. Crater population of western Mare Crisium. Note the slightly dark haloes around Picard and Peirce (see detail in Fig. 5). At this low sun illumination the rays of Proclus are barely visible.

photography and remote sensing data. Following is a selection of comments made by the CMP during the flight.

"I am just coming over Picard at the present time and wanted to make a comment that it looks like there are several ring structures inside the crater itself. They are all concentric, and I do not see a great deal of relief on those that look like they are in the bottom of the crater. But, looking at the scarps around the outer ring, Picard displays characteristics of caldera-type craters. The scarps look like fault planes along the outside. And I can see in the outer wall very distinct layering. For instance, right on the top is a very thin dark layer that runs all the way around. And there is a light-colored layer. And then there are alternating dark and light layers all within about the same distance from the top of the crater, all the way around." (LR26/05: 08: 03: 45 GET).

"Endeavour is coming up over Picard. Considering the color variations, Picard is a slightly different color than the rest of the mare basin. I would consider Crisium to be a light brownish gray. Picard itself is more of a brown tone and has a darker halo around it. I can see some of the brown material just on the outside of the rim, and outside of that is some darker material that gradually turns into the gray of Mare Crisium. Within Picard, I can see six distinct rings that go all around the crater interior. And the walls of Picard are very shallow. It looks like a very shallow dish-like basin. And I can see some definite layering, particularly in the upper boundary of the rim." (LR34/05: 23: 49: 24 GET).

The freshness of detail in the crater Picard and its layering (Fig. 5) was compared to the situation in other, nearby craters.

"We are looking down at Picard, Peirce, and Lick D. These craters display similar characteristics; they all look alike. They all have the same ring structure and show comparatively low rims. The rims look very shallow compared to the rims on the other craters in the area. Also, they all have a slightly darker halo effect around the entire crater. But the color difference is very subtle." (LR36/06: 01: 48: 01 GET).

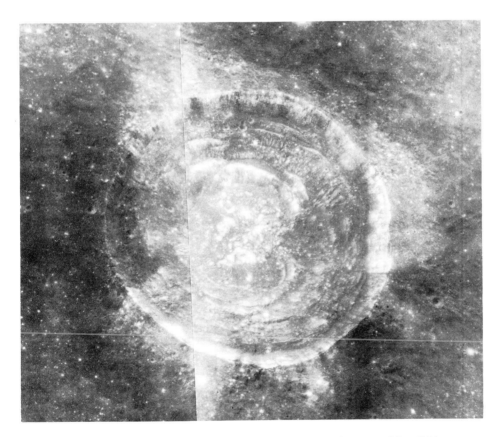

Fig. 5. Apollo 15 panoramic camera view of the crater Picard in western Mare Crisium. The width of the wall materials is about one-third of the crater diameter, and its heart-shaped floor is rather hummocky. Note that the wall is terraced, as many as 6 levels, and that there appears to be a number of layers within the terraces. The layers are visible especially on the northern part of the wall (mosaic of frames 9214 and 9216).

These characteristics contrast with those of other large craters in the same region, namely Lick and Yerkes, which display different features.

"Lick appears to be almost completely obscured. It looks very much like a collapse. All I can see is a small remnant ring, a color variation, with some positive relief. The interior of the crater appears to be similar to the surroundings, as far as the color and the texture are concerned. However, it does appear to slope gently in towards the center. Lick appears to be like a very large collapse feature, with the same kind of material both inside and outside the crater. And I would make the same comment about Yerkes." (LR34/05: 23: 49: 24 GET).

Ray-excluded zone of Proclus

The classical example of ray-excluded zones around impact craters is that of the crater Proclus on the western rim of Mare Crisium (El-Baz, 1969). Rays from Proclus extend in all directions except for a segment on the west-southwest (Fig. 6). Following are observations of the regional and local settings made from lunar orbit.

"The rays extending from Proclus are very light in color for about 240° to 260° around, and then there is a region of dark albedo. But the inner walls of Proclus are very light in color, almost white. The outer ring has a somewhat light gray appearance, and the difference in the rays is really between a light and a dark gray, as distinguished from the inner walls, which are quite white. The walls exhibit some debris on the upper slopes, maybe the upper 30 percent." (LR1/03: 07: 18: 37 GET).

"The excluded zone in the ray pattern is just very distinct at this point. And,

Fig. 6. At high sun illumination conditions the rays of the crater Proclus display a shadowed or excluded zone to the west-southwest. In an easterly direction, rays of Proclus extend over much of Mare Crisium. A detail is shown in Fig. 7.

from this angle looking at Proclus, about a crater diameter out to maybe a diameter and a half or so, you can see many small, bright, fresh craters, which appear to be in the general direction of a ray, like part of the ejecta blanket. They occur within a diameter to a diameter and a half of Proclus, and they are about the same brightness as the inner walls of Proclus, and they are small craters. I do see one which you might call a loop, which would suggest secondaries. They just seem to lie in the general direction of the rays of the ejecta from Proclus." (LR2/03: 09: 22: 54 GET).

During the debriefing, the CMP commented, "It is very strange the way the ejecta from, particularly, Proclus crosses Crisium. It is almost like flying above a haze layer and looking down through the haze at the surface. Ejecta from that crater does not look like it is resting on top of Crisium. It looks like it is suspended over it. It gives a very filmy, very gauzy appearance to the whole thing. It must be very thin. And I guess the reason it appears as if it is draped or suspended is that the ray material is visible no matter what it crosses. The ejecta forms lines when it goes through a crater, a wrinkle ridge or any topographic feature. It does not make any difference from what angle you view them; those lines of the ejecta material are straight. When the ray goes through a topographic prominence or a negative depression of some sort, you still see the ray through that." (NASA, 1971b, pp. 8–10).

The visual observation of Proclus was planned to determine the probable origin or cause of the ray-excluded zone. These zones occur around a number of lunar impact craters, and the cause for such zones may be one or more of the following:

1. The obliquity of the approach angle of the projectile would cause ejecta to be distributed all around the crater except for a segment directly below the path of the projectile. In some of these cases, a "rooster tail" pattern develops by the ejection of material in a direction opposite to that of the impact.

2. Topographic "shadowing" by a positive prominence would shield a segment around a crater against deposition of ejected ray material.

3. If differences exist in the materials in which the crater is situated, the two types of materials may respond differently to the impact pressure.

4. Deposition of younger units over the ray material.

Observations of Proclus from lunar orbit suggest a fifth class of reasons for ray exclusion; namely, structural control (Fig. 7). In the case of Proclus, a fault zone appears to have predated the crater. When the impact occurred, the fault plane formed part of the wall of the crater, and a broken-off segment of material may have been uplifted during the impact to inhibit the rays beyond it, later collapsing westward from the fault plane. This interpretation may, on first glance, appear complicated. Oblique missile impacts at White Sands show asymmetrical ejecta blankets (Moore, 1971). However, the probable role of structural control in the case of Proclus appears feasible and is based on the following comments:

"Something about Proclus was not obvious from the pictures we have seen before. The segment of the crater in the excluded zone seems to be discordant with the rest of the crater. In other words, if you made a circle to represent the crater, then this segment would lie outside of that circle. And I cannot see anything close to the rim that would account for any physical shadowing of the ray pattern. But I can see a diagonal fault zone that runs down into that little segment, and runs into

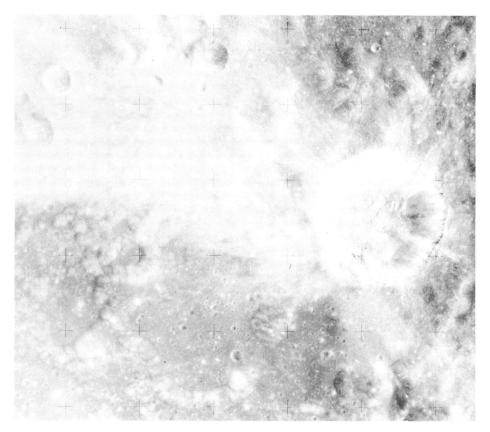

Fig. 7. Apollo 15 metric camera photograph of the crater Proclus and surrounding terrain. The lip on the southwestern corner of the rim corresponds with the ray-excluded zone. The straight segment of the wall nearest this lip may have been the result of a pre-crater fault which may have caused the ray-excluded zone as explained in the text (frame 0960).

it from the east side. I could not pick one out on the west side, but it is very distinct on the east side. And, in addition to that, I did not see a great deal of difference in the terrain or in the structure of the terrain across the excluded zone." (LR26/04: 08: 07: 19 GET).

"There is a tremendous variation in the wall [of Proclus] which does line up with the ejecta pattern. There is almost a straight wall on the side of Proclus that shows a ray-excluded zone. Also there is some breakthrough directly in the middle of that wall, which makes Proclus look like it is almost a circular crater. However, Proclus appears to be an elongate crater with one wall dipping quite steeply into the crater. Also the wall is oriented perpendicularly to a line bisecting the excluded zone, dipping into the crater. And then right in the middle of that portion, it looks like a small piece of that wall was also ejected, but it was only at the top part of the fault scarp. And so, if you look at it from the right angle, you can see almost a flat plane, which appears to have cut right into Proclus; and to the north and east of that flat plane is the crater Proclus, and to the south and west is a small chunk out of the top of it that coincides with the central part of the excluded zone." (LR34/05: 23: 52: 18 GET).

Cinder cones in the Littrow area

It was planned that the second author would study the geologic setting of the Littrow area. Special emphasis was to be placed on the nature and origin of the dark deposit on the southeastern rim of Mare Serenitatis between the upland massif units of the Taurus Mountains (Fig. 8).

"We are coming up over Serenitatis now. We are almost over Le Monnier at the present time, and we can see the Littrow area just out in front of us. And it is, in fact, about three different shades (Fig. 8). You can see in the upland area and particularly what looks like down in the valleys a darker color, and it does look like it is a light powdering or dusting over the entire area. And then, as you get out further into Mare Serenitatis, there is another zone which is somewhat lighter in color. And then, out at the last edge of the wrinkle ridge, out beyond that is the last zone, and the rest of Serenitatis looks fairly light in color. So it appears that the central part of Serenitatis is light, out beyond the first wrinkle ridge is a darker zone, and

Fig. 8. Southern part of Mare Serenitatis as displayed in this telescopic photograph (Lunar and Planetary Laboratory, C2544, September 1966). A dark annulus of mare material surrounds the lighter-colored interior zone. The basin is rimmed by the Sulpicius Gallus Rilles to the west, the Menelaus and Plinius Rilles to the south, and the Littrow and Chacornac Rilles to the east.

we are not up close enough to see what it is yet, and then as you get up into the high-lands around Le Monnier and Littrow, there is what appears to be a light dusting of dark material, and it certainly looks volcanic from here." (LR1/03: 07: 23: 11 GET).

"I am directly over Littrow at the present time. And I can see all the way around to the Apennine Front, encompassing all of Serenitatis between here and there except to the north over by Posidonius. So I have a very good view of Sulpicius Gallus. And the observation I wanted to make, in particular, pertains to the distinct way that the rilles do follow the old mare basin. And the fact that the second color band that we discussed in Littrow seems to be continuous across the basin into Tranquillitatis and on around—almost a shelf, a continental shelf appearance—into the Sulpicius Gallus area. There is a darker coloring in the uplands in Littrow and closer to the front or closer to the basin scarp." (LR23/05: 02: 18: 17 GET).

"If I had to give you the opinion right now, I would say the dark area in Littrow was some kind of ash. I am not sure it is a flow. But it certainly looks like a deposit over the entire surface. You can see it mostly in the upland areas, some in the mare areas, but mostly in valleys and in depressions. This mantle seems to have collected almost like there was some mass wasting down the hills, making the valleys darker in color and maybe a little thicker with this kind of material. But there are still at least three different distinctive color bands in the Littrow area, going from dark gray to a sort of brownish color. And it was the dark gray that looked like it was an ash fall to me." (LR25/05: 06: 17: 03 GET).

During the following revolution came the most important single observation of the entire flight: "I am looking right down on Littrow now, and a very interesting thing. I see the whole area around Littrow, particularly in the area of Littrow where I have noticed the darker deposits; there is a whole series of small, almost irregular-shaped cones, and there exists a very distinct dark mantling just around those cones. It looks like a whole field of small cinder cones in the area. And I say cinder cones because they are somewhat irregular in shape; they are not all round. These positive features display very dark haloes. The haloes which surround individual cones are mostly symmetric, but not always." (LR26/05: 08: 12: 46 GET).

Later during the mission, when asked whether the cinder cones were evenly distributed or whether they are concentrated in spots on the darkest unit, the CMP explained that they are concentrated in spots on the darkest unit, and that the darker the unit, the more cones: "Also, within the darker units, there are relatively high concentrations of these small cones, and then a few scattered ones in the out-lying areas. But I would say they were concentrated within the darker areas, more on the flat land side, in the valleys and in what looks like the lower areas. And within concentrations of cinder cones, there seems to be one locus of major activity, one locus of the greatest number of cones, and then they thin out beyond that." (LR37/06: 05: 58: 59 GET).

Discovery of the cinder cones in this area, which was later confirmed by the panoramic camera photographs (e.g. Fig. 9), and delineation of the dark deposit as an ash fall suggest late volcanism that postdates the major episodes of mare material extrusion. The dark deposit is relatively younger than Eratosthenian-age mare

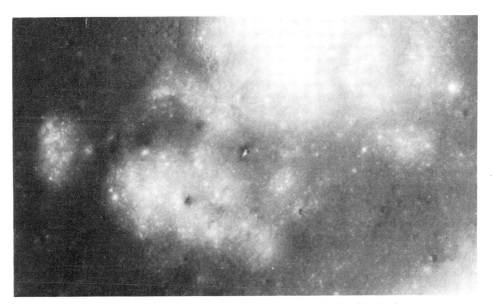

Fig. 9. Portion of an Apollo 15 panoramic camera photograph showing the largest symmetrical "cinder cone" in the Littrow area. The crater on the apex of the cone is about 75 m across and the symmetrical basal deposits form a dark halo, about 250 m across. A larger apron (as much as 1500 m in diameter) of ejecta can be seen as a thin mantle; it covers the light-colored highland material of the hill on which the cone is situated. Many other and smaller cones can be seen within the dark deposit, see for example, the lower right corner (frame 9554).

materials; that is, younger than about 3.2 billion years. The unit and the source vents appear much younger than any formation in the surrounding mare material. The cinder cones, as observed during the flight, "were very sharp, very distinct, and quite small. Most were orderly symmetrical structures. They were not broken down any way that I could see. They did not give the appearance of much degradation since they first formed. I did not see any of them partially obscured by an impact. Some of the aprons around them are symmetrical, some asymmetrical. Around the cones themselves, there is certainly a blanket that was laid down that is even darker than the surrounding material. The black spots are what drew my eye to the cinder cones." (NASA, 1971b, pp. 23–24).

Terminal portions of sinuous rilles

A problem of lunar sinuous rilles has been the uncertainty of definite terminal deposits or other indications that something had flowed out of the rilles onto the mare plains. The Apollo 15 groundtracks were well-suited for study of sinuous rilles because they covered innumerable rilles in Mare Imbrium and Oceanus Procellarum. Two rille complexes received particular attention: Rimae Prinz in the Harbinger Mountains region, and numerous rilles associated with the Aristarchus Plateau including Schroeter's Valley.

Rimae Prinz originate at circular depressions on the higher terrain near the rim of the old and particularly flooded crater Prinz. They terminate to the north in open-ended troughs as shown in Fig. 10. It was noted by the CMP that the rilles appear to have formed first in the higher terrain, and later been filled by mare lava flows. The latter appeared to have back-filled into the low terminal portions of the rilles: "Those flow fronts come back up into the rilles and chop off the terminus"; "It looked like this had rising water that went back in and filled up the termini of all the rilles." (NASA, 1971b, pp. 27–29). The interpretation was based on subtle differences in level of brightness, color-tone, texture, and changes in topographic expression. Some insight was probably gained by studying the rilles from various angles. This interpretation is a significant new input concerning the probable origin of sinuous rilles: it means that the probability of existence of terminal deposits prior to later flooding by younger mare material cannot be discarded.

A similar situation was encountered along the western boundaries of the Aristarchus Plateau. In this case two partly sinuous, partly linear rilles can be seen along the edge of the Plateau. They cut through the topographically higher plateau materials but not the surrounding mare of Oceanus Procellarum as illustrated in Fig. 11.

These features were discussed during the postmission visual observation debriefing, from which the following is excerpted: "On the west side of Aristarchus,

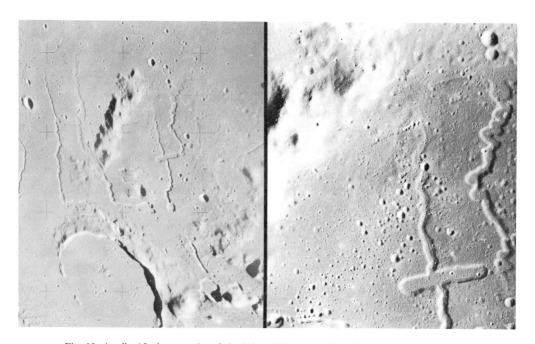

Fig. 10. Apollo 15 photographs of the Prinz Rilles. Left, Hasselblad 80 mm view of the rilles in the Harbinger Mountains region of Oceanus Procellarum (frame 11978). Right, panoramic camera view of the terminal portions showing the flooding of rille floors with mare material (frame 0314).

Fig. 11. Apollo 15 photographs of the western part of the Aristarchus Plateau. Left, Hassel-blad 80 mm lens view (looking southward) of the plateau materials and the mare materials of Oceanus Procellarum. The terminal portion of Schroeter's Valley can be seen on the left edge of the photograph (frame 12630). Right, Hasselblad 250 mm lens near-terminator view of the central part of the area shown to the left. Note that the mare material of Oceanus Procellarum fills the floors of the rilles that bisect the Aristarchus Plateau materials (frame 13345).

not only do you have the color to help you pick out the flows but you also have the surface texture. The surface texture of the top flow in the lower elevation flows is as shiny as a glass surface. It gives you a patent leather-like appearance and it is brown. Is it very different from the mare outside. It is a dirty brown color; inside the rille, it looks like a muddy river. And all the mare surface below the rille has been covered, too. It looks like a very viscous, muddy river that has somehow stopped in place. Those flow fronts come back up into the rilles and chop off the terminus. The basic mare around this area is gray." (NASA, 1971b, pp. 27–29).

In effect, this description is a characterization of the older Imbrian mare material in Mare Imbrium and farther east as grayish in color. The younger Eratosthenian mare material of Oceanus Procellarum, especially west of the Aristarchus Plateau, is brownish in color. "Imbrium was distinctly gray; Procellarum was distinctly brown."

A different situation, however, was encountered in the case of Schroeter's Valley. As illustrated in Fig.12 the main valley terminates in an enclosed area within the Aristarchus Plateau. The younger and more sinuous inner rille cuts across the valley rim and terminates in the mare materials of Oceanus Procellarum. A slight topographic elevation can be seen where this rille terminates. That however may be unrelated to the rille. From this it is concluded that there may be different modes of formation of sinuous rilles. Each rille or group of rilles should be treated separately

Fig. 12. Oblique (40° south) metric camera view of Schroeter's Valley. Note that the sinuous channel within the valley cuts across the wall and becomes progressively thinner until it terminates in the Oceanus Procellarum mare material (frame 2610).

based on the geomorphologic characteristics and the nature of the surrounding formations.

CONCLUSIONS

Before the Apollo 15 mission, observations from lunar orbit did not constitute an objective of the mission and, therefore, received only cursory interest. On the Apollo 15 mission, visual observations were treated as a formal mission objective: the task was conducted systematically, and targets were studied thoroughly. The extraordinary success of this undertaking proved the outstanding capabilities of man and his use in spaceflight. The unusual sensitivities of the human eye, when combined with the interpretive powers of the brain, constitute a combination that cannot be matched by one photographic system. However, this is not a competitive effort. Visual observations were made only to supplement the onboard photographic systems and are considered as a complement to other remotely sensed data.

The most significant results of performing the task on Apollo 15 include:

1. Contrary to photographic evidence, visual observations suggest that the mare fill of the crater Tsiolkovsky is not darker than the average (Eratosthenian) mare surface on the near side of the moon. The flow unit on the northwestern rim of the crater is interpreted as a landslide.

2. Alternating light and dark bands within the walls of the crater Picard are interpreted as discrete layers. The crater displays a brownish color-tone that is different from the surrounding mare materials and is believed to be volcanic in origin.

3. The ray-excluded zone of the crater Proclus appears to have been the result of a fault that predates the crater: when the impact occurred the western segment of the crater appears to have been displaced causing ray shadowing by structural control.

4. The dark deposit on the southeastern rim of Mare Serenitatis is interpreted as fine-grained or pyroclastic material related to explosive volcanic action that produced numerous cinder cones. Observations of cinder cones in the Taurus-Littrow area and subsequent study of the photography stimulated interest in the site which is now designated as the landing site of Apollo mission 17.

5. The termini of numerous sinuous rilles in the Harbinger Mountains—Aristarchus Plateau region appear to have been flooded with younger mare flows. This indicates that terminal deposits at the ends of sinuous channels could have existed prior to later flooding by mare materials.

The above examples illustrate that man does have special capabilities for observation that can be used to complement the orbital science return from a lunar mission. However, it is important to remember that man must be trained to be a good observer, and the task of looking must be planned before flight and conducted systematically. Otherwise, man will look but he may not see.

REFERENCES

El-Baz F. (1969) Crater characteristics. In *Analysis of Apollo 8 Photography and Visual Observations*, NASA SP–201, pp. 21–29.

El-Baz F. Worden A. M. and Brand V. D. (1972) Apollo 15 observations (abstract). In *Lunar Science—III* (editor C. Watkins), pp. 219–220, Lunar Science Institute Contr. No. 88.

Moore H. J. (1971) Lunar impact craters. In *Analysis of Apollo 10 Photography and Visual Observations*, NASA SP–232, pp. 24–26.

NASA (1971a) Apollo 15 technical air-to-ground voice transcription. Manned Spacecraft Center, Apollo Spacecraft Program Office, MSC–04558, July 1971, 1548 pp.

NASA (1971b) Apollo 15 debriefing for visual observations. Manned Spacecraft Center, Science Missions Support Division, MSC–04593, October 1971, 60 pp.

Worden A. M. and El-Baz F. (1971) Apollo 15 in lunar orbit: Significance of visual observations and photography (abstract). Geol. Soc. Amer. Ann. Meet., Abstracts with Program, pp. 757–758.

Proceedings of the Third Lunar Science Conference
(Supplement 3, *Geochimica et Cosmochimica Acta*)
Vol. 1, pp. 105–114
The M.I.T. Press, 1972

Mössbauer spectroscopy of lunar regolith returned by the automatic station Luna 16

T. V. Malysheva

V. I. Vernadsky Institute of Geochemistry and Analytical
Chemistry, USSR Academy of Sciences, Moscow, USSR

Abstract—Bulk fractions of lunar regolith from the surface layer (A) and the deep-seated layer (B) have been measured for the nuclide Fe^{57} by the method of Mössbauer spectroscopy. Troilite, metallic iron, ilmenite, olivine, pyroxene (mostly augite), and glasses have been detected. Iron distribution among the mineral phases demonstrates that the regolith of Luna 16 differs from the regolith of Apollo 11 by lesser ilmenite content and greater olivine content. The Luna 16 sample differs from the regolith of Apollo 12 by slightly greater olivine content. Within the limits of the analytical errors, the metallic iron does not contain nickel. No appreciable amount of iron in the trivalent state has been detected.

Introduction and Methods

The lunar soil sample returned by the automatic station Luna 16 consists of a 35 cm long column filled mostly with a powdery matter of black-brown color, the so-called lunar regolith (Fig. 1) (Vinogradov, 1971). The iron-bearing minerals have characteristic Mössbauer spectra (Sprenkel-Segal and Hanna, 1964; La Fleur *et al.*, 1969), that permit identification of individual minerals in the whole regolith sample without preliminary separation. Bulk regolith fractions from different layers of the column, as well as samples of different grain size were investigated by the technique of Mössbauer spectroscopy. Samples from the surface layer (3), the bulk sample 3–1, and the sample 3–3a, in the size range 0.2–0.4 mm were investigated. The bulk sample 8–2 and samples 9–4a, 9–5a, 9–6a and 9–1r in the size classes 0.200–0.450 mm, 0.127–0.200 mm, 0.083–0.127 mm and ⩽0.083 mm, respectively, were measured. These samples are from the deep-seated layers of the column. The measurements were carried out on a temporary Mössbauer spectrometer working in conditions of constant accelerations in combination with a 512-channel impulse analyzer, type LP 4050. The source was ^{57}Co in a palladium matrix with an activity 10 mci. One hundred to two hundred milligrams of lunar sample was placed in a 22 mm ϕ quartz

Fig. 1. Lower (deep-seated) part of the Luna 16 sample column with lunar soil.

vessel and the spectrum was measured for a period of ~40 hours to ensure the gathering of 10^6 impulses into each analyzer channel. The resulting data were refined by the least-squares method on electronic computers BESM–3 m and BESM–4.

RESULTS AND CONCLUSIONS

Mössbauer spectra of the sample 8–2 (Fig. 2) were measured at room temperature and at different relative velocities of the source to the absorber. Absorption spectra of sample 3–1 (Fig. 3) were measured at room temperature and at the temperature

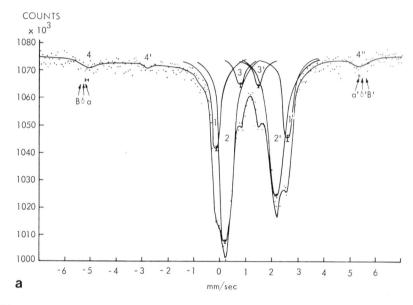

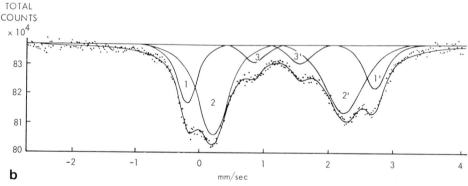

Fig. 2. Mössbauer spectra of sample 8–2, measured at (a) high velocity of the source relative to the absorber, (b) lesser velocity of relative motion (room temperature)

1 and 1′—peaks of olivine absorption
2 and 2′—peaks of pyroxene and glasses absorption
3 and 3′—peaks of ilmenite absorption
4, 4′ and 4″—peaks of metallic iron absorption.

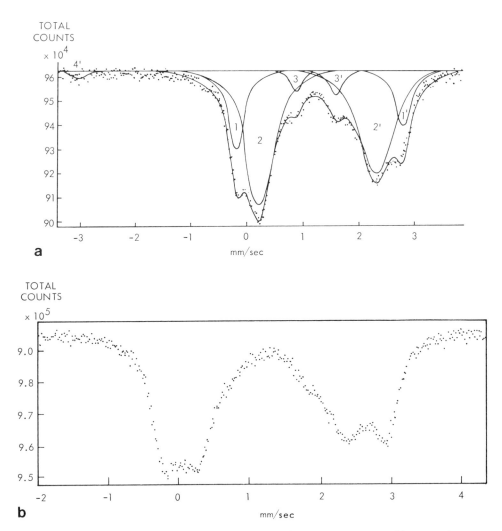

Fig. 3. Mössbauer spectra of samples 3–1 (a) measured at room temperature; (b) measured at liquid nitrogen temperature. Designations of peaks are the same as on Fig. 2.

of liquid nitrogen. The spectrum of sample 3–3a is identical with the spectrum of sample 3–1. The smooth line is the Lorenzian fit.

The peaks 1 and 1′ have been identified as a quadrupole doublet that was formed during resonance absorption of the Mössbauer radiation in olivine. The chemical shifts (δ), quadrupole slittings (Δ), and widths of lines (Γ) of olivines from lunar samples, terrestrial rocks and meteorites were compared (Table 1). It is apparent that the δ values coincide well with the values obtained for terrestrial and meteoritic olivines, and values of Δ are nearer to those obtained for olivines with a large value of $f = FeO/(FeO + MgO)\%$, which confirms the high iron content of lunar olivines (Makarov et al., 1971). A slightly lesser iron content has been detected in the bulk

Table 1. Experimental parameters of Mössbauer spectra of olivine.

Sample	$f(\%)$	Size of particles (mm)	Chemical shift (mm/sec)	Quadrupole splitting (mm/sec)	Reference
Earth	7		128 ± 0.02	3.02 ± 0.02	Malysheva et al. (1969)
Meteorite	10		130 ± 0.02	3.04 ± 0.02	Malysheva et al. (1969)
Earth	12.5.		1.26	3.02	Bancroft et al. (1967)
Earth	46.6		1.25	2.93	Bancroft et al. (1967)
Earth	50		1.28 ± 0.02	2.94 ± 0.02	Malysheva et al. (1969)
Earth	60.9		1.27	2.90	Bancroft et al. (1967)
Earth	96.1		1.27	2.90	Bancroft et al. (1967)
Luna 16					
8–2			1.28 ± 0.03	2.91 ± 0.02	
3–1			1.30 ± 0.02	2.99 ± 0.02	
3–3a		0.2 –0.4	1.26 ± 0.02	2.90 ± 0.02	
9–4a		0.200–0.450	1.27 ± 0.02	2.94 ± 0.01	
9–5a		0.127–0.200	1.26 ± 0.02	2.92 ± 0.01	
9–6a		0.083–0.127	1.26 ± 0.02	2.90 ± 0.01	
9–1r		0.083	1.29 ± 0.02	2.88 ± 0.02	

fraction of the surface layer (sample 3–1). A tendency toward increase in iron content olivine with the decrease of the grain size is distinctly apparent. The small olivine amount detected during petrological investigations of Luna 16 samples (Sellers et al., 1972) probably is connected with the fact that a great amount of olivine is present in breccias (Walter et al., 1972), glasses and also in a greatly crushed state.

The absorption lines 2 and 2′ are attributed to the pyroxene component of regolith. Parameters of Mössbauer spectra of pyroxene from meteoritic and terrestrial samples are presented in Table 2, together with the results of measurements of two samples of augites and glasses. Parameters of Mössbauer spectra of lunar samples are presented in Table 3. From comparison of Tables 2 and 3 it is seen that the parameters of lunar pyroxene Mössbauer spectra coincide best with the results obtained for augites and glasses. The slightly greater δ value in glasses may be due to depletion of alkalis. The lines of quadrupole splitting are broadened substantially. An attempt was made to dissociate quadrupole splitting lines into components corresponding to two unequivalent M1 and M2 positions in the lattice of the pyroxenes (Table 4, Fig. 4). It is seen that the division into components does not lead to satisfactory results. This apparently is associated with the presence of large amounts of glasses in the regolith. Strongly broadened lines of the Mössbauer spectra are noted for Apollo 14 samples (Battey et al., 1972).

In Table 5, δ and Δ are given for the quadrupole doublet with absorption lines 3 and 3′ pertaining to ilmenite. These values are in good agreement with the data presented by Gibb et al. (1969) on terrestrial ilmenites. However, the right peaks of the quadrupole doublets of lunar ilmenites are slightly broader and have greater intensities than the left peaks, which may indicate the presence of some unidentified components in the lunar regolith.

Measurement of samples at the temperature of liquid nitrogen does not improve the resolution of doublet lines that are characteristic of various iron-bearing minerals (Fig. 3).

Table 2. Parameters of pyroxene Mössbauer spectra at room temperature.

Samples	δ (mm/sec)		Δ (mm/sec)		Γ (mm/sec)		References
	M1	M2	M1	M2	M1	M2	
Rhombic pyroxenes	1.22 – 1.25	1.24 – 1.27	2.35 – 2.44	1.98 – 2.13	0.35 – 0.49	0.34 – 0.42	Bancroft et al. (1967)
Diogenite Johnston	1.28 ± 0.04	1.28 ± 0.04	2.31 ± 0.05	2.10 ± 0.04	0.35 ± 0.03	0.33 ± 0.03	Malysheva et al. (1970)
Howardite Yurtuk	1.27 ± 0.04	1.30 ± 0.02	2.62 ± 0.03	2.15 ± 0.01	0.35 ± 0.01	0.34 ± 0.01	Malysheva et al. (1970)
Eucrite Stannern	1.28 ± 0.02	1.27 ± 0.02	2.55 ± 0.03	2.05 ± 0.01	0.33 ± 0.02	0.34 ± 0.02	Malysheva et al. (1970)
Augite 11638	1.20 ± 0.03	1.21 ± 0.01	2.58 ± 0.03	2.02 ± 0.01	0.46 ± 0.03	0.40 ± 0.03	
Augite D × 379	1.22 ± 0.02	1.20 ± 0.02	2.68 ± 0.03	2.13 ± 0.02	0.46 ± 0.03	0.40 ± 0.03	
Alkaline ferrosilicate glasses	1.18 1.12 ± 0.05		2.11 2.00 ± 0.10		0.40		Gosselin et al. (1967) Belyustin et al. (1965)

Table 3. Parameters of Mössbauer spectra of the pyroxene component of lunar regolith.

Sample	Grain size (mm)	δ (mm/sec)	Δ (mm/sec)	Γ (mm/sec) left peak	right peak
3–1	Bulk sample	1.27 ± 0.01	2.10 ± 0.01	0.61 ± 0.01	0.74 ± 0.02
3–3a	0.2 –0.4	1.22 ± 0.01	2.02 ± 0.01	0.61 ± 0.02	0.64 ± 0.04
8–2	Bulk sample	1.23 ± 0.02	2.02 ± 0.01	0.56 ± 0.02	0.68 ± 0.03
9–4a	0.200–0.450	1.21 ± 0.03	2.06 ± 0.04	0.61 ± 0.01	0.83 ± 0.01
9–5a	0.127–0.200	1.24 ± 0.01	2.05 ± 0.01	0.61 ± 0.01	0.82 ± 0.01
9–6a	0.083–0.127	1.24 ± 0.01	2.05 ± 0.01	0.59 ± 0.01	0.68 ± 0.01
9–1r	≤0.083	1.26 ± 0.02	2.04 ± 0.02	0.62 ± 0.01	0.73 ± 0.01

Table 4. Parameters of Mössbauer spectra of the pyroxene component of lunar regolith.

Sample	Grain size (mm)	δ (mm/sec) I	II	Δ (mm/sec) I	II	Γ (mm/sec) left	right	left	right
3–1	Bulk sample	1.25 ± 0.07	1.27 ± 0.02	2.52 ± 0.07	1.92 ± 0.02	0.60± 0.07	0.76± 0.20	0.71± 0.02	0.79± 0.03
8–2	Bulk sample	1.18 ± 0.02	1.15 ± 0.02	2.31 ± 0.05	1.78 ± 0.05	0.36± 0.17	0.52± 0.03	0.69± 0.03	0.28± 0.05
9–4a	0.200–0.450	1.18 ± 0.01	1.16 ± 0.01	2.28 ± 0.01	1.77 ± 0.01	0.35± 0.03	0.45± 0.01	0.45± 0.02	0.45± 0.01
9–5a	0.127–0.200	1.20 ± 0.02	1.18 ± 0.02	2.33 ± 0.02	1.85 ± 0.03	0.35± 0.05	0.52± 0.01	0.58± 0.03	0.43± 0.04
9–1r	≤0.083	1.16 ± 0.01	1.17 ± 0.02	2.27 ± 0.01	1.79 ± 0.02	0.26± 0.02	0.52± 0.03	0.57± 0.01	0.53± 0.07

The absorption lines 4, 4′, and 4″ have been identified as three peaks of magnetic splitting from six peaks of metallic iron hyperfine structure. The ratio of the peak intensities is characteristic of finely dispersed metallic iron with particle size no less than 15 Å (Cleveland et al., 1970).

According to Johnston et al. (1962), the effective magnetic field on the nucleus of iron (H_{eff}) in the alloy iron-nickel differs from H_{eff} for pure iron. In order to elucidate the character of iron in the lunar regolith we have measured Mössbauer

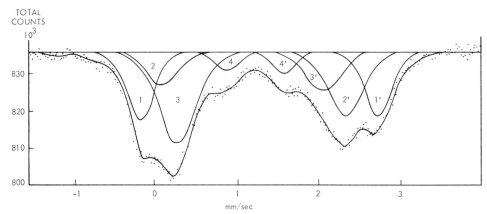

Fig. 4. Mössbauer spectrum of sample 8–2
1, 1'—peaks of olivine abs.
2, 2', 3, 3',—peaks of pyroxene abs. glasses abs.
4, 4'—peaks of ilmenite abs.

Table 5. Parameters of Mössbauer spectra of lunar ilmenites.

Sample	Grain size (mm)	δ (mm/sec)	Δ (mm/sec)	Γ (mm/sec) left peak	right peak
8–2	Bulk sample	1.22 ± 0.02	0.69 ± 0.02	0.29 ± 0.04	0.31 ± 0.04
3–3a	0.2 –0.4	1.23 ± 0.02	0.66 ± 0.03	0.21 ± 0.04	0.33 ± 0.04
3–1	Bulk sample	1.24 ± 0.02	0.70 ± 0.02	0.24 ± 0.03	0.36 ± 0.03
9–4a	0.200–0.450	1.24 ± 0.02	0.70 ± 0.02	0.22 ± 0.01	0.22 ± 0.01*
9–5a	0.127–0.200	1.21 ± 0.01	0.65 ± 0.01	0.22 ± 0.01	0.22 ± 0.01*
9–6a	0.083–0.127	1.21 ± 0.01	0.65 ± 0.01	0.21 ± 0.01	0.21 ± 0.01*
9–1r	0.083	1.20 ± 0.04	0.65 ± 0.01	0.21 ± 0.01	0.21 ± 0.01*

*Counted under conditions $\Gamma_r = \Gamma_e$.

Table 6. Magnetic hyperfine splitting measurements of Mössbauer absorbtion spectra for iron meteorites and lunar regolith samples.

Sample	Left peak No. of channel	Right peak No. of channel	$\dfrac{H_{eff}(\text{meteor.})}{H_{eff}(\text{iron})}$
Meteorite Chinge	144.0 ± 0.2	408.0 ± 0.3	1.026 ± 0.010
Meteorite Sikhote-Alin	147.0 ± 0.2	404.5 ± 0.2	1.036 ± 0.010
Reactive iron	149.0 ± 0.2	401.1 ± 0.2	1.000 ± 0.010
Luna 16 8–2	150.0 ± 1.4	401.5 ± 0.9	1.000 ± 0.010

spectra of the iron meteorites Sikhote-Alin and Chinge, containing 6% and 16% nickel, respectively, according to the chemical analysis data of Zavaritsky and Kvasha (1952), as well as pure metallic iron (Table 6, Fig. 5). It is apparent that H_{eff} in meteorites is greater than in pure iron. The meteorite-Sikhote-Alin consists almost entirely of kamacite and the meteorite Chinge is composed of a mixture of kamacite and taenite. The determination of nickel content from Mössbauer spectroscopy data in the measured meteorite samples is qualitative owing to the presence of

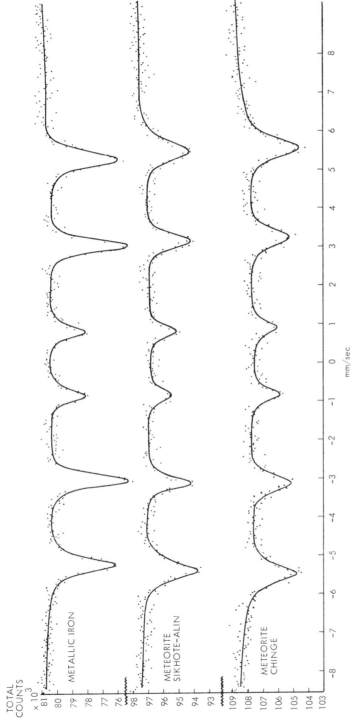

Fig. 5. Mössbauer spectra of the iron meteorites Sikhote-Alin, Chinge, and of pure metallic iron.

a small amount of cobalt (<1%). However, the difference in the positions of the
extreme peaks of the magnetic hyperfine structure in meteorites and in pure metallic
iron is clearly apparent. The positions of the 1st and 6th lines of the hyperfine
magnetic splitting for pure iron (a), the meteorite Sikhote-Alin (b) and the meteorite
Chinge (c) are indicated on Fig. 2. Pure metallic iron is present in sample 8–2, in so
far as can be determined within the limits of accuracy of the measurements. An
identical result has been obtained for all regolith samples analyzed.

Mössbauer spectra of the samples 9–4a (grain size 0.200–0.450 mm) and 9–5a
(0.127–0.200 mm) are identical with the spectra of samples 8–2 and 3–1. However,
spectra of the samples 9–6a (0.083–0.127 mm) and 9–1r (≤0.083 mm) contain lines
of magnetic superfine splitting that are characteristic of troilite (lines 5.5′ and 5″,
Fig. 6).

Trivalent iron has not been detected. According to the area of corresponding
peaks of hyperfine splitting, the mineralogical composition of the iron-bearing
portion of the regolith has been determined, assuming that the values of resonance
absorption are equal for all minerals, on the basis of the areas of the appropriate
hyperfine splitting peaks. Results obtained for samples of the lunar regolith returned
by Apollo 11 (Gay et al., 1970) and Apollo 12 (Bancroft et al., 1971) are listed for
comparison. The silicate component (olivines and pyroxene glasses) are the prevailing part of lunar soil (Table 7, Figs. 2–6), as was pointed out previously by Thiel
et al. (1971). The ilmenite content of the soil returned by Luna 16 is similar to the
soil sampled by Apollo 12, but the olivine content is slightly greater than the amount
reported in the regolith of Apollo 11 and 12. The increase of the area of quadrupole
doublets of the olivine component with column depth and with the decrease of grain
size probably is connected with an increase of the fayalite component in the olivine.

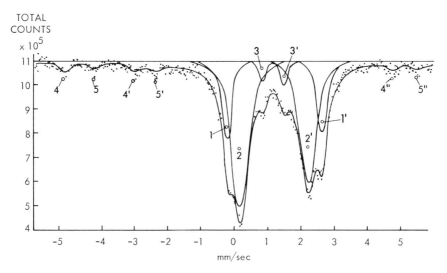

Fig. 6. Mössbauer spectrum of the sample 9–6a; 5, 5′ and 5″ are the absorption lines of
FeS, the other designations are the same as on Fig. 2.

Table 7. Phase distribution of iron in Luna 16 regolith.
Values are in percent of the total Mössbauer spectrum area.

Mineral	Luna 16							Apollo 11	Apollo 12
	3–1 0.2–0.4 mm	3–3a	8–2	9–4a	9–5a	9–6a	9–1r	10.084	12.051
Ilmenite	7.7	6.7	8.0	4.9	4.9	5.8	7.1	19.7	6.8
Pyroxene + glass	69.0	71.5	67.0	79.3	75.8	71.0	71.0	67.6	76.4
Olivine	18.3	16.7	25.1	15.8	16.2	22.6	21.9	4.4	11.6
Iron	4.5	5.1	4.1	4.2	3.3	3.1	3.1	5.8	4.3
Troilite	$\leqslant 1$	$\leqslant 1$	$\leqslant 1$	$\leqslant 1$	$\leqslant 1$	2.8	2.0	1.1	0.9

Thus, on the basis of iron distribution among the mineral phases. Mössbauer spectroscopy makes it possible to characterize the magma type in the region of lunar soil sampling.

Acknowledgments—I express my thanks to all of my laboratory collaborators who have participated in the measurements of samples, as well as to L. M. Satarova for aiding me to process spectra on the computer BESM–4.

REFERENCES

Bancroft G. M. Maddock A. G. and Burns R. G. (1967) Application of the Mössbauer effect to silicate mineralogy. *Geochim. Cosmochim. Acta* **31**, 2219–2246.

Battey M. H. Gibb T. C. Greatrex R. and Greenwood N. N. (1972) Mössbauer studies of Apollo 14 lunar samples (abstract). In *Lunar Science—III* (editor C. Watkins). p. 44, Lunar Science Institute Contr. No. 88.

Belustin A. A. Ostanevich Yu. M. Pisarevsky A. M. Tomilov S. B. Bai-Shu U. and Cher L. (1965) Mössbauer effect in alkaline ferrosilicate glasses. *Fisika tverdogo tela* **7**, No. 5, 1447–1461.

Boun M. G. Gay P. Muir J. D. Bancroft G. M. and Williams P. G. L. (1972) Mineralogical and petrographic investigation of some Apollo 12 samples. *Proc. Second Lunar Sci. Conf.*, *Geochim. Cosmochim. Acta* Suppl. 2, Vol. 1, pp. 377–392.

Cleveland B. Guarnicri C. R. and Walter J. C. (1970) Magnetic ordering in very thin iron films. *Bull. Amer. Phys. Soc.* **15**, 108.

Gay P. Bancroft G. M. and Boun M. G. (1970) Diffraction and Mössbauer studies of minerals from soil and rocks. *Proc. Apollo 11 Lunar Sci. Conf.*, *Geochim. Cosmochim. Acta* Suppl. 1, Vol. 1, pp. 481–497.

Gibb T. C. Greenwood N. N. and Twist W. (1969) The Mössbauer spectra of natural ilmenites. *J. Inorgan. Nucl. Chem.* **31**, 947–954.

Gosselin J. P. Chimony U. Grodzins L. and Cooper A. B. (1967) Mössbauer studies of iron in sodium trisilicate glasses. *Phys. Chem. Glasses* **8**, No. 2, 56–61.

Johnston K. Reedu M. Cranshow T. and Madson P. (1962) Field of superfine interaction and magnetic moment of iron atoms in ferromagnetic alloys. In *Mössbauer effect*, pp. 367–371, Moscow, Foreign Liter (IL).

Lafleur L. D. Goodman C. D. and King E. A. (1969) Mössbauer investigation of shocked and unshocked iron meteorites and fayalite. *Science* **162**, 1268.

Makarov E. S. Ilyin N. P. and Ivanov V. I. (1971) On the composition and crystalline structure of minerals of the regolith from the Sea of Fecundity. Report K-24 at the 14 Session of KOSPAR, June 1971, Seattle.

Malysheva T. V. Kurash V. V. and Ermakov A. N. (1969) Investigation of isomorphous Mg and Fe^{2+} substitution in olivines by the method of γ-resonance Mössbauer spectroscopy. *Geokhimiya* **11**, 1405–1408.

Malysheva T. V. Kurash V. V. Lavrukhina A. K. and Akolzina L. D. (1970) Investigation of pyroxenes in meteorites by the method of Mössbauer spectroscopy. Trans. 14 Meteorite Conf., December 14–19, 1970, Moscow.

Sellers G. A. Woo C. C. Finkelman R. B. Birol M. L. and Kirschenbaum H. Wash. D.C. (1972) Mineralogical and petrographic description of the sub-125 micrometer fraction of two Luna 16 core samples (abstract). *Lunar Science—III* (editor C. Watkins), p. 694, Lunar Science Institute Contr. No. 88.

Sprenkel-Segal E. L. and Hanna S. S. (1964) Mössbauer analysis of iron in stone meteorites. *Geochim. Cosmochim. Acta* **28**, 1913–1932.

Thiel K. Neugebauer F. W. Heimann M. and Herr W. (1972) Uranium distribution and Mössbauer studies of Apollo 11 to 14 samples (abstract). In *Lunar Science—III* (editor C. Watkins), p. 747, Lunar Science Institute Contr. No. 88.

Vinogradov A. P. (1971) Preliminary data of lunar soil, brought by the automatic station Luna 16. *Geokhimiya* **3**, 8–15.

Walter L. S. French B. M. Doan A. S. Jr. and Heinrich K. F. J. (1972) Petrographic analysis of lunar samples 14171 and 14305 (breccias) (abstract). In *Lunar Science—III* (editor C. Watkins), p. 773, Lunar Science Institute Contr. No. 88.

Zavaritsky A. N. and Kvasha L. G. (1952) *Meteorites of the USSR*. Moscow, Acad. Sci. USSR.

Proceedings of the Third Lunar Science Conference
(Supplement 3, *Geochimica et Cosmochimica Acta*)
Vol. 1, pp. 115–129
The M.I.T. Press, 1972

Petrology of Apollo 14 high-alumina basalt

Ikuo Kushiro

Geophysical Laboratory, Carnegie Institution of
Washington, Washington, D.C. 20008

Yukio Ikeda and Yasuo Nakamura*

Geological Institute, University of Tokyo,
Tokyo, Japan

Abstract—Melting experiments and microprobe analyses have been carried out on the Apollo 14 high-alumina basalt 14310 (Al_2O_3 21.7 wt% with MgO/FeO ratio higher than those of the Apollo 11 and 12 crystalline rocks). Another crystalline rock, 14053, and a breccia, 14321, have been analyzed with the microprobe. The liquidus phase of the rock 14310 is plagioclase up to about 10 kb, chromian spinel between about 10 and 20 kb, and pyrope-rich garnet above 20 kb. The composition of this rock, as well as those of feldspathic basalts that are considered to be highland materials, is separated from possible lunar "mantle" compositions by low-temperature cotectic boundaries at least up to 10 kb, suggesting that the magmas of these rocks are not products of direct partial melting of the lunar "mantle" or of fractional crystallization of a primary magma at least to the depth of 300 km under anhydrous conditions. It is most probable that these highland rocks were generated by partial or bulk melting of a plagioclase-cumulate rock that had been formed by a large-scale differentiation in the shallow parts of the lunar interior at an earlier stage. This argument is valid even if rock 14310 is a plagioclase-cumulate rock with less than 15% plagioclase enrichment. If considerable amounts (70% or more) of alkalis were lost before solidification, the original magma could have been formed by direct partial melting of the lunar "mantle" material with MgO/FeO ratio similar to that of the earth's upper mantle or by fractional crystallization of a primary magma at depths of about 100 km.

Introduction

Petrological studies including microprobe analysis, major element determination, and melting experiments at high pressures have been carried out on the Apollo 14 high-alumina basaltic rock 14310 (139 and 22). On the basis of these studies, a possible origin of the lunar high-alumina basalt is discussed. Another Apollo 14 crystalline rock (14053,6) and a breccia (14321,22) have been studied petrographically, and some of their constituent minerals have been analyzed with the microprobe.

Description of Rocks and Minerals

Crystalline rock 14310

The rock is basaltic and shows intersertal texture, although some pyroxenes show ophitic or subophitic texture and only minor amounts of glass are present. The most abundant minerals are plagioclase, orthopyroxene, and strongly zoned clinopyroxene. Minor constituents are K-feldspar containing Ba, ilmenite, ulvö-

---

*Present address: Department of Geology, University of Otago, Dunedin, New Zealand.

spinel, metallic iron, troilite, phosphate mineral (apatite or whitlockite), and Zr-rich minerals. The plagioclase consists of prismatic or needle-like crystals, which range in length from 0.1 to 1.0 mm. Some pyroxene crystals attain lengths of up to 2 mm. However, the grain size distribution appears to be continuous and a distinct phenocrystic texture is absent in the thin section examined. The mesostasis consists of very fine-grained crystals and a glass enriched in Si and K. The thin section contains one small patch (~ 1 mm across) of "cognate inclusion" (LSPET, 1971) consisting mainly of fine-grained plagioclase and pyroxene. The grain size of these minerals is one-fifth to one-tenth that of the main part. The boundary between the fine-grained part and the main part is sharp.

The abundances of major elements in rock 14310,139, obtained by the conventional wet-chemical analysis method, are given in Table 1. The rock is slightly undersaturated with silica, having 3.3 wt% normative olivine. The analysis shows high alumina and lime and relatively low iron. Alkalis, especially K_2O, and the MgO/FeO ratio are also higher than those of the Apollo 11 and 12 basaltic crystalline rocks. The composition is similar to those of some terrestrial high-alumina basalts, although Na_2O is still considerably lower and the alkali/alumina ratio of rock 14310 is much lower than those of the terrestrial high-alumina basalts (e.g., Kuno, 1960). It should be noted that the MgO/FeO ratios of the Apollo 15 gabbroic anorthosite 15415 (LSPET, 1972) and feldspathic basalts (Reid et al., 1972) are also higher than those of the Apollo 11 and 12 crystalline rocks.

Crystalline rock 14053

The rock is coarse-grained and gabbroic (most crystals range from 0.5 to 2.0 mm and some clinopyroxenes attain 5 mm in length); however, plagioclase and clinopyroxene commonly show ophitic texture, and the rock may be called coarse-grained dolerite. The constituent minerals are plagioclase, strongly zoned clinopyroxene, olivine, pyroxferroite, K-feldspar, cristobalite, chromite, ulvöspinel, ilmenite, troilite, metallic iron, and phosphate mineral (apatite or whitlockite). The rock also contains mesostasis with Si- and K-rich glass in the interstices. In some of them a myrmekitic intergrowth of metallic iron, silica mineral (probably cristobalite), phosphate mineral, and Si- and K-rich glass is observed. The abundances of major elements in rock 14053 (LSPET, 1971) are significantly different from those in rock 14310 and are closer to those in the Apollo 12 crystalline rocks of intermediate MgO/FeO ratio, although iron and Ti are lower and Al and Ca are higher in rock 14053.

Table 1. Major elements and CIPW norm of rock 14310,139.

SiO_2, 46.88; TiO_2, 1.19; Al_2O_3, 21.68; FeO, 8.22; MgO, 7.42; CaO, 12.55; Na_2O, 0.72; K_2O, 0.50; Cr_2O_3, 0.25; MnO, 0.13; P_2O_5, 0.17; Total, 99.71 (wt%)

CIPW Norm

Or, 2.96; Ab, 6.10; An, 54.61; Di (Wo, 2.80; En, 1.57; Fs, 1.11); Hy (En, 14.34; Fs, 10.21); Ol (Fo, 1.83; Fa, 1.43); Il, 2.27; Cm, 0.38; Ap, 0.39 (wt%)

Analyzed by H. Haramura by the conventional wet-chemical analysis method.

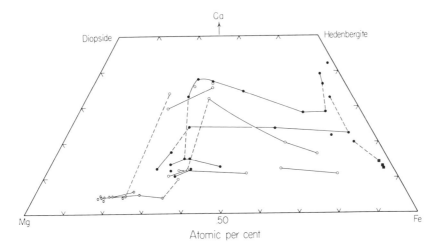

Fig. 1. Ca-Mg-Fe plot of the analyzed orthopyroxene, clinopyroxene, and pyroxferroite in rocks 14310,22 and 14053,6. Open and solid circles are clinopyroxenes in rocks 14310 and 14053, respectively; open circle with a vertical line and solid square are orthopyroxene (14310) and pyroxferroite (14053), respectively. Solid lines indicate the range of compositional zoning within single crystals, and dashed lines indicate discontinuous zoning (in the iron-rich side) and core-jacket relations.

Pyroxenes

Pyroxenes in rock 14310,22 are orthopyroxene, pigeonite, ferropigeonite, augite, and subcalcic ferroaugite. Their compositions are shown in the Ca-Mg-Fe diagram (Fig. 1), and selected microprobe analyses are given in Table 2. The method of microprobe analysis is the same as that described previously (Nakamura and Kushiro, 1970; Kushiro and Nakamura, 1970). Orthopyroxene (bronzite and hypersthene) occurs mostly as cores within pigeonite grains and is rarely in contact with augite. The Al content appears to decrease with increase of Fe/Mg ratio. Exsolution lamellae are not observed under the microscope in the orthopyroxene crystals. Pigeonite and ferropigeonite occur mostly as rims to the orthopyroxene. The Fe/(Mg + Fe) ratio ranges from 0.35 to 0.85, whereas the Ca content remains nearly constant (Wo, 10–12 mole %). Some pigeonite grains are zoned continuously to ferropigeonite; others are zoned discontinuously to (or surrounded by) augite, which is zoned continuously to subcalcic ferroaugite or ferropigeonite (Fig. 1). This may be due to the difference in the compositional zoning in different sectors of the pyroxene crystals, as shown by Hollister *et al.* (1971) and Boyd and Smith (1971) on the Apollo 12 pyroxenes. Augite occurs as rims to pigeonite or rarely to hypersthene. It is zoned to subcalcic ferroaugite near the margins. Some augite occurs as inclusions in pigeonite. The tie lines that connect the contacting Ca-poor pyroxenes and augite with core-jacket relations are shown in Fig. 1 (dashed lines). Although the analyses were made on the contacting pyroxenes within a range of 10 μ it is not certain that they are equilibrium tie lines.

Table 2. Selected microprobe analyses of pyroxene, olivine, pyroxferroite, feldspar, and glass in rocks 14310,22, 14053,6, and 14321,22.

14310,22

	Orthopyroxene		Pigeonite and ferropigeonite		Augite		Plagioclase		K-feldspar	Glass	
	1	2	3	4	5	6	7	8	9	10	11
SiO$_2$	52.2	51.8	50.8	47.1	49.6	48.4	44.0	50.9	62.0	76.8	30.5
Al$_2$O$_3$	3.44	1.58	1.35	0.95	2.81	1.91	34.8	30.0	18.9	11.1	3.08
TiO$_2$	0.69	0.65	0.89	0.74	1.53	1.85	—	—	—	0.89	14.6
Cr$_2$O$_3$	0.61	0.48	0.39	0.08	0.72	0.04	—	—	—	—	—
FeO	11.2	16.1	19.8	34.1	13.5	18.3	0.08	0.39	0.25	0.86	35.7
MnO	0.21	0.28	0.33	0.47	0.26	0.29	—	—	—	0.04	0.36
MgO	28.4	24.8	20.3	9.17	16.4	11.4	—	—	—	0.02	3.16
CaO	2.50	3.11	5.07	5.74	14.0	16.4	19.1	14.5	0.37	0.73	7.09
Na$_2$O	0.02	0.03	0.02	0.04	0.09	0.11	0.64	2.86	0.90	0.94	0.40
K$_2$O	—	—	—	—	—	—	0.05	0.73	14.2	7.39	0.52
TOTAL	99.3	98.8	99.0	98.4	98.7	98.7	98.6	99.4	96.7*	98.8	95.4
Ca	4.9	6.2	10.4	12.7	29.5	35.3	94.1	70.5	1.9		
Mg	77.9	68.8	57.8	28.3	48.3	34.1	5.7(Na)	25.2(Na)	8.6(Na)		
Fe	17.2	25.0	31.8	59.0	22.2	30.6	0.2(K)	4.2(K)	89.4(K)		

14053,6   14321,22

	Pigeonite		Subcalcic augite and subcalcic ferroaugite				Hedenberg-itic clino-pyroxene	Pyrox-ferroite	Oliv-ine	Plagio-clase	Glass	Oliv-ine	Orthopy-roxene
	12	13	14	15	16	17	18	19	20	21	22	23	24
SiO$_2$	51.9	50.3	49.7	46.0	50.2	47.0	44.6	45.0	30.1	45.1	76.3	40.0	54.5
Al$_2$O$_3$	1.80	1.83	1.82	0.87	3.01	2.08	1.52	0.37	0.02	33.9	9.95	0.01	0.85
TiO$_2$	0.84	0.60	1.39	0.96	1.62	1.00	1.78	0.55	0.10	—	0.54	0.05	0.39
Cr$_2$O$_3$	0.82	—	0.60	0.08	—	—	0.03	—	0.01	—	—	—	—
FeO	17.3	25.7	18.3	39.0	15.4	34.7	30.2	45.0	63.4	0.29	1.83	10.7	13.7
MnO	0.38	0.41	0.37	0.65	0.29	0.53	0.49	0.68	0.79	—	—	0.11	0.21
MgO	21.3	14.7	15.9	1.90	14.2	2.80	0.53	0.80	4.97	—	—	50.1	29.5
CaO	6.43	6.20	11.7	9.87	15.4	12.5	18.0	6.39	0.34	18.8	0.85	0.06	0.94
Na$_2$O	0.02	0.00	0.09	0.03	0.00	0.00	0.12	0.06	0.12	0.88	0.30	—	—
K$_2$O	0.00	0.00	0.01	0.00	0.00	0.00	0.00	0.00	0.00	0.03	8.40	—	—
TOTAL	100.8	99.7	99.9	99.3	100.1	100.6	97.2	98.9	99.9	99.0	98.2	101.0	100.1
Ca	12.9	13.2	24.3	22.7	32.6	28.4	42.1	14.8	—	92.0		—	1.8
Mg	59.4	43.5	45.6	6.1	41.6	8.9	1.7	2.6	12.1	7.8(Na)		89.2	77.7
Fe	27.7	43.3	30.1	71.2	25.9	62.7	56.2	82.6	87.9	0.2(K)		10.8	20.5

*Low total is due to presence of about 3% BaO.

Pyroxenes in rock 14053 are pigeonite, augite, subcalcic augite, subcalcic ferro-augite, and hedenbergitic clinopyroxene. Their compositions are also shown in Fig. 1. Very narrow compositional gaps are observed between contacting pigeonite and subcalcic augite; the gaps are different for different pairs, however, and none of them appears to represent the equilibrium solvus. The augite and subcalcic augite are strongly zoned to very iron-rich subcalcic ferroaugite, which is discontinuously zoned to ferroaugite or hedenbergitic clinopyroxene. The very iron-rich subcalcic ferroaugite is commonly in contact with pyroxferroite (Fig. 1). The two strongly zoned clinopyroxene crystals in Fig. 1 show a strong iron enrichment; however, their Ca contents are significantly different, indicating metastable crystallization of pyroxenes from residual liquids that were isolated from each other and had different compositions during crystallization.

Breccia 14321,22 contains relatively large (~ 0.5 mm) fragments of homogeneous orthopyroxene with a much lower Ca content (Table 2, No. 24) than those of 14310,22. In the same thin section a pyroxene grain consisting of very thin lamellae (2–3 μ wide) of pigeonite and hypersthene is found. Their compositions are $Ca_{11}Mg_{56}Fe_{33}$ and $Ca_5Mg_{57}Fe_{38}$, respectively. The bulk composition of this pyroxene, obtained by using a defocused electron beam (45 μm across) and by moving the sample at a speed of 50 μm/min, is $Ca_{8.2}Mg_{56.9}Fe_{34.9}$, which is the composition of a low-Ca pigeonite. This evidence indicates that pigeonite (or high-clinopyroxene with $C_{2/c}$ symmetry) exsolved hypersthene with lowering temperature, supporting the shape of the pigeonite field suggested in the join enstatite-diopside (Kushiro, 1969, 1972). Similar exsolution has been found in orthopyroxene grains from rock 14310 (Finger *et al.*, 1972).

The Al-Ti relations of the analyzed pyroxenes from rocks 14310 and 14053 are shown in Fig. 2. The Ti content in orthopyroxene from 14310 is relatively low and uniform (between 0.02 and 0.01 on the basis of 6 oxygens) and appears to be independent of Al content. Pigeonite in both 14310 and 14053 also has a relatively low Ti content (mostly below 0.03). The Ti contents of these analyzed pyroxenes and also those from the Apollo 12 crystalline rocks appear to increase systematically with an increase of the $Ca/(Ca + Mg + Fe)$ ratio (or wollastonite component) up to a value of 0.32 (Fig. 3) and are independent of the Fe/Mg ratio.

In the zoned clinopyroxenes, the Ti/Al ratio increases with increase of the Fe/Mg

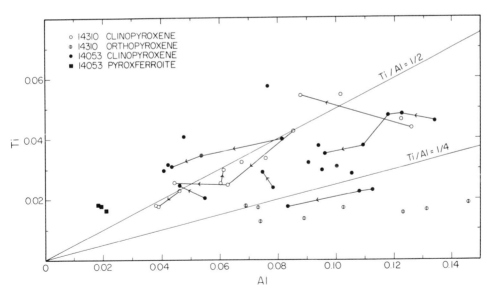

Fig. 2. Al-Ti relations of the analyzed clinopyroxenes in rocks 14310 and 14053. Symbols as in Fig. 1. Solid lines indicate the variations of Al and Ti in the zoned clinopyroxenes. Values based on 6 oxygens.

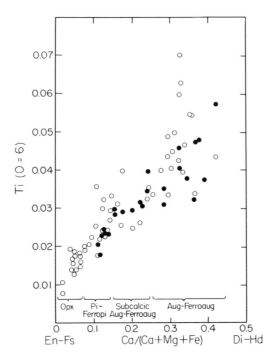

Fig. 3. The relations between Ti content and Ca/(Ca + Mg + Fe) ratio in the analyzed pyroxenes in rocks 14310, 14053, and 14321 (open circles) and those in the Apollo 12 crystalline rocks 12018, 12020, 12064, and 12065 (solid circles) (Kushiro and Nakamura, 1970).

ratio, and in very iron-rich clinopyroxenes and pyroxferroite this ratio is larger than 1/2 (Fig. 2). The Ti/(Al + Cr) ratio is also larger than 1/2 in most of these iron-rich clinopyroxenes and pyroxferroite. If all Ti is Ti^{4+}, the charge in the octahedral site is in excess of that in the tetrahedral site; that is, the charge balance is not maintained because it is unlikely that Fe^{3+} exists in these pyroxenes and pyroxferroite or that Ti^{4+} enters into the tetrahedral site. It seems most likely that a part of Ti is Ti^{3+}, at least in these very iron-rich pyroxenes and pyroxferroite. This possibility has been suggested previously by Boyd and Smith (1971) and Bence and Papike (1972).

Feldspar

Plagioclase is the most abundant mineral in rock 14310. Its composition ranges from $An_{94.1}Ab_{5.7}Or_{0.1}$ to $An_{70.5}Ab_{25.2}Or_{4.2}$. Selected analyses of plagioclase are given in Table 2. Plagioclase generally shows normal zoning, and in one case a single plagioclase crystal covers the above compositional range. The K content increases regularly with increasing Na content. Brown and Peckett (1971) distinguished phenocryst and groundmass plagioclase in two thin sections of 14310 and have shown that the phenocryst plagioclase is zoned to more sodic and potassic compositions, whereas the groundmass laths are as low in sodium and potassium as are the phenocryst cores. However, there are some exceptions; in the present analyses some of the very small plagioclase laths ("groundmass plagioclase") have compositions $An_{74}Ab_{26}$ and $An_{84}Ab_{16}$, which are considerably more sodic than those reported by Brown and Peckett (1971). Plagioclase in rock 14310 is often in contact with K-feldspar, which shows a compositional zoning, $An_{5.7}Ab_{12.8}Or_{81.5}$ to $An_{3.1}Ab_{6.4}Or_{90.5}$, and contains 3 to 6 wt% BaO. The K content of bytownite in contact with K-feldspar, but outside the effect of K emission from K-feldspar, is relatively high compared with that of bytownite of the same Na/Ca ratio in terrestrial basaltic rocks. This may be due to the fact that in the terrestrial igneous rocks more sodic plagioclase is in equilibrium with K-feldspar and bytownite is usually not saturated with K. Plagioclase in the "cognate inclusion" of rock 14310,22 has the composition $An_{92.3}Ab_{7.1}Or_{0.6}$, which is nearly the same as that of the calcic core of the zoned plagioclase in the main part.

Plagioclase in rock 14053,6 is relatively homogeneous ($An_{91}Ab_9$–$An_{93}Ab_7$) except near the rim, which is zoned to sodic bytownite and is commonly in contact with K-feldspar, cristobalite, and glass. The compositions of one pair of contacting plagioclase and K-feldspar are $An_{70.6}Ab_{24.9}Or_{4.5}$ and $An_{1.0}Ab_{1.8}Or_{97.2}$, respectively. The K-feldspar is slightly zoned from $An_{1.9}Ab_{2.0}Or_{96.0}$ in the core to $An_{1.0}Ab_{1.8}Or_{97.2}$ at the rim.

Plagioclase in an anorthositic rock fragment in breccia 14321,22 shows a compositional zoning from $An_{90.8}Ab_{8.6}Or_{0.6}$ in the core to $An_{82.4}Ab_{15.6}Or_{2.0}$ at the rim.

Olivine

Rock 14053,6 contains olivines of two distinctly different compositions. Inclusions of olivine in pigeonite crystals have the composition $Fo_{63}Fa_{37}$, whereas olivine that occurs in the interstices and is associated with cristobalite has the composition $Fo_{12}Fa_{88}$. Olivines of intermediate compositions have not been found, suggesting a reaction relation between magnesian olivine and pigeonite, and breakdown of the ferrosilite component in iron-rich pyroxene to iron-rich olivine and silica.

In breccia 14321,22 two fragments of zoned olivine crystals have cores with the compositions $Fo_{91}Fa_9$ and $Fo_{89}Fa_{11}$. They are more magnesian than those of the Apollo 11 and 12 crystalline rocks and nearly the same as the magnesian olivines crystallized from some terrestrial basic magmas, suggesting that at least some lunar basic magmas had MgO/FeO ratios similar to those of terrestrial basic magmas.

Glass

Pale brown to colorless glass is commonly found in the mesostasis of rock 14310,22. The composition determined with the microprobe is enriched in Si and K and is similar to those of the mesostasis of the Apollo 11 crystalline rocks (e.g., Roedder and Weiblen, 1970; Kushiro and Nakamura, 1970). In some of the mesostasis, colorless glass spherules are surrounded by a dark part, which consists of fibrous or skeletal ilmenite and very fine-grained pyroxenes. These appear to be quench crystals; that is, the dark mafic part was originally liquid. It seems likely that the colorless glass and the dark part were formed by liquid immiscibility, as first reported by Roedder and Weiblen (1970) in the Apollo 11 crystalline rocks.

The composition of the colorless glass (Table 2, No. 10) also is enriched in Si and K. The bulk composition of the quenched mafic part (Table 2, No. 11) is very rich in Fe and Ti, and its norm indicates that it is essentially a mixture of ilmenite and pyroxene. The ratio of the mafic part to the colorless glass is not large on the whole, and the bulk composition of the residual liquid is closer to that of the Si- and K-rich glass. These residual liquids would have been produced by small-scale fractionation of the magma, as suggested by Kushiro et al. (1971). However, the glass of such composition has considerably higher Or/Ab ratios than most terrestrial granites, and its composition is similar to those of some alaskite and aplite.

In rock 14053, glass patches also are found in the interstices and are commonly associated with pyroxferroite. Their compositions are very similar to those of rock 14310; i.e., of potassic granite or rhyolite (Table 2, No. 22).

MELTING EXPERIMENTS

Melting experiments have been carried out on high-alumina basalt 14310 at pressures of 5 to 25 kb and temperatures of 1100° to 1470°C in the piston-cylinder apparatus with graphite capsules. The piston-out technique, without correction, was applied to all the runs. The error of pressure in the low-pressure range (<10 kb) should be very large (probably up to 20%). Synthetic glass of this rock composition made from oxides (Table 1) was used as a starting material for preliminary determination of the phase relations. The results of the runs are shown in Table 3 and in a pressure-temperature diagram (Fig. 4). The results on both the natural rock and the synthetic glass are consistent with each other, except that plagioclase is finer

Table 3. Results of experiments on rock 14310,139
and a synthetic glass of the same composition.

Pressure (kb)	Temperature (°C)	Time (hr)	Results
			14310,139 (homogenized fines)
5	1300	1	Pl + Gl
5	1250	2	Pl + Gl
5	1200	$3\frac{1}{2}$	Pl + Px (Opx + Cpx?) + Sp + Gl (Opx: 4.3 wt% Al_2O_3)
7.5	1350	$\frac{1}{2}$	Gl + q-Cryst
7.5	1315	1	Gl
7.5	1290	$1\frac{1}{2}$	Pl + Gl
7.5	1250	2	Pl + Px (Opx + Cpx?) + Gl
10	1350	1	Gl
10	1325	1	Gl
10	1300	$1\frac{2}{3}$	Pl + Sp (rare) + Gl
10	1275	2	Pl + Sp + Px (trace) + Gl (Sp: 20.0 wt% Cr_2O_3, 0.4% TiO_2, and 17.5% MgO)
10	1250	2	Px (Opx + Cpx?) + Pl + Sp (rare) + Gl (Opx: Ca 5.1, Mg 78.5, Fe 16.4, with 6.6 wt% Al_2O_3)
10	1175	2	Pl + Px (Opx + Cpx) + Sp + Gl (trace)
15	1325	1	Sp + Gl
17.5	1300	$1\frac{1}{2}$	Gar + Cpx + Pl + Sp (trace) + Gl
20	1375	1	Gl (some Gl fragments contain very rare Sp)
20	1325	$1\frac{1}{2}$	Gar + Cpx + Pl + Sp (trace) + Gl (Cpx: Ca 40.5, Mg 47.1, Fe 12.4, with ~11% Al_2O_3; Gar: Ca 20.6 Mg 57.9, Fe 21.6)
25	1390	$\frac{2}{3}$	Gar + Cpx (rare) + Gl

Table 3. (continued).

Pressure (kb)	Temperature (°C)	Time (hr)	Results
			14310 synthetic glass
5	1325	1	Gl
5	1225	2	Pl + Px + Sp + Gl
5	1125	4	Pl + Px + Sp
5	1100	$4\frac{1}{6}$	Pl + Px + Sp
7.5	1360	$\frac{2}{3}$	Gl
7.5	1300	$1\frac{1}{2}$	Pl + Sp + Gl
7.5	1200	$2\frac{1}{2}$	Pl + Px + Sp + Gl
12.5	1325	1	Sp + Gl
12.5	1300	$1\frac{1}{2}$	Pl + Sp + Px? + Gl
15*	1350	$1\frac{1}{2}$	Sp (trace) + Gl
15	1325	$1\frac{1}{2}$	Sp + Gl
15	1300	$1\frac{2}{3}$	Pl + Cpx + Sp + Gl
15	1225	$2\frac{2}{5}$	Pl + Px + Gar + Sp
15	1175	$3\frac{1}{3}$	Pl + Px + Gar
17.5	1325	$1\frac{5}{6}$	Sp + Cpx + Gl
20*	1350	$1\frac{1}{2}$	Gar + Cpx + Sp (rare) + Gl
20	1250	2	Gar + Cpx + Sp + Gl
20	1200	3	Pl + Cpx + Gar + Sp
30†	1465	$\frac{1}{2}$	Gar (rare) + Gl
30†	1425	1	Gar + Gl

Abbreviations: Cpx, clinopyroxene; Gar, garnet; Gl, glass; L, liquid; Ol, olivine; Opx, orthopyroxene; Pl, plagioclase; Px, pyroxene (Opx and Cpx are not distinguished unless specifically mentioned); q, quench; Sp, spinel.
*Conducted in iron capsule. Others conducted in graphite capsule.
†Conducted by S. Akimoto in graphite capsule with tetrahedral anvil apparatus.

grained and the amount of spinel (a solid solution of spinel, hercynite and chromite) appears to be larger in runs made with the synthetic glass than in those made using the natural rock.

Below about 10 kb, calcic plagioclase is the liquidus phase and is followed by chromian spinel, aluminous orthopyroxene, and clinopyroxene with lowering temperature. Between about 10 and 20 kb, spinel is the liquidus phase, and above about 20 kb, pyrope-rich garnet is the liquidus phase. In these experiments spinel crystallizes over a wide P-T range. It is almost colorless to pale brown near the liquidus, becoming brown to dark brown with lowering temperature to near the solidus and dark brown to almost opaque below the solidus. Clearly its composition changes with temperature. Some investigators (e.g., Ford et al., 1972; Walker et al., 1972) reported that spinel crystallizes on the liquidus but in a much narrower pressure range (e.g., 12–17 kb). This discrepancy may be due to a difference in oxygen fugacity during the run. In the present experiments, made in graphite capsules, oxygen fugacity would have been higher than in those made with molybdenum and iron capsules (Heubner, 1971), and spinel would therefore be stable over a wider P-T range. The melting interval of rock 14310 is 150°–200°C in the pressure range studied.

Several runs were made at 1 atm on the synthetic glass in a CO_2 atmosphere by T. Katsura. Under these conditions, olivine crystallizes between 1213° and 1199°C. In the runs made at lower oxygen fugacity olivine also crystallizes near 1200°C (Muan et al., 1972; Ford et al., 1972; Ringwood et al., 1972). Since olivine is not

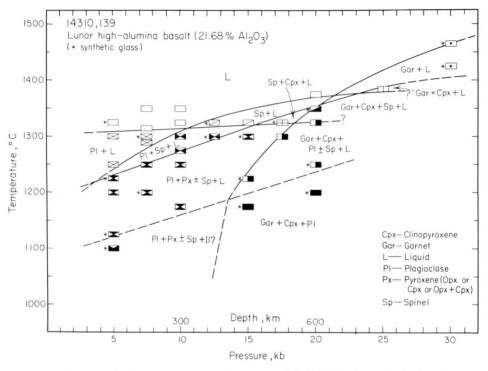

Fig. 4. Results of high-pressure experiments on rock 14310,139 and a synthetic glass of the same composition. The runs made with the latter starting material are shown by asterisks. Different box symbols correspond to different assemblages that are labeled in each field of the diagram.

observed at pressures higher than 5 kb, the P-T field for olivine crystallization terminates below 5 kb, as shown by Ford *et al.* (1972).

Orthopyroxene crystallized at 10 kb and 1250°C has the composition $Ca_{5.1}Mg_{78.5}Fe_{16.4}$ with 5.7 to 7.7 (average 6.6) wt% Al_2O_3 and 0.82 wt% Cr_2O_3. These amounts are greater than those of the orthopyroxene in rock 14310. At 5 kb and 1200°C the orthopyroxene contains 4.3 wt% Al_2O_3, which is still greater than that of the 14310 orthopyroxene. Spinel crystallized at 10 kb and 1275°C contains about 20.0 wt% Cr_2O_3, 48.5% Al_2O_3, 17.5% MgO, 13.5% FeO, and 0.4% TiO_2. Clinopyroxene crystallized at 20 kb and 1325°C has the composition $Ca_{40.5}Mg_{47.1}Fe_{12.4}$ with high alumina contents ($\sim 11\%$ Al_2O_3 and 0.35% Na_2O). The Ca:Mg:Fe ratio of this clinopyroxene is almost identical with that of a core augite ($\sim Ca_{40}Mg_{47}Fs_{13}$) found in orthopyroxene in rock 14310 by Hollister *et al.* (1972), although alumina is much higher and Ti is lower in the former than in the latter.

DISCUSSION

The experimental results indicate that the 14310 composition is within one of the volumes of alumina-rich phases in a pressure range from 1 atm to at least 30 kb;

that is, in the plagioclase volume at low pressures, in the spinel volume at intermediate pressures, and in the garnet volume at high pressures. At about 10 kb this composition lies on the boundaries between the plagioclase and spinel volumes, and at about 20 kb it lies between the spinel and garnet volumes. These relations are shown in the simple system anorthite-olivine-silica (Fig. 5), onto which the 14310 composition is projected from the apex of other components (normative Or, Ab, Wo, Il, Cm, and Ap). The total of these remaining components is about 15 wt% of the 14310 composition. At 1 atm the 14310 composition is in the plagioclase field, whereas at 10 kb it lies on or very close to the spinel-anorthite boundary. The phase boundaries at 10 kb are consistent with the runs made on 14310 at this pressure. The compositions of feldspathic basalts (or anorthositic gabbro), which Reid *et al.* (1972) have suggested to be lunar highland materials, also are plotted in this system. The compositions of rock 14310 and feldspathic basalts are on the An side of the piercing points at 1 atm, Sp + Ol + An + *L* (*C*) and Ol + An + Ca-poor Px + *L* (*D*), and those at 10 kb, Ol + Sp + Opx + *L* (*A*) and Sp + Opx + An + *L* (*B*), or the low-temperature cotectic boundaries at least to 10 kb. Walker *et al.* (1972) have indicated that the composition of the fines sample 14259 is close to the three-phase (olivine, pyroxene, and plagioclase) saturation boundary (or piercing point) at 1 atm. Addition of 20 wt% plagioclase (An$_{94}$Ab$_6$) to the fines gives a composition almost identical with that of rock 14310, indicating that the composition of the latter is away from the composition of the piercing point by about 20 wt% calcic plagioclase component. On the other hand, the possible lunar "mantle" materials (peridotite or pyroxenite with or without a small amount of plagioclase) would lie on the olivine or pyroxene side of these piercing points. If rock 14310 is not a plagioclase-cumulate rock, it cannot be a product of direct partial melting of the lunar "mantle" materials at least to 300 km under anhydrous conditions; nor is it a product of fractional crystallization of magma ("primary magma") that was formed by direct partial melting of the lunar "mantle" materials. Even if rock 14310 is a plagioclase-cumulate rock, these arguments would still be valid unless cumulate plagioclase is more than 15 (or close to 20) wt%, because the composition 14310, with less than 15 wt%

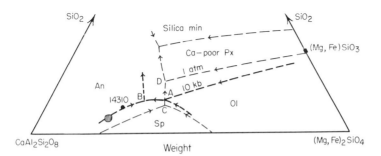

Fig. 5. System (Mg,Fe)$_2$SiO$_4$-CaAl$_2$Si$_2$O$_8$-SiO$_2$(Fe/[Mg + Fe] ratio, 0.3) at 1 atm and 10 kb. The liquidus boundaries are based on Andersen (1915), Roeder and Osborn (1966), and present experimental results. Solid circle, 14310 composition; dashed area, feldspathic basalts (Reid *et al.*, 1972). A and B are piercing points at 10 kb; C and D are those at 1 atm.

anorthite, would still be on the An side of these piercing points at pressures at least up to 20 kb.

Therefore, a two-stage model must be considered for the origin of rock 14310 and the feldspathic basalts. One possibility is that the magmas of these rocks were formed by bulk melting or partial melting of a plagioclase-cumulate rock, which had been formed in an earlier stage by a large-scale differentiation of magma in the shallow parts of the lunar interior and constituted a part of the lunar highland.

If rock 14310 is a plagioclase-cumulate rock with plagioclase enrichment by more than 15 (or close to 20) wt%, the rock could be formed either by partial melting of the lunar "mantle" material or by fractional crystallization of a more mafic magma, followed by plagioclase cumulation before the solidification of the magma. This possibility has been discussed by Walker et al. (1972). As described before, however, such a large amount of plagioclase phenocrysts was not observed in the thin section examined (14310,22). Brown and Peckett (1971) estimated about 10 volume % of plagioclase phenocrysts in two thin sections of rock 14310, although definition and estimation of plagioclase phenocrysts is extremely difficult in this rock, because the grain size distribution appears to be continuous. Because the discrepancy between the amount of plagioclase phenocrysts suggested and that required for the plagioclase-cumulation hypothesis appears to be significantly large, and because resorption of cumulate plagioclase was unlikely to occur during the ascent of magma, it is difficult to accept the plagioclase-cumulation hypothesis. The process involving garnet (at depths greater than 500 km) may not be related to the origin of this high-alumina basalt.

The above discussion is based on the assumption that there was no loss or gain of alkalis and volatile components during the magmatic stage. Brown and Peckett (1971) observed that "groundmass" plagioclase is more calcic than the outer part of "phenocryst" plagioclase, and to explain this observation they suggested significant loss of alkalis from the magma at or near the lunar surface. They estimated the loss of Na_2O and K_2O to be about 70 wt% of the amount originally present. If this is true, the discussion on the origin of rock 14310 may be subject to change. To examine

Table 4. Results of experiments on 14310 composition with alkalis added
(3.2 wt% Na_2O and $\sim 1\%$ K_2O)

Pressure (kb)	Temperature (°C)	Time (hr)	Results
2.5	1215	$2\frac{1}{3}$	Pl + Sp (rare) + Ol (rare) + Gl
3.5	1220	$2\frac{1}{3}$	Ol (rare) + Sp (rare) + Gl (Ol, $\sim Fo_{90}Fa_{10}$)
3.5*	1215	$3\frac{1}{2}$	Pl + Sp + Ol? + Gl
3.5*	1200	2	Pl + Sp + Px (mostly Opx) + Ol? + Gl
5	1240	$1\frac{5}{6}$	Sp + Gl
5	1225	$1\frac{1}{4}$	Sp + Pl (rare) + Gl
5*	1210	$2\frac{1}{3}$	Pl + Sp + Px + Ol + Gl
10	1300	$\frac{5}{6}$	Sp + Gl
10*	1275	2	Sp + Gl
10*	1250	1	Sp + Px + Gl
10*	1225	2	Sp + Pl + Px + Gl

*Synthetic glass with alkalis added. Others are 14310,139 with alkalis added.

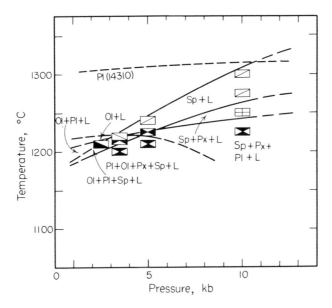

Fig. 6. Results of experiments on the alkali-enriched 14310 composition (total alkalis, Na_2O 3.2 and K_2O ~1.0 wt%). Dashed curve indicates the liquidus of plagioclase for the 14310 composition in Fig. 4. Symbols as in Fig.4 except presence of olivine in the low-pressure runs.

this possibility, melting experiments have been carried out on synthetic glass of the 14310 rock composition and on actual rock 14310,139 with addition of 2.5 wt% Na_2O and about 0.5% K_2O, making total alkali contents 3.2 wt% Na_2O and about 1% K_2O. These contents are similar to those of some terrestrial high-alumina basalts. The results are given in Table 4 and Fig. 6. The major changes are a drastic drop of the plagioclase liquidus (by 70°–90° between 2.5 and 10 kb) and crystallization of olivine to higher pressures (at least up to 5 kb). At 3.5 kb the liquidus temperature is slightly above 1220°C and four crystalline phases—olivine, spinel, plagioclase, and pyroxene—appear within a 20° interval, indicating that the alkali-enriched 14310 composition is very close to the boundary of the volumes of these four phases. This evidence suggests that the liquid of this alkali-enriched composition could be formed by direct partial melting of a plagioclase- and spinel-bearing peridotite at about 3 kb (~100 km) or by fractional crystallization of a more mafic magma. If significant alkali loss is accepted, a single-stage model may be possible for the origin of rock 14310; the original magma of rock 14310 was formed at depths near 100 km and erupted on the lunar surface, where alkalis were lost before the crystallization of most plagioclase and other minerals. The crystallization of plagioclase after alkali loss must have been very rapid, because the magma became supercooled with respect to plagioclase as soon as alkalis were lost. Most of the other crystals also would have crystallized rapidly.

The above discussion is based on the assumption that the lunar interior was completely or almost free of H_2O when the magmas were generated. If H_2O existed in the lunar interior, the magma would have been considerably enriched in An component, as shown by Yoder (1965), and the origin of An-rich magma could be explained by a single-stage process without considering a large amount of alkali loss. However, there is no positive evidence for the existence of a significant amount of water in the lunar interior.

As mentioned before and pointed out by other investigators (e.g., Ford *et al.*, 1972; Ringwood *et al.*, 1972), the MgO/FeO ratio of rock 14310 is greater than that of any of the mare-type crystalline rocks collected by the Apollo 11, 12, and 14 missions. If the magma of this rock were directly derived from the lunar "mantle," as in the single-stage model, the lunar "mantle" may have a MgO/FeO ratio similar to that of the earth's mantle. Olivine crystallized at 3.5 kb and 1220°C from the alkali-enriched 14310 composition is about $Fo_{90}Fa_{10}$, supporting this argument. Even for the double-stage model, the same conclusion would be obtained. As described before, the most magnesian parts of the zoned olivine fragments in breccia 14321 have the compositions $Fo_{91}Fa_9$ and $Fo_{89}Fa_{11}$. If they crystallized from magmas that were derived initially from the lunar interior, the lunar "mantle" should have a similar or higher MgO/FeO ratio, at least in the source regions of the high-alumina basalts of the lunar highlands.

In summary, if significant alkali loss did not occur before or during the solidification of magma of rock 14310, the rock could have been formed by remelting (either bulk melting or partial melting) of a plagioclase-cumulate rock that had been formed by a large-scale differentiation in the shallower part of the lunar interior in an earlier stage. Alternatively, if significant alkali loss ($\sim 70\%$ or more) occurred, the rock could have been formed either by partial melting of the lunar "mantle" materials or by fractional crystallization of a more primitive magma at depths of about 100 km.

Acknowledgments—The authors are grateful to Mr. H. Haramura for chemical analysis of rock 14310 and to Drs. S. Akimoto and T. Katsura for several runs made at 1 atm and 30 kb. They also thank Drs. F. R. Boyd, R. N. Thompson, D. Virgo, and H. S. Yoder, Jr., for their critical reading of the manuscript.

REFERENCES

Andersen O. (1915) The system anorthite-forsterite-silica. *Amer. J. Sci.* **39**, 407–454.

Bence A. E. and Papike J. J. (1972) Crystallization histories of pyroxenes from lunar basalts (abstract), *Lunar Science-III* (editor C. Watkins), p. 59, Lunar Science Institute Contr. No. 88.

Boyd F. R. and Smith D. (1971) Compositional zoning in pyroxenes from lunar rock 12021, *Oceanus Procellarum. J. Petrol.* **12** 439–464.

Brown G. M. and Peckett A. (1971) Selective volatilization on the lunar surface: evidence from Apollo 14 feldspar-phyric basalts. *Nature* **234**, 262–266.

Finger L. W. Hafner S. S. Schurmann K. Virgo D. and Warburton D. (1972) Distinct cooling histories and reheating of Apollo 14 rocks (abstract). *Lunar Science-III* (editor C. Watkins), p. 259, Lunar Science Institute Contr. No. 88.

Ford C. E. Humphries D. J. Wilson G. Dixon D. Biggar G. M. and O'Hara M. J. (1972) Experimental petrology of high alumina basalt, 14310, and related compositions (abstract). *Lunar Science-III* (editor C. Watkins), p. 274, Lunar Science Institute Contr. No. 88.

Heubner, J. S. (1971) Buffering techniques for hydrostatic systems at elevated pressures. In *Research techniques for high pressure and high temperature*. G. C. Ulmer, editor, pp. 123–177, Springer-Verlag.

Hollister L. Trzcienski W. Dymek R. Kulick C. Weigand P. and Hargraves R. (1972) Igneous fragment

14310,21 and the origin of the mare basalts (abstract). *Lunar Science-III* (editor C. Watkins), p. 386, Lunar Science Institute Contr. No. 88.

Hollister L. Trzcienski W. Hargraves R. and Kulick C. (1971) Petrogenetic significance of pyroxenes in two Apollo 12 samples. *Proc. Second Lunar Sci. Conf., Geochim. Cosmochim. Acta* Suppl. 2, Vol. 1, pp. 529–557. MIT Press.

Kuno H. (1960) High-alumina basalt. *J. Petrol.* **1**, 121–145.

Kushiro I. (1969) The system forsterite-diopside-silica with and without water at high pressures. *Amer. J. Sci., Schairer Volume* **267-A**, 269–294.

Kushiro I. (1972) Determination of liquidus relations in synthetic silicate systems with electron probe analysis: the system forsterite-diopside-silica at 1 atmosphere. *Amer. Mineral.* (in press).

Kushiro I. and Nakamura Y. (1970) Petrology of some lunar crystalline rocks. *Proc. Apollo 11 Lunar Sci. Conf., Geochim. Cosmochim. Acta* Suppl. 1, Vol. 1, pp. 607–626. Pergamon.

Kushiro I. Nakamura Y. Kitayama K. and Akimoto S. (1971) Petrology of some Apollo 12 crystalline rocks. *Proc. Second Lunar Sci. Conf., Geochim. Cosmochim. Acta* Suppl. 2, Vol. 1, pp. 481–495. MIT Press.

LSPET (1971) (Lunar Sample Preliminary Examination Team) Preliminary examination of lunar samples from Apollo 14, *Science* **173**, 681–693.

LSPET (1972) (Lunar Sample Preliminary Examination Team) A preliminary description of the Apollo 15 lunar samples, *Science* (in press).

Muan A. Hanck J. and Löfall T. (1972) Equilibrium studies with a bearing on lunar rocks (abstract). *Lunar Science-III* (editor C. Watkins), p. 561, Lunar Science Institute Contr. No. 88.

Nakamura Y. and Kushiro I. (1970) Compositional relations of coexisting orthopyroxene, pigeonite, and pigeonite in a tholeiitic andesite from Hakone Volcano. *Contrib. Mineral. Petrol.* **26**, 265–275.

Reid A. M. Ridley W. I. Warner J. Harmon R. S. Brett R. Jakeš P. and Brown R. W. (1972) Chemistry of highland and mare basalts as inferred from glasses in the lunar soils (abstract). *Lunar Science-III* (editor C. Watkins), p. 640, Lunar Science Institute Contr. No. 88.

Ringwood A. E. Green D. H. and Ware N. G. (1972) Experimental petrology and petrogenesis of Apollo 14 basalts (abstract). *Lunar Science-III* (editor C. Watkins), p. 654, Lunar Science Institute Contr. No. 88.

Roedder E. and Weiblen P. W. (1970) Lunar petrology of silicate melt inclusions, Apollo 11 rocks. *Proc. Apollo 11 Lunar Sci. Conf., Geochim. Cosmochim. Acta* Suppl. 1, Vol. 1, pp. 801–837. Pergamon.

Roeder P. L. and Osborn E. F. (1966) Experimental data for the system $MgO-FeO-Fe_2O_3-CaAl_2Si_2O_8-SiO_2$ and their petrologic implications. *Amer. J. Sci.* **264**, 428–480.

Walker D. Longhi J. and Hays J. F. (1972) Experimental petrology and origin of Fra Mauro rocks and soil (abstract). *Lunar Science-III* (editor C. Watkins), p. 770, Lunar Science Institute Contr. No. 88.

Yoder H. S. (1965) Diopside-anorthite-water at five and ten kilobars and its bearing on explosive volcanism, *Carnegie Inst. Washington Year Book* **64**, 82–89.

Proceedings of the Third Lunar Science Conference
(Supplement 3, *Geochimica et Cosmochimica Acta*)
Vol. 1, pp. 131–139
The M.I.T. Press, 1972

Petrography and crystallization history
of basalts 14310 and 14072

JOHN LONGHI, DAVID WALKER, and JAMES FRED HAYS

Department of Geological Sciences, Harvard University
Cambridge, Massachusetts 02138

Abstract—Rock 14310 is a high-alumina basalt containing plagioclase phenocrysts, grains of pigeonite with orthopyroxene cores, and a partially unmixed glassy residuum. Rock 14072 is a subophitic basalt with large resorbed olivine phenocrysts. Both rocks show evidence for strong post-crystallization reduction.

Consideration of Fe/Mg in plagioclase and other compositional and textural evidence leads to a rejection of the alkali-volatilization hypothesis.

INTRODUCTION

THE APOLLO 14 MISSION to the Fra Mauro Hills of the moon returned 43 kg of rocks and soils consisting chiefly of a complex assemblage of breccias, thus establishing the clastic nature of the Fra Mauro formation. Among the returned samples, however, were a few crystalline rocks showing basaltic textures. Similar clasts in many of the breccias are clear evidence for pre-Fra Mauro volcanic activity. This paper reports the results of petrographic and electron microprobe study of two polished thin sections of basaltic rocks 14310,30 and 14072,16 and our interpretation of their crystallization histories. These studies supplement phase equilibrium experiments on these rocks and related compositions; our interpretations of the origin and history of these rocks, are reported more fully in Walker *et al.* (1972).

DESCRIPTION

Samples 14310,30 and 14072,16 are fine to medium-grained basalts with rather different textures and mineralogies. Rock 14310 is a high alumina (20 wt% Al_2O_3) basalt with mixed subophitic and intergranular textures, a variable amount of plagioclase micro-phenocrysts, no olivine, and about 6% glassy residuum. Rock 14072 is a porphyritic basalt with medium-sized olivine phenocrysts, subophitic to ophitic texture and little glass in its residuum; it is similar in texture and mineralogy to rock 14053 (Bence and Papike, 1971) and several Apollo 12 basalts, though it possesses a lower Fe/Mg ratio. Bulk chemical and modal analyses are listed in Table 1.

Plagioclase in 14310 occurs in three habits: (1) more or less equidimensional, untwinned, subhedral crystals with blocky or rhomboid outlines (0.5 to 1.0 mm), interpreted as phenocrysts (Fig. 1); (2) lath-like crystals up to 1.0 mm long that show occasional albite or carlsbad twinning and which are intergrown with the larger pyroxenes; (3) fine to very fine lath to needle-like crystals commonly found in clusters with late pyroxenes or glasses and opaque minerals in the interstices. In 14072 the plagioclase grains commonly are ragged intergrowths of two or more

Table 1. Bulk chemical Compositions*
(wt %)

	14310,138	14072,3
SiO$_2$	48.27	44.94
TiO$_2$	1.27	2.56
Al$_2$O$_3$	20.26	11.31
Cr$_2$O$_3$	0.20	0.52
FeO	8.11	17.07
MgO	7.76	12.21
CaO	12.25	9.63
Na$_2$O	0.81	0.38
K$_2$O	0.55	0.11
TOTAL	99.48	98.73

Mode†
(vol. %)

	14310,30	14072,16
Orthopyroxene	13.8	—
Clinopyroxene	21.0	49.9
Olivine	—	2.5
Plagioclase	56.6	38.3
Opaques	2.9	7.7
Glass and Silica Minerals	5.9	1.7
Tranquillityite	tr	—

*Electron microprobe analyses of glasses formed
by quenching rock powders melted in graphite containers at 5 kb.
†Each mode based on 1000 counts.

Fig. 1. Photomicrograph of rock 14310,30 in plane light showing blocky, lath-like and needle-
like plagioclase. Larger laths are here intergrown with magnesian pigeonite, but augite
and more iron-rich pigeonite as well as opaque minerals and glasses are interstitial to the
smaller plagioclase laths and needles.

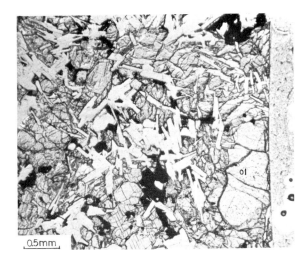

Fig. 2. Photomicrograph of rock 14072,16 in plane light. Plagioclase is more anhedral and ranges less in size and habit than that of 14310. Both opaque minerals and plagioclase commonly are included in large pigeonite crystals. Part of a medium-sized olivine pheno-cryst (Fa_{26}) occupies the lower right hand corner.

lath-like, subhedral crystals no more than 0.5 mm long. The equidimensional and acicular classes of 14310 are absent. Smaller laths are included in the larger pyroxenes, but other laths and composite grains appear pushed in between the pyroxenes, intruding the pyroxene margins (Fig. 2).

Systematic variations of Fe and Mg in the plagioclases provides a key to the interpretation of the crystallization history of rock 14310. Figure 3a shows that as An content decreases: K content increases, Fe increases (with some scatter), and Mg content ranges more or less randomly. These trends are similar to those shown by Apollo 11 plagioclase (Smith, 1971). The detailed mechanism of Fe and Mg incorporation in plagioclase is not known, but Quaide (1972) argues that under conditions of high temperature, rapid crystallization, Fe/Mg ratios in plagioclase may reflect Fe/Mg ratios in the magma at the time of crystallization. If so, the sequence of increasing Fe/Mg in plagioclase also should be the sequence of crystallization. Fig. 3c shows that Fe/Mg in 14310 plagioclase varies with crystal habit as follows: cores of phenocrysts and larger laths have lowest Fe/Mg; rims of phenocrysts and laths have intermediate ratios; and late needles have the highest. Because Fe/Mg ratio does not correlate closely with either K or An content (Fig. 3b, c), we have an additional clue to the alkali history of the 14310 magma.

Brown and Peckett (1971) observed a general decrease in An content from core (An_{94}) to rim (An_{67}) of phenocrysts, together with high An content (An_{92}) of ground mass laths. They interpret these analyses as a result of progressive increase of alkali content of the 14310 magma during crystallization of the phenocrysts followed by catastrophic volatilization of alkalis and rapid crystallization of calcic laths from

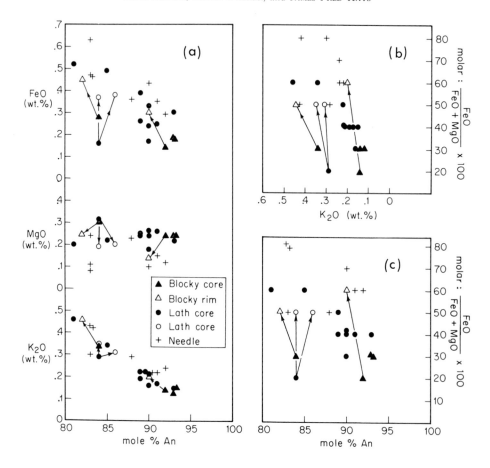

Fig. 3. Minor element variation in 14310 plagioclase. Solid lines connect core and rim of the same grain. (a) Weight percent FeO. MgO, and K_2O versus mol. percent An. (b) Molar Fe/Mg ratio versus weight percent K_2O. Grains strongly zoned with respect to Fe/Mg have essentially constant K_2O. (c) Molar Fe/Mg ratio versus mol. percent An. Zoning and large range in Fe/Mg are apparent. Variation in An and K_2O among grains of similar habit. Fe/Mg is attributed to local heterogeneity in alkali content of magma.

the alkali-depleted magma. We concur with Ridley, *et al.* (1972), who observe plagioclase laths intruding and included within phenocryst rims, variations in zoning of phenocrysts and laths, and considerable range of the An content of fine laths. There is no report of laths, large or small, with core compositions more sodic than An_{80} nor of reversely zoned phenocryst rims. Thus, chemical and textural evidence suggest that alkali content varied locally within the magma throughout crystallization, and that phenocryst rims grew later than many of the laths.

We, therefore, disagree with the scheme and magnitude of volatile loss proposed by Brown and Peckett. Small amounts of alkali loss may have occurred, but rapid crystallization of small Ca and Al depleted zones surrounding growing crystals of plagioclase (Lofgren, 1972) is a more satisfactory explanation for the observed pattern of zoning and variation.

Olivine appears in 14072 as large, subrounded phenocrysts ($Fa_{24\ 26}$) up to 2.0 mm across, as inclusions (Fa_{36}) in large pyroxenes, as anhedral members ($Fa_{39\ 42}$) of the general framework, and as part of the late stage assemblage (Fa_{65}) together with mosaic cristobalite and a spongy network of native iron. Our experimental work (Walker et al., 1972) suggests that the absence of olivine in 14310 results not from its failure to crystallize, but to crystallization over a very narrow temperature interval followed by efficient gravity settling and resorption.

Complex pyroxenes up to 1 mm in length with bronzite ($En_{74}Fs_{22}Wo_4$) cores and pigeonite ($En_{66}Fs_{27}Wo_7$) mantles, which may zone to augite, are prominent in 14310 (Fig. 4). Plagioclase laths commonly intrude the mantles and sometimes cut across the orthopyroxene cores (Fig. 5). In addition, there is much fine-grained, anhedral pigeonite and augite molded in between plagioclase laths and needles.

Rock 14072 lacks orthopyroxene, but 1 to 2 mm long grains of pigeonite ($En_{59}Fs_{29}Wo_{11}$) show complex patterns of zoning and rimming, with some rims reaching augite or ferroaugite compositions ($En_{15}Fs_{55}Wo_{31}$). A small amount of interstitial pyroxene also is present, but much less than in 14310. Table 2 gives selected olivine and pyroxene analyses.

At least three different glasses are present in 14310 (Table 2). One is a clear, rhyolitic type (76% SiO_2, 7% K_2O). The other two, high K_2O, low FeO, low SiO_2, and low K_2O, high FeO, low SiO_2, are intergrown on a very small scale and may be immiscible separates related to the lc-fa-SiO_2 system (Roedder, 1951). They appear dark and fuzzy in plane light because of small-scale divitrification products and swarms of minute metallic blebs (Fig. 6).

Fig. 4. Plagioclase lath intergrown with two pigeonite grains in rock 14310,30. The lower pigeonite grain has a bronzite core (dark). The core is unzoned, but exsolution lamellae ($2\ \mu$ thick) are detectable within the core by electron microprobe traverse. Crossed polarizers.

Fig. 5. Plagioclase laths cutting across bronzite core of complex pyroxene in rock 14310,30. Crossed polarizers.

Table 2. Selected electron microprobe analyses of minerals and glasses from 14310,30 and 14072,16 (wt %).

| | 14310,30 | | | | | | | | | 14072,16 | | | | |
	1–opx	2–cpx	3–cpx	4–plag	5–plag	6–plag	7–gl	8–gl	9–gl	10–plag	11–ol	12–cpx	13–cpx	14–cpx
SiO$_2$	54.59	53.39	50.31	45.40	45.77	45.44	75.90	55.93	45.18	45.62	38.68	52.34	47.04	47.46
TiO$_2$	0.47	0.60	2.09	0.03	0.05	0.01	0.72	2.14	2.33	0.01	0.0	1.07	1.16	0.91
Al$_2$O$_3$	1.26	1.52	2.43	34.93	34.35	35.02	12.67	12.93	2.90	34.56	0.20	2.15	1.31	1.14
Cr$_2$O$_3$	n.d.	n.d.	0.50	0.0	0.0	0.0	0.0	0.22	0.13	0.0	0.0	0.67	0.19	0.09
FeO	14.09	17.32	14.21	0.14	0.30	0.47	0.77	9.65	23.30	0.42	24.06	16.59	32.24	36.16
MgO	27.24	23.51	14.96	0.24	0.14	0.08	0.01	0.26	7.80	0.15	38.57	20.55	4.67	3.60
CaO	1.88	3.59	15.91	18.82	18.45	17.70	0.92	5.36	15.10	18.78	0.24	7.69	13.64	10.20
Na$_2$O	n.d.	n.d.	0.12	0.85	1.01	1.31	1.20	0.77	0.16	0.77	0.0	0.0	0.06	0.0
K$_2$O	n.d.	n.d.	0.11	0.14	0.20	0.30	7.79	12.68	0.65	0.11	0.09	0.09	0.11	0.11
TOTAL	99.43	99.93	100.64	100.55	100.27	100.33	99.98	99.94	97.55	100.42	101.84	101.15	99.42	99.67

n.d. = not determined.
  (1) and (2) Bronzite core and pigeonite rim of composite pyroxene in Fig. 4; (3) Interstitial augite; (4) and (5) Core and rim of large plagioclase phenocryst in Fig. 1; (6) Plagioclase needle; (7) Rhyolitic glass; (8) High K$_2$O—low FeO glass; (9) Low K$_2$O—high FeO glass; (10) Plagioclase core; (11) Olivine phenocryst with inclusions; (12), (13), and (14) Core, mantle and rim of zoned pyroxene.

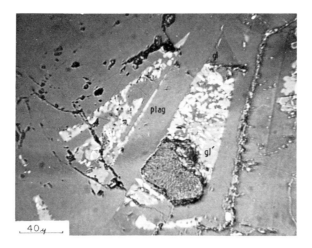

Fig. 6. Late stage immiscible (?) glasses occupying angular interstices between plagioclase laths in rock 14310,30 in reflected light. Glasses are crowded with ilmenite (?) blebs. gl = clear, colorless glass (high SiO_2); gl' = dark, fuzzy glass from which analyses 9 and 10 (Table 2) were obtained. Rough gray area is epoxy mounting medium.

Ilmenite, chrome-ulvöspinel, native iron and troilite are common to both rocks. Tranquillityite is present in 14310 only in the glassy residuum. Spinels from the 14310 residuum, are similar to those of Apollo 12 rocks, but 14072 spinels, some of which are embayed by silicates, are similar to Apollo 11 spinels (Haggerty and Meyer, 1971).

This difference results from higher crystallization temperatures of the 14072 spinels, and therefore, a narrower or closed miscibility gap along the (Fe, Mg) TiO_4–(Fe, Mg) $(Cr, Al)_2O_4$ join (Muan *et al.*, 1972). Chrome-ulvöspinel commonly is intergrown complexly with ilmenite. Native iron commonly is rimmed by troilite and associated with chrome-ulvöspinel in 14310 and 14072, but in 14072 it also forms a spongy network inside iron-rich olivines and pyroxenes associated with mosaic cristobalite (Fig. 7). In addition, the individual iron blebs are surrounded by another silica phase, which suggests that a late stage drop in the oxygen fugacity caused iron-rich silicates to break down to native iron plus silica (El Goresy *et al.*, 1972). El Goresy *et al.* have observed reduction of chrome-ulvöspinel to ilmenite plus native iron in both rocks.

We found no evidence in our sample of 14310 for the cognate inclusions reported by LPSET (1971).

CRYSTALLIZATION HISTORY

Both rocks contain well-developed vesicles with circular sections, such that a near-surface origin with some escape of volatiles seems probable. Furthermore, neither rock represents a pristine liquid, because both contain phenocrysts. It has been suggested that the orthopyroxene crystals, which commonly contain 1 to 3 wt% Al_2O_3, may also be phenocrysts or xenocrysts of high pressure origin (Hollister,

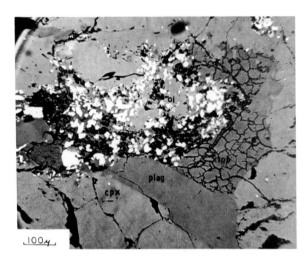

Fig. 7. Late stage assemblage of rock 14072,16 in reflected light. Dark gray, mosaic field is cristobalite; white grains are native iron surrounded by thin layer of a second silica mineral (tridymite?); plain dark gray field is plagioclase; medium gray is pyroxene and lightest gray is mostly iron-rich olivine and some glass (?). Black areas are defects in the thin section. Most of the iron grains seem to be developed within or bordering the olivine crystal which is optically continuous.

1972; Ridley *et al.*, 1972). Because we find plagioclase laths intergrown with ortho-pyroxenes in our thin section, and because we have been able to duplicate the plagioclase-orthopyroxene texture and high alumina content of the orthopyroxenes by laboratory crystallization in a vacuum, we believe that the orthopyroxenes formed near the surface. We also believe that the unusually high Al_2O_3 content in the ortho-pyroxenes results from high alumina activity in the early stages of crystallization, perhaps the result of very low initial concentrations of sodium and potassium. Later loss of volatiles may have been superimposed upon this original deficiency. Thus, we suggest the following crystallization sequences:

14310: Early addition of blocky phenocrysts to the original magma followed by slow crystallization of orthopyroxene, plagioclase, and possibly olivine; settling out or resorption of the olivine; reaction of the orthopyroxene and liquid to form pigeonite; more rapid crystallization of plagioclase, pigeonite, augite and perhaps ilmenite; finally, precipitation of plagioclase, clinopyroxene, ilmenite, ulvöspinel, tranquillityite, and partial unmixing and freezing of the residual liquid to form glasses.

14072: Early addition of olivine to a magma similar to, but more iron-rich than that of 14310; crystallization of plagioclase, ilmenite, and chrome-ulvöspinel; pigeonite joining the crystallizing phases together with inclusion, partial resorption, and making over of the smaller olivine grains; local separation of liquid from magnesian olivine yielding plagioclase–iron-rich clinopyroxene–cristobalite and finally plagioclase–iron-rich olivine–cristobalite assemblages.

Acknowledgments—This work was supported by the Committee on Experimental Geology and Geophysics. Harvard University, and by NASA grants NGR 22–007–175 and NGR 22–007–199. David Walker is an NSF predoctoral Fellow. We thank Cornelis Klein for his advice and comments.

REFERENCES

Bence A. E. and Papike J. J. (1971) Apollo 14 igneous rocks 14053 and 14310; phase petrology (abstract). In *Abstracts With Programs*, 1971 Annual Meetings, Geological Society of America, p. 503.

Brown G. M. and Peckett A. (1971) Selective volatilization on the lunar surface: evidence from Apollo 14 feldspar-phyric basalts. *Nature* **234**, 262–266.

El Goresy A. Ramdohr P. and Taylor L. A. (1972) Fra Mauro crystalline rocks: petrology, geochemistry, and subsolidus reduction of opaque minerals (abstract). In *Lunar Science—III* (editor C. Watkins), pp. 224–226, Lunar Science Institute Contr. No. 88.

Haggerty S. E. and Meyer H. O. A. (1970) Apollo 12: opaque oxides. *Earth Planet. Sci. Lett.* **9**, 379–387.

Hollister L. S. (1972) Implications of the relative concentrations of Al, Ti, and Cr in lunar pyroxenes. *Lunar Science—III* (editor C. Watkins), pp. 389–391, Lunar Science Institute Contr. No. 88.

Lofgren G. E. (1972) Crystallization studies on plagioclase (abstract). *Trans. Am. Geophys. U.* **253**, p. 549.

LSPET (1971) (Lunar Sample Preliminary Examination Team) Preliminary examination of lunar samples from Apollo 14, *Science* **173**, 681–693.

Muan A. Hauck J. and Löfall T. (1972) Equilibrium studies with a bearing on lunar rocks (abstract). In *Lunar Science—III* (editor C. Watkins), pp. 561–563, Lunar Science Institute Contr. No. 88.

Quaide W. (1972) Fe and Mg in bytownite-anorthite plagioclase in lunar basaltic rocks (abstract). In *Lunar Science—III* (editor C. Watkins), pp. 624–626, Lunar Science Institute Contr. No. 88.

Ridley W. I. Williams R. J. Brett R. and Takeda H. (1972) Petrology of lunar basalt 14310. *Lunar Science—III* (editor C. Watkins), pp. 648–650, Lunar Science Institute Contr. No. 88.

Roedder E. W. (1951) Low temperature liquid imiscibility in the system $K_2O–FeO–Al_2O_3–SiO_2$. *Am. Min.* **36**, 282–286.

Smith J. V. (1971) Minor elements in Apollo 11 and Apollo 12 olivine and plagioclase. *Proc. Second Lunar Sci. Conf.*, *Geochim. Cosmochim. Acta* Suppl. 2, Vol. 1, pp. 617–643, MIT Press.

Walker D. Longhi J. and Hays J. F. (1972) Experimental petrology and origin of Fra Mauro rocks and soil (abstract). In *Lunar Science—III* (editor C. Watkins), pp. 770–772, Lunar Science Institute Contr. No. 88. (also this volume).

Proceedings of the Third Lunar Science Conference
(Supplement 3, *Geochimica et Cosmochimica Acta*)
Vol. 1, pp. 141–157
The M.I.T. Press, 1972

Mineral-chemical variations in Apollo 14 and Apollo 15 basalts and granitic fractions

G. M. Brown, C. H. Emeleus, J. G. Holland,
A. Peckett, and R. Phillips

Department of Geology, Durham University, England

Abstract—Feldspar-phyric basalts from the Fra Mauro formation probably are plagioclase cumulates from a basalt rich in KREEP-type components. The residual phases are rhyolitic glass and minerals rich in K, Ba, REE, Zr, and P, similar to those in Apollo 11, 12 and 15 basalts. The pyroxenes are zoned from magnesian orthopyroxene cores to pyroxferroite rims, and Al:Ti values indicate initial crystallization in equilibrium with plagioclase. Calcic groundmass plagioclase compositions relative to phenocrysts indicate some loss of Na during late-stage crystallization. The four Apollo 15 basalt samples show three broadly contrasted pyroxene trends, suggestive of formation in three lava flows with different cooling rates. Al:Ti values range from 7:1 to $1\frac{1}{2}$:1 in zoned crystals, the rims probably containing Ti^{3+}. Apollo 14 breccias contain fairly abundant patches of rhyolite, of granophyre partly melted to rhyolite, and of troctolite (An_{95}, Fo_{85}) with harrisitic cumulate texture. This suggests high-level crystal fractionation, very early in the moon's history, to yield a granitic crustal differentiate. New analyses of tranquillityites in basalt and granophyre (Apollo 14 and 15) and of two new zirkelite-type minerals (one with 22.5% Y_2O_3 + RE_2O_3), suggest that increases in Y, Zr, Nb (as well as Rb, K and Ba) from the Apollo 11 through 12 to 14 soils could be associated with an increasing granitic component. As KREEP basalt compositions cannot be derived from addition of granite to basalt (or norite), the granite probably is a differentiate of ancient KREEP basalt (Meyer, 1972). Whitlockites range widely in RE_2O_3 and MgO:FeO ratios, and one crystal (Apollo 15 basalt) contains 1% SrO and the usual negative Eu anomaly. Variable contamination of mare basalts by granite-bearing phases (K-feldspar or Sr-whitlockite) could, therefore, produce anomalies in the Rb–Sr systematics.

Introduction

THE SAMPLING OF THE Fra Mauro formation by the Apollo 14 mission has yielded a highly feldspathic assemblage of breccias and soils and at least two isolated lava fragments that are high-alumina, plagioclase-phyric basalts. In contrast to the Apollo 11 and 12 suites, in which the nonmare-basalt and nonanorthositic lunar materials of the soils were referred to as a "magic" or "luny" component, here one can begin to classify much of the complex highlands material into different rocks of a primitive crustal assemblage. In particular, the concept of KREEP-rich material (Hubbard *et al.*, 1971) being an important lunar basalt-type (e.g., Meyer, 1972; Reid *et al.*, 1972) is of fundamental significance in explaining crust-mantle relationships on the moon. We have continued a study of the mineral composition ranges (by electron probe) in the basalts (Apollo 14 and 15), as a basis for understanding basalt generation, crystallization, and fractionation histories. Breccia and soil samples have been searched for clues as to the role of crystal fractionation on the moon, including studies of ultrabasic and granitic fragments. Rare minerals have been analyzed in particular detail, in association with x-ray fluorescence analysis of soils, to trace the behavior of elements such as P, REE, Ba, Rb, Zr, Hf, and Sr that are associated with problems in isotopic age determinations as well as the unique chemistry of the lunar basalts.

Apollo 14 Basalts, Breccias and Fines

The two basalt fragments we studied (14310,20 and 14073,9) are very similar, but sample 14310 was studied in greatest detail. It is a high-alumina basalt (20% Al_2O_3; LSPET, 1971), hence more feldspathic than mare basalts (Table 1), and is texturally plagioclase-phyric with 10% phenocrysts up to $1\frac{1}{2}$ mm long. We have described it elsewhere (Brown and Peckett, 1971) in regard to feldspar compositions and the evidence that a strong reversal from phenocryst rims (An67) to groundmass microlaths (An95) suggests loss of sodium during the final stages of crystallization. Further evidence presented at the Third Lunar Science Conference, however, points to heterogeneity within the rock (e.g., Hollister et al., 1972; Ridley et al., 1972), supported by more recent whitlockite analyses (Table 7). Even so, the well-developed normal zoning of the plagioclase phenocrysts indicates crystallization from a melt

Table 1. Modes (vol.%) of Apollo 14 and 15 basalts*.

	14310/20	15076/12	15085/14	15555/39
Plagioclase	54.1	35.7	31.3	26.8
Pyroxene	42.2	53.3	62.3	58.6
Olivine	—	—	1.3	8.4
Opaque minerals	1.8	3.9	3.5	4.2
Silica	—	6.4	1.6	2.0
Mesostasis	1.9	0.7	tr	tr

*Calculated void-free. Average of 3000 points per section.

Table 2. Pyroxene analyses showing some extreme compositions (see Figs. 1–3, 5–8)

	1	2	3	4	5	6	7	8	9	10
SiO_2	55.17	54.76	54.68	52.84	45.52	45.98	45.67	44.66	46.47	45.86
TiO_2	0.31	0.25	0.38	0.73	2.59	0.98	0.63	0.89	0.51	0.37
Al_2O_3	1.25	0.57	0.96	3.21	2.62	0.88	0.38	1.00	0.40	0.36
FeO	10.42	13.58	13.26	11.65	29.31	38.08	44.75	47.54	43.91	44.33
MnO	0.13	0.22	0.24	0.25	0.37	0.49	0.54	0.51	0.79	0.76
MgO	30.84	28.47	27.68	27.35	0.40	1.66	0.64	0.20	1.57	0.90
CaO	0.74	1.12	2.03	3.20	18.54	11.38	6.93	4.44	6.13	6.56
Cr_2O_3	0.57	0.28	0.30	0.60	—	0.09	0.08	0.08	0.06	0.08
TOTAL	99.43	99.25	99.53	99.83	99.35	99.54	99.62	99.32	99.84	99.22
Si	1.955	1.974	1.968	1.895	1.885	1.942	1.967	1.945	1.980	1.978
Ti	0.008	0.007	0.010	0.020	0.081	0.031	0.020	0.029	0.016	0.012
Al	0.052	0.024	0.041	0.136	0.128	0.044	0.019	0.051	0.020	0.018
Fe	0.309	0.409	0.399	0.349	1.015	1.345	1.612	1.732	1.565	1.599
Mn	0.004	0.007	0.007	0.008	0.013	0.018	0.020	0.019	0.029	0.028
Mg	1.629	1.530	1.485	1.462	0.025	0.105	0.041	0.013	0.100	0.058
Ca	0.028	0.043	0.078	0.123	0.823	0.515	0.320	0.207	0.280	0.303
Cr	0.016	0.008	0.009	0.017	—	0.003	0.003	0.003	0.002	0.003
A.R. { Fe	15.71	20.65	20.34	18.06	54.50	68.47	81.71	88.72	80.48	81.58
Mg	82.86	77.17	75.67	75.58	1.33	5.32	2.08	0.67	5.13	2.95
Ca	1.43	2.18	3.99	6.36	44.17	26.21	16.21	10.62	14.39	15.47

(1) Orthopyroxene (14161/20 fines), (2) orthopyroxene (14305/90 breccia), (3) low-alumina orthopyroxene and (4) high-alumina orthopyroxene (14310/20 basalt), (5) ferrohedenbergite (15555/39 basalt), (6) (7) (8) zoning to low calcium ferropyroxene (15076/12 basalt), (9) low-birefringent, probable pyroxferroite as wide sheath to ferroaugite (15475/125 basalt), (10) probable pyroxferroite grains with mesostasis silica (15475/125 basalt).

A.R., Atomic ratios. Analyses by electron microprobe.

with plagioclase on the liquidus, and the calcic plagioclase microlaths are inter-grown with the common lunar-basalt residual phases such as a potassic rhyolite glass (Table 4), tranquillityite (Table 5) and whitlockite (Table 7), as well as a new zirkelite-type phase (Table 5).

The pyroxenes are distinctive in containing cores of magnesian orthopyroxene (Table 2) zoned outwardly to subcalcic ferroaugite approaching pyroxferroite in composition (Fig. 1). The magnesian cores also are common in basalt 14073 and in soil fragments and breccias from this site (Fig. 2, Table 2), indicating closer affinity with norites and pyroxenic anorthosites of crustal derivation than with the mare basalts. The relatively high Al content and Al:Ti ratio of the orthopyroxenes is particularly significant because Al decreases with fractionation from the cores outward (Fig. 3). This is to be expected only if the orthopyroxenes crystallized in equilibrium with plagioclase, a stage reached fairly late in the crystallization sequence of mare basalts (e.g., Brown *et al.*, 1971, Fig. 4). Of equal significance is the fact that rock 14310 cannot be viewed purely as a norite (orthopyroxene-plagioclase rock). The pyroxene zoning (Fig. 1) differs from the noritic pattern of 14073 (Fig. 2) in overlapping the ferriferous field of mare-basalt pyroxenes. The trend shown is very similar to that observed by Meyer (1972, Fig. 2) in rock 15023, which he shows to have a similar composition to KREEP glasses, but the texture of a true basalt. Hence we view 14310 as a feldspathic KREEP basalt. We find this concept appealing, especially when linked with the hypothesis that there exists a high-K type of anortho-site that probably was derived from KREEP basalt Hubbard *et al.*, 1972). In that case, rock 14310 may have originated in part by some plagioclase floating in a KREEP-basalt magma chamber and is related to the formation of some anorthosites by local crystal fractionation. This does not preclude the fact that the parental KREEP basalt was itself quite high in alumina and derived from 1–3% partial

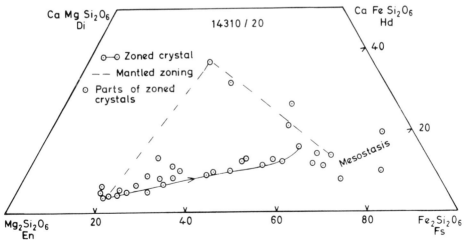

Fig. 1. Pyroxene trend for Apollo 14 plagioclase-phyric basalt. Cores of magnesian ortho-pyroxene zone through pigeonite and subcalcic augite to ferriferous rims and mesostasis grains, with augite rare (cf. KREEP-basalt 15023, Meyer, 1972).

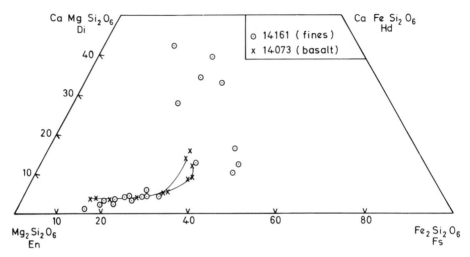

Fig. 2. Extreme Ca-poor magnesian orthopyroxenes in Apollo 14 fines (Table 2) and limited orthopyroxene–pigeonite trend in Apollo 14 plagioclase-phyric "basalt" (norite).

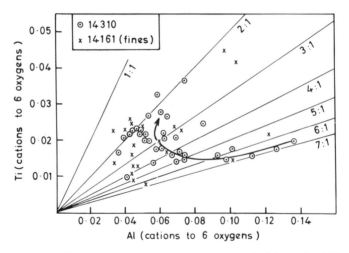

Fig. 3. Variable Al:Ti ratios in Apollo 14 pyroxenes. The trend of decreasing Al (arrowed line) indicates early crystallization of plagioclase with magnesian pyroxene (14310) in contrast to the mare-basalt crystallization sequences. Al:Ti < 2:1 implies some Ti^{3+} in the ferro-pyroxenes shown in Fig. 1.

melting of a more primitive feldspathic crust (Hubbard and Gast, 1971). If so, the development stage of the primitive crust must be >4.0 b.y. (the mineral age of 14310, Compston *et al.*, 1972).

Some of the Apollo 14 breccias contain an unusually aluminous pink or purple spinel, called chromian pleonaste (Haggerty, 1971). Zoning to more iron-rich rims (Table 3, Fig. 4) is not associated with Fe depletion in the matrix, as found by Roedder and Weiblen (1972), and suggests that the crystals are derived from a magmatic environment. If so, the primitive lunar crust may have consisted chiefly, related to accretion layering, of the assemblage calcic plagioclase–aluminous spinel–low-alumina magnesian orthopyroxene. The spinels would be refractory, and have not been found in the crustal-derived KREEP basalts. The breccias also contain a "magnesio" whitlockite similar to that in Apollo 12 fines but contrasted with "ferro" whitlockite in the KREEP and mare basalts (Table 7, discussed later). This also may be part of the primitive crustal assemblage.

The breccias and fines contain abundant glass fragments ranging from ultra-basic to granitic in composition (Table 4). The glasses referred to as picrite, olivine basalt, and olivine norite could be shock-melted either from crystal cumulates or from crystal-debris aggregates in the soil. However, we have seen fragments (2–3 mm) of an olivine–plagioclase intergrowth (14320,4), consisting of Fo86 and An95, very similar to the "crescumulate" texture of the terrestrial layered ultrabasic rocks of Rhum (Wager and Brown, 1968). Such a fragment, termed troctolite, has been shown by Compston *et al.* (1972) probably to have formed slightly earlier than 4.0 b.y. Separation of magnesian olivine and calcic plagioclase from basaltic magma

Table 3. Extreme compositions of oxides (see Fig. 4)

		1	2	3	4	5	6
TiO_2		2.46	11.68	37.01	55.27	0.21	0.21
Al_2O_3		10.84	7.23	0.02	0.11	60.50	59.24
FeO		26.38	44.08	61.69	36.67	13.17	17.68
MnO		0.41	0.49	0.40	0.30	0.08	0.17
MgO		6.15	0.70	0.76	6.29	18.21	15.60
Cr_2O_3		53.30	36.09	0.18	0.44	7.74	7.50
TOTAL		99.54	100.27	100.06	99.08	99.91	100.40
Ti		0.063	0.315	1.022	2.009	0.004	0.004
Al		0.434	0.306	0.001	0.006	1.844	1.836
Fe		0.750	1.323	1.893	1.482	0.285	0.389
Mn		0.012	0.015	0.012	0.012	0.002	0.004
Mg		0.312	0.038	0.042	0.453	0.702	0.611
Cr		1.433	1.024	0.005	0.017	0.158	0.156
O		4.000	4.000	4.000	6.000	4.000	4.000
A.R.	Ti	3.26	19.16	99.41	98.86	0.20	0.21
	Al	22.51	18.59	0.08	0.31	91.91	91.98
	Cr	74.23	62.25	0.51	0.83	7.89	7.81
A.R.	Fe	70.64	97.25	97.85	76.58	28.86	38.87
	Mg	29.36	2.75	2.15	23.42	71.14	61.13

(1) Chromite with high Mg/Fe ratio (15085/14 basalt), (2) titanochromite bridging the "gap" on the chromite-ulvöspinel join (part of zoned crystal, 15475/125 basalt), (3) extreme, near stoichiometric ulvöspinel (14161/20 fines), (4) magnesian ilmenite (14305/90 breccia), (5) core and (6) rim of zoned, pink, chromian pleonaste (14305/111 breccia).

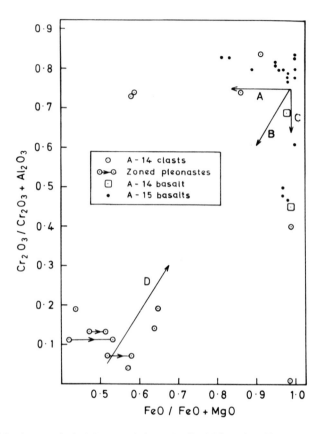

Fig. 4. Aluminous spinels (pleonastes) from Apollo 14 breccias. The compositions are plotted (D) on the diagram introduced by Haggerty (1971). Note the Fe/Mg zoning (Table 3). Trends A (Apollo 12 titanian chromites), B (Apollo 14 titanian chromites) and C (Apollo 12 aluminian ulvöspinels) from Haggerty. Additional data show extreme ulvöspinel, bottom right (Table 8); some Apollo 15 titanian chromites; and two Apollo 14 titanian chromites (top center) with low FeO/MgO.

is the mechanism necessary to produce free silica and K-feldspar (i.e., granite) as a residual fraction, and ancient crustal granite on the moon could therefore be of this localized origin. Fragments and lenses of rhyolite (up to 1 mm) are quite common in some of the breccias (e.g., 14305), but it is likely that many have been derived from melting of crystalline granophyre fragments (Table 4). Several cases of granophyre fragments ($1\frac{1}{2}$ mm) rimmed by brown rhyolite glass were observed, and their significance is discussed later.

APOLLO 15 BASALTS

Four basalts have been studied, and they are all mare-type, pyroxene-rich rocks (e.g., Table 1), relatively rich in silica (as cristobalite and tridymite) and containing a wide range of modal olivine. They are reminiscent of the Apollo 11 basalts except

Table 4. Analyses of selected basic and acid glasses.

	1	2	3	4	5	6	7	8	9	10	11
SiO$_2$	44.05	43.21	46.86	49.78	49.92	52.59	66.99	77.49	73.08	72.58	74.92
TiO$_2$	2.36	2.17	0.45	2.98	2.51	1.44	0.48	0.37	0.18	1.65	0.04
Al$_2$O$_3$	7.53	19.93	26.45	15.13	15.43	17.90	14.04	11.93	11.37	12.57	12.48
FeO	23.00	10.92	3.74	11.69	11.58	10.39	4.94	2.17	2.85	1.35	0.94
MnO	0.30	0.13	0.03	0.17	0.14	0.13	0.10	—	0.16	—	—
MgO	13.61	10.44	7.02	7.70	7.45	4.58	0.50	—	0.12	0.23	—
CaO	8.36	12.96	15.25	9.97	10.09	11.54	1.94	3.63	0.90	0.89	0.50
Na$_2$O	0.29	0.14	0.50	1.28	1.14	0.52	0.69	0.59	0.25	0.24	1.19
K$_2$O	0.15	0.06	0.04	1.34	0.89	0.60	9.48	3.93	8.29	10.07	9.47
Cr$_2$O$_3$	0.60	0.14	0.08	0.24	0.21	0.19	—	—	—	—	—
BaO	—	—	—	—	—	—	0.68	—	2.69	0.25	0.34
TOTAL	100.25	100.10	100.42	100.28	99.36	99.88	99.84	100.11	99.89	99.83	99.88
CIPW Norm:											
Q	—	—	—	0.64	2.73	10.78	18.30	49.65	34.78	30.60	30.27
Or	0.88	0.35	0.24	7.90	5.29	3.55	56.36	23.20	49.90	59.71	56.15
Ab	2.45	1.18	4.21	10.80	9.71	4.41	5.87	4.99	2.15	2.04	10.10
An	18.75	53.52	69.52	31.49	34.58	44.79	7.26	17.99	5.52	3.49	0.75
Di	18.79	8.73	4.67	14.68	13.04	10.50	3.20	—	3.54	0.68	2.11
Hy	28.40	14.69	18.18	28.49	29.54	22.95	8.10	3.37	3.77	0.26	0.55
Ol	25.37	17.19	2.21	—	—	—	—	—	—	—	—
Cm	0.88	0.21	0.12	0.35	0.31	0.28	—	—	—	Tn 0.37	—
Il	4.47	4.12	0.85	5.64	4.80	2.74	0.92	0.70	0.35	2.86	0.08

(1) to (6) Glasses selected from fines analyses (14161/20) to show range from (1) picrite, (2) olivine basalt and (3) olivine norite to (4,5) basalts with relatively high Na$_2$O and to (6) quartz basalt. (7) Potassic residual glass with relatively low silica (14310/20 basalt), (8) calcic rhyolite residuum (15076/12 basalt), (9) barian rhyolite residuum (15555/39 basalt), (10) low-barium rhyolite common as fragments in breccia (14305/90), (11) rhyolite patches from rim-melting of granophyre fragment in breccia (14305/111).

for the low content of Ti-bearing oxides. The pyroxene trends for three examples are shown in Figs. 5, 6, 7. The fourth basalt (15475,125) shows a trend like that in Fig. 7 except that pigeonite cores are abundant in the volume of each crystal. Hence, the four pyroxene suites are different in broad terms and may be a clue to the sampling of four different lava flows. Figure 5 shows a pigeonite-rich assemblage; the narrow rims zone towards a calcic pyroxferroite composition, with no detectable fayalite. Figure 6 shows coprecipitation of augite and pigeonite, and a continuous residuum zoning to ferrohedenbergite (with fayalite and silica). The latter trend is similar to that for pyroxenes in 14053 basalt (Gancarz et al., 1971) and two Apollo 12 basalts (Klein et al., 1971). Figure 7 shows a continuous zoning to "subcalcic pyroxferroite" with large, 0.5 mm fayalites (optical evidence for the pyroxenoid, in a similar trend, is better seen in 15475,125; Table 2). Clearly the extent to which the stable assemblage, ferrohedenbergite + fayalite + silica developed, varied according to the flow or flow portion from which each sample was derived.

The Al:Ti ratios of the Apollo 15 pyroxenes (Fig. 8) give an unusual pattern (the "hypersthenes" in 15085 may be subcalcic pigeonites, 2.2% CaO). The high Al:Ti ratio in the cores probably is due to low Ti-activity, because Al contents are also low. Unlike Apollo 12, there is no trend to Al-rich augite mantles and the main trend is one of Ti enrichment to the rim "ferropyroxenes", which show Al:Ti distinctly less than 2:1. This broad field (Fig. 8) implies low pO$_2$ such that appreciable Ti^{3+} is present (i.e., R^{2+} Ti^{3+} SiAlO$_6$, Bence and Papike, 1972). Note the very high Al and

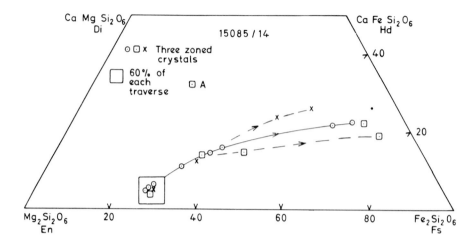

Fig. 5. Pyroxenes of an Apollo 15 basalt with large-volume pigeonite cores, augite very rare (A) and a distinctive trend in the subcalcic augite field (cf. Figs. 6, 7).

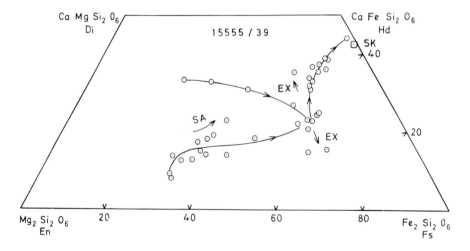

Fig. 6. Coexisting, intergrown augite and pigeonite zone to a common focus, Apollo 15 basalt. The subsequent, continuously zoned trend is to the stable assemblage ferrohedenbergite + fayalite + silica (cf. Figs. 5, 7). SA, subcalcic augites (one crystal). EX, intergrowths probably due to subsolidus exsolution. SK, extreme pyroxene of the Skaergaard fractionated layered series.

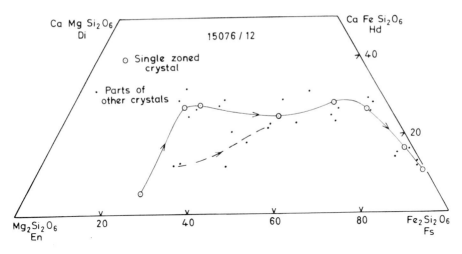

Fig. 7. Pyroxenes chiefly calcic (rare, small pigeonite core) trending to ferroaugite, Apollo 15 basalt (cf. Figs. 5, 6). The late-stage continuous trend is to a "subcalcic pyroxferroite" composition ("Fs" on Fig. 8; see Table 2).

Ti in the ferrohedenbergite and very low Al in the "ferrosilite" (ends of trends in Figs. 6 and 7, respectively; Table 2). Data on other phases are discussed later.

ZIRCONIUM-RICH MINERALS

Tranquillityite (Lovering *et al.*, 1971), phase ß (Haines *et al.*, 1971), and calcic phases referred to as "dysanalyte" (Ramdohr and El Goresy, 1970) but now analyzed and shown to be close to $CaZrTiO_5$ or zirkelite (Busche *et al.*, 1972) are members of a rare but important group of lunar minerals. We have analyzed two new crystals of this group in Apollo 14 and 15 basalts, referred to as Phase *X* and Phase *Y* (Table 5), and averaging 20–30 μ in size. Details are given elsewhere (Peckett *et al.*, 1972). New analyses of tranquillityite from Apollo 14 and 15 basalts show this phase to be almost ubiquitous and of nearly constant composition, differing from the other phases by having high silica (Table 5). Six crystals were analyzed from 14310, and over 12 crystals ($<10\ \mu$) occur within an interstitial cristobalite crystal in 15475. In contrast, the new minerals are nonsilicates and, like phase ß, each has at least one feature not shared by the other Zr-rich phases. Phase *X* could be viewed as a zirconian armalcolite, whereas phase *Y* contains 22.5 wt% Y_2O_3 + RE_2O_2. Optically, phase *Y* is similar to tranquillityite but phase *X* has a dark brown, rather than fox-red color in transmitted light. The abundant trivalent cations pose a problem in calculating a structural formula, but we have followed the method of Busche *et al.* (1972), arriving at a very close fit to the zirkelite or armalcolite formula of $A^{2+}B_2^{4+}O_5$ (Table 6 and footnote) for each phase.

The phases are plotted on Fig. 9 to show the variation in the three major oxides. Related lunar phases (baddeleyite, niobian rutile, zirconian ilmenite and armalcolite) and terrestrial zirkelite also are shown, together with ranges of the 4th and 5th most

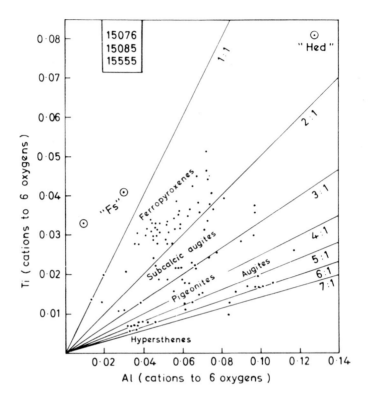

Fig. 8. Wide range in Al:Ti ratios for Apollo 15 basaltic pyroxenes. The zoning from hyperstenes or pigeonites to ferropyroxene residua (Figs. 5, 6, 7) is accompanied by increasing Ti, with nearly constant Al, once plagioclase began to crystallize (between pigeonite–augite and subcalcic augite fields). Hence the initially high Al:Ti ratios (7:1) reflect low Ti activity, because total Al is always low (cf. Apollo 12, Brown et al., 1971, Fig. 4). Strong evidence for Al:Ti much less than 2:1, implying Ti^{3+} and therefore low oxygen fugacities, in the late-stages of crystallization. Final stage "Hed" (Fig. 6) and "Fs" (Fig. 7) markedly different in Al and Ti contents.

abundant elements in the phases from Table 5. Phase ß is unique in its high content of U, Th, and Pb, but terrestrial zirkelite is also high in U and Th.

These Zr-rich phases commonly are richer in rare-earth oxides than the whit-lockites (Table 7) and also are more important as carriers of radiogenic heat sources. They are concentrated in the potassic, granitic residuum of mare and KREEP basalts, and we also found tranquillityite (50 µ) within a $1\frac{1}{2}$ mm granophyre fragment in breccia section 14305,111. Low totals in the Y + REE-rich phases may imply the presence of elements not detectable by the electron probe (Haines et al., 1971), of which lithium may well be the main one. Ion probe analysis (Fredriksson et al., 1971) showed up to 150 ppm Li in KREEP glasses. This could result from about 0.5% of the Li-bearing minerals (assuming 2–5% Li in phases Y and β, Table 5), which could include whitlockite (Lunatic Asylum, 1970).

Table 5. Two new Zr-rich minerals and tranquillityites, with comparisons.

	Phase X 14310/20	Phase Y 15555/39	Tranquillityite Apollo 11 and 12	Tranquillityite 14310/20*	Tranquillityite 15475/125	Phase β	Lunar zirkelite	Terrestrial zirkelite
MgO	1.7	0.1	—	1.3	0.6	—	0.7	0.2
CaO	3.1	3.2	1.3	1.1	1.4	3.0	8.8	10.8
MnO	0.2	0.3	0.3	0.3	0.3	—	—	—
FeO	13.4	11.4	42.5	42.1	43.5	11.6	6.1	7.7
PbO	—	—	—	—	—	4.1	0.5	—
Al₂O₃	0.9	0.5	1.1	1.2	1.4	—	1.4	—
Cr₂O₃	4.3	0.5	0.1	0.2	—	—	0.5	—
Y₂O₃	—	10.4	2.8	0.5	0.9	9.1	3.6	0.2
RE₂O₃	1.3†	12.1‡	0.2	—	—	5.0	1.8	2.5
SiO₂	0.2	—	14.0	14.0	14.4	2.1	0.4	—
TiO₂	68.8	27.1	19.5	21.9	21.8	22.1	34.5	15.0
ZrO₂	6.1	30.8	17.2	14.5	14.0	17.1	40.4	52.9
HfO₂	—	—	0.2	—	—	—	0.4	—
ThO₂	—	—	—	—	—	4.1	0.5	7.3
UO₂	—	—	72ppm U	—	—	3.4	0.2	1.4
Nb₂O₅	—	—	0.3	—	—	8.3	1.0	—
TOTAL	100.0	96.4	99.5	97.1	98.3	89.9	100.8	98.0

*Mean of 6 separate crystals. †0.2% La₂O₃, 0.9% Ce₂O₃, 0.1% Pr₂O₃, 0.1% Nd₂O₃. ‡0.6% La₂O₃, 1.9% Ce₂O₃, 0.7% Pr₂O₃, 3.3% Nd₂O₃, 1.7% Sm₂O₃, 0.4% Eu₂O₃, 2.1% Gd₂O₃, 0.3% Tb₂O₃, 0.9% Dy₂O₃, 0.2% Ho₂O₃. Tranquillityite in data column 3 is the mean of analyses from several laboratories (Lovering et al., 1971). Lunar zirkelite is the mean of 4 analyses from sample 14257/3 (Busche et al., 1972). Phase β is from 12013 (Haines et al., 1971) and terrestrial zirkelite from jacupirangite (Palache et al., 1944). (—, below detection).

Table 6. Structural formulae of two new zirconium-rich minerals (see Table 5).

	Phase X (14310/20)			Phase Y (15555/39)		
Mg	0.094	Si	0.006	Mg	0.008	Nd	0.058
Ca	0.125	Ti	1.923	Ca	0.173	Sm	0.030
Mn	0.008	Zr	0.110	Mn	0.013	Eu	0.007
Fe	0.418	O	5.000	Fe	0.475	Gd	0.035
Al	0.039	*Phase X:		Al	0.030	Tb	0.005
Cr	0.126	$(A^{2+})_{0.65}(B^{3+})_{0.18}(C^{4+})_{2.04}O_5$		Cr	0.018	Dy	0.014
La	0.002	equals $(A^{2+})_{1.00}(B^{4+})_{2.00}O_5$		Y	0.277	Ho	0.003
Ce	0.012	*Phase Y:		La	0.011	Ti	1.018
Pr	0.002	$(A^{2+})_{0.67}(B^{3+})_{0.54}(C^{4+})_{1.77}O_5$		Ce	0.035	Zr	0.748
Nd	0.002	equals $(A^{2+})_{1.00}(B^{4+})_{2.01}O_5$		Pr	0.012	O	5.000

*Transfer of all trivalent and the excess quadrivalent equivalents to the divalent site gives perfect fit for Phase X. For Phase Y, the divalent site is filled from equivalent trivalents, and the residue of trivalents allocated, as equivalents, to the quadrivalent site. The latter also gives perfect fit for the zirkelite-type formula.

WHITLOCKITES

Four new analyses show a great variation not only in total RE₂O₃ but also in the MgO:FeO ratios of lunar whitlockites (Table 7) with the general formula Ca₃(PO₄)₂. Contents of Y₂O₃ range by a factor of 4, and RE₂O₃ ranges from 1.9 to 8.8 wt%. Analyses by Lunatic Asylum (1970) and Keil et al. (1971) fall within this range. Magnesium and iron substitute for calcium, and the MgO/MgO + FeO values range from 0.04 to 0.82. Probably there is some significance in the fact that the "magnesio" whitlockites occur in the breccia and fines samples, and may be derived from the Mg-rich noritic crustal rocks. Whitlockite from 12013 (Lunatic Asylum,

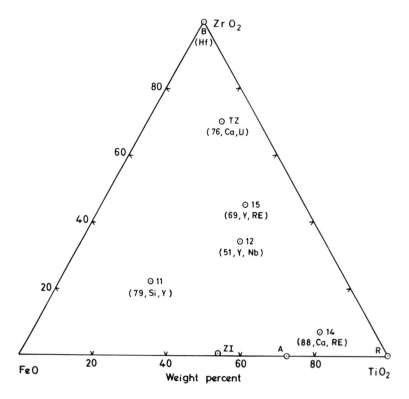

Fig. 9. Rare Zr-rich minerals (Table 5). Parentheses such as (76, Ca, U) = 76% ZrO_2 + FeO + TiO_2, with Ca and U next in abundance. Specimen numbers refer to main Apollo suite occurrence (11, tranquillityite; 12, phase β; 14, phase X; 15, phase Y). Associated lunar phases are baddeleyite (B, with Hf), R (niobian rutile), A (armalcolite), ZI (zirconian ilmenite). TZ, terrestrial zirkelite (Table 5).

1970) has an MgO/MgO + FeO ratio of 0.74. However, the mare basalt and KREEP-basalt whitlockites range from 0.75 (Keil *et al.*, 1971) to 0.04 (Table 7); therefore until more data are available, we can only use the whitlockite compositions as a broad guide to the source materials.

A more significant discovery is the presence of 1% of SrO in a mare basalt, residual-phase whitlockite (Table 7). This is by far the highest recorded concentration of strontium in any lunar mineral. It would be expected that Eu^{2+} would accompany Sr^{2+}, but europium is below detection (<0.05% as Eu_2O_3). Green *et al.* (1971) show experimentally that whitlockite–liquid partition coefficients for Sr^{2+} are close to unity (diopside–whitlockite liquids at 1280°C, atmospheric pressure), but ten times the diopside–liquid coefficients. Clearly, the unity partition ratio is not applicable to the lunar temperatures (?1000°C) of mesostasis crystallization in the Apollo 15 pyroxenic basalt. Source-rock partial melting may therefore have been the non-equilibrium process, suggested by Compston *et al.* (1972), which could perhaps separate radiogenic strontium in the mare basalts from common strontium in the

Table 7. Whitlockites ranging in RE_2O_3 and MgO:FeO ratios (wt%).

	14305/90	14310/20 A	14310/20 B	15475/125	Apollo 12 fines	Apollo 12 basalt
MgO	3.32	3.63	1.88	0.28	3.43	3.32
CaO	40.93	42.74	38.68	43.72	42.20	42.26
MnO	0.02	—	—	0.03	—	—
FeO	0.69	2.70	3.35	6.02	0.77	2.30
SrO	—	—	—	1.01	—	—
Al_2O_3	0.42	—	—	—	—	—
Y_2O_3	2.61	1.64	3.01	0.79	1.93	2.17
La_2O_3	0.53	0.47	1.02	0.20	n.a.	n.a.
Ce_2O_3	1.63	1.56	2.86	0.87	2.08	1.38
Pr_2O_3	0.20	0.39	0.74	0.20	0.54	0.44
Nd_2O_3	1.48	1.25	2.05	0.30	1.51	1.37
Sm_2O_3	0.50	0.62	1.02	0.35	0.66	0.30
Eu_2O_3	—	0.07	—	—	0.11	—
Gd_2O_3	0.98	—	0.84	—	0.67	0.60
Tb_2O_3	—	—	0.20	—	0.11	0.20
Dy_2O_3	—	—	—	—	0.43	0.51
Ho_2O_3	—	—	—	—	0.32	0.30
Er_2O_3	—	—	—	—	—	0.20
SiO_2	0.17	0.21	0.39	0.60	0.22	0.46
P_2O_5	45.71	45.46	43.22	45.82	44.81	44.27
TOTAL	99.19	100.74	99.26	100.19	99.79	100.08
MgO/MgO + FeO	0.82	0.57	0.36	0.04	0.82	0.59

14305/90, Breccia; 14310/20, feldspathic basalt (two separate grains); 15475/125, basalt; Apollo 12 fines (12003/10) and basalt (12040/39) from Brown et al. (table 8, 1971). Note the range in total RE_2O_3 (1.92 to 8.73) and MgO:FeO ratio (0.04 to 0.82). (n.a., not analysed; —, below detection).

residual, coarse-grained, Ca-poor pyroxenes of the mantle. However, subsequent fractionation of the basalts could lead to common strontium enrichment in the residual, whitlockite-bearing granitic liquids.

OTHER PHASES

Minerals analyzed, but not discussed here, have compositions that do not add materially to our knowledge of lunar mineralogy obtained from Apollo 11 and 12 data. These are plagioclase, alkali feldspar, olivine, ilmenite, ulvöspinel, chrome-spinel, troilite, nickel-iron, tridymite, cristobalite, F-apatite, zircon and baddeleyite. K-feldspar in granophyre (14305,111) contains 4.1% BaO, and a rectangular crystal (25 μ) in basalt 14310,20 contains 3.0% BaO. Barium increases $2V$ sharply (to 48° with BaO = 5%, Deer et al., 1963). The lunar basaltic sanidine crystal has $2V(-ve) < 20°$, implying a distinctly high-temperature structural state.

Some extreme oxides data are shown (Table 3, Fig. 4), and extreme fayalites occur (to Fo 1.5). Continuous zoning from Fo50 to Fo8 was observed in one olivine (15555,39). Troilite and silica-phases are near-stoichiometric. Baddeleyite contains 1.5 to 1.8% HfO_2. The iron grains range from 0.2% to 27.2% Ni, with no clear grouping into basaltic and fines minerals. One grain (15076,12 basalt) contains 5.9% cobalt (with 8% Ni), yet other grains in the same rock contain only 0.8% Co, 0.2% Ni.

Granitic Rocks

We have been interested for some time in the "granite problem" on the moon (e.g., Brown et al., 1971). This stemmed from the fact that analyses of mare basalt mesostasis areas always showed the presence of potassic rhyolite glass patches (Fig. 10) associated with KREEP-type minerals. These minerals are not only barian sanidine (K, Ba, Rb) and whitlockite (P, Y, REE) but also the ubiquitous tranquillityite and zirkelite minerals (Zr, Hf, Nb, Y, REE, U, Th). We preferred the term KREZP (op. cit.) but having made this point will now use the well-accepted term KREEP (Hubbard et al., 1971). More important is the fact that KREEP-granite can be derived from mare basalt, with attendant high Sr (whitlockite) and Rb concentrations, and high U and Th concentrations in the Zr-phases. Also, rhyolite and tranquillityite-bearing granophyre fragments are not uncommon in the Fra Mauro breccias (Table 4), with textural evidence for melting and hence probable ease of assimilation by basalts, whether of mare or KREEP type.

We formerly favored the idea that the few KREEP glasses previously analyzed from soils (Hubbard et al., 1971) were due to impact melting of granite fragments plus either mare basalt or highland norite fragments. It is now clear that the KREEP glasses have a nearly constant composition (Reid et al., 1972) and that actual basalt lavas can have the same composition (Meyer, 1972). Further, these compositions cannot be achieved by mixing any other basalt or norite with granite, in view of the SiO_2, FeO and other constraints. (P_2O_5 may be less significant because of whitlockite separation from the rhyolitic liquids.) Hence we are faced with the likelihood that

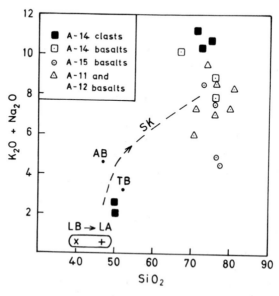

Fig. 10. Rhyolitic and dacitic glasses, from lunar basalt mesostasis areas and breccia fragments. Most of the alkali oxides are K_2O (Table 4). LB (lunar basalt), LA (lunar anorthosite) whole-rock analyses from the literature. AB, average alkali basalt; TB, average tholeiitic basalt; SK, Skaergaard trend (all terrestrial).

Table 8. X-ray fluorescence analyses of Apollo 14 fines samples (ppm).

Element	14003 7	14003 44	14141 28	14163 43	14190 3	14259 29	Mean
Ba	602	393	604	584	706	560	575
Nb	56	43	56	55	62	55	55
Zr	1089	914	1033	1069	1172	1068	1058
Y	246	209	260	243	261	241	243
Sr	209	179	198	209	201	205	200
Rb	15	13	19	15	20	15	16
Zn	32	37	57	41	34	26	38
Cu	9	12	6	10	7	7	9
Ni	350	349	175	319	279	383	309
Cr	2065	2135	2095	2022	2038	2040	2066
Mn	1518	1580	1644	1507	1402	1532	1531
La	78	76	85	82	80	75	79
K	4316	4185	6580	4785	6938	3736	5090

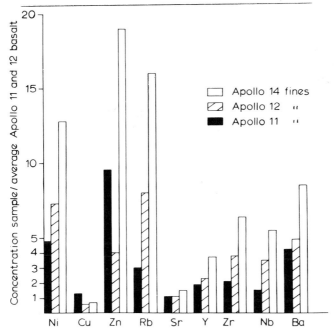

Fig. 11. Trace elements measured by x-ray fluorescence analysis of some Apollo 11, 12, and 14 fines samples (Table 8 and Brown et al., 1971). Enrichment relative to average Apollo 11 and 12 basalts (i.e., >1) for Ni, Cu and Zn not relevant here because of probable meteoritic contaminants. Rb and Ba could reflect increase in barian sanidine, and Y, Zr and Nb an increase in Zr-minerals (Table 5), towards a KREZP-rich source. Hubbard and Gast (1971, Fig. 11) show a similar pattern for Apollo 11 and 12 fines, including Eu, Ce, U, and K (but not Y, Zr, Nb).

granite can form either more than 4 b.y. ago (see earlier) by fractionation from KREEP-basalt generated by low-volume partial melting of a feldspathic crust, or can form in smaller volume by fractional crystallization of mantle-derived basalt (see section on Apollo 14 breccia troctolite). In either case, the granitic component will be detectable from Rb and Ba contents (sanidine) in the fines, but equally from

Y, Zr, and Nb in the "Z" component of the KREZP-granite. These concentrations are shown in Table 8, and the effect of this material, increasing from Apollo 11 to Apollo 14 fines, on Fig. 11. Compston *et al.* (1971) referred to these element enrichments in Apollo 12 fines. Whether this pattern can be attributed entirely to the effect of KREEP-basalt fragments is not yet clear, since the effect of KREEP-Zr-granite fragments would be an additional factor. The same trend is shown by Rb/Sr ratios in the soils (Wasserburg *et al.*, 1972, Fig. 3).

We must conclude that granite has been generated early in the evolution of the lunar crust, genetically related to the other lunar rocks in terms of O and Si isotopes (Epstein and Taylor, 1971). We have shown that zirkelite and whitlockite would be concentrated with granite, the whole assemblage not only carrying most of the K, U, and Th but also exerting a strong influence on Rb and Sr distributions. Contamination of mare basalts with either KREEP-granite or (less likely, in view of relative melting temperatures) with KREEP-basalt would chiefly result in an increased Rb/Sr ratio which, in turn, has influenced the measured I ($^{87}Sr/^{86}Sr$) values (Wasserburg *et al.*, 1972, p. 789). Further complications may result from the effect of Sr-whitlockite on these systematics.

The relationship between KREEP-basalt and KREEP-granite may now be the best explanation for rock 12013. Impact metamorphism of the two contrasted rocks would give rise to the partly mixed, dark and light lithologies. It is unlikely that a chance fragment of rare granite was available just for this event, but rather that a granitic unit exists with KREEP basalt in the lunar crust. The volume may be relatively small, but the chemical effects of redistribution would be very significant.

Acknowledgments—Financial support was from the Natural Environment Research Council (U.K.). Synthetic glasses for X-ray fluorescence standards were made by the Pilkington Research Laboratory (Lathom, England). Special technical assistance was from Mrs. A. W. Mines, Miss M. Lumley, and Messrs. G. Dresser, G. Randall, and A. Carr.

REFERENCES

Bence A. E. and Papike J. J. (1972) Crystallization histories of pyroxenes from lunar basalts (abstract). In *Lunar Science—III* (editor C. Watkins), pp. 59–61. Lunar Science Institute Contr. No. 88.

Brown G. M. Emeleus C. H. Holland J. G. Peckett A. and Phillips R. (1971) Picrite basalts, ferrobasalts, feldspathic norites and rhyolites in a strongly fractionated lunar crust. *Proc. Second Lunar Sci. Conf., Geochim. Cosmochim. Acta* Suppl. 2, Vol. 1, pp. 583–600. MIT Press.

Brown G. M. and Peckett A. (1971) Selective volatilization on the lunar surface: evidence from Apollo 14 feldspar-phyric basalts. *Nature* **234**, 262–266.

Busche F. D. Prinz M. Keil K. and Kurat G. (1972) Lunar zirkelite: a uranium-bearing phase. *Earth Planet. Sci. Lett.* (in press).

Compston W. Berry H. Vernon M. J. Chappell B. W. and Kaye M. J. (1971) Rubidium-strontium chronology and chemistry of lunar material from the Ocean of Storms. *Proc. Second Lunar Sci. Conf., Geochim. Cosmochim. Acta* Suppl. 2, Vol. 2, pp. 1471–1485. MIT Press.

Compston W. Vernon M. J. Berry H. Rudowski R. Gray C. M. Ware N. Chappell B. W. and Kaye M. (1972) Age and petrogenesis of Apollo 14 basalts (abstract). In *Lunar Science—III* (editor C. Watkins), pp. 151–153. Lunar Science Institute Contr. No. 88.

Deer W. A. Howie R. A. and Zussman J. (1963) *Rock-forming Minerals*, Vol. 4. Longmans.

Epstein S. and Taylor H. P. (1971) O^{18}/O^{16}, Si^{30}/Si^{28}, D/H, and C^{13}/C^{12} ratios in lunar samples. *Proc. Second Lunar Sci. Conf., Geochim. Cosmochim. Acta* Suppl. 2, Vol. 2, pp. 1421–1441. MIT Press.

Fredriksson K. Nelen J. Noonan A. Anderson C. A. and Hinthorne J. R. (1971) Glasses and sialic components in Mare Procellarum soil. *Proc. Second Lunar Sci. Conf., Geochim. Cosmochim. Acta* Suppl. 2, Vol. 1, pp. 727–735. MIT Press.

Gancarz A. J. Albee A. L. and Chodos A. (1971) Petrological and mineralogical investigation of some crystalline rocks returned by the Apollo 14 mission. *Earth Planet. Sci. Lett.* **12**, 1–18.

Green D. H. Ringwood A. E. Ware N. G. Hibberson W. O. Major A. and Kiss E. (1971) Experimental petrology and petrogenesis of Apollo 12 basalts. *Proc. Second Lunar Sci. Conf., Geochim. Cosmochim. Acta* Suppl. 2, Vol. 1, pp. 601–615. MIT Press.

Haggerty S. E. (1971) Compositional variations in lunar spinels. *Nature Phys. Sci.* **233**, 156–160.

Haines E. L. Albee A. L. Chodos A. A. and Wasserburg G. J. (1971) Uranium-bearing minerals of lunar rock 12013. *Earth Planet. Sci. Lett.* **12**, 145–154.

Hollister L. Trzcienski W. Jr. Dymek R. Kulick C. Weigand P. and Hargraves R. (1972) Igneous fragment 14310,21 and the origin of the mare basalts (abstract). In *Lunar Science—III* (editor C. Watkins), pp. 386–388. Lunar Science Institute Contr. No. 88.

Hubbard N. J. and Gast P. W. (1971) Chemical composition and origin of nonmare lunar basalts. *Proc. Second Lunar Sci. Conf., Geochim. Cosmochim. Acta* Suppl. 2, Vol. 2, pp. 999–1020. MIT Press.

Hubbard N. J. Meyer C. Jr. Gast P. W. and Wiesman H. (1971) The composition and derivation of Apollo 12 soils. *Earth Planet. Sci. Lett.* **10**, 341–350.

Hubbard N. J. Gast P. W. and Meyer C. Jr. (1972) Chemical composition of lunar anorthosites and their parent liquids (abstract). In *Lunar Science—III* (editor C. Watkins), pp. 404–406. Lunar Science Institute Contr. No. 88.

Keil K. Prinz M. and Bunch T. E. (1971) Mineralogy, petrology and chemistry of some Apollo 12 samples. *Proc. Second Lunar Sci. Conf., Geochim. Cosmochim. Acta* Suppl. 2, Vol. 1, pp. 319–341. MIT Press.

Klein C. Drake J. C. and Frondel C. (1971) Mineralogical, petrological and chemical features of four Apollo 12 lunar microgabbros. *Proc. Second Lunar Sci. Conf., Geochim. Cosmochim. Acta* Suppl. 2, Vol. 1, pp. 265–284. MIT Press.

Lovering J. F. Wark D. A. Reid A. F. Ware N. G. Keil K. Prinz M. Bunch T. E. El Goresy A. Ramdohr P. Brown G. M. Peckett A. Phillips R. Cameron E. N. Douglas J. A. V. and Plant A. G. (1971) Tranquillityite: A new silicate mineral from Apollo 11 and Apollo 12 basaltic rocks. *Proc. Second Lunar Sci. Conf., Geochim. Cosmochim. Acta* Suppl. 2, Vol. 1, pp. 39–45. MIT Press.

LSPET (1971) (Lunar Sample Preliminary Examination Team) Preliminary examination of lunar samples from Apollo 14, *Science* **173**, 681–693.

Lunatic Asylum (1970) Mineralogical and isotopic investigations on lunar rock 12013. *Earth Planet. Sci. Lett.* **9**, 137–163.

Meyer C. Jr. (1972) Mineral assemblages and the origin of nonmare lunar rock types (abstract). In *Lunar Science—III* (editor C. Watkins), pp. 542–544. Lunar Science Institute Contr. No. 88.

Palache C. Berman H. and Frondel C. (1944) *Dana's System of Mineralogy*, 7th edn., Vol. 1, Wiley.

Peckett A. Phillips R. and Brown G. M. (1972) New zirconium-rich minerals from Apollo 14 and 15 lunar rocks. *Nature* **236**, 215–217.

Ramdohr P. and El Goresy A. (1970) Opaque minerals of the lunar rocks and dust from Mare Tranquillitatis, *Science* **167**, 615–618.

Reid A. M. Ridley W. I. Warner J. Harmon R. S. Brett R. Jakes P. and Brown R. W. (1972) Chemistry of highland and mare basalts as inferred from glasses in the lunar soils (abstract). In *Lunar Science—III* (editor C. Watkins), pp. 640–642. Lunar Science Institute Contr. No. 88.

Ridley W. I. Williams R. J. Brett R. and Takeda H. (1972) Petrology of lunar basalt 14310 (abstract). In *Lunar Science—III* (editor C. Watkins), pp. 648–650. Lunar Science Institute Contr. No. 88.

Roedder E. and Weiblen P. (1972) Petrographic and petrologic features of Apollo 14, 15, and Luna 16 samples (abstract). In *Lunar Science—III* (editor C. Watkins), pp. 657–659. Lunar Science Institute Contr. No. 88.

Wager L. R. and Brown G. M. (1968) *Layered Igneous Rocks*. Oliver and Boyd.

Wasserburg G. J. Turner G. Tera F. Podosek F. A. Papanastassiou D. A. and Huneke J. C. (1972) Comparison of Rb–Sr, K–Ar and U–Th–Pb ages; lunar chronology and evolution (abstract). In *Lunar Science—III* (editor C. Watkins), pp. 788–790. Lunar Science Institute Contr. No. 88.

Proceedings of the Third Lunar Science Conference
(Supplement 3, *Geochimica et Cosmochimica Acta*)
Vol. 1, pp. 159–170
The M.I.T. Press, 1972

Petrology of Fra Mauro basalt 14310

W. I. Ridley, Robin Brett, Richard J. Williams, and Hiroshi Takeda

NASA Manned Spacecraft Center
Houston, Texas 77058

and

Roy W. Brown

Lockheed Electronics Corp.
Houston, Texas 77058

Abstract—A petrological study of rock 14310 indicates that it contains about 10% plagioclase pheno-crysts and may not necessarily represent a total melt composition. The plagioclase compositions can be explained by rapid crystallization in a closed system without requiring Na volatilization. Plagioclase crystallization was followed by orthopyroxene, pigeonite and augite, and ilmenite, chromian ulvöspinel and metal, leaving a siliceous, potassium-rich mesostasis. Unlike mare pyroxenes, metastable pyroxenes and augite pyroxene are rare.

The f_{O_2}, at least during the early portion of crystallization, was higher than that pertaining during the early crystallization of Apollo 12 basalts. The higher solidus and liquidus temperatures of 14310 com-pared to mare basalts, but greater enrichment in certain lithophile trace elements are inconsistent with 14310 and mare basalts being directly genetically related, and suggest chemically different source areas.

Introduction

Fra Mauro basalt 14310 is one of two crystalline rocks greater than 50 g weight returned by the Apollo 14 mission. It represents the only large crystalline rock at the Apollo 14 site with a bulk composition like that of the Fra Mauro breccias, and the KREEP component in soils from all landing sites (Meyer and Hubbard, 1970; Hubbard and Gast, 1971; Meyer *et al.*, 1971; Apollo Soil Survey, 1971; Reid *et al.*, 1972). If the rock is the product of a melt produced by the Imbrian impact, then its internal isochron age (3.89 ± 0.5 b.y., Wasserburg *et al.*, 1972) dates the time of the Imbrian event. If, on the other hand, the rock is pre-Imbrian, its internal isochron age dates either a period of nonmare igneous activity, or pre-Imbrian impact melting.

Additional attention has been focused on the rock by the work of Brown and Peckett (1971), who state that the rock shows indications of alkali loss by volatiliza-tion. This conclusion has broad implications on the low volatile content of all lunar samples returned to date.

Sections 14310,172 and 14310,80 have been examined by microscope and electron microprobe, and 14310,90 was used for single-crystal, x-ray diffraction studies. Several sections from the lunar sample curator's collection have also been examined petrographically.

Petrography

Rock 14310 has a wide range or textures from intersertal to subophitic to ophitic. The texture has been described in detail by Gancarz *et al.* (1971). The rock also contains a wide range of grain sizes, especially plagioclase, with some 10% large euhedral to subhedral plagioclase grains; the remainder of the plagioclase is lath-like and covers a range of grain sizes, down to felted groundmass laths about 0.5 mm long. Some small plagioclase laths are partly included in equant larger plagio-clase crystals, and there is no evidence of flow orientation of the feldspar. Local, finer grained patches in the rock grade into the coarser grained surroundings, and may represent cognate inclusions (LSPET, 1971) or reflect local variations in gas content during cooling. A mode (W. Melson and LSPET, unpublished data) calcu-lated from over 3,000 points is: plagioclase 66%, pyroxene 31%, opaque minerals 2%, mesostasis 0.5%. A mode of the opaque phases calculated from approximately 6000 points, averaged from six thin sections, confirms the presence of 2% opaque phases comprising ilmenite 72%, chromian ulvöspinel 8%, troilite 16%, Ni–Fe 4%. Trace amounts of tranquillityite and baddeleyite also occur. El Goresy et al. (1971) report more than 66% plagioclase in 14310, but Gancarz et al. (1971) report only 50% plagioclase in 14310,6 and Brown and Peckett (1971) report 53 ± 3% plagio-clase in 14310,20. The wide modal variation almost certainly confirms that 14310 is not a homogeneous rock. The variations do not seem to reflect only variations in phenocrystal plagioclase, but variation in all sizes of plagioclase grains from section to section. Consequently, major and trace element data should not be consistent throughout 14310.

Gancarz et al. (1971) report apatite, whitlockite, olivine and glass in the meso-stasis of 14310,6. El Goresy et al. (1971) report schreibersite, and "chalcopyrrhotite", and Brown and Peckett (1971) report pyroxferroite in 14310,13 and 14310,20.

Mineralogy

Plagioclase. Microprobe determinations of plagioclase phenocrysts in 14310,80 and 14310,172 indicate strong zoning (An_{94}–An_{58}) in most cases, and rare unzoned grains of An_{94}. Figures 1 and 2 indicate that the zoning is not necessarily symmetrical

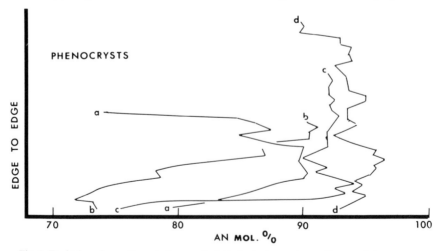

Fig. 1. Typical zoning patterns in large anhedral plagioclase. a-normally zoned; b, c-asym-metrically zoned, probably due to broken crystals; d-unzoned.

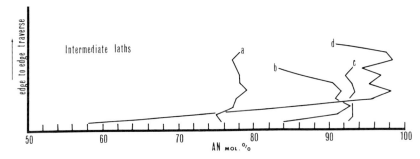

Fig. 2. Zoning patterns in intermediate sized plagioclases. a-unzoned solid plagioclase (xenocryst?); b-normally zoned; c-unzoned calcic plagioclase; d-strongly, but asymmetrically zoned.

across a grain, suggesting that many of the crystals are broken fragments. The zoning is essentially an albite substitution for anorthite with very little increase in K_2O with increasing Na_2O. The compositions of the groundmass laths and microlites also range widely (Fig. 3). Core and edge analyses of small laths indicate normal zoning from a calcic core (An_{95}) to sodic rim (An_{72-80}). The microlites range in composition from An_{94} to An_{80}.

Gancarz *et al.* (1971) have analyzed plagioclase of $An_{93}Ab_6Or_1$ composition, without pronounced zoning, in 14310,6 whereas Brown and Peckett (1971) have observed strongly zoned phenocrysts (An_{94-67}) but unzoned calcic groundmass feldspar (An_{93}) in 14310,13 and 14310,20.

Pyroxene. The pyroxenes of 14310 are intergranular to plagioclase but commonly include randomly oriented plagioclase laths. The cores of some pyroxene grains have no plagioclase inclusions, in contrast to their rims that have many included plagioclase laths. Small brownish grains of a discrete pyroxene-like phase (pyroxferroite?) occur in the mesostasis. Major and minor element zoning is evident in all grains

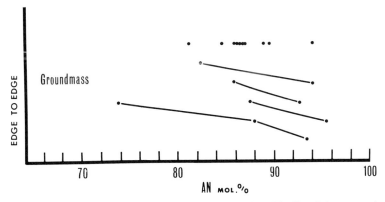

Fig. 3. Zoning in groundmass plagioclase laths and microlites. The lines join core analyses (calcic) to rim analyses (sodic). Single filled circles are analyses of microlites associated with residual glass.

examined but is different from grain to grain suggesting local chemical variations. Fe substitution for Mg is the most important major element variation, with a small degree of calcium enrichment (Fig. 4a). Low-calcium pyroxenes predominate, augitic pyroxenes being most common as fine lamellae (exsolution?) intergrown with pigeonite. Discrete augitic pyroxenes occur but are very rare.

The cores of some pyroxene grains are bronzitic ($Wo_4En_{80}Fs_{16}$) with up to 5% Al_2O_3. Aluminous bronzites are zoned outwards to low-alumina bronzites ($\sim 1\%$ Al_2O_3), but not all core bronzites are aluminous. The bronzites become enriched in Ca, Fe, and Ti and slightly depleted in Al outward, and pass into the pigeonite field. Single-crystal, x-ray diffraction studies supplemented by electron microprobe analyses (Takeda and Ridley, 1972) show that the orthopyroxene to clinopyroxene inversion composition is not easily identified. There is not the significant increase in Ca that commonly reflects this inversion in some terrestrial basaltic rocks and in some Apollo 14 coarse-fine pyroxenes (Fig. 4b).

The most iron-rich rim analyzed in 14310,80 is $Wo_{15}En_{40}Fs_{55}$, intergrown with lamellae of ferroaugite ($Wo_{32}En_{28}Fs_{40}$). Rare, discrete grains of zoned augite also occur. The cores are the most magnesian ($Wo_{37}En_{37}Fs_{25}$), zoning outward to ferro-augite with a composition similar to the lamellae.

Both the orthopyroxenes and clinopyroxenes show extensive Al solution, and the clinopyroxenes also contain significant Ti and Cr (Table 1). The alumina content of the orthopyroxenes ranges from 0.5 to 5 wt% and seems to be only weakly related

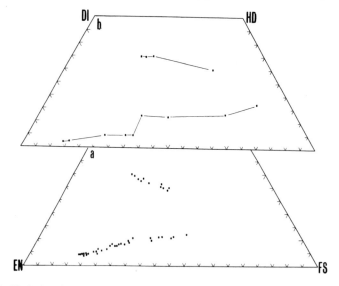

Fig. 4a. Variations in pyroxene compositions in 14310. Note the magnesian bronzites and the chemically ill-defined, orthopyroxene-pigeonite phase-change point. Calcic augites are single grains; ferroaugites are rim lamellae associated with intermediate pigeonites.

Fig. 4b. Pyroxene trend in a single crystal from coarse-fine sample 14259. Note the well-defined orthopyroxene to clinopyroxene phase-change point, and augite lamellae associated *only* with pigeonite.

Table 1. Microprobe analyses of pyroxenes in 14310.

	1	2	3	4	5	6	7	8
SiO_2	55.17	52.37	52.32	50.86	53.81	52.59	48.44	54.04
Al_2O_3	1.65	3.24	1.48	1.49	1.28	1.76	2.33	3.00
TiO_2	0.45	0.51	0.50	0.97	0.54	1.05	2.10	0.55
FeO	12.09	13.89	16.19	24.07	15.82	17.75	14.05	11.96
MnO	0.24	0.21	0.24	0.32	—	—	—	—
MgO	28.19	26.06	24.59	16.43	25.49	24.44	13.59	28.46
CaO	2.30	3.37	3.24	6.40	2.79	2.71	17.26	1.89
Na_2O	0.08	0.03	0.04	0.09	—	—	—	—
Cr_2O_3	0.48	0.45	0.39	0.11	—	—	—	—
TOTAL	100.65	100.13	98.99	100.74	99.73	100.30	97.77	99.90

1–4 = zoned pyroxene with bronzite core and intermediate pigeonite rim.
5–7 = zoned pigeonite (5, 6), and augite lamellae (7).
8 = ragged core of aluminous bronzite.
— = not determined.

to Ti solubility. In the low-alumina orthopyroxenes Al can be balanced by Ti and Cr but in the more aluminous bronzites, up to one-third Al must be octahedrally co-ordinated.

The Ti:Al ratio of the clinopyroxenes is about 1:3 and is roughly the same in magnesian pigeonites, intermediate pigeonites, augites and ferroaugites (Fig. 5). However, the absolute abundances of Al and Ti increase with iron enrichment, the ferroaugites and intermediate pigeonites being most enriched in Al and Ti. The trend can best be explained by Ti–Al and Cr–Al coupling without requiring any octahedrally coordinated Al.

Oxides. Lath-shaped ilmenite began to crystallize at about the onset of pyroxene crystallization, and crystallized continuously. It occurs as small laths and as dendritic crystals in the mesostasis. Rare ilmenite-ulvöspinel intergrowths also occur. Chromian ulvöspinel is the second most abundant oxide phase, mostly similar in composition and optical properties to the late-stage, khaki to buff ulvöspinel reported in the Apollo 12 basalts (Haggerty and Meyer, 1970; Gibb *et al.*, 1970). Oxide phases are discussed in detail by El Goresy *et al.* (1971).

In rare cases, ulvöspinel has undergone subsolidus reduction to ilmenite and metallic iron. Using the free-energy data of Taylor and Schmalzreid (1964) for this reaction, the reduction occurred at an f_{O_2} of less than $10^{-13.9}$ atm at 1100 C.

The characteristically foxy-red tranquillityite occurs as small irregular grains and rarely as hexagonal basal plates commonly associated with mesostasis. Tranquillityite is doubtlessly the Fe-Ti-Zr silicate discussed by Gancarz *et al.* (1971).

Nickel-iron and troilite. Nickel-iron occurs both as discrete, irregular grains and as inclusions in troilite, as it does in other crystalline lunar rocks. The metal-troilite aggregates range widely in Fe/FeS ratio. Metal is quite abundant in the mesostasis. Metal precipitation began at the onset of pyroxene crystallization and continued throughout the crystallization sequence, the last metal forming by subsolidus reduction of ulvöspinel. This is in agreement with El Goresy *et al.* (1971), who reported that metal precipitated middle to late in the paragenetic sequence.

Metal was precipitated throughout the crystallization sequence of most Apollo

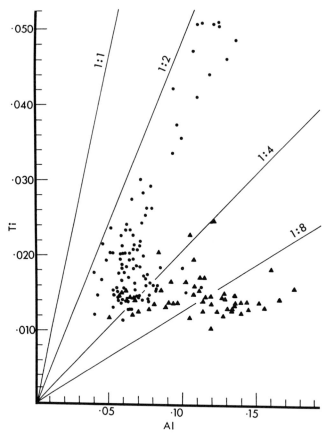

Fig. 5. Ti:Al plot, on the basis of 6 oxygens, for 14310 pyroxenes. Filled triangles are *chemically* defined orthopyroxenes (Fs < 30 mole %; Wo < 5 mole %). Filled dots are pigeonites, intermediate pigeonites and augites.

12 basalts (Reid *et al.*, 1970: Brett *et al.*, 1971). If one assumes that metal crystallization was an equilibrium process at least so far as the ratio Fe^0/Fe^{2+} is concerned, then the later-stage appearance of metal indicates that the redox state was higher during crystallization of 14310 than for Apollo 12 basalts, at least to the stage at which metal began to precipitate in 14310. If metal precipitation is caused by reactions such as:

$$2Cr^{2+}_{melt} + Fe^{2+}_{melt} = 2Cr^{3+}_{spinel} + Fe^0$$
$$\underset{pyroxene}{}$$

as suggested by Reid *et al.* (1970) and Brett *et al.* (1971), then the later precipitation of Cr-bearing spinels and pyroxenes in 14310, compared to the Apollo 12 rocks, might cause metal precipitation to be later.

Metal grains in rock 14310 range in composition from 23 to 0.7 wt% Ni. The mean is 5.5 ± 5.4 wt% Ni. Cobalt contents range from 2.2 to 0.3 wt% with a mean of 0.8 ± 0.4 wt%. There appears to be little, if any relationship between Ni and Co contents. We could find no relationship between the Ni content of a given metal grain and its position in the paragenetic sequence, unlike the relationship between Ni content and early crystallization of metal in some Apollo 12 rocks, as reported by Reid *et al.* (1970). This may be due to difficulty in determining the paragenesis of metal grains in this rock, because so few grains are included within other minerals. If one assumes that metal crystallization approached equilibrium conditions, then the Ni-rich metal grains precipitated earlier than the Ni-poor grains.

El Goresy *et al.* (1971) report that metal associated with troilite in 14310 is comparatively Ni-rich, unlike the metal included within troilite in Apollo 11 and 12 crystalline rocks. If the Fe–FeS melt separated from the silicate melt at a stage when the precipitating metal contained about 5.5 wt% Ni (the mean Ni content of metal grains in the rock) then such a sulfide melt would crystallize metal containing over 10 wt% Ni (Kullerud, 1963).

Schreibersite. El Goresy *et al.* (1971) report the presence of schreibersite in rock 14310 and state that this occurrence indicates conditions of formation more reducing than for the Apollo 11 and 12 rocks. The calculations of Olsen and Fuchs (1967) indicate that schreibersite is stable at 1200°C at a $\log f_{O_2}$ less than -10.5. This oxygen fugacity is near the wüstite-magnetite buffer so that highly reducing conditions are not required for schreibersite to be stable. The presence of schreibersite in rock 14310 probably is a function of the considerably higher phosphorus content of this rock (Rose *et al.*, 1972) compared to mare basalts. The presence of schreibersite *per se* cannot therefore be taken as any criterion of meteoritic contamination in rock 14310, as suggested by Dence *et al.* (1972).

Olsen and Fuchs (1967) have also calculated that the $\log f_{O_2}$ at 1100°C of the assemblage chlorapatite-metallic iron-pyroxene is -13.5 atm, which is in the range of oxygen fugacity of other lunar basalts at this temperature. Change in apatite composition would cause little change in the f_{O_2} at any given temperature. Therefore, the presence of the assemblage apatite-metallic iron-pyroxene in rock 14310 indicates normal redox conditions for this rock and not the highly reducing conditions suggested by El Goresy *et al.* (1972).

DISCUSSION

The texture of rock 14310 is that of a quickly cooled basalt. In hand specimen it appears to be a homogeneous, fine-grained basalt, but on closer examination it shows ranges of modal mineralogy, particularly in the distribution of plagioclase, and shows noticeable variations in grain size. We could find no petrographic criteria to decide between a true igneous and an impact-melt origin for 14310.

The texture is determined largely by the early and abundant crystallization of plagioclase. Plagioclase laths of many sizes are complexly intergrown into an open meshwork, suggesting initial rapid crystallization possibly due to supercooling. In this way, many of the small and large plagioclase laths with similar compositions may have crystallized simultaneously, so that decreasing grain size need not reflect the paragenetic sequence.

In some cases, sodic rims to plagioclases only develop where the plagioclase is in contact with alkali-rich mesostasis, so that the late-stage growth of plagioclase depends upon the availability of residual liquid. The meshwork texture, and probably increasing viscosity with crystallization, may have been important in determining the distribution of residual liquid.

Normal zoning of fine feldspar laths and microlites also is compatible with closed-system crystallization of a high-alumina basalt. More than 70% loss of Na_2O and K_2O from 14310, as suggested from experiments (Brown and Peckett, 1971), seems incompatible with the volatilization experiments of Gibson and Hubbard (1972) that require extreme Rb and K_2O loss prior to Na_2O loss. Such extreme loss of Rb should produce a K/Rb ratio significantly different from other KREEP basalts, which is not seen (Hubbard et al., 1972b). Extensive loss of K also seems to be partly at variance with the presence of a potassium-rich mesostasis in 14310.

The initial precipitation of orthopyroxene and subsequently of low-calcium clinopyroxenes, with only rare augitic pyroxenes, is unlike the pyroxene crystallization trends in mare basalts. The 14310 pyroxenes also differ from mare types in the rarity of metastable pyroxenes and the lack of erratic crystallization trends. Many of the differences probably derive from the quite different bulk compositions of mare basalts and 14310 basalt. Thus, initial crystallization of plagioclase in 14310 produced a residual liquid depleted in calcium, prior to pyroxene crystallization, and promoted the crystallization of calcium-poor pyroxenes. Other differences may result from the higher pyroxene liquidus temperatures for 14310 pyroxenes compared to mare basalts. Single crystals and electron microprobe data on a bronzite overgrown by twinned pigeonite with common (100) indicate that the orthopyroxene was succeeded by pigeonite at about 1200°C, using the En–Fs data of Kuno (1966).

Although both the orthopyroxene and clinopyroxene contain Al, Ti, and Cr as minor elements, the absolute abundances are lower than in mare basalts. The Ti–Al relationships in the clinopyroxenes most closely resemble those of Apollo 12 clinopyroxenes. In the Apollo 12 mare pyroxenes, the constancy of the Ti:Al ratio with increasing Fe content results from a gradual increase in Al in the melt due to the late crystallization of plagioclase. In contrast, plagioclase precipitates prior to pyroxene in 14310, and the similar trend of Ti:Al with Fe enrichment suggests either that plagioclase crystallization had almost ceased during pyroxene precipitation, or that the trend is dominated by the increase in Ti in the melt due to the late crystallization of Ti-bearing oxides.

The high alumina contents of some orthopyroxene crystals cores, necessitating some octahedrally coordinated Al, is unexpected in such a fine-grained basalt. The low-alumina orthopyroxenes, which are relatively common in 14310, are more consistent with a crystallization sequence in which much of the aluminum is removed from the melt as plagioclase crystallizes, prior to orthopyroxene crystallization. Hollister et al. (1972) suggest that some of the orthopyroxenes may be xenocrystic, derived from a higher pressure environment, which may account for the higher alumina contents. Bronzites in soil sample 14259 (Apollo Soil Survey, 1971) do not show the high alumina contents of bronzites from 14310 (Fig. 6). Because the Fra Mauro formation, as sampled at the Apollo 14 site, probably represents shallow

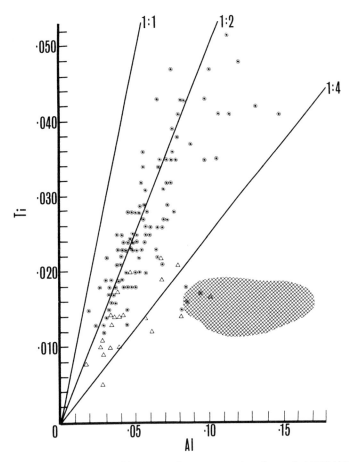

Fig. 6. Ti:Al plot, on the basis of 6 oxygens, for pyroxenes in soil sample 14259 (APOLLO SOIL SURVEY, 1971). Open triangles are *chemically* defined orthopyroxenes (See Fig. 5). Circles are pigeonites, intermediate pigeonite and augites. Shaded area is the field of many 14310 orthopyroxenes.

ejecta from the Imbrium Basin, the high-alumina bronzites may be from a deeper region.

Nonetheless, the fine-grained texture of 14310, in which even the aluminous bronzites are intergranular to plagioclase, as well as experimental evidence that orthopyroxene is an essential part of the crystallization sequence, suggests a low-pressure origin for the aluminous bronzites. The first orthopyroxene to crystallize may have been aluminous by a reaction similar to:

$$MgAl_2O_4 + SiO_2 \overset{reaction}{\rightleftharpoons} MgAl_2SiO_6$$

$$\underset{\text{spinel}}{} \quad \underset{\text{melt}}{} \quad \underset{\text{in orthopyroxene}}{}$$

Ford *et al.* (1972) report the crystallization of spinel after plagioclase in 14310. The spinel is later resorbed. Silica activity in the melt would also be increased by early precipitation of olivine and spinel (Ford *et al.*, 1972) thus driving the above reaction to the right. Experimental data (Kushiro, 1972) also suggests that aluminous orthopyroxenes could crystallize below 5 kb pressure.

The origin of the phenocrysts of plagioclase in 14310 still remains obscure. They do not show clear evidence of resorption, that might indicate a xenocrystric origin, but the textural evidence suggests that they were present when the bulk of the rock cooled. The normal zoning of many of the crystals would be consistent with an origin in which the phenocrysts were essentially in equilibrium with the high-alumina liquid. Some phenocrysts, however, have core compositions similar to the rims of the normally zoned grains and may be xenocrysts. Hubbard *et al.* (1972b) note that trace-element data are consistent with the addition of plagioclase to the 14310 melt.

If we assume some phenocrystic plagioclase, then removal of plagioclase from 14310 brings the major-element composition closer to the Type B glass (Apollo Soil Survey, 1971), a strongly preferred basaltic composition in the Apollo 14 soil. Type B glass corresponds generally to the major-element composition of KREEP basalt, that appears in crystalline, basaltic form in the Apollo 15 soil (Meyer, 1972). Gancarz *et al.* (1971) report the presence of plagioclase-rich basaltic fragments resembling 14310 in breccias 14321, 14305, 14083, and 14311. Because these breccias are most likely related to the Imbrian impact event, it follows that rocks similar to 14310, irrespective of their formation (igneous melt or impactite) must predate the Imbrian event, and may have been important rock types in the pre-mare Imbrium region.

The presence of liquidus plagioclase in 14310 suggests a close genetic relationship to lunar anorthosites, as suggested by Brown and Peckett (1971). Although 14310 may not be a pristine basaltic composition, high-alumina basalts of this general composition must be related to the more-aluminous lunar rocks. However, such a relationship on a regional scale seems most unlikely. KREEP basalts contain very high concentrations of radioactive elements, which on a regional scale would be inconsistent with thermal restraints (Wood *et al.*, 1971). Trace element data (Hubbard *et al.*, 1972b) also indicate that KREEP-like basalts cannot be parental to many of the anorthositic fragments examined to date. If highly feldspathic rocks formed an originally extensive lunar crust, their precursors, if any, still remain undetected.

The abundance of refractory trace elements in 14310 (LSPET, 1971), yet a high liquidus temperature ($\sim 1300°C$), precludes formation of a basaltic magma of 14310 composition, or any KREEP composition, by a small degree of partial melting of the source area of mare basalts. The aluminous nature of 14310 suggests an aluminous source region, probably plagioclase-rich, that is significantly different from the pyroxenitic source region of mare basalts.

Acknowledgments—W. I. Ridley and H. Takeda were supported by National Research Council Resident Research Associateships. We thank R. S. Harmon for modal counting of the opaque minerals.

References

Apollo Soil Survey (1971) Apollo 14: Nature of rock types in soil from the Fra Mauro Formation. *Earth Planet. Sci. Lett.* **12**, 49–54.

Brett R. Butler P. Jr. Meyer C. Jr. Reid A. M. Takeda H. and Williams R. (1971) Apollo 12, igneous rocks

12004, 12008, 12009, and 12022: A mineralogical and petrological study. *Proc. Second Lunar Sci. Conf., Geochim. Cosmochim. Acta* Suppl. 2, Vol. 1, 301–318.

Brown G. M. and Peckett A. (1971) Selective volatilization on the lunar surface: Evidence from Apollo 14 feldspar-phyric basalts. *Nature* **234**, 262–266.

Dence M. R. Plant A. G. and Traill R. J. (1972) Impact generated shock and thermal metamorphism in Fra Mauro lunar samples (abstract). *Lunar Science—III*, 174–176, Lunar Science Institute Contr. No. 88.

El Goresy A. Ramdohr P. and Taylor L. A. (1971) The opaque mineralogy of Apollo 14 crystalline rocks (abs.). *Meteoritics* **6**, 266.

El Goresy A. Ramdohr P. and Taylor L. A. (1972) Fra Mauro crystalline rocks: Petrology, geochemistry and subsolidus reduction of the opaque minerals. *Lunar Science—III*, 224–226, Lunar Science Institute Contr. No. 88.

Ford C. E. Humphries D. J. Wilson G. Dixon D. Biggar G. and O'Hara M. J. (1972) Experimental petrology of high alumina basalt, 14310, and related compositions. *Lunar Science—III*, 274–276, Lunar Science Institute Contr. No. 88.

Gancarz A. J. Albee A. L. and Chodos A. A. (1971) Petrologic and mineralogic investigation of some crystalline rocks returned by the Apollo 14 mission. *Earth Planet. Sci. Lett.* **12**, 1–18.

Gibb F. G. F. Stumpfl E. F. and Zussman J. (1970) Opaque minerals in an Apollo 12 rock. *Earth Planet. Sci. Lett.* **9**, 217–224.

Gibson E. K. Jr. and Hubbard N. J. (1972) Volatile element investigations on Apollo 11 and 12 lunar basalts via thermal volatilization (abstract). *Lunar Science—III*, 303–305, Lunar Science Institute Contr. No. 88.

Haggerty S. E. and Meyer H. (1970) Apollo 12: Opaque oxides. *Earth Planet. Sci. Lett.* **9**, 379–387.

Hollister L. S. Trzcienski W. I. Jr. Hargraves R. B. and Kulick C. G. (1971) Petrogenetic significance of pyroxenes in two Apollo 12 samples. *Proc. Second Lunar Sci. Conf., Geochim. Cosmochim. Acta* Suppl. 2, Vol. 1, 529–557.

Hollister L. S. Trzcienski W. I. Dymer R. Kulick C. Weigand P. and Hargraves R. (1972) Igneous fragment 14310,21 and the origin of the mare basalts. *Lunar Science—III*, 386–388, Lunar Science Institute Contr. No. 88.

Hubbard N. J. and Gast P. W. (1971) Chemical composition and origin of non-mare lunar basalts. *Proc. Second Lunar Sci. Conf., Geochim. Cosmochim. Acta* Suppl. 2, Vol. 2, 999–1020.

Hubbard N. J. Gast P. W. Meyer C. Jr. Nyquist L. E. Shih C. and Wiesmann H. (1972a) Chemical composition of lunar anorthosites and their parent liquids. *Earth Planet. Sci. Lett.* (in press).

Hubbard N. J. Rhodes J. M. Gast P. W. Bansal B. M. Wiesmann H. and Church S. E. (1972b) Nonmare basalts: Part II. (This volume).

Kullerud G. (1963) The Fe-Ni-S system. *Carnegie Inst. Wash. Yb.* **62**, 175–189.

Kuno H. (1966) Review of pyroxene relations in terrestrial rocks in the light of recent experimental work. *Mineral. J.* **5**, 21–43.

Kushiro I. (1972) Petrology of lunar high-alumina basalt. *Lunar Science—III*, 466–468, Lunar Sci. Institute Contr. No. 88, 466–468.

LSPET (1971) Preliminary examination of lunar samples from Apollo 14. *Science* **173**, 681–693.

Meyer C. Jr. (1972) Mineral assemblages and the origin of non-mare lunar rock types. *Lunar Science—III*, 543–545. Lunar Science Institute Contr. No. 88.

Meyer C. Jr. and Hubbard N. J. (1970) High potassium, high phosphorus glass as an important rock type in the Apollo 12 soil samples (abstract). *Meteoritics* **5**, 210–211.

Meyer C. Jr. Brett R. Hubbard N. J. Morrison D. A. McKay D. S. Aitken F. K. Takeda H. and Schonfeld E. (1971) Mineralogy, chemistry, and origin of the KREEP component in soil samples from the Ocean of Storms. *Proc. Second Lunar Sci. Conf., Geochim. Cosmochim. Acta* Suppl. 2, Vol. 1, 393–412.

Olsen E. and Fuchs L. H. (1967) The state of oxidation of some iron meteorites. *Icarus* **6**, 242–253.

Reid A. M. Meyer C. Jr. Harmon R. S. and Brett R. (1970) Metal grains in Apollo 12 igneous rocks. *Earth Planet. Sci. Lett.* **9**, 1–5.

Reid A. M. Ridley W. I. Warner J. Harmon R. S. Brett R. Jakes P. and Brown R. W. (1972) Chemistry of highland and mare basalts as inferred from glasses in the lunar soil. *Lunar Science—III*, 640–642, Lunar Sci. Inst. Contr. No. 88.

Rose H. J. Jr. Cuttitta F. Annell C. S. Carron M. K. Christian R. P. Dwornik E. J. and Ligon D. T. Jr.

(1972) Compositional data for fifteen Fra Mauro lunar samples. *Lunar Science—III*, 660–662, Lunar Science Institute, Contr. No. 88.

Takeda H. and Ridley W. I. (1972) Crystallography and chemical trends of orthopyroxene-pigeonite from rock 14310 and coarse-fine 12033. (This volume).

Taylor R. W. and Schmalzreid H. (1964) The free energy of formation of some titanates, silicates, and magnesium aluminate from measurements made with galvanic cells involving solid electrolytes. *Phys. Chem. J.* **68**, 2444–2449.

Wasserburg G. J. Turner G. Tera F. Posodek F. A. Papanastassiou D. A. and Huneke J. C. (1972) Comparison of Rb-Sr, K-Ar and U-Th-Pb ages; lunar chronology and evolution (abs.). *Lunar Science—III*, 788–790, Lunar Science Institute, Contr. No. 88.

Wood J. A. Marvin U. B. Reid J. B. Jr. Taylor G. J. Bower J. F. Powell B. N. and Dickey J. S. Jr. (1971) Mineralogy and petrology of the Apollo 12 lunar sample. *Smithsonian Astrophys. Obs. Spec. Rept. 333.*

Proceedings of the Third Lunar Science Conference
(Supplement 3, *Geochimica et Cosmochimica Acta*)
Vol. 1, pp. 171–184
The M.I.T. Press, 1972

Some textures in Apollo 12 lunar igneous rocks and in terrestrial analogs

H. I. Drever and R. Johnston

Dept. of Geology, University of St. Andrews, Scotland

P. Butler, Jr.

Manned Spacecraft Center, N.A.S.A., Houston, Texas

F. G. F. Gibb

Dept. of Geology, University of Manchester, England

Abstract—The interpretation of immature crystallization and some lunar textures characterized by it are the principal objectives of this investigation. A comparative and selective approach is adopted, and particular reference is made to the form and textural relations of olivine in 12009 and of pyroxenes and plagioclase in 12021, and to terrestrial analogs. The optic orientation of the olivines in 12009 is determined and their skeletal crystallization is illustrated and evaluated. Microprobe and optical data are associated in a textural analysis of an analog from the upper contact of a minor intrusion in Skye. The optic orientation of pyroxene enclosed in plagioclase cores is determined and the results plotted stereographically. The need for greater precision in the use of textural terms is stressed and a new term, intrafasciculate, introduced for textures in which pyroxene has crystallized within hollow, skeletal plagioclase. Apollo 12 pyroxene-phyric basalts are texturally reviewed and the crystallization of the phenocrysts discussed, emphasis being placed on size-independence of skeletal growth. On the basis of evidence that immature crystallization of pyroxene in a metastable field is not confined to quenched liquids, it is inferred that in mare basalts, both skeletal phenocrysts and skeletal groundmass crystals are the result of supersaturation, the earlier stage of lower nucleation density corresponding to a slightly lower degree of supersaturation. Similarities in the textural relationship of pyroxene and plagioclase, in the selected terrestrial analogs and in two Apollo 12 basalts, are correlated both optically and chemically. In the pyroxene quadrilateral, the clinopyroxene trend lines indicate, as in lunar basalts, rapid metastable crystallization. Some of the more promising lines of advancement in solving problems of texture and crystal growth in lunar basalts are summarized.

Introduction

In trying to comprehend the processes by which lunar igneous rocks were formed, some petrologists are compelled to place much reliance on concepts derived from their knowledge of terrestrial analogs. Lunar petrological research therefore requires that they ensure precision and reliability in those areas of terrestrial petrology that have greatest lunar relevance. In addition, an important side effect of the intensive research on lunar rocks is to stimulate new lines of enquiry in terrestrial petrology. The emergent situation is one in which progress can be achieved in both lunar and terrestrial petrology by cross-fertilization, and this is the view adopted in the present study of some critical textures in lunar igneous rocks. Undeniably, the factors determining lunar petrogenesis differ from those on earth. But, for very obvious reasons, it is necessary to lean heavily on our vastly greater knowledge of terrestrial rocks (Melson and Mason, 1971).

One of the most distinctive textural characteristics of lunar rocks of basaltic type is the crystallization of olivine, pyroxenes, and to a lesser degree plagioclase, in immature forms, the immaturity commonly extending to phenocrysts in porphyritic basalts. The interpretation of this immature crystallization, and the lunar textures it characterizes, are the principal objectives of this analogical investigation. Most of the interpretation has already been summarized (Drever *et al.*, 1972), and relevant information on immature crystallization has been expanded in another publication (Drever and Johnston, 1972).

The approach adopted is partly in the form of a review but mainly in that of a selective and comparative study of critical textures, particular attention being paid to Apollo 12 samples 12009 and 12021 together with those textures which, falling within our direct experience, represent close terrestrial analogs.

SKELETAL OLIVINES AND SKELETAL PYROXENES

Sample 12009 is the finest known rock portraying olivines in skeletal forms. It has been studied by Butler (1970) and Lofgren (1971) and is widely represented and figured in other publications (e.g., Plate 2, *Proc. Second Lunar Sci. Conf.*, Vol. 1, 1971; R. Brett *et al.* Fig. 2, ibid., p. 304; J. L. Warner, Fig 1, ibid., p. 471). Skeletal crystallization of olivine is also represented in other lunar samples or fragments but in a less spectacular manner (e.g., P. Gay *et al.*, ibid., p. 377; Wood *et al.*, 1971). Literature dealing with terrestrial examples has been reviewed by Drever and Johnston (1957), who developed the thesis that olivines in magnesium-rich magmas tend to crystallize very rapidly in skeletal forms. Hollow olivines or olivines with inclusions of groundmass (Roedder and Weiblen, 1971) are also typical products of rapid crystallization.

An exhaustive study of the optic orientation of the microphenocrysts in 12009,6 by Butler (1970) and R. Johnston (for this publication) yielded some evidence of preferred orientation in clusters, but not enough to suggest that movement of these crystals in relation to the liquid could be regarded as significant. Random optic orientation of a group of four olivines is demonstrated in Figs. 1 A,B. In size, there are all gradations from the largest microphenocrysts to the smallest skeletal microlites, and a greater range of skeletal shapes is demonstrated than in any one terrestrial rock. Because microphenocrysts with few recognizable skeletal features (in the plane of the section) are neither larger than, nor compositionally different from, those that are manifestly skeletal, they cannot be regarded as of significantly earlier formation.

The olivine microlites, although mainly in random orientation, may rarely develop as outgrowths from the equant microphenocrysts (Fig. 1B). They are rarely grouped in parallel and never so uniformly as in barred chrondrules or in certain samples of the Western Australian "spinifex" rock and the South African "peridotitic komatiite". Even more rarely, and in very small microlites, have epitaxial ilmenites developed at right angles to the elongation (Roedder and Weiblen, 1972a). Hitherto, such epitaxial ilmenite has been reported only in a Luna 16 sample (Roedder and Weiblen, 1972b). Also, skeletal pyroxenes, similar to those well-illustrated by Gansser (1950), develop patchily in 12009 and later than all the olivines. Some are markedly spherulitic (Lofgren, 1971).

The olivines in 12009 demonstrate, more convincingly than in any other known rock, that an inherent characteristic of their rapid crystallization in skeletal form is

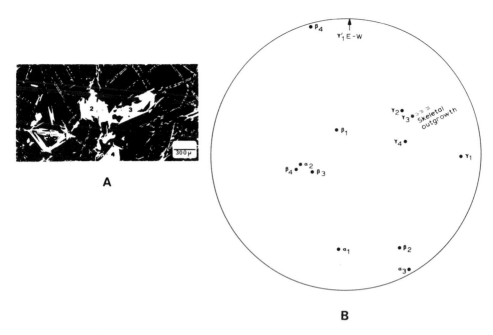

Fig. 1 (A). Two generations of skeletal crystals of olivine in Apollo 12 basalt 12009,6. Both tend to assume random orientations. The larger microphenocrysts exhibit forms that, in their variety, have no close counterpart in any known terrestrial rock. (B). Stereographic projection (upper hemisphere) of the optic orientation of microphenocrysts 1-4 in Fig. 1A. The arrow is approximately N-S.

the existence of embayments that cannot be ascribed to later resorption (Fig. 1A). Such embayments in skeletal-type olivines have led to uncertainty, in other instances, in postulating skeletal rather than resorption shapes. This range of shape could be ascribed to microscale environmental differences in the degree of supersaturation at the crystal-liquid interfaces, corresponding to different chemical gradients during rapid vectorial growth. At some points on a crystal there may be some cessation of growth, and even dissolution at other points, the contiguous liquid being no longer supersaturated (Chernov, 1963).

Because skeletal crystallization is most commonly observed in very small crystals, an undercooling model invoking rapid crystallization at the lunar surface is usually proposed. This may tend to obscure the fact that skeletal crystallization has no size limitations For instance, in certain of the ultrabasic, "harrisitic" rocks of Rhum, skeletal olivines of lamellar habit exceed half a meter in length (Drever and Johnston, 1972). More perfect skeletal types, several centimeters in length (Fig. 2), resemble those described by Naldrett and Mason (1968). A forsteritic olivine clast, reported by Brown *et al.* (1972) to exhibit "crescumulate" (or "harrisitic") texture, occurs in a sample of Fra Mauro breccia.

Skeletal or dendritic pyroxenes, several centimeters in length, also occur in terrestrial rocks (Drever and Johnston, 1972), and in some sections of 12009 it can even be observed that a later skeletal pyroxene has developed into larger crystals than

Fig. 2. Large dendritic crystals of forsteritic olivine in a specimen of eucritic "harrisite" from Harris Bay, Rhum, Scotland.

do any of the olivines (cf. Fig. A-67, Warner, 1970). Large, hollow pigeonite pheno-crysts, in Apollo 12 basalts, are reviewed later in the present paper.

Textural Analysis and Terminology

A specimen of a terrestrial rock with a texture relevant to the present study has been selected for rigorous textural analysis. This rock, recently discovered by one of the present authors, is located a few centimeters below the top of the horizontal western extension of the Gars-bheinn intrusion (Weedon, 1960) in the southern Cuillins of Skye, Scotland. Although it is predominantly ultrabasic, a more feldspathic facies is developed toward the top.

From a superficial assessment it might seem that the texture of this aphyric rock could be termed subophitic to subvariolitic, corresponding to a cotectic crystalliza-tion of the three principal minerals. The manner in which the olivine is interwoven with, and of similar size to, the other minerals might be regarded as ample justification for this assumption (Fig. 3).

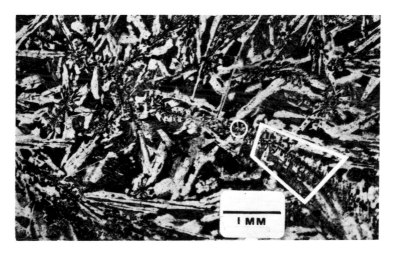

Fig. 3. Elongated skeletal olivine (O) intergrown with plagioclase. Radiating cross-sections of such olivines can be seen in the top left quadrant. Plagioclase with augite cores. Area in quadrilateral is sketched in Fig. 4A.

More careful observation and textural analysis (Figs. 4A,B,C,D) reveal that the elongated, and rarely curved, skeletal olivines have radiating outgrowths in cross-section, that some plagioclase has an epitaxial relationship with these outgrowths, and that the augite has a subophitic relationship with the plagioclase. In addition, an order of nucleation is clearly demonstrated, the olivine being followed by the plagioclase which is earlier than the slightly zoned augite. A metastable pyroxene trend (Muir and Tilley, 1964; Smith and Lindsley, 1971) is established by microprobe analysis (cf. Fig. 8, trend line a), and some range is evident in the composition of the forsteritic olivine.

On tracing back, in time, the sources and derivation of many of the textural terms still universally employed, it is disconcerting to discover how antiquated they are and how many of them have lost their original meaning (Johanssen, 1939). A case in point is the term *variolitic*, which is now employed so loosely as to have lost its original meaning entirely. Revision of textural terms in igneous petrology is long overdue, and every term should be based on a standard rock type. New terms such as *cumulate* (Wager *et al.*, 1960) or *crescumulate* (Wager and Brown, 1967), which are genetic in connotation, should be avoided in a purely descriptive context.

At different stages in crystallization, different textures can arise and, in order to achieve precision, three or even more textural terms may have to be employed in the textural analysis of a thin section of a homogeneous igneous rock. Only in this way can the changing composition of the mineral phases be correlated with the textural response at different temperatures and pressures. On the other hand, a single textural term, such as *poikilitic*, or *poikilophitic*, can be employed for a cumulate in which compositional zoning is absent, or a rock manifestly not a cumulate in which there is extreme compositional zoning in the minerals (e.g. Johnston, 1953). The four textural terms regarded here as appropriate to 12021, are porphyritic, subophitic (on

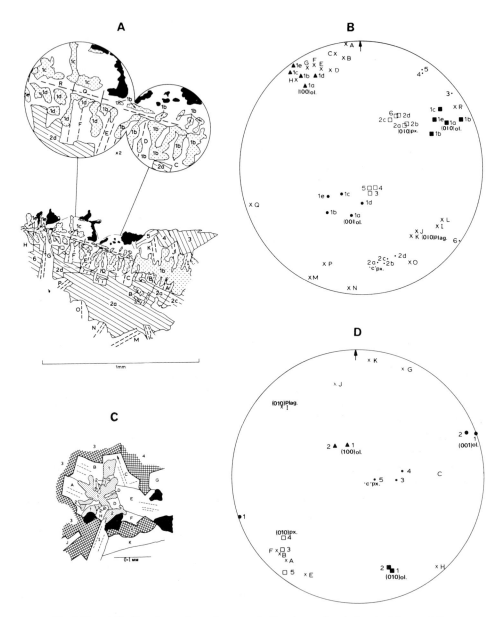

Fig. 4 (A and B). Drawing and equal area projection (upper hemisphere) of the crystallo-graphic directions in a longitudinal section (see Fig. 3) of an elongated, curved olivine (1) with plagioclase. A–R (An₇₄), in epitaxial relationship with the olivine and enclosed by augite (2–5). The reference arrow in projection (B) is the direction of the clinopyroxene (2a) cleavage. Locality: upper, western contact facies, Gars-bheinn ultrabasic intrusion, Skye, Scotland. (C and D). Drawing and equal-area projection (upper hemisphere) of a cross-section of an elongated, skeletal olivine (Fig. 3). Olivine (1 and 2) augite (3–5), plagioclase (A–K). The reference arrow in projection (D) is the (010) cleavage direction in plagioclase (C). Locality as above.

phenocryst margins), intrafasciculate, and radiate. The term subophitic is employed here for a texture characterized by a partial enclosure of plagioclase laths at pyroxene margins.

Intrafasciculate Texture in Lunar Basalts and Terrestrial Analogs

Apollo 12 sample 12021 had already been subjected to a very intensive petrological investigation. It has been called, on a textural and compositional basis, a porphyritic variolitic mare basalt (e.g., Dence *et al.*, 1971). The groundmass texture has also been referred to as subophitic (op. cit.), although this term is applicable only to the relationship of the plagioclase to the outer zones of the pyroxene phenocrysts.

A feature of the textural relations of the groundmass pyroxene and plagioclase in this rock, which is typical of a number of other Apollo 12 and Apollo 15 basalts, is the development within the plagioclase of elongated pyroxene in the form of "cores" (Figs. 5 and 6). This relationship has been well illustrated by Walter *et al.* (Fig. 6, 1971). The texture is sufficiently common and distinctive to justify a new name, and the term *intrafasciculate* is here suggested and employed. This may also help to focus more attention on it and lead to more definitive clarity.

This intrafasciculate texture is closely related to the "plumose" texture commonly referred to as variolitic (e.g. in 12002, 12052, 12053, 12065). In certain terrestrial rocks, which can be examined in a continuously exposed outcrop (e.g., the Gars-bheinn intrusion), the plumose aggregates are found to grade into a slightly coarser-grained facies, texturally typified by "cored" plagioclase crystals (Fig. 7). This seems to imply that such plumose (variolitic) textures are simply intrafasciculate on a finer scale. The texture of the rock illustrated in Figure 3 is mainly intrafasciculate.

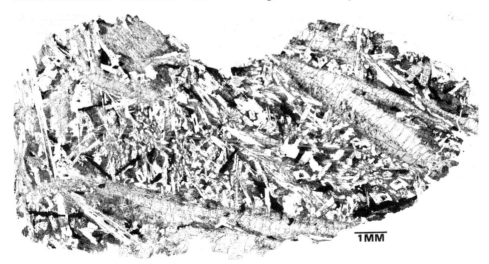

Fig. 5 Porphyritic, intrafasciculate, and radiate texture in Apollo 12 basalt 12021,3. Intrafasciculate texture is best seen on the far left (longitudinal section of hollow plagioclase containing pyroxene—see also Fig. 6); cross-sections can be seen on the far right, and radiate texture at upper center.

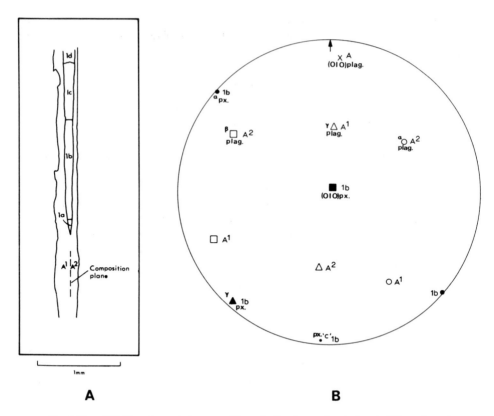

Fig. 6 (A and B). Drawing and stereographic projection (upper hemisphere) of the optic and crystallographic directions of plagioclase (A) enclosing augite (1a to 1d). The reference arrow in projection (B) is the (010) cleavage direction of the plagioclase. The drawing (A) is a representation of the plagioclase-pyroxene relationship in 12021,3 (see Fig. 5).

No optical or crystallographic relationship has been found to exist between the pyroxene cores and the plagioclase (Figs. 6A–B; Figs. 7A–D). In 12021, according to Walter *et al.* (1971), the pyroxene cores are richer in iron than similar-sized pyroxenes elsewhere in the groundmass, but no significant difference has yet been detected in the terrestrial analogs. From all the evidence assembled it is established that the pyroxene cores do not represent an intergrowth with the plagioclase but a development within a space formed in advance of the crystallization of the "core" pyroxene, this space being apparently in continuity with the interstitial liquid. Along the lines developed in this paper, the authors are at present investigating textures characterized by minerals, such as pyroxene and feldspar, radiating in all directions from a common nucleation center.

Porphyritic Texture in Lunar Basalts

Of all igneous rock textures, porphyritic is the one with which the interpretation of the major differentiation of lunar basalts is most concerned. Many lunar rocks

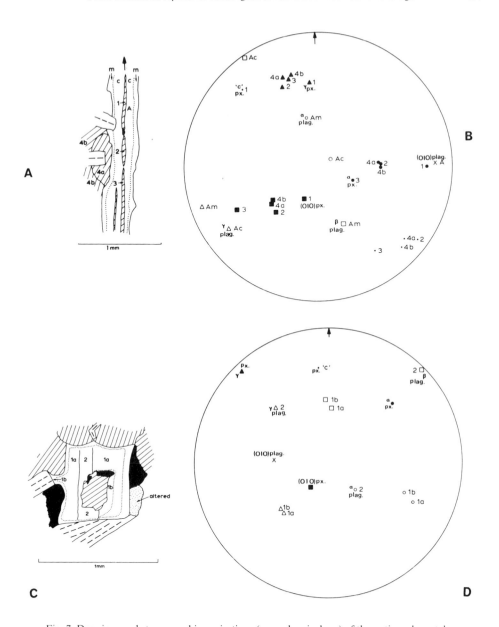

Fig. 7. Drawings and stereographic projections (upper hemisphere) of the optic and crystallographic directions of plagioclase enclosing augite, and also a little olivine and opaque iron ore in some cases. Locality: augite-plagioclase bands in the upper facies of the Ingia ultrabasic intrusion, Ubekendt Ejland, W. Greenland (Drever and Johnston, 1967). (A–B). Plagioclase, A (An$_{62}$) enclosing disoriented augites (1–3). The reference arrow is the (010) cleavage direction of the plagioclase. Augite (2) is nearly in optical continuity with augite (4a). (C–D). Plagioclase (1a, 1b, 2) enclosing augite and ore. The plagioclase is internally zoned (An$_{62/52-44}$) adjacent to the augite, and externally zoned (An$_{62/52-40}$). The reference arrow is the direction of the (010) cleavage of the plagioclase.

display this texture, particularly those from the Apollo 12 and Apollo 15 missions, the most typical phenocrysts being calcium-poor pyroxenes and, less commonly, olivine. Plagioclase-phyric rocks appear to be conspicuously lacking on the moon although they are very well represented on earth. The high-alumina Fra Mauro basalt 14310 is at least partly plagioclase-phyric (Brown and Peckett, 1971), but the polished thin section (14310,110) examined by the present writers appears to have a size gradation in the lamellar plagioclase from relatively large to relatively small, apart from rare little pools characterized by tiny plagioclase crystals of smaller size than the average. The texture in this thin section, however, may best be referred to as ophitic to poikilophitic rather than porphyritic, intersertal (Hollister *et al.*, 1972) or seriate. Variations in texture, not apparent in the one thin section (14310,110), have been observed when a number of thin sections were compared (Ridley *et al.*, 1972).

The phenocrysts in the pyroxene-phyric basalts commonly occur with hollow cores, and a number of investigators have examined this phenomenon in detail (e.g., Hollister *et al.*, 1971; Dence *et al.*, 1971; Bence *et al.*, 1971; Weill *et al.*, 1971), and the texture of the rocks has been well illustrated (Warner, 1970). It is generally agreed that such phenocrysts in 12021, 12052, 12053 and 12065, are immature or skeletal in form and crystallized rapidly. In most cases, the orientation is random, although a radiate arrangement has been observed (Hollister *et al.*, 1971) or lineate as, for example, in Apollo 15 basalt 15597 (Lunar Sample Information Catalog, 1971). To these pyroxene-phyric lunar basalts there appears to be no near terrestrial counterpart.

Textural Inferences

In many mare basalts the immature growth of the crystals, the compositional trend and complexity of the pyroxenes, and the abundant evidence of a lack of equilibrium have led many investigators of lunar samples to postulate quenching and crystallization in a metastable field (e.g. Brown *et al.*, 1970; Kushiro *et al.*, 1971).

The terrestrial rock-types, with clinopyroxene trends represented by a, b, c in Fig. 8, are not typical quench products. Yet the direction of these trends, corresponding as much to a reduction in Ca as an increase in Fe, is in accordance with the established quench or metastable trend (Smith and Lindsley, 1971). The clinopyroxenes defining these terrestrial trends are associated with facies that are not only richer in feldspar than the main body of rock in which they occur but appear to have crystallized somewhat later (cf. Aoki, 1962). The crucial factors, common to both the lunar and terrestrial clinopyroxene trends (a, b, c, d, e, Fig. 8), may be a lack of equilibrium and a relatively high degree of supersaturation (particularly of plagioclase) followed by rapid crystallization to which there is a characteristic textural response. Without more data it would be unrewarding at this stage, and largely irrelevant in this paper, to discuss trend differences in terms of differences in magmatic composition and oxygen fugacity.

Immaturity in crystal growth, although most common in small crystals, is not necessarily related to size, and if one were to reduce the average size of the hollow pyroxene phenocrysts in some mare basalts by a factor of ten, most petrologists

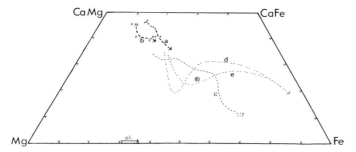

Fig. 8. Electron probe analyses illustrating the compositional trends (a, b, c) of clinopyroxenes in terrestrial igneous rocks and compositional trends of clinopyroxenes in Apollo 12 basalt 12021. Trend (a) represents intrafasciculate pyroxenes (Fig. 4) at the top of the Garsbheinn sill, Skye; (ol) represents the compositional range of the associated olivine. Trend (b) represents intrafasciculate pyroxenes (Fig. 7) in a band at the top of the Ingia intrusion, W. Greenland (Drever and Johnston, 1967). Trend (c) is the approximate compositional trend of clinopyroxenes in vein 96 in the upper part of the Makapohui lava (Evans and Moore, 1968, Fig. 3). K represents the clinopyroxene in a segregation vein in hypersthene-olivine basalt lava of the Okata Basalt Group, Izu, Japan (Kuno, 1955, Fig. 2). Approximate trends of clinopyroxenes in Apollo 12 basalt 12021 are represented by (d) Weill *et al.*, 1971 and (e) Brown *et al.*, 1971.

would interpret them as the result of rapid crystallization in situ. Intrafasciculate texture also appears to be the result of rapid crystallization.

The supersaturation responsible for rapid crystallization does not necessarily imply rapid cooling. Supersaturation, undercooling, and quenching are not synonymous, and a small degree of undercooling can yield a high degree of supersaturation (Wyllie, 1963).

The two-stage, rapid crystallization represented by the porphyritic olivine- and pyroxene-basalts referred to in this paper implies a sharp increase in the nucleation density, which could be ascribed simply to an increase in the degree of supersaturation. On the basis of chemical differences between pyroxene phenocrysts and groundmass pyroxenes, Bence *et al.* (1971) believe that the conditions governing early-stage pyroxene precipitation would only occur at some depth. On the other hand, Hollister *et al.* (1971) stress the immaturity in crystallization at both stages and prefer a surface or near-surface crystallization model.

The effects of the relatively low viscosity (Weill *et al.*, 1971) and low conductivity (Murase and McBirney, 1970) on the crystallization of lunar basaltic liquids are difficult to evaluate. But there is enough evidence to suggest that lunar lavas move over great distances and cool rapidly by surface radiation. Vesicularity and vug-formation indicate the removal of material in vapor form. According to Skinner and Winchell (1972), some of this material is relatively rich in alkalis. The effective removal of alkalis, favoring a lowering in viscosity, would entail some textural response.

Finally, it seems that further advances in understanding crystal growth and texture in lunar igneous rocks could depend, among other things, on (1) more definitive textural terminology and a closer integration of microprobe mineral analyses with their textural relationships; (2) more simulative experimentation, such as that of Brett *et al.* (1971); (3) more data on the cooling rate of lunar basalts (e.g., Provost and Bottinga, 1972); (4) growth rates of mineral phases determined experimentally for liquids of lunar composition (e.g., Sherer *et al.*, 1972); and (5) more information on the textural response to the removal of alkali-rich vapor.

Acknowledgments—For special assistance in a number of ways we thank Arch. M. Reid and Mary Reid, Neil Mackinnon, Sandra West, G. Tasker, S. Bateman, J. Allan, J. Zeeb, and J. Davie (for drafting the diagrams). Acknowledgment is also due to G. M. Brown and the referees for meticulous and effective critical reading of the manuscript.

References

Aoki K. (1962) The clinopyroxenes of the Otaki dolerite sill. *J. Japan. Mineral. Petrol. Econ. Geol.* **47**, 41–45.

Bence A. E., Papike J. J. and Lindsley D. E. (1971) Crystallization histories in two porphyritic rocks from Oceanus Procellarum. *Proc. Second Lunar Sci. Conf., Geochim. Cosmochim. Acta,* Suppl. 2, Vol. 1, pp. 559–574. M.I.T. Press.

Brett R. Butler P. Jr. Meyer C. Jr. Reid A. M. Takeda H. and Williams R. (1971) Apollo 12 igneous rocks 12004, 12008, 12009, and 12022: A mineralogical and petrological study. *Proc. Second Lunar Sci. Conf. Geochim. Cosmochim. Acta,* Suppl. 2, Vol. 1, pp. 301–317. M.I.T. Press.

Brown G. M. Emeleus C. H. Holland J. G. and Phillips R. (1970) Mineralogical, chemical and petrological features of Apollo 11 rocks and their relationship to igneous processes. *Proc. Apollo 11 Lunar Sci. Conf., Geochim. Cosmochim. Acta,* Suppl. 1, Vol. 1, pp. 195–219. Pergamon.

Brown G. M. Emeleus C. H. Holland J. G. Peckett A. and Phillips R. (1972) Mineral fractionation patterns between Apollo 14 primitive feldspathic rocks and Apollo 15 and other basalts (abstract). In *Lunar Science—III* (editor C. Watkins), pp. 95–97, Lunar Science Institute Contr. No. 88.

Brown G. M. and Peckett A. (1971) Selective volatilization on the lunar surface: Evidence from Apollo 14 feldspar-phyric basalts. *Nature* **234**, 262–266.

Butler P. Jr. (1970) Fabric and compositions of olivines in three Apollo 12 igneous rocks (abstract). Geol. Soc. Amer. Annual Meeting, 1970.

Chernov A. A. (1963) Crystal growth forms and their kinetic stability. *Kristallografiya* **8**, 63–67.

Dence M. R. Douglas J. A. V. Plant A. G. and Traill R. J. (1971) Mineralogical, petrological, and chemical features of four Apollo 12 lunar microgabbros. *Proc. Second Lunar Sci. Conf., Geochim. Cosmochim. Acta,* Suppl. 2, Vol. 1, pp. 285–299. M.I.T. Press

Drever H. I. and Johnston R. (1957) Crystal growth of forsteritic olivine in magmas and melts. *Trans. Roy. Soc. Edinburgh* **63**, 289–315.

Drever H. I. and Johnston R. (1967) Picritic minor intrusions. In *Ultramafic and Related Rocks* (editor P. J. Wyllie), Chap. 3, pp. 71–86, John Wiley.

Drever H. I. and Johnston R. (1972) Metastable growth patterns in some terrestrial and lunar rocks. *Repts., 24th Int. Geol. Congress,* Montreal (in press).

Drever H. I. Johnston R. Butler P. Jr. Gibb F. G. F. and Whitley J. E. (1972) Some textures in lunar igneous rocks and terrestrial analogs (abstract). In *Lunar Science—III* (editor C. Watkins), pp. 189–191, Lunar Science Institute Contr. No. 88.

Evans B. W. and Moore J. G. (1968) Mineralogy as a function of depth in the prehistoric Makapohui tholeiitic lava lake, Hawaii. *Contrib. Mineral. Petrol.* **17**, 85–115.

Gansser A. (1950) Geological and petrographical notes on Gorgona Island in relation to north-western S. America. *Schweiz. Mineral. Petrogr. Mitt.* **30**, 119–237.

Gay P. Bown M. G. Muir I. D. Bancroft G. M. and Williams P. G. L. (1971) Mineralogical and petrographic investigation of some Apollo 12 samples. *Proc. Second Lunar Sci. Conf., Geochim. Cosmochim. Acta,* Suppl. 2, Vol. 1, pp. 337–392. M.I.T. Press.

Hollister L. S. Trzcienski W. E. Jr. Hargraves R. B. and Kulick C. G. (1971) Petrogenetic significance of pyroxenes in two Apollo 12 samples. *Proc. Second Lunar Sci. Conf., Geochim. Cosmochim. Acta,* Suppl. 2, Vol. 1, pp. 529–557. M.I.T. Press.

Hollister L. Trzcienski W. Jr. Dymek R. Kulick C. Weigand P. and Hargraves R. (1972) Igneous fragment 14310,21 and the origin of the mare basalts (abstract). In *Lunar Science—III* (editor C. Watkins), pp. 386–388, Lunar Science Institute Contr. No. 88.

Johannsen A. (1939) *A Descriptive Petrography of the Igneous Rocks,* Vol. 1, Appendix 2, Univ. of Chicago Press.

Johnston R. (1953) The olivines of the Garbh Eilean sill, Shiant Isles. *Geol. Mag.* **90,** 161–171.

Kuno H. (1955) Ion substitution in the diopside-ferropigeonite series of clinopyroxenes. *Amer. Mineral.* **40,** 70–93.

Kushiro I. Nakamura Y. Kitayama K. and Akimoto S.-I. (1971) Petrology of some Apollo 12 crystalline rocks. *Proc. Second Lunar Sci. Conf., Geochim. Cosmochim. Acta,* Suppl. 2, Vol. 1, pp. 481–495. M.I.T. Press

Lofgren G. (1971) Spherulitic textures in glassy and crystalline rocks. *J. Geophys. Res.* **76,** 5635–5648.

Lunar Sample Information Catalog, Apollo 15 (1971) Lunar Receiving Laboratory, M.S.C., NASA.

Melson W. G. and Mason B. (1971) Lunar "basalts": Some comparisons with terrestrial and meteoritic analogs, and a proposed classification and nomenclature. *Proc. Second Lunar Sci. Conf., Geochim. Cosmochim. Acta,* Suppl. 2, Vol. 1, pp. 459–467. M.I.T. Press.

Muir I. D. and Tilley C. E. (1964) Iron enrichment and pyroxene fractionation in tholeiites. *Geol. J.* **4,** 143–156.

Murase T. and McBirney A. (1970) Thermal conductivity of lunar and terrestrial synthetic lunar rock in its melting range. *Science* **170,** 165–167.

Naldrett A. J. and Mason G. D. (1968) Contrasting Archaean ultramafic igneous bodies in Dundonald and Clergue Townships, Ontario. *Can. J. Earth Sci.* **5,** 111–143.

Provost A. and Bottinga Y. (1972) Cooling of lunar basalt flows (abstract). In *Lunar Science—III* (editor C. Watkins), pp. 622–623, Lunar Science Institute Contr. No. 88.

Ridley W. I. Williams R. J. Brett R. Takeda H. and Brown R. W. (1972) Petrology of lunar basalt 14310 (abstract). In *Lunar Science—III* (editor C. Watkins), pp. 648–650, Lunar Science Institute Contr. No. 88.

Roedder E. and Weiblen P. W. (1971) Petrology of silicate melt inclusions, Apollo 11 and Apollo 12 and terrestrial equivalents. *Proc. Second Lunar Sci. Conf., Geochim. Cosmochim. Acta,* Suppl. 2, Vol. 1, pp. 507–528. M.I.T. Press.

Roedder E. and Weiblen P. W. (1972a) Petrographic and petrologic features of Apollo 14, 15 and Luna 16 samples (abstract). In *Lunar Science—III* (editor C. Watkins), pp. 657–659, Lunar Science Institute Contr. No. 88.

Roedder E. and Weiblen P. W. (1972b) Silicate melt inclusions and glasses in lunar soil fragments from the Luna 16 core sample. *Earth Planet. Sci. Lett.* **13,** 272–285.

Scherer G., Hopper R. W. and Urlmann D. R. (1972) Crystallization behaviour and glass formation of selected lunar compositions (abstract). In *Lunar Science—III* (editor C. Watkins), p. 678, Lunar Science Institute Contr. No. 88.

Skinner B. J. and Winchell H. (1972) Vapor phase growth of feldspar crystals and fractionation of alkalis in feldspar crystals from 12038,22 (abstract). In *Lunar Science—III* (editor C. Watkins), pp. 710–711, Lunar Science Institute Contr. No. 88.

Smith D. and Lindsley D. H. (1971) Stable and metastable augite crystallization trends in a single basalt flow. *Amer. Mineral.* **56,** 225–233.

Wager L. R. Brown G. M. and Wadsworth W. J. (1960) Types of igneous cumulates. *J. Petrol.* **1,** 73–85.

Wager L. R. and Brown G. M. (1967) *Layered Igneous Rocks,* pp. 246–297. Oliver and Boyd, Edinburgh and London.

Walter L. S. French B. M. Heinrich K. J. F. Lowman P. D. Jr. Doan A. S. and Adler I. (1971) Mineralogical studies of Apollo 12 samples. *Proc. Second Lunar Sci. Conf., Geochim. Cosmochim. Acta,* Suppl. 2, Vol. 1, pp. 343–358. M.I.T. Press.

Warner J. L. (1970) Apollo 12 Lunar Sample Information. NASA Technical Report.

Warner J. L. (1971) Lunar crystalline rocks: Petrology and geology. *Proc. Second Lunar Sci. Conf.,* *Geochim. Cosmochim. Acta,* Suppl. 2, Vol. 1, pp. 469–480. M.I.T. Press.

Weedon D. S. (1960) The Gars-bheinn ultrabasic sill, Isle of Skye. *Quart. J. Geol. Soc. London* **116,** 37–54.

Weill D. F. Grieve R. A. McCallum I. S. and Bottinga Y. (1971) Mineralogy-petrology of lunar samples. Microprobe studies of samples 12021 and 12022; viscosity of melts of selected lunar compositions. *Proc. Second Lunar Sci. Conf., Geochim. Cosmochim. Acta,* Suppl. 2, Vol. 1, pp. 413–430. M.I.T. Press.

Wood J. A. Bower J. Dickey J. S. Jr. Marvin U. B. Powell B. N. Reid J. B. Jr. and Taylor C. J. (1971) Mineralogy and petrology of the Apollo 12 lunar sample. *Smithsonian Observatory Special Report 333,* pp. 1–272.

Wyllie P. J. (1963) Effects of the changes in slope occurring on liquidus and solidus paths in the system diopside-anorthite-albite. *Mineral. Soc. Amer., Spec. Paper* **1,** 204–212.

Proceedings of the Third Lunar Science Conference
(Supplement 3, *Geochimica et Cosmochimica Acta*)
Vol. 1, pp. 185–196
The M.I.T. Press, 1972

Equilibrium studies with a bearing on lunar rocks

A. Muan, J. Hauck,* and T. Löfall

The Pennsylvania State University,
University Park, Pa. 16802

Abstract—Equilibrium studies of lunar rocks and of synthesized oxide and silicate phases have been carried out in the temperature range of 1000–1400°C under strongly reducing conditions at a total pressure of 1 atm.

Liquidus and solidus temperatures and sequence of appearance of crystalline phases were determined for the two lunar rocks 14310 and 14259. The former has a liquidus temperature of approximately 1285°C with plagioclase as the primary crystalline phase, and a solidus temperature of approximately 1120°C. The latter sample has a liquidus temperature of approximately 1232°C, again with plagioclase as the primary crystalline phase, and a solidus temperature of approximately 1130°C.

Studies of synthesized mixtures were concentrated in two main areas, one dealing with subsolidus relations in titanate and spinel phases, the other with liquidus phase relations in simplified iron silicate systems. Equilibrium relations in the system $FeO–Al_2O_3–TiO_2$ at 1300°C were established, showing the existence of a miscibility gap in the $Fe_2TiO_4–FeAl_2O_4$ solid-solution series. A similar spinel miscibility gap is present in the system $MgO–Al_2O_3–TiO_2$ at 1300°C, whereas a continuous spinel solid-solution series is present at 1400°C. In each of the systems $Mg_2TiO_4–MgAl_2O_4–MgCr_2O_4$ and $Fe_2TiO_4–FeAl_2O_4–FeCr_2O_4$, at temperatures in the range of 1000–1300°C, a miscibility gap originates along the titanate-aluminate join and extends part-way toward the chromite end of the diagram. The model iron silicate system $CaMgSi_2O_6–Fe_2SiO_4–CaAlSi_2O_8$, displays phase relations involving the coexistence of liquids with olivine, pyroxene, anorthite, or spinel at temperatures similar to those of the liquidus-solidus range observed in lunar rocks (~1100–1300°C).

Introduction

Data on lunar rocks as well as on synthesized mixtures simulating phases or phase assemblages found in lunar rocks are needed in order to derive a better understanding of the petrogenesis of these rocks. In the present paper, we report results on solid-liquid relations in lunar rocks 14310 and 14259 at oxygen pressures corresponding to those of equilibrium with metallic iron at a total pressure of 1 atm. Furthermore, we present results of equilibrium studies involving spinel, ilmenite, and pseudobrookite phases in the systems $FeO–Al_2O_3–TiO_2$, $MgO–Al_2O_3–TiO_2$, $Fe_2TiO_4–FeAl_2O_4–FeCr_2O_4$, and $Mg_2TiO_4–MgAl_2O_4–MgCr_2O_4$, and results of equilibrium studies involving liquid and solid phases in the model silicate system $CaMgSi_2O_6–Fe_2SiO_4–CaAl_2Si_2O_8$ in contact with metallic iron.

Previously, we have shown that a miscibility gap exists in the solid-solution series between the two spinel end members Mg_2TiO_4 and $MgAl_2O_4$ in the system $MgO–Al_2O_3–TiO_2$ at 1300°C (Muan *et al.*, 1971). As part of the present study, we have investigated further the effect of temperature on the miscibility relations in this system, and we have delineated similar relations in the system $FeO–Al_2O_3–TiO_2$. Similar studies of the analogous Cr_2O_3-containing systems (Muan *et al.*, 1971; Shuart and Muan, unpublished data), on the other hand, have shown that continuous

---

*Now with Institut für Radiochemie der Kernforschungsanlage Jülich, West Germany.

solid-solution series exist between the spinel end members Mg_2TiO_4 and $MgCr_2O_4$, and between Fe_2TiO_4 and $FeCr_2O_4$ at the temperatures of the present investigation. Hence, if the spinel end members above are combined into the two ternary systems Mg_2TiO_4–$MgAl_2O_4$–$MgCr_2O_4$ and Fe_2TiO_4–$FeAl_2O_4$–$FeCr_2O_4$, a solid solubility gap will originate along the titanate-aluminate join in each system and gradually close as the contents of the chromite component increase. Delineation of the extent of the miscibility gap in these two systems as a function of temperature and determination of the direction of conjugation lines between coexisting solid-solution phases in the two-spinel fields constitute a major part of the present study.

Equilibrium relations in MgO-containing model silicate systems in which olivine, pyroxene, anorthite, or spinel, the main phases of lunar rocks, appear as crystalline phases at liquidus temperatures have been studied extensively in the past (see for instance $CaMgSi_2O_6$–Mg_2SiO_4–$CaAl_2Si_2O_8$; Osborn and Tait, 1952). Knowledge of the analogous iron oxide-containing systems, by comparison, is meager (e.g., $CaMgSi_2O_6$–Fe_2SiO_4–$CaAl_2Si_2O_8$). However, liquidus data for the latter system are now becoming available (Lofall and Muan, unpublished data), and are presented here as background information for further evaluation of crystallization phenomena in rocks of lunar compositions.

Experimental Methods

The equilibrium studies of the lunar rocks were carried out by experimental methods similar to those that have been described in detail previously (Muan and Schairer, 1970). The samples were contained in iron crucibles sealed into silica capsules and equilibrated for various periods of time (usually 1–3 days) at selected temperatures. The samples were then quenched to room temperature, the silica capsules were opened, and the iron crucibles containing the samples were sectioned longitudinally with a very thin diamond saw. One half of the crucible was mounted and examined by reflected-light microscopy, and the other half was ground for examination by transmitted-light microscopy and x-ray diffraction.

Conventional quenching techniques were used in the equilibrium studies of the synthesized mixtures. Reagent-grade oxides were used to make up starting materials of desired compositions, and these materials were heated at selected, carefully controlled temperatures in controlled atmospheres for sufficient lengths of time to attain equilibrium. The samples were then quenched rapidly to room temperature and the phases present were identified by microscopic examination (in transmitted as well as in reflected light), by x-ray diffraction, and in some instances by electron microprobe analysis.

Attainment of equilibrium was a problem in some of the experiments involving spinel phases (the systems Mg_2TiO_4–$MgAl_2O_4$–$MgCr_2O_4$ and Fe_2TiO_4–$FeAl_2O_4$–$FeCr_2O_4$). When dry oxides were used as starting materials, equilibrium usually was attained within a few days in runs at the highest temperatures used (1200–1300°C), particularly in iron oxide-rich mixtures. Reaction rates at the lowest temperatures used in the present investigation (1000–1100°C), and especially in mixtures low in iron oxide, were very sluggish. In these instances alternative methods were tried in order to attain equilibrium, namely, (a) runs under hydrothermal conditions, or (b) runs with a flux, and (c) runs of very long duration (approximately one month).

The hydrothermal runs were carried out at 1000°C. In each run, 60–100 mg of spinel solid solutions, pre-reacted at 1300°C, and approximately 20 mg water were sealed in platinum tubes and equilibrated for 6 days at an argon pressure of 2.7 kbar. In some of the hydrothermal runs involving mixtures of relatively high Cr_2O_3 contents hydrochloric acid (1 N) was used instead of water, but no significant beneficial effects on the rate of the reaction were observed.

In the equilibrations involving a flux, approximately 5% by weight of a 10% lead oxide (PbO) 90% potassium tetraborate ($K_2B_4O_7$, see Barks, 1966) mixture was added to each sample. The runs were carried out in sealed platinum tubes in order to prevent volatilization of the constituents of the flux.

In the iron oxide-containing spinels, "dry" runs of long duration (approximately one month) were the main method used for ascertaining spinel solubility relations at the lowest temperatures. The equilibrations were carried out in CO_2/H_2 atmospheres of ratio 1.38:1, corresponding to oxygen pressures of approximately $10^{-10.1}$, $10^{-11.3}$, and $10^{-14.4}$ atm at 1300, 1200, and 1000°C, in that order. A few hydrothermal runs were carried out with iron oxide-containing spinels sealed in 40%Ag60%Pd tubes that are permeable to hydrogen at elevated temperatures and in which the solubility of iron is fairly small even under strongly reducing conditions. In these runs a hydrogen pressure of approximately 100 bar was applied, by use of semipermeable Ag–Pd membrane, in addition to the argon pressure of 2.7 kbar. The resulting oxygen pressure inside the sample-containing capsules was approximately that of the SiO_2-fayalite-magnetite equilibrium.

The lattice parameters of the spinels were determined by x-ray diffractometer scans of the (440) and (311) reflections at a speed of 0.2° (2θ) per minute. The (220) reflection of MgO was used as a standard for these measurements.

RESULTS

Results of critical equilibration runs carried out on lunar rocks 14310 and 14259 are listed in Table 1. The liquidus temperatures in equilibrium with metallic iron were found to be 1285 and 1232°C, respectively, with plagioclase as the primary crystalline phase in both cases. Olivine is the second crystalline phase to appear in both cases, crystallizing at 1230°C in rock 14310 and 1225°C in rock 14259. Pyroxene follows at 1190°C in rock 14310 and 1215°C in rock 14259. The solidus temperatures are approximately 1120 and 1130°C, respectively, in the two rocks.

Phase relations in the system $FeO–Al_2O_3–TiO_2$ at 1300°C are shown in Fig. 1. Particularly noteworthy is the extensive miscibility gap in the spinel solid solution between the end members Fe_2TiO_4 and $FeAl_2O_4$. This is similar to observations previously made in the analogous MgO-containing system (Muan *et al.*, 1971), but in sharp contrast to the complete spinel solid-solution series observed in the systems $MgO–Cr_2O_3–TiO_2$ (Muan *et al.*, 1971) and $FeO–Cr_2O_3–TiO_2$ (Shuart and Muan, unpublished data).

The extent of the miscibility gap along the $Fe_2TiO_4–FeAl_2O_4$ join as a function

Table 1. Results of critical runs for determination of crystallization sequences in Apollo 14 rocks.

Rock No.	Temp. °C	Time (hrs.)	Phases Present*
14310	1290	16	Liq
	1281	16	An + Liq
	1239	16	An + Liq
	1220	16	An + Ol + Liq
	1209	16	An + Ol + Liq
	1182	16	An + Ol + Pyr + Liq
	1125	16	An + Ol + Pyr + Liq
	1110	16	An + Ol + Pyr
14259	1234	16	Liq
	1230	16	An + Liq
	1220	16	An + Ol + Liq
	1213	16	An + Ol + Pyr + Liq
	1136	16	An + Ol + Pyr + Liq
	1126	64	An + Ol + Pyr

*Abbreviations used have the following meanings: Liq = liquid; An = anorthite; Ol = olivine; Pyr = pyroxene.

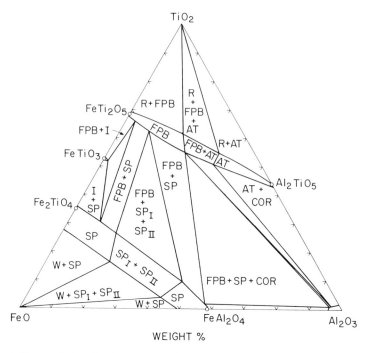

Fig. 1. Phase relations in the system FeO–Al$_2$O$_3$–TiO$_2$ in contact with metallic iron at 1300°C. Abbreviations used have the following meanings: R = rutile; FPB = ferro-pseudobrookite; AT = Al$_2$TiO$_5$; I = ilmenite; SP = spinel; COR = corundum; W = wüstite.

of temperature is shown in Fig. 2. It is seen that the miscibility gap does not close below solidus temperatures in this system. The compositions of the two coexisting phases within the spinel miscibility gap were determined from lattice-parameter measurements within each of the two terminal spinel solid solutions. The latter data are shown graphically in Fig. 3.

Similar data are shown in Figs. 4–6 for the system MgO–Al$_2$O$_3$–TiO$_2$. The first of these diagrams shows a comparison of phase relations in this system at 1400 and 1300°C. It is seen that only one spinel phase is present at the higher temperature (diagram on the left), whereas two spinel phases coexist at the lower temperature (diagram on the right). The extent of the miscibility gap along the spinel join as a function of temperature is shown in Fig. 5 and lattice parameters are shown in Fig. 6. It is seen that the extent of this miscibility gap is similar to that observed for the Fe$_2$TiO$_4$–FeAl$_2$O$_4$ system, but because of the much higher solidus temperatures in the MgO-containing system, the miscibility gap in Fig. 5 closes below the temperatures at which a liquid is present.

The extents of the spinel miscibility gaps in the systems Fe$_2$TiO$_4$–FeAl$_2$O$_4$–FeCr$_2$O$_4$ and Mg$_2$TiO$_4$–MgAl$_2$O$_4$–MgCr$_2$O$_4$ in the temperature range of 1000–1300°C are shown in Figs. 7 and 8, and lattice parameters of the ternary spinel solid solutions in the one-phase area of each system are shown in Figs. 9 and 10.

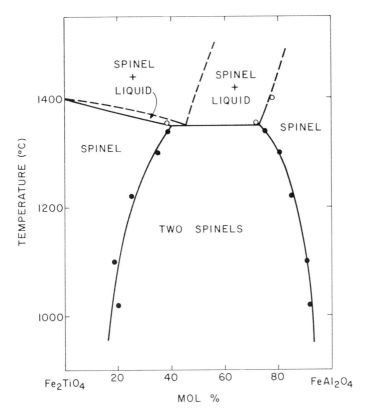

Fig. 2. Extent of miscibility gap along the Fe_2TiO_4–$FeAl_2O_4$ join of the system FeO–Al_2O_3–TiO_2, as a function of temperature.

The directions of conjugation lines between coexisting spinels in the two-phase areas of Figs. 7 and 8 are of particular interest in conjunction with reported compositions of spinel phases in lunar rocks (see for instance Agrell *et al.*, 1970, Champness *et al.*, 1971 and Haggerty and Meyer 1970). The directions of these conjugation lines are not likely to be greatly dependent on temperature, and hence their directions in the diagrams of Figs. 7 and 8 provide a valuable guide to the distribution of titanium, aluminum, and chromium among coexisting spinel phases under equilibrium conditions also at lower temperatures where the miscibility gap would be more extensive and perhaps would even extend to the titanate-chromite join (Fe_2TiO_4–$FeCr_2O_4$ or Mg_2TiO_4–$MgCr_2O_4$). A comparison between observed compositions of lunar spinels and the miscibility gaps experimentally determined for the temperature range 1000–1300°C is sketched in Fig. 11. A reasonable extrapolation of the experimental data to lower temperatures suggest that the compositions of coexisting lunar spinels are compatible with those predicted from equilibrium data.

The slight difference in behavior between the FeO- and the MgO-containing system has been ignored in projecting the spinel compositions in Fig. 11, which

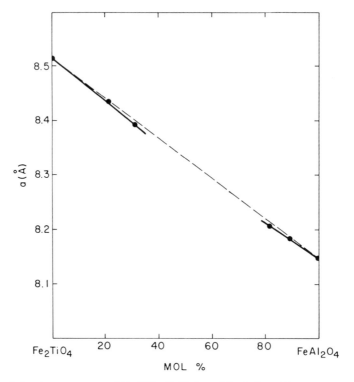

Fig. 3. Lattice parameters of Fe_2TiO_4–$FeAl_2O_4$ solid solutions. The samples were equilibrated at 1300°C and subsequently quenched to room temperature.

show only the ratios of TiO_2, Al_2O_3 and Cr_2O_3 in the spinel phase. The other
composition variables, FeO and MgO, are less critical than the other three composition variables, but can be taken into account by using three-dimensional repre-
sentations, for instance, as suggested by Haggerty, 1972. However, the experimental
data presently available are limited in extent and accuracy, and the present simplified
representation serves the purpose of establishing the trends of the immiscibility
relations.

Additional experimental data at lower temperatures (800–1000°C) and comparisons with additional data on lunar spinels that are now becoming available
are highly desirable for further checking of the inferences made regarding spinel
miscibilities in this paper.

Results of equilibration runs in a simplified model silicate system involving
olivine, pyroxene, anorthite, and spinel as crystalline phases in equilibrium with
liquids are shown in Fig. 12. Attention is called to the appearance of these crystalline
phases in equilibrium with liquid at temperatures similar to those prevailing in the
liquidus-solidus range in lunar rocks (~1100–1300°C). For comparison, it is noted
that the temperatures of coexistence of these crystalline phases in equilibrium with
liquid in the analogous MgO-containing model system $CaMgSi_2O_6$–Mg_2SiO_4–
$CaAl_2Si_2O_8$ (Osborn and Tait, 1952) are considerably (~100–200°C) higher.

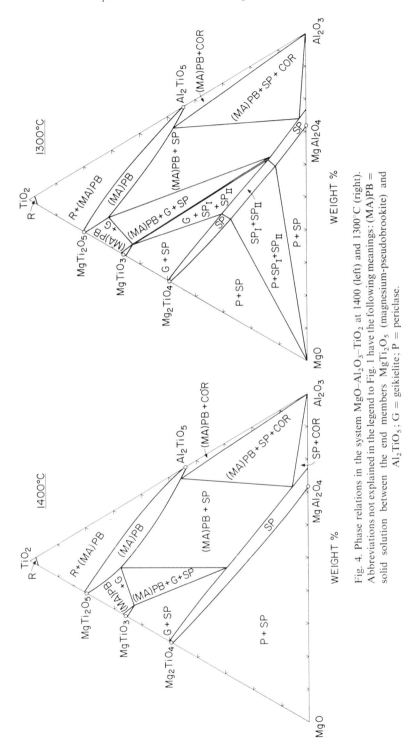

Fig. 4. Phase relations in the system MgO–Al$_2$O$_3$–TiO$_2$ at 1400 (left) and 1300°C (right). Abbreviations not explained in the legend to Fig. 1 have the following meanings: (MA)PB = solid solution between the end members MgTi$_2$O$_5$ (magnesium-pseudobrookite) and Al$_2$TiO$_5$; G = geikielite; P = periclase.

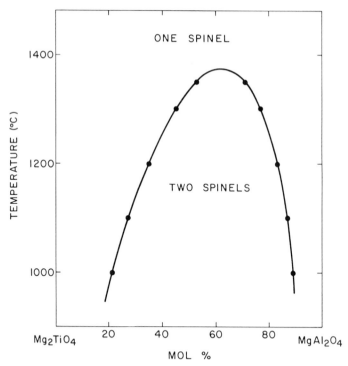

Fig. 5. Extent of miscibility gap along the Mg_2TiO_4–$MgAl_2O_4$ join of the system MgO–Al_2O_3–TiO_2, as function of temperature.

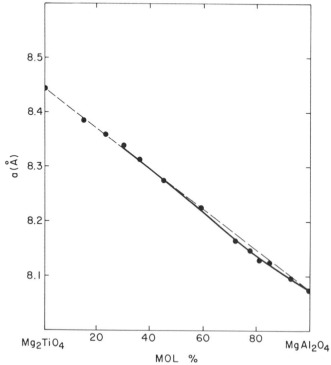

Fig. 6. Lattice parameters of Mg_2TiO_4–$MgAl_2O_4$ solid solutions. The samples were equilibrated at 1400°C and subsequently quenched to room temperature.

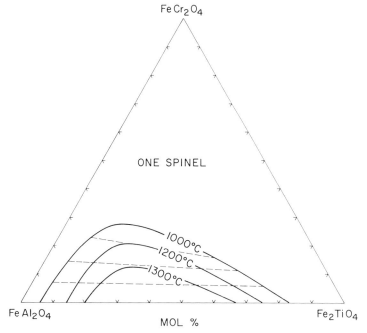

Fig. 7. Approximate extent of the spinel miscibility gap in the system Fe_2TiO_4–$FeAl_2O_4$–$FeCr_2O_4$ in the temperature range 1000–1300°C. Light dash curves are directions of conjugation lines between coexisting spinel phases within the two-phase area.

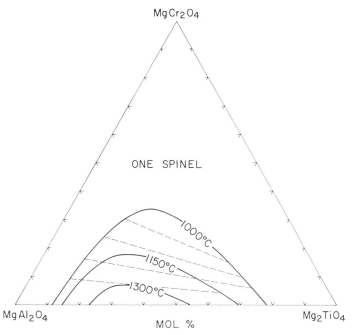

Fig. 8. Approximate extent of the spinel miscibility gap in the system Mg_2TiO_4–$MgAl_2O_4$–$MgCr_2O_4$ in the temperature range 1000–1300°C. Light dash curves are directions of conjugation lines between coexisting spinel phases within the two-phase area.

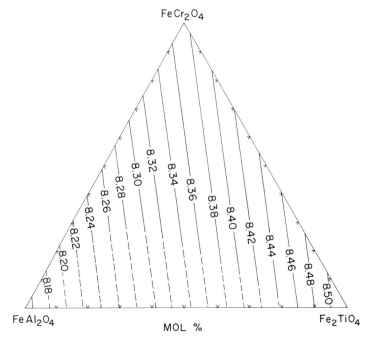

Fig. 9. Lattice parameters of ternary spinel solid solutions in the system Fe_2TiO_4–$FeAl_2O_4$–$FeCr_2O_4$. Samples were equilibrated at 1300°C and subsequently quenched to room temperature. The curves are dashed where the miscibility gap occurs.

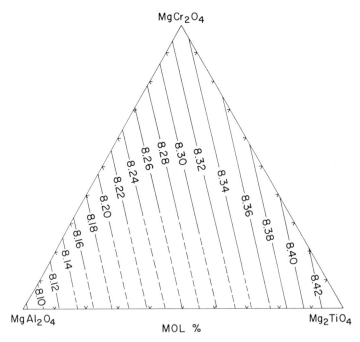

Fig. 10. Lattice parameters of ternary spinel solid solutions in the system Mg_2TiO_4–$MgAl_2O_4$–$MgCr_2O_4$. Samples were equilibrated at 1300°C and subsequently quenched to room temperature. The curves are dashed where the miscibility gap occurs.

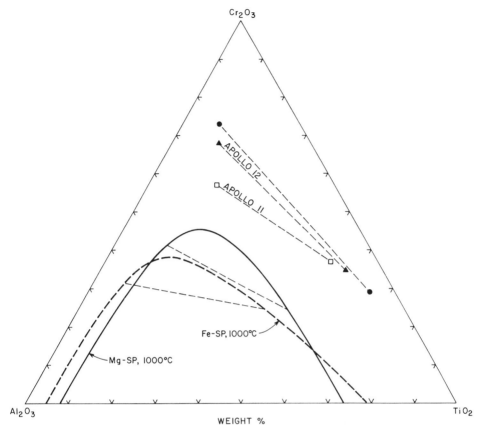

Fig. 11. Comparison of typical compositions of lunar spinels with miscibility gaps determined in the present investigation. The compositions have been projected into the plane
Al_2O_3–TiO_2–Cr_2O_3 for sake of simplicity (compare text). The curves labeled Fe–SP and
Mg–SP show the miscibility gaps along the spinel joins of the systems FeO–Al_2O_3–TiO_2
and MgO–Al_2O_3–TiO_2, respectively, at 1000°C.

The system shown in Fig. 12 is far from ternary because of the ready permutation
of iron and magnesium among the phases present in the system. However, by suitable
projections and supplementary diagrams it is possible to represent phase relations
in this system such that crystallization paths can be readily illustrated. Toward
this end, detailed experimental data are being obtained on the bounding system
$CaMgSi_2O_6$–$CaFeSi_2O_6$–Fe_2SiO_4 (Lofall and Muan, unpublished data). We
believe that phase relations in these model systems in equilibrium with metallic
iron will prove to be extremely valuable as a guide to the petrogenesis of lunar rocks.

Acknowledgments—The authors wish to thank C. W. Burnham and V. J. Wall for their assistance in the
hydrothermal runs and J. H. Puffer for help in microscopic examination of phases. This study was
supported by NASA Grant No. NGR39–009–184.

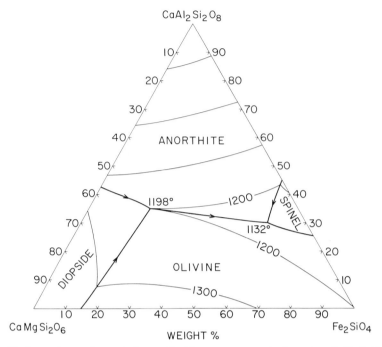

Fig. 12. Approximate phase relations at liquidus temperatures in the system $CaMgSi_2O_6$–Fe_2SiO_4–$CaAl_2Si_2O_8$ in contact with metallic iron. The system is not ternary, but is shown as a projection into the composition plane $CaMgSi_2O_6$–Fe_2SiO_4–$CaAlSi_2O_8$ for sake of simplicity.

REFERENCES

Agrell S. O. Boyd F. R. Bunch T. E. Cameron E. N. Dence M. R. Douglas J. A. V. Haggerty S. E. James O. B. Keil K. Peckett A. Plant A. G. Prinz M. and Traill R. J. (1970) *Proc. Apollo 11 Lunar Sci. Conf.*, *Geochim. Cosmochim. Acta* Suppl. 1, Vol. 1, pp. 81–86. Pergamon.

Barks R. E. (1966) Flux growth of single crystal R_2O_3 oxides with the corundum structure. Ph.D. Dissertation, The Pennsylvania State University.

Champness P. E. Dunham A. C. Gibb F. G. F. Giles H. N. MacKenzie W. S. Stumpfl E. F. and Zussman J. (1971) Mineralogy and petrology of some Apollo 12 lunar samples. *Proc. Second Lunar Sci. Conf.*, *Geochim. Cosmochim. Acta* Suppl. 2, Vol. 1, pp. 359–376. The MIT Press.

Haggerty S. E. (1972) Subsolidus reduction and compositional variations of lunar spinels (abstract). In *Lunar Science—III* (editor C. Watkins), pp. 347–349, Lunar Science Institute Contr. No. 88.

Haggerty S. E. and Meyer H. O. A. (1970) Apollo 12: Opaque phases. *Earth Planet. Sci. Lett.* **9**, 387–397.

Löfall T. and Muan A. (19) Phase equilibria in parts of the system CaO–MgO–FeO–Al_2O_3–SiO_2 under strongly reducing conditions. (Unpublished data).

Muan A. and Schairer J. F. (1970) Melting relations of materials of lunar compositions. *Carnegie Inst. Wash. Yearb. 1969–70*, pp. 243–245.

Muan A. Hauck J. Osborn E. F. and Schairer J. F. (1971) Equilibrium relations among phases occurring in lunar rocks. *Proc. Second Lunar Sci. Conf.*, *Geochim. Cosmochim. Acta* Suppl. 2, Vol. 1, pp. 497–505. The MIT Press.

Osborn E. F. and Tait D. B. (1952) The system diopside-forsterite-anorthite. *Amer. J. Sci.*, Bowen Volume, 413–433.

Shuart S. and Muan A. (19) Equilibrium relations among crystalline phases in the system FeO–Cr_2O_3–TiO_2 under strongly reducing conditions. (Unpublished data).

Proceedings of the Third Lunar Science Conference
(Supplement 3, *Geochimica et Cosmochimica Acta*)
Vol. 1, pp. 197–206
The M.I.T. Press, 1972

Experimental petrology and petrogenesis of Apollo 14 basalts

D. H. Green, A. E. Ringwood, N. G. Ware, and W. O. Hibberson

Department of Geophysics and Geochemistry,
Australian National University, Canberra, Australia

Abstract—Sample 14310 and a model KREEP composition have been studied experimentally as examples of the nonmare basaltic compositions that are represented at the Fra Mauro landing site.

Results of experimental melting studies at 1 atmosphere, compared with the observed mineralogy in 14310, show that the mineralogy of 14310 formed during crystallization at or near the lunar surface. Detailed study of the roles of plagioclase, olivine, pigeonite, and orthopyroxene crystallization in experiments and in the natural rock leads to the inference that 14310 cooled relatively slowly to $\sim 1180°C$ and then was more rapidly quenched, eliminating evidence for early transient olivine precipitation but preserving magnesian pigeonite (in part inverted to orthopyroxene) cores in zoned pyroxenes crystallized at $T < 1180°C$.

It is inferred that 14310 was an impact melt of pre-Imbrian regolith. High-pressure studies of 14310 and KREEP composition suggest primary derivation of nonmare basaltic compositions from a plagioclase-bearing source region in the outer 200 km of the moon. It is concluded that the moon is chemically zoned, the outer 200 km having significantly higher $\dfrac{MgO}{MgO + FeO}$ and U/K ratios together with a higher abundance of Al and perhaps Ca, relative to the region below 200 km which was the source of mare basalts. This chemical zonation appears to be a primary feature established during the accretion process.

Introduction

The breccia samples collected on the Apollo 14 mission contain rock fragments of mare-basalt type resembling examples of the Apollo 12 rock suite in texture and mineralogy. They include also fragments of plagioclase-rich basalt, distinctive in the textural dominance of plagioclase laths and in the common presence of orthopyroxene and absence or rarity of olivine. Sample 14310 is a large example of this rock type and, because of its differences from the Apollo 11 or 12 samples, has been studied experimentally to determine crystallization characteristics as functions of P and T. In addition, a model composition, designated KREEP (Meyer *et al.*, 1971) has been studied experimentally as a representative of the same class of nonmare basalts exemplified by 14310, but less rich in plagioclase. Compositions of the rocks studied are given in Table 1.

Subsolidus High Pressure Mineralogy

Experiments on 14310 composition used glass seeded with 10% Ga + Px + Qz + Rutile $\pm$ Kyanite (run at 25 kb, 1,100°C). High-pressure experiments used "piston-in" and a pressure correction of (-10%) of nominal load pressure. Techniques were the same as previously described (Ringwood and Essene, 1970; Green *et al.*, 1971). At 1,100°C, garnet disappeared from runs at 7 kb and 8 kb but occurred as minor but euhedral, well formed crystals at 9kb and 10 kb. Plagioclase remained the major phase at 9 and 10 kb but showed a large decrease at 15 kb, with accompanying

Table 1. Compositions of 14310* and KREEP† used in experimental study.

	14310 (wt%)	KREEP (wt%)		CIPW Norms 14310	KREEP
SiO$_2$	47.6	48.5	Qz	0.3	—
TiO$_2$	1.2	2.0	Or	3.0	7.2
Al$_2$O$_3$	20.7	16.7	Ab	5.9	5.2
FeO	8.2	11.1	An	52.0	39.2
MnO	0.1	0.0	Di	6.2	11.6
MgO	7.6	8.3	Hy	29.1	32.0
CaO	12.5	11.5	Ol	—	—
Na$_2$O	0.7	0.6	Ilm	2.3	3.8
K$_2$O	0.5	1.2	Ap	1.0	1.3
P$_2$O$_5$	0.4	0.6	Chr	0.3	—
Cr$_2$O$_3$	0.1	—	Plagioclase		
$\dfrac{100\ Mg}{Mg + Fe^{++}}$	62	56	An$_{85}$Ab$_{10}$Or$_5$	An$_{75.5}$Ab$_{10.5}$Or$_{14}$	

*B. W. Chappell, personal communication.
†Meyer et al., 1971.

increase in garnet and pyroxene. Ilmenite was present below 9 kb but at 10 kb was accompanied by rutile, which became more abundant at 15 kb and higher pressures. Plagioclase persisted to 21–22 kb but was absent at higher pressures. Quartz was present and kyanite tentatively identified as minor phases at >20 kb. At 1,300°C plagioclase was a minor phase at 20 kb but absent at 25 kb; garnet and pyroxene were the major phases at both pressures.

Subsolidus experiments on the KREEP composition were carried out using glass as starting material. At 1,100°C, 9 kb, rare garnet crystallized; garnet remained a very minor phase at 13 kb but increased markedly in a 15 kb run. Plagioclase crystallized as a minor phase at 17 kb but was absent at 18 kb. Minor quartz was present at 18 kb. The garnet which crystallized from KREEP composition at high pressures was lower in grossular content than that crystallized from 14310.

In comparison with Apollo 11 and 12 compositions, the pressure required for the first appearance of garnet resembles that of the olivine-bearing Apollo 12 basalts and may be attributed to reaction of anorthitic plagioclase with ferromagnesian minerals (ilmenite + hypersthene) in a *quartz-free* mineral assemblage. Very little garnet forms from this reaction and only above ∼14 kb, with hypersthene and plagioclase reacting to yield garnet + clinopyroxene + quartz, does garnet noticeably increase in amount. The very high anorthite content and high normative plagioclase/hypersthene + olivine ratio for these basalts cause plagioclase to coexist with increasing garnet and clinopyroxene to much higher pressures than for Apollo 11 and 12 compositions. These effects of chemical composition on the conditions of appearance and disappearance of garnet and plagioclase are consistent with compositional effects observed in terrestrial basalts (Ringwood and Green, 1966; Green and Ringwood, 1972). The depths within the moon over which the gabbro ($\rho \sim 2.95$ gm/cm^3) to eclogite ($\rho \sim 3.5$ gm/cm^3) transformation would occur in compositions such as 14310 or KREEP would allow persistence of plagioclase to a greater depth than was the case for Apollo 11 or Apollo 12 basalts. Nevertheless, consideration of mean lunar density and moment of inertia shows that a moon composed of rocks of compositions like 14310 or KREEP would have a substantially lower coefficient of moment of inertia than is observed (Ringwood and Essene, 1970), demonstrating that such compositions are not representative of the lunar interior.

MELTING RELATIONSHIPS AT 1 ATMOSPHERE

In experiments at one atmosphere, charges were held in iron capsules sealed in evacuated silica tubes. (See Figs. 1 and 2.) The starting material for most runs on 14310 was a finely crystalline (pyroxene + plagioclase + ilmenite) charge crystallized from glass at 1,050°C, 1 atmosphere. Runs were also carried out using glass as starting material particularly to obtain a reversal on the liquidus temperature and to grow phases sufficiently large for microprobe analyses. Starting glasses were prepared, by melting at 1,400–1,450°C for 1–2 mins. in an induction heater (argon atmosphere), from a homogenized 300 mg sample of the natural rock and from an oxide mix prepared to the 14310 composition provided by B. W. Chappell (personal

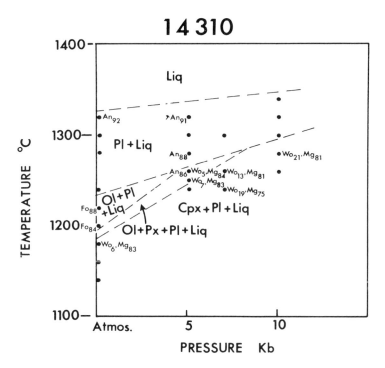

Fig. 1. Results of experimental crystallization studies at one atmosphere and up to 10 kb on sample 14310. Compositions of phases, where analyzable, are indicated in molecular

proportions: $\dfrac{100 \text{ An}}{\text{An} + \text{Ab} + \text{Or}}$ (plagioclase), $\dfrac{100 \text{ Fo}}{\text{Fo} + \text{Fa}}$ (olivine), and

$\dfrac{100 \text{ Wo}}{\text{Wo} + \text{Ens} + \text{Fs}}; \dfrac{100 \text{ Mg}}{\text{Mg} + \text{Fe}}$ (pyroxene).

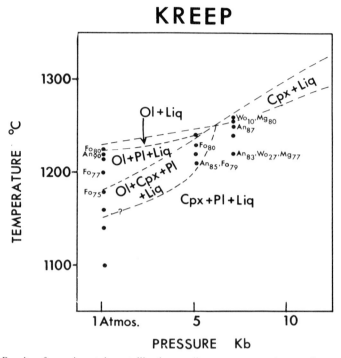

Fig. 2. Results of experimental crystallization studies at one atmosphere and up to 10 kb
on model KREEP basalt composition. Compositions of phases indicated as in Fig. 1.

communication). After preparation of the glasses and devitrification at 1,050°C and
1 atmosphere, it was found by analysis that FeO was 8.2% and the Fe_2O_3 was not
measurable (see Table 1). For runs on the KREEP composition, glass (see Table 1)
was used as the starting material.

For 14310 composition, plagioclase ($An_{92}Ab_{7.5}Or_{0.5}$) was the liquidus phase at
1,320°C, joined by olivine (Fo_{88}) at 1,230°C after approximately 10–15% plagioclase
had crystallized. Olivine continues to crystallize with plagioclase down to temperatures of <1,200°C, but between 1,200°C and 1,180°C olivine reacts with liquid to
precipitate low-calcium clinopyroxene (Table 2), and olivine is not detectable in
lower temperature runs. The solidus was not determined but lies above 1,050°C.

For the KREEP composition, olivine (Fo_{80}) occurs on the liquidus at 1,225°C
and is joined by plagioclase ($An_{89.5}Ab_{8.5}Or_{2.0}$) at very slightly lower temperature
(see Table 3). These minerals continue to precipitate together to 1,180°C, where
minor clinopyroxene appears. Runs below 1,180°C consisted of very fine-grained
pyroxene and plagioclase; the presence of olivine could not be confirmed and pyroxene and plagioclase were too fine grained for microprobe analysis.

Melting Relationships at High Pressures

For 14310 composition, plagioclase remains the liquidus phase to at least 10 kb.
(See Figs. 1 and 2.) At 5kb, the second phases to appear are olivine and orthopyroxene

Table 2. Compositions of natural and experimentally produced pyroxenes in 14310 and KREEP compositions (wt%).

	Natural Pyroxenes–14310				Experimentally crystallized pyroxenes–14310							Experimental pyroxenes–KREEP				
	Opx	Opx	Cpx	Cpx	Cpx 1 Atm. 1,180°C	Opx 5kb 1,260	Cpx 5kb 1,250	Cpx 6kb 1,260	Cpx 7kb 1,260	Cpx 7kb 1,240	Cpx 9kb 1,280	Cpx$_1$ 5kb 1,220	Cpx$_2$ 5kb 1,220	Cpx 5kb 1,210	Cpx 7kb 1,260	Cpx 7kb 1,220
SiO$_2$	54.1	53.8	54.6	53.2	53.3	50.8	53.3	51.3	50.8	49.7	49.5	50.3	53.0	50.1	54.2	47.9
TiO$_2$	0.5	0.7	0.5	0.7	0.8	0.4	0.7	1.0	0.6	1.1	0.9	0.9	1.1	1.5	0.5	1.7
Al$_2$O$_3$	2.6	2.2	2.1	1.4	2.4	8.5	3.4	7.4	8.0	8.2	9.2	5.8	3.4	7.6	3.2	9.8
FeO	11.0	14.1	12.5	14.8	10.4	9.4	9.9	9.5	10.3	12.1	9.1	11.2	12.2	9.9	11.4	9.7
MnO	0.2	0.2	0.2	0.3	0.2	0.2	0.2	0.3	0.3	0.3	0.2	—	—	—	—	—
MgO	28.0	26.8	27.7	25.6	28.3	28.4	27.9	23.8	23.9	19.7	21.4	23.9	26.2	20.7	25.2	19.1
CaO	2.5	2.3	2.5	2.9	2.8	2.4	3.3	6.0	6.0	8.9	9.3	8.3	4.7	10.7	5.0	12.3
Cr$_2$O$_3$	0.6	0.4	0.4	0.5	1.1	0.6	1.1	0.6	0.7	0.6	0.5	—	—	—	—	—
$\frac{100\,Mg}{Mg+Fe}$	81.5	77	80	75	83	84	83	82	81	75	81	79	79	78.5	80	77.5

Table 3. Compositions of natural and experimental phases from 14310 and KREEP compositions (wt%).

	Plagioclase 14310				KREEP				Ulvöspinel 14310	Experimentally crystallized spinel 14310	
	Cores of natural plagioclase laths	1 Atm. 1,320°C	5kb 1,280	5kb 1,260	1 Atm. 1,215	5kb 1,220	7kb 1,240			5kb 1,230	7kb 1,260
SiO$_2$	44.9	46.5	45.6	46.3	46.9	46.3	47.6	47.6	TiO$_2$ 29.4	0.7	0.4
Al$_2$O$_3$	35.1	33.3	34.2	33.2	32.0	33.0	31.6	32.0	Al$_2$O$_3$ 4.3	48.3	54.6
FeO	0.1	0.2	0.5	1.0	1.0	0.8	0.8	0.7	FeO 59.2	13.6	12.9
MgO	0.1	0.2	0.3	0.6	0.5	0.5	0.4	0.6	MgO 0.5	17.9	18.4
CaO	19.1	18.4	18.5	18.1	17.2	17.8	17.5	17.9	Cr$_2$O$_3$ 4.4	15.1	9.5
Na$_2$O	0.53	0.8	0.8	1.2	1.4	0.95	1.4	1.1	V$_2$O$_3$ 0.2	—	—
K$_2$O	0.05	0.12	0.09	0.16	0.29	0.31	0.6	0.5			
Or	0.3	0.7	0.5	0.9	1.7	1.8	3.3	3.1			
Ab	4.8	7.3	7.6	10.9	12.3	8.7	12.1	9.6			
An	94.9	92.0	91.9	88.2	86.0	89.5	84.6	87.3			

at 1,260°C, but the field of crystallization of orthopyroxene is very small, being replaced at 1,250°C by low-calcium clinopyroxene. Olivine is not detectable below 1,250°C and clinopyroxene + plagioclase are the major phases. Spinel is a minor phase below 1,250°C at 5 kb and at 1,260°C, 7 kb. At 7 kb neither orthopyroxene nor olivine appears, and the first clinopyroxene observed (at 1,260°C) is both more calcic and more aluminous than that crystallized at lower pressures. This trend continues at 10 kb (see Fig. 1 and Table 2).

At 5 kb for the KREEP composition, olivine, clinopyroxene, and plagioclase crystallize within 10°C of the liquidus, although olivine appears to be the actual liquidus phase. At 7 kb, clinopyroxene is the liquidus phase and plagioclase the second phase to appear. Olivine does not appear in the melting interval at this pressure. Clinopyroxene at 1 atmosphere was too fine grained for analysis, but at 5 kb, 1,220°C two coexisting clinopyroxenes, "pigeonite" and "subcalcic augite," were observed. Almost identical pairs were observed in two runs at 1,220°C, one of which was heated first to 1,260°C for 2 minutes and then rapidly cooled to 1,220°C and the other of which was heated to and held at 1,220°C. Quench clinopyroxenes observed in some runs were very different ($>10\%$ Al_2O_3, $>1.2\%$ TiO_2, $\dfrac{100Mg}{Mg + Fe}$ ~ 60) from either of the two clinopyroxenes at 5 kb, 1,220°C. The experiments are consistent with the existence of a miscibility gap *at low pressure* between pigeonite and subcalcic augite compositions (see Fig. 3). This is not to be confused with an apparent reaction relationship between *orthopyroxene* and pigeonite at low pressure (cf. 14310 composition with orthopyroxene at 5 kb, 1,260°C and "pigeonite" at 5 kb, 1,250°C). At 5 kb and 1,210°C only one pyroxene (subcalcic augite) is present in the KREEP composition system, and at 7 kb there appears to be a sudden change from a "pigeonite" composition in equilibrium with liquid at 1,260°C to a subcalcic augite composition in equilibrium with liquid at 1,220°C. No evidence for the coexistence of two pyroxenes was found at this pressure.

Interpretation of Crystallization History and Genesis of Basalt 14310 in the Light of Experimental Studies

The plagioclase observed on or very close to the liquidus at 1,320°C and 1 atmosphere is slightly less calcic (An_{92}) than the most calcium-rich natural plagioclase present in 14310 (An_{95}). The difference is not large enough for one to state with certainty that the most calcic plagioclases must be exotic or xenocrystal in relation to 14310 liquid, but the data are sufficient to show that the phenocrystal or larger plagioclase crystals did not precipitate from a liquid with a *higher Na/Ca ratio* than observed at present in 14310 (cf. Brown and Peckett, 1971). Olivine is the first ferro-

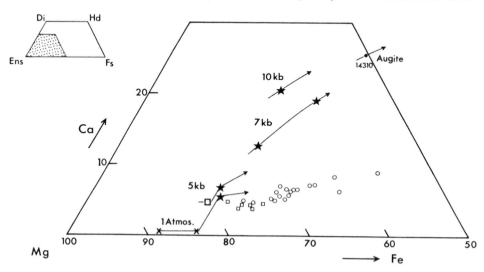

Fig. 3. Pyroxene diagram indicating compositions of natural and experimentally crystallized pyroxenes from 14310. □ Orthopyroxene cores in natural rock; o low-calcium clinopyroxenes in natural rock; ● high-calcium clinopyroxene in natural rock; ▫ orthopyroxene crystallized at 5 kb; ★ clinopyroxenes crystallized at 1 atm, 5 kb, 7 kb, and 10 kb; × olivines crystallized at 1 atm.

magnesian phase to appear in experiments at 1 atmosphere, and orthopyroxene does not appear. These observations appear to conflict with the natural rock in which the most Mg-rich phases are orthopyroxene and low-Ca, magnesian pigeonite. However, the olivine at 1,230°C has $\dfrac{100\text{Mg}}{\text{Mg} + \text{Fe}} = 88$ and that at 1,200°C has $\dfrac{100\text{Mg}}{\text{Mg} + \text{Fe}} = 84$, whereas the most magnesian orthopyroxenes observed in the natural rock have $\dfrac{100\text{Mg}}{\text{Mg} + \text{Fe}} = 81.5$. Between 1,200°C and 1,180°C olivine ($<\text{Mg}_{84}$) reacts with liquid to precipitate low-calcium pigeonite (Mg_{83} at 1,180°C). The low-calcium pigeonite is slightly more magnesian but otherwise almost identical in composition to ortho-pyroxene and magnesian pigeonite occurring in the natural rock (see Fig. 1 and Table 2). It is inferred that the orthopyroxene of 14310, at least that with the composition given in Table 2, has inverted from low-calcium pigeonite, the latter phase being in equilibrium with plagioclase + liquid (14310 bulk composition) at 1,180°C or slightly lower temperature. The zoned pigeonite and augite of the natural rock crystallized during relatively rapid cooling at T < 1,180 C. Thus, 14310 preserves a record of crystallization in which there is no evidence in the ferromagnesian phases of crystal-lization in the 1,230–1,180°C temperature interval, but in contrast the rock contains plagioclase that is too calcic to have been in equilibrium with liquid in 14310 *below 1,320°C* (plagioclase in 14310 composition at 1,180°C would be $<\text{An}_{88}$. See Fig. 2.).

These internal inconsistencies in the crystallization of 14310 cannot be explained by the hypothesis that 14310, at the time of crystallization of the larger euhedral plagioclase, contained $\sim 70\%$ higher $\text{Na}_2\text{O} + \text{K}_2\text{O}$ contents (Brown and Peckett, 1971). This would enlarge the crystallization field of olivine at the expense of pyroxene and plagioclase and would ensure that the earliest plagioclase to crystallize was much more sodic than even An_{92} (Green, Ware, and Hibberson, in preparation).

Two interpretations of the experimental results in relation to the observed petrology of the natural rock might be considered:

(a) 14310 was once completely liquid but cooled relatively slowly from $>1,320°C$ to 1,180°C, allowing time for all olivine precipitated between 1,230°C and 1,200°C to react with liquid but permitting only limited reequilibration of large, early-formed plagioclase crystals. This hypothesis is permissible whether 14310 was ultimately an impact melt or of internal lunar origin.

(b) 14310 is an impact melt of preexisting crystalline material in which calcic plagioclase and orthopyroxene were important and more refractory phases. Thus, 14310 may not have completely molten nor homogenized in the impact event, and some pyroxene and plagioclase may be "xenocrysts" (with overgrowths) in an inhomogeneous impact melt.

If the second interpretation of 14310 crystallization history is correct, then the rock allows no further conclusions about the nature of the lunar interior *except* insofar as the composition of various samples of the regolith* may be used to infer

---

* The term "regolith" is used in a broad sense to denote the outermost layer of the pre-Imbrian crust which probably was shattered, crushed, or shock melted and repeatedly mixed by the intense (meteoritic?) bombardment responsible for the highland craters. This "regolith" may be as much as 20 km thick.

the processes that operated in producing the pre-Imbrian regolith. If the first interpretation is correct, then either impact melting or internal partial melting to produce
14310 liquid remain possible processes. However, the chemical composition and
high-pressure crystallization characteristics of 14310 do not support an origin by
one-stage partial melting of the lunar interior. The high K and P and incompatible
trace element abundances of 14310 in comparison with Apollo 12 mare basalts or
relative to chondritic abundances, coupled with the high $\dfrac{Mg}{Mg + Fe^{++}}$ ratio and Cr
content, indicate that 14310 composition could only be a product of very small
degrees of melting, unless the source region possessed extremely high incompatible
trace element contents. However, the wide field of plagioclase crystallization below
the liquidus at 0–10 kb unaccompanied by other phases, shows that 14310, if postulated as a partial melt, would be derived by a sufficiently high degree of melting to
eliminate all phases but plagioclase from the residue. It has previously been argued
that plagioclase is absent from the source region (below approximately 200 km) of
mare basalts (Ringwood and Essene, 1970; Ringwood, 1971; Green et al., 1971). A
further difficulty with derivation of 14310 by partial melting of the lunar interior is
that temperatures in excess of 1,320°C in the outer 200 km of the moon would be
required at $T \sim 3.9$ b.y. (Turner et al., 1971), 0.6 b.y. after formation of the moon.
During this time interval, cooling and formation of a "rigid" crust make such high
temperatures, at shallow depths, rather improbable.

These constraints on the hypothesis of origin of 14310 as a direct partial melt of
the lunar interior can be reconciled only by postulating that 14310 was produced by
very small (probably $<5\%$) degrees of melting of an anorthosite source rock ($>95\%$
anorthite) as shallow depths in the moon. Our prejudice is that such a source rock
and origin for 14310 are improbable and we adopt as a preferred hypothesis that
14310 was produced as an impact melt of the pre-Imbrian regolith, the particular
regolith sample that was melted being enriched in plagioclase.

Nature of the Lunar Interior as Inferred from the Petrogenesis of Mare and Nonmare Basalts

Our preferred hypothesis interprets 14310 as an impact melt of pre-Imbrian
regolith. The composition of 14310 relates it to the component recognized in Apollo
11, 12, and 14 soils and breccias and referred to as KREEP, norite, or nonmare
basalt (Hubbard, et al., 1971; Meyer et al., 1971; Apollo Soil Survey, 1971; Fuchs,
1971; Lindsay, 1971). In the following discussion we will examine the significance of
this component with respect to the nature of the outer layers of the moon and the
early stages of lunar evolution.

The composition and mineralogy of nonmare basalt as defined by the above
investigators demonstrate its overall basaltic character (e.g. "Fra Mauro basalts"
Types B, C, and D of Apollo Soil Survey, 1971 or average KREEP basalt of Hubbard
et al., 1971). The nonmare basalt compositions possess high abundances of incompatible elements, including K, U, Th, Ba and the rare earths, coupled with high

Table 4. K/U ratios of mare and nonmare basalts.*

	Mare	Nonmare
Apollo 11	2800	—
Apollo 12	2200	1230 (KREEP)
Apollo 14	—	1200
Apollo 15	2700	1600

*Data from LSPET reports.

$\dfrac{Mg}{Mg + Fe^{++}}$ ratios and high Cr content. These characteristics may result from efficient fractionation in a partial melting process (implying a very small degree of partial melting) but are inconsistent with an origin by high degrees of crystal fraction-ation of some parental magma.

The relative abundance of plagioclase in the "KREEP" or "nonmare basalt" compositions, coupled with the experimental demonstration of the importance of near-liquidus plagioclase in 14310 and KREEP at high pressures, argues that plagioclase must have been present in the source region and left as a residual phase, accompanied by low-calcium, aluminous pyroxene(s) and possibly olivine. In com-parison with the source region for mare basalts, the relative composition and nature of the phases on the liquidi of 14310 and KREEP compositions show that the source region for nonmare or KREEP basalt had higher Al_2O_3/pyroxene ratio and higher $\dfrac{100Mg}{Mg + Fe^{++}}$ ratio (> 80) than the source region for mare basalts (with $\dfrac{100Mg}{Mg + Fe}$ < 80). Another significant difference between mare and nonmare basalts is the higher K/U ratios of the former (Table 4); this difference would also be characteristic of their respective source regions. To summarize, the source region of nonmare basalts is inferred to occur at depths < 200 km (because of the role of plagioclase in the source region; see also Ringwood, 1971) and to differ chemically from the source region for mare basalts in having higher Al_2O_3 content and $\dfrac{Mg}{Mg + Fe^{++}}$ and U/K ratios than the source region for mare basalts. This chemical inhomogeneity is inferred to be of primary origin and may be indicative of zonal chemical variation in the moon's outer layers established during accretion (Hubbard and Gast, 1971; Ringwood, 1972).

Evaluation of the significance of the nonmare or KREEP basaltic composition depends to a large extent on the abundance and distribution of this component on the lunar surface. The orbital gamma-ray spectrometer data (Metzger et al., 1972) and orbital x-ray fluorescence spectrometer data (Adler et al., 1972) both provide evidence that the characteristic or dominant (compositional) rock type of the highlands is related to gabbroic anorthosite and that KREEP is only locally (particularly near Mare Imbrium and Oceanus Procellarum) a major component. Although such con-clusions are highly tentative, it appears that the dominance of plagioclase-rich compositions, low in incompatible element contents, in the highland regions may result from a primitive stage of complete melting or high degrees of melting of the outermost layers in which pronounced crystal fractionation produced a dominance of near-surface plagioclase-rich rocks. The identification of a seismically fast horizon at depths >60 km (Toksöz *et al.*, 1972) may possibly be correlated with olivine-rich accumulated complementary to the plagioclase-rich near surface layers.

Acknowledgments—We wish to acknowledge the technical assistance of E. Kiss and E. H. Pedersen.

References

Adler I. Trombka J. Gerard J. Lowman P. Yen L. Blodgett H. Gorenstein P. and Bjorkholm P. (1972) Preliminary results from the S–161 x-ray fluorescence experiment (abstract). In *Lunar Science—III* (editor C. Watkins), p. 4, Lunar Science Institute Contr. No. 88.

Apollo Soil Survey (1971) Apollo 14: nature and origin of rock types in soil from the Fra Mauro formation. *Earth Planet. Sci. Lett.* **12**, 49–54.

Brown G. M. and Peckett A. (1971) Selective volatility on the lunar surface: evidence from Apollo 14 feldspar-phyric basalts. *Nature* **234**, 262–266.

Fuchs L. H. (1971) Orthopyroxene and orthopyroxene-bearing rock fragments rich in K, REE, and P in Apollo 14 soil sample 14163. *Earth Planet. Sci. Lett.* **12**, 170–174.

Green D. H. and Ringwood A. E. (1972) A comparison of recent experimental data on the gabbro-garnet granulite-eclogite transition. *J. Geol.* (in press).

Green D. H. Ringwood A. E. Ware N. G. Hibberson W. O. Major A. and Kiss E. (1971) Experimental petrology and petrogenesis of Apollo 12 basalts. *Proc. Second Lunar Sci. Conf., Geochim. Cosmochim. Acta,* Suppl. 2, Vol. 1, pp. 601–616. MIT Press.

Green D. H. Ware N. G. and Hibberson W. O. (in preparation) Experimental evidence against the role of selective volatility on the lunar surface.

Green D. H. Ware N. G. Hibberson W. O. and Major A. (1971) Experimental petrology of Apollo 12 basalts: part 1, sample 12009. *Earth Planet. Sci. Lett.* **13**, 85–96.

Hubbard N. J. and Gast P. W. (1971) Chemical composition and origin of non-mare lunar basalts. *Proc. Second Lunar Sci. Conf., Geochim. Cosmochim. Acta,* Suppl. 2, Vol. 2, pp. 999–1020. MIT Press.

Hubbard N. J. Meyer C. Gast P. W. and Wiesmann H. (1971) The composition and derivation of Apollo 12 soils. *Earth Planet. Sci. Lett.* **10**, 341–350.

Lindsay J. F. (1971) Mixing models and the recognition of end-member groups in Apollo 11 and 12 soils. *Earth Planet. Sci. Lett.* **12**, 67–72.

Metzger A. E. Trombka J. I. Peterson L. E. Reedy R. C. and Arnold J. R. (1972) A first look at the lunar orbital gamma-ray data (abstract). In *Lunar Science—III* (editor C. Watkins), p. 540, Lunar Science Institute Contr. No. 88.

Meyer C. Brett R. Hubbard N. J. Morrison D. A. McKay D. S. Aitken F. K. Jakeda H. and Schonfeld E. (1971) Mineralogy, chemistry and origin of the KREEP component in soil samples from the Ocean of Storms. *Proc. Second Lunar Sci. Conf., Geochim. Cosmochim. Acta,* Suppl. 2, Vol. 1, pp. 393–411. MIT Press.

Ringwood A. E. (1971) Petrogenesis of Apollo 11 basalts and implications for lunar origin. *J. Geophys. Res.* **75**, 6453–6479.

Ringwood A. E. (1972) Origin of the moon. *Phys. Earth Planet. Interiors* (submitted).

Ringwood A. E. and Essene E. J. (1970) Petrogenesis of Apollo 11 basalts, internal constitution and origin of the moon. *Proc. Apollo 11 Lunar Sci. Conf., Geochim. Cosmochim. Acta,* Suppl. 1, Vol. 1, pp. 769–800. Pergamon.

Ringwood A. E. and Green D. H. (1966) An experimental study of the gabbro-eclogite transformation and some geophysical implications. *Tectonophysics* **3**, 383–427.

Toksöz M. N. Press F. Anderson K. Dainty A. Latham G. Ewing M. Dorman J. Lammlein D. Sutton F. Duennebier F. and Nakamura Y. (1972) Velocity structure and properties of the lunar crust (abstract). In *Lunar Science—III* (editor C. Watkins), p. 758, Lunar Science Institute Contr. No. 88.

Turner G. Hunecke J. C. Podosek F. A. and Wasserburg G. J. (1971) ^{40}Ar–^{39}Ar ages and cosmic ray exposure ages of Apollo 14 samples. *Earth Planet. Sci. Lett.* **12**, 19–35.

Proceedings of the Third Lunar Science Conference
(Supplement 3, *Geochimica et Cosmochimica Acta*)
Vol. 1, pp. 207–229
The M.I.T. Press, 1972

Role of water in the evolution of the lunar crust; an experimental study of sample 14310; an indication of lunar calc-alkaline volcanism

C. E. Ford, G. M. Biggar, D. J. Humphries, G. Wilson, D. Dixon, and
M. J. O'Hara

Grant Institute of Geology,
West Mains Road,
Edinburgh EH9 3JW,
Scotland

Abstract—Intersertal feldspar-phyric high alumina basalt, 14310, from the Fra Mauro formation has anorthite as liquidus phase at approximately 1310°C at pressures to as much as 12 kb ($\sim$ 250 km depth) where it is replaced first by spinel, and then by garnet at pressures greater than approximately 17 kb ($\sim$ 350 km depth). Olivine is the next silicate to crystallize at low pressures ($\sim$ 1230°C), but is replaced in this role by one or more pyroxenes at pressures greater than 4 kb. Olivine reacts out on cooling at atmospheric pressure, calcium-poor pyroxene being the product at temperatures between 1200°C and 1160°C. The present bulk composition of 14310 contains too little oxygen to keep all the iron oxidized as FeO, or to maintain a Fe^{2+}/Mg^{2+} ratio in the liquid high enough to precipitate the relatively iron enriched pigeonites formed during the natural crystallization, and late stage loss of oxygen by volatilization is evident. The highly aluminous phases present on the liquidus are unsuitable as major residual constituents of the lunar mantle and 14310 cannot be a primary magma from that source. Water and alkalis may also have been lost by posteruption volatilization: addition of these components to 14310 produces a magma nearly cotectic with the minerals of spinel-troctolite at low pressures ($\sim$ 250 bars) and such rocks are observed in the breccias. We deduce that 14310 is a volatile depleted sample of liquid from a shallow seated fractionating magma chamber containing a liquid of calc-alkaline affinities.

Introduction

SIX GROUPS (Ford *et al.*, 1972; Kushiro, 1972; Muan *et al.*, 1972; Ridley *et al.*, 1972 (unrevised abstract); Ringwood *et al.*, 1972; Walker *et al.*, 1972) presented phase equilibria data for rock 14310 or related compositions. Atmospheric pressure results, though variable between groups, are shown to be broadly reconcilable, assuming differing oxygen contents of the charges recovered. The present composition of 14310 has suffered posteruption volatilization losses of O_2; significant losses of Na_2O and K_2O have also been claimed (Brown and Peckett, 1971, but challenged by Ridley *et al.*, 1972). We argue that water has also been lost; that water played a vital role in the origin of 14310 and the lunar crust in general, and re-emphasize that the reduced, volatile-depleted character of lunar surface rocks is no guide to lunar chemistry as a whole but is a consequence of exposure over many hours at $\sim$ 1200°C to a vacuum of $\sim 10^{-13}$ bars. Sample 14310 has not suffered significant near-surface fractional crystallization but may have evolved by fractional crystallization of spinel-bearing troctolites in magma chambers situated within 5 km of the lunar surface.

Data for 14053, 14162, 14321, and four preferred glass compositions in the soils, are also presented, both with and without restoration of alkalis that may have been volatilized.

Rock 14310: a feldspar-phyric endogenic high alumina lunar basalt

This rock contained about 8% plagioclase phenocrysts on eruption (Brown and Peckett, 1971). This rock may be an impact melt and as such may have no direct crystal-liquid connection with the lunar interior. However, the thermal histories of impact melted rocks are marked by superheating followed by rapid chilling under which conditions plagioclase *in particular* is very reluctant to nucleate or grow (Biggar *et al.*, 1972a, table 8, section C). Terrestrial impact melts crystallize to phenocryst-free rocks (Dence, 1971) and commonly retain inhomogeneous xenolithic textures. Brown and Peckett (1971) found no xenoliths in 14310, in contrast with other descriptions (LSPET, 1971; Walter *et al.*, 1972).

Clasts of this rock type occur in the Apollo 14 breccias (Wilshire and Jackson, 1972). Any resemblance between the compositions of 14310 and the subsequently formed soils at the Apollo 14 site merely testifies to the volumetric significance of high alumina, relatively magnesian basaltic rocks in the Imbrian target area. We assume 14310 to be a fragment of the solidified eruptive fraction of an endogenic magma, the problem of whose origin lies at the heart of a major question of lunar petrogenesis: *How is an early, relatively magnesian, and highly feldspathic crust derived from a lunar mantle that subsequently yielded mare basalts of relatively iron-rich, much less feldspathic character?*

Three hypotheses have been offered to explain a generally noritic-anorthositic lunar crust now thought to be about 50–65 km thick (Toksöz *et al.*, 1972). The first two hypotheses hold to a model of volatile-depleted moon. The third requires the presence initially of *substantial* amounts of water in the moon: (1) The last material accreted to the moon was more calcic and aluminous and poorer in Fe (Gast and McConnell, 1972) than all earlier material. The proponents of such hypotheses must explain convincingly how the anorthite rich material formed in space, or why such refractory material accumulated so late, and why such material is now so scarce among the meteorites. (2) The outermost layers of the moon were partly melted early in lunar history and during their recrystallization plagioclase floated to form the observed crust (Smith *et al.*, 1970; Wood *et al.*, 1970). Plagioclase is not observed to behave in this manner in terrestrial magmas, nor in experiments on lunar lavas (O'Hara *et al.*, 1970). The model presupposes either an outer layer of at least 200 km of the lunar mantle that is exceptionally rich in CaO and Al_2O_3 [see (1) above], or the formation of a surface layer, about 200 km thick, of *basaltic* melt followed by differentiation of much of its plagioclase to form a 60–70 km thick crust. In the latter case 130 km of complementary basic igneous rock (crystallized by some to eclogite, S.G. $\sim$ 3.7) would then underlie the anorthositic crust. (3) The lunar mantle underwent two major differentiating events: In the first, massive partial melting under high water vapor pressures produced magnesian melts whose condensed compositions were exceptionally rich in plagioclase. On eruption these outgassed, crystallized, and differentiated to form the lunar crust. Few hydrous minerals would have been stable in these magmas at depths less than 50 km (but see Mason *et al.*, 1972). In the second event, drier magnesian partial melts were derived from the largely dehydrated mantle, erupted into the maria, and there strongly fractionated to yield cumulus sequences of achondritic composition and iron-enriched late residual liquids (mare surface basalts). Enrichment of iron relative to magnesium was not pronounced during lunar crustal fractionation because plagioclase was the major phase separating. This model is summarized in Fig. 1. It interprets both 14310 and the mare surface basalts as late

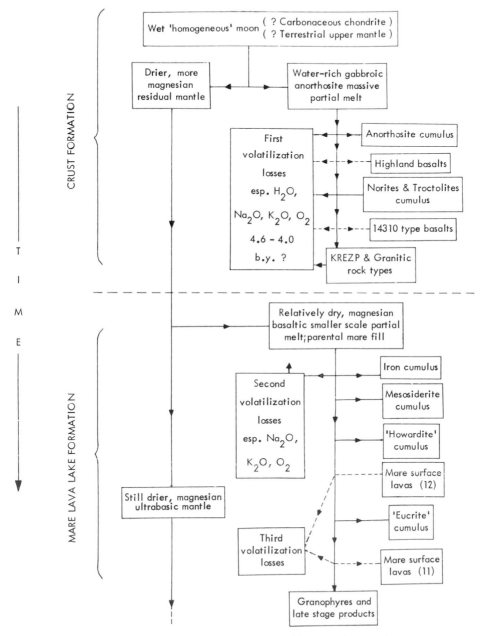

Fig. 1. Possible evolution and interrelationships of important lunar rock types.

stage residual liquids from high level fractional crystallization processes, thereby accommodating their high rare element concentrations without recourse to primary magmas derived from incipient partial melting events (an hypothesis leading to severe mass balance problems, Biggar *et al.* (1971, 1972a), and leaving unexplained how such magmas penetrate as much as two hundred kilometers of "cold" rocks without enforced differentiation en route to the surface).

Experimental Techniques

Atmospheric pressure techniques have been described, calibrated for temperature and oxygen fugacity, and checked for objectivity, reversibility, reproducibility, composition control, and attainment of equilibrium (Biggar *et al.*, 1972a, and references traceable thereby). Experiments between 1 and 10 kb were carried out in internally heated vessels (Ford, 1972). Dry experiments were carried out in sealed iron tubes using argon as the external pressure medium. Buffered water-bearing charges were contained in sealed $Ag_{70}Pd_{30}$, or Pt capsules inside larger sealed Pt tubes containing the buffer with sufficient water added to completely oxidize the buffer should all the hydrogen be lost, and run with an argon-hydrogen mixture as the external pressure medium to prolong the life of the buffer. Experiments in the range above 10 kb were carried out in piston-cylinder solid media equipment (Boyd and England, 1963) that has been extensively calibrated (O'Hara *et al.*, 1971; Herzberg and O'Hara, 1972) against five equilibria in the range 8–33 kb, 900–1700°C. No proven differences exist between our pressure/temperature scales for the solid media results and those quoted for similar work by the Geophysical Laboratory. Sealed platinum tubes were used in short exploratory experiments and sealed iron tubes in longer definitive experiments. Charges were dried in a nitrogen furnace before sealing.

Additions of alkalis to a 14310 sample were made by weighing and mixing in Analar Na_2CO_3 and K_2CO_3 followed by 4 hours heating at 1000°C at $f_{O_2} \sim 10^{-15}$ bars. In some of the wet melting experiments Na_2O was added in solution as NaOH (see Table 1).

Experimental Data

Conditions and results of quenching experiments on 4 natural and 4 synthetic samples, with and without added oxygen, water, and alkalis, are presented in Table 1, and interpreted in Figs. 2, 3, 4, 6, 7, and 8. Note that all atmospheric pressure data refer to charges that had an oxygen content imposed by the controlled oxygen fugacity. All dry melting data at higher pressures on 14310 refer to a composition too poor in oxygen to maintain all iron as FeO, except perhaps those in which Na_2O and K_2O had been added to the sample and the material preheated at the oxygen fugacity of the Fe/FeO equilibrium. All wet melting data were obtained at an oxygen fugacity determined by the FeO/Fe_3O_4 buffer.

Atmospheric Pressure Equilibrium Crystallization Data for 14310

The effect of variations in temperature and oxygen fugacity upon the phase assemblages stable in the 14310 composition is displayed in Fig. 4, which is based upon our own data points as shown, together with knowledge of the probable form of such diagrams from the work of Roeder and Emslie (1970). Quenching data obtained by other groups (Muan *et al.*, 1972; Ridley *et al.*, 1972; Ringwood *et al.*, 1972; Walker *et al.*, 1972) at unstated oxygen fugacities and low pressures are consistent with their charges having equilibrated at variable oxygen fugacities within the limited range $\sim 10^{-14}$ to 10^{-12} bars at about 1200°C, reflecting perhaps differing

Table 1

A. Runs in Atmospheric Pressure Equipment

14310, rock powder, 1 atm, Fe/FeO equilibrium (CO_2 + H_2), molybdenum capsules[1]

Run No.	Time (hr)	$-\log_{10} f_{O_2}$	Temp. (°C)	Result	Run No.	Time (hr)	$-\log_{10} f_{O_2}$	Temp. (°C)	Result
020	5	10.6	1310	(plag) 1	009	5	11.7	1211	plag crsp ol 1
016	5	10.8	1295	plag 1	004	11	11.8	1200	plag crsp ol (pig) 1
006	5	11.0	1274	plag 1	010	5	12.1	1189	plag crsp ol pig 1
007	5	11.3	1248	plag 1	005	9	12.2	1173	plag crsp ol pig 1
011	5	11.5	1237	plag crsp 1	008	5	12.6	1142	plag crsp pig (ol) 1
018	72	11.5	1231	plag crsp 1	017	5	13.2	1097	plag pig pk (1)
003	5	11.6	1226	plag crsp (ol) 1	002	7	13.7	1065	solid[2]

14310, glass starting material,[3,4] 1 atm, Fe/FeO equilibrium (CO_2 + H_2) molybdenum capsules

Run No.	Time (hr)	$-\log_{10} f_{O_2}$	Temp. (°C)	Result	Run No.	Time (hr)	$-\log_{10} f_{O_2}$	Temp. (°C)	Result
720	5	10.6	1310	(plag) 1	609	5	11.7	1211	plag crsp ol 1
716	5	10.8	1295	plag 1	504	11	11.8	1200	plag crsp pig 1
506	5	11.0	1274	plag 1	610	5	12.1	1189	plag crsp pig 1
507	5	11.3	1248	plag 1	505	9	12.2	1173	plag crsp ol pig 1
711	5	11.5	1237	plag crsp (ol) 1	508	5	12.6	1142	fine grained[2]
718	72	11.5	1231	plag crsp 1	717	5	13.2	1097	very fine grained[2]
503	5	11.6	1226	plag crsp ol 1					

14310, rock powder + $1\frac{1}{2}\%$ Na_2O + $\frac{1}{2}\%$ K_2O,[5] 1 atm, Fe/FeO equilibrium, molybdenum capsules

Run No.	Time (hr)	$-\log_{10} f_{O_2}$	Temp. (°C)	Result	Run No.	Time (hr)	$-\log_{10} f_{O_2}$	Temp. (°C)	Result
914	5	11.0	1276	(crsp) 1	912	5	11.6	1225	plag ol crsp 1
913	5	11.2	1255	plag crsp 1					

14310, rock powder, 1 atm, 90% H_2 10% CO_2, $f_{O_2} \approx 10^{-14}$, Fe and Mo capsules; Ni/NiO equilibrium AgPd capsules[6]

Run No.	Time (hr)	$-\log_{10} f_{O_2}$	Temp. (°C)	Result	Run No.	Time (hr)	$-\log_{10} f_{O_2}$	Temp. (°C)	Result
071	5	≈14(Mo)	1197	plag ol opx m 1	042	5	7.8	1207	plag ol sp 1
072	5	≈14(Fe)	1197	plag ol opx m 1	043	5	8.3	1161	plag ol sp opx 1

14321, breccia powder,[7] 1 atm, Fe/FeO equilibrium (CO_2 + H_2), molybdenum capsules

Run No.	Time (hr)	$-\log_{10} f_{O_2}$	Temp. (°C)	Result	Run No.	Time (hr)	$-\log_{10} f_{O_2}$	Temp. (°C)	Result
016	5	10.8	1295	1	010	5	12.1	1189	ol plag pk 1
006	5	11.0	1274	ol 1	005	9	12.2	1173	ol plag opx pk 1
007	5	11.3	1248	ol 1	021	5	12.3	1163	ol plag opx pk 1
011	5	11.5	1237	ol 1	008	5	12.6	1142	ol plag pig pk 1
003	5	11.6	1226	ol sp 1	017	5	13.2	1097	trace liquid[2]
009	5	11.7	1211	ol sp 1	002	7	13.7	1065	solid[2]
004	11	11.8	1200	ol plag sp 1					

14321, breccia powder, 1 atm, 90% H_2 10% CO_2, $f_{O_2} \approx 10^{-14}$, Fe and Mo capsules; NiNiO equilibrium Ag Pd capsules[6]

Run No.	Time (hr)	$-\log_{10} f_{O_2}$	Temp. (°C)	Result	Run No.	Time (hr)	$-\log_{10} f_{O_2}$	Temp. (°C)	Result
071	5	≈14(Mo)	1197	ol plag pig 1	042	5	7.8	1207	plag ol crsp 1
072	5	≈14(Fe)	1197	ol plag m 1	043	5	8.3	1161	plag pig ol crsp 1

Table 1 (continued)

A. Runs in Atmospheric Pressure Equipment (continued)

14053, powder, 1 atm, FeFeO equilibrium (CO₂ + H₂) molybdenum capsules

Run No.	Time (hr)	$-\log_{10} f_{O_2}$	Temp. (°C)	Result
011	5	11.5	1237	qu-pk 1
003	5	11.6	1226	crsp ol 1
009	5	11.7	1211	ol crsp 1
004	11	11.8	1200	ol crsp 1
010	5	12.1	1189	ol plag crsp 1
005	9	12.2	1173	ol plag crsp 1
021	5	12.3	1163	ol plag pig crsp 1
008	5	12.6	1142	ol plag pig crsp 1
019	5	12.9	1122	ol plag pig pk 1
017	5	13.2	1097	ol plag pig pk (1)[2]
002	7	13.7	1065	solid[2]

14053, powder, 1 atm, NiNiO buffer AgPd capsules;[6] 90% H₂ 10% CO₂ $f_{O_2} \approx 10^{-14}$ iron capsules

Run No.	Time (hr)	$-\log_{10} f_{O_2}$	Temp. (°C)	Result
042	4	7.8	1207	qu-pk 1
044	5	8.3	1150	plag pig ol pk 1
072	5	≈14	1197	ol plag pig m 1

14053, glass,[3] 1 atm, Fe/FeO equilibrium (CO₂ + H₂) molybdenum capsules

Run No.	Time (hr)	$-\log_{10} f_{O_2}$	Temp. (°C)	Result
511	5	11.5	1237	qu-sp 1
503	5	11.6	1226	qu-sp 1
509	5	11.7	1211	crsp (ol) 1
504	11	11.8	1200	crsp ol 1
510	5	12.1	1189	crsp ol (plag) 1
505	9	12.2	1173	pig plag ol crsp 1
521	5	12.3	1163	pig plag ol pk 1
508	5	12.6	1142	pig plag ol pk 1
719	5	12.9	1122	pig plag ol pk 1
717	5	13.2	1097	very fine grained (1)[2]

14162, soil powder, 1 atm, Fe/FeO equilibrium (CO₂ + H₂) molybdenum capsules

Run No.	Time (hr)	$-\log_{10} f_{O_2}$	Temp. (°C)	Result
007	5	11.3	1248	(plag) (crsp) 1
011	5	11.5	1237	plag crsp 1
018	72	11.5	1231	plag crsp 1
003	7	11.6	1226	plag crsp (ol) 1
009	5	11.7	1211	plag crsp ol 1
004	11	11.8	1200	plag crsp ol 1
010	5	12.1	1189	plag ol crsp 1
005	9	12.2	1173	plag ol crsp pig 1
008	5	12.6	1142[2]	
019	5	12.9	1122[2]	
017	5	13.2	1097[2]	
002	7	13.7	1065[2]	

Synthetic composition, Apollo Soil Survey (1971) Type E. 1 atm, Fe/FeO equilibrium (CO₂ + H₂) Mo capsules

Run No.	Time (hr)	$-\log_{10} f_{O_2}$	Temp. (°C)	Result
020	5	10.6	1310	plag 1
006	5	11.0	1274	plag 1
007	5	11.3	1248	plag ol 1
011	5	11.5	1237	plag ol 1
009	5	11.7	1211	plag ol 1
010	5	12.1	1189	plag sp ol 1
021	5	12.3	1163[2]	
008	5	12.6	1142	solid[2]

Synthetic composition, Apollo Soil Survey (1971) Type E, doped[5] with Na₂O, K₂O, 1 atm, Fe/FeO equilibrium

(i) plus 1% Na₂O plus ½% K₂O

Run No.	Time (hr)	$-\log_{10} f_{O_2}$	Temp. (°C)	Result
914	5	11.0	1276	plag sp 1
913	5	11.3	1255	plag ol sp 1
912	5	11.6	1225	plag ol sp 1

(ii) plus 2% Na₂O plus ½% K₂O

Run No.	Time (hr)	$-\log_{10} f_{O_2}$	Temp. (°C)	Result
914	5	11.0	1276	plag sp 1
913	5	11.3	1255	plag sp 1
912	5	11.6	1225	plag ol sp 1

Table 1 (continued)

A. Runs in Atmospheric Pressure Equipment (continued)

Run No.	Time (hr)	$-\log_{10} f_{O_2}$	Temp. (°C)	Result	Run No.	Time (hr)	$-\log_{10} f_{O_2}$	Temp. (°C)	Result
Synthetic composition, Apollo Soil Survey (1971) Type C, 1 atm, Fe/FeO equilibrium					Synthetic composition, Apollo Soil Survey (1971) Type C, doped[5] with Na$_2$O, K$_2$O, 1 atm, Fe/FeO equilibrium				
006	5	11.0	1274	l	010	5	12.1	1189	plag ol pk l
007	5	11.3	1248	(plag) l	021	5	12.3	1163	plag pig ol pk l
011	5	11.5	1237	plag l	008	5	12.6	1142	plag pig ol pk l
018	72	11.5	1231	plag l	019	5	12.9	1122	fine grained (l)[2]
009	5	11.7	1211	plag (ol) l	017	5	13.2	1097	fine grained (l)[2]
(i) plus 1% Na$_2$O plus ½% K$_2$O					(ii) plus 2% Na$_2$O plus ½% K$_2$O				
913	5	11.3	1255	l	913	5	11.3	1255	l
912	5	11.6	1225	(plag) l	912	5	11.6	1225	(plag) l
Synthetic composition, Apollo Soil Survey (1971) Type B, 1 atm, Fe/FeO equilibrium					Synthetic composition, Apollo Soil Survey (1971) Type B,[5] doped with Na$_2$O, K$_2$O, 1 atm, Fe/FeO equilibrium				
006	5	11.0	1274	l	009	5	11.7	1211	plag ol l
007	5	11.3	1248	plag (ol) l	010	5	12.1	1189	plag ol pig l
011	5	11.5	1237	plag ol l	008	5	12.6	1142	plag ol pig (l)
018	72	11.5	1237	plag ol l	019	5	12.9	1122	fine grained (l)[2]
(i) plus 1% Na$_2$O plus ½% K$_2$O					(ii) plus 2% Na$_2$O plus ½% K$_2$O				
914	5	11.0	1276	l	914	5	11.0	1276	(ol?) l
913	5	11.3	1255	ol l	913	5	11.3	1255	ol l
912	5	11.6	1225	ol plag l qu-sp	912	5	11.6	1225	ol l qu-sp
Synthetic composition, Apollo Soil Survey (1971) Mare type, 1 atm, Fe/FeO equilibrium									
011	5	11.5	1237	ol l	008	5	12.6	1142	ol pig plag pk l
009	5	11.7	1211	ol l	019	5	12.9	1122	ol pig plag pk l
010	5	12.1	1189	ol l	017	5	13.2	1097	ol pig plag pk l
021	5	12.3	1163	ol plag (pig) pk l					

Table 1 (continued)

14310, rock powder dried at 800°C and sealed in Fe tubes[6]

Run No.	Time (hr)	Pressure (kb)	Temp. (°C)	Result
143	4.5	2	1225	1 plag (ol) m qu-px
140	7	2	1200	1 plag ol m (sp) px
146	3	3	1220	1 plag ol qu-px
147	6	3	1175	1 plag cpx opx m
144	3.75	4	1221	1 plag opx cpx m
132	4.25	5	1311	1 v[8]
130	18.5	5	1290	1 plag m qu-px
126	17.5	5	1277	1 plag m qu-cpx
120	>0.25	5	1275	1 sp (plag) m qu-px[9]
129	4	5	1250	1 plag m qu-px
127	17.5	5	1223	1 plag cpx opx m

14310, rock powder dried at 110°C and sealed in Pt tubes

Run No.	Time (hr)	Pressure (kb)	Temp. (°C)	Result
111	0.17	10	1304	1 plag (sp) qu-cpx[4]

14310, rock powder + c. 10% H_2O sealed in $Ag_{70}Pd_{30}$ tubes and run at the FeO/Fe_3O_4 buffer[10]

Run No.	Time (hr)	Pressure (kb)	Temp. (°C)	Result
108	0.5	2	1126	1 plag (sp) qu-amph v[10]
124	21	2	1075	1 plag sp ol qu-amph v
110	0.5	2	1076	1 plag ol sp qu-cpx qu-mica v[11]
106	1	2	1025	1 plag (ol) cpx sp qu-mica v
105	1	2	952	plag cpx ilm v
119	16	5	1075	1 sp qu-amph v

14310, rock powder + c. 10% H_2O sealed in Pt tubes and run at FeO/Fe_3O_4 buffer

Run No.	Time (hr)	Pressure (kb)	Temp. (°C)	Result
107	0.17	2	1199	1 (sp) qu-amph v
102	0.17	2	1173	1 (sp) qu-amph v
112	0.17	2	1150	1 plag (sp) qu-amph v[11]
101	0.17	2	1100	1 plag (sp) qu-amph v

14310, rock powder + 1½% Na_2O + ¼% K_2O dried at 800°C and sealed in Fe tubes[5]

Run No.	Time (hr)	Pressure (kb)	Temp. (°C)	Result
148	21.5	5	1260	1 plag m[4]
155	1	7.5	1281	1 (plag) m qu-px

14310, rock powder + 1½% Na_2O + ¼% K_2O[5] + c. 10% H_2O sealed in $Ag_{70}Pd_{30}$

Run No.	Time (hr)	Pressure (kb)	Temp. (°C)	Result
149	18	2	1090	1 sp qu-amph qu-mica v
152	21	2	1075	1 ol sp qu-amph qu-mica v

B. Runs in Internally Heated Gas Media Equipment

Run No.	Time (hr)	Pressure (kb)	Temp. (°C)	Result
131	17	5	1201	1 plag cpx opx m
142	3.5	10	1336	1
138	3.5	10	1322	1 plag qu-px
137	3.5	10	1312	1 plag m qu-px
134	4	10	1290	1 plag m qu-px
136	3.5	10	1276	1 plag cpx m qu-px
135	4.25	10	1260	1 plag cpx opx qu-px
139	15.25	10	1226	1 plag cpx opx m
141	88	10	1175	1 plag cpx opx m pk
133	14	10→½	1250	1 plag m
109	0.17	10	1228	1 plag cpx sp pk qu-px[4]
118	22	5	1075	1 sp qu-amph qu-mica v
114	0.5	5	1074	1 plag sp qu-amph v[11]
115	0.17	5	1026	1 ol plag sp qu-amph v[11]
121	>0.25	5	1025	1 ol sp qu-amph v
117	17	5	1000	1 plag cpx ol sp qu-amph qu-mica v
122	>1.5	5	951	1 plag ol ilm cpx amph sp qu-mica v
103	0.33	2	1051	1 plag ol sp pk qu-amph v[12]
104	1	2	1000	1 plag ilm ol cpx (?qu-ru) v
116	0.17/ 0.17	2	1200/ 1025	1 sp plag qu-amph v[14]
150	17.75	10	1302	1 plag[4,13]
153	18.5	2	1060	1 ol sp qu-amph qu-mica v
154	16	2	1041	1 ol sp plag qu-amph qu-mica v

Table 1 (continued)

B. Runs in Internally Heated Gas Media Equipment (continued)

14310, rock powder + NaOH solution to give c. + 1½% Na$_2$O + c. 10% H$_2$O sealed in Ag$_{70}$Pd$_{30}$ tubes and run at the FeO/Fe$_3$O$_4$ buffer[4]

Run No.	Time (hr)	Pressure (kb)	Temp. (°C)	Result
123	22	2	1100	1 sp qu-amph v
124	21	2	1075	1 ol plag sp qu-amph v

14321, rock powder dried at 800°C and sealed in Fe tubes[7]

Run No.	Time (hr)	Pressure (kb)	Temp. (°C)	Result
143	4.75	2	1225	1 ol
140	7	2	1200	1 ol (plag) (cpx) sp m qu-px
146	3	3	1220	1 ol plag m qu-px
145	4.75	2.8	1210	1 ol opx plag m qu-px
144	3.75	4	1221	1 ol opx plag cpx m qu-px
132	4.25	5	1311	1 m (qu-px)
128	4	5	1301	1 (ol) m qu-px
130	18.5	5	1290	1 ol
126	17.5	5	1277	1 ol m
120	>0.25	5	1275	1 ol sp m[9]
129	4	5	1250	1 ol m

14321, rock powder dried at 110° and sealed in Pt tubes

Run No.	Time (hr)	Pressure (kb)	Temp. (°C)	Result
111	0.17	10	1304	1 qu-px

14321, rock powder + c. 10% H$_2$O sealed in Ag$_{70}$Pd$_{30}$ tubes and run at the FeO/Fe$_3$O$_4$ buffer

Run No.	Time (hr)	Pressure (kb)	Temp. (°C)	Result
108	0.5	2	1126	1 ol sp qu-amph qu-mica v
106	1	2	1025	1 plag ol cpx sp qu-mica v
105	1	2	952	plag ilm cpx ol sp v
113	0.17	5	1100	1 ol sp qu-amph qu-mica v
114	0.17	5	1074	1 ol sp qu-amph qu-mica v

14321, rock powder + c. 10% H$_2$O sealed in Pt tubes and run at the FeO/Fe$_3$O$_4$ buffer

Run No.	Time (hr)	Pressure (kb)	Temp. (°C)	Result
107	0.17	2	1199	1 sp qu-amph qu-pk v
102	0.17	2	1173	1 sp ol (pk) qu-amph v
101	0.17	2	1100	1 sp ol pk v

14321, rock powder + NaOH solution to give c. + 1½% Na$_2$O + 10% H$_2$O sealed in Ag$_{70}$Pd$_{30}$ capsules and run at the FeO/Fe$_3$O$_4$ buffer

Run No.	Time (hr)	Pressure (kb)	Temp. (°C)	Result
123	22	2	1100	1 ol sp qu-amph v
124	21	2	1075	1 ol sp qu-amph v

(right columns, continued)

Run No.	Time (hr)	Pressure (kb)	Temp. (°C)	Result
125	19	2	1049	1 plag ol sp qu-amph v
127	17.5	5	1223	1 plag opx cpx (ol) m
131	17	5	1201	1 plag opx cpx ol m
142	3.5	10	1336	1 opx qu-px m
138	3.5	10	1322	1 opx qu-px m
137	3.25	10	1312	1 opx qu-px m
134	4	10	1290	1 opx qu-px
136	3.75	10	1276	1 opx m
135	4.25	10	1260	1 ol plag opx cpx m qu-px
139	15.25	10	1229	1 cpx plag opx m
141	88	10	1175	plag cpx opx pk m
133	4	10 → ½	1250	1 ol m qu-px
109	0.17	10	1278	1 plag opx qu-cpx[11]
119	16	5	1075	1 ol sp qu-amph qu-mica v
121	>0.25	5	1025	1 ol sp qu-mica v
117	17.25	5	1000	1 ol plag sp (pk) qu-mica qu-cpx v
122	>1.5	5	951	1 ol plag ilm cpx amph sp qu-mica v
103	0.33	2	1051	1 sp ol qu-amph v
116	0.17/	2	1200/ 1025	1 sp ol cpx qu-cpx qu-amph qu-mica v
104	1	2	1000	1 sp ol plag cpx ilm (ru) v
125	19	2	1049	1 ol sp qu-amph qu-mica v

Table 1 (continued)

B. Runs in Internally Heated Gas Media Equipment (continued)

14053, rock powder + c. 10% H_2O sealed in $Ag_{70}Pd_{30}$ tubes and run at the FeO/Fe_3O_4 buffer

Run No.	Time (hr)	Pressure (kb)	Temp. (°C)	Result
108	0.5	2	1126	l sp qu-amph v[10]
110	0.5	2	1076	l ol sp (pk) cpx v
106	1	2	1025	l ol cpx ilm sp plag qu-amph qu-mica v
105	1	2	952	plag ilm cpx (sp) v
113	0.17	5	1100	l sp ol pk m qu-amph v
119	16	5	1075	l (ol) pk sp qu-amph v
114	0.17	5	1074	l sp ol pk cpx qu-amph v
115	0.5	5	1026	l ol cpx pk qu-amph v
117	17.25	5	1000	l ol cpx pk sp qu-amph qu-mica v
122	>1.5	5	951	l plag ol ilm cpx sp qu-cpx qu-mica v

14053, rock powder + c. 10% H_2O sealed in Pt tubes and run at the FeO/Fe_3O_4 buffer

Run No.	Time (hr)	Pressure (kb)	Temp. (°C)	Result
107	0.17	2	1199	l sp pk qu-amph v
102	0.17	2	1173	l sp ilm qu-cpx qu-amph v
103	0.3	2	1051	l ol cpx ilm sp qu-amph v
116	0.17/0.17	2	1200/1025	l ol cpx sp pk qu-amph v
101	0.17	2	1100	l ol pk sp qu-amph v
104	1	2	1000	l plag sp ilm cpx (ru) v

C. Runs in Solid Media Equipment

14310, rock powder dried at 800°C, sealed in platinum tubes

Run No.	Time (hr)	Pressure (kb)	Temp. (°C)	Result
205	0.17	15	1250	plag gr cpx m (l)
202	0.17	15	1225	plag gr cpx (l) (?opx)
206	0.17	20	1350	gr (cpx) m l
204	0.17	20	1275	gr cpx plag pk ru (l)[14]

14310, rock powder dried at 800°C, sealed in iron tubes

Run No.	Time (hr)	Pressure (kb)	Temp. (°C)	Result
254	2.5	10	1125	(gr) cpx plag opx m
235	1	15	1350	(sp) m l
238	1	15	1335	(sp) m l
228	1	15	1325	gr cpx plag m l (sp)
227	1	15	1300	gr cpx plag m l
215	1	15	1275	gr cpx plag m l
213	1	15	1250	gr cpx plag m (l) (?opx)
253	4	15	1175	gr cpx plag m (l)
240	1	17.5	1370	((gr)) m l
234	1	20	1400	l
239	1	20	1390	(gr) m l
231	1	20	1350	gr (cpx) m l
222	1	20	1317.5	gr cpx plag m l
223	1	20	1317.5	gr cpx plag m l
219	1.25	20	1310	gr cpx plag m l
236	1	25	1450	gr m l
230	1	25	1425	gr (cpx) m l
224	1	25	1400	gr cpx m l
220	1	25	1375	gr cpx m l
218	1	25	1350	gr cpx m l

Table 1 (continued)

C. Runs in Solid Media Equipment (continued)

14310, rock powder + $1\frac{1}{2}$% Na_2O + $\frac{1}{2}$% K_2O,[5] dried at 800°C, sealed in iron tubes

Run No.	Time (hr)	Pressure (kb)	Temp. (°C)	Result
243	1	15	1325	m l
244	1	15	1300	sp m l
246	1	15	1300	sp m l
248	1	15	1285	sp cpx m l

14321, rock powder, dried at 800°C, sealed in platinum tubes

Run No.	Time (hr)	Pressure (kb)	Temp. (°C)	Result
205	0.17	10	1300	opx (sp) l[4]
203	0.17	15	1350	opx (sp) l[4]
201	0.17	15	1325	opx (cpx) l

14321, rock powder, dried at 800°C, sealed in iron tubes

Run No.	Time (hr)	Pressure (kb)	Temp. (°C)	Result
231	1	15	1375	l
234	1	15	1365	opx m l
225	1	15	1335	opx qu-cpx l
222	1	15	1325	opx m l
232	1	15	1225	opx gr cpx plag (?sp)
230	3	15	1150	plag cpx opx gr (?ilm)

Run No.	Time (hr)	Pressure (kb)	Temp. (°C)	Result
249	1	17.5	1275	cpx plag gr m l
250	1	18	1300	cpx plag gr m sp l
247	1	20	1375	(cpx) m sp l
245	1	20	1360	cpx ((gr)) sp m l
242	1	20	1315	gr cpx plag m l (?sp)
204	0.17	20	1400	gr (opx) l
202	0.17	20	1375	gr opx cpx l qu-cpx
227	1	20	1425	l
233	1	20	1400	(gr) opx m l qu-cpx
215	1	20	1375	gr opx m l qu-cpx
236	1	25	1440	gr cpx l qu-cpx
223	1	25	1425	gr (opx) cpx m l qu-cpx
229	1	25	1250	gr cpx (opx) (ru) (?sp) (?qz) (pk)

(4) These results not used in constructing Figs. 2 or 3.
(5) Alkalis added as carbonates and sintered at 1000°C. Fe/FeO equilibrium in CO_2:H_2 atmosphere.
(6) Iron capsules for solid media work made up from stock containing 180 ppm C, 190 ppm S, 180 ppm Mn, 1230 ppm Cu, supplied by New Metals and Chemicals Ltd. $AG_{70}Pd_{30}$ capsules were used up to 1170°C, $Ag_{40}Pd_{60}$ at higher temperatures in all runs using silver-palladium alloy containers.
(7) Results presented for 14321 supersede those shown by Biggar et al.,1972b in a diagram containing some errors.
(8) Composition and source of vapor unknown.
(9) Dried at 110°C.
(10) Capsule partly melted.
(11) There may not have been sufficient time to dissolve plagioclase.
(12) Opaque phase assumed to be residual.
(13) Incompatible with solid media results if temperature scales are identical.
(14) Opaque phase and rutile fields not shown in Fig. 2.

Key (additional to that in Fig. 2):

m = metal (iron)
pig = pigeonite
pk = opaque phase (ilmenite, spinel)
px = pyroxene undistinguished
qu = quenching product

qz = quartz
ru = rutile
( ) = very small amount
? = doubtful identification

Where liquid is recorded, glass or quenching products were observed. Charges were dried in oxygen-free nitrogen at stated temperatures. Runs 120, 121, 122 in internally heated vessels were quenched by fuse blowing overnight. Durations were less than 18 hours.

(1) Molybdenum capsules were made by welding sheet supplied by Murex Ltd., Essex.
(2) Equilibrium probably not attained in these runs due to small proportion of liquid present.
(3) Glasses previously prepared at 1350–1400°C, 1–2 hours in molybdenum capsules at Fe/FeO equilibrium in CO_2:H_2 atmosphere.

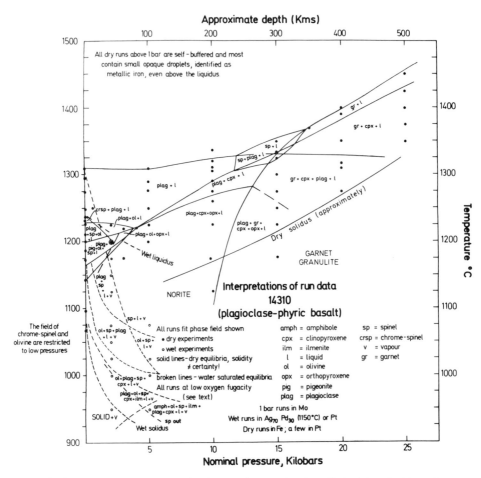

Fig. 2. Pressure–temperature section interpreting run data for 14310, with and without the addition of water and oxygen. Note that dry high-pressure runs on unmodified 14310 composition refer to a composition depleted in oxygen relative to *all* other experiments shown. (Revised from Ford *et al.*, 1972.)

degrees of success in controlling the charge composition as well as possible differences in oxygen content between subsamples ($\sim \pm 0.3\%$ O_2). Estimates of differences in oxygen content are based upon known, or assumed changes in the amounts of Fe metal, FeO, and Fe_2O_3 in the charges. Ringwood *et al.* (1972) seem to have attained oxygen fugacities intermediate between those achieved by Walker *et al.* (1972) and ourselves. Charges contain $\sim 1.1\%$ more O_2 at the highest f_{O_2} investigated by us than the samples do at the lowest f_{O_2}.

The compositions of 14310, its residual glasses and certain other investigated compositions lie almost in the plagioclase-olivine-silica plane, and are shown projected from diopside into that plane (Fig. 6). An internally consistent pseudo-ternary phase diagram can be drawn for the low pressure phase equilibria data, and

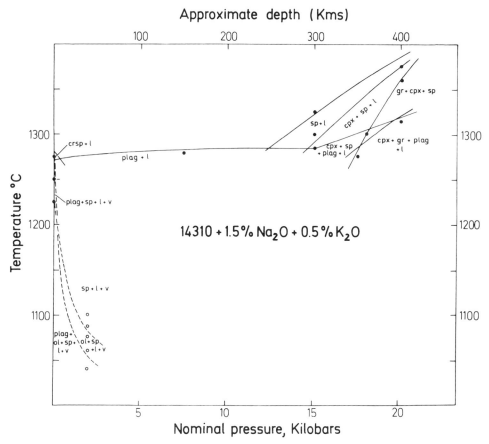

Fig. 3. Partial pressure–temperature section for 14310 plus alkalis, oxygen, and water.
Key as in Fig. 2.

even contoured for the plagioclase liquidus temperature. Note the implied high liquidus temperature, abundant plagioclase crystallization and subsequent lengthy olivine crystallization interval predicted for "highland basalts."

OXYGEN CONTENT OF 14310: LATE VOLATILIZATION LOSS

Our subsample 14310,141 when dried and sealed in iron or platinum capsules and run at high pressures almost always produced iron droplets. Microprobe analyses of glass in near-liquidus runs at 5 and 10 kb show substantial loss of iron by reduction to metal indicating about 0.7 % deficiency of oxygen relative to that necessary to keep the iron oxidized as FeO (Table 2). Iron did not, however, form early in the crystallization of 14310 (Ridley *et al.*, 1972).

Had iron ever formed in equilibrium with a liquid of 14310 composition, its rapid loss by sinking would be anticipated, and nickel should be severely depleted in this rock, but it is not. Moreover, the earlier pyroxenes observed to have crystallized from

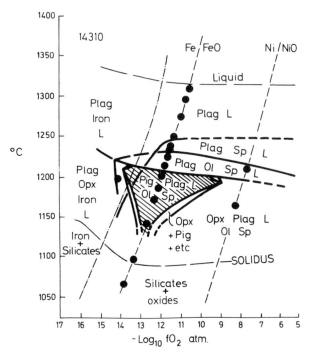

Fig. 4. Temperature versus log f_{O_2} plot of phases present in the 14310 composition at atmospheric pressure showing the critical effect of f_{O_2} upon the appearance of iron, chrome-spinel, and orthopyroxene versus pigeonite near 1200°C, $f_{O_2} \sim 10^{-13}$ bars in this composition.

the 14310 magma (Fig. 7) suggest that it then had a higher complement of FeO and therefore was not as oxygen depleted as the present bulk composition. The effects observed imply genuine loss of oxygen from the bulk rock, not merely a decline in oxygen fugacity with falling temperature (which also occurred). It follows that substantial losses of oxygen accompanied the eruption and consolidation of this rock, and its present oxygen content is not that appropriate to its previous evolution.

"Primitive" Character of the 14310 Magma

Rock 14310 originally crystallized with most of its iron as FeO in relatively iron-rich pigeonite. At atmospheric pressure substantial (~ 20–25%) plagioclase crystallization precedes the appearance of olivine, which is later joined by pigeonite, the plagioclase + olivine + pyroxene cotectic being reached at $\sim 1200°C$ when the charge is about 45–50 % crystalline (by construction from Fig. 6; visual estimate in Ford et al (1972) of 30 % crystallinity at olivine entry prior to pyroxene also is in accord with Fig. 6). The composition is not specially related to the cotectic liquids at atmospheric pressure. The probability of so distant a relationship to the cotectic liquid occurring by coincidence in a random liquid composition derived from some higher pressure is 0.94 (see Biggar et al., 1972a, Table 6, calculated using $C = 8$, $P^* = 3$, $X = 0.5$). This magma composition is, therefore, more primitive than those

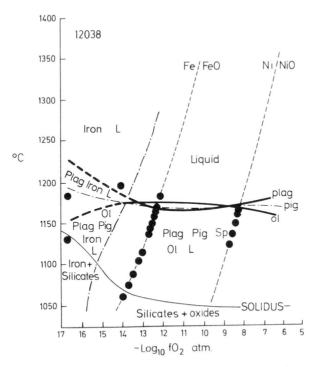

Fig. 5. Similar diagram to Fig. 4 but drawn for sample 12038, a near cotectic composition. The interchange of plagioclase and olivine entries is based upon visual observation of abundant plagioclase and very scarce olivine in the highest temperature run at the Ni/NiO equilibrium. The spinel entry boundary has not been included for the sake of clarity but lies close to the silicate liquidus between the Fe/FeO and Ni/NiO equilibria.

of the mare basalts, which have been established (Biggar *et al.*, 1971, 1972a) as being specially related to atmospheric pressure cotectic liquids. In the case of 14310, we differ from the interpretation of Walker *et al.* (1972), because the special relationship that they observe (plagioclase enrichment relative to the olivine + pyroxene + plagioclase cotectic) is a coincidence due to iron depletion in their closed system charges consequent on late stage oxygen loss from the natural rock (Figs. 6 and 7). Furthermore, if the 8% plagioclase phenocrysts are interpreted as equilibrium products of dry, near surface crystallization, the eruption temperature was about 60°C above that of the cotectic. (If the rock is interpreted as a wet magma of lower liquidus temperature whose groundmass was quenched by volatile loss, see below, further argument with regard to atmospheric pressure dry cotectic liquids is irrelevant).

SAMPLE 14310, THOUGH "PRIMITIVE" IS NOT A PRIMARY MAGMA

Posteruption oxygen loss has been demonstrated for 14310; substantial loss of alkalis has been claimed by Brown and Peckett (1971), and the case for substantial water loss will be argued below. Sample 14310, therefore, is not a primary magma from the lunar interior and any attempt to treat it as such leads inexorably toward

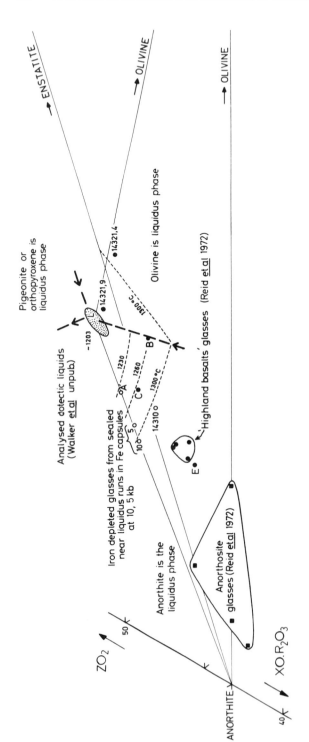

Fig. 6. Projection from "diopside" into the plagioclase-olivine-silica plane of the system XO-YO-R₂O₃-ZO₂ (Biggar et al., 1971. Fig. 1) showing preferred glass compositions B, C, E (Apollo Soil Survey, 1971) and the compositions of "highland basalt" and "anorthosite" glasses from Apollo 14 and other soils (Reid et al., 1972) and the analyses of two subsamples of breccia 14321 and of sample 14310 (shown as A; used by Ridley et al., 1972, unrevised abstract, for experiments) from LSPET (1971). A more recent analysis of 14310 (LSPET 1972) is also shown, together with the compositions of glasses from near liquidus experiments in equilibrium with iron droplets and plagioclase in sealed tube experiments at 5 and 10 kb (Table 2). Approximate boundaries between the liquidus fields of anorthite, olivine, and calcium-poor pyroxene are shown, meeting at the cotectic liquid compositions found by Walker et al. (1972) (field L). Approximate thermal contours on the plagioclase and olivine surface are shown.

Table 2. Phase compositions (wt%) from experiments on 14310.

Run No.	010	130	111	134	208
Container	Mo	Fe	Pt	Fe	Pt
Conditions	(1 bar, 1189°C, Fe/FeO)	(5 kb, 1290°C dry)	(10 kb, 1304°C dry)	(15 kb, 1200°C)	(15 kb, 1300°C)
Plagioclases:					
SiO_2	47.05	45.47		45.84	45.72
TiO_2	0.08	0.04		0.08	0.08
Al_2O_3	32.36	34.57		34.07	34.36
FeO	0.72	0.32		0.46	0.43
MgO	0.23	0.30		0.28	0.25
CaO	17.27	18.56		18.33	17.54
Na_2O	1.06	0.64		0.94	1.23
K_2O	0.46	0.18		0.29	0.22
TOTAL	99.23	100.08		100.29	99.83
An	87.6	93.2		90.0	87.6
Ab	9.7	5.8		8.3	11.1
Or	2.7	1.0		1.7	1.3
Glasses:					
SiO_2	50.13	51.13	50.47		46.96
TiO_2	2.26	1.46	1.45		1.24
Al_2O_3	13.97	19.87	20.18		20.37
Cr_2O_3	n.d.	n.d.	n.d.		n.d.
FeO	10.82	5.21*	5.22*		8.04
MnO	n.d.	n.d.	n.d.		n.d.
MgO	7.46	9.27	8.52		7.83
CaO	13.24	12.27	12.19		12.31
Na_2O	0.90	0.88	0.80		0.76
K_2O	0.64	0.67	0.64		0.47
TOTAL	99.42	100.76	99.47		97.97

* Iron droplets present.

hypotheses of derivation by partial melting of anorthosite at shallow depths, or of spinel or garnet rocks at greater depths (inadmissible on grounds of density).

EFFECT OF PRESSURE AND WATER VAPOR PRESSURE ON THE CRYSTALLIZATION OF 14310, 14321

Figure 6 locates the approximate boundary between the liquidus fields of plagioclase, olivine, and calcium-poor pyroxene in compositions close to the anorthite-olivine-silica plane, in which to a first approximation, Apollo 14 rock and soil compositions fall. Figure 8 reproduces this boundary in relation to the compositions of 14310 (LSPET, 1972, analysis), and breccia 14321 (LSPET, 1971, analyses). It shows, in sketch form, the effect of increased dry pressure in contracting the olivine liquidus field. Contrast this with the effect of addition of water to the system, where the olivine field expands into silica-saturated compositions of high normative plagioclase content, and a spinel-field also becomes prominent. Liquids in equilibrium with olivine and plagioclase at 1–3 kb water vapor pressure lie well inside the plagioclase liquidus field if pressure is released and the water lost. Similar effects are well-known from relevant synthetic systems (Yoder, 1968, 1969). These relationships may be rephrased in the statement that partial melting of plagioclase- or spinel-peridotites in the presence of water yields liquids of relatively high normative plagioclase content, which are quartz-normative if orthopyroxene was a phase in the crystalline residuum.

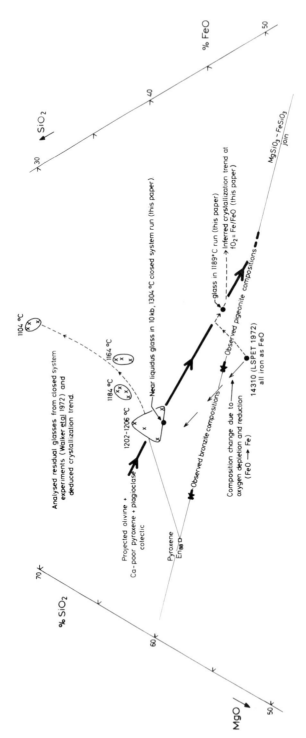

Fig. 7. Projection from CaSiO₃, normative feldspars, apatite, Cr₂O₃, and TiO₂ into the system MgO-FeO-SiO₂ (Kushiro et al., 1971. Fig. 5). showing (heavy line) the effective boundary between olivine and calcium-poor pyroxene as liquidus phase; the bulk composition of 14310 (LSPET, 1972) and (short dashed lines) the inferred evolution of its residual liquids during equilibrium crystallization in our experiments; and the oxygen and metallic iron depleted residual liquid compositions achieved or retained from the natural starting material in sealed system experiments on 14310 by ourselves and Walker et al. (1972) in relation to Walker et al. experimentally produced orthopyroxene and the ranges of pyroxene composition reported from the natural rock. The observed natural pigeonites appear to be more iron-rich than those that form during closed system crystallization of the present oxygen-depleted 14310 natural rock powder.

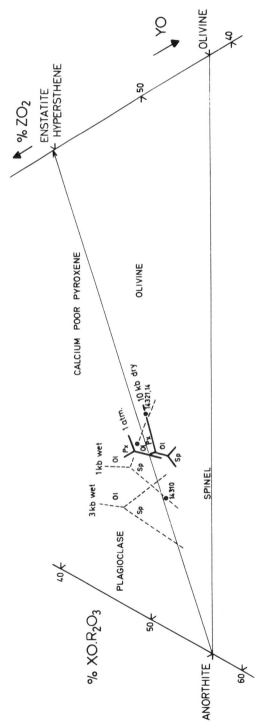

Fig. 8. Projection as in Fig. 6, but to smaller scale, showing the effect of pressure and water content upon the extent of the olivine, spinel, and calcic plagioclase liquidus fields in Apollo 14 compositions.

Affinities and Origin of 14310; The Complementary Spinel Trocolites

Brown et al. (1972), Drever et al. (1972), Roedder and Weiblen (1972) report spinel-troctolite or spinel and troctolite as clasts in the breccias and soils and comment on their cumulus textures. Brown and Peckett (1971) interpret 14310 as liquid extracted from a high-level fractionating magma chamber (perhaps elutriating some plagioclase phenocrysts). Addition of water and oxygen (at the FeO/Fe_3O_4 buffer) to 14310 brings olivine and plagioclase simultaneously into equilibrium with the liquid after trivial amounts of spinel precipitation (Fig. 2) at $\sim 1075°C$, 3 kb (~ 55 km). Restoration of lost alkalis to the sample changes this condition (Fig. 3) to $\sim 1175°C$, ~ 0.25 kb (~ 5 km). Some of the spinel observed in these experiments is of primary, equilibrium crystallization; some may, however, be a quenching product. A magma matching the deduced pre-eruption composition of 14310 could, evidently, have been in equilibrium with spinel-troctolite cumulus in a shallow-seated magma chamber.

The Natural Crystallization of 14310

Olivine has not been reported by groups studying 14310 (Hollister et al., 1972; Kushiro, 1972; Ridley et al., 1972; Walter et al., 1972), yet it was prominent in the atmospheric pressure equilibrium crystallization histories observed between 1230°C and $\sim 1190°C$ by four groups of experimentalists; below that temperature it is consumed by reaction with the liquid. At least three interpretations of the mineralogy are admissible:

(i) The dry magma cooled slowly enough after eruption to allow all olivine to react (unusual in terrestrial experience perhaps).

(ii) The dry magma was equilibrated at $\sim 1260°C$, with 8% plagioclase phenocrysts, erupted as a porphyritic liquid and quenched to $\sim 1150°C$, then annealed long enough for the liquid to crystallize to the plagioclase + pyroxene assemblage, and then further cooled.

(iii) The 14310 magma, containing about 1–2% dissolved water at, for example, 1150°C, 250 bars escaped from an approximately 5 km deep magma chamber, carrying with it the 8% plagioclase phenocrysts. Rapid boiling out of H_2O, Na_2O, K_2O, and O_2 led to crystallization of the relatively cool magma at and below 1150°C, where olivine would not be expected to appear.

Alkali Loss and Restoration; Plagioclase Compositions in 14310

Brown and Peckett (1971) with whom we concur, deduced significant posteruption loss of Na_2O, K_2O from 14310 from the zoning of plagioclase phenocrysts (An_{94-67}), the groundmass feldspar not having the expected alkali-rich bulk composition to represent potential feldspar in the liquid during the sodic rim crystallization. Ridley et al. (1972) however, interpret the relationships as indicating xenocrystal derivation for some of the plagioclase, yet this still leaves a problem regarding the more sodic liquid required to provide the sodic rims. Microprobe analyses of experimentally produced crystals in 14310 (Table 2) indicate that in the dry state near-liquidus plagioclase has composition An_{88} (15 kb), An_{94} (5 kb) and becomes more calcic with lower pressure. Even after plagioclase crystallization has proceeded for more than 100°C at atmospheric pressure and at 10 kb, feldspar compositions have advanced only to An_{88} and An_{90}, respectively. Simplistic interpretation is, however, impossible, because (i) the high pressure data refer to oxygen and iron depleted material and (ii) elevated water pressures are suspected to promote excessively calcic plagioclase crystallization from terrestrial calc-alkali magmas of relatively high Na/Ca ratio (Wager, 1962; Baker, 1968; Yoder, 1968, 1969).

We have restored Na_2O and K_2O to 14310 in lesser amounts than Kushiro (1972), finding this to lower liquidus temperatures, enhance olivine and spinel crystallization, and reduce the water vapor pressure required to suppress plagioclase as the liquidus phase (Fig. 3). The additions made were not, however, sufficient to alter the qualitative character of the phase equilibria.

If the analysis of 14310 is adjusted to restore Na_2O, K_2O, oxygen and water a composition remarkably similar to high alumina basalts of terrestrial calc-alkaline provinces is obtained (and vice versa; see Table 3).

Table 3

	SiO_2	TiO_2	Al_2O_3	FeO	MnO	MgO	CaO	Na_2O	K_2O	P_2O_5
I	49.38	0.86	19.89	9.14	0.18	6.79	12.81	0.52	0.35	0.08
II	47.82	1.26	20.41	8.49	0.11	7.98	12.45	0.64	0.50	0.34

I. Basalt cinders (14802) from Mansion pyroclastics, St. Kitts (Baker, 1968) recalculated to 100% after arbitrary "volatilization" of 1.81% Na_2O, 0.46% O_2 and all water. In its original alkali-rich state this type of material precipitated about Fo_{77} + An_{70} *after* eruption and water loss but was apparently precipitating Fo_{77} + An_{90-93} in a relatively shallow seated wet magma chamber immediately before eruption.

II. 14310 (LSPET, 1972) from Fra Mauro formation, normalized to 100%. Carries phenocryst cores of An_{94}, suggested here to form in water-bearing magma chamber when liquid contained about 1.5% more Na_2O; the groundmass plagioclase formed after water loss *and* Na_2O loss and is An_{93}. Liquidus plagioclase in experiments on present dry, Na_2O-poor composition is about An_{94}.

DIFFERENCES IN REPORTED RESULTS ON 14310 AT HIGH PRESSURES

Four groups (Ford, *et al.*, 1972; Kushiro, 1972; Ringwood *et al.*, 1972; Walker *et al.*, 1972) reported high-pressure data for 14310 composition that are in broad agreement on one point: Plagioclase is the liquidus phase to pressures in excess of 10 kb, then it is replaced by spinel, and at even higher pressures by garnet. Kushiro (1972) reported a much greater spinel field than shown in Fig. 2; from our own experience this *might* reflect the accession to the charge of a *small* amount of water through the permeable graphite capsules that he used; but it is also probable that his data refer to different oxygen contents of the charges. Most of our high pressure data can be reconciled with those of Walker *et al.* (1972), although our interpretation of the phase diagram differs somewhat, and we cannot accommodate 14310 as a primary melting product of any type of dry pyroxenite source region. Ringwood *et al.* (1972) found olivine present at 10 kb, but we observe no olivine above 4 kb.

SUMMARY AND CONCLUSIONS

Evidence presented or reviewed in this paper establishes that the present bulk composition of 14310 is not that of a primary magma from the lunar interior. Whether

14310 represents an endogenous partial melt or residual liquid from the lunar interior, or an exogenous impact melt, is largely immaterial to the major problem of the origin of anorthite-rich materials in the outer layer of the moon. We suggest that the anorthositic layer originated by early stage partial melting of a wet lunar interior, giving rise to liquids of calc-alkaline affinities (of which 14310 may represent a late stage derivative). Subsequent melting of the dehydrated residual mantle yielded the parental magmas of the maria suites.

REFERENCES

Apollo Soil Survey (1971) Apollo 14; nature and origin of rock types in soil from the Fra Mauro formation. *Earth Planet. Sci. Lett.* **12**, 49–54.

Baker P. E. (1968) Petrology of Mt. Misery volcano, St. Kitts, West Indies. *Lithos* **1**, 124–150.

Biggar G. M. O'Hara M. J. Peckett A. and Humphries D. J. (1971) Lunar lavas and the achondrites: Petrogenesis of protohypersthene basalts in the maria lava lakes. *Proc. Second Lunar Sci. Conf., Geochim. Cosmochim. Acta,* Suppl. 2, Vol. 1, pp. 617–643. M.I.T. Press.

Biggar G. M. O'Hara M. J. Humphries D. J. and Peckett A. (1972a). Maria lavas, mascons, layered complexes, achondrites and the lunar mantle. In *The Moon* (editors H. Urey and S. K. Runcorn, IAU. In press.

Biggar G. M. Ford C. E. Humphries D. J. Wilson G. and O'Hara M. J. (1972b) Melting relations of more primitive mare-type basalts 14053 and M (Reid 1971); and of breccia 14321 and soil 14162 (average lunar crust?) In *Lunar Science—III* (editor C. Watkins), pp. 74–76, Lunar Science Institute Contr. No. 88.

Boyd F. R. and England J. L. (1963) Effect of pressure on the melting of diopside, $CaMgSi_2O_6$, and albite, $NaAlSi_3O_8$, in the range up to 50 kb. *J. Geophys. Res.* **68**, 311–323.

Brown G. M. and Peckett A. (1971) Selective volatilization on the lunar surface: Evidence from Apollo 14 feldspar-phyric basalts. *Nature* **234**, 262–266.

Brown G. M. Emeleus C. H. Holland J. G. Peckett A. and Phillips R. (1972) Mineral fractionation patterns between Apollo 14 primitive feldspathic rocks and Apollo 15 and other basalts (abstract). In *Lunar Science—III* (editor C. Watkins), pp. 95–97, Lunar Science Institute Contr. No. 88.

Dence M. R. (1971) Impact melts. *J. Geophys. Res.* **76**, 5552–5565.

Drever H. I. Johnston R. and Gibb F. G. F. (1972) A note on three Imbrium spinels, and a twinned pigeonite in high alumina basalt 14310 (abstract). In *Lunar Science—III* (editor C. Watkins), pp. 186–188, Lunar Science Institute Contr. No. 88.

Ford C. E. (1972) Furnace design, temperature distribution, calibration and seal design in internally heated pressure vessels. *Progress in Experimental Petrology,* Second Rept. Natural Environment Research Council (in press).

Ford C. E. Humphries D. J. Wilson G. Dixon D. Biggar G. M. and O'Hara M. J. (1972) Experimental petrology of high alumina basalt, 14310, and related compositions (abstract). In *Lunar Science—III* (editor C. Watkins), pp. 274–276, Lunar Science Institute Contr. No. 88.

Gast P. W. and McConnell R. K. Jr. (1972) Evidence for initial chemical layering of the moon (abstract). In *Lunar Science—III* (editor C. Watkins), pp. 289–291, Lunar Science Institute Contr. No. 88.

Herzberg C. and O'Hara M. J. (1972) Temperature and pressure calibration and reproducibility of pressure in solid media equipment. *Progress in Experimental Petrology,* Second Rept. Natural Environment Research Council (in press).

Hollister L. Trzcienski W. Dymek R. Kulick C. Weigand P. and Hargraves R. (1972) Igneous fragment 14310,21 and the origin of the mare basalts (abstract). In *Lunar Science—III* (editor C. Watkins), pp. 386–388, Lunar Science Institute Contr. No. 88.

Kushiro I. (1972) Petrology of lunar high-alumina basalt (abstract). In *Lunar Science—III* (editor C. Watkins), pp. 466–468, Lunar Science Institute Contr. No. 88.

Kushiro I. Nakamura Y. Kitayama K. and Akimoto S. (1971) Petrology of some Apollo 12 crystalline rocks. *Proc. Second Lunar Sci. Conf., Geochim. Cosmochim. Acta,* Suppl. 2, Vol. 1, pp. 481–495. M.I.T. Press.

LSPET (1971) (Lunar Sample Preliminary Examination Team) *Apollo 14 Preliminary Science Report*, NASA SP-272, 309 pp., Washington, D.C.

LSPET (1972) (Lunar Sample Preliminary Examination Team) A preliminary description of the Apollo 15 lunar samples. *Science* (in press).

Mason B. Melson W. G. and Nelen J. (1972) Spinel and hornblende in Apollo 14 fines (abstract). In *Lunar Science—III* (editor C. Watkins), pp. 512–514, Lunar Science Institute Contr. No. 88.

Muan A. Hauck J. and Löfall T. (1972) Equilibrium studies with a bearing on lunar rocks (abstract). In *Lunar Science—III* (editor C. Watkins), pp. 561–563, Lunar Science Institute Contr. No. 88.

O'Hara M. J. Biggar G. M. Richardson S. W. Ford C. E. and Jamieson B. G. (1970) The nature of seas, mascons and the lunar interior in the light of experimental studies. *Proc. Apollo 11 Lunar Sci. Conf., Geochim. Cosmochim. Acta*, Suppl. 1, Vol. 1, pp. 695–710. Pergamon.

O'Hara M. J. Richardson S. W. and Wilson G. (1971) Garnet-peridotite stability and occurrence in crust and mantle. *Contr. Mineral. Petrol.* **32**, 48–68.

Reid A. M. Warner J. Harmon R. S. and Brett R. (1972) Chemistry of highland and mare basalts as inferred from glasses in the lunar soils (abstract). In *Lunar Science—III* (editor C. Watkins), pp. 640–642, Lunar Science Institute Contr. No. 88.

Ridley W. I. Williams R. J. Brett R. and Takeda H. (1972) Petrology of lunar basalt 14310 (abstract). In *Lunar Science—III* (editor C. Watkins), pp. 648–650, Lunar Science Institute Contr. No. 88.

Ringwood A. E. Green D. H. and Ware N. G. (1972) Experimental petrology and petrogenesis of Apollo 14 basalts (abstract). In *Lunar Science—III* (editor C. Watkins), pp. 654–656, Lunar Science Institute Contr. No. 88.

Roedder E. and Weiblen P. W. (1972) Petrographic and petrologic features of Apollo 14, 15 and Luna 16 samples (abstract). In *Lunar Science—III* (editor C. Watkins), pp. 657–659, Lunar Science Institute Contr. No. 88.

Roeder P. L. and Emslie R. F. (1970) Olivine-liquid equilibrium. *Contr. Mineral. Petrol.* **29**, 275–289.

Smith J. V. Anderson A. T. Newton R. C. Olsen E. J. Crewe A. V. Isaacson M. S. Johnson D. and Wyllie P. J. (1970) Petrologic history of the moon inferred from petrography, mineralogy and petrogenesis of Apollo 11 rocks. *Proc. Apollo 11 Lunar Sci. Conf., Geochim. Cosmochim. Acta*, Suppl. 1, Vol. 1, pp. 897–926. Pergamon.

Toksöz M. N. Press F. Anderson K. Dainty A. Latham G. Ewing M. Dorman J. Lammlein D. Sutton G. Duennebrier F. and Nakamura Y. (1972) Velocity structure and properties of the lunar crust (abstract). In *Lunar Science—III* (editor C. Watkins), pp. 758–760, Lunar Science Institute Contr. No. 88.

Wager L. R. (1962) Igneous cumulates from the 1902 eruption of Soufriere, St. Vincent. *Bull. Volcanol.* **29**, 93–99.

Walker D. Longhi J. and Hays J. F. (1972) Experimental petrology and origin of Fra Mauro rocks and soils (abstract). In *Lunar Science—III* (editor C. Watkins), pp. 770–772, Lunar Science Institute Contr. No. 88.

Walter L. S. French B. M. and Doan A. J. (1972) Petrographic analysis of lunar samples 14171 and 14305 (breccias) and 14310 (melt rock) (abstract). In *Lunar Science—III* (editor C. Watkins), pp. 773–775, Lunar Science Institute Contr. No. 88.

Wilshire H. G. and Jackson E. D. (1972) Petrology of the Fra Mauro formation at the Apollo 14 landing site (abstract). In *Lunar Science—III* (editor C. Watkins), pp. 803–805, Lunar Science Institute Contr. No. 88.

Wood J. A. Dickey J. S. Jr. Marvin U. B. and Powell B. N. (1970) Lunar anorthosites and a geophysical model of the moon. *Proc. Apollo 11 Lunar Sci. Conf., Geochim. Cosmochim. Acta*, Suppl. 1, Vol. 1, pp. 965–988. Pergamon.

Yoder H. S. Jr. (1968) Experimental studies bearing on the origin of anorthosites. In *Origin of Anorthosite and Related Rocks* (editor Y. W. Isachsen), pp. 13–22, Univ. State New York Mem. 18.

Yoder H. S. Jr. (1969) Calcalkalic andesites: Experimental data bearing on the origin of their assumed characteristics. In *Proc. Andesite Conf.* (editor A. R. McBirney), pp. 77–89, Oregon Dept. Geol. Mineral. Res., Bull. 65.

Proceedings of the Third Lunar Science Conference
(Supplement 3, *Geochimica et Cosmochimica Acta*)
Vol. 1, pp. 231–241
The M.I.T. Press, 1972

Electron petrography of Apollo sample 14310

Deane K. Smith, Peter A. Thrower, and Wesley P. Hoffman

College of Earth and Mineral Sciences, Pennsylvania State University,
University Park, Pennsylvania 16802

Abstract—Thin sections prepared from Apollo sample 14310 have been examined with 100 kV transmission electron microscopy. The sample is composed principally of anorthite and pyroxene. The anorthites fall into two categories. The transitional anorthites that show the characteristic "*c*" type diffraction maxima are characterized by very fine antiphase domain structures and some fine twinning. The body-centered anorthites that show *no* "*c*" reflections are characterized by an absence of fine twinning and a less well-defined antiphase domain structure.

Although both orthopyroxene and clinopyroxenes are reported in sample 14310, only clinopyroxenes were observed in this study and all show exsolution structures. The augite exsolved in pigeonite commonly is along lamellae parallel to (100), although (001) lamellae also are present. This unusual orientation ((001) lamellae commonly are most prominent) was verified by diffraction patterns. The pigeonite commonly shows antiphase domains, attributed to stacking errors when the sample inverted from its high temperature disordered form. In augite as the host phase, the pigeonite lamellae contain very complex structures including antiphase domains.

Introduction

Thin sections of Apollo sample 14310 from the Fra Mauro site have been examined both optically and by transmission electron microscopy. This rock is reported to be an aluminous basalt and has been shown by many investigators to have a marked range of bulk chemistry and modal mineral content in spite of its limited size. Several petrographic descriptions of this specimen are available from the preliminary examination team, Kushiro (1972) and Ridley *et al.* (1972), which are not in agreement with each other and are indicative of the heterogeneity of the sample. This range of texture and composition is also one of the puzzles surrounding the origin of this sample.

The texture of this rock is described as intergranular, subophitic, and ophitic. Mineralogically, the rock is dominated by plagioclase and pyroxene, which make up over 95% of the total. The plagioclase-pyroxene ratio, however, may vary from 1:1 to 2:1 depending on which portion of the rock is sampled. Figure 1 is an optical micrograph that is characteristic of the sample used in this study. Lath-shaped plagioclase with abundant twinning and grain sizes up to 1 mm dominate the field. A small portion of the plagioclase is found as equant anhedral essentially twin-free grains up to 2 mm in size. The pyroxene is anhedral and compositionally zoned and most commonly occurs interstitial to the plagioclase, although a few grains as large as 2 mm are present. Ridley *et al.* (1972) report microprobe data on both the feldspars and the pyroxenes. The plagioclase laths range from An_{95} to An_{72} with a few measurements to An_{58}. The equant grains may be unzoned An_{94} or zoned An_{91} to An_{75}. The pyroxenes commonly show a core of orthopyroxene zoning outward through a Mg pigeonite to intermediate pigeonite rims. The rims show exsolution

Fig. 1. Optical micrograph of sample 14310 showing subophitic texture of pyroxene and plagioclase, crossed nicols.

lamellae of subcalcic ferroaugite. Takeda and Ridley (1972) show that the ortho-pyroxenes contain little evidence of exsolution, but the pigeonites show twinning, stacking disorder and exsolution of augite both on (100) and (001). The exsolution is more evident in the intermediate pigeonites than in the magnesian pigeonites.

SAMPLE PREPARATION

Thin sections of sample 14310,109 were prepared by cutting thin slices with the aid of a microdiamond saw designed by Albrecht *et al.* (1969). Slices less than 250 μ thick were then mounted on microscope slides using Crystalbond 509 as the cement. Sections were ground to less than 50 μ thick using bonded diamond laps. These thinned sections were divided into areas less than 3 mm in diameter by scribing with a diamond pencil. The desired area was then floated free of the microscope slide using acetone. The resulting sample was mounted between thin brass washers and further thinned on the Commonwealth Scientific Ion Bean Thinning Device to the stage where small holes developed. No variation in thinning rate was evident for the plagioclase and pyroxene. The sample was then examined in a Philips EM300 Electron Microscope at 100 kV.

OBSERVATIONS

Feldspars

Feldspar is the most abundant phase in this rock. Ridley *et al.* (1972) showed that much compositional variation exists among these feldspars; however, because this study has not been coupled with microprobe analysis, it has not been possible to assign a specific composition to grains we have studied. Our observations result in classifying the feldspars into two groups based on the presence or absence of "c" type diffraction maxima, i.e., $h + k = 2n$, $\ell = 2n + 1$, when the feldspar pattern is indexed on a cell with $c = 14$ Å. All the feldspars studied showed strong, sharp "a" reflections with $h + k = 2n$, $\ell = 2n$, and weak, sharp "b" reflections with $h + k = 2n + 1$, $\ell = 2n + 1$. Where these reflections are the only ones observed, the structure has been called body-centered anorthite. Those structures which show the "c" class of reflections as diffuse spots are termed transitional anorthites. Both body-centered and transitional types have been observed and probably are a reflection of the range of composition.

The transitional anorthites generally are characterized by a lack of twinning and the presence of very fine domain texture. This domain structure probably is due to compositional variations of differing An contents, and the diffuseness of the "c" reflections is attributed to the fine domain size. Type "b" domains, Christie *et al.* (1971), have not been observed in this study.

The body-centered anorthites or those transitional anorthites requiring longer exposures of the diffraction pattern to reveal the "c" reflections showed an abundance of fine lamellar twinning, as well as the characteristic domain structure. Figure 2 shows the fine domain structure as revealed in a bright field image. It has the same general features as the "c" domains described above in spite of the absence or extreme weakness of "c" reflections. Figure 3 shows both bright field and dark field images of the twin lamellae. The dark field views show that with one exception alternate lamellae are in contrast and are related by the albite twin law. The exception is always out of contrast and its twin law has not been identified, but its presence does not disrupt the albite twin sequence. In all cases the twin interfaces are planar and show contrast effects similar to stacking faults.

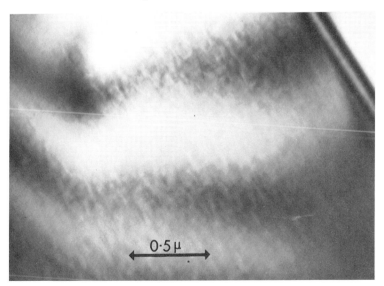

Fig. 2. Mottled texture of a body-centered anorthite attributed to antiphase domains. A twin lamella is evident in the upper right-hand corner.

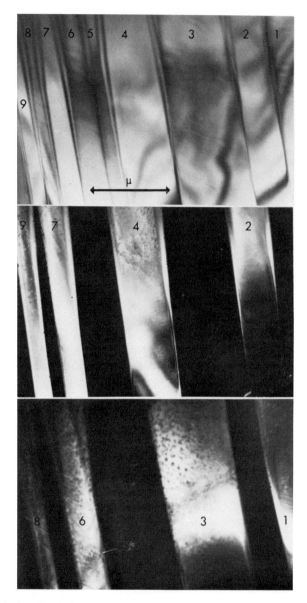

Fig. 3. Twin lamellae in body-centered anorthite. (Top) Bright field image showing all lamellae bounded by planar interfaces. (Middle) Dark field image showing alternate lamellae in contrast except for lamellae 5 and 6 which are both dark. (Bottom) Related dark field image of other lamellae. Note that lamella 5 is still dark.

Pyroxenes

Although both orthopyroxene and clinopyroxene have been reported from this sample, only the clinopyroxenes were observed in this study and all showed exsolution. In some cases the lamellae are simple and thin with very well-defined interfaces, but complex lamellae are equally common, in which the interfaces are irregular and the lamellae themselves show fine substructure. No orthopyroxene was identified in this study.

Figure 4 shows simple lamellae with widths of 50 to 200 Å. Their interfaces are sharp and planar. These lamellae, surprisingly, are parallel to (100). This unusual orientation has been determined from selected area diffraction patterns and was found to be true for all grains showing very thin lamellae. The diffraction spots from the lamellae are evident by their shapes, which are streaked parallel to a^*. In addition, the a^* direction of the lamellae and host are coincident. Both these features are apparent in the Fig. 4 insert. The host is pigeonite and the lamellae are augite, but their precise compositions are not known. The fineness of the lamellae may indicate the host is a magnesian pigeonite, but the accuracy of the electron diffraction pattern is not sufficient to verify this identification.

A second set of minor lamellae that is parallel to (001) is seen in the center of Fig. 4. It is more common to find this set as the dominant one with c^* of the augite lamellae coincident with c^* of the host pigeonite, as indicated by Ghose et al. (1972) and Takeda and Ridley (1972), but in this study the (100) set has proved more abundant.

Figure 5 shows a pigeonite that probably is more intermediate in composition than the one shown in Fig. 4. Augite lamellae are more abundant with exsolution having taken place on both (100) and (001). The thin, well-defined set appears similar to (100) lamellae observed in Fig. 4, in that they are thin with sharp planar boundaries, although it was not possible to verify this identification by selected area diffraction

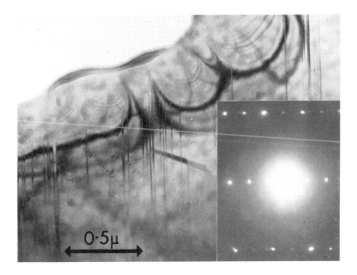

0·5μ

Fig. 4. Pigeonite with augite lamellae exsolved along (100). One lamella parallel to (001) is evident in the center of the field. The diffraction pattern insert shows coincidence of the augite and pigeonite a^* axes which are horizontal. Augite diffraction maxima are also streaked along a^*.

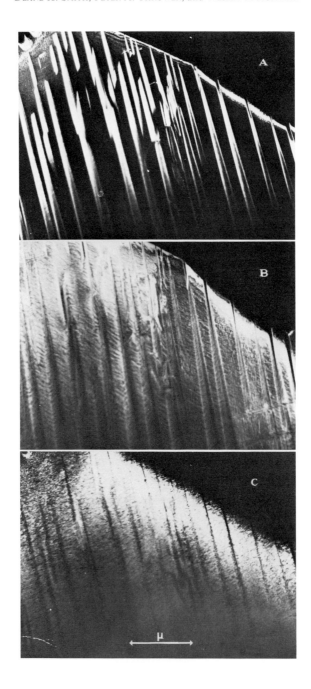

Fig. 5. Augite exsolved in pigeonite. (A) Discontinuous nature of augite lamellae parallel to (100) shown in dark field. (B) Indistinct augite lamellae parallel to (001) shown in dark field. (C) Antiphase domain texture in pigeonite revealed in dark field.

patterns. The dark field image, Fig. 5A, shows that these lamellae are composed of short discontinuous segments. The second set of augite lamellae have formed parallel to (001), as seen in Fig. 5B, but their interfaces are indistinct although they appear to cross over the sharper (100) lamellae. These latter lamellae actually may have been the first set to exsolve and may represent a spinodal decomposition in contrast to the (100) lamellae that formed subsequently by nucleation and growth and crossed over the pre-existing set. These features may indicate that the indistinct set, (001), formed initially as the sample was slowly cooling. Continued cooling at a more rapid rate quenched in the indistinct (001) set and initiated the sharp (100) exsolution.

The pigeonite host in Fig. 5C also shows a fine structure that can be attributed to anti-phase domains (Morimoto and Tokonami, 1969, and Christie *et al.*, 1971). These domains resulted when the pigeonite host, initially in its high-temperature disordered $C2/c$ form, developed short-range order of $P2_1/c$ domains on cooling. The originally disordered A and B tetrahedral layers were able to attain a state of short-range order, but the small size of these domains, generally less than 200 Å, suggests that the ordering process developed quickly and was restricted to very small domains due to rapid cooling of the rock.

In the lower right-hand corner of Fig. 5B intersecting exsolution lamellae and stacking faults may be faintly seen. A detailed photograph of this feature is given in Fig. 6 from which it can be seen that the two sets contain directions that appear perpendicular to each other.

Figure 7 shows a complex exsolution texture that appears to be pigeonite lamellae in an augite host. These exsolution bands show intricate branching patterns and evidently are not parallel to a rational lattice plane of the host. The two images in

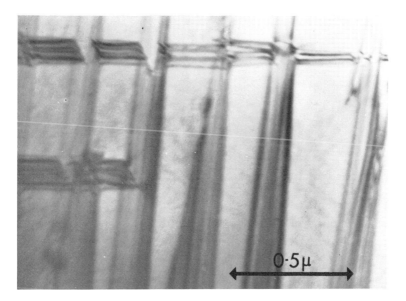

Fig. 6. Intersecting exsolution lamellae and stacking faults which may also be seen at lower right of Fig. 5B.

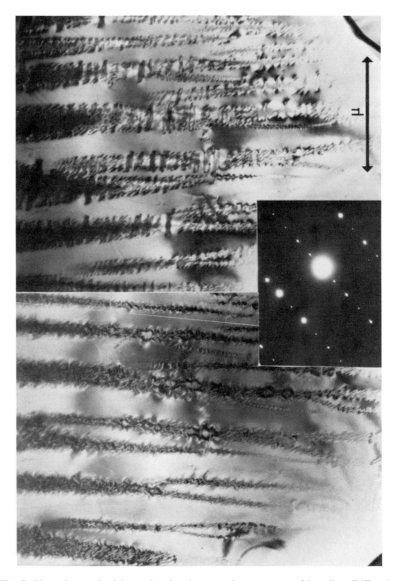

Fig. 7. Pigeonite exsolved in augite showing complex structure of lamellae. Diffraction insert is oriented with respect to the photographs and shows a streaking of spots parallel to a^*, which is sloped upper left to lower right. This streaking corresponds to the thin direction of the fine lamellae within the complex exsolution bands. The two photographs are taken with the specimen tilted 2° apart.

Fig. 7 are slightly tilted (2°) with respect to each other, although the structures are markedly different. In both views, the finest substructure within the complex bands are sets of indistinct, crudely planar lamellae, which make an angle near 45° with respect to the bands. These fine lamellae are parallel to (100), which is indicated by the elongated diffraction spots in the diffraction pattern insert. The remainder of the complex cross lamellae and bright-dark contrasting irregular areas is not as easily interpreted. The bright-dark regions probably are antiphase domains due to short-range order in stacking sequences of the pigeonite phase.

Radiation damage

Figure 8 is a higher magnification view of the same anorthite grain shown in Fig. 3. The area shows the typical black spot type of radiation damage. A few of the spots appear to have grown to sufficient size to resolve into complete loops.

The existence of radiation damage in our sample of 14310 is not surprising in that our sample did have one original surface as evidenced by glass-lined micro-craters. The particular anorthite grain that showed the black spot pattern could have been from very near the surface. Unfortunately, the distance to the surface was not preserved in preparing the specimen, and no other example of such damage was observed any grain. The structure is considered primary because electron beam damage observed in some areas shows entirely different features.

Conclusions

The value of electron petrography on lunar and terrestrial samples is demonstrated easily in the many and variable fine features that can be resolved, which are too small to interpret with the aid of only an optical microscope. Features observed in this study are similar to the results of other workers, Christie *et al.* (1971), Fernández-Morán *et al.* (1972). Exsolution, twinning, and antiphase domain structures are

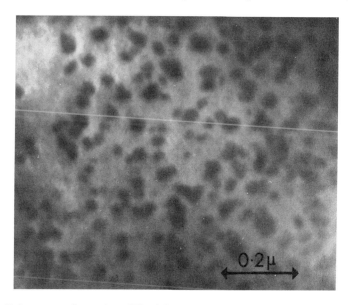

Fig. 8. Enlargement of a portion of Fig. 3 (Bottom) showing black spot radiation damage.

common in both pyroxenes and feldspars, and their sizes and mutual characteristics are related to compositional variations and thermal history. Evidence already cited seems to indicate that the cooling rates for sample 14310 may have changed during and after crystallization. Initially, cooling rates were slow enough for grain sizes up to 2 mm to develop as indicated by Finger *et al.* (1972). During this stage exsolution by spinodal decomposition in the pyroxenes and Al/Si ordering in the feldspars probably developed. A subsequent increase in cooling rate appears to have quenched in these features and superimposed the second thin set of exsolution lamellae parallel to (100) on the already existing (001) set.

The most surprising observation in this study is the abundance of (100) augite exsolution lamellae in the pigeonite. This orientation rarely has been reported before, and when it has, it is still less common than the (001) lamellae (Christie *et al.*, 1971; Fernández-Morán *et al.*, 1972). Because of its rarity, this identification has been carefully checked. Although electron diffraction patterns do have limitations with respect to accuracy, combined d-spacing and intervector angles on all diffraction patterns lead to unique indexings and provide the necessary evidence. Not all pyroxene grains studied yielded indexable patterns, but those that could be indexed yielded the quoted orientation.

Another feature generally apparent in this rock is the lack of dislocations and any evidence of microfracturing. Evidently, the specimen has undergone almost a pure thermal history and has not been subjected to the very strong forces associated with meteorite impact and ejecta effects. It is possible that a complex thermal history involving a second heating cycle could have obliterated evidence of earlier deformation, but the observed substructure does not show any apparent evidence of resorption of exsolution lamellae or other possible features that could be attributed to reheating. Finger *et al.* (1972) also show no evidence for reheating of this sample, although Hollister *et al.* (1972) indicate some reverse zoning does exist in the pyroxenes, and Ridley *et al.* (1972) indicate the same for feldspars. Thus, the conclusion must be drawn that this rock has seen little mechanical deformation during its history.

Acknowledgments—This study has been supported by Grant NGR–39–009–186 from the National Aeronautics and Space Administration.

References

Albrecht E. D. Stearns E. H. and Wittmayer F. J. (1969) Low speed diamond saw. *Proc. International Metallographic Soc.*, second annual mtg. pp. 8–10, UCRL 71611 Lawrence Radiation Laboratory, Livermore, Calif.

Christie J. M. Lally J. S. Heuer A. H. Fisher R. M. Griggs D. T. and Radcliffe S. V. (1971) Comparative electron petrography of Apollo 11, Apollo 12, and terrestrial rocks. *Proc. Second Lunar Sci. Conf.*, *Geochim. Cosmochim. Acta*, Suppl. 2, Vol. 1, pp. 69–89. M.I.T. Press.

Fernández-Morán H. Ohtsuki M. and Hough C. (1972) Correlated electron microscopy and diffraction studies of clinopyroxenes from Apollo 14 rocks (abstract). In *Lunar Science—III* (editor C. Watkins), p. 252, Lunar Science Institute Contr. No. 88.

Finger L. W. Hafner S. S. Schürmann K. Virgo D. and Warburton D. (1972) Distinct cooling histories and reheating of Apollo 14 rocks (abstract). In *Lunar Science—III* (editor C. Watkins), p. 252, Lunar Science Institute Contr. No. 88.

Ghose S. Ng. G. and Walter L. S. (1972) Clinopyroxenes from Apollo 12 and 14: Exsolution, cation order, and domain structure (abstract). In *Lunar Science—III* (editor C. Watkins), p. 300, Lunar Science Institute Contr. No. 88.

Hollister L. Trzcienski W. Jr. Dymek R. Kulick C. Weigand P. and Hargraves R. (1972) Igneous fragment 14310,21 and the origin of the mare basalts (abstract). In *Lunar Science—III* (editor C. Watkins), p. 386, Lunar Science Institute Contr. No. 88.

Kushiro I. (1972) Petrology of lunar high-alumina basalt (abstract). In *Lunar Science—III* (editor C. Watkins), p. 466, Lunar Science Institute Contr. No. 88.

Lally J. S. Fisher R. M. Christie J. M. Griggs D. T. Heuer A. H. Nord G. L. Jr. and Radcliffe S. W. (1972) Electron petrography of Apollo 14 and 15 samples (abstract). In *Lunar Science—III* (editor C. Watkins), p. 469, Lunar Science Institute Contr. No. 88.

Morimoto N. and Tokonami M. (1969) Domain structure of pigeonite and clinoenstatite. *Amer. Mineral.* **54**, 725–740.

Ridley W. I. Williams R. J. Brett R. Takeda H. and Brown R. W. (1972) Petrology of lunar basalt 14310 (abstract). In *Lunar Science—III* (editor C. Watkins), p. 648, Lunar Science Institute Contr. No. 88.

Takeda H. and Ridley W. I. (1972) Crystallography and mineralogy of pyroxenes from Fra Mauro soil and rock 14310 (abstract). In *Lunar Science—III* (editor C. Watkins), p. 738, Lunar Science Institute Contr. No. 88.

Proceedings of the Third Lunar Science Conference
(Supplement 3. *Geochimica et Cosmochimica Acta*)
Vol. 1, pp. 243–249
The M.I.T. Press, 1972

Mineralogical evidence for subsolidus vapor-phase transport of alkalis in lunar basalts

Brian J. Skinner and Horace Winchell

Department of Geology and Geophysics
Yale University, New Haven, Conn. 06511

Abstract—Thin, potassium-rich margins to feldspar grains, and along cracks within feldspar grains, in lunar basalts 12038 and 12040 provide evidence of late-stage and presumably subsolidus movement of material in the vapor phase. The compositions of the vapors are not known but are inferred to be alkali-rich relative to the parent rock. The same vapors are postulated to deposit perfect, whisker crystals of alkali-rich plagioclase in vugs and vesicles of 12038.

Introduction

The crystallization of igneous rocks is essentially a sequence of crystal nucleations within, and growth from, a melt. When volatile species with high vapor pressures are present in solution they may exsolve during cooling, and under suitable conditions the vapors may deposit phases either within cooler portions of the igneous rock or in adjacent rock units. The phenomenon is well known in terrestrial magmatic processes, but it is a particularly important process in H_2O-rich melts such as those forming the common salic igneous rocks.

The observations of DeMaria *et al.* (1971), Gibson and Hubbard (1972) and others concerning the amount of material volatilized from lunar basalts, even at temperatures below the solidus, indicate that vapor-phase transport also must be a factor in the cooling histories of lunar rocks. To the present time, however, no clear examples of vapor-phase transport and deposition have been reported from primary igneous rocks returned by the Apollo 11, 12 and 14 missions. It is the purpose of this paper to report occurrences of what appear to be vapor deposited feldspar crystals in vugs and vesicles of lunar basalt 12038, and evidence of subsolidus, presumably vapor transported, potassium-rich overgrowths on matrix feldspars in basalts 12038 and 12040.

Description of Samples 12038 and 12040

Rocks 12038 and 12040 are both ophitic basalts characterized by subhedral to euhedral crystals and with approximately equal grain size ranges of all major phases. Rock 12038 is classified as a Type 2 ophitic basalt by Warner (1971), meaning that euhedral lath-shaped crystals of plagioclase are enclosed in subhedral pyroxene crystals. Sample 12040 is classified as a Type 5 ophitic basalt, meaning that plagioclase laths are interstitial to the pyroxene crystals. The textures in both cases indicate that the liquids cooled and crystallized at such a rate that any differentiation processes would tend to proceed further in these rocks than they might in a porphyritic basalt of similar composition.

The compositions of both basalts are similar to those of other Apollo 12 basalts, but 12038 is the most aluminous of the ophitic basalts analysed by Kushiro and Haramura (1971), and 12040 is the most magnesian.

The major minerals in 12038 are plagioclase, pyroxene, and ilmenite, and the minor minerals are a chromium-rich spinel, olivine, troilite, and cristobalite. In 12040 olivine joins plagioclase, pyroxene, and ilmenite as a major mineral. The grain shapes of all major minerals are lath-like, but no obvious grain orientations are present. The random grain orientations suggest that most, if not all, crystallization occurred after the lavas had stopped flowing.

Rock 12038 is relatively coarse grained, with both feldspar and pyroxene grains present as lath-shaped crystals ranging up to 1 mm in length. In common with other coarse grained lunar igneous rocks, the minerals are not tightly bonded together, so that the rock is somewhat friable. In places, there are obvious openings between mineral grains, and in a small number of cases these irregularly shaped openings are sufficiently large to be called vugs. The vugs do not exceed 5 mm in diameter, and where the major minerals of the rock project into them, they display well-formed crystal terminations.

The origin of the vugs is problematic, but apparently they developed late in the cooling history of the basalt, after most of the melt had crystallized. Vesicles also are present in 12038, but are less common than vugs. The approximately 2.5 cm cube of sample available to us contained one well-developed vesicle 3 mm in diameter and a portion of a second vesicle that had been exposed during sample preparation at the Lunar Receiving Laboratory. The vesicles are spherical and so smooth-sided that, at first glance, they appear to be lined with glass. They actually are lined with in-numerable grain surfaces of the major minerals. The vesicles, unlike the vugs, must have formed early in the cooling history, when exsolved gas bubbles could form spherical surfaces against an essentially homogeneous liquid medium. The fact that they are present at all is good evidence that the lava cooled relatively rapidly.

Rock 12040 is also relatively coarse grained but contains fewer vugs than 12038 and apparently does not contain vesicles.

FELDSPARS IN THE MATRIX

The only feldspar in the matrix of 12038 is plagioclase and in common with all other lunar basalts. it is anorthite-rich. As observed by Christie et al. (1971), the crystals are slightly zoned, from An_{80-84} near the core to An_{75-78} near the margin. From 450 separate point analyses on 21 different grains, using the 4-channel Acton electron microprobe analyser, we obtained an average composition of $An_{79}Ab_{20}Or_{0.3}$ for the matrix feldspar. The zoning apparently is due solely to a change in the Ab and An contents, because the Or content remains constant at 0.3 throughout.

The feldspar in 12040 is even more anorthite-rich than 12038, reported by Walter et al. (1971) to fall in the range An_{92-95} but having a similar low Or content of about $Or_{0.3}$. Walter et al. (1971) report a slight enrichment of both sodium and potassium in the cores and towards the margins of matrix plagioclase. We have observed this phenomenon but find that it apparently is restricted to only a small percentage of crystals. The zoning is not so readily apparent as it is in 12038.

POTASSIUM-RICH SURFACES OF THE MATRIX FELDSPARS

When feldspar grains in 12038 and 12040 are scanned with an electron micro-probe it is observed that small, and at first seemingly random, potassium-rich areas occur in them. They are located either along the grain boundaries of the crystal or, less commonly, along minute fractures that apparently have formed from thermal stresses during cooling. Close examination shows these potassium-rich zones to be parallel to the grain boundary or fracture and to range markedly in width. The feature controlling the width of the zone is the angle at which the grain boundary or fracture intersects the upper surface of the thin section. If the angle of intersection is high, one observes either a very thin potassium-rich zone or, more commonly, no potassium enrichment. If the angle of intersection is low, however, a wide potassium-rich zone commonly is observed, the apparent width increasing as the angle of intersection decreases.

Two examples of this phenomenon are illustrated in Figs. 1 and 2, where the grain boundaries between feldspars and adjacent pyroxene crystals dip under the feldspar and intersect the surface of the thin section at less than 20°. We interpret our observations to arise from an extremely thin skin of a potassium-rich plagioclase.

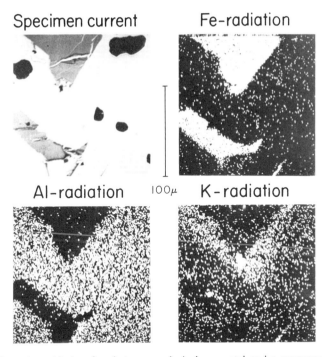

Fig. 1. Potassium-rich interface between a plagioclase crystal and a pyroxene grain in 12038, outlined by electron microprobe photographs. Pyroxene grains show dark gray in the specimen-current photograph, white in Fe-radiation, and black in Al-radiation. The interface between the large pyroxene crystal and the plagioclase dips under the plagioclase and irradiation of the interface shows a greater concentration of potassium compared to the body of the feldspar crystal.

Specimen current   Fe-radiation

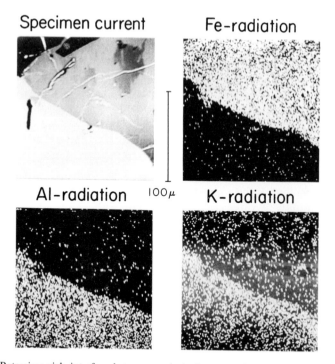

Al-radiation   100μ   K-radiation

Fig. 2. Potassium-rich interface between a plagioclase crystal and a pyroxene grain in 12038, outlined by electron microprobe photographs. Pyroxene grains show dark gray in the specimen-current photograph, white in Fe-radiation, and black in Al-radiation. The interface between the large pyroxene crystal and the plagioclase dips under the plagioclase and irradiation of the interface shows a greater concentration of potassium compared to the body of the feldspar crystal.

When a potassium-rich skin intersects the surface of the thin section at a low angle, the electron-beam penetrates the body of the feldspar crystal, and we can observe the skin. When the skin intersects at a high angle, we are unable to observe it because its width is apparently less than 0.1 microns, while the diameter of the electron beam we normally use for scanning feldspars is close to 1 μ. For the same reason we have been unable to get an analysis of the surface.

The phenomenon can be observed along any interface, but it is most marked along interfaces between plagioclase and pyroxene crystals. The examples shown in Figs. 1 and 2 are both of this kind. The phenomenon is also more marked in 12038 than in 12040. The time at which the external potassium-rich zones grew cannot be determined from textural observation. Those potassium-rich zones along fractures within the crystals. however, apparently developed as subsolidus features. Had they been formed by crystallization from a late-stage residual liquid, we would expect to find other mineral phases as well, and we might anticipate that the fractures them-selves would be healed over. These features are not observed, and we believe the potassium-rich rims must, therefore, have been formed after the last liquid crys-tallized. The only manner by which this could occur is through a vapor.

FELDSPAR CRYSTALS IN VUGS AND VESICLES

Within the vugs and vesicles of rock 12038 there are numerous thin, whisker-shaped crystals of feldspar (Fig. 3). Many crystals are extremely elongated, ranging from 2 to 5 μ wide and to as much as 800 μ long. The feldspar whiskers are not crystallographically continuous with the matrix phases, and appear to nucleate on each of the three major matrix minerals without preference.

The whiskers do not contain solid inclusions and do not overgrow, nor get overgrown by other phases, as is commonly observed for the matrix feldspars. They have the unusual property for natural minerals of being flexible and can be repeatedly flexed through as much as 90° until they abruptly lose their flexibility and the usual brittle properties of feldspars assert themselves with a consequent fragmentation of the crystal. We interpret this to mean that the crystals have low densities of dislocations and that breakage is associated with a sudden build-up of dislocations and a consequent loss of elastic properties.

The whiskers are all elongated parallel to the a axis, are untwinned and, despite x-ray exposure times of up to two weeks, do not show evidence of C-type reflections. The whiskers, therefore, appear to have the space group $I\bar{1}$, but the matrix feldspars clearly show C-type reflections and have space group $P\bar{1}$.

Microprobe analyses of the whisker crystals from 12038 give a wide range of compositions (from An_{70} to An_{81}), and show the crystals to be consistently richer in the Or component (from $Or_{0.8}$ to $Or_{3.8}$) than the matrix feldspars. Part of the apparent variability in composition is due to the difficulty of obtaining good analyses on small grains, but part too apparently is due to small compositional variations within the crystals. The remaining component in the whisker feldspars is Ab, and the relatively high Fe and Mg contents reported for the matrix feldspars (Keil *et al.*, 1971) apparently are not present in the whiskers. We obtain an average composition of $An_{74.0}Ab_{24.0}Or_{2.0}$ for the six whisker crystals that we have analysed. The whisker crystals are, therefore, consistently richer in the alkali components Ab and Or than the matrix feldspars.

The whisker crystals must have grown from vapors and the observed physical properties are consistent with this suggestion but are not consistent with growth

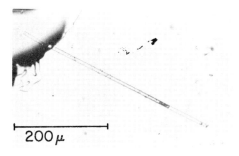

Fig. 3. Whisker crystal of plagioclase from vesicle of 12038. The crystal is approximately 400 μ long, 8μ wide and 4μ thick. The dark area, about one quarter of the length from the end, is due to a series of stepped faces on the crystal surface. The crystal is mounted on a slide and is cemented by an irregular mass of epoxy resin in the upper left-hand corner.

from a melt. While it may be argued that whisker crystals growing in vugs were simply deposited by late stage liquids that later drained away, those in the vesicles cannot be so explained. The vesicles are smooth-walled and have obviously remained empty of melt from the moment they were frozen in. The lengths of whisker crystals projecting into the vesicles may reach a third of the vesicle diameter. The only medium capable of transporting the components needed to grow such elongate crystals is a vapor.

The mineral textures of the matrix clearly are primary and do not allow the interpretation that the rock has been metamorphosed. Nor is there any evidence of alterations and depositions as might be expected if a postmagmatic vapor phase were introduced from the outside.

Conclusions

Two observations that we have presented for the feldspars in basalts 12038 and 12040 provide strong evidence of movement of material in the vapor phase during the cooling history of lunar basalts. The only evidence preserved suggests vapor phase transport during the subsolidus cooling history. The same phenomenon must proceed at even greater rates at higher temperatures while crystals are still growing from the liquid. We believe that it probably is not possible to find unequivocal mineralogical evidence within the rock to substantiate vapor phase transport during crystallization of the liquidus. Nevertheless, indirect arguments in favor of the process may be made on the basis of the bulk compositions of the rocks (Biggar *et al.*, 1971; Brown *et al.*, 1972).

Vapors forming from lunar basalts, eventually must be condensed on surfaces cooler than the matrix from which they were evaporated. In a crystallizing lava pile, the cooler marginal portions of the flows will serve to cold trap any vapors released from the hotter core and we suggest that very little, if any, material will actually escape from the rock unit once extrusion occurs. We have no direct evidence on the bulk composition of the transporting vapors, but they must have been alkali-rich relative to the parent rock. We suggest that although the moving vapor may have been composed largely of alkalis, the other components in the precipitated feldspars, i.e., aluminum, silicon and calcium, were locally derived, and were components within a tenuous vapor in direct equilibrium with minerals lining the vugs and vesicles. Addition of more alkalis to this tenuous vapor would naturally cause precipitation of a feldspar phase. Because feldspars are the principal repository of alkalis in lunar basalts, we do not believe that subsolidus crystal growth is a likely phenomenon within the other major silicate minerals, the pyroxenes and olivines. The process may well be important in the redistribution of volatile species such as sulfur.

Acknowledgments—We gratefully acknowledge the continuing expert assistance of Edward Weber with microprobe analyses. This work was supported by a grant from NASA for the study of lunar samples.

References

Biggar G. M. O'Hara M. J. Peckett A. and Humphries D. J. 1971. Lunar lavas and the achondrites: Petrogenesis of protohypersthene basalts in the lava lakes. *Proc. Second Lunar Sci. Conf., Geochim. Cosmochim. Acta* Suppl. 2, Vol. 1, pp. 617–643. MIT Press.

Brown G. M. Holland J. G. Peckett A. and Phillips R. (abstract) 1972. Mineral fractionation patterns between Apollo 14 primitive feldspathic rocks and Apollo 15 and other basalts. In *Lunar Science—III* (editor C. Watkins), pp. 95–98, Lunar Science Institute Contr. No. 88.

Christie J. M. Lally J. S. Hener A. H. Fisher R. M. Griggs D. T. and Radcliffe S. V. 1971. Comparative electron petrography of Apollo 11, Apollo 12 and terrestrial rocks. *Proc. Second Lunar Sci. Conf., Geochim. Cosmochim. Acta* Suppl. 2, Vol. 1, pp. 69–89. MIT Press.

DeMaria G. Balducci G. Guido M. and Piacente V. 1971. Mass spectrometric investigations of the vaporization process of Apollo 12 lunar samples. *Proc. Second Lunar Sci. Conf., Geochim. Cosmochim. Acta* Suppl. 2, Vol. 2, pp. 1367–1380. MIT Press.

Gibson E. K. Jr. and Hubbard N. J. 1972. Volatile element-depletion investigations on Apollo 11 and 12 lunar basalts via thermal volatilization. In *Lunar Science—III* (editor C. Watkins), pp. 303–305, Lunar Science Institute Contr. No. 88.

Keil K. Prinz M. and Bunch T. E. 1971. Mineralogy, petrology and chemistry of some Apollo 12 samples. *Proc. Second Lunar Sci. Conf., Geochim. Cosmochim. Acta* Suppl. 2, Vol. 1, pp. 319–341. MIT Press.

Kushiro I. and Haramura H. 1971. Major element variation and possible source materials of Apollo 12 crystalline rocks. *Science* **171**, pp. 1235–1237.

Walter L. S. French B. M. Heinrich K. F. J. Lowman P. D. Jr. Doan A. S. and Adler I. 1971. Mineralogical studies of Apollo 12 samples. *Proc. Second Lunar Sci. Conf., Geochim. Cosmochim. Acta* Suppl. 2, Vol. 1, pp. 343–358. MIT Press.

Warner J. L. 1971, Lunar crystalline rocks: Petrology and geology. *Proc. Second Lunar Sci. Conf., Geochim. Cosmochim. Acta* Suppl. 2, Vol. 1, pp. 469–480. MIT Press.

Proceedings of the Third Lunar Science Conference
(Supplement 3, *Geochimica et Cosmochimica Acta*)
Vol. 1, pp. 251–279
The M.I.T. Press, 1972

Petrographic features and petrologic significance of melt inclusions in Apollo 14 and 15 rocks*

EDWIN ROEDDER

U.S. Geological Survey
Washington, D.C. 20242

and

PAUL W. WEIBLEN

Univ. of Minnesota
Minneapolis, Minnesota 55455

Abstract—The occurrence and significance of silicate melt inclusions in a series of Apollo 14 and 15 igneous rocks, breccias, and soils are described. Electron microprobe analyses (114) are reported, giving the bulk composition of representative inclusions as well as their daughter and host minerals and other phases associated with the inclusions. Although many of the features seen are similar to those previously described, some novel ones were found.

Silicate melt inclusions in olivine are abundant and occasionally large (400 μ). Nucleation and growth (epitaxial or random) of daughter phases varies with inclusion size, bulk composition, and probably with cooling history. Solid inclusions in olivine consist of relatively large Cr-spinel euhedra (20–40 μ); the olivines in thermally metamorphosed rocks have, in addition, rows of very minute Cr-rich crystals ($\sim 1\ \mu$) that have decorated what apparently are otherwise invisible dislocations. Tiny melt inclusions in fractures in plagioclase of 15415 indicate the presence of at least a small amount of melt at the time of fracturing.

Melt inclusions related to the onset of immiscibility are common in all the samples, particularly 14310, and are like those in Apollo 11 and 12 and Luna 16 samples—potassic granite and ferropyroxenite in composition. There is an inverse relationship between the K_2O/Na_2O ratio in the bulk rock and its residual high-silica melt. A summary of all our melt inclusion data in the form of a single silica variation diagram suggests grossly similar liquid lines of descent for the various igneous rock types sampled at the five landing sites. All these lines appear to end at the same high-silica melt composition.

Numerous areas and fragments of glass of potassic granite composition (partly crystallized to K-feldspar) were found in the Apollo 14 breccias. The residual glass in these fragments is very similar in composition to that of the high-silica immiscible melt and could represent a phase of granitic composition from this process that occurs in the crust of the moon and was concentrated in the Apollo 14 source materials.

INTRODUCTION

IN THIS STUDY igneous rocks 14053, 14310, 15058, 15076, 15415, 15475, 15495, 15555; breccias 14303, 14305, 14319, 14320, 14321, 15405; and soils 14162 and 14163 were examined by microscope and electron microprobe. Our main goal was to obtain compositions of silicate melt inclusions and their hosts. Various other small-scale features were also observed that may be of interest to other workers. Results of studies on 12035 and 12036, that were not available previously, also are included.

The experimental techniques used were essentially as described earlier (Roedder and Weiblen, 1970, 1971, 1972a). Many details concerning the inclusions are similar to those described in our earlier papers and will not be repeated. The 114 analyses

---

*Publication authorized by the Director, U.S. Geological Survey.

for 8 to 12 elements reported here (consecutively numbered in Tables 1–8), and the many partial analyses referred to in the text all are previously unpublished data.

EARLY SILICATE MELT AND SOLID INCLUSIONS IN OLIVINE AND PLAGIOCLASE

Apollo 12 and 14 samples

Analyses of a silicate melt inclusion in olivine from 12036,8 (Fig. 1) include the residual melt after crystallization of daughter minerals, now a mass of dendritic crystals (Table 1, #1), the daughter crystal of plagioclase (Table 5, #94), and the host olivine (Table 5, #74). The opaque daughter crystal (Fig. 1) was too thin to analyze accurately but was verified as ilmenite, and a similar plate in another inclusion was analyzed (Table 5, #95). The CIPW norm for the residual melt (Table 7) shows plagioclase 43, calcic pyroxene 32, and quartz 20. The 20% quartz may seem surprising in a rock this mafic; it presumably represents metastable equilibrium on rapid cooling, because there is no evidence of a pyroxene reaction rim at the olivine walls, and such an amount of SiO_2 is almost certainly more than the amount representing stable incongruent melting.

Many clasts of olivine-rich rocks are found in the Apollo 14 breccias. Olivine phenocrysts in one of these, 14321, contain the largest silicate melt inclusions found in any of the lunar samples examined. The two largest of these inclusions (nearly 0.4 mm long; Fig. 2) occur in parallel array, in a single olivine crystal, yet have different compositions (Table 1, #s 2 and 3) and textures (Fig. 3). The inner inclusion (#3) unexpectedly has a composition representing a presumably later, more differentiated magma (see Table 7). Perhaps the crystal grew as a hollow tube, trapping the outer inclusion in the wall, and then closed off the tube at a later time to form the inner inclusion. A series of partial analyses of the host olivine show compositions ranging several percent on either side of $Fo_{65.4}Fa_{34.1}La_{0.5}$;[*] the variations are compatible with this interpretation of the growth patterns. It is also possible that kinetic disequilibrium was involved, due to rapid growth (Albarede and Bottinga, 1972). The cause of the obvious difference in textures is uncertain. The outer inclusion contains an oriented daughter (?) crystal of aluminian chromite (Table 5, #96), which was not included in the analysis of the melt (#2). Such oriented chromite crystals were also seen in other similar but smaller inclusions in the olivine of this sample (e.g., Table 5, #97) and frequently occur also as solid inclusions in most of the olivines; the absence of one in the inner inclusion could be merely a result of the plane of the section. This phenocryst has several inclusions of low-Ni metallic Fe (Fig. 2); the inner one also contains a mass of high-silica melt too small for a good analysis ($\sim 75\%$ SiO_2, 10% Al_2O_3, and 5% K_2O). Both inclusion types are rather inexplicable in olivine.

For comparison with analyses 2 and 3, the only previous analyses we have made are of laboratory homogenized inclusions, and hence have considerably lower silica contents (Roedder and Weiblen, 1970, p. 810; 1971, p. 515). Presumably the inclusions in 14321 had similar compositions when trapped, but have crystallized olivine onto

---

[*] All mineral element ratios and compositions in this report are on an atomic or molecular basis.

Table 1. Microprobe analyses (wt %) of melt inclusions in Apollo 12, 14, and 15 (Paul W. Weiblen, analyst). Notes: The photo areas listed are our identification numbers, and do not refer to the photographs presented. Averages are indicated by "A" after analysis number. Abbreviations: Inter. = interstitial inclusion (between crystals); Plag. = plagioclase; Pyrox. = pyroxene; Cristob. = cristobalite; Adjac. = adjacent; — = not determined.

	Apollo 12	Apollo 14		Apollo 15 (all from 15555,34, arranged in order of increasing silica content)						
Analysis	1	2	3	4	5	6	7	8	9	10
Sample	12036,8	14321,25	14321,25	15555,34	15555,34	15555,34	15555,34	15555,34	15555,34	15555,34
Photo area	18–1	29–2	29–3	7–3	8–5B	12–1	8–3	5–1	2–1	7–1
Host	Olivine	Olivine	Olivine	Olivine	Olivine	Ilmenite	Olivine	Olivine	Olivine	Olivine
SiO_2	55.4	51.6	54.7	47.0	48.3	50.3	55.7	57.5	59.4	61.4
Al_2O_3	15.3	15.0	17.5	14.1	11.9	10.4	17.5	12.0	16.3	15.8
FeO	7.3	9.42	10.6	15.3	12.2	18.5	8.47	8.06	6.74	7.02
MgO	3.4	3.47	1.85	2.85	5.09	1.22	1.77	1.40	2.17	1.03
CaO	15.3	16.1	12.1	18.3	17.3	12.7	13.6	14.1	13.0	11.4
Na_2O	0.16	0.15	0.51	—	0.0	0.0	0.07	0.41	0.45	0.94
K_2O	0.04	0.14	0.14	0.04	0.06	2.70	0.08	0.11	0.08	0.11
TiO_2	2.4	3.58	2.12	3.07	3.28	2.71	2.04	0.90	2.39	0.66
P_2O_5	0.14	0.05	0.07	0.05	0.13	0.39	0.12	0.12	0.11	0.05
MnO	0.2	0.21	0.19	0.44	0.14	0.02	0.08	0.13	0.08	0.10
Cr_2O_3	0.15	0.13	0.04	0.17	0.13	—	0.04	0.12	0.12	0.10
ARITHMETIC TOTAL	99.79	99.85	99.82	101.32	98.53	98.94	99.47	94.85	100.84	98.61

Table 1 (continued)

Apollo 15 (all from 15555,34, arranged in order of increasing silica content)										
11	12	13	14	15	16	17	18	19	20	20A
15555,34	15555,34	15555,34	15555,34	15555,34	15555,34	15555,34	15555,34	15555,34	15555,34	15555,34
6–1	15–1	8–5A	11–1	18–1	20–1	14–1	19–1	19–3	17–1	Avg. of
Olivine	Ilmenite	Olivine	Ilmenite	Ilmenite	Inter.	Inter.	Fayalite	Inter.	Ulvö-spinel	6 high silica
61.5	61.9	63.8	69.1	72.2	74.2	75.0	75.2	75.3	75.6	74.56
13.8	13.2	13.9	15.7	13.0	12.2	11.4	11.9	11.7	10.3	11.78
7.14	5.27	6.34	2.48	3.68	2.62	2.41	2.56	1.72	4.01	2.83
1.13	0.54	1.73	0.56	0.0	0.0	0.0	0.0	0.0	0.22	0.04
13.3	11.2	10.5	3.97	2.68	2.48	1.12	2.35	1.62	0.26	1.75
0.65	0.50	0.35	0.87	0.0	0.08	0.28	0.00	0.0	0.0	0.06
0.10	3.34	0.12	3.20	5.61	7.54	8.19	6.88	6.44	7.00	6.95
0.82	0.85	1.37	0.80	0.79	0.49	0.81	0.63	0.57	1.01	0.72
0.16	1.19	0.11	0.35	0.37	0.17	—	—	—	0.10	0.10
0.11	0.05	0.04	0.04	0.0	0.01	0.0	0.05	0.01	0.07	0.02
0.07	—	0.04	—	—	—	—	—	—	—	—
98.78	98.04	98.30	97.07	98.33	99.79	99.21	99.57	97.36	98.57	98.81

Sample notes:
1. Fine grained aggregate (defocused beam analysis). Norm in Table 7, Host crystal, analysis 74. (See Fig. 1).
2. Outer inclusion, no epitaxial ilmenite (defocused beam analysis). See Figs. 2 and 3. Norm in Table 7.
3. Inner inclusion, with epitaxial ilmenite. Defocused beam analysis includes epitaxial ilmenite. See Figs. 2 and 3. Norm in Table 7.
4. Partly nonluminescent (pyroxene?).
6. Host ilmenite verified by analysis.
8. Very small inclusion.
9. Crystalline inclusion.
12. Small inclusion.
14. Poor surface on inclusion.
15. Inclusion in ilmenite (analysis 92), with fayalite (analysis 75). (See Figs. 7 and 8).
18. Host is an interstitial crystal of fayalite (analysis 76).
20A. Average of 15–20; norm in Table 7.

Table 2. Microprobe analyses (wt %) of high-silica melt inclusions in Apollo 12, 14, and 15 samples (Paul W. Weiblen analyst). See also analysis 15 to 20, Table 1. Notes as in Table 1.

Analysis	21	22	23	24	25	25A	26	27	28	29	29A
Sample	12036,8	12036,8	12036,8	12036,8	12036,8	12036	14053,19	14053,19	14053,19	14053,19	14053,19
Photo area	15–2	17	16	15–1	13–1	Avg. of 5	D	C	E	E'	Avg. of 4
Host	Inter.	Inter.	Inter.	Ilmenite	Ilmenite		Inter.	Inter.	Ilmenite	Ilmenite	Ilmenite
SiO_2	75.6	72.5	75.5	74.0	71.4	73.80	73.68	74.5	73.49	73.7	73.84
Al_2O_3	11.2	13.6	12.3	11.9	8.4	11.48	15.0	12.2	14.2	13.8	13.80
FeO	1.3	2.0	1.7	2.2	5.8	2.60	0.91	1.1	1.5	1.5	1.23
MgO	0.08	—	0.0	0.45	1.7	0.56	0.0	0.08	0.0	0.0	0.02
CaO	0.92	0.82	2.5	2.0	5.4	2.33	1.3	3.8	1.2	0.89	1.80
Na_2O	0.60	0.55	0.8	0.92	0.3	0.63	0.21	0.64	0.48	0.26	0.40
K_2O	8.8	9.7	6.8	7.2	1.8	6.86	8.5	6.8	7.9	8.2	7.85
TiO_2	0.96	0.98	1.1	0.92	1.2	1.03	0.36	0.52	1.3	1.3	0.87
P_2O_5	—	—	—	—	—	0.00	—	—	—	—	—
MnO	—	0.02	0.0	0.04	—	0.02	0.04	0.0	0.03	0.01	0.02
Cr_2O_3	—	—	—	—	—	0.00	—	—	—	—	—
BaO	—	0.30	—	—	—	0.30	—	—	—	—	—
ARITHMETIC TOTAL	99.46	100.47	100.7	99.63	96.0	99.61	100.00	99.64	100.00	99.66	99.83

Table 2 (continued)

Analysis	30	31	32	33	34	35	36	37	37A	Std. Dev.	38
Sample	14053,57	14053,57	14053,57	14053,57	14053,57	14053,57	14053,57	14053,57	14053,57		14162,20
Photo area	1–2	3	7	12	9	11	10	15–11	Avg. of 8		10–2
Host	Inter.	Inter.	Inter.	Inter.	Inter.	Inter.	Ilmenite	Inter.			Plag.
SiO_2	77.7	73.2	73.3	79.2	81.2	74.8	77.3	78.5	76.90	2.883	71.4
Al_2O_3	13.6	14.8	12.8	11.2	9.9	11.6	12.2	10.9	12.12	1.575	13.6
FeO	2.2	1.6	3.6	0.64	1.1	2.5	1.30	0.89	1.73	0.968	1.43
MgO	0.2	0.0	0.05	0.0	0.02	0.0	0.0	0.02	0.04	0.068	0.72
CaO	1.2	1.2	1.7	0.95	1.9	1.2	2.8	2.4	1.67	0.660	2.37
Na_2O	0.0	0.2	0.42	0.13	0.0	0.17	0.0	0.19	0.14	0.144	0.83
K_2O	6.8	8.8	6.6	7.1	5.0	7.3	4.3	4.8	6.34	1.519	7.59
TiO_2	0.82	1.2	1.4	0.42	0.37	0.53	0.59	1.2	0.75	0.443	1.59
P_2O_5	—	—	—	—	—	—	—	—	—	—	0.52
MnO	0.17	0.08	0.19	0.09	0.02	0.0	0.03	0.0	0.07	0.074	0.01
Cr_2O_3	—	—	—	—	—	—	—	—	—	—	0.0
BaO	—	—	—	—	—	—	—	—	—	—	0.22
ARITHMETIC TOTAL	102.69	101.08	100.06	99.73	99.51	98.10	98.52	98.90	99.76		100.28

Table 2 (continued).

Analysis	39	40	41	42	43	44	45	45A	Std. Dev.	46	47
Sample	14303,55	14310,5	14310,5	14310,5	14310,5	14310,5	14310,5	14310,5		14310,13	14310,13
Photo area	8	1	3	7	13	15	18	Avg. of 6		2-1	1-1
Host	Pyrox.	Inter.	Inter.	Inter.	Inter.	Inter.	Inter.			Pyrox.	Inter.
SiO_2	73.0	76.1	76.9	77.0	77.8	76.5	74.3	76.43	1.189	73.7	73.4
Al_2O_3	13.8	11.0	11.9	12.5	12.7	13.1	12.2	12.23	0.731	14.1	12.4
FeO	0.5	1.96	2.17	1.46	1.49	0.98	2.94	1.83	0.684	1.39	1.40
MgO	0.2	0.76	0.24	0.67	0.52	0.42	0.61	0.54	0.187	0.41	0.04
CaO	0.6	1.72	1.22	0.96	0.86	0.93	1.13	1.14	0.316	1.93	0.68
Na_2O	—	0.74	0.49	0.94	0.55	0.37	0.11	0.53	0.288	1.01	0.99
K_2O	7.4	6.10	4.98	6.51	6.85	5.52	6.71	6.11	0.734	7.41	9.40
TiO_2	3.65	1.19	0.70	0.89	0.77	0.59	0.25	0.73	0.313	0.70	0.48
P_2O_5	—	0.0	0.07	0.0	0.00	0.00	0.00	0.01	0.029	0.20	0.15
MnO	—	0.0	0.06	0.0	0.0	0.0	0.0	0.01	0.024	0.15	0.01
Cr_2O_3	—	0.01	0.0	0.0	0.0	0.0	0.0	0.00	0.004	—	—
BaO	—	—	—	—	—	—	—	—	—	—	0.19
ARITHMETIC TOTAL	99.15	99.58	98.73	100.93	101.54	98.41	98.25	99.56		101.00	99.14

Table 2 (continued).

Analysis	48	48A	49	50	51	52	53	54	54A	Std. Dev.
Sample	14310,13	14310,13	15475,16	15475,16	15475,16	15475,16	15475,16	15475,16	15475,16	
Photo area	4-1	Avg. of 3	10-1	9-1	6-8	13-18	15-13	15-14	Avg. of 6	
Host	Inter.		Inter.	Inter.	Plag.	Ilmenite	Inter.	Inter.		
SiO_2	75.6	74.23	72.8	73.3	76.8	74.5	76.4	74.7	74.75	1.606
Al_2O_3	12.7	13.07	12.3	12.8	11.0	13.0	12.4	12.3	12.30	0.699
FeO	1.20	1.33	3.7	2.4	2.6	2.0	2.4	1.5	2.43	0.734
MgO	0.05	0.17	0.04	0.0	0.0	0.0	0.0	0.0	0.01	0.016
CaO	2.16	1.59	2.1	1.1	0.9	2.1	1.7	1.5	1.57	0.501
Na_2O	1.30	1.10	0.2	0.20	0.06	0.80	0.41	0.24	0.32	0.261
K_2O	6.19	7.67	6.3	7.8	7.3	6.3	6.9	8.2	7.13	0.781
TiO_2	0.37	0.52	0.45	0.54	0.28	0.72	0.37	0.24	0.43	0.178
P_2O_5	0.14	0.16	0.08	0.13	0.04	0.12	0.12	0.05	0.09	0.039
MnO	0.03	0.06	0.01	0.01	0.07	0.0	0.04	0.00	0.02	0.028
Cr_2O_3	—	—	—	—	—	—	—	—	0.00	—
BaO	0.03	0.11	0.19	0.47	0.15	0.13	0.15	0.16	0.21	0.130
ARITHMETIC TOTAL	99.77	100.01	98.17	98.75	99.20	99.67	100.89	98.89	99.26	

Sample notes:
25. Heterogeneous; average of high- and low-potassium parts.
25A. Norm in Table 7.
26. SiO_2 by difference.
27. Adjacent to plagioclase of An 95.6
28. SiO_2 by difference
29A. Norm in Table 7.
30. Heterogeneous.
33. Heterogeneous.
37A. Norm in Table 7.
38. Host plagioclase An 90 (analysis 79). Norm in Table 7.
39. Norm in Table 7. Host crystal analysis 85.
45A. Norm in Table 7.
46. Host crystal analysis 86.
47. Large uniform area.
48. Large uniform area.
48A. Norm in Table 7.
49. Heterogeneous; average of luminescent and nonluminescent parts.
53. Large area, so beam could be thoroughly defocused; Na value more reliable as a result.
54A. Norm in Table 7.

the walls on cooling. As we have reported earlier, these lunar (and terrestrial) olivine inclusions commonly show the ilmenite daughter crystal extending out into the olivine walls (as in Fig. 1), probably indicating subsequent crystallization of olivine after nucleation of the ilmenite. In Fig. 1 the volume of such crystallization can be crudely estimated at 15–20%; this amount of olivine crystallization would explain the difference in SiO_2 content.

Apollo 15 samples

The olivine in 15555 contains numerous tiny silicate melt inclusions. Many now consist of an epitaxially oriented ilmenite daughter crystal, glass, immiscible sulfide melt globule, and a shrinkage bubble (Fig. 4). Only some of the larger ones have nucleated plates of plagioclase or dendrites of pyroxene (Fig. 5).

The eight analyses of 15555 inclusions in early olivine (Table 1) range from 47.0 to 63.8% SiO_2, indicating a considerable amount of differentiation in the residual melt during crystallization of this mineral. The sample material precludes obtaining representative analyses, but inclusion percent SiO_2 varies sympathetically with the host olivine fayalite content (45.2 to 61.5%). Nine other inclusions also were analyzed (Table 1); the trends shown by these analyses are discussed later.

Inclusions in 15415,15 were studied to understand the complex history of this sample and to evaluate the extensive studies by others of this important sample. A large number of very tiny secondary inclusions, mostly $<1 \mu$, outline the many fractures through this rock (evidenced by slight offsets in the plagioclase twins and in segmented extinction positions). The inclusions are too small for positive identification, but some appear to be gas, whereas others presumably are pyroxene, formed from a melt that penetrated the crack during crushing.

Late Immiscible Silicate Melt Inclusions

Just as in Apollo 11 and 12 and Luna 16, many of the Apollo 14 and 15 crystalline rocks exhibit evidence of late immiscible high-silica and high-iron melts. Most of our

Table 3. Microprobe analyses (wt %) of high-iron melt inclusions in Apollo 12, 14, and 15 samples (Paul W. Weiblen, analyst). (Notes as in Table 1.)

Analysis	55	56	57	58	59	60	61	62	62A	Std. Dev.	63
Sample	12036,8	14053,57	14310,5	14310,5	14310,5	14310,5	14310,5	14310,5	14310,5		15475,16
Photo area	10-1	15-16	L-6	N-2	K-4	M-8	M-10	W-17	Avg. of 6		12-1
Host	Plag.	Inter.	Inter.	I.rter.	Inter.	Inter.	Inter.	Inter.			Plag.
SiO_2	43.5	33.5	44.7	36.2	40.0	33.8	33.3	37.7	37.62	4.268	39.9
Al_2O_3	2.5	0.28	5.62	3.10	3.46	3.88	3.05	4.47	3.93	0.983	2.5
FeO	27.1	47.2	23.4	32.1	32.2	33.4	33.8	28.9	30.63	3.940	42.6
MgO	4.8	15.5	8.35	3.51	5.50	6.48	6.97	7.13	6.32	1.661	0.34
CaO	11.9	0.12	9.56	10.5	10.4	9.93	8.15	17.2	10.96	3.174	9.9
Na_2O	0.08	0.69	0.23	0.21	0.0	0.04	0.35	0.08	0.15	0.134	0.0
K_2O	0.05	0.03	0.64	0.12	0.12	0.37	0.05	0.60	0.32	0.259	0.04
TiO_2	7.8	0.16	5.11	7.99	6.53	6.82	8.62	0.63	5.95	2.877	3.5
P_2O_5	0.30	0.34	1.41	5.2	1.66	3.58	3.64	3.23	3.12	1.405	0.64
MnO	0.64	1.0	0.16	0.30	0.36	0.41	0.36	0.16	0.29	0.108	0.40
Cr_2O_3	0.14	0.00	0.12	0.00	0.02	0.08	0.05	0.05	0.05	0.043	0.06
ARITHMETIC TOTAL	98.81	98.82	99.30	99.23	100.25	98.79	98.34	100.15	99.34		99.88

Analysis	64	65	66	66A	67	68	69	70	71	71A
Sample	15475,16	15475,16	15475,16	15475,16	15555,34	15555,34	15555,34	15555,34	15555,34	15555,34
Photo area	8-1	4-16	4-17	Avg. of 4	22-1	23-1	25-1	25-3	24-1	Avg. of 5
Host	Plag.	Plag.	Inter.		Cristob.	Cristob.	Plag.	Plag.	Plag.	
SiO_2	39.8	46.7	48.7	43.77	36.9	43.3	44.5	44.6	43.8	42.62
Al_2O_3	6.0	3.9	0.41	3.20	7.64	11.9	5.34	4.0	3.79	6.53
FeO	36.6	32.1	37.9	37.30	35.5	20.0	19.5	24.8	30.4	26.04
MgO	0.74	2.5	0.61	1.05	0.42	0.57	1.37	5.30	2.88	2.11
CaO	12.0	10.4	8.4	10.17	11.1	15.7	20.9	13.2	10.8	14.34
Na_2O	0.0	0.0	0.0	0.00	0.07	0.0	0.34	0.0	0.0	0.08
K_2O	0.04	0.01	0.0	0.02	0.51	1.52	0.11	0.38	0.03	0.51
TiO_2	6.1	4.6	0.59	3.70	3.76	5.49	5.12	5.35	7.10	5.36
P_2O_5	0.2	0.13	0.05	0.25	1.60	2.05	0.45	0.49	0.29	0.98
MnO	0.41	0.38	0.76	0.49	0.47	0.28	0.38	0.47	0.45	0.41
Cr_2O_3	0.03	0.18	0.03	0.07	0.03	0.0	0.14	0.18	0.14	0.10
ARITHMETIC TOTAL	101.92	100.90	97.45	100.02	98.00	100.81	98.15	98.77	99.68	99.08

Sample notes:
55. Norm in Table 7. Host crystal analysis 78.
56. Odd veinlet of birefringent material cutting mesostasis—see text, Table 4, and Fig. 12.
57. See Fig. 11.
62A. Norm in Table 7.
63–66. All heterogeneous; average of high and low values for Fe, Ca, and Si used.
66A. Norm in Table 7.
71. Plagioclase host An88.
71A. Norm in Table 7.

work was on 14053, 14310, 15475 and 15555, but similar features were also seen in 15058, 15076, and 15495, as well as in many of 44 additional sections examined at the Lunar Receiving Laboratory.

Excellent examples of immiscibility were also found in 12035,24 and 12036,8, as inclusions of dark, very finely crystalline high-iron melt in late plagioclase, many with globules of colorless high-silica melt (Fig. 6) and some with a shrinkage bubble. The plagioclase ranges from An_{93} to An_{83} away from immiscible melt inclusions (Table 5, # 77 and 78), and as low as An_{75} adjacent to high-silica melt. Five inclusions of high-silica melt (either interstitial or in ilmenite) were studied in 12036,8 (Table 2, # 21–25). Unfortunately, the bulk chemical composition of rock 12036 has not been determined (Warner 1972a).

As in Apollo 11 and 12 and Luna 16 rocks, most of the larger crystals of pyroxene from Apollo 14 and 15 (e.g., 14053, 14310, and 15475) have inclusion-free cores and more Fe-rich, more highly birefringent, inclusion-rich mantles; presumably this change signifies the onset of immiscibility (and hence heterogeneity) in the melt from which they crystallized. Although evidently high in Si and K, these inclusions generally are too small for accurate analysis. The two exceptions are Table 2, # 46 (and its pyroxene host, Table 5, # 86), and a similar inclusion in high-iron pyroxene from a breccia (Table 2, # 39, and its host, Table 5, # 85). This latter pyroxene also contains many globular inclusions of metallic Fe, and still other similar pyroxenes contain globular inclusions of troilite and of Fe, as well as abundant inclusions of high-silica melt, suggesting the simultaneous occurrence of *four* immiscible melts: sulfide, two silicate, and metal. Although some inclusions of sulfide and of metal obviously are in secondary healed fracture planes, many of these are densely spaced in parallel rows throughout the pyroxene crystals, implying a primary origin.

Most of the high-iron melt has crystallized to a fine-grained aggregate (Table 3), as in Apollo 11 and 12. The high-iron melt in 12036 (Table 3, # 55; in plagioclase, Table 5, # 78) has a composition very similar to that reported by Newton *et al.* (1971) for an otherwise very odd composition inclusion in plagioclase from 12040,44 (their analysis 18). Although not so interpreted by these authors, their photograph 1G almost certainly shows a bleb of high-silica melt in devitrified, high-iron melt. In more slowly cooled rocks, the high-iron melt now consists of single crystals of fayalite containing inter-grown, wormy inclusions of high-silica melt (Figs. 7 and 8; Table 1, # 15; Table 5, # 75 and 92). Others may have formed pyroxferroite. In addition, some of this late-stage material in several samples contains blebs of Fe metal and (or) troilite, which may also indicate the simultaneous occurrence of four immiscible melts. Other phases found in the late stage mesostasis include euhedral crystals of apatite low in rare earths ($<0.1\%Ce_2O_3$; $<0.1\%Y_2O_3$), K-Ba feldspar (Table 5, # 88), tridymite, and cristobalite, plus minor amounts of tranquillityite and other unidentified phases. It is difficult to ascribe the origin of some of these phases to one or the other immiscible melt with any certainty, except where there is a gross difference in the content of the element involved, e.g., phosphorus.

In 14053,19 and 14053,57 we analyzed 12 areas of high-silica melt (Table 2, # 26–37), nine interstitial and three in ilmenite. Although no significant differences were found between the two types, there are notable differences between the averages for inclusions in the two slides (# 29A and 37A). One analysis (# 27) shows 3.8% CaO, which is four times higher than the average for the other areas analyzed in this slide. It is adjacent to a relatively calcic plagioclase crystal (An 95.6), but this is not necessarily an equilibrium assemblage. It is important to note that this rock also has semi-euhedral crystals of cristobalite, indicating crystallization of some SiO_2 before the last of the pyroxene and plagioclase. Similar features were seen in Apollo 11 (Roedder and Weiblen, 1970, p. 816), as well as in Apollo 15.

Some of the analyzed melt inclusions compositionally are heterogeneous on a micro scale (e.g. # 30), probably from partial crystallization to fine-grained K-feldspar and SiO_2. This heterogeneity is also visible optically as a fibrous appearance and by gross differences in cathodoluminescence. This heterogeneity probably could account for the poor totals.

Apparently the high-iron melt in 14053 crystallized, as no high-iron glass was found. The mesostasis here consists of major amounts of metallic Fe, troilite, and SiO_2, together with lesser amounts of high-silica melt, apatite, ilmenite, sodic plagioclase, cristobalite, and unidentified phases including possible high-iron melt. (One small stringer of metallic Cu (Ni <0.1, Zn <0.1, and Cd 0.0%) was found in an adjoining area of mesostasis.) Most of the mass consists of a very fine-grained vermicular intergrowth of the first three phases, either metallic Fe and SiO_2 (Table 4, # 72) or troilite and SiO_2 (Table 4, # 73).

El Goresy et al. (1972) and Haggerty (1971 a,b; 1972) have proposed that the intergrowth of metallic Fe and SiO_2 in this sample arises from a reduction of original fayalite. Our analysis of several areas of this intergrowth (in the vicinity of Fig. 12 but more extensive and uniform) agrees fairly well with this interpretation, in that when recalculated to oxides (Table 4), it shows a composition similar to fayalite, but with a molecular ratio of Fe:Si of only 1.5. The other differences are in the minor constituents Mg, Ca, and Mn, all of which are lower than those found in other late-stage lunar fayalites (e.g., Table 5, # 75 and 76). Although we cannot accept Haggerty's suggestion of Fe metal vapor as an adequate reductant to reduce Fe silicate to Fe metal plus SiO_2, the evidence for reduction, both in analysis 72 and in the textures he and El Goresy et al. show, is convincing.

Among the areas of Fe metal-SiO_2 intergrowth, however, are a number of somewhat similar areas (darker in Fig. 12) that are a mixture of stoichiometric FeS and SiO_2 (analysis 73). Gancarz et al. (1971) report similar intergrowths in 14053,17, but with lower SiO_2 and much higher Al_2O_3 and K_2O in the siliceous phase.

Because Haggerty (1971b) has suggested the occurrence of both Fe and S in the reductant gas, we recalculated analysis 73 to an all-oxide, S-free basis (Table 4) under the assumption that this intergrowth merely represents sulfidation of the reduced metallic Fe-SiO_2 intergrowth. The ratio of Fe/Si is only 0.51, far below even ferrosilite (1.0). The suggestion that extra SiO_2 might come from the reduction and sulfidation of a mixture of fayalite plus high-silica melt, a mixture commonly found in other lunar samples (Fig. 8), is negated by the lack of K in this analysis.

Veinlets crosscutting both of these intergrowths seem to consist of round blebs of high-silica melt in high-iron melt (Fig. 12). The blebs are too small for a good analysis but are similar to high-silica melt: SiO_2, 71; Al_2O_3, 11; CaO, 2; MgO, 0;

Fig. 1. Silicate melt inclusion in olivine of 12036,8 that has formed an early epitaxially oriented daughter crystal of ilmenite (bright, in reflected light), then crystallized more olivine on the walls (note that the ilmenite appears embedded), then a parallel epitaxial plate of plagioclase (dark), and finally a dendritic mass of tiny pyroxene crystals. See analyses 1, 74, and 94. Bar = 10 μ.

Fig. 2. Olivine phenocrysts in olivine basalt in 14321,25, each with several large silicate melt inclusions (transmitted light; crossed polarizers). The arrow points to two apparently primary inclusions of metallic Fe and high-silica melt. See also Figs. 25 and 26. Bar = 500 μ.

Fig. 3. Detail on central part of Fig. 2, in reflected light. Note multiple epitaxially oriented ilmenites and fine grained matrix of inner inclusion (right), and coarse, random dendrites of outer inclusion (left). Bar = 10 μ.

Fig. 4. Three small melt inclusions in olivine of 15555,34, in transmitted plain light. Each contains a thin opaque epitaxially-oriented plate of ilmenite, a tiny opaque sphere of immiscible sulfide melt, a shrinkage bubble, and clear basaltic glass. Bar = 10 μ.

Fig. 5. Two small melt inclusions in olivine of 15555,34, in transmitted plain light. The lower one has nucleated ilmenite, plagioclase, and pyroxene (like Fig. 1), but the upper contains only ilmenite, glass, and a shrinkage bubble. Bar = 10 μ.

Fig. 6. Group of melt inclusions in plagioclase of 12035,24 in transmitted plain light. They contain globules of colorless high-silica melt in dark, granular, crystallized high-iron melt. Bar = 100 μ.

Figs. 7 and 8. Large melt inclusion in ilmenite in 15555,34, in plain transmitted (Fig. 7) and reflected light (Fig. 8). The major constituent is fayalite (light gray in Fig. 8), containing blebs of colorless, isotropic, high-silica glass (dark gray) and a bleb of metallic Fe and troilite (bright white). See analyses 15, 75, and 92. Bar = 100 μ.

Figs. 9 and 10. Interstitial inclusion of immiscible melts between plagioclase laths in 14310, 169, in plain transmitted (Fig. 9) and reflected light (Fig. 10). The globule of transparent high-silica melt contains a few acicular crystals. The high-iron melt has crystallized to a fine aggregate of several phases. Another similar but smaller inclusion is visible in the upper left. Bar = 10 μ.

Fig. 11. Similar to Fig. 10 but from 14310,5. High-iron melt (analysis 57) contains three globules of high-silica melt. Reflected light. Bar = 10 μ.

Fig. 12. Mesostasis in 14053,57, in reflected light. A euhedral plagioclase crystal occurs at the upper right. Metallic Fe, low in Ni (bright white) and troilite (very light gray) occur both in globules (as in lower left) and in fine myrmekitic intergrowth with silica (see Table 4). Two veinlets of fayalitic olivine (analysis 56), each with very dark gray globules of high-silica melt cross the intergrowth. Bar = 100 μ.

Fig. 13. Sample 14310, 169 in plain transmitted light with crossed polarizers. Note the radial group of thin plagioclase plates, apparently nucleated by some unknown nucleus. Bar = 100 μ.

Table 4. Microprobe analyses (defocused beam, wt %) of myrmekitic intergrowths of silicate phase with metal and sulfide phases in mesostasis of 14053,57. (See Fig. 12). (Paul W. Weiblen, analyst)

Analysis	72	Metal—silicate intergrowth recalculated analyses			73	Sulfide—silicate intergrowth recalculated analyses		
		Silicate	Metal	Oxide		Silicate	Sulfide	Oxides
SiO_2	39.63	93.5	—	34.45	58.8	97.85	—	61.13
Al_2O_3	1.04	2.5	—	0.90	0.92	1.53	—	0.96
Fe metal	56.09	—	99.57	62.89 (FeO)	28.0	—	65.88	37.53 (FeO)
MgO	1.28	3.0	—	1.11	0.32	0.53	—	0.33
CaO	0.13	0.31	—	0.11	0.01	0.02	—	0.01
Na_2O	0.00	0.00	—	0.00	0.00	0.00	—	0.00
K_2O	0.04	0.09	—	0.03	0.03	0.05	—	0.03
TiO_2	0.08	0.19	—	0.07	0.01	0.02	—	0.01
P_2O_5	0.18	0.42	—	0.16	—	—	—	
MnO	0.00	0.00	—	0.00	0.00	0.00	—	0.00
Cr_2O_3	0.02	0.05	—	0.02	—	—	—	
Ni metal	0.24	—	0.43	0.26 (NiO)	—	—	—	
S	—	—	—	—	14.5	—	34.12	—
ARITHMETIC TOTAL	98.73	100.06	100.00	100.00	102.59	100.00	100.00	100.00

Notes on analyses:
Column
1 Analysis 72 is the average of 4 areas each about 12.5 μ diameter. Range in Fe values 50.2–60.8%.
2 Refers to 42.4% of whole area of analysis 72. Total not 100.00 due to rounding off.
3 Refers to 56.3% of whole area of analysis 72.
4 Entire analysis 72 recalculated to an oxide basis; compare with pure fayalite: 70.5 FeO, 29.5 SiO_2.
5 Analysis 73 is the average of 2 areas each about 12.5 μ diameter, with Fe values of 26.6 and 29.4.
6 Refers to 60.09% of whole area of analysis 73.
7 Refers to 42.5% of whole area of analysis 73. Sulfide calculates to $Fe_{1.11}S$.
8 Entire analysis 73 recalculated to a sulfur-free oxide basis.

Table 5. Microprobe analyses (wt %) of host, associated, and daughter crystals for melt inclusions in Tables 1–4 (Paul W. Weiblen, analyst). (Notes as in Table 1.)

	Olivine				Plagioclase					Pyroxene		
Analysis	74	75	76	77	78	79	80	81	82	83	84	85
Sample	12036,8	15555,34	15555,34	12036,8	12036,8	14162,20	14310,13	14310,13	15555,34	14053,19	14053,19	14303,55
Photo area	18–2	18–2	19–2	17	10–2	10	1–2	4–2	25–2	E'	C	8
SiO_2	37.6	29.7	29.3	49.4	45.7	45.3	47.6	48.3	46.7	48.6	48.9	52.0
Al_2O_3	0.03	0.06	0.04	33.0	32.9	34.3	33.8	34.1	33.8	—	—	1.0
FeO	32.4	64.8	65.9	0.48	0.45	0.21	0.36	0.44	0.54	28.7	38.7	20.8
MgO	29.2	3.8	2.20	—	—	—	0.0	—	—	15.1	1.6	8.9
CaO	0.31	0.40	0.72	16.7	18.7	18.1	16.5	16.6	15.7	7.5	11.6	18.0
Na_2O	—	0.00	—	1.34	0.86	1.01	1.15	0.48	1.19	—	0.0	0.0
K_2O	—	0.00	—	0.36	0.07	0.09	0.36	0.36	0.02	—	—	0.85
TiO_2	0.03	0.80	0.12	0.08	0.08	0.04	—	0.03	0.03	—	—	—
P_2O_5	—	0.13	—	—	—	—	—	—	—	—	—	—
MnO	0.43	0.66	0.75	—	—	—	—	0.0	—	—	—	—
Cr_2O_3	0.13	0.04	0.0	—	—	—	—	—	—	—	—	—
ARITHMETIC TOTAL	100.13	100.39	99.03	101.36	98.76	99.05	99.77	100.31	97.98	99.9	100.8	101.55
Atomic ratios												
O	4.000	4.000	4.000	8.000	8.000	8.000	8.000	8.000	8.000	6.000	6.000	6.000
Si	1.025	0.977	0.988	2.231	2.138	2.108	2.185	2.198	2.176	1.922	2.024	1.987
Ti	0.001	0.020	0.003	0.003	0.003	0.001	—	0.001	0.001	—	—	0.024
Al	0.001	0.002	0.002	1.757	1.815	1.882	1.829	1.829	1.857	—	—	0.045
Cr	0.003	0.001	0.000	—	—	—	—	—	—	—	—	—
Fe	0.739	1.783	1.858	0.018	0.018	0.008	0.014	0.017	0.021	0.949	1.339	0.665
Mn	0.010	0.018	0.021	—	—	—	—	—	—	—	—	—
Mg	1.186	0.186	0.111	—	—	—	—	—	—	0.890	0.099	0.507
Ca	0.009	0.014	0.026	0.808	0.937	0.902	0.811	0.809	0.784	0.318	0.514	0.737
Ba	—	—	—	—	—	—	—	—	—	—	—	—
Na	—	—	—	0.117	0.078	0.091	0.102	0.042	0.108	—	—	—
K	—	—	—	0.021	0.004	0.005	0.021	0.021	0.001	—	—	—
TOTAL CATIONS	2.97	3.00	3.01	4.96	4.99	5.00	4.96	4.92	4.95	4.08	3.98	3.97

Table 5 (continued).

	Pyroxene		K-Spar	Ilmenite				Ulvöspinel	Daughter Crystals			
Analysis	86	87	88	89	90	91	92	93	94	95	96	97
Sample	14310,13	14310,13	12036,8	12036,8	12036,8	14053,19	15555,34	15555,34	12036,8	12036,8	14321,25	14321,25
Photo area	2-2	3-2	17	15-3	13-3	E'	18-3	17-2	18-3	2	29-4	2-1
									Plag.	Ilmenite	Spinel	Spinel
SiO_2	52.9	53.4	61.3	0.28	0.21	0.28	0.10	0.21	49.6	3.8	0.13	0.44
Al_2O_3	1.87	1.69	20.7	0.05	0.16	0.29	0.38	1.75	27.5	0.33	24.6	22.7
FeO	14.7	17.3	0.19	43.2	43.4	46.0	44.7	61.3	2.7	41.5	30.3	30.3
MgO	26.4	23.8	—	2.2	3.0	0.92	0.54	0.0	—	3.6	4.02	4.34
CaO	2.74	2.39	0.77	0.01	0.05	0.02	0.02	0.61	19.0	—	0.12	0.93
Na_2O	0.05	—	0.47	—	—	0.0	—	—	0.25	—	—	—
K_2O	—	—	14.0	—	—	0.11	—	—	0.01	—	—	—
TiO_2	0.72	0.54	—	52.7	53.0	53.4	52.7	33.1	0.55	49.7	3.37	3.38
P_2O_5	—	—	—	—	—	—	—	—	—	—	—	—
MnO	0.60	0.22	—	0.43	0.11	0.46	0.34	0.24	—	—	0.55	0.54
Cr_2O_3	1.92	1.24	—	0.49	0.54	—	0.36	2.64	—	0.34	36.0	36.7
ARITHMETIC TOTAL	101.90	100.58	100.73	99.36	100.47	101.48	99.14	99.85	99.61	99.27	99.09	99.33
Atomic ratios												
O	6.000	6.000	8.000	3.000	3.000	3.000	3.000	4.000	8.000	3.000	4.000	4.000
Si	1.894	1.945	2.870	0.007	0.005	0.007	0.003	0.008	2.313	0.094	0.004	0.014
Ti	0.019	0.015	—	0.989	0.980	0.990	0.999	0.917	0.019	0.920	0.082	0.083
Al	0.079	0.073	1.143	0.001	0.005	0.008	0.011	0.076	1.512	0.010	0.940	0.871
Cr	0.054	0.036	—	0.010	0.010	—	0.007	0.077	—	0.007	0.922	0.944
Fe	0.440	0.527	0.007	0.901	0.983	0.948	0.942	1.889	0.105	0.854	0.821	0.825
Mn	0.018	0.007	—	0.009	0.002	0.010	0.007	0.007	—	—	0.015	0.015
Mg	1.409	1.292	—	0.082	0.110	0.034	0.020	0.000	—	—	0.194	0.211
Ca	0.105	0.098	0.039	0.000	0.001	0.001	0.001	0.024	0.950	0.095	0.004	0.032
Ba	—	—	0.061	—	—	—	—	—	0.023	—	—	—
Na	0.003	—	0.043	—	—	0.003	—	—	0.001	—	—	—
K	—	—	0.836	—	—	—	—	—	—	—	—	—
TOTAL CATIONS	4.02	3.99	5.00	2.00	2.01	2.00	1.99	3.00	4.92	1.98	2.98	3.00

Sample notes:
74. Host for inclusion (analysis 1).
75. Fayalite in inclusion in ilmenite (analysis 92), with high-silica melt inclusions (analysis 15). (Figs. 7 and 8).
76. Fayalite with high-silica melt inclusions (analysis 18).
77. Away from high-silica melt inclusion; An 83. Adjacent to high-silica melt inclusion some is An 75.
78. An 93. Host for inclusion analysis 55.
79. An 90. Host for inclusion analysis 38.
80. An81.
81. An82.
82. An78.
85. Host for melt inclusion analysis 39.
86. Host for melt inclusion analysis 46.
88. BaO—3.3%.
92. Host for inclusion of fayalite (analysis 75) and high-silica glass (analysis 15). See Figs. 7 and 8.
96–97. Both crystals attached to walls of large inclusions in olivine; origin as daughter crystals uncertain.

FeO, 9.5; and K_2O, 3.0. The matrix for these blebs appears to be crystalline, fayalitic olivine, (Fo 37-Fa63; Table 3, #56) and tentatively is considered to be the crystalline equivalent of a high-iron melt. We do not have a satisfactory explanation at this time for this odd texture.

The best displays of immiscible melts were interstitial inclusions of late-stage residual melt in 14310 showing menisci between spherical blebs of colorless high-silica melt (now containing a few minute acicular crystals, presumably feldspar), protruding into almost opaque, high-iron melt (now a very fine grained crystalline aggregate) that fills the interstices between late plagioclase crystals (Figs. 9–11).

In view of the variety of theories that have been put forward to explain the origin of rock 14310, it is interesting to compare its immiscible melt inclusions with other lunar samples. First, the volume of residual immiscible melt is greater than in most other lunar samples examined. Point counts of representative areas of 14310,13 with reflected and transmitted light give the following mode (volume %): high-iron melt, 4.6; high-silica, 7.0; plagioclase, 53.0; pyroxene, 34.0; opaques, 1.4. Assuming the respective densities to be 3.1, 2.2, 2.75, 3.5, and 5.0 g/cm^3, and an average K_2O content of 0.32, 6.63, 0.36, 0.0, and 0.0% for the five phases (analyses 40–48, 57–62, 80–81, and 86–87), the rock should contain a total of 0.53% K_2O. (Approximately two-thirds of this is in the high-silica melt and one-third in the plagioclase.) This is to be compared with a preliminary value of 0.53% K_2O (LSPET, 1971); an analysis by Rose et al. (1972) showing 0.48% K_2O; and an analysis of 0.49% K_2O from a series of integrated microprobe scans on 14310,13. Thus the relatively large amount of residual immiscible melt agrees with the high content of K_2O. It is also important to note that at least 11.6 volume % melt was present when immiscibility occurred in this rock.

The lunar soil itself is relatively high in K, and in a previous paper (Roedder and Weiblen, 1971, p. 523) we suggested that the "KREEP" component proposed for some lunar soils could represent a concentration of the residual immiscible melts plus pyroxene and plagioclase. Thus, the high content of immiscible melts in 14310 is at least compatible with the theory of this rock being a remelted soil (Baedecker et al., 1972), or a KREEP basalt (Hubbard and Gast, 1972).

Second, both melts in this rock are surprisingly variable in composition (Table 2, #40–48 and Table 3, #57–62). Perhaps there is a bulk compositional heterogeneity in this rock on a millimeter scale comparable to the textural and mineralogical heterogeneity seen at a centimeter scale in some photographs of other sections of it. One might be tempted to suggest on this basis that there was only a partial melting of originally heterogeneous solid material, but some of the textures indicate that at least the finer matrix material was liquid. Thus the sunbursts of thin plagioclase crystals (Fig. 13) probably nucleated and grew in a (perhaps local) mass of liquid. It is also obvious that these particular plagioclase crystals must have formed after any mass flowage stopped, as this would be a delicate, easily broken structure. The nucleating agent is unknown, but apparently effective. Further, there are silicate melt inclusions in the outer zones of some of the pyroxene crystals, presumably indicating the onset of immiscibility in the melt from which they crystallized (Table 2, #46).

Third, in spite of the compositional variations mentioned, there are notable consistencies among the analyses, particularly in the MgO content. Both immiscible melts are considerably higher in MgO than equivalent melts in all other lunar samples, and Brown and Peckett (1971) refer to the very high-Mg, "primitive" pyroxenes in this rock, yet it is not high-Mg by lunar standards (MgO – 7.48%, Rose *et al.* 1972). These facts may merely reflect the much lower ratio of FeO/MgO (about 1.0 by weight, Hubbard and Gast, 1972) than most mare basalts. The high-iron melt also is unusually low in SiO_2; it shows 14.1% normative fayalitic olivine (Table 7), whereas other high-iron melts have only 5.7% olivine (Luna 16) or even 3.7% normative quartz (Apollo 11; Roedder and Weiblen 1972a). This also indicates that we are far from knowing yet what is the maximum compositional range of immiscible silicate melts.

The MgO/FeO ratios of the ilmenite host crystals and their melt inclusions suggest a sympathetic variation (data in Tables 1, 2, 5, and 6), indicating at least a local melt/ crystal equilibrium of the type found by Lovering and Widdowson (1968) for ilmenites in terrestrial rocks. Properly calibrated, such analysis pairs could yield temperature data, in analogy to the use of Mg-distribution between ilmenite and pyroxene (Anderson *et al.*, 1972).

One of the most striking features of the lunar magmas is their evolution to yield a high-K residue (immiscible, high-silica melt) by as much as a 300-fold enrichment from a parent magma that contained only a few hundred ppm K. Simultaneously, there are only small increases or decreases in Na (Table 8), so the net result is a large increase in the weight ratio K_2O/Na_2O. The amount of this increase, however, is approximately *inversely* proportional to the value of the ratio in the original bulk rock. Thus, the two samples with the lowest bulk rock ratios (15555 and Apollo 12 average) have the highest ratios in their respective high-silica melt; 14310 is at the other extreme. The variations in Na in the residual melts cause most of the change in ratio. The large differences in Na probably reflect the amount and the temperature at which plagioclase crystallized from the melt.

CLASTS OF GRANITIC COMPOSITION IN APOLLO 14 BRECCIAS

We have suggested (Roedder and Weiblen, 1970, p. 835) that immiscibility could have yielded areas of granitic composition ". . . in parts of the lunar highlands." This suggestion has been rejected by some other workers, and so we must point out that the presence of fragments of granitic composition in many of the Apollo 14 breccias, of a size range several orders of magnitude larger than any of the masses of residual immiscible melt found in the crystalline rocks, yields further corroboration of the concept that materials of granitic composition have been concentrated at or near the moon's surface and, hence, been available for brecciation.

Four of the Apollo 14 breccias available to us (14303, 14319, 14320, and 14321, all presumably throwout from Cone Crater) all contain some clasts enriched in K. Some, consisting of metamorphic assemblages of plagioclase and magnesian pyroxene (Fig. 14; Table 6, #110 and 112), have K-feldspar concentrated into intergranular layers (Fig. 15; Table 6, #105). The overall K content of such clasts would not be high.

Table 6. Microprobe analyses (wt %) of glasses of granitic compositions, and embedded crystals, from K-rich clasts and veins in Apollo 14 breccias. (Paul W. Weiblen, analyst) (Notes as in Table 1.)

Potassium-rich Glass

Analysis	98	99	100	101	102	103	103A
Sample	14320,6	14319,11a	14319,11a	14303,55	14303,55	14303,51	Avg. of 6
Photo area	7-1	2-3	2-1	11-1	12-1	A3	
SiO_2	73.6	74.7	71.7	74.3	74.0	72.2	73.42
Al_2O_3	13.5	14.3	12.7	12.5	11.9	14.3	13.20
FeO	0.80	0.95	2.69	0.84	1.20	1.9	1.40
MgO	0.19	0.0	0.08	0.24	0.08	—	0.12
CaO	0.44	0.43	1.29	0.62	0.58	0.58	0.66
Na_2O	0.51	0.96	0.32	0.34	0.12	1.1	0.56
K_2O	9.66	7.38	9.18	10.3	9.80	9.7	9.34
TiO_2	0.81	0.27	0.92	1.88	1.16	—	1.01
P_2O_5	0.11	0.06	0.20	—	—	—	0.12
MnO	0.0	0.04	0.04	0.0	0.0	—	0.02
Cr_2O_3	—	0.08	0.0	0.0	0.0	—	0.04
BaO	0.44	—	0.0	0.23	0.33	—	0.25
ARITHMETIC TOTAL	100.06	99.17	99.12	101.25	99.17	99.78	100.14

Table 6 (continued).

	Potassium Feldspar					Pyroxene		Ilmenite	Plagioclase		
Analysis	104	105	106	107	108	109	110	111	112	113	114
Sample	14320,6	14319,11a	14303,55	14303,51	14303,51	14319,11a	14319,11a	14319,11a	14319,11a	14303,51	14303,51
Photo area	7-2	8-1	12-2	A-2	A-4	2-2	8-2	2-4	8-3	A-1	A5
SiO_2	62.6	65.1	65.5	63.5	63.1	52.0	49.7	0.38	45.7	51.2	54.4
Al_2O_3	19.0	17.8	18.3	20.3	20.4	1.57	0.77	0.21	35.7	30.6	29.1
FeO	0.09	0.02	0.18	—	—	13.9	15.7	47.0	—	0.2	0.2
MgO	0.0	—	0.0	—	—	11.9	30.1	1.46	—	—	—
CaO	0.46	0.25	0.42	1.0	0.71	18.4	1.90	0.06	17.8	13.7	13.5
Na_2O	0.60	1.04	0.60	1.2	1.4	—	0.0	—	0.93	2.6	3.1
K_2O	14.1	14.8	14.3	14.2	14.0	0.59	—	—	0.20	0.92	1.1
TiO_2	—	—	0.44	—	—	1.07	0.91	50.9	0.11	—	—
P_2O_5	—	—	—	—	—	—	—	—	—	—	—
MnO	—	—	0.0	—	—	0.04	0.24	0.35	—	—	—
Cr_2O_3	—	—	—	—	—	0.19	0.29	0.66	—	—	—
BaO	3.37	1.72	1.04	—	—	—	—	—	—	—	—
ARITHMETIC TOTAL	100.22	100.73	100.78	100.2	99.61	99.66	99.61	101.02	100.44	99.22	101.4

Table 6 (continued).

	Potassium Feldspar					Pyroxene		Ilmenite		Plagioclase	
Atomic ratios	104	105	106	107	108	109	110	111	112	113	114
O	8.000	8.000	8.000	8.000	8.000	6.000	6.000	3.000	8.000	8.000	8.000
Si	2.941	3.009	2.999	2.916	2.912	1.973	1.833	0.009	2.092	2.348	2.437
Ti	—	—	0.015	—	—	0.031	0.025	0.954	0.004	—	—
Al	1.052	0.970	0.988	1.099	1.110	0.070	0.033	0.006	1.926	1.654	1.537
Cr	—	—	—	—	—	0.006	0.008	0.013	—	—	—
Fe	0.004	0.001	0.007	—	—	0.441	0.484	0.980	—	0.008	0.007
Mn	—	—	0.0	—	—	0.001	0.007	0.007	—	—	—
Mg	0.0	—	0.0	—	—	0.673	1.654	0.054	—	—	—
Ca	0.023	0.012	0.021	0.049	0.035	0.748	0.075	0.002	0.873	0.673	0.648
Ba	0.062	0.031	0.019	—	—	—	—	—	—	—	—
Na	0.055	0.093	0.053	0.107	0.125	0.029	0.0	—	0.083	0.231	0.269
K	0.845	0.873	0.835	0.832	0.824	—	—	—	0.012	0.054	0.063
TOTAL CATIONS	4.98	4.99	4.94	5.00	5.01	3.97	4.12	2.03	4.99	4.97	4.96

Sample notes:
113. Area within shell of skeletal ilmenite (analysis 111).
114. Main part of veinlike mass over 5 mm long (Fig. 19).
117. See analyses 107–108.
117A. Norm in Table 7.
119. Occurs as rims on pyroxene crystals (analysis 110) embedded in plagioclase crystals (analysis 112). Figs. 14–15. Rb_2O–0.0 %.
121–122. Rims on plagioclase crystals (analyses 113–114) embedded in glass of analysis 103. (Figs. 16–17).
123. Acicular crystals in glass of analysis 99.
124. Surrounded by K-feldspar (analysis 105) (Fig. 15).
125. Skeletal ilmenite mass in glass (analysis 99) (Fig. 19).
126. An87 (See analysis 105).
127. An67. Core under K-feldspar rims (analyses 107–108).
128. An65. Edge of plagioclase, immediately under K-feldspar rim.

Table 7. CIPW norms (wt %) for inclusion analyses from Tables 1, 2, 3, and 7. (Notes as in Table 1.)

	In olivine			High-silica melts				
Analysis	1	2	3	20A	25A	29A	37A	38
Sample	12036,8	14321,25		15555,34	12036,8	14053,19	14053,57	14162,20
	No. 18–1	Outer	Inner	Avg. of 6	Avg. of 6	Avg. of 4	Avg. of 8	No. 10–2
Q	19.8	13.9	18.5	42.8	37.6	37.3	47.4	32.8
C	—	—	—	1.3	—	1.4	2.0	1.0
or	0.2	0.8	0.8	41.5	40.8	46.4	37.5	44.8
ab	1.4	1.3	4.3	0.5	5.4	3.4	1.2	7.0
an	41.0	39.9	45.1	8.1	8.3	8.9	8.3	8.4
wo	14.3	16.6	6.1	—	1.4	—	—	—
en	8.5	8.7	4.6	0.1	1.4	—	0.1	—
fs	9.7	11.7	16.3	4.1	3.1	0.9	2.1	1.8
fo	—	—	—	—	—	—	—	—
fa	4.6	6.8	4.0	—	—	—	—	—
il	—	—	—	1.4	2.0	1.7	1.4	3.0
ru	0.2	0.2	0.1	—	—	—	—	—
cm	0.3	0.1	0.2	—	—	—	—	—
ap	—	—	—	0.2	—	—	—	1.2
TOTAL	100.0	100.0	100.0	100.0	100.08	100.0	100.0	100.0

Brackets in source: F = {or, ab, an}; P = {en, fs}; ol = {fo, fa}.

Table 7 (continued).

	High-silica melts					High-iron melts			
Analysis	39	45A	48A	54A	103A	55	62A	66A	71A
Sample	14303,55	14310,5	14310,13	15475,16		12036,8	14310,5	15475,16	15555,34
	No. 8	Avg. of 6	Avg. of 3	Avg. of 6	Avg. of 6	No. 10–1	Avg. of 6	Avg. of 4	Avg. of 5
					K-rich glass				
Q	43.5	46.0	34.5	41.2	32.8	4.8	—	0.9	1.6
C	4.7	2.7	0.5	1.4	1.3	—	—	—	—
or	44.1	36.2	45.3	42.6	55.2	0.3	1.9	0.1	3.0
ab	—	4.5	9.3	2.7	4.7	0.7	1.3	—	0.7
an	3.0	5.6	6.9	7.3	2.5	6.4	9.2	8.7	16.1
wo	—	—	—	—	—	21.5	10.5	16.8	20.6
en	0.5	1.4	0.4	—	0.3	12.1	11.2	2.6	5.3
fs	—	2.2	1.7	3.8	0.9	38.4	33.1	63.3	40.0
fo	—	—	—	—	—	—	3.3	—	—
fa	—	—	—	—	—	—	10.8	—	—
il	1.1	1.4	1.0	0.8	1.9	15.0	11.4	7.0	10.3
ru	—	—	—	—	—	—	—	—	—
cm	—	—	—	—	0.1	0.2	0.1	0.1	0.1
ap	3.1	—	—	0.2	0.3	0.6	7.2	0.5	2.3
TOTAL	100.0	100.0	100.0	100.0	100.0	100.0	100.0	100.0	100.0

Brackets in source: F = {or, ab, an}; P = {en, fs}; ol = {fo, fa}.

Table 8. Comparison of K_2O/Na_2O weight ratio in residual high-silica melts with that of the parent bulk rock. Figures in parentheses are the numbers of analyses averaged. Arranged in order of increasing melt ratio.

Sample	Residual Melt			Bulk Rock		
	K_2O	Na_2O	K_2O/Na_2O	K_2O	Na_2O	K_2O/Na_2O
Luna 16	5.0	0.85	5.9(19)[1]	0.15	0.38	0.39(1)[2]
14310,13	7.67	1.10	7[3]	0.53	0.63	0.84(1)[4]
12036,8	6.86	0.63	11(5)[3]		(unavailable)	
14310,5	6.11	0.53	12[3]	0.53	0.63	0.84(1)[4]
Apollo 14 K-rich glasses	9.34	0.56	17[3]		(unavailable)	
Apollo 11, avg.	6.5	0.37	18(33)[5]	0.14	0.55	0.25(8)[6]
14053,19	7.85	0.40	20[3]	0.14	0.38	0.37(1)[4]
15475,16	7.13	0.32	22[3]		(unavailable)	
14053,57	6.34	0.14	45[3]	0.14	0.38	0.37(1)[4]
Apollo 12, avg.	6.7	0.14	48(15)[7]	0.065	0.45	0.14(9)[8]
15555,34	7.26	0.09	81[3]	0.03	0.24	0.12(1)[9]

References: [1]Roedder and Weiblen, 1972a; [2]Vinogradov, 1971; [3]this report; [4]LSPET, 1971; [5]Roedder and Weiblen, 1970; [6]LSPET, 1969; [7]Roedder and Weiblen, 1971; [8]LSPET, 1970; [9]LSPET, 1972.

Other clasts essentially are glass of granitic composition, with much higher K. Six analyses of larger clasts are given in Table 6 (#98–103). Figure 16 is typical—a rounded mass of glass, with a composition very similar to the high-silica melt (analysis 103) embedded in a breccia matrix. Two plagioclase crystals (An 67; analysis 113) protruded into the glass from the edge of the clast. They presumably have reacted with the glass, as their outer surfaces have a narrow zone of K-feldspar (Fig. 17; analysis 107). The contact between the feldspars is extremely sharp, and the plagioclase immediately beneath the rim is still An65 (analysis 114). The glass contains numerous acicular crystals and has partly devitrified to form a brownish haze (in transmitted light); this is presumed to be K-feldspar from its similar cathodoluminescence (Fig. 17).

Other similar clasts (Fig. 18) contain even more sodic plagioclase (An56) and high-silica glass (SiO_2, 72.4; Al_2O_3, 13.2; and K_2O, 8.8%). The glass of these clasts has an unfortunate tendency to pluck out in sectioning (e.g., Fig. 18). As these particular breccia samples were metamorphosed (they are classed as "high facies" by Warner, 1972b), the clasts have had an opportunity to react with the surrounding breccia, yielding a network of crystals as a halo around the borders. Many have formed acicular crystals of augitic pyroxene (e.g., analysis 109), occasional crystals of apatite, and frequently larger tablets of new plagioclase (analysis 112) and K-feldspar (analyses 104–108). The composition of the glass, however, as given by analyses 98–103, was obtained on essentially crystal-free portions, so the high-K contents found are not attributable to accidentally included K-feldspar crystals.

Similar granitic clasts have been reported by numerous other workers and some give an indication of their abundance. Three groups report finding only one such clast each. Anderson et al. (1972) report 0.7 areal % granitic, rhyolitic, or granophyric fragments in 14036. In 14321 Grieve et al. (1972) report nine such fragments and Meyer (1972) reports six. However, none of these report analyses with as much K_2O as we found (avg. 9.34 wt%). Reid et al. (1972) report 1% of K-granite fragments in Apollo 14 samples, and Glass (1972) reports 1.6% such glass in 14257,26. The Apollo Soil Survey (1971) also reports 1.6% K-granite glass in soil 14259,26.

Because an immiscible, high-silica, high-K residual melt has been found at all the lunar sites, in a variety of rock types, we suggest that segregations of this late melt may be the source for the granitic clasts in the Apollo 14 breccias. Thus it is reasonable to expect the occurrence of late granophyric segregations in the anorthositic crust. The plagioclase in this type of segregation would be sodic, as was found. Similar plagioclase was also found in another high-K clast that is presumed to be a late segregation (Powell and Weiblen, 1972).

Metamorphism apparently was adequate in 14319 to cause a (presumed) clast of granitic composition material to flow into a veinlet over 5 mm long (Fig. 19), with an odd skeletal ilmenite in one end (analysis 111). The bulk of this veinlet has a composition given by analysis 100, with a rather high FeO content (2.69%). The glass within this ilmenite had a much lower FeO content (0.95%; analysis 99), so this ilmenite probably was quenched in the process of growing from the glass, rather than dissolving.

THERMAL METAMORPHISM OF THE APOLLO 14 BRECCIAS

Breccias 14303, 14319, and 14320 show much evidence of a complex thermal history, that must be involved in any evaluation of the silicate melt inclusions. At least some of the original rocks have apparently undergone slow cooling, as there are numerous very coarse exsolution textures in the pyroxene fragments, much coarser than we have seen in Apollo 11, 12, or Luna 16. Figure 20 is of a subcalcic augite with exsolution lamellae of intermediate calcium content broad enough to see with a handlens. (This crystal also shows another interesting feature visible in many of these breccias—a thorough rounding of individual mineral grains. Some plagioclase single crystal clasts are almost spherical, presumably by abrasion in transport). Figure 21 is of another pyroxene, with several different sets of coarse lamellae. The nearly horizontal set that is at extinction also has plates of a transparent, dark-purplish-red mineral, presumably ilmenite or spinel, marking part of their position. Figure 22 shows similar exsolved plates in a pigeonitic pyroxene that were sufficiently coarse to be verified as ilmenite by microprobe. These exsolution textures are common in slowly cooled terrestrial metamorphic and intrusive rocks; but as they are rare or lacking in the Apollo 11 and 12 basalts, it is reasonable to assume that the Apollo 14 materials have been excavated from deeper below the lunar surface. Evidence from the pleonaste spinels in the Apollo 14 breccias (Roedder and Weiblen, 1972b) confirms this idea.

These samples have been subjected to subsequent metamorphism, and most of the changes could be studied in the laboratory to provide rate data. Thus in many samples, all former glass fragments have crystallized except the glass of K-granite composition mentioned above; such anhydrous compositions have long been known to have extremely high viscosity and low crystallization rates. The flowage of such glass into veinlets (Fig. 19) might also be used to place some limits on the time-temperature conditions of metamorphism. The groundmass glass has been converted to a fine granular crystalline aggregate that varies in grain size from sample to sample. Fragments of shocked plagioclase have recrystallized to a mosaic of tiny plagioclase crystals, commonly elongated with a subparallel physical and optical orientation. Some twinned plagioclase crystals (e.g., Fig. 23) show gross differences in the amount and perfection of crystallinity in the two parts of multiply-twinned crystals after partial recrystallization. In this example the one set of twin bands (bright) extinguishes as a unit and has near normal birefringence, whereas the other set is a subparallel granular mass, reflecting different degrees of original shock damage.

Metamorphism also has resulted in reaction of the groundmass with some of the clasts. The groundmass of 14303, for example, must have excess (normative) SiO_2 in it, as all olivine grains in contact with it (other than fayalite) have reacted to form thin almost monomineralic rims of pyroxene.* In Fig. 24 the olivine ($Fo_{57}Fa_{42}La_{0.4}$)

*A bulk analysis of 14303,34 by Rose *et al.* (1972) yields 4.5 wt% normative olivine, but much of this is in ultramafic clasts.

Fig. 14. Relatively high-K clast in breccia 14319,11a in plain transmitted light. Pyroxene crystals are set in a matrix of plagioclase, with interstitial films of K-feldspar (see Fig. 15). Bar = 100 μ.

Fig. 15. K-Kα scanning photo of a small area from upper left of Fig. 14, showing rims of K-feldspar (analysis 105) surrounding square crystal of pyroxene (analysis 110) set in plagioclase (analysis 112). Bar = 10 μ.

Fig. 16. Rounded clast of glass of granitic composition from breccia 14303,51 (analysis 103), with two plagioclase crystals protruding from wall (analyses 113–114), in transmitted light. Each of these has a narrow rim of K-feldspar (see arrow) where in contact with the glass. Bar = 100 μ.

Fig. 17. Cathodoluminescence photograph of upper part of clast shown in Fig. 16; the K-feldspar rims luminesce bright blue, as do the minute devitrification crystals in the glass (brownish in transmitted light). Bar = 10 μ.

Fig. 18. Clast in breccia 14303,51 consisting of a mass of glass of granitic composition (K_2O–8.8%) containing acicular crystals of pyroxene and K-feldspar, attached to a crystal of rather sodic plagioclase (An 56). Partly crossed polarizers. A part of the glass has plucked out in sectioning. Bar = 10 μ.

Fig. 19. "Veinlet" of brown glass of granitic composition 5.4 mm long (analyses 99–100), in breccia 14319,11a, photographed in plain red light. A circular skeletal ilmenite (analysis 111) and many needles of pyroxene (analysis 109) are embedded in the glass. Bar = 1 mm.

Fig. 20. Very coarse exsolution lamellae in rounded pyroxene clast from breccia 14303,55 in transmitted light, crossed polarizers. Note that the formerly almost circular clast has been fractured in two places and offset. Bar = 100 μ.

Fig. 21. Very coarse exsolution lamellae in pyroxene clast from breccia 14319,11c, in transmitted light, crossed polarizers. The horizontal set (almost at extinction) is also the locus of an interrupted set of thin, purplish-red tablets (ilmentine or spinel?). Bar = 100 μ.

Fig. 22. Thin, transparent, purplish-red tabular exsolution platelets of ilmenite in pyroxene clast in breccia 14303,55, in plain transmitted green light. The trace of these at the surface is inclined 12° to the cleavage fractures. Bar = 10 μ.

Fig. 23. Twinned, shocked, and partly recrystallized plagioclase crystal in breccia 14303,55 in which one part of the twin (bright) has suffered less damage (or more recrystallization) than the other. Partly crossed polarizers. Bar = 100 μ.

Fig. 24. Olivine clast in breccia 14303,55, with a thin reaction rim of almost pure pyroxene, in transmitted light with crossed polarizers. Bar = 100 μ.

Fig. 25. Detail of olivine in center of Fig. 2, in transmitted plain light, showing decorated dislocations. Bar = 100 μ.

Fig. 26. Detail of part of Fig. 25, showing minute high-index, high-Cr crystals along dislocations. Some have dark centers (vacuoles?). Bar = 10 μ.

Fig. 27. Cr-pleonaste clast in breccia 14303,55 in plain reflected light (upper half) and transmitted light (lower half). Transmitted-light color is purplish-pink, becoming deeper at the margins due to compositional zoning. A corona of almost pure plagioclase, free of all pyroxene and opaque minerals (best seen in upper half) separates the grain from the breccia matrix. A tiny sphere rich in Ni, S, and Fe is embedded in the edge of the pleonaste at top; another pleonaste shows a similar sphere *within* the pleonaste (Roedder and Weiblen, 1972b). Bar = 100 μ.

is rimmed with low-calcium hypersthene having about equal weight % of MgO and FeO. The excess SiO_2 in the groundmass can be presumed to stem from a concentration of the late-stage, high-silica melt in these materials, as is expected from the areas of K concentration found in x-ray scans.

Many olivine grains in 14321 are more highly birefringent along their margins and along fractures; these areas have as much as 20 wt% more FeO than the unaffected parts. This apparently has diffused in from the groundmass. On the other hand, some high-Fe pyroxene crystals have lost iron to the matrix. Another evidence of heat treatment is seen within many olivine clasts where tiny (1μ) Cr-rich crystals with moderate Al (presumably a spinel) have precipitated out to decorate abundant dislocation loops. Thus, Figs. 25 and 26 show the olivine phenocryst of Fig. 2 at moderate and high magnification. Such decorations were made to form in Apollo 11 olivines by laboratory heating (Roedder and Weiblen, 1970, p. 810), and hence also may provide a tool for experimental time-temperature studies. Young (1969) has decorated dislocations in terrestrial olivine in the laboratory by diffusion of metal. Regardless of the mechanism of decoration, however, such dislocation studies may be useful in understanding the deformational history of the lunar olivines (Wegner and Grossman, 1969).

Perhaps the most striking reaction rims occur about grains of hercynitic or pleonaste spinels (Fig. 27). These have reacted with the groundmass, and there has been considerable transfer of materials as they not only have a zone of different composition spinel wherever they are in contact with the breccia matrix (Roedder and Weiblen, 1972b), but they have also developed a rim of almost pure plagioclase at this contact, free of pyroxene and opaque minerals. We believe that this spinel has been excavated from a deep source. It contains some minute spherical globules rich in Ni, Fe, and S; one of these is visible at the edge of the spinel in the upper part of Fig. 27.

THE LIQUID LINE OF DESCENT FOR LUNAR MAGMAS

Despite the complexity of lunar geology, a limited number of magma series currently are being implied or explicitly postulated by other workers (Hinners, 1971; Gast and McConnell, 1972): mare basalts, KREEP basalts, and magmas related to the formation of the anorthositic crust (Wood et al., 1970). At this point the genetic relationships in space and time can only be crudely outlined. Although differences among mare basalts, e.g., the high- and low-K Apollo 11 basalts, and the Apollo 11 versus the Apollo 12 mare basalts, have been recognized, their genetic relationships are not at all clear. However, some possible petrogenetic relationships have been suggested (Biggar et al., 1971).

In a cogenetic rock series, the early melt inclusions in olivine, the later melt inclusions in pyroxene, plagioclase, ulvöspinel, and ilmenite, and the late-stage interstitial immiscible melt inclusions should delineate a unique liquid line of descent (with its split into two lines at the onset of immiscibility). Thus melt inclusion data can provide valuable constraints on postulated magma series and their interrelationships. The ultimate usefulness of such data can only be realized when a relatively complete sequence of compositions along the liquid line of descent are found in a single rock or in several rocks that are genetically related. In addition, the difficulty of obtaining good data from very small melt inclusions with daughter minerals raises severe sampling requirements (Roedder and Weiblen, 1971).

Because we have only limited data at this point, we present a summary in the form of a single silica variation diagram (Fig. 28) for all inclusions studied, from the various igneous rocks sampled at the five landing sites. In a crude way, this summary suggests that there are common, dominant controls on the liquid line of descent for the wide variety of rocks sampled: olivine plus pyroxene crystallization in the range 40–55% SiO_2, plagioclase and pyroxene in the range 55–65, immiscibility 65–72, and K-feldspar 72–82%. The spread of points is much narrower, of course, if inclusions from a single rock are plotted (e.g., analyses 4–20 from 15555, but excluding those in ilmenite). Eventually, we hope to delineate unique lines of descent within the envelopes sketched in Fig. 28.

The most obviously deviant points in Fig. 28 are mainly from inclusions in ilmenite. In the previous work it was found that most melt inclusions in ilmenite are normal high-silica melt, but *some* are very high-K_2O compared to their SiO_2 content, and hence fall far above the bulk of the points on the K_2O plot. Anderson (1971) has proposed that similar high-K_2O inclusions in armalcolite in Apollo 11 rocks indicate the armalcolite to be exotic, having crystallized from a high-K_2O melt, and then been incorporated in this magma. As the viscosity of such an anhydrous magma would be high, removal of crystals of ilmenite (or armalcolite) would be difficult. As a large part of the parent magma would presumably be crystalline at this point, the only feasible mechanism would be to incorporate and disperse some of this crystalline mush in the new, lower silica magma. However, the compositions of some of these inclusions in ilmenite do not fit anywhere in the diagram. We have no satisfactory explanation of why this exotic high-K, low-SiO_2 parent magma, if that is indeed the source, has never been recognized in any of the lunar samples. Anderson and Wright (1972) report similar anomalously fractionated glass inclusions (but with far smaller alkali enrichment) in oxide phenocrysts in Kilauean lavas. They ascribe these to the mixing of magmas with different degrees of fractionation. Another possibility is that rapid growth of the ilmenite resulted in kinetic disequilibrium at the growing face (Albarede and Bottinga, 1972), thus causing the trapping of a non-representative sample of melt.

We have only two samples, both in ilmenite and presumably discordant, in the range of SiO_2 values where the K_2O content starts to rise abruptly. We believe that this lack of samples is a consequence of the initiation of immiscibility at this point; most melts in this range have presumably split to form high-silica melt and high-iron melt. The latter would have a composition falling off to the left of the diagram but is rarely preserved. The former commonly is preserved as inclusions, and the gap between its composition and that of the earlier melts ($<64\%$ SiO_2) is a logical consequence of removal from it of material of high-iron melt composition.

Particularly notable is the fact that the glass compositions for the granitic clasts in the breccias fall very close to those of the residual, immiscible high-silica melts. The differences are minor, except for K_2O, and even this may be simply a matter of degree of crystallization of K-feldspar prior to quenching to glass. Most of the final trend toward decreasing K_2O with higher SiO_2 visible in Fig. 28 may be a result of

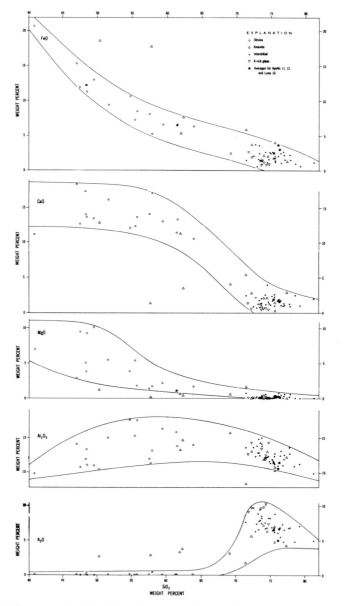

Fig. 28. Summary of melt inclusion data presented in the form of a silica variation diagram for all silicate melt inclusions analyzed by Roedder and Weiblen (1970, 1971, 1972a, and the present report), from all five sites. These include inclusions in host crystals of olivine and ilmenite, and interstitial inclusions, between crystals. Analyses of melt inclusions in olivine do not include large single daughter crystals of plagioclase or ilmenite (except when homogenized prior to analysis). Note that only the *averages* of Apollo 11, Apollo 12, and Luna 16 high-silica melts are plotted; the number of analyses averaged are 33, 15, and 19 respectively. K-rich glass refers to glass from Apollo 14 breccia clasts (Table 6). The envelopes are sketched in to include all points except a few aberrant inclusions in ilmenite (see text) and one low-calcium inclusion in olivine.

this crystallization. Hence we suggest that these clasts of granitic composition in the Apollo breccias could represent rhyolite, granophyre, or granite, produced by liquid immiscibility and concentrated in the Apollo 14 source rocks.

It is particularly important to note that the composition of all these late residual melts are very close to each other, and fall far from the "granite minimum" on "petrogeny's residual system" (albite-orthoclase-silica), as they rarely have even as much as 5% normative albite (Table 7). Apparently the bulk composition of these lunar magmas is such that crystallization of a large amount of plagioclase maintains a high K/Na ratio. This may assist in achieving immiscibility, as the low-temperature immiscibility volume in the quaternary system K_2O-FeO-Al_2O_3-SiO_2 (Roedder, 1953) has been found to be reduced or eliminated by almost all atomic substitutions tried, including that of replacing one half of the K ions with Na in a composition lying in the center of the volume (Roedder, unpublished data).

SUMMARY

1. Melt inclusions in olivine have compositions that presumably delineate a portion of the liquid line of descent for their parent magma.

2. Nucleation and growth of daughter phases (epitaxial ilmenite and plagioclase, pyroxene, and immiscible sulfide melt) in melt inclusions in olivine, similar to that described for Apollo 11 and 12, can make inclusion analyses difficult to interpret; such problems are best eliminated by homogenization before analysis.

3. Most Apollo 14 and 15 samples show late-stage silicate liquid immiscibility, as in Apollo 11 and 12 and Luna 16.

4. The onset of immiscibility commonly is marked by inclusion-rich mantles on otherwise almost inclusion-free pyroxenes, and by a gap with almost no data points in a SiO_2 variation diagram for the melt inclusion compositions.

5. Immiscibility occurred in 14310 while at least 11.6 volume % melt was still present, thus making feasible the possibility of a physical separation of the two immiscible melts; the compositions of these two melts are the most divergent from the averages of those in all other lunar samples.

6. There is an inverse relationship between the K_2O/Na_2O ratio of the bulk rock and its residual high-silica melt, probably related to the amount and the temperature at which plagioclase crystallized from the melt.

7. At the time of separation of the two immiscible silicate melts, there apparently were two other immiscible melts, metallic Fe and iron sulfide, making *four* simultaneously immiscible melts.

8. The MgO/FeO ratio of the ilmenite host crystals and their melt inclusions vary sympathetically, and could provide a measure of temperature of equilibration.

9. Clasts of granitic glass, in part crystallized to K-feldspar, occur in several breccias from near Cone Crater; the residual glass in these fragments has a composition similar to the high-silica melt and presumably originated from an earlier separation via immiscibility, possibly as late granophyric segregations in the anorthositic crust.

10. Some of the breccia clasts have undergone very slow cooling, and may have originated rather deep in the lunar crust; subsequent metamorphism of some of the

breccias has resulted in a variety of changes, many of which might be useful for rate studies in the laboratory.

11. In spite of the many variables involved, the compositions of all the silicate melt inclusions studied so far, in all minerals from all five sites, show grossly similar liquid lines of descent.

12. Numerous features remain puzzling or inexplicable. These include: metallic Fe and high-silica melt inclusions in early olivines; veinlets of fayalitic olivine and intergrowths of SiO_2 and troilite in 14053; gross textural differences in adjacent inclusions in one olivine; and high-K, low SiO_2 melt inclusions in some ilmenites.

Acknowledgments—We are indebted to James Kavenaugh for some of the data reduction, to H. E. Belkin for point counting and some excellent slide preparations and to several principal investigators for permission to examine their slides. We also want to thank the staff of the Lunar Receiving Laboratory for their expeditious handling of requests for exchanges of slides. The manuscript profited from technical review by Thomas L. Wright and Rosalind Helz.

REFERENCES

Albarede F. and Bottinga Y. (1972) Kinetic disequilibrium in trace element partitioning between phenocrysts and host lava. *Geochim. Cosmochim. Acta* **36**, 141–156.

Anderson A. T. Jr. (1971) Exotic armalcolite and the origin of Apollo 11 ilmenite basalts. *Geochim. Cosmochim. Acta* **35**, 969–973.

Anderson A. T. Jr. Braziunas T. F. Jacoby J. and Smith J. V. (1972) Breccia populations and thermal history: Nature of pre-Imbrian crust and impacting body (abstract). In *Lunar Science—III* (editor C. Watkins), pp. 24–26, Lunar Science Institute Contr. No. 88.

Anderson A. T. and Wright T. L. (1972) Phenocrysts and glass inclusions and their bearing on oxidation and mixing of basaltic magmas, Kilauea volcano, Hawaii. *Amer. Mineral.* **57**, 188–216.

Apollo Soil Survey (1971) Apollo 14: Nature and origin of rock types in soil from the Fra Mauro formation. *Earth Planet Sci. Lett.* **12**, 49–54.

Baedecker P.A. Chou C.-L. Kimberlin J. and Wasson J. T. (1972) Trace element studies of lunar rocks and soils (abstract). In *Lunar Science—III* (editor C. Watkins), pp. 35–37, Lunar Science Institute Contr. No. 88.

Biggar G. M. O'Hara M. J. Peckett A and Humphries D. J. (1971) Lunar lavas and the achondrites: Petrogenesis of protohypersthene basalts in the maria lava lakes. *Proc. Second Lunar Sci. Conf., Geochim. Cosmochim. Acta,* Suppl. 2, Vol. 1, pp. 617–643. M.I.T. Press.

Brown G. M. and Peckett A. (1971) Selective volatilization on the lunar surface: Evidence from Apollo 14 feldspar-phyric basalts. *Nature* **234**, 262–266.

El Goresy A Ramdohr P. and Taylor L. A. (1972) The geochemistry of the opaque minerals in Apollo 14 crystalline rocks. *Earth Planet. Sci. Lett.* (in press).

Gancarz A. J. Albee A. L. and Chodos A. A. (1971) Petrologic and mineralogic investigation of some crystalline rocks returned by the Apollo 14 mission. *Earth Planet. Sci. Lett.* **12**, 1–18.

Gast P. W. and McConnell R. K. Jr. (1972) Evidence for initial chemical layering of the moon (abstract). In *Lunar Science—III* (editor C. Watkins), pp. 289–290, Lunar Science Institute Contr. No. 88.

Glass B. P. (1972) Apollo 14 glasses (abstract). In *Lunar Science—III* (editor C. Watkins), pp. 312–314, Lunar Science Institute Contr. No. 88.

Grieve R. McKay G. Smith H. and Weill D. (1972) Mineralogy and petrology of polymict breccia 14321 (abstract). In *Lunar Science—III* (editor C. Watkins), pp. 338–340, Lunar Science Institute Contr. No. 88.

Haggerty S. E. (1971a) Compositional variations in lunar spinels. *Nature Phys. Sci.* **233**, 156–160.

Haggerty S. E. (1971b) Subsolidus reduction of lunar spinels. *Nature Phys. Sci.* **234**, 113–117.

Haggerty S. E. (1972) Subsolidus reduction and compositional variations of lunar spinels (abstract). In *Lunar Science—III* (editor C. Watkins), pp. 347–349, Lunar Science Institute Contr. No. 88.

Hinners N. W. (1971) The new moon: A view. *Rev. Geophys. Space Phys.* **9**, 447–522.

Hubbard N. J. and Gast P. W. (1972) Chemical composition of Apollo 14 materials and evidence for alkali volatilization (abstract). In *Lunar Science—III* (editor C. Watkins), pp. 407–409, Lunar Science Institute Contr. No. 88.

Lovering J. F. and Widdowson J. R. (1968) The petrological environment of magnesium ilmenites. *Earth Planet. Sci. Lett.* **4**, 310–314.

LSPET (1969) (Lunar Sample Preliminary Examination Team) Preliminary examination of lunar samples from Apollo 11. *Science* **165**, 1211–1227.

LSPET (1970) (Lunar Sample Preliminary Examination Team) Preliminary examination of lunar samples from Apollo 12. *Science* **167**, 1325–1339.

LSPET (1971) (Lunar Sample Preliminary Examination Team) Preliminary examination of lunar samples from Apollo 14. *Science* **173**, 681–693.

LSPET (1972) (Lunar Sample Preliminary Examination Team) The Apollo 15 lunar samples: A preliminary description. *Science* **175**, 363–375.

Meyer C. Jr. (1972) Mineral assemblages and the origin of non-mare lunar rock types (abstract). In *Lunar Science—III* (editor C. Watkins), pp. 542–544, Lunar Science Institute Contr. No. 88.

Newton R. C. Anderson A. T. and Smith J. V. (1971) Accumulation of olivine in rock 12040 and other basaltic fragments in the light of analysis and syntheses. *Proc. Second Lunar Sci. Conf., Geochim. Cosmochim. Acta*, Suppl. 2, Vol. 1, pp. 575–582. M.I.T. Press.

Powell B. N. and Weiblen P. W. (1972) Petrology and origin of rocks in the Fra Mauro formation. *Proc. Third Lunar Sci. Conf., Geochim. Cosmochim. Acta*, Suppl. 3, Vol. 0, p. 0000–0000. M.I.T. Press.

Reid A. M. Ridley W. I. Warner J. Harmon R. S. Brett R. Jakes P. and Brown R. W. (1972) Chemistry of highland and mare basalts as inferred from glasses in lunar soils (abstract). In *Lunar Science—III* (editor C. Watkins), pp. 640–642, Lunar Science Institute Contr. No. 88.

Roedder E. (1953) Liquid immiscibility in the system K_2O-FeO-Al_2O_3-SiO_2 (abstract). *Bull. Geol. Soc. Amer.* **64**, 1466.

Roedder E. and Weiblen P. W. (1970) Lunar petrology of silicate melt inclusions, Apollo 11 rocks. *Proc. Apollo 11 Lunar Sci. Conf., Geochim. Cosmochim. Acta*, Suppl. 1, Vol. 1, pp. 801–837. Pergamon.

Roedder E. and Weiblen P. W. (1971) Petrology of silicate melt inclusions, Apollo 11 and Apollo 12 and terrestrial equivalents. *Proc. Second Lunar Sci. Conf., Geochim. Cosmochim. Acta*, Suppl. 2, Vol. 1, pp. 507–528. M.I.T. Press.

Roedder E. and Weiblen P. W. (1972a) Silicate melt inclusions and glasses in lunar soil fragments from the Luna 16 core sample. *Earth Planet. Sci. Lett.* **13**, 272–285.

Roedder E. and Weiblen P. W. (1972b) Occurrence of chromian, hercynitic spinel ("pleonaste") in Apollo 14 samples and its petrologic implications. Accepted for publication by *Earth Planet. Sci. Lett.*

Rose H. J. Jr. Cuttitta F. Annell C. S. Carron M. K. Christian R. P. Dwornik E. J. and Ligon D. T. Jr. (1972) Compositional data for fifteen Fra Mauro lunar samples (abstract). In *Lunar Science—III* (editor C. Watkins), pp. 660–662, Lunar Science Institute Contr. No. 88.

Vinogradov A. P. (1971) Preliminary data on lunar ground brought to Earth by automatic probe "Luna 16." *Proc. Second Lunar Sci. Conf., Geochim. Cosmochim. Acta*, Suppl. 2, Vol. 1, pp. 1–16. M.I.T. Press.

Warner J. L. (1972a) Continuing compilation of Apollo chemical, age, and modal information. MSC Curator's Office (unpublished).

Warner J. L. (1972b) Apollo 14 breccias: Metamorphic origin and classification (abstract). In *Lunar Science—III* (editor C. Watkins), pp. 782–784, Lunar Science Institute Contr. No. 88.

Wegner M. W. and Grossman J. J. (1969) Dislocation etching of naturally deformed olivine (abstract). *Am. Geophys. Union Trans.*, EOS **50**, No. 11, 676.

Wood J. A. Dickey J. S. Jr. Marvin U. B. and Powell B. N. (1970) Lunar anorthosites and a geophysical model of the moon. *Proc. Apollo 11 Lunar Sci. Conf., Geochim. Cosmochim. Acta*, Suppl. 1, Vol. 1, pp. 965–988. Pergamon.

Young C. III (1969) Dislocations in the deformation of olivine. *Amer. J. Sci.* **267**, 841–852.

Proceedings of the Third Lunar Science Conference
(Supplement 3, *Geochimica et Cosmochimica Acta*)
Vol. 1, pp. 281–294
The M.I.T. Press, 1972

Uranium and potassium fractionation in pre-Imbrian lunar crustal rocks

J. F. Lovering,* D. A. Wark, A. J. W. Gleadow, and D. K. B. Sewell

School of Geology, University of Melbourne, Parkville
Victoria, Australia 3052

Abstract—Uranium and potassium abundance data from rock clasts in recrystallized Apollo 14 breccia 14305 indicate that three uranium-potassium fractionation groups can be recognized:

Group 1 Low U (<0.5 ppm)–low K ($<0.3\%$), composed of mare-type basalts (subgroup A) and anorthosites (subgroup B).

Group 2 Medium U (~1.6–4.8 ppm)–medium K (~0.4–2.3%), composed of nonmare type basalts and "norites."

Group 3 High U (~5–12 ppm)–high K (~3.9–6.2%), composed of "monzonitic" rocks (subgroup A) and glasses (subgroup B) probably derived from 3A rocks.

This rock association is suggested to be typical of the near-surface crust of the moon in pre-Imbrian times (i.e., $>3.9 \times 10^9$ years ago) and shows marked similarities to rock associations characteristic of Precambrian massif-type anorthosite terrains on earth.

Introduction

It seems generally agreed that most of the complex series of brecciated rocks returned from the Fra Mauro site of the Apollo 14 mission represent material ejected from the first 10 km of the pre-Imbrian lunar crust by the impact of the Imbrian planetesimal some 3.9×10^9 years ago (LSAPT, 1972). On this basis, the rock clasts found in the most texturally primitive of the Fra Mauro breccias should provide us with an irresistible opportunity to construct generalized petrological models for the near surface crust of the moon as it existed before 3.9×10^9 years ago.

The proposition that rock clast types observed in the Apollo 14 breccias may be used to construct such petrological models is complicated by uncertainties concerning the original source of the clasts. Some workers (e. g., Wilshire and Jackson, 1972) have stressed the complex nature of clasts in the breccias and the possibility that the Fra Mauro breccias were derived from an area that had already undergone a long history (over a half billion years) of impact in which the Imbrian event itself was simply the last large event. If this is the case, then the clasts themselves originally may have been derived from any portion of the moon's surface as it existed before the Imbrian event and their petrological and chemical variance can be taken as evidence of widespread differentiation of the lunar surface prior to 3.9×10^9 years ago. Another interpretation offered by von Engelhardt *et al.* (1972) and von Engelhardt (personal communication), largely on the basis of complex clast structures observed in breccias associated with the single terrestrial Ries event, is that the complex clasts can be derived by mixing two hot masses or melts produced from two different source rock units by the same impact.

---

*Also at Max-Planck-Institut für Kernphysik, 69 Heidelberg, Germany.

If this interpretation is correct then the variance of the clasts in the Fra Mauro breccias indicates the local differentiation present in the early lunar upper crust as it existed at the site of the Imbrian event. In either case a study of the clasts should provide considerable insight into the overall chemical differentiation present in the outer lunar crust before 3.9×10^9 years.

Several groups have proposed petrological models of the pre-Imbrian lunar crust on the basis of the clast types in the Apollo 14 soils and breccias (e.g., Anderson *et al.*, 1972; Meyer, 1972; Reid *et al.*, 1972). Although there is a common thread running through these models, each one proposed differs in detail depending on the criteria adopted by each group for the recognition of various clast types present in the breccias. Other difficulties arise in clast petrology studies with the overprinting of shock effects, and shock vitrification and recrystallization textures, which add considerable uncertainty to petrological recognition of important clast types.

Our approach has been to examine the clast types present in Apollo 14 breccias on the basis of important chemical criteria, and for this purpose we have chosen the geochemically important elements U and K. Both elements are known from terrestrial studies to be strongly fractionated during the evolution of the early terrestrial crust and also both will play an important part in establishing the thermal regime in the lunar crust. Having established groupings of rock clasts on the basis of U and K abundances, a number of well-crystallized, relatively unshocked representatives of each U–K grouping have been studied in further detail to characterise the petrographic character of each clast class and, by inference, the nature of the pre-Imbrian lunar crust.

The Fra Mauro breccias

The nature and evolution of the Fra Mauro breccias has been discussed by several groups (e.g., Chao *et al.*, 1972; von Engelhardt *et al.*, 1972; Jackson and Wilshire, 1972; Warner 1972) and a number of classification schemes have been proposed. These schemes differ in detail but basically all agree that the Fra Mauro breccias range from largely friable breccias rich in light colored and fragmental glassy clasts to highly coherent "recrystallized" or "metamorphosed" breccias rich in dark colored clasts that lack fragmental glass.

This series has been described by Jackson and Wilshire (1972) as being essentially discontinuous and indicative of at least three, and probably four, distinct statigraphic breccia units present in the Fra Mauro area. On the other hand Warner (1972) stressed the continuous nature of the breccia series and suggested they shared a common origin with the "medium" and "high" grade types being derived from "low" grade equivalents by a metamorphic process—"probably an autometamorphism in the thick, hot, Fra Mauro ejecta blanket." However, we have observed fragments of "high grade" breccias within breccia 14313,14 which Warner has classified as a Group 1 ("low grade") breccia. Clearly this breccia 14313 cannot be simply related to "high grade" breccias in a simple metamorphic sequence and 14313 must represent a later surface breccia formed in part from the fragmentation of pre-existing "high grade" breccias.

On the basis of any of the models proposed for the formation of the Fra Mauro breccias, the most likely types to be derived directly from the Imbrian event would be those showing substantial metamorphism or recrystallization. Those breccias showing the highest degree of metamorphism would also be unsuitable for a study of pre-Imbrian rock clasts because their original clasts would be severely modified by incipient melting processes. Consequently, we have studied breccia 14305,77 and 14305,12 that Warner (1972) classified as a Group 6 "high" grade type and which Jackson and Wilshire (1972) classified as an F4 coherent breccia with abundant dark clasts and rare fragmental glass clasts.

A "Lexan" plastic print study was first made to show the uranium distribution in 14305,77 and the first results showed that this, and all other Fra Mauro breccias studied, were highly enriched in uranium even compared to Apollo 12 breccias. It was also clear that rock clasts in this breccia showed a wide range in uranium abundance with some showing unusual enrichments up to 15 ppm U. Further chemical studies of these clasts showed a parallel enrichment in potassium so that it was clear that substantial fractionation of both uranium and potassium existed within the rocks of the pre-Imbrian lunar crust.

Analytical Methods

Uranium analyses. Lovering and Kleeman (1970) and Lovering and Wark (1971) have described the "Lexan" plastic rock-texture print method developed to record, and to locate precisely, fission tracks originating from U-enriched phases after irradiation of "Lexan" plastic-lunar rock couples in a thermal neutron flux. In determining overall U concentrations in individual rock clasts containing irregularly distributed U-enriched phases, errors arise from limitations of the sample size and for this reason uranium abundances in clasts are reported to two or less significant figures with individual uncertainties averaging about 6%.

Because most of the U-enriched sources in the clasts are grains $< 10 \mu$ across, U abundances in these phases are calculated by counting the total number of tracks arising from each grain and calculating the contributing volume of each (Wark, in preparation). The difficulties inherent in this method leads to uncertainties of around 20% for each determination. Other problems of analysis were discussed previously by Lovering and Wark (1971).

Overall clast compositions. Overall clast compositions (and , in particular, overall potassium compositions) were determined in the scanning beam mode either by electron microprobe (JEOL 5A with three computer controlled x-ray spectrometers) or scanning electron microscope (JEOL JSM-U3) equipped for x-ray analysis with a Nuclear Diode 165 eV Si (Li) solid state x-ray detector. Electron microprobe data were determined by standard procedures but the solid state x-ray spectra are converted to final analyses by a computer program for least squares fitting of a combination of standard mineral x-ray spectra to the unknown spectrum (Sewell and Lovering, to be published). Both the electron microprobe and scanning electron microscope analyses of rock clasts are subject to significant uncertainties in their final calculation, but these uncertainties are insignificant when compared to the problem of adequate sampling of the clasts. However, the overall potassium analyses reported are considered to have uncertainties of approximately $\pm 15\%$.

Results and Discussion

Apollo 14 breccias

"Lexan" plastic prints of breccia 14305,77 (Fig. 1) show rock clasts and individual mineral grains, some highly enriched in uranium, set in a recrystallized matrix of plagioclase ($\sim 45\%$) + pyroxene ($\sim 45\%$) + ilmenite ($\sim 8\%$) + high Si–K mesostasis material ($\sim 2\%$). This matrix material overall contains uranium abundances ranging from 2.3 to 11 ppm U but most of this U probably is concentrated in the mesostasis material, because individual plagioclase and pyroxene grains are impoverished in uranium. Plagioclase abundances are typically ≤ 15 ppb U and in pyroxene U abundances are ≤ 3 ppb. Individual U-enriched phases observed in the matrix, and listed in order of relative abundance are: apatite (20–700 ppm U) > zircon (70–800 ppm U) > whitlockite (40–160 ppm U) > a Zr–Ti–Fe–Ca phase B (400–>2800 ppm U), as defined by Lovering and Wark (1971). No tranquillityite (Lovering et al., 1971) or baddeleyite sources have been observed in the matrix.

The uranium distribution in the matrix commonly is quite inhomogeneous and areas relatively depleted in uranium are observed with distinct boundaries against more U-enriched areas (Fig. 1). The boundary itself commonly is made more obvious by a concentration of uranium around the margin of the depleted area. Similar

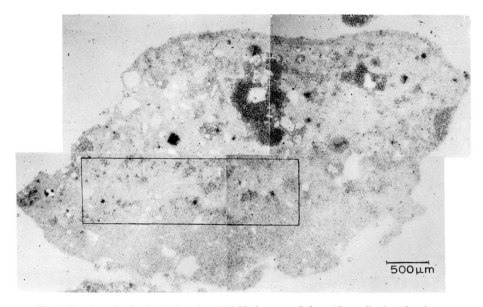

Fig. 1. Uranium distribution in breccia 14305,77 (fragment θ) from "Lexan" print, showing uranium enrichment in clasts. Area outlined shows uranium-depleted zone in breccia with marginal uranium enrichment similar to features observed in terrestrial ignimbrites.

concentrations of uranium have been observed by A. J. W. Gleadow (personal communication) at the margins of U-depleted glassy fragments in Upper Devonian rhyolitic ignimbrites from Robley's Spur in Victoria, Australia. These uranium concentrations at the margins of recrystallized glass shards in both impact and igneous generated base surge deposits suggests some transfer of uranium during initial cooling and recrystallization. Drake *et al.* (1972) have observed a similar marginal sodium enrichment of rhyolite in Apollo 14 breccias.

The rock clasts observed in breccia 14305,77 show overall uranium abundances ranging from ~0.2 ppm U in mare-type basalts to ~12 ppm U in certain red and colorless glass clasts. If overall K abundances for the clasts also are determined, Fig. 2 demonstrates how the clasts studied fall into three distinct chemical groups:
Group 1: Low U (<0.5 ppm U)–low K (<0.3% K)
Group 2: Medium U (~1.6–4.8 ppm U)–medium K (~0.4–2.3% K)
Group 3: High U (~5–12 ppm U)–high K (~3.9–6.2% K)
On the basis of this chemical classification it is then possible to further characterize each chemical group by certain petrological criteria as illustrated in Table 1 and discussed below.

Group 1 rock clasts (low U–low K)

Two subgroups can be distinguished on the basis of clast petrology. Sub group 1A clasts consist of fine and medium grained pyroxene-plagioclase assemblages with textures and U + K compositions characteristic of mare type basalts and gabbros. On this basis, they are called mare-type basalts. A few uranium-enriched sources are present in these clasts, but they are rare and have not yet been identified. Subgroup 1B clasts are essentially plagioclase aggregates (with minor amounts of pyroxene) and are characteristic of the anorthosite and anorthosite gabbro fragments

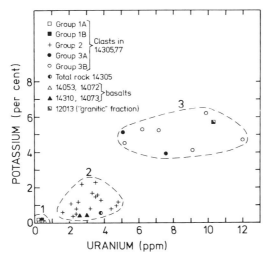

Fig. 2. Uranium-potassium distribution in clasts from breccia 14305,77 Apollo 14 basalts, and "granitic" fraction of 12013.

Table 1. Uranium and potassium abundances in clasts from breccia 14305,77, Apollo 14 basalts and the "granitic" fraction of 12013.

Clast group	Clast petrology	Clast number	Overall abundance	U-enriched phases* (rel. abund. + U-content)	Potassium (%)
				Uranium (ppm)	
Group 1 Low U– Low K	1A: Mare-type basalts	10	0.2	—	0.1
		22	0.2	—	0.2
		26	0.4	—	0.1
		38	0.3	—	0.1
	1B: Anorthositic rocks	33	0.4	—	0.2
Group 2 Medium U– medium K	Nonmare basalts and "norites"	7	2.7	—	2.2
		9	2.5	$\{$ B + Zc ($\sim$600 ppm) + Ap ($\sim$300 ppm) + M	1.2
		11	3.6	—	1.6
		12	3.5	M + B	1.5
		15	4.7	—	1.0
		17	3.3	—	1.7
		21	4.4	—	0.8
		23	3.8	M + Wt (70–125 ppm) + Ap ($\sim$100 ppm) + Zc ($\sim$300 ppm)	1.3
		28	2.4	Ap ($\sim$200 ppm)	0.5
		29	2.4	—	0.8
		31	4.8	—	1.2
		34	2.1	M + B	1.1
		35	1.6	M + B	0.6
		36	2.2	—	0.4
		45	3.5	M	2.3
		61	3.2	—	0.8
Group 3 High U– high K	3A: "Monzonitic" rocks	1	7.6	Ap ($\sim$50 ppm) + Wt ($\sim$75 ppm) + Zc (140–1300 ppm) + M (10–14 ppm)	3.9
		14	5.1		5.1
	3B: Red/Colorless glass	2	5.2	Ap (150–550 ppm) + G (5–14 ppm)	4.5
		3	6.2	—	5.3
		5	7.2	M + Ap	5.3
		6	9.9	Ap + Wt + B + G (13–17 ppm)	6.2
		8	9.1	M + Ap	4.1
		44	12.0	—	4.7

[Table continued opposite page.]

described by others (e.g., Wood *et al.*, 1970) in Apollo 11 soils and breccias. A few such clasts contain scattered U-enriched phases but they are very rare and have not yet been identified.

Group 2 rock clasts (medium U–medium K)

This group is characterized by clasts with opx/cpx + plag mineralogies and both basaltic and "type A norite" (after Marvin *et al.*, 1971) fragments are observed. Clasts in this group are very likely similar to the high Al, high K nonmare KREEP basaltic composition fragments described by other workers (e.g., Hubbard and Gast, 1971 and Meyer *et al.*, 1971) from the Apollo 12 soils and breccias.

Table 1 (continued)

Clast group	Rock petrology	Rock number	Overall abundance	Uranium (ppm) U-enriched phases* (rel. abund. + U-content)	Potassium (%)
Total rock 14305			3.8 ± 0.2 (1)†	Ap (36%, 20–700 ppm); Zc (34%, 70–800 ppm); Wt (24%, 40–160 ppm); B (6%, 400– >2800 ppm)	0.533 ± 0.10 (1)†
Apollo 14 basalts	Mare-type basalts	14072	0.56 ± 0.06	Ap (~60%, 100–600 ppm); Wt (~15%, ~50 ppm); Bd (~25%, ~170 ppm); Zc (<1%, ~500 ppm); M (3–4 ppm).	0.07 (2)
		14053	0.591 (9)	(5): Ap + Wt	0.09 (2)
	Nonmare type basalts	14310	3.0 ± 0.2 (1)	(3): Ap/Wt (20–200 ppm); Bd (400 ppm). (4): Wt + T + Bd + X? (5): Zc + Bd + Ap + Wt + "Zr–Ti phases"	0.414 ± 0.013 (1)
		14073	2.6 (10)	(5): Ap + Wt + "Zr–Ti phases"	0.41 (10)
"Granitic" fraction of 12013			10.3 (6)	(7): Zc (≤100 ppm); Ap (≤1000 ppm); Wt (≤46 ppm); β (≤3%)	5.7 (8)

*Ap: apatite; B: phase B (Zr–Ti–Fe–Ca); β: phase β (Ti–Zr–Fe–Ca) of Haines et al., 1971; Bd: baddeleyite; G: glass; M: mesostasis (usually K–feldspar and Si–glass); T: tranquillityite; X: phase X (Ti–Fe–Zr–Cr) of Brown et al., 1972; Zc: zircon.

†References

(1) Keith et al. (1972)	(6) O'Kelley et al. (1971)
(2) Compston et al. (1972)	(7) Haines et al. (1971)
(3) Crozaz et al. (1972)	(8) Schnetzler et al. (1970)
(4) Brown et al. (1972)	(9) Tatsumoto (1972)
(5) Gancarz et al. (1971)	(10) Laul et al. (1972)

Apollo 12 breccias have also been found to contain red-brown glass fragments, commonly forming accretionary lapilli with a core of anorthosite (<0.2 ppm U), that have U abundances of approximately 4 ppm and K abundances approximately 0.7% (Fig. 3). These glasses probably have been derived from Group 2 rocks.

Group 3 rock clasts (high U–high K)

The high U–high K group 3 rock clasts form two petrological subgroups that had not been recognized extensively as major lunar petrological rock types from material returned from Apollo 11 and 12 missions.

3A—"Monzonitic" rocks. These clasts commonly are recrystallized, but one clast (clast I) is relatively unaltered and has been studied in some detail to determine the normal mineralogy of these unusual rock types so highly enriched in both U and K (Table 2).

Clast I (~2 mm across) is a white fragment in which abundant plagioclase laths (~35 vol.%) and equigranular Ca-poor clinopyroxene grains (~18%) are associated

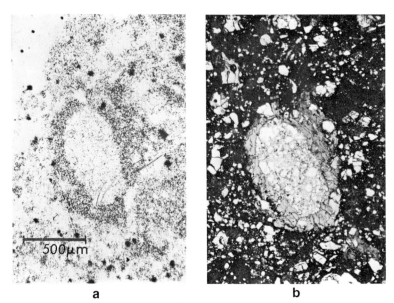

Fig. 3. Anorthositic clast in breccia 12073,33 with rim of red-brown glass forming accretionary lapilli structure (b). "Lexan" print (a) shows Group 2-type uranium-enrichment in border glass with uranium-depletion in anorthositic core.

with an abundant ($\sim 44.5\%$) interstitial mesostasis material rich in Si and K. Accessory phases present are apatite and whitlockite (0.8%) and zircon (0.5%). "Lexan" print studies indicate ~ 50 ppm U in the apatites, ~ 75 ppm U in the whitlockites and 140–1300 ppm U in the zircons, with the content of the abundant mesostasis areas approximately 10–14 ppm.

Scanning electron microscope examination of the mesostasis areas indicates that they are composed of Ba-rich K-feldspar intimately intergrown with a high-Si, high-K "glass" that contains minute, generally submicron grains of ilmenite, troilite, metallic iron, phosphates and possibly phase B. A survey shows that the composition of this "glass" appears to be identical with that described later in the mesostasis of Group 3B clasts (Table 2), and it is compositionally very similar to the late-stage high-Si immiscible melts described by Roedder and Weiblen (1971) in both Apollo 11 and 12 basaltic rocks.

Virtually all the relatively high-U content of the mesostasis areas is concentrated in the high-Si, high-K "glass", which thus becomes the major source of U in clast I. This "glass" is also relatively highly enriched in Li and/or B on the basis of the intensity of the "Lexan" print over these areas relative to the major phases present (Kleeman and Lovering, 1969).

The preliminary overall composition of clast I given in Table 3 and its calculated Niggli values suggest that this clast crystallized from a magma that most closely approaches the "leucomonzonitic" group of the potassic monzonitic magma type of Burri and Niggli (1945), although the clast I magma tends to be higher in Ca and depleted in Na. This Na depletion seems to be characteristic of all lunar crustal

Table 2. Preliminary average analyses* of mesostasis components in Group 3B clasts
in Breccia 14305,77.

	Mesostasis Components of Group 3B clasts 2 and 3 (Table 1)		Average high–Si interstitial glass in Apollo 12 basalts (Roedder and Weiblen, 1971)
	K–feldspar (average of 2)	High–Si, K "glass" (average of 4)	
SiO_2	62.5	75.8	76.3
TiO_2	0.30	1.3	0.68
Al_2O_3	19.5	13.8	11.5
FeO	0.23	0.85	3.0
MnO	<0.05	<0.05	0.05
MgO	0.02	0.10	0.07
CaO	<0.1	0.89	1.6
Na_2O	0.79	0.70	0.14
K_2O	15.2	7.9	6.7
P_2O_5	0.06	0.32	0.15
BaO	1.0	<0.1	0.62
TOTAL	99.7	101.8	100.81

*Analysis by 165 eV solid state x-ray detector on scanning electron microscope (see
section on analytical methods). Ti, Mn, Na, P, Ba, by electron microprobe.

magmas and makes comparison with terrestrial magma and rock types difficult. Mineralogical classification of the rock also is made difficult by the high proportion of mesostasis material, but assuming that the mesostasis contains about 50% crystallized K-feldspar and the associated high-Si, high-K "glass" could crystallize to ∼50% quartz and ∼50% K-feldspar, then the basic mineralogy of clast I would be quartz (∼11%) + clinopyroxene (∼18%) + plagioclase (∼35%) + K-feldspar (∼33%) + accessories (∼3%). According to the scheme of Streckeisen (1967), this rock would be classified as a "quartz-bearing monzonite" if it crystallized as a plutonic rock or as a "latite" if a volcanic rock.

Other Group 3A clasts have similar compositions and mineralogies, although quite commonly orthopyroxene also is present with or without clinopyroxene.

Many more mineralogical and chemical data on Group 3A clasts need to be acquired before the range in rock types present within Group 3A can be delineated. However, it seems that many of these 3A clasts represent rock and magma types that bear some relation to terrestrial "monzonitic" rocks.

3B—High U, K glass clasts. These clasts largely consist of patchy red-brown and colorless "glasses" in which relict plagioclase (and a few pyroxene grains) commonly can be observed. Relict apatite, whitlockite, baddeleyite, and phase B grains commonly are seen to be sources of U-enrichment in these clasts, and both the red and colorless "glasses" contain high U-enrichments (→17 ppm U). The "glasses" commonly show incipient crystallization of pyroxene blades (particularly around the margins of the clasts) and scanning electron microscope studies indicate that the "glasses" actually are fine intergrowths of both stumpy and feathery K-feldspar crystals and a high-Si, high-K "glass" (Table 2). A survey shows that the composition of this glass appears to be identical with that in the mesostasis of the Group 3A clasts. The observed chemical and mineralogical similarities between both 3A and 3B clasts suggests that the glassy 3B clasts simply are products of impact melting of the 3A "monzonitic" rocks.

Table 3. Composition of "monzonitic" clast I in breccia 14305,77.

	Preliminary electron microprobe analysis (wt%)		Modal Analysis (vol.%)
SiO$_2$	59.6	Plagioclase:	35.1
TiO$_2$	0.6	Clinopyroxene:	18.0
Al$_2$O$_3$	20.6	Mesostasis:	44.5
FeO	3.3	Opaques:	1.1
MgO	3.3	Apatite/whitlockite:	0.8
CaO	7.9	Zircon:	0.5
Na$_2$O	0.9		100.0
K$_2$O	4.7		
	100.9		

Norm			Niggli values	
Q	12.4		si	185.5
C	3.6		al	37.8
or	27.8		fm	23.9
ab	7.3		c	26.3
an	39.2		alk	12.0
hy	9.6		k	0.78
il	1.1		mg	0.64
			ti	1.4

Apollo 14 basalts

Basaltic rock fragments apparently are extremely rare at the Apollo 14 site (LSPET, 1971) and the few samples returned are considered by some (e.g., LSPET, 1971; Jackson and Wilshire, 1972) to be somewhat larger basalt clasts dislodged from breccias.

El Goresy *et al.* (1972) have studied four Apollo 14 basalts and report strong textural and mineralogical similarities between 14053 and 14072, on the one hand, and 14073 and 14310, on the other. A similar grouping is apparent from U and K abundance data reported in Table 1 with 14072 and 14053 depleted in both U and K, but 14310 and 14073 are both relatively enriched in U and K. Immediate similarities are obvious between these two basalt groups and the low U-low K group 1A clasts and the medium U-medium K Group 2 clasts in breccia 14305 (Fig. 2).

Low U–low K Apollo 14 basalts. Basaltic rocks 14053 and 14072 have relatively depleted U and K abundances comparable with the pre-Imbrian mare-type basalt clasts observed in breccia 14305 and the younger mare-fill basalts from the Apollo 11 and 12 sites.

Only basalt 14072,4 was available for the present study. This rock is a fine- to medium-grained porphyritic olivine basalt collected at Station C′ on the upper flanks of Cone Crater. "Lexan" print studies show that the U-enriched phases are all concentrated in interstitial areas between the major clinopyroxene, plagioclase, olivine, ilmenite phases that all contain negligible concentrations of uranium (i.e., $\lesssim 5$ ppb U). The late-stage U-enriched phases in the interstices are apatite ($\sim 60\%$ of observed U-enriched phases; 100–600 ppm U), whitlockite ($\sim 15\%$; ~ 50 ppm U); baddeleyite ($\sim 25\%$; ~ 170 ppm U) and zircon ($< 1\%$; ~ 500 ppm U) associated with a high-Si, K mesostasis material containing 3–4 ppm U.

The Zr-enriched phases associated with the high-Si, K mesostasis areas in this pre-Imbrian mare-type basalt (i.e., baddeleyite + zircon) are distinct from the baddeleyite + tranquillityite + phase B assemblages recorded by Lovering and Wark (1971) in the post-Imbrian (Apollo 11) and post-Eratosthenian (Apollo 12) mare-type basalts. However, there is a distinct similarity with the baddeleyite + zircon assemblages observed in eucritic achondrites (D. A. Wark, personal communication). It is not yet clear what parameters govern the observed associations in Zr-enriched phases in lunar basalts and achondrites but the overall Zr abundance would not seem to be critical on the basis of data given by Erlank *et al.* (1972) for total Zr abundances in Apollo 11, 12, and 14 basalts and eucritic achondrites.

The high-Si, K mesostasis areas, with significant uranium contents of 3 to 4 ppm U, make up about 0.5% of 14072,4. Scanning electron microscope studies indicate that these areas actually are composed of stumpy laths (averaging about 10 μ in the longest dimension) to feathery "crystallites", typically much less than 1μ across, of K-feldspar intergrown with a matrix composed almost exclusively of SiO_2 with <0.5% K. This submicroscopic intergrowth probably represents final crystallization of the late-stage high-Si and K glasses described by Roedder and Weiblen (1971) as immiscible silicate melts in Apollo 11 and 12 basalts. The high-Si matrix also contains submicron inclusions of troilite, iron, ilmenite, pyroxenes (?), Zr-rich phases and phosphates. Because the K-feldspar is unlikely to contain significant U abundances, the 3–4 ppm U observed in the mesostasis is most probably associated with the Si-rich matrix and the inclusions of phosphate and Zr-rich minerals.

On the basis of the development of the "Lexan" print over basalt 14072,4, the lowest Li and/or B abundances occur in ilmenite, troilite, and iron metal. Greater amounts occur in plagioclase and pyroxene with the pyroxene pit densities approximately twice those observed over plagioclase. The highest Li/B concentrations are associated with the mesostasis areas and indicate that Li and/or B are concentrated in the very late-stage liquids along with U, K, Zr, REE, etc.

Medium U–medium K Apollo 14 basalts. Both 14310 and 14073 are plagioclase-phyric basalts with medium enrichments in both U and K similar to the Group 2 basaltic and "noritic" clasts in breccia 14305,77 (Table 1). Neither 14310 nor 14073 were available for the present study, but other workers (Table 1) have reported the occurrence of the following suite of U-enriched phases in these basalts: apatite, whitlockite, zircon, baddeleyite, tranquillityite, phase X (?) of Brown *et al.* (1972) and various Zr–Ti phases (Gancarz *et al.*, 1971). This association is considerably more complex than that observed in the low U–low K Apollo 14 basalts and rather more reminiscent of the complex associations observed by us in the matrices of the Apollo 12 and 14 breccias. Morgan *et al.* (1972) have reported "meteoritic element" abundances in 14310 in exactly the same amounts and proportions as Apollo 14 soils and breccia and they speculate that 14310 represents remelted breccia (or fines). If this is the case then it is not surprising that 14310 exhibits chemical similarities to the "magic", "cryptic", and KREEP components of lunar breccias and fines, as pointed out by Gancarz *et al.* (1971).

Table 4. Relative abundance of U–K clast groups
in two samples of breccia 14305.

	U–K group	14305,77	14305,12	Average
Group	1A ⎫ 1B ⎬	22 ⎫ 30 8 ⎭	36 ⎫ 48 12 ⎭	29 ⎫ 39 10 ⎭
Group	2	43	42	42
Group Group	3A ⎫ 3B ⎬	4 ⎫ 27 23 ⎭	2 ⎫ 10 8 ⎭	3 ⎫ 19 16 ⎭
		100	100	100

Conclusions

On the basis of the observed chemical groupings of rock clasts in the recrystallized Apollo 14 breccia 14305, we propose that the pre-Imbrian crust was composed of the following rock associations:

Group 1 (low U–low K) rocks: mare-type basalts and anorthosites.

Group 2 (medium U–medium K) rocks: nonmare-type basalts and "norites" (opx-bearing basic rocks).

Group 3 (high U–high K) rocks: "monzonitic" rock types relatively enriched in plagioclase, K-feldspar and high-Si, K mesostasis "glass". It is very likely that the "granitic" fraction or component of the curious rock 12013 described by Drake *et al.* (1970) is a similarly fractioned member of this Group (Fig. 2).

The relative abundances of the three clast groupings measured on two sections of breccia 14305 are shown in Table 4. Clearly the distribution is inhomogeneous and the relationship with the relative abundances of these rocks in the source areas is impossible to determine. However, our data suggest that the high U–high K Group 3 rocks form a significant proportion (i.e., $\sim 20\%$) of rock clasts observed in breccia 14305. If this reflects a significant abundance of these "monzonitic" and "granitic" rocks in the early lunar crust, then their extreme U (and Th?) and K enrichments would pose severe constraints on the thermal history of the lunar crust.

There is a similarity between the rock association proposed here for the pre-Imbrian crust and the rock associations characteristic of the Precambrian massif-type anorthosite bodies on the earth as Hargraves and Buddington (1970) have proposed on the basis of other evidence. Terrestrial massif-type anorthosites are associated with a suite of more basic rocks (e.g., gabbroic anorthosites, anorthositic gabbros, noritic gabbros, iron-titanium ore bodies) and there is a further common association with more acidic rocks (e.g., syenites, granodiorites, adamellites, etc.) as Green (1969) has pointed out. The general similarity between the total terrestrial massif-anorthosite rock association and the proposed pre-Imbrian lunar crustal association is quite striking. Further examination of this analogy should provide parallel insight into the evolution of both important associations.

Acknowledgments—We wish to thank D. Campbell and R. Britten for their assistance in the project and one of us (J.F.L.) wishes to acknowledge the facilities made available by the Max-Planck-Institut für Kernphysik (Heidelberg) to prepare this paper. The work was supported by grants from the Australian Research Grants Committee and the Australian Institute for Nuclear Science and Engineering.

REFERENCES

Anderson A. T. Braziunas T. F. Jacoby J. and Smith J. V. (1972) Breccia populations and thermal history: Nature of pre-Imbrian crust and impacting body (abstract). In *Lunar Science—III* (editor C. Watkins), pp. 24–26, Lunar Science Institute Contr. No. 88.

Brown G. M. Emeleus C. H. Holland J. G. Peckett A. and Phillips R. (1972) Mineral fractionation patterns between Apollo 14 primitive feldspathic rocks and Apollo 15 and other basalts (abstract). In *Lunar Science—III* (editor C. Watkins), pp. 95–97, Lunar Science Institute Contr. No. 88.

Burri C. and Niggli P. (1945) Die jungen Eruptivgesteine des mediterranen Orogens I. Publ. herausgeg. v.d. Stiftung "Vulkaninstitut Immanuel Friedlaender" 3, Zurich. Reference in Burri C. (1959), Petrochemische Berechnungsmethoden auf äquivalenter Grundlage, 334 pages, Birkhauser Verlag, Basle, Switzerland.

Chao E. C. T. Minkin J. A. and Boreman J. A. (1972) The petrology of some Apollo 14 breccias (abstract). In *Lunar Science—III* (editor C. Watkins), pp. 131–132, Lunar Science Institute Contr. No. 88.

Compston W. Vernon M. J. Berry H. Rudowski R. Gray C. M. Ware N. Chappell B. W. and Kaye M. (1972) Age and petrogenesis of Apollo 14 basalts (abstract). In *Lunar Science—III* (editor C. Watkins), pp. 151–153, Lunar Science Institute Contr. No. 88.

Crozaz G. Drozd R. Graf H. Hohenberg C. M. Monnin M. Ragan D. Ralston C. Seitz M. Shirck J. Walker R. M. and Zimmermann J. (1972) Evidence for extinct Pu^{244}: Implications for the age of the pre-Imbrium crust (abstract). In *Lunar Science—III* (editor C. Watkins), pp. 164–166, Lunar Science Institute Contr. No. 88.

Drake M. J. McCallum I. S. McKay G. A. and Weill D. F. (1970) Mineralogy and petrology of Apollo 12 sample no. 12013: a progress report. *Earth Planet. Sci. Lett.* **9**, 103–123.

El Goresy A. Ramdohr P. and Taylor L. A. (1972) Fra Mauro crystalline rocks: Petrology, geochemistry, and subsolidus reduction of the opaque minerals (abstract). In *Lunar Science—III* (editor C. Watkins), pp. 224–226, Lunar Science Institute Contr. No. 88.

Von Engelhardt W. Arndt J. and Stöffler D. (1972) Apollo 14 soils and breccias, their compositions and origin by impacts (abstract). In *Lunar Science—III* (editor C. Watkins), pp. 233–235, Lunar Science Institute Contr. No. 88.

Erlank A. J. Willis J. P. Ahrens L. H. Gurney J. J. and McCarthy T. S. (1972) Inter-element relationship between the moon and stony meteorites with particular reference to some refractory elements (abstract). In *Lunar Science—III* (editor C. Watkins), pp. 239–241, Lunar Science Institute Contr. No. 88.

Gancarz A. J. Albee A. L. and Chodos A. A. (1971) Petrologic and mineralogic investigation of some crystalline rocks returned by the Apollo 14 mission. *Earth Planet. Sci. Lett.* **12**, pp. 1–18.

Green T. H. (1969) Experimental fractional crystallization of quartz diorite and its application to the problem of anorthosite origin. In "Origin of anorthosite and related rocks" (Y. W. Isachsen, ed.). N.Y. State Museum and Sci. Service, *Mem.* **18**, 23–29.

Haines E. L. Albee A. L. Chodos A. A. and Wasserburg G. J. (1971) Uranium-bearing minerals of lunar rock 12013. *Earth Planet. Sci. Lett.* **12**, 145–154.

Hargraves R. B. and Buddington A. F. (1970) Analogy between anorthosite series on the earth and moon. *Icarus* **13**, 371–382.

Hubbard N. J. and Gast P. W. (1971) Chemical composition and origin of nonmare lunar basalts. *Proc. Second Lunar Sci. Conf., Geochim. Cosmochim. Acta* Suppl. 2, Vol. 2, pp. 999–1020. MIT Press.

Jackson E. D. and Wilshire H. G. (1972) Classification of the samples returned from the Apollo 14 landing site (abstract). In *Lunar Science—III* (editor C. Watkins), pp. 418–420, Lunar Science Institute Contr. No. 88.

Keith J. E. Clark R. S. and Richardson K. A. (1972) Gamma ray measurements of Apollo 12, 14 and 15 lunar samples (abstract). In *Lunar Science—III* (editor C. Watkins), pp. 446–448, Lunar Science Institute Contr. No. 88.

Kleeman J. D. and Lovering J. F. (1969) Lexan plastic prints: how are they formed? In *Proc. Int. Conf. Nuclear Track Registration in Insulating Solids and Applications*, (editor M. Monnin). Clermont-Ferrand, 6–9 May, 1969.

Laul J. C. Boynton W. V. and Schmitt R. A. (1972) Bulk, REE and other elemental abundances in four Apollo 14 clastic rocks and three core samples, two Luna 16 breccias and four Apollo 15 soils (abstract).

In *Lunar Science—III* (editor C. Watkins), pp. 480–482, Lunar Science Institute Contr. No. 88.

Lovering J. F. and Kleeman J. D. (1970) Fission track uranium distribution studies on Apollo 11 lunar samples. *Proc. Apollo 11 Lunar Sci. Conf., Geochim. Cosmochim. Acta* Suppl. 1, Vol. 1, 627–631. Pergamon.

Lovering J. F. and Wark D. A. (1971) Uranium-enriched phases in Apollo 11 and Apollo 12 basaltic rocks. *Proc. Second Lunar Sci. Conf., Geochim. Cosmochim. Acta* Suppl. 2, Vol. 1, pp. 151–158. MIT Press.

Lovering J. F. Wark D. A. Reid A. F. Ware N. G. Keil K. Prinz M. Bunch T. E. El Goresy A. Ramdohr P. Brown G. M. Peckett A. Phillips R. Cameron E. N. Douglas J. A. V. and Plant A. G. (1971) Tranquillityite: A new silicate mineral from Apollo 11 and Apollo 12 basaltic rocks. *Proc. Second Lunar Sci. Conf., Geochim. Cosmochim. Acta* Suppl. 2, Vol. 1, pp. 39–45. MIT Press.

LSAPT (1972) (Lunar Sample Analysis Planning Team). Third Lunar Science Conference, *Science*.

LSPET (1971) (Lunar Sample Preliminary Examination Team). Preliminary examination of lunar samples from Apollo 14, *Science* **173**, 681–693.

Marvin U. B. Wood J. A. Taylor G. J. Reid J. B. Powell B. N. Dickey J. S. and Bower J. F. (1971) Relative proportions and probable sources of rock fragments in the Apollo 12 soil samples. *Proc. Second Lunar Sci. Conf., Geochim. Cosmochim. Acta* Suppl. 2, Vol. 1, pp. 679–699. MIT Press.

Meyer C. (1972) Mineral assemblages and the origin of nonmare lunar rock types (abstract). In *Lunar Science—III* (editor C. Watkins), pp. 542–544, Lunar Science Institute Contr. No. 88.

Meyer C. Brett R. Hubbard N. J. Morrison D. A. McKay D. S. Aitken F. K. Takeda H. and Schonfeld E. (1971) Mineralogy, chemistry and origin of the KREEP component in soil samples from the Ocean of Storms. *Proc. Second Lunar Sci. Conf., Geochim. Cosmochim. Acta* Suppl. 2, Vol. 1, pp. 393–411. MIT Press.

Morgan J. W. Krähenbühl U. Ganapathy R. and Anders E. (1972) Volatile and siderophile elements in Apollo 14 and 15 rocks (abstract). In *Lunar Science—III* (editor C. Watkins), pp. 555–557, Lunar Science Institute Contr. No. 88.

O'Kelley G. D. Eldridge J. S. Schonfeld E. and Bell P. R. (1971) Abundances of the primordial radionuclides K, Th, and U in Apollo 12 lunar samples by nondestructive gamma-ray spectrometry: Implications for origin of lunar soils. *Proc. Second Lunar Sci Conf., Geochim. Cosmochim. Acta* Suppl. 2, Vol. 2, pp. 1159–1168. MIT Press.

Reid A. M. Ridley W. I. Warner J. Harmon R. S. Brett R. Jakes P. and Brown R. W. (1972) Chemistry of highland and mare basalts as inferred from glasses in the lunar soils (abstract). In *Lunar Science—III* (editor C. Watkins), pp. 640–642, Lunar Science Institute Contr. No. 88.

Roedder E. and Weiblen P. W. (1971) Petrology of silicate melt inclusions, Apollo 11 and Apollo 12 and terrestrial equivalents. *Proc. Second Lunar Sci. Conf., Geochim. Cosmochim. Acta* Suppl. 2, Vol. 1, pp. 507–528. MIT Press.

Schnetzler C. C. Philpotts J. A. and Bottino M. L. (1970) Li, K, Rb, Sr, Ba and rare-earth concentrations, and Rb–Sr age of lunar rock 12013. *Earth Planet. Sci. Lett.* **9**, 185–192.

Streckeisen A. L. (1967) Classification and nomenclature of igneous rocks *Neues Jahrbuch für Mineral. Abh.* **107**, 144–214.

Tatsumoto T. (1972) U–Th–Pb and Rb–Sr measurements on some Apollo 14 lunar samples (abstract). In *Lunar Science—III* (editor C. Watkins), pp. 742–743, Lunar Science Institute Contr. No. 88.

Warner J. L. (1972) Apollo 14 breccias: Metamorphic origin and classification (abstract). In *Lunar Science—III* (editor C. Watkins), pp. 782–784, Lunar Science Institute Contr. No. 88.

Wilshire H. G. and Jackson E. D. (1972) Petrology of the Fra Mauro formation at the Apollo 14 landing site (abstract). In *Lunar Science—III* (editor C. Watkins), pp. 803–805, Lunar Science Institute Contr. No. 88.

Wood J. A. Dickey J. S. Marvin U. B. and Powell B. N. (1970) Lunar anorthosites and a geophysical model of the moon. *Proc. Apollo 11 Lunar Sci. Conf., Geochim. Cosmochim. Acta* Suppl. 1, Vol. 1, pp. 965–988. Pergamon.

Proceedings of the Third Lunar Science Conference
(Supplement 3, *Geochimica et Cosmochimica Acta*)
Vol. 1, pp. 295–303
The M.I.T. Press, 1972

Electron microprobe investigations of the oxidation states of Fe and Ti in ilmenite in Apollo 11, Apollo 12, and Apollo 14 crystalline rocks

M. Pavićević, P. Ramdohr, and A. El Goresy

Max-Planck-Institut für Kernphysik, 69 Heidelberg, Germany

Abstract—The L spectra of iron and titanium in members of the binary systems ilmenite-hematite and ulvöspinel-magnetite were studied using electron microprobe techniques. In compounds of a binary system the Fe L_β/L_α ratio decreases upon increase of the total amount of Fe $^{3+}$. Calibration curves for the L_β/L_α ratio at voltages between 2 and 25 kV are constructed for quantitative determination of the oxidation state of Fe. By applying such variation curves to ilmenites from the lunar samples 10047, 12063, and 14053 it was found that Fe in the studied lunar ilmenites is present as Fe $^{2+}$. The Ti L spectrum of ilmenite consists of three main bands: A, B, and C. These bands can be explained on the basis of the molecular orbital theory. Band B is due mainly to transitions from the $2t_{2g}$ molecular orbital level to the $2p_\frac{3}{2}$, whereas band C is due to transitions from $2t_{2g}$ molecular orbital level to $2p_\frac{1}{2}$. Band A is formed mainly by transitions from the $2e_g$ molecular orbital level to $2p_\frac{3}{2}$. The presence of one electron in the $2t_{2g}$ (in Ti $^{3+}$) level will cause an increase in the intensities of both bands B and C compared to band A. The Ti L spectra of ilmenites in the lunar samples 10047 and 12063 are identical to the spectrum of pure FeTiO$_3$ indicating that Ti in those ilmenites is present in the tetravalent state. In sample 14053 the Ti L spectra for ilmenite were found to show significant increase in the intensities of bands B and C. This feature is interpreted as due to the presence of some Ti $^{3+}$ in the structure of the ilmenite of this sample.

INTRODUCTION

THE OXIDATION STATES of iron and titanium in the Apollo samples have been a subject of concern in several lunar investigations. The presence of metallic iron in the Apollo 11 basalts and FeNi alloys in the Apollo 12 and 14 crystalline rocks suggests that iron is present in the divalent state in the Fe–Ti oxide minerals. In a detailed study of the Fra Mauro igneous rocks El Goresy *et al.* (1971, 1972) reported in rock 14053 a unique breakdown of the late stage fayalite to pure metallic Fe + SiO$_2$ glass. These findings demonstrate the extreme reducing conditions under which this rock was formed. However, Haggerty and Meyer (1970) suggested the presence of Fe $^{3+}$ in some members of the chromite-ulvöspinel series present in the Apollo 12 basaltic rocks. Recently Bell and Mao (1972) in a study of the crystal field spectra of Fe^{2+} in olivine and pyroxene crystals in samples from different Apollo missions found evidence for the presence of Ti^{3+} and Fe^{3+}. Hafner *et al.* (1971) did not detect any trivalent iron in clinopyroxene separates from their Apollo 11 and Apollo 12 samples. It should be noted that it is difficult to gain information on the oxidation state of titanium in rocks or mineral separates by conventional chemical analytical methods.

Albee and Chodos (1970), Pavićević (1971) and O'Nions and Smith (1971) demonstrated that a commercial electron microprobe can provide information on the oxidation states of the 3*d* transition elements in many rock forming minerals. However, in order to apply the technique on a quantitative basis, measurements should be restricted to members of simple binary systems like the ilmenite-hematite or the ulvöspinel-magnetite series. In such a simple binary system the L emission spectra will be influenced mainly by the oxidation state of the transition element under investigation. Ilmenite, the major opaque mineral in many lunar samples contains two transition elements. Fe and Ti, as major cations. The major aim if this study was to investigate the oxidation states of both iron and titanium in lunar ilmenites by studying the L spectra of those elements with the electron microprobe and comparing those spectra with the L spectra of standards from the system ilmenite-hematite with well-known oxidation states.

Method

The equipment used in the present investigations are an ARL-EMX and a Siemens electron microprobe. The ARL was used for the investigations on the Fe L spectra using a KAP crystal, whereas the Siemens probe was used for the study of the Ti L spectra using a clinochlore crystal (28.4 Å) focused for the wave length range 26.5 to 28.0 Å. Operations were carried out under vacuum conditions between 2.10^{-6} and 10^{-5} Torr. A liquid nitrogen cold trap was employed to minimize the effect of carbon contamination on the surface of the studied sample under the stationary electron beam. Measurements of the Fe L spectra were carried out at excitation potentials from 2 to 25 kV. Measurements of the Ti L spectra were made at 6 kV only and not at higher voltages because of the severe absorption with the low take-off angle (30°) of the spectrometers in the Siemens probe. Accurate determinations of the peak positions were made at every voltage employed, because it is known that there is an apparent shift of the peak positions with increasing excitation voltage due to the overlap of the L absorption edge with the high energy side of the L_{III} band (Pavićević, 1971; O'Nions and Smith, 1971).

Results and Discussion

Fe L Spectra

Our microprobe investigations of the Fe L_α and L_β spectra indicate that any change in the oxidation state of iron among members of a solid solution series influences the spectra in the following manner:

1. As the total amount of Fe^{3+} increases, the peak position shifts continuously, but nonlinearly, towards shorter wavelengths.
2. The L_β/L_α intensity ratio decreases almost linearly with the increase in the oxidation state.
3. There is an apparent wavelength shift of the spectra to higher wavelengths upon increase of the excitation potential.

Figure 1 displays the Fe L_α and L_β peak positions for pure synthetic ilmenite, synthetic Ilm_{96}–Ht_4, and Elba hematite (which was found to be pure stoichiometric Fe_2O_3) at excitation potentials between 2 and 25 kV. Figure 2 shows the peak positions for pure synthetic ulvöspinel (Usp_{100}), synthetic Usp_{80}–Mt_{20}, and synthetic pure magnetite (Mt_{100}) respectively at the same operating conditions. Two important phenomena can be seen in Figs. 1 and 2. In compounds of both binary systems, increase in the total amount of Fe^{3+} causes a wavelength shift of both L_α and L_β towards shorter wavelengths. This wavelength shift is by no means quantitatively proportional to the increase in the amount of Fe^{3+}. Furthermore, there is an apparent wavelength shift of the spectra to higher wavelengths upon increase of the excitation potential. An additional interesting phenomenon in both figures is the complicated overlap of the L_α wavelength variation curves of all studied compounds at excitation energies below 5 kV. The L_β wavelength variation curves do not show such a phenomenon. We cannot offer a satisfactory explanation for this observation at present.

The nonlinearity of the wavelength shift with change in the oxidation state of iron disqualifies this phenomenon from further quantitative evaluation. The wavelength variation curves for lunar ilmenites in samples 10047, 12063, and 14053 were found to agree within the experimental error with the variation curve for pure synthetic ilmenite.

The Fe L_β and L_α intensity ratios were measured first using conventional peak intensity measurements and later using the integrating procedure as described by Smith and O'Nions (1971). Measurements of the peak height intensities rather than using the integrating technique is subject to error, because the peak heights are not proportional to the energy under the peak.

The L_β/L_α variation curves for members of a solid solution series by applying the integration technique were found to shift to higher values compared to the variation curves using the peak intensity ratios. However, the general relationships of the curves for both methods remain unchanged. This is mostly due to the fact that

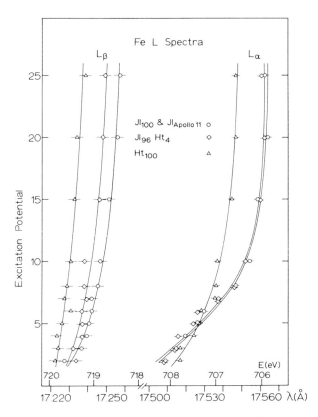

Fig. 1. L_α, L_β wavelength variation with excitation potential for the ilmenite-hematite series.

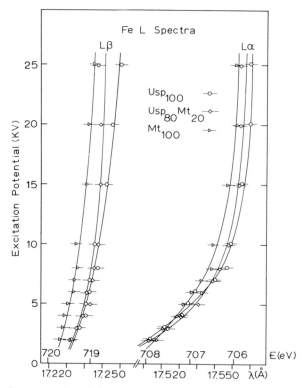

Fig. 2. L_α, L_β wavelength variation with excitation potential for the ulvöspinel-magnetite series.

minerals of the same series possess similar stereochemical properties and the change in the Fe–O distances is negligible. The only difference is the amount of Fe^{3+} present. For these reasons we employed the peak intensity ratio method since it is less time consuming compared to the integrating technique. No absorption corrections were applied.

Smith and O'Nions (1971) report that the theoretical ratio Fe $L_{II}:L_{III}$ of 0.5 for pure iron is never likely to be observed by simple measurement of peak heights even after applying absorption corrections. Pavićević (1971) carried out L_β/L_α peak height measurements on numerous thin iron layers ranging in thickness from 65 Å to 1060 Å coated on highly polished graphite plates. His measurements demonstrate that such a ratio is indeed obtainable for Fe thin layers, in the thickness range of 160 Å, and that the ratio increases upon increase of the excitation potential. Furthermore, very thin iron layers of 65 Å thickness, in which negligible self-absorption takes place, give an L_β/L_α ratio of 0.65 at 2 kV. This ratio increases linearly upon increase of the excitation potential to a value of 0.8 at 25 kV. The integration method revealed even higher values. These results demonstrate that self-absorption is the most severe problem. On the other hand, the theoretical ratio of 0.5 is very probably in error and needs revision.

O'Nions and Smith (1971) and Pavićević (1971) have shown that certain features of the L_β and L_α spectra from Fe–Ti oxides, and variations in the L_β/L_α ratio with operating voltage and bond character, can be explained on the basis of molecular orbital theory. The Fe L band originates from mixed metal and ligand orbitals. If the assumption is made that the L_{II} band is due to transitions from antibonding molecular orbitals and the L_{III} band to transitions from both bonding and anti-bonding molecular orbitals (O'Nions and Smith, 1971), certain aspects of the L_β/L_α intensity variation in oxides may be explained. In iron oxides and iron-titanium oxides the band shape and the intensity ratio depend on the oxidation state of iron (and hence the bond type), the degree of delocalization of the bonding electrons, the stereochemistry of the oxygen polyhedra, and the metal-ligand distances. In order to be able to use the L spectra for quantitative determination of the oxidation state, measurements should be confined to compounds with similar stereochemical proper-ties and belonging to the same solid solution series in which only the oxidation state changes gradually from one end member to the other.

Deduction of the oxidation state of iron from quantitative electron microprobe analyses with the assumption of stoichiometry is *indeed unrealistic* because members of ulvöspinel-magnetite and ilmenite-hematite solid solution series are known to show cation vacancies. Though unlikely, iron-titanium oxides may be oxygen deficient. Any slight deviation from stoichiometry is likely to cause changes in the oxidation states of iron and titanium. In members of the ilmenite-hematite series both iron and titanium are octahedrally coordinated. Members of the ulvöspinel-magnetite solid solution series possess an inverse spinel type of structure in which both octahedral and tetrahedral coordinations are present. In an oxygen octahedron with O_h symmetry the Fe^{2+}–O bond is more ionic in character than the Fe^{3+}–O

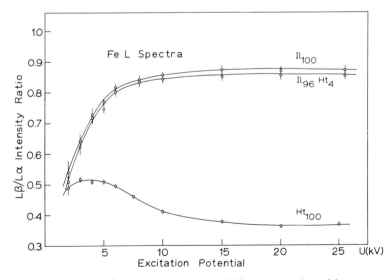

Fig. 3. Fe L_β/L_α intensity variation with excitation potential for some members of the system Fe TiO_3–Fe$_2O_3$.

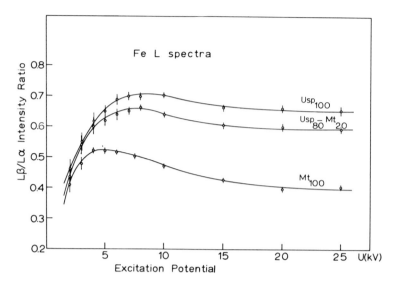

Fig. 4. Fe L_β/L_α intensity variation with excitation potential for members of the system Fe_2TiO_4–Fe_3O_4.

bond. According to O'Nions and Smith (1971) this factor together with the additional t_{2g} electron in Fe^{2+} should increase the transition probabilities from the anti-bonding molecular orbitals to Fe L_{II} and L_{III} relative to the compound with more Fe^{3+}. O'Nions and Smith argue that if the L_{II} band arises from the $L_{II} \leftarrow t_{2g}$ transition, the rate of increase of intensity of the L_{II} relative to the L_{III} band may be faster for a compound with Fe^{2+} than for another with Fe^{3+}.

The L_β/L_α (L_{II}/L_{III}) intensity variations for members of the ilmenite-hematite and ulvöspinel-magnetite solid solution series with the excitation potential are shown in Figs. 3 and 4, respectively. It is apparent that the L_β/L_α ratio increases with increase of the total amount of Fe^{2+}. The distance between any two curves at a given excitation potential in the range from 5 to 25 kV was found to be a direct function of the increase in the amount of the Fe_2O_3 molecule present. These curves can only be applied to members of the binary systems. Our investigations indicate that the increase in the Mg/Fe ratio causes a decrease in the L_β/L_α ratios. For this reason we have chosen for this study lunar ilmenites with low Mg concentrations and with almost the same composition. Unfortunately, the calibration curves shown in Fig. 4 cannot be applied for the majority of the lunar ulvöspinels, because they contain appreciable amounts of the $(Fe, Mg) Cr_2O_4$ molecule.

The measured L_β/L_α intensity variation curves for lunar ilmenites in samples 10047, 12063, and 14053 were found to coincide with the curve for pure synthetic ilmenite. The L_β/L_α ratios for the lunar ilmenites together with the values for the standards used, as measured at 10 kV are compiled in Table 1. The intensity ratios for lunar ilmenites agree within the experimental error with the values for pure synthetic $Fe^{2+}Ti^{4+}O_3^{-6}$. These results indicate that the iron in ilmenites in the studied lunar samples is present as Fe^{2+}.

Table 1. L_β/L_α ratios for lunar ilmenites and standards.

Composition	L_β/L_α ratio	δ_s
1. Ilm_{100}	0.857	± 0.024
2. Ilm_{96}–Hem_4	0.842	± 0.024
3. Hem_{100}	0.413	± 0.006
4. 10047	0.867	± 0.028
5. 12063	0.855	± 0.026
6. 14053	0.857	± 0.030

Ti L spectra

Microprobe investigations of the Ti L spectra were carried out only at an excitation potential of 6 kV. Because of the severe absorption due to the low take-off angle of the Siemens spectrometers we failed to get good counting rates at higher voltages. The Ti L spectra for pure ilmenite (Il_{100}) and for ilmenites in the studied lunar samples are shown in Fig. 5. Every spectrum consists of 3 major bands designated A, B, and C. Like iron Ti in ilmenite is octahedrally coordinated with oxygen. These 3 bands can be explained if molecular orbital theory is applied. In a comprehensive study of the Ti L spectra in the octahedrally coordinated titanium oxides TiO, Ti_2O_3, and TiO_2 under conditions of minimum self-absorption Fischer (1970) demonstrated that the Ti $3d$ and ligand $2p$ orbitals interact to form the $2e_g$, $1t_{2g}$, $2t_{2g}$, and $3e_g$ molecular orbital levels. The first two levels being bonding and the latter two antibonding orbitals. There is also some slight interaction between ligand $2s$ and metal $3d$, $4s$, and $4p$ orbitals giving rise to the $1a_{1g}$, $1t_{1u}$, $1e_g$ molecular orbital levels (Fischer, 1970). All of the molecular orbitals levels below the $2t_{2g}$ level are completely filled with electrons. The major difference in the electron structure of Ti_2O_3 and TiO_2 is that Ti_2O_3 will have one electron on the $2t_{2g}$ level but TiO_2 will have none. Because Ti in the ilmenite structure has the same octahedral coordination as in Ti_2O_3 and TiO_2, Fischer's concept holds also for ilmenite. The L spectra for Ti_2O_3 and TiO_2 contain the three major bands also found for ilmenite. One of the main differences between the L spectra for Ti_2O_3 and TiO_2 is that band A is significantly stronger for TiO_2 than for Ti_2O_3. Furthermore, bands B and C are more intense for Ti_2O_3 than

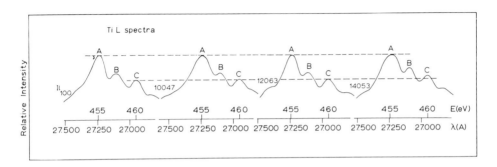

Fig. 5. Ti L_α and L_β spectra of pure ilmenite, Apollo 11, 12, and 14 ilmenites at 6 kV.

for TiO_2. Fischer (1970) suggested that band B is mostly due to transitions from the $2t_{2g}$ molecular orbital level to the $2p_{\frac{3}{2}}$ (Ti) level and that band A is due to transitions from the $2e_g$ molecular orbital level to the $2p_{\frac{3}{2}}$ (Ti) level. Oxidizing Ti^{3+} to Ti^{4+} will result in electron loss from the $2t_{2g}$ level while the $2e_g$ level remains filled. The intensity of band B will hence decrease compared to A as the oxidation state increases. Band C is mostly due to transitions from the $2t_{2g}$ level to the $2p_{\frac{1}{2}}$ (Ti), while the transition $2e_g \rightarrow 2p_{\frac{1}{2}}$ will contribute to the intensity of band B. The oxidation of Ti^{3+} to Ti^{4+} will cause a decrease of the intensity of band C compared to bands B and A, respectively. Assuming that all Ti in pure synthetic ilmenite is present as Ti^{4+}, the presence of Ti^{3+} in an ilmenite with the same chemical composition will cause the increase in the intensities of bands B and C compared to A and a relative increase in the intensity of band C compared to band B.

As can be seen from Fig. 5 the Ti L spectra for pure synthetic ilmenite and for ilmenites in the lunar samples 10047 and 12063 are identical. The Ti L spectrum for ilmenite in the Apollo 14 sample 14053 shows a significant increase in the intensities of bands B and C with respect to band A as compared with the spectra of the other three ilmenites. The relative intensities of the three bands for ilmenites of the same chemical composition should be identical if, and only if, Ti is solely present as Ti^{4+}. The higher intensities for B and C bands for the Apollo 14 sample 14053 is thus interpreted as due to the presence of some Ti^{3+}. However, it would be difficult in our present state of development to estimate the amount of Ti^{3+} present. On the other hand the presence of Ti^{3+} in ilmenite is not surprising when other mineralogical features of rock 14053 are considered. El Goresy et al. (1971, 1972) have recently described several observations made on this rock that led them to conclude that rock 14053 crystallized under more reducing conditions than any Apollo 11 or 12 rocks. That is, crystallization in an environment conducive to the formation and preservation of Ti in the trivalent state.

Acknowledgments—We wish to thank J. Janicke and H. Weber for their help in the preparation of the figures. We are grateful to Don Lindsley and E. Woermann for supplying us with synthetic Fe–Ti oxides.

References

Albee A. A. and Chodos (1969) Semiquantitative electron microprobe determinations of Fe^{2+}/Fe^{3+} and Mn^{3+} in oxides and Silicates and its application to petrologic problems. *Amer. Miner.* **55**, 491–501.

Bell P. M. and Mao H. K. (1972) Initial findings of a study of chemical composition and crystal field spectra of selected grains from Apollo 14 and 15 rocks, Glasses and fine fractions (less than 1mm). In *Lunar Science—III*, (editor C. Watkins), pp. 55–57, Lunar Science Institute Contr. No. 88.

El Goresy A. Ramdohr P. and Taylor L. A. (1971) The geochemistry of the opaque minerals in the Apollo 14 crystalline rocks. *Earth Planet. Sci. Lett.* **13**, 121–129.

El Goresy A. Ramdohr P. and Taylor L. A. (1972) Fra Mauro crystalline rocks: petrology, geochemistry, and subsolidus reduction of the opaque minerals. In *Lunar Science—III*, (editor C. Watkins), pp. 224–226, Lunar Science Institute Contr. No. 88.

Fischer D. W. (1971) Molecular-orbital interpretation of the soft X-ray L_{II}, L_{III} emission and absorption spectra from some titanium and vanadium compounds. *Journ. Appl. Phys.* **41**, 9, 3561–3569.

Hafner S. S. Virgo D. and Warburton D. (1971) Cation distributions and cooling history of clinopyroxenes from Oceanus Procellarum. *Proc. Second Lunar Sci. Conf., Geochim. Cosmochim. Acta* Supple. 2, Vol. 3, pp. 91–108. MIT Press.

Haggerty S. E. and Meyer H. O. A. (1970) Apollo 12: Opaque oxides, *Earth Planet. Sci. Lett.* **9**, 379–387.

O'Nions R. K. and Smith D. G. W. (1971) Investigation of the $L_{II, III}$ X-ray emission spectra of Fe by electron microprobe. Part 2. The Fe $L_{II,III}$ spectra of Fe and Fe-Ti oxides. *Amer. Miner.* **56**, 1452–1463.

Pavićević M. K. (1971) Der Einfluß der chemischen Bindung in Mineralien des Systems Eisen-Titan-Sauerstoff auf die Röntgenemissions-Spektren im ultraweichen Bereich. Dissertation, University of Heidelberg, 1–45.

Smith D. G. W. and O'Nions R. K. (1971) Investigation of the $L_{II,III}$ X-ray emission spectra of Fe by the electron microprobe. Part I: Some aspects of the Fe $L_{II,III}$ spectra from metallic iron and hematite. *Brit. J. Appl. Phys.* **4**, 147–159.

Proceedings of the Third Lunar Science Conference
(Supplement 3, *Geochimica et Cosmochimica Acta*)
Vol. 1, pp. 305–332
The M.I.T. Press, 1972

Apollo 14: Subsolidus reduction and compositional variations of spinels

STEPHEN E. HAGGERTY

Department of Geology, University of Massachusetts, Amherst, Mass. 01002

Abstract—Reflection microscopy and electron microprobe analyses undertaken on spinels of diverse origin and composition, in samples from the Apollo 14 site, show that these spinels are structurally of two classes and constitute two separate solid solution series: one is between ($FeAl_2O_4 - FeCr_2O_4$) and Fe_2TiO_4, the normal-inverse series; and a second is between $MgAl_2O_4$ and $FeAl_2O_4$, the normal series. Members of the normal spinel series are defined compositionally as pleonaste; the normal-inverse series is a continuous crystallization series from early Mg-aluminian-chromite, to intermediate Al-titanian-chromite and late stage Cr-Al-ulvöspinel. Compositional discontinuities are identified within and between these series. Early formed chromites are characterized by high Al_2O_3 (18 %–20 %) contents comparable only to those in the Luna 16 samples, and by intermediate compositions in the normal-inverse series comparable only to spinels from Apollo 11. Pleonastes are zoned compositionally with Mg-Al cores and Fe-Cr-rich mantles.

Subsolidus reduction of Cr-Al-ulvöspinel is identified as a widespread phenomenon and, based on the relative stabilities of chromite and ilmenite, the instability of Fe_2TiO_4-enriched spinels and fayalite, and the absence of TiO_2 polymorphs, it is concluded that reduction takes place at high temperatures ($> 600°C$) and low fO_2's ($< 10^{-15}$ atm at 1000°C). Cr-Al-ulvöspinel decomposes to Fe + $FeTiO_3$ and by coupled precipitation of this metal-oxide assemblage the host spinel is reconstituted, attains a higher degree of normal-type coordination, recedes along the initial upward path of crystallization, and progressively is enriched in Cr + Al + Mg and depleted in Fe + Ti with increasing reduction intensity.

The probability that B site coordinated Fe^{+2} and Ti^{+4} are mobilized selectively during subsolidus reduction is demonstrated, and because of the high octahedral site preference energies of Cr^{+3} and Al^{+3} these properties have led to a definitive quantitative index referred to as the TAC reduction index $TiO_2/(TiO_2 + Al_2O_3 + Cr_2O_3)$, which permits intercorrelation and comparison of intensities of reduction within and between samples. The TAC index reveals that low intensity reduction is heterogeneous (TAC = 0.6–0.9), whereas high intensity (TAC = 0.35–0.50) results in a more uniformly reduced sample, and this is interpreted to indicate that low intensity is associated with deuteric cooling and that high intensity results at high temperatures. Furthermore the widespread occurrence of the low intensity assemblage at all five lunar landing sites suggests a commonly occurring process and is in contrast to the high intensity assemblage which because of its rarity is more likely linked to sporadic selenological events.

INTRODUCTION

A DETAILED EXAMINATION of the opaque minerals by reflection microscopy, and electron microprobe analyses of spinels, ilmenite, and a group of zirconium-titanates has been carried out on samples from the Apollo 14 site. The analytical data described in this report will be restricted to the spinels and their intergrowths. Subsolidus reactions, phase intergrowths, compositional variations, and the systematics of cationic substitution are outlined, with a view to determining the environments of subsolidus reduction, evaluating the factors that may control the discontinuities that are observed in spinel solid solutions, and contrasting the behavior of primary crystallization trends with subsolidus re-equilibration trends.

Two crystalline basalts (14053 and 14310), three microbreccias (14063, 14066, and 14321) and two soil samples (14191, 14258) were examined. Results now completed on samples from the Apollo 15 site, and to be published elsewhere (Haggerty, 1972a), are called upon and used for comparative purposes.

The Apollo 14 spinels are diverse in composition, textural characteristics, and origin. The effects of subsolidus reduction are widespread, and a detailed description of these effects is presented before the properties of primary spinels are considered. The paper is divided into five sections: (1) chemistry and characteristics of lunar spinels; (2) subsolidus reduction and related subsolidus reactions; (3) spinel types, compositions, and oxide variations; (4) cationic substitution; and (5) solid solutions and compositional discontinuities. A number of arguments that relate to subsolidus reduction can only be evaluated with respect to the evidence presented for the non-reduced primary spinels and for this reason some aspects of the chemistry and refinement of the mechanism of reduction are discussed under topics 3 and 4 rather than in section 2.

Chemical Characteristics of Lunar Spinels

Two compositional groups of spinels are recognized in the lunar material and these may be structurally classified into (1) the normal-inverse spinel series and (2) the normal spinel series.

The normal-inverse series is typified broadly by compositions within the ternary system chromite ($Fe_8^{IV}Cr_{16}^{VI}O_{32}$ normal)—hercynite ($Fe_8^{IV}Al_{16}^{VI}O_{32}$ normal)—ulvö-spinel ($Fe_8^{IV}Fe_8^{VI}Ti_8^{VI}O_{32}$ inverse); the superscripts IV and VI indicate tetrahedral and octahedral site coordination, respectively, in the spinel structure. The series is complete between approximately ($0.75\ FeCr_2O_4 - 0.25\ FeAl_2O_4$) and Fe_2TiO_4 (Haggerty, 1972b), and solid solution is accomplished by the substitution of Fe + Ti for Cr + Al (Agrell et al., 1970). In detail the ternary system is extended to a six-component system because there also is substantial substitution of Mg for Fe. The degree of divalent substitution however is variable; it is at a maximum in the early crystalliza-tion products ($FeCr_2O_4 - FeAl_2O_4$), falls off rapidly with increasing TiO_2 content, and essentially is absent in phases highly enriched in Fe_2TiO_4. Thus $MgCr_2O_4$ and $MgAl_2O_4$ are significant components of the low-Ti, and paragenetically early, phases enriched in $FeCr_2O_4 - FeAl_2O_4$, while concentrations of Mg_2TiO_4 in Fe_2TiO_4 tend to be minor or totally absent, and the system therefore is effectively reduced to five components. Manganese and vanadium are present in variable but minor concentrations and are strongly fractionated in the ulvöspinel-rich and chromite-rich members of the series respectively (El Goresy et al., 1971a).

Members of the normal-inverse spinel series are typically present in mare basalts and are identified in samples from the five lunar landing sites. The spinels from each site have a unique compositional characteristic: The Apollo 11 spinels are restricted to intermediate members of the series and the Apollo 12 spinels, with few exceptions, show a discontinuity in the series, referred to as the Apollo 12 miscibility *gap* (Hag-gerty and Meyer, 1970); the compositional range of the Apollo 14 spinels determined in this study show a more extensive discontinuity between ($FeCr_2O_4 - FeAl_2O_4$) and Fe_2TiO_4. However, this discontinuity is displaced with respect to the Apollo 12 *gap* towards Fe_2TiO_4, and compositions comparable to those of Apollo 11 fill the

Apollo 12 *gap*. Although the Apollo 14 spinels are similar compositionally to those of Apollo 11, with respect to their TiO_2 contents and therefore with respect to their relative position in the normal-inverse series, the former are products re-equilibrated by subsolidus reduction and are in contrast to the latter, which are primary crystalline phases. Apollo 14 spinels are enriched in Al_2O_3 to the degree that the series is extended from 0.25 $FeAl_2O_4$ to 0.45 $FeAl_2O_4$. The Apollo 15 samples contain basalts with spinels that span the entire series, basalts with spinels that reflect the Apollo 12 *gap*, isolated spinels in breccia fragments that show high Al enrichment and spinels that have partially re-equilibrated (Haggerty, 1972a). Spinels in the Luna 16 material have compositional characteristics that are identical to those from the Apollo sites with two major differences: (1) high Al concentrations for intermediate members of the normal-inverse series and (2) extensive substitution of Cr for Al as a function of Fe/Mg (Haggerty 1972b).

The normal spinel series contains members of the solid solution series between spinel ($Mg_8^{IV}Al_{16}^{VI}O_{32}$) and hercynite ($Fe_8^{IV}Al_{16}^{VI}O_{32}$). Substitution of Al by Cr is the only other major ionic substitution that is encountered in this series and the binary $MgAl_2O_4 - FeAl_2O_4$ is extended therefore to include $FeCr_2O_4 - MgCr_2O_4$; minor amounts of Mn and Ti also are present. Members of the series are termed pleonastes or chromium-pleonastes, and of the spinels examined in this study are restricted to compositions between approximately (0.5 $MgAl_2O_4$ − 0.75 $MgAl_2O_4$) and between (0.05 $FeCr_2O_4$ − 0.3 $FeCr_2O_4$).

The normal spinel series are identified in typical non-mare material (Roedder and Weiblen, 1972; Steele, 1972). Members of the series are present in rare discrete grains from the Luna 16 site (Jakes *et al.*, 1972), and $MgAl_2O_4$ (3.33% FeO) is reported by Keil *et al.* (1970) as a single grain in a felspathic-rich fragment from Apollo 11. Pleonastes and chromian-pleonastes are abundant in the Apollo 14 samples (LSPET, 1971; Haggerty, 1971a; Drever *et al.*, 1972; Brown *et al.*, 1972; Michel-Levy *et al.*, 1972; Roedder and Weiblen, 1972; Steele, 1972), but have not yet been identified in the Apollo 12 or 15 material.

In summary, mare basalts are typified by the normal-inverse series with the following observed paragenetic sequence: Mg-aluminian-chromite,* aluminian-chromite, Ti-aluminian-chromite, Al-titanian-chromite, Al-chromian-ulvöspinel, chromian-ulvöspinel, and ulvöspinel. The normal-spinel series are identified in non-mare material and are defined compositionally as spinel ($MgAl_2O_4$), pleonaste ($FeMgAl_2O_4$) or chromian-pleonaste.

SUBSOLIDUS REDUCTION

Cr-Al-ulvöspinel-ilmenite-metallic iron intergrowths are a characteristic feature of many lunar basalts and the assemblage now has been identified in samples from the five lunar landing sites. The origin of these intergrowths is ascribed either to late

---

*Chromian, titanian, aluminian prefixed by the abbreviated symbols Cr, Ti, Al, Mg indicates that the latter is subordinate; where both elements are abbreviated, e.g., Cr-Al-ulvöspinel, both elements are present in approximately equal concentrations.

stage crystallization re-equilibration (Brown *et al.*, 1971), or to the effects of sub-solidus reduction of a primary spinel highly enriched in Fe_2TiO_4 (Haggerty and Meyer, 1970; Haggerty, 1971b; El Goresy *et al.*, 1971b). Definitive evidence in support of either magmatic re-equilibration, deuteric or post-crystallization altera-tion is absent, but it is nevertheless evident that these metal-oxide intergrowths must have resulted by solid state diffusion at high temperatures and low oxygen fugacities. It will be shown that in those basalts containing spinels of contrasting paragenesis and composition (early Mg-aluminian-chromite and late Cr-Al-ulvöspinel) that it is the *later* precipitated and more TiO_2-rich spinels that are metastable; these spinels dissociate into three compositionally distinct phases, with the composition of the ilmenite and the iron remaining unchanged but the spinel becoming progressively enriched in Cr + Al + Mg (Haggerty, 1971b); the initial trend of crystallization is thus inverted, and the concept of re-equilibration as suggested by Brown *et al.* (1971) therefore is valid.

Reduction textures

Three phase intergrowths are encountered and these are classified according to the composition of the host spinel and on whether metallic iron is present or absent. The assemblages are as follows: (A) Cr-Al-ulvöspinel + ilmenite (Fig. 1a); (B) Cr-Al-ulvöspinel + ilmenite + Fe (Fig. 1b); and (C) Al-titanian-chromite + il-menite + Fe (Fig. 1c).

Type A and B intergrowths are identified in basalts from Apollo 11 (Cameron, 1970), Apollo 12 (Haggerty and Meyer, 1970; El Goresy *et al.*, 1971a; Brown *et al.*, 1971; Bowie and Simpson, 1971; Brett *et al.*, 1971) and Luna 16 (Haggerty, 1972b); and are identified in this study in basalt samples 14310 and 15555. Type A and B intergrowths also are present in basaltic fragments in soils 14191 and 14258. Type C intergrowths are restricted to basalt sample 14053 and to two fragments of basalt in breccia 14321.

The ilmenite in these intergrowths generally is lamellar in form and is oriented along {111} planes in the spinel host; internal and external composite textures (Buddington and Lindsley, 1964) also are present. In types B and C, metallic Fe is peppered along the ilmenite-spinel contact as fine discrete blebs (Fig. 1b), or as larger nucleated droplets dispersed throughout, but intimately associated with, the oxide assemblage (Fig. 1c).

Apart from the compositional differences of the host spinel, the following addi-tional noteworthy features are associated with these intergrowths: (1) Type A and B assemblages invariably coexist within the same specimen; Type B occurs alone, but Type A without Type B is not observed; (2) spinel crystals that have reacted to the reducing environment do not react with equal intensity; the reducing environment must be a nonequilibrium environment and a wide range of ilmenite/spinel, and ilmenite + Fe/spinel ratios result. These variations apply particularly to Type A and B assemblages; the variation of ilmenite + Fe/spinel between grains is not as extensive in the Type C assemblage, and this is interpreted to indicate that intense reduction takes place at elevated temperatures and is therefore more pervasive; (3)

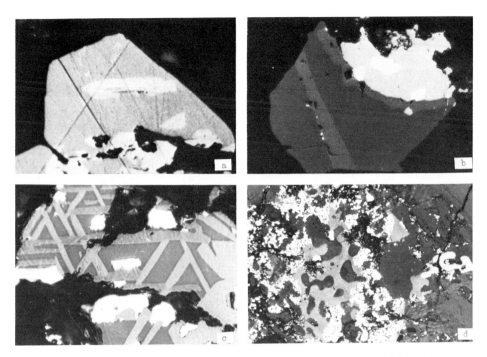

Fig. 1. *Subsolidus reduction textures.* (Photomicrographs were taken under oil-immersion objectives and in reflected light.)

(a) Type A assemblage. TAC index = 0.79. Cr-Al-ulvöspinel ($Fe_{15.08}Mg_{0.18}Mn_{0.11}$ $Cr_{1.25}Al_{0.87}Ti_{6.60}O_{32}$) + internal and external composite ilmenite. Iron and troilite are white, ilmenite is light gray and Cr-Al-ulvöspinel is medium gray (15555). Field width = 230 μ.

(b) Type B assemblage. TAC index = 0.78. Cr-Al-ulvöspinel ($Fe_{14.99}Mg_{0.18}Cr_{1.45}$ $Al_{0.74}Ti_{6.61}O_{32}$) + lamellar and external composite ilmenite + Fe. Blebs of Fe are present along the ilmenite-spinel contact and are also associated in a pseudo-eutectoid with troilite. Note that the composite ilmenite follows the grain contour of the adjacent troilite, suggesting that reduction may have been induced by the late stage precipitation of the Fe + FeS assemblage. (15555). Field width = 470 μ.

(c) Type C assemblage. TAC index = 0.39. Al-titanian-chromite ($Fe_{11.65}Mg_{0.36}Mn_{0.19}$ $Cr_{5.52}Al_{1.85}Ti_{4.18}O_{32}$) + oriented reduction lamellae of ilmenite along $\{111\}$ spinel planes + Fe. Symplectic Fe is associated with SiO_2 in the ovoid metostasis adjacent to the metal-oxide assemblage. (14053). Field width = 400 μ.

(d) Decomposition of fayalite to Fe + SiO_2. The decomposition assemblage is present in the symplectic intergrowth. Unaltered fayalite is light gray, and the darker gray lobate inclusions are Al-rich glass. The mottled areas are SiO_2 (cristobalite), the white phase is Fe, and anchor-shaped crystal is Mg-Al-chromite ($Fe_{7.42}Mg_{1.48}Mn_{0.10}Cr_{8.22}Al_{5.81}Ti_{0.93}$ O_{32}). TAC index = 0.08. (14053). Field width = 240 μ.

the modal ratio of ilmenite + Fe/spinel in Type C is markedly greater than it is in the Type B assemblage; this modal increase corresponds to a substantial increase in the Cr + Al + Mg/Fe + Ti ratio of the host spinel and is interpreted to reflect a greater degree of dissociation in Type C; (4) of the Fe-Ti-Cr oxides present in any one basalt sample it is generally only within the most Fe_2TiO_4-enriched spinels that new phases are generated. Thus in spite of the fact that both Fe and Ti are removed from the spinel host, these spinels continue to be somewhat enriched in TiO_2. Primary ilmenite and primary Mg-aluminian-chromite or mantles of Al-Ti-chromite (14053) are totally unaffected as far as can be determined microscopically; (5) the ilmenite lamellae in Type C, and to a lesser degree the lamellar and composite ilmenite in Types A and B, are mantled by parallel Cr-rich spinel borders (Fig. 1b, c); the ilmenite-border contacts are sharp, whereas the border-host contacts are diffuse; and (6) Type B intergrowths commonly contain troilite in addition to metallic iron.

Chemistry and the TAC-reduction index

Type A and B assemblages are practically indistinguishable in terms of major element abundances. There is some suggestion, however, for slight decreases in the ratios of Fe/Fe + Mg and Ti/Ti + Al + Cr in spinels that also contain free metallic Fe; the reality of these subtle differences is difficult to substantiate because the ratios more obviously are related to the abundance of ilmenite rather than to the presence or absence of metallic Fe.

The differences between Type A and B on the one hand, and Type C on the other, are gross as exemplified by a comparison of basalts 15555 (Types A and B) and 14053 (Type C). To illustrate these differences a comparison of the composition of the spinels in Type B and C assemblages and calculated oxide ratios are summarized in Table 1 for typical grains in these samples. While accepting that the primary compositions of the reaction assemblages in the two rocks may have been very different to start with, the fact remains that the progressive re-equilibration trends to be discussed at a later point are still restricted (by the confining limits set by a major proportion of many hundreds of primary lunar spinel analyses) to compositions between $(0.65 \, FeCr_2O_4 - 0.75 \, FeCr_2O_4) - (0.35 \, FeAl_2O_4 - 0.25 \, FeAl_2O_4)$ and

Table 1. Ulvöspinel, chromite, and ilmenite analyses (wt %)*

Anal.	1	2	3	4	5	6	7	8	9	10	11	12
TiO_2	30.23	29.53	22.77	22.60	20.55	20.10	19.14	18.71	18.57	13.11	11.58	9.76
MnO	0.40	0.44	0.63	0.74	0.76	0.72	0.66	0.52	0.51	0.44	0.46	0.46
FeO	60.53	60.21	47.16	48.33	49.23	48.00	47.32	48.27	47.59	37.06	39.51	38.26
MgO	0.87	0.40	3.03	3.22	0.59	0.90	1.00	0.57	0.70	3.16	2.70	2.74
Cr_2O_3	4.52	6.16	20.68	18.99	21.60	23.81	25.08	25.59	26.06	32.41	34.37	37.00
Al_2O_3	2.21	2.11	4.94	4.52	5.99	5.26	5.97	5.99	6.00	11.86	9.80	10.86
SiO_2	0.30	0.30	0.33	0.29	0.40	0.33	0.42	0.16	0.31	0.31	0.57	0.31
CaO	0.07	0.07	0.11	0.13	0.05	0.10	0.20	0.04	0.04	0.00	0.00	0.00
TOTAL	99.13	99.22	99.65	98.82	99.17	99.22	99.79	99.85	99.78	98.35	98.99	99.39

Oxide ratios

	1	2	3	4	5	6	7	8	9	10	11	12
Divalent	0.986	0.993	0.940	0.938	0.988	0.982	0.979	0.988	0.986	0.921	0.936	0.933
Trivalent	0.672	0.745	0.807	0.808	0.783	0.819	0.808	0.810	0.813	0.732	0.778	0.773
TAC	0.818	0.781	0.471	0.490	0.427	0.409	0.381	0.372	0.367	0.228	0.208	0.169

Atomic proportions based on 32 oxygen atoms

	1	2	3	4	5	6	7	8	9	10	11	12
R^{+2}	15.531	15.285	12.654	13.208	12.346	12.176	11.918	11.999	11.858	9.967	10.505	10.162
R^{+3}	1.836	2.191	6.301	5.860	6.980	7.238	7.692	7.861	7.951	10.942	10.860	11.717
R^{+4}	6.657	6.614	4.889	5.021	4.600	4.499	4.302	4.111	4.114	3.010	2.603	2.131
SUM	24.024	24.090	23.944	24.089	23.926	23.913	23.912	23.971	23.923	23.919	23.968	24.010

Table 1 (continued)

Anal.	13	14	15	16	17	18	19	20	21	22	23	24
TiO_2	19.67	4.32	16.72	16.26	15.37	10.03	7.80	5.80	5.63	4.86	4.60	3.87
MnO	0.36	0.44	0.52	0.49 .	0.48	0.54	0.51	0.45	0.47	0.43	0.44	0.48
FeO	48.58	37.31	42.73	44.82	44.89	35.77	35.48	32.47	31.82	33.38	33.13	29.29
MgO	1.60	0.83	3.39	0.75	0.84	4.02	3.53	4.41	4.69	4.58	3.71	6.49
Cr_2O_3	23.71	45.27	27.55	29.83	30.43	35.18	38.45	39.84	41.23	38.46	38.83	38.25
Al_2O_3	5.74	11.41	8.68	7.18	7.49	13.46	13.09	16.13	15.66	17.00	18.39	20.55
SiO_2	0.21	0.19	0.21	0.26	0.38	0.39	0.41	0.29	0.27	0.27	0.15	0.33
CaO	0.04	0.01	0.05	0.06	0.05	0.14	0.04	0.06	0.13	0.04	0.05	0.08
TOTAL	99.91	99.77	99.85	99.65	99.93	99.53	99.31	99.45	99.90	99.02	99.30	99.34

Anal.	25	26	27	28	29	30	31	32	33	34	35	36
TiO_2	6.59	4.24	2.78	2.86	2.05	4.79	12.72	11.53	9.82	14.61	13.93	12.54
MnO	0.44	0.39	0.42	0.25	0.29	0.20	0.28	0.31	0.27	0.33	0.33	0.25
FeO	33.86	32.64	31.66	27.72	27.50	29.04	36.37	37.13	35.80	38.43	38.81	36.61
MgO	3.65	3.72	4.06	7.50	7.38	6.14	4.32	4.14	4.14	5.23	5.09	5.34
Cr_2O_3	38.94	38.93	39.50	42.72	43.77	43.52	35.11	35.11	36.98	32.24	32.22	33.30
Al_2O_3	15.26	18.69	21.03	16.72	16.90	14.80	9.95	10.39	11.73	8.42	8.62	10.78
SiO_2	0.34	0.13	0.22	0.43	0.44	0.43	0.30	0.34	0.32	0.27	0.25	0.38
CaO	0.01	0.02	0.01	0.07	0.07	0.49	0.21	0.18	0.14	0.11	0.09	0.22
TOTAL	99.09	98.76	99.68	98.27	99.40	99.41	99.26	99.11	99.20	99.64	99.34	99.42

Anal.	37	38	39	40	41	42	43	44	45	46	47	48
TiO_2	0.47	12.08	3.65	1.35	1.00	0.19	0.56	0.41	0.24	0.20	0.47	0.64
MnO	0.04	0.28	0.26	0.09	0.22	0.15	0.27	0.19	0.19	0.19	0.16	0.24
FeO	21.74	33.95	19.99	21.25	17.78	17.82	21.22	18.94	17.07	18.18	17.34	20.32
MgO	14.71	5.73	15.79	14.01	16.00	15.57	12.39	14.47	15.56	14.81	16.30	12.98
Cr_2O_3	7.86	37.13	8.93	9.52	8.31	12.56	18.82	11.92	12.18	14.30	7.88	18.71
Al_2O_3	53.86	9.49	50.49	52.18	56.48	53.44	46.34	52.65	53.61	51.06	57.64	45.87
SiO_2	0.37	0.34	0.10	0.41	0.28	0.17	0.13	0.30	0.22	0.25	0.25	0.35
CaO	0.14	0.31	0.06	0.13	0.18	0.06	0.07	0.14	0.07	0.11	0.10	0.16
TOTAL	99.19	99.31	99.27	98.94	100.25	99.96	99.80	99.02	99.14	99.26	100.14	99.27

Anal.	49	50	51	52	53	54	55	56	57	58	59	60
TiO_2	51.17	50.70	51.98	48.19	51.26	50.91	53.18	53.57	52.78	52.52	53.24	52.71
MnO	0.38	0.52	0.38	0.33	0.36	0.39	0.62	0.54	0.58	0.47	0.47	0.54
FeO	46.86	46.84	46.07	50.37	47.24	47.02	46.01	45.44	46.04	46.39	45.30	45.67
MgO	0.47	0.27	0.70	0.27	0.29	0.78	0.43	0.41	0.43	0.69	0.72	0.74
Cr_2O_3	0.39	0.23	0.35	0.45	0.42	0.45	0.35	0.42	0.42	0.32	0.35	0.79
Al_2O_3	0.17	0.09	0.09	0.60	0.34	0.11	0.14	0.09	0.09	0.07	0.09	0.07
SiO_2	0.29	0.50	0.16	0.27	0.06	0.23	0.15	0.16	0.16	0.02	0.16	0.16
CaO	0.10	0.16	0.01	0.02	0.08	0.08	0.07	0.06	0.07	0.05	0.05	0.04
TOTAL	99.82	99.32	99.75	100.51	100.06	99.97	100.98	100.69	100.57	100.53	100.38	100.72

* Analyses 1–12 are reduced spinels. Anal. 1—Type A Cr-Al-ulvöspinel in 14310; Anal. 2—Type B Cr-Al-ulvöspinel in 15555; Anals. 3–9—Type C Al-titanian-chromite in 14053.

Analyses 13–27 are primary nonreduced spinels. Anal. 13—Al-titanian-chromite in 15555; Anal. 14—Mg-aluminian-chromite in 15555; Anal. 15–24—Al-Ti-chromite mantles and Mg-aluminian-chromites in cores and in olivine in 14053; Anal. 25–27—Mg-aluminian-chromite in 14321.

Analyses 28–30—Mg-aluminian-chromite veinlet in pyroxene (14258). Analyses 31–36—purple Al-Ti-chromite-ilmenite intergrowths in 14191 and 14258. Analyses 37–38—noritic spinels in 14258. Analyses 39–41—pleonaste-ilmenite intergrowths in 14063. Analyses 42 and 43—zoned pleonaste with core (42)-mantle (43) relationship in 14066. Analyses 44–48—are pleonaste and chromian-pleonaste in 14063 and 14066.

Analyses 49–60 are for ilmenite. Analyses 49–51 (15555) and 55–57 (14053) are discrete ilmenite; Anal. 52–54 (15555) and 58–60 (14053) are lamellar ilmenite in reduced spinels.

Fe_2TiO_4; thus any initial differences in parental composition will tend to change the starting point of the reaction and influence its relative stability but not the partitioning trend of re-equilibration.

The chromian-ulvöspinel in 15555 (Fig. 1b) is weakly anisotropic and tan in color, whereas the Al-titanian-chromite in 14053 (Fig. 1c) is optically isotropic and gray in color; the modal ratios of the reaction products (ilmenite + Fe)/nonreacted hosts (Cr-Al-ulvöspinel or Al-titanian-chromite) are approximately 1:10 in 15555 and approximately 10:1 in 14053. Sample 14053 is enriched in Cr_2O_3 by a factor of four, and in Al_2O_3 and MgO by factors of three and two, respectively; whereas sample 15555 shows a 13% increase in FeO content and a 10% increase in TiO_2 content. Apart from these differences, which may be due in part to their respective parental compositions, the divalent (FeO/FeO + MgO) and trivalent ($Cr_2O_3/Cr_2O_3 + Al_2O_3$) ratios do not reflect the overall effects of the reaction. However the ($TiO_2/TiO_2 + Al_2O_3 + Cr_2O_3$) ratio (abbreviated as the TAC ratio) shows a twofold decrease between 15555 and 14053. This decrease is compatible with the increase in ilmenite + Fe/spinel ratio and because modal analyses of intergrowths are difficult to determine and inherently subjective, the TAC ratio becomes a potentially useful quantitative reduction index. Alternatively, the TAC ratio also may be viewed as reflecting increasing degrees of crystallization, so that early formed Mg-aluminian-chromites typically have TAC ratios <0.1; this value increases progressively to 0.95 in primary Cr-Al-ulvöspinel and the theoretical maximum is 1.0 for pure Fe_2TiO_4. By induced phase separation of ilmenite + Fe the receding subsolidus path of the spinel very nearly is identical to the initial crystallization path, so that the TAC ratios conveniently may be applied to both processes. Furthermore, since Ti, Al, and Cr all have strong octahedral site preference energies (Navrotsky and Kleppa, 1967) the TAC ratio is restricted to the changes in one coordination site and circumvents the structural complexities that probably exist for intermediate members of the normal-inverse spinel series.

Closer examination reveals that the TAC ratio is in fact a sensitive re-equilibration parameter for the five major elements common in lunar spinels. For example, in the Type B assemblage illustrated in Fig. 1b the TAC ratio of the Cr-Al-ulvöspinel shows a steady decrease from 0.80 to 0.75 with increasing proximity to ilmenite; this difference is even more pronounced in traverses in 14053 where the TAC ratio may decrease from 0.35, in the center of one of the triangular or rectangular areas of the spinel, to <0.20 in the Cr-enriched border zone adjacent to ilmenite. The mean TAC ratio for ten spinel intergrowths (25 analyses) in 15555 is 0.79 (range = 0.6–0.9) and the mean for six spinel grains (21 analyses) in 14053 is 0.38 (range = 0.3–0.5). Sample 14053 is, therefore, more intensely and more uniformly reduced than 15555. By applying a similar analytical approach to Type B intergrowths in samples from Apollo 12 and Luna 16 we find that the TAC ratios are comparable with those of 15555 and that very little variation exists for an extensive survey of Apollo 15 samples (Haggerty, 1972a). Two fragments of basalt in breccia 14321 contain the Type C (14053) assemblage, both fragments have lower TAC ratios (range = 0.11–0.25; mean = 0.19) and both are therefore more intensely reduced than 14053. In summary, Apollo 12, 15, and Luna 16 samples have variable but high TAC ratios ($\simeq 0.8$),

Apollo 14 samples (14053 and 14321) have considerably lower TAC ratios ($\simeq 0.3$) and are more intensely reduced.

Further discussion of the chemistry and the systematics of individual oxide or element variations are left to a later section so that the reduction products may be contrasted with the crystallization trends exhibited by primary crystalline spinels.

With respect to the associated ilmenite (Table 1) little or no difference is observed between primary and secondary ilmenite nor are any major differences observed between Types B and C, although a critical property of these ilmenites is the valency state of Ti (Pavicevic et al., 1972) and the possibility of slight anionic deficiency ($FeTiO_{3-x}$).

Reduction environment and mechanism

The presence of metallic iron alone, but also the association of Fe + $FeTiO_3$, in Type B and C spinel assemblages is indicative of subsolidus reduction at low oxygen fugacities, and this is based on the following factors: (1) the reduction of Fe from ulvöspinel and ilmenite, determined experimentally at 1200°C, takes place at $10^{-12.6}$ and $10^{-13.0}$ atm, respectively (Webster and Bright, 1961). (2) Calculated univariant curves for the dissociation of Fe from ulvöspinel, ilmenite, hercynite, and chromite (Fig. 2) show that this is the relative order of stability—with ulvöspinel being the least stable and chromite the most stable (Haggerty, 1971b). These curves show, furthermore, that the initial stage ($Fe_2TiO_4 = Fe + FeTiO_3 + \frac{1}{2}O_2$) as well as the more advanced dissociation of ulvöspinel ($Fe_2TiO_4 = 2Fe + TiO_2 + O_2$) takes place at values of fO_2 that are higher than the decomposition of ilmenite ($FeTiO_3 = Fe + TiO_2 + \frac{1}{2}O_2$); these curves are no more than 0.5 orders of magnitude fO_2 apart, and the two latter curves in fact become coincident at lower temperatures ($<800°C$). Because TiO_2 is not observed, and the fact that ilmenite as well as chromite remain unaltered, these factors indicate that the maximum values of fO_2, for sub-solidus reduction, must lie between the stability curves for ulvöspinel and ilmenite. An estimate of these values would bracket fO_2 approximately between 10^{-15} and 10^{-16} atm at 1000°C. (3) In basalt sample 14053 intense subsolidus reduction of chromian-ulvöspinel (Type C) is accompanied in the same sample by the decomposition of fayalite to Fe + SiO_2 (El Goresy et al., 1971b; Haggerty, 1971b); the decomposition assemblage is illustrated in Fig. 1d. The fayalite-quartz-iron (FQI) buffer curve falls between the two ulvöspinel curves (Fig. 2), confirming therefore that the estimates of fO_2 cannot be grossly in error. These values of fO_2 are well below those determined for bulk rock fO_2 analyses ($10^{-10.8}$ to $10^{-12.8}$ atm) on Apollo 12 samples (Sato and Helz, 1971) and are also well below those required for the stable crystallization of ulvöspinel-rich members of the normal-inverse series (Muan et al., 1972), and finally (4) Reid et al. (1970) have presented convincing evidence for the progressive reduction of Fe from the melt with increasing crystallization, and although this evidence is not directly applicable to the reduction of spinels and does not apply to all samples, their proposed mechanism strongly supports the contention that processes of late stage reduction may occur in lunar basalts.

In Type A intergrowths metallic Fe is absent at the maximum resolving power of the reflecting microscope ($<1\,\mu$), and although Types A and B have comparable

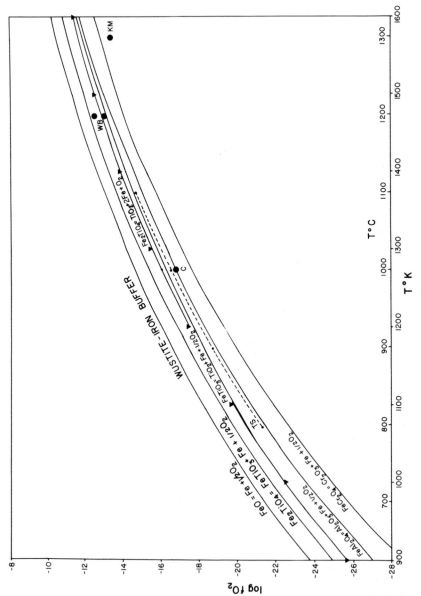

Fig. 2. Equilibrium dissociation curves for the Fe^{+2}-bearing spinels and ilmenite determined from basic thermodynamic data (Haggerty, 1971b). The curve marked TS is an alternative curve for $FeTiO_3$ (Taylor and Schmalreid, 1964). The full circles are specific experimental determinations for the decomposition of ulvöspinel and ilmenite—WB (Webster and Bright, 1961); chromite—KM (Katsura and Muan, 1964); and hercynite—C (McLean and Ward, 1966).

TAC-reduction ratios, the assemblage Cr-Al-ulvöspinel + ilmenite poses a slightly different genetic problem. Does this assemblage represent an even lower intensity of reduction than those containing metallic Fe, or are the oriented lamellae the products of exsolution (sensu-stricto)? Exsolution is not supported (a) by experimental determinations at low oxygen fugacities, insofar as spinel^I coexists with spinel^{II} in the normal-inverse series (Muan *et al.*, 1972); (b) there is no evidence in support of extensive FeTiO$_3$ solubility in the cubic spinel structure (Verhoogen, 1962; Bud-dington and Lindsley, 1964); and (c) the possibilities of an initially exsolved spinel inverting to ilmenite, or the transformation of cubic FeTiO$_3 \rightarrow$ rhombohedral FeTiO$_3$ seems unlikely.

The observation that the Type A assemblage does not occur unless Type B is also present suggests that although metallic Fe is absent, the assemblage is still a product of reduction, but is considered to be one of slightly lower intensity. If this interpretation is correct, and experimental verification is required, then the reduction process must proceed in two distinct stages, re-equilibrating initially to accommodate the metastable excess TiO$_2$ (as FeTiO$_3$), and subsequently the metastable excess FeO (as FeTiO$_3$ and Fe). The inference is not that Ti is more susceptible to mobilization than Fe^{+2} but rather that the more stable elements (Cr, Al, Mg) must maintain their balance of Fe^{+2} in the spinel structure as (FeMg)(CrAl)$_2$O$_4$, but once satisfied the system becomes saturated in Fe^{+2} and Fe^0 precipitates.

The mechanism proposed to account for the spinel-ilmenite assemblages, with and without metallic iron, therefore is as follows: With low intensity reduction, Fe and Ti selectively are mobilized, ions diffuse to {111} spinel parting planes and ilmenite is precipitated. The diffusion mechanism must be somewhat analogous to that of ulvöspinel in a mildly oxidizing environment (O'Reilly and Banerjee, 1967), and to ilmenite in a highly oxidizing environment (Haggerty, 1971c) because in both cases Fe and Ti exhibit selective although contrasting mobility rates. A process of selective diffusion,* induced by subsolidus reduction, or a process akin to it, must furthermore be invoked to account for the prominent Cr-enriched borders that mantle the ilmenite lamellae (Fig. 1c). With more intense reduction, the spinel host becomes highly enriched in Mg, Cr, Al, and is depleted in Fe + Ti; enrichment in the spinel is balanced by the concomitant development of proportionately larger concentrations of ilmenite, but at the same time discrete metallic Fe is liberated. Because rutile is not observed (see Fig. 2), the iron must be exclusively derived from the decomposition of the spinel by coupled precipitation of Fe + FeTiO$_3$, and this reaction must proceed at values of T and fO_2 below the dissociation curve for ulvö-spinel but within the stability field of ilmenite.

The widespread occurrence of the low intensity Type A, B assemblages suggests that the process responsible for reduction is an all-embracing one to which many basalts are subjected. The rarity of the intensely reduced Type C assemblage, however, suggests that this assemblage is more likely related to a sporadic selenological event. Common factors to both are (1) the relatively coarse grained nature of the inter-growth; (2) similar reaction products; (3) textural evidence for solid state reactions;

---

* See section on (Cr + Al) − Ti substitution for further discussion based on site preference energies.

(4) reaction only of late stage Fe_2TiO_4-enriched spinels; and (5) reaction at high temperature and low fO_2.

If we assume that crystallization takes place close to or above the QFI buffer curve then the only condition by which crystalline Fe_2TiO_4 can remain within its stability field are the constraints of having the cooling fO_2 curve parallel or remain above the buffer curve. If the cooling curve falls off more rapidly than the buffer curve the slope increment would not need to be substantially different before the ulvöspinel dissociation curve is crossed. If this condition arises at high temperatures the reaction would be spontaneous, if at intermediate temperatures more sluggish, and at lower temperatures would proceed with difficulty. Thus, the major differences between Types A, B and Type C may be a function of temperature with Type C occurring at elevated temperatures and Types A, B at somewhat lower temperatures. If this is the case then the widespread occurrences of Types A, B may be accounted for by deuteric cooling such that the rate at which fO_2 decreases with cooling is not drastically greater than the curve defined by the slope of the QFI buffer. To account for the Type C assemblage as a function of the rate of cooling would necessitate a rapid decrease in fO_2 with correspondingly small changes in temperature. With increasing concentrations of Cr + Al + Mg in the reacted spinel the stability regime increases to lower values of fO_2 (Fig. 2), so that the fO_2 cooling curve must be considerably steeper than the QFI buffer curve. This situation is unlikely to occur during normal cooling but could arise as a consequence of volatilization and oxygen loss at high temperatures. The evidence for reheating of sample 14053 presented by Finger *et al.* (1972), which is based on the natural cation distribution in pigeonite, could reasonably account for the observed intense reduction assemblage; whether reduction ensued as a result of oxygen loss during reheating or was induced by a gaseous reduction agent is unknown, but perhaps may be related to the closely associated impact event at Cone Crater.

Related subsolidus reactions

Subsolidus reactions involving spinels are identified in terms of five basic types: (1) subsolidus reduction of Cr-Al-ulvöspinel; (2) *exsolution** of ilmenite from chromian-pleonaste; (3) *exsolution* of ilmenite from a distinctively colored purple Al-titanian-chromite (abbreviated hereafter as purple-chromite to distinguish it from the more usual bluish-gray varieties); (4) exsolution of Mg-aluminian chromite from pyroxene; and (5) symplectic intergrowths of ulvöspinel and chromite in olivine or pyroxene.

Subsolidus reduction is widespread as outlined in the above discussion. The four remaining occurrences have not been observed previously in lunar samples and these assemblages, although rare in the Apollo 14 material, are considered sufficiently important for brief discussion.

---

* *Exsolution* is used in the textural sense to denote exsolution-like intergrowths and is not intended to be used in the genetic sense of exsolution (*sensu-stricto*) from solid solution.

Pleonaste-ilmenite and chromite-ilmenite intergrowths

Subsolidus reduction of Cr-Al-ulvöspinel is associated classically, on decomposition, with the reaction products Fe + FeTiO$_3$. Discrete metallic Fe is absent in the purple-chromite and the pleonaste assemblages (Fig. 3a–c), and also is absent in the Type A Cr-Al-ulvöspinel assemblage discussed earlier (Fig. 1a); although, because the latter is closely associated with Cr-Al-ulvöspinel that does contain metallic Fe, it is assumed that the reaction proceeds in two distinct stages in environments of low intensity reduction. The same mechanism of low intensity reduction and induced diffusion perhaps can be argued in favor of these Ti-aluminian and Mg-Al-rich spinel-ilmenite intergrowths, but for the same reasons as were quoted

Fig. 3. *Subsolidus reactions involving spinels.* (Photomicrographs were taken under oil-immersion objectives and in reflected light.)

(a) Chromian-pleonaste (Fe$_{3.68}$Mg$_{4.98}$Mn$_{0.05}$Cr$_{1.56}$Al$_{13.12}$Ti$_{0.61}$O$_{32}$) with thin ilmenite lamellae along {111} planes. (14063). Field width = 270 μ.

(b) Purple Al-Ti-chromite (Fe$_{8.34}$Mg$_{2.24}$Mn$_{0.06}$Cr$_{7.12}$Al$_{3.47}$Ti$_{2.59}$O$_{32}$) containing fine oriented lamellae of ilmenite along {111} planes. (14191). Field width = 205 μ.

(c) High magnification of (b) under partial oblique illumination. Although the ilmenite lamellae appear to be elevated, the Al-Ti-chromite is in fact the harder of the two phases. (14191). Field width = 110 μ.

(d) Thin parallel lamellae of chromite and a veinlet of Mg-aluminian-chromite (Fe$_{6.73}$ Mg$_{2.49}$Mn$_{0.06}$Cr$_{9.29}$Al$_{4.46}$Ti$_{0.87}$O$_{32}$) in pyroxene (En$_{88.4}$Fs$_{11.0}$Wo$_{0.7}$). The veinlet is compositionally heterogeneous along its length as shown by two typical analyses in Table 1. (14258). Field width = 460 μ.

earlier, exsolution (*sensu-stricto*) or high intensity oxidation cannot have been responsible. The ilmenite lamellae in these spinel assemblages are well-oriented along
{111} spinel parting planes, are present in small concentrations ($<5\%$), and in thin
(<1–$5\,\mu$) discontinuous plates (Fig. 3b). The low ilmenite concentration suggests
that although these spinels must originally have had small contents of titanium in
solid solution (Table 1), and the purple-chromites still do (10%–15% TiO_2), the
TiO_2 content could not have been very large and certainly cannot have been initially
comparable in composition to Cr-Al-ulvöspinel. The mean TAC ratios of the purple-
chromites and the chromian-pleonastes are 0.21 and 0.01, respectively. Thus, although the TAC ratio of chromian-pleonaste is lower, and that of the purple-chromite
higher, than the most intensely reduced examples observed in this study, these ratios
must be strongly influenced by their respective parental compositions. From a
thermodynamic point of view, Mg-Al-Cr spinels will tend to be considerably more
stable than their Fe or Fe-Ti-bearing counterparts (Fig. 2), so that while the ilmenite
may have resulted by subsolidus reduction, the energy required to liberate free
metallic Fe would be considerably greater and the environment of fO_2 would almost
certainly have been well below the stability field of ilmenite.

Although the origin of these spinel-ilmenite assemblages remains enigmatic the
fact that they do occur, albeit in minor concentrations, possibly may be the initial
clue to the elusive occurrence of Mg_2TiO_4 in the lunar samples.

Spinel-silicate intergrowths

Fine (<1–$5\,\mu$) platelets of gray isotropic chromite were observed in four discrete
crystals of pyroxene ($En_{88.4}Fs_{11.0}Wo_{0.7}$) in sample 14258. These lamellae are
parallel sided, 10–15 μ in length and are generally oriented in one prominent direction.
A veinlet of heterogeneous Mg-aluminian chromite (Table 1) is associated with one
of these pyroxene crystals (Fig. 3d) and although accurate analyses of individual
lamellae are impossible the lamellae and the veinlet have comparable optical properties; it is concluded therefore that although the two textural forms of chromite
may have had quite different origins, they are probably not substantially different
in composition. The mean TAC ratio of the veinlet is 0.04, and of the analyses
determined in this study, comparably low TiO_2 contents (2%–4.8%) are exhibited
only by early formed primary Mg-aluminian-chromites, but the latter rarely show
the same degree of Mg-enrichment (7.5%). The FeO/(FeO + MgO) ratio of primary
chromite in 14053 and 14321 is approximately 0.9, the veinlet is 0.8 and the chromian-
pleonastes range from 0.5 to 0.65. The chromite lamellae are considered to be
products of exsolution but the veinlet is more probably the result of primary coprecipitation, or alternatively a metamorphic product; the variation in the composition of the veinlet (Table 1) perhaps would argue more favorably for a solid state
reaction; the nature of the redox environment is of considerable interest but is
unknown.

The spinels in the second group of silicate intergrowths are at present only qualitatively identified and these results will not be presented until more detailed work has
been carried out.

SPINEL TYPES, COMPOSITIONS AND OXIDE VARIATIONS

A representative selection of electron microprobe analyses of spinels are given in Table 1 and all data are shown in Figs. 4–8. It is important to note that these data embrace not only compositions of primary spinels, but also spinels that have undergone varying degrees of subsolidus reduction; analyses of the chromite veinlet in pyroxene, the purple-chromites and the pleonaste-ilmenite assemblages, described above are also included.

The multicomponent spinel prism employed in previous studies (Haggerty, 1971a, 1972b) with spinel analyses from the Apollo 14 samples is shown in Fig. 4. The base and a rectangular face projection for the prism are shown in Figs. 5a and 5b, res-

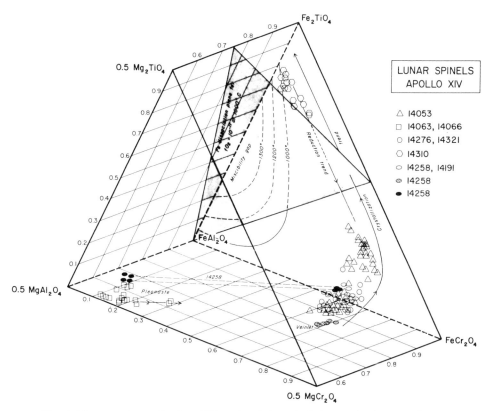

Fig. 4. Electron microprobe data (125 analyses) for spinels in samples from Apollo 14, showing crystallization and reduction trends, experimentally determined miscibility gaps for the face Fe_2TiO_4–$FeAl_2O_4$–$FeCr_2O_4$ at 1000, 1200, and 1300°C and 10^{-10} atm (Muan *et al.*, 1972), and one Fe-dissociation plane at 1000°C and 10^{-17} atm (Haggerty, 1971b). The tie-lines join pleonaste and Al-Ti-chromite in a noritic fragment (full ellipses). The arrow in the chromian-pleonaste region indicates the compositional trend for core-mantle zoning. Reduced spinels in 14053 (Δ) and 14321 (◯) are above the join $FeAl_2O_4$–$FeCr_2O_4$ and are close to the ternary Her-Chr-Usp face. Nonreduced spinels in the same samples fall close to the prism base. The chromite veinlet (shaded ellipses) is shown in Fig. 3d. Purple-chromite (Fig. 3b–c) are open ellipses; hexagons are Cr-Al-ulvöspinel.

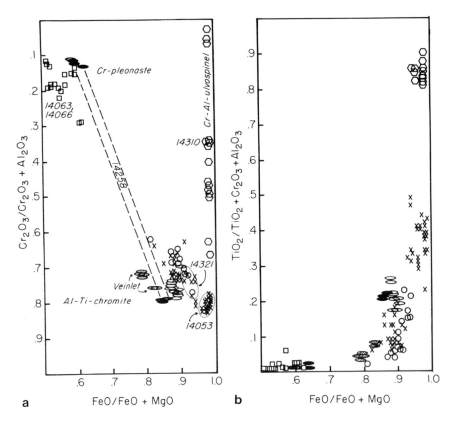

Fig. 5. (a) A base projection of the multicomponent spinel prism shown in Fig. 4. The reduced spinels in 14053 (✗) and 14321 (○) are enclosed by large open ellipses, and the noritic spinels are joined by tie lines.

(b) A front rectangular face projection of the multicomponent prism shown in Fig. 4. The vertical axis represents the TAC-reduction or TAC-recrystallization index. Symbols are the same as those used in Figs. 4 and 5a.

pectively. The base projection is defined by the trivalent ratio $Cr_2O_3/(Cr_2O_3 + Al_2O_3)$ as a function of the divalent ratio $FeO/(FeO + MgO)$, and the face projection is similarly defined by the TAC ratio $TiO_2/(TiO_2 + Al_2O_3 + Cr_2O_3)$ versus the divalent ratio. These projections permit a less integrated but more detailed examination of crystallization and re-equilibration trends; individual analyses can be more accurately located and the discontinuities, within and between the normal-inverse and the normal spinel series, also become more apparent. Furthermore, these discontinuities conveniently are defined in terms of the divalent, trivalent, or TAC ratios.

The Apollo 14 samples contain the largest variety of spinels identified in any of the lunar material to date. These spinels fall into three discrete compositional groups: (1) chromian-pleonastes with low concentrations of titanium (1% TiO_2); (2) Mg-aluminian-chromite grading into Al-titanian-chromite; and (3) Cr-Al-ulvöspinel.

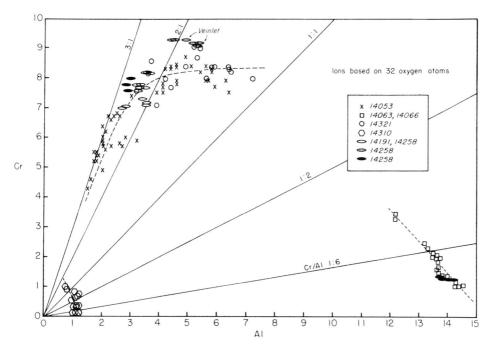

Fig. 6. Cr-Al substitution in Apollo 14 spinels. Substitutional trends are discussed in the text and the symbols are the same as those used in Figs. 4 and 5a.

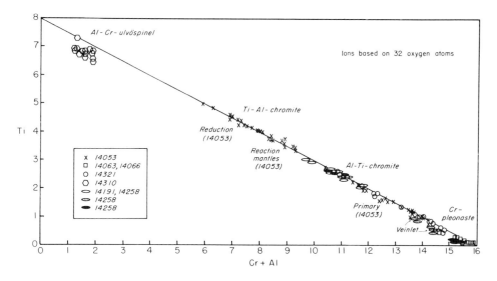

Fig. 7. (Cr + Al) − Ti substitution for Apollo 14 spinels. The straight line represents the ideal octahedral site concentrations between normal and inverse spinels; the significance of intermediate members (Ti = 4; Cr + Al = 8) are discussed in the text. (See captions to Figs. 4 and 5a for symbol identification.)

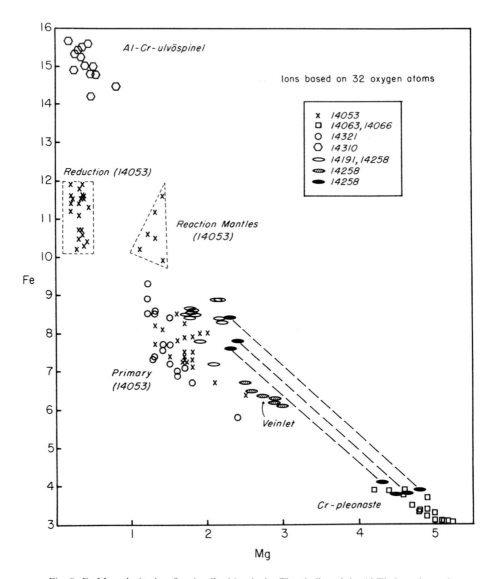

Fig. 8. Fe-Mg substitution for Apollo 14 spinels. The tie-lines join Al-Ti-chromite and pleonaste in a noritic fragment. (See captions to Figs. 4 and 5a for symbol identification.)

Group 1. Pleonaste and chromian-pleonaste are compositionally located close to or on the base of the spinel prism between $MgAl_2O_4$ and $FeAl_2O_4$ (Fig. 4). These spinels are present in discrete mineral grains in the breccias (14063 and 14066) and also are present in a single fragment of norite in soil sample 14258. The spinels are optically isotropic, transparent or semi-opaque of low reflectivity (5%–10%), and are brown in reflected light but pink in transmitted light. Grains are subhedral or angular in shape. Two of the fifteen grains observed in the breccias are zoned compositionally with Mg-Al-rich cores ($Fe_{3.22}Mg_{5.01}Mn_{0.03}Al_{13.59}Cr_{2.14}Ti_{0.03}O_{32}$) and Fe-Cr-rich mantles ($Fe_{3.89}Mg_{4.24}Mn_{0.05}Al_{12.28}Cr_{3.42}Ti_{0.1}O_{32}$). Four grains contain oriented ilmenite lamellae (Fig. 3a) and both of the zoned crystals contain anhedral ilmenite in the peripheral mantle. Both textural forms of ilmenite are fine grained and accurate analyses were not possible, but strong reflection anisotropy, positive high Ti content, and the absence of Cr confirm the identification of ilmenite.

The chromian-pleonaste in the noritic fragment is unzoned and the associated opaque minerals are ilmenite (4.1 % MgO), Al-titanian-chromite (Table 1), troilite, and metallic Fe. Both spinels are present as discrete fragments (joined by tie-lines in Fig. 4) and their paragenetic relationship is therefore indeterminate; both are earlier than either troilite or Fe and both appear also to be earlier than either plagioclase or pyroxene. The coexistence of these two spinels of such widely contrasting compositions suggests possible magmatic contamination. Ilmenite is the dominant opaque mineral and the spinels are present in small concentrations ($<1\%$). An exotic origin for one or the other of these spinels cannot be ruled out but it is important to note that the zoned pleonastes do show Fe-Cr enrichment towards $FeCr_2O_4$ (Fig. 5a) and that this enrichment trend closely parallels the direction that is subsequently followed by the early formed Mg-Al-rich chromites in the basalts (Figs. 4 and 5a).

Group 2. (a) Mg-aluminian-chromites are present in three textural forms (1) as discrete crystals (14053, 14321); (2) in cores mantled by Al-Ti-chromite (14053); and (3) in veinlets and lamellae in pyroxene (14258). Discrete crystals and the core-mantle relationship are common to most lunar basalts, but this is the first occurrence of chromite veinlets and lamellae to be reported in the lunar samples. This spinel group is characterized by high Al_2O_3 contents; 14053 varies from 16%–20% Al_2O_3; 14321 displays a range from 18% to 23%, and the veinlet varies from 14% to 16% Al_2O_3 (Table 1). These values are considerably higher than those determined for Apollo 11 and 12 and are matched only by those from the Luna 16 site (Haggerty, 1972b; Jakeš et al., 1972), where values of up to 25% Al_2O_3 were determined. The veinlet has the highest MgO content (7.5%) in this group and the MgO contents of 14053 and 14321 are in the 4%–6% range. The trivalent ratio of the veinlet is 0.74 and this value is intermediate between those of 14053 and 14321 (Fig. 5a); the TAC ratios of all three types are comparable (0.05); the divalent ratio of the veinlet is 0.8 and is slightly lower than that of the spinels in the basalts (0.9), although there is some degree of overlap (Figs. 5a–b).

(b) Al-Ti-chromites have almost equal contents of Al_2O_3 and TiO_2 (8%–14%), and these spinels have TAC ratios between 0.1 and 0.25 (Fig. 5a). Spinels that fall within this subclass are: (1) mantles on Mg-aluminian-chromite in basalt 14053; (2) the noritic spinel that coexists with pleonaste in 14258; (3) the intensely reduced spinels (Fe + $FeTiO_3$) in 14321; and (4) the purple-chromite-ilmenite assemblages in 14258 and 14191. These spinels have remarkably similar compositions in spite of their diverse origins and associated mineralogies and are located along the $FeAl_2O_4$-$FeCr_2O_4$ join but slightly above the base of the prism (Fig. 4). The divalent ratio centers around 0.9; the trivalent ratio varies from 0.7 to 0.8 with the noritic spinel showing the highest value, and the remaining types tightly clustered around 0.75 (Fig. 5a, b).

(c) Al-titanian-chromite is characterized by its higher TAC ratio (0.3–0.5) and is present in the intensely reduced basalt 14053. Compositions are located close to the ternary face (FeAl$_2$O$_4$ – FeCr$_2$O$_4$ – FeTiO$_4$) but well above the prism base (Fig. 4). As noted earlier this sample is less intensely reduced than the fragments in breccia 14321, as defined by the TAC-reduction ratio, although both samples contain approximately equal concentrations of Fe + FeTiO$_3$. The Al$_2$O$_3$ content of the Al-titanian-chromite in 14053 varies from 3% to 5%; this value contrasts significantly with the 16%–20% range determined for the early formed Mg-aluminian-chromites and is also very different to the 12%–14% range determined for the Al-Ti-chromite mantles. Apart from the spinels in the Luna 16 material, intermediate members of the normal-inverse series are characterized by relatively low Al$_2$O$_3$ contents regardless of the overall aluminum content of the rock; so that in this respect 14053 appears to follow the primary crystallization trend exhibited by many other lunar basalts. The re-equilibrated Al-titanian-chromites contain approximately 1% MgO, and the coexisting ilmenite contains < 1% MgO, which substantiates the absence of significant Mg$_2$TiO$_4$ in the parent spinel. Thus, although the primary composition of the late stage spinel in 14053 is unknown, and furthermore is difficult to recast from the reaction assemblage, there is compelling evidence in support for a parental spinel with high TAC values (0.8–0.9) and this is provided collectively by the Al$_2$O$_3$ content, the MgO content, the TAC ratio, and the apparent similarity between the initial path of crystallization and the receding trend of re-equilibration. The composition of this spinel cannot be far removed from the solid solution series (0.7 FeCr$_2$O$_4$ – 0.3 FeAl$_2$O$_4$) – Fe$_2$TiO$_4$, and because the width of the series narrows with increasing Fe$_2$TiO$_4$ content (Haggerty, 1972b), the spinel in 14053 must have been somewhat similar to the *initial* composition of the Cr-Al-ulvöspinel in 15555, cited in an earlier discussion.

A comparison of the re-equilibrated spinels in 14053 with the primary spinels shows that the former have high divalent ratios (0.96), high trivalent ratios (0.82), and the highest TAC ratios (0.3–0.5) of the chromites classified in this group.

Group 3. Cr-Al-ulvöspinel is restricted to sample 14310, and although optically similar spinels were observed in the soil samples these were not analyzed in the present survey. These spinels have high TAC ratios (0.8–0.9) with Al$_2$O$_3$ contents in the range 2.2%–3.5% and Cr$_2$O$_3$ contents between 1.7% and 4.5%. The trivalent ratios vary between 0.02 and 0.68, whereas the divalent ratio is almost constant at 0.98.

CATIONIC SUBSTITUTION

The oxide ratios discussed above delineate in a general way both the path of crystallization and the trend of subsolidus re-equilibration. Because of the apparent and surprising similarity in these trends and because the end members represent the extremes of two spinel structural classes, a more detailed examination of ionic substitution is warranted; firstly, to evaluate the nature of intermediate coordination types and secondly, to determine the possible presence of Cr^{+2} and Ti^{+3} in the reduced spinels.

The site locations of Cr, Al, and Ti in the spinel structure are critical to confirmation of the normal-inverse character of the lunar spinel series. Cr^{+3} has the largest octahedral (B) site preference energy of the transition metals found in spinels (McClure, 1957; Dunitz and Orgel, 1957) and is of the order of -20 kcal (Navrotsky and Kleppa, 1967). Al^{+3} also has a large B site coordination preference which is approximately -10 kcal (Navrotsky and Kleppa, 1967) and is surpassed only by Ni^{+2} and Mn^{+3}. Al^{+3} has been shown to occupy the A site only in the nickel spinel aluminates (McClure, 1957), and in the synthesis of $MgAl_2O_4$ below $1000°C$, where a small degree of inversion is observed by tetrahedrally coordinated Al^{+3} (Dalta and Roy, 1967). Ti^{+4} also has a large octahedral site preference energy (Stephenson, 1970) and is probably lower or equal to that of Al^{+3}, except in Ni-bearing spinels where Ti^{+4} is tetrahedrally coordinated (Gorter, 1954). Fe^{+2} and Mg^{+2} have small octahedral and tetrahedral site preference energies respectively (Navrotsky and Kleppa, 1967). The site preference of Ti^{+3} is unknown and Cr^{+2} probably favors tetrahedral coordination (Ulmer and White, 1966).

In the following empirical analysis the electron microprobe data have been recalculated on the basis of the spinel formula with 32 oxygen atoms; the substitution of Cr for Al; Cr + Al for Ti; and Fe for Mg are considered separately.

Cr-Al substitution

Three contrasting trends of Al-Cr substitution are shown for the Apollo 14 spinels (Fig. 6). The Group 1 pleonastes show a simple linear trend of negative slope with high Al contents grading to intermediate Cr contents. Group 2 chromites have Cr/Al ratios between 1:1 and 3:1. Values between these limits fall approximately along a hyperbolic curve as shown in Fig. 6. Cr/Al ratios between 1:1 and 2:1 are accompanied by small decreases in Cr content, but significantly larger decreases in Al content result in a relatively flat positive slope. Between 2:1 and 3:1 however there is a dramatic decrease in the Cr content and the curve steepens significantly to low values of Al and intermediate values of Cr. This rapid decrease is correlated with the reduction-equilibrated spinels in 14053 and to a lesser extent to the intensely reduced spinels in 14321. The primary Mg-aluminian-chromites fall on the flat portion of the curve, the chromite veinlet falls above the curve and the noritic and purple chromites occupy the region of maximum slope increment. Group 3 Cr-Al-ulvöspinels in 14310 have small contents of Cr and Al but the Cr/Al ratio varies from 2:1 to <1:6; and the slope like that of the pleonastes is negative although the increase in the Cr content of the ulvö-spinel is significantly smaller than that of the pleonastes.

The character of the hyperbolic curve is of particular interest with respect to subsolidus reduction, and the possible presence of Cr^{+2}, which has been shown to be present in the series $FeCr_2O_4 - MgCr_2O_4$ at low fO_2 (Ulmer and White, 1966). Furthermore if the purple coloration of the Ti^{+3} glasses referred to by Dunitz and Orgel (1957) is indicative of Ti^{+3} then the chromium in the purple-chromites may also be in the Cr^{+2} state rather than Cr^{+3}. The property responsible for the unusual color of these chromites is unknown as in the nature of the hyperbolic curve but for the substitution of (Cr + Al) $-$ Ti it would appear that Cr^{+2} and Ti^{+3} probably are absent but spectral data are obviously required. The "compressional" effects associated with the substitution of the larger Cr^{+3} ion in replacing the smaller Al^{+3} ion (Dunitz and Orgel, 1957), and the relative volume expansions associated with redox reactions in spinels are of interest but are beyond the scope of this investigation.

$(Cr + Al) - Ti$ substitution

In contrast to the variable trends shown for Al-Cr substitution (Fig. 6) the trivalent-tetravalent relationship is a straight line relationship (Fig. 7). The TAC ratio is related to the substitution of $Cr + Al$ for 2Ti and this substitution is effectively limited to the octahedrally coordinated B sites in the spinel structure. In the structurally normal chromites and pleonastes (Fig. 7) the B site is occupied by a maximum of 16 trivalent cations ($Cr + Al$); and in the structurally inverse ulvöspinels (Fig. 7) the B site is replaced by a maximum of 8 divalent (Fe, Mg) + 8 tetravalent (Ti) cations. The Apollo 14 data (Fig. 7) demonstrate that the principle of a normal-inverse spinel series is satisfied insofar as with decreasing Ti content the octahedrally coordinated Fe + Ti ions are replaced progressively by Cr + Al. This replacement must be rigidly effected because except for the Cr-Al-ulvöspinels in 14310 there are no large deviations from ideality. The amount by which the ulvöspinels are displaced negatively is considered to be larger than that due to analytical error. This deviation from ideality is considered to be real (it may for example account for the reflection anisotropy in ulvöspinel caused by lattice distortions) and the displacement is confirmed by ulvöspinels in the Luna 16 samples.

The intermediate structural regime (Ti = 4; Cr + Al = 8) is occupied predominantly by the reduced spinels in 14053, and a specific example of one of the electron microprobe analyses (Anal. 8, Table 1) is sufficient to show that the expected coordination type must, at least empirically, be between $Fe_8[Fe_8Ti_8]O_{32}$ and $Fe_8[(Cr + Al)_{16}]O_{32}$ and should therefore conform to the intermediate formula $Fe_8[Fe_4Ti_4(Cr + Al)_8]O_{32}$; square brackets indicate B site coordination. The analysis shows $\Sigma R^{+2} = 11.99$; $\Sigma R^{+3} = 7.86$ ($Cr^{+3} = 5.83$; $Al^{+3} = 2.03$) and $Ti^{+4} = 4.05$. Thus, the normal-inverse character of this spinel is confirmed in terms of tetravalent and trivalent ions, with $\frac{2}{3}$ of the R^{+2} ions predictably occupying the A sites, because of the coincidence of this analysis with the B site spinel *line* in Fig. 7.

The plot in Fig. 7 shows furthermore that the paragenetic sequence Mg-aluminian-chromite, Ti-aluminian-chromite, and Al-titanian-chromite is exactly the reverse of the subsolidus re-equilibration trend, so that regardless of whether ionic substitution is accomplished by initial crystallization or by subsequent subsolidus reduction, the content of Ti + Cr + Al is maintained and balanced in the octahedrally coordinated sites. In calculating the atomic proportions it is assumed that titanium and chromium are in the Ti^{+4} and Cr^{+3} states, respectively, and because the data fall close to or exactly on the ideal octahedral spinel *line*, defined by the slope of $Cr + Al = 2:1$, it is concluded that if Ti^{+3} or Cr^{+2} are present, their respective concentrations must be very small.

In the ideal case of a normal $(FeCr_2O_4)$—inverse (Fe_2TiO_4) spinel series the divalent ions are exclusively in the A sites in normal spinels, and are equally distributed between the A and B sites in the inverse spinels; and it follows therefore that intermediate members have intermediate divalent distribution. With respect to subsolidus reduction the question now arises as to whether it is the A or B site Fe^{+2} that is mobilized. As the spinel is reconstituted a higher degree of the normal type co-ordination is achieved (Fig. 7) and this is accomplished in part because Fe^{+2} does not have a strong coordination preference for either A or B sites (Navrotsky and Kleppa, 1967), but more importantly the increased concentrations of Cr + Al will tend to replace the B site Ti^{+4} and presumably in doing so also a percentage of B site Fe^{+2}. The increase in Cr + Al content is twice as large as the Ti content that is removed to form $FeTiO_3$, so that most of the B site Fe^{+2} must be removed, and since the potential for B site coordination by Fe^{+2} is now decreased, Fe^{+2} occupying a B site may migrate either to an A site position, which is effectively an $Fe^{+2} - Fe^{+2}$ exchange or alternatively combine with the available Ti as $FeTiO_3$. Because Ti^{+4} is the critical ion that is mobilized by subsolidus reduction this may provide permissive evidence for B site instability of Fe^{+2} under strongly reducing conditions. Thus, by preferential B site activity, in the presence of Cr + Al (which both have higher octahedral site preference energies), the Ti is effectively forced to relinquish its B site position, in favor of Cr + Al, and follow the Fe^{+2}. It is also perhaps at this stage of instability that part of the Fe^{+2}, which is already in excess, is reduced to the metallic state.

If the reduction process can be equated to that of oxidation, then the above model is very similar to that described by O'Reilly and Banerjee (1966). They con-clude that the tetrahedral site population in $Fe_2TiO_4 - Fe_3O_4$ solid solutions remains unchanged because tetrahedrally sited ions in spinels have very low diffusion rates, and that oxidation in this series occurs largely at the expense of octahedral Fe^{+2}. However the presence of Cr^{+3} and Al^{+3}, in the lunar spinel system, are con-sidered to be an equally important property and perhaps the overriding factor, because of their high octahedral site preference energies.

Fe-Mg substitution

It was argued in an earlier paper (Haggerty, 1972b) that the incoherence of Fe-Mg substitution in the normal-inverse series probably was a crystallochemical effect related to the changes in the A or B site positions of the divalent cations between the two end members of the spinel series. Part of the scatter that is observed can be removed if the Ti content is simultaneously considered, and when this is implemented three distinct linear slopes with negative inclination emerge: one for Cr-Al-ulvöspinel (steep), a second for intermediate members (intermediate slope), and a third for the Mg-aluminian-chromites (flat). The Apollo 14 data (Fig. 8) show much the same pattern. At the two extremes Cr-Al-ulvöspinel and the pleonastes show good negative linear correlations, but those falling within the interim show little or no correlation whatsoever unless the groups of spinels are separately considered. For example, the chromite veinlet, the noritic chromite and the purple-chromite (all ellipses in Fig. 8) fall approximately along a common straight line; a second curve could account for the spinels in 14321 (circles in Fig. 8), but the spinels in 14053 do not form a coherent unit and spatially are separated into three groups. This separation correlates with whether the spinel is primary Mg-aluminian-chromite, mantle Al-Ti-chromite or reduced Al-titanian-chromite. The reduced spinels contain the lowest Mg contents (comparable to Cr-Al-ulvöspinel in 14310), the mantles are intermediate, and the primary chromites contain the highest Mg contents which are similar to the purple-chromites and the Mg-aluminian-chromites in 14321. Although there is considerable scatter for the intermediate region, the overall tendency is for Mg to increase as Fe decreases; and in this respect it is of interest to note that the tie-lines joining the noritic spinels would closely approximate in slope to a curve drawn from Cr-Al-ulvöspinel through the chromite regime to pleonaste. The trend between chromite and pleonaste is discussed more fully in the following section.

SOLID SOLUTIONS AND COMPOSITIONAL DISCONTINUITIES

The compositional separation of the three spinel groups determined in this study is shown ideally in the multicomponent prism in Fig. 4. The low-Ti pleonastes and the Mg-aluminian-chromites are close to or on the base of the prism, but are separated by a prominent compositional discontinuity defined by trivalent ratios between 0.3 and 0.6 (Fig. 5a). This discontinuity obviously is not reflected in the TAC ratios because of the uniformly low Ti content, but is apparent in the divalent ratio for values between 0.65 and 0.77 (Fig. 5b).

The small but overall increases in both the trivalent and divalent ratios shown by the pleonastes (Fig. 5b) and the corresponding crystallization trend from Mg-aluminian-chromite to Al-titanian-chromite (Fig. 5a) suggests that the similarity in the direction of these enrichment trends may in fact represent a second normal-spinel solid solution series between $(0.5\ MgAl_2O_4 - 0.5\ FeAl_2O_4)$ and $FeCr_2O_4$ (Fig. 4). The Mg-Al-spinels are typically associated with non-mare material, whereas the Mg-aluminian, aluminian and Al-titanian-chromites are typical of mare basalts, so that although the basaltic chromites show early enrichment of Mg and Al, inter-mediate members of this fractionation series have not yet been observed; or, at least, if these are present this fractionation is apparently not reflected in the compositions of the spinels. The fractionation is specifically one of early Mg depletion and to a lesser extent Al. It would seem that at least part of an Al-fractionation trend is observed in the Luna 16 samples, but once again high Mg concentrations are absent (Haggerty, 1972b).

A second discontinuity is present between the base of the prism and Fe_2TiO_4. Spinels on or close to the prism base include the chromite veinlet, the noritic chromite and primary Mg-aluminian chromites in 14053 and 14321. Moving away from the prism base and towards Fe_2TiO_4 we encounter the purple-chromites, the Al-Ti-chromite mantles in 14053, the reduced spinels in 14321 and finally the less intensely reduced spinels in 14053. No compositions were determined between this latter group and the Cr-Al-ulvöspinels in 14310.

This discontinuity is defined by TAC ratios between 0.5 and 0.8, and forms an interesting comparison with the Apollo 12 miscibility *gap*. This *gap* is defined by TAC ratios between 0.30 and 0.55 and with one exception all of the Apollo 11 spinels fall precisely within the Apollo 12 *gap* (Haggerty, 1971a). Thus, the discontinuity in the Apollo 14 series is displaced with respect to the Apollo 12 *gap* and although this may be significant it may be due also to insufficient data. The important property is the fact that the Apollo 12 *gap* is occupied by the reduced spinels in 14053. These spinels and those of Apollo 11 span comparable TAC values (0.3–0.5 and 0.3–0.55, respectively), and have identical trivalent (mean = 0.78) and divalent ratios (mean = 0.96), but the two groups of spinels have had totally contrasting histories and this suggests that the precipitation of immediate compositions, in the normal-inverse series, may be T and fO_2 dependent. Alternatively, the Apollo 11 spinels may have totally re-equilibrated and if this is the case the lower prevailing fO_2 would account in part for the larger proportion of ilmenite and the absence of Fe_2TiO_4. One of the striking features of lunar basalts, other than Apollo 11, is the occurrence of two spinels of contrasting composition (Mg-aluminian-chromite and Cr-Al-ulvöspinel) or the occurrence of only one spinel, which is always Cr-Al-ulvöspinel but never just Mg-aluminian-chromite. Thus, if two spinels are present with an intervening discontinuity in composition it is possible that this reflects not only a paragenetic discontinuity, but a radical change in the environment of late stage crystallization (Cameron, 1971). If only one spinel is present then either the Apollo 11 situation arises or alternatively the liquid is highly fractionated and early formed (FeMg) $(CrAl)_2O_4$ is removed by gravitational settling.

Muan and his co-workers (1971, 1972) have added significantly to our understanding of the complex spinel system and their experimental studies have delineated two major miscibility gaps; one between Fe_2TiO_4 and $FeAl_2O_4$ and a second between Mg_2TiO_4 and $MgAl_2O_4$. The first of these is shown schematically in Fig. 4 and is taken from Muan *et al.* (1972); the second is not shown and is of less interest because Mg_2TiO_4 is not observed in lunar spinels. The extent of the miscibility gap increases towards the join $Fe_2TiO_4 - FeCr_2O_4$ with decreasing temperature and perhaps accounts for the lack of more aluminous-rich intermediate members of the series but does not explain the two spinel occurrences that are so common in lunar basalts. There is little doubt that the Apollo 12 *gap* exists as shown by the summarized data from 11 different laboratories (Haggerty, 1971a) and furthermore is confirmed beyond all doubt by the Apollo 15 samples which show that approximately 10% of 356 spinel analyses have intermediate compositions in the normal-inverse series (Haggerty, 1972a). A number of possibilities have been suggested to account for the lack of intermediate compositions (Brett *et al.*, 1971; Weill *et al.*, 1971; Haggerty, 1972b) and these will not be repeated here, except to add that T, fO_2 and the crystallochemical properties of a complex normal-inverse spinel series have not been fully explored.

Summary and Conclusions

Subsolidus reduction of paragenetically late Fe_2TiO_4-enriched spinels decompose at high temperatures and low fO_2 to the reaction assemblage Cr-Al-ulvöspinel + ilmenite $\pm$ Fe, and with more intense reduction to Al-titanian-chromite + ilmenite + Fe. Estimates of T and fO_2 for subsolidus reduction are based on the observed stabilities of ilmenite and Mg-Al-chromite, the observed instabilities of Cr-Al-ulvöspinel and fayalite, and the absence of TiO_2. Calculated dissociation curves, as a function of T and fO_2, suggest that subsolidus reduction of Fe_2TiO_4-rich phases takes place at 10^{-15} to 10^{-16} atm at 1000°C, and between 10^{-20} and 10^{-21} atm at

800°C. Low intensity reduction is widespread, is identified in basalts from the five lunar landing sites and is considered to be due to a deuteric process on cooling. High intensity reduction is identified only in the Apollo 14 samples and the rarity of these occurrences suggests that reduction is induced by reheating, volatilization, and oxygen loss.

Oxide and cationic substitutional relationships demonstrate that the initial trend of spinel crystallization is indistinguishable from the receding trend of spinel subsolidus re-equilibration. These relationships confirm the notion of a continuous structural spinel series between normal and inverse spinel types. The formation of intermediate structural coordination can be brought about either as a result of initial crystallization or by subsolidus reduction. Based on the large octahedral site preference energies of Cr^{+3} and Al^{+3}, and on the selective mobility of B sited Ti^{+4} it is concluded that the site population of A sited Fe^{+2} probably remains unchanged in accordance with the O'Reilly-Banerjee model (1966) and that because of the higher mobility of B site Fe^{+2} the liberated iron is derived essentially from the octahedral site. Low intensity reduction probably proceeds in two distinct steps with B site Fe^{+2} and Ti^{+4} selectively activated to precipitate $FeTiO_3$ initially, and with saturation of the reconstituted spinel in Fe^{+2}, further precipitation of Fe^{+2} (in $FeTiO_3$) and Fe^0 are induced.

The preferential and radical exchange of ions that take place in the B site during the sequential stages of crystallization or subsolidus reduction, allow these changes to be defined in terms of the TAC reduction index or the TAC crystallization index. In the normal-inverse series this index simultaneously equates the variations of the five major elements present in lunar spinels and provides a simple numerical value for intercorrelative purposes, for relative intensities of reduction, and for paragenetic stages of spinel crystallization.

In conclusion, the implications of subsolidus reduction as identified in the spinels are far greater than magnetic oxide (Banerjee, 1972) decomposition and the reduction process may well have a substantial influence on at least five major properties of lunar geochemistry. These are (1) oxygen loss and associated volatilization; (2) lower ionic valency states; (3) trace element migration and repartitioning; (4) lattice distortions associated with vacancies and with anionic deficiency; and (5) silicate decomposition and the secondary formation of discrete Fe.

Acknowledgments—I wish to acknowledge H. S. Yoder, Director of the Geophysical Laboratory for continued use of the electron microprobe, and to thank F. R. Boyd, L. W. Finger and particularly C. Hadidiacos for their endless advice during the course of this study. Earlier reviews of some portions of this paper benefited through extensive discussion with R. Brett, R. Williams, and G. C. Ulmer; this manuscript was reviewed by A. El Goresy and L. A. Taylor. I am grateful to these colleagues for their comments and criticisms. This work was supported by NASA under contract NGR-22-010-089 and by a University of Massachusetts Faculty Research Grant FR-J36-72.

REFERENCES

Agrell S. O. Peckett A. Boyd F. R. Haggerty S. E. Bunch T. E. Cameron E. N. Dence M. R. Douglas J. A. V. Plant A. G. Traill R. J. James O. B. Keil K. and Prinz M. (1970) Titanian chromite aluminian chromite and chromian ulvöspinel from Apollo 11 rocks. *Proc. Apollo 11 Lunar Sci. Conf., Geochim. Cosmochim. Acta*, Suppl. 1, Vol. 1, pp. 81–86. Pergamon.

Banerjee S. K. (1972) Iron-titanium-chromite. A possible new carrier of remanent magnetization in lunar rocks (abstract). In *Lunar Science—III* (editor C. Watkins), pp. 38–40, Lunar Science Institute Contr. No. 88.

Bowie S. H. U. and Simpson P. (1971) Opaque phases in Apollo 12 samples. *Proc. Second Lunar Sci. Conf., Geochim. Cosmochim. Acta,* Suppl. 2, Vol. 1, pp. 207–218. M.I.T. Press.

Brett R. Butler P. Meyer C. Reid A. M. Takeda H. and Williams R. (1971) Apollo 12 igneous rocks 12004, 12008, 12009, and 12022: A mineralogical and petrological study. *Proc. Second Lunar Sci. Conf., Geochim. Cosmochim. Acta,* Suppl. 2, Vol. 1, pp. 301–317. M.I.T. Press.

Brown G. M. Emeleus C. H. Holland J. G. Peckett A. and Phillips R. (1971) Picrite basalts, ferrobasalts, feldspathic norites, and rhyolites in a strongly fractionated lunar crust. *Proc. Second Lunar Sci. Conf., Geochim. Cosmochim. Acta,* Suppl. 2, Vol. 1, pp. 583–600. M.I.T. Press.

Buddington A. F. and Lindsley D. H. (1964) Iron-titanium oxide minerals and synthetic equivalents. *J. Petrol.* **5**, 310–357.

Cameron E. N. (1970) Opaque minerals in certain lunar rocks from Apollo 11. *Proc. Apollo 11 Lunar Sci. Conf., Geochim. Cosmochim. Acta,* Suppl. 1, Vol. 1, pp. 221–245. Pergamon.

Cameron E. N. (1971) Opaque minerals in certain lunar rocks from Apollo 12. *Proc. Second Lunar Sci. Conf., Geochim. Cosmochim. Acta,* Suppl. 2, Vol. 1, pp. 193–206. M.I.T. Press.

Datta R. K. and Roy R. (1967) Order-disorder in spinels. *J. Amer. Ceram. Soc.* **50**, 578–583.

Drever H. I. Johnston R. and Gibb F. G. F. (1972) Chromian pleonaste and aluminous picotite in two Apollo 14 microbreccias 14306 and 14055. *Nature Phys. Sci.* **235**, 30–31.

Dunitz J. D. and Orgel L. E. (1957) Electronic properties of transition-metal oxides—II. Cation distribution amongst octahedral and tetrahedral sites. *Phys. Chem. Solids.* **3**, 318–323.

El Goresy A. Ramdohr P and Taylor L. A. (1971a) The opaque minerals in the lunar rocks from Oceanus Procellarum. *Proc. Second Lunar Sci. Conf., Geochim. Cosmochim. Acta,* Suppl. 2, Vol. 1, pp. 219–236. M.I.T. Press.

El Goresy A. Ramdohr P. and Taylor L. A. (1971b) The geochemistry of the opaque minerals in Apollo 14 crystalline rocks. *Earth Planet. Sci. Lett.* **13**, 121–129.

Finger L. W. Hafner S. S. Schürmann K. Virgo D. and Warburton D. (1972) Distinct cooling histories and reheating of Apollo 14 rocks (abstract). In *Lunar Science—III* (editor C. Watkins), pp. 259–261, Lunar Science Institute Contr. No. 88.

Gorter E. W. (1954) Ionic distribution deduces from the g-factor of a ferrimagnetic spinel: Ti^{+4} in fourfold coordination. *Nature* **173**, 123–124.

Haggerty S. E. (1971a) Compositional variations in lunar spinels. *Nature Phys. Sci.* **233**, 156–160.

Haggerty S. E. (1971b) Subsolidus reduction of lunar spinels. *Nature Phys. Sci.* **234**, 113–117.

Haggerty S. E. (1971c) High temperature oxidation of ilmenite in basalts. *Carnegie Institute of Washington Year Book* **70**, 165–176.

Haggerty S. E. (1972a) Apollo 15: Compositional variations of spinels. A test of the normal-inverse spinel series. *Meteoritics,* to appear.

Haggerty S. E. (1972b) Luna 16: An opaque mineral study and a systematic examination of compositional variations of spinels from Mare Fecunditatis. *Earth Planet. Sci. Lett.* **13**, 328–352.

Haggerty S. E. and Meyer H. O. A. (1970) Apollo 12: Opaque oxides. *Earth Planet. Sci. Lett.* **9**, 379–387.

Jakeš P. Warner J. Ridley W. I. Reid A. M. Harmon R. S. Brett R. and Brown R. W. (1972) Petrology of a portion of the Mare Fecunditatis regolith. *Earth Planet. Sci. Lett.* **13**, 257–271.

Katsura T. and Muan A. (1964) Experimental study of equilibria in the system $FeO-Fe_2O_3-Cr_2O_3$ at 1300°C. *Trans. Met. Soc. AIME* **230**, 77–83.

Keil K. Bunch T. E. and Prinz M. (1970) Mineralogy and composition of Apollo 11 lunar samples. *Proc. Apollo 11 Lunar Sci. Conf., Geochim. Cosmochim. Acta,* Suppl. 1, Vol. 1, pp. 561–598. Pergamon.

LSPET (1971) (Lunar Sample Preliminary Examination Team) Preliminary examination of lunar samples from Apollo 14. *Science* **173**, 681–693.

McClure D. S. (1957) The distribution of transition metal cations in spinels. *J. Phys. Chem. Solids.* **3**, 311–317.

McLean M. C. and Ward R. G. (1966) Thermodynamics of hercynite formation. *J. Iron Steel Inst.* **204**, 8–11.

Michel-Levy M. C. and Levy C. (1972) Mineralogical aspects of Apollo 14 samples: Lunar chondrules, pink spinel bearing rocks, ilmenite (abstract). In *Lunar Science—III* (editor C. Watkins), pp. 136–138, Lunar Science Institute Contr. No. 88.

Muan A. Hauck J. Osborn E. F. and Schairer J. F. (1971) Equilibrium relations among phases occurring in lunar rocks. *Proc. Second Lunar Sci. Conf., Geochim. Cosmochim. Acta*, Suppl. 2, Vol. 1, pp. 497–505. M.I.T. Press.

Muan A. Hauck J. and Lofall T. (1972) Equilibrium studies with a bearing on lunar rocks (abstract). In *Lunar Science—III* (editor C. Watkins), pp. 561–564, Lunar Science Institute Contr. No. 88.

Navrotsky A. and Kleppa O. J. (1967) The thermodynamics of cation distribution in simple spinels. *J. Org. Nucl. Chem.* **29**, 2701–2714.

O'Reilly W. and Banerjee S. K. (1966) Oxidation of titanomagnetites and self-reversal. *Nature* **211**, 26–28.

O'Reilly W. and Banerjee S. K. (1967) The mechanism of oxidation in titanomagnetites: A magnetic study. *Mineral. Mag.* **36**, 29–37.

Pavicevic M. Ramdohr P. and El Goresy A. (1972) Microprobe investigations of the oxidation state of Fe and Ti in ilmenite in Apollo 11, Apollo 12, and Apollo 14 crystalline rocks (abstract). In *Lunar Science—III* (editor C. Watkins), pp. 596–598, Lunar Science Institute Contr. No. 88.

Reid A. M. Meyer C. Harmon R. S. and Brett R. (1970) Metal grains in Apollo 12 igneous rocks. *Earth Planet. Sci. Lett.* **9**, 1–5.

Roedder E. and Weiblen P. W. (1972) Petrographic and petrologic features of Apollo 14, 15, and Luna 16 samples (abstract). In *Lunar Science—III* (editor C. Watkins), pp. 657–659, Lunar Science Institute Contr. No. 88.

Sato M. and Helz R. T. (1971) Oxygen fugacity studies of Apollo 12 basalts by the electrolyte method. Second Lunar Science Conf. (unpublished proceedings).

Steele I. M. (1972) Chromian spinels from Apollo 14 rocks. *Earth Planet. Sci. Lett.* (in press).

Stephenson A. (1969) The temperature dependent cation distribution in titanomagnetites. *Geophys. J. Roy. Astron. Soc.* **18**, 199–210.

Taylor R. W. and Schmalreid H. (1964) The free energy of formation of some titanates, silicates, and magnesium aluminate from measurements made with galvanic cells involving solid electrolytes. *J. Phys. Chem.* **68**, 2444–2449.

Ulmer G. C. and White W. B. (1966) Existence of chromous ion in the spinel solid solution series $FeCr_2O_4 - MgCr_2O_4$. *J. Amer. Ceram. Soc.* **49**, 50–51.

Verhoogen J. (1962) Oxidation of iron-titanium oxides in igneous rocks. *J. Geol.* **70**, 168–181.

Webster A. H. and Bright N. F. H. (1961) The system iron-titanium-oxygen at 1200°C and oxygen partial pressures between 1 atm and 2×10^{-14} atm. *J. Amer. Ceram. Soc.* **44**, 110–116.

Proceedings of the Third Lunar Science Conference
(Supplement 3, *Geochimica et Cosmochimica Acta*)
Vol. 1, pp. 333–349
The M.I.T. Press, 1972

Fra Mauro crystalline rocks: Mineralogy, geochemistry and subsolidus reduction of the opaque minerals

AHMED EL GORESY, LAWRENCE A. TAYLOR*

and

PAUL RAMDOHR

Max-Planck-Institut für Kernphysik, Heidelberg, Germany

Abstract—The opaque minerals observed in Apollo 14 crystalline rock samples (14053,2; 14072,12; 14073,7; 14310,101) are ilmenite, chromian ulvöspinel, aluminian-titanian chromite, picotite, rutile, baddeleyite, schreibersite, native FeNi metal, troilite, mackinawite, chalcopyrrhotite and tranquillityite. Textural and mineralogical similarities exist between rocks 14073 and 14310 versus 14053 and 14072.

The Ni content of the native FeNi metals in 14310 and 14073 ranges between 1.5 and 37wt% and 5.5 and 24 wt% Ni, respectively; there is a preference of Ni for the metal phase associated with troilite that may be a result of the prevailing a_{S_2}. The Co content is close to 0.5% and the metal grains have meteorite-type compositions. In 14053 and 14072, the Ni content of the metal phase ranges from <0.01 to 4.8 wt% and between 0.2 and 6.1 wt% Ni, respectively, and is independent of the mineral assemblage. Two coexisting schreibersites were found in FeNi metal grains of 14310, one Ni-rich (28 wt%) and the other $\sim Fe_3P$. Rocks 14310 and 14073 may contain more baddeleyite than any lunar rock examined to date. It commonly occurs with ilmenite (containing up to 0.57 wt% ZrO_2) and chromian ulvöspinel (up to 0.25 wt% ZrO_2). Two tranquillityites were observed in 14310; one is similar to the type material and the other contains ~ 4 wt% less ZrO_2 and ~ 4 wt% more $(TiO_2 + SiO_2)$. Only chromian ulvöspinel was observed in 14073 and 14310, whereas, in addition, aluminian-titanian chromite and picotite occur in 14053 and 14072. The ulvöspinel in the latter two rocks is much more Cr rich than in the former.

Subsolidus reduction of chromian ulvöspinel to ilmenite + aluminian chromite + native Fe occurs in 14053 and 14072, and the compositional trends of ulvöspinel during primary crystallization and subsolidus reduction are in an opposite sense. MgO partitionings between primary ilmenite and ulvöspinel versus "exsolved" ilmenite and coexisting ulvöspinel indicate that the reduction process represents a closer approach to equilibrium than initial crystallization. In addition, fayalite in 14053 shows breakdown to native Fe + tridymite + SiO_2-rich glass. Rocks 14053 and 14072 have undergone more reducing conditions than any other Apollo 11, 12, or 14 lunar rock.

INTRODUCTION

THERE IS A DISTINCT paucity of rocks that are wholly igneous in origin among the samples returned by the Apollo 14 mission (LSPET, 1971). Of these samples, only two crystalline rocks (14053 and 14310) weigh more than 50 gm; however, smaller samples of crystalline rock were also returned and are present as clasts in the numerous microbreccias.

In this report, observations are presented concerning the opaque mineralogy of samples 14053,2; 14310,101; 14073,7; and 14072,12, the last two samples are from crystalline rocks weighing less than 50 gm. Some preliminary findings were reported by El Goresy *et al.* (1971a, c, d; 1972).

Textural and mineralogical similarities were noted between rocks 14310 and 14073, on the one hand, and 14053 and 14072 on the other. This subdivision has

---

*Department of Geosciences, Purdue University, Lafayette, Indiana 47907.

recently been substantiated by oxygen isotope (Clayton *et al.*, 1972) and additional petrological studies (Melson *et al.*, 1972). Rock 14310 consists of euhedral plagioclase laths (~60%) with intersertal to interstitial anhedral pyroxene (~35%); the remainder being the opaque minerals and a late-stage mesostasis (fayalite, glass, tridymite, a phosphate mineral, and some additional opaque minerals). In contrast, rock 14053 is more mafic than 14310. It has an ophitic texture, and consists of pyroxene (~50%) and plagioclase (~40%). The remainder consists of olivine, opaque minerals, fayalite displaying an unusual texture, and mesostasis.

The opaque minerals make up but a few percent of the Fra Mauro crystalline rocks, much less than in either Apollo 11 or 12 rocks. These minerals include ilmenite, chromian ulvöspinel, aluminian-titanian chromite, picotite, tranquillityite, rutile, baddeleyite, schreibersite, native FeNi metal, troilite, mackinawite, and chalcopyrrhotite. The last two minerals were previously reported on by El Goresy *et al.* (1971d).

The mineralogy and geochemistry of these opaque minerals are the subjects of this investigation. In addition, the Apollo 14 crystalline rocks, particularly 14053 and 14072, contain more abundant evidence for subsolidus reduction reactions than are present in any of the rocks returned by the earlier Apollo missions. These observations emphasize the extreme reducing environment present during the formation of these Fra Mauro rocks.

FeNi METAL

The association of native Fe metal with troilite that was so pronounced in Apollo 11 samples, is not common in the Apollo 14 rocks. In this respect, the Apollo 12 and 14 rocks are similar. The metal phase in all Apollo crystalline rocks collected to date, regardless of the Ni content of this phase, consists entirely of kamacite (i.e., no taenite has been optically observed). However, some interesting geochemical observations were made that set the Apollo 14 metal phases apart from those occurring in rocks collected during the previous lunar missions.

The petrographic differences between rocks 14053 and 14072 versus 14310 and 14073 are further emphasized by the composition of the native FeNi metal. Electron microprobe analyses were performed on many of the FeNi grains in these rocks. As can be seen in Fig. 1, the Ni content of the metals in samples 14310 and 14073 ranges between 1.5 and 37 wt% and 5.5 and 24 wt% Ni, respectively. The composition of the metal phase is dependent upon the mineral assemblage; there is a preference of Ni for the metal phase associated with troilite, especially in 14073. In samples 14053 and 14072, the Ni content of the metal phase ranges from <0.01 to 4.8 wt% and between 0.2 and 6.1 wt% Ni, respectively, and there is no strong correlation between the Ni content of a given metal grain and the mineral assemblage in which it occurs. Also, no evidence was found for the high Ti and Cr contents of the metal phase in 14053, as reported by Melson *et al.* (1972). These low Ni compositions, as compared with metals in 14310, reflect the generally low Ni content of the rocks (50 ppm Ni in 14053 versus 630 ppm in 14310; LSPET, 1971). The Ni versus Co contents of approximately half of the metal grains analyzed are within the meteoritic composition range

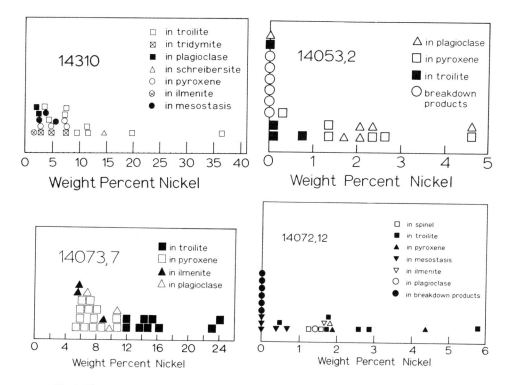

Fig. 1. Histograms of Ni concentrations in native Fe grains in Fra Mauro crystalline rocks.

as defined by Goldstein and Yakowitz (1971). This would appear to support the theory of a melting of the lunar soil, containing meteoritic particles, to form rock 14310 (e.g., Dence *et al.*, 1972). This would necessitate that not all of the metal grains melted or reequilibrated with the magma. However, the compositions of the other metal grains would seem to indicate that 14310 is a typical lunar igneous rock.

In certain Apollo 12 and 14 rocks, the Ni content of the metal grains is dependent upon the mineral assemblage (e.g. El Goresy *et al.*, 1971b, d; Reid *et al.*, 1970; Taylor *et al.*, 1971). The paragenesis of native Fe appears to be a function of the prevailing oxygen fugacity, as well as sulfur activity. The troilite + native Fe assemblage generally is considered to result from crystallization of an immiscible sulfide (Fe–FeS) liquid at 988°C (Skinner, 1970). If the oxygen fugacity is low enough, native metal will appear early in the crystallization sequence and continue precipitation throughout cooling. In this case, the early metal grains will contain relatively large amounts of the highly siderophile Ni. However, if the oxygen fugacity is higher, the Fe metal will not crystallize until later, and some of Ni in the magma may become concentrated in the immiscible sulfide liquid. This would result in high Ni contents for the metal grains associated with troilite as proposed by Brett (personal communication) such as are present in 14310. This explanation may also hold for 14073 (Fig.1).

Table 1. Selected Electron Microprobe Analyses.

| | Troilite* | | | | Schreibersite† | Metal |
	14310	14073	14072	14053	14310 Coexisting	
Fe	63.4	63.4	63.0	63.4	56.1	82.2
S	36.3	36.3	36.3	36.5	—	—
Ti	0.01	<0.01	0.75	0.05	—	—
Ni	0.02	0.04	<0.01	<0.01	28.0	15.6
Co	—	—	—	—	0.46	1.11
P	—	—	—	—	11.4	0.09
TOTAL	99.7	99.7	100.1	100.0	96.0	99.0

*Average of ∼10 analyses in each specimen.— = not analyzed.
†Less than 4 μ grain.

TROILITE

Troilite is associated with FeNi metal in a eutectic texture where it is late in the crystallization sequence of the Apollo 14 rocks. However, it more commonly occurs without the metal phase and is paragenetically related to the oxide minerals ilmenite, spinels, and baddeleyite.

Microprobe analyses of troilite grains in the various rocks revealed that, without exception, it is always stoichiometric (Table 1). There is a measurable Ti partitioning (up to 0.8 wt% Ti) between troilite and coexisting ilmenite or ulvöspinel. This observation was reported also for Apollo 12 rocks by El Goresy et al. (1971b) and Taylor et al. (1971). The geochemical significance of this partitioning currently is being investigated.

SCHREIBERSITE

The presence of schreibersite in the Apollo 11 and 12 breccias and fines was used as a criterion for meteoric contamination (e.g., Goldstein and Yakowitz, 1971). El Goresy et al. (1971d) stated that this criterion is suspect because schreibersite is observed as small rounded inclusions within some of the FeNi metal grains in rock 14310. Recent investigations (e.g., Dence et al., 1972) have indicated that rock 14310 may have formed as a result of shock melting of a lunar soil or breccia. In this case, the schreibersite occurrence in 14310 may be due to initial meteoritic contamination, although it is doubtful that schreibersite would have survived the remelting.

A schreibersite assemblage and the chemistry of the association are shown in Fig. 2. Two different compositions of schreibersite were found to coexist. One is almost pure iron phosphide and was too fine grained for accurate microprobe analysis, whereas the other contains 28 wt% Ni in addition to the Fe and P (Table 1). Also, the FeNi metal grain in which these schreibersites occur contains, besides ∼16 wt% Ni, only a trace of P (i.e., 0.09 wt%).

From the x-ray scans shown in Fig. 2, it can be seen that there is a high P-bearing phase in contact with the schreibersite. This is a Ca phosphate (whitlockite or apatite) associated with the surrounding pyroxene. It may have formed as a result of a reaction such as:

$$Fe_3P + pyx_1 + O_2 \rightleftharpoons Ca_3(PO_4)_2 + pyx_2$$

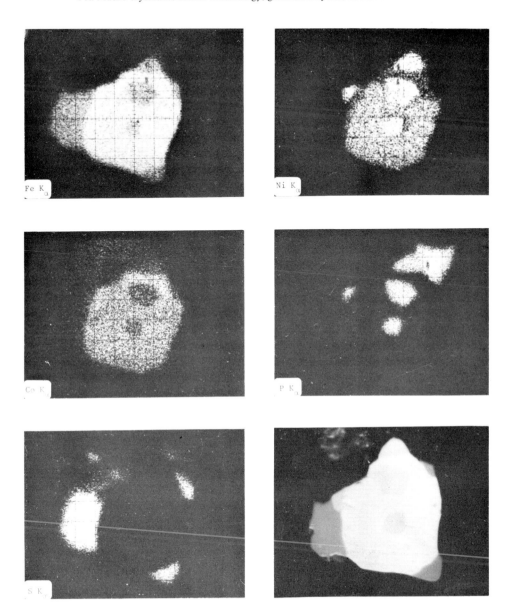

Fig. 2. X-ray microprobe scans of an assemblage of schreibersite (light gray), FeNi metal (white) and troilite (dark gray) in sample 14310,101.

Ridley *et al.* (this volume), on the basis of calculations by Olsen and Fuchs (1967), emphasize that the univariant curve involving this reaction lies at f_{0_2} below the iron-wüstite curve in the approximate range of f_{0_2} generally thought to have prevailed during crystallization of these basaltic rocks. Ridley *et al.* also mention that the presence of the assemblage schreibersite + native Fe does not necessarily indicate extreme reducing conditions as we suggested previously (El Goresy *et al.*, 1971d).

ZrO_2-Bearing Minerals

Rock 14310 has a somewhat KREEP-like composition and contains one of the highest amounts of ZrO_2 (0.13%, LSPET, 1971) of the Apollo 14 samples. This ZrO_2 content is contributed mainly by the mineral *baddeleyite*, which occurs with ilmenite, troilite, and ulvöspinel, and the amount of this mineral is greater than in any sample from the Apollo 11 or Apollo 12 missions that we have examined.

The ilmenites and ulvöspinels associated with baddeleyite in rocks 14310 and 14073 contain noteworthy amounts of ZrO_2 (Fig. 3): ilmenites = 0.17 to 0.57 wt%

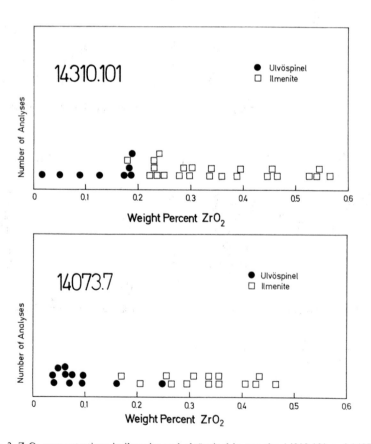

Fig. 3. ZrO_2 concentrations in ilmenite and ulvöspinel in samples 14310,101 and 14073,7.

ZrO_2; chromium ulvöspinel $= <0.02$ to 0.25 wt% ZrO_2. Obviously, there is a partitioning of ZrO_2 in favor of the ilmenites (the possible geochemical significance of this presently is under experimental investigation). These are the highest values of Zr reported to date for ulvöspinels and ilmenites and obviously represent some degree of substitution of Zr^{+4} for Ti^{+4}.

Baddeleyite was not observed in rocks 14053 or 14072, and the ilmenites and ulvöspinels contain only traces of ZrO_2 (i.e., maximum of 0.1 wt% in ilmenite; <0.02 wt% in ulvöspinel).

TRANQUILLITYITE

Two optically and compositionally distinct varieties of tranquillityite were observed in rock 14310. The normal variety was not analyzed quantitatively because of the small grain size. However, semiquantitative comparison with Apollo 11 and 12 tranquillityites, using a solid state detector, revealed the same approximate composition. As shown in Table 2, the other variety of tranquillityite contains ~ 4 wt% more combined $TiO_2 + SiO_2$ than the type material. The Zr/Hf ratios are low, i.e., 19–24.

ILMENITE

In the rocks collected during the Apollo 11, 12, and 14 missions, ilmenite is the major opaque mineral. This mineral, in the Fra Mauro crystalline rocks, is present in amounts of 1–3 modal %, as compared to 16–21% in Apollo 11 and 5–10% in Apollo 12 samples. However, the Fra Mauro ilmenites possess minor element chemistry that is even more interesting than those of ilmenites from earlier Apollo missions. Microprobe analyses of numerous grains of ilmenite have shown their compositions to consist mainly of $FeTiO_3$, with minor to trace amounts of MgO, MnO, Al_2O_3, Cr_2O_3, V_2O_3, and ZrO_2. The amounts of MgO in these ilmenites is noteworthy (Fig. 4).

In the Fra Mauro rocks studied, the MgO contents of some of the ilmenites are relatively high (up to 3.9 wt% MgO; Fig. 4). El Goresy *et al.* (1971b) have reported

Table 2. Selected Electron Microprobe Analyses.

Tranquillityite in 14310		Ba-Rich Glass		
(avg. of 3 grains)			14053	14072
SiO_2	15.2	SiO_2	60.6	61.2
TiO_2	23.0	TiO_2	0.12	0.09
Al_2O_3	1.76	Al_2O_3	19.6	19.7
FeO	41.4	FeO	0.54	0.51
MgO	1.33	MgO	0.01	0.01
CaO	1.31	CaO	0.20	0.31
ZrO_2	12.6	Na_2O	0.05	0.13
HfO_2	0.45	K_2O	12.5	13.0
Y_2O_3	1.60	BaO	7.32	6.51
TOTAL	98.7		100.9	101.5
Zr/Hf	19–24	K/Ba	1.59	1.85
$Fe_{6.92} Mg_{0.42} Ca_{0.40} (Zr_{1.23} Hf_{0.03} Y_{0.17} Al_{0.42}) Ti_{3.46}$				
$Si_{3.04} O_{24}$				

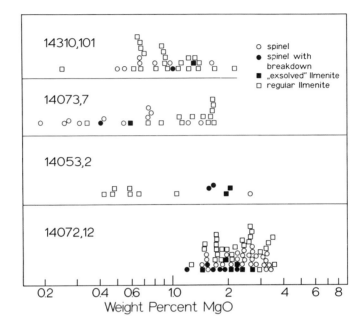

Fig. 4. MgO concentrations in ilmenite and spinel in Fra Mauro crystalline rocks. The "exsolved" ilmenite refers to ilmenite formed as a subsolidus reduction product from the breakdown of the ulvöspinel host (labelled "spinel with breakdown"). "Regular ilmenite" refers to the primary crystallization phase.

MgO contents of ilmenites in 12018 of up to 5.14 wt%, and they state that this may reflect the high magnesia of the whole rock (15.1 wt%; LSPET, 1970). Apollo 14 crystalline rocks are low in MgO (14053 = 8.4 wt%; 14310 = 8.0 wt%; LSPET, 1971), yet the ilmenites, relative to those from most other Apollo rocks, contain abundant MgO. Thus, this minor element content of ilmenite is not always simply a reflection of the whole rock chemistry.

In addition, the ilmenites in sample 14053,2 have been found to contain Ti^{+3} (Pavićević et al., 1972) which is thought to reflect the extreme reducing conditions present during formation of this rock (see Subsolidus Reduction Reactions). In contrast, Ti^{+3} has not been observed in ilmenites from rock 14072.

SPINEL

The spinels present in the lunar rocks are quite varied in composition (e.g., see Haggerty, 1971a). We observed chromian ulvöspinel, aluminian-titanian chromite, and picotite in the Apollo 14 crystalline rock samples examined. Based on microprobe analyses of more than 160 spinel grains, the distinctions and similarities between the four crystalline rocks are again emphasized.

The only spinel phase present in rocks 14310 and 14073 is chromian ulvöspinel (Fig. 5 and Table 3) whose composition and optical properties (i.e., reflectivity,

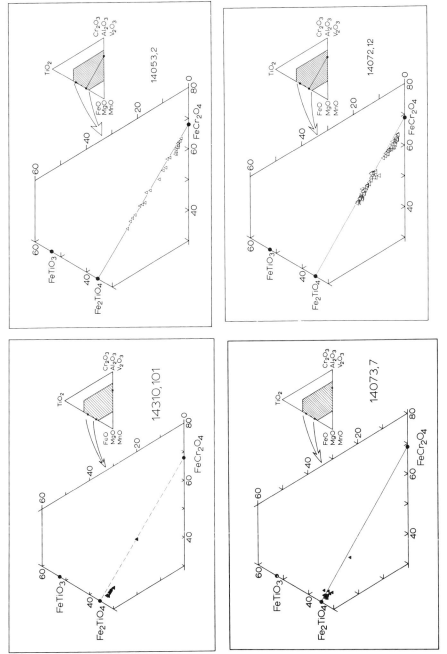

Fig. 5. Compositional diagram of +2, +3, and +4 cation oxides illustrating the relative compositions of analyzed spinels in the Fra Mauro crystalline rocks.

Table 3. Selected Electron Microprobe Analyses of Spinels in the Fra Mauro Crystalline Rocks.
All values in weight percent.

	Chr 14053	Chr 14072	Usp 14053	Usp 14072	Usp 14310	Usp 14073	Chr 14053	Pic Coexisting	Usp
TiO$_2$	2.66	3.31	23.0	16.8	32.4	32.3	3.67	4.28	18.3
FeO	28.8	31.7	52.0	42.1	62.5	63.5	28.5	24.9	43.7
MnO	0.19	0.24	0.71	0.38	0.40	0.47	0.21	0.08	0.44
MgO	6.18	4.62	1.55	3.28	0.66	0.26	6.75	11.3	5.37
Cr$_2$O$_3$	38.2	39.1	17.3	27.0	0.97	0.32	38.0	18.8	21.9
Al$_2$O$_3$	21.7	20.1	4.38	8.03	2.10	1.33	20.7	38.2	10.4
V$_2$O$_3$	0.56	0.63	0.45	0.52	0.01	0.04	0.56	0.22	0.57
ZrO$_2$	—	0.04	0.04	0.05	0.19	0.08	—	—	—
TOTAL	98.3	99.7	99.4	98.2	99.2	98.3	98.4	97.8	100.7

Atomic Proportions Based on 32 Oxygens

Ti	0.5251	0.6565	5.0456	3.5846	7.2605	7.3869	0.7259	0.7673	3.7488
Fe	6.3274	7.0028	12.6790	10.0246	15.5848	16.1386	6.2564	4.9699	9.9440
Mn	0.0431	0.0540	0.1761	0.0927	0.1011	0.1210	0.0471	0.0153	0.1006
Mg	2.4205	1.8180	0.6708	1.3919	0.2954	0.1164	2.6417	4.0194	2.1780
Cr	7.9390	8.1583	3.9808	6.0684	0.2298	0.0766	7.8839	3.5412	4.7227
Al	6.7244	6.2507	1.5026	2.6935	0.7381	0.4751	6.4099	10.7357	3.3400
V	0.1186	0.1341	0.1061	0.1178	0.0025	0.0093	0.1174	0.0421	0.1249
Zr	0.0000	0.0054	0.0052	0.0064	0.0275	0.0114	0.0000	0.0000	0.0000
TOTAL	24.0981	24.0798	24.1662	23.9799	24.2397	24.3553	24.0823	24.0909	24.1590

color, anomalous anisotropism) are similar to those Apollo 12 ulvöspinels which are found in the late-stage groundmass. The single analysis in each sample that plots approximately midway between the ulvöspinel (Fe$_2$TiO$_4$) and chromite (FeCr$_2$O$_4$) end points in Fig. 5, is from an ulvöspinel associated with a subsolidus breakdown texture and will be discussed later.

In contrast, samples 14053,2 and 14072,12 contain chromian ulvöspinel, aluminian-titanian chromite, and picotite (Table 3). Figure 5 shows a projection of the compositions of these spinels. Compared to titanian chromites from Apollo 12 rocks, these chromites (Table 3) are distinctly higher in Al$_2$O$_3$ contents (13 to 22 wt% versus 9 to 12 wt% in Apollo 12). The chromian ulvöspinels are very rich in Cr$_2$O$_3$ (Table 3 and Fig. 5) and actually contain larger amounts of FeCr$_2$O$_4$ (chromite) molecule than Fe$_2$TiO$_4$ (ulvöspinel) and have compositions essentially identical with many of the chromian ulvöspinels occurring in rock 12018 (El Goresy et al., 1971b). Although Haggerty (1971a; 1972) refers to these as aluminian-titanian chromites, we prefer to consider them as chromian ulvöspinels because they closely resemble other less chrome-rich ulvöspinels in color, reflectivity, and anomalous anisotropism and are not optically similar to the bluish chromite with which they are associated. There obviously is a need for revision and clarification of the spinel nomenclature, because the chemical criteria alone do not appear to be wholly adequate.

The MgO contents of the chromian ulvöspinels in these rocks are quite variable (Fig. 4). The ulvöspinels coexisting with ilmenites in rocks 14053 and 14072, in general, contain higher MgO than the same minerals in 14310 and 14073. The MgO contents of these minerals are, obviously, not a reflection of result of the total magnesia content of the rocks, because the rocks are all low in MgO.

In light of the experimental work of Johnson *et al.* (1971), the equilibrium partitioning of MgO between ilmenite and ulvöspinel is in favor of ilmenite. In the Apollo 14 rocks, ilmenite formed mainly as a primary phase but also as a product of subsolidus reduction of ulvöspinel. In 14310 and 14073 (Fig. 4), both in the primary ilmenite (and ulvöspinel) and in the two analyzed breakdown assemblages, the MgO partitionings are in general agreement with the phase equilibria studies. In contrast, the MgO in the primary ilmenites and ulvöspinels in 14053 and 14073 are in an opposite sense, i.e., $MgO_{il} < MgO_{usp}$, representing disequilibrium; however, the secondary ("exsolved") ilmenite and coexisting ulvöspinel host have MgO contents in favor of ilmenite. It would appear that the partitioning of MgO between ilmenite and ulvöspinel *more closely approaches equilibrium during subsolidus reduction processes* than during initial crystallization.

An unusual assemblage of spinels was observed in sample 14053 (Table 3). It consists of coexisting aluminian-titanian chromite, chromian ulvöspinel, and picotite (as defined by Deer *et al.*, 1962). The physico-chemical significance of this assemblage is unknown at present.

SUBSOLIDUS REDUCTION

The Fra Mauro crystalline rocks, particularly 14053 and 14072, contain more examples of reactions involving extensive subsolidus reduction than any of the Apollo 11 or 12 lunar rocks examined to date.

Fayalite breakdown

As first reported by El Goresy *et al.* (1971a), rock 14053 contains a unique texture (Fig. 6a) due to the breakdown of fayalite as a result of subsolidus reduction processes. The fayalite (Fa_{86-96}) crystallized during the end-stages of crystallization; it subsequently was broken down to a spongy mass of pure Fe metal (Fig. 1) + tridymite + SiO_2-rich glass. All degrees of incipient to complete breakdown occur. This SiO_2-rich glass contains only a small amount of MgO ($\sim 1\%$) and does not account for all the MgO from the fayalite. However, crystallites of a nonopaque phase, tentatively identified as a pyroxene, commonly are associated with this assemblage and may have formed as a result of the breakdown process. Lindsley *et al.* (1972) have reported fayalite + clinopyroxene + tridymite as breakdown products of pyroxferroite at 990°C. Although these phases are all present in this spongy texture, the proportions are entirely different from what one would expect if this texture resulted from the breakdown of pyroxferroite. Locally, within the spongy texture of sample 14053, there are all gradations between fresh native Fe and native Fe that has been converted to troilite. This is interpreted as evidence for a very *late-stage period of sulfurization*.

This fayalite breakdown texture and assemblage is unique. It has never been reported as occurring in any other rock, either terrestrial or extraterrestrial. Rock 14072, which is similar to 14053 in many ways, also contains fayalite in the mesostasis; however, this fayalite shows no evidence of breakdown.

In addition to the glass associated with the fayalite, 14053 and 14072 also contains

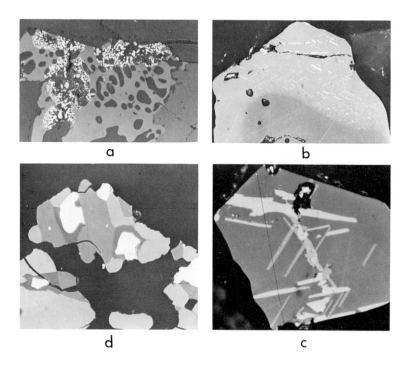

Fig. 6. Examples of subsolidus reduction reactions in rock 14053 (a) and 14072 (b, c, and d). (a) spongy texture of native Fe + tridymite + SiO_2-rich glass resulting from the breakdown of fayalite (lt. gray) occurring in a sieve texture. Photo width ~ 300 μ. (b) ulvöspinel (lt. gray) breakdown to ilmenite + native Fe (white) along {111} planes. Note that the breakdown does not occur in the chromite (dark gray) but is confined to the ulvöspinel. Photo width ~ 45 μ. (c) a more advanced stage of ulvöspinel (dark gray) breakdown. Here, in addition to the ilmenite (lt. gray) + native Fe (white), thin rims (< 1 μ) of chromite occur around the ilmenite blades. The ilmenite also contains rutile + native Fe due to subsolidus reduction. Photo width ~ 45 μ. (d) An example of complete breakdown of ulvöspinel (no longer present) to ilmenite (lt. gray) + native Fe (white) + chromite (dark gray). Width of field of view is ~ 45 μ.

interstitial glass that occurs in the groundmass. This primary glass contains large amounts of K_2O and BaO (Table 2) and has K/Ba ratios of 1.5–1.9 versus the whole rock ratio of ~ 6.

Ulvöspinel breakdown

The subsolidus reduction of chromian ulvöspinel to ilmenite + native Fe has been observed in all Apollo 11 through 15 rocks (see, for example, El Goresy *et al.*, 1971b and d; Haggerty, 1971b); the native Fe contains no detectable Ni or Co and is pure Fe (Fig. 1). This breakdown of chromian ulvöspinel is developed to a greater extent in the Apollo 14 crystalline rocks 14053 and 14072, with all stages of reduction to complete breakdown.

Figure 6b shows an excellent example of the breakdown texture present in a chromian ulvöspinel in rock 14072. The ilmenite, together with accompanying fine grains of native Fe, have "exsolved" along the {111} planes of the host. Note that the breakdown is confined entirely to the ulvöspinel. A more advanced stage of this breakdown is shown in Fig. 6c. On the borders of the ilmenite in this grain, thin rims of bluish aluminian-titanian chromite are present. This represents a more complicated reaction in which the chromian ulvöspinel reduction products are aluminian-titanian chromite + ilmenite + native Fe. The complete reduction of the ulvöspinel to these products is shown in Figure 6d. A similar type of breakdown reaction was recently discussed by Haggerty (1971b).

In Fig. 7, the compositions are plotted for the chromian ulvöspinels that show evidence of reduction. It should be noted that in this type of projection these compositions plot approximately half-way between the ulvöspinel and chromite end members. In general, the closer the ulvöspinel composition is to the chromite end, the more highly reduced it is. Study of the involved phases has resulted in the determination of compositional trends that have revealed certain systematic changes in the spinel chemistry during the reduction processes.

The compositional variations of spinels during crystallization of lunar magmas, as discussed for example, by El Goresy et al. (1971b; 1972), Haggerty (1971a) and Taylor et al. (1971), are shown in Fig. 8a. In general, the spinels, both chromites and ulvöspinels, increase in the percentage of Fe_2TiO_4 molecule as crystallization progresses, with the chromite precipitating first. However, during subsolidus reduction, a reverse trend is evident for the ulvöspinel compositions.

If the initial ulvöspinel composition contained a large amount of Fe_2TiO_4 molecule, partial reduction resulted in ilmenite + native Fe metal. In these cases, the ulvöspinel "exsolves" minerals (i.e., ilmenite and native Fe) containing Fe, FeO, and TiO_2, and thereby becomes progressively enriched an Al_2O_3 and Cr_2O_3 (i.e., the chromite molecule $Fe(Cr,Al)_2O_4$). This is depicted diagrammatically by the trend of compositions shown by the upper arrow in Fig. 8b. However, if the initial composition of the ulvöspinel contains sufficient $Fe(Cr,Al)_2O_4$ molecule, or if the reduction process proceeds to an advanced stage, the ulvöspinel composition changes approximately as shown by the lower arrow in Fig. 8b. A point is reached (approximately Usp 32–Chr 68 for 14072 on Fig. 7) where the ulvöspinel becomes so supersaturated in the chromite molecule that aluminian-chromite nucleates and is precipitated as distinct rims around the ilmenite and native Fe grains (Fig. 6d).

The one chromian ulvöspinel in each of 14310 and 14073 (Fig. 5), whose composition plots near the midpoint on the join between ulvöspinel and chromite, shows this subsolidus reduction breakdown. All of the others analyzed in these two specimens do not. It is probable that the composition of these reduced ulvöspinels was initially within the compositional range shown by the pristine ulvöspinels.

Thus, the *compositional change of ulvöspinel during subsolidus reduction* is, indeed, in an *opposite direction to that occurring during crystallization* from the melt. This subsolidus reduction is a disequilibrium process; however, as discussed earlier in this paper, the MgO partitioning between the ilmenite reduction product and the coexisting ulvöspinel appears to indicate a closer approach to equilibrium than

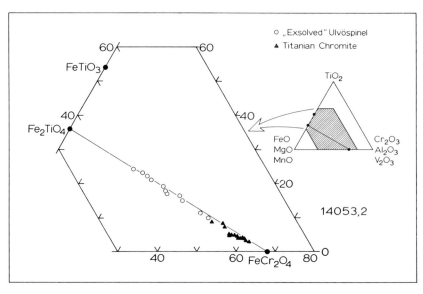

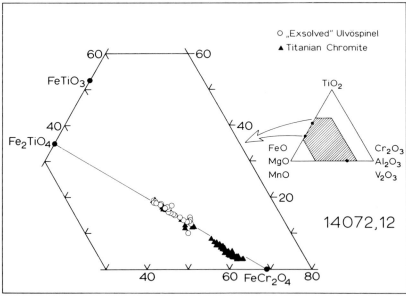

95 Complete Analyses

Fig. 7. Compositional diagram of +2, +3 and +4 cation oxides illustrating the relative compositions of ulvöspinels with breakdown products and chromites.

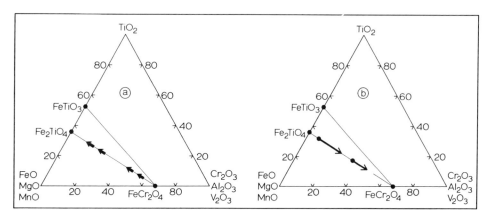

Fig. 8. Spinel compositional variations during (a) crystallization from a melt and (b) subsolidus reduction.

occurred during primary crystallization of these coexisting phases from the melt.

In addition to the breakdown assemblages discussed above, *ilmenite*, formed by the reduction of ulvöspinel in 14072 (as well as in 12018), has been reduced to rutile + native Fe (Fig. 6). It would appear that the reduction of ulvöspinel can occur in two steps with ilmenite as an intermediate product.

Based on the observation of complete reduction of chromian ulvöspinel to aluminian chromite + ilmenite + Fe metal and the additional breakdown of this ilmenite to rutile + Fe, it would appear that rock 14072 has undergone more reducing conditions than 14053. However, whereas both rocks contain fayalite in the mesostasis, only the fayalite in 14053 has broken down during reduction processes. The conclusions to be reached from these observations seem contradictory. A possible explanation can be found in considerations of the kinetics involved. Differences in the initial ulvöspinel compositions of the two rocks may impart sufficiently different breakdown rates to these spinels so that even though ulvöspinel reduction apparently has progressed further in 14072, the actual environment was, in reality, not as reducing. The fact that the fayalite compositions in the two rocks are identical, but fayalite breakdown only occurs in 14053, supports our conclusion that rock 14053 has undergone more severe reducing conditions. This is also supported by the presence of Ti^{+3} in ilmenites of 14053 as discussed earlier in this paper.

How reducing was this environment? It is possible to calculate the data for and plot various curves associated with these breakdown reactions (i.e., Haggerty, 1971b), namely the fayalite, ulvöspinel and ilmenite reductions discussed above. However, these curves only give upper limits of f_{O_2} because the kinetics and mechanisms of the breakdown processes are unknown at present. In other words, it is not known how far removed to lower oxygen fugacities the particular mineral must be before the reduction reaction will proceed at a geologically feasible rate. This is particularly true at the temperatures at which these subsolidus reactions probably occurred (i.e., 800–1100°C).

Thus, with the present state of knowledge, it is not possible to state lower values of f_{O_2} for these breakdown reactions. For example, Haggerty (1971b) states that because ulvöspinel breaks down to ilmenite + native Fe, the reaction must occur above the ilmenite stability curve and thereby gives lower f_{O_2} boundary conditions. However, we do not really understand this breakdown reaction well enough to rule out a stepwise formation of ilmenite which in turn breaks down to TiO_2 + native Fe. In fact, our observations of rutile + native Fe as ilmenite reduction products appears to indicate such a sequence of steps. The presence of this assemblage gives an upper limit of f_{O_2}, but no lower boundary can be stated. This, as well as many other kinetic problems, remains to be investigated.

Acknowledgments—The authors would like to express their appreciation to Drs. R. Brett, S. E. Haggerty, R. H. McCallister, Henry O. A. Meyer and M. Prinz for constructive criticisms and suggestions during the preparation of this manuscript. The drafting and photographic assistance of H. Weber is also acknowledged. One of us (L.A.T.) would like to thank Professor Dr. Gentner, Director of the Max-Planck-Institut für Kernphysik at Heidelberg, for arranging for financial support during this study.

References

Clayton R. N. Hurd J. M. and Mayeda T. K. (1972) Oxygen isotope abundances in Apollo 14 and 15 rocks and minerals (abstract). In *Lunar Science—III* (editor C. Watkins), pp. 141–143, Lunar Science Institute Contr. No. 88.

Deer W. A. Howie R. A. and Zussman J. (1962) *Rock-forming Minerals*, Vol. 5, pp. 61–62. John Wiley.

Dence M. R. Plant A. G. and Traill R. J. (1972) Impact-generated shock and thermal metamorphism in Fra Mauro lunar samples (abstract). In *Lunar Science—III* (editor C. Watkins), pp. 174–176, Lunar Science Institute Contr. No. 88.

El Goresy A. Ramdohr P. and Taylor L. A. (1971a) The opaque mineralogy of Apollo 14 crystalline rocks. Meteoritics, p. 266, Abst. of 34th Ann. Mtg. of Meteor. Soc. (Aug.).

El Goresy A. Ramdohr P. and Taylor L. A. (1971b) The opaque minerals in the lunar rocks from Oceanus Procellarum. *Proc. Second Lunar Sci. Conf.*, *Geochim. Cosmochim. Acta* Suppl. 2, Vol. 1, 219–236.

El Goresy A. Ramdohr P. and Taylor L. A. (1971c) The opaque mineralogy of Apollo 14 crystalline rock 14310. Geol. Soc. Amer. Ann. Mtg., Washington, Abst., 555.

El Goresy A. Ramdohr P. and Taylor L. A. (1971d) The geochemistry of the opaque minerals in Apollo 14 crystalline rocks. *Earth Planet. Sci. Lett.* **13**, 121–129.

El Goresy A. Ramdohr P. and Taylor L. A. (1972) Fra Mauro crystalline rocks: petrology, geochemistry, and subsolidus reduction of the opaque minerals (abstract). In *Lunar Science—III* (editor C. Watkins), pp. 224–226, Lunar Science Institute Contr. No. 88.

Goldstein J. I. and Yakowitz H. (1971) Metallic inclusions and metal particles in the Apollo 12 lunar soil. *Proc. Second Lunar Sci. Conf.*, *Geochim. Cosmochim. Acta* Suppl. 2, Vol. 1, 177–191.

Haggerty S. E. (1971a) Compositional variations in lunar spinels. *Nature Phys. Sci.* **233**, 156–160.

Haggerty S. E. (1971b) Subsolidus reduction of lunar spinels. *Nature Phys. Sci.* **234**, 113–117.

Haggerty S. E. (1972) Luna 16: an opaque mineral study and a systematic examination of compositional variations of spinels from Mare Fecunditatis. *Earth Planet. Sci. Lett.* **13**, 328–352.

Johnson R. E. Woermann E. and Muan A. (1971) Equilibrium studies in the system MgO–"FeO"–TiO_2. *Amer. J. Sci.* **271**, 278–292.

Melson W. G. Mason B. Nelen J. and Jacobson S. (1972) Apollo 14 basaltic rocks (abstract). In *Lunar Science—III* (editor C. Watkins), pp. 535–537, Lunar Science Institute Contr. No. 88.

Lindsley D. H. Papike J. J. and Bence A. E. (1972) Pyroxferroite: breakdown at low pressure and temperature (abstract). In *Lunar Science—III* (editor C. Watkins), pp. 483–485, Lunar Science Institute Contr. No. 88.

LSPET (1970) (Lunar Sample Preliminary Examination Team) Preliminary examination of the lunar samples from Apollo 12. *Science* **167**, 1325–1339.

LSPET (1971) (Lunar Sample Preliminary Examination Team) Preliminary examination of lunar samples from Apollo 14, *Science* **173**, 681–693.

Olsen E. and Fuchs L. H. (1967) The state of oxidation of some iron meteorites. *Icarus* **6**, 242–253.

Pavićević M. Ramdohr P. and El Goresy A. (1972) Microprobe investigations of the oxidation state of Fe and Ti in ilmenite in Apollo 11, Apollo 12, and Apollo 14 crystalline rocks (abstract). In *Lunar Science—III* (editor C. Watkins), pp. 596–598, Lunar Science Institute Contr. No. 88.

Reid A. M. Meyer C. Jr. Harmon R. S. and Brett R. (1970) Metal grains in Apollo 12 igneous rocks. *Earth Planet. Sci. Lett.* **9**, 1–5.

Ridley W. I. Williams R. J. Brett R. Takeda H. and Brown R. W. (1972) Petrology of lunar basalt 14310 (abstract). In *Lunar Science—III* (editor C. Watkins), pp. 648–650, Lunar Science Institute Contr. No. 88.

Skinner B. J. (1970) High crystallization temperatures indicated for igneous rocks from Tranquillity Base. *Science* **167**, 652–654.

Taylor L. A. Kullerud G. and Bryan W. B. (1971) Opaque mineralogy and textural features of Apollo 12 samples and a comparison with Apollo 11 rocks. *Proc. Second Lunar Sci Conf., Geochim. Cosmochim. Acta* Suppl. 2, Vol. 1, 855–871.

Walter L. S. French B. M. and Doan A. S. Jr. (1972) Petrographic analysis of lunar samples 14171 and 14305 (breccias) and 14310 (melt rock) (abstract). In *Lunar Science—III* (editor C. Watkins), pp. 773–775, Lunar Science Institute Contr. No. 88.

Proceedings of the Third Lunar Science Conference
(Supplement 3, *Geochimica et Cosmochimica Acta*)
Vol. 1, pp. 351–362
The M.I.T. Press, 1972

Mineralogical and petrographic features of two Apollo 14 rocks

P. Gay, M. G. Bown, and I. D. Muir

Department of Mineralogy and Petrology,
University of Cambridge, England

Abstract—Petrographic descriptions are given of 14310, a feldsparphyric high-alumina basalt rich in pigeonite, and of 14321, a complex microbreccia containing many different types of clasts and igneous rock fragments. Textures and very unusual compositional zoning in the plagioclase of the basalt indicate that it completed crystallization at or near the surface with a loss of soda to a presumed vapor phase. A detailed study of 14321 reveals some unexpected postcrystallization reactions, including the exsolution of oxide phases from pyroxenes and olivines.

Comparison with analogous terrestrial occurrences suggests that some fragments of 14321 were thermally metamorphosed before incorporation into the microbreccia. Oxide phases could have been produced, even under near vacuum conditions, in metastable initial crystals compositionally different from their terrestrial counterparts; alternatively, the metamorphism could have occurred under weakly oxidizing conditions, which could be important in the detailed crystallization history of the lunar surface.

Introduction

IN THIS PAPER WE DESCRIBE the petrography and general mineralogy of two Apollo 14 samples, 14310, a feldsparphyric olivine-free basalt, and 14321, a breccia containing xenoliths of basalt fragments, anorthosite, and micronorite with many large mineral clasts. Some unusual postcrystallization reactions have been found in minerals of the breccia 14321; these are described separately and their possible implications discussed in a later section.

Petrography and Mineralogy

Rock 14310

A polished thin section (27) and rock chips from the interior (198) and exterior (105) of the sample were examined by standard optical, x-ray diffraction, and electron microprobe methods. The thin section (Fig. 1a) reveals that this rock appears to be homogeneous with an intergranular to subophitic texture. The texture is unusual for a lunar basalt, for there are some phenocrysts of calcic plagioclase that may reach 2 mm in length, as well as other phenocrysts of orthopyroxene and pigeonite; these phenocrysts are accompanied by more numerous microphenocrysts of the same minerals. They are all set in an intergranular, or more rarely, an intersertal matrix composed of pyroxene (ferropigeonite and ferroaugite) and a felted mass of plagioclase laths. Minor amounts of opaque phases (ilmenite, troilite and metallic iron) also are present, and in the mesostasis sanidine and other phases, thought to be whitlockite and baddeleyite, were observed together with an iron-rich pyroxene (hedenbergite?) and possibly fayalite.

The plagioclase phenocrysts and microphenocrysts display normal zoning from cores of An_{94} to margins of An_{80} (determined by electron microprobe analyses),

Fig. 1a. High alumina basalt 14310. Note phenocrysts of calcic plagioclase, An_{93} zoned at extreme margins to An_{66}. Groundmass feldspar is also zoned, but its mean composition is near An_{92}. At extreme left, beyond the end of the long plagioclase (anals. 1 and 2, Table 1), a gray resorbed core of low-calcium pigeonite is mantled by a more magnesian pigeonite. (anal. 5, Table 1).

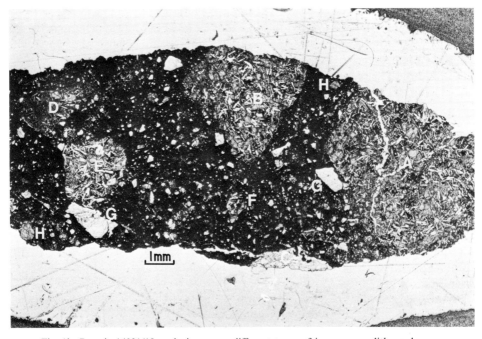

Fig. 1b. Breccia 14321/19 enclosing many different types of igneous xenoliths and xeno-crysts. A picrite basalt; B pigeonite-rich basalt; C anorthosite with ilmenite and zircon (anal. 4 Table 1); D basalt richer in plagioclase, part of this has recrystallized from a melt; E basalt transitional to high alumina type; F micronorite; G clasts of calcic plagioclase; H clasts of hypersthene with prominent exsolution features.

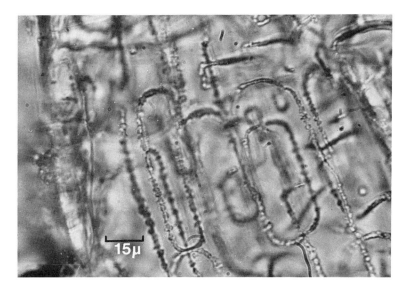

Fig. 1c. Breccia 14321/19 olivine in xenolith A (anals. 7 and 8, Table 1) with curved and looped threads and beads of a phase believed to be spinel developed during reheating and slight oxidation. These structures are about 1 μ in width.

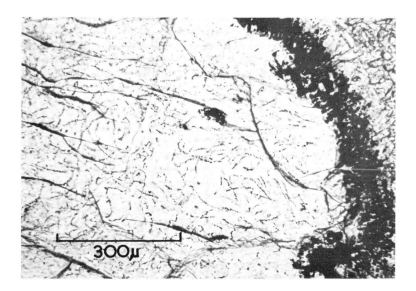

Fig. 1d. Thermally metamorphosed picrite basalt 63454, Kilauea. Ejected block from 1924 explosion of Halemaumau. Phenocryst of olivine mantled by magnetite and surrounded by a dense sheath of new hypersthene. Inside the olivine, note the development of curved loops and beads of magnetite.

though at the extreme margins of some grains compositions ranging down to An_{61} were detected optically. These compositions contrast with that of the groundmass feldspar, where maximum extinction angles in the (010) zone of up to 58°, (determined by universal stage), suggest compositions of An_{92-90}, in agreement with those determined by Brown and Peckett (1971). Most of the phenocrysts are twinned on the albite or albite-carlsbad laws, with vicinal composition planes, but single pericline cross lamellae are common. Some optical measurements with the universal stage were made on different parts of twinned phenocrysts whose compositions were known to range from An_{94-92} (Fig. 1a) but the results were not consistent; the variations are thought to be due to slight changes in structural state in different parts of the same crystal or, less likely, to the effects of shock. Diffraction studies of crystals removed from the rock chips confirm the widespread twinning and show transitional anorthite patterns with some range of diffuseness of the type (c) reflections. In general, our examination of this plagioclase is in accord with the results of other workers (Wenk et al., 1972), but, as noted for Apollo 12, the optical measurements consistently suggest slightly more sodic compositions than do probe analyses; such optical deviations may be due to the significant amounts of Fe and Mg reported in lunar plagioclases (Table 1). One crystal removed from the surface of the rock chip (105) contained a number of thin hexagonal plates of ilmenite, apparently in a regular orientation with respect to the plagioclase host; further work is in hand to assess the significance of this association.

Turning to pyroxenes, a few early phenocrysts of hypersthene (Of_{15-20}) are

Table 1. Selected mineral analyses (wt%) from 14310/27 and 14321/19.

	1	2	3	4	5	6	7	8
SiO_2	44.50	46.70	44.06	32.41	53.56	37.76	36.58	37.88
ZrO_2	—	—	—	66.93	—	—	—	—
TiO_2	—	—	0.02	0.19	0.53	0.02	0.11	0.02
Al_2O_3	35.24	32.36	35.56	0.20	2.68	1.12	0.38	0.19
Cr_2O_3	—	—	0.03	0.03	—	0.27	0.92	0.11
FeO	0.02	0.37	0.51	0.02	14.93	21.31	31.48	26.86
NiO	0.02	0.02	0.02	0.06	0.20	0.05	0.07	0.06
CoO	—	—	—	0.02	—	0.09	0.12	0.12
MnO	—	—	0.03	nil	n.d.	0.29	0.29	0.29
MgO	0.19	0.20	0.44	0.19	25.68	38.73	31.16	34.92
CaO	18.77	16.45	18.13	0.26	2.04	0.54	0.19	0.39
Na_2O	0.76	1.81	0.91	—	n.d.	—	—	—
K_2O	0.12	0.27	0.07	—	n.d.	—	—	—
BaO	0.01	0.01	0.07	—	n.d.	—	—	—
TOTAL	99.63	98.19	99.85	100.31	99.62	100.18	101.30	100.85
Composition (mole%)	An93	An81	An90		Ca4.2 Mg72.4 Fe23.4	Fa_{23}	Fa_{38}	Fa_{33}

1. Anorthite, basalt 14310/27: core of phenocryst (Fig. 1a).
2. Bytownite 14310/27: near margin of same crystal.
3. Anorthite 14321/19: small crystal in picrite basalt xenolith A.
4. Zircon 14321/19: crystal in anorthosite xenolith C (Fig. 1b).
5. Pigeonite 14310/27: margin to composite phenocryst shown in Fig. 1a.
6. Olivine 14321/19: clast with threads of spinel and a distinct inclusion of green spinel.
7. Olivine 14321/19: zoned crystal with threads of spinel in picrite basalt xenolith A.
8. Olivine 14321/19: another crystal with threads of spinel in picrite basalt xenolith A.

found, commonly mantled by pigeonites displaying the curved fractures seen in the cores of zoned Apollo 12 pyroxenes and suggesting protohypersthene (see Fig. 5 of Gay *et al.*, 1971). Pigeonites ($2V = 20°–10°$) commonly display prominent polysynthetic twinning; they are commonly zoned outwards into more iron-rich margins, though in some cases they are jacketed by a subcalcic augite ($2V = 36°–40°$). The late pyroxene is distinctly yellow in color and appears to include both ferroaugite and nearly uniaxial pigeonite; pyroxferroite is thought to be present in the residuum (identified optically but not microprobed). In this rock, pigeonite greatly predominates over augite, a feature demonstrated by the composition of the normative pyroxene (Fig. 2). A curious feature revealed by microprobe analysis of two grains shown in Fig. 1a is that a resorbed core of a more iron-rich pigeonite ($Ca_{4.4}Mg_{67.5}Fe_{28.1}$) is enclosed by a more magnesian overgrowth ($Ca_{4.2}Mg_{72.4}$ $Fe_{23.4}$); similar observations have been reported by Hollister *et al.* (1972), who have pointed out the abnormally low calcium contents of these pigeonites whose compositions lie in the field normally occupied by hypersthene. The presence of the more magnesian recurrences suggests the presence of xenocrysts.

Chemical analyses of this rock (Kushiro, 1972) show it to be a high alumina basalt with more than 20% alumina. Although the textures show that this rock has some phenocrysts of calcic plagioclase and orthopyroxene, as well as xenocrysts,

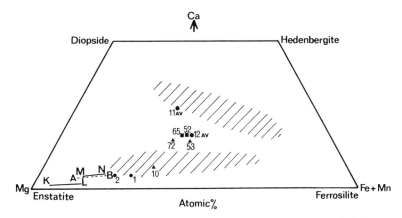

Fig. 2. Pyroxene quadrilateral showing composition of normative pyroxene of high alumina basalt 14310 in relation to others of the Apollo 14 suite and to the averages for the mare basalts of the Apollo 11 and 12 suites, expressed in terms of atomic % Ca, Mg, and Fe + Mn. ▲₁₀ normative pyroxene of 14310; ▲₇₂, ▲₅₃ normative pyroxenes of picrite basalts (mare type) 14072 and 14053, respectively.

●11ᴀᵥ, ●12ᴀᵥ, normative pyroxenes of average Apollo 11 and Apollo 12 basalts respectively. ■₅₂, ■₆₅, normative pyroxenes of 12052 and 12065 respectively.

A B; 14310, range of zoning determined in orthopyroxene. 1 and 2, compositions of core and margin respectively of composite pigeonite grain illustrated in Fig. 1a. Also shown is the approximate range of compositions reported by various authors for the pigeonites and augites in this rock.

K–L, M–N Composition ranges of phenocrysts of clinoenstatite and bronzite respectively from the magnesia-rich tholeiite of Cape Vogel, Papua. (Dallwitz, Green and Thompson, 1966).

basalt liquid. Brown and Peckett (1971) discuss the evidence for selective loss of alkalis, particularly of sodium during groundmass crystallization (based on zoning in the plagioclase), and interpret this observation in terms of a two stage magma eruptive process. The textural evidence, and mineral compositions reported here are in full accord with their conclusions.

In Fig. 2 the normative pyroxene of this rock is compared with those of the mafic Apollo 14 basalts 14053 and 14072, and with the mean compositions of the mare basalts of the Apollo 11 and 12 suites. Also shown is the range of actual compositions reported from 14310 along with points indicating our own determinations. From this plot the very different nature of the plagioclase-phyric basalts of the Apollo 14 suite is apparent: they are characterized by high-magnesium, low-calcium pigeonites; augite is very subordinate. Nor, in view of the high magnesia-iron ratios of these rocks, can they be derived from the same parents as the picrite basalts of the Apollo 14 suite, even if a considerable portion of the plagioclase is accepted as a cumulate phase.

Rock 14321

Two thin sections (19 and 240), a polished butt end (96) and a rock chip (106) of the sample were studied by standard optical, x-ray diffraction and electron microprobe methods. The thin sections (Fig. 1b) reveal that this rock is an extremely heterogeneous microbreccia composed of angular rock fragments as much as 5 mm in diameter with broken crystals of calcic plagioclase, pyroxene and olivine, set in a dark matrix containing further mineral clasts, which is cut by thin lighter colored veins.

Most of the rock fragments are of igneous origin and may be classified as microanorthosites, micronorites and various types of basalt. Of these the basalt and micronorite fragments are the most abundant; there are also fragments of preexisting breccias.

Two types of microanorthosite occur, one with an igneous habit characterized by a marked parallelism of the plagioclase, and a second with a granoblastic texture. Micronorite fragments grade into hypersthene basalts, with textures that range from hypidiomorphic granular to microgranulitic. Plagioclase or hypersthene may occur as "phenocrysts" set in a fine base composed of plagioclase, hypersthene, pale clinopyroxene and potash feldspar or interstitial glass. Textures range from igneous to those that suggest recrystallized and reconstituted breccias; this micronorite group predominates among the smaller fragments. In one section (240), the largest xenolith is notable in that it is composed in part of recrystallized micronorite with a dark strip resembling the matrix of the breccia, but the remainder appears to be a crushed micronoritic microbreccia. There are three main types of basalt. The picrite basalts (as much as 20% olivine and 50% clinopyroxene, approximately 5% of Fe–Ti oxide minerals and a variable amount of plagioclase) contain a few microphenocrysts of olivine (Table 1, Nos. 7 and 8) but the textures are essentially intergranular to sub-ophitic. There are also olivine basalts with similar mineralogy to the picrite basalts but containing much less olivine and more plagioclase. Other basalt fragments are transitional to the 14310 type, but fragments of others are in part granular and in part subvariolitic in aspect; parts of these may have recrystallized from a melt.

the specimen provides clear evidence of the existence on the moon of a high alumina

Mineral clasts set in the partially recrystallized fine-grained matrix range from 500 μ down to a few microns in size; below this limit they are difficult to resolve because of marginal recrystallization. Most are plagioclase, though pyroxene, olivine, ilmenite plus spinel, and metallic iron also occur, in order of decreasing abundance. Glassy fragments and spheres characteristically are absent, although a few small pools of pale brown glass occur interstitially in the matrix.

The plagioclase consists mainly of weakly shocked fragments; a few single crystals are greater than 400 μ long, suggesting very coarse-grained source rocks. Maskelynite has not been identified, but recrystallized or thermally annealed plagioclase, which once may have been maskelynite, is common; such fragments commonly show a distinct recrystallization rim around the margins. Compositions range through bytownite and anorthite. These observations are consistent with the diffraction studies which, although commonly showing slight crystal disorientations probably due to shock effects, indicate transitional anorthite structures for all plagioclase crystals removed from this rock; the diffuseness of type (c) reflections is variable. Dusty inclusions, provisionally identified as rutile or ilmenite, have occasionally been observed optically, but they have not yet been recognized in the x-ray studies. These may be compared with exsolution structures observed in some terrestrial anorthosites (viz. those from Egersund, S. Norway). Augite (rather pink in color, similar to the augite from the basalts), hypersthene and pigeonite have been identified and their exsolution textures will be described in a later section, as will those in olivines. Metallic iron occurs as small angular to subrounded crystal fragments as much as 100 μ in diameter. In (19) the metal compositions are unusually high in Ni compared to other breccia samples; from ten fragments, eight had taenite compositions (Ni_{11-17}) and only two lay in the kamacite range (Ni_7). In the polished butt end (96), only three of the thirty grains analyzed were taenite and the rest were kamacite (Ni_{3-8}), presumably reflecting again the heterogeneity of this specimen. In no case was an iron oxide phase observed as an alteration product.

Although at low magnification about 40% of the matrix appears to be glassy, at the highest magnifications it can be seen to be recrystallized and composed mainly of colorless pyroxene with hypersthene predominating over augite; there are also rather shapeless grains of plagioclase dusted with ilmenite. Recrystallization is seen best at the margins of plagioclase clasts into which minute pyroxene crystals project. There are also a few vugs lined with inward pointing crystals of pyroxene and plagioclase.

At an extremity of one section (240) it appears that the microbreccia has been invaded by an irregular mass of basaltic material. This material is an olivine-pigeonite-augite basalt with randomly oriented plagioclase laths and a considerable amount of dark interstitial late-stage products. The basalt also appears to include rounded areas of the microbreccia at whose contacts the basalt becomes a little finer grained, and a less lathy habit characterizes its plagioclase. In general, the textures suggest that although the basalt was the later member, the breccia was hot and plastic at the time it was invaded. It seems likely that there was no significant time interval between the formation of the two components, which are essentially the products of the same period of activity. Superficially the relations are similar to those observed in some terrestrial mixed rocks, but the inspection of such a small area could provide quite misleading comparisons.

Topotactic Reactions in Rock 14321

Oriented products developed within host single crystals can provide evidence of postcrystallization reactions. At the earliest stages of a reaction sequence, such products commonly are detectable only by single crystal diffraction studies. We have established techniques for the examination of substructures in terrestrial feldspars, pyroxenes, and olivines and have already reported on their application to lunar minerals from Apollo 11 and 12 rocks. Continuing such studies on Apollo 14 rocks, we find some unexpected reaction products in rock 14321 that are reported here; these may be of some significance in the interpretation of this particular specimen and more generally in lunar surface processes.

The petrographic description of this rock in the preceding section is in general accord with other accounts which emphasize the extremely complex structure. The rock chip (106) from which crystals for x-ray study were removed has a very heterogeneous surface texture. We wish to emphasize, therefore, that it is impossible for us to correlate the single crystals from the chip with any particular surface feature or with any particular occurrence of the same minerals in thin section.

Feldspars

In plagioclase we have been unable to find any evidence for antiperthitic structures (Bown and Gay, 1971) or any other oriented product. In the anorthosite xenolith C of Fig. 1b a few subparallel, pale yellow lamellae were observed in a few crystals; these appeared to be lower in refractive index than their host, but because they are less than 1 μ wide their nature cannot be determined with the probe. There are, however, a few feldspar crystals with fine yellowish lamellar inclusions that could be parallel and in optical continuity with each other; unfortunately it has not been possible to identify similar occurrences in this section. X-ray examination shows a homogeneous monoclinic host (which is presumably an alkali or bariun feldspar) together with a second as yet unidentified phase. The orientation of the product phase appears to be consistent in at least two of the crystals, but unfortunately is not in accordance with the intergrowth symmetry principle. Further study of the phase and its relationship to the feldspar host is now in progress.

Pyroxenes

By x-ray diffraction, clinopyroxene crystals are shown to contain a second oriented pyroxene, though this is rarely visible optically. In most of the crystals examined, the host, whether an augite (C2/c) or pigeonite (P2$_1$/c) structure type, has exsolution of various amounts of the other monoclinic structure sharing (001), and nearly all crystals exhibit additional complexities as already described for other lunar specimens (Papike *et al.*, 1971). A few crystals of augite show no pigeonite but carry only orthopyroxene sharing (100), commonly rather poorly developed.

Orthopyroxene crystals show exsolution of twinned augite-type structures on (100), commonly in such small amounts that considerable over-exposure is necessary for their detection.

Further symmetry-related diffraction maxima can be seen on the overexposed orthopyroxene photographs. These can be attributed to ilmenite and spinel structures in their normal topotactic orientations (Bown and Gay, 1959); some crystals contain both oxides, but in others, only one or other of the oxide phases is present. Re-examination of the clinopyroxene photographs revealed faint traces of similar oxide phases, but only in some of the crystals that carry exsolved orthopyroxene. The spinel phase is only present in the m.1 orientation, suggesting that it (and probably the ilmenite) is associated with orthopyroxene either as host or exsolution product. Similar oxide phases are reported in pyroxenes from the anorthosite 15415 (Stewart *et al.*, 1972).

Examined in thin section, some orthopyroxene single crystal clasts reveal, in addition to fine pale green lamellae of augite and fine droplets of a green spinel (probably incorporated during growth), Schiller lamellae of ilmenite of variable thickness; other fine lamellae of an opaque phase (ilmenite or a dark spinel?) are commonly observed in a separate orientation. Fuchs (1971) has reported oriented ilmenite lamellae in orthopyroxenes from Apollo 14 soils.

Olivines

Although some olivine crystals show normal diffraction patterns typical of the composition range 20–40% Fa with rare traces of superposed powder arcs due to misoriented inclusions (as is common in many lunar olivines), some show additional weak symmetry-related diffraction maxima on heavily exposed photographs; these maxima are not sharp in every case and commonly are extended around constant θ loci. Unfortunately these extra maxima are not the same in each different crystal which, coupled with their general feebleness, makes interpretation in terms of particular developing structures very difficult; indeed it seems likely that different products develop in different crystals. However, one set of extra spots occurs with more regularity than any others, and these are in accord with the occurrence of a spinel structure in the common topotactic orientation, (Champness and Gay, 1968). In one crystal there are also additional spinel orientations recognized by these authors in heat-treated olivines. The other reaction products cannot be identified with any certainty, though in one or two crystals there may be traces of a wüstite-like structure. Rarely, in addition to the various extra weak maxima, there are broadened and diffuse regions contiguous to the sharp spots of the host olivine; these are the same as the diffraction effects developed in terrestrial olivines at the earliest stages of the "iddingsitization" process (Gay and Le Maitre, 1961).

In some of the single crystals that have been x-rayed it is possible to see optical inhomogeneity in the form of erratic trails. These are more clearly seen in many of the olivine clasts in thin section, though they are more common in one section (19) than the other (240). These pale inclusions (Fig. 1c) are slightly darker in color than the olivine, have a higher refractive index, and occur as micronwide threadlike loops;

occasionally they have a beaded appearance. They can be compared to the loops of spinel crystals present in the cores of olivine phenocrysts in the picrite basalt ejected from Halemaumau, Kilauea (Fig. 1d) (Muir and Tilley, 1957). Similar features are shown in the peridotite blocks from the volcano of San Quentin, Baja California, where dark loops present in the olivine may be an oxide phase. Furthermore, Gay and Le Maitre (1961) noted minute hair-like growths at the transition from olivine to iddingsite in a basalt from Vogelsberg, Hesse. However, the paler character of loops in the lunar olivines compared to these terrestrial examples make it uncertain whether complete correspondence is established. There may be a relationship to the curving rod-like inclusions described in some olivines from rock 12018 that Roedder and Weiblen (1971) tentatively have identified as chrome-spinel. Another type of exsolution is seen in some olivines from 14321, in which micron-sized olive-brown platelets (probably biconvex discs) are developed, apparently lying in the (010) plane. At high magnification these display a dendritic structure similar to that illustrated by Harker (1954, Fig. 29a) in some of the allivalite rocks from Rhum; Brown (1956), following Harker, suggests that in the Rhum rocks these may be magnetite, though tabular inclusions, identified as brown chromite, may be present as well.

Possible Significance of Topotactic Reactions in 14321

From the petrographic description given earlier (and from others presented at the Third Lunar Science Conference) it is clear that 14321 is a fragmental rock of considerable complexity in formation and history. Some of the postcrystallization reactions that have been described are unique among the lunar specimens that we have examined. They are only weakly developed and localized in occurrence so that the following discussion of their significance in a lunar environment must be regarded as speculative at this stage.

We have already drawn attention to terrestrial analogs in which similar structures and reaction products can be interpreted as due to oxidation and possibly hydration mechanisms. But there is overwhelming evidence from other lunar specimens that the conditions for such mechanisms do not obtain generally in lunar processes; indeed Haggerty (1971) cites evidence from the subsolidus decomposition of spinels in 14321 that suggest a process of postcrystallization thermal reduction. In view of the sporadic occurrence of the reactions observed, the most plausible alternative explanation is that the various topotactic products were exsolved when some of the rock fragments were reheated locally before incorporation into the microbreccia. Even in the high vacuum conditions of the lunar surface, this could possibly produce exsolved oxide phases, because the quenched and metastable minerals have some-what abnormal compositions by comparison with terrestrial analogs. Provisional microprobe data confirm compositional variations of the kind reported in other lunar minerals (e.g. olivines can contain up to 0.4–0.5 wt% of Cr_2O_3), but so far there is no positive correlation with crystals containing visible or sub-microscopic products. It could be argued less plausibly that the various phases that have been detected are formed when random foreign inclusions caught up during the crystallization of a mineral are recrystallized in some subsequent process under the topotactic control of the host.

Although we have suggested these explanations, and indicated which seem most plausible, we must also point out that the postcrystallization reactions in 14321, deduced from x-ray diffraction and optical evidence, could alternatively be explained in a straight-forward manner by very localized thermal metamorphism under weakly oxidizing or hydrothermal conditions, or both together. Such a possibility is admittedly against the general run of evidence, but there are some pointers in its favor, which are worth listing. Bulk chemical analysis of another sample of this rock shows 0.05 wt% H_2O^+ (Scoon, 1972), and in a similar fragmental rock 14301, rims of a hydrous mineral, goethite, have been found around metallic iron particles (Agrell *et al.*, 1972). No conclusive result rewarded the search for similar evidence in 14321. In the thin section of basalt xenolith A, Fig. 1b, there is a local late stage reaction that has produced a reddish-brown region (surrounding ilmenite) that is very similar in appearance to the stained areas around the goethite-rimmed particles in 14301 described by Agrell *et al.* (1972). Nearby there are patches of an olive-green low birefringence material that would be identified as secondary amphibole if it were observed in a terrestrial basalt. One most unusual crystal was found among material removed from the surface of the rock chip; it was clear at one end, shading to a reddish-brown at the other, and gave a composite diffraction pattern consistent with olivine and chondrodite crystals in a reaction relationship.

The present position is inconclusive and there is no unassailable evidence for any of these (or any other) mechanisms. It undoubtedly would be helpful to have more data on the composition of product phases and the host crystals. Apart from the identified phases in certain olivines, the characterized spinel and ilmenite structures are recognized only by a few weak and ill-defined diffraction maxima. The rough cell constants that can be deduced (e.g. for the spinel phase a $\sim 8.40 \pm 0.15\text{Å}$) embrace wide compositional ranges in complex solid solutions including, for instance, the Cr and Al bearing exsolved spinel (presumably a chromous aluminate) reported by Stewart *et al.* (1972). The problems posed by the postcrystallization reactions in 14321 are probably not unique to this specimen, and their resolution is important if only because they might indicate that the lunar surface has contained water, which if substantiated could be of significance in the detailed history of its formation (see the discussion of Agrell *et al.*, 1972).

Acknowledgments—We are grateful to our colleagues, in particular Dr. S. O. Agrell, in the Department of Mineralogy and Petrology for their continued help in both practical aspects of the work and discussion of the results. We are greatly indebted to Mr. K. O. Rickson and Mr. N. F. Smith for their assistance in the diffraction studies, Mr. N. F. Smith and Mr. A. R. Arnold for electron microprobe determinations, and Mr. J. A. F. Fozzard for photographic reproductions.

REFERENCES

Agrell S. O. Scoon J. H. Long J. V. P. and Coles J. N. (1972) The occurrence of goethite in a microbreccia from the Fra Mauro formation (abstract). In *Lunar Science—III* (editor C. Watkins) pp. 7–9. Lunar Science Institute Contr. No. 88.

Bown M. G. and Gay P. (1959). Identification of oriented inclusions in pyroxene crystals. *Amer. Mineral.* **44**, 592–602.

Bown M. G. and Gay P. (1971) Lunar antiperthites. *Earth Planet. Sci. Lett.* **11**, 23–27.

Brown G. M. (1956) The layered ultrabasic rocks of Rhum, Inner Hebrides; *Phil. Trans. Roy. Soc. Series B*, **240**, 1–53.

Brown G. M. and Peckett A. (1971) Selective volatilization on the lunar surface: evidence from Apollo 14 feldspar-phyric basalts. *Nature* **234**, 262–266.

Champness P. E. and Gay P. (1968) Oxidation of olivines. *Nature* **218**, 157–158.

Dallwitz W. B. Green D. H. and Thompson J. E. (1966) Clinoenstatites in a volcanic rock from Cape Vogel, Papua. *J. Petrol.* **7**, 375–403.

Fuchs L. H. (1971) Orthopyroxene and orthopyroxene-bearing rock fragments rich in K, REE, and P in Apollo soil sample 14163. *Earth Planet. Sci. Lett.* **12**, 170–174.

Gay P. Bown M. G. Muir I. D. Bancroft G. M. and Williams P. G. L. (1971) Mineralogical and petrographic investigation of some Apollo 12 samples. *Proc. Second Lunar Sci. Conf., Geochim. Cosmochim. Acta* Suppl. 2, Vol. 1, 377–392. MIT Press.

Gay P. and Le Maitre R. W. (1961) Some observations on "iddingsite". *Amer. Mineral.* **46**, 92–111.

Haggerty S. E. (1971) Sub-solidus reduction of lunar spinels. *Nature* **234**, 113–117.

Harker A. (1954) *Petrology for Students*, 8th Edn. Cambridge University Press.

Hollister L. Trzcienski W. Jr. Dymek R. Kulick C. Weigand P. and Hargraves R. (1972) Igneous fragment 14310,21 and the origin of the mare basalts (abstract). In *Lunar Science—III* (editor C. Watkins), pp. 386–388, Lunar Science Institute Contr. No. 88.

Kushiro I. (1972) Petrology of lunar high-alumina basalt. In *Lunar Science—III* (editor C. Watkins) pp. 466–468, Lunar Science Institute Contr. No. 88.

Muir I. D. and Tilley C. E. (1957) Contributions to the petrology of Hawaiian basalts: I, the picrite basalts of Kilauea; *Amer. J. Sci.* **255**, 241–253.

Papike J. J. Bence A. E. Brown G. E. Prewitt C. T. and Wu C. H. (1971) Apollo 12 clinopyroxenes: exsolution and epitaxy. *Earth Planet. Sci. Lett.* **10**, 307–315.

Roedder E. and Weiblen P. W. (1971) Petrology of silicate melt inclusions, Apollo 11 and Apollo 12 and terrestrial equivalents. *Proc. Second Lunar Sci. Conf., Geochim. Cosmochim. Acta* Suppl. 2, Vol. 1, pp. 507–528. MIT Press.

Scoon J. H. (1972) Chemical analyses of lunar samples 14003, 14311 and 14321 (abstract). In *Lunar Science—III* (editor C. Watkins) pp. 690–691. Lunar Science Institute Contr. No. 88.

Stewart D. B. Ross M. M. Morgan B. A. Appleman D. E. Huebner J. S. and Commeau R. F. (1972) Mineralogy and petrology of lunar anorthosite 15415 (abstract). In *Lunar Science—III* (editor C. Watkins) pp. 726–728, Lunar Science Institute Contr. No. 88.

Wenk H. R. Ulbrich M. and Muller W. (1972) Lunar plagioclase (a mineralogical study) (abstract). In *Lunar Science—III* (editor C. Watkins), pp. 797–799. Lunar Science Institute Contr. No. 88.

Proceedings of the Third Lunar Science Conference
(Supplement 3, *Geochimica et Cosmochimica Acta*)
Vol. 1, pp. 363–378
The M.I.T. Press, 1972

The major element compositions of lunar rocks as inferred from glass compositions in the lunar soils

Arch M. Reid, Jeff Warner, W. I. Ridley,
Dennis A. Johnston, Russell S. Harmon

NASA Manned Spacecraft Center,
Houston, Texas 77058

Petr Jakeš

Lunar Science Institute,
Houston, Texas 77058

and

Roy W. Brown

Lockheed Electronics Corporation,
Houston, Texas 77058

Abstract—Data on the major element composition of glasses in Apollo 11, 12, and 14 and Luna 16 soils have been classified into groupings of preferred compositions by cluster analysis techniques. These preferred compositions are interpreted as being representative of the composition of rock types contributing to the various soils.

Nonmare rock types are characterized by high weight percent Al_2O_3 (>14), low FeO (<14), low Cr_2O_3 (<0.2) and low CaO/Al_2O_3 (<0.7). Two major nonmare rock types are recognized. Fra Mauro basalts (KREEP) predominate at the Apollo 14 site, are abundant in some Apollo 12 soils, but are uncommon in the Apollo 11 and Luna 16 soils. Highland basalt, or anorthositic gabbro, is abundant at all four sites and is interpreted as a major rock type in the lunar highlands.

Apollo 11 and 12 basalts are higher in Fe, Cr and Ca/Al, and lower in Al than the Luna 16 basalts. Ilmenite pyroxenites form a minor group at each mare site. Mare basalts probably are partial melts of a pyroxenitic mantle whereas Highland basalts and possibly Fra Mauro basalts originate from a shallower, more aluminous source region. The outer regions of the moon apparently are heterogeneous and layered, with a Ca-, Al-rich feldspathic crust overlying a more mafic pyroxenitic mantle at depth.

Introduction

THE BUILDING BLOCKS required to construct models of the petrologic evolution of the moon are the rock types discernible in the lunar samples. In order to define these rock types, evidence from soils, breccias and igneous rocks must be combined. The advantage in studying well-mixed soils is that they contain (1) locally derived materials that may be more representative of a specific area of the moon than the few large rock samples that can be returned, and (2) material derived from more distant sources. Much of the material present in the lunar soils has been so altered by brecciation, mixing, recrystallization, and deformation that the nature of the parent rocks is difficult to decipher. Two types of material, lithic fragments and glasses, provide clues to the compositions of these parent rocks. Lithic fragments of primary igneous character offer the most direct evidence, but not all of the contributing rock types are present as primary igneous fragments. Rock fragments are transported from distant sources by major impacts that may destroy the original texture and phase chemistry. Thus lithic fragment studies must include non-igneous fragments, many of which may be mixed rocks derived from more than one parent rock. To characterize such lithic fragments properly requires a large number of very time-consuming analyses (e.g., Prinz *et al.*, 1971).

We have attempted to study the rock types contributing to the lunar soils by analyzing glasses and using glass compositions as a guide to the composition of these rocks. Melting of the parent rock tends to homogenize the material so that analysis of a small amount of glass may be used to infer the composition of a heterogeneous crystalline rock. Several objections can be raised to the assumption that glass compositions represent the composition of the parent rocks from which the glasses are derived. In small impacts, the portion of the parent rock melted may not be representative of the bulk rock composition An extreme case would be the production of glass by melting a single mineral constituent in the parent rock. This factor undoubtedly contributes to the spread of compositions that we find in glass analyses. However, it should be noted that, with the exception of feldspar grains rendered isotropic by shock effects (maskelynite), we have not found any monomineralic glasses. A second major problem is that glasses form not only from primary rocks but from secondary mixed rocks such as breccias and soils. This factor must also contribute to the spread in composition found in the lunar soil glasses. In addition we must consider the possibility that the glass-forming process causes fractionation so that the glass does not have the same composition as the parent. Of the elements considered here, only the alkalis appear likely to be affected by high temperature fractionation (Gibson and Hubbard, 1972).

The practical advantage of studying glasses is that they can be analyzed much more rapidly than lithic fragments so that a large number of analyses can be accumulated in a relatively short time. We believe that because of the factors outlined above, no single glass analysis can be taken as representative of the composition of the parent rock. However, if a sufficiently large number of glass analyses are considered, preferred compositions can be distinguished. Such preferred compositions, we believe, provide the best guide to the composition of the parent rocks from which the glasses are derived. Various factors in the glass-making process contribute towards scatter in the analytical data, but by considering a sufficiently large number of analyses we hope to deduce the parent rock compositions.

In this paper we summarize the data, using the above approach, on two Apollo 14 soils (Apollo Soil Survey, 1971; Reid *et al.*, 1972a) and Luna 16 soils (Jakeš *et al.*, 1972). In addition, glass data from several authors (Warner, 1972) are summarized for the Apollo 11 and 12 soils. The data are interpreted in terms of the major rock compositions contributing to the soils at each site.

For each site, the glasses that have compositions like the mare basalts are discussed separately from those with compositions indicating a nonmare origin. Different mare-type glass groups at each site are given a letter code (e.g. Tranquillitatis A, Tranquillitatis B). Two major nonmare types are recognized: (1) Fra Mauro basaltic glasses (KREEP) that are characteristic of the Apollo 14 site and (2) Highland basaltic glasses that are common at all four sites.

METHODS

All analyses were made with the electron microprobe as reported in Apollo Soil Survey (1971, 1972). The actual analyses for the Apollo 14 soils are given in Brown *et al.* (1971a) and Reid *et al.* (1971a), and for the Luna 16 soils in Brown *et al.* (1971b). Various binary plots and histograms were studied in order to establish a classification based on natural breaks or minima between preferred groups. For each group, averages were calculated and checked against plotted data to establish preferred compositions. In order to minimize subjectivity, the data were regrouped by using a cluster analysis procedure to determine the groupings. The two approaches yield very similar groupings. In the following discussion, data from the cluster analysis treatment are used and average glass compositions may thus differ in detail from the averages published elsewhere.

The cluster analysis method starts with an input set of cluster centers and improves on that set in an iterative manner. During each interation, new clusters may be defined and existing clusters may be combined, if certain input test values are exceeded. The N parameters used (Ca, Fe, Mg, Si, Ti, Al, K, and Na) are not weighted. They are normalized by dividing each value by the standard deviation of the appropriate parameter, i.e. the standard deviations of the normalized values are all 1.0. Distance measures are calculated using an N-dimensional Euclidian distance formula. Each interation consists of 10 steps:

1. Assign each point to one of the current clusters by minimizing the distance calculation.

2. For each cluster, calculate (a) cluster mean, (b) average distance from the cluster mean to each point in that cluster, (c) standard deviation of each N parameter. Report mean, average distance, and the maximum standard deviation.

3. If this is the final pass, report the cluster assignment for each data point and stop.

4. Test each cluster for splitting; split a cluster if the average distance is greater than the input value TEST-1 and the maximum standard deviation is greater than the input value TEST-2.

5. If a cluster is to be split, (a) find the parameter n that displays the maximum standard deviation, (b) define new cluster 1 that has values of all parameters equal to the old values except that parameter n has a value of old value plus input value TEST-3, (c) define new cluster 2 as with cluster 1 except that parameter n has a value of old value minus TEST-3.

6. Assign each data point to one of the current clusters as in step 1.

7. If any cluster has 2 or fewer points, eliminate that cluster.

8. For each cluster, calculate and report as in step 2.

9. Test each cluster with every other cluster for combining; combine two clusters if the distance between the cluster mean is less than the input value TEST-4.

10. Combine old clusters to be combined into a new cluster, calculate new cluster mean, and delete old clusters.

The same results are achieved by starting with random input cluster centers and iterating 50 times, or by starting with a standard set of estimated input cluster centers and iterating 20 times. In practice, the latter procedure was used. The technique is similar to that described by Ball and Hall (1967). This approach yields major groupings and several minor groups. Some of the minor groups have been combined where their unique composition marks them as a separate but somewhat variable group. Thus all the high-silica glasses have been combined into a single "granite" group despite the fact that the range of compositions in the high-silica group causes the cluster analysis routine to divide this group into several clusters.

PREFERRED GLASS COMPOSITIONS IN LUNAR SOILS

Apollo 11

The Apollo 11 data considered here are culled from the literature (Warner, 1972). Not all of the data are from random analyses of glasses in the soil and thus the proportions of the various glass types given in Table 1 are only very crude estimates of relative abundance.

Table 1. Chemical compositions of glass types in the Apollo 11 soils (wt%).

	SiO_2	TiO_2	Al_2O_3	FeO	MgO	CaO	Na_2O	K_2O	Approximate abundances
Tranquillitatis A	40.2	7.2	13.2	16.5	8.5	12.0	0.3	0.10	40
Tranquillitatis B	39.1	8.7	6.2	23.2	14.3	7.6	0.3	0.08	19
Tranquillitatis C	45.0	3.0	16.9	12.3	9.2	12.1	0.7	0.13	2
Highland basalt	44.6	0.9	24.4	6.6	8.5	14.5	0.3	0.05	21
Anorthositic	45.2	0.2	32.1	1.9	2.3	17.1	0.8	0.03	6
Fra Mauro basalt	50.4	2.5	17.2	10.5	6.0	10.1	1.1	0.87	5
Granitic	75.6	0.5	12.0	2.2	0.3	1.9	0.4	6.1	7

Mare-type glasses. The Mare Tranquillitatis soils are very complex and do not show such well-defined groupings, on binary plots, as do the Fra Mauro soils (Fig. 1). The majority of the glasses have compositions like mare basalts, i.e. high in Fe and Ti, low in Al and K. Few glasses in the soil correspond in composition to the large samples of mare basalt from the Apollo 11 mission. The high iron, mare-type glasses split into two clusters. The larger cluster, Tranquillitatis basalt A, is a large very diffuse group that in comparison with the Apollo 11 mare basalts is richer in Al and poorer in Fe (Fig. 1). The composition and the large amount of scatter in the analyses suggest that these glasses may be mixtures of mare basalts and more aluminous material such as the Highland basalt glasses (described below). The Ti content of the glasses is, however, too high to be compatible with such simple mixing. The

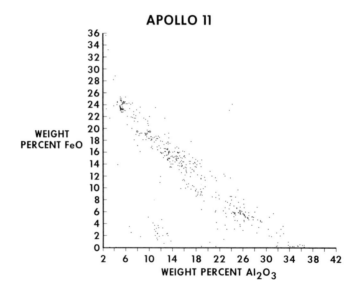

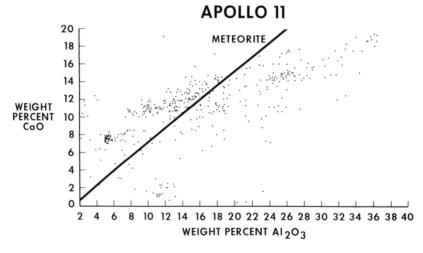

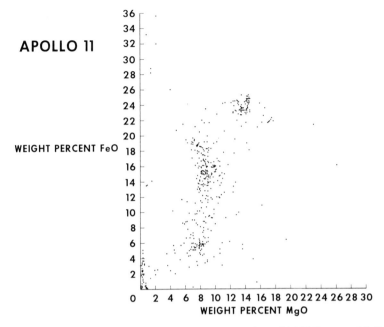

APOLLO 11

WEIGHT PERCENT FeO

WEIGHT PERCENT MgO

Fig. 1. Plots of (a) FeO versus Al_2O_3, (b) CaO versus Al_2O_3 and (c) FeO versus MgO for glasses in Apollo 11 soils. Open circles represent the compositions of Apollo 11 mare basalts.

group may represent a very complex mixture, such as would be generated by melting of the Apollo 11 soil. Bulk analyses of the soil are consistent with this hypothesis. We are then left with the problem of explaining the near absence, in the soil, of glasses that have the composition of the larger igneous rocks. A third possibility is that the large Apollo 11 igneous rocks are not representative of average Mare Tranquillitatis basalt. The present sampling coverage allows the possibility that many basalts in the region are lower in Fe and Ti and higher in Al than the large basalt samples. The origin of Tranquillitatis basalt A glasses remains unknown in view of the very large scatter in the data. Absence of strongly preferred compositions may indicate extensive mixing in a mature soil.

The second mare-basalt glass group (Tranquillitatis B) forms a much tighter grouping with very high Fe and Ti, and low Al. The grouping of glasses around a specific composition (Fig. 1) suggests that these glasses formed by total melting of a single rock unit. No large crystalline rocks of this composition have been recovered. Tranquillitatis B glasses commonly have a distinctive red-brown color. Several workers have analyzed glasses from this Ti-rich group: Prinz *et al.* (1972) discuss the glasses and report that no lithic fragments of comparable composition were found. Glasses of similar composition occur in less abundance in the Apollo 12, Apollo 14, and Luna 16 soils. The composition (Table 1) corresponds to that of an ilmenite pyroxenite with 46% pyroxene, 19% ilmenite, 17% plagioclase feldspar and 16% olivine. The composition resembles the Apollo 11, low-K basalts but has higher Fe and Mg, and lower Al. Compared with the mare basalts, the source rock would be expected to contain less feldspar and be very rich in pyroxene, olivine and ilmenite. The higher Mg/Fe ratio is consistent with formation of ilmenite pyroxenite as a cumulate.

A third glass group, Tranquillitatis C, has lower Fe and higher Al than the other mare-type glasses. This small group is not an obvious preferred composition, in plots of the data, and may not be representative of a parent rock composition. The composition is similar to that of Fra Mauro basalt (KREEP) but with a lower K content.

Nonmare glasses. The major nonmare glass type is a well-defined group with higher Al and Ca, and lower Fe, Mg and Ti than mare-type glasses. The average composition corresponds to a rock with 70% plagioclase, 20% pyroxene, 9% olivine, and 1% ilmenite. These glasses constitute a major group at each site, where they outweigh the more anorthositic glasses in abundance. The average composition is essentially the same in Mare Tranquillitatis, Oceanus Procellarum, Fra Mauro, and Mare Fecunditatis. The presence of the same nonmare component at each site suggests that it is derived from a parent rock that predominates in the lunar highlands (Reid *et al.*, 1972b). The parent rock may be a fine-grained basalt (Highland basalt, Reid *et al.*, 1972b) formed by partial or total melting of an extremely aluminous source (Al_2O_3, 25 wt% or higher) or may be an anorthositic gabbro cumulate from a high-alumina basalt parent.

A less abundant but more aluminous group includes glasses that are compositionally equivalent to gabbroic anorthosite and anorthosite. These glasses probably derive from highly feldspathic cumulates. Some portion of the anorthositic glasses (normative feldspar greater than approximately 95%) probably are feldspar grains melted or rendered isotropic by shock processes.

A small number of glasses in the Apollo 11 soil have compositions similar to the dominant glass compositions at the Apollo 14 site and are thus Fra Mauro or KREEP basaltic glasses. This group is characterized by high Al and K, and low Fe and Mg. Higher K values provide the major discriminant of the group but the entire major element chemistry is very similar to Fra Mauro basaltic glasses (Apollo Soil Survey, 1971) suggesting that there is a minor component of Fra Mauro material in the Mare Tranquillitatis regolith.

The remaining glass group encompasses the high-silica glasses that are rich in Si, Al, K and Ba, low in Fe, Ca, Mg, and have high Fe/Mg ratios. The individual glasses have a wide range of compositions but the average high-silica glass from each site is remarkably similar. The dangers involved in using non-random data from the literature and calculating abundances is well-illustrated by the high calculated abundance of "granitic" glasses in the Apollo 11 soils (Table 1).

Two possible sources can be proposed for these high-silica, "granitic" glasses. They may be derived from the late-stage residual liquids shown to form in the crystallization of lunar basalts (e.g. Roedder and Weiblen, 1970). Alternatively, they may be derived from "granitic" source rocks. Some of these glasses are homogeneous, inclusion-free grains probably that formed by total melting of a pre-existing "granitic" parent rock. The existence of separate "granitic" bodies of small size is evidenced by the work on rock 12013 (e.g. Drake *et al.*, 1971) or by the presence of "granitic" lithic fragments in the soil (e.g. Mason *et al.*, 1971; Meyer, 1972). There does not appear to be any direct evidence from the returned lunar samples for the presence at the lunar surface of extensive bodies of "granitic" rock.

Apollo 12

The discussion here is based on glass data from the literature (Warner, 1972). The data show large variations in the constitution of different Apollo 12 soils. For example, the proportions of Fra Mauro-type material (KREEP) vary widely for different soils (e.g. Meyer *et al.*, 1971). The data base used in this work is an indiscriminant mixture of analyses from several soil samples. These data are not extensive enough to provide a thorough insight into the glass chemistry of the various soils.

Mare-type glasses. The major glass grouping, Procellarum basalt A (Fig. 2), contrasts in composition with the average Apollo 12 igneous rock, being higher in Al and lower in Fe and Mg (Table 2). The relationship between the major glass type and the large igneous rocks is analogous to the relationship at the Apollo 11 site and the discussion need not be repeated here. Compared to the Apollo 11 glasses, this group has higher Si and Mg, and much lower Ti.

As in the Apollo 11 soils, there is a small, distinctive group of glasses, Procellarum basalt B, with compositions resembling Ti-rich mare basalts. This group resembles Tranquillitatis B in composition (Tables 1 and 2) but has higher Al and lower Fe and Mg. These glasses may be derived from an ultramafic mare rock.

Nonmare glasses. The major nonmare contribution to the Apollo 12 soils is material of the type that predominates at the Fra Mauro site (Fra Mauro basalt or KREEP). The equivalent glasses have a distinctive composition (e.g. Meyer *et al.*, 1971) with high Al and K, and low Fe in comparison with mare-type glasses (Table 3). These glasses are similar to those found at the Fra Mauro site (Apollo Soil Survey, 1971).

Highly aluminous glasses occur in the Apollo 12 soil and a single group is dis-

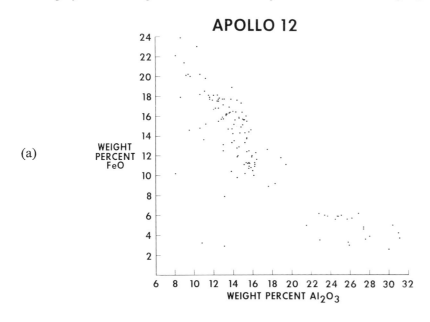

(a)

APOLLO 12

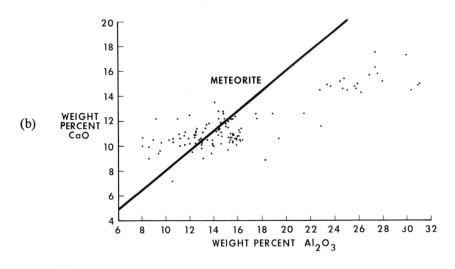

(b)

APOLLO 12

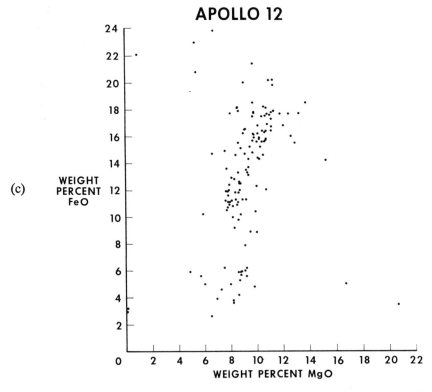

(c)

Fig. 2. Plots of (a) FeO versus Al_2O_3, (b) CaO versus Al_2O_3 and (c) FeO versus MgO for glasses in Apollo 12 soils.

Table 2. Chemical compositions of glass types in the Apollo 12 soil (wt%).

	SiO$_2$	TiO$_2$	Al$_2$O$_3$	FeO	MgO	CaO	Na$_2$O	K$_2$O	Approximate abundances
Procellarum A	43.7	3.3	13.2	16.5	9.9	11.1	0.2	0.16	53
Procellarum B	41.1	9.9	9.6	14.4	8.9	9.8	0.1	0.02	4.5
Highland basalt	43.9	0.4	26.2	4.9	8.9	15.0	0.2	0.04	16
Fra Mauro basalt	48.5	2.2	15.2	11.7	8.4	10.7	0.7	0.78	25
Granitic	76.1	0.6	12.0	3.1	0.1	1.5	0.2	6.6	1.5

Table 3. Chemical compositions of glass types in the Apollo 14 soil (wt%).

	SiO$_2$	TiO$_2$	Al$_2$O$_3$	FeO	MgO	CaO	Na$_2$O	K$_2$O	Approximate abundances
Fra Mauro basalt	48.0	2.1	17.0	10.9	8.7	10.7	0.7	0.54	60
Highland basalt A	45.3	0.4	25.8	5.6	7.9	14.8	0.2	0.07	29
Highland basalt B	48.7	0.3	28.5	3.0	2.1	15.0	1.4	0.81	1
Anorthositic	38.0	0.2	34.5	1.2	5.6	20.4	0.0	0.0	< 1
Granitic	71.5	0.4	14.25	1.8	0.7	2.0	0.9	6.5	1
Mare-type	45.4	2.7	9.6	18.4	12.9	9.4	0.4	.23	8
Mare-type	41.2	8.3	10.5	16.4	11.8	10.6	1.1	.66	1

tinguished. The average composition of this group is that of Highland basalt (anortho-sitic gabbro) with 26 wt% Al$_2$O$_3$. This composition is essentially the same as at other sites. True anorthosite or gabbroic anorthosite compositions are very rare at the Apollo 12 site. "Granitic" glasses have been reported in minor amounts.

Apollo 14

Some 500 glasses have been analysed (Fig. 3) in two Apollo 14 soils, 14259 and 14156. The resultant glass groupings, almost identical for the two soils, are reported

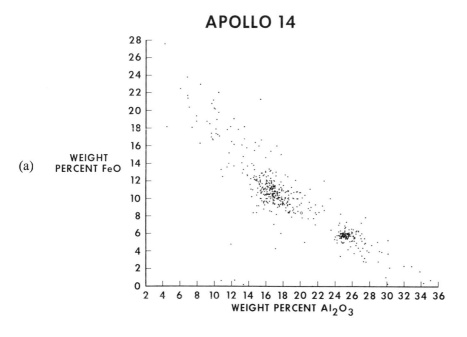

(a)

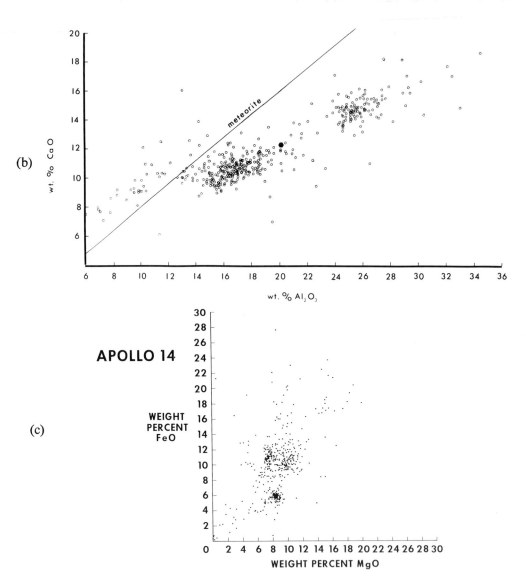

Fig. 3. Plots of (a) FeO versus Al_2O_3, (b) CaO versus Al_2O_3 and (c) FeO versus MgO for glasses in Apollo 14 soils. Circle and hexagon represent the composition of rock 14310.

and discussed in detail elsewhere (Apollo Soil Survey, 1971, 1972; Reid *et al.*, 1972a). *Nonmare glasses.* The major glass type, Fra Mauro basaltic glass or KREEP, has higher Al and K, and lower Fe than mare basalt and resembles terrestrial tholeiites in major element composition (Table 3). Three subgroups with different MgO contents (Types B, C and D) can be distinguished (Table 4). The Apollo Soil Survey (1971) suggested that Fra Mauro basalts and anorthositic gabbro (Highland basalt) might be complimentary differentiates from a magma with the composition of rock 14310.

Table 4. Chemical compositions of Fra Mauro basaltic glasses
(wt%)

	TYPE B	TYPE C	TYPE D
SiO_2	47.3	48.2	51.6
TiO_2	1.9	2.2	1.8
Al_2O_3	17.0	17.1	16.9
FeO	10.6	10.5	11.0
MgO	10.2	7.6	4.7
CaO	10.5	10.9	10.9
Na_2O	0.6	0.8	0.8
K_2O	0.5	0.6	0.8
TOTAL	98.6	97.9	97.9
Relative abundance	47.3	48.2	4.5

Later work (e.g. Ridley *et al.*, 1972) has shown that 14310 probably does not represent a primary magma composition. Also it has become apparent that Highland basalt is abundant in Apollo 11, 12 and 14 and Luna 16 soils, whereas Fra Mauro basalt is abundant only in the Apollo 12 and 14 soils. These data are inconsistent with the complimentary differentiate hypothesis and the parents of these two glass types are probably not related by a simple one-stage process. The experimental data of Walker *et al.* (1972) and the very high concentrations of certain trace elements in Fra Mauro basalts (e.g. Hubbard and Gast, 1971) indicate that they are highly fractionated compositions. Such compositions could arise as residues from extensive crystal-liquid fractionation or as liquids formed from a small degree of partial melting. The Cr content of Fra Mauro basalts (Type B, 0.16 weight percent Cr_2O_3; Reid *et al.*, 1972a) may be inconsistent with an origin involving extensive crystal-liquid fractionation (Hubbard and Gast, 1971). Some fractionation of this type probably has occurred, however, and Types B, C and D may represent the composition of successive residual liquids produced by crystal-liquid fractionation (Walker *et al.*, 1972).

The second largest glass grouping (30% of the glasses) have Highland basalt (anorthositic gabbro) compositions, essentially the same as at other sites. A subgroup comprises very few glasses with a similar major element chemistry but higher K. There is thus some indication that there may be two types of highly feldspathic material, with different K and REE contents, as suggested by Hubbard *et al.* (1972). The higher K feldspathic rocks may be cumulates from Fra Mauro (KREEP) basalt. The much more abundant, low K, Highland basalts have a separate origin. Anorthositic and gabbroic anorthosite glasses form a much smaller group than the Highland basalts. High-silica glasses are present but also rare. Their abundance is high relative to mare-derived glasses, indicating a nonmare origin. These non-mare "granites" have very similar average composition to the "granite" glasses in mare soils.

Mare type glasses. A portion of the Apollo 14 glasses have high Fe and low Al, and closely resemble mare glasses. These glasses are probably mare-derived and in average composition they resemble Oceanus Procellarum soil glasses. The group is compositionally diffuse and contains a few high-Ti glasses similar to Tranquillitatis B. Mixed glasses intermediate between mare-type and Fra Mauro type can be recognized.

Luna 16

The Luna 16 soils from Mare Fecunditatis have been studied intensively (e.g. Jakeš *et al.*, 1972). The major glass group, Fecunditatis basalt A (Fig. 4), has an average composition that resembles mare basalts (Table 5). Fe is lower, however, and Al substantially higher than in Apollo 11 and 12 mare basalts. There is thus indirect evidence that the major basalt type in Mare Fecunditatis is a mare basalt with less Fe and more Al than the other known mare basalts. Again, a minor Ti-rich composition is present in the glasses. Five percent of the glasses are rich in Fe and Ti (Fecunditatis B) and resemble Tranquillitatis B compositions.

Cluster analysis yielded a third glass group that was included within Fecunditatis A in the paper by Jakeš *et al.* (1972). This group, Fecunditatis C, has the composition of a high-alumina basalt with 21 wt% Al_2O_3 and only 10% FeO (Table 5). The composition is similar to that of Fra Mauro basalt (KREEP) but without the higher K content. The status of this group is uncertain in that it is not a strongly preferred composition and these glasses may represent mixtures of Fecunditatis A and Highland basalt. If indeed they do derive from a single rock type, then this composition would be an important component in constructing petrologic models of lunar evolution.

As at the other sites, Highland basalt is an important glass type (16%) and more abundant than the more anorthositic glasses (6%). Fra Mauro basalt glass is not abundant (less than 2%) and "granitic" glasses are rare (less than 1%).

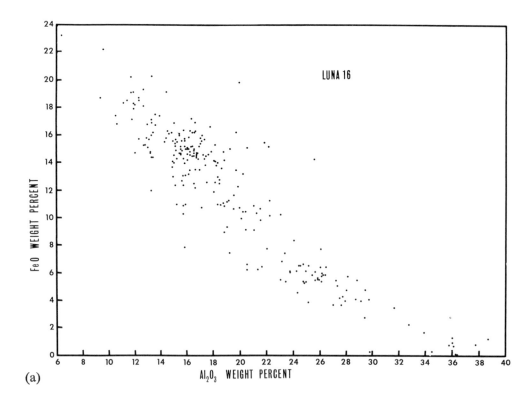

(a)

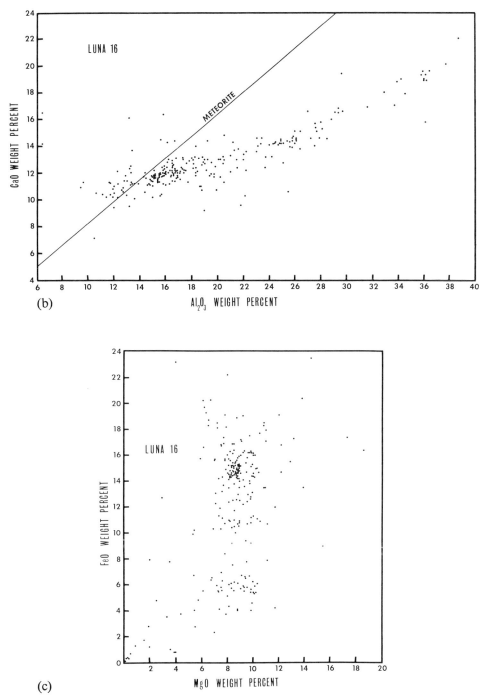

Fig. 4. Plots of (a) FeO versus Al_2O_3, (b) CaO versus Al_2O_3 and (c) FeO versus MgO for glasses in Luna 16 soils.

Table 5. Chemical compositions of glass types in the Luna 16 soil (wt%).

	SiO_2	TiO_2	Al_2O_3	FeO	MgO	CaO	Na_2O	K_2O	Approximate abundances
Fecunditatis A	44.1	3.0	15.9	14.9	9.1	11.8	0.3	0.09	52.5
Fecunditatis B	41.6	6.9	12.2	18.5	7.7	11.0	0.4	0.16	5
Fecunditatis C	46.0	1.0	20.7	9.8	8.8	13.0	0.4	0.12	17
Highland basalt	45.0	0.3	26.6	5.3	8.2	14.9	0.1	0.02	16.5
Anorthositic	43.6	0.1	35.0	1.0	1.4	18.9	0.4	0.14	6.5
Fra Mauro basalt	46.8	1.9	14.5	14.3	7.3	12.4	0.3	0.88	1.5
Granitic	77.6	n.d.	11.3	1.4	n.d.	1.0	0.5	6.3	<1

SUMMARY OF ROCK TYPES DEDUCED FROM GLASS ANALYSES

The major element characteristics of the lunar glasses can be seen in Figs. 1 to 4 and in the Tables 1 to 5. Over the entire range of glasses, Fe and Al vary more than any other major elements, but a striking feature is the near constancy (with the exception of the "granitic" glasses) of Fe plus Al. In part in this trend there is a mixing line between Fe-rich, Al-poor mare rocks and Fe-poor, Al-rich nonmare rocks. The major glass types also have very similar Mg contents, the exceptions being the anorthosites, gabbroic anorthosites, "granites", and a few high-Mg mare glasses. Si values are also similar for all the glasses except for the "granites" and the Ti-rich glasses.

Mare types

Three mare sites have been studied and in each case the glasses in the soil are dominated by a single, diffuse glass group. The compositions of these groups (tranquillitatis A, Procellarum A, Fecunditatis A) are characteristic for each mare site and do not match the compositions of the larger igneous rock samples from these sites. Either the larger rock samples are not representative of the regional mare basalt population, or the glasses represent mixtures of various components in the mare regolith (J. B. Reid et al., 1972). Some mixing of materials in the thoroughly reworked, mature soils undoubtedly has occurred. Nevertheless, the mare-basalt component is substantial and the unique chemical characteristics of the different mare are reflected in the glasses. Mare-type glasses are a minor component in the Apollo 14 soils and resemble Apollo 12 basaltic glasses.

At each site there is a distinctive group of Ti-rich glasses that are Fe-rich and Al-poor, and correspond in composition to ilmenite pyroxenites. They may in part represent cumulates from mare-basalt liquids.

Mare basalts are characterized by high Fe, Cr and Ca/Al, and low Al, in comparison with the non-mare samples. These "mare" characteristics become increasingly prominent in the sequence Luna 16 → Apollo 11 → Apollo 12 → Apollo 15 (Reid et al., 1971b). Luna 16 basalts have some characteristics, such as Al content, that are intermediate between nonmare and the other mare samples. Mare basalts appear to have formed at depth by partial melting of a dominantly pyroxenitic mantle (Ringwood, 1970).

Nonmare types

The nature of the nonmare lunar rocks can be determined from a study of the one nonmare site (Apollo 14) and from a study of the exotic fragments in the mare soils. Nonmare samples are characterized by high Al ($Al_2O_3 > 14\%$), low Fe (FeO < 14%), low Cr ($Cr_2O_3 < 0.2\%$; Reid *et al.*, 1972a, b), and low Ca/Al ($CaO/Al_2O_3 < 0.7$). Two major types of nonmare material are recognized in mare soils and at the Apollo 14 site. Fra Mauro basalts predominate at the Apollo 14 site, are prominent at the Apollo 12 site, but are rare in the Apollo 11 and Luna 16 soils. The other nonmare component is present in significant amounts in soils at all four sites and has essentially the same composition (anorthositic gabbro) at each site. These data are consistent with this component being derived from the lunar highlands and we have called it Highland basalt. In comparison, glasses with gabbroic anorthosite and anorthosite compositions are much less abundant. If, as seems probable, there is a highland-derived component common to all four sites, the best candidate for this component is the Highland basalt. There are some indications in both Luna 16 and Apollo 11 soils of the existence of high-alumina basalts similar to Fra Mauro basalts, but lacking their high K contents.

The highly feldspathic composition and the high Mg/Fe ratio of Highland basalt is suggestive of a cumulative origin. Whether Highland basalts are cumulates or not, they apparently are derived from source material that is more aluminous, more feldspathic, and at a shallower depth than the source material for mare basalts (Reid *et al.*, 1972b). Fra Mauro basalts have compositions that lie near the cotectic in the system silica-olivine-anorthite (Walker *et al.*, 1972). According to Ford *et al.* (1972), feldspar is the liquidus phase for Types B, C, and D. Feldspar-rich cumulates from Fra Mauro basalts do occur (Hubbard *et al.*, 1972; Apollo Soil Survey, 1972). The parent rocks for the Fra Mauro basalts apparently lie to the feldspar-rich side of the cotectic, also implying a source rock more aluminous than the proposed source for mare basalts.

Lunar surface igneous rocks are derived from both a feldspathic outer layer (non-mare basalts) and a deeper pyroxenite source (mare basalts). Present models call for a heterogeneous layered moon, high in Ca and Al near the surface and more mafic at depth, with both layers depleted in volatile and siderophile elements. This layering may be due to differentiation early in lunar history, or may partly reflect an accretionary stratigraphy.

Acknowledgments—We thank Robin Brett for substantial contributions to this project and P. Richardson and A. Kenworthy for their help in data handling. W. I. Ridley was supported by an NRC Resident Research Associateship. P. Jakeš was supported by contract number NSR 09–051–001 between the National Aeronautics and Space administration and the Universities Space Research Association. This paper constitutes Lunar Science Institute Contribution 91.

REFERENCES

Apollo Soil Survey (1971) Apollo 14: Nature and origin of rock types in soil from the Fra Mauro formation. *Earth Planet. Sci. Lett.* **12**, 49–54.
Apollo Soil Survey (1972) Phase chemistry of Apollo 14 soil sample 14259. In preparation.

Ball G. H. and Hall D. J. (1967) A clustering technique for summarizing multivariate data. *Behavioral Sci.* **12**, 153–155.

Brown R. W. Reid A. M. Ridley W. I. Warner J. L. Jakeš P. Butler P. Williams R. J. and Anderson D. H. (1971a) Microprobe analyses of glasses, and minerals from Apollo 14 sample 14259. NASA Tech. Memo. TMX–58080.

Brown R. W. Harmon R. S. Jakeš P. Reid A. M. Ridley W. I. and Warner J. L. (1971b) Microprobe analyses of glasses and minerals from Luna 16 soil. NASA Tech. Memo. TMX–58082.

Drake M. J. McCallum I. S. McKay G. A. and Weill D. F. (1970) Mineralogy and petrology of Apollo 12 sample No. 12013: A progress report. *Earth Planet. Sci. Lett.* **9**, 103–123.

Ford C. E. Humphries D. J. Wilson G. Dixon D. Biggar G. M. and O'Hara M. J. (1972) Experimental petrology of high alumina basalt 14310 and related compositions (abstract). In *Lunar Science—III* (editor C. Watkins) pp. 274–276, Lunar Science Institute Contr. No. 88.

Gibson E. K. and Hubbard N. J. (1972) Volatile element depletion investigations on Apollo 11 and 12 lunar basalts via thermal volatilization. In *Lunar Science—III* (editor C. Watkins), pp. 303–305, Lunar Science Institute Contr. No. 88.

Hubbard N. J. and Gast P. W. (1971) Chemical composition and origin of nonmare lunar basalts. *Proc. Second Lunar Sci. Conf., Geochim. Cosmochim. Acta* Suppl. 2, Vol. 2, pp. 999–1020.

Hubbard N. J. Gast P. W. and Meyer C. Jr. (1972) Chemical composition of lunar anorthosites and their parent liquids. In *Lunar Science—III* (editor C. Watkins), pp. 404–406, Lunar Science Institute Contr. No. 88.

Jakeš P. Warner J. Ridley W. I. Reid A. M. Harmon R. S. Brett R. and Brown R. W. (1972) Petrology of a portion of the Mare Fecunditatis regolith. *Earth Planet. Sci. Lett.* **13**, 257–276.

Mason B. Melson W. G. Henderson E. P. Jarosewich E. and Nelen J. (1971) Mineralogy and petrology of some Apollo 12 samples. *Second Lunar Sci. Conf.* (unpublished proceedings).

Meyer C. Jr. Brett R. Hubbard N. J. Morrison D. A. McKay D. S. Aitken F. K. Takeda H. and Schonfeld E. (1971) Mineralogy, chemistry and origin of the KREEP component in soil samples from the Ocean of Storms. *Proc. Second Lunar Sci. Conf., Geochim. Cosmochim. Acta* Suppl. 2, Vol. 1, pp. 393–411.

Meyer C. Jr. (1972) Mineral assemblages and the origin of non-mare lunar rock types. In *Lunar Science—III* (editor C. Watkins), pp. 542–544, Lunar Science Institute Contr. No. 88.

Prinz M. Bunch T. E. and Keil K. (1971) Composition and origin of lithic fragments and glasses in Apollo 11 samples. *Contrib. Mineral. Petrol.* **32**, 211–230.

Reid A. M. Ridley W. I. Jakeš P. and Warner J. L. (1971a) Microprobe analyses of glasses from Apollo 14 sample 14156. NASA Tech. Memo. TMX–58081.

Reid A. M. Ridley W. I. Warner J. L. Harmon R. S. Brett R. and Jakeš P. (1971b) Lunar basalts (abstract). *Trans. Amer. Geophys. Union* **52**, 857.

Reid A. M. Ridley W. I. Jakeš P. and Harmon R. S. (1972a) Compositions of glasses lunar soil 14156. In preparation.

Reid M. A. Ridley W. I. Harmon R. S. Warner J. L. Brett R. Jakeš P. and Brown R. W. (1972b) Highly aluminous glasses in lunar soils and the nature of the lunar highlands. Submitted to *Geochim. Cosmochim. Acta*.

Reid J. B. Jr. Taylor G. J. Marvin U. B. and Wood J. A. (1972) Luna 16: Relative proportions and petrologic significance of particles in the soil from Mare Fecunditatis. *Earth Planet. Sci. Lett.* **13**, 286–298.

Ridley W. I. Williams R. J. Brett R. Takeda H. and Brown R. W. (1972) Petrology of lunar basalt 14310. In *Lunar Science—III* (editor C. Watkins), Lunar Science Institute Contr. No. 88.

Ringwood A. E. (1970) Petrogenesis of Apollo 11 basalts and implications for lunar origin. *J. Geophys. Res.* **75**, 6453–6479.

Roedder E. and Weiblen P. (1970) Lunar petrology of silicate melt inclusions, Apollo 11 rocks. *Proc. Apollo 11 Lunar Sci. Conf., Geochim. Cosmochim. Acta* Suppl. 1, Vol. 1, pp. 801–837.

Walker D. Longhi J. and Hays J. F. (1972) Experimental petrology and origin of Fra Mauro rocks and soil. In *Lunar Science—III* (editor C. Watkins), pp. 770–772, Lunar Science Institute Contr. No. 88.

Warner J. L. (1972) Unpublished compilation of Apollo chemical, age and modal information on file in the Curator's Office, NASA, MSC, Houston, Texas.

Proceedings of the Third Lunar Science Conference
(Supplement 3, *Geochimica et Cosmochimica Acta*)
Vol. 1, pp. 379–399
The M.I.T. Press, 1972

Analysis of Fra Mauro samples and the origin of the Imbrium Basin

M. R. DENCE AND A. G. PLANT

Earth Physics Branch, Department of Energy,
Mines and Resources, Ottawa, Canada

Geological Survey of Canada, Department of Energy,
Mines and Resources, Ottawa, Canada

Abstract—Apollo 14 soils and rocks are predominantly fragmental, and comprise glasses, glass-bearing breccias, annealed fragmental rocks, and a small proportion of igneous rocks of basaltic texture. The textures and composition of glasses and rock fragments in soils 14258,34 and 14167,7 are reported and are shown to be mainly of four types: alkali-rich (Fra Mauro) basalts, feldspathic (Highland) basalts and anorthosites, mare basalts, and potassic granites. The same components are found in fragmental rocks. Glass-bearing breccia 14315,11 contains many chondrule-like bodies and devitrified glasses of feldspathic basalt composition. Annealed fragmental rocks 14305,5 and 14314,13 contain a wide variety of clasts including mineral fragments, mare basalt, and annealed glass predominantly of Fra Mauro basalt composition enclosing smaller amounts of potassic granite. Fragmental rock 14311,88 is more thoroughly annealed and contains many vug linings and inclusions of potassic granite glass, some partly devitrified. Crystalline rock 14310,4 is a feldspathic basalt with a wide range of texture and, by analogy with rocks from terrestrial impact craters, is interpreted as an impact melt, in which coarser crystals nucleated around relict xenocrysts. From an impact model for the Imbrium Basin, based on studies of terrestrial craters, the argument is developed that the ejecta at the Fra Mauro site were probably only weakly shocked and heated by the Imbrian event and were derived from the uppermost crust in the Imbrium Basin area. The multiple shock and thermal events recorded in the fragmental rocks and feldspathic basalts are considered to result from intensive meteorite bombardment prior to the Imbrian impact. Some mare basalts also predate this event, but most glass-bearing breccias probably were of later derivation.

INTRODUCTION

THERE HAS BEEN general acceptance of the interpretation by Eggleton (1963, 1964) that the Fra Mauro formation, sampled at the Apollo 14 site, is part of the thick ejecta blanket from the Imbrium Basin. The hummocky terrain in the vicinity of the site has been compared with deposits around other large lunar craters and basins, notably those around the Orientale Basin (McCauley, 1967), from which it has been suggested that the dominant mode of transport of these deposits was by flow along the ground. Such flows have been compared to base surge deposits from volcanoes, and the assumption commonly has been made that they were necessarily hot (McKay and Morrison, 1971). Certainly, impact-produced breccias with a high content of shock melted components will be sufficiently hot to undergo welding and annealing of some constituents. However, the study of terrestrial impact structures demonstrates that large quantities of breccia both inside and outside the craters do not contain strongly shocked materials (Engelhardt, 1971) and are only mildly heated unless adjacent to large bodies of impact melt (Dence, 1971).

We have studied samples of Apollo 14 soils 14258,34 and 14167,7, fragmental rocks 14305,5, 14311,88, 14314,13, 14315,11, and crystalline rock 14310,4.

Preliminary examination (LSPET, 1971) demonstrated that the Apollo 14 rocks and minerals are similar to those of Apollo 11 and 12 samples, although in strikingly different proportions, and that conditions of crystallization, low f_{O_2} and f_{H_2O}, essentially were identical at all three sites. We have therefore given particular attention to textural relationships and the characterization of the bulk composition of glasses and fragmental rocks, with the aim of relating them to the history of the Fra Mauro formation and its relationship to the Imbrium Basin.

REGOLITH FINES, 14258,34 AND 14167,7

Fines samples were washed and sorted, and polished thin section mounts were made of 44 selected 1–2 mm grains from 14258,34 and six 2–4 mm grains from 14167,7. The fragments analyzed contain representatives of all the main constituents in 14258,34 except glassy agglomerates that comprise about half the sample and are interpreted, like Apollo 11 and 12 agglomerates, as being the splash products of micrometeorite bombardment of the regolith. The remainder include massive glasses (5%), basaltic rocks (1%), glass-bearing fragmental rocks (11%) and crystalline fragmental rocks (33%), which range from those with extremely fine-grained dark matrices, to rocks of lighter, even-grained granular texture.

Glasses, both as individual fragments up to 1mm across and small inclusions in glass-bearing breccias, show textures and forms typical of impact products. They include homogeneous and heterogeneous varieties, the latter containing schlieren of different refractive indices (Fig. 1) and inclusions of olivine, pyroxene, plagioclase, and ilmenite showing different degrees of shock and reaction with the enclosing glass. A few glasses are vesicular and approximately 10% contain fine crystallites, though none of these samples show general devitrification textures. Most are irregular in form but there are a few spheres, broken or intact. Analyses of 23 of these glasses are given in Table 1. We recognize four main groups: the most abundant (column 1) has the composition of feldspathic basalt with relatively high alkali and phosphorus content; the next most numerous (column 2) are distinctly aluminous but lower in alkalis and correspond to an anorthositic gabbro in composition; the remainder include two glasses of basaltic composition similar to the first group but lower in alkalis and phosphorus and two basalts with high FeO, TiO_2, and low Al_2O_3 and alkalis typical of mare basalts. These compositional groupings have been recognized by others including the Apollo Soil Survey (1971), Reid *et al.* (1972), Glass (1972) and Engelhardt *et al.* (1972), the name Fra Mauro basalt having been assigned to the first group, and Highland basalt to the second.

Glass-bearing breccias are distinguished easily from other fragmental rocks not only by their glassy clasts but also by their fragmental, glass-rich matrices, which have a yellow-brown hue in thin section. Other clasts include small fragments of crystalline rocks and mineral fragments: plagioclase, pyroxene (both low and high calcium varieties), rare olivine (Fo 64 ± 5), ilmenite, rare iron (up to 50 μ) and rarer troilite. A few grains show distinct evidence of shock deformation. The mean composition of two such breccias is given in Table 2, column 1. It shows the dominant

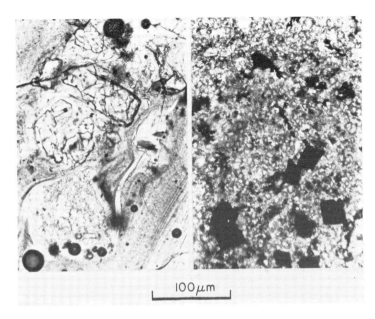

Fig. 1. 14258,34 (left): Massive greenish glass fragment showing schlieren and inclusions of olivine (Fo_{70}), pyroxene ($Wo_{38}En_{46}Fs_{16}$) and plagioclase (An_{85} and An_{95}) reacting with matrix. Glass analysis is included in Table 1, column 1. Plane polarized light. (Right): Detail in plane polarized light of annealed compound fragmental rock; brownish central area (Table 2, analysis 3) with euhedral ilmenite; intermediate zone (Table 2, analysis 4) with irregular ilmenite; light colored edge (Table 2, analysis 5) richer in plagioclase.

influence of the main types of glass and is virtually identical to the mean composition of the regolith samples (LSPET, 1971).

The second group of fragmental rocks belong to the type we have referred to as annealed breccias (Dence *et al.*, 1972) and correspond to F4 breccias of Jackson and Wilshire (1972). They are the most abundant group of the regolith grains we have analyzed, comprising 65% of those from 14258,34 and all six from 14167,7. They are characterized by fine-grained granular matrices, ranging from dark to light gray with increasing matrix grain size. Most 1 mm regolith grains show one matrix texture but in larger grains textural variations are seen with areas of dark, fine-grained texture, interpreted as annealed glasses enclosed in a matrix of less regular, generally lighter and coarser texture. One type of unrecrystallized glass is found as irregular patches in several of these rocks. It is a highly siliceous glass rich in potash, an analysis of which is given later in Table 6, analysis 1. Matrix minerals predominantly are plagioclase, low calcium pyroxene, ilmenite and rare olivine, spinel, iron and troilite. In many cases ilmenite is a useful index to the textural range in the matrices, being fine $(1-2\ \mu)$ and uniformly distributed in the rocks of extremely fine grain size, and commonly clustered in coarser-grained rocks, in some cases assuming euhedral habit (Fig. 1). The same minerals form larger clasts up to 0.5 mm across, along with fragments of crystalline rocks up to 2 mm across, including olivine basalts and clasts

Table 1. Electron microprobe analyses and CIPW norms (wt %) of some glasses from a sample of 1–2 mm fines, 14258,34.

	1	Range	2	Range	3	4	5	6
SiO$_2$	49.1	47.5 –51.1	45.5	44.6 –46.6	48.5	45.4	48.0	43.4
TiO$_2$	1.70	1.20– 2.38	0.25	0.16– 0.33	1.68	1.93	3.64	4.57
Al$_2$O$_3$	15.4	12.1 –17.4	25.1	23.0 –29.3	16.8	17.9	9.2	9.1
Cr$_2$O$_3$	0.21	0.11– 0.27	0.11	0.06– 0.14	0.21	0.12	0.29	0.35
FeO	11.1	8.7 –15.1	5.7	3.6 – 6.4	11.8	11.0	20.8	21.1
MgO	9.4	7.4 –11.7	7.4	3.4 – 8.8	9.8	10.0	7.6	10.6
CaO	10.6	9.7 –11.6	15.3	14.3 –16.9	11.2	11.8	9.0	9.6
Na$_2$O	0.72	0.48– 0.95	0.25	0.0 – 0.46	0.14	0.0	0.16	0.39
K$_2$O	0.56	0.43– 0.69	0.03	0.01– 0.05	0.18	0.05	0.14	0.18
P$_2$O$_5$	0.35	0.2 – 0.5	0.0	0.0 – 0.0	0.05	0.0	0.05	0.15
TOTAL	99.13	—	99.64	—	100.36	98.20	98.88	99.44
Q	2.0	—	—	—	2.4	—	6.1	—
Or	3.3	—	0.2	—	1.1	0.3	0.8	1.1
Ab	6.2	—	2.1	—	1.2	—	1.4	3.3
An	37.5	—	67.6	—	44.6	49.6	24.3	22.7
Di	6.6	—	4.8	—	5.2	5.2	7.1	10.0
He	4.3	—	2.2	—	3.4	3.0	10.3	10.1
En	20.6	—	10.3	—	21.9	21.8	15.9	14.4
Fs	15.1	—	5.5	—	16.6	14.7	26.6	16.6
Fo	—	—	4.2	—	—	0.8	—	5.3
Fa	—	—	2.5	—	—	0.6	—	6.8
Il	3.3	—	0.5	—	3.2	3.7	7.0	8.7
Cr	0.3	—	0.2	—	0.3	0.2	0.4	0.5
Ap	0.8	—	—	—	0.1	—	0.1	9.4

(1) Average composition of 12 pale green to pale yellow-brown glasses; (2) Average composition of 7 colorless to very pale yellow glasses; (3) very pale yellow; (4) very pale yellow; (5) pale yellow; (6) bright yellow.

similar to the coarser-grained fragmental rocks. Many of the clasts, both minerals and rocks, show evidence of shock deformation, though many plagioclase grains have recrystallized in part or in whole, largely obliterating primary shock metamorphism textures. Only one case of shock deformation affecting every matrix crystal was noted in rocks in this category.

Bulk analyses (Table 2) were obtained for 13 of these fragmental rocks. The majority have compositions (column 2) differing little from those of the glass-bearing breccias. They appear to be slightly richer in alkalis and correspondingly less calcic. Two fragments (analyses 7 and 8), though not noticeably different in texture, have distinctly greater contents of ferromagnesian minerals, but two others (analyses 3–5 and 9) are enriched in both plagioclase and alkali feldspar. One of the latter is a composite grain (Fig. 1) in which a feldspar-rich brownish central region is surrounded by zones alternately richer and poorer in alkali feldspar (analyses 3, 4, and 5). The intermediate zone, about 50–100 μ wide, may result from diffusion between the central region and the marginal zone.

Fragmental Rocks

Four fragmental rocks were examined, one containing abundant glass, the other three being of the crystallized type.

Table 2. Electron microprobe analyses (using a defocussed beam) and CIPW norms (wt %) of breccia fragments from a sample of 1–2 mm fines, 14258,34.

	1	2	Range	3	4	5	6	7	8	9
SiO$_2$	47.6	49.4	46.4–50.1	49.3	49.4	49.8	47.2	46.6	48.9	50.8
TiO$_2$	1.4	1.2	0.7– 1.8	0.7	0.3	0.7	1.4	2.3	0.8	0.1
Al$_2$O$_3$	18.8	17.6	15.8–20.4	23.3	28.4	18.9	19.7	14.4	15.3	28.1
Cr$_2$O$_3$	0.2	0.1	0.1– 0.2	0.1	0.1	0.3	0.4	0.2	0.3	0.0
FeO	9.9	9.5	8.1–11.5	3.7	1.8	5.7	4.8	13.7	13.9	0.4
MgO	8.2	9.5	7.4–11.0	2.6	2.0	5.9	6.6	11.2	11.8	0.1
CaO	11.4	10.1	8.8–12.3	16.3	14.7	15.7	17.7	9.8	6.9	13.0
Na$_2$O	0.7	1.0	0.9– 1.3	1.2	1.4	1.1	0.5	0.5	0.8	0.9
K$_2$O	0.5	0.6	0.2– 1.0	1.3	1.6	0.8	0.2	0.2	0.5	3.5
P$_2$O$_5$	0.4	0.6	0.3– 1.0	1.5	1.6	0.5	0.1	0.3	0.2	0.0
TOTAL	99.1	99.6	—	100.0	101.3	99.4	98.6	99.2	99.4	96.9
Q	0.6	2.1	—	3.9	3.9	1.9	1.2	—	—	4.5
C	—	—	—	—	1.5	—	—	—	1.4	—
Or	3.0	3.7	—	7.7	9.3	4.8	1.2	1.2	3.0	21.4
Ab	6.0	8.5	—	10.2	11.7	9.4	4.3	4.3	6.8	7.9
An	47.1	42.3	—	54.4	61.7	44.6	51.7	36.8	33.2	64.4
Di	3.6	1.2	—	7.9	—	16.4	22.6	5.1	—	0.6
He	2.4	0.6	—	5.8	—	8.5	6.9	3.3	—	1.1
En	18.9	22.8	—	2.8	4.9	7.2	6.2	23.0	29.1	—
Fs	14.4	14.9	—	2.4	2.6	4.3	2.2	17.3	23.4	—
Fo	—	—	—	—	—	—	—	1.9	0.4	—
Fa	—	—	—	—	—	—	—	1.6	0.3	—
Il	2.7	2.3	—	1.3	0.6	1.3	2.7	4.4	1.5	0.2
Cr	0.3	0.1	—	0.2	0.2	0.5	0.6	0.3	0.5	—
Ap	0.9	1.4	—	3.5	3.7	1.2	0.2	0.7	0.5	—

(1) Average composition of two glass-bearing breccia fragments; (2) average composition of eight non-glass bearing, annealed breccias; (3), (4) and (5) analyses of core, intermediate and edge zones respectively of fragment in Fig. 1 (right); (6), (7), (8), and (9) four fragments.

1. Rock 14315,11 Station H

The rock has been described as a coherent, glass-rich breccia (F2 of Jackson and Wilshire, 1972; group 3 of Warner, 1972). Our slice shows a distinctive texture of closely packed fragments that are of two main types: glasses and fine-grained feld-spar-phyric basaltic rocks, many of which have an irregular interstitial or intersertal texture (Fig. 2). Mineral fragments, notably plagioclase, are common but there is a dearth of all types of coarser grained rock fragments. Both glasses and fine-grained rock fragments show irregular tabloid to ovoid or spherical cross sections giving the rock a strikingly uniform texture. This is enhanced by an approximate alignment of the long axes of most tabular and ovoid fragments (Fig. 2) suggesting crude bedding. Most glass fragments are devitrified in whole or in part, the devitrification having taken place since assuming their present form (Lofgren, 1971). The commonly ovoid shape 0.3 to 0.7 mm across of the fine-grained feldspathic basalt fragments has prompted comparisons with chondrules (e.g. Fredriksson et al., 1972 and King et al., 1972). However, irregular masses up to 2 mm long of the same texture also are present. Despite their considerable range in texture, they show a remarkable uniformity of composition (Table 3), which corresponds to that of the feldspathic or Highland basalt (Table 1, column 2). The majority of the glasses have compositions in the same range, though showing greater variation. The most feldspathic glass (Table 4, analyses 1 and 2) shows some color and compositional differences between

Fig. 2. 14315,11 Photomicrograph in plane polarized light, field of view 1.01 by 0.79 mm, of glass-bearing breccia showing large, devitrified orange-brown glass (Table 4, analysis 3), smaller glass and mineral fragments and two chondrule-like clasts typical of those included in Table 3. Long axis of glass fragment is subparallel to crude bedding.

Table 3. Average composition, range and CIPW norm (wt %) for 10 chondrule-like clasts in breccia 14315,11 by electron microprobe analysis using a defocussed beam.

	Av	Range
SiO_2	46.7	45.0 –47.6
TiO_2	0.2	0.15– 0.5
Al_2O_3	26.0	23.3 –28.3
Cr_2O_3	0.1	0.05– 0.2
FeO	6.5	5.4 – 8.0
MgO	5.6	4.7 – 7.3
CaO	14.0	13.4 –14.3
Na_2O	0.7	0.5 – 1.1
K_2O	0.2	0.10– 0.31
P_2O_5	0.02	0.0 – 0.1
TOTAL	100.02	

Or	1.2
Ab	5.9
An	67.2
Di	1.0
He	0.7
En	11.1
Fs	9.1
Fo	1.7
Fa	1.5
Il	0.4
Cr	0.1
Ap	0.1

Table 4. Electron microprobe analyses and CIPW norms (wt %) of some glasses in breccia 14315,11.

	1	2	3	4	5	6	7	8	9	10	11
SiO_2	42.8	42.5	47.8	46.4	50.9	49.3	51.1	51.5	48.0	53.9	47.8
TiO_2	0.12	0.16	0.18	0.16	0.15	0.26	0.22	0.44	0.09	0.25	2.11
Al_2O_3	31.4	30.6	26.3	26.2	23.7	22.5	22.4	20.8	20.7	18.9	12.2
Cr_2O_3	0.10	0.11	0.11	0.18	0.24	0.24	0.18	0.23	0.17	0.24	0.31
FeO	2.1	4.2	4.1	2.5	5.0	4.7	5.4	7.1	4.7	6.1	16.6
MgO	6.7	5.0	5.4	8.5	6.5	9.0	6.3	6.5	13.5	7.5	8.8
CaO	16.7	16.4	15.5	15.3	13.4	13.5	13.9	13.3	12.1	12.1	10.7
Na_2O	0.00	0.29	0.12	0.15	0.32	0.03	0.22	0.03	0.09	0.77	0.57
K_2O	0.00	0.03	0.01	0.00	0.03	0.00	0.03	0.02	0.00	0.08	0.26
P_2O_5	0.0	0.0	0.0	0.0	0.0	0.0	0.0	0.0	0.0	0.0	0.2
TOTAL	99.92	99.29	99.52	99.39	100.24	99.53	99.75	99.92	99.35	99.84	99.55
Q	—	—	2.9	—	7.4	4.5	8.3	9.7	—	9.9	0.7
C	1.0	0.3	—	—	—	—	—	—	—	—	—
Or	—	0.2	0.1	—	0.2	—	0.2	0.1	—	0.5	1.5
Ab	—	2.5	1.0	1.3	2.7	0.3	1.9	0.3	0.8	6.5	4.9
An	83.0	82.0	72.3	71.3	63.1	61.5	60.2	56.7	56.5	48.0	30.1
Di	—	—	3.2	3.5	1.9	3.6	4.8	4.7	2.6	6.7	8.9
He	—	—	1.5	0.6	0.8	1.1	2.5	3.0	0.6	3.2	9.4
En	3.5	0.4	12.2	17.8	15.3	21.0	13.5	14.0	28.5	15.6	17.9
Fs	0.7	0.2	6.4	3.4	8.1	7.3	8.0	10.3	7.0	8.7	21.6
Fo	9.2	8.5	—	1.4	—	—	—	—	2.9	—	—
Fa	2.1	5.5	—	0.3	—	—	—	—	0.8	—	—
Il	0.2	0.3	0.3	0.3	0.3	0.5	0.4	0.8	0.2	0.5	4.0
Cr	0.1	0.2	0.2	0.3	0.4	0.4	0.3	0.3	0.3	0.4	0.5
Ap	—	—	—	—	—	—	—	—	—	—	0.5

(1) and (2) pale orange-brown core and dark brown rim respectively; (3) orange-brown—see Fig. 2, devitrified; (4) and (5) both colorless; (6) colorless center and very pale yellow-brown devitrified margin; (7) and (8) orange-brown, devitrified; (9) colorless; (10) pale orange-brown; (11) pale greenish-yellow.

rim and core, apparently from reaction with the matrix during devitrification. Another (Table 4, analysis 6) shows no difference between clear core and devitrified rim. The majority of the glasses, however, are distinctly lower in alkalis and somewhat lower in iron than the chondrule-like bodies, suggesting either that the glasses have undergone alkali loss or that the two assemblages have been derived from associated but different source areas. In either case, the source area was notably lacking in either the common Fra Mauro high alkali basalt type or typical iron-rich mare basalts. The nearest approach to the latter are a distinct group of small greenish-yellow glasses represented by analysis 11 of Table 4, which resembles the reported analysis of basalt 14053 (LSPET, 1971).

The texture of 14315,11, if representative of the rock, suggests deposition in a dune-like base-surge deposit, incorporating relatively slowly cooled droplets of impact melt (chondrule-like bodies) and chilled, possibly pre-existing glasses from the same distinctive source area. Four samples, 14301, 14313, 14315, and 14318, have been reported as having the chondrule-like fragments in abundance, suggesting that their source is the Triplet Crater group, southeast of the landing site.

2. Rocks 14305,5 and 14314,13 from west of LM and Station H (Turtle Rock), respectively

These rocks and 14311,88, described below, are members of the heterogeneous, glass-deficient fragmental rocks, group F4 of Jackson and Wilshire (1972), Type III of Engelhardt *et al.* (1972), which are the most distinctive group of samples from the Fra Mauro site. Their heterogeneity and textural variability are well shown in Figure 3. In both slices the largest clasts (3 to 5 mm) are light colored crystalline rocks with fragmental textures, in which plagioclase inclusions are held in a granular matrix of plagioclase, low calcium pyroxene, minor olivine, ilmenite, iron, troilite, and phosphate minerals (Fig. 4). Less clearly defined are dark, fine-grained clasts of irregular to ovoid form that contain scattered inclusions of plagioclase, pyroxene, olivine, and patches of highly siliceous potassic glass or its recrystallized equivalents (Fig. 5). Small clasts, generally 0.5 to 2 mm long, include basaltic rocks of both feldspar-rich and olivine-clinopyroxene varieties (Fig. 6, left), ophitic plagioclase-pyroxene rocks and single crystals of plagioclase, pyroxene, olivine and ilmenite. It is notable that the largest (0.5 to 1 mm) plagioclase grains in 14305,5 are unzoned and have compositions in the range An_{96-98}. Also in 14305,5 are two low calcium pyroxene grains ($Wo_6En_{51}Fs_{43}$ and $Wo_3En_{53}Fs_{44}$) with exsolution lamellae of augitic composition ($Wo_{36}En_{41}Fs_{23}$ and $Wo_{38}En_{39}Fs_{23}$, respectively). Smaller

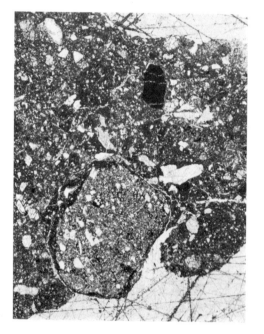

Fig. 3. 14305,5 Photomicrograph in plane polarized light giving general view of annealed fragmental texture. Field of view 7.2 by 5.6 mm. Clasts include large, light gray annealed breccia with plagioclase clasts (An_{97} to An_{83}), pyroxene, olivine (Fo_{65}) and dark, fine-grained annealed glass matrix (Table 5, analysis 1). Clast is rimmed by annealed glass (Table 5, analysis 3). Immediately above the large clast is an ill-defined clast with dark fine-grained matrix (Table 5, analysis 2) containing inclusions of recrystallized K and Ba-rich glass, plagioclase (An_{95}), and olivine (Fo_{59}). Other clasts include ilmenite, a large, recrystallized, shocked plagioclase (An_{96}), feldspathic basalt and a grain of pink spinel (top right).

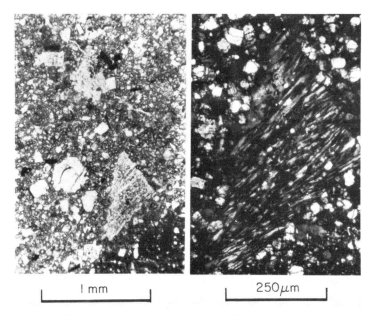

Fig. 4. 14314,13 Photomicrographs in plane polarized light (left) and crossed polarizers (right) of light gray annealed fragmental rock clast. Triangular plagioclase (An_{91}) shows texture of recrystallized maskelynite and evidence of reaction with matrix. Pigeonite ($Wo_{10}En_{56}Fs_{34}$) occurs as inclusions in plagioclase as well as in the matrix.

clasts include potassic feldspar and a distinctive pink spinel that has been reported by several groups (e.g. Drever *et al.*, 1972 and Haggerty, 1972). Analysis of one grain in 14305,5 gave a core composition of

$$(Fe_{3.82}Mg_{4.40}Mn_{0.04}Cr_{1.55}Al_{14.25}Ti_{0.04})_{24.10}O_{32},$$

a low-Ti chromian pleonaste; the dark red rim is slightly enriched in titanium

$$(Fe_{3.97}Mg_{4.33}Mn_{0.04}Cr_{1.56}Al_{14.02}Ti_{0.15})_{24.07}O_{32}.$$

A distinctive group of dark, fine-grained inclusions in 14314,13 have an abundance of fine acicular olivine crystallites (Fig. 6, right) in keeping with their olivine-troctolite compositions (Table 5, column 5). They show some resemblance to fragmental rock 14068 (Hubbard *et al.*, 1972 and Warner, 1972).

Both rocks show the effects of shock and thermal metamorphism. In the case of 14314,13 the latest event in the history of the rock prior to ejection to the lunar surface was the injection of small veinlets of glass (Fig. 7, left) that compositionally (Table 5, column 6) resemble fused soil. In the same figure and in detail (Fig. 7, right) is shown a shocked clinopyroxene with well-developed shock lamellae. This is one of the few clasts in either rock in which shock features are present and unaffected by later thermal events. The large feldspar in Fig. 3, for example, shows relics of original twin lamellae and shock-produced planar features, both largely obliterated by recrystallization to a fine-grained mosaic. A similar effect is seen in Fig. 4. Clasts

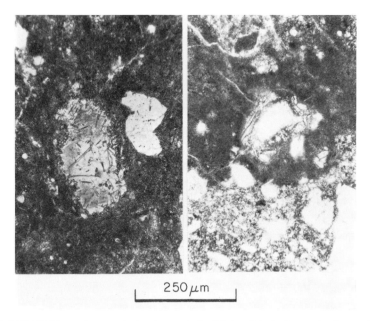

Fig. 5. 14314,13 Photomicrographs in plane polarized light of dark, fine-grained granular clasts containing: (left) clear and brown glasses (Table 6, analyses 2 and 3, respectively) with crystallites of alkali feldspar and a border zone of granular pyroxene $(Wo_7En_{67}Fs_{26})$; (right) silica grains, probably quartz, mantled by rhyolitic glass, alkali feldspar, and pyroxene $(Wo_4En_{69}Fs_{27})$.

of olivine (Fo_{57}) in 14305,5 have developed rims of pigeonite $(Wo_8En_{62}Fs_{30})$ in reaction with the matrix, but similar rims are rare in 14314,13 except around a strongly shocked olivine fragment. The most widespread evidence for thermal metamorphism, however, is the presence of dark, extremely fine-grained granular aggregates as clasts (Figs. 3 and 5) and in the matrix. These are clearly former glasses, which in some cases reacted extensively with inclusions of quartz and alkali feldspar to produce potassic siliceous glasses (Fig. 5), but chilled rapidly through the temperature range of reaction with plagioclase. The equigranular texture of these rocks is taken to be the result of thermal solid state metamorphism rather than direct crystallization from the molten state. Compositionally most of the recrystallized dark granular areas compare closely with the Fra Mauro basalt type, being relatively rich in alkalis and phosphorus. As illustrated by Fig. 5, the larger clasts of this type most commonly contain inclusions of plagioclase and glasses of potassic rhyolite composition (Table 6, analyses 2 and 3) suggesting that the source area for Fra Mauro basaltic rocks was also relatively rich in rocks of granitic composition.

3. Rock 14311,88, Station Dg.

A further example of the recrystallized fragmental rock suite, this specimen is distinguished from the two previously described by an even greater irregularity of matrix texture and above all by the development of vugs in the matrix, as seen in

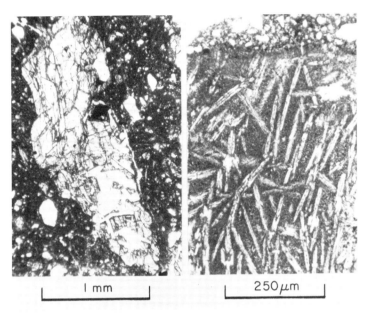

Fig. 6. 14314,13 Photomicrographs in plane polarized light of (left) moderately shocked olivine basalt with zoned olivine (core Fo_{67} to rim Fo_{60}), pyroxene, ($Wo_{30}En_{48}Fs_{22}$ to $Wo_9En_{56}Fs_{35}$), plagioclase (An_{93}), and ilmenite; (right) olivine-phyric clast (Table 5, analysis 5) with skeletal olivine (Fo_{82}) and rare plagioclase inclusions.

both rock chip and thin section. The vugs are as much as 5 mm across and are lined by siliceous, potassic glasses similar to those included in rock 14314,13. Three compositional varieties were noted in the glass lining (Fig. 8, left), a clear highly siliceous glass (Table 6, analysis 5), a clear less siliceous but more potassic barium-rich glass (Table 6, analysis 4) and a brownish area, possibly partly devitrified and intermediate in composition (Table 6, analysis 6) between the two. Cutting the glass and growing into the vug are acicular crystals of ilmenite, equant calcic pyroxene and plagioclase. McKay et al. (1972) have discussed crystallization in these vugs in terms of growth from the vapor phase.

There are several examples in the rock of siliceous glasses not obviously associated with vugs. An example is shown in Fig. 8, right. Again, brown and clear low and high silica phases are present (Table 6, analyses 7–10), with indications of devitrification to form alkali feldspar intergrowths. The clast population is dominated by fragments of plagioclase up to 1 mm across and rarer grains of orthopyroxene and ilmenite. Olivine clasts are rare and are mantled by reaction rims of pigeonite. Clinopyroxene also is rare as clasts. Rock fragments up to 2 mm across are also dominated by plagioclase-rich varieties including feldspathic basalt, plagioclase breccia with alkali feldspar and phosphate minerals in the matrix, ophitic plagioclase-pyroxene rock and fine-grained recrystallized fragmental rocks with small plagioclase clasts. Shock effects again are obscured by subsequent thermal overgrowths.

Compositions were obtained for five of the finer-grained, uniformly granular areas, whether clearly distinguished as clasts or apparently part of the matrix. The

Table 5. Electron microprobe analyses (using a defocussed beam) and CIPW norms (wt %) of matrix, clasts and a glass vein from three breccias.

	14305,5				14314,13		14311,88				
	1	2	3	4	5	6	7	8	9	10	11
SiO_2	45.5	46.8	48.9	49.3	43.1	48.3	44.7	48.6	49.8	50.6	52.1
TiO_2	5.1	1.8	1.7	1.9	0.8	1.3	0.8	0.6	1.9	1.2	0.9
Al_2O_3	15.1	16.2	16.9	16.1	20.1	20.0	24.3	21.0	17.5	14.2	16.2
Cr_2O_3	n.d.	n.d.	n.d.	n.d.	n.d.	n.d.	0.1	0.1	0.1	0.2	0.1
FeO	9.9	11.6	10.1	9.7	7.0	8.3	9.2	9.2	11.0	13.1	10.9
MgO	8.5	11.6	9.3	9.1	17.1	8.9	6.8	7.8	8.2	11.3	7.8
CaO	9.9	9.4	9.1	13.7	11.3	11.2	12.4	10.9	9.1	7.7	9.0
Na_2O	1.1	1.0	1.4	0.8	0.6	0.9	1.1	0.9	1.2	1.2	1.1
K_2O	1.3	0.2	0.5	0.9	0.1	0.9	0.1	0.5	0.6	0.6	0.9
P_2O_5	1.0	0.5	0.5	0.0	0.1	0.4	0.3	0.7	0.8	0.4	0.8
TOTAL	97.4	99.1	98.4	101.5	100.2	100.2	99.8	100.3	100.2	100.5	99.8
Q	0.8	—	0.5	—	—	—	—	1.3	3.4	0.1	5.9
C	—	—	—	—	—	—	0.6	0.8	0.3	—	—
Or	7.9	1.2	3.0	5.2	0.6	5.3	0.6	3.0	3.5	3.5	5.3
Ab	9.6	8.5	12.0	6.7	5.1	7.6	9.3	7.6	10.1	10.1	9.3
An	33.3	39.5	39.0	37.1	51.8	47.8	59.7	49.4	39.8	31.5	36.7
Di	6.0	2.2	1.8	15.6	2.3	2.7	—	—	—	2.0	1.3
He	2.4	1.2	1.1	8.8	0.5	1.4	—	—	—	1.3	1.0
En	19.0	20.8	22.7	11.0	0.9	17.7	7.2	19.4	20.4	27.1	18.9
Fs	8.7	13.2	15.4	7.1	0.2	10.4	6.5	15.7	16.9	20.9	17.8
Fo	—	5.2	—	2.9	28.4	2.2	6.9	—	—	—	—
Fa	—	3.6	—	2.1	8.5	1.5	6.9	—	—	—	—
Il	9.9	3.5	3.3	3.6	1.5	2.5	1.5	1.1	3.6	2.3	1.7
Cr	—	—	—	—	—	—	0.1	0.1	0.1	0.3	0.1
Ap	2.4	1.2	1.2	—	0.2	0.9	0.7	1.6	1.9	0.9	1.9

n.d. = not determined.

├─── 250 μm ───┤   ├─── 100 μm ───┤

Fig. 7. 14314,13 Photomicrographs (left) in plane polarized light of matrix cut (at top) by a glass vein (Table 5, analysis 6); (right) in crossed polarizers, detail of unzoned clino-pyroxene ($Wo_{38}En_{46}Fs_{16}$) with well-developed shock induced planar elements.

Table 6. Electron microprobe analyses of potash-rich glasses.

	14258,34	14314,13		14311,88							14310,4
	1 Brown	2 Clear	3 Brown	4 Clear	5 Clear	6 Brown	7 Clear	8 Brown	9 Clear	10 Clear	11 Brown
SiO$_2$	75.9	76.4	72.3	60.9	77.7	73.6	61.0	74.0	64.0	76.3	75.0
TiO$_2$	0.72	0.78	1.12	0.26	0.75	0.74	0.21	0.70	0.26	0.89	0.88
Al$_2$O$_3$	12.5	12.2	13.1	19.0	12.6	12.7	19.5	11.9	18.0	12.4	12.9
FeO	1.22	0.71	0.82	0.24	0.42	0.60	0.23	0.91	0.38	0.56	1.82
MgO	0.0	0.14	0.19	0.12	0.14	0.23	0.14	0.17	0.12	0.17	0.15
CaO	0.64	0.83	0.78	0.43	0.54	0.28	0.97	0.45	0.36	0.62	0.88
Na$_2$O	0.42	1.22	1.17	0.75	0.80	0.50	0.77	0.50	0.84	1.11	0.44
K$_2$O	8.6	7.7	9.4	13.7	7.3	10.0	13.5	10.4	13.6	7.3	8.0
P$_2$O$_5$	0.1	0.03	0.10	0.01	0.06	0.09	0.00	0.11	0.04	0.15	n.d.
BaO	0.05	0.07	0.49	4.40	0.08	0.74	3.95	0.71	2.70	0.06	0.37
TOTAL	100.15	100.08	99.47	99.81	100.39	99.48	100.27	99.85	100.30	99.56	100.44

n.d. = not determined.

majority of the results (Table 5, analyses 8–11) show a general conformity to the aluminous, alkali basaltic compositions of Fra Mauro type, and an apparent absence of the low alkali feldspathic Highland basalt, so abundant in 14315,11.

Fig. 8. 14311,88 Photomicrographs in plane polarized light: (left) detail of crystals and glass lining a vug, with calcic pyroxene (Wo$_{39}$En$_{32}$Fs$_{29}$) in center of field above acicular ilmenite. At top of field pyroxenes are Wo$_{41}$En$_{37}$Fs$_{22}$, below which is a patch of brown glass (Table 6, analysis 6) adjacent to clear glass (Table 6, analysis 5). At center left edge is a small patch of high potash glass (Table 6, analysis 4); (right) patch of brown glass (Table 6, analysis 8) with crystals of potash feldspar surrounded by clear glass (Table 6, analysis 7) in a matrix of high and low calcium pyroxenes, plagioclase, and ilmenite.

Crystalline Rock 14310

The rock chip 14310,99 and thin section 14310,4 we examined are of a plagioclase-rich basalt alternating in texture between intergranular patches about 2 to 3 mm across, with equant 0.4 to 1.0 mm grains of subhedral plagioclase and anhedral pyroxene, and regions of interstitial texture, with boxworks of thin blades of plagioclase intergrown with pyroxene, ilmenite and lesser amounts of alkali feldspar, troilite, metallic iron, tranquillityite, spinel and brown glass. The larger, tabular grains of plagioclase are clear of inclusions, show few twin lamellae and are highly calcic (An_{90-98}). Zoning is rare and generally is of simple normal type, typically developed adjoining iron-rich pyroxene. Irregular normal zoning also is seen (Fig. 9) in which the composition on one side is as sodic as An_{79}. The finer plagioclase laths show some evidence of normal zoning and have cores as calcic as An_{95}. The plagioclase laths are intergrown with pyroxene extending virtually to the low calcium core of the pyroxenes (Fig. 10). In a typical case, pyroxene ranges in composition from a core of $Wo_4En_{75}Fs_{21}$ to $Wo_{35}En_{38}Fs_{27}$ to an outer zone $Wo_{14}En_{22}Fs_{64}$. Ilmenite and other accessory minerals precipitate late and are intergrown with the marginal pyroxene phases. At the margins of some pyroxenes, compositions appropriate to pyroxferroite were recorded but no undoubted crystals of the latter were detected. Among the minor constituents are iron with a range of Ni content, troilite, a few poorly formed grains of tranquillityite, an interstitial residual brown glass (Table 6,

Fig. 9. 14310,4 Photomicrograph in crossed polarizers of rare, irregular zoning in equant plagioclase crystal. Main part of the grain is An_{93} dropping to An_{79} in corner. Field of view 0.44 × 0.34 mm.

Fig. 10. 14310,4 Photomicrograph in crossed polarizers of pyroxene with orthopyroxene core and iron-rich pigeonite rim, intergrown with laths of plagioclase. Field of view 1.01 × 0.79 mm.

analysis 11) similar to those in the fragmental rocks, and a potash- and barium-rich feldspar. No silica phases or fayalite (reported by Gancarz *et al.*, 1971) or evidence of shock were recorded from our section.

Many workers have reported anomalous features in rock 14310. It is unique among returned Apollo basalts in that it contains inclusions (LSPET, 1971) that are of virtually the same mineralogy as the host but of much finer grain size (Walter *et al.*, (1972). Hollister *et al.* (1972) note anomalous relationships in the cores of pyroxenes indicating the presence of at least two structural varieties of hypersthene, with some suggestion of resorbtion. These relationships have also been discussed by Ridley *et al.* (1972), who noted that plagioclase relationships suggested that xenocrysts of plagioclase were present when the melt was emplaced. According to this view coarser "phenocrystic" plagioclase and groundmass plagioclase would crystallize simultaneously, the larger crystals being overgrown on preexisting nuclei. This would remove the argument for extensive volatile loss presented by Brown and Peckett (1971), which is difficult to reconcile with the appreciable potash content of the rock (Hubbard *et al.*, 1972). Indeed, the major and minor element composition of the rock (LSPET, 1971 and Morgan *et al.*, 1972a) is compatible with formation by impact melting. A close terrestrial analogy is presented by impact melts from the 20 km Mistastin Lake crater, Labrador (Taylor and Dence, 1969) where plagioclase-rich melt rocks formed from anorthosite have developed microporphyritic textures by overgrowth of partly resorbed xenocrysts in similar fashion to that proposed for rock 14310.

Discussion

Our analyses of glasses in soil 14258,34 support the observations of Reid *et al.*
(1972) and many others that pale yellow-green glasses with the composition of
alkali-rich basalts are predominant. Glasses of similar composition had been noted
previously in Apollo 12 soils and breccias and have been called KREEP, norite or
Fra Mauro basalt. We note, however, that a distinction can be made between the
pale yellow-green glasses common to both sites and the distinctive brown glasses
found in abundance in soils 12032 and 12033 to which the name KREEP was first
given (Meyer *et al.*, 1971). The distinction was recognized by Marvin *et al.* (1971)
and Dence *et al.* (1971) who recorded the absence of brown KREEP in soil 12070.
Chemically, the two varieties differ principally in the ratio of normative orthoclase to
normative albite which is greater than 1:1 in brown KREEP but less in yellow-green
glasses (Fig. 11). There are also differences in normative apatite to albite. Lithic
fragments, mainly annealed fragmental rocks, analyzed by Meyer *et al.* (1971) were
put forward as possible source materials for the KREEP glasses. They have Or:Ab
ratios typical of the yellow-green glasses and correspond closely to the compositions

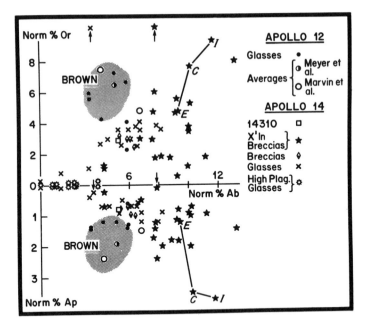

Fig. 11. Plot of normative components orthoclase (Or) and apatite (Ap) against albite (Ab)
for Apollo 12 glasses and Apollo 14 glasses and fragmental rocks in 14258,34 and 14311,88.
Brown refers to shaded fields of brown KREEP glass in Apollo 12 samples. C, I, and E refer
to center, intermediate, and edge zones of the fragment shown in Fig. 1 (right), analyses 3,
4, and 5, respectively, in Table 2. X^IIn breccias refers to fragmental rocks of annealed
texture or clasts in rock 14311,88. Breccias are glass-bearing fragmental rocks in 14258,34.
High plag. glasses have more than 60% normative anorthite. Arrows point to highly potassic
siliceous glass (Table 6, analysis 1) and feldspathic fragment in Table 2, analysis 9, with
50.8% and 21.4% normative orthoclase, respectively.

of lithic fragments we have analyzed from 14258,34 and 14311,88. There appear to be no lithic fragments equivalent to the brown glasses, which appear to have the restricted distribution and exposure ages appropriate to ray material from Copernicus (Eberhardt *et al.*, 1972). The more widespread yellow-green glasses and their lithic equivalents may well be characteristic of the Imbrian ejecta blanket for which Fra Mauro basalt is the most apt designation.

It is also apparent from Fig. 11 that our analyses suggest that most lithic fragments are richer in total alkalis and phosphorus than the associated glasses. Possibly there has been partial loss of volatiles during impact melting. The lithic fragments also include a group that have normative Or:Ab less than 1:4. In some cases these are dark fine-grained granular masses enclosing blebs of highly potassic siliceous glass, suggesting that in these cases diffusion of K, and possibly P, has taken place during annealing and reaction with the inclusion.

The second major component of 14258,34 glasses, the feldspathic basalt composition or highland basalt (Reid *et al.*, 1972), which is apparently ubiquitous to all sampled lunar sites, may be representative of the pre-Fra Mauro materials exposed east of the Apollo 14 site. We have already noted that components of this composition are abundant in glass-bearing breccia 14315,11 and apparently also in the texturally similar breccia 14318 (Kurat *et al.*, 1972), but are relatively rare in the annealed fragmental rocks. It is therefore unlikely that the glass-bearing breccias are simply unmetamorphosed members of the same rock suite as the annealed rocks, as has been suggested (Warner, 1972).

The origin of the annealed fragmental rocks remains a problem of considerable significance. We agree with those who interpret the granular fabric of the matrix and many clasts in these rocks as being the result of annealing recrystallization of rocks previously in a glassy or clastic state. Grieve *et al.* (1972) and Anderson *et al.* (1972) propose temperatures between 700°C and 1000 ± 100°C for various components, the higher temperatures relating to the lighter clasts and potassic granite inclusions, the lower temperatures to some of the dark, very fine-grained recrystallized glasses and the clastic matrix. The lower temperatures are not unreasonable for a relatively thin layer of breccia such as the Ries suevite in which hot, highly shocked material is intermixed with cool, moderately to weakly shocked components. The higher temperatures would seem to require the proximity of a substantial body of impact melt. We may question whether such substantial quantities of hot Imbrian ejecta are likely to be present at a distance of more than 500 km from the margin of the Imbrium Basin.

To make such an evaluation it is necessary to argue from terrestrial analogs. Consider an impact model for the Imbrian structure based on comparisons with terrestrial craters of complex type 20 or more km in diameter (Dence, 1968, 1971). The moon is depicted with a crust of feldspar-rich rocks about 50 km thick overlying a mafic mantle (Toksöz *et al.*, 1972). In Fig. 12a are sketched two stages in the initial excavation of a lunar crater with maximum dimensions of about 180 km depth and 450 km radius, considered to result from the hypervelocity impact of a planetesimal with kinetic energy of about 10^{33} ergs. At an early stage in the excavation, only strongly shocked and heated rocks, accelerated to escape velocities, are involved,

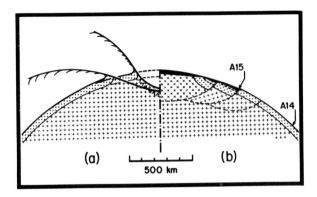

Fig. 12. Impact model of the Imbrium Basin, based on hypervelocity impact of a planetesimal onto a moon with a 50 km thick feldspathic crust overlying a mafic mantle. (a) Two stages in the excavation of the initial cavity; an early stage with high velocity ejection of strongly shocked material and a later stage in which most ejecta is weakly shocked crustal material. Inverted v s indicate limit of significant shock deformation of target materials. (b) Final configuration after completion of overturning, collapse of rim, rapid uplift of mantle in the center (cf. Wise and Yates, 1970) and flooding by mare basalt flows. The relative positions of Apollo 14 and 15 landing sites are indicated.

but in the later stages of cavity growth massive excavation of weakly shocked, and hence cool, crustal materials takes place. These form the great bulk of the ejecta blanket and are overlain by a relatively thin veneer of hot material of suevite type, rich in shock melted fragments. The closest terrestrial analog is that of the Ries crater, where weakly shocked Bunte breccia is widespread around the crater and is overlain by patches of strongly shocked suevite (Engelhardt, 1971). As shown in Fig. 12b the Apollo 14 site is about $2\frac{1}{2}$ crater radii from the center of the Imbrium Basin, a greater relative distance than any patch of Ries suevite. The analogy therefore suggests that the great bulk of the Fra Mauro formation is a Bunte breccia deposit only weakly shocked and heated by the Imbrian event. By this argument, if the Fra Mauro formation is no more than 76 m thick (Kovach and Watkins, 1972) any Imbrian suevite breccia would have been only a few meters thick at most, and would have been completely disrupted by regolith-forming events.

According to the Bunte breccia interpretation, the Fra Mauro materials would have formed the uppermost levels of the pre-Imbrian crust, as suggested by Anderson *et al.* (1972) and would have been subjected to a heavy bombardment prior to the Imbrian event. It is to these early impacts that we attribute the multiple brecciation and thermal events recorded in the annealed Fra Mauro fragmental rocks, and also their high content of meteoritic components (Morgan *et al.*, 1972b). A suevite deposit, on the other hand, would contain material from deeper levels in the lunar crust or mantle. The Fra Mauro samples, with their relatively high content of radioactive elements and potassic granite components, are much more likely to be representative of the upper rather than the deeper lunar crust.

Concluding Remarks

The lunar history deduced from the analysis of Apollo 14 samples is dominated by large scale impact events. Annealed fragmental rocks and feldspathic basalt 14310 are interpreted as the products of direct or indirect heating by multiple impact cratering events on the scale of tens to hundreds of kilometers prior to the Imbrian impact. It follows that ages determined from crystalline rocks of the Fra Mauro suite (Wasserburg *et al.*, 1972) predate the Imbrian excavation. Mare basalt extrusion also pre-dated that event, as olivine-clinopyroxene basalts form clasts in some annealed breccias.

We agree with the distinction, most clearly apparent in glasses, between alkali-rich (Fra Mauro) basalt components and low alkali, feldspathic (Highland) basalts. Potassic granites are close associates of the former, anorthosites and troctolites of the latter. The alkali-rich rocks appear to be concentrated in the vicinity of the Imbrian Basin, but the Highland basalts are more widespread and may form the pre-Imbrian surface at the Fra Mauro site. Glass-bearing breccias and soils of late formation are derived from both rock types in differing proportions.

The Fra Mauro formation has been interpreted as a Bunte breccia equivalent that was only weakly shocked and heated by the Imbrian event and so contains little direct information on the age and nature of that event. Such information may be forthcoming from Apollo 15 samples.

Acknowledgments—We thank Dr. R. J. Traill for access to the lunar samples and preparation of the regolith fines for study. F. J. Cooke, E. J. Gelinas and R. Wirthlin provided assistance with photography, and Carole Dence greatly lightened the task of completing the manuscript.

References

Anderson A. T. Jr. Braziunas T. F. Jacoby J. and Smith J. V. (1972) Breccia populations and thermal history: nature of pre-Imbrium crust and impacting body (abstract). In *Lunar Science—III* (editor C. Watkins), pp. 24–26, Lunar Science Institute Contr. No. 88.

Apollo Soil Survey (1971) Apollo 14: nature and origin of rock types in soil from the Fra Mauro formation. *Earth Planet. Sci. Lett.* **12**, 49–54.

Brown G. M. and Peckett A. (1971) Selective volatilization on the lunar surface: evidence from Apollo 14 feldspar-phyric basalts. *Nature* **234**, 262–266.

Dence M. R. (1968) Shock zoning at Canadian craters: petrography and structural implications. In *Shock Metamorphism of Natural Materials*, (editors B. M. French and N. M. Short), pp. 169–184. Mono.

Dence M. R. (1971) Impact melts. *J. Geophys. Res.* **76**, 5552–5565.

Dence M. R. Douglas J. A. V. Plant A. G. and Traill R. J. (1971) Mineralogy and petrology of some Apollo 12 samples. *Proc. Second Lunar Sci. Conf., Geochim. Cosmochim. Acta* Suppl. 2, Vol. 1, pp. 285–299. MIT Press.

Dence M. R. Plant A. G. and Traill R. J. (1972) Impact-generated shock and thermal metamorphism in Fra Mauro lunar samples (abstract). In *Lunar Science—III* (editor C. Watkins), pp. 174–176, Lunar Science Institute Contr. No. 88.

Drever H. I. Johnston R. and Gibb F. G. F. (1972) Chromian pleonaste and aluminous picotite in two Apollo 14 microbreccias 14306 and 14055. *Nature* **235**, 30–31.

Eberhardt P. Eugster O. Geiss J. Grögler N. Schwarzmüller J. Stettler A. and Weber L. (1972) When was the Apollo 12 KREEP ejected? (abstract). In *Lunar Science—III* (editor C. Watkins), pp. 206–208. Lunar Science Institute Contr. No. 88.

Eggleton R. E. (1963) Thickness of the Apenninian series in the Lansberg region of the Moon. Astrogeol. Studies, Ann. Prog. Rept., Aug. 1961–Aug. 1962. Pt. A, U.S. Geol. Survey open-file Rept. pp. 19–31.

Eggleton R. E. (1964) Preliminary geology of the Riphaeus Quadrangle of the Moon and definition of the Fra Mauro Formation. Astrogeol. Studies, Ann. Prog. Rept., Aug. 1962–July 1963. Pt. A, U.S. Geol. Survey open-file Rept. pp. 46–63.

Engelhardt W. von (1971) Detrital impact formations. *J. Geophys. Res.* **76**, 5566–5574.

Engelhardt W. von Arndt J. and Stöffler D. (1972) Apollo 14 soils and breccias, their compositions and origin by impacts (abstract). In *Lunar Science—III* (editor C. Watkins), pp. 233–235, Lunar Science Institute Contr. No. 88.

Fredriksson K. Nelen J. and Noonan A. (1972) Apollo 14: glasses, breccias, chondrules (abstract). In *Lunar Science—III* (editor C. Watkins), pp. 280–282, Lunar Science Institute Contr. No. 88.

Gancarz A. J. Albee A. L. and Chodos A. A. (1971) Petrologic and mineralogic investigation of some crystalline rocks returned by the Apollo 14 mission. *Earth Planet. Sci. Lett.* **12**, 1–18.

Glass B. P. (1972) Apollo 14 glasses (abstract). In *Lunar Science—III* (editor C. Watkins), pp. 212–214, Lunar Science Institute Contr. No. 88.

Grieve R. McKay G. Smith H. and Weill D. (1972) Mineralogy and petrology of polymict breccia 14321 (abstract). In *Lunar Science—III* (editor C. Watkins), pp. 338–340, Lunar Science Institute Contr. No. 88.

Haggerty S. E. (1972) Subsolidus reduction and compositional variations of lunar spinels (abstract). In *Lunar Science—III* (editor C. Watkins), pp. 347–349, Lunar Science Institute Contr. No. 88.

Hollister L. Trzcienski W. Jr. Dymek R. Kulick C. Weigand P. and Hargraves R. (1972) Igneous fragment 14310, 21 and the origin of the mare basalts (abstract). In *Lunar Science—III* (editor C. Watkins) pp. 386–388, Lunar Science Institute Contr. No. 88.

Hubbard N. J. Gast P. W. Rhodes M. and Wiesmann H. (1972) Chemical composition of Apollo 14 materials and evidence for alkali volatilization (abstract). In *Lunar Science—III* (editor C. Watkins), pp. 407–409, Lunar Science Institute Contr. No. 88.

Jackson E. D. and Wilshire H. G. (1972) Classification of the samples returned from the Apollo 14 landing site (abstract). In *Lunar Science—III* (editor C. Watkins), pp. 418–420, Lunar Science Institute Contr. No. 88.

King E. A. Butler J. C. and Carman M. F. (1972) Chondrules in Apollo 14 breccias and estimation of lunar surface exposure ages from grain size analyses (abstract). In *Lunar Science—III* (editor C. Watkins), pp. 449–451, Lunar Science Institute Contr. No. 88.

Kovach R. L. and Watkins J. S. (1972) The near-surface velocity structure of the moon (abstract). In *Lunar Science—III* (editor C. Watkins), pp. 461–462, Lunar Science Institute Contr. No. 88.

Kurat G. Keil K. Prinz M. and Bunch T. E. (1972) A "chondrite" of lunar origin: textures, lithic fragments, glasses and chondrules (abstract). In *Lunar Science—III* (editor C. Watkins), pp. 463–465, Lunar Science Institute Contr. No. 88.

Lofgren G. (1971) Devitrified glass fragments from Apollo 11 and Apollo 12 lunar samples. *Proc. Second Lunar Sci. Conf., Geochim. Cosmochim. Acta* Suppl. 2, Vol. 1, pp. 949–955. MIT Press.

LSPET (1971) (Lunar Sample Preliminary Examination Team) Preliminary examination of lunar samples from Apollo 14. *Science* **173**, 681–693.

Marvin U. B. Wood J. A. Taylor G. J. Reid J. B. Jr. Powell B. N. Dickey J. S. Jr. and Bower J. F. (1971) Relative proportions and probable sources of rock fragments in the Apollo 12 soil samples. *Proc. Second Lunar Sci. Conf., Geochim. Cosmochim. Acta* Suppl. 2, Vol. 1, pp. 679–699. MIT Press.

McCauley J. F. (1967) Geologic results from the lunar precursor probes. *Amer. Inst. Aeronaut. Astronaut.* Ann. Meet., 4th, Anaheim, Calif., Oct. 23–27, Pap. 67–862, 9 pp.

McKay D. S. and Morrison D. A. (1971) Lunar breccias. *J. Geophys. Res.* **76**, 5658–5669.

McKay D. S. Clanton U. S. Meiken G. H. Morrison D. A. Taylor R. M. and Ladle G. (1972) Vapor phase crystallization in Apollo 14 breccias and size analysis of Apollo 14 soils (abstract). In *Lunar Science—III* (editor C. Watkins), pp. 529–531, Lunar Science Institute Contr. No. 88.

Meyer C. Jr. Brett R. Hubbard N. J. Morrison D. A. McKay D. S. Aitken F. K. Takeda H. and Schonfeld E. (1971) Mineralogy, chemistry, and origin of the KREEP component in soil samples from the Ocean of Storms. *Proc. Second Lunar Sci. Conf., Geochim. Cosmochim. Acta* Suppl. 2, Vol. 1, pp. 393–411. MIT Press.

Morgan J. W. Krähenbühl U. Ganapathy R. and Anders E. (1972a) Volatile and siderophile elements in Apollo 14 and 15 rocks (abstract). In *Lunar Science—III* (editor C. Watkins), pp. 555–557, Lunar Science Institute Contr. No. 88.

Morgan J. S. Laul J. C. Krähenbühl U. Ganapathy R. and Anders E. (1972b) Major impacts on the moon: chemical characterization of projectiles (abstract). In *Lunar Science—III* (editor C. Watkins), pp. 552–554. Lunar Science Institute Contr. No. 88.

Reid A. M. Ridley W. I. Warner J. Harmon R. S. Brett R. Jakeš P. and Brown R. W. (1972) Chemistry of Highland and Mare basalts as inferred from glasses in the lunar soils (abstract). In *Lunar Science—III* (editor C. Watkins), pp. 640–642, Lunar Science Institute Contr. No. 88.

Ridley W. I. Williams R. J. Brett R. Takeda H. and Brown R. W. (1972) Petrology of lunar basalt 14310 (abstract). In *Lunar Science—III* (editor C. Watkins), pp. 648–650, Lunar Science Institute Contr. No. 88.

Taylor F. C. and Dence M. R. (1969) A probable meteorite origin for Mistastin Lake, Labrador. *Can. J. Earth Sci.* **6**, 39–45.

Toksöz M. N. Press F. Anderson K. Dainty A. Latham G. Ewing M. Dorman J. Lammlein D. Sutton G. Duennebier F. and Nakamura Y. (1972) Velocity structure and properties of the lunar crust (abstract). In *Lunar Science—III* (editor C. Watkins), pp. 758–760, Lunar Science Institute Contr. No. 88.

Walter L. S. French B. M. Doan A. S. Jr. and Heinrich K. F. J. (1972) Petrographic analysis of lunar samples 14171 and 14305 (breccias) and 14310 (melt rock) (abstract). In *Lunar Science—III* (editor C. Watkins), pp. 773–775, Lunar Science Institute Contr. No. 88.

Warner J. L. (1972) Apollo 14 breccias: metamorphic origin and classification (abstract). In *Lunar Science —III* (editor C. Watkins), pp. 782–784, Lunar Science Institute Contr. No. 88.

Wasserburg G. J. Turner G. Tera F. Podosek F. A. Papanastassiou D. A. and Huneke J. C. (1972) Comparison of Rb-Sr, K-Ar and U-Th-Pb ages; lunar chronology and evolution (abstract). In *Lunar Science —III* (editor C. Watkins), pp. 788–790, Lunar Science Institute Contr. No. 88.

Wise D. U. and Yates M. T. (1970) Mascons as structural relief on a lunar "Moho." *J. Geophys. Res.* **75**, 261–268.

Proceedings of the Third Lunar Science Conference
(Supplement 3, *Geochimica et Cosmochimica Acta*)
Vol. 1, pp. 401–422
The M.I.T. Press, 1972

Electron petrography of Apollo 14 and 15 rocks

J. S. LALLY, R. M. FISHER

U.S. Steel Research Laboratories, Monroeville, Pennsylvania 15146

J. M. CHRISTIE, D. T. GRIGGS

Department of Geology and
Institute of Geophysics and Planetary Physics
University of California, Los Angeles, California 90024

A. H. HEUER, G. L. NORD, JR., AND S. V. RADCLIFFE

Division of Metallurgy and Materials Science
Case Western Reserve University
Cleveland, Ohio 44106

Abstract—Comparative optical and high-voltage (800–1000 kV) electron petrographic analyses have been conducted on igneous rocks and several types of breccias from Apollo 14 and 15. The principal results can be summarized as follows:

Breccias: The matrix of rock 14321, although optically isotropic, is shown to consist principally of submicroscopic crystals exhibiting highly heterogeneous deformation and recrystallization substructures, which are particularly prominent at grain contacts. These characteristics are believed to result from the high pressures and temperatures accompanying the shock waves associated with meteoritic impact. It is proposed that the cohesion of this breccia is due mainly to deformation ("pressure welding") associated with recrystallization; this mechanism is termed "shock sintering."

Another common lunar breccia, the glass-welded type, was represented by two breccia fragments from the 14161 rake sample. One extreme of this type, in which the glass was the major component ($\sim 90\%$), exhibited sharp glass-crystal interfaces, indicative of rapid cooling. In the other extreme, the glass is a minor phase ($\sim 10\%$) and exists as an intercrystalline film, with diffuse glass-crystal interfaces. Despite this indication of slow cooling, an unusual feature of many of the crystal fragments in this breccia was a high density (3–$6 \times 10^{10}/\mathrm{cm}^2$) of fossil particle tracks which apparently survived the heating associated with consolidation.

Laboratory experiments on lunar dust have established that consolidation *can* be effected by simple thermal sintering. However, none of the limited number of breccias examined to date exhibited substructures characteristic of this mechanism.

Crystalline Rocks: On the basis of the presence of very fine-scale exsolution and the size of the (b) type domains in plagioclase, and on the scale of the M–T APB structure in pigeonite, rock 14310 is thought to have cooled more slowly than Apollo 12038 and 12053 basalts.

Anorthite in 15415 is somewhat more deformed than the plagioclase in Apollo 11 and 12 basalts and contains numerous microtwins, on which submicroscopic pyroxene has precipitated. Type (c) domain structures are present and are considerably larger than in lunar bytownites; this is believed to be a consequence of the high An ($\sim 95\%$) content.

INTRODUCTION AND TECHNIQUES

HIGH-VOLTAGE (800–1000 kV) electron petrography has been applied successfully to the study of the structural evolution and geologic history of igneous rocks from the Apollo 11 and 12 missions (Radcliffe *et al.*, 1970; Christie *et al.*, 1971). This work has been continued with samples of igneous and brecciated rocks from the Fra Mauro and Apennine sites of the Apollo 14 and 15 landings.

The lunar samples examined here include the crystalline rocks 14310,159 and 15415,25 and the breccias 14321,42, 14161,38–4, and 14161,38–7; the latter two specimens are fines from a rake sample. The techniques for the preparation of suitably thin electron-transparent sections and for electron transmission analysis have been described previously (Radcliffe *et al.*, 1970; Christie *et al.*, 1971; Heuer *et al.*, 1971).

RESULTS AND DISCUSSION

Crystalline rocks

Optical Petrography. Sample 14310,159 is from one of the two crystalline rocks with masses greater than 250 gm returned by the Apollo 14 mission (LSPET, 1971). The sample is a feldspathic basalt containing plagioclase and pyroxene with an intersertal to intergranular texture. The pyroxenes are zoned, with small orthopyroxene cores and pigeonite rims, with fine augite exsolution lamellae present in the latter. The plagioclase composition ranges from An_{90} in unzoned grains to grains with normal zoning in the range An_{90} to An_{65}. Minor amounts of ilmenite and metallic particles of Fe–Ni alloy are also present. This sample (159) was cut from the upper surface of the rock to permit structural examination at and adjacent to this surface, with particular respect to radiation damage.

Rock 15415 was collected from the N–NW rim of Spur Crater and is the first large specimen of anorthosite returned from the moon. It is thus one of several "unique" specimens collected at the Apennine front (LSPET, 1972). The sample described in this paper consists of fine fragments, apparently from the rock sample bag. As their small size (~ 1 mm in largest dimension) precluded direct sectioning, the fragments first had to be embedded in a larger block of a medium appropriate for sectioning and ion thinning. A mixture of 50% cassiterite (SnO_2) powder and 50% polystyrene resin was found suitable for this purpose. Optically, the several plagioclase fragments in these fines were found to be homogeneous with a composition of An_{93-95}. Fine albite and pericline twins are present with no optical evidence of deformation. All the plagioclase fragments examined contained small amounts (between 1 and 2%) of small subhedral to euhedral inclusions of augite. A chip sample of 15415,46, received later and currently under study, shows fractures and deformation twins in thin section, in keeping with the variety of cataclastic textures reported elsewhere for this rock (LSPET, 1972).

Electron petrography. In the igneous sample 14310,159 the plagioclase contains few dislocations ($< 10^7/cm^2$), and is thus similar to the plagioclase from the Apollo 11 and 12 basalts previously studied (Radcliffe *et al.*, 1970; Christie *et al.*, 1971). However, unlike the Apollo 11 and 12 plagioclases examined by us, fine scale exsolution is present (Fig. 1), confirming the recently discovered miscibility gap in bytownite (Nissen, 1968). The presence of this feature in 14310 suggests a slower cooling history than for the Apollo 11 and 12 basalts. This is consistent with Mössbauer studies of pyroxene in this rock (Finger *et al.*, 1972). However, the possibility that exsolution in plagioclase in 14310 arises because the composition of the small area sampled differed significantly from the plagioclase studied earlier cannot be

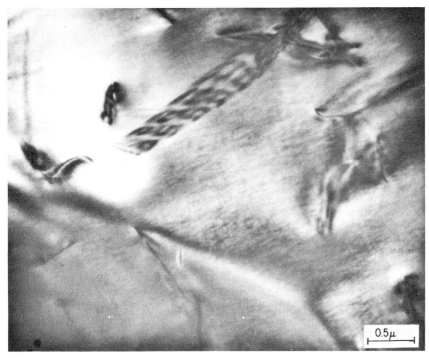

Fig. 1. "Tweed" background structure in plagioclase from 14310,159 indicating fine-scale exsolution. Transmission electron micrograph, 800 kV.

excluded completely at this time. Also consistent with relatively slow cooling is the observation that type (b) antiphase domains, due to the $C\bar{1}$–$I\bar{1}$ transformation (Christie *et al.*, 1971; Heuer *et al.*, 1972) are larger (Fig. 2a) than in Apollo 12 plagioclase (they are similar in size to those in 10029). In the other plagioclase grains in this sample of 14310, substantial differences occur in the contrast of the (b) type domain structure in regions separated by only 25 μ (Figs. 2b and c); this effect undoubtedly is due to changes in stoichiometry or in the Al–Si ordering, as the cooling rate must have been very similar in both regions. It is of interest that the parallel antiphase boundaries in Fig. 2b all terminate at a twin boundary. The twin boundary apparently assists in nucleation of the $I\bar{1}$ "superlattice" from the $C\bar{1}$ matrix.

In the clinopyroxene of rock 14310, the substructure resulting from exsolution and APB formation appears to be more complex than that in Apollo 11 and 12 basalts (Christie *et al.*, 1971). The host pigeonite studied here contains fine-scale augite exsolution lamellae, but only on (001) (Fig. 3a) [(100) exsolution, although not observed here, has been found in x-ray studies of 14310 pigeonite (Takeda and Ridley, 1972)]; the Morimoto-Tokonami (M–T) APB's also tend to be on a fine scale. Rock 14310 also differs from Apollo 11 and 12 basalts in that the pyroxene crystallized with bronzite cores (0 in Fig. 3b) and pigeonite rims. Coarse and fine-scale exsolution lamellae of augite in pigeonite (see A and B, Fig. 3b) were observed

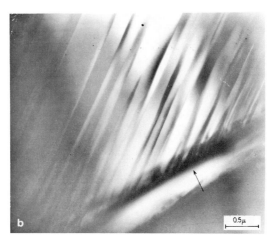

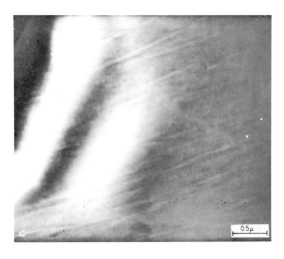

near the ortho-clinopyroxene boundary. The coarse augite exsolution lamellae on (001) were restricted to the region within 5–15 μ from the orthopyroxene boundary (Fig. 4a), probably due to local enrichment of Ca in these regions. In addition to the coarse and fine-scale (001) exsolution, Fig. 3c shows a second set of lamellae (arrowed) not on (100), which apparently nucleated on the augite-pigeonite interfaces. This set of lamellae is believed to have formed under more nonequilibrium conditions than the conventional (001) and (100) lamellae and will be discussed further in a later publication. In addition, the two-pyroxene interfaces appear to assist in nucleating the M–T APB's (Fig. 3d).

The M–T antiphase domains are approximately 600 Å in diameter near the two-phase interfaces but have an average size of only 150 Å away from them. In regions where the domain size is unaffected by heterogeneous nucleation (that is, away from the coarse augite exsolution lamellae) the domain size in 14310 is about twice as large as that found in sample 12038, again consistent with a slower cooling history for rock 14310.

It is clear from this evidence of heterogeneous nucleation, as well as variations of domain size in micrographs in our earlier publications (Christie *et al.*, 1971), that care must be taken in interpreting the significance of domain size in pyroxenes. As in many order-disorder transformations, the size of the M–T APB's in pigeonite is dependent on the ratio of the rate of nucleation of the new phase to its growth rate. In host pigeonites of C2/c symmetry, in which augite exsolved above the C2/c–P2$_1$/c pigeonite transformation, there will be preferential nucleation sites for the P2$_1$/c phase on the augite-pigeonite interfaces, leading to the formation of large APB's (Fig. 8a below; see also Fig. 15 of Christie *et al.*, 1971). The small domain size observed in other pigeonites is due to the homogeneous nucleation of these domains prior to exsolution (see Figs. 13 and 14 of Christie *et al.*, 1971).

The interfacial region between the ortho and clinopyroxene (Fig. 4) does not exhibit strain contrast and appears to be incoherent. However, the diffraction patterns are matched closely on opposite sides of the boundary (see insets in Fig. 4a), suggesting that an epitaxial relationship may have existed at high temperatures. The subboundary (A in Fig. 4a) running through the interface is consistent with this idea. The orthopyroxene itself contains a low density of dislocations, and a few isolated small-angle subboundaries. Although large areas of the orthopyroxene are completely free from crystalline defects, small regions containing a high density of growth faults on (100) (Fig. 5) are present. The planar stacking faults and the partial dislocations bounding them show weak contrast using diffraction vectors perpendicular to the fault plane. Invisibility conditions are consistent with a fault vector of $\frac{1}{2}(\vec{a} + \vec{c})(\equiv \frac{1}{2}[101])$. Thus, these faults represent a stacking disorder, due to

Fig. 2. Type "b" antiphase domain structures in plagioclase in 14310,159. The domain structure ranges in size even within a single grain. Large domains are shown in (a) and are comparable in size to those in 10029. The domain boundaries are parallel in (b) and appear to have nucleated at a twin boundary (arrowed). The boundary contrast is very weak in (c) which is due either to a low degree of long-range Al–Si order or to a change in stoichiometry. Dark field electron micrographs, type b reflection, 800 kV.

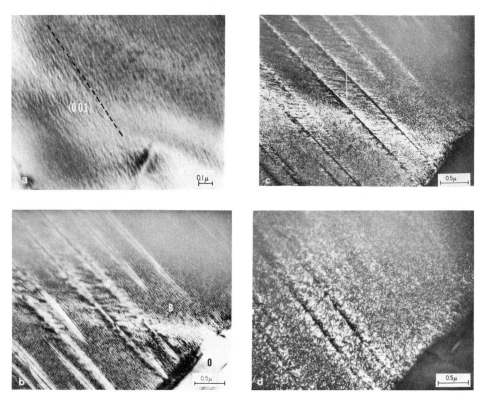

Fig. 3. Augite exsolution lamellae and M-T APB's in a pigeonite rim in 14310,159. The scale of these structures range widely with position in this crystal. (a) shows a fine exsolution structure on (001), typical of regions exhibiting homogeneous nucleation. [The dashed line is the trace of (001).] Near the orthopyroxene core [marked 0 in (b)], both coarse and fine-scale lamellae (at A and B respectively) are present. The internal interfaces between two pyroxenes serve as heterogeneous nucleation sites for subsequent secondary exsolution lamellae [arrowed in (c)], and for the M-T APB's, as seen in (d). (b), (c) and (d) are the identical areas taken under different diffraction conditions. Transmission electron micrographs: (a) dark field, $h + k = 2n$; (b) bright field; (c) dark field, $h + k = 2n$; (d) dark field, $h + k = 2n + 1$. 800 kV.

the accidental removal (or insertion) of two layers of the silicate chains during growth and have some similarity to faulted loops in metals (Fourdeux *et al.*, 1961). Two closely spaced growth faults of this type will restore the long-range perfection of the crystal and in such regions the fault image will appear to be very weak, as at A in Fig. 5. The (100) faults in the orthopyroxene generally are widely separated, but do overlap in a few regions.

Etching studies on thin sections of 14510,159 indicate* that the intensity of radiation damage is relatively low. Optical examinations reveal a particle track density of 10^7 to $10^8/cm^2$ at the exposed surface, which falls to $10^6/cm^2$ or less at a

---

*Conducted at UCLA by our colleague M. Wegner.

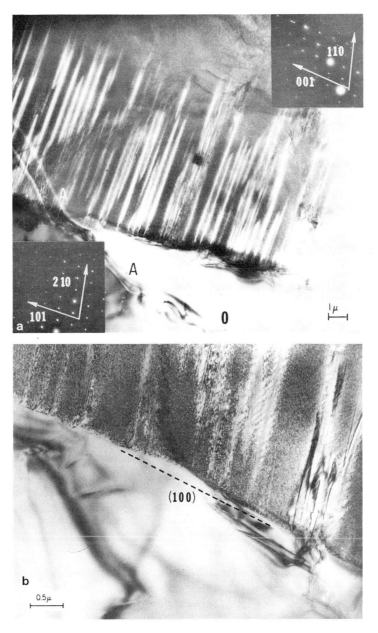

Fig. 4. Bronzite-pigeonite interface in 14310,159. The interface is inclined ~35° to the plane of the foil and exhibits contrast typical of an incoherent boundary; its irregularity is believed to be due to uneven foil thickness. Coarse augite exsolution lamellae [bright linear features in (a)] occur in the pigeonite adjacent to the interface. A is a subboundary that runs through both phases (0 indicates the orthopyroxene phase). The insets in (a) show the similarity of the diffraction patterns in the two pyroxenes. The bronzite-pigeonite interface is roughly parallel to (100), as can be seen in (b). Transmission electron micrographs. 800 kV.

Fig. 5. Stacking faults on (100) in bronzite. Diffraction contrast experiments indicate that these faults correspond to the removal or insertion of a layer of crystal a/2[100] thick; both individual faults and clusters of faults can be seen. At A the contrast is weak because of the overlapping of two faults. Transmission electron micrograph, 800 kV.

2 mm depth. Several of the plagioclase grains contain isolated fission track bursts due to entrapped fissionable minerals. Transmission electron microscope studies of the surface and near-surface regions of this rock fail to disclose any regions with particle track densities $> 10^7/cm^2$. If the outermost portion of the rock do contain a higher density of tracks (say, $10^9/cm^2$) that was not detected in the optical study, this material did not survive the ion-thinning foil preparation.

In the anorthosite 15415,25 fragments, many of the regions examined are free from evidence of deformation, but in some regions deformation and recovery substructures were found. For example, Fig. 6a shows several dislocation subboundaries and dislocation "debris", which appear to have arisen from annealing of a deformed region. Other regions contain dense tangles of dislocations. This diversity of deformation substructures is consistent with a complex deformation-recovery history (James, 1972). Our general impression, based only on the fines examined to date in the electron microscope, is that this sample, on the average, is more deformed than any previous lunar plagioclase we have studied from an igneous rock. Numerous submicroscopic twins were also present (see Figs. 6a, b, and c). As would be expected from the high anorthite content, the domain structures in this anorthosite are markedly different from those in the lunar bytownites previously examined in the electron microscope. No type (b) APB's could be imaged, possibly because crystallization may have occurred directly to the phase with $I\bar{1}$ symmetry. The diffraction patterns from the anorthite show strong c and d type diffraction spots, indicating $P\bar{1}$

Fig. 6. Substructures in anorthite from sample 15415,25. The dislocation walls and dislocation "debris" in (a) suggests deformation followed by recovery (T is a twin). Large type (c) domains adjacent to a pericline twin are shown in (b) and illustrate heterogeneous nucleation of these domains at the twin. Submicroscopic pyroxene precipitates (arrowed in (c)), have nucleated preferentially on the twin walls [note also the (c) domains in weak contrast in this micrograph]. The diffraction patterns in (d) and (e) illustrate the $P\bar{1}$–$I\bar{1}$ transformation; (d) was taken at 20°C and (e) at 400°C. The latter diffraction patterns have been rotated with respect to the micrographs (a), (b), and (c). The diffraction spots in the vestical row missing in (e) (the spots in this row are actually present but have very low intensity) have h + k even, l odd. (a), (b) and (c) are bright field transmission electron micrographs.

symmetry. Dark-field micrographs, using both (c) and (d) type reflections, reveal a domain structure (Fig. 6b) larger than the type (c) structures observed in lunar bytownite (Heuer *et al.*, 1972). Furthermore, the domain size adjacent to the twin boundary is much greater than in the remainder of the crystal. Hence, the twin boundary appears to provide a preferential site for nucleation of the $P\bar{1}$ phase in the $I\bar{1}$ matrix; the subsequent growth from these nuclei gives rise to a larger-than-average domain size. This is similar to the observations noted above in the pyroxenes. An additional significant observation is that small pyroxene grains (identified by qualitative chemical analysis using a SEM and by electron diffraction) are present on some twin boundaries (Fig. 6c), indicating that the twin boundaries act as preferential sites for precipitation of a pyroxene phase in the anorthite.

Apollo sample 15415 has been variously called a complex polymetamorphic anorthosite (James, 1972) (a result of recrystallization of the deformed parent rock) and a shock metamorphosed fragment derived from a gabbroic anorthosite magma (Hargraves and Hollister, 1972), the included augite grains arising from droplets of magma trapped within the growing feldspar. The observed association of sub-microscopic pyroxene precipitates and twin boundaries tends to support the mechanism of formation suggested by James. It is not likely that entrapped droplets of magma would be associated with twin boundaries, and precipitation rather than inclusion during crystallization appears to be the likeliest explanation.

In situ heating experiments were conducted on foils of the 15415 plagioclase in the electron microscope to observe the $I\bar{1}$–$P\bar{1}$ phase transformation. Diffraction patterns taken at 20°C and 400°C (Figs. 6d and e) show that the (c) spots have lost most of their intensity at 400°C but are still present. The domain structure is no longer visible, however, because of the weakness of these reflections. The domain contrast reappears on cooling to 20°C; the domain structure after cooling is similar in scale to the initial structure but the domains are not identical in shape. It should be noted here that these experiments were hindered by the tendency of plagioclase to become amorphous due to electron radiation damage during heating in the electron microscope. This effect is not troublesome, provided the area under study is irradiated with electrons only for short periods when recording images and diffraction patterns.

Breccias

Apollo 14161 is a rake sample and our allocation of fines contains a variety of breccia types as well as a few igneous rock fragments. Only two individual fragments, on which the electron petrographic study has been completed, will be discussed here. Detailed observations have also been made on a sample of breccia 14321. The remaining breccia types available to us from Apollo 14 and 15 will be discussed in a future publication.

Optical petrography. Sample 14321,42 is a shocked nonporous breccia already described in detail by Grieve *et al.* (1972), Duncan *et al.* (1972), and others and hence will only be briefly described here. Because of the extreme heterogeneity of this sample, a general description does not complement the electron petrographic study to the same degree as with igneous rocks. In general, the rock consists of crystalline fragments from 1 cm to submicroscopic size. The finest material (referred to as *matrix*) is optically isotropic. Shock-induced deformation structures, such as twinning and bending in the plagioclase and pyroxene crystals, are abundant. There also is extensive optical evidence of recovery in the larger plagioclase and pyroxene clasts; this includes partial resorption of twin lamellae in both minerals, and rims of different birefringence on the boundaries of some pyroxenes. Large (>1 cm) basaltic fragments in the sample show little internal deformation, but near the boundaries of these fragments there is considerable crushing, intermingling with the dark, fine-grained breccia, and internal deformation (fracture, bending, and twinning) of the crystals.

Sample 14161,38–4 is a brown, highly vesicular heterogeneous and flow-banded glass (see Fig. 11 below), containing 7 to 10% angular crystalline fragments of pyroxene and plagioclase, which range from 0.2 mm to less than the resolution of the optical microscope. The glass contains 1 to 2% of fine iron or iron-nickel spheres of less than 0.01 mm in diameter.

Sample 14161,38–7 (see Fig. 14a below) is a glass-welded aggregate. Several large vesicles (0.4 mm) are present on the edge of the sample, and small vesicles are abundant in the interior. The presence of glass is easily seen on the exterior; the interior groundmass is charged with fine crystal fragments and much of it appears optically isotropic. Approximately two-thirds of the large crystal fragments >0.2 mm are plagioclase, the remainder being pyroxene. The groundmass contains more pyroxene than plagioclase and a substantial proportion of glassy fragments and spheres. Mineral and glass fragments appear to be distributed according to size in the following way: >0.2 mm, 15 to 20%; 0.2 to 0.01 mm, 60%; and the rest a ground-mass of less than 0.01 mm.

Electron petrography. The high resolution of the electron microscope can be used to advantage with breccias to examine the structure of the submicroscopic groundmass and the interfaces between crystal fragments. Examination of the substructure of the grains themselves give additional information on deformation and thermal history.

Before discussing our results on the breccias, it is useful to describe a preliminary experiment conducted in order to ascertain the type of bonding or cohesion that might occur in lunar soil as a result of thermal sintering. Specimens of lunar dust (from sample 10084,89) were compacted at room temperature in a die under 3-kbar load and heated for 1 hour at several different temperatures. Only at 1100°C, the highest temperature studied, did cohesion of the particles occur; the resulting sintered material was sufficiently well bonded to allow sectioning for electron micro-scopy. The grain boundary regions were studied in detail; a typical example is shown in Fig. 7. The boundary is very irregular and contains many small crystallites

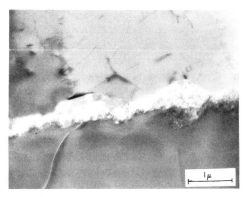

Fig. 7. Interface region (bright feature) in thermally sintered specimen of lunar dust, 10084,89. The dust was compacted at room temperature at 3 kb, and then heated in vacuum at 1100°C. Dark-field electron micrograph, 800 kV.

which can be made to diffract strongly in dark field; it is free from glass and cavities. Such boundaries have not been observed in any of the lunar breccias so far examined. We can therefore eliminate this type of thermal sintering as a significant mechanism in their consolidation. However, the fact that such a mechanism is effective at sufficiently high temperature suggests that some of the lunar breccias may have formed in this way. The failure to obtain sintering in an earlier attempt at 900°C (Chao *et al.*, 1971) probably was due to too low a temperature.

Because of the wide diversity of microstructures and complex history of rock 14321, we cannot give a complete account but will concentrate on the substructure of the matrix. Its optical isotropy has been attributed to either a glassy or submicroscopic crystalline structure. In electron transmission, the groundmass is seen to be a composite structure of submicroscopic crystals (a few microns or less in diameter) and fine rock fragments with only a minor amount (< 10%) of glass; thus the isotropic optical character is due to small crystal size. Within the crystalline fragments there is a great diversity of substructures, illustrated in Fig. 8. Some pyroxene grains are undeformed (Fig. 8a) showing only large exsolution lamellae and M–T APB's, but others (Fig. 8b) are heavily deformed and contain recrystallized areas (Christie *et al.*, 1972). Plagioclase (Figs. 8c and d) shows a similar diversity: some grains (Fig. 8c) show an extremely high dislocation density; others (Fig. 8d) exhibit regions of extensive deformation separated by only 1 to 2 μ from areas that have recovered and/or recrystallized (the bright areas in this micrograph are recrystallized regions that have fallen out during specimen preparation). Olivine in this sample, shows a substantial density of slip dislocations (Fig. 8e), but no evidence of internal recrystallization.

Recrystallization is also a prominent feature in the vicinity of grain contacts, both within the fine-grained matrix and at the matrix-clast interfaces, and at many internal boundaries in the larger crystals; it is invariably associated with regions of heavy deformation (Fig. 9). The heating that caused the recrystallization either accompanied or followed the deformation. It is consistent with shock wave effects associated with meteoritic impact on the lunar surface, in that the deformation and temperature rise are greatest in porous materials, or in porous regions of heterogeneous aggregates like the lunar soils or breccias. The mechanism of cohesion in this breccia, which has been termed "shocked sintering" (Christie *et al.*, 1972), involves deformation ("pressure welding") associated with recrystallization. It should be noted that some regions of the matrix in this breccia do appear to be bonded by glass (Fig. 10). However, although much of this glass may be produced by the shock events, grain bonding by glass is minor in this breccia, compared with "shock sintering."

Breccia 14161,38–4 (Fig. 11) is similar to the type 4 breccia of Chao *et al.* (1972) and consists of vesicular and partially devitrified glass surrounding crystal fragments. The interface (arrowed in Fig. 12a) between the large crystal fragments and the glass does not involve a gradual loss of crystallinity but is quite abrupt. The plagioclase adjacent to the glass boundary in this micrograph has many small subgrains and exhibits considerable lattice strain, in contrast to the interior regions of the crystal. This suggests that the surface regions of this crystal were more highly deformed

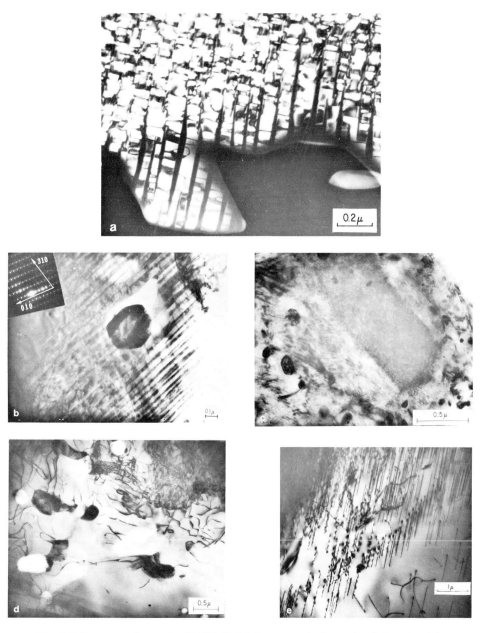

Fig. 8. Substructures in matrix of 14321,42. (a) shows a strain-free grain of pigeonite with exsolved augite; M–T APB's are present in the pigeonite. By way of contrast, (b) shows a deformed pyroxene grain also containing exsolution lamellae with an out-of-contrast recrystallized grain. Plagioclase grains exhibit a range of substructures, (c) showing heavy deformation while (d) indicates that some recovery has occurred; both grains contain small recrystallized regions. Olivine (e) contains slip dislocations but no recrystallized regions. Transmission electron micrographs; (a) is taken in dark field, 800 kV.

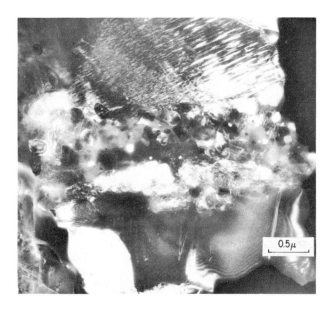

Fig. 9. Grain contact between pyroxene (top) and plagioclase (bottom). Extensive deformation and recrystallization occurs in the vicinity of the interface and leads to breccia consolidation via "shock sintering". Dark-field electron micrograph, 800 kV.

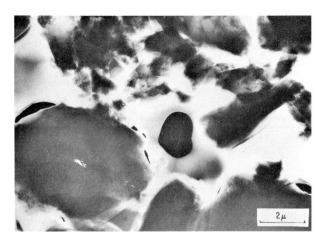

Fig. 10. Groundmass of 14321,42 containing micron-sized crystalline fragments embedded in glass (light featureless regions). The total glass content of the groundmass is < 10%. The black area in the center is a hole. Dark-field electron micrograph, 800 kV.

Fig. 11. Optical micrograph of 14161, 38-4. This breccia is a vesicular glass containing
7–10% angular crystal fragments. Plane polarized light.

than the interior and were heated sufficiently during consolidation to permit sub-grain formation (recovery). Metal spheres are present within the plagioclase grains but were too thick to be transparent in the electron microscope, even at 1 MEV.

The glass portions of this sample generally are featureless but do contain a few micron-sized tabular crystals of plagioclase (not visible optically), which represent the early stages of devitrification (Fig. 12b). Widely separated spheres of fcc Fe–Ni (positively identified by electron diffraction and with Ni contents from 7 to 20% as measured on a SEM but also containing some Cr and Co) also are observed (Fig. 13), usually within glass-filled vesicles in the parent glass. Within the alloy spheres, fine particles (arrowed) ∼200 Å in diameter were observed to have been precipitated; their morphology suggests that they are phosphides rather than carbides, although more positive identification by electron diffraction was unsuccessful.

In the remaining sample to be discussed here, 14161,38-7, crystal fragments range from 10 to 500 μ in diameter and are surrounded by vesicular glass (Fig. 14a). Submicron-sized metallic particles are also present (Fig. 14b). The latter micrograph was taken in dark field, under which conditions both the vesicles and crystal fragments appear darker than the glassy matrix. The total glass content of this sample is approximately 10%. The interfaces between the glass and crystalline regions (Fig. 15) appear gradational, in contrast to those in 14161,38-4. The plagioclase shown in the right half of Fig. 15a contains two type (b) antiphase domain boundaries that gradually become diffuse as they approach the glassy region. This observation indicates that the glass-crystal boundary is itself quite diffuse, extending over approximately 1 μ. The glass adjacent to this crystal contains small spherical particles ∼200 Å in diameter (Figs. 15a and b; see also Fig. 14b). The particles

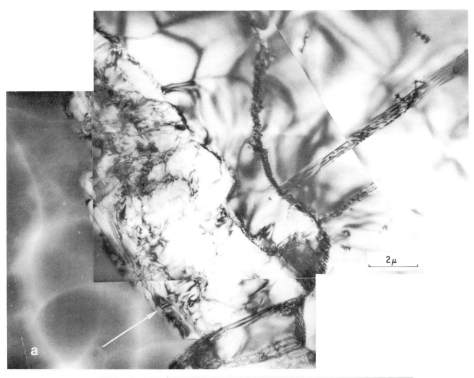

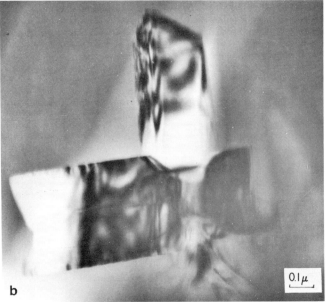

Fig. 12. Interface between plagioclase and glass in 14161,38–4. (a) shows that subgrains are present in the plagioclase fragment adjacent to the interface (arrowed). The onset of devitrification (the formation of small tabular plagioclase crystals) is shown in (b). Transmission electron micrographs, 800 kV.

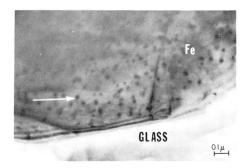

Fig. 13. Interface between fcc iron alloy (top) and glass (bottom) in 14161,38–4. Micro-analysis of this alloy using a scanning electron microscope and a non-dispersive spectro-meter indicated a Ni content of $\sim 15\%$ and Cr of $\sim 5\%$ with some cobalt present. The particles in the metal alloy (arrowed) are thought to be phosphides. Transmission electron micrograph, 800 kV.

were shown by electron diffraction to be bcc iron, the polymorph normally stable at room temperature. The glass composition, obtained by microanalysis using a scanning electron microscope in conjunction with a solid-state nondispersive x-ray microanalyzer, has been shown (see inset of Fig. 15a) to be rich in Fe, K, and to a lesser degree Ti and is possibly of KREEP composition (Meyer *et al.*, 1971). A phosphorous peak is not present but the minor P content of KREEP glass is below the resolution of the analyzer.

The most striking and significant feature observed in sample 14161,38–7 is a high density of radiation tracks in the mineral clasts studied. The high-energy particle tracks observed in the plagioclase crystal marked A in Fig. 14a are shown in Fig. 16a. The density of tracks in this region is $6 \times 10^{10}/cm^2$. Tracks in an ortho-pyroxene crystal (located near B in Fig. 14a) are shown in Fig. 16b; the track density in this crystal is $3 \times 10^{10}/cm^2$. Measurements of track densities were made from the other arrowed regions in Fig. 14a and were in the range from 3 to $6 \times 10^{10}/cm^2$. Tracks were observed clearly in all crystals examined in this breccia, with the exception of clinopyroxene, where the very strong contrast due to exsolution obscures the much weaker track contrast. The very high track densities observed in all the grains in this rock suggest that the crystallites were irradiated as individual fragments prior to consolidation into the breccia. High-energy cosmic rays, whose range would allow particle registration in the interior of a consolidated breccia of this size, have too low a flux to be the source of the tracks observed (Crozaz *et al.*, 1971). Similarly, the range of solar flare protons is too shallow for the tracks to be visible in the interior of 25 μ grains. We therefore believe that the tracks were caused by Fe nuclei from solar flares, and were accumulated while the fragments were lying on the lunar surface prior to consolidation into the breccia. It was surprising that the tracks did not anneal out during the process of breccia consolidation, which involved bonding with a molten glass; in this process a substantial rise in temperature would be expected. As can be seen from Fig. 15b, the tracks extend right up to the plagioclase-glass interface. Because it has been demonstrated that tracks will anneal

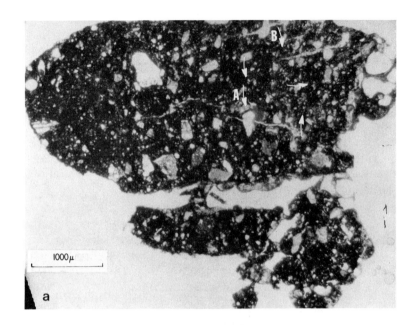

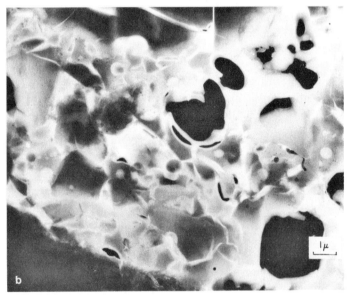

Fig. 14. (a) Optical micrograph (plane polarized light) of 14161,38–7 showing crystal fragments <0.5 mm embedded in vesicular glass. (b) Transmission electron micrograph of 14161,38–7 showing mineral fragments, including iron particles (which appear as spherules) embedded in vesicular glass. 800 kV.

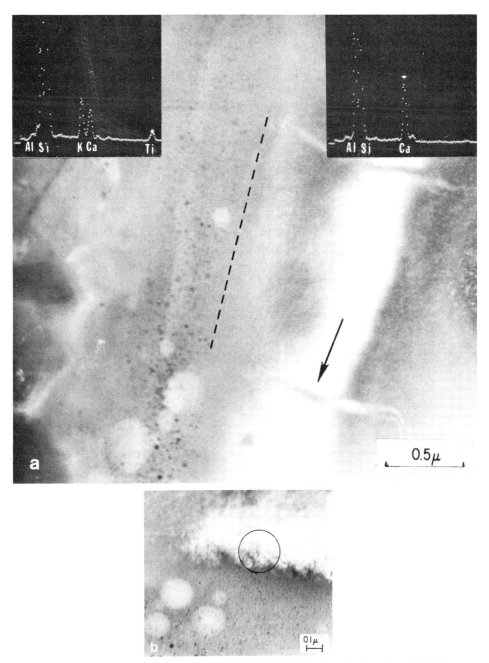

Fig. 15. Plagioclase-glass interfaces in 1416,38–7. (a) shows that the boundary (dashed) between the plagioclase [marked A in Figure 14(a)], and the surrounding glass is diffuse. Antiphase domain boundaries (arrowed) are visible in the plagioclase. The inserts show the composition of the crystalline and glass regions, as determined by microanalysis using a scanning electron microscope. Particle tracks adjacent to the glass boundary can be seen (circled area) in (b). Dark-field electron micrographs, 800 kV. Type (b) reflection used for (a).

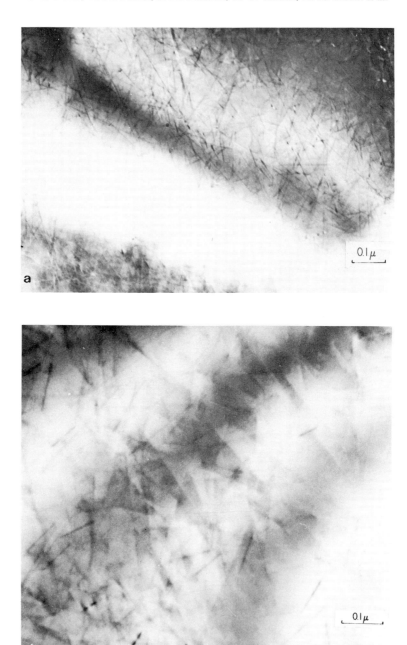

Fig. 16. Particle tracks in crystalline grains in 14161,38–7. (a) shows a track density of approximately $6 \times 10^{10}/cm^2$ in a plagioclase grain (marked A in 14(a)), while the orthopyroxene in (b) (marked B in Fig. 14(a)) has a track density of $3 \times 10^{10}/cm^2$. Dark field electron micrographs, 800 kV.

out during heat treatments of 1 hour at 800°C (Borg *et al.*, 1971), and possibly at lower temperatures, this observation suggests only minimal heating during consolidation. In one crystal, the contrast was not continuous along the whole length of the tracks. This condition perhaps represents the earliest stage of track recovery from annealing. However, it can be seen in Figs. 14 and 15 that the glass is vesicular, contains Fe precipitates and appears to have dissolved the surface contacts of the grains, suggesting that the grains were in contact with a glass originally at high temperatures and which cooled slowly enough for those features to develop. The apparent paradox between evidence for minimal heating (persistence of tracks) and evidence for slow cooling has not been resolved to date.

CONCLUSIONS

The crystalline rocks 14310 and 15415 differ substantially from the Apollo 11 and 12 basalts. Sample 14310 appears to have cooled more slowly than previous Apollo 12 basalts studied. Anorthite from 15415 is more heavily deformed than plagioclase from the Apollo 11 and 12 basalts. (c) type domains are relatively large and easily imaged in the electron microscope. Submicroscopic pyroxene has precipitated on twins in the anorthite.

Substructures in the Apollo 14 breccias are complex. At least two bonding mechanisms can be discerned from the electron petrographic examination so far conducted: (a) "shock sintering," due to high pressures and temperatures associated with the shock waves of meteoritic impacts, resulting in highly deformed and partially recrystallized regions at grain contacts; and (b) lithification by glass bonding. One breccia of the latter type exhibited a high density (3 to $6 \times 10^{10}/cm^2$) of particle tracks, thought to have been accumulated by the grains prior to their consolidation in the glass-welded breccia. The glass which formed breccia is thought to have been quite hot initially; the survival of these tracks is thus paradoxical.

Acknowledgments—This research was supported through the Manned Spacecraft Center, NASA, Houston. We wish to thank the following colleagues for their assistance and advice: W. R. Duff and G. P. Wray (U.S. Steel), J. DeGrosse and W. A. Dollase (UCLA), and A. R. Cooper and R. F. Hehemann (CWRU).

REFERENCES

Borg J. Maurette M. Durrieu L. and Jouret C. (1971) Ultramicroscopic features in micron-sized lunar dust grains and cosmophysics. *Proc. Second Lunar Sci. Conf., Geochim. Cosmochim. Acta* Suppl. 2, Vol. 3, pp. 2027–2040. MIT Press.

Chao E. C. T. Boreman J. A. and Desborough G. A. (1971) The petrology of unshocked and shocked Apollo 11 and Apollo 12 microbreccias. *Proc. Second Lunar Sci. Conf., Geochim. Cosmochim. Acta* Suppl. 2, Vol. 1, pp. 797–816. MIT Press.

Chao E. C. T. Minkin J. A. and Boreman J. A. (1972) The petrology of some Apollo 14 breccias (abstract). In *Lunar Science—III* (editor C. Watkins), p. 131, Lunar Science Institute Contr. No. 88.

Christie J. M. Lally J. S. Heuer A. H. Fisher R. M. Griggs D. T. and Radcliffe S. V. (1971) Comparative electron petrography of Apollo 11, Apollo 12 and terrestrial rocks. *Proc. Second Lunar Sci. Conf., Geochim. Cosmochim. Acta* Suppl. 2, Vol. 1, pp. 69–90. MIT Press.

Christie J. M. Griggs D. T. Fisher R. M. Lally J. S. Heuer A. H. and Radcliffe S. V. (1972) Deformation of lunar and terrestrial minerals. In *Electron Microscopy and Structure of Materials* (editors G. Thomas R. M. Fulrath and R. M. Fisher), University of California Press.

Crozaz G. Walker R. and Woolum D. (1971) Nuclear track studies of dynamic surface processes on the moon and the constancy of solar activity. *Proc. Second Lunar Sci. Conf., Geochim. Cosmochim. Acta* Suppl. 2, Vol. 3, pp. 2543–2558. MIT Press.

Duncan A. R. Lindstrom M. M. Lindstrom O. J. McKay S. M. Stoeser J. W. Goles G. G. and Fruchter, J. S. (1972) Comments on the genesis of breccia 14321 (abstract). In *Lunar Science—III* (editor C. Watkins), p. 192, Lunar Science Institute Contr. No. 88.

Finger L. W. Hafner S. S. Schurman K. Virgo D. and Warburton D. (1972) Distinct cooling histories and reheating of Apollo 14 rocks (abstract). In *Lunar Science—III* (editor C. Watkins), p. 259, Lunar Science Institute Contr. No. 88.

Fourdeaux A. Berghezan A. and Webb W. W. (1960) Stacking faults in zinc. *J. Appl. Phys.* **31**, 918.

Grieve R. McKay G. Smith M. and Weill D. (1972) Mineralogy and petrology of polymict breccia 14321 (abstract). In *Lunar Science—III* (editor C. Watkins), p. 338, Lunar Science Institute Contr. No. 88.

Hargraves R. B. and Hollister L. S. (1972) Mineralogic and petrologic study of lunar anorthosite slide 15415,18. *Science* **175**, 430–431.

Heuer A. H. Firestone R. F. Snow J. D. Green H. W. Howe R. G. and Christie J. M. (1971) An improved ion-thinning apparatus. *Rev. Sci. Inst.* **42**, 1177.

Heuer A. H. Lally J. S. Christie J. M. and Radcliffe S. V. (1972) Phase transformations and exsolution in lunar and terrestrial calcic plagioclase. *Phil. Mag.*, in press.

James O. B. (1972) Lunar anorthosite 15415: texture, mineralogy and metamorphic history. *Science* **175**, 432–435.

LSPET (1971) (Lunar Sample Preliminary Examination Team) Preliminary examination of the lunar samples from Apollo 14. *Science* **173**, 681–693.

LSPET (1972) (Lunar Sample Preliminary Examination Team) Preliminary examination of the lunar samples from Apollo 15. *Science* **175**, 363–374.

Meyer C. Jr. Brett R. Hubbard N. J. Morrison D. A. McKay D. S. Aitken F. K. Takeda H. and Schonfeld E. (1971) Mineralogy, chemistry, and origin of the KREEP component in soil samples from the Ocean of Storms. *Proc. Second Lunar Sci. Conf., Geochim. Cosmochim. Acta* Suppl. 2, Vol. 1, p. 393–411. MIT Press.

Nissen M.-V. (1968) A study of bytownite in amphibolites of the ivrea-zone (Italian Alps) and in anorthosites: a new unmixing gap in the low plagioclases. *Schweiz. Mineral. Petrogr. Mitt.* **48**, 53.

Radcliffe S. V. Heuer A. H. Fisher R. M. Christie J. M. and Griggs D. T. (1970) High voltage (800 kV) electron petrography of type B rock from Apollo 11. *Proc. Apollo 11 Lunar Sci. Conf., Geochim. Cosmochim. Acta* Suppl. 1, Vol. 1, pp. 731–748. Pergamon.

Takeda H. and Ridley W. I. (1972) Crystallography and mineralogy of pyroxenes from Fra Mauro soil and 14310 (abstract). In *Lunar Science—III* (editor C. Watkins), p. 738, Lunar Science Institute Contr. No. 88.

Proceedings of the Third Lunar Science Conference
(Supplement 3, *Geochimica et Cosmochimica Acta*)
Vol. 1, pp. 423–430
The M.I.T. Press, 1972

Crystallography and chemical trends of orthopyroxene-pigeonite from rock 14310 and coarse fine 12033

HIROSHI TAKEDA* and W. I. RIDLEY

NASA Manned Spacecraft Center,
Houston, Texas 77058

Abstract—X-ray single crystal diffraction studies, supplemented by electron microprobe analyses of orthopyroxenes and pigeonites from rock chip 14310,90, have identified the following Fra Mauro pyroxenes: bronzite with minor exsolution of augite on (100); twinned magnesian pigeonite with minor exsolution of augite on (100) and (001), with or without core bronzite sharing (100); more Fe-rich pigeonite, and augite that resembles some eucritic pyroxenes. The cation distribution coefficient, $(Fe/Mg)_{M1}/(Fe/Mg)_{M2}$, of a 14310 bronzite refined by x-ray methods is 0.10, which is similar to that of a bronzite of terrestrial volcanic origin. The features of overgrowth and exsolution, which are distinct from those of mare pyroxenes, indicate that crystallization was not as metastable as that of many mare rocks but was much more rapid than that of plutonic or intrusive rocks.

INTRODUCTION

PYROXENES IN APOLLO 11 and 12 mare basalts are dominantly augites and pigeonites. Orthopyroxene, an indicator of nonmare rocks, has been reported in trace amounts in Apollo 11 and 12 soil and breccia samples (Fuchs, 1970). The character of ortho-pyroxene and pigeonite from a typical KREEP fragment (12033,97), as revealed by x-ray single crystal methods (Meyer *et al.*, 1971), suggests a crystallization trend in Fra Mauro rocks from orthopyroxene to pigeonite. In the Apollo 14 soil, low-Ca pyroxene represents as much as 41% of all analyzed pyroxenes (Apollo Soil Survey, 1971). The Mg/(Mg + Fe) ratios range from 0.85 to 0.61. Orthopyroxene is common in crystalline rock 14310 (Papike and Bence, 1971; Ridley *et al.*, 1971), in which it is succeeded by magnesian and intermediate pigeonite and augite. As part of a study of rock 14310 (Ridley *et al.*, 1972), the exsolution and overgrowth patterns, cation distributions and structures of the orthopyroxenes, together with those of the co-existing pigeonites and augites, were studied by single crystal x-ray diffraction methods, combined with microprobe analysis, to provide information on crystallization and cooling conditions. The cation distribution of the orthopyroxene from KREEP fragment 12033,97 is also given.

EXPERIMENTAL TECHNIQUE

Three crystals of brownish Fe-rich pyroxene (SB1–3) and one orthopyroxene crystal (SY1) were selected from the small grains produced when a part of rock 14310,90 was chipped. Nine pyroxene grains were separated under a microscope from three small rock fragments (R1, R3, R7), each about 3 mm in diameter. The sizes of crystals used are given in Table 1. The crystals were mounted along c^*, and precession photographs of $h0l$ and $0kl$ nets were taken using Zr-filtered, MoKα radiation. If the patterns were

---

*On leave from Mineralogical Institute, Faculty of Science, University of Tokyo, Hongo, Tokyo, Japan.

Table 1. Crystallographic and chemical data on pyroxenes from rock 14310 and coarse fines 12033.

| Sample | 12033,97 | 14310,90 | | |
Mineral	Bronzite	Bronzite	Magnesian pigeonite	Intermediate pigeonite	Augite*
a, Å	18.304(3)	18.301(3)	9.673(3)	9.715(1)	9.713(2)
b, Å	8.887(2)	8.869(2)	8.896(2)	8.963(1)	8.964(3)
c, Å	5.215(1)	5.215(1)	5.228(3)	5.239(1)	5.266(2)
β, deg.	90.00	90.00	108.65 (4)	108.64 (2)	105.93 (2)
Space group	$Pbca$	$Pbca$	$P2_1/c$	$P2_1/c$	$C2/c$
Crystal size, (mm)	$0.12 \times 0.05 \times 0.10$	$0.15 \times 0.15 \times 0.17$	$0.15 \times 0.25 \times 0.10$	$0.14 \times 0.28 \times 0.25$	
SiO_2†	52.98	53.84	53.07	50.07	49.24
TiO_2	0.52	0.73	0.69	0.54	1.03
Al_2O_3	0.60	1.74	1.25	0.55	1.26
Cr_2O_3	0.00	0.41	0.30	0.05	0.09
FeO	18.37	14.87	19.16	33.04	20.25
MnO	0.31	0.22	0.30	0.57	0.46
MgO	24.33	26.11	22.25	13.21	11.00
CaO	1.55	2.41	3.72	2.48	15.58
Na_2O	0.02	0.00	0.03	0.00	0.03
TOTAL	98.66	100.33	100.76	100.52	98.95

*Lamellae in the intermediate pigeonite.
†All values in weight percent.

complicated due to twinning, or stacking faults, overexposed $h0l$ photographs of the same crystals were taken using Ni-filtered, CuKα, radiation. Subsequently, each crystal was studied with the electron microprobe.

The cell dimensions of selected crystals (Table 1) were obtained by least-squares refinements using the diffraction angles of 12 reflections measured on a Picker FACS–1 system with MoKα_1 radiation. The diffuseness of class b reflections ($h + k$:odd) was estimated for pigeonite crystals by measuring the 2θ angles at half peak height. The nearby a reflections ($h + k$:even) were used as a standard for mechanical line broadening.

X-ray diffraction intensities of an orthopyroxene from a KREEP-type glass-matrix breccia fragment (12033,97,2B, Meyer *et al.*, 1971), and an orthopyroxene (R7–1) and its coexisting magnesian pigeonite (R7–5) from rock 14310,90, were measured on a Picker FACS–1 diffractometer system in the ω–2θ mode using MoKα radiation, a graphite monochrometer and a scintillation detector. To detect weaker reflections for the orthopyroxenes, a fixed count method with longer counting time also was employed. The x-ray diffraction intensities of a coexisting intermediate pigeonite and augite pair from rock 14310,90, also were measured in the same manner.

Structure Refinements

The measured intensities were corrected for Lorentz, polarization and absorption factors and were reduced to structure factors $|F_o|$ by the method used for previous pyroxene work (Takeda, 1972a, b). Only 389 (A–12033) and 887 (A–14310 R7–1) reflections were observable because of the small size of the orthopyroxene crystals (Table 1). 651 observable reflections of the magnesian pigeonite (A–14310 R7–5) are free from the overlapping due to twinning.

Refinements of the atomic coordinates, anisotropic temperature factors and site occupancy factors for Mg and Fe in the M1 and M2 sites of the orthopyroxenes and pigeonites were carried out on a UNIVAC 1108 by using the full-matrix least-squares program RFINE (Finger, 1969) and methods similar to those described in

Takeda (1972a, b). The positional parameters and isotropic temperature factors of the Takasima bronzite (Takeda, 1972b) were used as starting parameters for the orthopyroxenes, and those of the 12052 pigeonite (Takeda, 1972a) were used for the 14310 pigeonite. The conventional unweighted residuals for the final refinements of anisotropic temperature factors are 0.060 (A–12033) and 0.039 (A–14310) for the orthopyroxenes and 0.057, for the pigeonite (A–14310, R7–5). The final parameters for the orthopyroxene (A–14310) are given in Table 2. Parameters for the other structures refined, using smaller numbers of reflections, presently are being evaluated.

RESULTS

Based on their chemical compositions, exsolution patterns, twinning and color, Ridley *et al.* (1971, 1972) classified the 14310 pyroxene grains into three major groups. The present results are discussed in terms of these groups.

(1) Orthopyroxene (4 crystals), pale yellow or almost colorless but not transparent. The compositions are bronzitic ($Wo_4En_{80}Fs_{16}$–$Wo_5En_{72}Fs_{23}$). Diffuse streaks along a^*, indicative of the disordered orthopyroxene commonly found in meteorites (Pollack, 1968), were not detected within usual exposure times. Longer exposed photographs showed submicroscopic exsolution of augite on (100). Two doublets of very weak orthopyroxene-augite reflections: 202 (aug) and 602 (opx), and 402 (aug) and 10.0.2 (opx), were observable. The intensities of both reflections of a doublet are nearly equal. One grain (SY1) is composed of several slightly misoriented crystals.

(2) Twinned magnesian pigeonite (5 crystals), has almost the same appearance as the orthopyroxenes, and invariably is twinned on (100) or $[001]_{180}$. The intensities of paired reflections due to twinning vary from grain to grain indicating that the twinning in these pigeonites is not as fine as that of twinned clinoenstatite transformed from protoenstatite. The class *b* reflections ($h + k$: odd) of these pigeonites are sharp. Exsolution of augites on both (100) and (001) are observed on films exposed for longer times. The reflection intensities of exsolved augites in a pigeonite twin are proportional to the intensities of each twin individual of the host (pigeonite).

Table 2. Atomic coordinates and equivalent isotropic temperature factor $B(\text{Å}^2)$, derived from anisotropic temperature factors of the 14310 bronzite.*

Atom	Atomic coordinates			B (equiv.)
	x	y	z	
M1	0.37556(6)	0.6544(1)	0.8698(2)	0.75(3)
M2	0.37825(4)	0.4820(1)	0.3643(2)	1.01(2)
SiA	0.27127(5)	0.3412(1)	0.0482(2)	0.62(2)
SiB	0.47400(5)	0.3375(1)	0.7981(2)	0.63(2)
O1A	0.1833(1)	0.3389(3)	0.0390(5)	0.69(4)
O1B	0.5628(1)	0.3376(3)	0.7992(5)	0.76(4)
O2A	0.3105(1)	0.5011(3)	0.0478(5)	0.79(4)
O2B	0.4339(1)	0.4856(3)	0.6966(6)	0.98(4)
O3A	0.3021(1)	0.2303(3)	−0.1773(5)	0.92(4)
O3B	0.4472(1)	0.2001(3)	0.5977(5)	0.98(4)

*The standard deviations given in parentheses refer to the estimated error in the last digit.

The β angle of pigeonite (Table 1) is the same as for a fully exsolved one, e.g. 12021 (Papike *et al.*, 1971), and the β angles of exsolved augites, ($\beta = 106°23'$ for (100) augite, and $\beta = 106°12'$ for (001) augite) are not as large as those of rapidly cooled pyroxenes, e.g. 12052 (Takeda and Reid, 1971). One of the twinned pigeonites ($Wo_8En_{71}Fs_{20}$) shares a common (100) with an orthopyroxene core ($Wo_4En_{80}Fs_{16}$). Another pigeonite, without orthopyroxene, is more iron-rich ($Wo_8En_{58}Fs_{34}$).

(3) Colored intermediate pigeonites (3 crystals, SB1–3) are pink to brownish. All of the grains that we studied are untwinned and more Fe-rich than the twinned pigeonites. The diffraction patterns show predominant exsolution of augite on (001). The intensity ratio, pig/aug is roughly 3/1. The lamellae or patches of augite ($Wo_{31}En_{32}Fs_{37}$) barely are detectable by the electron microprobe in the matrices of pigeonites ($Wo_7En_{35}Fs_{58}$). The separations of the a^* axes of pigeonite and augite are close to the maximum values observed for Apollo 12 clinopyroxenes (Papike *et al.*, 1971). The class b reflections of pigeonite are sharp. Very weak diffuse streaks along a^* in some parts of the $h02$ rows were observed on films exposed for longer periods. One augite grain ($Wo_{27}En_{44}Fs_{27}$) has exsolved an amount of pigeonite equal to the host augite.

In many photographs of the orthopyroxene and magnesian pigeonites, weak diffraction patterns of oxide phases (spinel and possibly ilmenite with certain common crystallographic orientation, Gay *et al.*, 1972; Ross, personal comm., 1972) are observed.

The site populations and mean M–O bond distances for the 12033 and 14310 orthopyroxenes are given in Table 3. They are compared with those of other orthopyroxenes determined by the x-ray method, and also those of the 14310 orthopyroxene determined by the Mössbauer method (Finger *et al.*, 1972). The (Si, Al)–O distances, together with those of other bronzites (Takeda, 1972b) are given in Table 4.

Discussion

A characteristic of the 14310 pyroxenes is the presence of orthopyroxene overgrown by and sharing (100) with the twinned magnesian pigeonite, as first reported by Ridley *et al.* (1971). The orthopyroxene is more Mg-rich and more Ca-poor than the pigeonite. This pigeonite coexisting with orthopyroxene is, to our knowledge, the most Mg-rich ever reported (Kushiro and Ross, private communication).

The possibility that this Mg-rich twinned pigeonite is twinned clinoenstatite inverted from protoenstatite is unlikely, because the pigeonite contains more Fe and Ca than the bronzite, and the twin lamellae are coarse enough to be observed optically. In addition, the diffuse streaks along a^*, commonly found in clinoenstatite inverted from protoenstatite (Sadanaga *et al.*, 1969), were not observed in our Mg-rich pigeonite, and the exsolved augite found in 14310 Mg-rich pigeonite has not yet been found in clinoenstatite. The possibility that the core bronzite is inverted pigeonite is excluded by the compositional relation of the bronzite to the twinned pigeonite and the presence of a little exsolved augite in the orthopyroxene on (100).

The above orthopyroxene-pigeonite characteristics may be explained as follows: Bronzite was the first pyroxene to crystallize from the 14310 magma. When the

Table 3. Site occupancy factors and the intracrystalline cation distribution coefficients, defined by $k = (Fe/Mg)_{M1}/(Fe/Mg)_{M2}$, and the mean M1–O and M2–O distances of the bronzites from lunar sample 12033, lunar rock 14310 and Takasima, Japan.

Octahedral site	12033,97 bronzite	14310,90 bronzite	Takasima bronzite
M1 occupancy*			
Fe	0.07	0.064	0.03
Mg	0.91	0.89	0.86
Ti, Al, Cr, Mn	0.02	0.05	0.21†
M1–O length, Å	2.097	2.083	2.070
M2 occupancy*			
Fe	0.50	0.39	0.29
Mg	0.44	0.52	0.64
Ca	0.06	0.09	0.07
M2–O length, Å	2.204	2.200	2.176
k	0.07	0.10	0.08

*The errors in site occupancy are approximately 0.01.
†0.07 Fe^{3+} is included.

Table 4. Si–O bond lengths in Å of the 14310 bronzite and the Takasima bronzite*
(Takeda, 1972b)

	SiA Chain 14310	Takasima		SiB Chain 14310	Takasima
SiA–O1A	1.611(3)†	1.613(1)	SiB–O1B	1.626(3)	1.630(1)
SiA–O2A	1.590(3)	1.595(1)	SiB–O2B	1.595(3)	1.600(1)
SiA–O3A	1.634(3)	1.642(1)	SiB–O3B	1.671(3)	1.671(2)
SiA–O3A′	1.664(3)	1.658(2)	SiB–O3B′	1.679(3)	1.680(1)
Mean	1.622	1.627	Mean	1.639	1.645

*7.5% Al in the tetrahedral sites.
†The standard deviation given in parentheses refer to the estimated error in the last digit.

magma reached the ortho- to high clino-pyroxene inversion temperature (ca 1200°C in accordance with known composition-temperature relations of pyroxenes, for example, Kuno, 1966) pigeonite crystallization began and continued until final solidification. The x-ray pattern for the twinned pigeonite sharing (100) with ortho-pyroxene is similar to that for an orthopyroxene heated in the stability field of high pigeonite (Ross et al., private communication). Monoclinic pigeonite growing on orthopyroxene or growing in orthopyroxene by inversion may have one of two orientations, thus producing the twinning. A similar mechanism of twinning may be postulated when the pyroxene crystallization path crosses the orthopyroxene-high pigeonite inversion boundary. Another explanation is that the cell dimensions of high-pigeonite within the above composition and temperature range may give lower "twin-obliquity".

The exsolution of augite on (100) of magnesian pigeonite in rock 14310 is like the common mode of exsolution in Apollo 12 magnesian pigeonites (Papike et al., 1971). Such exsolution was not observed in more Fe-rich pigeonites that occur with

the 14310 magnesian pigeonites. Some (100) x-ray patterns do not represent exsolution, but overgrowth of augite on pigeonite (Takeda and Reid, 1971; Brett *et al.*,
1971). These observations do not support the explanations proposed by Morimoto
and Tokonami (1969) for the formation of augite in pigeonite with common (100).

The diffuse streaks observed in x-ray patterns of the intermediate pigeonites
have not been reported in the patterns of terrestrial pyroxenes or of those from mare
basalts. Similar diffuse streaks have been found in the x-ray patterns of pigeonites,
with or without coexisting secondary orthopyroxene, from some eucrites, and from
the eucritic portions of some mesosiderites (Takeda and Reid, 1972). The intensities
of the streaks are stronger where the strong reflections of orthopyroxene are expected
in reciprocal space, e.g., between 102 and 202, and between $\bar{2}02$ and $\bar{3}02$, and thus
indicate the presence of stacking faults. If the coexisting secondary orthopyroxene
in some eucrites is an inverted pigeonite, as suggested for the Moore County
pyroxenes (Hess and Henderson, 1948), the pigeonite with diffuse streaks may be in
an intermediate inversion stage in which orthopyroxene-like sequences occur at
random intervals. Similarities between Fe-rich 14310 pigeonites and eucritic
pigeonites suggest similar thermal histories of the eucrite source rock and parts of
the Fra Mauro formation.

The cation distribution coefficient, $k = (Fe/Mg)_{M1}/(Fe/Mg)_{M2}$, of the 14310
orthopyroxene (Table 3) is 0.10, as compared with 0.097 obtained by the Mössbauer
method (Finger *et al.*, 1972) for powdered orthopyroxene separated from rock
14310,116. Considering that the Mössbauer data refer to many grains of different
chemical compositions and our data to a single crystal, the agreement is good. The
values are similar to those for orthopyroxenes from volcanic rocks (Virgo and
Hafner, 1969). The distribution coefficient of the 12033 orthopyroxene is 0.07,
which is lower than expected for an orthopyroxene from a glass-matrix breccia
(12033,97–2B). The cation distribution appears to have been reequilibrated on or in
a hot ejecta blanket, because the cation distribution would be more disordered if the
fragment cooled quickly following impact.

The exsolution patterns give more information on the cooling history at high
temperatures than the cation distribution (Takeda, 1972a). The fully exsolved
pigeonites and augites of rock 14310 indicate that it cooled more slowly than many
mare rocks such as 12052 (Brett *et al.*, 1971), although zoning within single grains
indicates that crystallization was much more rapid than that of plutonic or intrusive
rocks. The presence of discrete grains of orthopyroxene, magnesian pigeonite,
augite, and intermediate pigeonite with detectable lamellae of augite indicates a
much smaller degree of metastable crystallization than for many mare pyroxenes.
The quite different liquidus paths of 14310 and mare basalts probably account for
some of the differences in the pyroxene crystallization (Ridley *et al.*, 1972).

Some 14310 orthopyroxenes are aluminous (Ridley *et al.*, 1971), but the orthopyroxene used for the structure refinements (1.7% Al_2O_3) is not. In an aluminan
orthopyroxene structure of high pressure origin from Takasima, Japan (Takeda,
1972b), tetrahedral aluminum is concentrated in the Si(B) tetrahedra. The M1
octahedron of the Takasima bronzite, due to its Al and Fe^{3+} content, is smaller than
the octahedron with the same Fe/Mg ratio, but we could find no evidence for such

a small octahedron in the 14310 orthopyroxene (Table 3). The Si–O distances (Table 4) suggest the concentration of Al in the Si(B) tetrahedra.

Conclusions

(1) The crystallographic features of 14310 (Fra Mauro) pyroxenes, such as twinned magnesian pigeonite overgrown on bronzite with common (100), are distinct from those of mare pyroxenes and suggest that such features may be useful in characterizing lunar rocks and their crystallization trends. (2) The mode of exsolution and the chemical trends of the 14310 pyroxenes indicate that crystallization was much more rapid than that of plutonic or intrusive rocks but was not as metastable as that of many mare rocks. (3) Similarities between Fe-rich 14310 pigeonites and eucritic pigeonites, such as diffuse streaks along a^* and exsolution patterns, suggest similar thermal histories for these pyroxenes. (4) The cation distribution and crystal structure of the Fra Mauro orthopyroxenes are similar to those of terrestrial orthopyroxenes of volcanic origin.

Acknowledgments—We thank Dr. Robin Brett (NASA–MSC) for the lunar samples, Dr. Ikuo Kushiro (Geophysical Lab.), Dr. Malcolm Ross (USGS) and Prof. S. S. Hafner (Univ. of Chicago) for discussion, Dr. Arch M. Reid (NASA–MSC) for some of the microprobe analyses, and Mr. Grover Moreland (Smithsonian Institution) for the microprobe sections of the single crystals. The authors were supported by National Research Council Resident Research Associateships.

References

Apollo Soil Survey (1971) Apollo 14: Nature and origin of rock types in soil from the Fra Mauro formation. *Earth Planet. Sci. Lett.* **12**, 49–54.

Brett R. Butler P. Jr. Meyer C. Jr. Reid A. M. Takeda H. and Williams R. (1971) Apollo 12 igneous rocks 12004, 12008, 12009, and 12022: A mineralogical and petrological study. *Proc. Second Lunar Sci. Conf., Geochim. Cosmochim. Acta* Suppl. 2, Vol. 1, pp. 301–317. MIT Press.

Finger L. W. (1969) Determination of cation distribution by least-squares refinements of single crystal x-ray data. *Carnegie Inst. Wash. Yearb.* **67**, 216–217.

Finger L. W. Hafner S. S. Schürmann K. Virgo D. and Warburton D. (1972) Distinct cooling histories and reheating of Apollo 14 rocks (abstract). In *Lunar Science—III* (editor C. Watkins), p. 259, Lunar Science Institute Contr. No. 88.

Fuchs L. (1970) Orthopyroxene-plagioclase fragments in the lunar soil from Apollo 12. *Science* **169**, 866–867.

Gay P. and Bown M. G. (1972) Topotactic reactions in some lunar pyroxenes and olivines (abstract). In *Lunar Science—III* (editor C. Watkins), p. 291, Lunar Science Institute Contr. No. 88.

Hess H. H. and Henderson E. P. (1948) The Moore County meteorite: A further study with comment on its primordial environment. *Amer. Mineral.* **33**, 494–507.

Kuno H. (1966) Review of pyroxene relations in terrestrial rocks in the light of recent experimental works. *Mineral. Jour.* **5**, 21–43.

Meyer C. Jr. Brett R. Hubbard N. J. Morrison D. A. McKay D. S. Aitken F. K. Takeda H. and Schonfeld E. (1971) Mineralogy, chemistry, and origin of the KREEP component in soil samples from the Ocean of Storms. *Proc. Second Lunar Sci. Conf., Geochim. Cosmochim. Acta* Suppl. 2, Vol. 1, pp. 393–411.

Morimoto N. and Tokonami M. (1969) Oriented exsolution of augite in pigeonite. *Amer. Mineral.* **54**, 1101–1117.

Papike J. J. Bence A. E. Brown G. E. Prewitt C. T. and Wu C. H. (1971) Apollo 12 clinopyroxenes: Exsolution and epitaxy. *Earth Planet. Sci. Lett.* **10**, 307–315.

Papike J. J. and Bence A. E. (1971) Apollo 14 pyroxenes: Subsolidus relations and implied thermal histories. *Geol. Soc. Am. Abstr.*, 1971 Annual Meeting, pp. 666–667.

Pollack S. S. (1968) Disordered pyroxenes in chondrites. *Geochim. Cosmochim. Acta* **32**, 1209–1217.

Ridley W. I. Williams R. J. Takeda H. Brown R. W. and Brett R. (1971) Petrology of Fra Mauro basalt 14310. *Geol. Soc. Am. Abstr.*, 1971 Annual Meeting, pp. 682–683.

Ridley W. I. Williams R. J. Brett R. Takeda H. and Brown R. W. (1972) Petrology of lunar basalt 14310 (abstract). In *Lunar Science—III* (editor C. Watkins), p. 648, Lunar Science Institute Contr. No. 88.

Sadanaga R. Okamura F. P. and Takeda H. (1969) X-ray study of the phase transformations of enstatite. *Mineral. J.* **6**, 110–130.

Takeda H. (1971) Silicon-aluminum substitution in some aluminan orthopyroxenes and micas. *Amer. Crystallogr. Ass. Prog. Abstr.*, Columbia, S.C., p. 45.

Takeda H. (1972a) Structural studies of rim augite and core pigeonite from lunar rock 12052. *Earth Planet. Sci. Lett.* **14**, No. 3.

Takeda H. (1972b) Crystallographic studies of coexisting aluminan orthopyroxene and augite of high-pressure origin. *J. Geophys. Res.* (in press).

Takeda H. and Reid A. M. (1971) Euhedral clinopyroxenes in a vug from an Apollo 12 rock. *Trans. Amer. Geophys. Union* **52**, 271.

Takeda H. and Reid A. M. (1972) Crystallography and chemical trends of pigeonites in some basaltic achondrites (abstract). *Trans. Amer. Geophys. Union* **53**, 437.

Virgo D. and Hafner S. S. (1969) $Fe^{2+}-Mg^{2+}$ order-disorder in heated orthopyroxenes. *Mineral. Soc. Amer. Spec. Pap.* **2**, 67–81.

Proceedings of the Third Lunar Science Conference
(Supplement 3, *Geochimica et Cosmochimica Acta*)
Vol. 1, pp. 431–469
The M.I.T. Press, 1972

Pyroxenes as recorders of lunar basalt petrogenesis: Chemical trends due to crystal-liquid interaction

A. E. Bence and J. J. Papike

Department of Earth and Space Sciences
State University of New York
Stony Brook, New York 11790

Abstract—Pyroxenes from basalts collected on the Apollo 11, 12, 14, 15, and Luna 16 missions have experienced a diverse range of crystallization histories as indicated by their chemical, crystallographic, morphological, and paragenetic relationships. Although the final stages of lunar basalt crystallization appear to be rapid near-surface events, the initial stages vary considerably among the different basalt types. Differences in basalt bulk rock compositions, emplacement histories, and intensive parameters (T, P, f_{O_2}) are recorded in the paragenetic sequence of the basalt crystallization, and the pyroxene crystallization trends on the quadrilateral, Ti/Al plots, Ti–Cr–Al^{VI} plots, and Al/Si versus Fe/Mg plots.

Four broad types of basalts can be differentiated on the basis of their pyroxene chemical trends and textures. Basalts with high TiO_2/Al_2O_3 ratios, relatively early crystallization of plagioclase, one-stage, near-surface crystallization, and $f_{O_2} \simeq 10^{-13}$ atmospheres at 1000°C (Apollo 11, Luna 16, 12022) have pyroxenes with Ti/Al (atomic) $\simeq \frac{1}{2}$, a continuous crystallization trend on the pyroxene quadrilateral, compositions on or near the Ti–Cr join in the ternary plot Ti–Cr–Al^{VI}, and Al/Si versus Fe/Mg trends showing a sharp decrease in Al/Si at low Fe/Mg followed by a more gradual decrease at higher Fe/Mg ratios. Pyroxenes from basalts with lower TiO_2/Al_2O_3 ratios, delayed plagioclase crystallization, and a two-stage history show sharp breaks in the crystallization trends on the quadrilateral, and on the Ti/Al, Ti–Cr–Al^{VI}, Al/Si versus Fe/Mg diagrams when plagioclase starts to crystallize. Basalts of essentially the same composition but low f_{O_2} ($\simeq 10^{-16}$ atm at 1000°C) show a marked difference on the Ti/Al plot with late-stage pyroxenes approaching Ti/Al = 1, which we interpret as addition of component $R^{2+}Ti^{3+}SiAlO_6$. High alumina basalts with early crystallization of plagioclase and a one-stage near-surface history show significant Al^{VI} enrichment in the early pyroxenes (orthopyroxenes) and Ti/Al ratios approaching $\frac{1}{2}$ in the late stages (clinopyroxenes).

Introductory Statement

THE AMERICAN APOLLO missions and the Russian Luna 16 unmanned probe have returned a diversity of basalt types exhibiting a wide range of bulk compositions, textures, and mineral parageneses, suggesting that they have experienced different crystallization histories. Although the final stages of lunar basalt crystallization appear to be rapid, near-surface events, the initial stages appear to vary considerably among the different basalt types. Differences in basalt bulk-rock compositions, emplacement histories, and intensive parameters (P, T, f_{O_2}) are recorded in their mineralogical variations. Sufficient data are now available to permit comparison of the chemical and mineralogical differences among the basalt types and to make some statements concerning the details of their crystallization histories.

Detailed investigations correlating chemical, textural, and crystallographic relationships of the basaltic pyroxenes reveal that these minerals are sensitive to variants in their cooling histories (Bence *et al.*, 1970, 1971; Bence and Papike, 1972; Boyd and Smith, 1971; Hollister *et al.*, 1971; Papike *et al.*, 1971; Champness and Lorimer, 1971). In particular, the pyroxenes appear most sensitive to the position of plagioclase in the paragenetic sequence (Bence *et al.*, 1970; Ross *et al.*, 1970; Bence *et al.*, 1971; Hollister *et al.*, 1971). Furthermore, pyroxene textural relationships for all of the lunar basalts indicate that they grew throughout much of the crystallization history of the basaltic magmas and, consequently, should record changes in mineralogy, chemistry, and intensive thermodynamic properties (T, P, f_{O_2}) that took place around them. If compositional and textural variants in the pyroxenes can be correlated with variations in basalt crystallization, then it should be possible to examine basalt fragments and even pyroxene single-crystal fragments from the soil or breccias and infer the crystallization histories of the rocks from which they were derived. It should be remembered, however, that most of the crystallization trends observed in the lunar pyroxenes are metastable—resulting from extremely rapid late-stage cooling histories in an anhydrous environment. Nevertheless, these metastable trends can be correlated with differences in bulk chemistry, emplacement histories, and intensive parameters.

Analytical Procedures

Pyroxene compositional variations were determined by electron microprobe step scans. The data were obtained on a four-spectrometer ARL–EMX–SM electron microprobe and were reduced following the modified techniques of Bence and Albee (1968). Calcium-poor pyroxenes from 14310 were separated from the rock and x-rayed by single-crystal precession techniques to verify whether or not they were ortho- or clinopyroxenes. These crystals were then analyzed to correlate compositions with the x-ray diffraction data. Compositional trends for the composite phenocrysts were subsequently obtained from polished thin sections.

The following samples have been studied in this investigation: 12022,30,113; Luna 16; 12052,5,68; 12021,20,132; 15016,9; 15555,33,250; 14310,8,12,101; 14053, 16,20,34; 15058,15; 15499,6; 14276,12.

Basalt Classification

To demonstrate the validity of utilizing pyroxene crystallization trends as petrogenetic recorders, we have subdivided the lunar basalts into seven types (Table 1) by correlating textures, mineralogy, bulk compositions, and paragenetic sequences.

I. Fine-grained intersertal to subophitic mare basalts are the predominant basalts at the Apollo 11 landing site (see James and Jackson, 1970). In general, they contain abundant calcic pyroxene, calcic plagioclase, and ilmenite, lesser olivine,

Table 1. Basalt classification followed in text.

Type	
I	Apollo 11, Luna 16
II	Olivine cumulate 12022
III	Pyroxene porphyritic 12052, 12021, 15499, 15058
IV	Vesicular 15016
V	Poikilitic plagioclase 15555
VI	14053
VII	Aluminous basalt 14310, 14276

and minor interstitial glass, SiO_2, and pyroxferroite. Primary pigeonite is rare, although its presence has been reported in several rocks (Kushiro and Nakamura, 1970; Weill et al., 1970; Bence and Papike, 1971). Similar (but with significantly less ilmenite) mare basalts were found in the soil at the Luna 16 landing site in Mare Fecunditatis (Albee et al., 1972; Bence et al., 1972; Grieve et al., 1972; Jakes et al., 1972; Reid et al., 1972). Chemically, the Apollo 11 and Luna 16 basalts are characterized by high TiO_2 and low SiO_2 and MgO (Table 2). The paragenetic sequence of these basalts is, with few exceptions, Fe–Ti opaque oxides + olivine → olivine + augite → augite + plagioclase → pyroxferroite + plagioclase + glass + SiO_2. Silicate parageneses are shown schematically in Fig. 1.

II. Oceanus Procellarum basalts (Apollo 12) include additional textural types— one of them, pyroxene porphyritic, is also common to the Apollo 15 landing site at the Apennine Front. These basalts contain complex composite (commonly skeletal) clinopyroxene phenocrysts with pigeonite cores and augite rims in a groundmass that ranges from aphanitic (15497) to variolitic (15499, 12052, 12021) to subophitic (15058). A portion of this textural range is illustrated in Fig. 2.

Chemical analyses of these porphyritic basalts (Table 2) reveal that they are all very similar and are characterized by uniformly high SiO_2 (47–48 wt%), moderate Al_2O_3 (9–11 wt%), uniform FeO, CaO, and MgO, and variable, but low, TiO_2 (3.7–1.7 wt%). Ratios of titania to alumina vary from 0.34 to 0.19. The silicate paragenetic sequence (Fig. 1) in all of these basalts is quite similar with the only major variation being the presence or absence of olivine. The general sequence is: ± olivine → ± olivine + pigeonite → pigeonite + augite → pigeonite + augite + plagioclase. Pyroxferroite and either tridymite or cristobalite are late-stage minerals. The pyroxene subsolidus relations for several basalts from this textural group are discussed by Papike et al. (1971).

III. An olivine porphyry (12022), believed to be a cumulate (Brett et al., 1971), was found at the Apollo 12 landing site. This basalt contains large (2–3 mm) anhedral olivine phenocrysts set in a matrix of fine-grained olivine, calcic clinopyroxene,

Table 2. Chemical analyses (wt%) and TiO_2/Al_2O_3 weight ratios of lunar basalts discussed in text.

	10057 Engel et al. (1971)	12022,56 Engel et al. (1971)	Luna 16 Vino-gradov (1971)	12021,51 Engel et al. (1971)	12052 Maxwell & Wiik (1971)	15016 LSPET (1972)	15555 LSPET (1972)	15058 LSPET (1972)	15499 LSPET (1972)	14053 M. Rhodes, Pers. comm.	14310 LSPET (1971)
SiO_2	39.79	43.20	43.8	47.05	46.6	43.97	44.24	47.81	47.62	46.3	47.19
Al_2O_3	10.84	9.04	13.65	10.97	10.24	8.43	8.48	8.87	9.27	13.60	20.14
FeO	19.35	21.44	19.35	19.04	19.82	22.58	22.47	19.97	20.26	16.80	8.38
CaO	10.08	9.56	10.40	11.34	10.70	9.40	9.45	10.32	10.40	11.15	12.29
MgO	7.65	10.43	7.05	7.08	8.14	11.14	11.19	9.01	8.94	8.72	7.87
TiO_2	11.44	5.16	4.90	3.74	3.30	2.31	2.26	1.77	1.81	2.64	1.24
Zr	520 ppm	180 ppm	0.04	180 ppm	150 ppm	95 ppm	78 ppm	98 ppm	112 ppm	—	847 ppm
Cr_2O_3	2400 ppm	3800 ppm	0.28	2400 ppm	0.54	—	0.70	—	—	—	0.18
MnO	0.20	0.25	0.20	0.25	0.26	0.33	0.29	0.28	0.28	—	0.11
Na_2O	0.54	0.47	0.33	0.29	0.27	0.21	0.24	0.28	0.29	—	0.63
K_2O	0.32	0.07	0.15	0.08	0.067	0.03	0.03	0.03	0.06	0.12	0.49
S	—	—	0.17	—	—	0.07	0.05	0.07	0.07	0.14	0.02
P_2O_5	0.17	0.13	511 ppm	0.09	0.083	0.07	0.06	0.08	0.08	0.09	0.34
$\dfrac{TiO_2}{Al_2O_3}$	1.055	0.571	0.359	0.341	0.322	0.274	0.267	0.1995	0.1953	0.193	0.062

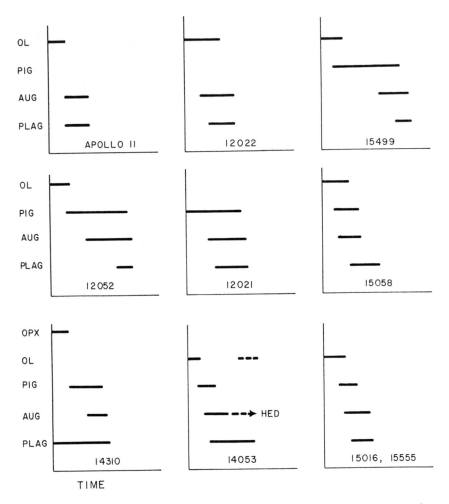

Fig. 1. Schematic representation of (major mineral) silicate paragenetic sequences for
basalts discussed in text.

abundant oriented ilmenite plates, and plagioclase (Fig. 3). Primary pigeonite is
absent. The chemical analysis of this rock (Table 2) reveals that it is comparably
low in SiO_2 as the Apollo 11 and Luna 16 basalts, and high in FeO and MgO.
Titania is intermediate to the TiO_2 values of Apollo 11 and Luna 16. The paragenetic
sequence of this basalt is: olivine → olivine + augite → augite + plagioclase →
augite + plagioclase + ilmenite (Fig. 1).

IV. A fourth textural type is represented by Apollo 15 mare basalt 15016 (Fig.
4). It is a highly vesicular equigranular basalt containing abundant weakly-zoned
composite clinopyroxenes (augite more abundant than pigeonite) and olivine, lesser
plagioclase, and ilmenite plates which line the vesicles. Its texture is intergranular
and its silicate paragenetic sequence (Fig. 1) is olivine → olivine + pigeonite →

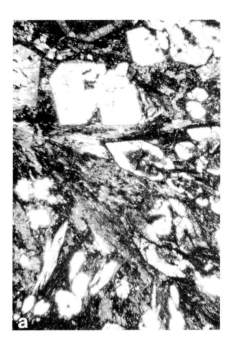

Fig. 2. (a) Photomicrograph of 15499 (X nicols). Field of view is 3.8 × 2 mm. (b) Photomicrograph of 15058 (X nicols). Field of view is 3.8 × 2 mm.

pigeonite + augite → pigeonite + augite + plagioclase. This basalt is characterized
by intermediate silica, low Al_2O_3, and high FeO (Table 2). Its TiO_2/Al_2O_3 ratio is
slightly lower than the pyroxene porphyritic basalts.

V. A fifth textural type is represented by basalt 15555 that is in some ways similar
to the pyroxene porphyritic basalts. It contains early composite clinopyroxene
crystals and olivine phenocrysts; however, the late-stage minerals are dominated
by large poikilitic plagioclases enclosing relatively small euhedral olivines and
augites (Fig. 5). Chemically, this rock is identical to 15016. Its silicate paragenetic
sequence is olivine → olivine + pigeonite → pigeonite + augite → pigeonite +
augite + plagioclase (Fig. 1).

VI. Additional textural types are observed in the two large basalts returned by
the Apollo 14 mission from the Fra Mauro formation. Both appear to have been
clasts in breccias. Rock 14053 (Fig. 6) is a medium-grained plagioclase-rich basalt
containing strongly zoned, anhedral, composite clinopyroxenes. The boundary
between the core-pigeonites and rim-augites in these composite crystals is extremely
diffuse. Late-stage minerals in the interstitial regions include the association
hedenbergite-fayalite-tridymite. These are the stable breakdown products of pyroxferroite as experimentally documented by Lindsley *et al.* (1972). The texture ranges
from subophitic in the vicinity of the pyroxenes to intersertal in the interstitial
areas. The observed breakdown of fayalite to iron metal, silica glass, and tridymite
(El Goresy *et al.*, 1972) indicates that this basalt had the lowest f_{O_2} yet observed.
Studies of subsolidus oxide reduction reactions (Haggerty, 1972) indicate oxygen

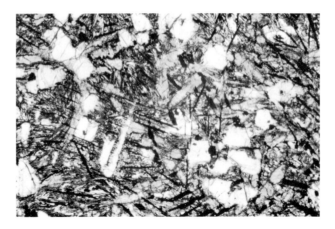

Fig. 3. Photomicrograph (plain light) of basalt 12022. Field of view is 3.8 × 2 mm.

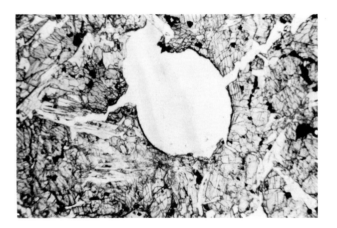

Fig. 4. Photomicrograph (plain light) of basalt 15016. Field of view is 3.8 × 2 mm.

fugacities of the order of 10^{-15} to 10^{-16} atm at 1000°C were attained. Chemically, 14053 is characterized by higher Al_2O_3 and lower FeO than the previous basalt groups (Table 2). Its TiO_2/Al_2O_3 ratio is 0.193. Silicate paragenesis (Fig. 1) is olivine → pigeonite + olivine → pigeonite + augite → pigeonite + augite + plagioclase → hedenbergite + fayalite + SiO_2 + plagioclase.

VII. Nonmare basalts 14310 and 14276 are very high alumina basalts (Table 2) and contain abundant plagioclase (60–70%, LSPET, 1971) occurring in euhedral laths with a wide range in grain size (Fig. 7). This type is extremely fine-grained and contains complex pyroxene phenocrysts with orthopyroxene cores, mantled by pigeonite which, in turn, is mantled by a discontinuous augite rim (Fig. 7). Olivine is absent, and ilmenite has a wide range of concentration. The texture is intersertal. The high Al_2O_3 relative to TiO_2 gives this basalt the lowest TiO_2/Al_2O_3 ratio of all the basalts (Table 2). In addition, it is enriched in CaO and extremely depleted in

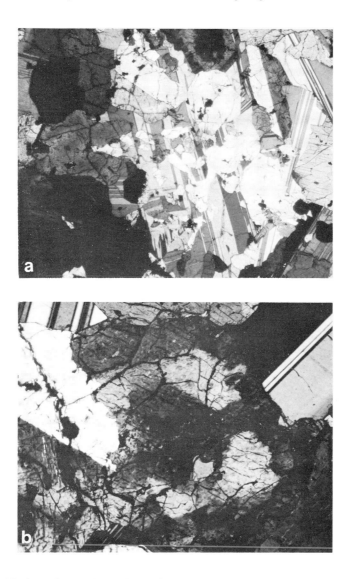

Fig. 5. (a) Photomicrograph (X nicols) of basalt 15555. Field of view is 2.5 × 2 mm. (b) Photomicrograph (X nicols) of composite pyroxene from basalt 15555. Field of view is 0.8 × 0.7 mm.

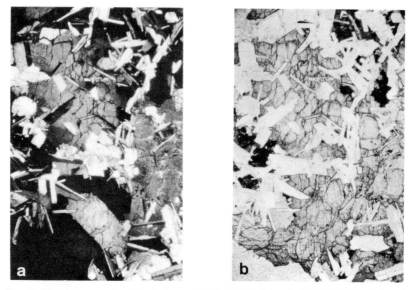

Fig. 6. Photomicrographs (X nicols and plain light) of basalt 14053. Field of view is 3.8 × 2 mm.

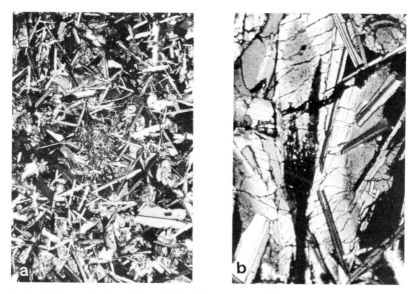

Fig. 7. (a) Photomicrograph (X nicols) of basalt 14310. Field of view is 3.8 × 2 mm. (b) Photomicrograph (X nicols) of composite pyroxene from basalt 14310. Orthopyroxene core is at extinction. Field of view is 1.2 × 0.7 mm.

FeO. Its silicate paragenesis is: plagioclase → orthopyroxene + plagioclase → pigeonite + plagioclase → pigeonite + augite + plagioclase (Fig. 1). Pyroxene subsolidus studies for basalt 14310 have been reported by Takeda and Ridley (1972).

Correlation of Pyroxene Crystallization Trends with Basalt Chemistry and Paragenetic Sequences

Four types of data displays are used for the purpose of correlating the pyroxene crystallization trends with basalt chemistry and paragenetic sequences. These are the familiar pyroxene quadrilateral, titanium-aluminum relationships, ternary plots of the proportions of $R^{2+}TiAl_2O_6$, $R^{2+}CrSiAlO_6$, and $R^{2+}AlSiAlO_6$, and plots of Al/Si versus Fe/Mg (atomic) in the pyroxenes. The order of calculation of the Ti, Cr, and Al components in the pyroxenes is: (1) $R^{2+}TiAl_2O_6$, (2) $R^{2+}Cr^{3+}SiAlO_6$, and (3) $R^{2+}AlSiAlO_6$. Sodium is insignificant and, assuming it is present as $NaCrSi_2O_6$ or $NaAlSi_2O_6$, its omission does not measurably affect calculations for the above components. These calculations assume Ti^{4+} and Cr^{3+}.

Apollo 11 and Luna 16 basalts

Pyroxenes from the Apollo 11 and Luna 16 mare basalts have very similar textural and compositional relationships. They appear to have crystallized, for the most part, in the presence of plagioclase (Ross et al., 1970). The extremely high TiO_2/Al_2O_3 (weight) ratios in the Apollo 11 basalts (Table 2), moderate Al_2O_3 concentrations, the coprecipitation of augite and plagioclase, and f_{O_2} at the time of crystallization all combine to give pyroxenes with Ti/Al (atomic) ratios of $\frac{1}{2}$ (see, for example, Ross et al., 1970; Hargraves et al., 1970; Kushiro and Nakamura, 1970). Similar observations can be made for groundmass clinopyroxenes from other types (see, for example, Bence et al., 1970).

If Ti is quadrivalent, and the component that incorporates Ti in the pyroxene is $R^{2+}TiAl_2O_6$, then the maximum Ti/Al ratio is $\frac{1}{2}$. Therefore, at high TiO_2/Al_2O_3 ratios in the melt the pyroxenes will obtain this limiting value. In view of the very low observed values for f_{O_2} for the Apollo 11 basalts (Brown et al., 1970), it is possible that some of the Ti may be trivalent and, if this is the case, some of the aluminum present in the pyroxenes must be octahedrally coordinated. Trivalent titanium in the Apollo 11 pyroxenes tentatively has been identified by spectral techniques (Burns et al., 1972). However, considering the restricted nature of the Ti/Al ratios, if Ti^{3+} is present, it must be in relatively minor concentrations.

The crystallization trends on the pyroxene quadrilateral (Fig. 27), are continuous, i.e., no discontinuities in Ca, to the late stages. The late-stage assemblages of the unrecrystallized basalts which include interstitial glasses and metastable pyroxferroite, suggest very rapid cooling of the Apollo 11 basalts—presumably at the lunar surface. Anderson et al. (1970), however, observed both pyroxferroite and the assemblage hedenbergite-fayalite-silica in two Apollo 11 microgabbros. We interpret the above observations to indicate that the Apollo 11 basalts crystallized in one near-surface event and that plagioclase and clinopyroxene coprecipitated.

Luna 16 basalts compositionally are quite similar to those from Apollo 11 but have lower TiO_2/Al_2O_3 (weight) ratios and their pyroxenes have a larger range of Ti/Al (atomic) ratios (1/2–1/3). High Al relative to Ti (Ti $<$ 2Al) in the presence of low Cr suggests that the pyroxenes have some octahedral aluminum present in a presumed component $R^{2+}AlSiAlO_6$ (Bence *et al.*, 1972). The presence of octahedral aluminum in these pyroxenes can be explained by lower TiO_2/Al_2O_3 ratios in the basalts and, therefore, the pyroxenes are not saturated with respect to Ti^{4+} and have Ti/Al ratios less than $\frac{1}{2}$. We conclude that these basalts crystallized in one rapid, near-surface event in the same manner as the Apollo 11 basalts and the presence of Al^{VI} is a function of different bulk chemistry.

Olivine cumulate basalt (12022)

Olivine cumulate 12022 has a lower TiO_2/Al_2O_3 (weight) ratio and lower total TiO_2 (Table 2) than the Apollo 11 basalts, although it is still quite high. As a consequence, plagioclase, augite, and ilmenite coprecipitated following the removal of olivine and some early augite. The Ti/Al (atomic) ratios of the first-formed augites (Fig. 8) fall below the $\frac{1}{2}$ line. Ratios less than $\frac{1}{2}$ (assuming all Ti as Ti^{4+}) can be interpreted as a consequence of the presence of the additional components $R^{2+}Cr^{3+}SiAlO_6$ and $R^{2+}AlSiAlO_6$ in the pyroxene structure. However, $R^{2+}Cr^{3+}SiAlO_6$ is in too low concentrations to account for the extra Al observed and Al^{VI} must be present.

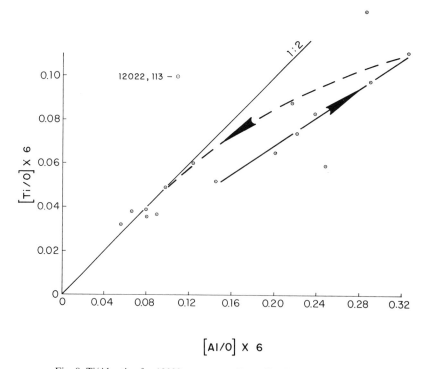

Fig. 8. Ti/Al ratios for 12022 pyroxenes. Crystallization directions as shown.

With crystallization, Ti/Al decreases as Al increases and Fe/Mg remains essentially constant. A rapid increase in Fe/Mg coincides with a sharp break Ti/Al back to $\sim\frac{1}{2}$ with significantly lowered Al concentrations. These breaks probably reflect the near simultaneous crystallization of augite, plagioclase, and ilmenite.

Proportions of Ti, Cr, and Al^{VI}, shown schematically in Fig. 29 show that initially, the pyroxenes have approximately equal proportions of Ti and Cr and very little Al^{VI}. The amounts of octahedral aluminum and chromium decrease with crystallization.

From the textural and chemical observations, we interpret this rock to have crystallized in one near-surface event; however, it is possible that the olivines and the early augites may have started to form at depth.

Poikilitic plagioclase basalt (15555)

Basalt 15555 contains two different pyroxene associations: early, relatively coarse composite pigeonite-augite crystals and, later, fine-grained, euhedral augites enclosed in poikilitic plagioclase. Unlike the porphyritic basalts discussed below, the pigeonite-augite composites do not have hollow cores. In the paragenetic sequence, plagioclase crystallized very late relative to most of the pyroxenes. Representative microprobe analyses taken across the composite phenocrysts and of the augites enclosed within the plagioclases are given in Table 3. There are no compositional differences between the augite rims and augites enclosed within plagioclase.

Titanium-aluminum ratios for the composite pyroxenes (Fig. 9) reveal that the early pigeonites and augites have significant Al^{VI} as indicated by Ti/Al (atomic) ratios of $\frac{1}{4}$. Late-stage pigeonites and augites, including the augites enclosed within plagioclase crystals, have Ti/Al (atomic) ratios approaching $\frac{1}{2}$. A compositional discontinuity in this diagram is abrupt—occurring over a distance of a micron in the pyroxene crystal. A correlation of Al/Si (atomic) versus Fe/Mg (atomic) (Fig. 15a) in these pyroxenes shows that the earliest pyroxenes (low Fe/Mg pigeonites) are depleted in Al relative to Si but that this value increases very quickly to a maximum as augite forms. With a slight increase in Fe/Mg, the Al/Si ratio decreases rapidly. This decrease coincides with the change of Ti/Al from $\frac{1}{4}$ to $\frac{1}{2}$ (Figure 9). With increased differentiation (higher Fe/Mg), the Al/Si ratio remains relatively constant at a low value which is comparable to the initial value. Two trends are observed on this diagram. One which shows the pronounced Al/Si maximum and a discontinuous increase from pigeonite to augite and a second trend corresponding to the profile normal to (010) (Fig. 5), which has been previously shown to be a direction of continued pigeonite growth (Bence et al., 1970, 1971; Boyd and Smith, 1971; Hollister et al., 1971). In the second case, the Al/Si maximum is not nearly so pronounced.

Crystallization trends corresponding to both of these plots can be recognized on the familiar pyroxene quadrilateral (Fig. 10). The low-Ca trend corresponds to the profile normal to (010) along the pigeonite channel and the pigeonite-augite trend corresponds to a profile normal to (110) in sections cut parallel to (001) (Fig. 5). From the size of the triangles, which represent the amounts of components other

Table 3. Representative electron microprobe analyses (wt%) of pyroxenes from basalt 15555.
Columns 1–6: Analyses obtained in direction normal to (010) along low calcium trend. Columns 7–18: Analyses obtained along high calcium (pigeonite → augite) trend. (Normalized to 6 oxygens.)

	1	2	3	4	5	6	7	8	9	10	11	12	13	14	15	16	17	18
SiO$_2$	51.9	51.3	51.0	50.1	48.7	47.0	52.5	51.9	51.7	51.8	51.4	50.7	50.4	51.9	52.0	48.2	48.5	48.1
Al$_2$O$_3$	1.59	1.71	1.89	1.53	0.88	1.16	1.60	1.81	1.53	2.11	2.15	2.83	2.75	1.64	1.29	1.72	1.57	1.49
TiO$_2$	0.47	0.48	0.74	0.77	0.83	0.92	0.61	0.76	0.76	0.57	0.83	0.87	1.11	0.74	0.66	1.35	1.25	1.15
FeO	19.1	20.3	21.4	21.6	31.6	36.6	18.5	18.5	17.4	14.5	13.6	13.1	15.3	20.8	21.0	29.8	30.5	33.5
MgO	20.4	18.9	17.3	16.9	9.15	5.59	20.4	20.0	19.4	18.0	16.8	15.1	14.4	17.4	18.0	6.52	5.54	5.59
CaO	5.40	6.35	6.60	7.31	7.85	7.81	4.68	6.26	7.89	13.0	14.4	14.9	13.5	6.78	5.86	11.3	11.8	9.04
Na$_2$O	0.0	0.0	0.0	0.0	0.0	0.0	0.0	0.0	0.0	0.0	0.0	0.0	0.0	0.0	0.0	0.0	0.0	0.0
Cr$_2$O$_3$	0.68	0.63	0.51	0.48	0.12	0.00	0.81	0.69	0.81	0.86	0.75	0.97	0.93	0.68	0.69	0.29	0.18	0.25
	99.5	99.7	99.4	98.7	99.1	99.1	99.1	99.9	99.5	100.8	99.9	98.5	98.4	99.9	99.5	99.2	99.3	99.1
Si	1.942	1.935	1.939	1.931	1.963	1.948	1.962	1.935	1.936	1.919	1.920	1.922	1.924	1.957	1.965	1.947	1.963	1.965
Al^{IV}	0.058	0.065	0.061	0.069	0.037	0.052	0.038	0.065	0.064	0.081	0.080	0.078	0.076	0.043	0.035	0.053	0.037	0.035
Al^{VI}	0.012	0.011	0.024	0.001	0.005	0.004	0.032	0.014	0.003	0.011	0.015	0.049	0.048	0.030	0.022	0.029	0.038	0.036
Ti	0.013	0.014	0.021	0.022	0.025	0.029	0.017	0.021	0.021	0.016	0.023	0.025	0.032	0.021	0.019	0.041	0.038	0.035
Fe	0.598	0.640	0.680	0.696	1.065	1.269	0.579	0.577	0.546	0.448	0.424	0.417	0.488	0.654	0.665	1.005	1.032	1.145
Mg	1.139	1.064	0.981	0.969	0.550	0.345	1.134	1.111	1.084	0.992	0.937	0.853	0.819	0.977	1.013	0.392	0.334	0.340
Ca	0.217	0.257	0.269	0.302	0.339	0.347	0.187	0.250	0.316	0.514	0.578	0.604	0.554	0.274	0.238	0.489	0.510	0.395
Na	0.0	0.0	0.0	0.0	0.0	0.0	0.0	0.0	0.0	0.0	0.0	0.0	0.0	0.0	0.0	0.0	0.0	0.0
Cr	0.020	0.019	0.015	0.015	0.004	0.00	0.024	0.020	0.024	0.025	0.022	0.029	0.028	0.020	0.021	0.009	0.006	0.008

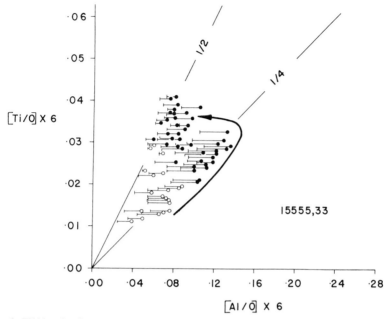

Fig. 9. Ti/Al ratios for 15555 pyroxenes. Bars represent substraction of Al in an assumed component $R^{2+}CrSiAlO_6$. Crystallization trend as shown. Open circles: pigeonites. Solid circles: augites.

than Wo-En-Fs for point analyses, it can be seen that the amounts of these other components (primarily $R^{2+}Ti^{4+}Al_2O_6$, $R^{2+}Cr^{3+}SiAlO_6$, and $R^{2+}AlSiAlO_6$) remain essentially constant with crystallization.

Correlations of the pyroxene crystallization trends in the quadrilateral, the Ti/Al relationships, and Al/Si versus Fe/Mg relationships indicate that the Ti/Al (atomic) break from $\frac{1}{4}$ to $\frac{1}{2}$, the Al/Si drop from the maximum and breaks in the crystallization trends on the quadrilateral (particularly pronounced in the augite trend) all occur at the same point in the crystal.

Plots resulting from calculations of the proportions of $R^{2+}Ti^{4+}Al_2O_6$, $R^{2+}Cr^{3+}SiAlO_6$, and $R^{2+}AlSiAlO_6$ for compositions across the zoned composite crystals are shown in Fig. 11 with the crystallization trend as indicated. Note the buildup of Al^{VI} in the early stages followed by an abrupt turnaround to lower Al^{VI} and Cr. This discontinuity in the trend coincides exactly with the observed breaks discussed above.

Our interpretation of these crystallization trends is: (1) crystallization of pigeonite followed by pigeonite + augite in a melt having a high plagioclase activity following the removal of olivine from the melt; (2) simultaneous crystallization of clinopyroxene and plagioclase.

Vesicular basalt (15016)

This basalt has an identical bulk composition to 15555 and a similar paragenetic

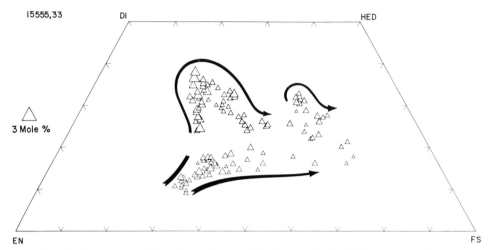

Fig. 10. Pyroxene quadrilateral showing crystallization trends for 15555 pyroxenes. Size of the triangle is proportional to the abundance of components other than $CaSiO_3$–$MgSiO_3$–$FeSiO_3$.

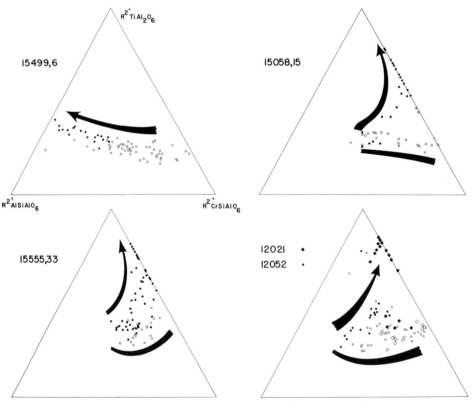

Fig. 11. Ti–Cr–Al^{VI} plot showing proportions of $R^{2+}TiAl_2O_6$, $R^{2+}CrSiAlO_6$ and $R^{2+}AlSiAlO_6$. Crystallization trends as indicated. Open circles: pigeonites. Solid circles: augites.

sequence; however, the textures are very different and, consequently, differences in crystallization history must be invoked. Composite pyroxenes occur in anhedral grains with the demarcation between pigeonite and augite barely recognizable optically. Microprobe analyses (Table 4) across these composite grains reveal a crystallization trend from high-calcium pigeonite to subcalcic augite (Fig. 12) followed by a sharp reversal to a low Ca-pyroxene at higher Fe/Mg ratios. This sharp reversal correlates with the break in Ti/Al ratios in Fig. 13, with the reversal in Ti–Cr–Al^{VI} relationships in Fig. 14, and with the decrease in Al/Si on the Al/Si–Fe/Mg diagram (Fig. 15a). Relative to 15555 pyroxenes, the crystallization trends for 15016 pyroxenes on the pyroxene quadrilateral are truncated at both the high and low Mg ends. In every respect but texture, 15016 and 15555 are identical. Basalt 15016 evidently cooled very quickly in a low-pressure regime and released gases formed vesicles whereas 15555 cooled relatively slowly in the late stages to produce the observed poikilitic feldspars. It is possible that 15555 and 15016 came from the same flow unit with 15555 representing a deeper portion.

Pyroxene porphyritic basalts (15499, 12052, 12021, 15058)

These basalts have similar bulk chemistry and similar paragenetic sequences. The only differences between them are the nature of the groundmass and the presence or absence of early olivine. Their crystallization trends on the pyroxene quadrilateral are, with minor variations, quite similar.

Skeletal pyroxene phenocrysts from basalt 15499 have two crystallization trends on the pyroxene quadrilateral—a low Ca trend normal to (010) and the pigeonite to augite to Fe-augite trend normal to (110) and (001) (Fig. 16). The total amount of $R^{2+}TiAl_2O_6 + R^{2+}Cr^{3+}SiAlO_6 + R^{2+}AlSiAlO_6$ as indicated by the size of the triangles increases right up to the peripheries of the phenocrysts. On the

Table 4. Representative electron microprobe analyses of pyroxenes (wt%) from basalt 15016. Columns 1–3: Pigeonite trend. Columns 4–6: Early augite trend. Columns 7–10: High iron augite trend. (Normalized to 6 oxygens.)

	1	2	3	4	5	6	7	8	9	10
SiO_2	52.3	51.9	52.1	51.8	50.4	50.6	51.8	50.4	49.9	48.8
Al_2O_3	2.18	2.74	3.30	2.88	2.86	3.55	1.66	1.52	1.66	1.63
TiO_2	0.71	0.97	1.05	0.90	1.15	1.16	0.99	1.00	1.21	1.31
FeO	18.1	17.8	16.0	14.5	12.2	14.4	25.0	25.5	25.8	33.0
MgO	18.4	16.5	16.0	16.3	16.3	14.9	15.5	13.8	11.4	8.68
CaO	7.88	8.68	10.9	12.8	16.5	14.3	5.33	8.39	10.6	7.22
Na_2O	0.0	0.0	0.0	0.0	0.0	0.0	0.0	0.0	0.0	0.0
Cr_2O_3	0.87	0.92	0.95	0.86	1.00	1.05	0.41	0.23	0.22	0.26
	100.4	99.5	100.3	100.0	100.4	100.0	100.7	100.8	100.8	100.9
Si	1.942	1.946	1.936	1.928	1.880	1.898	1.964	1.935	1.932	1.938
Al^{IV}	0.058	0.054	0.064	0.072	0.120	0.102	0.036	0.065	0.068	0.062
Al^{VI}	0.037	0.067	0.080	0.045	0.006	0.055	0.038	0.004	0.008	0.015
Ti	0.020	0.027	0.029	0.025	0.032	0.033	0.028	0.010	0.035	0.039
Fe	0.563	0.560	0.495	0.452	0.381	0.450	0.794	0.820	0.836	1.098
Mg	1.018	0.922	0.884	0.904	0.905	0.831	0.875	0.793	0.663	0.513
Ca	0.314	0.349	0.432	0.510	0.658	0.575	0.217	0.345	0.442	0.307
Na	0.0	0.0	0.0	0.0	0.0	0.0	0.0	0.0	0.0	0.0
Cr	0.026	0.027	0.028	0.025	0.029	0.031	0.012	0.007	0.007	0.008

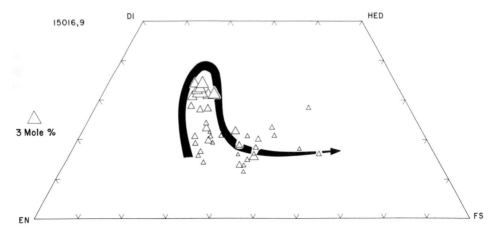

Fig. 12. Crystallization trend for pyroxenes from 15016 on pyroxene quadrilateral. Triangles as for Fig. 10.

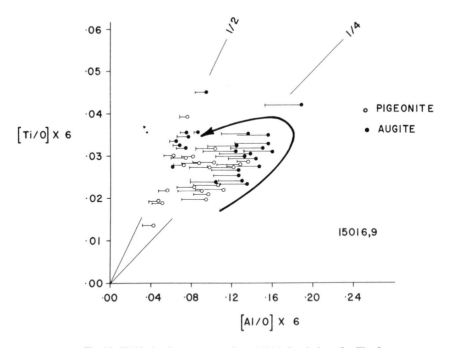

Fig. 13. Ti/Al plot for pyroxenes from 15016. Symbols as for Fig. 9.

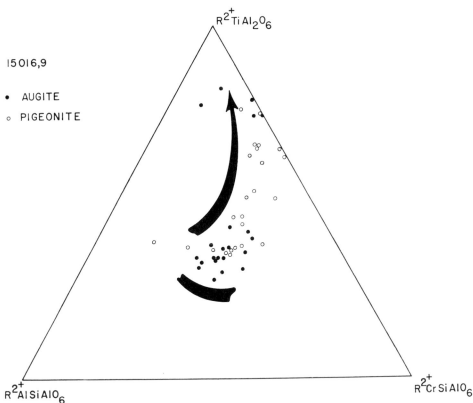

Fig. 14. Ti–Cr–Al^{VI} plot for pyroxenes from 15016. Symbols as for Fig. 11.

Ti–Al plot (Fig. 17), this trend is one of progressively increasing Al at Ti/Al ratios less than $\frac{1}{4}$. The outer rims of these crystals have Al_2O_3 concentrations of nearly 10 weight percent (Table 5). There is no dramatic jump back to Ti/Al ratios of $\frac{1}{2}$ such as has been observed for rim and groundmass pyroxenes from 12052 and 12021 (Bence *et al.*, 1970). When the proportions of Ti, Cr, and Al^{VI} are plotted together (Fig. 11), the trend is one of essentially uniform Ti, but continually increasing Al^{VI} at the expense of Cr. Again it should be remembered that the sum of these components is continually increasing along this trend.

Finally, Al/Si–Fe/Mg plots (Fig. 15a) show progressively increasing Al/Si ratios with increasing Fe/Mg in the pyroxenes. No decrease in Al/Si can be seen right to the outer few microns of the crystal. Textural relationships indicate that plagioclase did not precipitate until the pyroxene phenocrysts had, to all intents and purposes, stopped adding material.

A single skeletal olivine phenocryst in the slide studied is extremely zoned with respect to Fe/Mg and ranges in composition from Fa_{12} to Fa_{67} from core to rim (Fig. 16).

Basalt 15058, which is included in the same textural and compositional group as 15499, 12052, and 12021, represents one extreme of that group. It has a relatively

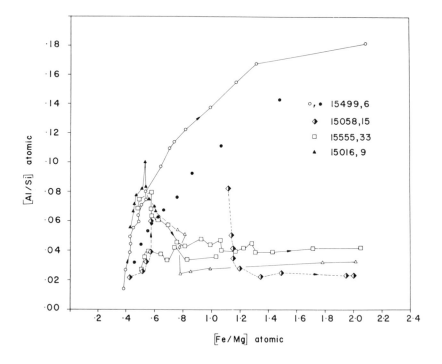

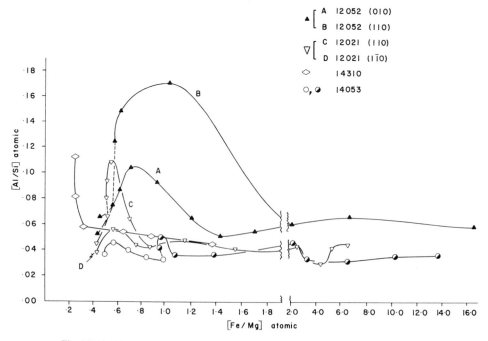

Fig. 15. (A, B) Al/Si versus Fe/Mg crystallization trends for some lunar pyroxenes.

Table 5. Representative electron microprobe analyses of pyroxenes (wt%) from basalt 15499.
Columns 1–11: High calcium (pigeonite-augite) trend. Columns 12–15: Low calcium trend.
(Normalized to 6 oxygens.)

	1	2	3	4	5	6	7	8	9	10	11	12	13	14	15
SiO_2	52.4	52.9	52.1	50.1	50.1	49.4	49.4	49.3	48.5	46.7	43.2	47.8	53.2	52.9	53.2
Al_2O_3	1.89	1.73	1.96	3.55	3.96	4.25	4.35	4.59	5.05	6.67	9.66	4.52	1.75	1.97	1.46
TiO_2	0.36	0.39	0.41	0.88	0.98	1.16	1.30	1.32	1.45	2.01	3.99	0.98	0.43	0.37	0.37
FeO	16.9	17.4	17.5	18.1	18.6	17.5	16.2	16.5	17.4	19.7	20.1	25.0	18.0	16.4	16.5
MgO	23.9	22.6	22.6	18.9	16.5	15.2	14.4	13.1	11.9	9.43	8.58	13.3	21.4	23.1	23.0
CaO	2.62	3.40	3.71	6.51	8.52	10.6	12.2	13.3	14.2	14.3	13.8	6.20	2.96	2.46	2.37
Na_2O	0.03	0.00	0.00	0.00	0.02	0.01	0.05	0.04	0.03	0.02	0.05	0.02	0.00	0.03	0.00
Cr_2O_3	1.04	1.09	1.15	1.18	1.26	1.39	1.45	1.36	1.11	0.71	0.66	0.27	1.08	1.15	1.05
	99.1	99.5	99.4	99.2	99.9	99.5	99.4	99.5	99.6	99.5	100.0	98.1	98.8	98.4	98.0
Si	1.934	1.952	1.932	1.884	1.887	1.872	1.872	1.873	1.854	1.809	1.676	1.884	1.997	1.973	1.995
Al^{IV}	0.076	0.048	0.068	0.116	0.113	0.128	0.128	0.127	0.146	0.191	0.324	0.110	0.003	0.027	0.005
Al^{VI}	0.006	0.027	0.018	0.041	0.063	0.062	0.066	0.079	0.081	0.113	0.118	0.100	0.074	0.060	0.059
Ti	0.010	0.011	0.011	0.025	0.028	0.033	0.037	0.038	0.042	0.059	0.117	0.029	0.012	0.010	0.010
Fe	0.523	0.536	0.541	0.571	0.584	0.555	0.513	0.525	0.557	0.639	0.653	0.824	0.565	0.512	0.518
Mg	1.316	1.244	1.246	1.061	0.923	0.859	0.814	0.742	0.677	0.543	0.496	0.781	1.197	1.284	1.286
Ca	0.104	0.134	0.147	0.262	0.343	0.430	0.495	0.542	0.582	0.594	0.572	0.262	0.119	0.098	0.095
Na	0.002	0.0	0.0	0.0	0.001	0.000	0.004	0.003	0.002	0.002	0.004	0.002	0.0	0.002	0.0
Cr	0.030	0.032	0.034	0.035	0.038	0.042	0.044	0.041	0.034	0.022	0.020	0.008	0.032	0.034	0.031

coarse-grained groundmass exhibiting a subophitic texture surrounding rather poorly defined composite clinopyroxene phenocrysts (Fig. 2). The crystallization trends (Table 6, Fig. 18) of the phenocrysts are very similar to those from 15555. Both the low- and high-calcium trends are observed and the augite trend has a pronounced calcium discontinuity that occurs relatively early. Plots of Ti/Al (atomic) ratios (Fig. 19) show that the early pigeonites have Ti/Al ratios less than $\frac{1}{4}$ and relatively low total Al. Initially, the aluminum concentrations of the 15058 pigeonites are comparable to the early pigeonites from 15499. With crystallization, Al and Ti increase in approximately the same ratio. Shortly after the first appearance of augite, there is a sharp increase in the Ti/Al ratio to about $\frac{1}{2}$, which is due to a decrease in Al while Ti remains essentially constant. Except for the absence of a break between pigeonite and augite and lower total Al in the first augites, the Ti/Al trends for 15058 pyroxenes are, in the early stages, comparable to those of 15499. Ti–Cr–Al^{VI} trends (Fig. 11) reveal, as in 15499, the enrichment of Cr relative to both Ti and Al^{VI} in the early stages, and the increase in Al^{VI} at the expense of Cr with crystallization. A discontinuity in the trend back to the Ti–Cr join at higher Ti/Cr ratios occurs shortly after the appearance of augite.

Apollo 12 basalts 12052 and 12021, which are texturally intermediate between 15499 and 15058, have been described extensively elsewhere (Bence et al., 1970, 1971; Boyd and Smith, 1971; Brett et al., 1971) and only comparisons with the Apollo 15 porphyritic basalts will be made here. Composite pyroxene phenocrysts from both basalts exhibit Fe–Ca–Mg, Ti–Al, Ti–Cr–Al^{VI} and Al/Si–Fe/Mg crystallization trends generally similar to those in 15058 and 15499 (Figs. 27, 28, 11, 15b) and all but 15499 have late-stage assemblages that include pyroxferroite. However, compositional variations can be seen in their late-stage pyroxene crystallization trends.

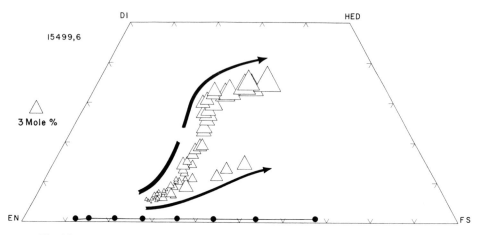

Fig. 16. Pyroxene quadrilateral crystallization trend for pyroxenes from 15499. Triangles as for Fig. 10. Point compositions for a skeletal olivine crystal shown by solid circles. Rim composition at high Fe end.

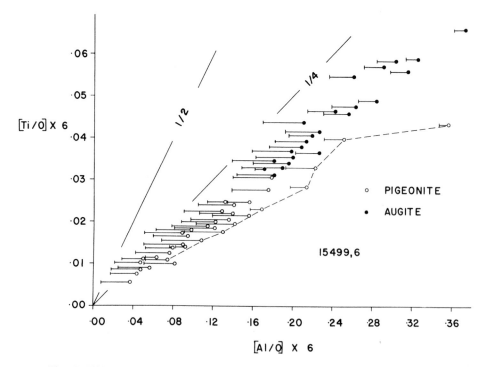

Fig. 17. Ti/Al relationships on pyroxenes from 15499. Symbols as in Fig. 9. Dashed line joins points along profile normal to (010).

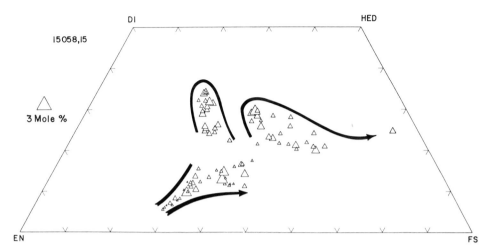

Fig. 18. Crystallization trends on pyroxene quadrilateral for pyroxenes from 15058. Triangles as for Fig. 10.

Crystallization trends on the pyroxene quadrilateral reveal sharp discontinuities with respect to Ca near the inner and outer margins of the skeletal crystals as shown schematically in Fig. 27. These discontinuities range from a few microns to tens of microns in width, have considerable sectoral control (Bence *et al.*, 1971), and their positions within the phenocrysts differ from rock to rock. In 12052, approximately 95% of the phenocryst grew before the Ca variations were incorporated, whereas in 12021 only 60–70% of the phenocrysts as now constituted had crystallized. In 15058, the comparable figure is 50%.

Pyroxene titanium-aluminum relationships for 12052 and 12021 are similar to those in 15499, i.e., considerable Al^{VI} in both the early pigeonites and epitaxial augites (Fig. 28). However, peripheral and groundmass pyroxenes have Ti/Al ratios that increase dramatically from 1/4–1/8 to 1/2 with most of the change being due to a sharp decrease in aluminum. As in the case of the Ca discontinuity in the Fe–Ca–Mg crystallization trends, the breaks in the Ti/Al trends occur at different positions within the phenocrysts of the various basalts. Approximately 50% of each composite crystal in 15058 has Ti/Al ratios of $\sim\frac{1}{2}$, in 12021 the figure is about 25–33%, in 12052 it is less than 10% and in 15499 it is zero. This is the same relative arrangement as noted for the Ca discontinuities.

The differences are carried over on the Ti–Cr–Al^{VI} plots. Basalts 12021 and 12052 (Fig. 11) show Al^{VI} increasing in the early pigeonites and augites at the expense of Cr much the same as 15499 and 15058. As in 15058 the peripheral pyroxenes change direction abruptly on this plot and return to the Ti–Cr join with much higher Ti/Cr ratios.

Plots of Al/Si versus Fe/Mg for the composite phenocrysts from 15499, 15058 (Fig. 15a) and 12052, 12021 (Fig. 15b) show that the early pigeonites and augites are relatively depleted in Al but, as in 15555, the Al/Si (atomic) ratio increases very quickly from the core outward. At higher Fe/Mg ratios, an Al/Si maximum is

Table 6. Representative electron microprobe analyses of pyroxenes (wt%) from basalt 15058.
Columns 1–3: Low calcium trend. Columns 4–10: High calcium trend. (Normalized to 6 oxygens.)

	1	2	3	4	5	6	7	8	9	10	11
SiO_2	52.0	50.9	51.4	52.5	51.7	51.3	50.7	50.4	48.2	51.6	51.5
Al_2O_3	2.01	3.95	1.81	3.07	2.41	2.81	2.07	1.31	1.43	2.47	2.77
TiO_2	0.53	0.58	0.40	0.61	0.65	0.71	0.66	0.84	1.11	0.73	0.63
FeO	23.4	22.1	24.2	18.8	18.1	23.7	23.8	27.0	30.2	15.4	14.5
MgO	15.3	15.6	15.4	17.9	14.6	14.0	13.9	10.4	7.41	14.4	14.2
CaO	5.24	6.50	5.37	6.14	11.1	7.40	7.46	9.80	10.8	14.1	14.7
Na_2O	0.0	0.0	0.0	0.05	0.06	0.02	0.01	0.01	0.03	0.04	0.03
Cr_2O_3	0.86	0.58	0.66	1.18	1.14	0.93	0.74	0.54	0.45	1.05	1.06
	99.3	100.2	99.2	100.3	99.8	100.9	99.3	100.3	99.6	99.8	99.4
Si	1.982	1.921	1.973	1.946	1.953	1.943	1.956	1.969	1.942	1.943	1.941
Al^{IV}	0.018	0.079	0.027	0.054	0.047	0.057	0.054	0.031	0.058	0.057	0.059
Al^{VI}	0.072	0.097	0.055	0.080	0.060	0.068	0.040	0.029	0.010	0.053	0.064
Ti	0.015	0.017	0.012	0.017	0.018	0.020	0.019	0.025	0.034	0.021	0.018
Fe	0.745	0.698	0.778	0.584	0.571	0.752	0.769	0.882	1.017	0.486	0.458
Mg	0.872	0.875	0.879	0.990	0.824	0.790	0.798	0.606	0.444	0.805	0.798
Ca	0.214	0.263	0.221	0.244	0.447	0.300	0.309	0.410	0.464	0.568	0.594
Na	0.000	0.000	0.000	0.004	0.004	0.002	0.001	0.000	0.002	0.003	0.002
Cr	0.026	0.017	0.020	0.035	0.034	0.028	0.022	0.017	0.014	0.031	0.031

reached and a sharp decrease in Al/Si occurs. The height of the Al/Si maximum and the Fe/Mg position of the drop off varies considerably among these basalts. In 15499 (Fig. 15a), a drop off is never observed and extensive Al-enrichment occurs right to the crystal margins. Pyroxenes from 12052 show a lesser Al-enrichment (Al/Si = 0.1–0.22) and a maximum occurring at Fe/Mg $\simeq$ 1. The range in Al/Si is due to sectoral control. Pyroxenes from 12021 have Al/Si maxima of 0.06 in one sector and 0.11 in a second both occurring at Fe/Mg $\simeq$ 0.6 (Fig. 15b), whereas 15058 pyroxenes have a poorly defined Al/Si maximum exceeding 0.08 somewhere between Fe/Mg = 0.6 to 1.0. In the later stages of 12052, 12021, and 15058, the pyroxenes have relatively uniform Al/Si ratios of 0.04 to 0.06. A small but well-defined depression in Al/Si for 12021 pyroxenes occurs at Fe/Mg $\simeq$ 4.

The existence of strong correlations among groundmass grain size, the proximity of the compositional discontinuities in all of the crystallization trends to the margins of the phenocrysts, and the extent to which the phenocrysts exhibit their crystal forms are unmistakable. Phenocrysts from 15499 have regular and distinct forms with all the faces clearly recognizable. Those from 12052 have less regular crystal faces and those from 12021 are defined only in part. Phenocrysts from 15058 exhibit no regular forms.

The correlation of groundmass grain size with the degree to which the pyroxene phenocrysts exhibit their crystal forms might be accounted for either by late-stage resorption or to increased interference effects (Boyd, personal communication) that might be expected as the groundmass plagioclases became coarser. Of these two, we prefer the resorption although both could have been operative. Resorption is to be expected because the Al content of the pyroxene rims is exceedingly high and forms a metastable assemblage with the melt relative to plagioclase. When plagioclase precipitates, the activity of the Al_2O_3 in the melt drops drastically, and the pyroxene rim becomes even further out of equilibrium with the melt and would start to dissolve.

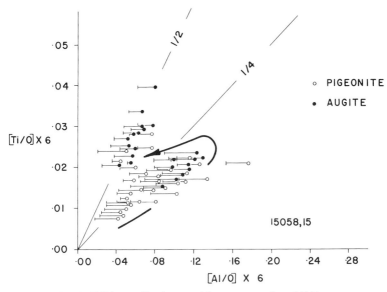

Fig. 19. Ti/Al crystallization trend for pyroxenes from 15058.

In such a situation pyroxene crystals would be expected to show a degree of resorption proportional to their residence time in the melt.

The relative arrangement of the chemical discontinuities in terms of percent of phenocryst crystallized is: 15499—100%, 12052 ~ 95%, 12021 ~ 75%, 15058 ~ 50%. In view of the similarities in bulk compositions of these basalts, the variations in texture and pyroxene chemistry must reflect differences in the late-stage cooling rate. This suggests a two-stage model for pyroxene porphyritic basalt petrogenesis: the first stage involving crystallization at a relatively slow rate in a melt with a low TiO_2/Al_2O_3 ratio; a second stage, at a relatively fast rate, at which time plagioclase is crystallizing. As we have suggested previously (Bence *et al.*, 1971), the first stage of a two-stage basalt crystallization model can be explained in one of two ways: (1) Rapid crystallization in a melt super-saturated with respect to plagioclase, or (2) early clinopyroxene formation at depth with pressure contributing to the suppression of plagioclase precipitation. Both of these models can account for the observed textural and compositional relationships in the porphyritic basalts. Model (1) might account for the observed composition and textural breaks and the range of textures from sharply defined phenocrysts with a glassy matrix to irregular resorbed phenocrysts in a granular matrix by differences in depths in a lava flow or in a near-surface intrusive. If this model holds for Apollo 15 basalts, then 15058 and 15499 could have originated at different depths within the same lava flow. Model (2) involves initial crystallization at depth where pressures could have contributed to the suppression of plagioclase crystallization. The onset of plagioclase crystallization would occur in the second, eruptive stage. The arguments for and against these two models were discussed by Bence *et al.* (1971), Boyd and Smith (1971), and Hollister *et al.* (1971).

Basalt type 14053

Basalt 14053, characterized by extremely low TiO_2/Al_2O_3 (weight) ratios, high Al_2O_3, and low FeO is unusual in that it appears to have had an extremely low f_{O_2} in its late stages as indicated by the breakdown of fayalite to iron and silica (El Goresy *et al.*, 1972; Gancarz *et al.*, 1971; Haggerty, 1972). Additional evidence for low f_{O_2} is the suggestion of Ti^{3+} in the pyroxenes from this rock (Bence and Papike, 1971; 1972) and Ti^{3+} in ilmenite (El Goresy *et al.*, 1972).

Composite clinopyroxenes from 14053 have, for the most part, similar crystallization trends on the pyroxene quadrilateral (Fig. 20, Table 7) as the phenocrysts from the porphyritic basalts. The major difference is in the late-stage trend. In 12021, 12052, and 15058, the late-stage trend is towards pyroxferroite (Fig. 20), whereas in 14053, a sharp discontinuity in the trend occurs in the vicinity of the "forbidden field" of Lindsley and Munoz (1969) and late-stage pyroxenes approach hedenbergite compositions. A discontinuity in the augite trend (high-to-low Ca), occurs at some depth in these crystals. This sharp decrease in Ca corresponds to the crystallization of plagioclase. Prior to the incoming of plagioclase, the pyroxenes crystallized in a melt with a low TiO_2/Al_2O_3 ratio.

Titanium-aluminum relationships of 14053 pyroxenes (Fig. 21) show that the early pigeonites have ratios less than $\frac{1}{4}$ but with very low total Al and Ti. This indicates that some Al^{VI} is present. With crystallization both Ti and Al increase but the Ti/Al ratios increase as well and approach the 1/1 line. Ratios greater than $\frac{1}{2}$ can be explained by the presence of Ti^{3+} in a component $R^{2+}Ti^{3+}SiAlO_6$ when the amount of univalent cations is insignificant (Bence and Papike, 1971). Because an additional component (Ti^{3+}) is assumed to be present, a plot of $Ti–Cr–Al^{VI}$ is unique for this rock and cannot be compared with similar plots from other basalts. In addition to trivalent titanium, the possibility that divalent chromium is present in the pyroxenes has been considered. On the basis of spectral observations, Haggerty *et al.* (1970) conclude that Cr^{2+} is present in olivines from the Apollo 11 basalts. If Cr^{2+} is in the pyroxene structure, it would be present in the M-octahedral positions and would be expressed as a component $Cr^{2+}SiO_3$. All of the Al must then be present in the components $R^{2+}TiAl_2O_6$, $R^{2+}Ti^{3+}SiAlO_6$, and $R^{2+}AlSiAlO_6$ (Na is in very low concentrations) and the removal of an amount of Al equal to Cr for the component $R^{2+}CrSiAlO_6$ (represented by the bar on Figure 21) is no longer valid. However, in view of the very small amounts of chromium in the late-stage pyroxenes of 14053, its presence as either Cr^{2+} or Cr^{3+} will not affect the arguments for Ti^{3+}.

It is of interest to note that the pigeonites never attained significant Al-enrichment prior to the crystallization of plagioclase (Figs. 15b, 30, Table 7). This may be due to similar sectoral control as observed for 12021, crystallization at low pressure, or the very low f_{O_2} indicated for this basalt. The low f_{O_2} may be responsible for Al_2O_3 being incorporated in the pyroxene component $R^{2+}Ti^{3+}SiAlO_6$; addition of this component rather than $R^{2+}Ti^{4+}Al_2O_6$ would have the effect of reducing the Al/Si ratio.

The Al/Si–Fe/Mg (atomic) relationships for these pyroxenes (Fig. 15b) are quite distinctive from those of the porphyritic type. There is very little Al-enrichment

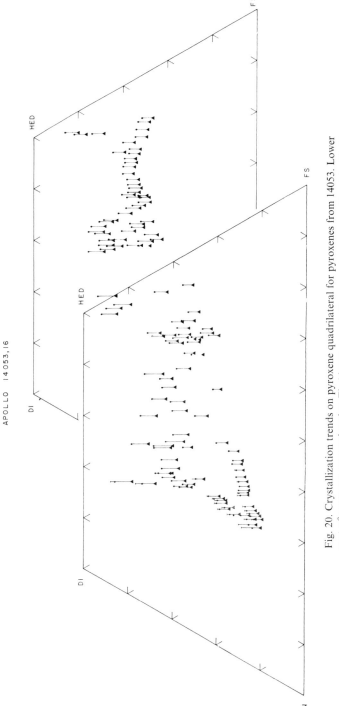

Fig. 20. Crystallization trends on pyroxene quadrilateral for pyroxenes from 14053. Lower part of posts are proportional to Ti–Al component. Upper parts are proportional to Al–Al component.

Table 7. Representative electron microprobe analyses of pyroxenes (wt%) from basalt 14053.
(Normalized to 6 oxygens.)

	1	2	3	4	5	6	7	8	9	10	11	12	13	14
SiO_2	51.3	51.9	50.5	51.1	51.2	50.3	50.6	49.2	48.9	47.7	48.0	46.9	46.3	46.4
Al_2O_3	1.73	1.04	1.46	1.74	1.77	1.95	1.08	1.51	1.61	1.40	1.36	1.22	1.62	1.52
TiO_2	0.84	0.68	1.11	0.75	0.99	1.39	0.93	1.25	1.32	1.23	1.12	1.04	1.43	1.40
FeO	18.6	21.7	23.5	18.6	18.5	17.1	21.4	24.9	26.7	30.8	32.3	35.2	29.8	30.5
MnO	—	—	—	—	—	—	—	—	—	—	—	—	—	—
MgO	21.2	18.2	16.0	20.6	18.8	13.8	12.9	10.2	8.11	6.69	5.56	4.0	1.24	1.03
CaO	4.37	5.06	7.09	5.06	7.32	14.0	12.0	12.5	13.5	11.3	11.8	10.5	18.7	18.1
Na_2O	0.02	0.06	0.04	0.04	0.04	—	0.07	0.04	0.11	0.09	0.07	0.07	0.10	0.09
Cr_2O_3	0.73	0.42	0.41	0.71	0.73	0.65	0.38	0.42	0.32	0.33	0.27	0.22	0.21	0.20
	98.8	99.1	100.1	98.6	99.4	99.2	99.4	100.0	100.6	99.5	100.5	99.2	99.4	99.2
Si	1.929	1.971	1.934	1.928	1.929	1.924	1.960	1.931	1.930	1.933	1.941	1.947	1.920	1.928
Al^{IV}	0.071	0.029	0.066	0.072	0.071	0.076	0.040	0.069	0.070	0.067	0.059	0.053	0.080	0.072
Al^{VI}	0.006	0.017	—	0.005	0.008	0.012	0.009	0.001	0.005	—	0.006	0.007	—	0.002
Ti	0.024	0.020	0.032	0.021	0.028	0.040	0.627	0.037	0.039	0.038	0.034	0.033	0.045	0.044
Fe	0.584	0.691	0.756	0.589	0.584	0.547	0.694	0.818	0.881	1.043	1.092	1.220	1.032	1.061
Mn	—	—	—	—	—	—	—	—	—	—	—	—	—	—
Mg	1.188	1.031	0.910	1.159	1.055	0.790	0.742	0.594	0.477	0.404	0.335	0.249	0.076	0.064
Ca	0.176	0.206	0.291	0.205	0.296	0.573	0.497	0.527	0.571	0.492	0.510	0.468	0.829	0.807
Na	0.002	0.004	0.003	0.003	0.003	0.000	0.005	0.003	0.008	0.007	0.006	0.005	0.008	0.008
Cr	0.022	0.013	0.012	0.021	0.022	0.020	0.012	0.013	0.010	0.010	0.009	0.007	0.007	0.007

in the pigeonite cores and Al/Si drops off gradually prior to the crystallization of augite. A sharp decrease in Al/Si soon after augite crystallizes coincides with the calcium decrease on the quadrilateral trend. A marked Al/Si depression occurs at high Fe/Mg ratios.

The Al/Si depression in the late-stage pyroxenes from 14053 can be accounted for by the introduction of a new aluminous phase—possibly an aluminous spinel (El Goresy et al., 1972). From the textural and compositional data, we conclude that 14053 crystallized at a relatively shallow depth in the lunar crust and that its crystallization rate was slow enough to produce the stable late-stage assemblage hedenbergite-fayalite-silica. Recent experimental evidence (Lindsley et al., 1972) indicates that this cooling rate could have been of the order of days.

Aluminous basalts (14310, 14276)

This type is unusual both in its abundance of plagioclase and in the nature of its pyroxenes. Composites composed of three pyroxenes in this fine-grained rock contain calcium-rich orthopyroxene cores mantled by pigeonite, which in turn is mantled by much less abundant augite. The crystallization trend in the pyroxene quadrilateral is shown on Fig. 22 and representative analyses are given in Table 8.

Titanium-aluminum relationships (Fig. 23) show aluminous-rich orthopyroxenes in the early stages. With crystallization the Ti/Al ratios increase due to a drop in total Al. At Fe/Fe + Mg $\simeq$ 0.3 in the pyroxene, pigeonite crystallizes and the Ti content of the pyroxene increases very rapidly with only a minor increase in Al and the Ti/Al ratio approaches $\frac{1}{2}$.

On the Ti–Cr–Al^{VI} plot (Fig. 24), the trend is from a high proportion of octahedral Al to a composition in which the aluminum is present primarily in the component $R^{2+}TiAl_2O_6$.

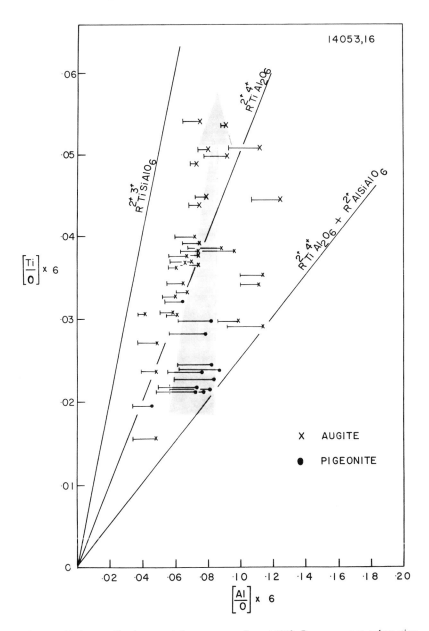

Fig. 21. Ti/Al crystallization trend for pyroxenes from 14053. Bars represent subtraction of Al in assumed component $R^{2+}CrSiAlO_6$.

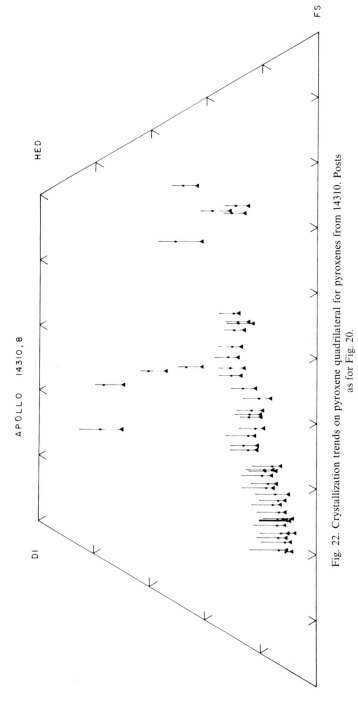

Fig. 22. Crystallization trends on pyroxene quadrilateral for pyroxenes from 14310. Posts as for Fig. 20.

Table 8. Representative electron microprobe analyses of pyroxenes (wt%) from nonmare basalt 14310. (Normalized to 6 oxygens.)

	1	2	3	4	5	6	7	8	9	10	11	12	13	14	15	16	17
SiO_2	53.8	54.3	52.8	53.8	52.9	51.9	52.1	52.7	52.2	50.7	49.0	50.3	50.2	49.4	49.3	49.2	46.6
Al_2O_3	1.50	1.82	2.91	1.90	1.66	2.58	1.40	1.32	1.30	2.28	2.84	2.01	2.19	2.37	0.93	1.56	2.23
TiO_2	0.44	0.67	0.80	0.55	0.71	0.71	1.01	0.68	0.58	1.86	2.05	1.25	1.50	1.74	0.69	1.21	0.97
FeO	13.3	13.6	14.9	14.1	15.6	17.9	19.6	19.9	20.4	13.9	14.3	17.8	19.5	22.4	29.4	33.7	37.2
MnO	—	—	0.23	—	—	0.37	—	—	—	—	—	—	—	—	—	—	0.61
MgO	27.5	27.3	26.2	27.3	24.6	22.3	19.8	20.5	21.7	16.5	12.5	13.2	14.0	11.2	12.7	6.14	6.46
CaO	2.35	2.04	2.19	2.38	3.66	3.85	6.37	4.66	3.13	13.8	18.6	14.7	12.8	13.1	5.64	7.55	5.42
Na_2O	0.01	0.03	0.03	0.0	0.0	0.04	0.02	0.00	0.0	0.0	0.08	0.18	0.0	0.02	0.10	0.02	0.07
Cr_2O_3	0.51	0.50	0.56	0.44	0.45	0.57	0.45	0.41	0.41	0.24	0.59	0.41	0.37	0.23	0.54	0.14	0.06
	99.4	100.3	100.6	100.5	99.6	100.2	100.8	100.2	99.7	99.3	100.0	99.9	100.6	100.5	99.3	99.5	99.6
Si	1.942	1.943	1.902	1.929	1.936	1.914	1.938	1.959	1.948	1.921	1.868	1.923	1.908	1.907	1.952	1.988	1.917
Al^{IV}	0.058	0.057	0.098	0.071	0.064	0.086	0.062	0.041	0.052	0.079	0.132	0.077	0.092	0.093	0.048	0.012	0.083
Al^{VI}	0.006	0.020	0.026	0.009	0.008	0.026	—	0.017	0.005	0.023	—	0.014	0.006	0.015	—	0.062	0.025
Ti	0.012	0.018	0.022	0.015	0.020	0.020	0.028	0.019	0.016	0.053	0.059	0.036	0.043	0.050	0.021	0.037	0.030
Fe	0.402	0.407	0.448	0.424	0.478	0.553	0.608	0.618	0.638	0.440	0.456	0.569	0.620	0.723	0.974	1.137	1.280
Mn	—	—	0.007	—	—	0.012	—	—	—	—	—	—	—	—	—	—	0.021
Mg	1.482	1.455	1.403	1.457	1.340	1.223	1.096	1.135	1.205	0.780	0.711	0.750	0.791	0.646	0.748	0.370	0.397
Ca	0.091	0.078	0.084	0.092	0.144	0.152	0.254	0.186	0.125	0.669	0.761	0.603	0.523	0.542	0.239	0.327	0.239
Na	0.001	0.002	0.002	0.0	0.0	0.002	0.002	0.0	0.0	0.0	0.006	0.013	0.0	0.001	0.007	0.001	0.006
Cr	0.014	0.014	0.016	0.013	0.013	0.017	0.013	0.012	0.012	0.007	0.018	0.013	0.011	0.007	0.017	0.005	0.002

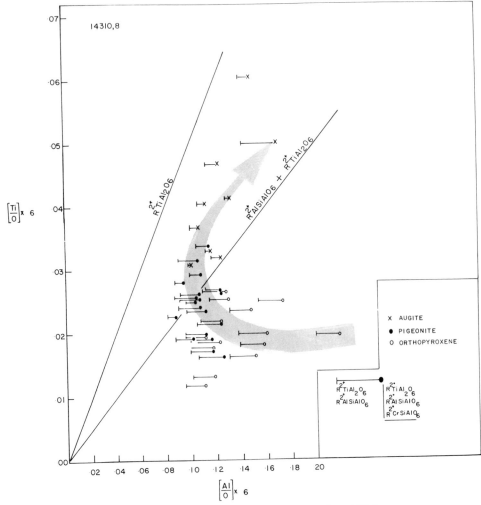

Fig. 23. Ti/Al crystallization trend for pyroxenes from 14310.

The Al/Si versus Fe/Mg crystallization trend for these pyroxenes (Fig. 15b) shows extreme Al-enrichment in the early orthopyroxenes. This Al-enrichment drops off very rapidly prior to the crystallization of pigeonite. As pigeonite and, subsequently, augite crystallize, the Al/Si ratio decreases gradually with increasing Fe/Mg.

Similar crystallization trends are observed for pyroxenes from 14276 (Figs. 25, 26).

From its unusual bulk composition, non-mare basalt type 14310 appears to have been produced by melting of the lunar crust (Walker *et al.*, 1972), possibly by meteorite impact, and there is some evidence to suggest that there are unmelted relicts of the source material present (Hollister *et al.*, 1972). Crystallization experiments (Walker *et al.*, 1972) support the textural evidence that plagioclase was on

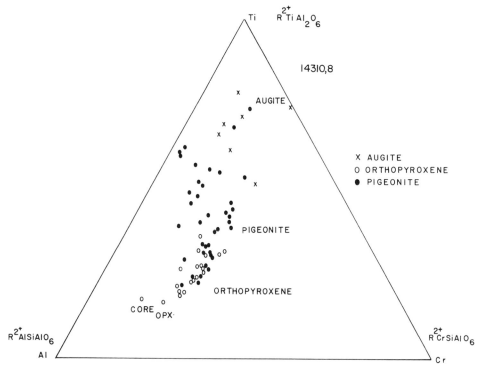

Fig. 24. Ti–Cr–Al^{VI} relationships in pyroxenes from 14310.

the liquidus. This is consistent with the pyroxene chemistry, which suggests that the earliest pyroxenes (orthopyroxenes) crystallized in the presence of a very high activity of alumina, but as they grew, the alumina activity in the melt dropped off— presumably due to the continued removal of plagioclase. This removal of plagioclase resulted in an increase in the TiO_2/Al_2O_3 ratio in the melt and the Ti/Al ratios of the pyroxenes eventually approached the limiting value of $\frac{1}{2}$. The change in direction on the Ti/Al plot (Fig. 28) with the appearance of pigeonite might reflect the increased Ca content of the clinopyroxenes. A correlation between Ca content and the Ti/Al couple has already been demonstrated by Bence and Papike (1971). Similar relationships are observed for 14276.

We interpret these observations to indicate that 14310 and 14276 cooled very quickly near the lunar surface.

SUMMARY

The above discussion demonstrates that one can correlate the chemical trends in lunar pyroxenes with the bulk composition of the host rock, paragenetic sequence, emplacement history, and variations in f_{O_2}.

Basalts with high TiO_2/Al_2O_3 ratios, early crystallization of plagioclase, one-stage near-surface crystallization, and oxygen fugacities in the range 10^{-13} atmos-

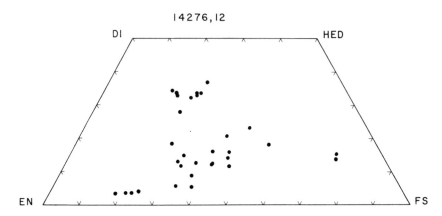

Fig. 25. Pyroxene compositions from 14276 projected on pyroxene quadrilateral.

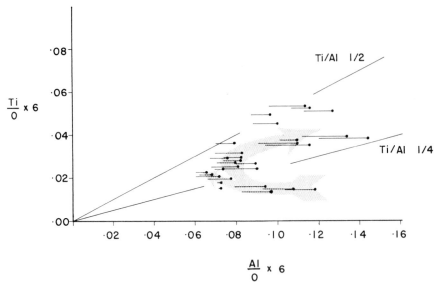

Fig. 26. Ti/Al crystallization trend for pyroxenes from 14276. The trends for pigeonite
(constant Al) and augite (increasing Al) are shown.

pheres (Apollo 11 basalts, Luna 16 basalts and 12022) show the following pyroxene relationships:

1. Ti/Al (atomic) ratios approaching $\frac{1}{2}$ (Fig. 28).
2. A continuous trend on the pyroxene quadrilateral from augite to pyroxferroite (Fig. 27).
3. Compositions close to the Ti–Cr join on the ternary plot Ti–Cr–Al^{VI} (Fig. 29).
4. A sharp decrease in Al/Si with increasing Fe/Mg at low Fe/Mg ratios followed by a very gradual decrease over a wide range in Fe/Mg (Fig. 30).

Pyroxenes from rocks with lower TiO_2/Al_2O_3 ratios, delayed crystallization of

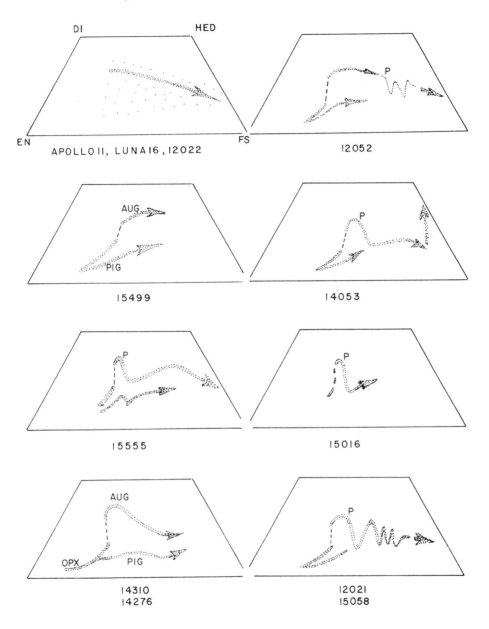

Fig. 27. Schematic of crystallization trends on pyroxene quadrilateral discussed in text.

plagioclase, oxygen fugacities of $\sim 10^{-13}$ atmospheres, and a two-stage crystallization history (12021, 12052, 15016, 15058, 15499, 15555) show the following:

(1) Clinopyroxenes commence crystallization with low Ti–Al ratios, have significant Al^{VI}, and show an enrichment of total Al with crystallization (Fig. 28). The
incoming of plagioclase is reflected by a sharp break in the Ti/Al trend to Ti/Al $= \frac{1}{2}$.
In rocks 15555, 15016, 15058, this break occurs relatively early in the crystallization
sequence and a relatively small enrichment of Al in the clinopyroxenes prior to
plagioclase precipitation is noted. In 12052 and 12021 significant Al-enrichment is
observed (up to $\sim 8\%$ before plagioclase begins to crystallize).

In 15499, the Al_2O_3 content of the clinopyroxene reaches $\sim 10\%$ before the
plagioclase crystallized and no sharp discontinuity in Ti/Al is observed, presumably
due to the narrow width of this zone on the rims of the phenocrysts.

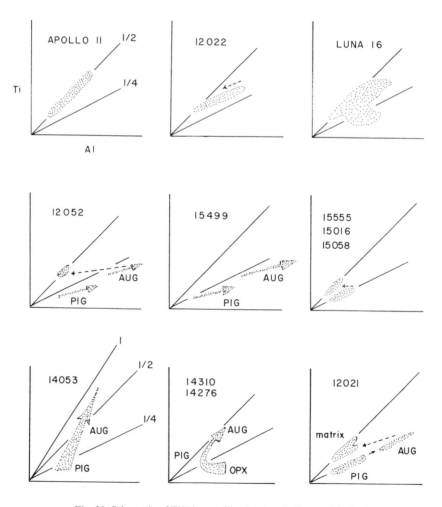

Fig. 28. Schematic of Ti/Al crystallization trends discussed in text.

(2) On the pyroxene quadrilateral, two trends are observed: a calcium-rich trend with a discontinuity between pigeonite and augite in microprobe traverses normal to (110), a calcium-poor trend in traverses normal to (010) in which this discontinuity is not observed. The incoming of plagioclase is reflected by a decrease in calcium in both trends and is indicated by the letter P in Fig. 27. The position of this Ca decrease relative to the percentage of growth of the phenocrysts as presently observed correlates with the grain size of the groundmass and the degree to which the phenocrysts display euhedral outlines.

(3) On the Ti–Cr–Al^{VI} diagrams (Fig. 29) early clinopyroxenes start out near the Ti–Cr join with Cr/Ti > 1. With crystallization Al^{VI} increases at the expense of Cr. At the onset of plagioclase crystallization, a discontinuity occurs with a rapid decrease in Al^{VI} and a preferential enrichment of the component R^{2+}TiAl$_2$O$_6$. No discontinuity in the trend is observed for pyroxenes from 15499.

(4) On the Al/Si–Fe/Mg diagram early clinopyroxenes show a rapid increase in Al/Si (Fig. 30). At the onset of plagioclase crystallization there is a reversal in this trend. In the late stages a relatively low Al/Si ratio obtains. The positions and the magnitude of the Al/Si maximum correlate with the degree to which the basalt was fractionated prior to the onset of plagioclase crystallization.

Rock 14053 with a low TiO$_2$/Al$_2$O$_3$ ratio, delayed plagioclase crystallization, and an extremely low f_{O_2} (approximately 10^{-16} atm, Haggerty, 1972) has the following pyroxene trends:

(1) Ti/Al (atomic ratios $\leq \frac{1}{4}$ in the early stages and gradually increasing ratios due almost entirely to the addition of Ti, at least some of which is in a component R^{2+}Ti^{3+}SiAlO$_6$ (Fig. 28). Final Ti/Al ratios trending towards Ti/Al = 1 are obtained.

(2) The crystallization trend on the pyroxene quadrilateral shows a Ca decrease at the onset of plagioclase crystallization shortly after the appearance of augite (Fig. 27).

(3) Although this rock has a low TiO$_2$/Al$_2$O$_3$ ratio comparable to porphyritic rocks and a moderately high Al$_2$O$_3$ content, it does not show a high Al/Si maximum in the Al/Si–Fe/Mg diagram (Fig. 30). Three possible explanations for this low Al/Si maximum come to mind: (a) Crystallization took place entirely at the lunar surface and pressure was not a contributing factor to the incorporation of Al^{VI} in the pyroxene structure. (b) The low f_{O_2} is responsible for a significant amount of the Al$_2$O$_3$ being incorporated in the pyroxene component R^{2+}Ti^{3+}SiAlO$_6$. Addition of this component rather than R^{2+}Ti^{4+}Al$_2$O$_6$ would have the effect of diminishing the magnitude of the Al/Si maximum. (c) Sectoral control similar to that observed in 12021.

Rocks 14310 and 14276 with extremely high Al$_2$O$_3$ ($\sim 20\%$), low TiO$_2$/Al$_2$O$_3$ ratios, a paragenetic sequence in which plagioclase is on the liquidus, which crystallized entirely near the lunar surface, and which had f_{O_2} approximately 10^{-14} atm at 1100 C (Ridley *et al.*, 1972) have pyroxenes with the following chemical relationships:

(1) Initial low Ti/Al ratios and high Al in the orthopyroxenes followed by a gradual drop off in Al and concomittant Ti/Al increase as the Fe/Mg ratio in the

A. E. Bence and J. J. Papike

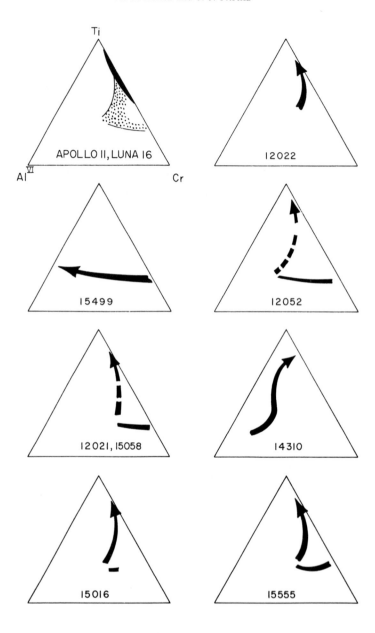

Fig. 29. Schematic of Ti–Cr–Al^{VI} crystallization trends discussed in text.

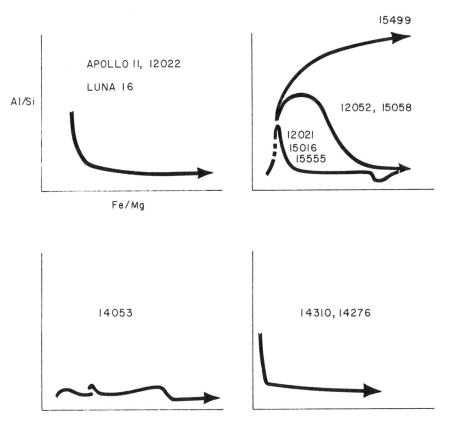

Fig. 30. Schematic of Al/Si versus Fe/Mg crystallization trends discussed in text.

orthopyroxenes increases. With the introduction of pigeonite and, subsequently, augite, a sharp increase in Ti/Al occurs (Fig. 28).

(2) The quadrilateral (Fig. 27) reflects the three pyroxenes and does not show a sharp Ca-decrease in the augite trend since plagioclase was on the liquidus.

(3) On the Ti–Cr–Al^{VI} plot (Fig. 29) the orthopyroxenes are highly enriched in Al^{VI} relative to Ti and Cr. With crystallization Al^{VI} drops off while Cr remains essentially constant. Late-stage pyroxenes have a high proportion of the component $R^{2+}Ti^{4+}Al_2O_6$.

(4) The Al/Si–Fe/Mg diagram for these pyroxenes (Fig. 30) shows Al/Si enrichment in the early orthopyroxenes followed by a sharp decrease as the activity of Al_2O_3 is decreased in the melt due to the removal of plagioclase.

The observed compositional and textural differences for the pyroxenes in the various basalt types can be explained by differences in bulk composition, paragenetic sequence, f_{0_2}, and emplacement histories. In particular the position of plagioclase in the paragenetic sequence is a function of its activity in the melt and its appearance is recorded in the crystallization trends of the pyroxenes by compositional discontinuities. We conclude that lunar pyroxenes are effective recorders of their detailed evolutionary history and single pyroxene crystals (providing that they contain core-to-rim portions) should be capable of distinguishing the four groupings considered in this summary.

Acknowledgments—We wish to acknowledge the assistance of W. Holzwarth in the electron microprobe laboratory. This research has been funded by NASA Grant No. NGR 33-015-123.

References

Albee A. L. Chodos A. A. Gancarz A. J. Haines E. L. Papanastassiou D. A. Ray L. Tera F. Wasserburg G. J. and Wen T. (1972) Mineralogy, petrology and chemistry of a Luna 16 basaltic fragment, sample B-1. *Earth Planet. Sci. Lett.* **13**, 353–367.

Anderson A. T. Jr. Crewe A. V. Goldsmith J. R. Moore P. B. Newton J. C. Olsen E. J. Smith J. V. and Wyllie P. J. (1970) Petrologic history of the moon suggested by petrography, mineralogy, and crystallography. *Science* **167**, 587–590.

Bence A. E. and Albee A. L. (1968) Empirical correction factors for the electron microanalysis of silicates and oxides. *J. Geol.* **76**, 382–403.

Bence A. E. Papike J. J. and Prewitt C. T. (1970) Apollo 12 clinopyroxenes: Chemical trends. *Earth Planet. Sci. Lett.* **8**, 393–399.

Bence A. E. and Papike J. J. (1971) A martini-glass clinopyroxene from the moon. *Earth Planet. Sci. Lett.* **10**, 245–251.

Bence A. E. and Papike J. J. (1971) Apollo 14 igneous rocks 14053 and 14310; phase petrology (abstract). Abstracts with programs, 1971 Annual Meeting, Geological Society of America **3**, 503.

Bence A. E. Papike J. J. and Lindsley D. H. (1971) Crystallization histories of clinopyroxenes in two porphyritic rocks from Oceanus Procellarum. *Proc. Second Lunar Sci. Conf., Geochim. Cosmochim. Acta* Suppl. 2, Vol. 1, pp. 559–574. MIT Press.

Bence A. E. Holzwarth W. and Papike J. J. (1972) Petrology of basaltic and monomineralic soil fragments from the Sea of Fertility. *Earth Planet. Sci. Lett.* **13**, 299–311.

Bence A. E. and Papike J. J. (1972) Crystallization histories of pyroxenes from lunar basalts (abstract). In *Lunar Science—III* (editor C. Watkins), pp. 59–61, Lunar Science Institute Contr. No. 88.

Boyd F. R. Jr. and Smith D. (1971) Compositional zoning in pyroxenes from lunar rock 12021, Oceanus Procellarum. *J. Petrol.* **12**, 439–464.

Brett R. Butler P. Jr. Meyer C. Jr. Reid A. M. Takeda H. and Williams R. (1971) Apollo 12 igneous rocks 12004, 12008, 12009, and 12022: A mineralogical and petrological study. *Proc. Second Lunar Sci. Conf., Geochim. Cosmochim. Acta* Suppl. 2, Vol. 1, pp. 301–317. MIT Press.

Brown G. M. Emeleus C. H. Holland J. G. and Phillips R. (1970) Petrographic, mineralogic, and x-ray fluorescence analysis of lunar igneous-type rocks and spherules. *Science* **167**, 599–601.

Burns R. G. Abu-Eid R. M. and Huggins F. E. (1972) Crystal field spectra of lunar silicates (abstract). In *Lunar Science—III* (editor C. Watkins), pp. 108–109, Lunar Science Institute Contr. No. 88.

Champness P. E. and Lorimer G. W. (1971) An electron microscopic study of a lunar pyroxene. *Contrib. Mineral. Petrol.* **33**, 171–183.

El Goresy A. Ramdohr P. and Taylor L. A. (1972) Petrology and geochemistry of the opaque minerals in the crystalline rocks of the Fra Mauro region (abstract). In *Lunar Science—III* (editor C. Watkins), pp. 224–226, Lunar Science Institute Contr. No. 88.

Gancarz A. J. Albee A. L. and Chodos A. A. (1971) Petrographic and mineralogic investigation of some crystalline rocks returned by the Apollo 14 mission. *Earth Planet. Sci. Lett.* **12**, 1–18.

Grieve R. A. F. McKay G. A. and Weill D. F. (1972) Microprobe studies of three Luna 16 basalt fragments. *Earth Planet. Sci. Lett.* **13**, 233–242.

Haggerty S. E. (1972) Subsolidus reduction and compositional variations of lunar spinels (abstract). In *Lunar Science—III* (editor C. Watkins), pp. 347–349, Lunar Science Institute Contr. No. 88.

Haggerty S. E. Boyd F. R. Bell P. M. Finger L. W. and Bryan B. W. (1970) Opaque minerals and olivine in lavas and breccias from Mare Tranquillitatis. *Proc. Apollo 11 Lunar Sci. Conf., Geochim. Cosmochim. Acta* Suppl. 1, Vol. 1, pp. 513–538. Pergamon.

Hargraves R. B. Hollister L. S. and Otalara G. (1970) Compositional zoning and its significance in pyroxenes from three coarse-grained lunar samples. *Science* **167**, 631–633.

Hollister L. S. Trzcienski W. E. Jr. Hargraves R. B. and Kulick C. G. (1971) Petrogenetic significance of pyroxenes in two Apollo 12 samples. *Proc. Second Lunar Sci. Conf., Geochim. Cosmochim. Acta* Suppl. 2, Vol. 1, pp. 529–557. MIT Press.

Hollister L. S. Trzcienski W. Jr. Dymek R. Kulick C. Weigand P. and Hargraves R. B. (1972) Igneous fragment 14310,21 and the origin of the mare basalts (abstract). In *Lunar Science—III* (editor C. Watkins), pp. 386–388, Lunar Science Institute Contr. No. 88.

Jakeš P. Warner J. Ridley W. I. Reid A. M. Harmon R. S. Brett R. and Brown R. W. (1972) Petrology of a portion of the Mare Fecunditatis regolith. *Earth Planet. Sci. Lett.* **13**, 257–271.

James O. B. and Jackson E. O. (1970) Petrology of the Apollo 11 ilmenite basalts. *J. Geophys. Res.* **75**, 5793–5824.

Kushiro I. and Nakamura Y. (1970) Petrology of some lunar crystalline rocks. *Proc. Apollo 11 Lunar Sci. Conf., Geochim. Cosmochim. Acta* Suppl. 1, Vol. 1, pp. 607–626. Pergamon.

Lindsley D. H. and Munoz J. L. (1969) Subsolidus relations along the join hedenbergite-ferrosilite. *Amer. J. Sci. Schairer* **267**, sec. A, 295–324.

Lindsley D. H. Papike J. J. and Bence A. E. (1972) Pyroxferroite: Breakdown at low pressure and high temperature (abstract). In *Lunar Science—III* (editor C. Watkins), pp. 483–485, Lunar Science Institute Contr. No. 88.

LSPET (1971) (Lunar Sample Preliminary Examination Team) Preliminary examination of lunar samples from Apollo 14. *Science* **173**, 681–693.

LSPET (1972) (Lunar Sample Preliminary Examination Team) The Apollo 15 lunar samples: A preliminary description. *Science* **175**, 363–375.

Papike J. J. Bence A. E. Brown G. E. Prewitt C. T. and Wu C. H. (1971) Apollo 12 clinopyroxenes: Exsolution and epitaxy. *Earth Planet. Sci. Lett.* **10**, 307–315.

Reid J. B. Jr. Taylor G. J. Marvin U. B. and Wood J. A. (1972) Luna 16: Relative proportions and petrologic significance of particles in the soil from Mare Fecunditatis. *Earth Planet. Sci. Lett.* **13**, 286–298.

Ridley W. I. Williams R. J. Brett R. Takeda H. and Brown R. (1972) Petrology of lunar basalt 14310 (abstract). In *Lunar Science—III* (editor C. Watkins), pp. 648–650. Lunar Science Institute Contr. No. 88.

Ross M. Bence A. E. Dwornik E. J. Clark J. R. and Papike J. J. (1970) Lunar clinopyroxenes: Chemical composition, structural state, and texture. *Science* **167**, 628–631.

Takeda H. and Ridley W. I. (1972) Crystallography and mineralogy of pyroxenes from Fra Mauro soil and rock 14310 (abstract). In *Lunar Science—III* (editor C. Watkins), pp. 738–740, Lunar Science Institute Contr. No. 88.

Walker D. Longhi J. and Hays J. F. (1972) Experimental petrology and origin of Fra Mauro rocks and soil (abstract). In *Lunar Science—III* (editor C. Watkins), pp. 770–772, Lunar Science Institute Contr. No. 88.

Weill D. F. MaCallum I. S. Bottinga Y. Drake M. J. and McKay G. A. (1970) Mineralogy and petrology of some Apollo 11 igneous rocks. *Proc. Apollo 11 Lunar Sci. Conf., Geochim. Cosmochim. Acta* Suppl. 1, Vol. 1, pp. 937–955. Pergamon.

Proceedings of the Third Lunar Science Conference
(Supplement 3, *Geochimica et Cosmochimica Acta*)
Vol. 1, pp. 471–480
The M.I.T. Press, 1972

Pyroxenes from breccia 14303

P. W. Weigand and L. S. Hollister

Department of Geological and Geophysical Sciences
Princeton University
Princeton, New Jersey 08540

Abstract—Adjacent polished thin sections 47 and 53 of breccia 14303 have been studied. A basaltic lithic clast, composed mainly of plagioclase, olivine, and pigeonite zoned to sub-calcic augite, occupies the same relative position in both sections and is assumed to represent a single lithic chip. Similarly a fragmental clast, characterized by mineral clasts of plagioclase, opaque minerals, and orthopyroxene (En69–82), is found in both sections and is probably the same breccia chip. A second fragmental clast, characterized by shocked plagioclase, Mg-rich bronzite (En86), and essential lack of opaque minerals >0.01 mm, is found only in section 53. Groundmass pyroxenes are mainly pigeonites and augites, and are interpreted as being fragments of quickly cooled surface basalts.

Orthopyroxenes from 14303 and anorthosite 15415 exhibit a linear coherence of $Ti/(Ti + Cr)$ and $Fs/(Fs + En)$. This relationship, plus the observations that 14303 orthopyroxene clasts generally are larger than pyroxenes from typical lunar basalts, essentially are unzoned chemically, and are distinct chemically from meteoritic analogs, suggests that they originally crystallized in a plutonic-metamorphic environment beneath the pre-Imbrian crust.

Study of lithic and fragmental clasts in breccias probably will yield information on lunar history between 4.6 and 3.9 b.y. ago.

Introduction

Breccia sample 14303 is the largest rock returned by Apollo 14 for which a location has not been reasonably well established by photographs or crew comments (Swan *et al.*, 1971). It has been assigned tentatively to the comprehensive sample collected between the LM and Doublet Crater (Swan *et al.*, 1971, p. 71), although Wilshire and Jackson (1972) indicate that it was collected at station H on a boulder-ray deposit associated with Cone Crater. Before splitting for distribution, the rock weighed 898 g and measured about $16 \times 9 \times 7$ cm. It has been described in hand-specimen as a blocky, subrounded fragmental rock, fine-grained, and having less than 1% of sub-rounded leucocratic clasts in a medium gray matrix (Swan *et al.*, 1971). Sample 14303 has been placed into group 6 of the high metamorphic facies of Warner's (1972) classification. This group is characterized by having no matrix glass, a few glass clasts, and an annealed to euhedral matrix texture. Jackson and Wilshire (1972) place 14303 into subdivision F4 of their breccia classification, which represents coherent breccias with dark clasts dominant over light ones. The whole-rock chemistry of this rock has been studied by Brunfelt *et al.* (1972).

Examination of polished thin sections 47 and 53 (Fig. 1) showed that 14303 is a breccia consisting of mineral, lithic, and fragmental clasts. The mineral clasts range in size up to 2 mm across and are mainly plagioclase, pyroxene, olivine, and small disseminated opaque minerals. Plagioclase clasts predominate and are up to 1.5–2.0 mm in length; pyroxene clasts are 0.5 mm across or less, and one olivine grain is 1.0 mm across.

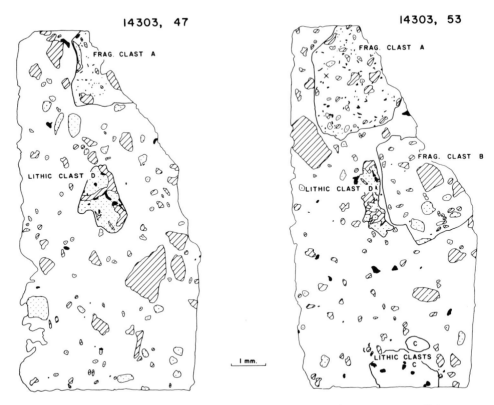

Fig. 1. Sketches of thin sections studied. Lines: plagioclase; dots: pyroxene + olivine; solid: opaque minerals; × : spinel. Lithic clasts C have a fine, granulitic texture of plagioclase and orthopyroxene.

Three distinct fragmental clasts occur. The largest (here designated A) in section 53 (Fig. 1b) measures $>2.5 \times >3.5$ mm and consists of plagioclase and pyroxene in a matrix much lighter than the bulk matrix; opaque grains, generally 0.05–0.10 mm across, are numerous and conspicuous. This same texture is found in a clast ($>1.0 \times >2.0$ mm) located in the same relative position in section 47 (Fig. 1a); because the two sections were cut adjacent to each other (R. Laughon, pers. comm. 1972), we assume that these two clasts represent a single fragmental chip. A second fragmental clast (B; $>3.5 \times >2.5$ mm) occurs only in section 53, has an intermediate gray matrix, and contains a few opaque grains ~ 0.05 mm across and many disseminated ones less than 0.01 mm across. The main constituent mineral clasts are plagioclase and pyroxene. Some of the pyroxene clasts are up to 0.5 mm long, which is larger than pyroxene grains in typical lunar basalts (Grieve *et al.*, 1972; Powell and Weiblen, 1972). The plagioclase grains have a fine mosaic texture, probably an effect of shock, and generally are unlike those elsewhere in the two slides. There are a few very indistinct dark breccia clasts less than 0.5 mm across in each section.

The lithic clasts are of several types. The most prominent lithic clast (D, Fig. 1), which occupies the central portion in both sections and is assumed to represent a single lithic chip, is an angular fragment of basaltic rock with an intergranular texture consisting of plagioclase, pyroxene, olivine, and opaque minerals. The clast measures approximately 2.5 × 0.5 mm in section 53 and 2.2 × 1.6 mm in section 47. Two fragments (C, the larger approximately >1.0 × 2.5 mm, Fig. 1b), showing a hornfelsic or granulitic texture composed predominantly of plagioclase and less pyroxene and opaque grains, occur in section 53. The texture suggests the fragments were recrystallized. Several other lithic clasts, <0.5 mm across, consist of plagioclase only or ophitic to subophitic intergrowths of plagioclase and pyroxene.

Section 47 consists of approximately 85% matrix, 5% fragmental clast A, 4% lithic clast D, and 6% plagioclase mineral clasts >0.5 mm in size; clast 47-A contains approximately 79% matrix, and mineral clasts >0.01 mm across of plagioclase (15%) and orthopyroxene (6%). Section 53 has the modal composition fragmental clasts A (13%) and B (12%), lithic clasts C (4%) and D (2%), plagioclase clasts >0.5 mm in size (3%), and the rest matrix (66%); clast 53-A is composed of matrix (87%), and plagioclase (10%) and orthopyroxene (2%) mineral clasts with less than 1% olivine and clinopyroxene clasts, and clast 53-B is comprised of plagioclase (16%), orthopyroxene (8%) and olivine (<1%) mineral clasts and matrix (76%). These figures probably are not representative of the rock as a whole.

For reference in the following discussion, the numbering scheme for mineral analyses is: first number represents section (47 or 53), letter represents clast designation (if any), and second number refers to mineral clast. For instance, pyroxene 53-A-4 refers to grain 4 in clast A from section 53.

All mineral analyses were made on an ARL-EMX electron microprobe using analytical procedures similar to those described in Hollister and Hargraves (1970) and in Hollister and Gancarz (1971).

RESULTS

Pyroxenes

Several distinct pyroxene groups are evident from the data plotted on the pyroxene quadrilateral (Fig. 2a). The augites are magnesian, more so than the most magnesian augites reported from Apollo 11 rocks; pyroxene 47-13 (a mineral clast) is the most magnesian augite as yet reported from any lunar material (Walter *et al.*, 1972, refer to diopsides of $\sim Di_{90}Fs_{10}$ from clasts of plagioclase-rich rocks from 14305). The augite and sub-calcic augite analyses are of mineral clasts with the exception of one grain from basaltic clast 47D.

The pigeonites form a fairly discrete group, and are similar to those reported for many Apollo 12 basalts. These analyses represent mineral clasts as well as pyroxenes in gabbro clast D. One pyroxene grain from the gabbroic clast (D) in section 47 ranges continuously in composition from pigeonite to sub-calcic clinopyroxene (arrow, Fig. 2a).

The hypersthenes and bronzites (distinguished from the Mg-rich bronzites from clast 53-B discussed below) exhibit a rather wide range in En content and appear to be concentrated in the fragmental clast (A) that extends through both sections.

The Mg-bronzites form a unique group and exhibit virtually no range of chemical composition. This field represents analyses of 9 separate grains that all are enclosed

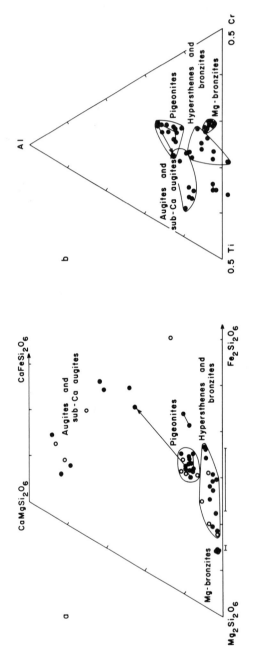

Fig. 2. (a) Pyroxene quadrilateral showing groupings of pyroxenes from 14303 (mol. %). Lines show zoning within single crystals, arrow shows direction to rim where known. Open circles represent 12 pyroxene analyses from 14303, 48, and 49 (W. L. Griffin, unpub. microprobe data). Range in olivine composition is indicated along base. (b) Ti-Cr-Al (atomic %) diagram for same pyroxene analyses.

within fragmental clast B in section 53. (A single mineral clast of the same composition occurs in section 47 not associated with a fragmental clast.) They are more Mg-rich than any reported from Apollo 11 or 12 basaltic or fragmental rocks, and are among the most Mg-rich of those reported from Apollo 14. Other Apollo 14 samples containing orthopyroxenes of $\sim En_{85}$ include soil samples 14163 (Fuchs, 1971) and 14259 (Apollo Soil Survey, 1971), and breccias 14313 (Floran et al., 1972) and 14305 (Walter et al., 1972); they have also been reported from an undesignated soil sample by Brown et al. (1972).

The groupings evident from the pyroxene quadrilateral also are shown on the Ti-Cr-Al diagram (Fig. 2b). The orthopyroxene fields are distinct from fields or trends of Apollo 11, 12, or 14 basaltic pyroxenes (Hollister, 1972).

Representative analyses are given in Table 1 (additional analyses may be obtained from the authors).

Other minerals

Olivines associated with the Mg-bronzites in fragmental clast 52-B range from Fo_{68} to Fo_{77}, those in fragmental clasts 47-A and 53-A are Fo_{66}. Olivine mineral clasts in the bulk matrix are Fo_{64-78}, and olivines associated with the basaltic clasts are Fo_{65}. Mg-rich olivines (Fo_{85}) occur as inclusions in a large plagioclase mineral clast in section 47 (see Table 1).

The An content of plagioclase ranges from 88 to 96; plagioclases in fragmental clast 53-B are among the highest in An content and show little chemical range (An_{94-96}).

A dark red isotropic spinel grain was found in fragmental clast 53-A with dimensions of about $40 \times 70 \mu$. It is continuously zoned with a relatively homogeneous center and a thin rim of lower Al and higher Cr. Representative analyses are given in Table 1, along with an analysis of a spinel from 14393,49 (W. L. Griffin, unpub. microprobe data).

DISCUSSION

Fragmental clasts A and B contrast with the clasts and bulk matrix in the rest of the two slides with respect to matrix color, size and distribution of opaque phases, and composition of pyroxene mineral clasts. They also differ from each other with regard to these same features; compared to clast A, clast B has a darker matrix and contains much smaller opaque mineral grains and more Mg-rich orthopyroxenes. The high Mg content of the bronzites and shocked nature of the plagioclase mineral clasts in fragmental clast 53-B both suggest that it has had a complex history.

The Mg-bronzite mineral clasts from fragmental clast 53-B are distinct chemically from terrestrial and meteoritic analogs as well. They are significantly lower in Al and Ca, and higher in Ti, than bronzites from terrestrial layered intrusions (c.f. Atkins, 1969; Hess, 1960); Cr contents are similar. Similarly, the bronzites and hypersthenes from clasts 47-A and 53-A have lower Al and Ca and higher Ti than cumulate hypersthenes (ibid). 14303 ortho-pyroxenes also are higher in Ca and lower in Al than those from anorthositic bodies (c.f. Griffin, 1971).

Table 1. Results of microprobe analyses of minerals in breccia 14303, 47 and 53 (wt %). Mineral components in mol. %.

	1	2	3	4	5	6	7	8	9	10
SiO$_2$	51.8	52.3	55.8	53.2	52.3	56.2	40.3	—	—	0.00
TiO$_2$	1.72	0.54	0.33	0.41	0.81	0.38	0.01	1.08	1.64	0.00
Al$_2$O$_3$	1.94	1.50	0.63	0.54	0.83	0.68	0.02	52.2	49.0	62.9
Cr$_2$O$_3$	0.33	0.75	0.42	0.34	0.31	0.52	0.04	11.8	14.5	5.61
FeO	6.15	17.7	11.7	19.3	21.0	9.19	13.4	22.2	22.8	13.4
MnO	0.14	0.34	0.20	0.34	0.30	0.17	0.12	0.15	0.16	0.00
MgO	17.0	22.3	30.0	25.1	22.2	33.5	44.7	12.4	11.8	18.2
CaO	20.3	4.31	0.79	1.10	2.20	0.79	0.14	—	—	0.00
Na$_2$O	0.14	0.02	0.01	0.01	0.02	0.01	0.03	—	—	—
Sum	99.5	99.8	99.9	100.3	100.0	101.4	98.8	99.8	99.9	100.1
En	48.3	62.8	80.6	68.0	62.2	85.2	Fo 85.5			
Wo	41.6	8.7	1.5	2.2	4.4	1.4	Fa 14.5			
Fs	10.1	28.5	17.9	29.8	33.4	13.4				

1. Fe-poor augite (47–13).
2. Average of 10 pigeonites from both sections.
3. 47-A-10A ⎫ Extreme compositions of orthopyroxenes from clasts A.
4. 53-A-14 ⎭
5. Average of 3 hypersthenes from clast 53-C.
6. Average of 9 Mg-bronzites from clast 53-B.
7. Average of two Mg-rich olivine inclusions in plagioclase from section 47.
8. Center ⎫ Spinel from 53-A (SiO$_2$ ~ 0.05, CaO ~ 0.06, Na$_2$O ~ 0.00).
9. Rim ⎭
10. Spinel from 14303,49 (W. L. Griffin, unpub. microprobe data).

Both groups of 14303 orthopyroxenes can be distinguished from those in olivine-bronzite and olivine-hypersthene chondrites on the basis of higher Ca content (cf. Keil and Fredriksson, 1964), and higher contents of Ti, Cʀ, and Al (F. D. Busche, pers. comm. 1972). Similarly, Al contents appear higher in the 14303 orthopyroxenes than in achondritic orthopyroxenes (Duke and Silver, 1967). It appears from the available data that the orthopyroxenes in 14303 are lunar in origin, not meteoritic.

Analyses of orthopyroxenes from Apollo 11, 12, and 15 igneous and fragmental rocks have been compiled (generally from Proceedings of the Lunar Science Conferences) and plotted on the pyroxene quadrilateral and Ti-Cr-Al diagram (Fig. 3) for comparison with 14303 orthopyroxenes. The exceptionally high En component of the Mg-bronzites is thus emphasized (Fig. 3a) and a striking relation of progressive depletion of Cr is exhibited in the Ti-Cr-Al diagram (Fig. 3b). Orthopyroxenes from 14303 and 15415, a cumulate anorthosite of probable plutonic origin (Hargraves and Hollister, 1972), fall on linear trends in plots of absolute concentrations of TiO_2, Cr_2O_3, and Al_2O_3 versus relative Fs content. They also show a positive correlation of $Ti/(Ti + Cr)$ with relative Fs content (Fig. 4), which results from slightly decreasing Cr contents with Ti remaining essentially constant. Hyperstenes from 15597 (a pyroxene vitrophyre) and 12065, which have crystallized in the absence of plagioclase (Hollister *et al.*, 1971), fall well below the latter trend, but those from 14310, a rock of complex history, and those from Apollo 12 soil samples fall above it.

The linear coherence of $Ti/(Ti + Cr)$ with $Fs/(Fs + En)$ exhibited by orthopyroxenes from 14303 and anorthosite 15414 therefore appears distinctive from other lunar pyroxene groups. This relationship, plus the observations that the orthopyroxene clasts chemically are unzoned and that some are larger than pyroxene grains in typical lunar basalts, suggests that the 14303 orthopyroxenes crystallized in a plutonic-metamorphic environment beneath the pre-Imbrian crust. Later sampling by the impacting body that produced Mare Imbrium brought these fragments to the Apollo 14 landing site.

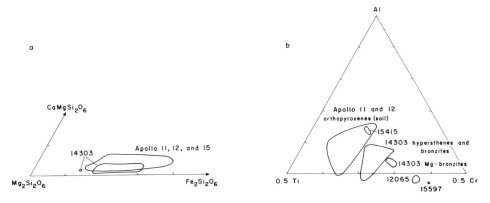

Fig. 3. (a) Portion of pyroxene quadrilateral comparing orthopyroxenes from clasts 47A, 53-A, and 53-B in 14303 with those from Apollo 11, 12, and 15 rocks. (b) Same comparison on Ti-Cr-Al diagram. Data for 15415 from Hargraves and Hollister (1972), that for 12065 from Hollister *et al.* (1971).

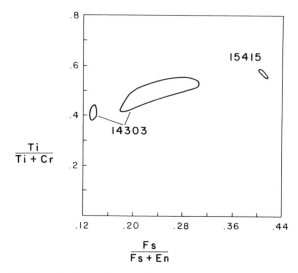

Fig. 4. Plot of Ti/(Ti + Cr) (atomic) versus relative Fs content of orthopyroxenes of probable plutonic origin.

The 14303 pigeonites and sub-calcic augites are represented by both mineral clasts and pyroxenes in the basaltic clast D. They are similar to pyroxenes from Apollo 12 quench basalts and we conclude that they represent fragments of quickly cooled, surface or near-surface basalts.

Although possibly more complex, our study of breccia sample 14303 suggests the following simple history:

1. brecciation and lithification of deep lunar crust
2. volcanism producing flows and near-surface intrusions similar to Apollo 11 and 12 basaltic rocks
3. brecciation and lithification of 1 and 2.

If 3 represents the Imbrian event dated at about 3.9 b.y. (e.g. Papanastassiou and Wasserburg, 1971), then events 1 and 2 suggest one or more pre-Imbrian volcanic events subsequent to at least one major earlier impact event. This is consistent with ideas of Cliff *et al.* (1972) and Wilhelms and McCauley (1972) based on age and geological relations. It seems probable that study of fragmental and lithic clasts in breccia samples will give information on lunar history between 4.6 and 3.9 b.y. ago.

Acknowledgments—R. B. Hargraves and C. G. Kulick are thanked for their assistance. Financial support was provided by NASA contract NGR 31-001-205

References

Apollo Soil Survey (1971) Apollo 14: Nature and origin of rock types in soil from the Fra Mauro formation. *Earth Planet. Sci. Lett.* **12**, 49–54.
Atkins F. B. (1969) Pyroxenes of the Bushveld intrusion, South Africa. *J. Petrol.* **10**, 222–249.
Brown G. M. Emeleus J. G. Holland A. and Phillips R. (1972) Mineral fractionation patterns between

Apollo 14 primitive feldspathic rocks and Apollo 15 and other basalts (abstract). In *Lunar Science—III* (editor C. Watkins), pp. 95–97, Lunar Science Institute Contr. No. 88.

Brunfelt A. O. Heier K. S. Nilssen B. Steinnes E. and Sundvoll B. (1972) Distribution of elements between different phases of Apollo 14 rocks and soil (abstracts). In *Lunar Science—III* (editor C. Watkins), pp. 99–101, Lunar Science Institute Contr. No. 88.

Cliff R. A. Lee-Hu C. and Wetherill G. W. (1972) K, Rb, and Sr measurements in Apollo 14 and 15 material (abstract). In *Lunar Science—III* (editor C. Watkins), pp. 146–147, Lunar Science Institute Contr. No. 88.

Duke M. B. and Silver L. T. (1967) Petrology of eucrites, howardites, and mesosiderites. *Geochim. Cosmochim. Acta* **31**, 1637–1666.

Floran R. J. Cameron K. Bence A. E. and Papike J. J. (1972) The 14313 consortium: A mineralogical and petrologic report (abstract). In *Lunar Science—III* (editor C. Watkins), pp. 268–269, Lunar Science Institute Contr. No. 88.

Fuchs L. H. (1971) Orthopyroxene and orthopyroxene-bearing rock fragments rich in K, REE, and P in Apollo 14 soil sample 14163. *Earth Planet. Sci. Lett.* **12**, 170–174.

Grieve R. McKay G. Smith H. and Weill D. (1972) Mineralogy and petrology of polymict breccia 14321 (abstract). In *Lunar Science—III* (editor C. Watkins), pp. 338–340, Lunar Science Institute Contr. No. 88.

Griffin W. L. (1971) Genesis of coronas in anorthosites of the Upper Jotun Nappe, Indre Sogn, Norway. *J. Petrol.* **12**, 219–244.

Hargraves R. B. and Hollister L. S. (1972) Lunar anorthosite 15415: Texture, mineralogy, and metamorphic history. *Science* **175**, 430–432.

Hess H. H. (1960) Stillwater igneous complex, Montana: A quantitative mineralogical study. *Geol. Soc. Amer. Mem.* **80**, 230 pp.

Hollister L. S. (1972) Implications of the relative concentrations of Al, Ti and Cr in Lunar pyroxenes (abstract). In *Lunar Science—III* (editor C. Watkins), pp. 389–391, Lunar Science Institute Contr. No. 88.

Hollister L. S. and Gancarz A. J. (1971) Compositional sector-zoning in clinopyroxene from the Narce area, Italy. *Amer. Mineral.* **56**, 959–979.

Hollister L. S. and Hargraves R. B. (1970) Compositional zoning and its significance in pyroxenes from two coarse-grained Apollo 11 samples. *Proc. Apollo 11 Lunar Sci. Conf., Geochim. Cosmochim. Acta*, Suppl. 1, Vol. 1, pp. 541–550. Pergamon.

Hollister L. S. Trzcienski W. E. Jr. Hargraves R. B. and Kulick C. G. (1971) Petrogenetic significance of pyroxenes in two Apollo 12 samples. *Proc. Second Lunar Sci. Conf., Geochim. Cosmochim. Acta*, Suppl. 2, Vol. 1, pp. 529–557. M.I.T. Press.

Jackson E. D. and Wilshire H. G. (1972) Classification of the samples returned from the Apollo 14 landing site (abstract). In *Lunar Science—III* (editor C. Watkins), pp, 418–420, Lunar Science Institute Contr. No. 88.

Keil K. and Fredriksson K. (1964) The iron, magnesium, and calcium distribution in coexisting olivines and rhombic pyroxenes of chondrites. *J. Geophys. Res.* **69**, 3487–3515.

Papanastassiou D. A. and Wasserburg G. J. (1971) Rb-Sr ages of igneous rocks from the Apollo 14 mission and the age of the Fra Mauro formation. *Earth Planet. Sci. Lett.* **12**, 36–48.

Powell B. N. and Weiblen P. W. (1972) Petrology and origin of rocks in the Fra Mauro formation (abstract). In *Lunar Science—III* (editor C. Watkins), pp, 616–618, Lunar Science Institute Contr. No. 88.

Swann G. A. Bailey N. G. Batson R. M. Eggleton R. E. Hait M. H. Holt H. E. Larson K. B. McEwen M. C. Mitchell E. D. Schaber G. G. Schafer J. P. Shepard A. B. Sutton R. L. Trask N. J. Ulrich G. E. Wilshire H. G. and Wolfe E. W. (1971) Preliminary geologic investigations of the Apollo 14 landing site. NASA Interagency Rpt. **29**, 124 pp.

Walter L. S. French B. M. and Doan A. S. Jr. (1972) Petrographic analysis of lunar samples 14171 and 14305 (breccias) and 14310 (melt rock) (abstract). In *Lunar Science—III* (editor C. Watkins), pp. 773–775, Lunar Science Institute Contr. No. 88.

Warner J. L. (1972) Apollo 14 breccias: Metamorphic origin and classification (abstract). In *Lunar Science—III* (editor C. Watkins), pp. 782–784, Lunar Science Institute Contr. No. 88.

Wilhelms D. E. and McCauley J. F. (1972) Geological setting of the Apollo 14 and Apollo 15 landing sites (abstract). In *Lunar Science—III* (editor C. Watkins), pp. 800–801, Lunar Science Institute Contr. No. 88.

Wilshire H. G. and Jackson E. D. (1972) Petrology of the Fra Mauro formation at the Apollo 14 landing site (abstract). In *Lunar Science—III* (editor C. Watkins), pp. 803–805, Lunar Science Institute Contr. No. 88.

Proceedings of the Third Lunar Science Conference
(Supplement 3, *Geochimica et Cosmochimica Acta*)
Vol. 1, pp. 481–492
The M.I.T. Press, 1972

Fe²⁺-Mg site distribution in Apollo 12021 clinopyroxenes: Evidence for bias in Mössbauer measurements, and relation of ordering to exsolution

ERIC DOWTY,* MALCOLM ROSS, AND FRANK CUTTITTA

U.S. Geological Survey, Washington, D.C. 20242

Abstract—Mössbauer-derived "cation distribution" numbers, and exsolution relationships from single-crystal x-ray studies are reported for three pyroxene fractions from rock 12021: Mg-pigeonite, $Wo_{11}En_{60}Fs_{29}$; Mg-augite, $Wo_{30}En_{43}Fs_{27}$; and Fe-augite, $Wo_{26}En_{24}Fs_{50}$. Each of these fractions has been analyzed by semi-micro wet-chemical methods. For the Fe-augite, the Mössbauer measurements indicate an anomalous "excess" of $M2$ cations. Error in chemical characterization can be eliminated as a cause of this anomaly, because the same sample was used in both Mössbauer measurements and chemical analysis. Consideration of exsolution relations suggests an even greater anomaly for the Fe-augite. Because it contains 50% exsolved pigeonite, which we expect to be disordered by analogy with the result for the Mg-pigeonite, this fraction should be partially disordered in bulk, i.e., have some magnesium in $M2$.

The anomaly can be explained as a result of a bias in the Mössbauer measurements due to (1) differing local environments for iron atoms in $M1$, which "splits" the $M1$ absorption, causing some of it to lie under the $M2$ doublet and thereby enhancing the latter; and (2) nonsuperposition of the $M1$ doublets for the exsolved pigeonite and augite in each fraction (x-ray studies show that all the fractions are unmixed). Both these types of bias have been documented in synthetic clinopyroxenes. The bias is probably present in all Mössbauer measurements of clinopyroxenes and may be large enough to account for the disagreement between Mössbauer and x-ray results—previous Mössbauer studies have suggested that ordering increases with the amount of calcium in clinopyroxenes, whereas x-ray refinements show just the opposite.

INTRODUCTION

As A RESULT of extensive Mössbauer work, and to a lesser extent crystal structure analysis, temperature dependent Fe^{2+}-Mg distribution patterns between the $M1$ and $M2$ sites of orthopyroxenes have been established (e.g., Ghose, 1965; Virgo and Hafner, 1969, 1970; Saxena and Ghose, 1970, 1971). This work has recently been extended to clinopyroxenes, particularly those from lunar rocks (Hafner and Virgo, 1970; Gay *et al.*, 1970; Hafner *et al.*, 1971; Ghose *et al.*, 1972; Finger *et al.*, 1972) but here there are complicating factors: the presence, in addition to Fe and Mg, of large amounts of Ca, as well as lesser amounts of Al, Ti, and Cr in the M sites; and the fact that the clinopyroxenes augite and pigeonite commonly are exsolved from each other, resulting in intergrowths too fine even to be detected by ordinary microscope or electron microprobe techniques. Single-crystal x-ray study of pigeonites and sub-calcic augites almost invariably shows the presence of crystallographically oriented exsolved phases, sometimes in large amounts (for example, up to 50% exsolved pigeonite in augite host and up to 35% augite in pigeonite host; see below).

---

*Present address: Department of Geology and Institute of Meteoritics, University of New Mexico, Albuquerque, New Mexico 87106.

In the following, we wish to explore some of the problems encountered in Mössbauer analysis of cation order-disorder in clinopyroxenes, particularly those from Apollo 12 rocks. These problems include the accuracy of Mössbauer-determined site occupancies, and the relation of cation ordering to exsolution.

SAMPLE CHARACTERIZATION

We have carried out wet-chemical, x-ray diffraction, and Mössbauer analysis of three pyroxene separates from Apollo rock 12021. Modal analysis based on a complete mineral separation of this rock shows: Silica, 2%; plagioclase ($\sim An_{90}$), 27%; pigeonite, 10%; Mg-rich augite, 13%; moderate- to high-iron augite, 41%; pyroxferroite, 2%; and metal oxides, sulfides, and iron-nickel, 5%. Compositions and unit-cell parameters of many pyroxenes from this rock have been previously reported (Bence *et al.*, 1970; Papike *et al.*, 1971; Ross *et al.*, 1971a, b). The exsolution relationships in these pyroxenes are relatively well understood, because careful x-ray study has been carried out on them in connection with phase-petrologic heating experiments (Ross *et al.*, 1971a, b).

The three pyroxene concentrates were separated from rock chips 12021,103 and 12021,29 by means of magnetic fractionation, centrifuging in heavy liquid, and by hand picking. The concentrates were characterized as to exsolution relations by taking single-crystal x-ray precession photographs of several grains from each fraction. Some idea of the chemical variation in each separate was obtained from the cell-parameters, and from electron microprobe analyses, which were carried out on several of the grains from each separate.

Fraction (A) contains 95% or more of Mg-rich pigeonite mostly of the bulk composition $Wo_9En_{63}Fs_{28}$ with the rest being more calcium-rich pigeonite grains with a bulk composition ranging from $Wo_{10}En_{61}Fs_{28}$ to $Wo_{17}En_{54}Fs_{29}$. The x-ray precession photographs show that the low-calcium (Wo_{9-10}) pigeonite grains contain about 10% augite exsolved on the (001) and (100) planes of the host. The high-calcium ($Wo_{>10}$) pigeonite grains contain 15–35% augite exsolved on (001). Fraction (B) consists of Mg-rich augite grains with an approximate bulk composition range from $Wo_{33}En_{45}Fs_{22}$ to approximately $Wo_{26}En_{44}Fs_{30}$. This augite contains about 35% pigeonite exsolved on (001). Fraction (C) is not so well characterized as to range in composition and undoubtedly has a wider range in Mg/Fe ratio than the other two separates, but most grains are approximately Wo_{26-23} and are probably less than about 15% En-Fs on either side of the average composition of the fraction (from the wet-chemical analysis) of $Wo_{26}En_{24}Fs_{50}$. The augite grains in Fraction (C) invariably contain about 50% pigeonite exsolved on (001). These grains are classified as augite because, when they are homogenized by heating, their symmetry is $C2/c$ (Ross *et al.*, 1971a, b).

Mössbauer spectra of the three fractions were taken at liquid-nitrogen temperature (77°K) by standard techniques. Four absorption peaks, corresponding to a doublet for each of the sites $M1$ and $M2$, were fitted by least-squares methods to each spectrum. After recovery from the Mössbauer absorbers, the greater part of each concentrate (about 0.050 gram) was analyzed by semi-micro, wet-chemical methods. As a control, samples of the terrestrial Kakanui augite (Mason, 1966) were analyzed simultaneously. The results of the chemical analyses are shown in Table 1.

We can use the chemical analyses and the observed amount of exsolution to derive approximate average compositions for the exsolved phases in each separate. We assume that fractions (A) and (B) have on the average 15 and 35%, respectively, of exsolution, and that host pigeonite and exsolved augite in Fraction (A) have the same respective calcium contents as exsolved pigeonite and host augite in Fraction (B). We also assume that the exsolved phases in each separate have the same Fe/Mg ratio as the bulk separate. We have not calculated the composition of the exsolved phases in Fraction (C) but assume that exsolved augite and pigeonite in this fraction have the same calcium content as the exsolved phases in the other two fractions. The resultant compositions are shown by the circles in Figure 1. These compositions agree fairly well with those derived from the cell parameters of the exsolved phases taken from precession photographs (Ross *et al.*, 1971a, b). The latter compositions were obtained by use of determinative nomograms developed from unit-cell parameter (b, β)-composition plots (Papike *et al.*, 1971).

METHODS OF MÖSSBAUER ORDER-DISORDER STUDIES AND THEIR VALIDITY

Previous Mössbauer studies of the cation ordering in lunar pyroxenes (Hafner and Virgo, 1970; Gay *et al.*, 1970; Hafner *et al.*, 1971; Ghose *et al.*, 1972; Finger *et al.*, 1972) utilized separation procedures and Mössbauer techniques very similar to those of the present work. The principal differences in method and scope are that in the previous studies chemistry of the pyroxenes was characterized only by micro-probe analyses, and that the Mössbauer results were not related in any systematic way to exsolution. Despite the difference in chemical characterization and minor differences in Mössbauer methods, the "cation distribution" numbers obtained are very similar, as will be shown below.

Two principal types of conditions must be fulfilled if Mössbauer measurements of cation ordering in pyroxenes are to be meaningful. First, the ratios of the peak areas of the doublets (or peak heights, if these are used) must represent accurately the ratios of iron site occupancies in $M1$ and $M2$. In fact, it has never been established that this is the case in clinopyroxenes, although there is some evidence that Mössbauer measurements are at least approximately accurate in orthopyroxenes (Virgo and Hafner, 1968; Burnham *et al.*, 1971). The present study will show that Mössbauer measurements of lunar clinopyroxenes are *not* accurate, although the exact extent of the bias cannot be established from our data alone.

Second, it is necessary to evaluate the influence of exsolution on the interpretation of Mössbauer data. It has been assumed (Hafner and Virgo, 1970; Gay *et al.*, 1970; Hafner *et al.*, 1971; Ghose *et al.*, 1972; Finger *et al.*, 1972) that even when exsolution is present, as it is in most lunar pyroxenes, the *average* cation distribution of the two exsolved phases in a sample can be obtained by Mössbauer measurements which are made as if the sample were homogeneous. The errors introduced by this assumption

Table 1. Chemical analyses of lunar pyroxenes from rock 12021.

| | (A) Mg-pigeonite | | (B) Mg-augite | | (C) Fe-augite | | Kakanui augite | | Mason (1966) |
	(wt %)	cations* (6 oxygens)	(wt %)	cations* (6 oxygens)	(wt %)	cations* (6 oxygens)	(1) (wt %)	(2) (wt %)	(wt %)
SiO_2	50.1	1.910	48.0	1.852	46.6	1.911	48.6_8	50.6	50.73
Al_2O_3	1.88	0.082	3.97	0.177	1.94	0.092	8.41	8.10	7.86
FeO	18.0	0.561	15.9	0.505	29.3	0.983	6.84	n.d.	6.77
MgO	21.1	1.172	14.4	0.814	7.70	0.461	16.4_0	16.1_3	16.65
CaO	5.57	0.222	14.2	0.577	11.8	0.507	16.5_9	16.5_0	15.82
Na_2O	0.058	—	0.11	—	0.10	—	1.65	0.15_8	1.27
K_2O	0.086	—	0.11	—	0.12	—	0.70	0.12_4	0.00
TiO_2	0.71	0.020	1.26	0.036	1.20	0.036	0.81	n.d.	0.74
Cr_2O_3	1.15	0.039	1.11	0.033	0.10	0.002	0.060	n.d.	n.d.
MnO	(0.35)†	0.011	(0.35)†	0.011	(0.35)†	0.012	—	—	0.13
TOTAL	98.65		99.06		98.86		100.14		100.34
Wo	11.4		30.4		26.0				
Fs	28.7		26.7		50.4				
En	59.9		42.9		23.6				

* In deriving the structural formulas shown, silica values obtained by difference were used, and alkalies were ignored. The difference between these formulas and those derived from the numbers shown is not significant.

† Manganese was not determined. The values shown are estimates taken from microprobe analyses (e.g., Bence et al., 1970).

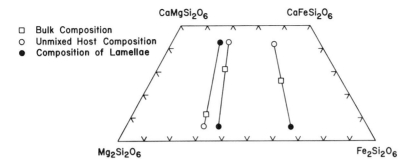

Fig. 1. Projection onto the pyroxene quadrilateral of the compositions of three pyroxene fractions from rock 12021.

alone probably are small. The assumption is not precisely correct, but even if it were valid, and if the Mössbauer measurements are accurate, it is still necessary to rationalize the "average" cation distribution obtained in terms of the exsolved phases in the sample. This has not been done in the past. Lack of consideration of unmixing has led to interpretations that are not internally consistent and has contributed to the overlooking of situations in which the first type of condition (i.e., the accuracy of the Mössbauer measurements) is not fulfilled.

In addition to these problems of the interpretation of "average" cation distribution values, we must also consider the possible interrelations of the cation ordering and exsolution processes. It has been assumed that the two processes occur in essentially distinct temperature ranges, and that the effects of one on the other could be ignored in practice. However, recent experimental results (Ross *et al.*, 1971a, b) have shown that crystallization and exsolution of some lunar pyroxenes occurred in a temperature range that may overlap to a considerable extent with the temperature range of cation ordering processes.

Apparent Cation Distribution

The cation distribution numbers (Table 2) were obtained from the chemical (Table 1) and Mössbauer data. It is assumed in deriving these numbers that the relative areas of the two doublets in the Mössbauer spectrum accurately represent the "average" distribution of iron between sites $M1$ and $M2$ in each fraction. All calcium is allocated to $M2$, and all aluminum, titanium, and chromium to $M1$. The equilibrium constant, $k(T)$ is calculated according to the formula

$$k(T) = \left(\frac{\text{Fe}}{\text{Mg}}\right)^{M1} \bigg/ \left(\frac{\text{Fe}}{\text{Mg}}\right)^{M2}.$$

Ordering will be discussed in terms of this equilibrium constant. The cation distribution values and $k(T)$ in Table 2 are similar to those obtained by Hafner *et al.* (1971) for clinopyroxenes from rock 12021 (shown also in Table 2) and are comparable to those obtained for clinopyroxenes from several other lunar rocks by Mössbauer methods (Hafner and Virgo, 1970; Hafner *et al.*, 1971; Ghose *et al.*, 1972).

Table 2. Site occupancies of 12021 pyroxenes from Mössbauer spectra.

	Present study (rock 12021, 103,29)				
	Composition		Site occupancies		
			$M1$	$M2$	
(A)	$Wo_{11.4}$	Ca	—	0.22	
Mg-pigeonite	$Fs_{28.7}$	Fe	0.11	0.45	$k(T) = 0.096$
	$En_{59.9}$	Mg	0.84	0.33	
(B)	$Wo_{30.4}$	Ca	—	0.58	
Mg-augite	$Fs_{26.7}$	Fe	0.15	0.36	$k(T) = 0.033$
	$En_{42.9}$	Mg	0.75	0.06	
(C)	$Wo_{26.0}$	Ca	—	0.51	
Fe-augite	$Fs_{50.4}$	Fe	0.43	0.55	$k(T) = 0.0$
	$En_{23.6}$	Mg	0.46	—	

	Hafner et al., 1971 (rock 12021,150)				
	Composition		Site occupancies		
			$M1$	$M2$	
P1	$Wo_{8.8}$	Ca	—	0.18	
pigeonite	$Fs_{32.1}$	Fe	0.12	0.52	$k(T) = 0.081$
	$En_{59.2}$	Mg	0.88	0.30	
P2	$Wo_{26.7}$	Ca	—	(0.53)	
Fe-augite	$Fs_{42.4}$	Fe	(0.34)	(0.51)	
	$En_{30.9}$	Mg	(0.62)	—	
P3	$Wo_{27.0}$	Ca	—	(0.54)	
Fe-augite	$Fs_{56.3}$	Fe	(0.58)	(0.54)	
	$En_{16.7}$	Mg	(0.34)	—	

Fraction (A), the Mg-rich Ca-poor pigeonite concentrate, appears to contain some magnesium in $M2$ and some iron in $M1$ and is, therefore, not completely ordered; $k(T)$ is relatively high for lunar pyroxenes. This concentrate contains about 15% exsolved augite. If ordering in clinopyroxenes at a given temperature increases with calcium content, as may be inferred from previous Mössbauer studies, this exsolved augite should be more ordered than the "bulk" separate, and therefore the pure unmixed pigeonite host should be slightly more disordered than the result for the "bulk" exsolution intergrowth indicates.

Fraction (B), the Mg-rich augite, would appear from the results in Table 2 to be somewhat disordered, although $k(T)$ is much smaller than that for the pigeonite (A). This concentrate contains 35% exsolved pigeonite. If this exsolved pigeonite has approximately the same calcium content and state of ordering as the exsolved pigeonite host in fraction (A), it would by itself account for all of the apparent disorder in fraction (B). This would require that the exsolved augite host in fraction (B) be *completely* ordered, in order to obtain the average result for the "bulk" fraction (B). Actually, there is a hint of an anomaly in these numbers, because with this much exsolved disordered pigeonite present, we would expect the "bulk" fraction (B) to be slightly more disordered than the results indicate. However, the difference is within the precision of the chemical and Mössbauer determinations.

In fraction (C), the iron-rich sub-calcic augite, there is a definite anomaly. This fraction is composed of 50% host augite and 50% exsolved pigeonite. There is good reason to believe, from their cell parameters (Ross et al., 1971a, b), that the exsolved phases have calcium contents comparable to those of the unmixed augite and pigeon-ite in fractions (A) and (B). We assume for the moment that the Fe/Mg ratio of pyroxenes has little effect on cation ordering in terms of $k(T)$. Because the results for fraction (A) and (B) indicated that exsolved pigeonite is significantly disordered, and exsolved augite is completely ordered, we would predict that fraction (C) in bulk would have a state of ordering intermediate to exsolved augite and pigeonite, that is, (C) should be about half as disordered as the pure exsolved pigeonite host in fraction (A), or slightly more than half as disordered as the "bulk" fraction (A). Instead of indicating partial disorder in (C), i.e., the presence of some magnesium in $M2$, the results show an *excess* of iron and calcium alone in $M2$.

We must note here that the result for fraction (C) is not unique to our study, but is completely consistent with the results of Hafner *et al.* (1971) for similar augite fractions from the same rock, 12021, as shown in Table 2. The fractions P2, (C) and P3 have increasing iron content and also increasing apparent excess of $M2$ cations in that order. It appears to be the rule, rather than the exception, that Mössbauer measurements indicate an excess of cations in $M2$ of iron-rich augites. Hafner *et al.* (1971) explained this as a result of inaccuracy in chemical characterization of the pyroxene fractions, because of sampling problems in microprobe analysis. The method used in the present study, whereby the same sample was used in the chemical analysis as in the Mössbauer absorber, and the sample was analyzed in bulk, allows us to eliminate this possibility.

It is extremely unlikely that the apparent excess of cations in $M2$ could be caused by the presence of calcium in $M1$. Calcium in $M1$ has not been indicated in any x-ray structural refinement of natural pyroxenes. Also, there is no reason to expect calcium in $M1$ only in the iron-rich pyroxenes. Magnesian augites such as fraction (B) have a higher calcium content than the iron-rich augites and crystallized at higher tempera-tures (Ross *et al.,* 1971a, b), which suggests that they would be much more likely to have calcium in $M1$. The only reasonable explanation for the apparent excess of cations in $M2$ is an overestimation of the $M2$ iron site occupancy, because of a bias in the Mössbauer measurements.

CAUSES OF THE MÖSSBAUER MEASUREMENT BIAS

It has recently been shown (Williams *et al.,* 1971) that the Mössbauer spectra of $C2/c$ pyroxenes are anomalous—the relative area for the $M2$ doublet in room tempera-ture spectra is often far larger than it is in liquid-nitrogen temperature spectra. It was furthermore found in spectra of synthetic hedenbergite-ferrosilite pyroxenes (Dowty and Lindsley, 1971, 1972) that the relative area of $M2$ is larger than expected even at liquid-nitrogen temperature (Fig. 2).

To explain this anomaly it must be supposed that even in seemingly homogeneous clinopyroxenes the $M1$ absorption is somehow split, so that part of it lies effectively under the $M2$ doublet, enhancing the apparent area of the latter. Williams *et al.* (1971) postulated a domain structure in $C2/c$ pyroxenes, but because there is no independent evidence for domains, a better explanation is that the splitting is a result of differing next-nearest neighbor configurations for iron atoms in $M1$ (Dowty and Lindsley,

1972). The $M1$ coordination octahedron shares edges with three $M2$ coordination polyhedra. Varying occupancies of these $M2$ positions, particularly occupancy by calcium versus iron or magnesium, can lead to distinct types of local environments for iron atoms in $M1$. The $M2$ absorption is not subject to this kind of splitting, primarily because it has only $M1$ cations as next-nearest neighbors, and $M1$ does not contain calcium.

The anomaly in room temperature spectra has been reduced by fitting more than one $M1$ doublet; two $M1$ doublets were assumed by Williams $et\ al.$ (1971), and up to four by Dowty and Lindsley (1972). Fitting multiple $M1$ doublets to the liquid-nitrogen temperature spectra of the synthetic hedenbergite-ferrosilite pyroxenes was not successful, apparently because the $M1$ doublets are closer to each other at low temperature. Our best estimate of the iron site occupancies must still come from the two-doublet (one $M1$ doublet) fits of liquid-nitrogen temperature spectra, but these give biased results, as illustrated in Fig. 2.*

We should add that regardless of whether it is the next-nearest neighbor hypothesis, the domain hypothesis, or some other hypothesis which is accepted, a bias has been experimentally demonstrated. The room temperature bias is certainly undeniable, and it would be indeed fortuitous if the causative factors were completely absent at lower temperatures.

The type of bias just discussed is a function of composition, particularly calcium content, and is expected to be present whether or not the particular specimen under

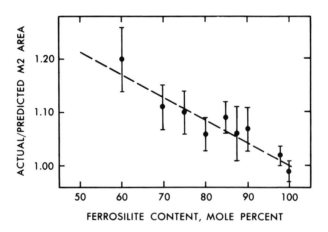

Fig. 2. Bias in liquid-nitrogen temperature Mössbauer measurements of synthetic hedenbergite-ferrosilite pyroxenes (Dowty and Lindsley, 1972). Actual $M2$ area is that measured with two-doublet (one $M1$ doublet) fits to the spectra. Predicted $M2$ area assumes that calcium is in $M2$ only. Ferrosilite content of 50 mole % represents hedenbergite, $Wo_{50}Fs_{50}$.

*The discrepancies in the liquid-nitrogen temperature spectra shown in Fig. 2 were originally ascribed to the presence of calcium in $M1$ (Dowty and Lindsley, 1971). However, in view of the much larger discrepancies subsequently verified in the room temperature spectra, and other considerations discussed by Dowty and Lindsley (1972) and in the text above, the presence of calcium in $M1$ is now considered to be very unlikely.

consideration has unmixed. There may be a second cause of bias in natural pyroxenes, which can be attributed directly to exsolution. Even if we assume that a single doublet for each site could accurately represent the site occupancies of homogeneous pyroxenes, there is some reason to believe that the respective $M1$ doublets of each phase in an exsolution intergrowth would not be exactly coincident. To demonstrate this, we compare the Mössbauer parameters of two specimens in the synthetic hedenbergite-ferrosilite series that are analogous to the exsolved phases in a natural intergrowth: $Fs_{60}Wo_{40}$, which has calcium content comparable to that of the augite in an exsolution intergrowth; and $Fs_{90}Wo_{10}$, which is comparable to exsolved pigeonite. When only two doublets (one $M1$ doublet) were fitted to liquid-nitrogen temperature spectra of these specimens, quadrupole splittings of 2.88 and 3.07 mm/sec, respectively, were obtained for the $M1$ doublets in $Fs_{60}Wo_{40}$ and $Fs_{90}Wo_{10}$. On the other hand, the $M2$ doublets in the two specimens were found to have almost exactly the same quadrupole splitting, about 1.96 mm/sec. The isomer shift of the $M1$ doublet was about the same in the two specimens, as was that of the $M2$ doublet. Thus, the two $M2$ doublets of a pigeonite-augite intergrowth would be almost exactly superimposed, but the $M1$ doublets would be slightly separated. This further "splitting" of the $M1$ absorption would probably lead to additional underestimation of the $M1$ site occupancy. This type of bias may be less serious than the first type, because some measurements (Hafner and Virgo, 1970; Hafner et al., 1971) have suggested that quadrupole splitting does not vary significantly with calcium content in natural magnesian lunar pyroxenes, in contrast with the results for the synthetic series. However, these measurements themselves represent averages of augite and pigeonite, because they were made on material which was, in most cases, exsolved.

It has often been assumed in the past that if the line widths obtained from least-squares fits of Mössbauer spectra are not excessively large, there are no unaccounted-for components in the spectra, that is, extra peaks or "split" absorptions. This may be a valid criterion when only one doublet is expected, or when the peaks in question are almost completely resolved, but when multidoublet spectra with only partial resolution are considered, it may be misleading. In the latter case some of the "split" absorption of one doublet may lie effectively under a different doublet, which would tend to minimize the apparent broadening of the lines. Such is apparently the case in clinopyroxenes.

In natural pyroxenes it is impossible to separate the effects of the two kinds of bias or to predict the exact extent of either. The first effect might well be greater in natural pyroxenes than in synthetic hedenbergite-ferrosilite pyroxenes with the same calcium content, because of the presence of Mg and especially such cations as Al, Ti and Cr, which would increase the number of possible next-nearest neighbor configuration for iron atoms in the $M1$ site. It is out of the question to try to fit separate doublets for all these configurations in both exsolved phases, as we have no idea what their Mössbauer parameters might be. The effect of these biases on absolute site occupancies would depend of course on the amount of iron present and its actual site distribution between $M1$ and $M2$ sites (and therefore on the state of ordering of the specimen) and on the relative proportions of exsolved phases.

In the case of the 12021 pyroxenes, the effect is probably different for each of the different fractions. The pigeonite, fraction (A), has low iron and low calcium content, and also has very little exsolution. Because the next-nearest neighbor effect is least in low-calcium pyroxenes, as shown in Fig. 2, and because the total error introduced by a bias in the Mössbauer spectra is proportional to the amount of iron present, the results for fraction (A), and also for fraction P1 of Hafner *et al.* (1971) probably are close to accurate.

For the augite fractions (B) and (C), and fractions P2 and P3 of Hafner *et al.* (1971), we cannot assume that the results are close to accurate. Of course, only a relatively small bias is definitely proven, by means of the "excess" of cations in $M2$, but here it is pertinent to examine the data on cation ordering in low- and high-calcium pyroxenes obtained independently by x-ray structure analysis. X-ray refinements of coexisting lunar pigeonite and augite (Takeda, 1972a) and coexisting terrestrial orthopyroxene and augite (Takeda, 1972b) indicate that the augite in both cases is more *disordered* than the low-calcium pyroxene. This directly contradicts the inference, which one would make from previous Mössbauer studies that high-calcium pyroxenes tend to be more *ordered* than low-calcium pyroxenes. There also are possible sources of error in the x-ray method, but in lieu of more concrete evidence, we suggest that this disagreement could be accounted for by a large bias in the Mössbauer measurements.

If we do not accept a large bias in the Mössbauer measurements, there are some very serious difficulties in the interpretation of cation ordering data. Let us assume for the moment that the bias in the iron-rich augites is only large enough to account for the "excess" of iron in $M2$. This would mean that the bulk fraction (C) is fully ordered, and that *both* the augite and pigeonite, which are unmixed from this fraction, are fully ordered. Because the Mg-rich pigeonite (A) is not fully ordered, we would have to conclude that the difference in ordering between the unmixed pigeonites in (A) and (C) is due to either (1) an increase in cation ordering with increasing Fe/Mg ratio, or (2) an increase in cation ordering with the amount of exsolution—the fractions (A), (B), and (C) contain respectively 15, 35, and 50% exsolved phases, and have increasing apparent cation order in that sequence.

Even if we do accept a large bias in the Mössbauer measurements, hypothesis (2) should not be dismissed lightly, because homogenization experiments (Ross *et al.*, 1971a, b) have shown that iron-rich augites such as fraction (C) crystallized in the temperature range of 1050°–950°C, and probably were in the process of exsolution at temperatures well below 900°C. This is within the temperature range of cation ordering processes in pyroxenes. If calcium, iron, and magnesium are mobile between exsolution lamellae, this might well cause a closer approach to equilibrium with respect to intralamellar cation distribution, and therefore greater ordering. Mueller (1969) discusses some of the kinetic aspects of multiple distribution equilibria like this.

It might seem that the determination of equilibration temperatures by means of heating experiments combined with Mössbauer measurement is still a valid procedure. There are complications, however, if exsolution occurred contemporaneously with or after the last significant cation exchange between $M1$ and $M2$ on original cooling. In this case, it would be necessary to cause an appropriate remixing of lamellae at the same time as the $M1$–$M2$ distribution is disordered by heating. This might be difficult in a laboratory experiment if the rates for the two processes are different.

Acknowledgments—Publication authorized by the Director, U.S. Geological Survey. Research supported by NASA. E. D. was NRC-USGS Research Associate during this work. We appreciate critical comments by P. Toulmin III and J. J. Papike.

REFERENCES

Bence A. E. Papike J. J. and Prewitt C. T. (1970) Apollo 12 clinopyroxenes: Chemical trends. *Earth Planet. Sci. Lett.* **8**, 393–399.

Burnham C. W. Ohashi Y. Hafner S. S. and Virgo D. (1971) Cation distribution and atomic thermal vibrations in an iron-rich orthopyroxene. *Amer. Mineral.* **56**, 850–876.

Dowty E. and Lindsley D. H. (1970) Mössbauer spectroscopy of synthetic Ca-Fe pyroxenes. *Carnegie Inst. Wash. Year Book* **69**, 190–193.

Dowty E. and Lindsley D. H. (1972) Mössbauer spectra of synthetic hedenbergite-ferrosilite pyroxenes. Submitted to *Amer. Mineral.*

Finger L. W. Hafner S. S. Schürmann K. Virgo D. and Warburton D. (1972) Distinct cooling histories and reheating of Apollo 14 rocks (abstract). In *Lunar Science—III* (editor C. Watkins), pp. 259–261, Lunar Science Institute Contr. No. 88.

Gay P. Bancroft G. M. and Bown M. G. (1970) Diffraction and Mössbauer studies of minerals from lunar soils and rocks. *Proc. Apollo 11 Lunar Sci. Conf., Geochim. Cosmochim. Acta* Suppl. 1, Vol. 1, pp. 481–497. Pergamon.

Ghose S. (1965) Mg^{2+} – Fe^{2+} order in an orthopyroxene, Mg$_{0.93}$Fe$_{1.07}$Si$_2$O$_6$. *Z. Kristallogr.* **122**, 81–89.

Ghose S. Ng. G. and Walter L. S. (1972) Clinopyroxenes from Apollo 12 and 14: Exsolution, cation order and domain structure (abstract). In *Lunar Science—III* (editor C. Watkins), pp. 300–302, Lunar Science Institute Contr. No. 88.

Hafner S. S. and Virgo D. (1970) Temperature-dependent cation distributions in lunar and terrestrial pyroxenes. *Proc. Apollo 11 Lunar Sci. Conf., Geochim. Cosmochim. Acta,* Suppl. 1, Vol. 3, pp. 2183–2198. Pergamon.

Hafner S. S. Virgo D. and Warburton D. (1971) Cation distributions and cooling history of clinopyroxenes from Oceanus Procellarum. *Proc. Second Lunar Sci. Conf., Geochim. Cosmochim. Acta,* Suppl. 2, Vol. 1, pp. 91–108. M.I.T. Press.

Mason B. (1966) Pyrope, augite and hornblende from Kakanui, New Zealand. *N. Z. J. Geol. Geophys.* **9**, 474–480.

Mueller R. F. (1969) Kinetics and thermodynamics of intracrystalline distributions. *Mineral Soc. Amer. Spec. Pap.* **2**, 83–93.

Papike J. J. Bence A. E. Brown G. E. Prewitt C. T. and Wu C. H. (1971) Apollo 12 clinopyroxenes: Exsolution and epitaxy. *Earth Planet. Sci. Lett.* **10**, 307–315.

Ross M. Huebner J. S. and Dowty E. (1971a) Melting and subsolidus phase relations of augite and pigeonite from lunar rock 12021. Second Lunar Science Conference unpublished proceedings.

Ross M. Huebner J. S. and Dowty E. (1971b) Delineation of the one atmosphere augite-pigeonite solvus and orthopyroxene-pigeonite reaction curve, and pigeonite melting relations (abstract). *Geol. Soc. Amer. Ann. Meeting,* 1971, Abstracts with Programs.

Saxena S. K. and Ghose S. (1970) Order-disorder and the activity-composition relation in a binary crystalline solution. Part 1. Metamorphic orthopyroxene. *Amer. Mineral.* **55**, 1219–1225.

Saxena S. K. and Ghose S. (1971) Mg^{2+} – Fe^{2+} order-disorder and the thermodynamics of the ortho-pyroxene crystalline solution. *Amer. Mineral.* **56**, 532–559.

Takeda H. (1972a) Structural studies of rim augite and core pigeonite from lunar rock 12052. *Earth Planet. Sci. Lett.* (in press).

Takeda H. (1972b) Crystallographic studies of coexisting aluminian orthopyroxene and augite of high-pressure origin. In preparation.

Virgo D. and Hafner S. S. (1968) Re-evaluation of the cation distribution in orthopyroxenes by the Mössbauer effect. *Earth Planet. Sci. Lett.* **4**, 265–269.

Virgo D. and Hafner S. S. (1969) Fe^{2+} − Mg order-disorder in heated orthopyroxenes. *Mineral. Soc. Amer. Spec. Pap.* **2**, 67–81.

Virgo D. and Hafner S. S. (1970) Fe^{2+}, Mg order-disorder in natural orthopyroxenes. *Amer. Mineral.* **55**, 201–223.

Williams P. G. L. Bancroft G. M. Bown M. G. and Turncock A. C. (1971) Anomalous Mössbauer spectra of $C2/c$ clinopyroxenes. *Nature Phys. Sci.* **230**, 149–151.

Proceedings of the Third Lunar Science Conference
(Supplement 3, *Geochimica et Cosmochimica Acta*)
Vol. 1, pp. 493–506
The M.I.T. Press, 1972

Distinct subsolidus cooling histories of Apollo 14 basalts

K. Schürmann and S. S. Hafner

Department of the Geophysical Sciences
The University of Chicago
Chicago, Illinois 60637

Abstract —The Mg^{2+}, Fe^{2+} exchange reaction between the $M1$ and $M2$ sites is analyzed in natural and heated samples of pigeonite from basalt 14053,47 and orthopyroxene from basalt 14310,116. The distribution coefficient k_n in orthopyroxene 14310 is smaller than the critical coefficient k_c. It corresponds to an equilibrium temperature of approximately 600°C. The subsolidus cooling of basalt 14310 occurred at the depth of several meters in a coherent body of appreciable size.

k_n of pigeonite 14053 is significantly greater than k_c. It corresponds to an equilibrium temperature of approximately 840°C. 14053 must have been quenched extremely rapidly from a temperature higher than T_n probably by impact at Fra Mauro as a fragment of a larger body from the Imbrian ejecta.

Introduction

The distribution of magnesium and ferrous iron over the $M1$ and $M2$ positions in pyroxenes from the basaltic rocks 14053 and 14310 has been studied in order to analyze the subsolidus cooling history.

The principles of this approach have been described elsewhere (Hafner *et al.*, 1971, particularly in the section "cooling history"). Here, we restrict ourselves to a summary of the relevant points. The kinetics of the Mg^{2+}, Fe^{2+} exchange between $M1$ and $M2$ may be described by three characteristic temperature intervals. Between the melting point and a certain critical temperature T_c, the exchange reaction is exceedingly rapid. Disordering as well as ordering can be studied by heating crystals at constant temperatures in the laboratory, and the reaction rates and equilibrium conditions can be analyzed experimentally. The data obtained in this range are apparently consistent with the ideal solution model, i.e., assuming ideal distribution of Mg^{2+} and Fe^{2+} at each M position, at least to a first approximation. A steady state of maximum disorder is attained in crystals quenched at temperatures higher than T_h. The distribution coefficient (equilibrium constant) k approaches an apparent maximum value k_h, which is smaller than one and cannot be exceeded at higher temperatures until the melting point is reached. Here, the distribution coefficient (equilibrium constant) is

$$k = (Fe^{2+}/Mg^{2+})_{M1}/(Fe^{2+}/Mg^{2+})_{M2},$$

and Fe^{2+}/Mg^{2+} is the ratio of the iron and magnesium ions at the position in question. The critical point T_c is the temperature at which a critical degree of Mg^{2+}, Fe^{2+} ordering described by the critical distribution coefficient k_c is at equilibrium. Here k_c determines a well-defined steady state in an ordering experiment at constant temperature (Hafner and Virgo, 1970, Fig. 8). Excess ordering beyond k_c cannot be attained in the laboratory because the rate constants for ordering are too small below T_c; geological times are required. The second temperature interval is limited by T_c and the quench-in temperature T_q. Only disordering can be studied in the laboratory at temperatures in this range. The Mg^{2+}, Fe^{2+} distribution equilibrated at T_q yields the minimum distribution coefficient, $k = k_q$, which represents the maximum degree of ordering. Ordering in excess of k_q cannot be attained even over exceedingly long geological times. In the third temperature interval, i.e., at temperatures below T_q, the exchange reaction is terminated. No change in site occupancy occurs, even for a disordered pyroxene, no matter how long the time may be.

At temperatures above T_c the kinetics appear to be, to a first approximation, independent of the concentrations of minor elements and lattice defects in the pyroxenes, as concluded from the heating experiments that have been carried out until this time. Below T_c this may not be so, particularly not for the ordering reaction. T_c (and k_c) depends somewhat on the temperature of the experiment (Hafner and Virgo, 1970) and varies approximately between 600 and 800°C. It depends critically on shock effects (Dundon and Hafner 1971), and probably also on the presence of microexsolution (Hafner et al., 1971) and on chemical composition (Schürmann et al., 1972). It should be determined individually for each pyroxene. The T_q appears to be close to 480°C for orthopyroxenes and calcium-poor pigeonites, but more data will be needed to analyze its dependence on space group, chemical composition, etc.

Distribution coefficients k_n observed in natural pyroxenes may be divided into three categories:

$$k_h > k_n > k_c.$$

In this range the natural distribution can be approached experimentally from higher or lower temperatures either by ordering or by disordering. The cooling history can be determined quantitatively by numerical integration of the rate equation (Hafner and Virgo 1970, equations 1 and 2) with respect to temperature and time, assuming certain heat-flow models. Natural pyroxenes with $k_n > k_c$ are rare. They occur in extremely rapidly cooled rocks, e.g., in spatters of lava vents quenched in the atmosphere, in spines at the surface of lava flows, or in lava flows at depths of not more than a few meters.

$$k_c > k_n > k_q$$

The cooling history of pyroxenes with Mg, Fe distributions in this range cannot be analyzed quantitatively, because the rate constants of ordering are not known below T_c. However, some qualitative information is obtained from inspection of the terms $T_c - T_n$ or $k_c - k_n$ (Hafner et al., 1971). Distributions within this range require fairly slow cooling. The crystals must have been located at least at a depth of several meters in a large size body during cooling. Pyroxenes from volcanic rocks commonly will fall into this range.

$$k_n = k_q.$$

Pyroxenes with Mg^{2+}, Fe^{2+} distributions equilibrated at the quench-in temperature T_q are found in exceedingly slowly cooled plutonic igneous or regional metamorphic rocks.

DESCRIPTION OF ROCK SPECIMENS

Two Apollo 14 rocks have been studied: the medium to coarse-grained basalt 14053,47 and the fine-grained basalt 14310,116.

Rock 14053 (total weight 251 gm) was collected at station C2, 110 m south of the rim of the Cone Crater (NASA PSR 1971). It is probably a clast from "filleted rock," a 1–2 m diameter boulder that is a very rounded and friable-looking coarse breccia (NASA PSR 1971). The moderately inhomogeneous rock shows an ophitic texture with a grain size of 0.5–2 mm and contains 40–45% plagioclase ($Ab_{10-20}Or_{1-5}An_{75-90}$), 45–50% clinopyroxene, 3% ilmenite, and minor amounts of olivine, cristobalite, troilite, ulvöspinel, metallic iron, glass, etc. (Gancarz *et al.*, 1971). In thin section 14053,56, olivine is relatively abundant and occurs commonly as cores of pigeonite phenocrysts that may have augite rims. However, in our chip 14053,47 large phenocrysts were not present, and the amount of olivine was less than 1%.

Rock 14310 (total weight 3439 gm), was collected as a grab sample from the "smooth region" close to the landing site, probably at Station G (NASA PSR 1971). It is a fine-grained basaltic rock with an intergranular to intersertal structure and a grain size between 0.1 and 0.5 mm.

It consists of 50–55% plagioclase ($Ab_{5-20}Or_{1-4}An_{75-95}$), 35–40% ortho- and clinopyroxene and minor components such as ilmenite, olivine, troilite, ulvöspinel, glass, metallic iron, etc. (Gancarz *et al.*, 1971). In the section 14310,175 two different generations of plagioclase and orthopyroxene crystals with small rims of pigeonite and/or augite are present.

EXPERIMENTAL TECHNIQUES

Mineral separations

The rock chips 14053,47 (3.83 gm) and 14310,116 (4.30 gm) were crushed and sieved for a yield of fractions $> 149\ \mu$, $74–149\ \mu$, and $44–74\ \mu$. Pyroxene and feldspar were concentrated by heavy liquid separations (Table 1). All separates except 14053,47-P9 were finally purified by hand picking. A pigeonite concentrate 14053,47-P9 (Table 1) was purified by use of a FRANTZ-Magnetic separator and the final

Table 1. Separated pyroxene fractions.

Number	Pyroxene	Grain size	Specific gravity range
14053,47-P1	pigeonite	$> 149\ \mu$	3.45–3.56
14053,47-P4	pigeonite	$74–149\ \mu$	3.45–3.56
14053,47-P9	pigeonite[a]	$44–74\ \mu$	3.45–3.56
14310,116-P1	orthopyroxene[b]	$74–149\ \mu$	3.30–3.45
14310,116-P4	orthopyroxene[b]	$44–74\ \mu$	3.30–3.45

[a] Contaminated by approximately 5–7% of subcalcic augite crystals.
[b] Contaminated by some pigeonite crystals (less than 20%; the contamination could not be detected in a Guinier diffraction pattern). The contamination of fraction P4 was probably higher than that of P1.

specimen was contaminated by about 5–7% of subcalcic augite crystals. Ortho-pyroxene of 14310 was separated from clinopyroxene using diluted Clerici solutions and the concentrate was tested by taking an x-ray diffraction pattern with a Guinier camera.

Chemical composition of the pyroxenes

X-ray emission microanalyses for Na, Mg, Al, Si, Ca, Ti, Cr, Mn, and Fe were carried out by Dr. D. Virgo at the Geophysical Laboratory. For a detailed description of analytical technique, see Virgo (1971). The composition of our pyroxene separates are shown in Table 2 and plotted in Figs. 1 and 2.

Heating experiments

Clino- and orthopyroxene samples were heated in evacuated quartz tubes $(10^{-4} - 10^{-5}$ mm Hg) under controlled temperatures in a calibrated platinum furnace for different time periods. Temperatures were continuously recorded on a calibrated potentiometer. The maximum errors are for $500°C: \pm 3°$, for $1000°C: \pm 5°$, and for $1100°C: \pm 10°$. For the rate constant determination the heat flow characteristics of the furnace were examined by calibrated test runs. After the heating, the quenched samples were crushed for optical examination. No decomposition or oxidation could be observed visually or in the Mössbauer spectra.

Mössbauer spectroscopy

The resonant absorption spectra of ^{57}Fe were taken with a constant acceleration system using a 1024 channel analyzer and a symmetric velocity wave form (500 channels per spectrum). The absorbers consisted of 18–32 mg of pyroxene, which were distributed homogeneously over an area of 1.27 cm^2. This corresponded to densities of 2–3 mg natural iron per cm^2. No thickness corrections were employed. Least-squares fits were made assuming four Lorentzian lines to the uncorrected analyzer counts without employing constraints. The positive and negative parts of the spectra were fitted independently.

RESULTS

The Mössbauer spectra were analyzed on the basis of the same assumptions as reported previously (Hafner and Virgo, 1970, pp. 2186–2192; Hafner et al., 1971, pp. 96–98). Our pyroxene separates were, of course, not homogeneous. The standard

Table 2. Chemical composition of pyroxene separates.

Number	Pyroxene	Average composition			Standard deviation[a]			Number of spot analyses[b]
		Wo	En	Fs	Wo	En	Fs	
14053,47-P1	pigeonite	11.6	58.1	30.3	2.9	6.1	3.9	50
14053,47-P4	pigeonite	12.8	57.2	30.0	4.0	5.6	2.7	31
14310,116-P1	orthopyroxene	6.6	69.7	23.7	3.6	7.2	4.3	25

[a] $\sigma = ([vv]/(n - 1))^{\frac{1}{2}}$.
[b] Na, Mg, Al, Si, Ca, Ti, Cr, Mn, Fe determined (Virgo 1971).

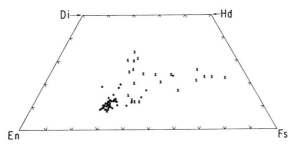

Fig. 1. Scatter in the chemical compositions of pigeonite separate 14053,47-P1 (dots). The subcalcic augite separate (crosses) was not studied.

deviations of 30–50 spot analyses on each separate are shown in Table 2. An electron microscopy study by Fernández-Morán *et al.* (1972) revealed the common presence of exsolution lamellae with widths of 200–400 Å in a large number of pigeonite crystals from separate 14053,47-P1. The total amount of augite is difficult to estimate because commonly no exsolution lamellae could be detected over large areas of a selected crystal, whereas in other areas of the same crystal lamellae occur in high concentrations. Pigeonite crystals from this separate were heated at 1000°C for 7 days in evacuated tubes and examined. Approximately 40–60% of the lamellae disappeared. After heating at 1100°C for 6 days the total number of lamellae was drastically reduced, but a few, sporadic lamellae were still present. A large number of crystals were apparently homogeneous. We believe that in our pigeonite heating experiments for the Mössbauer study the duration of the heating was not long enough to substantially homogenize the crystals. An electron microscopy study of orthopyroxene crystals from the separate 14310,116-P1 showed that most of the selected crystals were free of exsolution lamellae (Dr. M. Ohtsuki, personal communication).

Site occupancy

The site occupancy was obtained from the Fe^{2+} distribution numbers that were referred to the average ferrosilite component of the absorber. Ca^{2+} was assumed to be located exclusively at $M2$ sites. The remaining unoccupied fraction of the sites

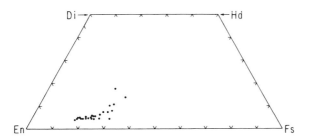

Fig. 2. Scatter in the chemical compositions of orthopyroxene separate 14310,116-P1. The figure shows some contamination by foreign pyroxenes, predominantly pigeonite (approximately 6 out of 25 spots).

was assigned to Mg^{2+}. Al, Ti, Cr, and Mn concentrations at the M sites were ignored. This seemed reasonable, because the data were interpreted in terms of an idealized binary Mg^{2+}, Fe^{2+} exchange reaction and the distribution of the minor elements over the M sites at elevated temperatures is not known. The spectrum data and site occupancy are reported in Tables 3 and 4. The ^{57}Fe data of a pigeonite separate $(Wo_{11}En_{60}Fs_{29})$ from 14053,44 and a mixed pigeonite-orthopyroxene separate $(Wo_9En_{65}Fs_{26})$ from 14310,115 reported by Ghose et al. (1972) are in good agreement with our values.

The least-squares fits of the pigeonite and orthopyroxene spectra (assuming four Lorentzian lines with no constraints) were considered as acceptable from a statistical point of view. Typical χ^2 values of the orthopyroxene spectra varied between 1.1 and 1.3 units per channel for counting rates of approximately 0.7×10^6. The χ^2 of the pigeonite spectra was generally somewhat higher (1.3–1.5). A spectrum of pigeonite 14053,47-P1 is shown in Fig. 3.

Brown et al. (1972b) recently reported distribution coefficients $k = 0.07$ for a natural pigeonite crystal of the composition $Wo_5En_{65}Fs_{30}$ from basalt 12021, and $k = 0.16$ for the same crystal after heating at 1100°C for 9 days (homogenized composition $Wo_9En_{61}Fs_{30}$). These values which were obtained from crystal structure refinements using x-ray diffraction, may be compared with $k = 0.08$ for our pigeonite separate 12021,150-Pl and $k = 0.18$ for the same separate heated at 1000°C for one day (Hafner et al., 1971). The maximum differences in occupancy occur for the

Table 3. Pyroxenes from Apollo 14: Peak heights and widths at 77°K.*

Specimen	A_1		B_1		B_2		A_2	
	I	FWHH (mm/sec)	I	FWHH (mm/sec)	I	FWHH (mm/sec)	I	FWHH (mm/sec)
Pigeonite								
14053,47-P1 natural	0.0270	0.325	0.0868	0.314	0.0758	0.359	0.0264	0.305
14053,47-P1 550°C, 12h	0.0150	0.300	0.0675	0.318	0.0588	0.345	0.0153	0.284
14053,47-P1 550°C, 3d	0.0188	0.326	0.0829	0.305	0.0714	0.341	0.0196	0.303
14053,47-P1 1000°C, 1d	0.0213	0.322	0.0587	0.317	0.0491	0.363	0.0213	0.320
14053,47-P4 natural	0.0170	0.321	0.0707	0.309	0.0598	0.344	0.0178	0.299
14053,47-P4 1000°C, 1 min	0.0182	0.336	0.0515	0.340	0.0448	0.376	0.0187	0.303
14053,47-P4 1000°C, 5 min	0.0103	0.304	0.0273	0.314	0.0226	0.333	0.0102	0.322
Orthopyroxene								
14310,116-P1 natural	0.0129	0.322	0.0719	0.321	0.0612	0.365	0.0126	0.290
14310,116-P4 550°C, 3d	0.0129	0.267	0.0691	0.326	0.0577	0.364	0.0123	0.302
14310,116-P4 1000°C, 21d	0.0212	0.354	0.0664	0.359	0.0563	0.403	0.0215	0.342

* The nuclear quadrupole splittings at $M1$ and $M2$ are for pigeonite 14053: 3.03 and 2.12, and for orthopyroxene 14310: 3.01 and 2.15, respectively; the isomer shifts (referred to a metallic iron absorber at 295°K) at $M1$ and $M2$ are for pigeonite 14053: 1.29 and 1.27 and for orthopyroxene 14310: 1.29 and 1.27, respectively (numbers in mm/sec). No changes in splittings and shifts were observed after the heat treatments.

Table 4. Apollo 14 pyroxenes: site occupancies and distribution coefficients*.

Specimen	Ratio Area $\dfrac{A(M1)}{A(M1+M2)}$	Peak height $\dfrac{I(M1)}{I(M1+M2)}$	M1 Mg	M1 Fe	M2 Mg	M2 Fe	Ca	Distribution coefficient k
Pigeonite								
14053,47-P1 natural	0.236(5)	0.247(2)	0.850	0.150(3)	0.312	0.456(4)	0.232	0.121(2)
14053,47-P1 550°C, 12h	0.175(6)	0.193(3)	0.883	0.117(3)	0.279	0.489(4)	0.232	0.076(2)
14053,47-P1 550°C, 3d	0.196(5)	0.199(3)	0.879	0.121(3)	0.283	0.485(4)	0.232	0.080(2)
14053,47-P1 1000°C, 1d	0.273(5)	0.283(3)	0.829	0.171(3)	0.333	0.435(4)	0.232	0.158(3)
14053,47-P4 natural	0.202(6)	0.210(3)	0.874	0.126(3)	0.268	0.474(4)	0.258	0.082(2)
14053,47-P4 1000°C, 1min	0.256(6)	0.277(3)	0.834	0.166(3)	0.308	0.434(4)	0.258	0.141(3)
14053,47-P4 1000°C, 5min	0.28(1)	0.291(5)	0.825	0.175(3)	0.317	0.425(4)	0.258	0.158(3)
Orthopyroxene								
14310,116-P1 natural	0.146(6)	0.160(3)	0.924	0.076(3)	0.470	0.398(8)	0.132	0.097(4)
14310,116-P1 550°C, 3d	0.141(7)	0.166(4)	0.921	0.079(4)	0.473	0.395(8)	0.132	0.103(5)
14310,116-P1 1000°C, 21d	0.242(5)	0.258(3)	0.877	0.123(5)	0.517	0.351(7)	0.132	0.206(9)

* The values are average values of two spectra (positive and negative acceleration) that were fitted independently. The numbers between parentheses are standard deviations (σ) determined from the error matrix of the least-squares fit of a single spectrum (positive or negative acceleration) and refer to the last decimal place shown. σ of columns 5, 7, and 9 was computed from column 3 and Table 2 using the error propagation law. Systematic errors are *not* accounted for in this Table and may be significantly higher than the listed standard deviations (see text). The total error of the site occupancy numbers for Fe^{2+} is estimated to be at least ± 0.02 and for k at least ± 0.01. The *differences* between two absorbers of the same pyroxene separate, e.g., in the heating experiments, probably are more accurate.

Mg^{2+} and Ca^{2+} numbers at the $M2$ sites (~ 0.08). In view of the somewhat different chemical compositions and applied simplifications, the data are within the experimental error. The k values observed in natural pigeonites until this time range between 0.09 and 0.04 (Table 5).

The site occupancy of our orthopyroxene separate 14310,116-P1 (Table 4) may be compared with a recent x-ray diffraction structure refinement by Takeda and Ridley (1972). The occupancy numbers are in accord within 0.02–0.04 for each cation, and the differences result primarily from the somewhat lower Ca^{2+} concentration of the crystal used by Takeda and Ridley (wollastonite component 4.5%). It should be noted that structure refinements of three pyroxenes from lunar basalts can be directly compared with Mössbauer data at this time. The Fe^{2+} occupancy at $M1$ and $M2$ deviates by not more than 0.04 in each case, and the major part of this deviation is almost certainly due to somewhat different chemical compositions of the crystals in question.

Mg^{2+}, Fe^{2+} *distribution coefficients and exchange energies*

The distribution coefficients of our natural and heated 14053 pigeonite and 14310 orthopyroxene separates are consistent with previously determined k values of

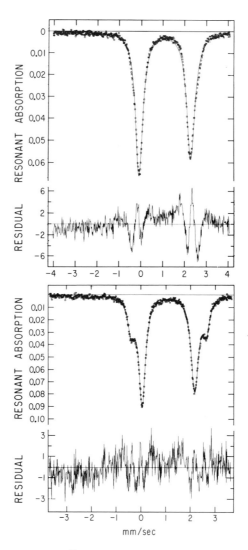

Fig. 3. Mössbauer spectrum of ^{57}Fe in pigeonite separate 14053,47-P1 at 295°K (upper spectrum, least-squares fit without constraints assuming two Lorentzian lines) and 77°K (lower spectrum, least-squares fit without constraints assuming four Lorentzian lines). Outer peaks: Fe^{2+} doublet at $M1$; inner peaks: Fe^{2+} doublet at $M2$. The solid line is a least-squares fit to the uncorrected data (one variable for background). The deviations of the solid line from the data (divided by the square root of the background) are plotted as "residual" below the spectra. The residual for the spectrum at 77°K is reasonably small. The χ^2 of this spectrum was 1.3 per data point (500 points).

pyroxenes with similar compositions. They are shown in Table 5. It would be interesting to know the effect of the low-clinopyroxene (P2$_1$/c) to high-clinopyroxene (C2/c) phase transition on ΔG_E^0, because some of the distribution coefficients in Table 5 refer to the high-clinopyroxene phase. It is assumed that the effect is small. The

Table 5. Distribution coefficients and Mg^{2+}, Fe^{2+} exchange energies of lunar and terrestrial pigeonites and orthopyroxenes.

Pyroxene	Composition			Temperature	k	ΔG^0	Reference
	En	Fs	Wo	(°C)		kcal mole	
Pigeonites							
12021,150-P1	59	32	9	natural	0.081		[a]
12021,150-P1	59	32	9	555,3d	0.077		[a]
12021,150-P1	59	32	9	800,1d	0.15	4.1	[a]
12021,150-P1	59	32	9	1000,1d	0.18	4.3	[a]
12021	65	30	5	natural	0.07		[b]
12021	61	29	10	1100	0.16		[b]
12053,79	57	32	11	natural	0.091		[a]
14053,47-P1	58	30	12	natural	0.12		this work
14053,47-P1	58	30	12	550,3d	0.080		this work
14053,47-P1	58	30	12	1000,1d	0.16	4.7	this work
14053,47-P4	57	30	13	natural	0.082		this work
14053,47-P4	57	30	13	1000,5 min	0.16		this work
Mull	39	52	9	natural	0.04		[c]
	39	52	9	960	0.14	4.9	[c]
Orthopyroxenes							
14310,116-P1	70	24	6	natural	0.097		this work
14310,116-P1	70	24	6	550,3d	0.10		this work
14310,116-P1	70	24	6	1000,21d	0.21	4.0	this work
14310,90	72	23	5	natural	0.11		[d]

[a] Hafner *et al.* (1971).
[b] Brown *et al.* (1972b).
[c] Brown *et al.* (1972a).
[d] Takeda and Ridley (1972).

difference between the unheated samples 14053-P1 and 14053-P4 probably reflects a trend due to distinct chemical differences between the two separates.

It has been suggested that the standard Gibbs free energy difference ΔG_E^0, for the Mg^{2+}, Fe^{2+} exchange between $M1$ and $M2$ in clinopyroxenes depends on the calcium concentration at $M2$ and probably increases in proportion to the wollastonite component (Hafner *et al.*, 1971). ΔG_E^0 has been estimated for orthopyroxenes with ratios $Fe/(Mg + Fe)$ less than 0.5 using the ideal solution model: $\Delta G_E^0 \approx 3.6$ kcal per molecular unit $M_2Si_2O_6$. It is not yet known whether the Mg^{2+}, Fe^{2+} distribution in pigeonites with some wollastonite component behaves as ideally as in magnesium-rich orthopyroxenes. In the following we assume that this is the case. At any rate, it seems that in pigeonites with wollastonite components of approximately 10%, ΔG_E^0 is approximately one kcal per mole higher.

Rate constants for the Mg^{2+}, Fe^{2+} reaction

Two heating experiments at 1000°C were carried out to study the rate constant K_{21} for the *disordering reaction* in pigeonite 14053,47-P4 at that temperature (rate equation and technique, Virgo and Hafner, 1969). Heating a pigeonite sample at 1000°C for one minute yielded an intermediate, nonequilibrated site occupancy that could be fitted to the integrated rate equation. The rate constant K_{21} was found to be ≤ 0.28 (min^{-1}), referred to two sites per volume unit. It is difficult to estimate an error limit for this value in view of the various theoretical assumptions (e.g., ideal solution model) and technical difficulties (e.g., temperature control and cooling characteristics of the furnace). However, after heating for 5 minutes at 1000°C the equilibrated state is almost completely attained as an additional experiment showed, which indicates that K_{21} must be larger than one-fifth of 0.28. Using K_{21} (1000°C) = 0.28 and K_{12} (500°C) = 6×10^{-5} (min^{-1}), as estimated for orthopyroxene (Virgo and Hafner, 1969), yields an activation energy for the disordering reaction of approximately 33 kcal per mole. This value, which seems to be higher than that for orthopyroxene, is a very rough estimate in view of scarcity of data on Mg^{2+}, Fe^{2+} kinetics in pigeonites. The rate constant for the *ordering reaction* can be estimated for the relationship

$$K_{21}/K_{12} = k(1000°C).$$

In summary, we believe that the present data on the kinetics of the exchange reaction in pigeonite 14053 are consistent with the previous values obtained from an orthopyroxene (Virgo and Hafner, 1969). The experimental errors are large and more heating experiments are needed for a precise analysis of the rate constants.

DISCUSSION

An attempt is made to analyze the subsolidus cooling history of orthopyroxene 14310 and pigeonite 14053 from the characteristics of the Mg^{2+}, Fe^{2+} exchange reaction and in terms of the distribution coefficients k_n, k_c and the corresponding equilibrium temperatures T_n, T_c. The practical range of temperature in which the Mg^{2+}, Fe^{2+} distribution over $M1$ and $M2$ depends critically on the cooling rate is 500–1000°C. The cooling histories of the two basalts are, in fact, significantly different. This is reflected in their natural distribution coefficients: in orthopyroxene 14310 k_n is lower than k_c whereas in pigeonite 14053 k_n is higher than k_c.

Cooling history of orthopyroxene 14310

The natural distribution coefficient, $k_n = 0.097$ (Table 5), of orthopyroxene 14310 is considerably higher than k_n values observed in slowly cooled terrestrial orthopyroxenes ($k_n = k_q < 0.05$, Virgo and Hafner 1970). However, a heating experiment at 550°C for 3 days ($k_{550} = 0.10$) did not change the cation distribution, within the experimental error. Therefore, k_n is smaller than k_c and, correspondingly, T_n is lower than T_c. If an exchange energy $\Delta G_E^0 = 4.0$ kcal per mole (Table 5) is assumed and if it is assumed that ΔG_E^0 is nearly invariant over the temperature interval 500–1000°C, one obtains an estimate of the equilibrium temperature, $T_n = 600°C$, for k_n This model temperature T_n is consistent with the heating experiment at 550°C and indicates that the various assumptions (ideal solution model, invariant ΔG_E^0) are approximately correct. If T_n is lower than T_c, but higher than the furnace temperature T_{exp}, the Mg^{2+}, Fe^{2+} distribution is in a steady state and no change in the distribution can be detected experimentally. This is confirmed by the experiment at $T_{exp} = 550°C$ as expected. The total error of T_n is estimated to be within

$\pm 50°$C. The T_n is certainly lower than T_c, which was generally found to be higher than 600°C in terrestrial orthopyroxenes (Hafner and Virgo, 1970) and in a lunar pigeonite (Hafner *et al.*, 1971).

The coefficient k_n of orthopyroxene 14310 is close to typical values observed in orthopyroxenes from terrestrial volcanic rocks. However, the cooling rate was *significantly lower* than rates on the surface of certain Hawaiian lava flows (Hafner and Virgo, 1970; Schürmann *et al.*, 1972), or at depths of approximately one meter from the surface. Mineralogical and petrological studies of basalt 14310 suggest that this rock crystallized under near-surface conditions (Bence and Papike 1972; Dence *et al.*, 1972; Finger *et al.*, 1972). A near-surface crystallization would also be in accord with the alkali-loss models of Brown and Peckett (1971) and the suggestion of Morgan *et al.* (1972) that 14310 may have crystallized from a molten pocket in the ejecta blanket at Fra Mauro. But the history of basalt 14310 may well be more complicated because small cognate, but finer grained inclusions in 14310 (Walter *et al.*, 1972) are indicative of some remelting process.

In summary, it is concluded that the natural Mg^{2+}, Fe^{2+} distribution in ortho-pyroxene 14310 indicates a subsolidus cooling rate of a fairly large body (i.e., of *at least* several meters diameter, cf. second part of discussion) under near-surface conditions at a depth of at least several meters. The cation distribution reflects the *most recent* cooling event from temperatures *higher* than T_q, i.e., approximately 480°C. It is consistent with the suggestion that 14310 was part of a molten pocket in the Fra Mauro ejecta blanket and was excavated without substantial reheating *after* cooling to at least $T_n = 550°$C or lower.

Cooling history of pigeonite 14053

The natural Mg^{2+}, Fe^{2+} distribution over $M1$ and $M2$ in pigeonite P1 from basalt 14053 is considerably disordered: $k_n = 0.12$. This distribution coefficient is significantly higher than $k_c = 0.076$, which was determined by the heating experiments at 550°C. Thus, k_n is greater than k_c and consequently T_n is higher than T_c. If it is assumed that the Mg^{2+}, Fe^{2+} exchange energy $\Delta G_E^0 = 4.7$ kcal per mole is nearly invariant at temperatures lower than 1000°C an equilibrium temperature of $T_n \sim 840°$C is obtained for k_n. The total error in T_n is probably within $\pm 50°$C. The natural Mg^{2+}, Fe^{2+} distribution of a different separate, P4, from the same rock is lower ($k_n = 0.082$) than that of P1, but no heating experiment at 550°C was carried out. We believe that this apparent difference is due to the distinct chemical composi-tions of the two separates (Table 2).

Values T_n higher than T_c have been reported for pyroxene crystals from ejected glassy spatters which were rapidly quenched in the atmosphere, but not for pyroxene crystals from holocrystalline rocks. Then $T_n > T_c$ is indicative of extremely rapid cooling, which cannot occur in a lava flow at some depth. But grain size and texture of 14053, as well as the presence of exsolution lamellae in the pigeonite crystals show that the original cooling rate of basalt 14053 was fairly slow (cf. also Papike and Bence, 1971). Two possible explanations for this discrepancy should be considered: (1) After its solidification and cooling down to lower temperatures, this rock was reheated and then cooled extremely rapidly. In this case the reheating temperature must have been somewhat higher than T_c and was probably associated with an impact event. One can now assume that the reheating and successively rapid cooling of 14053 may have been associated with a more recent event at the Fra Mauro region, for example the impact that formed Cone Crater. The exposure age of 14053 (≤ 30 m.y., Turner *et al.*, 1971) suggests that this rock was excavated by the Cone Crater impact. Substantial reheating by the Cone Crater impact is not likely. The position of 14053 at station C2, about 120 m from the rim of Cone Crater as a clast of a larger boulder, indicates a ballistic trajectory under low-energy conditions and an excavation from some depth of the crater. A shock intensity of about 100 kbar would not have been able to increase the temperature by more than about 100–150°C (Dr. W. Müller, written communication), and a substantially higher shock effect must be excluded, because rock 14053 does not exhibit any sign of shock effects. (2) The rock originated from a hot source due to some magmatic activity shortly before the Imbrian event. An additional heating in a range from not more than 100–150°C could have been added by shock heat during the Imbrian impact and some heating may have resulted from the impact of the debris at Fra Mauro. Extremely rapid cooling could have occurred by heat loss due to radiation during the flight. But cooling from an assumed initial temperature of 900°C by radiation during a projection of 15–30 minutes would only affect the Mg^{2+}, Fe^{2+} distribution in crystals from the surface or in the outermost rim of approximately one centimeter thickness (Gray *et al.*, 1972). A more likely quench-in process would be due to conduction after an early impact in the cold regolith at Fra Mauro region. This suggestion is supported by the fact that 14053 had been excavated as a part of a larger breccia boulder from the deeper parts of Cone Crater. It must have been located in the deeper parts of the Fra Mauro formation. Reheating of 14053 to temperatures less than T_q (i.e., $\sim$480°C) by an overlaying, hot ejecta blanket would not have disturbed the quenched Mg^{2+}, Fe^{2+} distribution.

In summary, we conclude that the observed natural Mg^{2+}, Fe^{2+} distribution in pigeonite of rock 14053 cannot be derived from the original cooling rate of a lava flow. It is the result of a more recent process that included reheating of rock 14053, probably by impact events at the Imbrium Basin and/or the Fra Mauro region and successive rapid cooling by radiation or conduction. The simplest and most likely process would be quenching the hot fist-size rock fragment in a cold regolith at Fra Mauro: hot debris, of which 14053 was a part, were deposited after the projection from the Imbrium Basin; some smaller fragments, including 14053, cooled very rapidly in the cold Fra Mauro regolith by conduction.

Erratum

In Table 6 (p. 98) of S. S. Hafner, D. Virgo, and D. Warburton, *Proc. Second Lunar Sci. Conf., Geochim. Cosmochim. Acta,* Suppl. 2, Vol. 1, (1971) the following numbers should be corrected: The composition of clinopyroxene 12021,150-P1 (1000°C, 600°C) is $Wo_9En_{59}Fs_{32}$ (and not $Wo_6En_{61}Fs_{33}$); the $M2$ site occupancy of clinopyroxene 12021,150-P1 (1000°C) for Mg is 0.368 (and not 0.470); the $M1$ site occupancy numbers of clinopyroxene 12021,150-P3 for Mg and Fe should be between parentheses. Numbers between parentheses in Table 6 indicate model numbers and should be interpreted with caution (site occupancies in excess of one; error estimates of text of p. 96–98 and reference mentioned therein).

Acknowledgments— We thank Dr. Gordon E. Brown and Dr. Charles T. Prewitt for kindly communicating results on crystal structure refinements of pigeonites, Dr. Alfred T. Anderson Jr., Dr. Norman Gray, Dr. Ian Steele, Dr. David Virgo, and Mr. David Warburton for their interest in our work, Miss Barbara Janik for carefully performing the computations, and the National Aeronautics and Space Administration (Grant NGR 14-001-173) for financial support.

REFERENCES

Bence A. E. and Papike J. J. (1972) Crystallization histories of pyroxene from lunar basalts. *Lunar Science— III* (editor C. Watkins), pp. 59–61, Lunar Science Institute Contr. No. 88.

Brown G. E. Prewitt C. T. Papike J. J. and Sueno S. (1972a) A comparison of the structures of low and high pigeonite. *J. Geophys. Res.,* in press.

Brown G. E. *et al.* (1972b) written communication; manuscript in preparation.

Brown G. M. and Peckett A. (1971) Selective volatilization on the lunar surface: evidence from Apollo 14 feldspar-phyric basalts. *Nature* **234,** 262–266.

Dence M. R. Plant A. G. and Traill R. G. (1972) Impact-generated shock and thermal metamorphism in Fra Mauro lunar samples. *Lunar Science—III* (editor C. Watkins), pp. 174–176, Lunar Science Institute Contr. No. 88.

Dundon R. W. and Hafner S. S. (1971) Cation disorder in shocked orthopyroxenes. *Science* **174,** 581–583.

Fernández-Morán H. Ohtsuki M. and Hough C. (1972) Correlated electron microscopy and diffraction studies of clinopyroxenes from Apollo 14 rocks. *Lunar Science—III* (editor C. Watkins), p. 252, Lunar Science Institute Contr. No. 88.

Finger L. W. Hafner S. S. Schürmann K. Virgo D. and Warburton D. (1972) Distinct cooling histories and reheating of Apollo 14 rocks. *Lunar Science—III* (editor C. Watkins), pp. 259–261, Lunar Science Institute Contribution No. 88.

Gancarz A. J. Albee A. L. and Chodos A. A. (1971) Petrologic and mineralogic investigation of some crystalline rocks returned by the Apollo 14 mission. *Earth Planet. Sci. Lett.* **12,** 1–18.

Ghose S. Ng G. and Walter L. S. (1972) Clinopyroxenes from Apollo 12 and 14: Exsolution, cation order and domain structure. *Lunar Science—III* (editor C. Watkins), pp. 300–302, Lunar Science Institute Contr. No. 88.

Gray N. Hafner S. S. Schürmann K. and Virgo D. (1972) Distinct cooling histories of Apollo 14 basalts. *Nature Physical Science* **236,** 71–73.

Hafner S. S. and Virgo D. (1970) Temperature-dependent cation distributions in lunar and terrestrial pyroxenes. *Proc. Apollo 11 Lunar Sci. Conf., Geochim. Cosmochim. Acta,* Suppl. 1, Vol. 3, pp. 2183–2198.

Hafner S. S. Virgo D. and Warburton D. (1971) Cation distributions and cooling history of clinopyroxenes from Oceanus Procellarum. *Proc. Second Lunar Sci. Conf., Geochim. Cosmochim. Acta,* Suppl. 2, Vol. 1, pp. 91–108.

Morgan J. W. Krähenbül U. Ganapathy R. and Anders E. (1972) Volatile and siderophile elements in Apollo 14 and 15 rocks. *Lunar Science—III* (editor C. Watkins), pp. 555–557, Lunar Science Institute Contr. No. 88.

NASA Apollo 14 Preliminary Science Report (1971) Preliminary examination of lunar samples. Apollo 14 Preliminary Science Report pp. 109–131, NASA, Washington D.C. 1971.

Papike J. J. and Bence A. E. (1971) Apollo 14 pyroxenes subsolidus relations and implied thermal histories. GSA Annual Meeting Abstracts, p. 666.

Schürmann K. Anderson A. T. and Hafner S. S. (1972) Mg,Fe order-disorder in ortho- and clinopyroxenes from a Hawaiian lava flow. *Trans. Amer. Geophys. Union* **53,** 541.

Takeda H. and Ridley W. I. (1972) Crystallography and mineralogy of pyroxenes from Fra Mauro soil and
    rock 14310. *Lunar Science—III* (editor C. Watkins), pp. 738–740, Lunar Science Institute Contr.
    No. 88.
Turner G. Huneke J. C. Podosek F. A. and Wasserburg G. J. (1971) ^{40}Ar-^{39}Ar ages and cosmic ray
    exposure ages of Apollo 14 samples. *Earth Planet. Sci. Lett.* **12,** 19–35.
Virgo D. (1971) Written communication, manuscript in preparation.
Virgo D. and Hafner S. S. (1969) Fe^{2+}, Mg order-disorder in heated orthopyroxenes. *Mineral. Soc. Amer.
    Spec. Paper* **2,** 67–81.
Virgo D. and Hafner S. S. (1970) Fe^{2+}, Mg order-disorder in natural orthopyroxenes. *Amer. Mineral.*
    55, 201–223.
Walter L. S. French B. M. and Doan A. S. Jr. (1972) Petrographic analysis of lunar samples 14171 and
    14305 (breccias) and 14310 (melt rock). *Lunar Science—III* (editor C. Watkins), pp. 773–775, Lunar
    Science Institute Contr. No. 88.

Proceedings of the Third Lunar Science Conference
(Supplement 3, *Geochimica et Cosmochimica Acta*)
Vol. 1, pp. 507–531
The M.I.T. Press, 1972

Clinopyroxenes from Apollo 12 and 14:
Exsolution, domain structure, and cation order

Subrata Ghose,* George Ng,† and L. S. Walter

Planetology Branch, Goddard Space Flight Center, Greenbelt, Maryland 20771

Abstract—Core pigeonites from rock 12053 show very small amounts of fine-scale exsolution lamellae of augite, presumably parallel to (001). The *b*-type (h + k = odd) x-ray reflections are very diffuse and are stretched parallel to *a**. Rim subcalcic augite from the same rock shows the smallest separation in β-angles ($\Delta\beta = 1°35'$) between the host augite and exsolved pigeonite phases. Subcalcic-augite from rock 12038 shows larger $\Delta\beta$ ($2°17'$) and sharper exsolved pigeonite spots. In both rocks the exsolved pigeonite spots are connected to the corresponding augite spots by strong diffuse streaks. A second generation of very fine-scale (001) exsolution lamellae in subcalcic augites from both rocks are rotated about 1° (*b* as the rotation axis) from the orientation of the first generation (001) lamellae. Cation order determined by Mössbauer resonance spectroscopy in pigeonites from both rocks is high, the K_D values for pigeonites from rocks 12053 and 12038 being 0.086 and 0.030, respectively. Pigeonites from rock 12040 contain only one set of very fine-scale augite exsolution lamellae parallel to (001), rotated by $\sim 1°$ with respect to the host pigeonite. These second-generation lamellae indicate that these rocks must have been subjected to reheating for a considerable period of time, following initial cooling after crystallization from the melt.

In contrast, pigeonites from rock 12021 show two sets of exsolved augite lamellae parallel to (001) and (100), with sharp augite spots, sharp *b*-type reflections and a fairly high degree of cation order ($K_D = 0.097$). This indicates a continuous cooling of this rock at a slow rate, with no evidence for a subsequent reheating.

The exsolution behavior in pigeonites from rocks 14310 and 14053 are similar, both showing large separation in β angles from exsolved augite and pigeonite ($\Delta\beta - 2°30'$ to $2°50'$). The corresponding augite and pigeonite spots are connected by faint diffuse streaks. The *b*-type reflections are fairly sharp. However, the degree of cation disorder in pigeonite from rock 14053 ($K_D = 0.127$) is much higher than that in pigeonite from rock 14310 ($K_D = 0.094$). Both rocks have cooled fairly slowly after crystallization from the melt. Rock 14053 has subsequently been reheated to a temperature higher than 840°C for a short duration.

The size of the domains in pigeonite (as observed through the electron microscope by imaging through the *b*-type reflections) increases progressively in the order: 12053 (50 to 100 Å), 12038 (~ 500 Å), 12021 (~ 1000 Å), 14310 (~ 1500 Å). Viewed down the *b*-axis the shape of the domains are blocky and are slightly elongated parallel to the *c*-axis.

Introduction

The subsolidus cooling history of lunar rocks 12021,21, 12038,72, 12040,24, 12053,72, 14310,115, and 14053,44 has been elucidated through the study of fine-scale exsolution, domain structure (in pigeonite), and cation order in clinopyroxenes by means of x-ray diffraction, high-voltage electron microscopy and ^{57}Fe Mössbauer resonance. Considered together, these phenomena reveal a multievent cooling history for a number of these lunar rocks.

---

* Present address (academic year 1971–72): Department of Geology and Geophysics, University of California, Berkeley, California 94720
† Permanent address: Department of Chemistry, Federal City College, Washington, D.C. 20001

Description of the Rocks

Rock 12021 is a porphyritic basalt with elongate pyroxene phenocrysts as much as 2 cm long in a variolitic groundmass of pyroxene and plagioclase. The phenocrysts are commonly hollow, consisting of pigeonite, which are mantled successively by augite, Fe-rich pigeonite, and ferro-augite. The groundmass consists of fine-grained elongate plagioclase, high-Fe clinopyroxene, pyroxferroite, and ilmenite plus small amounts of chromite, ulvöspinel, metallic iron, troilite, tridymite, and cristobalite (Weill *et al.*, 1971; Bence *et al.*, 1971; Klein *et al.*, 1971). Sector zoning and chemical variations in the pyroxenes have been described in detail by Bence *et al.* (1971) and Boyd *et al.* (1971). From melting experiments on this rock, Green *et al.* (1971) estimated the highest temperature of crystallization of the pigeonite to be 1140°C, followed by rapid cooling at ~ 1120°C.

Rock 12053 is very similar to rock 12021 except that its grain size is much smaller, indicating a much more rapid cooling. The phenocrysts in this rock are about 5 mm long in a groundmass of small equant pyroxenes (~ 0.1 mm), plates and rods of plagioclase (up to 1 mm long), and ilmenite. The rock consists of 80% pyroxene, 10% each of plagioclase and ilmenite with minor amounts of cristobalite. The pyroxene phenocrysts are in radiating groups consisting of a pigeonite core and augite rim (Christie *et al.*, 1971).

Rock 12038 is a medium-grained sub-ophitic basalt with about 55% pyroxene, 30% plagioclase, 10% ilmenite and < 5% cristobalite and other minor phases. Pyroxenes are elongate (up to 1 cm long), subhedral to anhedral, and consist of pigeonite cores and augite rims. Exsolution lamellae in pyroxenes can be resolved optically. Lath-shaped plagioclase crystals ranging up to 3 mm long occur locally in radiating clusters (Christie *et al.*, 1971).

Rock 12040 is a coarse-grained olivine basalt consisting of 45% pyroxene, 20% each of olivine and plagioclase, 10% ilmenite, and minor amounts of troilite, spinel, metallic iron, and devitrified glass. Olivines occur as large inclusions in pigeonite-augite megacrysts and as clusters associated with smaller pyroxene crystals. The glass inclusions are commonly even-textured and coarsely devitrified, consisting of ilmenite, plagioclase, and glass. Inclusions in plagioclase consist of tiny droplets of high-silica glass in high-iron devitrified glass (Newton *et al.*, 1971). From melting experiments, Green *et al.* (1971) estimated the temperature at which pigeonite appeared to be 1190°C.

Rock 14310 is a fine-grained plagioclase-rich basalt with scattered small cognate inclusions showing a sharp interface, that are finer-grained than the body of the rock. The texture is intergranular to intersertal, consisting of euhedral plagioclase laths (66%) and anhedral pyroxene crystals (31%) with minor amounts of ilmenite, troilite, metallic iron, chromium spinel, and ulvöspinel (LSPET, 1971; Gancarz *et al.*, 1971). This rock has been termed a "crystalline melt rock" (Walter *et al.*, 1972), which preserves mineral phases indicative of a prior pre-melt history (Hollister *et al.*, 1972). The pyroxenes are primarily magnesian hypersthene cores, rimmed by pigeonite and augite. The anomalous compositional variations within the pyroxene grains indicative of a pre-melt history have been described in detail by Hollister *et al.* (1972).

Rock 14053 is a coarser-grained basalt with ophitic texture and is more mafic than rock 14310. The rock is moderately inhomogeneous and consists mostly of pyroxene, plagioclase, and a small amount of olivine, commonly occurring as anhedral grains in the core of pigeonite. Minor phases are ilmenite, metallic iron, ulvöspinel, and cristobalite (LSPET, 1971; Gancarz et al., 1971). The pyroxene crystals consist mostly of pigeonite with some rim augite.

EXSOLUTION

Submicroscopic exsolution phenomena in lunar clinopyroxenes have been studied by x-ray diffraction using the precession technique with MoK_α radiation. Single-crystal fragments with average diameters of 0.1 to 0.2 mm were selected under the optical microscope. Because some of these pyroxenes show very fine-scale exsolution lamellae causing diffuse spots, the exposure times varied between 2–3 days to one week (45 kV, 10 mA). Through these long exposures we have been able to bring out secondary diffuse spots due to very fine-scale second-set of (001) exsolution lamellae, which can be correlated with electron micrographs showing the same features. From the measured unit-cell parameters (Table 1), the compositions of the exsolved phases were obtained using the b-β nomogram (Papike et al., 1971) and are shown in Fig. 1a, b.

Core pigeonite from 12053,72 shows practically no exsolution. Very long-exposure photographs show very diffuse augite-type spots with a^* common and $c_A^* \wedge c_p^* \simeq 1°$, indicating the presence of a small amount of very fine exsolution lamellae (Fig. 2a). These lamellae are believed to be parallel to (001), but are rotated by about 1° (with b as the rotation axis) with respect to the host pigeonite (see below). Two sets of (001) exsolution lamellae are clearly evident in subcalcic augites from

Table 1. Cell dimensions and estimated compositions of lunar pigeonites (space group $P2_1/c$) and subcalcic augites (space group $C2/c$).

Sample	Pyroxenes	a(Å)	b(Å)	c(Å)	β(°)	Wo–En–Fs
14310,115	Pigeonite	9.67	8.90	5.23	108°40′	6, 64, 31
	(001) Augite	9.69	8.96	5.26	106°10′	42, 42, 16
14053,44	Augite	9.74	8.96	5.27	106°05′	40, 30, 30
	(001) Pigeonite	9.76	8.96	5.25	108°55′	2, 45, 53
	Augite	9.78	8.99	5.27	106°00′	40, 20, 40
	(001) Pigeonite	9.78	8.99	5.25	108°25′	7, 31, 62
	Pigeonite	9.71	8.94	5.22	108°50′	3, 52, 45
	(001) Augite	9.71	8.94	5.26	106°00′	45, 34, 21
14162,34	Augite	9.76	8.88	5.30	106°45′	35, 52, 13
12053,72	Pigeonite	9.67	8.90	5.23	108°25′	10, 63, 27
	Augite	9.74	8.88	5.27	106°40′	38, 51, 11
	(001) Pigeonite	9.74	8.88	5.25	108°15′	10, 70, 20
12038,72	Augite	9.71	8.90	5.31	106°35′	38, 45, 18
	(001) Pigeonite	9.71	8.90	5.23	108°50′	2, 66, 32
12040,24	Pigeonite	9.67	8.88	5.21	108°30′	7, 71, 22
	(001) Augite	9.74	8.88	5.26	106°15′	34, 53, 13
	Augite	9.68	8.89	5.25	105°25′	40, 46, 14
	(001) Pigeonite	9.68	8.89	5.20	108°50′	4, 68, 28
12021	Pigeonite	9.66	8.88	5.21	108°45′	5, 72, 13
	(001) Augite	9.66	8.88	5.25	106°05′	46, 42, 13
	(100) Augite	9.70	8.88	5.25	106°00′	46, 42, 13

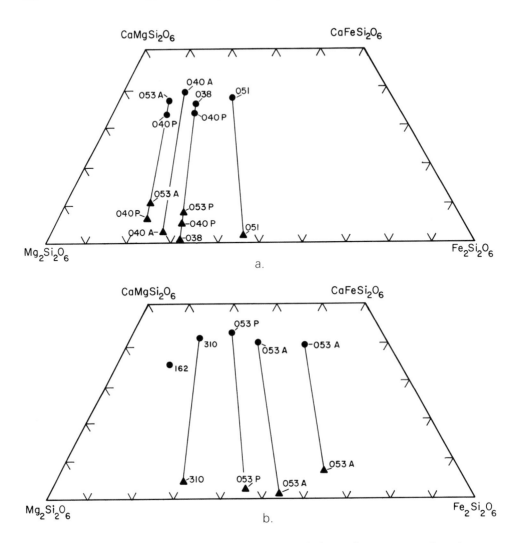

Fig. 1. Chemical compositions of the exsolved phases in lunar clinopyroxenes estimated on the basis of cell dimensions using the b-β nomogram (Papike *et al.*, 1971) (a) Apollo 12 clinopyroxenes, (b) Apollo 14 clinopyroxenes.

12053 and 12038 (Figs. 2b, c); the very fine-scale second-generation lamellae are rotated about 1° with reference to the first set of (001) lamellae. These features are clearly shown by a transmission electron micrograph of augite from rock 12053 (Fig. 3) taken by Christie *et al.* (1971).

Rim augite from 12053,72 (Fig. 2b) shows the smallest $\Delta\beta$ (1°35′), which is comparable to the value found in a pigeonite from 12052 (Papike *et al.*, 1971); the exsolved pigeonite spots are diffuse and connected by diffuse streaks with corresponding augite spots. Subcalcic augite from 12038,72 (Fig. 2c, d) shows similar behaviour, but has a larger $\Delta\beta$ (2°17′) and sharper exsolved pigeonite spots.

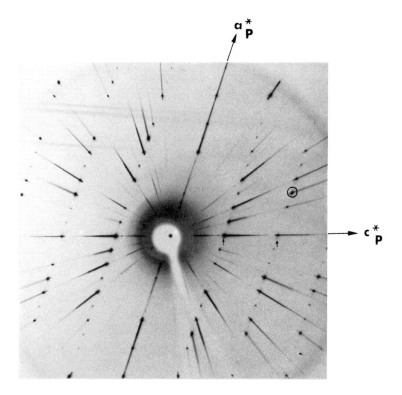

Fig. 2a. X-ray precession photographs of the h0l nets of pyroxenes. The reciprocal axes for pigeonite are indicated as a_p^*, c_p^*, for (001) augite as a_A^* (001), c_A^* (001) and for (100) augite as a_A^* (100), c_A^* (100). Where (100) augite lamellae are absent, the reciprocal axes from augite parallel to (001) are indicated as a_A^*, c_A^*. Reciprocal axes in orthopyroxene are indicated as a^* and c^*. Mo rad., Zr filter, 45 kV, 10 mA. (a) Pigeonite from rock 12053,72. Very diffuse reflections from exsolved augite lamellae are indicated by arrows. $c_A^* \wedge c_p^* \simeq 1°$. The b-type reflections are very diffuse. The circled b-type reflection is 304.

In contrast, 12021 pigeonites show two sets of augite lamellae parallel to (001) and (100) with sharp augite spots and a large $\Delta\beta$ (2°55'; Fig. 2e). In another pigeonite crystal from rock 12021, the spots from augite exsolved parallel to (001) are sharp, whereas those from augite lamellae parallel to (100) are diffuse (Fig. 2f). This means that the (100) augite lamellae are much finer in scale than the (001) set. Augite shows only one set of exsolved pigeonite lamellae parallel to (001). Exsolution and epitaxy in pyroxenes from 12021 have been studied extensively by Papike *et al.* (1971) through x-ray diffraction. Fernández-Morán *et al.* (1971) determined by electron microscopy the width of the augite lamellae in pigeonite to be about 200 to 300 Å.

Diffuse spots for augite lamellae parallel to (100) would mean a lamellae width of ~ 100 Å or less. A characteristic of 12021 pigeonites is the absence of the diffuse streak connecting exsolved augite and pigeonite spots; furthermore, $a_A \neq a_p$, in addition to $c_A \neq c_p$. In pyroxenes from 12053 and 12038 $a_A = a_p$, $c_A \neq c_p$, indicating

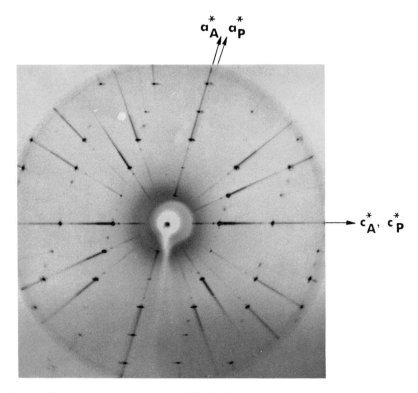

Fig. 2b. Subcalcic augite from rock 12053,72. The exsolved pigeonite and augite spots are connected by diffuse streaks. The pigeonite spots are very diffuse indicating the presence of very thin (~ 50 Å) lamellae.

a semicoherent interface between the exsolved phases. This is indicative of an earlier stage of exsolution than that in pyroxenes from rock 12021.

Pigeonites from 12040 show diffuse spots from exsolved augite lamellae parallel to (001). As in pigeonite from 12021, $a_A \neq a_p$ and $c_A \neq c_p$. However, the exsolved augite lamellae are rotated by about 1° with respect to the host pigeonite (Fig. 2g).

We have examined by x-ray diffraction an orthopyroxene and a pigeonite from rock 14310,115. The orthopyroxene, with cell dimensions $a_0 = 18.30$, $b_0 = 8.85$, $c_0 = 5.21$ Å, does not show any exsolved augite (Fig. 2h). The pigeonite shows two sets of exsolved augite lamellae parallel to (001) and (100), the set parallel to (100) being much finer than that parallel to (001) (Fig. 2i). The difference in the β angles of the exsolved phases ($\Delta\beta = 2°30'$) is large. The corresponding pigeonite and augite spots are connected by weak diffuse streaks. The b-type ($h + k$ = odd) pigeonite reflections are sharp, indicating the presence of large domains. Extensive exsolution of pigeonite lamellae in augite has been observed by us (Fig. 4), whereas exsolution in plagioclase in this rock has been observed by Lally *et al.* (1972) through the electron microscope. All these features indicate a fairly slow cooling rate for the rock 14310, which is comparable to that for rock 12021.

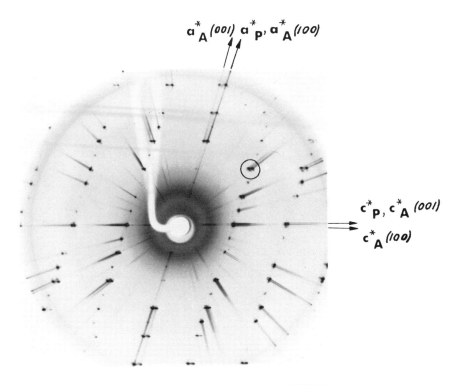

Fig. 2c. Subcalcic augite from rock 12038,72.

Takeda and Ridley (1972) have observed a twinned magnesian pigeonite with an orthopyroxene core that have the (100) face in common. We have observed widespread twinning in pigeonite by electron microscopy (Fig. 5). Takeda and Ridley (1972) considered the twinning in pigeonite to be a growth feature, because monoclinic pigeonite growing on orthopyroxene may have one of two twin orientations. Although twinning due to crystal growth definitely is possible, in view of the fairly slow cooling rate of this rock some of the twinning may represent the first stage of the phase transformation of pigeonite to the stable orthopyroxene phase.

Both pigeonite and augite (Fig. 2j) from rock 14053 show extensive exsolution. $\Delta\beta$ ranges from 2°25' to 2°50' and the b-type reflections in pigeonite range from diffuse to sharp.

A pink subcalcic augite from the soil sample 14162 does not show any exsolution lamellae. The cell dimensions indicate a composition, $Wo_{35}En_{52}Fs_{13}$. This pyroxene probably is derived from a basalt quenched very quickly above the pyroxene solvus. Alternatively, it may have been homogenized by shock heating.

The strong diffuse streaks connecting exsolved augite and pigeonite reflections in pyroxenes from rocks 12038 (Fig. 2c, d) and 12053 (Fig. 2b) indicate the presence of continuous structural and chemical variation from one lamellar region to the next. The structurally and chemically intermediate regions will presumably be restricted

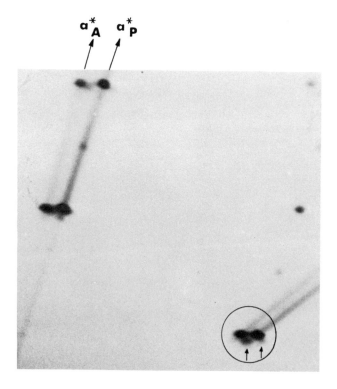

Fig. 2d. Detail of the upper right quadrant of Fig. 2c. The reflections within the circle are from (402) planes in augite and pigeonite. Note the strong diffuse streak connecting the augite and pigeonite spot. Diffuse subsidiary reflections, indicated by arrows are due to a second-generation exsolution lamellae rotated by $\sim 1°$ with respect to the first set.

to the interface between two exsolved lamellae. Because the lamellae are thin, the disordered interface region will form a sizable part of the whole crystal. When the lamellae grow thicker, the interface volume and the accompanying disorder decreases and the diffuse streaks disappear. For example, in the pigeonite from rock 12021, the exsolved augite lamellae are about 300 Å thick (Fernández-Morán et al., 1971) and the pigeonite and augite diffraction spots are sharp without any connecting diffuse streak.

Through electron microscopy, the existence of thin exsolved lamellae (50-100 Å) parallel to (001) and a thinner set parallel to (100) in lunar pyroxenes have been observed (Champness et al., 1971; Christie et al., 1971). This evidence suggests the spinodal decomposition mechanism for the very early stages of exsolution in some clinopyroxenes. In the early stages of the exsolution process, $c_A \neq c_p$ $(a_A = a_p)$ and composition fluctuations develop parallel to (001), followed by fluctuations parallel to (100) when $a_A \neq a_p$. The spinodal mechanism has been suggested by Bailey et al. (1970), Champness et al. (1971), and Christie et al. (1971) to be operative in lunar clinopyroxenes. However, to be able to prove that this mechanism is operative, as opposed to nucleation and growth, it is necessary to observe the compositional

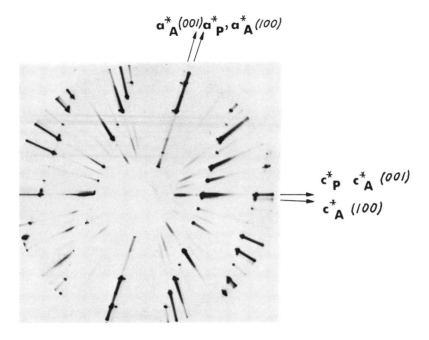

Fig. 2e. Pigeonite from rock 12021,21 showing augite exsolution lamellae parallel to (001) and (100). Augite reflections are all sharp.

fluctuations *before* the solid solution breaks down into two recognizably separate phases. X-ray diffraction studies on pyroxenes from 12052 (Papike *et al.*, 1971) and 12053 indicate the presence of two distinct phases, pigeonite and augite with $\Delta\beta \simeq 1°30'$. These two phases are only semicoherent, with $a_A = a_p, c_A \neq c_p$. In view of the existing evidence, it cannot be considered that the spinodal mechanism is conclusively proven in the case of lunar clinopyroxenes.

The rotation of the second set of exsolved (001) lamellae with respect to the first set is caused by the misfit of the respective cell dimensions of augite and pigeonite phases, which are continuously changing as a function of cooling and changing composition. The observation of two distinct sets of (001) lamellae rotated by 1° with respect to each other indicates an interruption in the cooling of the pyroxene. Because the rotated second set is always finer than the first set and does not cut across the first set (Fig. 3), the second set must have originated later than the first. It is proposed here that the second set of (001) lamellae has been generated during a reheating of the rock, subsequent to the crystallization and initial cooling, which generated the first set of (001) lamellae.

DOMAIN STRUCTURE IN PIGEONITE

On the basis of diffuse *b*-type (h + k = odd) reflections in pigeonite, first observed by Bown and Gay (1957), Morimoto and Tokonami (1969) postulated the

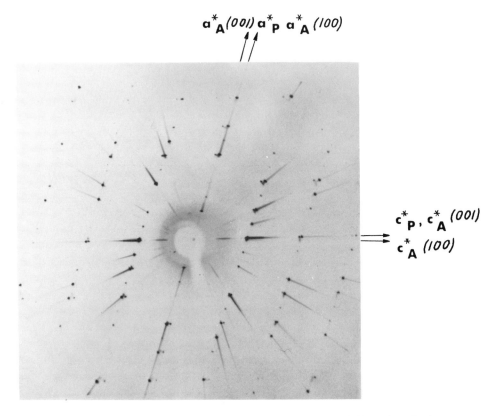

Fig. 2f. Same as Fig. 2e. (001) augite reflections are sharp, whereas (100) augite reflections
are diffuse.

presence of domains in the pigeonite structure that are shifted with respect to each other by $(\mathbf{\bar{a}} + \mathbf{\bar{b}})/2$. These domains presumably originate during the C2/c → P2₁/c transition in pigeonite. The diffuseness of the b reflections in pigeonite is a measure of the size of these domains. Provided the chemical composition is the same, a quick quenching will cause small domains to form (diffuse b reflections), whereas slow cooling will result in large domains (sharp b reflections). Hence domain size in pigeonite can be used as an indicator of the cooling history. These domains have been directly observed through transmission electron microscopy by imaging the pigeonite crystal using the b-type reflection (Bailey et al., 1970; Christie et al., 1971; Champness et al., 1971).

The b-type reflections in pigeonite from 12021 are sharp (Figs. 2e, f), whereas those in pigeonite from 12053 are quite diffuse and are stretched parallel to a^* (Fig. 2a, b); the b reflections in pigeonite from 12038 show an intermediate degree of diffuseness (Fig. 2c). Pigeonites from 14310 show sharp b reflections (Fig. 2i). The domains in pigeonites from rocks 12021, 12038, 12053, and 14310 have been observed directly through high-voltage electron microscopy (Figs. 6, 7). Indeed, the

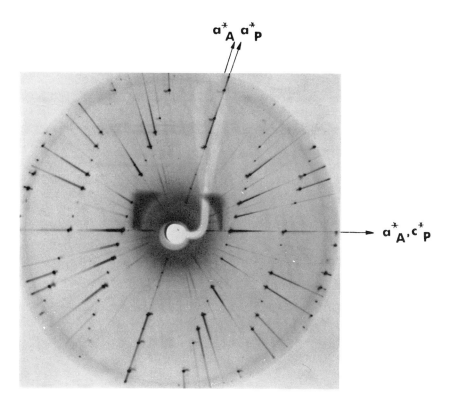

Fig. 2g. Pigeonite from rock 12040,26. Note the diffuse reflections from exsolved augite;
a_A^* and c_A^* are rotated clockwise ($\sim 1°$) with respect to a_p^* and c_p^*.

correlation of the observed domain size with the diffuseness of the b-type reflection
is good. The size of the domains in pigeonites increases progressively in the order
12053 (50 to 100 Å), 12038 (~ 500 Å), 12021 (~ 1000 Å), and 14310 (~ 1500 Å).
Viewed down the b-axis, the domains appear roughly blocky in shape and slightly
elongated parallel to the c-axis. This observation confirms the prediction of Morimoto and Tokonami (1969) that the domains in pigeonite are columnar parallel to
the c-axis. The temperature of phase transition in pigeonite from $C2/c$ to $P2_1/c$ space
group depends on the chemical composition, ranging from $\sim 1000°C$ for magnesian
pigeonite to $\sim 200°C$ for ferropigeonite ($Fs_{85}Wo_{15}$) (Prewitt *et al.*, 1971). The
domain size presumably also will depend on composition as well as on cooling rate
(Christie *et al.*, 1971). To be able to compare domain sizes in pigeonites from rocks
12021, 12053, and 12038 in terms of cooling history, we have chosen as far as possible
core pigeonites that are Mg-rich and are comparable in composition.

CATION ORDER

Experimental

Mineral separation. The rock chips were individually crushed into coarse (0.5 mm) grains. The coarse
powder was passed through the magnetic separator and the coarse pyroxene-rich fractions thus obtained

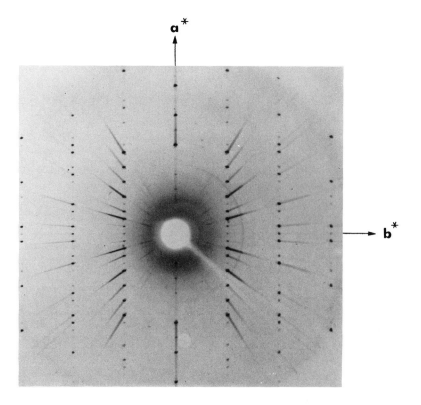

Fig. 2h. Orthopyroxene from rock 14310,115. Note the absence of spots from exsolved augite. Extra spots appear on the photograph, indicating the presence of other unidentified exsolved phases.

were crushed in boron carbide mortars and sieved, using nylon cloth, into 200 and 325 mesh fractions. Where the powder retained on 200 mesh showed aggregate grains, it was pulverized further to pass through 200 mesh. Heavy liquid separation was carried out in stages until a nearly complete macroscopic separation of pink augite and yellow pigeonite was obtained. Each of these augite and pigeonite phases contains extensive exsolution lamellae of the other phase and show considerable chemical variation. The final separation was achieved through hand-picking under a binocular microscope, resulting in a final purity of about 98–99 % for each sample.

Chemical analysis. Microprobe analyses of Ca, Fe, Mg, Al, Ti, and Si were made on grains from the same powder sample used for the Mössbauer experiment. Natural pyroxene standards were used and the data were processed by the Bence and Albee (1968) correction method. More than 20 grains in each sample were analysed and the results were averaged. The chemical analyses of pyroxene powders used for Mössbauer experiments are shown in Table 2.

[57]*Fe Mössbauer Resonance Spectroscopy.* The Mössbauer resonance spectrometer consists of an electro-mechanical drive of the Kankeleit type and a multichannel analyzer operated in the time mode. Fifty mCi of [57]Co diffused in a Pd foil was used as the source for the 14.4 keV gamma rays. Absorbers were prepared by encapsulating 50 to 100 mg of the finely powdered pyroxene between sticky mylar sheets. All spectra were recorded with the absorbers cooled to liquid-nitrogen temperature. The transmitted gamma-ray pulses were counted by means of a proportional counter filled with 90 % krypton and 10 % methane at 1 atm. The inner four lines from a standard Fe foil of 99.999 % purity (provided by Dr. J. J. Spijkerman of the National Bureau of Standards) were used for the velocity calibration. The measured line widths of

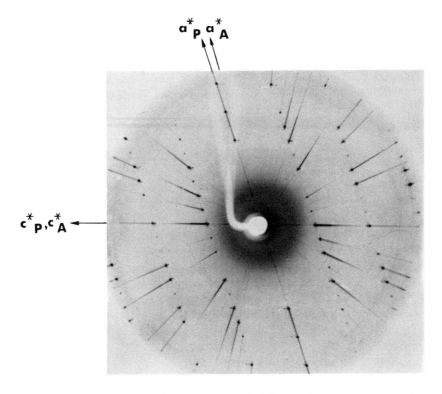

Fig. 2i. Pigeonite from rock 14310,115. Very weak diffuse streaks connect corresponding augite and pigeonite spots.

the inner pair is 0.26 mm/sec. A least-squares program was used on the IBM 360/75 computer to fit Lorentzian curves to the two overlapping doublets without any constraints.

The inner pair of lines (A2-B2) originates from Fe^{2+} at the M2 site, whereas the outer pair (A1-B1) is from Fe^{2+} at the M1 site. The site-occupancy factors were calculated assuming the recoilless fraction at M1 and M2 sites to be the same. The site occupancies were calculated using the procedure of Hafner *et al.* (1970), assuming all the Ca to be in the M2 site. The Mössbauer resonance spectra of pigeonite and augite from rock 12021 were kindly recorded by Dr. P. A. Pella at the National Bureau of Standards, Gaithersburg, Maryland, using the NBS spectrometer. Thirty mCi of ^{57}Co diffused into Pd foil was used as the source. The data were computer-fitted, using a nonlinear least-squares program. For these two samples, two least-squares fitting procedures were used. In procedure A (Figs. 8, 9), two overlapping Lorentzian line shapes constrained to the same half-width to each component of the doublets (a total of eight lines) were fitted. This particular fitting model was chosen to obtain an estimate of the range of the quadrupole splitting for ^{57}Fe within the M1 and M2 sites. For example, Δ_1 and Δ_{11} correspond to quadrupole splittings for Fe^{2+} at the M1 site and likewise Δ_2 and Δ_{22} correspond to quadrupole splittings for Fe^{2+} at the M2 site. In procedure B (Figs. 8, 9), one Lorentzian line was fitted to each component of the doublets, resulting in a four-line fit. The Fe^{2+} site-occupancy factors in augite from 12021 using both fitting procedures indicate agreement within the experimental error. However, comparison of the results on pigeonite shows that the M1 site-occupancy factors calculated by procedure A was about 50% higher than that calculated using procedure B. This difference appears to be due to the poorer resolution of the quadrupole split doublets from Fe^{2+} at the M1 site in comparison with that observed in the spectra from augite. The results of the four- and eight-line fit with the Mössbauer parameters and site-occupancy factors for

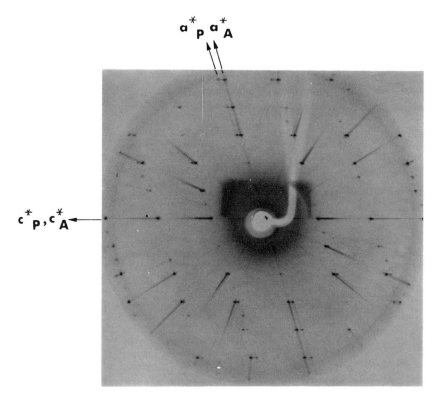

Fig. 2j. Subcalcic-augite from 14053,44. Note weak diffuse streaks connecting augite and pigeonite spots.

augite and pigeonite from rock 12021 are listed in Table 3a. The Mössbauer parameters for the other samples are listed in Table 3b. The chemical composition and site occupancies in terms of Ca, Fe, and Mg are listed in Table 4.

Results

The cation distribution in augite and pigeonite phases, as determined by the Mössbauer resonance spectroscopy, is a statistical average because both augite and pigeonite phases show extensive exsolution as well as a wide range in chemical composition. The cation-ordering parameters, determined this way, are bulk parameters and should be interpreted with caution. In general, they may not be comparable to the ordering parameters of pyroxene single crystals, as determined by x-ray diffraction. Furthermore, there may be a bias in this technique, resulting in an apparent higher degree of cation order, depending on the amount of Ca present (Dowty *et al.*, 1972). If present, this effect would be much less in pigeonite, which contains much less Ca and commonly shows much smaller amounts of exsolved augite lamellae. Hence, ordering parameters of pigeonite determined by the Mössbauer technique are probably more reliable than those of augite. However, using

Fig. 3. Dark field transmission electron micrograph (800 kV) (taken by Christie *et al.*, 1971) showing two generations of exsolved augite lamellae in pigeonite from rock 12053. First-generation augite lamellae parallel to (001) are thick. The second generation consists of two sets of augite lamellae parallel to (001) and (100). Note that the first and second-generation augite lamellae parallel to (001) are rotated by about 1° with respect to each other. (Courtesy of Christie *et al.*, 1971.)

the Mössbauer technique, these bulk ordering parameters can be determined with a high degree of precision. Hence, they can be used for comparative purposes to determine the relative cooling histories of lunar rocks.

The degree of cation order in the lunar pyroxenes can be expressed in terms of the distribution coefficient K_D, where

$$K_D = \frac{X_{Fe}^{M1}(1 - X_{Fe}^{M2})}{X_{Fe}^{M2}(1 - X_{Fe}^{M1})}$$

and

$$X_{Fe} = \frac{Fe}{(Fe + Mg)}$$

The K_D values for the pigeonite from rock 12021 and the mixed pigeonite-ortho-pyroxene phases from rock 14310 are 0.097 and 0.094, respectively, indicating a

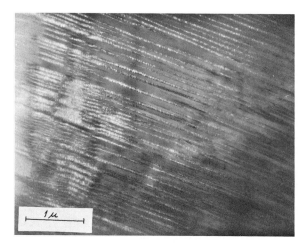

Fig. 4. Dark-field transmission electron micrograph (800 kV) showing pigeonite exsolution lamellae in augite from rock 14310. Note the presence of the antiphase domains in the pigeonite lamellae.

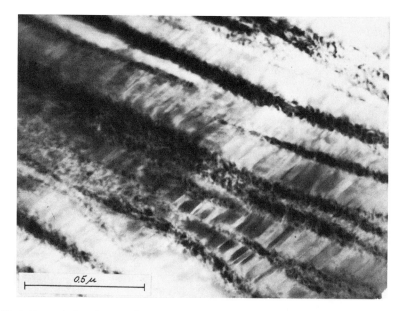

Fig. 5. Transmission electron micrograph (800 kV) showing widespread twinning in pigeonite from rock 14310.

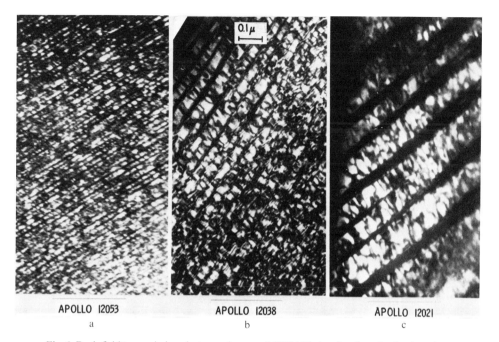

APOLLO 12053   APOLLO 12038   APOLLO 12021

a   b   c

Fig. 6. Dark-field transmission electron micrograph (800 kV) showing domains in pigeonites from rocks 12053, 12038, and 12021. View is down the b-axis. Imaged through a b-type reflection in pigeonite. The augite lamellae are out of contrast and appear dark. (6a, b taken by Christie *et al.*, 1971; reproduced here by permission.)

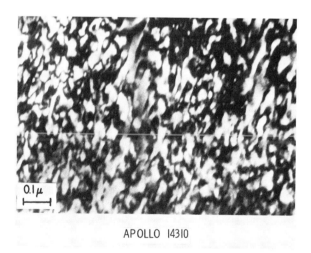

APOLLO 14310

Fig. 7. Dark-field transmission electron micrograph (800 kV) (taken by Lally *et al.*, 1972.) showing domains in a pigeonite from rock 14310. Imaged through a b-type reflection. (Courtesy of Lally *et al.*, 1972.)

Table 2. Microprobe analyses (wt %) of pyroxene powders used for Mössbauer resonance spectroscopy.

| | 12021,21 | | 12053 | | 12038 | 14310,115 | 14053 | |
	Pig.	Aug.	Pig.	Aug.	Pig.	Pig + Orpx	Pig.	Aug.
CaO	5.39	10.06	4.84	13.40	6.46	4.42	5.57	14.28
MgO	21.72	11.08	20.26	15.30	19.25	22.61	21.35	13.01
FeO	18.35	24.60	17.43	15.05	20.25	16.63	18.52	16.55
Al_2O_3	1.93	1.34	2.02	3.88	2.19	1.75	1.60	1.73
TiO_2	0.81	1.40	0.78	1.60	1.17	0.81	0.92	1.55
SiO_2	52.46	48.91	52.22	48.84	51.20	53.29	52.86	50.35

fairly high degree of cation order. For a pigeonite from rock 12021 with a slightly different composition ($Wo_9En_{59}Fs_{32}$), Hafner et al. (1971) obtained a K_D value of 0.081, whereas for the orthopyroxene-pigeonite sample from rock 14310, Finger et al. (1972) obtained a K_D value of 0.097. Based on heating of lunar and synthetic clinopyroxenes, Hafner et al. (1971) presented a plot of temperature against the distribution coefficient (Fig. 8, Hafner et al., 1972). Using this plot we obtain an equilibration temperature of $\sim 700°C$ for both pigeonites from rocks 12021 and 14310. The K_D value for pigeonite from rock 12053 is slightly lower, namely 0.086 (Hafner et al., 1971, reports a K_D value of 0.091 for this sample), but not significantly different from those in pigeonites from 12021 and 14310. Though rock 12053 had been initially quenched quickly from close to the temperature of crystallization from the melt ($\sim 1200°C$), the high degree of cation order can only be explained by assuming a reheating of the rock at $\sim 700°C$, which also causes the second-generation exsolution lamellae to appear. This interpretation is different from that given by Hafner et al. (1971), who suggested a quick quenching from the melt without subsequent reheating. The pigeonite from rock 12038 is also highly ordered ($K_D = 0.030$); this can also be explained by assuming a subsequent reheating of the rock at a temperature lower than 700°C.

The pigeonite from rock 14053 shows the highest degree of cation disorder ($K_D = 0.127$). Finger et al. (1972) reported a K_D value of 0.120 for this sample and an equilibration temperature of 840°C. Since the exsolution behavior in pigeonites from rocks 14053 and 14310 are comparable, the significantly higher degree of cation disorder in pigeonite from 14053 can only be explained by later reheating (Finger et al., 1972) to a temperature above 840°C. However, as no second-generation exsolution lamellae appear, the duration of the reheating must have been very short. Because the cation exchange rate in pyroxenes is high at high temperatures, the very fact that such a degree of cation disorder has been retained indicates rapid quenching.

The three augites from rocks 12021, 12053, and 14053 all show a very high degree of cation order, the K_D values being 0.031, 0.014, and 0.052, respectively. The higher K_D value for augite from rock 14053 is consistent with the high K_D value for the pigeonite from the same rock.

Summary and Conclusions

Combination of several experimental techniques, namely x-ray diffraction, electron microprobe, Mössbauer resonances, and high-voltage electron microscopy,

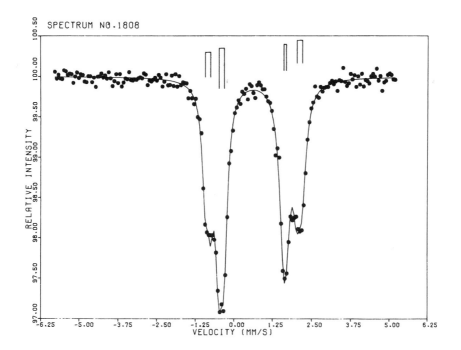

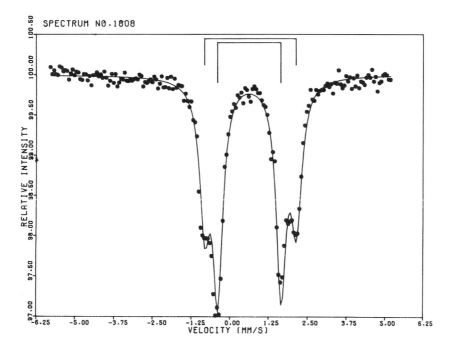

Fig. 8. ^{57}Fe Mössbauer resonance spectra of subcalcic augite from rock 12021. (a) top, 8-line fit (b) bottom, 4-line fit.

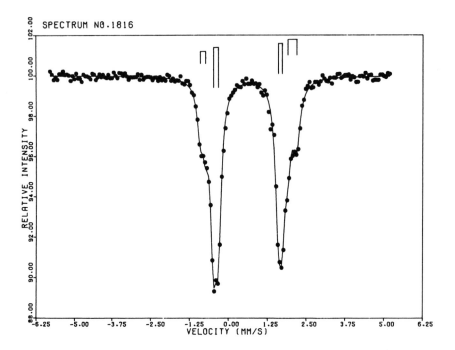

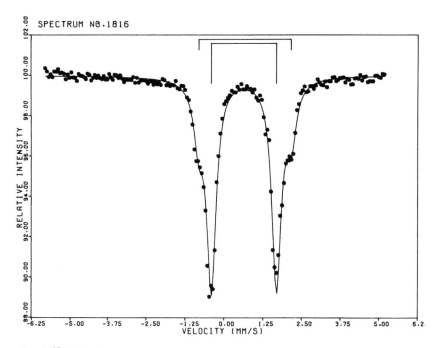

Fig. 9. ^{57}Fe Mössbauer resonance spectra of pigeonite from rock 12021. (a) top, 8-line fit (b) bottom, 4-line fit.

Table 3A. Mössbauer parameters on pyroxenes from rock 12021.

Pyroxene	Fe²⁺ site Occupancy factors[1]	Isomer shift δ, mm/s[2]	Quadrupole splitting, Δ, mm/s	Γ = FWHM	Temp, °K	Fe²⁺ site Occupancy factors[3]
Augite 12021,21 # 1 (1808) ($Wo_{22.5}Fs_{43}En_{34.5}$)	$M_1 = 0.395 \pm .01$ $M_2 = 0.605 \pm .01$	$\delta_1 = 1.179 \pm .02$ $\delta_2 = 1.164$	$\Delta_1 = 3.172 \pm .06$ $\Delta_{11} = 2.806$ $\Delta_2 = 2.215$ $\Delta_{22} = 1.934$	$M_1 = .52 \pm .02$ $M_2 = .58$	80	$M_1 = 0.391 \pm .01$ $M_2 = 0.609 \pm .01$
Augite 12021,21 # 2 (1814) ($Wo_{22.5}Fs_{43}En_{34.5}$)	$M_1 = 0.398 \pm .01$ $M_2 = 0.602 \pm .01$	$\delta_1 = 1.203$ $\delta_2 = 1.183$	$\Delta_1 = 3.163$ $\Delta_{11} = 2.837$ $\Delta_1 = 2.221$ $\Delta_{22} = 1.911$	$M_1 = .50 \pm .02$ $M_2 = .62 \pm .02$	80	$M_1 = 0.402 \pm .01$ $M_2 = 0.598 \pm .01$
Pigeonite* 12021,14 (1816) ($Wo_{11}Fs_{28.5}En_{60.5}$)	$M_1 = 0.295 \pm .02$ $M_2 = 0.705 \pm .01$	$\delta_1 = 1.310 \pm .01$ $\delta_{11} = 1.129$ $\delta_2 = 1.180$ $\delta_{22} = 1.163$	$\Delta_1 = 3.145 \pm .01$ $\Delta_{11} = 2.681$ $\Delta_2 = 2.236$ $\Delta_{22} = 1.971$	$M_1 = .54 \pm .05$ $M_1 = .44$ $M_2 = .48$ $M_2 = .48$	80	$M_1 = 0.202 \pm .01$ $M_2 = 0.789 \pm .01$

[1] Used procedure A for computer fitting the spectra: eight-line fit.
[2] Isomer shift is with respect to NBS SRM Fe foil # 1541.
[3] Used procedure B for computer fitting the spectra: four-line fit.
* Resolution of the two sites is poor.

Table 3b. Mössbauer parameters of lunar clinopyroxenes based on fitting of four Lorentzian lines.

Clinopyroxene	Composition			Quadrupole splitting		Isomer shift*		Area Ratio $(A_2 + B_2)$
	Wo	Fs	En	M1	M2	M1	M2	$\overline{(A_1 + B_1 + A_2 + B_2)}$
				mm/sec	mm/sec	mm/sec	mm/sec	
14310,115	9	26.5	64.5	2.99	2.14	1.29	1.27	0.819
14053,44 Pigeonite	11	29	56	3.00	2.12	1.28	1.25	0.756
12053,72 Pigeonite	10.5	29	60.5	2.98	2.10	1.29	1.27	0.806
12038,72 Pigeonite	13	32.5	54.5	2.96	2.10	1.28	1.26	0.755
14053,44 Augite	32	28.5	39.5	3.01	2.08	1.28	1.26	0.569
12053,72 Augite	29	25.5	45.5	2.98	2.06	1.28	1.25	0.724

* Referred to metallic iron at room temperature.

Table 4. Cation distribution in lunar pigeonites and augites determined by ^{57}Fe Mössbauer resonance spectroscopy.

a. Pigeonites

Sample No.	Composition*	Site occupancy					K_D
		M1			M2		
		Fe	Mg	Ca	Fe	Mg	
14310,115	$(Ca_{.18}Fe_{.53}Mg_{1.29})Si_2O_6$	.096	.904	.181	.435	.384	.094
14053,44	$(Ca_{.22}Fe_{.58}Mg_{1.12})Si_2O_6$	.142	.858	.224	.439	.337	.127
12053,72	$(Ca_{.21}Fe_{.58}Mg_{1.21})Si_2O_6$	.112	.888	.208	.471	.321	.086
12038,72	$(Ca_{.26}Fe_{.65}Mg_{1.09})Si_2O_6$	.159	.841	.263	.486	.251	.030
12021,21	$(Ca_{.22}Fe_{.57}Mg_{1.21})Si_2O_6$	.118	.882	.216	.456	.328	.097

b. Augites

14053,44	$(Ca_{.64}Fe_{.57}Mg_{.79})Si_2O_6$	.253	.747	.638	.314	.048	.052
12053,72	$(Ca_{.58}Fe_{.51}Mg_{.91})Si_2O_6$	.140	.860	.577	.366	.057	.014
12021,21	$(Ca_{.45}Fe_{.86}Mg_{.69})Si_2O_6$	.342	.658	.450	.519	.031	.031

* Determined by microprobe analysis on more than 20 grains and averaged.

applied to the same sample, can yield a wealth of information about the subsolidus cooling history of lunar rocks. Exsolution and (probably) antiphase domain formation starts soon after crystallization of the pyroxenes, but cation ordering continues to fairly low temperatures. These data, combined with other data on chemistry, petrography, rare-earth fractionation, and so on should yield a fairly complete history of the lunar rocks.

On the basis of the observed subsolidus phenomena in pyroxenes from rocks 12021, 12053, 12038, 12040, 14310, and 14053 we have been able to reconstruct the thermal history of these rocks.

Rock 12021, following crystallization out of the melt at ~ 1200°C, cooled slowly at a fairly uniform rate that is comparable to that of terrestrial volcanic rocks. This is evidenced by the formation of fairly extensive exsolution lamellae, large domains (~ 1000 Å) in pigeonite and a fairly high degree of cation order.

Rock 12053, following crystallization, was quenched rapidly at ~ 1200°C. This is evidenced by the fine-scale first-generation exsolution lamellae and very small domains (~ 50 Å) in pigeonite. A second generation of very fine-scale exsolution lamellae and a fairly high degree of cation order indicate a subsequent reheating at or above 700°C.

Pyroxenes in rock 12038 show similar phenomena, as observed in pyroxenes from rock 12053, but the second-generation exsolution lamellae are thicker and the degree of cation order is higher than that in pyroxenes from rock 12053. Hence the reheating of this rock must have taken place at a temperature < 700°C.

Because the core pigeonite does not show any first-generation exsolution lamellae, rock 12040 must have been quenched very quickly immediately after crystallization above the pyroxene solvus. This rock was subsequently reheated; this presumably caused the devitrification of the residual glass inclusions.

Rock 14310, presumably formed by the impact melting of the lunar soil, cooled after crystallization at a rate comparable to that of rock 12021, as evidenced by large domains (~ 1500 Å) and a fairly high degree of cation order in pigeonite. No evidence for a subsequent reheating has been found.

Rock 14053 must have cooled initially at approximately the same rate as rock 14310. Subsequently this rock was subjected to a reheating above 840°C for a short duration.

Acknowledgments—We are indebted to C. W. Kouns and Mrs. C. Inman for mineral separations, F. Wood for help with the microprobe analysis, E. Tidy for taking some of the x-ray diffraction photographs, and W. R. Riffle for assistance with the Mössbauer experiments. Dr. P. A. Pella of the National Bureau of Standards kindly recorded and processed the Mössbauer spectra of pigeonite and augite from rock 12021. The electron micrographs were taken at the million-volt electron microscope facility of the U.S. Steel Research Center, Monroeville, Pennsylvania. We are grateful to Drs. R. M. Fisher and J. S. Lally for assistance and helpful discussions. We are also indebted to Dr. R. M. Fisher and his colleagues for permission to reproduce Figs. 3, 6a, b, and 7. This research has been supported by NASA grant NGR 05003486.

References

Bailey J. C. Champness P. E. Dunham A. C. Esson J. Fyfe W. S. MacKenzie W. S. Stumpfl E. F. and Zussman J. (1970) Mineralogy and petrology of Apollo 11 lunar samples. *Proc. Apollo 11 Lunar Sci. Conf., Geochim. Cosmochim. Acta,* Suppl. 1, Vol. 1, pp. 169–194. Pergamon.

Bence A. E. and Albee A. L. (1968) Empirical correction factors for the electron microanalysis of silicates and oxides. *J. Geol.* **76**, 382–403.

Bence A. E. Papike J. J. and Prewitt C. T. (1970) Apollo 12 clinopyroxenes: Chemical trends. *Earth Planet. Sci. Lett.* **8**, 393–399.

Bown M. G. and Gay P. (1957) Observations on pigeonite. *Acta Crystallogr.* **10**, 440–441.

Boyd F. R. Bell P. M. and Smith D. (1971) Compositional variation in pyroxenes and olivines from Oceanus Procellarum. Second Lunar Science Conference unpublished proceedings.

Champness P. E. Dunham A. C. Gibb F. G. F. Giles H. N. MacKenzie W. S. Stumpfl E. F. and Zussman J. (1971) Mineralogy and petrology of some Apollo 12 lunar samples. *Proc. Second Lunar Sci. Conf., Geochim. Cosmochim. Acta,* Suppl. 2, Vol. 1, pp. 359–376. M.I.T. Press.

Christie J. M. Lally J. S. Heuer A. H. Fisher R. M. Griggs D. T. and Radcliffe S. V. (1971) Comparative electron petrography of Apollo 11, Apollo 12, and terrestrial rocks. *Proc. Second Lunar Sci. Conf., Geochim. Cosmochim. Acta,* Suppl. 3, Vol. 1, pp. 69–89. M.I.T. Press.

Dowty E. Ross M. and Cuttita F. (1972) Fe^{2+}-Mg site distribution in Apollo 12021 clinopyroxenes: Evidence for bias in Mössbauer spectra, and relation of ordering to exsolution. Unpublished manuscript.

Fernández-Morán H. Ohtsuki M. and Hibino A. (1971) Correlated electron microscopy and diffraction of lunar clinopyroxenes from Apollo 12 samples. *Proc. Second Lunar Sci. Conf., Geochim. Cosmochim. Acta,* Suppl. 2, Vol. 1, pp. 109–116. M.I.T. Press.

Finger L. W. Hafner S. S. Schürmann K. Virgo D. and Warburton D. (1972) Distinct cooling histories and reheating of Apollo 14 rocks (abstract). In *Lunar Science—III* (editor C. Watkins), pp. 259–261, Lunar Science Institute Contr. No. 88.

Gancarz A. J. Albee A. L. and Chodos A. A. (1971) Petrologic and mineralogical investigation of some crystalline rocks returned by the Apollo 14 mission. *Earth Planet. Sci. Lett.* **12**, 1–18.

Green D. H. Ringwood A. E. Ware N. G. Hibberson W. O. Major A. and Kiss E. (1971) Experimental petrology and petrogenesis of Apollo 12 basalts. *Proc. Second Lunar Sci. Conf., Geochim. Cosmochim. Acta,* Suppl. 2, Vol. 1, pp. 601–615. M.I.T. Press.

Hafner S. S. and Virgo D. (1970) Temperature dependent cation distributions in lunar and terrestrial pyroxenes. *Proc. Apollo 11 Lunar Sci. Conf., Geochim. Cosmochim. Acta,* Suppl. 1, Vol. 3, pp. 2183–2198. Pergamon.

Hafner S. S. Virgo D. and Warburton D. (1971) Cation distributions and cooling history of clinopyroxenes from Oceanus Procellarum. *Proc. Second Lunar Sci. Conf., Geochim. Cosmochim. Acta,* Suppl. 2, Vol. 1, pp. 91–108. M.I.T. Press.

Hollister L. Trzcienski W. Jr. Dymek R. Kulick C. Weigand P. and Hargraves R. (1972) Igneous fragment 14310,21 and the origin of the Mare basalts (abstract). In *Lunar Science—III* (editor C. Watkins), pp. 386–388, Lunar Science Institute Contr. No. 88.

Klein C. Jr. Drake J. C. and Frondel C. (1971) Mineralogical, petrological, and chemical features of four Apollo 12 lunar microgabbros. *Proc. Second Lunar Sci. Conf., Geochim. Cosmochim. Acta,* Vol. 1, Vol. 1, pp. 265–284. M.I.T. Press.

Lally J. S. Fisher R. M. Christie J. M. Griggs D. T. Heuer A. H. Nord G. L. and Radcliffe S. V. (1972) Electron petrography of Apollo 14 and 15 samples (abstract). In *Lunar Science—III* (editor C. Watkins), pp. 469–471, Lunar Science Institute Contr. No. 88.

LSPET (1971) (Lunar Sample Preliminary Examination Team) Preliminary examination of lunar samples. In *Apollo 14 Preliminary Science Report,* pp. 109–131, NASA SP-272, Washington, D.C.

Morimoto N. and Tokonami M. (1969) Domain structure of pigeonite and clinoenstatite. *Amer. Mineral.* **54**, 725–740.

Newton R. C. Anderson A. T. and Smith J. V. (1971) Accumulation of olivine in rock 12040 and other basaltic fragments in the light of analysis and synthesis. *Proc. Second Lunar Sci. Conf., Geochim. Cosmochim. Acta,* Suppl. 2, Vol. 1, pp. 575–582. M.I.T. Press.

Papike J. J. Bence A. E. Brown G. E. Prewitt C. T. and Wu C. H. (1971) Apollo 12 clinopyroxenes: Exsolution and epitaxy. *Earth Planet. Sci. Lett.* **10**, 307–315.

Prewitt C. T. Brown G. E. and Papike J. J. (1971) Apollo 12 clinopyroxenes: High temperature x-ray diffraction studies. *Proc. Second Lunar Sci. Conf., Geochim. Cosmochim. Acta,* Suppl. 2, Vol. 1, pp. 59–68. M.I.T. Press.

Takeda H. and Ridley W. I. (1972) Crystallography and mineralogy of pyroxenes from Fra Mauro soil and rock 14310 (abstract). In *Lunar Science—III* (editor C. Watkins), pp. 738–740, Lunar Science Institute Contr. No. 88.

Walter L. S. French B. M. and Doan A. S. Jr. (1972) Petrographic analysis of lunar samples 14171 and 14305 (breccias) and 14310 (melt rock) (abstract). In *Lunar Science—III* (editor C. Watkins), pp. 773–775, Lunar Science Institute Contr. No. 88.

Weill D. F. Grieve R. A. MacCallum I. S. and Bottinga Y. (1971) Mineralogy-petrology of lunar samples. Microprobe studies of samples 12021 and 12022; viscosity of melts of selected lunar compositions. *Proc. Second Lunar Sci. Conf., Geochim. Cosmochim. Acta*, Suppl. 2, Vol. 1, pp. 413–430. M.I.T. Press.

Proceedings of the Third Lunar Science Conference
(Supplement 3, *Geochimica et Cosmochimica Acta*)
Vol. 1, pp. 533–543
The M.I.T. Press, 1972

Crystal field spectra of lunar pyroxenes

ROGER G. BURNS, RATEB M. ABU-EID, and FRANK E. HUGGINS

Department of Earth and Planetary Sciences
Massachusetts Institute of Technology
Cambridge, Massachusetts 02139

Abstract—Absorption spectra in the visible and near infrared regions have been obtained for pyroxene single crystals in rocks from the Apollo 11, 12, 14, and 15 missions. The polarized spectra are compared with those obtained from terrestrial calcic clinopyroxenes, subcalcic augites, pigeonites, and orthopyroxenes. The lunar pyroxenes contain several broad, intense absorption bands in the near infrared, the positions of which are related to bulk composition, Fe^{2+} site occupancy and structure-type of the pyroxene. The visible spectra contain several sharp, weak peaks mainly due to spin-forbidden transitions in Fe^{2+}. Additional weak bands in this region in Apollo 11 pyroxenes are attributed to Ti^{3+} ions. Spectral features from Fe^{3+}, Mn^{2+}, Cr^{3+}, and Cr^{2+} were not observed. The spectral evidence for Ti^{3+} ions suggests that cation substitution in the lunar pyroxenes may be more complex than the $R_{oct}^{2+} + 2Si_{tet}^{4+} \rightleftharpoons Ti_{oct}^{4+} + 2Al_{tet}^{3+}$ coupled substitution mechanism suggested previously.

INTRODUCTION

THIS PAPER DESCRIBES measurements of the polarized absorption spectra in the region 360–2200 nm made on single crystals of pyroxene minerals in rocks from the Apollo 11, 12, 14, and 15 missions. The goal of the investigation was to identify the valencies of iron and titanium cations in the pyroxene phases.

First reports of the Apollo 11 rocks (LSPET, 1969) not only revealed the highly titaniferous composition of the rocks but also suggested that the pyroxene phases might be a titaniferous variety. Spectral measurements of terrestrial titanaugites in the visible region (Chesnokov, 1959; Burns *et al.*, 1964; Manning and Nickel, 1969) had demonstrated that these minerals contain Ti^{3+} ions. Therefore, one facet of our investigation was to search for Ti^{3+} ions in the lunar pyroxenes.

Numerous spectral measurements have been made of Fe^{2+} ions in terrestrial pyroxenes (Burns, 1965, 1970; White and Keester, 1966, 1967; Bancroft and Burns, 1967; Lewis and White, 1972). It is apparent that the crystal field spectra depend not only on the bulk chemical composition of the pyroxene phases, but also on the structure-type and site occupancy of Fe^{2+} ions in the pyroxene structure. Some groups have reported spectral evidence for Fe^{3+} ions in the lunar pyroxenes (Cohen, 1970; Mao and Bell, 1971; Bell and Mao, 1972). Therefore, a second facet of our work has been to characterize the iron cations and to correlate peak maxima with pyroxene composition in the lunar pyroxenes.

EXPERIMENTAL METHODS

Specimens. Polarized absorption spectra were recorded for pyroxenes in thin sections of the following Apollo rocks: 10045, 10047, 10058, 10072, 12021, 12052, 14053, 14310, and 15555. Initial measurements were made on petrographic thin sections 30–50 μ thick, but data from more recent missions are being recorded on 100 μ thick rock sections mounted on glass slides.

Technique. Measurements were made using the "microscope technique" described previously (Burns, 1966) with minor modifications. Microscopes with calcite Nicol polarizers were mounted horizontally in a specially constructed light-tight compartment built into a Cary model 17 recording spectrophotometer between the light sources and detectors. Pyroxene single crystals had been selected beforehand that completely covered the field of view of the microscope when viewed through 62X or lower powered objectives. The general fine-grained texture of most Apollo rocks made it difficult to use the universal stage in order to orient the pyroxene single crystal in three dimensions as in previous measurements. Therefore, crystals were selected that had two optical indicatrix axes lying approximately in the plane of the flat stage microscope. They were aligned parallel to the plane of polarization of the lower calcite Nicol prism. This experimental set-up enabled us to measure the spectra of single crystals as small as 0.5 mm in diameter, provided that they filled the field of view of the microscope. Although it was impossible to avoid working with compositionally zoned and finely exsolved crystals, we were able to obtain separate spectra from core pigeonite and mantle augite in some pyroxenes.

Summary of Terrestrial Pyroxene Spectra

The numerous absorption spectral measurements of pyroxenes in crustal rocks and in meteorites by several groups (Burns, 1965, 1970; White and Keester, 1966; Lewis and White, 1972; Adams, personal communication) may be summarized as follows. In light polarized along the b axis, orthopyroxenes give at least two broad and intense bands at 890–940 nm and 1800–2100 nm arising from spin-allowed transitions in Fe^{2+} ions. The peak maxima depend on the Fe/Mg ratio of the orthopyroxene and the ranges are 890–920 nm and 1800–1950 nm for enstatite-hypersthene compositions, Fs_{5-50}, in which Fe^{2+} ions largely are concentrated in the pyroxene M2 positions. Clinopyroxenes of the diopside-hedenbergite series, containing Fe^{2+} ions in the M1 positions, give rise to absorption bands at 1035–1045 nm and 2250 nm, the positions of which are not affected by changing the Fe/Mg ratio. These bands move to shorter wavelengths in augites, ferroaugites, and subcalcic augites, as a result of Fe^{2+} ions filling calcium deficiencies in the M2 positions. In augites and ferroaugites, the peak maxima occur at 990–1030 nm and 2000–2300 nm, and in subcalcic augites the bands span the ranges 970–1000 nm and 1900–2100 nm. Pigeonites have absorption spectra analogous to orthopyroxenes with peak maxima dependent on Fe/Mg ratios, but the bands occur at slightly longer wavelengths, 920–960 nm and 1850–2100 nm, respectively. The trends for the absorption band centered around 890–1040 nm are summarized in Fig. 1. The decrease in band maxima from 1040 nm in calcic clinopyroxenes to 890–920 nm in orthopyroxenes is a qualitative measure of the increasing Fe^{2+} site occupancy of the pyroxene M2 positions.

Terrestrial pyroxenes also show weak, sharp peaks in the visible region arising from spin-forbidden transitions in Fe^{2+} ions (Furlani, 1957). These occur typically at about 550, 505, and 425 nm (Burns, 1965). Pyroxenes containing Fe^{3+} ions commonly contain a broad intense band between 700–800 nm and very sharp peaks around 440 nm and 370 nm. The former band originates from charge transfer between adjacent Fe^{2+} and Fe^{3+} ions in the pyroxene structure, whereas the latter peaks represent spin-forbidden transitions in Fe^{3+} ions. Titanaugites also contain a broad intense band around 450 nm, which is attributed to Ti^{3+} ions.

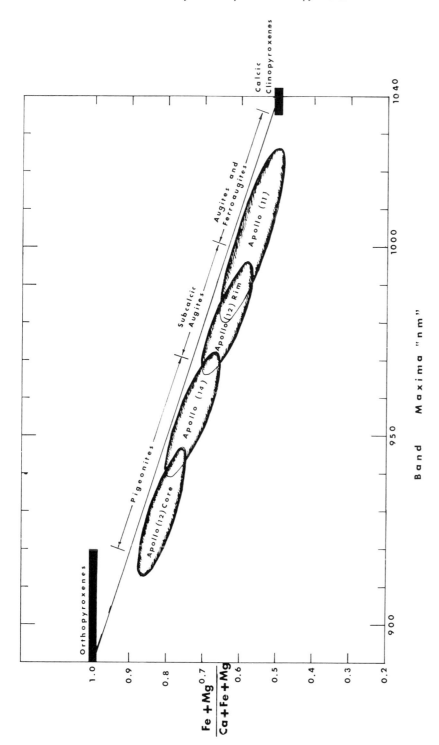

Fig. 1 Ranges of absorption maxima for terrestrial and lunar pyroxenes. The data are for spectra measured with light polarized along the b crystallographic axis.

Spectral data for minerals in certain Apollo 11 rocks are described elsewhere (Burns, Huggins, and Abu-Eid, 1972). We shall summarize here important features in the Apollo 11 pyroxene spectra and compare them with results obtained for pyroxenes from later missions.

Apollo 11 pyroxenes

The polarized absorption spectra of pyroxenes in rock 10047 are shown in Fig. 2. Numerous microprobe analyses have been reported for these pyroxenes (Dence *et al.*, 1970; Lovering and Ware, 1970; Ross *et al.*, 1970). Although compositionally zoned, the average composition of the pyroxenes may be approximated to be:

$$Ca_{0.56}Na_{0.01}Mn_{0.01}Fe_{0.51}Mg_{0.86}Ti_{0.04}Cr_{0.01}Al_{0.07}Si_{1.89}O_6$$

The broad intense bands centered around 1000 nm and 2000 nm in Fig. 2a represent spin-allowed transitions in octahedrally coordinated Fe^{2+} ions in the pyroxene structure. By analogy with terrestrial pyroxenes, the most intense band occurs in the Y spectrum, corresponding to light polarized along the *b* crystallographic axis. In the Apollo 11 pyroxenes that we have examined, the maxima of the intense band in the Y spectra span the range 980–1030 nm. This may be compared on Fig. 1 with values found for terrestrial augites and ferroaugites (990–1030 nm) and subcalcic augites (970–1000 nm).

The visible spectrum of pyroxene 10047 shown in Fig. 2b displays several weak sharp peaks arising from spin-forbidden transitions in Fe^{2+} ions. However, certain broad bands centered around 465 nm and 660 nm have been interpreted as spin-allowed transitions in Ti^{3+} ions (Burns, Huggins, and Abu-Eid, 1972). These bands are most pronounced in the Apollo 11 pyroxenes, including 10045, 10047, 10058, and 10072, in which the Ti/Fe ratios exceed 0.1 in certain areas.

Apollo 12 pyroxenes

Pyroxenes from rock 12021 contain prominent cores of pigeonite mantled by augite. Microprobe analyses of these phases (Bence *et al.*, 1970; Hafner *et al.*, 1971; Klein *et al.*, 1971; Prewitt *et al.*, 1971; Weill *et al.*, 1971; Boyd and Smith, 1971). indicate the following average compositions of the pigeonite cores

$$Ca_{0.19}Mn_{0.01}Fe_{0.56}Mg_{1.21}Cr_{0.03}Ti_{0.02}Al_{0.08}Si_{1.90}O_6$$

and of the augite mantles

$$Ca_{0.50}Mn_{0.02}Fe_{1.06}Mg_{0.40}Cr_{0.01}Ti_{0.04}Al_{0.07}Si_{1.92}O_6$$

Thus, the Ti/Fe ratios of these pyroxenes are smaller than those in the pyroxenes from Apollo 11 rocks that we examined.

Polarized absorption spectra of two pyroxene crystals in different orientations from rock 12021 are shown in Fig. 3. The spectra in Fig. 3a represent light polarized along the X and Y axes of the pigeonite cores; the corresponding spectra for the

Fig. 2. Polarized absorption spectra of clinopyroxene single crystals in rock 10047.

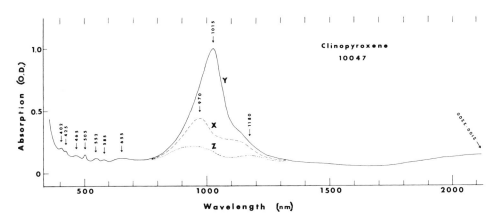

(a) Visible and near infrared regions.

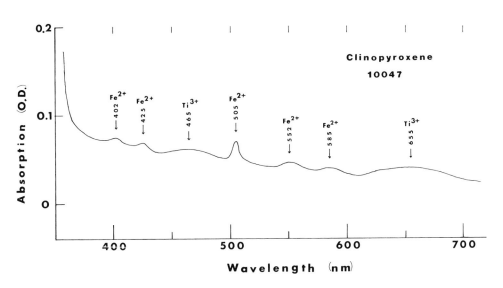

(b) Enlargement of visible spectrum with peak assignments.

Fig. 3. Polarized absorption spectra of Apollo 12 pyroxenes in rock 12021.

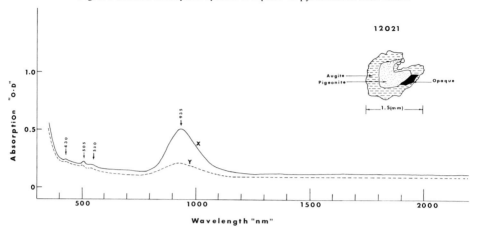

(a) Core pigeonite, X and Y spectra.

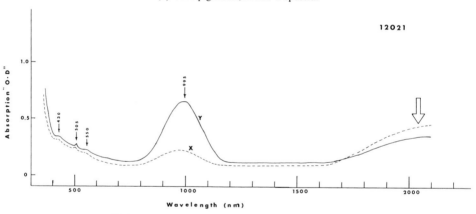

(b) Mantle augite. X and Y spectra on the same crystal.

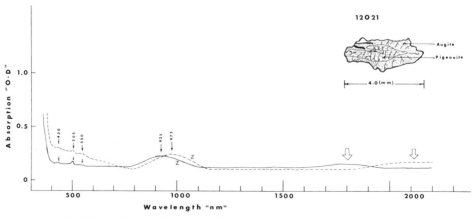

(c) Z Spectra for core pigeonite and mantle augite. —— pigeonite, – – – – augite.

augite mantles are shown in Fig. 3b. The Z polarized spectra for the core pigeonite and mantle augites of a different crystal are shown in Fig. 3c. Intense broad absorption bands occur around 1000 nm and 2000 nm, which are due to spin-allowed transitions in Fe^{2+} ions. The pigeonite spectra resemble those illustrated by other groups (Mao and Bell, 1971; Bell and Mao, 1972). In these more magnesian pigeonites, the absorption maxima in the Y spectra occur at 915–935 nm and at about 1850 nm by analogy with terrestrial Mg pigeonites. The peak maxima for the augite Y spectra span the range 970–990 nm found also for terrestrial subcalcic augites. There is a notable reduction in spectral features in the visible region of Apollo 12 pyroxenes compared with the pyroxenes from Apollo 11 rocks. In particular the bands centered around 465 nm and 660 nm are virtually absent from the pigeonite and augite in Apollo 12 rocks. This correlates with the lower Ti/Fe ratios in these minerals, and suggests that the Ti^{3+} ion concentration is lower in the Apollo 12 pyroxenes. Other weak sharp peaks arising from spin-forbidden transitions in Fe^{2+} ions occur in the visible region. The increased intensity of these peaks in the augite spectra compared to pigeonite conforms with the higher iron concentration in the augite mantles. The iron enrichment also manifests itself in increased charge transfer (metal to O) absorption in the ultraviolet region below 400 nm.

Apollo 14 pyroxenes

The clinopyroxenes in rock 14053 have very complex textures, structures and compositions. Gancarz et al. (1971) report microprobe analyses across zoned pigeonite-augite-ferroaugite crystals, the average composition of which may be approximated as

$$Ca_{0.54}Na_{0.01}Mn_{0.01}Fe_{0.89}Mg_{0.51}Cr_{0.01}Ti_{0.03}Al_{0.07}Si_{1.93}O_6$$

Also in this case the Ti/Fe ratio is smaller than that for the Apollo 11 pyroxenes.

The absorption spectra of the 14053 pyroxenes reflect their complex crystal chemistries. Spectra measured from the centers of crystals give band maxima at 930–940 nm and about 1800 nm, but the margins yield bands centered around 965 nm and 1800 nm. Weak spin-forbidden peaks occur at 550, 505, and 425 nm, but spectral features at 465 and 660 nm were not observed. Thus, although Ti^{3+} ions have been suggested in Apollo 14 pyroxenes from crystal chemical arguments (Bence and Papike, 1972), their concentration is below the detection limits of the spectral measurements.

Apollo 15 pyroxenes

Representative polarized spectra of pyroxene single crystals in rock 15555 are shown in Fig. 4. These measurements were made on 100 μ-thick sections, with the result that absorption bands are more intense than those recorded for Apollo 12 and 14 pyroxenes. Measurements were made on the central regions and margins of each crystal. Absorption maxima occur around 950 nm and 2000 nm for the cores and around 985 nm and 2100 nm for the margins. By analogy with terrestrial pyroxenes.

Fig. 4. Polarized absorption spectra of Apollo 15 pyroxenes in rock 15555.

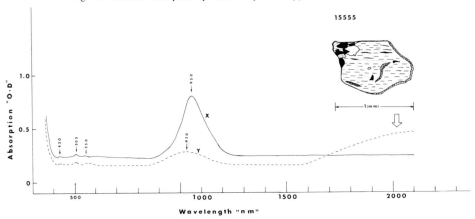

(a) Spectra measured at the center of a crystal.

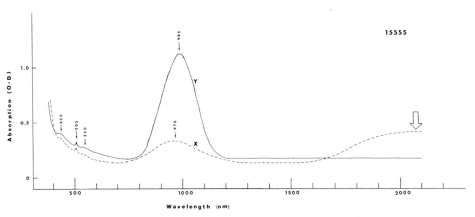

(b) Spectra measured at the margin of the same crystal.

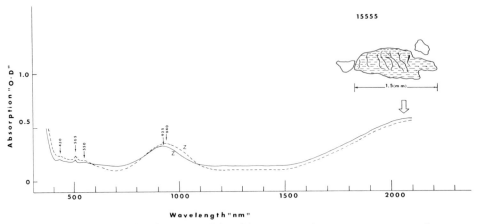

(c) Z Spectra measured at the center and rim of the same crystal; ——— center, ----- rim.

the band centered around 2000 nm is most intense in Z polarized light. The omnipresent spin-forbidden Fe^{2+} bands occur at 550, 505, and 430 nm in the visible region of each spectrum. Other bands at 465 nm and 660 nm attributed to Ti^{3+} ions were not observed.

Although microprobe analyses of pyroxenes in rock 15555 are not yet available to us, we can deduce the compositions of these pyroxenes from the absorption spectral measurements. The crystals are compositionally zoned and probably contain a pigeonite core which is more ferrugenous than the core pigeonites of rock 12021. The margins of the pyroxenes from rock 15555 probably are ferroaugite with average compositions intermediate between the pyroxenes in rock 10047 and the mantle augite in rock 12021.

DISCUSSION

The spectral data reported here provide further information in the search for Ti^{3+} ions in the lunar pyroxenes. The presence of this reduced oxidation state of titanium in the moon rocks and minerals has been postulated by several groups (for example Agrell et al., 1970; Lovering and Ware, 1970; Dence et al., 1970; Bence and Papike, 1972; Boyd and Smith, 1971; and Bell and Mao, 1972). The identification of Ti^{3+} ions in the crystal field spectra of lunar pyroxenes, although a plausible direct approach, is fraught with difficulties arising from the low Ti concentration relative to Fe. The predicted and observed Ti^{3+} spectral features occur among the spin-forbidden peaks of Fe^{2+}, the locations and assignments of which are poorly understood. However, the presence of weak broad bands at 465 nm and 600 nm in the more titaniferous pyroxenes from Apollo 11 rocks and the failure to resolve similar bands in the Apollo 12 and 14 pyroxenes, which have lower Ti/Fe ratios, are consistent with the hypothesis that Ti^{3+} ions only occur in the Apollo 11 pyroxenes to any measurable extent. The Ti^{3+} ions may also be present in other lunar pyroxenes, but at the present time their spectral features are below the detectibility of our spectrophotometer.

Contributions to the spectra from transition metal ions besides Fe^{2+} $(3d^6)$ and Ti^{3+} $(3d^1)$ appear to be unlikely. The Ti^{4+} ion with zero $3d$ electrons does not absorb in the visible region. Both Mn^{2+} and Fe^{3+} with $3d^5$ configurations give rise only to spin-forbidden peaks in oxide and silicate minerals. Because these peaks are one-tenth to one-hundredth the intensity of spin-allowed transitions, the concentrations of Mn^{2+} and Fe^{3+} in the lunar pyroxenes would have to be comparable to that of Fe^{2+} in order for these $3d^5$ ions to contribute to the crystal field spectra. This is clearly at variance with chemical and Mössbauer spectral data for the pyroxenes from the moon. Cr^{2+} $(3d^4)$ and Cr^{3+} $(3d^3)$ ions give rise to spin-allowed transitions. Broad bands centered around 430 nm and 620 nm are well-known for Cr (III) compounds and minerals, including chrome-diopside (Neuhaus, 1960). However, the Cr content of the lunar pyroxenes is generally very small and similar in the Apollo 11, 12, and 14 pyroxenes, but only the Apollo 11 pyroxenes show the spectral features at 465 nm and 660 nm.

Assuming that Ti^{3+} ions do, indeed, occur in the Apollo 11 pyroxenes, then their presence is inconsistent with one of the mechanisms for coupled substitution in the lunar pyroxenes. Several groups have noted a strong 1:2 correlation between the Ti and Al concentrations of the Apollo 11 pyroxenes, and have suggested the following coupled substitution scheme (for example, Ross *et al.*, 1970; Hollister and Hargraves, 1970):

$$R_{oct}^{2+} + 2Si_{tet}^{4+} \rightleftharpoons Ti_{oct}^{4+} + 2Al_{tet}^{3+}$$

Clearly this scheme cannot account for the crystal chemistry of titanium if a portion exists as Ti^{3+} ions.

We suggest that the cation substitution in pyroxenes from Apollo 11 rocks is more complex than has been suggested previously. Published microprobe analyses (Ross *et al.*, 1970, p. 845) show a deficiency of (Si + Al) in tetrahedral sites, and the cation sum (Cr + Mn + Mg + Fe + Ca + Na), excluding Ti, may or may not exceed 2.00 units in the chemical formulae. If the Ti contents are included, there is generally a surplus of cations in either the tetrahedral or Ml sites. It has been tacitly assumed that all the Al occurs in tetrahedral coordination. If this is true, then the titanium must be distributed among the Si and Ml sites. We propose that not only does some of the titanium occur as Ti^{3+} ions, but also that some titanium (perhaps Ti^{4+}) replaces silicon in the linked SiO_4 tetrahedra of pyroxenes from Apollo 11 rocks.

Note added: Dr. E. Dowty, who reviewed this paper, has drawn our attention to his spectral data for a titaniferous pyroxene from the Pueblito de Allende meteorite, in which most of the titanium is Ti^{3+}. Bands occur at 21,000 cm^{-1} and 16,500 cm^{-1}, in good agreement with the Ti^{3+} spectral features described here. Dowty's results and interpretations are to be submitted and published elsewhere.

Acknowledgments—This work is supported by a grant from the National Aeronautics and Space Administration (Grant No. NGR 22–009–551). We thank Dr. J. B. Adams and Professor W. B. White for spectral data in advance of publication. Mrs. Virginia Mee Burns assisted in bibliographic research and Sue Hall helped in preparation of the manuscript.

References

Agrell S. O. Scoon J. H. Muir I. D. Long J. V. P. McConnell J. D. C. and Peckett A. (1970) Observations on the Chemistry, Mineralogy and Petrology of Some Apollo 11 Lunar Samples. *Proc. Apollo 11 Lunar Sci. Conf., Geochim. Cosmochim. Acta* Suppl. 1, vol. 1, pp. 93–128. Pergamon.

Bancroft G. M. and Burns R. G. (1967) Interpretation of the electronic spectra of pyroxenes. *Amer. Mineral.* **52**, 1278–1287.

Bell P. M. and Mao H. K. (1972) Initial findings of a study of chemical composition and crystal field spectra of selected grains from Apollo 14 and 15 rocks, glasses and fine fractions (less than 1 mm). (abstract). In *Lunar Science—III* (editor C. Watkins) pp. 55–57, Lunar Science Institute Contr. No. 88.

Bence A. E. and Papike J. J. (1972) Crystallization histories of pyroxenes from lunar basalts (abstract). In *Lunar Science—III* (editor C. Watkins), pp. 59–61, Lunar Science Institute Contr. No. 88.

Bence A. E. Papike J. J. and Prewitt C. T. (1970) Apollo 12 clinopyroxenes: chemical trends. *Earth Planet. Sci. Lett.* **8**, 393–399.

Boyd F. R. and Smith D. (1971) Compositional zoning in pyroxenes from lunar rock 12021. Oceanus Procellarum. *J. Petrol.* **12**, 439–464.

Burns R. G. (1965) Electronic spectra of silicate minerals: applications of crystal field theory to aspects of geochemistry. Ph.D. thesis, Univ. of Calif., Berkeley.

Burns R. G. (1966) Apparatus for measuring polarized absorption spectra of small crystals. *J. Sci. Instr.* **43**, 58–60.

Burns R. G. (1970) *Mineralogical Applications of Crystal Field Theory*. Cambridge Univ. Press.

Burns R. G. Clark R. H. and Fyfe W. S. (1964) Crystal field theory and applications to problems in geochemistry. In. Vinogradov, A.P., ed., *Chemistry of the Earth's Crust. Proc. Vern. Centen. Symp.* **2**, 88–106 (transl. *Israel Progr. Sci. Transl., Jerusalem,* 1967, **2**, 93–112).

Burns R. G. Huggins F. E. and Abu-Eid R. M. (1972) Polarized absorption spectra of single crystals of lunar pyroxenes and olivines. *The Moon*, in press.

Chesnokov B. V. (1959) Spectral absorption curves of some minerals colored by titanium. *Dokl. Akad. Nauk S.S.S.R.* **129**, 647–649.

Cohen A. J. (1970) Origin of trace Fe^{3+} and Ti^{3+} in Apollo 11 and 12 crystalline rocks. *Trans. Amer. Geophys. Union* **52**, 272.

Dence M. R. Douglas J. A. V. Plant A. G. and Traill R. J. (1970) Petrology, mineralogy and deformation of Apollo 11 samples, *Proc. Apollo 11 Lunar Sci. Conf., Geochim. Cosmochim. Acta* Suppl. 1, vol. 1, pp. 315–340. Pergamon.

Furlani C. (1957) Spettri di assorbimento di complessi electrostatic del Fe^{2+}. *Gazz. Chim. Ital.* **87**, 371–379.

Gancarz A. J. Albee A. L. and Chodas A. A. (1971) Petrologic and mineralogic investigation of some crystalline rocks returned by the Apollo 14 mission. *Earth Planet. Sci. Lett.* **12**, 1–18.

Hafner S. S. Virgo D. and Warburton D. (1971) Cation distributions and cooling history of clinopyroxenes from Oceanus Procellarum. *Proc. Second Lunar Sci. Conf., Geochim. Cosmochim. Acta* Suppl. 2, vol. 1, pp. 91–108, M.I.T. Press.

Haggerty S. E. Boyd F. R. Bell P. M. Finger L. W. and Bryan W. B. (1970) Opaque minerals and olivine in lavas and breccias from Mare Tranquillitatis. *Proc. Apollo 11 Lunar Sci. Conf., Geochim. Cosmochim. Acta* Suppl. 1, vol. 1, pp. 513–538. Pergamon.

Hollister L. S. and Hargraves R. B. (1970) Compositional zoning and its significance in pyroxenes from two coarse grained Apollo 11 samples. *Proc. Apollo 11 Lunar Sci. Conf., Geochim. Cosmochim. Acta* Suppl. 1, vol. 1, pp. 541–550. Pergamon.

Klein C. Jr. Drake J. C. and Frondel C. (1971) Mineralogical, petrological and chemical features of four Apollo 12 lunar microgabbros. *Proc. Second Lunar Sci. Conf., Geochim. Cosmochim. Acta* Suppl. 1, vol. 1, pp. 265–284. M.I.T. Press.

Lewis J. F. and White W. B. (1972) Electronic spectra of iron in pyroxenes. *J. Geophys. Res.*, in press.

Lovering J. F. and Ware N. G. (1970) Electron probe microanalysis of minerals and glasses in Apollo 11 lunar samples. *Proc. Apollo 11 Lunar Sci. Conf., Geochim. Cosmochim. Acta* Suppl. 1, vol. 1, pp. 633–654. Pergamon.

LSPET (1969) (The Lunar Sample Preliminary Examination Team) Preliminary examination of lunar samples from Apollo 11. *Science* **167**, 631–633.

Manning P. G. and Nickel E. H. (1969) Optical absorption and electron microprobe studies of some high–Ti andradites. *Canad., Mineral.* **10**, 71–83.

Mao H. K. and Bell P. M. (1971) Crystal field spectra. *Ann. Rept. Geophys. Lab., Year Book* **70**, 207–215.

Neuhaus A. (1960) Uber die Ionenfarben der Kristalle und Minerale am Beispiel der Chromfarbungen. *Z. Kristallogr.* **113**, 195–233.

Prewitt C. T. Brown G. E. and Papike J. J. (1971) Apollo 12 clinopyroxenes: high temperature X-ray diffraction studies. *Proc. Second Lunar Sci. Conf., Geochim. Cosmochim. Acta* Suppl. 2, vol. 1, pp. 59–68. M.I.T. Press.

Ross M. Bence A. E. Dwornik E. J. Clark J. R. and Papike J. J. (1970) Mineralogy of the lunar clinopyroxenes, augite and pigeonite. *Proc. Apollo 11 Lunar Sci. Conf. Geochim. Cosmochim. Acta* Suppl. 1, vol. 1, pp. 839–848. Pergamon.

Weill D. F. Grieve R. A. McCallum I. S. and Bottinga Y. (1971) Mineralogy, petrology of lunar samples. Microprobe studies of samples 12021 and 12022; viscosity of melts of selected lunar compositions. *Proc. Second Lunar Sci. Conf., Geochim. Cosmochim. Acta* Suppl. 2, vol. 1, pp. 413–430. M.I.T. Press.

White W. B. and Keester K. L. (1966) Optical absorption spectra of iron in the rock-forming silicates. *Amer. Mineral.* **51**, 774–791.

White W. B. and Keester K. L. (1967) Selection rules and assignments for the spectra of ferrous iron in pyroxenes. *Amer. Mineral.* **52**, 1508–1514.

Proceedings of the Third Lunar Science Conference
(Supplement 3, *Geochimica et Cosmochimica Acta*)
Vol. 1, pp. 545–553
The M.I.T. Press, 1972

Crystal-field effects of iron and titanium in selected grains of Apollo 12, 14, and 15 rocks, glasses, and fine fractions*

P. M. BELL and H. K. MAO

Geophysical Laboratory, Carnegie Institution of
Washington, Washington, D.C. 20008

Abstract—Polarized crystal-field absorption bands in the near infrared caused by iron and titanium indicate their atomic site coordination and oxidation state in lunar pyroxenes and olivines from samples 12040,18, 12040,49, 12063,79, 14306,6, and 15601,94. The energies of the crystal-field transitions, and the resulting bands, shift with concentration.

In lunar glasses from samples 14163,33 and 15601,94, a near infrared band caused by ferrous iron shifts only in intensity with concentration. Interaction of this band with part of a strong absorption in the ultraviolet from titanium causes the colors of these glasses. The visible "window," a minimum between the iron and titanium absorptions, shifts as a function of the total concentration of titanium and of the ratio of trivalent to quadrivalent titanium.

The constancy of energy of the iron absorption in lunar glasses implies the following: (a) The structural symmetry of the atomic site of iron in lunar glasses is independent of bulk composition. (b) The thermal radiative properties of lunar soils composed primarily of lunar glass are remarkably uniform. (c) Spectra of soils composed of glass measured on the lunar surface should include values of the amplitude as well as wavelength of absorption if they are to be used to estimate the iron content.

INTRODUCTION

CRYSTALS AND GLASS grains from Apollo samples 12040,18, 12040,49, 12063,79, 14306,6, 14163,33, and 15601,94 were measured in this study, and conclusions concerning relationships between their optical properties and the concentrations and states of Fe and Ti were drawn. The properties of Fe and Ti are determined from polarized spectra of single crystals of pyroxene and olivine. Individual grains of glass have similar but broader absorption bands that correlate with the same properties. The positions and splitting patterns of optical absorption bands are highly sensitive to the states of transition elements, even if they are present in low concentration. Crystal-field effects have yielded evidence of unusually low oxidation states in lunar samples (Haggerty *et al.*, 1970). It should be emphasized that assignments of bands are often difficult and not always unique. The greatest difficulties arise at energies above 14,000 wave numbers.

PYROXENE SPECTRA

Measurements of single crystals were made on oriented grains so as to obtain polarized spectra (Mao and Bell, 1972a). The polarized bands indicate the transition element, its oxidation state, and certain characteristics of its coordination in the crystal structure. Figure 1 shows polarized spectra of a pyroxene (pigeonite; average dimension, 225 μ) from rock sample 12040,49, whose composition is given in Table 1. The bands at 11,000 and 5.300 cm^{-1} are caused by Fe^{2+} in distorted M2 octa-

*The authors wish to acknowledge the following grants for financial support of this research: NASA, NGL–09–140–012; NSF, GP–4384 and GA–22707.

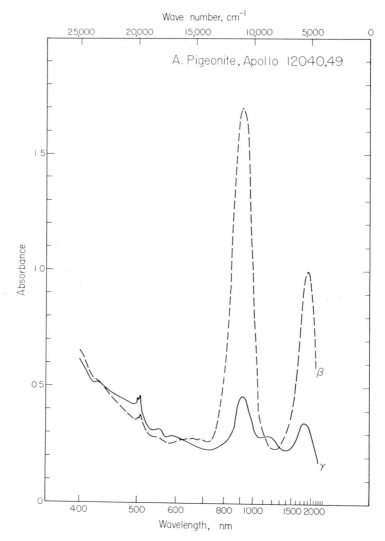

Fig. 1. Gamma and beta polarized crystal-field spectra of a pigeonite crystal from 12040,49. (L. W. Finger, coinvestigator, selected the crystal and identified it as pigeonite, $P2_1/c$, by x-ray diffraction.)

hedral sites. Fe^{2+} is also present in the less distorted M1 site, causing absorption at 8,500 cm^{-1}. The bands at 15,000 (doublet), 17,000, 18,200, and 19,700 (doublet) cm^{-1} are in the charge-transfer region. Their assignment is uncertain, but they are probably caused by Ti^{3+} and perhaps Fe^{3+} as well. At energies above 21,000 cm^{-1} the poorly resolved fine structure is caused almost entirely by Ti^{3+} and Ti^{4+}. The spectra in Fig. 1 and the band assignments are typical of lunar pyroxenes, changing slightly in augites, pigeonites, and hypersthenes (Mao and Bell, 1971). The shifts of bands with changes in Fe concentration are consistent with those described by Burns (1970) for terrestrial counterparts. Data on pyroxenes from samples 12040,49, 12063,79, and 14306,6 are listed in Table 1.

Table 1. Compositional* and spectral data

Sample	Phase	Optical directions measured	Cr_2O_3	MnO	FeO	Na_2O	MgO	Al_2O_3	SiO_2	CaO	TiO_2	K_2O	TOTAL
12040,18	Olivine† (loose coarse-grained rock)	α,β,γ											
12040,49	Pigeonite	β,γ		0.32	15.75	n.a.	23.63	1.27	52.55	3.85	0.52	n.a.	98.67
12063,79	Pyroxene‡	α,β,γ											
14306,6	Hypersthene (breccia)	α,β,γ	0.28	0.36	18.72	0.01	23.90	0.90	54.15	1.69	0.52		100.53
14306,6	Olivine§ (breccia)	α,β,γ											
14163,33–1	Glass	n	0.14	0.15	9.07	0.09	12.69	17.37	47.53	10.28	1.83		99.15
14163,33–2¶	Glass	n											
14163,33–3	Glass	n	0.57	0.29	17.01	0.22	14.72	9.94	48.54	9.20	0.60		101.09
14163,33–4¶	Glass	n											
14163,33–5¶	Glass	n											
14163,33–6	Glass	n	0.15	0.31	20.75	0.21	5.86	9.41	44.23	10.19	7.36	0.08	98.55
14163,33–7	Glass	n	0.60	0.28	17.02	0.24	14.60	9.86	46.42	9.38	0.64	0.06	99.10
14163,33–8	Glass	n	0.18	0.05	5.54	0.24	7.42	22.82	48.12	13.33	0.46	0.17	98.33
14163,33–9	Glass	n	0.13	0.10	5.08	0.03	9.01	24.34	44.07	14.00	0.29	0.00	97.05
14163,33–10	Glass	n	0.60	0.32	21.91	0.18	16.45	6.71	44.56	7.83	0.99	0.03	99.58
14163,33–11	Glass	n	0.28	0.18	11.16	0.44	10.54	16.85	47.15	10.75	1.58	0.18	99.11
14163,33–12	Glass	n	0.34	0.26	16.68	0.28	8.73	11.89	47.00	9.94	3.36	0.12	98.60
14163,33–13	Glass	n	0.47	0.33	20.31	0.21	8.07	10.03	45.85	10.17	2.93	0.01	98.38
14163,33–14	Glass	n	0.00	0.00	0.41	3.70	0.15	30.23	52.38	13.14	0.05	0.12	100.18
15601,94	Olivine (soil)	α,β,γ	0.15	0.37	31.13	0.00	31.00	0.00	36.09	0.27	0.06	0.00	99.07
15601,94	Feldspar	α,β,γ	0.00	0.01	0.56	1.07	0.19	32.13	46.61	17.56	0.06	0.03	98.22
15601,94–1	Glass	n	0.42	0.26	17.65	0.18	9.49	10.10	41.42	10.42	9.19	0.00	99.13
15601,94–2	Glass	n	0.31	0.24	15.39	0.03	12.12	14.47	43.17	11.54	2.13	0.00	99.40
15601,94–3	Glass	n	0.27	0.15	9.56	0.49	9.02	16.65	50.47	10.50	1.52	0.50	99.13
15601,94–4	Glass	n	0.27	0.19	11.55	0.14	11.55	17.11	45.10	11.08	2.44	0.03	99.46
15601,94–5	Glass	n	0.17	0.17	9.44	0.03	10.77	22.19	41.19	13.63	1.37	0.00	98.96
15601,94–6	Glass	n	0.59	0.29	18.66	0.10	16.83	7.72	45.98	8.57	0.41	0.00	99.15
15601,94–7	Glass	n	0.60	0.30	19.00	0.15	18.16	7.74	46.82	8.17	0.45	0.00	101.39
15601,94–8	Glass	n	0.26	0.16	9.03	0.57	9.14	16.62	50.78	10.25	1.38	0.48	98.67
15601,94–9	Glass	n	0.12	0.10	5.76	0.49	6.53	25.40	47.57	14.59	0.82	0.22	101.60
15601,94–10	Glass	n	0.62	0.30	19.72	0.14	17.93	7.58	46.92	8.27	0.43	0.00	101.91
15601,94–11	Glass	n	0.34	0.23	14.50	0.04	11.11	15.28	46.37	11.14	1.05	0.01	100.07

*Analyses were performed by electron microprobe.
†Analysis in Bell (1971).
‡Analysis in Adams and McCord (1972).
§Analyses are given in Bell and Mao (1972); discussions of the olivines of breccia 14306 and rock 15555 are in preparation for publication elsewhere.
¶Grain not suitable for microprobe analysis.

OLIVINE SPECTRA

The olivine spectra shown in Fig. 2 (A and B) were measured from an oriented single crystal from Apollo 12040,18. They show most of the features observed in similar crystals from samples 14306,6 and 15601,94. The bands caused by Fe^{2+}, both in M1 and M2 octahedral sites, are designated in Fig. 2A. Figure 2B shows detail in the visible and near ultraviolet regions between 14,000 and 25,000 wave numbers. The weak band located between 15,500 and 17,500 cm^{-1} changes intensity with optical orientation (α, β, γ directions) of the crystal. This band probably is caused by Fe^{3+}, although the assignment is not unique because of the possible

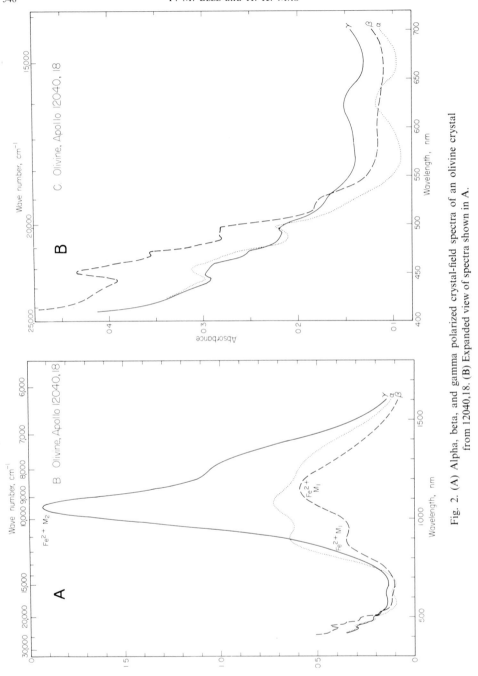

Fig. 2. (A) Alpha, beta, and gamma polarized crystal-field spectra of an olivine crystal from 12040,18. (B) Expanded view of spectra shown in A.

interference of a spin-forbidden band in Fe^{2+} (Burns *et al.*, 1972). The fine structure and strong absorption starting at 19,000 cm^{-1} also could be caused by a charge-transfer transition in Fe^{3+} (Shankland, 1966). This is not an intrinsic absorption edge caused by the band gap of olivine (Mao and Bell, 1972).

Data for the olivine of Fig. 2 and other olivines from samples 14306,6 and

15601,94 are given in Table 1. The iron concentrations were sufficiently high to obscure absorption bands of other transition elements in these crystals.

PLAGIOCLASE SPECTRA

Spectra were measured of single crystals of plagioclase and fragments of plagioclase glass from samples 15601,94 and 14163,33 (Table 1). Both crystals and glass are transparent, showing no strong absorption bands in the visible region. Iron as Fe^{2+} (some Fe^{3+} may be present as well; see Weeks and Kolopus, 1972) causes a polarized band in the near infrared. The polarized α spectrum shows an intense peak at 8,000 cm^{-1} (absorption coefficient, 14.3 cm^{-1}), but no peaks were observed in the β and γ orientations. The Fe absorption in plagioclase glass was observed as a broad band at 9,000–10,000 cm^{-1}, much like those in the other glasses described below.

GLASS SPECTRA

Analysis of the data on single crystals provided the basis for interpretation of the broad spectral bands of lunar glasses. The bands in these glasses are caused almost entirely by Fe and Ti; the effects of Cr and other transition elements are minor. Spectra of typical glasses are shown in Fig. 3A, 3B, and 3C. The first of these is a nearly colorless spheroid of glass, low in both Fe (5.09% FeO) and Ti (0.29% TiO_2), and might have formed from a contaminated melt, approximately equivalent to a low-iron, plagioclase-rich basalt. Figure 3B shows the spectrum of a typical green lunar glass fragment, containing low Ti (0.6% TiO_2) and high Fe (17.01% FeO), which would be classified as a "mare" type (Apollo Soil Survey, 1972). The spectrum plotted in Fig. 3C is representative of red lunar glass, characteristically high in Fe (20.84% FeO) and Ti (7.30% TiO_2), and also would be classified as a "mare" type. Brown glasses contain intermediate amounts of Ti.

Positions of the broad maximum of the absorption band at 9,000–10,000 cm^{-1} do not shift with the concentration of total iron, but the amplitude changes in proportion to the amount of iron. Data shown in Fig. 5 and discussed below suggest that absorption caused by titanium is sensitive to the total titanium content and to the Ti^{3+}/Ti^{4+} ratio as well. The maximum absorption for Ti, located in the ultraviolet region, is beyond the sensitivity of the present technique (Mao and Bell, 1972b), but its low-energy edge slopes into the visible region, intersecting the high-energy edge of the Fe band. It is postulated that the minimum of the resultant absorption correlates with the Ti^{3+}/Ti^{4+} ratio and the total amount of Ti present. The concentration of iron in the range 5–22 wt% FeO does not affect this absorption minimum. This is a visible "window," the energy of which actually determines the color of lunar glass.

The intensity of the Fe absorption band correlates moderately well with the total amount of iron. The weight percentage of FeO of several lunar glasses listed in Table 1 is plotted against the absorption coefficient (absorbance per unit thickness, in cm^{-1}) in Fig. 4. The scatter of points probably is caused by the errors introduced in measurement of the height of the maximum of the broad band in irregularly shaped grains.

The position, in wavelength, of the absorption minimum (λ min, nm) versus

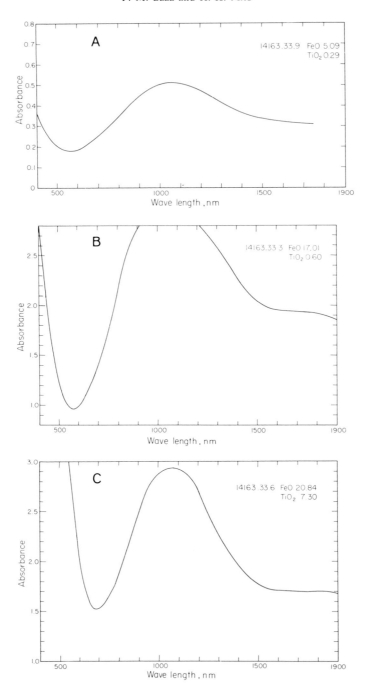

Fig. 3. Crystal-field spectra of three glasses from 14163,33. (Compositions given on the diagrams.)

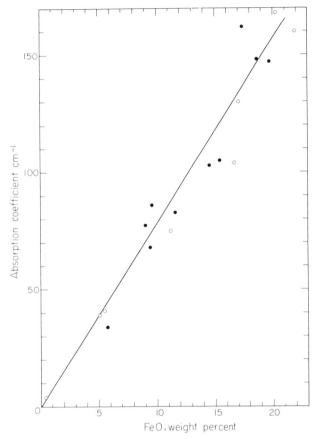

Fig. 4. Graph of the absorption coefficient (absorbance per centimeter thickness measured at 1000–1100 nm) of several glasses versus their weight percentage of FeO measured by electron microprobe. (Analyses given in Table 1.) Open circles, 14163; solid circles, 15601.

weight percentage of TiO_2 is plotted in Fig. 5. Data are listed in Table 1. The scatter of the points within the upper left half of Fig. 5 appears to be related to a combination of the Ti^{3+}/Ti^{4+} ratio of the glasses and their concentrations of Ti. Experimental data on the optical properties, oxidation states, and fugacities of oxygen as a function of the Ti^{3+}/Ti^{4+} ratio currently are only crudely known (White *et al.*, 1971). Eventually it should be possible to construct a family of curves for the diagram in Fig. 5, each corresponding to a fixed oxygen fugacity. Two such curves can be placed approximately. The first would be located near the vertical axis, corresponding to all Ti present as Ti^{3+} or as Ti^{4+} (White *et al.*, 1971). The second is approximated by the dashed line, which corresponds to the limiting value of oxygen fugacity in these lunar glasses. The estimated range of oxygen pressure for the two lines and for the glasses plotted between them is $10^{-7}–10^{-14}$ atm (Haggerty *et al.*, 1970).

IMPLICATIONS OF THE DATA

The present measurements are summarized by the data listed in Table 1. The

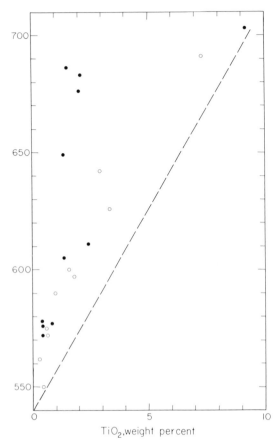

Fig. 5. Plot of wavelengths (λ) of the absorption minimum or "window" of several glasses
versus their weight percentage of FeO measured by electron microprobe. (Analyses given
in Table 1.) Open circles, 14163; solid circles, 15601.

values show that the variations in color of lunar glasses are caused by a crystal-field
effect of Ti, originating in the ultraviolet region, that extends to the visible. These
colors, and the Ti^{3+}/Ti^{4+} ratio that governs them, relate to partial pressures of
oxygen that characterized the lunar environment when they were formed. This
basically is true for lunar pyroxenes and amphiboles as well, although the iron
absorption also is sensitive to concentration and crystal structure.

If the constancy of band position in glasses and its variability in crystals are
generally true of all materials on the lunar surface, it is of interest to note the implications. The iron band is in the near infrared, where it strongly influences thermal
radiation. Soils composed mainly of lunar glasses would, therefore, possess remarkably uniform properties in the near-infrared.

The fact that the iron band is constant in energy suggests that the average atomic
coordination site for iron is nearly identical in all of the lunar glasses studied (Cotton,
1963). A practical consequence is that the Fe content in lunar glasses can be estimated

only from the peak height, not from the peak position as is possible in crystals. This makes the compositional determinations of soil mixtures uncertain unless both peak height and wavelength are known. The wide range of compositions and related optical absorptions appears to strengthen this conclusion. The wide range also points to the fact that the Apollo 14 and 15 soil samples studied are mixtures of highland material contaminated with mare material.

The results are valuable as indicators of the lunar atmosphere and the compositions of the bulk samples as well as single grains. Some of the present data have been used by other investigators to study reflectivity spectra measured by telescope and in the laboratory (Adams and McCord, 1972). There is a strong influence of Fe and Ti on the color, albedo, and radiative properties of the lunar surface.

References

Adams J. B. and McCord T. (1972) *Proc. Third Lunar Sci Conf., Geochim. Cosmochim. Acta* Suppl. 3, Vol. 3. MIT Press.

Apollo Soil Survey (Aitken F. K. Anderson D. H. Bass M. N. Brown R. W. Butler P. Jr. Heiken G. Jakeš P. Reid A. M. Ridley W. I. Takeda H. Warner J. and Williams R. J.) (1971) Apollo 14: Nature and origin of rock types in soil from the Fra Mauro formation. *Earth Planet. Sci. Lett.* **12**, 49–54.

Bell P. M. (1971) Analysis of olivine crystals in Apollo 12 rocks. *Carnegie Inst. Washington Year Book* 69, 228–229.

Bell P. M. and Mao H. K. (1972) Initial findings of a study of chemical composition and crystal field spectra of selected grains from Apollo 14 and 15 rocks, glasses and fine fractions (less than 1 mm) (abstract). In *Lunar Science—III* (editor C. Watkins), pp. 55–57, Lunar Science Institute Contr. No. 88.

Burns R. G. (1970) *Mineralogical Application of Crystal Field Theory.* Cambridge University Press, Cambridge.

Burns R. G. Abu-Eid R. E. and Huggins F. E. (1972) Crystal-field spectra of lunar pyroxenes (abstract). In *Lunar Science—III* (editor C. Watkins), pp. 108–109, Lunar Science Institute Contr. No. 88.

Cotton F. A. (1963) *Chemical Applications of Group Theory.* Interscience.

Haggerty S. E. Boyd F. R. Bell P. M. Finger L. W. and Bryan W. B. (1970) Opaque minerals and olivine in lavas and breccias from *Mare Tranquillitatis. Proc. Apollo 11 Lunar Sci. Conf., Geochim. Cosmochim. Acta* Suppl. 1, Vol. 1, pp. 513–538. Pergamon.

Mao H. K. and Bell P. M. (1971) Crystal-field spectra. *Carnegie Inst. Washington Year Book* 70, 207–215.

Mao H. K. and Bell P. M. (1972a) Polarized crystal-field spectra of microparticles of the moon. In "Analytical methods developed for application to lunar sample analyses," *Amer. Soc. Test. Mater. Spec. Tech. Publ.,* in press.

Mao H. K. and Bell P. M. (1972b) Electrical conductivity and the red shift of absorption in olivine and spinel at high pressure. *Science* **176**, 403–406.

Shankland T. J. (1966) Synthesis and optical properties of forsterite. *Office Naval Res. Tech. Rept.* **HP–16**, 136 pp.

Weeks R. A. and Kolopus J. L. (1972) Magnetic phases in lunar material and their electron magnetic resonance spectra: Apollo 14. In *Lunar Science—III* (editor C. Watkins), p. 791, Lunar Science Institute Contr. No. 88.

White W. B. White E. W. Gorz H. Henisch H. K. Fabel G. W. Roy R. and Weber J. M. (1971) Physical characterization of lunar glasses and fines. *Proc. Second Lunar Sci. Conf., Geochim. Cosmochim. Acta* Suppl. 2. Vol. 3, p. 2215. MIT Press.

Proceedings of the Third Lunar Science Conference
(Supplement 3, *Geochimica et Cosmochimica Acta*)
Vol. 1, pp. 555–568
The M.I.T. Press, 1972

X-ray investigations of lunar plagioclases and pyroxenes

H. Jagodzinski and M. Korekawa
Institut für Kristallographie und Mineralogie der Universität München,
8 München 2, Theresienstr. 41, Germany

Abstract—The theory of x-ray diffraction of lamellar systems with an incongruent and a congruent plane of intergrowth is developed. For small thicknesses of lamellae considerable deviations from the usual "incoherent" calculation occur. Effects of this kind are observed in plagioclases showing twinning and exsolution and in pigeonites with a beginning exsolution of augite. Twins according to the "Karlsbad law," albite law, and a combined "Karlsbad-albite law" are commonly observed in terrestrial and lunar plagioclases. This twinning is partly submicroscopic. Some lunar plagioclase specimens with bytownite composition show an "incoherent" two-phase diffraction pattern and are interpreted as submicroscopically intergrown phases with different An contents. Disorder effects in orthopyroxenes are most probably due to the observed zonal structure of the crystals.

Introduction

Electron micrographs of feldspars and pyroxenes commonly show a typical lamellar structure, which may be explained by either exsolution processes or sub-microscopical twinning (Ross *et al.*, 1970a, b; McConnell and Fleet, 1963). Because of the similarity of the structures, the decision as to which of these two possibilities is realized can commonly be given only by a detailed x-ray analysis. Difficulties arise if the lattice constants of the different types of lamellae are very similar. In this particular case the assumption of "incoherent" scattering of various types of lamellae, which is normally used, is no longer valid.

Preliminary x-ray studies of lunar feldspars and pyroxenes showed marked differences between lunar and terrestrial specimens, which seem to be due to the different prehistory of the samples. For this reason we felt it necessary to generalize the diffraction theory of lamellar systems. It will be shown that the theory of one-dimensional disorder, which has been developed in the most general form by Jagodzinski (1948, 1953) and Kakinoki and Komura (1952, 1954), can be applied to systems of lamellae having the same lattice vectors at the common plane of inter-growth. The theory fails to be applicable for the more general problem, with different lattice constants of the lamellae at the plane of intergrowth.

Diffraction Theory of Lamellar Intergrowth

Congruent planes of intergrowth

In the following, we consider only two types of lamellae differing in their structure. At the common plane of intergrowth, the lamellae shall have the translation vectors a and b, but it shall be admitted that one of the lattices may be described by a smaller plane unit cell (fractional translation vectors); this implies extinction rules in the diffraction problem. The third noncoplanar vector c is different for both types of lamellae (c_1, c_2). In case of twins the direction of c only is different for both types of lamellae, but we wish to deal with the more general case of completely different lattice constants c, which is typical for two different phases.

Let us assume that the lamellae are completely ordered in the a and b directions, but there shall be some disorder of the following kind in the c direction: The number of unit cells varies from lamella to lamella (compare Fig. 1). Consequently, the resulting scattering amplitude of the nth layer consisting of $N_1 N_2$ cells in the corresponding a and b directions and one cell in the c direction is:

$$L_n = RF_i\, e^{-2\pi inl}\, e^{-i\phi_n}, \tag{1}$$

where

$$R = \frac{\sin \pi N_1 h}{\sin \pi h} \frac{\sin \pi N_2 k}{\sin \pi k},$$

$F_i = F_1$ or F_2 (scattering amplitude of one layer of the lamellae of type 1, 2 as shown on Fig. 1), ϕ_n = phase factor, originating from the displacement of the nth layer, and $h, k,$ and l are coordinates in reciprocal space (indices of reflections).

The intensity of the observed x-ray diffraction pattern is described by

$$I(hkl) = R^2 \sum_{i,k} \sum_{m=-N_3}^{+N_3} (N_3 - |m|)\, p_i P_{ik}(m,\,\alpha,\,\beta,\,\phi_m)\, F_i F_k^*\, e^{2\pi iml}, \tag{2}$$

where l refers to the averaged lattice constant c, which is defined by $c_1 = c + \Delta_1 a + \Delta_2 b + \Delta_3 c$, $c_2 = c - \Delta_1 a - \Delta_2 b - \Delta_3 c$, N_3 = number of unit cells in the c direction, p_i = a priori probability of finding a layer of type i ($i = 1, 2$ in our case), P_{ik} = complex probability function of finding a layer of type k with the phase factor ϕ_m, if the original layer was of type i, m = summation index of two cells at a distance m, α = probability that a layer of type 1 be continued, and β = corresponding probability for layer 2.

The outstanding problem of solving equation (2) is the evaluation of $P_{ik}(m,\,\alpha,\,\beta,\,\phi_m)$. The solution is given by characteristic equations with values λ_v, the number of which reduces to 2 in our problem (2 types of lamellae).

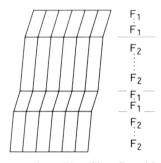

Fig. 1. Model for calculating x-ray intensities of lamellae with a congruent plane of intergrowth.

The most simple solution (next neighbors interaction) is given by

$$I(hkl) = R^2 \sum_{v=1}^{2} \left[A_v \frac{1 - |\lambda_v|}{1 - 2|\lambda_v|\cos 2\pi(l + \phi_v) + |\lambda_v|^2} \right.$$

$$\left. - 2B_v|\lambda_v| \frac{\sin 2\pi(l + \phi_v)}{1 - 2|\lambda_v|\cos 2\pi(l + \phi_v) + |\lambda_v|^2} \right], \quad (3)$$

where $\lambda_v = |\lambda_v| e^{2\pi i \phi_v}$ are the two solutions of the quadratic equation

$$\lambda^2 - \lambda(\alpha e^{2\pi i \psi} + \beta e^{-2\pi i \psi}) - 1 + \alpha + \beta = 0_1 \quad (4)$$

with $\psi = \Delta_1 h + \Delta_2 k + \Delta_3 l$. The constants A_v, B_v may be calculated in a cumbersome way; they depend on α, β, F_i, and ϕ_m.

Each λ_v in equation (3) describes one reflection, which is as sharp as a Bragg reflection if $|\lambda_v| = 1$, and represents a diffuse streak in the c^* direction if $|\lambda_v| = 0$. ϕ_v gives the position of the maximum. If $\Delta_1 = \Delta_2 = \Delta_3 = 0$, the problem reduces to the solution given by Jagodzinski (1948) with $\lambda_1 = 1$ and $\lambda_2 = \alpha + \beta - 1$ (it should be noted that α, β correspond to $1 - \alpha$, $1 - \beta$ in the paper cited). According to the previous discussion, only one reflection is observed as long as $\alpha + \beta > 0$, but a diffuse streak (λ_2) is superimposed by a sharp reflection ($\lambda_1 = 1$). This behavior is observed in labradorites with Schiller effect, as reported by Korekawa and Jagodzinski (1967). The diffuse reflection is resolved into supersatellite reflections, due to a more regular thickness of lamellae, when compared with our statistical model. A more detailed discussion of equation (3) will be given later.

Incongruent plane of intergrowth

The solution of the diffraction problem becomes more complicated if the condition of equal translation vectors at the plane of intergrowth is no longer valid. Fig. 2 gives some idea of the difficulties: The difference in lattice constants causes a strained area between the lamellae of the two phases. Sticking to the condition that both types of lamellae are strictly ordered in the a and b directions, and introducing

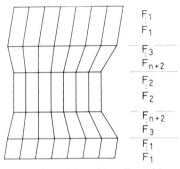

Fig. 2. Model for calculating x-ray intensities of lamellae with an incongruent plane of intergrowth.

structure amplitudes $F_3, \ldots, F_{n+2}$ for the strained areas of intergrowth according to Fig. 2, we have to define lattice amplitudes R_i correlated with the summation over the cells in the a and b direction. These amplitudes are considered to be sharp functions of the type

$$\frac{\sin \pi N_1 h_v}{\sin \pi h_v} \frac{\sin \pi N_2 k_v}{\sin \pi k_v},$$

as defined in equation (1) for both kinds of lamellae, but h_v, k_v are now different. The lattice amplitudes $R_3, \ldots, R_{n+2}$ are no longer functions of comparable sharpness because of the fact that the misfit of layers causes strains, which are partly reduced by dislocations. Their strain field has a strong influence on the shape of $R_3, \ldots, R_{n+2}$. Thus the solution of this diffraction problem is determined largely by the number of, the profiles of $R_3, \ldots, R_{n+2}$, and the F_i involved. The characteristic equation may be given in the general form:

$$[\lambda^2 - \lambda(\alpha e^{2\pi i\psi} + \beta e^{-2\pi i\psi} + \alpha\beta)] \lambda^{2n} - (1-\alpha)(1-\beta) = 0. \tag{5}$$

The solution of equation (5) shows that under certain circumstances two eigenvalues are similar to those of equation (4), but in general at least one of them (the one for which $|\lambda_v|$ differs appreciably from unity) is different. It may also be shown that the remaining $2n$ eigenvalues have λ_v with $|\lambda_v| \ll 1$; consequently, they generate very diffuse reflections. Because they are similar, interference effects are possible, which may be demonstrated by discussion of the intensity equation (6).

$$I(hkl) = \sum_{i,k} \sum_{m=-\infty}^{+\infty} p_i P_{ik}(m, \alpha, \beta, \phi_m) R_i R_k^* F_i F_k^* e^{2\pi i ml}. \tag{6}$$

The solution of equation (6) is now strongly dependent on the products $R_i R_k^*$, which may become zero if $i \neq k$. Thus the general discussion of the solution may best be done by considering 3 different cases, which shall be described qualitatively only:

(1) All products $R_i R_k^* (R_i R_k^* \neq 0)$ are strictly "coherent" (this is valid for h, k, $l \rightarrow 0$ or for very small differences of lattice constants a and b). The qualitative diffraction pattern is shown in Fig. 3a. The two spots corresponding to the Bragg reflections of the lamellae and all accompanying diffuse streaks are on the same reciprocal lattice row. The diffuse streak vanishes if both types of lamellae become very thick (> 1000 Å).

(2) Only R_1 and R_2 behave "incoherently" ($R_1 R_2^* = 0$), but the remaining R_i

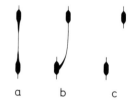

a   b   c

Fig. 3. X-ray intensities (qualitative) for a lamellae system (two phases or two twins) with an incongruent plane of intergrowth.

are still coherent. In this case the diffraction pattern depends on the R_i and the F_i. Fig. 3b shows one of the various possibilities: The diffuse streak may have the direction of the normal of the plane of intergrowth, especially the one which points from the reflection of the first phase opposite to the reflection position of the second phase. The connecting diffuse streak may be curved. The conditions of an eventual curvature of the streaks will be discussed in more detail in a later paper.

(3) All R_i are incoherent ($R_i R_k^* = 0$, if $i \neq k$). Only streaks vertical to the plane of intergrowth accompanying the reflections of the two lamellae may be observed (compare Fig. 3c). A very diffuse background between both reflections may rarely be observed, but this diffuse scattering also may be obscured by the general background of the diffraction pattern.

It should be pointed out here that the intensity distribution on the streaks, their symmetrical or asymmetrical behavior, and their diffuseness can only be discussed if a specific model of the structure of the strained intermediate area is given. For a general solution of our problem it seems to be more convenient to derive the model from the diffraction pattern. A possible method of obtaining this information from the diffuse scattering will be described in a later paper.

PLAGIOCLASES

Crystals from Apollo 11 (10050,45), Apollo 12 (12038,79) and Apollo 14 lunar sample (14310,106) were studied and compared with some chemically corresponding terrestrial samples by single-crystal x-ray diffraction methods. Besides Weissenberg and Buerger precession cameras, a monochromatically focussing single-crystal camera with high resolution power (Jagodzinski and Korekawa, 1965; Korekawa and Jagodzinski, 1967) was used.

Some examples of twin formation are shown in Fig. 4. Specimen No. 2 of sample 14310,106 consists essentially of two individuals described by a complex "Karlsbad-albite" twin law (Fig. 5), and specimens No. 5 and No. 15 of the same sample show only pairs of twins after the albite law. With these exceptions the other specimens investigated show three or four individuals after albite and "Karlsbad" twin laws. The same has been reported by Czank *et al.* (1972). The volume compositions of twin individuals are different (Fig. 4). Total symmetry of reflection positions (regardless their intensities) is mm2.

On precession and Weissenberg photographs of the lunar crystals, specimens 12038,79, No. 2 and 14310,106, No. 5 excepted, (a) and (b) or (a), (b), and (c) reflections were observed. The (c) reflections are diffuse, and the diffuseness is approximately parallel to b^* direction. Crystal 12038,79, No. 2 shows weak (e) reflection-satellite reflections as well as (a), (b), and diffuse (c) reflections. Crystal 14310,106, No. 5 shows (e) and (f) reflection-satellite reflections as well as (a) and (b) reflections (Fig. 6, c and d).

Recently one of the authors (Korekawa, to appear) has interpreted the Weissenberg photographs of some terrestrial specimens $An_{70} \sim An_{75}$ as "incoherent" two-phase patterns of submicroscopically intergrown phases with different An contents. Approximately $An_{65 \pm 3}$ and $An_{80 \pm 3}$ were assumed as the end members of exsolution.

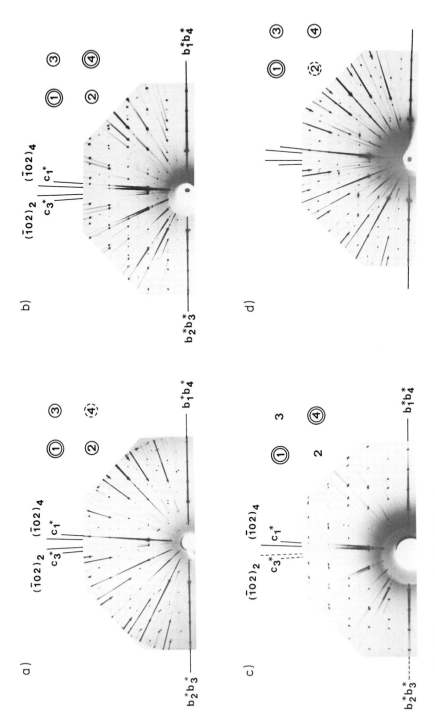

Fig. 4. Some examples of twin formation in lunar and terrestrial plagioclases with bytownite composition. Precession photographs $(0kl)$, MoK$_\alpha$ radiation. (a) Terrestrial specimen 103, No. 9, An$_{69.7}$ (Sognefjord, Norway): (b) lunar specimen 10050,45, No. 2 (Apollo 11) (c) lunar specimen 14310,106. No. 2 (Apollo 14) (d) lunar specimen 14310,106, No. 6 (Apollo 14). Nos. 1–4: twin individuals. Double circle: predominant; single circle: moderate; broken circle: small amount in volume.

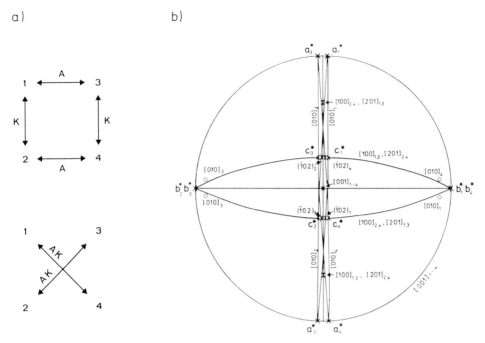

Fig. 5. Twinning of plagioclases with bytownite composition. (a) Schematic representation of twin relations. A: albite twin law, mirror plane (010). K: "Karlsbad" twin law, rotation axis [001]. AK: complex "Karlsbad-albite" twin law, (010) and [001]. (b) Stereographic projection of twin individuals 1–4; the index of 1 refers to $c = 14 \text{Å}$.

Examples of $0kl$ and $4kl$ Weissenberg photographs of the terrestrial specimen 103, No. 3 with bulk composition $An_{69.7}$ (Bambauer *et al.*, 1967) taken with better collimated incident x-rays and with extremely long exposure times are given in Figs. 6a and b. The following characteristics of the (a), (b), and (e) reflections were observed.

On the $0kl$ photograph the position of (e) reflections is symmetric with respect to (b) reflections (like satellites). On the $4kl$ photograph the same is not true (compare the reflection groups $01\bar{3}_1$ and $41\bar{5}_1$), additionally (a) reflections are split into two reflections (see reflection groups $40\bar{2}_1$ and $40\bar{4}_1$ in Figs. 6a and 6b). This is characteristic for an "incoherent" two-phase pattern of submicroscopically intergrown phases with different anorthite contents, one phase with intermediate structure having, besides (a) reflections, typical (e) and (f) reflections (satellites), and the other phase with the body-centered anorthite structure deduced from (a) and (b) reflections (Fig. 7). On the other hand, the lunar specimen 12038,79, No. 2 is exsolved into two phases, one with intermediate structure and the other with anorthite-like structure having primitive lattice ((c) reflection). The a and c axes of both phases are parallel to each other and the boundary between two phases is most probably (010). Whether the formation of twins and the exsolution are correlated or not is not yet known.

As can be seen in Fig. 6, $0kl$ and $4kl$ photographs of both specimens 103, No. 3 (terrestrial) and 14310,106, No. 5 (lunar) are identical with respect to reflection

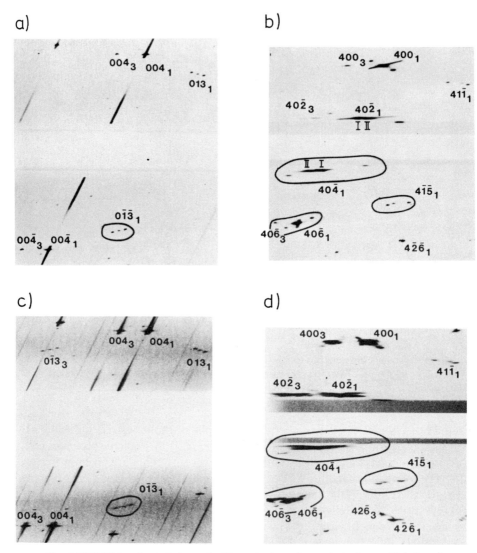

Fig. 6. [100] Weissenberg photograph of terrestrial and lunar plagioclases with bytownite composition (parts enlarged). CuK$_\alpha$ radiation. (a) 0-layer (100^h for 70°) and (b) fourth-layer photograph (150^h for 70°) of terrestrial specimen 103, No. 3 An$_{69.7}$ bulk composition), Sognefjord, Norway. (c) 0-layer (240^h for 60°) and (d) fourth-layer photograph (150^h for 60°) of lunar specimen 14310,106, No. 5. hkl_1 and hkl_3: reflections of twin individuals 1 and 3, respectively. I, II: reflections of phase I (An$_{65\pm3}$) and of phase II (An$_{80\pm3}$), respectively.

positions, but the $4kl$ reflections in particular of the lunar specimen are much more diffuse, indicating a higher degree of disorder of the lunar specimen because of its different prehistory. Thus, the splitting of (a) reflections is indicated as elongated reflections (see reflection group $40\bar{2}_1$ and $40\bar{4}_1$ in Fig. 6d). The curved streaks at the reflection groups $40\bar{6}$ and $42\bar{6}$ of the lunar plagioclases (Fig. 6d) are characteristic

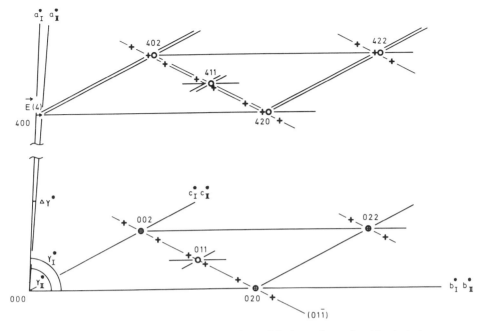

Fig. 7. Schematic representation of the reciprocal lattices of two An-rich plagioclases.
$+$: phase I (An$_{65\pm3}$) with (a), (e), and (f) reflections; $\bigcirc$: phase II (An$_{80\pm3}$) with (a) and (b)
reflections. E(4) $\approx 4a^*\tan\Delta\gamma^*$.

for an incongruent plane of intergrowth of lamellae, which is demonstrated qualitatively in Fig. 3b. The corresponding streaks are missing on the zero-layer photograph in spite of the extremely long exposure time (Fig. 6c). This behavior may be
interpreted with the aid of the discussion in (1) of the section entitled "Incongruent
plane of intergrowth," where all products $R_iR_k^*$ in equation (6) are assumed to be
different from zero. The chemistry of the two plagioclases differs only slightly (An$_{65}$
and An$_{80}$). For small values of h the F_i and R_i are very similar because of the small
differences in lattice constants of the two An-rich plagioclases. Therefore the diffraction theory simplifies to that of a congruent plane of intergrowth. Consequently the
x-ray photographs are essentially described by two eigenvalues λ_1 and λ_2, the amounts
of which are near unity. Therefore, diffuse streaks may be visible only in the neighborhood of Bragg reflections of the two phases. This discussion is qualitative insofar as
four types of lamellae (1$_I$, 1$_{II}$, 3$_I$, and 3$_{II}$) exist in our particular case. Consequently
an extended theory has to be applied which leads to four different eigenvalues in the
case of a common congruent plane of intergrowth of all four kinds of lamellae. For
the zero-layer line pair of 2 reflections of the exsolution lamellae coincide.

Single-crystal photographs of high resolution power of specimen No 6 of the
sample 14310,106 were taken in a focussing camera. The (a) reflections observed
were more diffuse than the corresponding reflections of terrestrial specimens. No
supersatellites were observed.

Pyroxenes

Clinopyroxenes

The pigeonites of the lunar sample 14310,106 were partly microscopically well-defined single crystals without any indication of unmixing. Consequently, the Bragg reflections were fairly sharp. The lattice constants, determined by oscillation and Weissenberg photographs, indicate rather high Fe and low Ca contents (see LSPET, 1971). Some single crystals showed diffuse streaks originating from reflections (*hkl*) with $h + k$ even, without any indication of reflections of a second phase; a pair of reflections was observed in some specimens for specific reflections, but it was impossible to find all corresponding pairs necessary to describe a diffraction pattern of a second phase. Streaks connected with the same extinction rule have been reported by Morimoto and Tokonami (1969), Ross *et al.* (1970a, b), Takeda (1972), and Ghose (1972), but contrary to our results, in all cases the reflections of augite were already visible. Thus, we arrived at the conclusion that the streaks observed in our specimens should be correlated with a beginning exsolution of augite.

Both planes of intergrowth (100) and (001) could be observed, but their x-ray diffraction patterns were completely different: Although the (100) exsolution lamellae clearly belong to the case of a congruent plane of intergrowth, the same is not true for the (001) exsolution lamellae.

As the usual (*h*01) precession photographs are not adequate to demonstrate this effect distinctly, we show some magnified parts of Weissenberg photographs of the (*hk*0) and (*hk*2) plane of the reciprocal lattice. Fig. 8a shows the area (*hh*0) of the (*hk*0) plane, indicating a more or less clear separation of reflection of the two kinds of lamellae. The behavior of (*hk*2) reflections is completely different. A diffuse streak (Fig. 8b) accompanies the reflections without any indication of a maximum, although the extension of the diffuse streak is even larger than the separation of the pair of

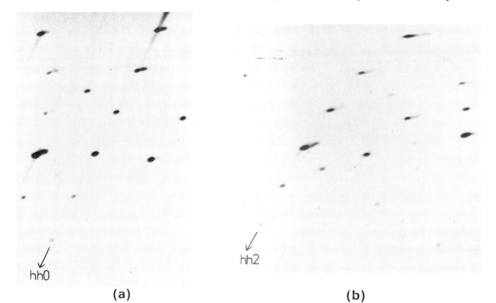

hh0

hh2

(a)   (b)

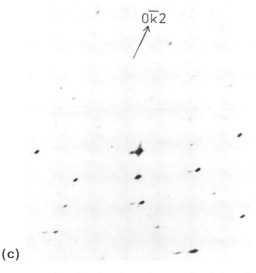

(c)

Fig. 8. Magnified areas of [001] Weissenberg photographs of pigeonite from lunar sample 14310,106, crystal No. 50104. CuK$_\alpha$ radiation. (a) Zero layer, area ($hh0$); (b) second-layer, area ($hh2$); (c) second-layer, area ($0k2$).

reflections in the ($hk0$) plane. The intensity distribution of the ($0k2$) area of the ($hk2$) plane is again different, as shown on Fig. 8c. A clear separation of reflections is visible, but the diffuse streaks accompanying them are asymmetric. All streaks have the (100) direction in reciprocal space, which corresponds to the normal of the plane of intergrowth (100). This behavior of diffuse intensities is typical for equation (4). The line profiles for diffuse reflections ($|\lambda_v| < 1$) are given on Fig. 9 for the three typical cases $B < 0$, $B = 0$, $B > 0$. Apparently the reflections ($hh0$) are very near to the case $B = 0$, while the reflections ($hh2$) belong to $B > 0$ and $0\bar{h}2$) to $B < 0$. The form of the line profile may be used to determine the ratio $B_v : A_v$ (compare equation (4)). On the other hand, the diffuseness and the position may be used to measure $|\lambda_v|$ and ϕ_v (equation (3)) experimentally.

A typical diffraction picture of exsolution lamellae with the (001) plane as plane

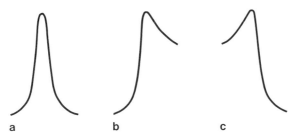

a   b   c

Fig. 9. Intensity distribution of diffuse reflections according to equation (6). (a) B = 0; (b) B > 0 (c) B < 0.

of intergrowth is shown in Fig. 10, which shows magnified parts of a precession photograph of the ($h0l$) plane. It should be pointed out that these pictures are strongly overexposed in order to show the diffuse effect more distinctly. Thus, the (004) reflection especially is nothing else but the reflection and a diffuse halo pointing into the c^* direction. Because the precession photograph was taken with the usual filter-monochromatized radiation, a streak of white radiation accompanies the reflections, thus indicating the reciprocal directions a^* and c^*, respectively. The diffraction picture of the (004) reflection shows a clear asymmetry into the a^* direction, which can be due to a curvature of the streak that is typical for and incongruent plane of intergrowth. There is now doubt that these diffuse effects are completely different from the effects described above for the (100) exsolution lamellae, because all diffuse streaks pointed from the host reflections into the (100) direction. It would be desirable to study the kinetics of these two exsolution processes. Whether the precise measurement of lattice constants leads to different unit cells has not yet been checked.

Orthopyroxenes

Single crystals of orthopyroxene are rare in sample 14310,106; the same observation has been reported by Takeda (1972). Thus, only two single crystals of orthopyroxene could be studied by x-ray diffraction methods to date. Oscillation photographs taken in a high-resolution camera with crystal-monochromatized radiation showed that the reflections have a typical "structure" at large diffraction angles which may be due to a zonal structure. Some additional reflections were observed in both crystals, but it is improbable that they may be attributed to pyroxenes. Although the reflections are considerably better defined than the corresponding reflections of clinopyroxenes, a clear anisotropic behavior of the reflection profiles could be observed. The reflections were more diffuse in the reciprocal (010) directions than in the other two, which may be due to the zonal structure of the crystals.

Another curious effect is shown in Fig. 11, which gives a magnified reflection in the third-layer line of an oscillation photograph. The reflection splits into two well-

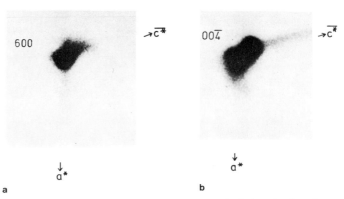

Fig. 10. Magnified areas of a precession photograph ($h01$) of pigeonite from lunar sample 14310,106, crystal No. 13. CuK$_\alpha$ radiation. (a) Reflection (600); (b) reflection (004).

Fig. 11. Oscillation photograph of orthopyroxene in a focussing camera, CuK$_{\alpha 1}$ radiation.

defined lines, connected by diffuse bands. It should be stressed that no effect of this kind may be seen on the second-layer line at the bottom of the figure. There are three possible explanations of this effect, which will be studied in more detail in a subsequent paper: (1) anisotropical polygonization of very small crystals, (2) different lattice constants of various zones, or (3) bent areas with different radii of curvature. Which of these possible explanations is correct will be shown after a second series of measurements.

Acknowledgments—We thank Profs. Laves and Ribbe, Drs. Nissen and Ross for stimulating discussions of the diffraction problem, Dr. Stewart for providing the lunar specimens 10050,45 and 12038,79. The Deutsche Forschungsgemeinschaft has assisted our investigations by lending the x-ray equipment for diffraction photographs. We also thank the Bundesministerium für Bildung und Wissenschaft for meeting the travel expenses for our participation in the Third Lunar Science Conference.

References

Bambauer H. U. Eberhard E. and Viswanathan K. (1967) The lattice constants and related parameters of "plagioclase (low)." *Schweiz. Mineral. Petrogr. Mitt.* **47**, 351–364.

Czank M. Girgis K. Harnik A. B. Laves F. Schmid R. Schulz H. and Weber L. (1972) Crystallographic studies of some Apollo 14 plagioclases (abstract). In *Lunar Science—III* (editor C. Watkins), p. 171, Lunar Science Institute Contr. No. 88.

Ghose S. (1972) Clinopyroxenes from Apollo 12 and 14: Exsolution, cation distribution and domain structure (abstract). In *Lunar Science—III* (editor C. Watkins), p. 738, Lunar Science Institute Contr. No. 88.

Jagodzinski H. (1949) Eindimensionale Fehlordnung und ihr Einfluss auf die Röntgeninterferenzen I, II und III. *Acta Cryst.* **2**, 201–207, 208–214, 298–304.

Jagodzinski H. (1954) Der Synnetrieeinfluss auf den allgemeinen Lösungsansatz eindimensionaler Fehlord-
    nungsprobleme. *Acta Cryst.* **6**, 17–25.
Jagodzonski H. and Korekawa M. (1965) Supersatelliten im Beugungsbild des Labraderits. *Naturwiss.*
    52, 640.
Kakinoki J. and Komura Y. (1952) Intensity of x-ray diffraction by a one-dimensionally disordered
    crystal I. *J. Phys. Soc. Japan* **7**, 30–35.
Kakinoki J. and Komura Y. (1954) Intensity of x-ray diffraction by a one-dimensionally disordered
    crystal II. *J. Phys. Soc. Japan* **9**, 169–176.
Korekawa M. and Jagodzinski J. (1967) Die Satellitenreflexe des Labraderits. *Schweiz. Mineral. Petrogr.
    Mitt.* **47**, 269–278.
LSPET (19   ) (Lunar Sample Preliminary Examination Team) Preliminary examination of lunar samples
    from Apollo 14. *Science* **173**, 681–693.
McConnell S. D. C. and Fleet S. G. (1963) Direct electron optical resolution of anti-phase domains in a
    silicate. *Nature* **109**, 586.
Morimoto N. and Tokonami M. (1969) Oriented exsolution of augite in pigeonite. *Amer. Mineral.* **54**,
    1101–1117.
Ross M. Bence A. E. Dwornik E. J. Clark J. R. and Papike S. J. (1970a) Lunar clinopyroxenes: Chemical
    composition, structured state and texture. *Science* **167**, 628–630.
Ross M. Bence A. E. Dwornik E. J. Clark J. R. and Papike S. J. (1970b) Mineralogy of the lunar clino-
    pyroxenes, augite and pigeonite. *Proc. Apollo 11 Lunar Sci. Conf., Geochim. Cosmochim. Acta* Suppl. 1,
    Vol. 1, pp. 839–848. Pergamon.
Takeda H. and Ridley W. I. (1972) Crystallography and mineralogy of pyroxenes from Fra Mauro soil
    and rock 14310 (abstract). In *Lunar Science—III* (editor C. Watkins), p. 738, Lunar Science Institute
    Contr. No. 88.

Proceedings of the Third Lunar Science Conference
(Supplement 3, *Geochimica et Cosmochimica Acta*)
Vol. 1, pp. 569–579
The M.I.T. Press, 1972

Lunar plagioclase: A mineralogical study

H.-R. Wenk and M. Ulbrich

Department of Geology and Geophysics
and Space Sciences Laboratory,
University of California at Berkeley

and

W. F. Müller

Department of Materials Science and Engineering,
University of California at Berkeley

Abstract—Mineralogical properties of calcic plagioclase have been analyzed using U-stage, microprobe, x-ray precession cameras, and a 650 kV electron microscope. The orientation of the optical indicatrix in lunar and eucrite anorthites is described with Euler angles. All crystals, except one, show strong b- and diffuse c-reflections in precession photographs. In 10017, b-split-reflections have been found. Dark-field electron micrographs of 14310 anorthite show both large and small b-antiphase domains, and an exsolution structure in crystals that display b-split reflections in the diffractogram. Diffuseness of c-reflections in x-ray photographs and the inability to resolve c-domains in electronmicrographs in An 94 anorthite of 14310 indicate relatively rapid cooling of this rock compared to plutonic rocks.

Introduction

It was the purpose of this study to describe the properties of lunar plagioclase and to investigate, by comparison with appropriate terrestrial feldspars, whether the optics and structure of anorthite are indicative of the thermal history of the crystals. Documentation on anorthites is sparse; therefore, lunar rocks provide excellent material to study calcic plagioclase and these new data contribute to a better understanding of this important mineral series.

Techniques used include determination of optical properties on the universal stage, single crystal x-ray photography by the precession method, chemical microprobe analyses, and 650–kV transmission electron microscopy. Procedures are similar to those described by Wenk and Nord (1971). First, a polished thin section was prepared and examined on the petrographic microscope. U-stage measurements were made on several multiply twinned plagioclase crystals (preferentially twinned after albite and albite-Carlsbad laws). Then the chemical composition (Ca, Na, K) was determined on the same spots by microprobe. One crystal was picked either from the thin section or from the rock for x-ray study and later analyzed chemically. Electron microscope samples were prepared from thin sections using the ion thinning technique.

Most of our data were obtained on the only two available thin sections, 14310 (basalt) and 14319 (breccia). Fines were usually not satisfactory for this analysis. Some measurements have been done on small fragments of 10017, 12021, and 14162. The results reported apply only to these few specimens and not to lunar rocks in general.

Petrography

Basalt 14310 (Fig. 1a) consists mainly of clear, inclusion-free crystals of calcic plagioclase (An 85–95), clinopyroxene ($2V_\gamma = 12$–$20°$), orthopyroxene ($2V_\alpha = 72°$),

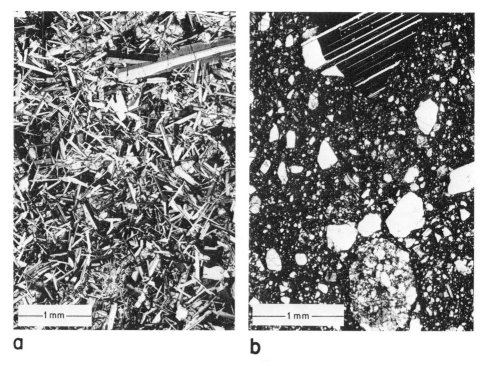

Fig. 1. (a) Photomicrograph of subophitic basalt 14310. Crossed nicols. Notice the range in grain size of plagioclase, and the areas with extremely fine-grained anorthite needles. (b) Photomicrograph of breccia 14319. Crossed nicols. Notice a large anorthite crystal and a breccia fragment.

and opaque minerals. The texture is subophitic. Plagioclase ranges greatly in size. The largest crystals in the thin section are 2 mm long, but most of them range between 0.3 and 0.5 mm. Chemical analysis indicates that there are two groups of plagioclase present; these, however, are not apparent by grain size, shape, or texture. An average of microprobe analyses on many crystals gives a composition An 86.9, Or 1.44 for one group and An 94.0, Or 0.42 for the other. Of special interest is the large difference in potassium. These two groups are quite distinct, without intermediate compositions. No zoning has been found except for one very large crystal, which shows a thin rim of the potassium-rich phase; therefore, this An-poor and Or-rich plagioclase may be younger than the other crystals. There are small areas of very fine-grained plagioclase needles chemically indistinguishable from the Or-poor, large ones that show polysynthetic twinning. These "clots" have almost no interstitial pyroxene. Twin laws identified on the U-stage are albite, Carlsbad, albite-Carlsbad (common), pericline (less common), Baveno-r (one crystal only). Cruciform intergrowth is quite common. Attention has been given to the commonly occuring pseudotwins and peculiar intergrowths.

   In one group of crystals, the plane between two intergrown crystals looks like a "composition plane" of a twin (Fig. 2a). The two indicatrices are related by a single rotation as in a regular twin, but neither axis nor composition plane is a rational direction. The composition plane is in the vicinity but distinctly different from $(0\bar{2}1)$ (Fig. 2b). These "irrational" intergrowths are not uncommon in lunar and meteoritic plagioclase (Ulbrich, 1971; Wenk and Nord, 1971).

In another case (Fig. 2c), three crystals are intergrown. Crystal 1–2 is a regular albite-Carlsbad twin with (010) as composition plane; 2–3 is a pericline twin, whose composition plane should be the rhombic section which is close to (001) for anorthite. The actual plane of intergrowth in these crystals is a rough surface close to (010) (Fig. 2d).

Pyroxenes have not been studied in detail. Routine checks showed that clino-pyroxenes with small $2V_\gamma$ (12–22°) twinned on (100) are common. No exsolution lamellae were seen in the petrographic microscope. Orthopyroxene (hypersthene) is less common and frequently mantled by pigeonite. A similar type of intergrowth as described for plagioclase (Figs. 2c, d) has been found in pyroxene. Fig. 2e is an example of polysynthetically twinned pigeonite (1–2) ([010] common axis = Y), yet the composition plane 1–2 is not (100) but an irrational and slightly curved surface (C.P. 1–2, Fig. 2f). This twinned crystal is partly rimmed by another clino-pyroxene (3), which appears to be more iron-rich, because it has the same optical orientation as 2 ($X_2 = X_3$, $Y_2 = Y_3$, and $Z_2 = Z_3$), but has a higher birefringence and higher $2V_\gamma$.

A fragment from 14162 coarse fines has properties very similar to that of 14310 basalt. Textures are identical and two groups of plagioclase are present. We assume that it is a fragment of the same rock.

A large number of plagioclase fragments appear in 14319 breccia (Fig. 1b). Plagioclase is heterogeneous: many crystals are twinned, some are undeformed, some are fractured, some have patchy extinction and bent lamellae. One crystal of plagioclase shows very thin platelets of an opaque mineral on (010). Other com-ponents of the breccia are ortho- and clinopyroxene with exsolution and twin lamellae, perovskite, and an unidentified small fragment of a yellow biaxial positive crystal. Apart from crystal fragments, the breccia contains many lithic fragments. Euler angles of plagioclase have been determined in anorthositic and gabbroic fragments. A basalt fragment appears to be closely related to basalt 14310. In the breccia there also are fragments of an older breccia. Brown glass inclusions show beginning crystallization. Noteworthy are spherical aggregates of pyroxene with radial crystallites, resembling meteoritic chondrules. Many lithic and crystallite fragments in the breccia are rounded; others, however, have sharp corners.

Sample 12021 is a basalt of ophitic texture. Some of the large plagioclase crystals are skeletal with pyroxene and opaque inclusions.

Calcic plagioclase from terrestrial rocks and from eucrite meteorites with similar mineralogical composition has been analyzed and compared with the lunar crystals. Some results on meteoritic plagioclase are included here, because, to our knowledge, this paper gives the first description of optical properties of plagioclase in such meteorites. The eucrites are composed of calcic plagioclase, and pigeonite with exsolution lamellae of subcalcic augite to augite. Cachari (Argentina) is brecciated, Serra de Magè (Brazil) shows a beautiful equigranular texture, Ibitira (Brazil) has

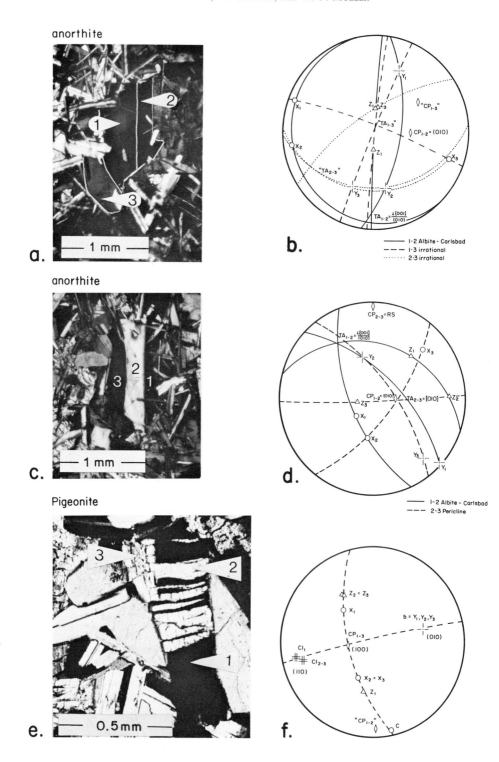

few large crystals and a ground mass of fine annealed grains with polygonal outlines. Plagioclase in all of them is twinned after the albite, albite-Carlsbad, Carlsbad, and pericline laws (Ulbrich, 1971).

U-Stage Analysis of Plagioclase

Euler angles are used to describe the orientation of the optical indicatrix in the crystal. Euler I angles (Θ, Ψ, and Φ) have been derived from measurements of the albite composition plane (010) and the albite-Carlsbad twin axis $\perp[001]$ in (010), and in some cases, from the cleavage (001). The results are listed in Table 1. In order to evaluate the use of the orientation of the indicatrix as an indicator for the thermal history, we plot the angles Φ and Ψ, which show the largest variation in calcic plagioclase, as a function of the anorthite content (Fig. 3). For reference we also plot all data on calcic plagioclase found in the literature and add new measurements. This gives a better measure of the significance of an interpretation than if we compare the new data points with averaged determinative curves (Burri et al., 1967). Above An 75 there are no longer two distinct curves for plutonic and volcanic plagioclase; there is, instead, a diffuse band of scattering points. To make any statement about the thermal history of the rock, a statistical number of high-precision measurements is necessary. It is difficult also to predict the chemical composition in this range of the plagioclase series with accuracy greater than ± 5 to 10% An. The scatter in the data is larger than the accuracy of the measurements and we expect that in addition to the chemical composition and the thermal history, submicroscopic features such as twins and domains may account for it. Looking at all anorthites, Φ and Ψ appear to be slightly larger for plutonic than for volcanic feldspars, and the rather large angles in lunar and meteoritic plagioclase agree with plutonic optics (Wenk and Nord, 1971; Ulbrich, 1971; E. Wenk et al., 1972). As has been shown by E. Wenk et al. (1972), in $\Phi\Psi$ plots our data scatter in the same field as theirs. The $\Phi\Psi$ plots also indicate that lunar plagioclase may have slightly different optical properties than terrestrial ones. To prove this, very pure anorthites have to be studied.

Fig. 2. Unusual twins in lunar basalt 14310. Photomicrographs and stereographic projection in upper hemisphere are in the same orientation. (a), (b): anorthite showing irrational intergrowth between crystals 1 and 3. The pseudocomposition plane, CP 1–3, is close to but different from (0$\bar{2}$1). The "twin axis" 1–3 is also irrational. Crystals 1 and 2 are in albite-Carlsbad relation. (c), (d): anorthite with albite-Carlsbad (1–2) and pericline twin (2–3). The 2–3 composition plane (rhombic section) should be close to (001). The actual plane of intergrowth is close to (010). (e), (f): Polysynthetically twinned pigeonite (1–2). The common axis between 1 and 2 is $[010] = Y$. The plane of intergrowth (C.P. 1–2) is not the theoretical composition plane (100), but an irrational surface. The twinned crystal is rimmed by another clinopyroxene, which is in the same optical orientation as 2 but has higher $2V_\gamma$ (3).

Table 1. Euler I angles and chemical composition of plagioclase from lunar rocks and eucrite meteorites. The accuracy of Euler angles is $\pm 1/2°$.

Lunar Rocks	Twin laws	$2V_\alpha$	Φ	Ψ	Θ	An	Or	Ab
12021, 118C basalt fragment	Ab, Ca Ab-Ca; Pe	77°	$25\frac{1}{4}$	-6	36	92.9	0.34	6.7_4
14310, 23 & 95 plagioclase-pyroxene basalt	Ab, Ca Ab-Ca	—	$25\frac{1}{2}$	-2	38	85.3_0	1.72	12.9_8
	Pe (rare)	—	25	-4	38	87.6_9	1.22	11.0_9
	Baveno-r (only one)	81°	$23\frac{3}{4}$	$-5\frac{1}{2}$	38	87.6_0	1.38	11.0_2
		77°	22	$-5\frac{1}{2}$	37	93.0_9	0.48	6.4_3
		78°	24	-6	34	94.3_3	0.39	5.2_8
		—	$23\frac{1}{2}$	-6	38	93.4_1	0.45	6.1_4
14319, 6 Breccia	Ab, Ca							
(A) plag.-pyr. aggr. (anor.)	Ab-Ca Pe	79° 78	$22\frac{3}{4}$ $21\frac{1}{2}$	-4 -7	38 37	93.8_9 94.6_1	0.37 0.54	5.7_4 4.8_5
(B) plag.-pyr. aggr. (gab.)	Ab, Ca Ab-Ca, Pe	—	20	-6	37	92.6_9	0.63	6.6_8
(C) plag. aggregate	Ab, Ca Ab-Ca	—	26	-1	34	84.2_3	1.56	14.2_1
(D) plag. frag.	Ab	—	23	-8	41	96.7_0	0.39	2.9_1
14162, 13 coarse fines plag.-pyrox. in the fines	Ab, Ca Ab-Ca Pe (rare)	— —	23 17	$-6\frac{1}{2}$ -10	38 $37\frac{1}{2}$	93.5_6 93.7_2	0.44 0.31	6.0_0 5.9_6
Eucrite Meteorites								
Serra de Magè	Ab, Pe	—	20	$-8\frac{1}{4}$	$37\frac{1}{2}$	94.8_3	0.09	5.0_8
	Ca, Ab-Ca	77°	23	-6	38	95.3_7	0.04	4.5_9
Ibitira	Ab-Ca,	81°	$22\frac{3}{4}$	-6	37	95.3_0	0.21	4.4_9
Cachari	Ab, Ca, Pe	—	28	-6	42	88.2_9	0.45	11.2_5

Ab: Albite; Ca: Carlsbad; Pe: Pericline.

Lattice Constants and Structure

Lattice constants were measured on x-ray precession photographs, which were used to determine the structural state from the presence and diffuseness of indicative b- and c-reflections. The lattice parameters along with other structural information for four crystals are listed in Table 2. The variations are within the standard deviation and no conclusions can be drawn. Except for crystals from 10017, all analyzed crystals showed a transitional anorthite structure with strong b-reflections and diffuse c-reflections streaking parallel to b^* in 0kl sections. As has been pointed out by Gay (1953), Gay and Taylor (1953), and Laves and Goldsmith (1954a, b), the diffuseness of c-reflections is a function of the chemical composition and of the thermal history. Comparing lunar crystals with terrestrial and eucrite anorthites *of*

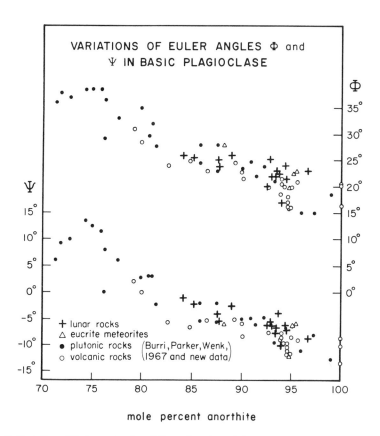

Fig. 3. Variation of Euler angles Φ and Ψ as a function of chemistry in anorthites. Data from
the literature and new measurements.

Table 2. X-ray data of anorthites from precession photographs.

Sample	a	b (Å)[1]	c	α	β (degrees)[2]	γ	Structure	An	Or	Ab
10017 (A)	8.19	12.87	14.18	93.1	116.1	90.9	b weak			
10017 (B)							e (b-split) and b			
12021 (A)	8.18	12.87	14.18	93.3	115.9	90.9	b sharp, strong c diffuse, streaks along b^*	92.0_6	0.42	7.5_2
12021 (B)							b sharp, strong c diffuse, streaks along b^*			
12021 (C)							b sharp, strong c diffuse, streaks along b^*			
14310	8.18	12.87	14.19	93.3	116.2	91.0	b sharp, strong c diffuse, streaks along b^*	93.4_4	0.16	6.4_0
Serra de Magè eucrite	8.19	12.9	14.18	92.9	116.1	91.4	b sharp, weak c sharp, strong	95.5_7	0.05	4.3_9

[1]error ± 0.01 Å   [2]error ± 0.1°

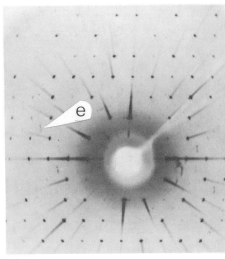

10017

Fig. 4. X-ray single-crystal precession photograph (0kl). (Mo-radiation, Zr filter) of a plagioclase crystal in rock 10017. Notice asymmetric *e*-reflections and sharp *b*-reflections in the same crystal.

the same chemical composition, the structure of lunar plagioclase is similar to that of plagioclase from volcanic rocks. All comparative metamorphic, plutonic and meteoritic feldspars that have been studied by the authors show much stronger and sharper *c*-reflections (Wenk and Nord, 1971; Müller *et al.*, 1972). Of special interest is a crystal from 10017 that shows *e(b*-split)-reflections and sharp *b*-reflections in the same crystal (Fig. 4). Similar patterns have been observed by Jagodzinski and Korekawa (1972) in terrestrial An 70–77 plagioclase and have been interpreted as exsolution of an An 65 and an An 80 end member. We did not find that lunar plagioclase is "intimately twinned" on a submicroscopic scale according to the albite-Carlsbad law (Czank *et al.*, 1972). The x-ray patterns of 14310 plagioclase and other crystals analyzed indicate perfect single crystals.

Electron Microscopy

High voltage (650 kV) transmission electron microscopy of plagioclase was done on specimens from 14310. Isolated submicroscopic twin lamellae were observed (Fig. 5a). They commonly occur singly or as two to three parallel lamellae, approximately 1 micron wide, with a few lamellae as small as 0.2 microns. Twin laws identified were albite and Baveno-r. It is not known what influence these submicroscopic twins have on the U-stage determined orientation of the indicatrix. Possibly they account for the difficulty in obtaining perfect extinction in the optical measurements. Selected-area electron diffraction patterns showed sharp *a*- and *b*-, diffuse *c*-, and very weak, diffuse *d*-reflections. The *c*- and *d*-reflections were streaked in

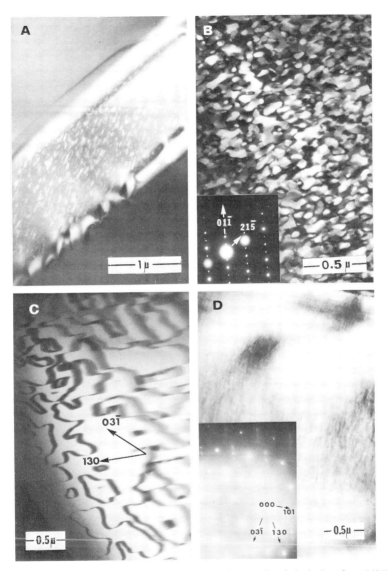

Fig. 5. High voltage (650 kV) transmission electron micrographs of plagioclase from 14310. (a) Baveno-r twin lamellae with b-antiphase-domain boundaries in contrast. Dark-field image with $\tilde{g} = 011$ as the operating beam. (b) Small b-antiphase domains. Dark-field image with $\tilde{g} = 31\bar{5}$. The selected area electron diffraction pattern is inserted. (c) Boundaries of b-antiphase domains. Dark-field image with $\tilde{g} = 03\bar{1}$. Note the larger domain size compared to (b). (d) Cross-hatched structures. Dark-field image with $\tilde{g} = 130$. The corresponding selected area diffraction pattern (insert) displays "b-split" reflections. Different part of the same crystal as in (c). Note radiation tracks.

directions perpendicular to about ($2\bar{3}\bar{1}$) in a diffractogram normal to [211]. This agrees with x-ray studies by Ribbe and Colville (1968) and electron diffraction results by Appleman *et al.* (1971). Dark-field images using *b*-reflections revealed smoothly curved antiphase-domain boundaries (Figs. 5a, b and c). Most of these *b*-antiphase domains were 500–1000 Å wide (Fig. 5b) but in some cases larger *b*-domains have been observed in the same specimen (3000–10,000 Å, Fig. 5c; compare also Christie *et al.*, 1971). No *c*-domains could be resolved in dark-field images exposed for as much as two minutes. Of special interest was a crystal with a range of chemical composition, probably similar to the one described in the section on petrography, which had a composition of An 94 in the core and had a rim of An 87 bytownite. This crystal shows in one part very large *b*-domains (Fig. 5c). No *c*-domains could be imaged. The crystal is bordered by a 2.5 micron broad band of a plagioclase phase that displays strong *b*-split reflections and very small *b*-domains in dark-field pictures. In this crystal a cross-hatched structure resembling peristerite was seen (Fig. 5d). The same structure has been observed by Lally *et al.* (1972) and may represent an exsolution in the bytownite composition range (Nissen, 1968; Jagodzinski and Korekawa, 1972). Radiation tracks can be seen in the same picture (Fig. 5d) with a maximum track density of about 2×10^8 cm^{-2}. The presence of *b*-domains and the absence of *c*-domains has been seen in at least 15 of the crystals that were examined. This makes it very probable that many of these anorthites have the far more common An 94–95 composition. The diffraction patterns agree with the x-ray precession photographs of the crystal that had a chemical composition of An 93.4.

Müller *et al.* (1972) found that anorthites (An 95–97) from a metamorphic calc-silicate rock and a eucrite meteorite contained *c*-antiphase-domain boundaries, the domain walls being as much as a few microns apart; whereas an anorthite (An 95) from a volcanic tuff displayed only small *c*-domains of the order of 70 to 100 Å in diameter. In view of these observations, it appears that the fact that no *c*-domains could be imaged in anorthite from rock 14310, which showed more diffuse *c*-reflections than all comparative terrestrial anorthites of similar chemical composition, can be interpreted as an indication for relatively rapid cooling of 14310. This characterization is not in disagreement with the results of other investigators who describe 14310 as having been more slowly cooled than other lunar basalts (e.g. Finger *et al.*, 1972). The distinction suggested by structural variations of anorthite so far is merely between a volcanic and a plutonic/metamorphic geological history. Other lunar igneous rocks may have had a different history than 14310. Large *c*-antiphase domains were observed in the anorthite of 15415 anorthosite (Lally *et al.*, 1972), which indicates that this rock may have formed under plutonic conditions.

Acknowledgments—Support from NASA grants NGR 05–002–414 and NGR 05–003–410 and from AEC (G. Thomas, electron microscope at Berkeley) is acknowledged. R. Heming and J. Donnelly helped with the microprobe analyses. H.-R.W. thanks the Miller Institute for basic research for a professorship that relieved him of teaching during the year 1971–1972. W.F.M. thanks the Deutsche Forschungsgemeinschaft for support and Drs. W. L. Bell, M. Bouchard, P. Phakey and G. Thomas for discussions.

References

Appleman D. E. Nissen H. -U. Stewart D. B. Clark J. R. Dowty E. and Heubner J. S. (1971) Studies of lunar plagioclases, tridymite, and cristobalite. *Proc. Second Lunar Sci. Conf., Geochim. Cosmochim. Acta* Suppl. 2, Vol. 1, pp. 927–930. MIT Press.

Burri C. Parker R. L. and Wenk E. (1967) *Die optische Orientierung der Plagioklase.* Birkhäuser, Basel.

Christie J. M. Lally J. S. Heuer A. H. Fisher R. M. Griggs D. T. and Radcliffe S. V. (1971) Comparative electron petrography of Apollo 11, Apollo 12 and terrestrial rocks. *Proc. Second Lunar Sci. Conf., Geochim. Cosmochim. Acta* Suppl. 2, Vol. 1, pp. 69–89. MIT Press.

Czank M. Girgis K. Harnik A. B. Laves F. Schmid R. Schulz H. and Weber L. (1972) Crystallographic studies of some Apollo 14 plagioclases (abstract). In *Lunar Science—III* (editor C. Watkins), pp. 171–173, Lunar Science Institute Contr. No. 88.

Finger L. W. Hafner S. S. Schurmann K. Virgo D. and Warburton D. (1972) Distinct cooling histories and reheating of Apollo 14 rocks (abstract). In *Lunar Science—III* (editor C. Watkins), pp. 259–261, Lunar Science Institute Contr. No. 88.

Gay P. (1953) The structures of the plagioclase feldspars: III. An x-ray study of anorthites and bytownites. *Mineral. Mag.* **30**, 169–177.

Gay P. and Taylor W. H. (1953) The structures of the plagioclase feldspars. IV. Variations in the anorthite structure. *Acta Cryst.* **6**, 647–650.

Jagodzinski H. and Korekawa M. (1972) X-ray studies of plagioclases and pyroxenes (abstract). In *Lunar Science—III* (editor C. Watkins), pp. 427–429, Lunar Science Institute Contr. No. 88.

Lally J. S. Fisher R. H. Christie J. M. Griggs D. T. Heuer A. H. Nord G. L. Jr. and Radcliffe S. V. (1972) Electron petrography of Apollo 14 and 15 samples (abstract). In *Lunar Science—III* (editor C. Watkins), pp. 469–471, Lunar Science Institute Contr. No. 88.

Laves F. and Goldsmith J. R. (1954a) Long-range short-range order in calcic plagioclases as a continuous and reversible function of temperature. *Acta Cryst.* **7**, 465–472.

Laves F. and Goldsmith J. R. (1954b) On the use of calcic plagioclases in geologic thermometry. *J. Geol.* **62**, 405–408.

Müller W. F. Wenk H. -R. and Thomas G. (1972) Structural variations in anorthite. *Contr. Mineral. Petrol.* **34**, 304–314.

Nissen H. U. (1968) A study of bytownites in amphibolites of the Ivrea Zone (Italian Alps) and in anorthosites: A new unmixing gap in the low plagioclases. *Schweiz. Mineral. Petrogr. Mitt.* **48**, 53–55.

Ribbe P. H. and Colville A. A. (1968) Orientation of the boundaries of out-of-step domains in anorthite. *Mineral. Mag.* **36**, 814–819.

Ulbrich M. (1971) Systematics of eucrites and howardite meteorites and a petrographic study of representative individual eucrites. M.A. thesis, Univ. of Calif., Berkeley.

Wenk E. Glauser A. Schwander H. and Trommsdorff V. (1972) Optical orientation, composition and twin laws of plagioclase from rocks 12051, 14053 and 14310 (abstract). In *Lunar Science—III* (editor C. Watkins), pp. 794–796, Lunar Science Institute Contr. No. 88.

Wenk H. -R. and Nord G. L. (1971) Lunar bytownite from sample 12032, 44. *Proc. Second Lunar Sci. Conf., Geochim. Cosmochim. Acta* Suppl. 2, Vol. 1, pp. 135–140. MIT Press.

Proceedings of the Third Lunar Science Conference
(Supplement 3, *Geochimica et Cosmochimica Acta*)
Vol. I, pp. 581–589
The M.I.T. Press, 1972

Twin laws, optic orientation, and composition of plagioclases from rocks 12051, 14053, and 14310

E. Wenk, A. Glauser, H. Schwander, and V. Trommsdorff

Mineralogisch-Petrographisches Institut der Universität Basel,
CH–4000 Basel, Switzerland

Abstract—Universal stage studies show that lunar plagioclases are twinned according to laws found in terrestrial feldspars. Interpenetrant twin groups with symmetrical and pseudosymmetrical mutual orientation are common. The frequency of the different twin laws corresponds to the evidence known from volcanic rocks of the earth. For lunar bytownites and anorthites sixty new sets of Euler I angles are presented. These Euler I angles relate the mutual positions of the three chief vibration directions $[n\alpha]$, $[n\beta]$, and $[n\gamma]$ of the indicatrix to the rectangular cartesian system $X = \perp[001]$ in (010), $Y = (010)$ and $Z = [001]$. Optical data for lunar bytownites are consistent with those of terrestrial bytownites. Lunar anorthites, however, show distinctly higher negative ψ and higher positive ϕ Euler angles than anorthites from young volcanic rocks of the earth. Optic measurements by different operators on one and the same twin group agree within 0.5° to 1°. Chemical inhomogeneity of the plagioclases to which our optical data refer is proven by microprobe analyses.

Plagioclase Twinning and Crystal Optics

From thin sections 12051,57, 14053,11, 14053,19 14053,61, and 14310,14, plagioclases were selected for this study.

Nearly all feldspars of these rocks are twinned with the great majority having (010) or vicinal faces as composition planes. Simple or polysynthetic albite twins predominate (90% of all) and none of the other laws represents more than about 5% of the total twins. The general sequence in frequency appears to be: albite > Roc Tourné, Carlsbad, pericline-acline, Baveno > Ala, albite-Ala > Manebach, which is very rare. But there is some variation from rock to rock; in sample 14310 simple, never polysynthetic, pericline twins form conspicuous squares and are second in importance after albite, third being Baveno. In samples 12051 and 14053, the laws Ala B and albite Ala, which are easily overlooked unless a full U-stage study is made, are combined with albite and Roc Tourné twins, and pericline and Baveno are the least common. Thus, on the moon a set of twin laws known from igneous rocks of the earth is found. No law other than albite forms polysynthetic lamellae. In composite (010) twin groups, Carlsbad as well as Roc Tourné laws rarely may be repeated, but never in adjacent lamellae.

For a composition of An 80 high or of An 82 low, the directions $[n\beta]$ of Carlsbad twins coincide and occupy the position of pole (1̄10). This plane contains the optic axes and becomes an additional plane of symmetry of the twin (Fig. 1), as does very nearly also (130), which is virtually perpendicular to (1̄10). Therefore, at this very special composition, a Carlsbad twin can be interpreted also as a twin according to the prism law (130) as far as the symmetry of the indicatrices is concerned. However, in all cases examined, (010) was determined as composition plane and also the

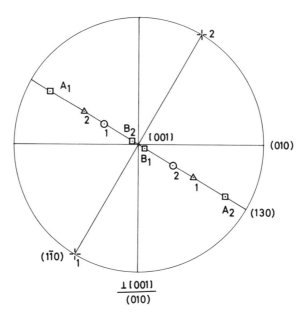

Fig. 1. Stereographic projection perpendicular to [001] showing the position of the chief optic directions for a Carlsbad twin whose optic vibration directions β_1 and β_2 coincide and are perpendicular to [001]. Composition near An 80. Open circles vibration direction α, crosses β, triangles γ, squares optic axes A and B for the two twinned individuals.

combination with albite and Roc Tourné proves that the interpretation as Carlsbad twin is correct. For lunar feldspars studied to date, (110) and (130) were ascertained only as faces of interpenetrant crystals, not as twin-planes.

Interpenetrant twin groups show regular and irregular mutual orientation. Apart from pseudosymmetry, which appears to be quite common, some examples of interpenetrant twins were found in which individuals of the two intergrown groups show the symmetry of the Baveno law.

The optic orientation was derived from 61 complex (010) twin groups by constructing the twin axes and symmetry planes between the indicatrices. As with feldspars of terrestrial igneous rocks, the composition planes of twins are highly irregular. Therefore, we did not rely upon morphological data. Cleavage is poor, though the crystals are sometimes bent or broken. Imperfect optic extinction, caused by chemical inhomogeneity of the crystals, as proven by microprobe analysis, is common.

In triplets there is little uncertainty in locating the three mutually perpendicular twin axes of the Carlsbad (Z-axis), albite (Y-axis) and Roc Tourné (X-axis) laws within one degree, and reading the Eulerian angles from the stereogram (diameter 40 cm) (Fig. 2). In this way two sets of optic data were obtained, one referring to individual crystals, the other giving the average for each thin section. Subsequently, the crystals selected for the optic study were analyzed by electron microprobe.

The optic analysis was done with special care because of the restricted migration of the indicatrix of basic plagioclases in response to change in composition, thermal

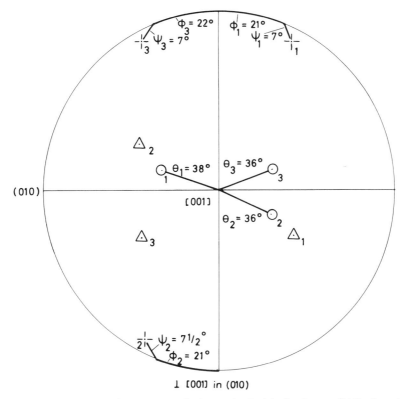

Fig. 2. Stereographic projection perpendicular to the Carlsbad twin axes [001] of a twin group with 3 individuals. Open circles vibration direction α, crosses β, triangles γ. The indicatrices of twinned individuals show the symmetry of the albite law (1/3), Carlsbad law (1/2) and Roc Tourné law (2/3). Euler angles ϕ, Θ, ψ for the three individuals are read as shown in the figure. Mean Euler I angles for this twin group: ϕ 21.3°, Θ 36.7°, $\psi - 7.2°$.

history, and ordering. In several instances the same crystal was examined by three operators with different microscopes and U-stages. The angles measured by various observers agreed very well (within $\pm$ 0.5°) in homogeneous parts. Only when influenced by zonal structure do the measured angles differ more than 1°. The determination of $2V$ by measurement of both optic axes requires high tilting angles, at which birefringence of the cement used for the moon slides interferes. Accordingly, $2V$ data show more scatter.

We are concerned about a strange experience and feel obliged to report it, even though we cannot offer a satisfactory explanation.

In all instances the universal-stage determinations were made in advance of the microprobe analysis. But, for various reasons, the optic orientation of some feldspar crystals in PTS 14053,11 was checked again after repeated microprobe analysis. The slide was then thinner (ca. 20 μ), but the grains measured could be recognized immediately on the previous photographs. No optic anomalies were noted in anorthite crystals. However, the universal-stage work on two, potassium-rich acid bytownites, adjacent to silica minerals, now became tantalizing. Extinctions were vague and inhomogeneities in the crystal more pronounced. Accordingly, the construction of the twin axes became inaccurate and the optic data scattered widely (see Fig. 3). The postmicroprobe data show a definite trend and indicate more calcic plagioclase and/or disordering. Possible effects of zonal structure within the reduced thickness of the PTS, of domain structures becoming more conspicuous in the thinner section and thus disturbing extinctions, of grain deformation by repeated polishing, or of damage caused by electron beam, are difficult to assess.

Microprobe Analyses

Feldspar reference samples, used for Ca, Na, and K calibration are those described by Schwander (1967). Anorthite contents are calculated irrespective of Si and Al.

For the microprobe analyses done in Basel, the following technique was applied: The PTS was polished, coated with carbon (a few Å) and subsequently with gold (~ 200 Å). Excitation conditions for plagioclase analysis by Jeol JXA-3A: Ca, Na 15 kV, 90 nA; K 20 kV, analyzing crystal quartz, KAP; measuring time (point analysis) 30 sec, 4 to 6 points per crystal; beam diameter 20 μ.

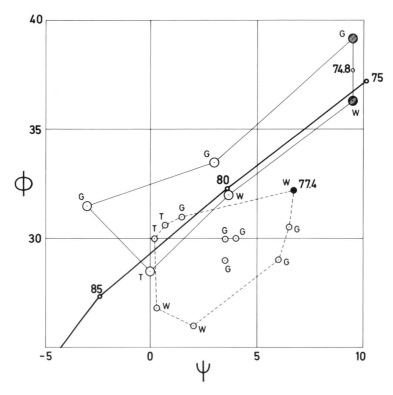

Fig. 3. $\phi \, \psi$-diagram. Data of the two acid bytownites An 74.8 and An 77.4, measured on the universal stage prior (dots) and after (open circles) the microprobe analyses.

The slide 14053,11 was cleaned mechanically after the first measurements, repolished, recoated again in the same way, and reexamined by Jeol JXA-5 in Munich with higher sensitivity and higher take off angle using the same set of standards as before. Excitation conditions: Ca, Na: 15 kV, 15 nA; Si, Al: 10 kV, 15 nA; analyzing crystal PE, KAP, measuring time: 40 sec, 3 points per grain, beam diameter 10 μ. This second series of determinations resulted in slightly higher Ca- and lower Na-contents. The irregular optical properties of the bytownites of this slide are mentioned in the preceding section.

The RMS data of the analyzed plagioclases are listed in Table 1. Point analyses (up to 27) along several traverses across one and the same crystal show that the broad centers of the plagioclases, to which our optical data refer, are not homogeneous. The wide cores have a compositional range An 86 to 94; in the narrow marginal zones An content can be as little as 70%.

In slide 14053,11 bytownites An 78–81, with the exceptional Or content of 3%, are surrounded by silica minerals and can hardly be distinguished morphologically from the predominant anorthites. The nuclei of these bytownites have compositions similar to the marginal zones of the anorthites.

Table 1. Euler angles and microprobe analyses of lunar plagioclases.

Sample	Twin laws	ϕ	Θ	ψ	$2V\gamma$	Opt. analyst	An, calculated from Ca determination
12051,57	A + C + RT	22.0	38.8	−9.8	105	G	
		23.5	36.0	−6.3	101.7	W	88, 90, 90
		24.3	37.3	−7.3	99.7	W	
		23.2	37.4	−7.8	102.1	x̄	89.3
	A + RT	27.5	37.2	−0.7	96	G	87, 90, 93
		26.8	35.5	−3.0	98	W	
		27.2	36.4	−1.9	97	x̄	90
	A + RT + Ala	24.4	37.6	−3	101	W	89.7, 88, 89, 92
	A + RT	26.5	36.0	+0.3	—	W	82.3, 80, 83, 84
	A + RT	23.5	35.7	−2.3	99.3	W	
	A + RT	25.2	37.3	−6.0	103	G	
	A + C + RT	28.0	35.0	−1.5	91.5	G	
	A + RT	26.0	37.0	−2.5	—	G	
	A + RT	24.0	37.5	−5.0	103	G	
	A + RT	26.0	35.0	−3.0	—	G	
	A + RT	25.0	38.0	−5.0	101	G	
		25.2	36.5	−4.0	99.5	x̄	88.2 (80–93) overall mean
14053,11	A + C + RT	26.0	39.0	−2.0	100	G	
		25.8	38.7	−2.3	99	W	
		25.5	37.8	−3.3	—	T	90.8, 89.7, 92.3, 84.6, 82.5, 86.6
		25.8	38.5	−2.5	99.5	x̄	87.7 (82–92)
	A + C + RT	39.0	36.0	+9.5	—	G	
		36.3	33.5	+9.5	—	W	81.5, 72.8, 73.3, 71.2
		37.7	34.8	+9.5	—	x̄	74.8 (Ab. 21.3, Or 3.0)
	A + C + RT	21.2	36.8	−6.8	—	G	
		23.2	37.3	−4.8	—	G	
		24.0	35.7	−5.5	—	T	
		20.3	37.0	−5.7	—	W	89.7, 90.8, 96.0, 92.3, 86.1
		22.2	36.7	−5.7	—	x̄	90.8 (86–96)
	A + C + RT	32.2	35.2	+6.7	—	W	72.2, 78.4, 78.9, 78.9
						x̄	77.4 (72–79)
	Ala + M − Ala	27.0	36.5	+3.2	101	W	81.5, 83.5, 82.5, 85.1, 79.9, 81.5
						x̄	82.5 (80–85)
	A + RT	21.0	38.2	−10.5	—	W	
	A + RT	31.0	38.0	+2.0	—	G	
	A + RT	27.0	37.0	+1.0	86	G	
	A + C + RT	23.5	38.0	−8.5	—	G	
	A + C + RT	22.5	36.5	−6.0	108	G	
	A + C + RT	20.5	35.5	−12.0	102	G	

Table 1 (continued)

Sample	Twin Laws	ϕ	Θ	ψ	$2V\gamma$	Opt. Analyst	An, calculated from Ca determination
14053,11	A + C + RT	41.5	32.0	+17.0	99	G	
	A + C + RT	20.5	37.0	−13.0	—	G	
	A + C + RT	21.0	38.5	−8.0	97	G	
14053,19	A + Ala	25.2	33.2	−4.5	103	W	
	RT + Ala	22.5	34.0	−7.3	105	W	85, 85, 89, 91, 91, 93, 94, 94 94, 93, 94, 91, 92, 91, 91 91, 91, 90, 94, 92 94, 94, 91, 91
						$\bar{x}$	91.5 (85–94)
	RT	23.5	35.0	−5.0	94	W	
	A + C + RT	23.8	37.0	−5.2	98	G	90, 90, 89, 86, 91, 91, 90, 89
		20.3	37.3	−10.5	104	W	90, 90, 89, 88
		22.1	37.2	−7.9	101	$\bar{x}$	90.2 (86–91)
	A + RT	22.0	38.0	−6.2	99	W	
		22.0	36.0	−5.0	104	G	
		22.0	37.0	−5.6	101.5	$\bar{x}$	
	C + P	21.3	37.0	−10.7	99.5	W	
	A + C + RT	23.7	38.0	−6.8	95	W	
		24.0	35.0	−6.3	101	T	
		23.8	36.5	−6.5	98	$\bar{x}$	
	C	22.8	38.0	−9.7	98.5	W	
		20.0	37.0	−7.8	—	T	
		21.4	37.5	−8.8	98.5	$\bar{x}$	
	A + C + RT	22.0	37.0	−10.5	106	G	
		22.6	36.3	−7.3	100.5	$\bar{x}$	91.3 overall mean
14053,61	A + RT	21.7	38.0	−6.8	100	W	91, 86, 91
						$\bar{x}$	89.3
14310,14	A + RT	21.5	36.8	−7.5	103	G	91, 86, 93
		22.7	36.8	−6.7	100.7	T	90, 92, 91, 91, 91, 91, 87, 90
		22.3	38.0	−7.8	101.3	W	91, 93, 91, 93, 91, 90, 90, 91
		22.2	37.2	−7.3	101.7	$\bar{x}$	90.7 (86–93)
	A + C + RT	24.0	38.0	−9.0	106	G	91, 93, 90, 86
		25.5	37.0	−8.8	102	T	90, 91, 93, 92, 93, 93, 92, 94, 91, 93 (82, 79, 86) 89, 90, 93, 92, 93, 92, 92 94, 92, (83, 92)
		24.7	37.5	−8.9	104	$\bar{x}$	91.7 (86–94), resp. 90.4 (79–94)
	A + C + RT	20.0	35.7	−10.6	96	G	91, 94, 92
		21.5	37.0	−9.3	100	T	
		20.7	36.8	−10.3	100	T	
		18.5	35.5	−10.3	99	W	
		20.2	36.2	−10.1	98.8	$\bar{x}$	92.3 (91–94)
14310,14	A + RT	24.0	38.0	−8.0	101	G	
	RT	24.0	37.0	−5.7	100.5	W	
	A + RT	20.5	37.2	−7.2	—	G	
	A + RT	20.6	38.3	−10.3	101	T	
							91, 93, 91, 91
		22.0	37.1	−8.6	100.8	$\bar{x}$	91.3 (86–94) marginal zones excluded, this mean value refers only to cores optically analyzed

Microprobe analyses for 14053,19 by R. T. Helz, U.S. Geological Survey, all others by H. Schwander. A = albite. C = Carlsbad. P = pericline. RT = Roc Tourné = complex albite-Carlsbad.

Discussion and Conclusions

Our data (Table 1) can be compared with those of the Berkeley team (H. R. Wenk and G. L. Nord, 1971; H. R. Wenk *et al.*, 1972). In calcic plagioclases the angle Θ is almost constant, while ϕ and ψ change distinctly with composition and structural state. Therefore, the $\phi\,\psi$ diagram (Fig. 4) was chosen for graphical presentation.

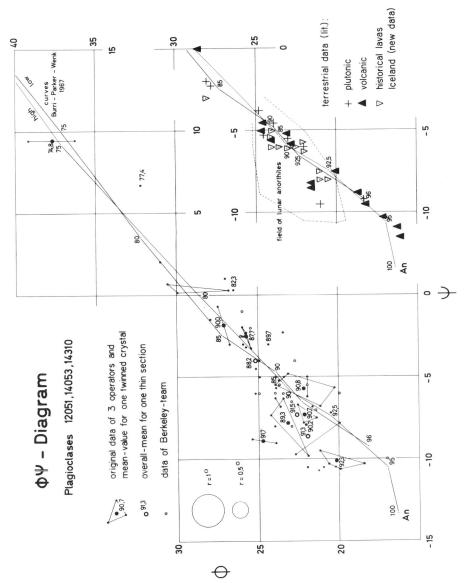

Fig. 4. $\phi\psi$-diagram. Data from Table 1 and from literature.

In the *bytownite* range, the relations between the optic orientation and the chemical composition of lunar feldspars are consistent with the terrestrial evidence compiled by Burri *et al.* (1967). Though the differences between high and low optics are small in the range An 75–90, the anorthite content determined by microprobe is a better fit to the low curve. This holds not only for selected twin groups, but also for the average values of sample 12051,57 (Mean of 14 series of Euler angles ϕ 25.2°, Θ 36.5°, ψ 4.0°, $2V\gamma$ 99.5; mean value of 15 determinations by microprobe An 88.2). Towards the calcic end of the plagioclase series the distinction between high and low optics has little basis.

In the *anorthite* range, however, the situation is different. Here only a minority of the lunar data agrees with the meager data for plutonic anorthites from earth, or with the more ample evidence on terrestrial feldspars from very young volcanoes (Vesuvius, Etna, Soufrière, Gunung Guntur). The majority shows higher negative ψ angles, scatters towards the left of the curves and occupies the field indicated in Fig. 4. The results of the Berkeley group show a similar trend. However, in our data this tendency is more pronounced, and careful cross checks of our cooperative study prove that the trend is consistent and must be significant. No information is yet available for An-contents greater than 92. Apparently lunar anorthites in the narrow compositional range 90–92 can show quite different and reproducably distinct optic orientations.

To test the possibility of operator bias, two of us measured, with the same universal stages and microscopes, anorthite phenocrysts of historic Tungnaa lavas from Iceland. The results are shown in the inset in the lower right hand corner of Fig. 4. The data follow remarkably well the terrestrial high curve and agree with published Euler angles for young volcanic feldspars. Thus we contend that most of the anorthites known from the very old basalts of the moon differ in optic orientation from those of young volcanic rocks of the earth. However, caution is indicated because in the world-literature of the past 50 years, optic-chemical information only for 22 synthetic and natural anorthites of the earth is found. In contrast, about three times as many data exist for lunar feldspars and these results have been accumulated during the past 6 months.

Before any firm conclusions can be drawn, optical-chemical data are required on *lunar anorthites An 93–100* and on *anorthites from metamorphic and plutonic rocks of the earth.*

Acknowledgments—We are indebted to D. B. Stewart for discussions and for opportunity to examine some of his lunar specimens, to R. T. Helz, U.S. Geol. Survey, Washington, for microprobe analyses and to H. R. Wenk and W. F. Müller for discussions. We also wish to thank H. Hänni who skillfully prepared the polished sections and to H. H. Klein for photographs and drafts. Our investigations were supported by Schweizerischer Nationalfonds zur Förderung der wissenschaftlichen Forschung (grant 2.438.71).

REFERENCES

Appleman D. A. Nissen H. U. Stewart D. B. Clark J. R. Dowty E. and Huebner J. S. (1971) Studies of lunar plagioclases, tridymite, and cristobalite. *Proc. Second Lunar Sci. Conf., Geochim. Cosmochim. Acta*, Suppl. 2, Vol. 1, pp. 117–133. M.I.T. Press.

Burri C. Parker R. L. and Wenk E. (1967) *Die optische Orientierung der Plagioklase*. Birkhäuser, Basel.

Schwander H. and Wenk E. (1967) Studien mit der Röntgen-Mikrosonde an basischen Plagioklasen alpiner Metamorphite. *Schweiz. Mineral. Petrogr. Mitt.* **47**, 225–234.

Stewart D. B. Appleman D. E. Huebner J. S. and Clark J. R. (1970) Crystallography of some lunar plagioclases. *Proc. Apollo 11 Lunar Sci. Conf., Geochim. Cosmochim. Acta,* Suppl. 1, Vol. 1, pp. 927–930.
Pergamon.

Wenk E. Schwander H. and Trommsdorff V. (1967) Optische Orientierung zweier Anorthite aus meta-
morphen Gesteinen. *Schweiz. Mineral. Petrogr. Mitt.* **47**, 219–224.

Wenk E. and Trommsdorff V. (1967) The optical orientation of synthetic anorthite. *Schweiz. Mineral.
Petrogr. Mitt.* **47**, 213–218.

Wenk H. R. and Nord G. L. (1971) Lunar bytownite from sample 12032,44. *Proc. Second Lunar Sci.
Conf., Geochim. Cosmochim. Acta,* Suppl. 2, Vol. 1, pp. 135–140. M.I.T. Press.

Wenk H. R. Ulbrich M. and Müller W. (1972) Lunar plagioclase (A mineralogical study) (abstract). In
Lunar Science—III (editor C. Watkins), pp. 797–799, Lunar Science Institute Contr. No. 88.

Proceedings of the Third Lunar Science Conference
(Supplement 3, *Geochimica et Cosmochimica Acta*)
Vol. 1, pp. 591–602
The M.I.T. Press, 1972

Plagioclase and Ba–K phases from Apollo samples 12063 and 14310

W. E. Trzcienski, Jr.,* and C. G. Kulick

Department of Geological and Geophysical Sciences
Princeton University
Princeton, N.J. 08540

Abstract—Plagioclases in Apollo 12 sample 12063 began growth as hollow crystals commonly enclosing a late crystallizing pyroxene. With continued growth away from the initial shell of the hollow plagioclase crystal, sector zoning as well as normal zoning (An_{93}–An_{80}) patterns developed. Relative enrichment in potassium (0.06 to 1.89% K_2O), iron (0.63 to 1.12% FeO), and barium (0.00 to 0.04% BaO) occurs toward the outer grain margins. Also in 12063 late-stage mesostases containing two distinct K–Ba phases exist: One phase contains 12.2% BaO and is a celsian-orthoclase feldspar whereas the other phase is a potassium-rich silica glass. In contrast, plagioclases from Apollo 14 sample 14310 are more anorthitic (An_{96}–An_{88}) than those in 12063, and they occur in two distinct habits. The two morphologically different plagioclases may be related to liquid immiscibility. A late-stage potassium feldspar also occurs in sample 14310. The high potassium phases and liquid immiscibility suggest that "granite" may exist on an unsampled portion of the lunar surface.

Introduction

Feldspars, especially the plagioclases, are nearly ubiquitous to all the returned lunar samples. Optical and electron microprobe studies of plagioclases and K–Ba phases from lunar samples 12063,15 and 14310,21 presented in this paper show: (1) that the feldspars have important morphological and chemical characteristics and (2) that the feldspars are critical to and play an important role in the petrogenetic history of each rock. Both of these samples are fine-grained lunar basalts and have been petrographically described by Hollister *et al.* (1971b, 1972).

Feldspars in Sample 12063,15

Plagioclases

The plagioclases in rock 12063 compose 25 modal percent of the rock and occur generally as euhedral to subhedral phenocrysts. Some phenocrysts (Fig. 1a) have core areas that contain pyroxenes that have been interpreted by Hollister *et al.* (1971a, p. 549) to have crystallized after the onset of plagioclase crystallization. The majority of plagioclases, however, occur as subhedral laths (Fig. 1b) commonly possessing albite twinning. Step scanning across five representative crystals (the analytical procedure is similar to that described in Hollister *et al.*, 1971a) indicates that the chemical zoning pattern is similar for each plagioclase grain. For all crystals there is a general increase in Na, K, and Fe with a complementary depletion in Ca and Mg as a crystal's outer rim is approached. This chemical trend has also been described for other Apollo samples (Brown *et al.*, 1970; Keil *et al.*, 1970, 1971; Lovering and Ware, 1970; Ridley *et al.*, 1972).

*Present address: Département de Génie géologique, Ecole Polytechnique, Montréal 250, Canada.

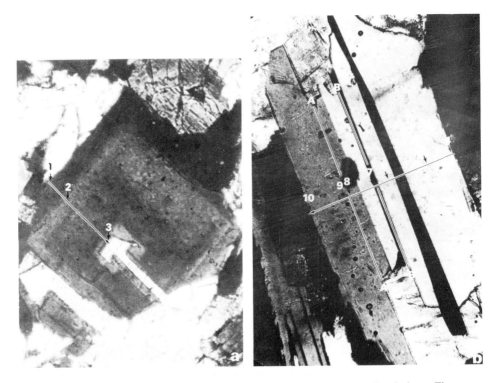

Fig. 1. Photomicrograph of plagioclase crystals in rock 12063: crossed polarizers. The arrows represent the direction of microprobe traverses and the numbered dots are point analyses listed in Table 1. (a) A "hollow" plagioclase crystal showing optical zoning outward toward the rim and inward toward the center from the area (light band) of highest anorthite content. The length of traverse is 100 μ. The pyroxene within the plagioclase is a group 3 pyroxene. (b) A subhedral plagioclase lath. AA' and BB' form boundaries within which a homogeneous core "sector" exists bounded on either side by normally zoned rim "sectors". The length of the traverse is 250 μ.

The plagioclase in Fig. 1a is one of the hollow plagioclases found in section 12063,15 and is unique in that it possesses the maximum amount of zoning (Fig. 2a) found in any of the plagioclases studied from this section. This zoning, however, is not the maximum for lunar samples (Agrell *et al.*, 1970; Brown and Peckett, 1971). Three analyses (Table 1) corresponding to the three numbered points in Figs. 1a and 2a indicate the extent of the zoning; the highest An content occurs in the area of analysis No. 2. The optical variations parallel the chemical changes. It should be noted that the 11% orthoclase content of the outermost rim is unusual for a plagioclase with so high an anorthite content and represents the highest orthoclase content from lunar plagioclases (Weill *et al.*, 1970b; Appleman *et al.*, 1971).

The zoning profile (Fig. 2b) for the subhedral plagioclase shows the more typical zonation found in the plagioclases. Representative analyses along this profile are given in Table 1. The An content varies from about An_{94} to An_{88} with a maximum Or component of 2% at the grain rim. Fe increases approximately two-fold (0.68 to

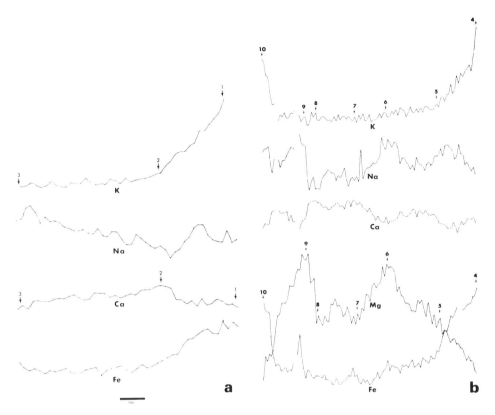

Fig. 2. Element variation along microprobe traverses across grains illustrated in Fig. 1. Traverse directions are from right to left. The numbers refer to the points in Fig. 1 and to the analyses given in Table 1. Electron beam diameter was less than 1 μ, and the analysis interval is 2 μ. (a) Microprobe traverse across grain in Fig. 1a. (b) Microprobe traverse across grain in Fig. 1b.

1.12 wt% FeO) from center to rim, and Ba (0.04 wt% BaO) becomes detectable, by microprobe analysis, only at the rim. The relatively sharp breaks in the Na, Ca, and Mg profiles (Fig. 2b) occur at the boundaries AA' and BB' (Fig. 1b). Zoning profiles across several other plagioclase grains exhibit similar discontinuities.

All plagioclase analyses (Table 1) indicate a deficiency of aluminum (Weill *et al.*, 1970a) in the tetrahedral site. This deficiency possibly may be compensated for in part by iron, for most analyses show an excess of cations in the alkali metal position. The oxidation state of iron is indeterminate from microprobe analyses and other techniques are now being employed (Appleman *et al.*, 1971; Virgo *et al.*, 1971) to determine ferrous-ferric ratios.

K–Ba phases

Associated with the pyroxenes and plagioclases that compose the major part of rock 12063 is a late-stage mesostasis (2 modal percent) consisting of fayalitic olivine

Table 1. Microprobe analyses of feldspars in sample 12063 (wt%).

Analysis No.	1	2	3	4	5	6	7	8	9	10	11	12
Na_2O	1.04	0.97	1.33	0.87	0.99	1.17	0.89	0.74	1.19	1.12	0.13	0.16
MgO	0.01	0.17	0.32	0.02	0.17	0.34	0.17	0.18	0.36	0.07	0.00	0.00
Al_2O_3	31.74	33.40	32.04	32.67	33.12	32.63	34.10	32.38	32.36	31.64	21.35	10.31
SiO_2	48.52	46.71	48.18	47.73	47.05	47.66	46.04	48.94	47.66	49.05	54.45	79.99
K_2O	1.89	0.13	0.06	0.37	0.07	0.05	0.04	0.05	0.04	0.26	8.83	7.65
CaO	16.67	18.09	16.93	17.32	17.85	17.50	18.47	18.79	17.69	17.09	0.55	0.00
FeO	1.12	0.68	0.63	1.48	0.77	0.66	0.54	0.63	0.66	1.33	1.00	0.00
BaO	0.04	0.01	0.00	0.02	0.02	0.00	0.01	0.01	0.01	0.04	12.15	0.72
TOTAL	101.03	100.15	99.48	100.48	100.03	100.01	100.26	101.73	99.97	100.60	98.47	98.83

Cation Proportions

	1	2	3	4	5	6	7	8	9	10	11	12
Na	0.093	0.087	0.119	0.078	0.089	0.105	0.079	0.065	0.106	0.099	0.013	0.014
Mg	0.000	0.011	0.022	0.001	0.011	0.023	0.012	0.012	0.025	0.005	0.000	0.000
Al	1.719	1.814	1.742	1.770	1.799	1.769	1.851	1.728	1.757	1.707	1.270	0.530
Si	2.228	2.151	2.222	2.193	2.167	2.191	2.120	2.214	2.194	2.244	2.746	3.486
K	0.111	0.007	0.003	0.022	0.004	0.003	0.002	0.003	0.002	0.015	0.568	0.425
Ca	0.821	0.893	0.837	0.853	0.881	0.862	0.911	0.911	0.873	0.838	0.030	0.000
Fe	0.043	0.026	0.024	0.057	0.029	0.025	0.021	0.024	0.025	0.051	0.042	0.000
Ba	0.001	0.000	0.000	0.000	0.000	0.000	0.000	0.000	0.000	0.001	0.215	0.011
O	8	8	8	8	8	8	8	8	8	8	8	8
An	80	90	87	90	90.5	89	92	93	89	88	—	—
Ab	9	9	12.5	8	9	11	8	7	11	10	—	—
Or	11	1	0.5	2	0.5	0	0	0	0	2	—	—

(Fa_{95}), cristobalite, and two distinct K–Ba rich phases. These mesostasis areas commonly are less than 150 μ in diameter.

An electron microprobe sample current scan (Fig. 3a) over one of these areas shows that the olivine has a sieve texture. The areas within and around the high iron olivine (Fig. 3b) have high concentrations of potassium (Fig. 3c), some parts of which are distinct, for they also have a high barium concentration (Fig. 3d). The high potassium areas probably are glass phases, because a short period of exposure to the electron beam results in a rapid decrease of the potassium x-ray intensity. Similar time exposures to the electron beam of a crystalline orthoclase standard resulted in a slower volatilization of potassium. An analysis (No. 11, Table 1) for the high barium area (Fig. 3d) yields a BaO content of 12.2 wt%, the greatest concentration of barium yet reported in any phase from the returned lunar samples. The next highest BaO concentration is 9.3 wt% reported from sample 12039 (Keil et al., 1971).

Recast into a feldspar formula based on eight oxygens, the barium phase closely approximates a celsian-orthoclase feldspar (Cn_{27}) with the formula:

$$(K_{0.57}Na_{0.01}Ca_{0.03}Ba_{0.22}Fe_{0.04})_{0.89}Al_{1.27}Si_{2.75}O_8.$$

If the low analysis total (Table 1) is due to partial volatilization of about 1.50 wt% K_2O, a new formula based on 10.3 K_2O yields:

$$(K_{0.66}Na_{0.01}Ca_{0.03}Ba_{0.24}Fe_{0.04})_{0.98}Al_{1.26}Si_{2.73}O_8.$$

This latter formula approaches that of an alkali feldspar containing 24 mole % celsian; it therefore merits consideration as a new phase in the lunar rocks.

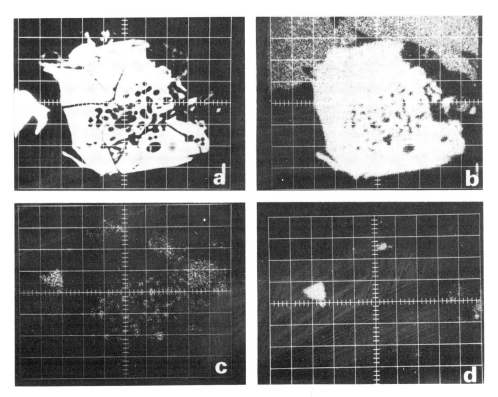

Fig. 3. X-ray scans from sample 12063; the grid boxes are 15 μ square. (a) Sample current scan over a mesostasis area. Note seive texture in olivine. (b) Scan for Fe across mesostasis. White area is fayalitic olivine; gray area is pyroxene. (c) Scan for K across mesostasis area. (d) Scan for Ba across mesostasis area. The analysis for the bright triangular area is No. 11 in Table 1.

The barium "hot spots" (Fig. 3d) are rare in the mesostasis. An average analysis (No. 12, Table 1) from several typical high potassium (Fig. 3c) mesostasis areas indicates that the predominant composition is high in silica and is similar to the high silica immiscible glasses described by Roedder and Weiblen (1971) in other Apollo 12 samples.

DISCUSSION OF FELDSPARS IN 12063,15

The crystal forms (Fig. 1a and 1b) and the zoning profiles across the plagioclases are consistent with a model that indicates that many of the grains probably initially were skeletal or hollow crystals. This type of crystal has been described elsewhere for pyroxenes (Bence *et al.*, 1971; Boyd, 1971; Hollister *et al.*, 1971a) and for olivines (Drever and Johnston, 1957; Brett *et al.*, 1971). As the crystal grew, growth progressed both outward and inward from the initial "shell." This is most clearly manifested by the crystal in Fig. 1a, where the high An-zone (the inferred initial shell) is bounded inward toward the crystal center and outward toward the crystal rim by successively lower An containing layers. The difference between the outer and inner rim compositions (Nos. 1 and 3, Table 1), especially in the proportion of K, Ba, and Fe, reflects the fact that at the point of final crystallization each rim was in communication with a different liquid. The lower K, Ba, and Fe concentrations of the interior rim suggest that it was most likely isolated from the last liquid (the mesostasis) before the outer rim crystallized. A similar growth history was postulated for the plagioclases in sample 12021 by Walter *et al.* (1971).

Distribution coefficients calculated for K and Ba between the outer rim plagioclase and the mesostasis suggest that when the plagioclase began crystallization, 60–75% of the original liquid was crystalline. This is consistent with an earlier conclusion (Hollister *et al.*, 1971a) that plagioclase appeared late on the liquidus during the crystallization sequence of rock 12063. The appearance of plagioclase on the liquidus coincides with a sharp drop in the calcium and aluminum concentrations of the zoned pyroxenes in this rock.

The discontinuities in the profile (Fig. 2b) of the lath-like plagioclase (Fig. 1b) are similar in some respects to discontinuities found in sector-zoned pyroxenes (Albee and Chodos, 1970; Bence *et al.*, 1970; Hollister and Hargraves, 1970; Hollister *et al.*, 1971). The core area of this grain lying between AA' and BB' is relatively homogeneous and probably represents a single growth face. Because the twin plane is (010), this growth face is (100) assuming the long direction of the grain parallels (001). The continuous zoning of the bounding outer sectors results from the section cutting at an angle the (100) and/or the (110) growth directions. The encroachment of the center sector by late crystallizing pyroxene is interpreted to be morphologically equivalent to the late pyroxene in the center of the crystal in Fig. 1a. The outer sectors show normal zoning: Fe, Na, and K increase and Ca and Mg decrease from core to rim. In all profiles, there is little to no oscillation in the concentration of elements; this indicates that during crystallization, the change in chemical concentrations was monotonic.

Feldspars in Sample 14310,21

Plagioclases

The abundant (55 modal percent) plagioclase in rock 14310 occurs in two distinct habits. Large subhedral crystals account for less than one quarter of the plagioclase; the remainder are small, polysynthetically twinned laths, the smallest being microlites. Textural relations suggest that the large phenocrysts are an early generation of plagioclase and that the laths are a later generation.

Microprobe step scan analysis across one of the largest crystals in section 14310,21 shows that the grain is asymmetrically zoned with a maximum An content of An_{96} (No. 1, Table 2) zoning out to rims of An_{90} and An_{88} (Nos. 2 and 3, Table 2). The asymmetrical pattern also was observed by Ridley *et al.* (1972). The asymmetry is most clearly manifested by magnesium, which increases monotonically from 0.13 to 0.23 wt% MgO from one edge to the other edge of the grain.

Analyses of the smaller plagioclase laths scattered throughout the section show a variation in anorthite content similar to that found within the large grain. Some

Table 2. Microprobe analyses of feldspars in sample 14310.

Analysis No.	1	2	3	4	5	6	7	8
Na$_2$O	0.39	0.97	1.31	0.76	1.45	0.58	1.06	0.83
MgO	0.16	0.13	0.23	0.19	0.17	0.13	0.14	0.01
Al$_2$O$_3$	35.34	34.56	33.58	34.73	33.18	35.16	34.33	19.24
SiO$_2$	43.77	45.32	46.35	45.30	47.16	44.95	45.80	60.79
K$_2$O	0.00	0.19	0.00	0.21	0.43	0.20	0.31	13.56
CaO	19.36	18.12	17.11	18.45	17.03	19.01	17.87	0.58
TiO$_2$	—	—	—	0.03	0.06	0.03	0.03	—
FeO	0.14	0.34	0.38	0.27	0.48	0.27	0.39	0.24
BaO	0.00	0.00	0.00	0.02	0.04	0.00	0.03	3.84
TOTAL	99.16	99.63	98.96	99.96	100.00	100.33	99.96	99.09

Cation Proportions

Na	0.036	0.087	0.117	0.068	0.130	0.051	0.095	0.077
Mg	0.011	0.009	0.016	0.013	0.012	0.009	0.009	0.000
Al	1.944	1.888	1.838	1.891	1.802	1.911	1.869	1.084
Si	2.041	2.099	2.151	2.092	2.172	2.072	2.115	2.905
K	0.000	0.011	0.000	0.012	0.025	0.012	0.018	0.827
Ca	0.968	0.899	0.851	0.913	0.841	0.939	0.884	0.030
Ti	—	—	—	0.001	0.002	0.001	0.001	—
Fe	0.006	0.013	0.015	0.010	0.019	0.010	0.015	0.010
Ba	0.000	0.000	0.000	0.000	0.001	0.000	0.000	0.064
O	8	8	8	8	8	8	8	8
An	96	90	88	92	84	94	89	—
Ab	4	9	12	7	13	5	9	—
Or	0	1	0	1	3	1	2	—

of these laths, however, are greater in anorthite than the rim composition of the large phenocryst. Four representative analyses (Nos. 4–7, Table 2) show a variation from An$_{94}$ to An$_{84}$, but each individual grain is nearly homogeneous.

In general, the plagioclases from 14310 are significantly lower in their Fe, Ba, and K contents than those in 12063. The high anorthite cores (An$_{96}$) of the large phenocrysts are quite close in composition to the plagioclases in the "genesis rock" (An$_{96-97}$) (Steele and Smith, 1971; Hargraves and Hollister, 1972). Nowhere in the section was there found optically or by electron microprobe step scanning any suggestion of hollow or sector-zoned plagioclases.

Ba–K phase

As does sample 12063, sample 14310 contains a late-stage potassium phase. The K-bearing phase in 14310 commonly occurs within an encircling mesh of small plagioclase laths. One of the larger grains of this phase is outlined in Figure 4a. The area shows a high concentration of potassium (Fig. 4b) and barium (Fig. 4c) and essentially is devoid of calcium (Fig. 4d). *Extremely* high calcium areas (Fig. 4d), commonly associated with the K-phase, also are high in phosphorus and probably are whitlockite (Gancarz *et al.*, 1971).

In contrast to the potassium glass in sample 12063, the K-phase in sample 14310 appears to be crystalline K-feldspar. Five different K-rich areas were analyzed at

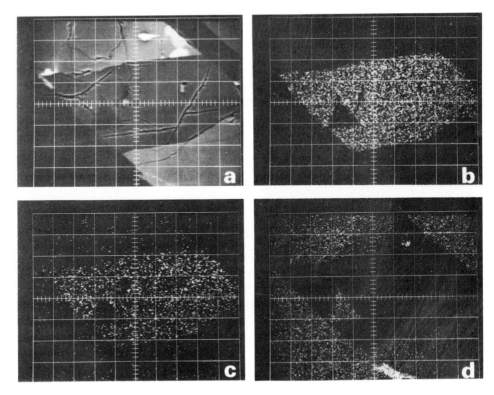

Fig. 4. X-ray scans from sample 14310; the grid boxes are 20 μ square. (a) Sample current scan over K-feldspar crystal (beneath subdivided grid lines) and surrounding plagioclase. (b) Scan for K across crystal. (c) Scan for Ba across crystal. (d) Scan for calcium across crystal. Note triangular inclusion of plagioclase. The bright elongate area at the bottom center is a small grain of whitlockite.

different locations in the section. One microprobe spectrometer was used as a moniter for potassium in order that only the high potassium areas would be represented in the total analysis. Little variation in element x-ray intensity occurred from spot to spot and from grain to grain suggesting that the grains are homogeneous. Based on the rate of volatilization of potassium from the sample compared with that from a crystalline standard, and based on the fact that all analyses are nearly the same, it is concluded that the K-feldspar is, indeed, a crystalline phase and not a glass.

An average analysis (No. 8, Table 2) for the high potassium phase shows significant barium and sodium concentrations. However, the BaO content (3.8 wt%) is only about one-third that of the crystalline (?) high potassium, high barium feldspar in 12063. Recast into a feldspar structural formula based on eight oxygens, the mineral analysis results in a nearly perfect fit for an alkali feldspar with seven % celsian component. The cation amount of sodium in the K-feldspar is the same as in the associated plagioclase laths.

Discussion of Feldspars in 14310,21

One possibility for the two generations of plagioclase crystals in sample 14310 is that it is the result of a shock melt in which all crystals had not undergone complete resorption. By this mechanism one generation of plagioclases represent premelt crystals and the other generation of crystals result from the crystallization of the shock melt. Pyroxene cores indicative of a premelt history for sample 14310 have been described by Hollister *et al.* (1972). Experimental work (Ford *et al.*, 1972; Ringwood *et al.*, 1972; Ridley *et al.*, 1972; Walker *et al.*, 1972) also suggest that sample 14310 is not a primary magma.

The fact that some of the plagioclase lath compositions have higher anorthite content than the large phenocryst rims, have led Brown and Peckett (1971) to speculate that appreciable alkali loss occurred in the time between crystallization of the two generations of plagioclases. If this were the case, however, it would seem improbable to find alkali feldspar intimately associated between many of the late plagioclase laths.

An alternative speculation calls upon liquid immiscibility, which has been found in other Apollo samples (Roedder and Weiblen, 1970, 1971), as a mechanism. Assuming that a primitive lunar "basalt" liquid originally began to fractionally crystallize (as the zoned crystals suggest) from a single liquid, the early crystals, possibly high anorthite plagioclases, floated upward to form an anorthositic rock (Smith *et al.*, 1970; Wood *et al.*, 1970). At the same time, the remaining liquid became relatively enriched in the nonplagioclase components. At some point in the crystallization sequence the liquid would have separated into two liquids: one a basic, phosphorus-rich liquid and the other a potassium, silica-rich liquid.* The latter liquid would be lighter than the basic one, would also float upward (probably carrying with it inclusions of the basic liquid), and would crystallize as a rock approximating a granite in composition. In fact, lunar sample 12013 (Drake *et al.*, 1970) suggests that a granitic fraction does exist on the lunar surface. An immiscible granitic fraction such as this also was proposed by Roedder and Weiblen (1970) in connection with their work on immiscible glasses. The remaining basic liquid, which was still a major portion of the original liquid, then continued to crystallize into lunar basalt. This basalt could retain small isolated blebs of the granitic component, as well as remnant phenocrysts of the anorthositic plagioclases.

Liquid immiscibility in a simplified, schematic, multicomponent system might develop along a path such as XX' in Fig. 5 where A and B represent two combinations of a number of elements. The works of both Roedder and Weiblen (1971) and Philpotts (1971), the latter on terrestrial rocks, indicate that the composition of immiscible liquids need not be too different in their constituent components. From X to X' solids crystallize from one liquid with zoning evident in the crystals. At X' the original liquid separates into two liquids with compositions Y and Z. If an original mineral formed at X were to form again as new crystals after separation into two liquids, it could at Z, have a composition more "primitive" (for example a

---

*Immiscibility has been shown experimentally to exist in a number of phosphorus-bearing systems (Levin *et al.*, 1964).

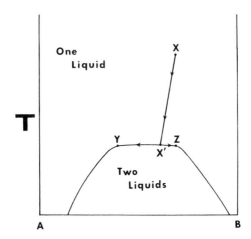

Fig. 5. Schematic diagram showing liquid immiscibility. Along the path XX′ there is crystallization from one liquid only. At X′ two immiscible liquids form: One has a composition at Y and the other a composition at Z.

more anorthitic plagioclase) than the original phenocryst core. The composition for the new mineral habit would definately be more "primitive" than that at the rim of the early formed phenocryst. Because the structure of the liquid at Z is probably different from that at X, the habit of the mineral at Z may be different from that at X. By this mechanism two generations of the same mineral species could develop such as the plagioclases in sample 14310. There could also be a difference in the trace element distribution pattern between the phenocryst and the small lath plagioclases from this sample.

SUMMARY

The chemical and morphological characteristics of feldspars in Apollo samples 12063 and 14310 support the following contentions: (1) Plagioclases in Apollo sample 12063 initially formed as hollow crystals, when the rock was 60–75% crystalline, with continued growth sector-zoning as well as normal zoning developed. (2) Plagioclases from Apollo sample 14310 occur in two distinct habits suggestive of two generations of plagioclase growth which may be related to liquid immiscibility. (3) In both samples late-stage Ba–K phases occur approximating the composition of alkali feldspars. These Ba–K phases may be trapped remnants, within basalt, of a larger granitic fraction that occurs on some unsampled portion of the lunar surface.

Acknowledgments—R. B. Hargraves, L. S. Hollister, and D. R. Waldbaum have provided helpful discussions during various parts of this work. The comments and criticisms on the manuscript by L. S. Hollister are appreciated. Comments by the reviewers, J. J. Papike and J. F. Lovering, were also helpful. This work was supported by NASA contract NGR 31–001–205.

References

Agrell S. O. Scoon J. H. Muir I. D. Long J. V. P. McConnell J. D. C. and Peckett A. (1970) Observations on the chemistry, mineralogy, and petrology of some Apollo 11 lunar samples. *Proc. Apollo 11 Lunar Sci. Conf., Geochim. Cosmochim. Acta* Suppl. 1, Vol. 1, pp. 93–128. Pergamon.

Albee A. L. and Chodos A. A. (1970) Microprobe investigations of Apollo 11 samples. *Proc. Apollo 11 Lunar Sci. Conf., Geochim Cosmochim. Acta* Suppl. 1, Vol. 1, pp. 135–157. Pergamon.

Appleman D. E. Nissen H. -U. Stewart D. B. Clark J. R. Dowty E. and Huebner J. S. (1971) Studies of lunar plagioclases, tridymite, and cristobalite. *Proc. Second Lunar Sci. Conf., Geochim. Cosmochim. Acta* Suppl. 2, Vol. 1, pp. 117–133. MIT Press.

Bence A. E. Papike J. J. and Lindsley D. H. (1971) Crystallization histories of clinopyroxenes in two porphyritic rocks from Oceanus Procellarum. *Proc. Second Lunar Sci. Conf., Geochim. Cosmochim. Acta* Suppl. 2, Vol. 1, pp. 559–574. MIT Press.

Boyd F. R. (1971) Anatomy of a mantled pigeonite from Oceanus Procellarum. *Carnegie Inst. Wash. Yearb.* **69**, 216–228.

Brett R. Butler P. Myer C. Reid A. M. Takeda H. and Williams R. (1971) Apollo 12 igneous rocks 12004, 12008, 12009, and 120022: A mineralogical and petrological study. *Proc. Second Lunar Sci. Conf., Geochim. Cosmochim. Acta* Suppl. 2, Vol. 1, pp. 301–317. MIT Press.

Brown G. M. Emeleus C. H. Holland J. G. and Phillips R. (1971) Mineralogical, chemical, and petrological features of Apollo 11 rocks and their relationship to igneous processes. *Proc. Apollo 11 Lunar Sci. Conf., Geochim. Cosmochim. Acta* Suppl. 1, Vol. 1, pp. 195–219. Pergamon.

Brown G. M. and Peckett A. (1971) Selective volatilization on the lunar surface: Evidence from Apollo 14 feldspar-phyric basalts. *Nature* **234**, 262–266.

Drake M. J. McCallum I. S. McKay G. M. and Weill D. F. (1970) Mineralogy and petrology of Apollo 12 sample 12013: A progress report. *Earth Planet. Sci. Lett.* **9**, 103–123.

Drever H. I. and Johnston R. (1957) Crystal growth of forsteritic olivine in magmas and melts. *Trans. Roy. Soc. Edinburgh* **63**, 289–315.

Ford C. E. Humphries D. J. Wilson G. Dixon D. Biggar G. M. and O'Hara M. J. (1972) Experimental petrology of high alumina basalt 14310, and related compositions (abstract). In *Lunar Science—III* (editor C. Watkins), pp. 274–276, Lunar Science Institute Contr. No. 88.

Gancarz A. J. Albee A. L. and Chodos A. A. (1971) Petrologic and mineralogic investigation of some crystalline rocks returned by the Apollo 14 mission. *Earth Planet. Sci. Lett.* **12**, 1–18.

Hargraves R. B. and Hollister L. S. (1972) Mineralogic and petrographic study of lunar anorthosite slide 15415,18. *Science* **175**, 430–432.

Hollister L. S. and Hargraves R. B. (1970) Compositional zoning and its significance in pyroxenes from two course grained Apollo 11 samples. *Proc. Apollo 11 Lunar Sci. Conf., Geochim. Cosmochim. Acta* Suppl. 1, Vol. 1, pp. 541–550. Pergamon.

Hollister L. S. Trzcienski W. E. Jr. Hargraves R. B. and Kulick C. G. (1971a) Petrogenetic significance of pyroxenes in two Apollo 12 samples. *Proc. Second Lunar Sci. Conf., Geochim. Cosmochim. Acta* Suppl. 2, Vol. 1, pp. 529–557. MIT Press.

Hollister L. S. Trzcienski W. E. Jr. Hargraves R. B. and Kulick C. G. (1971b) Crystallization histories of two Apollo 12 basalts. In *Hess Memorial Volume* (editor R. Shagam), Geological Society of America Memoir 132 (in press).

Hollister L. S. Trzcienski W. E. Jr. Dymek R. Kulick C. G. Weigand P. and Hargraves R. B. (1972) Igneous fragment 14310,21 and the origin of the mare basalts (abstract). In *Lunar Science—III* (editor C. Watkins), pp. 386–388, Lunar Science Institute Contr. No. 88.

Keil K. Bunch T. E. and Prinz M. (1970) Mineralogy and composition of Apollo 11 lunar samples. *Proc. Apollo 11 Lunar Sci. Conf., Geochim. Cosmochim. Acta* Suppl. 1, Vol. 1, pp. 561–598. Pergamon.

Keil K. Prinz M. and Bunch T. E. (1971) Mineralogy, petrology, and chemistry of some Apollo 12 samples. *Proc. Second Lunar Sci. Conf., Geochim. Cosmochim. Acta* Suppl. 2, Vol. 1, pp. 319–341. MIT Press.

Levin E. M. Robbins C. R. and McMurdie H. F. (1964) *Phase Diagrams for Ceramists.* American Ceramic Society.

Lovering J. F. and Ware N. G. (1970) Electron probe microanalysis of minerals and glasses in Apollo 11 lunar samples. *Proc. Apollo 11 Lunar Sci. Conf., Geochim. Cosmochim. Acta* Suppl. 1, Vol. 1, pp. 633–654. Pergamon.

Philpotts A. R. (1971) Immiscibility between feldspathic and gabbroic magmas. *Nature* **229**, 107–109.

Ridley W. I. Williams R. J. Brett R. Takeda H. and Brown R. W. (1972) Petrology of lunar basalt 14310 (abstract). In *Lunar Science—III* (editor C. Watkins), pp. 648–650, Lunar Science Institute Contr. No. 88.

Ringwood A. E. Green D. H. and Ware N. G. (1972) Experimental petrology and petrogenesis of Apollo 14 basalts (abstract). In *Lunar Science—III* (editor C. Watkins), pp. 654–656, Lunar Science Institute Contr. No. 88.

Roedder E. and Weiblen P. W. (1970) Lunar petrology of silicate melt inclusions, Apollo 11 rocks. *Proc. Apollo 11 Lunar Sci. Conf., Geochim. Cosmochim. Acta* Suppl. 1, Vol. 1, pp. 801–837. Pergamon.

Roedder E. and Weiblen P. W. (1971) Petrology of silicate melt inclusions, Apollo 11 and Apollo 12 and terrestrial equivalents. *Proc. Second Lunar Sci. Conf., Geochim. Cosmochim. Acta* Suppl. 2, Vol. 1, pp. 507–528. MIT Press.

Smith J. V. Anderson A. T. Newton R. C. Olsen E. J. Wyllie P. J. Crewe A. V. Isaacson M. S. and Johnson D. (1970) Petrologic history of the moon inferred from petrography, mineralogy, and petrogenesis of Apollo 11 rocks. *Proc. Apollo 11 Lunar Sci. Conf., Geochim. Cosmochim. Acta* Suppl. 1, Vol. 1, pp. 897–925. Pergamon.

Steele I. M. and Smith J. V. (1971) Mineralogy of Apollo 15415 "Genesis rock": Source of anorthosite on moon. *Nature* **234**, 134–140.

Virgo D. Hafner S. S. and Warburton D. (1971) Cation distribution studies in clinopyroxenes, olivines, and feldspars using Mossbauer spectroscopy of ^{57}Fe. Second Lunar Science Conference (unpublished proceedings).

Walker D. Longhi J. and Hays J. F. (1972) Experimental petrology and origin of Fra Mauro rocks and soil (abstract). In *Lunar Science—III* (editor C. Watkins), pp. 770–772, Lunar Science Institute Contr. No. 88.

Walter L. S. French B. M. Heinrich K. F. J. Lowman P. D. Jr. Doan A. S. and Adler I. (1971) Mineralogical studies of Apollo 12 samples. *Proc. Second Lunar Sci. Conf., Geochim. Cosmochim. Acta* Suppl. 2, Vol. 1, pp. 343–358. MIT Press.

Weill D. F. McCallum I. S. Bottinga Y. Drake M. J. and McKay G. A. (1970a) Petrology of a fine-grained igneous rock from the Sea of Tranquility. *Science* **167**, 635–638.

Weill D. F. McCallum I. S. Bottinga Y. Drake M. J. and Gay G. A. (1970b) Mineralogy and petrology of some Apollo 11 igneous rocks. *Proc. Apollo 11 Lunar Sci. Conf., Geochim. Cosmochim. Acta* Suppl. 1, Vol. 1, pp. 937–955. Pergamon.

Wood J. A. Dickey J. S. Marvin U. B. and Powell B. N. (1970) Lunar anorthosites and a geophysical model of the moon. *Proc. Apollo 11 Lunar Sci. Conf., Geochim. Cosmochim. Acta* Suppl. 1, Vol. 1, pp. 965–988. Pergamon.

Proceedings of the Third Lunar Science Conference
(Supplement 3, *Geochimica et Cosmochimica Acta*)
Vol. 1, pp. 603–613
The M.I.T. Press, 1972

Crystallographic studies of lunar plagioclases from samples 14053, 14163, 14301, and 14310

M. Czank, K. Girgis, A. B. Harnik, F. Laves, R. Schmid,
H. Schulz and L. Weber

Institut für Kristallographie und Petrographie, Eidg. Techn. Hochschule,
Sonneggstr. 5, 8006 Zürich, Switzerland

Abstract—We studied plagioclases from Apollo 14 basalts, fines, and a breccia. The samples have been investigated optically, by microprobe, and x-ray precession methods.

The chemical composition of the crystals investigated by x-rays lies between An 75 and An 95. Basaltic plagioclases predominantly have An contents between 85% and 94% in the weakly zoned cores whereas the An contents of four breccia crystals range in an interval of 76% to 93%. Small zones of high potassium concentrations have been detected in these margins.

Most plagioclase crystals are intimately twinned. We optically observed in decreasing order of frequency albite, carlsbad, albite-carlsbad, pericline, and baveno twins. The volume ratios of the different twin individuals present in single crystals were determined both optically and by x-rays. Discrepancies were noted between the results found by these two methods. Submicroscopic carlsbad and/or albite-carlsbad as well as albite twinning might be the cause of the observed differences.

The c-reflections of the investigated crystals are weak and diffuse. Only one crystal from breccia 14301 has sharp and strong c-reflections.

The c-reflections of all heat-treated crystals were less intense and slightly more diffuse after than before heating. Therefore, we propose that these plagioclases have been cooled from about 1000°C in a much longer time than the cooling time in our experiments (i.e., from 1000°C down to 300°C in one hour).

INTRODUCTION

THE CHARACTERISTICS of calcic plagioclases and particularly the appearance of their c-reflections ($h + k + l$ odd) as a function of their thermal history have been the subject of much work; for references, see Laves and Goldsmith (1951, reprinted 1954a; 1954b); Gay (1953); Gay and Taylor (1953); and Gay (1954).* Our intention has been to contribute to a better understanding of these still puzzling phenomena in the lunar plagioclases.

We have examined calcic plagioclases from samples 14053,45 and 14310,107 (basalts), 14301,93 (breccia) and 14163,166 (fines); and from thin sections 14053,61, 14301,14, and 14310,14.

EXPERIMENTAL METHODS

Apparently untwinned or weakly twinned plagioclase grains showing (010) and/or (001) cleavage planes were selected from the crushed rock material. A preliminary

---

*One must realize that the diffuseness of the c-reflections at room temperature depends, among other things, on the quenching temperature. In addition, if the x-ray photographs are taken *at elevated temperatures* the diffuseness is also a function of such temperatures. This means that the c-reflections (they may be sharp or diffuse at room temperature) become continuously more diffuse with increasing temperature and virtually disappear at approximately 240°C. This process is instantaneously reversible. Details and earlier work are described by Laves *et al.* (1970).

test by the Tsuboi (1923) method yielded their An contents. By planimetric analysis of photographs taken in four directions normal to b^*, the *volume ratio* of the main individual plus its albite-law-related lamellae to that of the carlsbad plus albite-carlsbad lamellae was determined (Table 1). A further distinction between the main individual and its albite-law-related lamellae was possible only in two grains.

Buerger precession photographs of the $(0kl)$ reciprocal lattice plane were registered, and only that plane was used for measurements of reflection intensities. The b^*c^* planes of the twin individuals showing the strongest reflections were chosen as zero-level photographs. We shall call these strongest reflections "reflections I" and the corresponding lamellae "individuals I" (Fig. 2b). Those lamellae that are related to I by the albite law are designated as Ia, and the lamellae related to I and Ia by the carlsbad law are designated as Ic and Iac. Because of the albite law, the $(b^*c^*)_I$ plane coincides with the $(b^*c^*)_{Ia}$ plane and (due to the special metrics of plagioclases) nearly coincides with the $(b^*, [\overline{1}02]^*)_{Ic, Iac}$ planes of the individuals twinned according to the carlsbad law.

Some of the x-rayed crystals were heat treated. These crystals were wrapped in platinum foil and placed in a quartz tube. The quartz tubes were heated to approximately 1000°C for ~ 40 h (for details see Table 1). Vacuum ($\sim 10^{-7}$ Torr) was used to avoid any contamination during heat treatment. After heating, the samples were cooled to 300 C during one hour, and then the furnace was turned off.

The density of the reflection spots along reciprocal lattice rows parallel to b^* was measured with a scanning photometer. Only measurements with a density ≤ 1.5 were used. In the following only integrated intensities are considered. The influence of heat treatment was investigated by comparing the intensities of the c-reflections $(0\overline{4}5)$ and $(0\overline{6}5)$ taken from scaled precession photographs recorded before and after heat treatment. In order to evaluate the volume of the individuals I, Ia, Ic, and Iac it

Table 1. X-ray, chemical, and optical data for eight Apollo 14 plagioclases.

No.	Specimen	Type	Reflections *before* heat treatment		Reflections *after* heat treatment	
			b-reflections	c_1-reflections	b-reflections	c_1-reflections
1	14310,107 -A3	basalt	sharp weak	diffuse very weak	—	—
2	14310,107 -B1	basalt	sharp weak	very diffuse very very weak	sharp weak	somewhat more diffuse very very weak
3	14053,45 -A1	basalt	sharp weak-medium	very diffuse weak	sharp weak-medium	extremely diffuse very weak
4	14053,45 -A3	basalt	sharp weak	very diffuse very weak	—	—
5	14053,45 -B1	basalt	somewhat diffuse weak-medium	very diffuse very weak	—	—
6	14053,45 -B2	basalt	sharp very weak	very diffuse very weak	—	—
7	14163,166 -1	fines	sharp weak	diffuse weak	sharp slightly stronger	slightly more diffuse weak
8	14301,93 -A1	breccia	sharp medium	sharp[a] strong	sharp medium	sharp strong

Table 1 (continued)

No.	Heat treatment	Microprobe analysis (wt. %)			Twin volume percentage determined by	
		An	Ab[b]	Or	x-ray intensities[c]	planimetric analysis[d]
1	—				I	50
					Ia	30
					Ic	} 20
					Iac	
2	1000° ± 10°C, 48[h]				I	
					Ia	
					Ic	
					Iac	
3	1015° ± 10°C, 36[h]	88,6	7,1	0,59	I 71 } 83	86 } 94
					Ia 12	8
					Ic 14 } 17	} 6
					Iac 3	
4	—	74,5	18,4	2,77	I 73 } 77	} 85
					Ia 4	
					Ic 15 } 23	} 15
					Iac 8	
5	—				I	} 95
					Ia	
					Ic	} 5
					Iac	
6	—				I 44 } 87	} 84
					Ia 43	
					Ic 2 } 13	} 16
					Iac 11	
7	1015° ± 10°C, 36[h]	87,5	7,9	0,35	I	} 99
					Ia	
					Ic	P:1
					Iac	
8	1015° ± 10°C, 36[h]	94,8	3,0	0,27[e]	I	62
					Ia	
					Ic	} 20 MP:6
					Iac	G:12

[a] Weak and sharp c_2 reflections have been observed in precession photographs of the crystal No. 8.

[b] Ab content probably too low, cf. text.

[c] For explanation of symbols see text.

[d] P: Pericline twins
  MP: Twins with compositional planes //[010]
  G: Not according to a twin law associated second grain
  Rel. error of percentages, determined by planimetric analysis: ±30%.

[e] Complete analysis (wt. %): K 0.037, Ca 13.651, Na 0.260, Al 18.685, Si 20.157, Ti 0.051, Fe 0.128, O (not measured) (46.140); total 99.109.

was necessary to use two scaled precession photographs: A $(b^*c^*)_I$ photograph showing b^* of all four individuals coinciding and c^* axes of individuals I and Ia; and a $(b^*, [\bar{1}02]^*)_I$ photograph showing the same four coinciding b^* axes and the c^* axes of individuals Ic and Iac. The volume ratios should be equal to the intensity ratios of corresponding reflections.

Some x-rayed crystals were analyzed by electron microprobe (15 kV, 0.15 μA electron beam; approximate beam diameter = 1 μ).

Results of Preliminary Optical Investigations

In nearly all cases, the plagioclases are intimately twinned. Albite twinning is predominant; carlsbad and albite-carlsbad twins are less common. Other twin laws occur more rarely. We rarely observed baveno twins (in a basalt crystal) and pericline twins (in a fines crystal).

Index of refraction determinations of 11 plagioclase grains of sample 14310,107 and of 4 grains of sample 14053,45 and U-stage measurements of eight plagioclases in the thin sections 14310,14 and 14053,61 led to the following results concerning the plagioclases in these basalt samples:

(a) The thickness of the twin lamellae ranges from 0.1 mm to the order of the resolution limit of the microscope. The weakly zoned inhomogeneous cores of the grains are surrounded by a thin margin of lower An content that in itself is very strongly zoned.

(b) The refractive indices $n\alpha'_D$ and $n\gamma'_D$ in the cores of plagioclase plates parallel (010) and (001) range with only a few exceptions, from 1.574 to 1.577 and 1.579 to 1.583, respectively, with an estimated error ± 0.0015.

(c) Referring to the calibration curves of Burri et al. (1967) the optical measurements give An contents that lie in the following range:

U-stage: An 89–94 (± 1.5)
Index of refraction: An 85–93 (± 2.5).

In the *breccia* 14301 the An concentration in 9 plagioclase grains has been determined, in 4 cases by the electron microprobe (Table 2) and in 5 cases by index of refraction determinations. The average An contents in the cores in 7 of these 9 grains range from 88 to 95. The other two grains contain only 76.4 and 78 An, respectively. The variation of An content in the cores may be as high as $\pm 5.5\%$ An (Table 2). The An content in the margins lies between 85 and 65. Within the margins of grains no. 3 and 4 and in the core of grain no. 1 (Table 2) small lamellae (about 0.01 mm thick) of lower Ca concentration but much higher K content (corresponding to more than 15 wt.% Or) could be recognized. The Na concentration in these cases is not much different from that in the adjoining plagioclase.

Electron Microprobe Studies of X-Rayed Crystals

The composition of x-rayed feldspars after heat treatment as determined by electron microprobe methods is given in Table 1. It was difficult to measure element

Table 2. Electron microprobe analyses (wt. %) of four plagioclase grains in thin section 14301,14 (breccia).

| | Grain No. | | | |
| Origin | 1 | 2 | 3 | 4 |
	weakly recrystallized subbreccia	moderately recrystallized subbreccia	strongly recrystallized subbreccia	weakly recrystallized subbreccia
Average composition of the grain core				
An	76.4 ($\pm$5.5)*	89.2 ($\pm$2.5)*	89.2 ($\pm$2)*	93.3 ($\pm$3.5)*
Ab	18.0	9.2	9.0	4.9
Or	3.0_6	0.6_3	0.7_6	0.6_4
Total of analyzed weight percents of CaO, Na$_2$O, K$_2$O, and calculated Al$_2$O$_3$, SiO$_2$	97.5	99.0	99.0	98.8
Approximate range of An content in the margin	no margin	no margin	71–87	65–85
Average diameter of the grain in mm	0.28	0.18	0.43	0.46
Shape	xenomorphic	xenomorphic	xenomorphic	xenomorphic (shows shock phenomena)

* In parentheses: indication of the range of irregular variation of An content in the cores.

concentrations because even on the polished surface (typical area about 0.1 × 0.2 mm) there were numerous cracks and holes. The sum (An + Ab + Or) for all specimens studied is less (about 4%) than 100 wt.%. On the other hand, similarly analyzed feldspars in the thin sections did not show such striking discrepancies. One might suspect that during the heating of these crystals under vacuum a certain amount of elements may have evaporated from the surface. Of the elements present in the plagioclases the most volatile would be sodium, but calcium should not have been influenced by our experiments. Therefore we assume that the measured An concentrations are reliable. In this connection the interesting observations on the non-stoichiometry of lunar feldspars reported by Weill *et al.* (1970) should be noted.

X-Ray Investigation Results

The *c*-reflections of most of the investigated crystals are weak and diffuse. We found only one crystal from the breccia 14301 (Fig. 2a) with sharp *c*-reflections. Table 1 gives a survey of our observations.

The integrated intensities of the *a*- and *b*-reflections did not show any significant changes by the heat treatment. However, the intensities of the *c*-reflections of all heat treated crystals were lower (e.g. 20% lower for crystal no. 3) and the shapes slightly more diffuse than before the heating (Fig. 1).

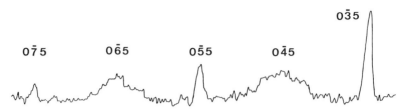

Fig. 1. Part of a photometer plot from crystal no. 3 (unheated). The density is shown in a scan across *b*- and *c*-reflections on the reciprocal lattice row $(0\bar{k}5)_1$. *b*-reflections are $(0\bar{3}5)$, $(0\bar{5}5)$ and $(0\bar{7}5)$; *c*-reflections are $(0\bar{4}5)$ and $(0\bar{6}5)$.

As a tentative quantitative measure for the diffuseness of *c*-reflections the ratio of the halfwidths of *c*- and *b*-reflections photometered parallel to *b** is introduced. On all precession photographs the *c*-reflections appear more diffuse than the *b*-reflections, and the corresponding halfwidth ratio therefore is greater than 1. Diffuseness has been calculated in the following range:

$\sim$ 5 (basalt, crystal no. 3);
$\sim$ 3 (fines, crystal no. 7);
$\sim$ 1.1 (breccia, crystal no. 8).

There are only slight differences, if any, in the diffuseness of reflections before and after heat treatment (Table 1); accordingly the calculated halfwidth ratio does not differ significantly.

None of the 8 crystals investigated by x-rays was untwinned. All of them showed twinning according to one or more of the abovementioned twin laws. In some cases there were distinct differences observed between volume ratios of the several twin lamellae determined by optical and x-ray methods (see Table 1). This discrepancy led us to the assumption that albite as well as carlsbad and albite-carlsbad twins might exist on a submicroscopic scale. To test this assumption we did cut crystal no. 3 (Fig. 3). In this small fragment only a very small part of the individual Ia was visible beside the main individual I. However, we clearly observed spots corresponding to individuals Iac in a *(0kl)* precession photograph of this fragment (Fig. 2c).

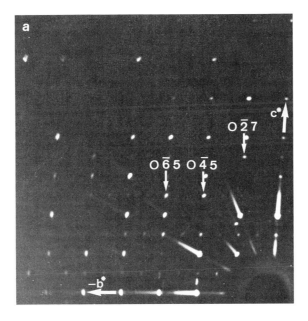

Fig. 2. Precession photographs of *(0kl)* reciprocal lattice plane of lunar plagioclase crystals. Zr-filtered Mo-radiation.

(a) Crystal no. 8 (breccia), unheated. Exposure 20 hours. Arrows indicate 3 strong *c*-reflections.

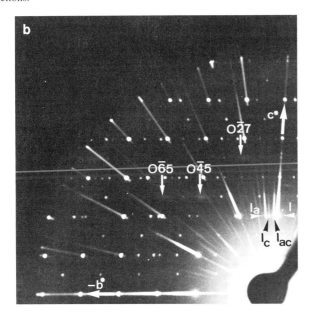

(b) Crystal no. 3 (basalt), heated. Exposure 356 hours. Arrows indicate 3 diffuse *c*-reflections. Arrow-*heads* only indicate reflections (004) of I; (00$\bar{4}$) of Ia; ($\bar{2}$04) of Ic; and (20$\bar{4}$) of Iac.

(c) Fragment of crystal no. 3, cut off after making the photograph shown in Fig. 2b. Exposure 336 hours. For notation see Fig. 2b.

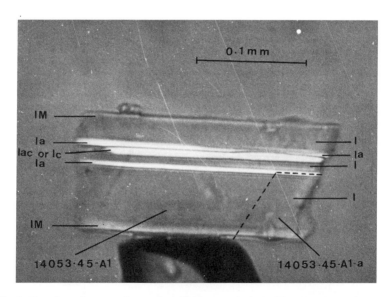

Fig. 3. Crystal no. 3 mounted on a glass fiber; crossed nicols. The (001) faces of the main mass individuals (I) lie in the plane of the photograph, with the *a* axis horizontal. Lamellae in albite (Ia), carlsbad (Ic) and albite-carlsbad (Iac) twin relation to the main mass are indicated, as well as inhomogeneous margins (IM) of lower An content. Afterwards the lower right corner has been cut off for further X-ray work (see Fig. 2c).

DISCUSSION

We have already presented a short account of the thermal history, as concluded from diffraction photographs of lunar Ca-rich plagioclase crystals before and after heat treatment under laboratory conditions (Czank *et al.*, 1972).

Thermal history

As previously stated, the *c*-reflections of all heat treated crystals were less intense and slightly more diffuse after heat treatment than before. Thus we came to the conclusion that these crystals must have been cooled down from about 1000°C in a much longer time than the cooling time provided in our experiment (i.e., from 1000°C down to 300°C in one hour).

It can be noted that a difference exists between the intensities of *c*-reflections of crystals nos. 3 and 7 both before and after heat treatment (Table 1), although their chemical composition appears to be almost the same. This may be due to some errors in the composition determination, which was especially difficult for crystal no. 7. The cause might also be a different thermal history.

Similar conclusions on the cooling rate of the basaltic rocks 14053 and 14310 also were presented by other authors. Ghose *et al.* (1972) studied clinopyroxenes of samples 14053,44 and 14310,115. They state that rock 14053 "has been quenched quickly above 1000°C" whereas the results for rock 14310 "indicate a slow cooling rate."

Finger *et al.* (1972) came to the conclusion, on the basis of pyroxene studies, that the original cooling rate of sample 14053,116 must have been quite slow; and that it must have been reheated afterwards to more than 840°C and then cooled down extremely rapidly. On the other hand the cooling rate of sample 14310,116 was fairly slow compared with rock 14053.

H. R. Wenk *et al.* (1972) conclude, on the basis of the diffuseness of *c*-reflections of anorthites from samples 14310,23 and 14310,95, that the plagioclases "have all been cooled rapidly" and that their structure is similar to that of plagioclases from terrestrial volcanic rocks.

The statements of Finger *et al.* and H. R. Wenk *et al.* are not in contradiction to our conclusions. It may appear that the results of Ghose *et al.* on rock 14053 are different from ours. However, they do not give any quantitative definition of the term "quenched quickly" and therefore a direct comparison with our results is not possible.

Twinning

The twin lamellae volume and intensity percentages of crystals nos. 3, 4, and 6 which were measured by optical and x-ray methods are summarized in Table 1. Even considering possible errors, it seems that for crystal no. 3 the x-ray and optical data are definitely not equal. This result easily could be interpreted by assuming sub-microscopic twinning according to the carlsbad and albite-carlsbad laws. Only in that case one would expect the ratio (Ic + Iac)/(I + Ia) to appear larger in the x-ray analysis as compared with the optical analysis.

As described above, the precession photograph of an almost untwinned fragment of crystal no. 3 (in which, besides the big lamellae I, only a small lamellae Ia was visible) shows reflection spots I, Ia, and Iac (Fig. 2c). Again we are tempted to try to explain these findings by making allowance for carlsbad and/or albite-carlsbad twinning on a submicroscopic scale. This is a surprising result. As far as we know submicroscopic twinning according to the carlsbad and/or albite-carlsbad law has not yet been reported for terrestrial plagioclases. Further work is in progress with respect to this phenomenon. A submicroscopic carlsbad and/or albite-carlsbad twinning may contribute to our understanding of the puzzling results of the optical investigations reported by E. Wenk *et al.* (1972).

Acknowledgments—We wish to thank all who have so kindly assisted us in carrying out this study. The preparation of samples for the microprobe analysis was most skillfully done by E. Schärli. All microprobe data were supplied by R. A. Gubser, who always was very patient with our various desires. We are indebted to C. Schäfer, L. Schultz, and P. Signer for their help with the heating of crystals. We wish to express our gratitude to H. Jagodzinski and M. Korekawa (München) as well as to E. Wenk and his coworkers (Basel) for valuable discussions. We are very grateful to NASA for the lunar basalt and breccia samples and to P. Signer (Zürich) for the fines.

References

Burri C. Parker R. L. and Wenk E. (1967) *Die optische Orientierung der Plagioklase.* Verlag Birkhäuser, Basel and Stuttgart.

Czank M. Girgis K. Harnik A. B. Laves F. Schmid R. Schulz H. and Weber L. (1972) Crystallographic studies of some Apollo 14 plagioclases (abstract). In *Lunar Science—III* (editor C. Watkins), pp. 171–173, Lunar Science Institute Contr. No. 88.

Finger L. W. Hafner S. S. Schürmann K. Virgo D. and Warburton D. (1972) Distinct cooling histories and reheating of Apollo 14 rocks (abstract). In *Lunar Science—III* (editor C. Watkins), pp. 259–261, Lunar Science Institute Contr. No. 88.

Gay P. (1953) The structures of the plagioclase felspars: III. An x-ray study of anorthites and bytownites. *Min. Mag.* **30,** 169–177.

Gay P. (1954) The structures of the plagioclase felspars. V. The heat-treatment of lime-rich plagioclases. *Min. Mag.* **30,** 428–438.

Gay P. and Taylor W. H. (1953) The structures of the plagioclase felspars. IV. Variations in the anorthite structure. *Acta Cryst.* **6,** 647–650.

Ghose S. Ng G. and Walter L. S. (1972) Clinopyroxenes from Apollo 12 and 14: Exsolution, cation order and domain structure (abstract). In *Lunar Science—III* (editor C. Watkins), pp. 300–302, Lunar Science Institute Contr. No. 88.

Laves F. and Goldsmith J. R. (1954a) Discussion on the anorthite superstructure. *Acta Cryst.* **7,** 131–132.

Laves F. and Goldsmith J. R. (1954b) Long-range-short-range-order in calcic plagioclases as a continuous and reversible function of temperature. *Acta Cryst.* **7,** 465–472.

Laves F. Czank M. and Schulz H. (1970) The temperature dependence of the reflection intensities of anorthite ($CaAl_2Si_2O_8$) and the corresponding formation of domains. *Schweiz. Mineral. Petrogr. Mitt.* **50,** 519–525.

Tsuboi S. (1923) A dispersion method of determining plagioclases in cleavage-flakes. *Min. Mag.* **20,** 108–122.

Weill D. F. McCallum I. S. Bottinga Y. Drake M. J. and McKay G. A. (1970) Mineralogy and petrology of some Apollo 11 igneous rocks. *Proc. Apollo 11 Lunar Sci. Conf., Geochim. Cosmochim. Acta,* Suppl. 1, Vol. 1, pp. 937–955. Pergamon.

Wenk E. Glauser A. Schwander H. and Trommsdorff V. (1972) Optical orientation, composition and twin-laws of plagioclases from rocks 12051, 14053, and 14310 (abstract). In *Lunar Science—III* (editor C. Watkins), pp. 794–796, Lunar Science Institute Contr. No. 88.

Wenk H. R. Ulbrich M. and Muller W. (1972) Lunar plagioclase (a mineralogical study) (abstract). In *Lunar Science—III* (editor C. Watkins), pp. 797–799, Lunar Science Institute Contr. No. 88.

Proceedings of the Third Lunar Science Conference
(Supplement 3, *Geochimica et Cosmochimica Acta*)
Vol. 1, pp. 615–621
The M.I.T. Press, 1972

On the amount of ferric iron in plagioclases from lunar igneous rocks

K. Schürmann and S. S. Hafner

Department of the Geophysical Sciences
The University of Chicago
Chicago, Illinois 60637

Abstract—Mössbauer spectra of ^{57}Fe in plagioclases from rocks 14053, 14310, and 15415 and from terrestrial anorthosites and basalts have been studied. The center of gravity of the total resonance absorption area has been tentatively interpreted in terms of the Fe^{3+}/Fe_{tot} ratio. The significantly different values can be correlated with oxygen partial pressure conditions during crystallization.

Introduction

IRON IS THE most abundant transition element in plagioclases of lunar rocks. Its distribution within individual crystals commonly is inhomogeneous and its concentration may range over more than an order of magnitude. Typical average values in plagioclases from different rocks range between approximately 0.01 and 0.04 iron atoms (predominantly as Fe^{2+}) per 8 oxygen atoms, which corresponds to approximately 0.03 and 1.2 wt% FeO. The Mössbauer spectra of ^{57}Fe exhibit a considerable complexity (Appleman *et al.,* 1971, Hafner *et al.,* 1971) that is undoubtedly related to the particular conditions of crystallization and the subsequent subsolidus cooling history of this mineral. Unfortunately, a precise analysis of the spectra is hampered because of the small concentrations of iron.

In this paper, we present a tentative simplified estimate of the Fe^{3+}/Fe^{2+} ratio in terms of the average isomer shift obtained from the center of gravity of the total resonant absorption area and add a few qualitative comments on the crystallization and subsequent cooling history of the plagioclase.

Technique

The resonant absorption effect of the strongest peak in the ^{57}Fe spectra of lunar plagioclases ranged between 0.002 (15415 anorthite) and 0.007 (these values should be compared with an effect of 0.20–0.22 for the strongest peak of a 0.001 inch thick metallic iron foil obtained with our apparatus under equivalent conditions). The count rates were 1.5–2.0 × 10⁷ counts per channel of a 400-channel analyzer that was operated in conjunction with a linear velocity generator (symmetric velocity wave form). Generally, our absorbers comprised 60–70 mg. plagioclase per cm². The density of the thinnest absorber studied (15415 anorthite) was 4 × 10⁻⁵ gm. natural iron per cm². The plagioclase separates were carefully handpicked to avoid superimposed peaks from foreign inclusions (e.g. pyroxene or ilmenite). The observed ^{57}Fe effects definitely were due to iron in the plagioclase crystal structures.

The spectra were fitted assuming a number N of Lorentzian lines with arbitrary positions, where N (usually 2 or 3) was increased until the fits were considered acceptable from a statistical point of view. The center of gravity C of the total resonance areas, interpreted in terms of an average isomer shift, is then

$$C = \sum_{i=1}^{N} V_i A_i / \sum A_i$$

where V_i is the position and A_i the area of peak i. The standard deviations of C were calculated from the standard deviations of V_i and A_i obtained from the least-squares fits of the spectra, using the error propagation law.

Results

Terrestrial calcium-rich plagioclases

The ^{57}Fe spectra of plagioclases from lunar basalts are significantly different from those of terrestrial plagioclases. The spectrum of plagioclase (An_{74}) of an anorthosite from the Stillwater Complex reveals one single Fe^{2+}-doublet, which is assigned to iron at Ca^{2+} sites. The resonant absorption effect of ^{57}Fe in this plagioclase is sufficiently large so that the spectrum can be analyzed with reasonable accuracy. The data are presented in Table 1 and the spectra at 77 and 295°K are shown in Fig. 1. The area ratio and the widths of the doublet peaks are independent of temperature between 77 and 295°K. This indicates that the recoil-free fraction of ^{57}Fe in this plagioclase cannot be significantly anisotropic. The thermal shift over

Table 1. ^{57}Fe hyperfine data of ($Ab_{26}An_{74}$) from Stillwater anorthosite[a]

Absorber temperature (°K)	Number of spectra[b]	Heat treatment (°C)	Absorption effect[c]	Quadrupole splitting (mm/sec)	Isomer shift[d] (mm/sec)
295	4	natural	0.0064	2.63 [0.02]	1.14 [0.02]
77	2	natural	0.0095	2.78 [0.02]	1.26 [0.02]
295	4	1000	0.0051	2.12 [0.02]	1.13 [0.02]

Absorber temperature (°K)	Heat treatment (°C)	Low velocity peak			High velocity peak		
		height[e]	width[f] (mm/sec)	area[e]	height[e]	width[f] (mm/sec)	area[e]
295	natural	0.54 (0.02)	0.44 (0.02)	0.60 (0.03)	0.46 (0.02)	0.35 (0.01)	0.40 (0.02)
77	natural	0.55 (0.01)	0.47 (0.02)	0.60 (0.03)	0.45 (0.02)	0.39 (0.02)	0.40 (0.02)
295	1000	0.60 (0.02)	0.70 (0.10)	0.62 (0.03)	0.40 (0.02)	0.65 (0.03)	0.38 (0.03)

[a] Numbers between parentheses: standard deviations obtained from the least-squares fits; numbers between brackets: estimated total errors.
[b] Positive and negative acceleration counted (and fitted) separately.
[c] Height of low velocity peak referred to background counting rate.
[d] Referred to metallic iron at 295°K.
[e] Referred to the sum of the two peaks.
[f] Full width at half height.

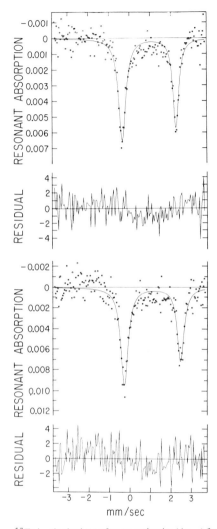

Fig. 1. *Upper spectrum:* ^{57}Fe in plagioclase of an anorthosite (An$_{74}$) from Stillwater Complex (Montana) at 295°K. The observed Fe^{2+} doublet is assigned to Ca^{2+} sites. The "residual" is the normalized difference between the two-line least-squares fit (solid curve) and the data. *Lower spectrum:* Same absorber at 77°K. Note that the low-velocity peak is more intense in both spectra.

this temperature range is $(-5.3 \pm 1.0) \times 10^{-4}$mm/sec°K. This value is in accord with typical thermal shifts for iron in silicates and oxides. The isomer shift at 77°K is consistent with Fe^{2+} in a regular octahedral or a distorted 7-fold oxygen coordination in silicates.

It should be noted that iron in the Stillwater plagioclase is predominantly in the ferrous state. Fe^{3+} peaks could not be resolved. However, the low velocity peak of the observed doublet is more intense and also exhibits an apparently broader width at 77°K as well as at 295°K. This is interpreted in terms of a superimposed weak Fe^{3+} doublet with a small quadrupole splitting, which cannot be resolved.

Spectra of plagioclase (An_{66}) from a Lake Superior anorthosite (Duluth Gabbro Complex) and of plagioclase (An_{56}) from the Kii lava flow (Hawaii), also were studied. These spectra are complicated by the superimposition of strong Fe^{3+} peaks in addition to the Fe^{2+} patterns. The resonant absorption effect was too low to allow analysis of the complex spectra in detail, but the significant increase of the ratio Fe^{3+}/Fe^{2+} compared to the Stillwater plagioclase is obvious in these spectra.

Two samples of the Stillwater plagioclase were heated at 1000 and 1123°C in evacuated quartz tubes (10^{-5} mm Hg) for two days and successively measured at room temperature. After the heat treatment, the spectra were significantly different and showed a remarkable similarity to the plagioclase spectra of lunar basalts. The high velocity peak apparently was broadened by the generation of various overlapping peaks, although the effect of broadening was less marked than in lunar plagioclases (e.g., 3-line fits were not successful). The center of gravity of the total resonant absorption area was *not* shifted in the heated samples (cf. Table 2). Therefore, the changes in the spectra cannot be attributed to oxidation of Fe^{2+}. They probably result from a redistribution of Fe^{2+} among distinct positions in the crystal structure.

Table 2. ^{57}Fe data and estimated $Fe^{3+}/(Fe^{2+} + Fe^{3+})$ ratios (spectra at 295 K)

Plagioclase	Anorthite (mol. %)	Wt % Fe absorption effect	Heat treatment (°C)	Center of gravity[a] (C)	Fe^{3+}/Fe_{tot}[b] (a)
Stillwater anorthosite	74	0.41 0.0064	natural 1000 1123	0.90(3) 0.90(2) 0.97(3)	0.25(3) 0.25(2) 0.18(3)
Lake Superior anorthosite	66	0.42 0.0015	natural	0.76(6)	0.40(6)
Hawaii Kii lava flow	56	0.66 0.0028	natural	0.59(4)	0.57(4)
10044,25 basalt	88–92	0.31 0.0065	natural	1.04(7)	0.10(7)
12021,150 basalt	88–90	0.41 0.0070	natural	1.02(3)	0.12(3)
14053,47 basalt	77–90[c]	0.25–0.65[c,d] 0.0055	natural 1123	1.05(3) 1.05(3)	0.09(3) 0.09(3)
14310,116 basalt	92–93[c]	0.28–0.47[c,e] 0.0040	natural	1.03(3)	0.11(3)
15415,48 anorthosite	97–99[f]	0.08–0.09[f] 0.0024	natural	1.10(4)	0.04(4)

[a] Referred to a metallic iron absorber; numbers between parentheses are standard deviations.

[b] Tentative estimate using equation (1): 1.14 mm/sec for Fe^{2+} and 0.18 mm/sec for Fe^{3+} as isomer shifts were assumed; in Table 3 of Finger *et al.* (1972) a shift of 0.22 mm/sec was assumed for Fe^{3+} (the effect of this difference is very small). The *a* values give solely the trend. For 10044,25 *a* was estimated to be between 0.02 and 0.1 (Hafner *et al.*, 1971); numbers between parentheses are standard deviations.

[c] Gancarz *et al.* (1971).

[d] Quaide (1972).

[e] Walter *et al.* (1972).

[f] Hargraves and Hollister (1972), Steele and Smith (1971), Stewart *et al.* (1972).

Lunar plagioclases (14053, 14310, 15415)

In contrast to the unheated Stillwater plagioclase, the lunar plagioclases studied thus far show at least two fairly distinct Fe^{2+} peaks in the high velocity region that were assigned to two distinct Fe^{2+} positions with different coordination number (Appleman *et al.*, 1971, Fig. 1; or Hafner *et al.*, 1971, Fig. 1). One sample was heated at 1000°C for two days, but no significant change was observed in the spectrum after the heat treatment (cf. Table 2). Hafner *et al.* (1971) suggested, on the basis of a two Fe^{2+} doublet analysis of spectra taken at 77 and 295°K, that the recoil-free fractions at the two assumed positions are probably anisotropic. In addition a significant contribution of Fe^{3+} to the spectrum was indicated, and the ratio $Fe^{3+}/(Fe^2 + Fe^{3+})$ was estimated to be between 0.02 and 0.1. However, the resonant absorption effects in lunar plagioclases studied until this time were generally too poor to allow an analysis of the spectrum in full detail. Fe^{3+} may be present in the form of a complex pattern that is unknown, and if its concentration is high it must be accounted for in a precise analysis of the Fe^{2+} splittings. At any rate, the lunar plagioclase spectra are significantly different from the unheated Stillwater plagioclase, but they show a certain resemblance to the latter heated at 1000°C or higher. The line broadening in the heated samples from Stillwater, of course, cannot be due to Fe^{2+} at tetrahedral sites; it seems to indicate positional disorder of iron after the heat treatment. The Fe^{2+} positions in plagioclase seem not to be restricted to regular atomic sites. They may be accidental sites, probably associated with dislocations, vacancies, or defects in the crystal structures. Anisotropic recoil-free fractions at some of the iron positions almost certainly are present, but their intrinsic properties are not known.

Ferric to ferrous iron ratio

The center of gravity C of the total resonant absorption area (at 295°K) tentatively is interpreted in terms of an average isomer shift c that results from the superimposition of Fe^{2+} and Fe^{3+} doublets using the relationship

$$C = ac_{Fe^{3+}} + (1 - a)c_{Fe^{2+}} \tag{1}$$

where a is the ratio $Fe^{3+}/Fe^{2+} + Fe^{3+}$. The presence of distinctly anisotropic recoil-free fractions and distinct thermal shifts at the different iron positions are neglected in this simplified estimate. The center of gravity C of the spectrum of plagioclase from basalt 10044 was not significantly affected by temperature between 77 and 295°K. Its thermal shift was approximately -2×10^{-4} mm/sec°K. In equation (1), the correct correlation between C and the oxidation state of iron in plagioclases is not yet known and precise ratios for $Fe^{3+}/(Fe^{2+} + Fe^{3+})$ cannot be obtained. However, the differences between samples of different origin should exhibit the correct trend. As isomer shift for Fe^{2+}, the value of 1.14 mm/sec obtained from the Stillwater plagioclase was used. For Fe^{3+}, which most probably is substituted for Al^{3+} at tetrahedral sites, an isomer shift of 0.18 mm/sec was assumed. This value was accurately determined for Fe^{3+} in ferri-diopside (Hafner and Huckenholz, 1971), and it is known that the Fe^{3+} shifts at tetrahedral sites in oxides are not very sensitive to the crystal structures of silicates. Besides, a change of 20%, for example, in the Fe^{3+}

shift would affect the $Fe^{3+}/(Fe^{2+} + Fe^{3+})$ ratio in the lunar plagioclases only slightly (cf. Table 2, footnote b). If it is assumed that approximately one third of ferrous iron is in fact located at tetrahedral sites with an isomer shift of 0.95 mm/sec as suggested previously (Hafner *et al.*, 1971), the estimated $Fe^{3+}/(Fe^{2+} + Fe^{3+})$ ratios in Table 2 would be reduced by approximately 50%. However, the spectra of the heated Stillwater samples (cf. also Hafner *et al.*, 1971) seem to indicate that tetrahedral Fe^{2+} may be less than one third of total iron in the plagioclase.

DISCUSSION

It is difficult to estimate reliably the error limits to the absolute values in Table 2. We believe that the total errors are within approximately ± 0.05. Fe^{3+} in lunar plagioclases has been positively identified in electron paramagnetic resonance spectra (Kolopus *et al.*, 1971, Weeks *et al.*, 1972). Weeks (1972) reported a value for $Fe^{3+}/Fe_{tot} = 0.008 - 0.12$ for a plagioclase separate from 14053,47, which is in agreement with our own value within the experimental error. Weeks' study of the Stillwater sample yielded an approximately 5 times greater ratio than that of 14053.

Despite the probably large systematic error in our estimates, the trends in Table 2 appear reasonable. For example, in anorthosites from the Stillwater Complex (Montana), magnetite is absent in the exposed lower levels (Hess 1960, McCallum 1968) whereas anorthosites from the Duluth Gabbro Complex commonly have magnetite as an interstitial mineral phase (Phinney 1969). McCallum (1968) suggested $p(O_2) \sim 10^{-12}$ at 1000°C for the Stillwater anorthosites, whereas for lunar mare basalts $p(O_2)$ was estimated to be 10^{-15} to 10^{-16} at same temperature (Haggerty, 1972). For Hawaiian lava flows values of 10^{-8} to 10^{-10} were reported for 1100–1000°C (Anderson and Wright 1972). Thus it appears that the ratio Fe^{3+}/Fe_{tot} in plagioclases indeed reflects the partial pressure of oxygen during crystallization. It should be noted that we find no appreciable difference in the Fe^{3+}/Fe_{tot} ratios in the plagioclases from various mare basalts, whereas Bence and Papike (1972) suggested a significantly lower partial oxygen pressure during the final crystallization stages of basalt 14053 compared to 14310.

In addition, the Mössbauer spectra of plagioclase from the Stillwater anorthosites show that, in the case of very slow cooling and mostly undisturbed crystallization from a magma, Fe^{2+} can be located primarily at Ca^{2+} sites. But the plagioclase spectrum of lunar anorthosite 15415 is indistinguishable from the plagioclase spectra of lunar basalts. They show the same pattern of broad lines, in contrast to the Stillwater plagioclase, and one can assume that the crystallization of 15415 was more of a "basaltic nature", i.e., rapid crystallization and rapid cooling. The crystallization was, of course, also drier. Thus, although the petrology of 15415 indicates a "plutonic" provenance (Hargraves and Hollister, 1972), there may have been some later process, which destroyed the originally ordered Fe^{2+} location as expected in a slowly cooled and undisturbed plutonic or metamorphic complex.

Acknowledgments—We wish to thank Drs. A. T. Anderson, W. L. Griffin and R. A. Weeks for exchange of samples, Miss Barbara Janik for her assistance, and NASA for financial support.

REFERENCES

Anderson A. T. and Wright T. L. (1972) Phenocrysts and glass inclusions and their bearing on oxidation and mixing of basaltic magmas, Kilauea Volcano, Hawaii. *Amer. Mineral.* **57,** 188–216.

Appleman D. E. Nissen H.-U. Stewart D. B. Clark J. R. Dowty E. and Huebner J. S. (1971) Studies of lunar plagioclases, tridymite, and cristobalite. *Proc. Second Lunar Sci. Conf., Geochim. Cosmochim. Acta,* Suppl. 2, Vol. 1, pp. 117–133. M.I.T. Press.

Bence A. E. and Papike J. J. (1972) Crystallization histories of pyroxenes from lunar basalts (abstract). In *Lunar Science—III* (editor C. Watkins), pp. 59–61, Lunar Science Institute Contr. No. 88.

Finger L. W. Hafner S. S. Schürmann K. Virgo D. and Warburton D. (1972) Distinct cooling histories and reheating of Apollo 14 rocks (abstract). In *Lunar Science—III* (editor C. Watkins), pp. 259–261, Lunar Science Institute Contr. No. 88.

Gancarz A. J. Albee A. L. and Chodos A. A. (1971) Petrologic and mineralogic investigation of some crystalline rocks returned by the Apollo 14 mission. *Earth Planet. Sci. Lett.* **12,** 1–18.

Hafner S. S. and Huckenholz H. G. (1971) Mössbauer spectrum of synthetic ferri-diopside. *Nature Phys. Sci.* **233.** 9–10.

Hafner S. S. Virgo D. and Warburton D. (1971) Oxidation state of iron in plagioclase from lunar basalts. *Earth Planet. Sci. Lett.* **12,** 159–166.

Haggerty S. E. (1972) Subsolidus reduction and compositional variations of lunar spinels (abstract). In *Lunar Science—III* (editor C. Watkins), pp. 347–349, Lunar Science Institute Contr. No. 88.

Hargraves R. B. and Hollister L. S. (1972) Mineralogic and petrologic study of lunar anorthosite slide 15415, 18. *Science* **175,** 430–432.

Hess H. H. (1960) Stillwater igneous complex, Montana: A quantitative mineralogical study. *Mem. Geol. Soc. Amer.* **80,** 230 pp.

Kolopus J. L. Kline D. Chatelain A. and Weeks R. A. (1971) Magnetic resonance properties of lunar samples: mostly Apollo 12. *Proc. Second Lunar Sci. Conf., Geochim. Cosmochim. Acta,* Suppl. 2, Vol. 3, pp. 2501–2514. M.I.T. Press.

McCallum I. S. (1968) Equilibrium relationships among the coexisting minerals in the Stillwater complex, Montana. Ph.D. thesis, University of Chicago, 175 pp.

Phinney W. C. (1969) Anorthosite occurrences in Keweenawan rocks of northeastern Minnesota. *New York State Mus. Sci. Serv.—Mem.* **18,** 135–147.

Quaide W. (1972) Fe and Mg in bytownite-anorthite plagioclase in lunar basaltic rocks (abstract). In *Lunar Science—III* (editor C. Watkins), pp. 624–626, Lunar Science Institute Contr. No. 88.

Steele I. M. and Smith J. V. (1971) Mineralogy of Apollo 15415 ("genesis rock"): Source of anorthosite on moon. *Nature* **234,** 138–140.

Stewart D. B. Ross M. Morgan B. A. Appleman D. E. Huebner J. S. and Commeau R. F. (1972) Mineralogy and petrology of lunar anorthosite 15415 (abstract). In *Lunar Science—III* (editor C. Watkins, pp. 726–728, Lunar Science Institute Contr. No. 88.

Walter L. S. French B. M. and Doan A. S. Jr. (1972) Petrographic analysis of lunar samples 14171 and 14305 (breccias) and 14310 (melt rock) (abstract). In *Lunar Science—III* (editor C. Watkins), pp. 773–775, Lunar Science Institute Contr. No. 88.

Weeks R. A. (1972) Paramagnetic resonance spectra of Ti^{3+}, Fe^{3+} and Mn^{2+} in lunar plagioclases. *Transact. Amer. Geophys. Union* **53,** 551.

Weeks R. A. Kolopus J. L. and Kline D. (1972) Magnetic phases in lunar material and their electron magnetic resonance spectra: Apollo 14 (abstract). In *Lunar Science—III* (editor C. Watkins), pp. 791–793, Lunar Science Institute Contr. No. 88.

Proceedings of the Third Lunar Science Conference
(Supplement 3, *Geochimica et Cosmochimica Acta*)
Vol. 1, pp. 623–643
The M.I.T. Press, 1972

Metamorphism of Apollo 14 breccias

Jeffrey L. Warner

National Aeronautics and Space Administration
Manned Spacecraft Center, Houston, Texas 77058

Abstract—All Apollo 14 breccias for which thin sections are available (27 rocks) were studied petrographically. The samples display a range in matrix mineralogy and texture from glassy breccias with a detrital matrix, through intermediate members with little glass and an annealed matrix, to glass-free breccias with an interlocking, euhedral matrix. Based on the abundance of matrix glass, the abundance of glass clasts, and the matrix texture, the 27 samples are classified into eight groups fitting into three metamorphic grades. Other petrographic features are correlated with this series.

From one sample in each of six groups, electron microprobe analyses were performed on matrix (less than 25 μ) plagioclase, pyroxene, and olivine and of rims and cores of plagioclase and pyroxene clasts (50–300 μ).

Matrix plagioclase, pyroxene, and olivine in the high-grade rocks display a narrow range of compositions (approximately An_{75-80}; $Wo_5En_{60-80}Fs_{15-35}$; and Fo_{65-80}), suggesting an equilibrated or metamorphic origin. However, in the low-grade rocks a wide range of compositions, suggesting a detrital or unequilibrated origin, is observed. Partially equilibrated states are observed in medium-grade samples.

In high-grade rocks, essentially all pyroxene and plagioclase clasts display a homogeneous core and a 1–5 μ rim. The rim has the "equilibrated composition" regardless of the core composition. In the low-grade rocks no rims were found, and about half of the clasts in the medium-grade rocks display rims.

It is concluded that the Apollo 14 breccias formed as part of the Imbrium Basin ejecta blanket—the Fra Mauro formation. The continuous nature of the metamorphic series suggests that the medium- and high-grade breccias were derived from unmetamorphosed equivalents by autometamorphism in the thick, hot Fra Mauro formation at the Apollo 14 site.

Introduction

The surface of the moon is dominated by craters and their continuous ejecta blankets. Breccias, samples of the ejecta blankets, contain a wide variety of fragments. In order to interpret the history of the fragments, and thus of the moon, the breccias must be understood. This study concerns the metamorphism of Fra Mauro breccias and considers the history of the Imbrium Basin ejecta blanket.

All Apollo 14 breccias for which thin sections exist (27 samples) have been studied petrographically, concentrating on the matrix mineralogy and texture. As is the case with Apollo 11 and 12 breccias, Apollo 14 breccias are a complex aggregate of glass, mineral, breccia, and igneous rock fragments. The fragments range in size from lithic clasts, more than 10 cm across down to glass and mineral grains less than 1μ. Those fragments less than 25 μ across are designated matrix (following Chao *et al.*, 1971); larger fragments are considered clasts.

The Apollo 14 breccias form a continuous series from unrecrystallized breccias, through partly recrystallized breccias, to fully recrystallized breccias. The breccias recrystallized, or metamorphosed, in the Imbrium Basin ejecta blanket. Those breccias that display no recrystallization are referred to as glassy or low-grade breccias; they are similar to Apollo 11 and 12 breccias. Partly and fully recrystallized breccias are referred to as medium-grade breccias and high-grade breccias, respectively. This matrix metamorphic classification of Apollo 14 breccias is in agreement with the coherency-clast color index classification of Jackson and Wilshire (1972).

Microprobe analyses of plagioclase, pyroxene, and olivine were performed to determine the reactions involved in breccia metamorphism. Chemical data were gathered from one low-grade breccia (14307), one medium-grade breccia (14318) and four high-grade breccias (14006, 14066, 14068, and 14311). As the metamorphism increased, first mineral clasts developed rims, second matrix olivine and pyroxene equilibrated, and third matrix plagioclase equilibrated.

Lunar breccias also interpreted as having metamorphic textures have been described from the KREEP fragments of Apollo 12 by Meyer *et al.* (1971) and Anderson and Smith (1971).

This study of matrix metamorphism should be considered a companion to the clast population studies of Jackson and Wilshire (1971) and Wilshire and Jackson (1971). Pre-Imbrian history and local stratigraphy may be interpreted from the clasts, whereas the thermal regime within the Imbrian ejecta blanket (the Fra Mauro formation) may be interpreted from the metamorphic reactions.

Description of Petrologic Groups

Each of the 27 breccias has been assigned to one of eight mineralogic-textural groups that are defined on (1) amount of matrix glass, (2) amount of glass clasts, and (3) texture of the matrix. The groups, designated 1 to 8, form a series in order of increasing differences from Apollo 11 and 12 breccias (excluding 12013). The chief characteristics of each group, along with sample assignments, are set out in Table 1.

Table 1. Mineralogic-textural characteristics of the breccia groups.

Grade	Group	Matrix glass amount[1]	Glass clasts amount[2]	Matrix texture	Sample numbers (14XXX)			
Low	1	Abundant	Abundant	Detrital	042 313	047	055	307
	2	Intermediate	Abundant	Detrital	049	301		
Medium	3	Low	Intermediate	Detrital	063	082	315	318
	4	None	Low	Partly annealed	171	306	321	
High	5	None	None	Equant	311			
	6	None	Low	Equant to euhedral	006 320	303	304	305
	7	None	Low	Euhedral	(065) 314	066 319	270	312
	8	None	None	Sheath-like	068			

[1] In matrix glass amount:
  Present—dark brown, >50% isotropic material.
  Intermediate—light brown, <50% isotropic material.
  Low—gray, <5% isotropic material.
  None—gray, trace of isotropic material.
[2] In glass clast amount:
  Present—>10% glass clasts, little devitrification.
  Intermediate—<10% glass clast, about half are devitrified.
  Low—<2% glass clasts, show major devitrification.
  None—no isotropic material present in clasts.

Group 1 samples are similar to Apollo 11 and 12 breccias in that they contain abundant glass, both as clasts and in the matrix, and a detrital matrix texture. Progressively higher numbered grades differ first by less and less matrix glass, second by fewer glass clasts, third by displaying increasingly more recrystallized matrix textures, and fourth by a sheath-like texture. Group 1 and 2 samples are referred to as low grade or glassy, Group 3 and 4 samples are referrred to as medium grade, and Group 5, 6, 7, and 8 samples are referred to as high grade or recrystallized.

Definition of textural terms

The term "detrital" or "unrecrystallized matrix texture" (a in Figs. 1–3) refers to rocks in which the matrix contains abundant glass and displays no evidence of recrystallization. There is a continuum of grain sizes, for each phase and for the total rock, ranging from less than a micron to more than 10 cm. Individual grains are angular to subrounded and have irregular shapes. This texture suggests an origin by comminution and accumulation.

The term "equant matrix texture" (c in Figs. 1–3) refers to rocks in which the matrix contains no glass and the silicate and oxide phases form a mosaic pattern. There are distinct discontinuities in the grain size distribution: Ilmenite crystals are approximately 1μ across; plagioclase and pyroxene crystals are approximately 3 to 6μ across. There are no crystals in the 1 to 3 and 7 to 25μ size ranges. Ilmenite crystals have the shape of rounded prisms, and the silicate crystals tend to form regular polygons. This texture suggests an origin by solid state recrystallization, presumably from a detrital texture.

The term "euhedral matrix texture" (e in Figs. 1–3, Fig 5) refers to rocks in which the matrix contains no glass and the minerals form a lath-shaped pattern. A euhedral matrix differs from an equant matrix in the shape and size of the silicate minerals. Plagioclase and pyroxene form lath-shaped crystals ranging in size from 2×10–20μ to $5 \times 30 \mu$. In any one rock, however, the range of sizes is small. As is the case with the equant texture, a euhedral texture suggests an origin by solid state recrystallization, presumably from an equant texture.

The term "sheath-like matrix texture" (f in Figs. 1–3, Fig. 6) refers to rocks in which the matrix contains no glass and plagioclase and pyroxene form a sheath-like or spherulitic network of lath-shaped crystals ($5 \times 150 \mu$). Skeletal olivine prisms ($20 \times 40 \times 150$–200μ) are scattered throughout the spherulitic network. Opaque minerals form 1μ rounded blebs in pyroxene, and there is no mesostatsis. By analogy with experimental work on devitrification (Lofgren, 1971), a sheath-like texture suggests an origin by rapid crystallization from a supercooled melt containing olivine crystallites.

Groups 1, 2, and 3

Group 1 breccias, similar to the Apollo 11 and 12 breccias, are dark honey-brown in thin section; the color is caused by the abundance of brown glass in the matrix (more than 50%). Irregular glass clasts and yellow, red, and brown glass spheres are present ($> 10\%$); few show devitrification. Lithic and mineral clasts display sharp

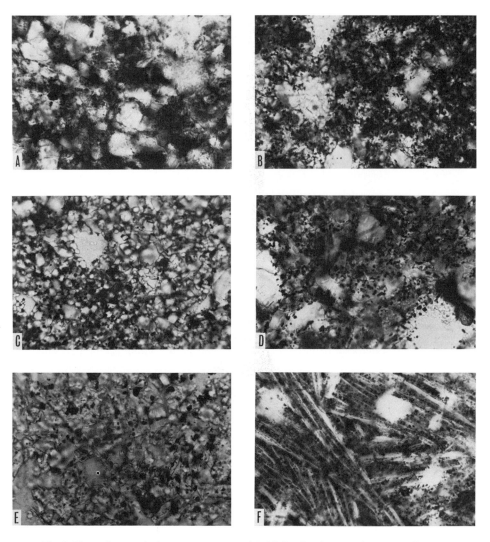

Fig. 1. Photomicrographs in convergent transmitted light of various matrix textures. Longer dimension of image is 220 μ. (a) Sample 14047, Group 1, detrital; (b) sample 14092, Group 4, partly annealed; (c) sample 14311, Group 5, equant; (d) sample 14303, Group 6, equant to euhedral; (e) sample 14066, Group 7, euhedral; (f) sample 14068, Group 8, sheath-like.

borders in a detrital matrix. Pore space, making up about 25 % of the rock, is in the form of intergranular cavities (the dark parts in Fig. 2a) approximately 1 μ across.

Group 2 breccias differ from Group 1 breccias only by a lesser content of matrix glass (between 15 and 50%), resulting in a lighter brown color in thin section. Because counting a mode of the matrix is so difficult, the assignment of a rock to Group 1 or 2 is subjective and prone to error; thus the Group 1–2 boundary is relatively indistinct.

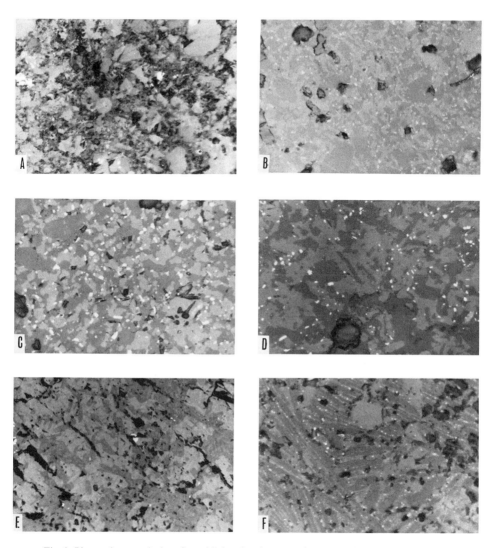

Fig. 2. Photomicrographs in reflected light of various matrix textures. Longer dimension of image is 220 μ. (a) Sample 14042, Group 1, detrital; (b) sample 14321, Group 4, partly annealed; (c) sample 14311, Group 5, equant; (d) sample 14303, Group 6, equant to euhedral; (e) a clast in sample 14319, Group 7, euhedral; (f) sample 14068, Group 8, sheath-like.

Group 3 breccias have a pale brown matrix, the color being due to about 5% brown glass. Irregular glass clasts, glass spheres, and mineral clasts are as in Groups 1 and 2. Lithic clasts display ragged borders. Rocks 14315 and 14318 contain an unusually high content of > 1 mm lithic clasts (over 50%), and the large clasts display a faint foliation. Group 3 rocks display a detrital matrix texture. Pore space forms inter-granular cavities as in Group 1 and 2 rocks.

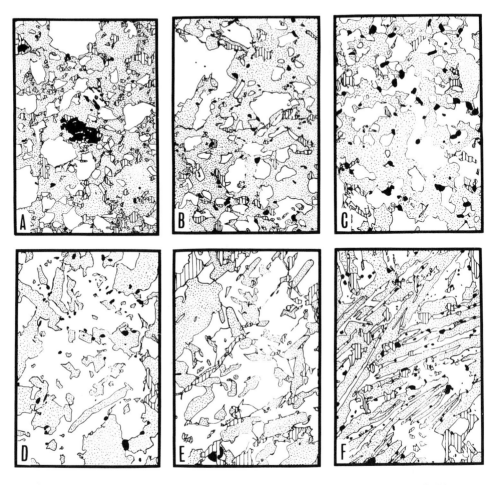

Fig. 3. Sketches of the various textures. Longer dimension of sketch is 190 μ. (a), (c), (d), (e), and (f) as for Fig. 2; (b) sample 14171, Group 4. Clear areas are pyroxene plus olivine, dotted areas are feldspar plus glass; black areas are opaque minerals; and lined areas are vugs and polishing artifacts.

The white rocks (14063, 14082) do not fit well into the proposed classification. They contain matrix glass (1 to 3%), but it is colorless rather than brown. Thus, the rocks are light gray rather than brownish. The rocks contain a few (2–3%) lithic and glass clasts (1–2%). The glass clasts are all pale brown. However, these rocks have a detrital matrix texture (Fig. 4). Plagioclase, orthopyroxene, and pink spinel are the most abundant clast types; the feldspar commonly displays ragged boundaries. The white rocks are assigned to Group 3 even though they do not match the other members of this group in all aspects. Although the white rocks must be chemically different from the other breccias, a major element analysis is not yet available.

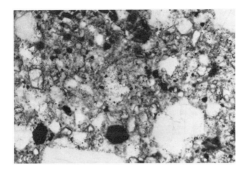

Fig. 4. Photomicrograph illustrating the detrital matrix of sample 14063, a white rock. Longer dimension of image is 560 μ.

Group 4

Group 4 breccias contain no matrix glass and less than 5 % irregular glass clasts, many of which show some devitrification. No glass spheres were observed, but some spherical objects that have been interpreted as "chondrules" by others (King *et al.*, 1972a, b; Kurat *et al.*, 1972) may be devitrified glass spheres. Mineral and lithic clasts display both sharp and diffuse borders. Pink spinel occurs in these rocks. The matrix has a partially annealed texture (b in Figs. 1–3), that is, the texture is not detrital, yet it is not as organized as an equant or euhedral texture. Pore space is present both as micron-sized intergranular cavities and as > 30 μ vugs.

A detailed study of 14321 is now in progress. A re-evaluation of the matrix texture in light of the Grieve *et al.* (1972) description of microbreccia 3 and light matrix tentatively suggests that the texture is more organized than stated previously. Perhaps more detailed work on the rocks in this group will demonstrate that groups 4 and 5 should be switched.

Groups 5, 6, and 7

Groups 5, 6, and 7 rocks contain no matrix glass. Irregular glass clasts are rare (<2%) and are highly devitrified. "Chondrules" are present. Mineral and

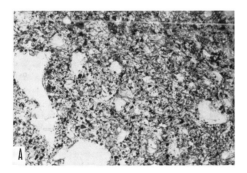
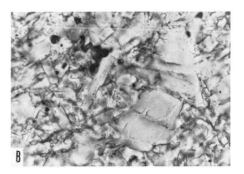

Fig. 5. Photomicrographs of the euhedral matrix texture. (a) General view of sample 14066, longer dimension of photo equals 560 μ; (b) a coarse-grained part of sample 14303, longer dimension of image is 220 μ.

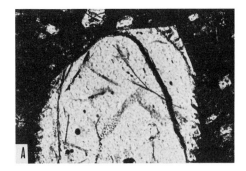
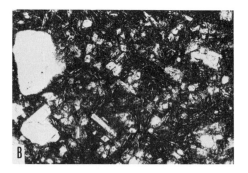

Fig. 6. Photomicrographs of sample 14068. (a) Olivine clast with a euhedral overgrowth, longer dimension of photo equals 560 μ; (b) general view of rock showing skeletal olivine prisms, longer dimension of image is 1.1 mm.

lithic clasts commonly display ragged borders. The matrix tends to be patchy under low magnifications—i.e., the matrix consists of lighter and darker regions. These regions are approximately $\frac{1}{2}$ mm across and are irregular in shape. The lighter regions are coarser grained, the lighter color being due to fewer, but larger, ilmenite crystals (Fig. 1c). This phenomenon is especially well developed in sample 14311. Pore space in the form of intergranular cavities occurs sporadically, but vugs are common. The pink spinel occurs in most samples in these groups.

The difference between groups 5, 6, and 7 is based entirely on matrix texture:

(1) Group 5 has equant texture (c in Figs. 1–3).

(2) Group 6 has equant to euhedral texture (d in Figs. 1–3).

(3) Group 7 has euhedral texture (e in Figs. 1–3).

The assignment of a rock to one of these groups is subjective and prone to error; thus the Group 5–6 and the Group 6–7 boundaries are relatively indistinct.

Group 8

Group 8 is represented by one sample (14068) that contains 17.6 wt% MgO (Hubbard *et al.*, 1972), twice as much as the other breccias. This chemical difference shows up in the mode—14068 contains approximately 20% olivine both as skeletal crystals in the matrix and as clasts. No matrix glass is present, and partly devitrified glass clasts are rare. Mineral and lithic clasts have ragged borders; in one case there is an obvious overgrowth on an olivine clast (Fig. 6a). Group 8 displays a sheath-like texture (e in Figs. 1–3, Fig. 6b) and contains vugs.

Discussion

As pointed out previously, some of the group boundaries are indistinct, but the general position of a sample in the sequence is accurate. For example, sample 14318 was originally assigned to group 4 (Warner, 1972), but restudy of 14318 has led to reassigning that sample to Group 3. Many thin sections of several samples were studied to determine if the metamorphic group was the same for the entire rock. Sections were studied that covered about 75% of a slice through 14307, a small rock; petrographic examination of all sections indicated that 14307 belongs in Group 1. Sections were studied that formed a strip through 14311, a large rock; all sections indicated that 14311 belonged to either Group 5 or 6. As many as 15 sections of several other rocks were studied, but not in a systematic way. In each case, all sections implied the same or adjacent groups. It was concluded that the samples are homogeneous as to group assignment.

Clast stratigraphy has been extensively studied by Grieve *et al.* (1972) in sample 14321 and by Wilshire and Jackson (1972) for all samples. My observations in this area were limited, but the generalization that clasts have a higher or equal group assignment than their host is true. This holds if the host is the matrix of the sample or if the host is a clast itself. This generalization is contradicted by only one clast-in-clast observation of Wilshire and Jackson.

METAMORPHIC ORIGIN

The mineralogic-textural groups 1 to 8 define a series from glassy, unrecrystallized breccias through nonglassy, recrystallized breccias, to a devitrification breccia. It is suggested that the different groups represent samples that have been metamorphosed to various degrees. Because the groups are based on matrix characteristics, it is believed that this metamorphism took place in the hot ejecta blanket resulting from the formation of the Imbrium Basin, i.e. in the Fra Mauro formation. Using various geothermometers, Grieve *et al.* (1972) have calculated a temperature of 700°C for a Group 4 rock (14321), and Grieve *et al.* (1972) and Anderson *et al.* (1972) have calculated a temperature of about 1000°C for breccia clasts that would fit into Groups 6 and 7. The concept that the Apollo 14 breccias were subjected to an autometamorphism in the Imbrian ejecta blanket is supported by the sample homogeneity and clast stratigraphy outlined above. The isochemical character of the majority of the Apollo 14 breccia suite is also consistent with the concept that glassy, unrecrystallized breccias were metamorphosed into recrystallized breccias. Thus, the Apollo 11 and 12 breccias and the Apollo 14 Group 1 and 2 (low grade) breccias are unmetamorphosed; the Group 2 and 3 (medium grade) breccias are partly metamorphosed at perhaps 700°C; the Group 5, 6 and 7 (high grade) breccias are fully metamorphosed at about 1000°C, and the Group 8 rock probably was melted at a still higher temperature.

There is a possibility that the temperature order of Groups 5–8 may be reversed. Experimental work on crystallization from a glass (e.g., Lofgren, 1971) shows that the higher the temperature of crystallization, the more equant the crystals; the lower the temperature, the more feathery or spherulitic the crystals. This data would apply only if the matrix texture is the result of direct crystallization from a glass, that is, a higher-grade rock did not go through the medium-grade states. Then the suggestion would be that the equant texture rock (Group 5) represents higher temperature than the lath-like rocks (Group 6 and 7), which in turn represent higher temperatures than the spherulitic rock (Group 8).

There are several petrologic, mineralogic, and chemical features besides content of matrix glass, amount of glass clasts, and matrix texture that correlate with metamorphic grade. Chemical equilibrium during metamorphism involving equilibration of matrix silicates and growth of rims on plagioclase and pyroxene clasts is set out in detail in a later section of this paper. Other phenomena are outlined in the next few paragraphs, though I have not studied most of them in detail.

Metal particle grain size

By studying the viscous remanent magnetization, Gose *et al.* (1972) were able to determine the size of metal particles in the matrix. Low-grade rocks are dominated by very small sizes (mostly $<500\,\text{Å}$), medium-grade rocks show medium grain sizes (as much as $1\,\mu$), and high-grade rocks show large grain sizes (several microns). This systematic increase in grain size may be explained by progressive metamorphic recrystallization.

Pore space

All breccias have between $\frac{1}{5}$ to $\frac{1}{3}$ volume $\%$ pore space. In the low-grade breccias it takes the form of intergranular porosity, whereas most of the high-grade rocks display vugs that contain a vapor deposited assemblage (McKay *et al.*, 1972). The change in the form of the pores may be interpreted as "recrystallization" of the pores with increasing metamorphic grade.

K-rich glass blebs

K-rich glass blebs exist in medium and high grade rocks (Anderson *et al.*, 1972; Dence *et al.*, 1972). It is suggested that these blebs are the lowest melting fraction of the rocks. The metamorphic temperature was high enough to produce these blebs by a partial melt or "sweating-out" process. This process has been considered for the "granitic" component of sample 12013 by Drake *et al.* (1970). The glass did not devitrify completely during cooling because reactions are very slow in a lowering temperature environment.

Olivine reaction rims

In the high-grade rocks some olivine clasts display a rim of orthopyroxene plus ilmenite. This may result from a reaction between the matrix and the olivine clasts.

Pink spinel

The spinel described by Mason *et al.* (1972) is found in the white rocks of Group 3 and the high-grade rocks. This mineral may be an indicator of metamorphic grade as staurolite is in terrestrial pelitic schists.

Total carbon and solar wind gases

Total carbon abundances (Moore *et al.*, 1972) and noble gas contents (Bogard and Nyquist, 1972; LSPET, 1971) correlate with metamorphic grade; low-grade rocks have large amounts, medium-grade rocks have moderate amounts, and high-grade rocks have very low amounts. These relations are only approximate. There is also a correlation between the noble gas content and coherency; the friable samples tend to have more gas than the coherent samples. It is suggested that these constituents were released during the high temperature recrystallization of the breccias. The very high gas samples, the friable ones, may be later surface breccias.

CORRELATION TO OTHER STUDIES

Jackson and Wilshire (1972) have classified the Apollo 14 breccias into four groups on the basis of coherency and color index of visible clasts. They called their groups F1 (friable, light clasts), F2 (coherent, light clasts), F3 (friable, dark clasts), and F4 (coherent, dark clasts). The coherency parameter was determined by visual observation, and the color index of clasts is based on visual and microscopic study.

There is a very good fit between the matrix groups of the present study and the Jackson-Wilshire groups:

Groups 1, 2, 3 (excluding the white rocks)—F1, F2
Group 3 (white rocks)—F3
Groups 4, 5, 6, 7—F4
Group 8—not studied

The low-grade samples are both friable and coherent and contain mostly light colored clasts. The high-grade samples are coherent and contain mostly dark clasts. The white rocks are friable and contain mostly dark clasts.

All of the friable samples contain matrix glass and large amounts of intergranular porosity. All of the recrystallized samples are coherent. It is suggested that coherency may be explained mostly by intergranular porosity and matrix texture.

The color index data presents a problem. The low-grade rocks contain predominantly light clasts and the high-grade rocks contain predominantly dark clasts (disregarding the white rocks which are not in the main sequence). This macroscopic clast color difference with grade, which is confirmed by Jackson and Wilshire's triangular plots (1972, Figs. 1, 2, and 3) of microscopic clast colors, suggests a chemical difference between the low-grade and high-grade breccias. This suggestion is not borne out by the available chemical data (LSPET, 1971; Hubbard et al., 1972; and Rose et al., 1972). If all the breccias (except for 14063, 14082, and 14068) are essentially isochemical, as appears to be the case, then, in going from low to high grade rocks, either (1) some light clasts must disappear, (2) some dark clasts must appear, or (3) some light clasts must change color. The origin of the clast color systematics remains a problem.

Many other workers who studied Apollo 14 breccias reach somewhat different conclusions from those contained in this paper. For example, Engelhardt et al. (1972) classify the breccias into (I) Glass-rich breccias due to impact into the regolith (my low-grade breccias); (II) Glass-poor breccias due to impact into solid rock (my medium-grade breccias); and (III) Glass-poor breccias with crystalline matrix due to either recrystallization of a base surge or an impact melt (my high-grade breccias).

CHEMICAL TRENDS DURING METAMORPHISM

Methods

Samples representing six metamorphic groups were analyzed with the electron microprobe: 14307 (group 1), 14318 (group 3), 14311 (group 5), 14006 (group 6), 14066 (group 7), and 14068 (group 8). Plagioclase was analyzed for Ca, K, and Na; pyroxene for Ca, Fe, and Mg; and olivine for Fe and Mg. Five types of experiments were performed on each sample:

1. Analysis of 20 to 60 matrix plagioclase crystals.
2. Analysis of ≤ 20 rims and cores of plagioclase clasts.
3. Analysis of 20 to 50 matrix pyroxene crystals.
4. Analysis of ≤ 20 rims and cores of pyroxene clasts.
5. Analysis of matrix olivine crystals.

In addition, olivine analyses were performed on the skeletal prisms and rounded clasts in sample 14068.

For each sample studied, three areas ($2-3$ mm^2) in widely separate parts of the polished section were chosen. Approximately one-third of the analyses of each type were performed in each area. No differences were noted among the three subsets of data, thus suggesting chemical phase homogeneity on the scale of 1 cm.

The overall quality of the microprobe data reported herein is low. Matrix corrections were not performed. More important is the very small grain size of the analyzed phases; the grain size: beam diameter ratio was commonly between 2 and 4.

Results

Plagioclase—matrix Fig. 7 contains histograms of the An content of matrix plagioclase for each sample. The lower-grade rocks display a continuum of compositions similar to the feldspar distribution in a soil (Reid *et al.*, 1971); that is, these feldspars have a detrital origin (comminution and accumulation). The close grouping of An values in higher-grade rocks probably results from plagioclase equilibrium during metamorphic recrystallization. The average compositions of equilibrated plagioclases are:

Grade 6 (Sample 14006)—An$_{83}$; Or$_1$
Grade 7 (Sample 14066)—An$_{76}$; Or$_2$
Grade 8 (Sample 14068)—An$_{73}$; Or$_7$

The increase in Or content with grade apparently is produced by the dissolution of the tiny K-feldspar blebs, present in the matrix of low- and medium-grade breccias, into the plagioclase.

Plagioclase clasts A traverse across a typical plagioclase clast in a high-grade rock is shown in Fig. 8. The cores are homogeneous but range in composition from clast to clast. However, the rims seem to develop identical compositions throughout each sample, but edge effects and resolution limitations of the microprobe make the observation somewhat uncertain. No correlation was noted between rim thickness

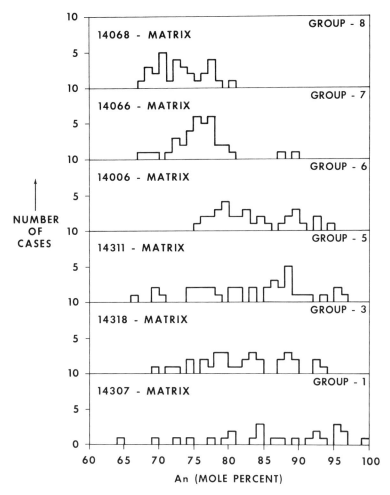

Fig. 7 Histograms of the An content of matrix plagioclase.

and metamorphic grade. The rims within one thin section range in thickness from
1 to 5 μ. The rims follow the clast border regardless of clast shape, and most rims are
complete, but others, especially in 14068, are not. The boundaries between rim and
core are smooth and parallel to the outer border of the clast.

Fig. 9 contains Or-Ab-An plots that show rim and core compositions for each
sample. The low-grade sample (14307) does not show a systematic picture. In the
high-grade rocks almost every plagioclase clast shows a rim of about the same
composition, i.e., the composition of the equilibrated matrix plagioclase. In sample
14318, approximately half of the clasts display rims and half are unrimmed. Cores of
rimmed clasts range in composition from more sodic (An_{59}) to more calcic (An_{97})
than the rims. As is to be expected, clasts that have An_{75-80} cores show no dis-
cernable rims. Note that rims first appear in a Group 3 sample, but the matrix
plagioclase does not show even a tendency towards equilibrium except in Group 5

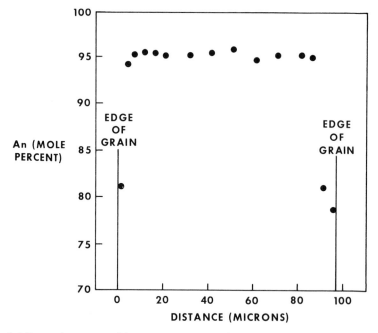

Fig. 8. Microprobe traverse of An content across a typical plagioclase clast in sample 14006.

and higher group samples. It is suggested that the rims formed by precipitation of new plagioclase, the material of which came from devitrifying matrix glass. This implies that the rims are not due to reaction between the original clast and the matrix. Core compositions show a continuum, similar to the 14307 matrix continuum, suggesting that the clast cores have a detrital origin and that they did not take part in the metamorphic reactions. This lack of major diffusion and equilibration in the clasts implies

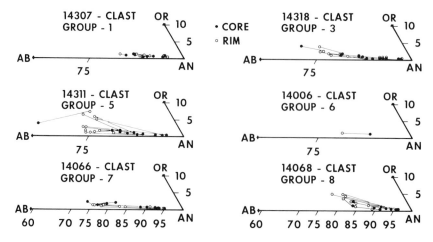

Fig. 9. Core and rim compositions of plagioclase clasts in terms of Or-Ab-An.

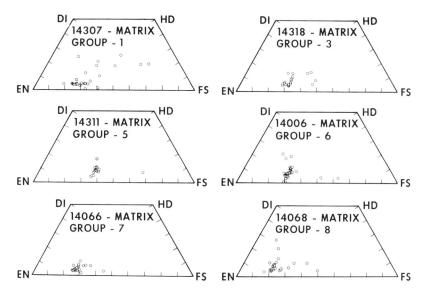

Fig. 10. Wo-En-Fs diagrams of the compositions of matrix pyroxene.

that these Fra Mauro rocks were not subjected to high temperatures for long times. *Pyroxene—matrix* Pyroxene tetrahedra showing matrix compositions for each sample are plotted in Fig. 10. The situation is similar to the plagioclase story. Low-grade sample 14307 shows a distribution of compositions similar to a soil (see Apollo Soil Survey, 1971), that is, consistent with a detrital origin. The higher-grade pyroxene distributions display a preferred orthopyroxene composition in all cases. The range in CaO contents is due to contamination from adjacent plagioclase crystals. The average En values of the equilibrated pyroxenes are:

Grade 3 (Sample 14318)—En_{61}
Grade 5 (Sample 14311)—En_{55}
Grade 6 (Sample 14006)—En_{63}
Grade 7 (Sample 14066)—En_{68}
Grade 8 (Sample 14068)—En_{72}

If one assumes that 14318 is not completely equilibrated, then there is a trend in composition towards higher Mg content. This is consistent with changing pyroxene equilibrium at increasing metamorphic grade, probably in response to declining oxygen fugacity.

Pyroxene clasts Rim and core compositions for pyroxene clasts from each sample are plotted in Fig. 11. Here again, the plagioclase story is repeated. Low-grade sample 14307 does not show rimming, medium-grade sample 14318 displays rims about most of the clasts, and all the high-grade samples display rims with compositions closely matching the equilibrium matrix pyroxene composition. Those "rim" analyses that are not on the preferred composition are due to analysis of a volume of part rim and part core. In these cases the tie line from core to rim composition forms a vector

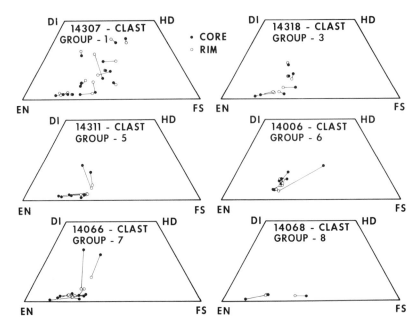

Fig. 11. Core and rim compositions of pyroxene clasts in terms of Wo-En-Fs.

pointing to the true rim composition. Because lunar pyroxenes normally display extensive zoning, they are more difficult to interpret. However, pyroxenes zone outwards towards Fe-rich subcalcic augite, but the rims are Mg-rich orthopyroxene. It is suggested that, as is the case for the plagioclase rims, these rims were produced by the precipitation of new pyroxene.

Olivine Matrix olivine compositional data for Group 1–7 samples is:

Grade 1 (Sample 14307)—Fo_{38} to Fo_{77} (4 analyses)
Grade 3 (Sample 14318)—Fo_{64} (1 analysis)
Grade 5 (Sample 14311)—No data
Grade 6 (Sample 14006)—Fo_{65} to Fo_{67} (3 analyses)
Grade 7 (Sample 14066)—Fo_{67} to Fo_{68} (3 analyses)

Preliminary study of olivine compositions suggests relationship to grade as observed for plagioclase and pyroxene. The lowest-grade sample shows a broad spread in compositions, whereas high-grade samples display a trend toward uniform compositions. As with the pyroxenes, the olivine equilibrium composition becomes more Mg rich with increasing grade.

Sample 14068 is a special case in that the matrix contains skeletal olivine prisms that may have been phenocrysts. Olivine compositions are:

Sample 14068 (prisms)—Fo_{77} to Fo_{79} (14 analyses)
Sample 14068 (clast cores)—Fo_{67} to Fo_{84} (7 analyses)

The clasts display a detrital distribution, whereas the skeletal prisms are equilibrated. The composition of the prisms (Fo_{78}) continues the above trend to more Mg-rich values, indicative of higher grades of metamorphism, even though 14068 has excep-tionally high Mg content ($MgO = 17.6$ wt %, Hubbard *et al.*, 1972).

Discussion

The chemical relations are summarized in Table 2.

The first chemical event observed is the rimming of pre-existing pyroxene and plagioclase clasts with newly formed pyroxene and plagioclase, respectively; the new material has the equilibrium composition. This event is correlated with the disap-pearance of matrix glass. The rims effectively isolate the clasts from taking part in succeeding reactions and equilibria. An alternate interpretation that the rims formed by reaction between clast and matrix has not been disproved.

At a specific point in the metamorphic series, each silicate phase in the matrix reaches equilibrium, i.e., analyses of that phase display a strongly preferred com-position. The point in the series where equilibrium is attained is specific for each phase: pyroxene at Group 4 or 5; plagioclase at Group 7; and olivine at Group 6 or below.

With increasing grade, compositions of the equilibrated silicates, both rims and matrix, change systematically in response to a lower oxygen fugacity or a higher temperature of equilibration, e.g., olivine and pyroxene becomes more Mg rich. Textural evidence indicates that matrix of the grade 8 rock was partially melted.

An alternate interpretation of the high-grade textures (discussed previously) suggests that the matrix of the high-grade breccias crystallized directly from glass and implies that the order of Groups 5 to 8 is reversed, i.e., Group 5 would represent the highest temperature. In terms of the mafic silicates, this reversal would present the unlikely situation of the higher-grade rocks containing more Fe-rich pyroxenes and olivines. However, the plagioclase data can fit the reversal. Assume the bulk feldspar composition is about An_{75}. The Group 7 and 8 rocks that would have crystallized below the solidus should form feldspars of approximately the bulk composition; the more equant textural rocks of Groups 5 and 6, which would have crystallized between the solidus and liquidus, should attempt to form equilibrium feldspars that would range from the bulk composition to higher An values.

The metamorphism of lunar breccias is similar to the metamorphism of chondritic meteorites (see King *et al.*, this volume) as reviewed by Dodd (1969), Van Schmus

Table 2. Chemical characteristics of the breccia groups.

Grade	Group	Sample	Rims on plag. and pyrox. clasts	Equilibrated matrix		
				Pyroxene	Plagioclase	Olivine
Low	1	14301	None	No	No	No
	2		None			
Medium	3	14318	Sporadic	Partial	No	?
	4		None			
High	5	14311	Present	Yes	Some	?
	6	14006	Present	Yes	Partial	Yes
	7	14066	Present	Yes	Yes	Yes
	8	14068	Present	Yes	Yes	Yes

(1969), Wood (1969), and others in that both unmetamorphosed lunar rocks and meteorites have a broad range of silicate compositions, but metamorphosed breccias and chondrites have an equilibrium distribution of compositions. Breccia and chondrite metamorphism differs in that in chondrites, once the mafic silicates achieve chemical equilibrium at type 4, changes to type 5 and 6 are only textural, whereas in breccias, after the initial equilibrium is achieved, the composition of both olivine and pyroxene (besides the texture) continues to evolve.

The metamorphism of lunar breccias is similar to terrestrial regional metamorphism in that equilibrium and observed composition of various phases changes as a function of metamorphic grade. However, terrestrial metamorphic rocks commonly, at least metastably, achieve textural equilibrium at low grade, whereas textural equilibrium of breccias is not achieved until the highest grades.

POSTMETAMORPHIC EVENTS

Repeated impacts on the Fra Mauro formation should have fractured and jostled the rocks before they were excavated and thrown out on the lunar surface. Hait (1972) observed results of this process on the meter scale at the Apollo 14 site.

In studying sample 14321, Grieve et al. (1972) describe regions several cm across of "microbreccia-3" (a medium-grade breccia) plus basaltic clasts that are loosely bound together by a "light matrix" to form 14321. The light matrix is a mixture of ground-up basalt and microbreccia-3. Most basalt clasts, however, are immersed directly in microbreccia-3. It is suggested that postmetamorphic fracturing of microbreccia-3 about some included basalt clasts, followed by grinding, mixing, and consolidation without recrystallization, may explain the origin of the light matrix material. This would be an example of a postmetamorphic breccia on the cm scale.

LSPET (1971) reported fractures and veins in some breccias. The veins are filled with brown glass that contains spherules of metallic iron. The veins are as much as 1 mm thick, commonly occur in parallel or subparallel swarms, and cut glass, mineral, and lithic fragments. The vein-filling material has penetrated the "wall rock" on both sides of the veins, reducing porosity to zero. In sample 14307, veins form two sets that intersect at about 90 degrees. In sample 14306 one vein thinned out to less than 0.1 μ, becoming a microfault with a displacement of approximately 1 μ. It is suggested that these veins and microfaults are the result of postmetamorphic fracturing on the scale of a few μ to mm.

CONCLUSIONS

The Apollo 14 breccias have been metamorphosed to various metamorphic grades in response to increased temperature. Each metamorphic group is characterized texturally and mineralogically in Table 1 and chemically in Table 2. The reactions between groups are set out in Table 3.

Apollo 11 and 12 breccias (excluding 12013) are surface breccias that were probably formed as relatively local base surge deposits (McKay and Morrison, 1971). Low-grade Apollo 14 breccias are texturally and mineralogically similar to Apollo 11

Table 3. Processes involved during breccia metamorphism.

Transition	Process
1 to 2 to 3	Glass devitrifies; new plagioclase and pyroxene forms as rims on pre-existing clasts
3 to 4 to 5	Matrix pyroxene starts to equilibrate; pyroxene forms equant texture
4 to 5 to 6	Matrix pyroxene reaches equilibration, matrix plagioclase starts to equilibrate; plagioclase forms euhedral texture
6 to 7	Matrix plagioclase reaches equilibration
7 to 8	Matrix melts, texture due to devitrification

and 12 breccias. The Apollo 14 site lies in the Fra Mauro formation, which is a large ejecta blanket from the Imbrium Basin. Because the postulated base surge breccias are all low grade, and ejecta breccias are to be expected at the Apollo 14 site, it is concluded that at least the medium- and high-grade Apollo 14 breccias formed as part of the Imbrian ejecta blanket that is the Fra Mauro formation.

However, the continuous nature of the Apollo 14 breccia metamorphic series suggests a common origin for all samples. Medium- and high-grade ejecta breccias may be derived from low-grade equivalents by a metamorphic process—an auto-metamorphism in the thick, hot, Fra Mauro ejecta blanket at the Apollo 14 site. The data in this paper concerning the matrix of the breccias is in accord with the conclusion that the Apollo 14 breccias were metamorphosed in the Fra Mauro formation as a single cooling unit. Dence *et al.* (1972) have pointed out that by terrestrial standards, an ejecta blanket 500 km from the edge of the crater should not be as hot as is deduced from the observed recrystallization and equilibrium in the Fra Mauro formation. The origin of the energy necessary for the *in situ* thermal metamorphism of the breccia matrix is a problem.

The Fra Mauro ejecta breccias underwent a postmetamorphic history—they were fractured and jostled, causing microfaulting, glass veining, and post meta-morphic brecciation.

The presence of later surface breccias among the Apollo 14 suite is not denied; the low-grade breccias may be either later surface breccias or unmetamorphosed ejecta breccias. However, at least one of the unmetamorphosed breccias (14307) displays postmetamorphic glass veins; this particular rock cannot be a later surface breccia. The high noble gas content rocks (14047, 14049; LSPET, 1971), however, are probably surface breccias. It is concluded that surface breccias grade into, and are indistinguishable from, low-grade ejecta breccias that lack postmetamorphic features. Perhaps the coherent, low-grade breccias (F2 of Jackson and Wilshire, 1972) such as 14307 are all unmetamorphosed ejecta breccias and the high noble gas, friable, low-grade rocks (F1 of Jackson and Wilshire, 1972) are all later surface breccias.

Acknowledgments—I thank the Curator's Office at MSC for the unlimited use of thin sections. I first recognized the series nature of the Apollo 14 breccias during PET with the help of A. M. Reid. The idea for the microprobe portion of this study is the product of discussions with D. B. Stewart. I have benefited from discussions with my colleagues at MSC, especially C. Simonds and W. Phinney.

References

Anderson A. T. Braziunas T. F. Jacoby J. and Smith J. V. (1972) Breccia populations and thermal history: Nature of pre-Imbrium crust and impacting body (abstract). In *Lunar Science—III* (editor C. Watkins), pp, 24–26, Lunar Science Institute Contr. No. 88.

Anderson A. T. and Smith J. V. (1971) Nature, occurrence, and exotic origin of "grey mottled" (Luny Rock) basalts in Apollo 12 soils and breccias. *Proc. Second Lunar Sci. Conf., Geochim. Cosmochim. Acta*, Suppl. 2, Vol. 1, pp. 431–438. M.I.T. Press.

Apollo Soil Survey (1971) Apollo 14: Nature and origin of rock types in soil from the Fra Mauro formation. *Earth Planet. Sci. Lett.* **12,** 49–54.

Bogard D. D. and Nyquist L. E. (1972) Noble gas studied on regolith materials from Apollo 14 and 15 (abstract). In *Lunar Science—III* (editor C. Watkins), pp. 89–91, Lunar Science Institute Contr. No. 88.

Chao E. C. T. Boreman J. A. and Desborough G. A. (1971) The petrology of unshocked and shocked Apollo 11 and Apollo 12 microbreccias. *Proc. Second Lunar Sci. Conf., Geochim. Cosmochim. Acta*, Suppl. 2, Vol. 1, pp. 797–816, M.I.T. Press.

Dence M. R. Plant A. G. and Traill R. J. (1972) Impact-generated shock and thermal metamorphism in Fra Mauro lunar samples (abstract). In *Lunar Science—III* (editor C. Watkins), 174–176, Lunar Science Institute Contr. No. 88.

Dodd R. T. (1969) Metamorphism of the ordinary chondrites: A review. *Geochim. Cosmochim. Acta*, **33,** 161–203.

Drake M. J. McCallum I. S. McKay G. A. and Weill D. F. (1970) Mineralogy and petrology of Apollo 12 sample no 12013: A progress report. *Earth Planet. Sci. Lett.* **9,** 103–123.

Engelhardt W. von Arndt J. and Stoffler D. (1972) Apollo 14 soils and breccias, their compositions and origin by impacts (abstract). In *Lunar Science—III* (editor C. Watkins), 233–235, Lunar Science Institute Contr. No. 88.

Gose W. A. Pearce G. W. Strangway D. W. and Larson E. E. (1972) On the magnetic properties of lunar breccias (abstract). In *Lunar Science—III* (editor C. Watkins), pp. 332–334, Lunar Science Institute Contr. No. 88.

Grieve R. McKay G. Smith H. and Weill D. (1972) Mineralogy and petrology of polymict breccia 14321 (abstract). In *Lunar Science—III* (editor C. Watkins), pp. 338–340, Lunar Science Institute Contr. No. 88.

Hait M. H. (1972) The white rock group and other boulders of the Apollo 14 site: A partial record of Fra Mauro history (abstract). In *Lunar Science—III* (editor C. Watkins), p. 353, Lunar Science Institute Contr. No. 88.

Hubbard N. J. Gast P. W. Rhodes M. and Wiesmann H. (1972) Chemical composition of Apollo 14 materials and evidence for alkali volatilization (abstract). In *Lunar Science—III* (editor C. Watkins), pp. 407–409, Lunar Science Institute Contr. No. 88.

Jackson E. D. and Wilshire H. G. (1972) Classification of the samples returned from the Apollo 14 landing site (abstract). In *Lunar Science—III* (editor C. Watkins), pp. 418–420, Lunar Science Institute Contr. No. 88.

King E. A. Butler J. C. and Carman M. F. (1972b) Chondrules in Apollo 14 samples: Implications for the lunar surface exposure ages from grain size analyses (abstract). In *Lunar Science—III* (editor C. Watkins), pp. 449–451, Lunar Science Institute Contr. No. 88.

King E. A. Carman M. F. and Butler J. C. (1972b) Chondrules in Apollo 14 samples: Implications for the origin of chondritic meteorites. *Science* **175,** 59–60.

Kurat G. Keil K. Prinz M. and Bunch T. E. (1972) A "chondrite" of lunar origin: Textures, lithic fragments, glasses and chondrules (abstract). In *Lunar Science—III* (editor C. Watkins), pp. 463–465, Lunar Science Institute Contr. No. 88.

Lofgren G. (1971) Spherulitic textures in glassy and crystalline rocks. *J. Geophys. Res.* **76,** 5635–5648.

LSPET (1971) (Lunar Sample Preliminary Examination Team) Preliminary examination of lunar samples from Apollo 14. *Science* **173,** 681–693.

Mason B. Melson W. G. and Nelen J. (1972) Spinel and hornblende in Apollo 14 fines (abstract). In *Lunar Science—III* (editor C. Watkins), 512–514, Lunar Science Institute Contr. No. 88.

McKay D. S. Clanton U. S. Heiken G. H. Morrison D. A. and Taylor R. M. (1972) Vapor phase crystalli-
    zation in Apollo 14 breccias and size analysis of Apollo 14 soils (abstract). In *Lunar Science—III* (editor
    C. Watkins), pp. 529–531, Lunar Science Institute Contr. No. 88.
McKay D. S. and Morrison D. A. (1971) Lunar breccias. *J. Geophys. Res.* **76**, 5658–5669.
Meyer C. Jr. Brett R. Hubbard N. J. Morrison D. A. McKay D. S. Aitken F. K. Takeda H. and Schonfeld E.
    (1971) Mineralogy, chemistry and origin of the KREEP component in soil samples from the Ocean of
    Storms. *Proc. Second Lunar Sci. Conf., Geochim. Cosmochim. Acta.,* Suppl. 2, Vol. 1, pp. 393–411,
    M.I.T. Press.
Moore C. B. Lewis C. F. Cripe J. Kelly W. R. and Delles F. (1972) Total carbon, nitrogen, and sulfur
    abundances in Apollo 14 lunar samples (abstract). In *Lunar Science—III* (editor C. Watkins), pp. 550–
    551, Lunar Science Institute Contr. No. 88.
Reid A. M. Frazer J. Z. Fujita H. and Everson J. E. (1971) Apollo 11 samples: Major mineral chemistry.
    Proc. Apollo 11 Lunar Sci. Conf., Geochim. Cosmochim. Acta, Suppl. 1, Vol. 1, pp. 749–761. Pergamon.
Rose H. J. Cuttitta F. Annell C. S. Carron M. K. Christian R. P. Dwornik E. J. and Ligon D. T. (1972)
    Compositional data for fifteen Fra Mauro lunar samples (abstract). In *Lunar Science—III* (editor C.
    Watkins), pp. 660–662, Lunar Science Institute Contr. No. 88.
Simonds C. H. (1972) Sintering of Fra Mauro composition glass (abstract). *Trans. Amer. Geophys. Union*
    53, 428–429.
Van Schmus R. (1969) The mineralogy and petrology of chondritic meteorites. *Earth Sci. Rev.* **5**, 145–184.
Warner J. L. (1972) Apollo 14 breccias: metamorphic origin and classification (abstract). In *Lunar Science
    —III* (editor C. Watkins), pp. 782–784, Lunar Science Institute Contr. No. 88.
Wilshire H. G. and Jackson E. D. (1972) Petrology of the Fra Mauro formation at the Apollo 14 landing
    site (abstract). In *Lunar Science—III* (editor C. Watkins), pp. 803–805, Lunar Science Institute Contr.
    No. 88.
Wood J. A. (1968) *Meteorites and the Origin of Planets.* McGraw-Hill.

Proceedings of the Third Lunar Science Conference
(Supplement 3, *Geochimica et Cosmochimica Acta*)
Vol. 1, pp. 645–659
The M.I.T. Press, 1972

Apollo 14 breccias: General characteristics and classification*

E. C. T. Chao, Jean A. Minkin, and Judith B. Best

U.S. Geological Survey, Washington, D.C. 20242

Abstract—The Apollo 14 breccias are complex and are characterized by a wide range of clast types and textures. The principal fragment types in the breccias are (1) hornfelsed noritic microbreccias; (2) micronorite hornfels; (3) devitrified glass; and (4) ophitic and other kinds of basalt. Dark vitrophyric fragments, fine-grained anorthositic rocks and anorthositic breccias, olivine-rich rocks, pyroxene-rich fragments, and breccias with inclusions of earlier breccias are also present, but they are comparatively rare.

Based on fragment population, nature of the matrix, grain size and porosity, metamorphic history, and bulk chemical composition, the Apollo 14 breccias can be classified chiefly into regolith microbreccias, Fra Mauro breccias and spherule-rich microbreccias. Further subdivisions are based principally on the interpretation of metamorphic history. Twenty-six Apollo 14 breccias have been so classified.

The complex fragment types contained in Apollo 14 breccias represent multiple episodes of heating and fragmentation. They probably are pre-Imbrian and were not produced by the impact effects of a single event. The variability of their fragment populations is important in the interpretation of pre-Imbrian geologic history.

Introduction

The Apollo 14 Fra Mauro site was chosen because it was interpreted to be a highland area underlain by ejecta from the Imbrium Basin (Swann *et al.*, 1971). Among the hand specimens returned from this site, two are basalts and the rest are breccias (LSPET, 1971). The Apollo 14 breccias have a wide range of clast types and textures, indicating complex histories of fragmentation and metamorphism. They represent the best suite of breccias returned from the moon to date. Hence, a serious effort should be made to determine their origin, age, and geologic significance.

For the sake of brevity we present in this paper only the general characteristics of the various types of Apollo 14 breccias and their classification. To understand their complex history thoroughly, it is necessary to have detailed knowledge of the fragment or clast types they contain. More than 20 fragment types have been commonly observed. We do not yet have adequate petrographic and chemical information for all of these fragment types, nor have we had an opportunity to compare the small breccia chips we have studied (generally <1 cm) with the large breccia samples. The information presented here must thus be regarded as preliminary.

The criteria for classifying the Apollo 14 breccias are the same as those applied to Apollo 11 and 12 breccias (Chao *et al.*, 1971). The classification is genetic.

Lithic Fragment Types, General Occurrence and Association

Several categories of lithic fragments can be recognized on the basis of gross mineral composition and textures. They are (1) fine-grained to very fine-grained

---

*Publication authorized by the Director, U.S. Geological Survey.

Fig. 1. The annealed textures and minerals of a type 1a fragment (right) and an adjacent type 1b fragment (left) from a polished thin section of 14303,34. Note the difference in grain size between types 1a and 1b, and the similarity of interlocking grain boundaries between pyroxene (light gray) and plagioclase (medium gray); disseminated ilmenite is white. Reflected light. Bar scale 100 μ.

Fig. 2. A well-annealed type 1a fragment (upper part) in contact with the less well-annealed matrix (lower part) in a polished thin section of 14303,34. The matrix is more plucked and has a more random fabric than the type 1a fragment. Reflected light. Bar scale 100 μ.

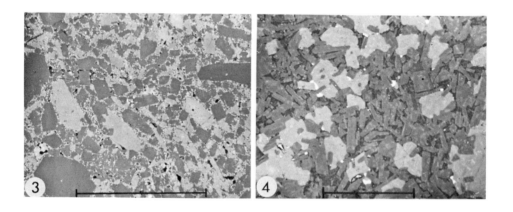

Fig. 3. A type 1c fragment in 14321,241. Note the abundance of angular xenocrysts embedded in devitrified glass (mottled areas) which contains numerous tiny ilmenite grains (white). Pyroxene and olivine xenocrysts are light gray; plagioclase xenocrysts, medium gray. Reflected light. Bar scale 100 μ.

Fig. 4. A polished thin section of a type 2 fragment showing the holocrystalline texture with orthopyroxene (light gray), plagioclase (medium gray, lath-shaped), Si–K-rich meso-stasis (dark gray) and ilmenite (white). Fragment is from 14063,13, a friable, largely unannealed, feldspathic microbreccia. Reflected light. Bar scale 100 μ.

hornfelsed noritic microbreccias and extremely fine-grained annealed fragment-laden glass, referred to by Carr and Meyer (1972) as light and dark hornfels breccias; (2) micronorite hornfels, referred to by Floran et al. (1972) and by Marvin et al. (1972) as micronorite; (3) extremely fine-grained dark gray to black devitrified glass with xenocrysts and a few xenoliths; (4) ophitic mare basalts (Jackson and Wilshire, 1972) and other types of basalts; (5) fine-grained dark vitrophyric fragments; (6) anorthositic igneous and metamorphic rocks and anorthositic breccias; (7) troctolites and brecciated troctolites, olivine-rich microbreccias and metamorphic rocks; (8) pyroxene-rich microbreccias and metamorphic rocks; and (9) breccias containing earlier breccia fragments. Many of these categories may be subdivided into types on the basis of grain size, variability in mineral proportions, or evidence of metamorphism. We recognize more than 20 types of lithic fragments that are commonly observed in the Apollo 14 breccias. Some of these types are not yet adequately characterized because of the ambiguity of textures and the lack of information on the bulk chemical composition of the fragments.

For brevity, only characteristics of the dominant lithic fragment types of the various Apollo 14 nonregolith (Fra Mauro) breccias are summarized in Table 1. Among these, 1a, 1b, and 2 are more common than types 1c, 3, and 4a in the breccias and are also more common as individual fragments in the coarse fines.

Types 1a (Figs. 1 and 2), 1b (Fig. 1), and 1c (Fig. 3) have a granular texture. They probably are fragment-laden glasses (glass-welded aggregates, Chao et al., 1971) that have been annealed by thermal metamorphism. We refer to types 1a and 1b as hornfelsed noritic microbreccias (Table 1). Type 1c consists largely of angular mineral fragments and a few lithic fragments rimmed by minute amounts of devitrified glass that contains numerous tiny blebs of ilmenite.

Type 2 (Table 1) is vesicular or vuggy. It is coarser in grain size than type 1a and more thoroughly recrystallized (Fig. 4). These fragments have fine-grained textures with inclusion-laden orthopyroxenes, laths of plagioclase, and a Si–K-rich mesostasis. The vesicular variety is holocrystalline and relatively rare. The vuggy variety contains abundant xenocrysts and some xenoliths. We refer to type 2 fragments as micronorite hornfels. In breccia 14311, type 2 fragments occur as inclusions in type 1a.

The mineral modes of types 1a, 1b, and 2 are shown in Table 2 and their bulk chemical compositions plus that of type 1c are included in Table 3. All are feldspar-rich and are remarkably similar in bulk chemical composition. Minor differences, however, may be significant. Type 2 fragments are generally slightly higher in silica, alumina, and potash and lower in iron, magnesia, and titania. This is reflected by the slightly lower pyroxene content and higher Si–K-rich mesostasis in type 2 than in types 1a and 1b. Further studies are needed to test the significance of these apparently consistent minor differences.

Types 1a, 1b, 1c, and particularly type 2 all contain appreciable amounts of alkalis (Table 3); hence, it is possible that they are the principal contributors to the KREEP components in the Apollo fines and regolith microbreccias (see Chao et al., 1972, this issue).

Minute euhedral crystals occur in vugs in types 1a, 1b, and 2. These suggest vapor deposition and are beautifully illustrated by McKay *et al.* (1972). It should be noted, however, that the vugs and pore openings of their associated unannealed and partly annealed breccias contain none of these euhedral crystals.

Lithic fragment type 3 consists largely of devitrified glass with sparse xenocrysts and xenoliths. Chemically, type 3 fragments contain high alumina and lime, and low iron, magnesia and potash (Table 3).

The most common type of basalt that occurs in millimeter- to centimeter-size fragments is ophitic mare type basalt similar to 14053 (type 4a, Tables 1 and 3). Such fragments are abundant in 14321, but are rare in most other Apollo 14 breccias.

There appear to be several kinds of dark, fine-grained vitrophyric fragments that may differ in bulk composition. One type, common in breccias 14304 and 14319,

Table 1. Summary of principal lithic

Type	Examples in:	Color
1a Fine-grained, hornfelsed, noritic microbreccia.	14303, 14311, and in coarse fines 14161,3 and 14166,4.	Medium gray in chips; light gray in thin sections.
1b Very fine-grained, hornfelsed, noritic microbreccia.	14303 and in coarse fines.	Dark gray to black in chips; black in thin sections.
1c Extremely fine-grained, annealed, fragment-laden glass.	14321, 14306, and in coarse fines.	Dark gray to black in chips; black in thin sections.
2 Micronorite hornfels.	14063, 14311 and in coarse fines.	Yellowish white to light gray in chips; colorless in thin sections.
3 Devitrified glass.	14063, 14082.	Brownish black in chips; purplish-grayish-brown in thin sections.
4a Ophitic mare basalt (similar to 14053).	14321.	Dark brown to gray in chips; colorless to pale brown in thin sections.
4b Other basalts (some similar to 14310).	14303, 14063.	Dark gray in chips; colorless in thin sections.

*See text for description or reference to minor lithic fragment types.

has radiating crystals of pyroxenes and olivine and minor plagioclase. Another rare type has microphenocrysts of olivine and forms the bulk of sample 14068. The origin of these finely devitrified glasses is uncertain. Information on their bulk chemical composition is available only for 14068 (R. T. Helz, 1972, this issue).

Anorthositic rocks (both igneous and thermally metamorphosed types), brecciated anorthositic rocks, and plagioclase-rich polymict microbreccias have been observed as fragments in the Apollo 14 breccias. They are not abundant, although they are widespread in occurrence. They commonly occur as inclusions within annealed fragments of types 1a, 1b, 1c, and 2, indicating that they were formed by episodes of brecciation and metamorphism prior to the events that formed the type 1a, 1b, 1c, and 2 fragments.

fragment types in Fra Mauro breccias*.

Xenocrysts and xenoliths (types 1 to 3) or principal minerals (types 4a and 4b)	Matrix minerals	Texture
Xenocrysts abundant: Mainly calcic plagioclase, also pigeonite-subcalcic augite, olivine, orthopyroxene, chrome-spinel and ilmenite. Xenoliths: Mainly anorthositic (some thermally metamorphosed).	Predominantly calcic plagioclase and pyroxenes. Ilmenite finely disseminated. Blebs of metallic iron and troilite also present. Si-K-rich mesostasis rare.	Vuggy, otherwise nonporous. Annealed granular, interlocking grain boundaries (Fig. 1). Olivine, pyroxene and spinel xenocrysts often surrounded by reaction rims (Wilshire and Jackson, 1972).
Similar to type 1a.	Similar to type 1a.	Finer grain size than type 1a (Fig. 1), otherwise similar.
Xenocrysts abundant, variable: Mainly calcic plagioclase, pyroxenes and olivine. Xenoliths sparse: Mainly devitrified glass, and anorthositic and olivine-rich metamorphic rocks and microbreccias.	Devitrified glass, generally in minute amounts, consisting of plagioclase, pyroxene, finely disseminated blebs of ilmenite, and some metallic iron and troilite.	Vuggy, otherwise non-porous. Minute amounts of extremely fine-grained devitrified glass matrix rimming angular xenocrysts and xenoliths (Fig. 2).
Xenocrysts sparse to abundant: Mainly calcic plagioclase, also calcic augite, olivine, spinel and ilmenite. Xenoliths: Anorthositic rocks and brown devitrified glass.	Calcic plagioclase and orthopyroxenes dominant, with Si-K-rich mesostasis (silica phase, probably tridymite, and K-feldspar). Scattered ilmenite and a few blebs of metallic iron and troilite. Accessory phosphate mineral (probably whitlockite or apatite) and zircon.	Vesicular or vuggy. Holocrystalline, "microporphyritic" (Fig. 3). Orthopyroxenes subhedral, inclusion-laden; calcic plagioclase granular to lath-shaped. Coarser-grained than type 1a.
Xenocrysts sparse: Mainly plagioclase, some pyroxene and chrome-spinel. Xenoliths sparse: Anorthositic rocks.	Predominantly crystallites of calcic plagioclase, some crystallites of pyroxene. No ilmenite.	Sparsely vesicular, otherwise nonporous. Mostly extremely fine-grained devitrified glass with scattered xenocrysts and rare xenoliths.
Principal early-formed minerals: Pigeonite-augite pyroxenes dominant over calcic plagioclase. Accessory olivine and ilmenite.	Groundmass: Si-rich mesostasis with blebs of troilite and metallic iron; tranquillityite sometimes present.	Subophitic to ophitic. Zoned pyroxenes common. Fine-to medium-grained.
Principal early-formed minerals: Calcic plagioclase dominant over pigeonite pyroxenes. Accessory ilmenite.	Groundmass: Si-rich mesostasis.	Intersertal to sub-ophitic. Generally fine-grained.

Table 2. Mineral modes of lithic fragment types (vol.%) in Fra Mauro breccias.

	Fine-grained hornfelsed noritic microbreccia (type 1a)			Very fine-grained hornfelsed noritic microbreccia (type 1b)		Micronorite hornfels (type 2)		
	14161,3–12	14161,3–10	14166,4–1	14303,34	14161,3–13	14161,3–1	14161,3–2	14161,3–5
Plagioclase								
Matrix	37.7	38.7	37.2	37.0	41.5	40.4		41.5
Xenocryst	13.2	9.7	13.2	15.8	8.9	11.6		9.2
TOTAL	50.9	48.4	50.4	52.8	50.4	52.0	49.9	50.7
Pyroxene								
Matrix	39.7	45.7	34.8	32.7	37.0	36.3		36.9
Xenocryst	1.4	0.7	4.4	10.1	4.0	0.4		1.6
TOTAL	41.1	46.4	39.2	42.8	41.0	36.7	43.4	38.5
Olivine	3.6	2.1	5.2	1.3	4.4	2.9	2.5	0.5
Ilmenite	3.8	2.8	2.8	1.7	3.0	2.2	1.0	1.0
Native iron	0.3	0.3	1.2	0.1	0.2	tr.	tr.	0.3
Troilite	0.3	tr.	0.4	0.4	tr.	0.3	0.4	tr.
Mesostasis								
K-feldspar	—	—	0.8	0.2	0.5	5.1	2.3	7.9
Silica	—	—	tr.	—	0.5	0.8	0.4	0.3
Other*	—	—	—	0.7	—	—	—	0.8

*Includes glass, spinel, chromite.

Table 3. Bulk chemical compositions of lithic fragment types (in weight %) in Fra Mauro breccias*.

Fragment type	1a		1b	1c		2		3	4a	Ortho-pyroxene olivine rock
Sample no.	14161,3 –10	14161,3 –11	14161,3 –13	14321,241 –5	14063,46 –5	14161,3 –5	14161,3 –2	14063,46 –4	14321,241 –2	14161,3 –3
SiO$_2$	47.6	48.2	48.1	48.8	49.1	50.7	48.7	44.6	48.0	46.8
TiO$_2$	1.96	1.78	1.62	1.79	1.18	1.59	1.68	1.20	2.29	1.86
Al$_2$O$_3$	15.6	16.3	15.2	15.6	22.1	16.0	16.0	24.8	13.5	13.2
FeO	10.4	10.0	10.7	10.0	5.7	9.3	9.5	4.5	14.5	13.0
MgO	12.5	11.6	13.3	11.6	6.9	10.3	12.7	9.0	7.9	15.1
CaO	9.5	9.4	8.8	9.6	12.4	9.2	9.0	14.2	10.9	7.7
Na$_2$O	0.91	0.94	0.88	0.87	0.85	0.92	0.91	0.95	0.77	0.99
K$_2$O	0.36	0.45	0.54	0.65	0.61	1.02	0.48	0.16	0.26	0.52
MnO	0.11	0.09	0.09	0.02	0.09	0.07	0.07	—	0.20	0.09
Cr$_2$O$_3$	0.18	0.12	0.12	0.11	0.06	0.09	0.13	—	0.33	0.06
TOTAL	99	99	99	99	99	99	99	99	99	99

—Below limit of detection.
*Electron microprobe analyses of 4- to 15-mg chips fused with equal amounts of LiB$_4$O$_7$. The chips are splits of the same samples studied petrographically; Analyst Jean A. Minkin.

There are also several types of olivine-rich fragments. These include troctolites (in 14321, Compston *et al.*, 1972), olivine-rich microbreccias (in 14063,13), and metamorphic orthopyroxene-olivine rocks (in 14161,3 coarse fines, see 14161,3–3 in Table 3).

General Characteristics and Classification of Apollo 14 Breccias

In order to describe, characterize, and genetically classify lunar breccias, several aspects need to be studied. These are (1) the nature of lithic, mineral, and glass fragment types, their bulk compositions, and their relative abundance; (2) the nature of the matrix; (3) textures, including grain size, fractures, and porosity; (4) deformational and metamorphic effects; and (5) bulk chemical composition of the whole breccia.

In our study of Apollo 11 and 12 breccias (Chao *et al.*, 1971), we distinguished four types of lunar breccias. These were (1) porous, unshocked (regolith) micro-breccias; (2) shock-compacted or compressed (regolith) microbreccias; (3) glass-welded aggregates; and (4) thermally metamorphosed breccias. The first two types were dominant among the hand specimens, the third was common among coarse fines, and the fourth was very rare. In contrast, among the Apollo 14 breccias, the fourth type is dominant and, in addition, two new types can be recognized that are not evident among the Apollo 11 and 12 breccias. They are friable feldspathic microbreccias and spherule-rich transported regolith breccias. Table 4 shows the classification of Apollo 14 breccias, based on consideration of the general charac-teristics listed above. This classification is roughly comparable to that of Jackson and Wilshire (1972). Their F1 is equivalent to our unshocked, porous regolith microbreccias; F2 (except 14315 and 14318) to our compact or shocked regolith microbreccias; F3 to our unannealed (Fra Mauro) friable feldspathic microbreccias; and F4 (except 14321) to our strongly annealed (Fra Mauro) breccias. Detailed comparison with Jackson and Wilshire (1972) cannot be made without resolution of differences in terminology applied to fragment types and to the nature of the matrix. We are unable to compare our classification with that proposed by Warner (1972) because he emphasizes only matrix textures.

Porous, unshocked, regolith microbreccias

These samples consist of poorly compacted, weakly lithified lunar fines. They have several basic characteristics: (1) uniform fine fragment size (mostly less than 200 μ although a few large fragments may be present); (2) high porosity ($>20\%$); (3) a very high content (about 50 vol.%) of glass particles with a wide range of

Table 4. Classification of Apollo 14 breccias based on the study of polished thin sections.

Classification	Samples
1. Regolith microbreccias	
a. Unshocked, porous	14042, 14047, 14049, 14055, 14301*
b. Compact, nonporous	14313
c. Shocked	14307
2. Fra Mauro breccias	
a. Unannealed or slightly annealed, feldspathic	
breccias	14063†, 14082†
b. Moderately annealed breccias	14321†
c. Strongly annealed (thermally metamorphosed)	
breccias	
Unshocked	14006, 14270, 14311
Shocked	14066, 14171, 14303, 14304, 14305, 14306†, 14308, 14312†, 14314, 14319†, 14320
3. Spherule-rich, transported microbreccias	14315†, 14318

*Contains large lithic clasts (up to 1 cm) that are common in strongly annealed Fra Mauro breccias.
†Samples with highly distinctive fragment population.

chemical composition; (4) most lithic fragments apparently derived from the under-lying (Fra Mauro) formation, with a small but significant amount of exotic lithic and glass particles also present; and (5) bulk chemical composition nearly identical to the fines of the regolith (Table 5).

Samples of porous, unshocked regolith microbreccias are listed in Table 4. Sample 14301 (Fig. 5) has been classified here with the porous regolith microbreccias because it has the general characteristics listed above. However, it contains a larger proportion of coarse clasts of type 1a and 2 (Table 1) and a smaller amount of undevitrified glass particles than a typical regolith microbreccia, such as 14047. The alumina and lime of this sample are slightly lower than in the regolith breccia 14047 and the fines (Table 5).

Compact and/or shocked regolith microbreccias

Shocked regolith microbreccias (14307, Table 4) are characterized by very low porosity and by the presence of a set of subparallel to parallel, discontinuous tension microfractures that cause these microbreccias to break across grain boundaries. In addition, most fragments have extensive microfractures, and most ilmenite grains show shock-induced fine lamellar twinning. Small chips of shocked regolith micro-breccias are abundant among the coarse fines (<1 cm), and they are commonly coated with glass. Compact regolith microbreccias that show no evidence of shock also have been observed (14313, Table 4). The porosity of these microbreccias is less than 20%.

Spherule-rich microbreccias

These samples are characterized by abundant devitrified spherules and spheroids of varied composition and by subrounded to rounded lithic fragments of both breccia and crystalline rocks. The long dimensions of fragments (in 14315 especially) are aligned in a subparallel manner (Fig. 6). Sample 14318 is similar in bulk com-position to most of the regolith microbreccias, but 14315 is much richer in alumina, lower in iron and magnesia, and very low in titania (Table 5). Most of the spheroids and spherules in 14315 are devitrified, although some of them only on the exterior. Others are completely undevitrified.

Unannealed to slightly annealed Fra Mauro microbreccias

These samples (Figs. 7a and 7b) are white to grayish-white, friable, highly porous, and largely free of thermal metamorphic effects. Plagioclase feldspar fragments predominate. The subordinate lithic fragments are of two main types: yellowish-white to light gray fine-grained micronorite hornfels (type 2); and purplish to grayish-brown devitrified glass containing scattered xenocrysts of plagioclase, spinel, and plagioclase-rich xenoliths (type 3). The bulk compositions of these microbreccias are characterized by high alumina and lime and low iron and potash, (14063,46, Table 5). Thus, in both fragment population and bulk composition, these breccias are distinct from the rest of the Apollo 14 breccias. No undevitrified glass particles were found in this type.

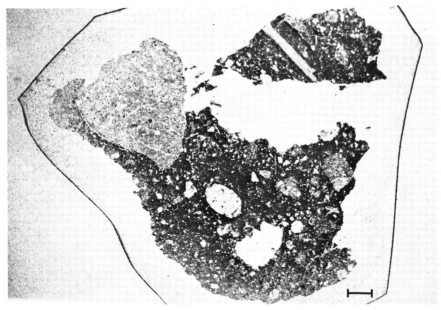

Fig. 5. A thin section of porous regolith microbreccia 14301,84 containing glass spheroids (light-colored elliptical fragment, center, and an egg-shaped particle in the lower portion of the detached part of the thin section; note also the glass veinlet just above it). The largest triangular fragment is a micronorite hornfels. The fragment population of this thin section is given in Table 6. Plane polarized light. Bar scale 1 mm.

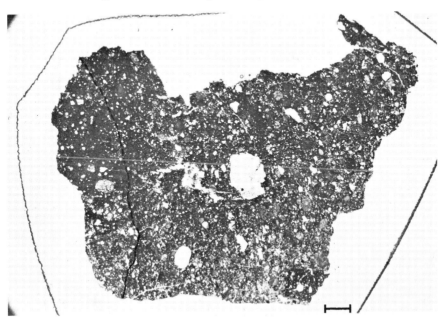

Fig. 6. A thin section of a shocked, transported regolith microbreccia 14315,8. It contains a large number of devitrified spheroids (Table 6) and a few undevitrified glass particles (white elliptical grain, central lower part, and white circle, upper right). Left of the marked dashed line is a complex annealed breccia fragment. Note the alignment of the long axes of fragments in the NNW direction. Plane polarized light. Bar scale 1 mm.

Table 5. Chemical composition of centimeter-size Apollo 14 breccias in weight %.*

	Friable feldspathic microbreccia 14063,46	Spherule-rich microbreccias		Annealed breccia 14303,34	Regolith breccias		Fines 14259,12
		14315,4	14318,27A		14301,62	14047,27	
SiO_2	44.69	47.76	47.97	47.49	48.26	47.45	48.16
TiO_2	1.48	0.80	1.48	1.98	2.06	1.48	1.73
Al_2O_3	22.31	21.31	17.80	16.05	16.52	17.75	17.60
FeO	6.71	7.82	9.62	10.96	10.29	10.36	10.41
MgO	10.80	8.28	9.79	10.99	9.98	9.35	9.26
CaO	12.70	12.77	11.16	10.03	10.29	11.19	11.25
Na_2O	0.76	0.76	0.79	0.87	0.84	0.75	0.61
K_2O	0.15	0.35	0.60	0.46	0.75	0.49	0.51
P_2O_5	0.22	0.23	0.56	0.56	0.64	0.39	0.53
MnO	0.08	0.11	0.13	0.15	0.14	0.13	0.14
Cr_2O_3	0.21	0.23	0.19	0.21	0.21	0.22	0.26
TOTAL	100.11	100.42	100.09	99.75	99.98	99.56	100.46

*Analyzed by combined microchemical and x-ray fluorescence methods. Analysts H. Rose, Jr., et al., U.S. Geological Survey.

Moderately annealed Fra Mauro breccias

The only large sample of this type is 14321, but other smaller fragments are a significant part of coarse fines 14161,3 and 14166,4. Sample 14321 has a moderately coherent, white, feldspathic matrix that contains two main fragment types: (1) ophitic mare basalt similar to 14053; and (2) black fine-grained, annealed, fragment-laden glass (type 1c). Among the Apollo 14 breccias studied, this sample contains the most basalt fragments (Table 6).

Similar smaller breccia fragments in 14161,3 and 14166,4 are gray to light gray, moderately coherent, and contain lithic fragment types 1a, 1b, and 2 in a less coherent, porous feldspathic matrix. Neither 14321 nor these smaller microbreccias contain undevitrified glass particles.

Strongly annealed Fra Mauro breccias

This category contains the largest number of hand specimen-size Apollo 14 breccias. The samples are light gray to gray, coherent, nonporous (or with very low porosity) and contain a wide range of fragment types (Fig. 8). Several generations of metamorphosed breccias are present, each with a range of bulk composition, texture, and perhaps annealing conditions. The matrices are, in general, more feldspathic than the bulk of the annealed fragments. Basalt fragments are present in some samples, but are sparse to very sparse in others. The bulk composition of a chip of one of these breccias, 14303,34 (Table 5), is similar to that of 14301. No undevitrified glass particles are present in these breccias.

DISCUSSION

Only partial results of investigations of many millimeter-size and a few centimeter-size Apollo 14 breccias are reported here. The main task remains to extend the detailed petrographic, x-ray, and chemical studies to large samples of Apollo 14

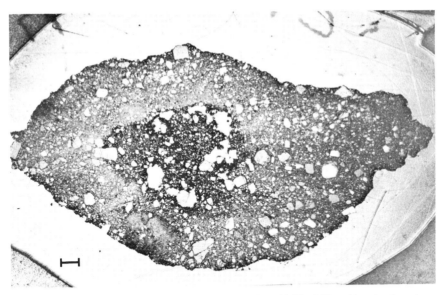

Fig. 7a. A polished thin section of largely unannealed, friable, feldspathic microbreccia 14063,13. Note the high degree of porosity. Two fragment types are dominant in this section (see Table 6): white micronorite hornfels (type 2) and gray devitrified glass (type 3). A large olivine-rich microbreccia fragment in the lower left cannot be distinguished in this photo-micrograph. Dark central area is plucked. Reflected light. Bar scale 1 mm.

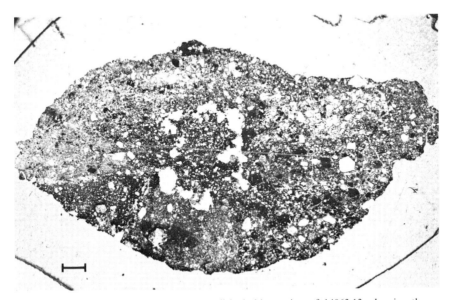

Fig. 7b. Photomicrograph of the same polished thin section of 14063,13, showing the dominant black devitrified fragments (type 3) and an olivine-rich microbreccia fragment in the lower left. The outlines of most of the light-colored micronorite hornfels (type 2) are indistinct. Plane polarized light. Bar scale 1 mm.

Table 6. Fragment population of centimeter-size Apollo 14 breccias in volume percent.

	Friable, white feldspathic		Moderately annealed	Annealed	Spherule-rich	Unshocked regolith
	14063,13	14063,46	14321,241	14303,2	14315,8	14301,84
Lithic and glass fragments						
Fine-grained hornfelsed noritic microbreccia (1a)	—	—	—	44.4	0.2	9.4
Very fine-grained hornfelsed noritic microbreccia (1b)	0.8	3.1	—	11.1	16.8	1.3
Extremely fine-grained annealed fragment-laden glass (1c)	—	—	59.3	—	2.6	—
Micronorite hornfels (2)	21.4	33.0	0.3	—	0.8	44.8
Devitrified glass (3)	16.0	19.0	3.3	—	—	—
Basalts (4a and 4b)	0.6	0.3	27.5	8.2	0.4	2.8
Anorthositic rocks	1.1	1.8	0.5	3.6	3.6	1.4
Olivine-rich rocks	9.9	—	0.5	—	—	0.1
Devitrified glass spheroids	—	—	—	—	38.9	7.2
Undevitrified glass particles	—	—	—	—	1.9	8.7
Mineral fragments						
Plagioclase	24.2	28.6	3.1	7.1	0.8	6.4
Pyroxene	20.7	10.4	5.0	6.5	0.6	5.4
Olivine	3.8	1.6	0.1	0.5	0.3	—
Ilmenite	1.2	0.5	—	0.6	—	0.2
Spinel	0.4	1.8	—	0.2	—	—
Ni-Fe	tr.	tr.	tr.	0.1	—	—
Matrix ($<25\,\mu$)	*	*	*	17.6	33.1	12.2
Pores	(51)	(38)	(18)	—	—	(20)

*Matrix was not determined for these samples; it is included in the count of mineral fragments.

breccias, even though it is impossible to extend these studies directly to the large boulders at the rim of the Cone Crater. The relationship, however, may be inferred as discussed below.

Like the Apollo 11 and 12 porous, unshocked regolith microbreccias, those of Apollo 14 are very fine-grained. They apparently are the products of cold compaction by multiple small impact events on the surface of the regolith (Chao *et al.*, 1971). The major source material, in the case of Apollo 14, is the underlying Fra Mauro formation.

Sample 14301 is porous and unshocked and has the general characteristics of a regolith microbreccia, but it contains a large number of coarse clasts of type 1a and 2 fragments that are abundant in the strongly annealed Fra Mauro breccias. It is possible that 14301 was formed in the lower part of the regolith, which is the most likely source of coarse fragments from the underlying Fra Mauro formation.

We have classified samples 14315 and 14318 as spherule-rich, transported micro-breccias because they contain abundant devitrified spherules and rounded lithic fragments and exhibit some evidence of "bedding" or "foliation," as demonstrated by the subparallel alignment of the long dimensions of fragments, particularly in 14315. It is not clear what type of process transported the materials. Possibly mass wasting, base surge, or the ejection of material traveling long distances may be involved. Because 14315 contains a large number of devitrified feldspathic spheroids (Table 6), it is possible that it is a regolith microbreccia that has been transported from a highly feldspathic terrain. Devitrification suggests that the spherules in this sample have undergone a high-temperature thermal history. However, it is not clear whether the spherules were devitrified before or after incorporation into the breccia. Regolith microbreccias of Apollo 11 and 12 also contain devitrified spherules and devitrified glass-coated or bonded grains (Duke *et al.*, 1970; Lovering and Ware, 1970; McKay *et al.*, 1970; and Chao *et al.*, 1971), but they are scattered and un-aligned. Although we have tentatively classified 14318 together with 14315 because it contains rounded grains, 14318 also resembles regolith microbreccia 14301 in that both samples contain undevitrified glass particles and both have similar bulk chemical compositions (Table 5). Both 14315 and 14318 show evidence of shock compression at low to moderate shock pressures. It is not clear at present whether these spherule-rich microbreccias are clasts in the Fra Mauro ejecta.

In the millimeter- to centimeter-size Apollo 14 breccia samples, strongly annealed fragments such as types 1a, 1b, 1c, 2, and 3 occur as inclusions in unannealed or partly annealed breccias. Hence the strongly annealed fragments have an early thermal metamorphic history (pre-Imbrian) which precedes their incorporation into the surrounding breccia matrix.

The large breccia samples, some boulder size, presumably thrown out by the Cone Crater event. are probably our best representatives of the Fra Mauro formation. The white, largely unannealed, feldspathic microbreccia 14082 was collected from a large boulder near the rim. Strongly annealed gray breccias 14312 and 14319 were collected from the top of a large gray boulder, and moderately annealed sample

Fig. 8. Photomicrograph of annealed breccia 14303,2, consisting essentially of type 1a fine-grained hornfelsed noritic microbreccia fragments (mottled gray areas), and very fine-grained type 1b fragments (black areas). Note the elliptical-shaped fragment, consisting of two generations of breccias, in the lower central part of the section. Plane polarized light. Bar scale 1 mm.

14321 was partially buried near the rim of the crater. The coexistence of these types of breccias is also suggested in photographs of the large boulders in the vicinity of the crater. The moderately to strongly annealed light to medium gray breccias may correspond to the gray to dark gray parts of the boulders, and the white friable feldspathic microbreccias may correspond to the white parts of these boulders.

The coexistence of unannealed and annealed breccias indicates that the Fra Mauro formation is essentially a low-temperature, porous ejecta blanket which suffered little compaction or annealing after deposition at the Fra Mauro site. This interpretation is consistent with the data from the seismic refraction experiment carried out during the Apollo 14 mission. On the basis of the low compressive wave velocity, Watkins and Kovach (1972) conclude that the Fra Mauro formation must be very porous. Hence the nonporous, strongly annealed hand-specimen size breccias must be clasts in the porous Fra Mauro formation. From all this evidence we infer that the geochronologic clock of the Imbrian ejecta at the Fra Mauro site very probably has not been reset by the Imbrian impact event.

Furthermore, Table 6 shows that the fragment populations of 14063, 14321,241, and 14303,2 are different. Preliminary petrographic studies also show that among strongly annealed samples there is a wide range of fragment populations. Examples of this are seen in samples 14311, 14306, and 14319 (Table 4). This variation in fragment population, accompanied by a wide range of annealing textures, is a significant observation which will be studied in greater detail. It signifies that a multiple, intimately mixed fragment source probably existed in the upper part of the Mare Imbrium site before the Imbrian event took place.

It seems clear that the rock suites in the Fra Mauro formation, many of which show evidence of multiple episodes of thermal and impact events, might reveal not only those fragment types which may be derived from earlier multi-ring basin events, such as the formation of Serenitatis, but also a suite of rocks from the upper crust of the moon. A clearer understanding of the various episodes of heating would also give us a better foundation for the dating of the Imbrian event.

Acknowledgments—We are grateful to our colleagues G. A. Desborough, USGS Denver, and P. M. Bell and C. G. Hadidiacos, Geophysical Laboratory, Carnegie Institution of Washington, Washington, D.C., for making electron microprobe facilities available. We thank our colleagues E. D. Jackson and H. G. Wilshire, USGS Menlo Park, and LSAPT for their recommendations and assistance in selecting a suite of breccias for this study. N. G. Benjamin, USGS, Washington, D.C., deceased, prepared many of the fine doubly polished thin sections studied. We thank our colleague O. B. James for critical review of this paper. This work was performed under NASA Contracts T–75412 and W13,130.

References

Carr M. H. and Meyer C. E. (1972) Petrologic and chemical characterization of soils from the Apollo 14 landing site (abstract). In *Lunar Science—III* (editor C. Watkins), pp. 116–118, Lunar Science Institute Contr. No. 88.

Chao E. C. T. Boreman J. A. and Desborough G. A. (1971) The petrology of unshocked and shocked Apollo 11 and Apollo 12 microbreccias. *Proc. Second Lunar Sci. Conf., Geochim. Cosmochim. Acta* Suppl. 2, Vol. 1, pp. 797–816. MIT Press.

Chao E. C. T. Best J. B. and Minkin J. A. (1972) Apollo 14 glasses of impact origin and their parent rock types. *Proc. Third Lunar Sci. Conf., Geochim. Cosmochim. Acta* Suppl. 3, Vol. 1. MIT Press.

Compston W. Vernon M. J. Berry H. Rudowski R. Gray C. M. and Ware N. (1972) Age and petrogenesis of Apollo 14 basalts (abstract). In *Lunar Science—III* (editor C. Watkins), pp. 151–153, Lunar Science Institute Contr. No. 88.

Duke M. B. Woo C. C. Sellers G. A. Bird M. L. and Finkelman R. B. (1970) Genesis of lunar soil at Tranquillity Base. *Proc. Apollo 11 Lunar Sci. Conf., Geochim. Cosmochim. Acta* Suppl. 1, Vol. 1, pp. 347–361. Pergamon.

Floran R. J. Cameron K. Bence A. E. and Papike J. J. (1972) The 14313 consortium: A mineralogic and petrologic report (abstract). In *Lunar Science—III* (editor C. Watkins), pp. 268–270, Lunar Science Institute Contr. No. 88.

Helz R. T. (1972) Rock 14068: An unusual lunar breccia. *Proc. Third Lunar Sci. Conf., Geochim. Cosmochim. Acta* Suppl. 3, Vol. 1. MIT Press.

Jackson E. D. and Wilshire H. G. (1972) Classification of the samples returned from the Apollo 14 landing site (abstract). In *Lunar Science—III* (editor C. Watkins), pp. 418–420, Lunar Science Institute Contr. No. 88.

Lovering J. F. and Ware N. G. (1970) Electron microprobe analyses of minerals and glasses in Apollo 11 lunar samples. *Proc. Apollo 11 Lunar Sci. Conf., Geochim. Cosmochim. Acta* Suppl. 1, Vol. 1, pp. 633–654. Pergamon.

LSPET (1971) (Lunar Sample Preliminary Examination Team) Preliminary examination of lunar samples from Apollo 14. *Science* 173, 681–693.

Marvin U. B. Reid J. B. Jr. Taylor G. J. and Wood J. A. (1972) A survey of lithic and vitreous types in the Apollo 14 soil samples. *Proc. Third Lunar Sci. Conf., Geochim. Cosmochim. Acta* Suppl. 3, Vol. 1. MIT Press.

McKay D. S. Greenwood W. R. and Morrison D. A. (1970) Origin of small lunar particles and breccias from the Apollo 11 site. *Proc. Apollo 11 Lunar Sci Conf., Geochim. Cosmochim. Acta* Suppl. 1, Vol. 1, pp. 673–694. Pergamon.

McKay D. S. Clanton U. S. Heiken G. H. Morrison D. A. and Taylor R. M. (1972) Vapor phase crystallization in Apollo 14 breccias and size analysis of Apollo 14 soils (abstract). In *Lunar Science—III* (editor C. Watkins), pp. 529–531, Lunar Science Institute Contr. No. 88.

Swann G. A. Trask N. J. Hait M. H. and Sutton R. L. (1971) Geologic setting of the Apollo 14 samples. *Science* 173, 716–719.

Warner J. L. (1972) Apollo 14 breccias: Metamorphic origin and classification (abstract). In *Lunar Science—III* (editor C. Watkins), pp. 782–784, Lunar Science Institute Contr. No. 88.

Watkins J. S. and Kovach R. L. (1972) Apollo 14 active seismic experiment. *Science* 175, 1244–1245.

Wilshire H. G. and Jackson E. D. (1972) Petrology of the Fra Mauro formation at the Apollo 14 landing site (abstract). In *Lunar Science—III* (editor C. Watkins), pp. 803–805, Lunar Science Institute Contr. No. 88.

Proceedings of the Third Lunar Science Conference
(Supplement 3, *Geochimica et Cosmochimica Acta*)
Vol. 1, pp. 661–671
The M.I.T. Press, 1972

Apollo 14 breccia 14313: A mineralogic and petrologic report

R. J. FLORAN, K. L. CAMERON, A. E. BENCE, AND J. J. PAPIKE

Department of Earth and Space Sciences
State University of New York, Stony Brook, New York 11790

Abstract—Lunar sample 14313 is a coherent polymict breccia that has had a complex history of comminution and reagglomeration. The dominant types of clasts are (1) noritic rock fragments, (2) monomineralic fragments, (3) microbreccia clasts, and (4) glassy fragments including glass spherules. Mare-type basalt clasts are rare. The matrix of the breccia is composed primarily of fine particles of brownish glass. Petrographic examination reveals varying degrees of shock damage of the clasts ranging from unshocked through shock-melted fragments.

Both thermally recrystallized and unrecrystallized microbreccia clasts are present in 14313. The former lack glass and have an annealed matrix, whereas the latter contain abundant glass both as clasts and in their matrices. Based on the criterion of breccia-within-breccia, a sequence of four unrecrystallized microbreccias has been recognized. The number of impact events needed to form this sequence cannot be determined uniquely, but the high glass content and high average level of shock features of the clasts in 14313 suggest many events.

We believe that breccia 14313 was formed by shock lithification of regolith developed on the Fra Mauro formation. The abundance of micronorite clasts suggests that noritic rocks were an important pre-Imbrian rock type in the area now occupied by the Imbrium Basin. Breccia 14313 appears to be a multi-impact soil breccia with great textural variation and clast diversity reflecting a mature regolith deposit.

INTRODUCTION

SPECIMEN 14313 is a coherent polymict breccia that is being studied by a consortium of 9 laboratories. These studies include mineralogy-petrology, stable light isotopes, rare gases, geochronology, shock deformation, and major trace-element chemistry. The purpose of this paper is to describe the mineralogy and petrology of 14313 and to consider its origin. We have examined 5 thin sections of the breccia (14313,7; 14313,39; 14313,40; 14313,41; 14313,42) that have a total rock area of approximately 1370 mm^2.

The Apollo 14 landing site was on the Fra Mauro formation (Wilhelms and McCauley, 1971), believed to be an ejecta blanket thrown out of the Imbrium Basin approximately 3.9×10^9 years ago. The Fra Mauro formation is characterized by a series of discontinuous parallel ridges that are radial to the Imbrium Basin. In the vicinity of the landing site, the Fra Mauro formation has been divided into two geologic units: smooth terrain material and ridge material (Eggleton and Offield, 1970). Specimen 14313 was collected from the smooth terrain material at sample station G1 approximately 150 m east of the Lunar Module landing point (Swann *et al.*, 1971b). Station G1 was on the north rim of North Triplet Crater, interpreted to be one of the oldest craters on the geologic traverse. The smooth terrain material in this area is not covered by the Cone Crater ray deposit, that contains ridge material (Fig. 1).

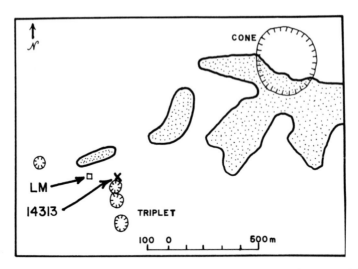

Fig. 1. Apollo 14 locality map (after Swann *et al.*, 1971b). Stippled area is ray material from Cone Crater.

This specimen has been briefly characterized and classified in two separate studies of the Apollo 14 breccias. In a study of the thermal metamorphism of the breccias, Warner (1972) classified 14313 as essentially unmetamorphosed (Low Facies, group 1). The low-grade breccias contain abundant glass both as clasts and as matrix material. In contrast, high-grade breccias contain few, if any, undevitrified glass clasts, and their matrices are recrystallized and have a fine granoblastic texture. Warner concluded that the high-grade rocks were annealed within the hot Fra Mauro ejecta blanket, whereas the low-grade rocks were soil breccias or unmetamorphosed Imbrian ejecta. Jackson and Wilshire (1972) placed 14313 with their F2 type breccias that generally correspond with Warner's low-grade rocks and the shock-compressed microbreccias of Chao *et al.* (1971).

PETROGRAPHIC DESCRIPTION

Breccia 14313 is a poorly sorted aggregate that contains a complete range in fragment size to as much as 10 mm. In the following discussion, we refer to particles larger than 25 μ as clasts and material less than 25 μ as matrix. The dominant clast types in 14313 are glass fragments, noritic rock fragments, microbreccia clasts, and monomineralic fragments of plagioclase and pyroxene. The matrix is composed primarily of fine particles of brown glass with lesser amounts of crystalline fragments.

Glass clasts

Irregular glass fragments are very abundant in 14313 and they have a wide range of color, size, and optical homogeneity. They range from colorless to very dark brown and from fragments free of crystalline inclusions to crystalline aggregates cemented by fused glass. Glass spherules also are common, and most are optically homogeneous with no devitrification and/or quench crystals.

Representative analyses of glasses are presented in Table 1. Glass clasts 1 and 2, Table 1, have KREEP affinities. Their K and P contents are slightly lower than that of typical KREEP (Hubbard *et al.*, 1971; Meyer *et al.*, 1971); however, they are within the range of KREEP analyses reported by Meyer *et al.* (1971, Table 1). The composition of a common type of glass in 14313 that occurs as yellowish brown fragments, generally containing irregular stringers of dark brown dust-like inclusions, is presented in Analysis 2.

Micronorite clasts

The most abundant type of "igneous rock" clast in 14313 is an orthopyroxene-plagioclase-rich rock that we call micronorite. Fragments of similar mineralogy and texture have been called orthopyroxene-plagioclase fragments (Fuchs, 1970), norite (Wood *et al.*, 1970), KREEP (Meyer *et al.*, 1971), gray mottled basalt (Anderson and Smith, 1971), Fra Mauro basalt (Apollo Soil Survey, 1971), and more recently KREEP basalt (Meyer, 1972, and other workers).

We have used the term micronorite to emphasize the very fine-grained nature of these rocks. They contain phenocrysts of plagioclase and more rarely orthopyroxene that may be as much as $500\,\mu$ long; nevertheless, the average grain size of most micronorites is in the range of 100 to $300\,\mu$. The fine-grained texture of these rocks suggests they are hypabyssal and/or volcanic rather than plutonic in origin. There is no textural evidence for their being cumulates.

The texture of the micronorites may be different from clast to clast, and is somewhat variable within individual clasts. Some clasts have to a large degree retained their original igneous intergranular to subophitic textures (Fig. 2). Most, however, have been brecciated and/or thermally annealed, and a few have been strongly shocked.

Table 1. Electron microprobe analyses* of glasses from 14313. All values in weight percent.

	1	2	3	4	5	6
SiO_2	48.8	48.7	46.9	45.2	44.8	43.0
TiO_2	2.14	2.41	2.12	0.78	0.86	3.42
Al_2O_3	15.3	17.1	18.4	7.94	6.98	15.14
FeO	12.1	11.3	11.1	20.0	23.0	16.4
MgO	9.24	6.86	9.62	16.6	15.1	8.95
MnO	0.17	0.16	0.19	0.29	0.31	1.11
CaO	9.02	9.97	11.1	8.74	7.76	11.9
Na_2O	1.08	1.11	0.13	0.22	0.27	1.13
K_2O	0.78	0.62	0.05	0.01	0.06	0.0
P_2O_5	0.62	0.59	0.04	0.11	0.12	0.05
TOTAL	99.3	98.8	99.7	99.9	99.3	100.2

* Electron microprobe data were reduced using the method of Bence and Albee (1968).

1 dark brown glass fragment.
2 yellowish brown fragment.
3 very light brown sphere.
4 light brown sphere.
5 light brown sphere.
6 light brown sphere.

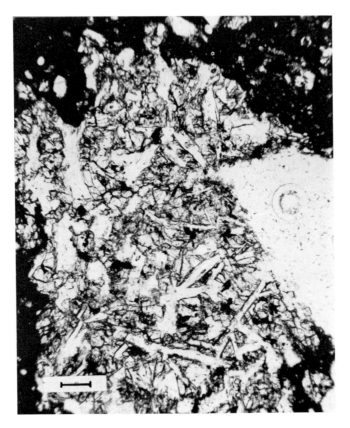

Fig. 2. Micronorite clast with intergranular texture. Scale bar is 100 μ

There does not appear to be any substantial difference in the major element chemistry of the minerals from the various textural types of micronorites. Pyroxene compositions from about 15 different noritic clasts and one basaltic clast are shown in Fig. 3, and selected analyses are presented in Table 2. Most pyroxenes with ≤ 5 mole % Wo have low birefringence and parallel extinction, indicating that they are orthopyroxenes. These orthopyroxenes have a small compositional range, from about 15 to 35 mole % Fs. The pyroxenes are zoned and it appears that the range within single crystals (Fig. 3) may approach the range in composition of ortho-pyroxenes from different noritic fragments.

Olivine in the micronorites ranges in composition from about 25 to 40 mole % Fa (Fig. 3). The plagioclase compositions (Fig. 4, Table 3) range from approximately 75 to 95 mole % An and many crystals are zoned over a range of 10 to 15 mole % An.

Monomineralic fragments

The most common monomineralic fragments are plagioclase and pyroxene, and most of these clasts are single crystals. The compositions of several of these fragments are shown in Figs. 3 and 4. The monomineralic feldspar crystals are similar in

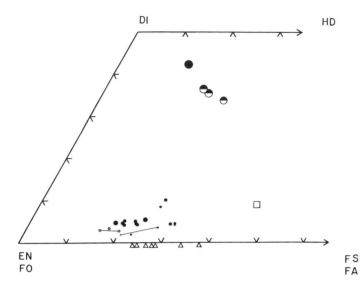

Fig. 3. Pyroxene and olivine analyses. Circles indicate the compositions of pyroxenes, and areas of the circles are proportional to their content of elements other than Ca, Fe, Mg, and Si. See Table 2 for representative analyses. Solid circles = micronorites; open circles = monomineralic fragments; half-solid circles = basalt; square = bulk composition of a pyroxene fragment with exsolution lamellae. Triangles indicate the Fe–Mg content of olivines from micronorites.

Table 2. Electron microprobe analyses* of pyroxenes (weight percent).

	1	2	3	4	5	6
SiO$_2$	55.2	54.0	52.8	54.6	51.7	51.8
TiO$_2$	1.23	0.98	0.79	0.63	1.62	1.53
Al$_2$O$_3$	1.36	1.54	0.77	2.40	2.70	1.77
FeO	12.8	14.3	19.7	15.3	8.46	13.1
MgO	27.7	26.2	23.4	25.4	14.5	14.4
MnO	n.d.	0.21	n.d.	0.17	0.11	0.25
CaO	2.22	2.14	2.16	2.60	19.8	16.7
Na$_2$O	n.d.	0.03	n.d.	0.10	0.23	0.14
K$_2$O	n.d.	0.00	n.d.	0.04	0.01	0.01
TOTAL	100.5	99.4	99.6	101.1	99.1	99.7
Si	1.96	1.96	1.96	1.95	1.93	1.95
Al	0.06	0.07	0.03	0.10	0.12	0.08
Ti	0.03	0.03	0.02	0.02	0.05	0.04
Fe	0.38	0.43	0.61	0.46	0.26	0.41
Mg	1.47	1.41	1.29	1.35	0.81	0.81
Mn	—	0.01	—	0.01	0.00	0.01
Ca	0.09	0.08	0.09	0.10	0.79	0.67
Na	—	0.00	—	0.01	0.02	0.01
K	—	0.00	—	0.00	0.00	0.00

* Electron microprobe data were reduced using the method of Bence and Albee (1968). Formulae were normalized to 6 oxygens.

1 orthopyroxene, micronorite.
2 orthopyroxene, micronorite.
3 orthopyroxene, brecciated micronorite.
4 orthopyroxene, shocked noritic fragment.
5 clinopyroxene, brecciated micronorite.
6 clinopyroxene, basalt.

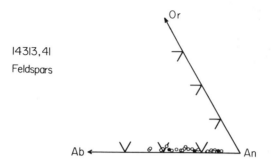

Fig. 4. Plagioclase compositions from 14313,41. Open circles = micronorites; solid circles = monomineralic fragments.

composition to the feldspars in the noritic clasts. Some of the monomineralic pyroxene clasts contain exsolution lamellae, a feature that was not observed in the pyroxenes from the micronorite clasts. The bulk composition, determined by defocused beam electron-probe analysis, of a single crystal pyroxene clast containing exsolution lamellae is shown in Fig. 3. This pyroxene appears to be significantly more Fe-rich than the pyroxenes from the micronorites.

Recrystallized microbreccia clasts

Some clasts in 14313 appear to have been thermally metamorphosed. These include the annealed micronorites and a rather heterogeneous group of clasts

Table 3. Electron microprobe analyses* of feldspars from clasts in 14313 (weight percent).

	1	2	3	4
SiO_2	47.6	48.5	45.3	47.0
TiO_2	0.09	0.05	0.05	0.04
Al_2O_3	33.9	32.9	35.8	34.2
FeO	0.50	0.31	0.31	0.15
MgO	0.26	0.10	0.08	0.10
CaO	16.3	15.0	18.1	16.5
Na_2O	1.46	2.50	0.70	1.71
K_2O	0.16	0.12	0.03	0.12
TOTAL	100.3	99.5	100.4	99.8
Si	2.18	2.22	2.08	2.16
Al	1.83	1.78	1.93	1.85
Ti	0.00	0.00	0.00	0.00
Fe	0.02	0.01	0.01	0.01
Mg	0.02	0.01	0.01	0.01
Ca	0.80	0.74	0.89	0.81
Na	0.13	0.22	0.06	0.15
K	0.01	0.01	0.00	0.01

* Electron microprobe data were reduced using the method of Bence and Albee (1968). Formulae were normalized to 8 oxygens.

1 micronorite.
2 brecciated micronorite.
3 brecciated micronorite.
4 monomineralic fragment.

referred to as recrystallized microbreccias. The latter do not contain glass or recognizable clasts of other rock types, but they have a recrystallized matrix, generally contain
more olivine and clinopyroxene than the typical annealed micronorites, and some
contain small vugs. There is no clear compositional or textural division between the
annealed micronorites and the recrystallized microbreccias. These recrystallized
microbreccia clasts would be classified as Warner's (1972) high-grade breccias.

Unrecrystallized microbreccia clasts

Unrecrystallized microbreccia clasts that contain abundant glass both as clasts
and matrix material are common in 14313. Based on the textural criterion of "breccia
enclosing breccia," a sequence of four unrecrystallized breccias has been recognized,
and they have been numbered in the following manner: breccia 4 is the whole rock
14313; breccia 3 is within breccia 4; breccia 2 is within breccia 3; and finally breccia 1
is within breccia 2. Breccia clasts of the same number were not necessarily formed
contemporaneously, and it is not possible to give a unique value to the number of
events necessary to form the observed sequence of four breccias.

Unrecrystallized microbreccia clasts are most abundant in thin section 14313,42.
The most conspicuous ones in this slide are four dark, rounded clasts (Fig. 5) that
range from 5 to 10 mm in diameter. Only one clearly defined unrecrystallized breccia
2 clast, approximately 2.5 × 1 mm, was observed, and it contains a single unrecrystallized microbreccia clast (breccia 1), approximately 1 × 0.5 mm. The types and
relative proportions of clasts appear to be similar in all four breccias of the sequence.
The various types of clasts in each breccia are shown in Table 4.

Separate modal analyses were made of two regions within 14313,42. One area
consisted of the four large dark breccia 3 clasts (Fig. 5); the second area was the

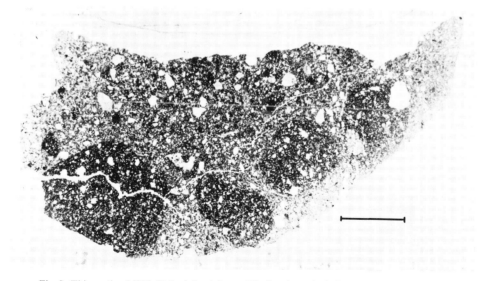

Fig. 5. Thin section 14313,42. Scale bar is 5 mm. The four large dark clasts are unrecrystallized
microbreccias.

Table 4. Clast types in each breccia. Breccia 4 encloses breccia 3; breccia 3 encloses breccia 2; etc.

Clast $>25\,\mu$	Breccia 4	Breccia 3	Breccia 2	Breccia 1
Micronorite	X	X	X	
Glass fragments	X	X	X	X
Basalt	X	X		
Plagioclase fragments	X	X	X	X
Pyroxene fragments	X	X	X	X
Recrystallized microbreccia	X	X		

remainder of the rock. Table 5 demonstrates that the two areas contain similar types and proportions of clasts. The darker color of the breccia 3 clasts probably is caused by a higher content of matrix material composed primarily of fine particles of dark brown glass.

Shock features

Petrographic examination reveals varying degrees of shock damage ranging from unshocked through shock-melted fragments. This complete range in shock features indicates that the highly shocked clasts underwent shock deformation prior to incorporation within 14313. Shock effects in 14313 are discussed in greater detail by Avé Lallemant and Carter (1972, this volume).

The Preliminary Examination Team reported that sample 14313 contained two sets of intersecting fractures (Swann *et al.*, 1971a). In thin section the fractures can be seen to pass around some clasts but through others (Fig. 5). They commonly consist of groups of discontinuous fractures that closely follow the boundaries between the large dark clasts (breccia 3) and the remainder of 14313. However, these

Table 5. Size distributions and modes of two areas in 14313,42 (area percent).

	Size distribution	
	Large dark breccia 3 clasts	Remainder of 14313
Matrix $<25\,\mu$	68.7%	60.1%
Fragments $25-250\,\mu$	20.7	17.7
Fragments $250-1000\,\mu$	8.0	14.9
Fragments $>1000\,\mu$	2.6	7.3
	100.0%	100.0%
	Relative proportions of clasts $>25\,\mu$	
Glass fragments	32.9%	40.1%
Micronorites	37.6	34.9
Basalt	2.4	0.4
Microbreccias	12.8	6.4
Opaque minerals	0.0	0.4
Monomineralic plagioclase fragments	8.8	11.5
Monomineralic pyroxene fragments	4.8	6.3
Monomineralic olivine fragments	0.7	0.0
	100.0%	100.0%
Total number of clasts $>25\,\mu$	125	252

fractures break across clast-matrix interfaces, indicating similar coherency. Except for fractures, the rock has very low porosity ($<1\%$). Sample 14313 appears to be best classified as a shock compressed breccia following Chao *et al.* (1971).

DISCUSSION AND INTERPRETATION

We believe that shock lithification of regolith material formed breccia 14313. Evidence supporting this interpretation is (1) the large content of glass both as clasts and in the matrix, (2) similarity in lithology between Apollo 14 soils and breccia 14313, (3) low porosity, and (4) the fractured nature of the specimen.

Wilshire and Jackson (1972) consider their F2 type breccias, of which 14313 is an example, to be representative of the top depositional unit of the Fra Mauro formation. Their conclusion was based on the tacit assumption that the lithification of the breccias dates from the deposition of the Fra Mauro. We, in contrast, believe that the material in 14313 has been extensively reworked by meteoritic impact since the deposition of the Fra Mauro formation and that the specimen was shock lithified at or near the lunar surface by a post-Imbrian event. The high average level of shock features of clasts in 14313 suggests multiple periods of deformation. If the recrystallized microbreccia clasts were thermally metamorphosed in the Fra Mauro ejecta blanket, then breccias 3 and 4, that contain these fragments, must have been lithified *after* the Imbrian event. Alternatively, these recrystallized fragments may have been annealed in pre-Imbrian times, either in the area now occupied by the Imbrium Basin or in an older underlying ejecta blanket at the Apollo 14 site. Although we consider 14313 to be lithified regolith, other F2 type breccias may well be unreworked upper Fra Mauro formation.

The compositional and textural similarities of all the unrecrystallized microbreccia in 14313 suggest that all could have been soil breccias. The absence of basalts and recrystallized microbreccias in breccias 1 and 2 (Table 5) probably is not significant. They are rather uncommon rock types in 14313, and breccias 1 and 2 are represented by only very small fragments. The latter reason also may explain the absence of micronorite clasts in breccia 1. Although mare-type basalt clasts are rare in 14313, they are abundant in some other Apollo 14 breccias (e.g., 14321, see Grieve *et al.*, 1972).

Unfortunately it is not possible to formulate a unique number of impact events to explain the sequence of breccia-within-breccia texture. However, the presence of recrystallized microbreccias within 14313 strongly suggests that more than one impact event was involved. All of the unrecrystallized microbreccia clasts are rounded. Rounding could occur by abrasion in a rapidly moving and turbulent base surge (Moore, 1967; O'Keefe *et al.*, 1969) or by repeated micrometeorite impact on the lunar regolith. Sample 14313, however, lacks accretionary lapilli, high porosity, and other features that have been ascribed to base surge deposits (McKay and Morrison, 1971; Waters and Fisher, 1971).

The abundance of micronorite clasts in 14313 suggests that noritic rocks were an important rock type in the pre-Imbrian impact area. It was predicted from cratering mechanics that, because of the great distance from the impact, the Fra Mauro ejecta in the Apollo 14 vicinity would be composed of shallow rock types from the Imbrium Basin. The very fine-grained texture of the micronorites is consistent with an extrusive and/or hypabyssal origin. However, coarse-grained igneous rock clasts in the lunar regolith probably would be reduced to monomineralic fragments relatively quickly, making it very difficult to detect their former presence, especially in the fine-grained breccias. The visible exsolution lamellae in some of the monomineralic pyroxene fragments in 14313 suggest that these grains were derived from igneous rocks that cooled relatively slowly (Papike and Bence, 1972); although it is possible that exsolution could have taken place within a thick hot ejecta blanket prior to incorporation into 14313.

Summary and Conclusions

(1) Sample 14313 is a coherent polymict breccia containing clasts of plagioclase, pyroxene, glass, and micronorite set in a matrix composed primarily of very fine particles of brown glass.

(2) Microbreccia clasts are of two types: recrystallized and unrecrystallized. The recrystallized microbreccias are believed to have been thermally metamorphosed. The unrecrystallized microbreccias contain abundant glass and, based upon breccia-within-breccia texture, a sequence of four unrecrystallized microbreccias has been recognized.

(3) The abundance of micronorite clasts in 14313 suggests that noritic rocks were an important pre-Imbrian rock type in the area now occupied by the Imbrium Basin. Most micronorites have undergone a complicated history as a result of shock and/or thermal reheating, and there is a range in texture from intergranular to hornfelsic.

(4) Shock lithification of a well-mixed regolith formed 14313. Its glassy matrix is incompatible with formation deep within a hot ejecta blanket but suggests rapid quenching at the lunar surface.

(5) It is not possible at the present time to delineate clearly the number of separate events that led to the formation of 14313. However, the presence of *recrystallized* microbreccia within 14313 strongly suggests that more than one event was involved. It is quite possible that a large number of events affected some of the clasts, but direct evidence is lacking.

References

Anderson A. T. Jr. and Smith J. V. (1971) Nature, occurrence and exotic origin of "grey mottled" (Luny Rock) basalts in Apollo 12 soils and breccias. *Proc. Second Lunar Sci. Conf., Geochim. Cosmochim. Acta*, Suppl. 2, Vol. 1, 431–438. M.I.T. Press.

Apollo Soil Survey (1971) Apollo 14: Nature and origin of rock types in soil from the Fra Mauro Formation. *Earth Planet. Sci. Lett.* **10**, 49–54.

Avé Lallemant H. G. and Carter N. L. (1972) Deformation of silicates in some Fra Mauro breccias (Abstract). In *Lunar Science—III* (editor C. Watkins), p. 33, Lunar Science Institute Contr. No. 88.

Bence A. E. and Albee A. L. (1968) Empirical correction factors for the electron microanalysis of silicates and oxides. *J. Geol.* **76**, 382–403.

Chao E. C. T. Boreman J. A. and Desborough G. A. (1971) The petrology of unshocked and shocked Apollo 11 and Apollo 12 microbreccias. *Proc. Second Lunar Sci. Conf., Geochim. Cosmochim. Acta*, Suppl. 2, Vol. 1, pp. 797–816. M.I.T. Press.

Eggleton R. E. and Offield T. W. (1970) Geologic maps of the Fra Mauro region of the moon, U.S. Geol. Surv. Misc. Geol. Invest. Map I-708.

Fuchs L. H. (1970) Orthopyroxene-plagioclase fragments in the lunar soil from Apollo 12. *Science* **169**, 866–867.

Grieve R. McKay G. Smith H. and Weill D. (1972) Mineralogy and petrology of polymict breccia 14321 (Abstract). In *Lunar Science—III* (editor C. Watkins), p. 339, Lunar Science Institute Contr. No. 88.

Hubbard N. J. Meyer C. Jr. Gast P. W. and Weismann H. (1971) The composition and derivation of Apollo 12 soils. *Earth Planet. Sci. Lett.* **10**, 341–350.

Jackson E. D. and Wilshire H. G. (1972) Classification of the samples returned from the Apollo 14 landing (Abstract). In *Lunar Science—III* (editor C. Watkins), p. 418, Lunar Science Institute Contr. No. 88.

McKay D. S. and Morrison D. A. (1971) Lunar breccias. *J. Geophys. Res.* **76**, 5658–5669.

Meyer C. Jr. (1972) Mineral assemblages and the origin of non-mare lunar rock types (Abstract). In *Lunar Science—III* (editor C. Watkins), p. 542, Lunar Science Institute Contr. No. 88.

Meyer C. Jr. Brett R. Hubbard N. J. Morrison D. A. McKay D. S. Aitken F. K. Takeda H. and Schonfeld E. (1971) Mineralogy, chemistry, and origin of the KREEP component in soil samples from the Ocean of Storms. *Proc. Second Lunar Sci. Conf., Geochim. Cosmochim. Acta*, Suppl. 2, Vol. 1, pp. 393–411. M.I.T. Press.

Moore J. C. (1967) Base surge in recent volcanic eruptions. *Bull. Volcanol.* **30**, 337–363.

O'Keefe J. Cameron W. S. and Masursky H. (1969) Hypersonic gas flow in Analysis of Apollo 8 Photographs and Visual Observations, NASA SO-201, p. 30.

Papike J. J. and Bence A. E. (1972) Apollo 14 inverted pigeonites: Possible samples of lunar plutonic rocks. *Earth Planet. Sci. Lett.* **14**, 176–182.

Swann G. A. Bailey N. G. Batson R. M. Eggleton R. E. Hait M. H. Holt H. E. Larson K. B. McEwen M. C. Mitchell E. D. Schaber G. G. Schafer J. P. Shepard A. B. Sutton R. L. Trask N. J. Ulrich G. E. Wilshire H. G. and Wolfe E. W. (1971a) Preliminary geologic investigations of the Apollo 14 landing site. Apollo 14 Preliminary Science Report, NASA SP-272, Chapter 3, pp. 39–85.

Swann G. A. Trask N. J. Hait M. H. and Sutton R. L. (1971b) Geologic setting of the Apollo 14 samples. *Science* **173**, 716–719.

Warner J. L. (1972) Apollo 14 breccias: metamorphic origin and classification (Abstract). In *Lunar Science —III* (editor C. Watkins), p. 782, Lunar Science Institute Contr. No. 88.

Waters A. C. and Fisher R. V. (1971) Base surges and their deposits: Capelinhos and Taal volcanoes. *J. Geophys. Res.* **76**, 5596–5614.

Wilhelms D. E. and McCauley J. F. (1971) Geologic map of the near side of the moon. U.S. Geol. Surv. Misc. Geol. Invest. Map I-703.

Wilshire H. G. and Jackson E. D. (1972) Petrology of the Fra Mauro Formation at the Apollo 14 landing site (Abstract). In *Lunar Science—III* (editor C. Watkins), p. 803, Lunar Science Institute Contr. No. 88.

Wood J. A. Dickey J. S. Jr. Marvin J. B. and Powell B. N. (1970) Lunar anorthosites and a geophysical model of the moon. *Proc. Apollo 11 Lunar Sci. Conf., Geochim. Cosmochim. Acta*, Suppl. 2, Vol. 1, pp. 965–988. M.I.T. Press.

Proceedings of the Third Lunar Science Conference
(Supplement 3, *Geochimica et Cosmochimica Acta*)
Vol. 1, pp. 673–686
The M.I.T. Press, 1972

Chondrules in Apollo 14 samples and size analyses of Apollo 14 and 15 fines

Elbert A. King, Jr., John C. Butler, and Max F. Carman

Department of Geology, University of Houston, Houston, Texas 77004

Abstract—Chondrules have been observed in several breccia samples and one fines sample returned by the Apollo 14 mission. The chondrules are formed by at least three different processes that appear to be related to large impacts: (1) crystallization of shock-melted spherules and droplets; (2) rounding of rock clasts and mineral grains by abrasion in the base surge; and (3) diffusion and recrystallization around clasts in hot base surge and fall-back deposits. In the case of the Apollo 14 samples, the large impact almost certainly is the Imbrian event. Some of the chondrules in chondritic meteorites may have formed by similar processes on other planetary surfaces. Many of the differences between petrologic types of chondritic meteorites can be accounted for by this model. Chondrules may be an inevitable result of the terminal stages of accretion of silicate planetary bodies.

Grain size analyses of undisturbed fines samples from the Apollo 14 site and from the Apollo 15 Apennine Front are almost identical, indicating that the two localities have similar meteroid bombardment exposure ages, approximately 3.7×10^9 yr. This observation is consistent with the interpretation that both the Fra Mauro formation and the Apennine Front material originated as ejecta from the Imbrian event. Size analyses of fines from under, on and around the boulder that was sampled near St. George Crater indicate that the boulder has been exposed at that site for a short period of time, probably less than 1×10^5 yr.

Lunar Chondrules: Introduction and Previous Work

Numerous chondrules and chondrule-like bodies occur in at least three Apollo 14 breccia samples. Many of these lunar chondrules are identical, in texture and general mineralogy, to common types of meteoritic chondrules. These chondrules are not the relatively simple glass and glassy spherules that have been described previously by virtually all mineralogy and petrology investigators (King *et al.*, 1970; Chao *et al.*, 1971; Fredriksson *et al.*, 1971; and others), although some of the chondrules are partly glass and others are devitrified glass in part. Roedder (1971) noted crystallized spherules in some Apollo 11 soil and breccia samples and stated that there were some interesting textural similarities with meteoritic chondrules. Fredriksson *et al.* (1971) emphasized the textural similarities of some glassy lunar particles and some types of meteoritic chondrules. However, the first *convincing* occurrence of lunar chondrules was in the Apollo 14 breccia and soil samples.

Lunar chondrules were first reported in Apollo 14 samples by Butler *et al.* (1971), Kurat *et al.* (1971), Fredriksson *et al.* (1971a) and von Engelhardt *et al.* (1971) at the 34th Annual Meeting of the Meteoritical Society in Tübingen, Germany. A brief article illustrating lunar chondrules that discussed the possible implications for the origin of chondritic meteorites was published by King *et al.* (1972). Four papers that discussed lunar chondrules were presented at the Third Lunar Science Conference (King *et al.*, 1972a; Kurat *et al.*, 1972; Fredriksson *et al.*, 1972 and Lévy and Lévy, 1972).

CHONDRULE OCCURRENCE AND DISCUSSION

Apollo 14 breccia samples 14313, 14318, and 14301 contain abundant chondrules and chondrule-like bodies, as much as 10% by volume of the portions of the samples that have been examined. Rare to moderately abundant chondrules have been observed in breccia samples 14305, 14306, and 14311. In addition, a fragment containing lunar chondrules was identified in fines sample 14162. Many of these lunar chondrules have the textures and mineralogies of common types of meteoritic chondrules. These textures include: euhedral olivine and pyroxene crystals in brown transparent glass (Figs. 1 and 2); radiating groups of pyroxene and plagioclase crystals that appear to have nucleated on the surface of a spherical body and crystallized into the interior (Fig. 3); and olivine-rich cores surrounded by recrystallized aureoles or diffusion halos of pyroxene and opaque minerals (Fig. 4). Also, Fredriksson *et al.* (1972) have observed a barred olivine-glass chondrule in sample 14259,33. The textural and mineralogical similarity of these lunar chondrules to meteoritic chondrules is apparent to meteorite researchers. There are differences between the compositions of lunar chondrules and most meteoritic chondrules, as has been noted by Kurat *et al.* (1971), but this difference is most likely a function of target composition.

The important conclusion is that chondrules are formed on the moon by several mechanisms that are related to large impacts (Butler *et al.*, 1971; King *et al.*, 1972, 1972a). These mechanisms include the following:

(1) Crystallization of shock-melted silicate spherules and droplets (Fig. 3). This

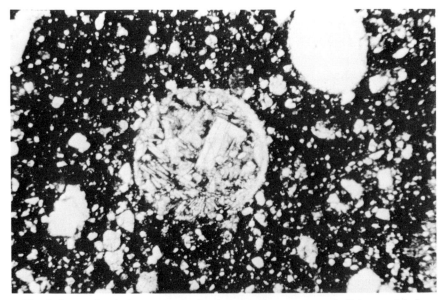

Fig. 1. Lunar chondrule composed of euhedral orthopyroxene and olivine crystals in brown transparent glass in sample 14313. Note that there is a glass spherule in the field of view (upper right) that shows no signs of devitrification or crystallization. Plane polarized light, length of field of view is approximately 3 mm.

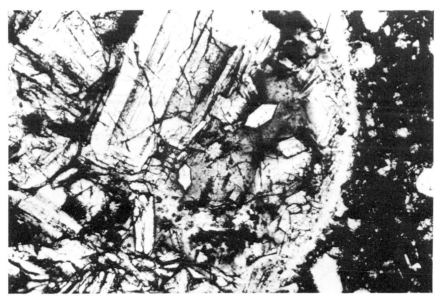

Fig. 2. A portion of the orthopyroxene-olivine-glass lunar chondrule in 14313 from Fig. 1, showing small euhedral olivine crystals in matrix of light brown transparent glass. Plane polarized light, length of field of view is approximately 0.8 mm.

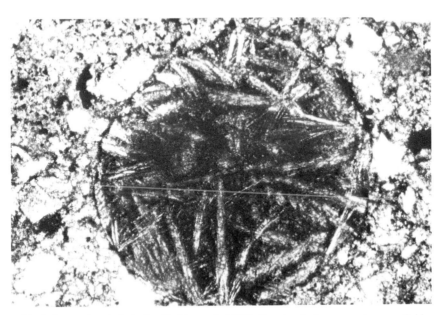

Fig. 3. Lunar chondrule in Apollo 14 sample 14318. The chondrule apparently was a fluid drop that assumed a spherical shape due to surface tension, then crystals of plagioclase and pyroxene began to nucleate at the surface of the sphere, and crystallization proceeded into the interior. The crystals are surrounded by dark brown turbid glass. Plane polarized light, diameter of chondrule is 0.5 mm. [Courtesy of *Science*]

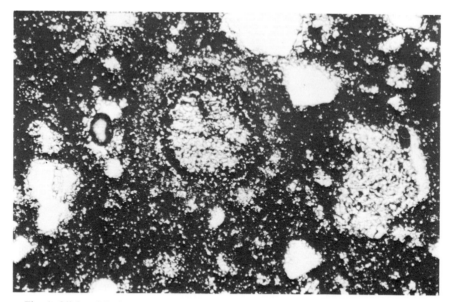

Fig. 4. Olivine-rich clast (dunite?) surrounded by diffusion halos (aureoles) of pyroxene and opaque minerals in an Apollo 14 sample. This texture is found in many Apollo 14 breccias and is similar to the textures of some meteoritic chondrules in slightly recrystallized meteorites. Plane polarized light, length of field of view is 0.8 mm. [Courtesy of *Science*]

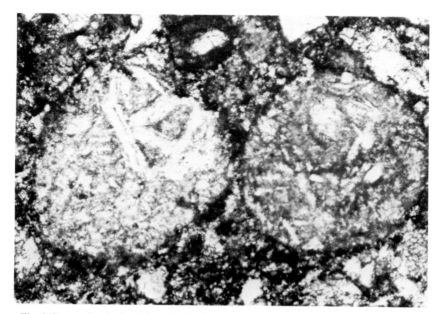

Fig. 5. Lunar chondrules believed to be formed by rounding of rock clasts. This type of chondrule is abundant in samples 14318 and 14313. Similar chondrules have been described by Tschermak (1885) in the Soko-Banja chondrite. Plane polarized light, length of field of view is 0.8 mm.

mechanism will produce chondrules by the free silicate melt forming a roughly spherical shape, because of surface tension, followed by supercooling and rapid crystallization as has been demonstrated by Nelson *et al.* (1972). Slower devitrification of the glassy body may occur if it is buried in hot fall-back and/or base surge deposits ejected by the cratering event. Sorby (1864) first observed that some meteoritic chondrules are glass; he interpreted many other meteoritic chondrules as partially devitrified glass. Impact as the process to produce the fluid silicate droplets, from which this type of chondrule forms, was first postulated by Urey (1952) and was discussed again by Urey and Craig (1953).

(2) Rounding of rock clasts and mineral grains in impact generated base surge deposits. As in the case of the Imbrian event, virtually all large impacts must produce base surges that travel hundreds of kilometers across planetary surfaces. There should be ample opportunity for particle interactions to produce rounding of some of the clasts and grains by impact and abrasion (Fig. 5). Roedder (1971) noted "rounded masses of rock or mineral that have apparently been shaped by abrasion" in Apollo 11 samples 10046 and 10019. We recently have observed similar bodies that appear to be rounded clasts and mineral grains in the eucrite Pasamonte. Tschermak (1885) observed chondrules that appeared to be rounded mineral grains and rock fragments, particularly in the Soko-Banja chondrite.

(3) Diffusion around rock clasts and mineral grains that are of markedly different composition from the surrounding detritus in hot base surge and fall-back deposits (Fig. 4). Certainly much of the recrystallization of matrix in chondritic meteorites tends to be concentric to chondrules. Wilshire and Jackson (1972) and Jackson and Wilshire (1972) have found numerous examples of olivine-rich clasts with concentric diffusion halos of pyroxene and opaque minerals (Fig. 4) in the Apollo 14 breccias.

The observation of lunar chondrules that are related to impact processes on the moon leads to the possibility that some, perhaps many, meteoritic chondrules have formed by similar processes. Warner (1972), Wilshire and Jackson (1972), and Jackson and Wilshire (1972) have interpreted the petrologic types of breccia found at the Apollo 14 landing site in terms of a thick, single cooling unit of impact-implaced rock and mineral fragments, probably base surge from the Imbrian event. This kind of setting on a planetary surface may relate to several systematic variations in chondritic meteorites. In the Van Schmus and Wood (1967) classification of chondritic meteorites, there is the underlying assumption that the variations between chondritic meteorites of similar bulk chemistry are the result of progressive thermal metamorphism. Such criteria as distinctness of chondrules from matrix, homogeneity of olivine compositions, volatile content, etc., are used to distinguish the various petrologic types. However, it is attractive to consider some of these variations as functions of the possible position of the chondritic material in a vertical section of an impact-implaced cooling unit (Fig. 6). Anderson *et al.* (1972) and Grieve *et al.* (1972) estimate that the cooling of parts of the Fra Mauro formation may have started from temperatures in the range of from 700° to 1050°C. Initial temperatures in this range, or even substantially lower, would be sufficient to explain the homogenization of mineral compositions, high degree of recrystallization, lack of volatiles, and indistinctness of chondrules in the lower portions of the cooling unit.

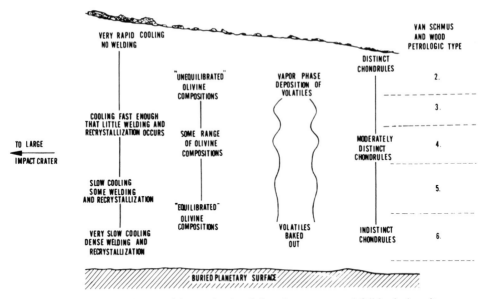

Fig. 6. Schematic diagram of impact-implaced, hot, base surge, and fall-back deposits. Possible relationships to some of the petrologic and chemical variations between compositionally similar chondrites that have been used to classify chondritic meteorites by Van Schmus and Wood (1967) are indicated.

Table 1. Weight in grams of size fractions from Apollo 14

Micron sieve size*	14003,23[a]	14141,29[b]	14148,22[c]	14149,38[d]	14156,22[e]	14163,120[f]	14258,88[g]	14421,22[h]
841	0.0201	0.0170	0.0020	0.0063	0.0106	0.0571	0.0127	0.0532
420	0.1348	0.0682	0.0291	0.0460	0.0414	0.1289	0.1075	0.1063
250	0.1404	0.0510	0.0372	0.0417	0.0383	0.1275	0.1247	0.1200
177	0.1212	0.0350	0.0289	0.0293	0.0377	0.1058	0.1128	0.1089
150	0.0600	0.0161	0.0178	0.0141	0.0190	0.0553	0.0598	0.0800
125	0.0715	0.0179	0.0181	0.0178	0.0210	0.0649	0.0735	0.0688
105	0.0829	0.0188	0.0257	0.0218	0.0259	0.0756	0.0835	0.0900
74	0.1350	0.0277	0.0372	0.0317	0.0383	0.1291	0.1349	0.1435
53	0.1150	0.0218	0.0282	0.0297	0.0322	0.1076	0.1094	0.1028
44	0.0895	0.0172	0.0232	0.0202	0.0255	0.0948	0.0920	0.0886
37	0.0927	0.0185	0.0229	0.0274	0.0321	0.0865	0.1057	0.0946
30	0.1596	0.0262	0.0388	0.0317	0.0407	0.1306	0.1525	0.1359
20	0.2154	0.0421	0.0773	0.0850	0.0702	0.1923	0.3241	0.2112
10	0.1906	0.0210	0.0425	0.0470	0.0569	0.1802	0.1424	0.2919
pan	0.3245	0.0241	0.0682	0.0136	0.0295	0.3410	0.2915	0.2373
Total weight	1.9532	0.4226	0.4971	0.4633	0.5193	1.8772	1.9270	1.9330

*U.S. Standard Sieves, woven mesh, square aperture (37–841 μ); 10–30 μ sieves are electroformed, square aperture.
[a]Contingency sample   [e]Middle of trench
[b]Cone Crater   [f]Bulk fines
[c]Top of trench   [g]Comprehensive fines
[d]Bottom of trench   [h]Bulk

If this concept of the genesis of chondritic meteorites is correct, then the study of large numbers of lithic fragments in the unequilibrated chondrites should provide an excellent opportunity to identify the parent materials (target materials) of the different chemical groups of chondritic meteorites. The unequilibrated chondrites would be the best material for this type of study, because the included lithic fragments should have suffered the least amount of baking and recrystallization. Achondritic fragments, of course, would be of greatest interest.

SIZE ANALYSES: APOLLO 14 AND 15 FINES

Grain size analyses have been performed on 13 samples from the Apollo 14 mission and on 4 samples from Apollo 15 in the less than 1 mm size fraction (Table 1) by sieving (for technique, see King *et al.*, 1971). These grain size analyses have approximately the same characteristics (Fig. 7) that have been observed in previous lunar sample size analyses (King *et al.*, 1971; Butler *et al.*, 1971; and others), except that they are mostly finer than the Apollo 11 and 12 samples, reflecting the older exposure ages of the Apollo 14 and 15 (Apennine Front) sample collection sites. In the comparison of fines size analyses from the different lunar landing sites, care must be taken to avoid coarse ejecta from relatively recent impact craters. For example, the Cone Crater ejecta at the Apollo 14 site is quite coarse and does not have similar size frequency distribution parameters to the relatively undisturbed plains samples. Quantitative statistical parameters of the grain size frequency distributions are presented in Table 2. The "average" of the size frequency distributions from the Apollo 11, 12, and 14 sample collection sites was determined

and 15 samples retained on sieves after grain size analysis.

14230,67[i]	14230,93[j]	14230,76[k]	14230,80[l]	14230,83[m]	15101,103[n]	15231,67[o]	15201,17[p]	15221,57[q]
n.d.	n.d.	n.d.	n.d.	n.d.	0.0059	0.0031	0.0048	0.0049
n.d.	n.d.	n.d.	n.d.	n.d.	0.0241	0.0171	0.0248	0.0201
0.0177	0.0203	0.0196	0.0290	0.0331	0.0300	0.0277	0.0331	0.0337
n.d.	n.d.	n.d.	n.d.	n.d.	0.0279	0.0280	0.0285	0.0302
n.d.	n.d.	n.d.	n.d.	n.d.	0.0174	0.0154	0.0172	0.0156
n.d.	n.d.	n.d.	n.d.	n.d.	0.0214	0.0190	0.0183	0.0197
n.d.	n.d.	n.d.	n.d.	n.d.	0.0225	0.0241	0.0226	0.0224
n.d.	n.d.	n.d.	n.d.	n.d.	0.0392	0.0395	0.0351	0.0384
n.d.	n.d.	n.d.	n.d.	n.d.	0.0302	0.0300	0.0285	0.0312
0.0413	0.0501	0.0339	0.0424	0.0321	0.0271	0.0290	0.0312	0.0238
0.0089	0.0107	0.0074	0.0053	0.0067	0.0192	0.0265	0.0319	0.0272
0.0120	0.0070	0.0059	0.0087	0.0084	0.0465	0.0402	0.0437	0.0427
n.d.	n.d.	n.d.	n.d.	n.d.	n.d.	n.d.	n.d.	n.d.
0.0182	0.0175	0.0121	0.0141	0.0053	0.0788	0.1071	0.0792	0.0992
0.0057	0.0046	0.0108	0.0042	0.0052	0.0565	0.0492	0.0490	0.0715
0.1038	0.1102	0.0897	0.1037	0.0908	0.4467	0.4559	0.4479	0.4806

n.d. = not determined.
[i]Core sample (Drive tube 1) 11.6–13.0 cm
[j]Core sample (Drive tube 1) 15.3–16.7 cm
[k]Core sample (Drive tube 1) 19.0–19.8 to 20.0–20.8 cm
[l]Core sample (Drive tube 1) 21.5–22.6 cm
[m]Core sample (Drive tube 1) 23.6–24.0+ cm

[n]St. George: Rake
[o]St. George: under large rock
[p]St. George: broken from large rock
[q]St. George: Crater

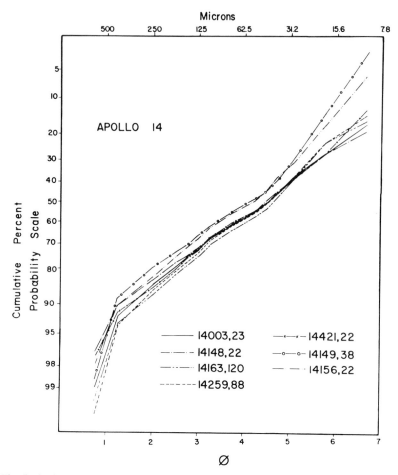

Fig. 7. Grain size frequency distributions of the less than 1 mm size fractions of seven Apollo 14 fines samples. The shapes of these curves are similar to those from Apollo 11 and 12, but the mean size is finer. The finer sizes almost certainly are a function of the older meteoroid bombardment exposure age of the Apollo 14 sample collection site.

by fitting a third order polynomial equation to all of the size data from relatively *undisturbed* samples from each site. Only directly comparable data, collected by our investigating team, were used. Coarse ejecta from relatively recent impacts in each site was excluded. The quality of the numerical fits (Fig. 8) is excellent, and the residuals are extremely small. The mean value of the size distribution from each of these curves was plotted linearly against the estimated site exposure ages. The Apollo 11 and 12 sites were used to establish the slope of the line, because both of these sites are in relatively uncomplicated settings, away from large impact craters that might influence the size distributions in a major way. King *et al.* (1971) have suggested that the mean size of lunar sample fines material decreases with exposure to additional comminution by micrometeoroids. If the linear approximation of the

Table 2. Quantitative statistical parameters of Apollo 14 and 15 size frequency distributions.

Sample number*	Weight (grams)	Graphic mean	Graphic standard deviation	Graphic skewness	ϕ_{50}
14003,23[a]	1.9532	4.45ϕ	2.33ϕ	−0.034	4.52ϕ
14141,29[b]	0.4226	3.20ϕ	2.10ϕ	+0.071	3.19ϕ
14148,22[c]	0.4971	4.37ϕ	2.10ϕ	−0.010	4.50ϕ
14149,38[d]	0.4633	3.80ϕ	1.77ϕ	−0.380	4.14ϕ
14156,22[e]	0.5193	3.92ϕ	1.90ϕ	−0.234	4.13ϕ
14163,120[f]	1.8772	4.58ϕ	2.60ϕ	+0.018	4.51ϕ
14258,88[g]	1.9270	4.53ϕ	2.20ϕ	−0.050	4.60ϕ
14421,22[h]	1.9330	4.33ϕ	2.15ϕ	−0.120	4.50ϕ
14230,67[i]	0.1038	n.d.	n.d.	n.d.	4.10ϕ
14230,93[j]	0.1102	n.d.	n.d.	n.d.	3.80ϕ
14230,76[k]	0.0897	n.d.	n.d.	n.d.	3.95ϕ
14230,80[l]	0.1037	n.d.	n.d.	n.d.	3.40ϕ
14230,83[m]	0.0908	n.d.	n.d.	n.d.	2.91ϕ
15101,103[n]	0.4467	4.30ϕ	2.10ϕ	0.000	4.30ϕ
15231,67[o]	0.4559	4.45ϕ	1.98ϕ	−0.012	4.45ϕ
15201,17[p]	0.4479	4.13ϕ	2.00ϕ	−0.100	4.32ϕ
15221,57[q]	0.4806	4.43ϕ	2.10ϕ	−0.047	4.50ϕ

*For sample identities see Table 1.
n.d. = not determined.

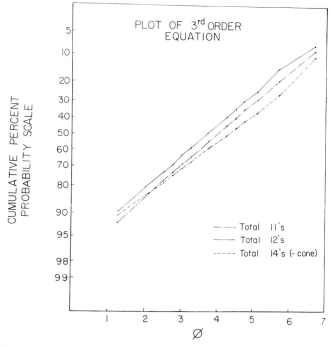

Fig. 8. Plot of third order polynomial equations fitted to the size frequency distribution data for the less than 1 mm regolith samples from the Apollo 11, 12, and 14 sample collection sites. Note that the curves are finer for the older sites.

decrease of grain size with exposure age is extrapolated to the sizes of samples measured for the Apollo 14 and 15 (Apennine Front), we find a meteoroid bombardment exposure age of approximately 3.7×10^9 yr (Fig. 9). However, this should be a *minimum* age for the Imbrian event, and the radiometric ages from unmetamorphosed clasts in Apollo 14 Fra Mauro breccias will approximate the crystallization ages of rocks prior to the Imbrian event. The linear relationship probably is not the best approximation to use. The relationship is almost certainly curvilinear, but at this time the shape of the curve is not exactly known, thus our estimates of lunar surface exposure ages are subject to an unknown amount of error.

The great similarity of the size distributions from fines collected at the Apollo 14 and Apollo 15 (Apennine front) sites, although only four Apennine front samples have been analyzed, indicates that the Fra Mauro formation and the Apennine Front have very similar, if not identical, meteoroid bombardment exposure ages. Thus, both materials probably are Imbrian ejecta, and it seems unlikely that the Fra Mauro formation at the Apollo 14 site could be ejecta from the older Mare Serenitatis Basin, as has been speculated by some workers.

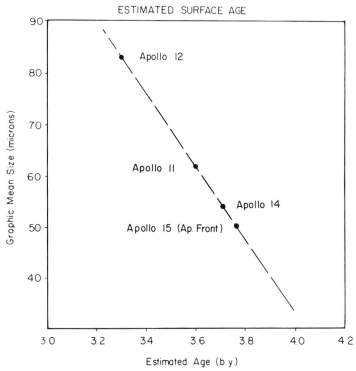

Fig. 9. Plot of the average graphic mean size of fines samples from each of the Apollo landing sites versus estimated meteoroid bombardment exposure age. The average radiometric ages of the Apollo 11 and 12 samples were used to establish the slope of the curve as a first approximation. Extrapolation of the line to the mean sizes of the Apollo 14 and Apollo 15 (Apennine Front) samples leads to the conclusion that the Imbrian event occurred approximately 3.7×10^9 yr ago.

The extreme similarity of the size analyses of the four samples analyzed from on, under and around the boulder at St. George Crater (Fig. 10) indicates that the boulder has been in that location for a relatively short period of time, probably less than 1×10^5 yr, which we believe to be the approximate limit of sensitivity of the grain size method for distinguishing surfaces of different ages. Otherwise, we should observe some effect of coarse detritus shed by the boulder onto the nearby surface. We conclude that the fines on the boulder were picked up from the older surface on which the boulder rests when it rolled to its present location.

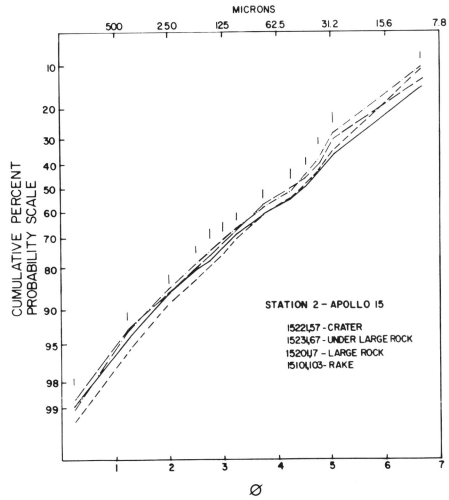

Fig. 10. Grain size frequency distributions of the less than 1 mm fines samples from on, under and around the boulder at St. George Crater (Apollo 15, Apennine Front). There is virtually no difference between the four curves. Hence, we conclude that the exposure age to meteoroids of the boulder at its present site is very short, probably less than 1×10^5 yr. The fines on the boulder probably were picked up during rolling to its final location. Vertical hashes indicate the actual sieve sizes used.

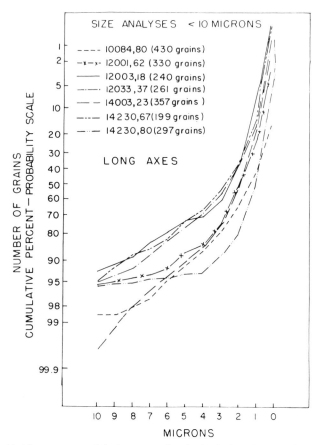

Fig. 11. Measurements of the long axes of grains in the less than 10 μ fraction of lunar fines samples. Note that more than half of the grains have long axes of less than 2 μ in almost all samples. (Measurements by Mike Daily and Frank Kramer.)

Size analyses of Apollo fines samples in the less than 10 μ size range are now in progress. These measurements are performed after mounting small volumes of particles in this size range in an immersion slide mount with a strong dispersant. Several fields of view are photographed for each sample at several different focal planes, and the scale of the photographs is precisely determined. The measurements of long axes and maximum dimension perpendicular to the long axes are made on individual grains with a calibrated scale. The preliminary results are presented in Figs. 11 and 12. Although the general similarity of the size distributions of the less than 10 μ fraction of most samples is striking, there are real differences between some samples. At the present time, the significance of these differences is not known.

Note Added in Proof: Chondritic samples also occur in the Apollo 15 material, *i.e.*, Sample 15426.

Acknowledgments—The authors thank Mike Daily and Frank Kramer for measurements of the less than 10 μ grains. Mike Daily also made the photographic prints. This work was supported by a lunar sample analysis grant from the National Aeronautics and Space Administration.

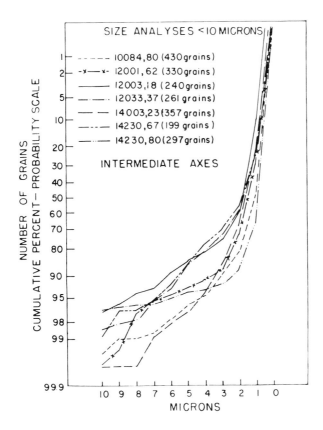

Fig. 12. Measurements of the intermediate axes of grains in the less than 10 μ fraction of lunar fines samples. Note that more than half of the grains have intermediate axes whose lengths are less than 2 μ in all samples. (Measurements by Mike Daily and Frank Kramer.)

References

Anderson A. T. Jr. Braziunas T. F. Jacoby J. and Smith J. V. (1972) Breccia populations and thermal history: Nature of the pre-Imbrian crust and impacting body (abstract). In *Lunar Science—III* (editor C. Watkins), pp. 24–26, Lunar Science Institute Contr. No. 88.

Butler J. C. King E. A. Jr. and Carman M. F. (1971) Grain size frequency distributions of lunar samples and estimation of lunar surface ages (abstract). *Meteoritics* 6, No. 4, 254–255.

Chao E. C. T. James O. B. Minkin J. A. Boreman J. A. Jackson E. D. and Raleigh C. B. (1971) Petrology of unshocked crystalline rocks and evidence of impact metamorphism in Apollo 11 returned lunar sample. *Proc. Apollo 11 Lunar Sci. Conf., Geochim. Cosmochim. Acta* Suppl. 1, Vol. 1, pp. 287–314. Pergamon.

Engelhardt W. von Arndt J. Müller W. F. and Stöffler D. (1971) Evidence of meteorite impacts found in lunar soil and breccias of the Apollo landing sites (abstract). *Meteoritics* 6, No. 4, 267. Note: Although lunar chondrules are not mentioned in the abstract, the find of a chondrule-like object in Apollo 14 samples was discussed in the presentation.

Fredriksson K. Nelen J. and Noonan A. (1971) Lunar chondrules (abstract). *Meteoritics* 6, No. 4, 270–271.

Fredriksson K. Nelen J. and Melson W. G. (1971a) Petrography and origin of lunar glasses and breccias. *Proc. Apollo 11 Lunar Sci. Conf., Geochim. Cosmochim. Acta* Suppl. 1, Vol. 1, pp. 419–432. Pergamon.

Fredriksson K. Nelen J. Noonan A. and Kraut F. (1972) Apollo 14: Glasses, breccias, chondrules (abstract). In *Lunar Science—III* (editor C. Watkins), pp. 280–282, Lunar Science Institute Contr. No. 88.

Grieve R. McKay G. Smith H. and Weill D. (1972) Mineralogy and petrology of polymict breccia 14321 (abstract). In *Lunar Science—III* (editor C. Watkins), pp. 338–340, Lunar Science Institute Contr. No. 88.

Jackson E. D. and Wilshire H. G. (1972) Apollo 14 rock classification (abstract). In *Lunar Science—III* (editor C. Watkins), pp. 419–420, Lunar Science Institute Contr. No. 88.

King E. A. Jr. Carman M. F. and Butler J. C. (1970) Mineralogy and petrology of coarse particulate material from lunar surface at Tranquillity Base. *Science* **167**, 650–652.

King E. A. Jr. Carman M. F. and Butler J. C. (1970a) Mineralogy and petrology of coarse particulate material from the lunar surface at Tranquillity Base. *Proc. Apollo 11 Lunar Sci. Conf., Geochim. Cosmochim. Acta* Suppl. 1, Vol. 1, pp. 599–606. Pergamon.

King E. A. Jr. Butler J. C. and Carman M. F. (1971) The lunar regolith as sampled by Apollo 11 and Apollo 12: Grain size analyses, modal analyses, and origins of particles. *Proc. Second Lunar Sci. Conf., Geochim. Cosmochim. Acta* Suppl. 2, Vol. 1, pp. 737–746. MIT Press.

King E. A. Jr. Carman M. F. and Butler J. C. (1972) Chondrules in Apollo 14 samples: Implications for the origin of chondritic meteorites. *Science* **175**, No. 4017, 59–60.

King E. A. Jr. Butler J. C. and Carman M. F. (1972a) Chondrules in Apollo 14 breccias and estimation of lunar surface exposure ages from grain size analyses (abstract). In *Lunar Science—III* (editor C. Watkins), pp. 449–451, Lunar Science Institute Contr. No. 88.

Kurat G. Keil K. Prinz M. and Nehru C. E. (1971) Chondrules of lunar origin (abstract). *Meteoritics* **6**, No. 4, 285–286.

Kurat G. Keil K. and Prinz M. (1972) A "chondrite" of lunar origin: Textures, lithic fragments, glasses and chondrules (abstract). In *Lunar Science—III* (editor C. Watkins), pp. 463–465, Lunar Science Institute Contr. No. 88.

Levy M. C.-M.- and Levy C. (1972) Mineralogical aspects of the Apollo XIV samples: Lunar chondrules, pink spinel bearing rocks, ilmenites (abstract). In *Lunar Science—III* (editor C. Watkins), pp. 136–138, Lunar Science Institute Contr. No. 88.

Nelson L. S. Blander M. Skaggs S. R. and Keil K. (1972) Use of a CO_2 laser to prepare chondrule-like spherules from supercooled molten oxide and silicate droplets. *Earth Planet. Sci. Lett.* (in press).

Roedder E. (1971) Natural and laboratory crystallization of lunar glasses from Apollo 11. *Min. Soc. Japan*, Spec. Paper 1, Proc. IMA-IAGOD Mtg., 1970, IMA Vol., pp. 5–12.

Sorby H. C. (1864) On the microscopic structure of meteorites. *Phil. Mag.* **28**, 157–159.

Tschermak G. (1885) *Die mikroskopische Beschaffenheit der Meteoriten erläutert durch photographische Abbildungen.* Schweizerbart'sche Verlagshandlung, Stuttgart, 24 pp.

Urey H. C. (1952) *The Planets.* Yale Univ. Press. New Haven, Conn.

Urey H. C. and Craig H. (1953) The composition of the stone meteorites and the origin of the meteorites. *Geochim. Cosmochim. Acta* **4**, 36–82.

Van Schmus W. R. and Wood J. A. (1967) A chemical-petrologic classification for the chondritic meteorites. *Geochim. Cosmochim. Acta* **26**, 739–749.

Warner J. L. (1972) Apollo 14 breccias: Metamorphic origin and classification (abstract). In *Lunar Science—III* (editor C. Watkins), pp. 782–784, Lunar Science Institute Contr. No. 88.

Wilshire H. G. and Jackson E. D. (1972) Petrology of the Fra Mauro formation at the Apollo 14 landing site (abstract). In *Lunar Science—III* (editor C. Watkins), pp. 803–805, Lunar Science Institute Contr. No. 88.

Proceedings of the Third Lunar Science Conference
(Supplement 3, *Geochimica et Cosmochimica Acta*)
Vol. 1, pp. 687–705
The M.I.T. Press, 1972

Petrology and chemistry of some Apollo 14 lunar samples

V. C. Juan, J. C. Chen, C. K. Huang, P. Y. Chen, and C. M. Wang Lee

Institute of Geology,
National Taiwan University

Abstract—The petrography of 14310,178; 14066,37; 14305,103 and 14313,55 and the chemistry of 14310,197 have been studied by means of optical, x-ray diffraction and atomic absorption techniques. The three welded lunar breccias (14066,37, 14305,103, and 14313,55) consist essentially of lithic and mineral clasts set in a fine-grained matrix. The occurrence of pre-existing regoliths, crystalline lithic fragments, basaltic rock fragments, and chondritic lunar samples in the breccias may have resulted from large-scale cratering events with thick, hot ejecta blanket. The minor amounts of anorthositic fragments found in the three breccias may have originated from the highlands. The mechanism of base surge deposition is perhaps the best hypothesis proposed to produce thermal sintering in the welded breccias. 14310 (178, 197) is a fine-grained, diabasic to subophitic nonmare basalt. Its petrographic characteristics are: (1) ratio of plagioclase (highly calcic) to mafics, approximately 65:35, (2) some pigeonites are rimmed by subcalcic augites, (3) the grain size of all phases is less than 1 mm (4) there are very few vugs or vesicles, and (5) the opaque minerals are in low concentrations ($\sim 2\%$). Sample 14310,197 has higher SiO_2, Al_2O_3, Na_2O, K_2O, Rb, Sr, and Li contents and Al_2O_3/CaO, MgO/FeO, Rb/Sr and Ni/Co ratios but lower FeO, TiO_2, MnO, and Cr_2O_3 contents and Na_2O/K_2O ratio than previously analyzed lunar samples. The noritic fragments (enriched in KREEP) found in Apollo 12 soils chemically are similar to 14310,197. The high Si, Al, Na, K, Rb, Sr, and Li contents in 14310,197 and other Fra Mauro basalts indicate that basaltic rocks from the maria could not have been derived from materials of Fra Mauro composition by partial melting. We suggest that 14310 may represent a primary lunar high-alumina basalt magma generated by a small percentage of partial melting of plagioclase-bearing peridotite at a depth of less than 200 km. Primary high alumina basalt magma may play an extremely important role in the early stage of lunar evolution, because it may be the parental liquid for a whole spectrum of differentiates, such as feldspathic basalts, feldspathic gabbro, and anorthosite cumulates.

Introduction

Four polished thin sections, 14310,178; 14066,37; 14305,103; and 14313,55 were obtained from NASA for optical petrographic studies. In addition, a 1.2 gram chip of 14310,197 was received for chemical and x-ray diffraction analyses. The purposes of the present investigations are (1) to study the petrographic and chemical characteristics of the lunar samples (2) to compare the major and trace element data with previously analyzed lunar rocks, meteorites and terrestrial basalts and (3) to draw conclusions as to the petrological and chemical evolution of the lunar samples investigated.

The limited sample available for the present study greatly restricts the available data concerning the complex petrogenesis of Fra Mauro rocks. However, the data presented could provide some basic information as to the dynamothermal history of the lunar samples.

Petrography

Four polished thin sections were investigated under polarized light microscopes, and universal stage techniques were used to determine textures, modes and optical characters of discrete mineral phases. The areas of the sections are 1.0 cm^2, 0.7 cm^2, 2.0 cm^2 and 0.3 cm^2 for 14313,55; 14066,37; 14305,103; and 14310,178, respectively. Among them, 14310,178 is the only basaltic rock studied. The rock was called "Fra Mauro basalt" by Reid *et al.* (1971). The other three rocks tentatively are classified by the authors as welded lunar breccia (McKay and Morrison, 1971). However, Chao (1972, personal communication) considered 14066 and 14305 to be thermally metamorphosed breccias and 14313 to be shocked regolith microbreccia.

All of these four thin sections have igneous textures (from pyroclastic to diabasic) indicating that they represent the products of igneous activities on the moon. The mineral phases in 14310, that have been identified by optical and x-ray analyses, include plagioclase, pigeonite, subcalcic augite, hypersthene, ilmenite, troilite, metallic iron, rutile, and apatite. In the three breccias (14066,37; 14305,103; and 14313,55), besides lithic clasts, the following minerals have been recognized optically: plagioclase, clinopyroxene, orthopyroxene, olivine, ilmenite, apatite, ulvöspinel, and possibly potash feldspar. Neither magnetite nor hydrous minerals were found. Petrographic summaries of these sections are as follows:

Polished thin section 14310,178 nonmare basalt

The sample 14310 is a grab sample with the location tentatively identified as Station G (APSR 1971, p. 112). It is one of the nonmare basalts (or "Fra Mauro basalts" of Reid *et al.*, 1971) returned by the Apollo 14 Mission. The rock is a medium-gray, fine-grained, diabasic to subophitic basalt (Fig. 1) consisting mostly of plagioclase (63%), pigeonite (18%), subcalcic augite (9%), and hypersthene (3%) with minor amounts of ilmenite (2%), interstitial glass (3%), troilite (0.5%) and traces of metallic iron, rutile, and apatite. The grain size is generally between 0.2 and 0.6 mm with no grain larger than 1 mm. Very few vugs (1%) or vesicles were found in the section, suggesting a near surface rather than surface crystallization. In addition, the absence of preferred orientation of plagioclase laths precludes the possibility of flow banding.

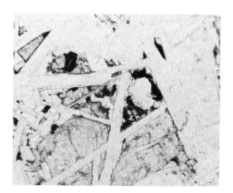

Fig. 1. Photomicrograph of 14310,178 showing diabasic texture. Light gray plagioclase, dark gray clinopyroxene (with 90° cleavage), black ilmenite. Plane polarized light. Length of field of view is approximately 0.55 mm.

Under microscopic observation, clear, euhedral laths of plagioclase commonly are twinned after the Albite law. The maximum symmetrical extinction angle determined by universal stage measurements is $55° \pm 2°$, indicating a composition of An_{88}. However, x-ray diffraction data of 14310,197 indicate that the plagioclase composition is $An_{82 \pm 2}$. This value is determined by means of the curve published by Jackson (1961) and cross-checked with the diagrams constructed by Smith and Gay (1958), based on $\gamma[2\theta(131) + 2\theta(220) - 4\theta(13\bar{1})]$ and $\beta[2\theta(1\bar{1}1) - 2\theta(\bar{2}01)]$. Agreement among the values is excellent. Ophitic subcalcic augite is pale yellow to pale brown and subhedral. Lamellar twins on (100) and (001) occasionally are found in augite. The weakly pleochroic augite (from pale yellowish green to pale brown) has the following optical properties: $2V_z = 46°$, $c \wedge Z = 37°$ and O.A.P. parallel to (010). The pale greenish pigeonite has a small 2V (approximately 25°) and occurs as single grains or as cores rimmed by augite. This texture indicates that precipitation of pigeonite occurred at an early, slow cooling stage followed by epitaxial growth of augite on selected pigeonite phases while pigeonite continued to grow. There apparently was little subsolidus exsolution (Papike *et al.*, 1971). It is interesting to note that hypersthene is parallelly intergrown with augite, a common feature found in terrestrial basalts.

Ilmenite blades commonly less than 0.1 mm long are found as inclusions in clinopyroxenes. Interstitial minute grains of troilite also are present. We have systematically searched for cristobalite but it has not been positively identified. However, interstitial glass ($1.50 > n > 1.48$), probably rich in SiO_2 (Gancarz *et al.*, 1971 reported 77% SiO_2 in glass of 14310,6), has been found. It should be noted that both plagioclase and pyroxenes exhibit wavy extinction suggesting some sort of shock effects.

14310,178 is distinctly different from the mare basalts in several petrographic aspects:

(1) The ratio of plagioclase to mafic minerals is about 65:35, whereas in mare basalts it is closer to 30:70 (see Brown *et al.*, 1970).

(2) The pyroxene is pale yellow, whereas in mare basalt it is deep cinnamon brown.

(3) The grain size of all phases is less than 1 mm, whereas in the mare suite some grains may be as much as 1 cm.

(4) There are very few vugs or vesicles, whereas the mare basalts are slightly to highly vesicular and/or vuggy.

(5) The opaques are present in small amounts ($\sim 2\%$), whereas in mare basalt they are generally greater and may reach 26% by volume (Kushiro and Nakamura, 1970, p. 608, Table 1).

Polished thin section 14066,37 breccia

The rock was collected at station F about 900 meters from Cone Crater. The polymict fragmental rock is light gray with medium gray clasts. The grain size of the breccia ranges from <0.01 mm to 0.10 mm in matrix and 0.11 mm to 1.50 mm in clasts.

The breccia consists essentially of subangular to subrounded lithic clasts (7%), mineral clasts (9%, including 6% plagioclase, 2% clinopyroxene and orthopyroxene, 0.3% ilmenite, 0.7% olivine and trace amounts of ulvöspinel and apatite) set in a fine-grained matrix (84%). The lithic clasts commonly are composed of plagioclase and pyroxenes with pc/px ratios ranging from 80/20 to 20/80, suggesting that they were derived from different parent rocks. Minor lithic clasts (0.5% by volume) consist entirely of plagioclase (An 93) and tecturally are similar to the anorthositic rock fragment in Apollo 11 bulk sample described by Wood *et al.* (1970b, p. 967). The plagioclase grains are equant and polygonal (Fig. 2) which may be due to some degree of recrystallization.

Among the mineral clasts, plagioclase occurs as single grains and commonly also as fine-grained mosaics that probably were produced by shock from single crystals. The single-crystal plagioclase grains, commonly twinned after the Albite and/or Carlsbad law, have an An percentage ranging from 62 to 67. Clinopyroxene ($2V_z = 64°$, probably a high Al and Ca variety) grains are anhedral and show no exsolution lamellae in this particular section. Optically negative orthopyroxene crystals have good cleavage and parallel extinction. Minor amounts of euhedral olivine (Fo_{78}) crystals (Fig. 3) have been found. They have $2V_z = 93.5°$ and optic axial plane parallel to (001). The olivine crystals have cracks but no serpentinization, suggesting an anhydrous environment of formation. Ulvöspinel occurs as anhedral brownish-red isotropic grains. Some ilmenite has skeletal texture similar to that described by Haggerty *et al.* (1970, p. 515). However, the trapped phase is plagioclase instead of pyroxene. The texture may be formed in immiscible silicate-rich and metal-rich liquids that have differentiated on cooling. The fine-grained matrix consists essentially of plagioclase with minor amounts of pyroxenes, olivine, and opaques.

Many lithic clasts and mineral clasts within the rock are fractured. In addition, many plagioclase grains and lithic clasts show wavy extinction, suggesting that they have been shocked. Extreme mosaicism also is present. Some lithic fragments and plagioclase grains show evidence of melting (transitional and/or embayed contacts

Fig. 2. Anorthositic rock fragment in 14066,37. The plagioclase (An_{93}) is characteristically equant and polygonal suggesting considerable recrystallization. Crossed nicols. Length of field of view is approximately 0.55 mm.

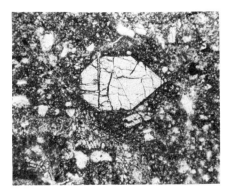

Fig. 3. Euhedral olivine (Fo_{78}) crystal in 14066,37. The olivine has cracks but no serpenti-nization indicating an anhydrous environment. Plane polarized light. Length of field of view is approximately 0.55 mm.

with the matrix and occurrence of glassy material in cracks), suggesting that they have reacted with the matrix that may have been remobilized during a shock event.

Polished thin section 14305,103 breccia

The rock was sampled on EVA–1 (APSR, 1971, p. 112). The rock has pyro-clastic texture and consists of lithic clasts (16%, including 5% fine-grained basaltic rock, 2% gabbroic anorthosite, 3% welded breccias and 6% glassy breccias) and mineral clasts (6%, including 3% plagioclase, 0.5% olivine, 2% pyroxenes, 0.5% opaques and trace amount of microcline) set in a fine-grained (<0.05 mm) crypto-crystalline to microcrystalline matrix (78%) composed mainly of plagioclase with subordinate olivine, pyroxenes and opaque minerals.

Among the lithic clasts, the fine-grained basaltic rocks consist of euhedral plagioclase (An_{86}, 55%) subhedral pigeonite (40%), olivine (Fo_{92}, 2%) and opaque minerals (3%). On the basis of their relatively greater plagioclase and lesser opaque mineral contents, these basaltic rocks are similar in composition to 14310,178 and may be classified as nonmare type. The gabbroic anorthosite (Wood, 1970b) clasts contain coarse grained angular to subangular crystals of plagioclase (An_{94}, 75–90%), pyroxene (10–20%) and opaque minerals (5% or less). The plagioclase crystals in the gabbroic anorthosite characteristically are equant and show strong wavy ex-tinction suggesting shock effects. It is suggested that these gabbroic anorthosite fragments might have been derived from the lunar highlands. Subrounded to irregular glassy breccias and welded breccias (1–2 mm) also are among the lithic clasts. The former consists mainly of plagioclase (0.2–0.6 mm long) and olivine set in a glassy to cryptocrystalline matrix, and the latter contains plagioclase, pyroxene, and olivine set in a fine-grained crystalline matrix.

Plagioclase and pyroxenes predominate among the mineral clasts. Some well-rounded plagioclase grains have strong wavy extinction suggesting that they have experienced shock events. Extreme mosaicism also was found in plagioclase (Fig. 4).

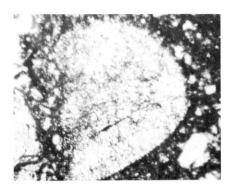

Fig. 4. Extreme mosaicism in plagioclase in 14305,103. The roundness of the clast may be due to abrasion. Crossed nicols. Length of field of view is approximately 0.65 mm.

Extreme mosaicism and glassy rims found in some plagioclases may be the result of granulation and melting produced by shock metamorphism.

The large variety of lithic clasts found in the rock indicates that the breccia has had a complex history. The lithic clasts may have originated from the bedrock, the pre-existing regoliths and possibly some clasts from the lunar highlands.

Polished thin section 14313,55 breccia

This sample was taken at station G1 in known orientation (APSR, 1971, p. 112). It is a polymictic fragmental rock consisting of lithic clasts (10%), glass fragments and spherules (7%) and mineral clasts (5%) set in a fine-grained partly glassy matrix (77%). It should be mentioned that glass spherules have not been found in the two breccias (14066,37 and 14305,103) described above.

The clasts commonly are angular and are rather uniformly distributed in the matrix, without appreciable preferred orientation. The lithic clasts include 3% 3% basaltic rocks, 3% welded breccias and 4% pyroxene-plagioclase rock with a range of px/pc ratios. The basaltic rocks consist mainly of minute lath-shaped plagioclase (33%) with minor intersertal pyroxenes (10%) and opaque minerals (2%) set in a fine-grained matrix (55%, containing essentially plagioclase with or without glass). The texture and mineral components in the rocks suggest that they belong to the nonmare basalts. A small lithic fragment composed entirely of a polygonal mosaic of plagioclase (An_{95}) has been recognized. It is texturally and mineralogically similar to anorthosite described by Wood *et al.* (1970b) and may have been derived from the lunar highlands.

A bronzite chondrule (1 mm in diameter) has been found in the thin section. This round, equant chondrule (Fig. 5) is composed mostly of euhedral to subhedral bronzite with a minor amount of minute euhedral olivine and glassy material. The colorless bronzite has $2V_z = 73°$ (En_{78} Fs_{22}) and contains some tear-drop and dumbell-like bubble inclusions. Pale brown glassy materials fill the interstices between crystals. King *et al.* (1972) have presented an excellent discussion on the origin of lunar chondrules observed in breccia samples returned by the Apollo 14

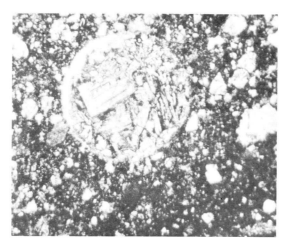

Fig. 5. Bronzite chondrule composed mostly of bronzite (En_{78}) with minute euhedral olivine (section parallel to 100) in 14313,55. Plane polarized light. Length of field of view is approximately 2 mm.

mission. They concluded that these chondrules were formed by impact processes such as crystallization after shock melting and abrasion and diffusion in base-surge and fall-back deposits generated by large impacts on the lunar surface.

The glass fragments have a range of size, shape, color, and probably also a range of chemical composition. The pale yellow to brown irregularly shaped glass fragments rarely show flow texture and some contain vesicles or tiny crystal inclusions. Some purplish brown glass fragments are partly devitrified and contain plagioclase microlites with radiated or feather-like texture. In addition, some well-rounded, smooth and undeformed glass spherules (0.05–0.6 mm in diameter) (Fig. 6) have a range of colors (clear, yellow, or brown) and possibly are of different compositions.

The mineral clasts in the rock include plagioclase (An_{85}), olivine (Fo_{60}, $2V_z = 103°$), augite, and ilmenite, with minor amounts of pinkish brown spinel. The somewhat turbid matrix of the breccia is composed mostly of minute plagioclase, pyroxene, olivine (?), dust-like opaque minerals and glassy materials. Recrystallization textures generally are lacking in the matrix.

The lithic clasts found in the breccia may have been derived from different parent rocks including crystalline bedrocks, pre-existing regoliths and highland materials. The glass spherules may be formed during the base surge event; however, some partly devitrified glass fragments with radiating texture may have been formed at an earlier stage than the spherules.

CHEMISTRY

14310,197 (nonmare basalt or Fra Mauro basalt) is the only lunar sample chemically analyzed in the present investigation. The basalt chip weighed 1.2198 gm when received. The texture, mineral mode and general mineralogy of 14310 have already been described. The sample was prepared for chemical analysis in the following

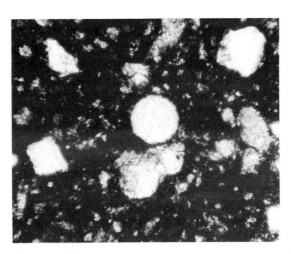

Fig. 6. Nearly spherical undeformed glass spherule in 14313,55. Plane polarized light. Length of field of view is approximately 0.55 mm.

steps: (1) one part of the chip was ground into powder in an agate mortar, (2) 0.2000 gm of rock powder was dissolved in a mixture of ultrapure HNO_3 and HF. Special care was taken to avoid contamination during the analytical procedures.

With the exception of Si, Al, and Ti, all of the analyses for major and trace elements were performed with a Perkin-Elmer Model 303 atomic absorption spectro-photometer using standard solutions of various dilutions. Twelve artificial standards of basaltic composition containing Al, Ti, Fe, Mn, Ca, Na, K, Cr, Ag, Au, Co, Cu, Ga, Li, Ni, Rb, Sr, and Zn were prepared from Johnson and Matthey spectro-graphically pure chemicals and were used to construct the working curves. U.S. Geological Survey rock standards W–1 and BCR–1 were used as references. The instrumental settings for the analyses are those listed in Perkin-Elmer (1971) Ana-lytical Methods for Atomic Absorption Spectrophotometry. Precisions for AAS determinations are generally $\pm 2\%$ of the amounts present.

SiO_2, Al_2O_3, and TiO_2 in the rock were determined by colormetric methods described by Shapiro and Brannock (1962) using U.S. Geological Survey rock standards W–1, BCR–1, AGV–1, PCC–1, DTS–1 and artificial standards to construct the working curves. The precision for SiO_2, Al_2O_3, and TiO_2 determinations is $\pm 3\%$ of the amounts present.

The results of the chemical analyses together with chemical data from Apollo 11 and 12 rocks and terrestrial basalts are listed in Table 1. The normative com-position of 14310,197 is presented in Table 2.

Reid *et al.* (1971) classified lunar basalts into five categories: They are: (1) feldspathic basalts (or anorthositic gabbro) (with approximately 70% normative plagioclase), (2) Fra Mauro basalts (high-alumina basalt with approximately 50% normative plagioclase), (3) Luna 16 basalts, (4) Apollo 11 basalts and (5) Apollo 12 basalts. Sample 14310,197 is considered here to be in the second category. It is interesting to note that 14310,197 resembles some terrestrial high-alumina basalts

Table 1. Chemical data for 14310,197 (nonmare Fra Mauro basalt) compared with Apollo 11 and 12 samples and terrestrial basalts. All values are in weight percent except as indicated.

	14310,197[1]	10072[2]	10045[3]	12038[4]	12021–86[5]	12070[6]	High-Al[7] basalt	Hawaii[8] alkali basalt	Oceanic[9] tholeiite
SiO_2	48.33	40.49	39.04	49	46.46	45.91	50.11	46.46	49.85
TiO_2	1.26	11.99	11.32	3.2	3.44	2.81	1.84	3.01	2.50
Al_2O_3	20.45	7.74	9.51	12	10.55	12.50	19.45	14.64	12.47
Cr_2O_3	0.16	0.33	0.35	0.32	0.40	0.43	n.d.	0.018	0.07
ΣFeO	8.40	19.38	19.40	17	19.68	16.40	7.75	12.05	11.70
MnO	0.13	0.24	0.27	0.26	0.26	0.22	0.03	0.14	n.d.
MgO	8.00	7.45	7.73	6.5	7.60	10.00	6.53	8.19	8.79
CaO	11.70	10.56	11.28	11	11.37	10.43	9.39	10.33	10.91
Na_2O	0.628	0.50	0.36	0.60	0.35	0.41	2.60	2.92	2.56
K_2O	0.520	0.29	0.05	0.057	0.07	0.25	1.46	0.84	0.60
H_2O^+	0.00	n.d.	n.d.	n.d.	n.d.	n.d.	0.00	n.d.	n.d.
H_2O^-	0.00	n.d.	n.d.	n.d.	0.00	n.d.	0.00	n.d.	n.d.
Ag(ppm)	0.04	n.d.	n.d.	n.d.	n.d.	n.d.	n.d.	n.d.	n.d.
Au(ppm)	0.008	n.d.	n.d.	n.d.	n.d.	n.d.	n.d.	n.d.	n.d.
Co(ppm)	45	34	23	23	n.d.	n.d.	n.d.	39	55
Cu(ppm)	16	22	20	n.d.	n.d.	n.d.	n.d.	n.d.	n.d.
Ga(ppm)	10	4	3	n.d.	n.d.	n.d.	n.d.	22	25
Li(ppm)	25	n.d.	n.d.	5.5	n.d.	n.d.	n.d.	13	10
Ni(ppm)	205	<20	<20	14	n.d.	276	n.d.	33	350
Rb(ppm)	16	5.61	0.62	0.70	n.d.	6.9	n.d.	61	n.d.
Sr(ppm)	258	168.4	137.7	230	n.d.	136	n.d.	1200	350
Zn(ppm)	13	34	14	n.d.	n.d.	n.d.	n.d.	n.d.	n.d.
Na_2O/K_2O	1.21	1.72	7.20	10.53	5.00	1.64	1.78	3.48	4.27
$MgO/\Sigma FeO$	0.95	0.38	0.40	0.38	0.39	0.61	0.84	0.68	0.75
K/Rb	270	429	669	676	n.d.	301	n.d.	114	n.d.
Ni/Co	4.56	<0.59	<0.87	0.61	n.d.	n.d.	n.d.	0.85	6.36
GaX1000/Al	0.092	0.097	0.596	n.d.	n.d.	n.d.	n.d.	0.284	0.379
Rb/Sr	0.062	0.033	0.0045	0.0030	n.d.	0.051	n.d.	0.0508	n.d.
Fe/Ni	318	>7143	>7540	9439	n.d.	462	n.d.	2838	260

(1) This work (analyst, J. C. Chen)
(2) Apollo 11 Type A fine-grained vesicular igneous rock (Compston et al., 1970, Table 1)
(3) Apollo 11 Type B medium-grained vuggy crystalline igneous rock (Compston et al., 1970, Table 1)
(4) Apollo 12 crystalline rock (LSPET, 1970, Table 2)
(5) Apollo 12 crystalline rock (Kushiro and Haramura, 1971)
(6) Apollo 12 soil (LSPET, in press, Table 4)
(7) High-Aluminum Basalt (Juan et al., 1963, Table 1), H_2O-free basis
(8) Hawaiian alkali basalt (Major element data from MacDonald and Katsura 1964, Table 10; Trace element data from Prinz, 1967, p. 285)
(9) Oceanic tholeiite (Nockolds and Allen, 1956, Table 32)

Table 2. Normative composition of 14310,197.

Component		Norm
Chromite		0.22
Ilmenite		2.43
Orthoclase		3.34
Plagioclase	Albite	5.24
	Anorthite	51.43
Diopside	Wo	2.78
	Fs	1.06
	En	1.60
Hypersthene	En	18.40
	Fs	12.41
Quartz		0.84
An % in Pc		90.75%

(Table 1 (7)) in major element content, but the former has relatively low Na₂O owing to the occurrence of highly calcic plagioclase in the rock.

The chemical composition of 14310,197 (Table 1 (1)) and the compositions of other Apollo 14 samples (LSPET, 1971, Table 1) are clearly different from those of Apollo 11 and 12 samples. Sample 14310,197 and other Apollo 14 rocks generally have higher Si, Al, Na, K, Rb, Sr, and Li but lower Fe, Ti, Mn, and Cr contents. In addition, 14310,197 and other Apollo 14 samples generally are characterized by relatively higher MgO/FeO and Ni/Co ratios but have lower Na₂O/K₂O ratio as compared with Apollo 11 and 12 rocks. The occurrence of normative quartz (Table 2) and highly silicic glass in the mode of 14310 suggests that the basaltic magma is saturated with silica. The virtual absence of H₂O in 14310,197 indicates that it was formed in, and has remained in, a low oxygen fugacity environment as have the Apollo 11 and 12 rocks.

Reid *et al.* (1971) noted that in the five categories of lunar basalts, there are compositional trends of increasing Fe, Cr, Ca/Al, Cr/Ni, and decreasing Al and Ni in the order mentioned above. However, the Ni content of 14310,197 (205 ppm) is the highest obtained to date for a lunar basaltic rock and is higher than most terrestrial oceanic basalts (av 120 ppm, Prinz, 1967, Table IV). The Fe/Ni ratio of 14310,197 (i.e. 318) is among the lowest values yet obtained for lunar material. The Ni/Co ratios of Apollo 14 samples (including 14310,197) generally are higher than those of previously analyzed lunar rocks. Whether the high nickel content is related to meteoritic admixture or not is still uncertain.

It should be noted that 14310,197 has a considerably higher Rb/Sr ratio (i.e. 0.062) than the Apollo 11 Type A and Type B rocks (Compston *et al.*, 1970, Table 1) and Apollo 12 crystalline rocks (LSPET, 1970, Table 2).

The chromium contents of Apollo 14 samples (including 14310,197) mostly are less than 0.2% by weight. These values are less than in previously analyzed lunar samples, but they are considerably greater than those of terrestrial oceanic basalts (Cr av 220 ppm). A review of Table 1 (this paper) and Table 1 of LSPET (1971) indicates that Apollo 14 samples generally have lesser gallium contents than oceanic basalts (av 20 ppm, Prinz, 1967 p. 285). The latter are also characterized by relatively greater Al/Ga ratios.

A ternary plot of FeO–TiO₂-alkali for Apollo 11, 12, and 14 (including 14310,197) samples and terrestrial basalts is presented in Fig. 7. Sample 14310,197 and other Apollo 14 samples are higher in alkali content than Apollo 11 and 12 rocks. The fact that Apollo 14 (including 14310,197) and Apollo 12 samples have considerably lower TiO₂ contents than Apollo 11 samples is clearly apparent. Samples of 14310, 14053, and one terrestrial high-alumina basalt, lying on a line with nearly constant TiO₂ content, may indicate that a magma of 14310 composition could differentiate into two different suites of rocks: (1) with iron enrichment and (2) with alkali enrichment, depending possibly on the amount of volatile component present and the oxygen fugacity. It should be noted that some Apollo 14 sample fields and terrestrial basalt fields (especially oceanic ridge basalts) overlap in the ternary plot.

The MgO versus FeO plot for Apollo 14 samples (including 14310,197), Apollo 11 and 12 rocks, eucrites and terrestrial basalts is presented in Fig. 8. It is apparent

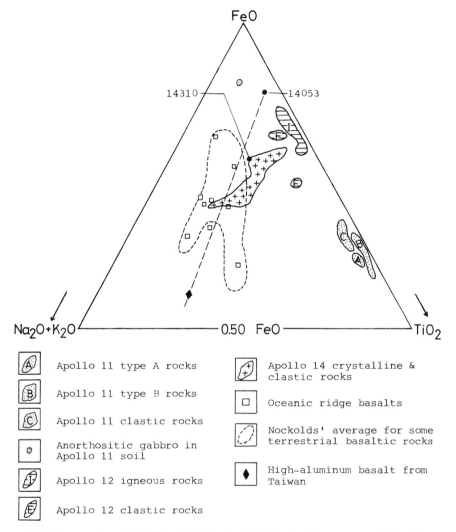

Fig. 7. Ternary plot of FeO–TiO$_2$–alkali showing the distinction between terrestrial basaltic rocks and the Apollo 11, 12, and 14 materials. Data are from this work, Goles (1971) and LSPET (1971).

that the mare basalts (Apollo 11 and 12 samples) are higher in FeO (i.e. >17%) than the Fra Mauro basalts (including 14310,197) and the oceanic ridge and Islandic basalts. Apollo 14 samples generally have FeO contents less than 14% by weight. Some Apollo 11 and 12 samples are similar in MgO and FeO contents to eucritic meteorites. The MgO/FeO ratio in Apollo 14 samples is close to one (MgO/FeO = 0.95 for 14310,197) or greater, but the ratio is approximately 0.5 in Apollo 11 and 12 samples and in eucrites. Many Apollo 14 samples have MgO and FeO contents similar to those of oceanic ridge and Islandic basalts. Sample 14310 particularly is related to terrestrial high-alumina basalt and noritic rocks found in Apollo 12 soil, as shown by a systematic change in the MgO/FeO ratio (Fig. 8).

The CaO versus Al_2O_3 plot for 14310,197 and other lunar rocks is presented in Fig. 9. Various meteoritic and terrestrial basalt data also are plotted for comparison. Sample 14310,197 and other Apollo 14 samples are characterized by high Al_2O_3 (12–20%) and low CaO (7.5–12%) contents and high Al_2O_3/CaO ratios ($\sim$1.0–2.0) as compared to Apollo 11 and 12 rocks (Al_2O_3/CaO < 1). However, sample 14310, noritic rocks from Apollo 12 soil and terrestrial high-alumina basalts (Fig. 9) occupy an area that could be designated as a group representing a distinct magma type characterized essentially by Al_2O_3 in excess of 17%, as defined by Tilley (1950) and Kuno (1960) for terrestrial basalts. It is apparent that the Al_2O_3/CaO ratio for Apollo

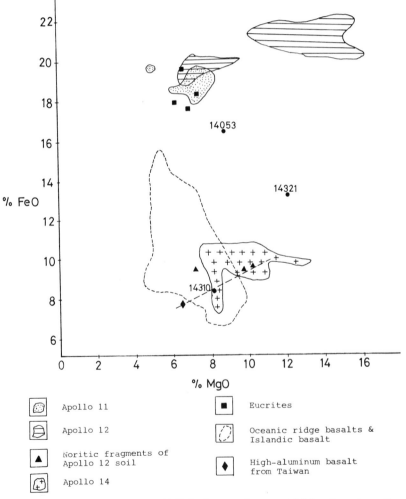

Fig. 8. Weight percent MgO versus FeO for lunar rocks, terrestrial basalts and eucrites. Data are from this work, LSPET (1971) and LSPET (in press).

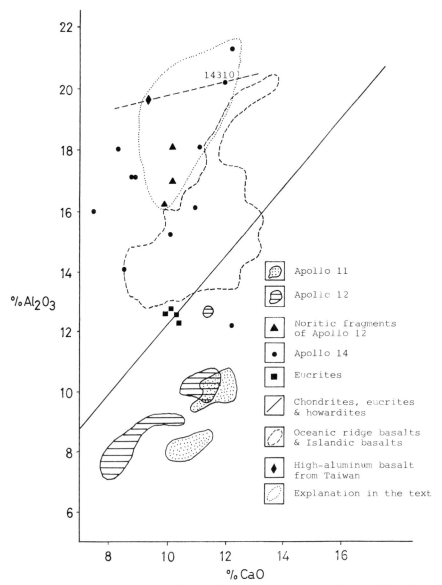

Fig. 9. Weight percent CaO versus Al_2O_3 for lunar rocks, meteorites and terrestrial basalts. Data are from this work, LSPET (1971) and LSPET (in press).

14 samples and previously analyzed lunar rocks has a wide range, whereas the ratio is essentially constant (around 1.2) for most meteorites. This fact does not support the hypothesis that achondrites come from the lunar surface.

It should be noted that the noritic fragments enriched in **KREEP** (Hubbard *et al.*, 1971) found in the Apollo 12 soil, the dark-colored portion of rock 12013 (Wakita and Schmitt, 1970), the coarse fractions of Apollo 12 soil 12033 (Baedecker *et al.*, 1971) and Luny Rock 1 from the Apollo 11 site (Albee and Chodos, 1970) all are chemically similar to Apollo 14 samples. These facts further emphasize the importance of the 14310 magma type on the moon, a point that will be discussed further in a later section.

LSPET (1971) concluded that the mineralogical and chemical similarity of the Fra Mauro materials and noritic (KREEP) fragments in Apollo 12 soil and Luny Rock 1 from the Apollo 11 site suggest that material of this composition is widespread on the lunar surface. They suggested also that, because this material is further removed in composition from chondritic abundances than are the mare basalts, a profound premare lunar differentiation history must have occurred that produced a lunar crust.

The K versus Rb plot for 14310,197 and other lunar samples is presented in Fig. 10. Apollo 14 samples (including 14310,197) have relatively greater K and Rb contents than Apollo 11 and 12 rocks and achondrites (K = 12.9–657 ppm, Rb = 0.0315–0.696 ppm, Tera et al., 1970, p. 1639). The K/Rb ratios are quite constant for Apollo 14 samples, averaging 325 ± 50 in the fragmental rocks and 330 ± 25 in the fines (LSPET, 1971). The high K, Rb, Sr, Li, Na, Al, and Si contents in 14310,197 and other Fra Mauro basalts indicate that basaltic rocks from the maria could not have been derived from materials of Fra Mauro composition by partial melting.

Discussion

It is evident from the petrographic observations stated above that the welded breccias (14066,37; 14305,103; and 14313,55) had a complex history. The occurrence of pre-existing breccias, crystalline lithic fragments, basaltic rock fragments and

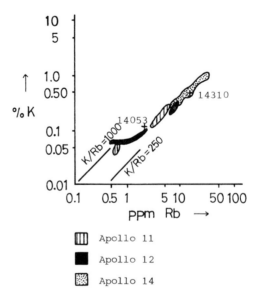

Fig. 10. The K/Rb relationship for 14310 (this work) and other Apollo 14 rocks and fines compared with some Apollo 11 and 12 rocks and soils (after LSPET, 1971).

chondritic lunar samples may have resulted from large-scale cratering events with thick, hot ejecta blankets. The minor amounts of anorthositic rock fragments found in the three breccias may have originated at an area remote from the present sampling location. Although Reid *et al.* (in press) suggested that feldspathic basalt (with approximately 70% normative plagioclase, and low in K and Ba content) is the prime candidate for lunar highland material, Smith *et al.* (1970), Wood *et al.* (1970a) and Short (1970) concluded that the lunar highlands are composed mainly of anorthosite that may be the parent rock of the anorthositic fragments found in the breccias under consideration.

Welded breccias were first assumed by LSPET (1969) to be lithified lunar soil by shock. However, Smith *et al.* (1970) suggested that these breccias were thermally sintered and they proposed welding in a hot glowing cloud. Frederiksson *et al.* (1970) pointed out the similarities between lunar breccias, terrestrial impactites, and some ignimbrites and proposed that "impact-ignimbrite" was a suitable name for the welded lunar breccias. Chao *et al.* (1971) suggested that the welded breccias were formed by low level shock compaction of soil located some distance from the point of impact and near the base of the lunar regolith.

The preservation of delicate structure (i.e. smooth, well-rounded, and undeformed glass spherules in 14313,55) in the welded breccia indicates that matrix grains probably were not welded to larger fragments by shock lithification. Perhaps the mechanism of base surge deposition, proposed by McKay and Morrison (1971), is the best hypothesis to produce thermal sintering in the welded breccias. The base surge probably was produced by a large meteorite impact in what is now the Imbrium Basin. Hot gas from the volatilized meteorite and from the heated lunar materials in the vicinity of the impact drove radially outward from the site of the impact carrying ejecta from the crater. The ejecta includes both pre-existing regolith (such as welded breccias and glassy breccias found in 14305,103 and 14313,55) and crystalline bed rock (such as pyroxene-plagioclase rocks and basaltic rocks found in 14066,37; 14305,103; and 14313,55). Some residual meteoritic materials also may occur in the ejecta. The velocity of the base surge may be rather high and the flow may be turbulent and may thoroughly mix the regolith with bedrock ejecta. The base-surge flow will deposit a layer or blanket that thins radially outward from the impact site. It is suggested that the base surge might have been hot enough to sinter or weld after deposition in the Fra Mauro area. The preservation of perfectly rounded and undeformed glass spherules in 14313,55 indicates that the rock was not subjected to significant post-depositional compressive stress.

The petrogenesis of 14310, a nonmare basalt, deserves further consideration. The rock appears to be high in Al_2O_3 and low in TiO_2 and FeO and is highly enriched in K, Rb, Sr, REE, U, and Th. Both from the modal and chemical compositions, we may infer that the parental liquid for 14310 lunar basalts could have been much like the terrestrial high-alumina basalts. Textural evidence in 14310,178 indicates a sequence of crystallization with plagioclase first, followed by ilmenite and pyroxenes. Ophitic augites and inclusions of ilmenite in pyroxene infer that plagioclase had a very wide field of occurrence, confirming to the experimental work with synthetic mixtures of lower TiO_2 content (O'Hara *et al.*, 1970). A moderate amount of separation of ilmenite from an alumina-rich liquid would produce a high rubidium liquid (Gast *et al.*, 1970, p. 1156). The chemical characteristics clearly indicate that plagioclase is the prevailing phase in the crystallization of this type of lunar basaltic liquid.

Both Tilley (1950) and Kuno (1960) recognized that terrestrial high-alumina basalts constituted a new magma type. The essential feature of these high-alumina basalts is Al_2O_3 in excess of 17%, which is expressed as plagioclase in the norm. Kuno also considers high-alumina basalt a primary magma, transitional in its chemistry and mineralogy between the tholeiitic and alkali olivine basalt magmas, and with a fractionation trend generally similar to tholeiite. However, Yoder and Tilley (1962) pointed out that aphyric high-alumina basalts form important members of both the tholeiites, as chilled marginal facies, and the alkali olivine basalts as variants within their lineage. Apart from the critical feature of tholeiitic rocks, pyroxenes that consist of augite zoned to subcalcic augite with pigeonite as defined by Tilley (1961), the present samples, 14310,197 and 14310,178 could have experienced extreme differentiation, indicated by residual quartz in the norm, a consequence of the reaction relation between olivine and Ca-poor pyroxene in the early stages of crystallization.

Two aspects of the origin of basaltic magma must be considered. The first aspect concerns the nature of the material at depth. The second aspect concerns the nature of the process by which the basaltic liquid is produced. In dealing with terrestrial basalts, Clark and Ringwood (1964) indicated that temperatures necessary to produce basaltic magmas are reached below the M discontinuity, a few tens to a few hundred kilometers below the earth's surface in the upper mantle. Kushiro and Haramura (1971) suggested that the interior of the moon at depths less than 200 km is most likely plagioclase-bearing peridotite. Haskin *et al.* (1970) concluded that the trace element abundances, and especially the europium anomaly in Apollo 11 basalts, were more readily explained in terms of partial melting of a source material containing plagioclase. However, Ringwood (1970), based on experimental data, advocated that an unfractionated pyroxenite at depths of 200–600 km is capable of explaining both the major element chemistry and the trace-element abundances of Apollo 11 basalts.

One of the many interesting features of Apollo 11 and Apollo 12 rocks is that their average density, approximately 3.3, is close to that of the mean density of the moon, which is 3.31. However, these rocks cannot be regarded as representing the bulk composition of the moon. Their composition, when subjected to moderate pressures, such as those that exist only a few hundred kilometers below the surface of the moon, would be transformed from pyroxene-plagioclase rock to pyroxene-garnet rock. A pyroxene-garnet rock would have a density of 3.6 or 3.7, a value much too high to be compatible with the overall density of the moon.

Yoder and Tilley (1962, p. 507) suggested garnet peridotite as a probable source for terrestrial basalt magma. They concluded that this source would yield, by partial melting at low pressure, a tholeiitic magma. However, many more data are needed in order to apply our knowledge of terrestrial magmatic processes directly to the interpretation of lunar rocks.

On the other hand, the 14310 collected from Apollo 14 mission indicates extensive chemical fractionation. If the density of the melt were approximately 3, plagioclase would tend to float and other mineral would sink. This would produce a plagioclase-rich crust floating on a denser substratum. A small percentage (probably less than 5%) of partial melting of plagioclase-bearing peridotite (presumably heated by long-lived radioactive elements such as U and Th) could produce a liquid enriched in Al, Na, K, Rb, Sr, REE, U, and Th, that compositionally is similar to 14310. The liquidus temperature may be considerably higher than that of terrestrial high-alumina basalt owing to the anhydrous environment. If such primary high-alumina basaltic magma did exist on the moon, then this liquid is extremely important in the early stage of lunar evolution. It may be the parental liquid for a whole spectrum of differentiates, such as feldspathic basalts of the highland (Reid *et al.*, in press), feldspathic gabbro and anorthosite cumulates.

Sample 14310 was dated at $3.88 \pm 0.04 \times 10^9$ years by the Rb^{87}–Sr^{87} method (Papanastassiou and Wasserburg, 1971). Husain *et al.* (1971) reported that the age of Fra Mauro crystalline rocks (including 14310) as $3.77 \pm 0.15 \times 10^9$ years, which is the same as the age obtained for Apollo 14 fragmental rocks, $3.75 \pm 0.15 \times 10^9$ years. These dates suggest that some of the pre-Imbrian basalts are no more than 150 million years older than the Imbrian event itself. In addition, the possibility that the basalts were generated by the Imbrian impact event cannot be ruled out.

Acknowledgments—We are indebted to U.S. National Aeronautics and Space Administration for sample allocation, and to Academia Sinica for sponsorship. The work reported here was supported by the research fund of the National Taiwan University. We would like to thank Messrs. W. L. Jeng and C. Y. Kuo for laboratory assistance.

REFERENCES

Albee A. L. and Chodos A. A. (1970) Microprobe investigations on Apollo 11 samples. *Proc. Apollo 11 Lunar Sci Conf., Geochim. Cosmochim. Acta* Vol. 1, 135–157. Pergamon.

APSR (1971) Apollo 14 Preliminary Science Report NASA SP–272 309 p.

Apollo Soil Survey (1971) Apollo 14 nature and origin of rock types in soil from the Fra Mauro formation. *Earth Planet. Sci. Lett.* **12**, 49–54.

Baedecker P. A. Cuttitta F. Rose H. J. Jr. Schaudy R. and Wasson J. T. (1971) On the origin of lunar soil 12033. *Earth Planet. Sci. Lett.* **10**, 361–364.

Brown G. M. Emeleus C. H. Holland J. G. and Phillips R. (1970) Mineralogical, chemical and petrological features of Apollo 11 rocks and their relationship to igneous processes. *Proc. Apollo 11 Lunar Sci. Conf., Geochim. Cosmochim. Acta* Vol. 1, 195–219. Pergamon.

Chao E. C. T. (1972) personal communication.

Chao E. C. T. Boreman J. A. and Desborough G. A. (1971) The petrology of unshocked and shocked Apollo 11 and Apollo 12 microbreccias. *Proc. Apollo 12 Lunar Sci. Conf., Geochim. Cosmochim. Acta* 797–816. Pergamon.

Clark S. P. and Ringwood A. E. (1964) Density distribution and constitution of the mantle. *Rev. Geophys.* **2**, 35–85.

Compston W. Chappel B. W. Arriens P. A. and Vernon M. J. (1970) The chemistry and age of Apollo 11 lunar material. *Proc. Apollo 11 Lunar Sci. Conf., Geochim. Cosmochim. Acta* Vol. 2, 1007–1027. Pergamon.

Fredriksson K. Nelen J. and Melson W. G. (1970) Petrography and origin of lunar breccias and glasses. *Proc. Apollo 11 Lunar Sci. Conf., Geochim. Cosmochim. Acta* Vol. 1, 419–432. Pergamon.

Gancarz A. J. Albee A. L. and Chodos A. A. (1971) Petrologic and mineralogic investigation of some crystalline rocks returned by the Apollo 14 mission. *Earth Planet. Sci. Lett.* **12**, 1–18.

Gast P. W. Hubbard N. J. and Wiesmann H. (1970) Chemical composition and petrogenesis of basalts from Tranquillity Base. *Proc. Apollo 11 Lunar Sci. Conf., Geochim. Cosmochim. Acta* Vol. 2, 1143–1163. Pergamon.

Goles G. G. (1971) A review of the Apollo Project. *Amer. Sci.* **59**, pp. 326–331.

Haggerty S. E. Boyd F. R. Bell P. M. Finger L. W. and Bryan W. B. (1970) Opaque minerals and olivine in lavas and breccias from Mare Tranquillitatis. *Proc. Apollo 11 Lunar Sci. Conf., Geochim. Cosmochim. Acta* Vol. 1, 513–538. Pergamon.

Haskin L. Allen R. Helmke P. Paster T. Anderson M. Krotev R. and Zweifel K. (1970) Rare earths and other trace elements in Apollo 11 lunar samples. *Proc. Apollo 11 Lunar Sci. Conf., Geochim. Cosmochim. Acta* Vol. 2, 1213–1231. Pergamon.

Hassain L. Sutter J. F. and Schaeffer O. A. (1971) Ages of crystalline rocks from Fra Mauro. *Science* **173**, 1235–1236.

Hubbard N. J. Meyer C. Jr. Gast P. W. and Wiesmann H. (1971) The composition and derivation of Apollo 12 soils. *Earth Planet. Sci. Lett.* **10**, 3, 341–350.

Jackson E. D. (1961) X-ray determinative curves for some natural plagioclase of composition An_{60-85}. U.S.G.S. Prof. Paper 424 C, 286–288.

Juan V. C. Hsu L. C. and Yao T. S. (1963) High-alumina basalt from northern Taiwan. *Proc. Geol. Soc. China* **6**, 67–71.

King E. A. Jr. Carman M. F. and Butler J. C. (1972) Chondrules in Apollo 14 samples: Implications for the origin of chondritic meteorites. *Science* **175**, 59–60.

Kuno H. (1960) High-alumina basalt. *J. Petrol.* **1**, 121–145.

Kushiro I. and Nakamura Y. (1970) Petrology of some lunar crystalline rocks. *Proc. Apollo 11 Lunar Sci. Conf., Geochim. Cosmochim. Acta* Vol. 1, 607–626. Pergamon.

Kushiro I. and Haramura H. (1971) Major element variation and possible source materials of Apollo 12 crystalline rocks. *Science* **171**, 1235–1237.

LSPET (1969) Preliminary examination of lunar samples from Apollo 11. *Science* **165**, 1211–1227.

LSPET (1970) Preliminary examination of lunar samples from Apollo 12. *Science* **167**, 1325–1339.

LSPET (1971) Preliminary examination of lunar samples from Apollo 14. *Science* **173**, 681–693.

LSPET (in press) A preliminary description of the Apollo 15 lunar samples.

MacDonald G. A. and Katsura T. (1964) Chemical composition of Hawaiian lavas. *J. Petrol.* **5**, 82–133.

McKay D. S. and Morrison D. A. (1971) Lunar breccias. *J. Geophys. Res.* **76**, 5658–5669.

Nockolds S. R. and Allen R. (1956) The geochemistry of some igneous rock series III. *Geochim. Cosmochim. Acta* **9**, 34–77.

O'Hara M. J. Bigger G. M. and Richardson S. W. (1970) Experimental petrology of lunar material: The nature of mascons, seas and lunar interior. *Science* **167**, 605–607.

Papanastassiou D. A. and Wasserburg G. J. (1971) Rb–Sr ages of igneous rocks from the Apollo 14 mission and the age of the Fra Mauro formation. *Earth Planet. Sci. Lett.* **12**, 36–48.

Papike J. J. Bence A. E. Brown G. E. Prewitt C. T. and Wu C. H. (1971) Apollo 12 clinopyroxenes: Exsolution and Epitaxy. *Earth Planet. Sci. Lett.* **10**, 307–315.

Perkin-Elmer (1971): Analytical methods for atomic absorption spectrophotometry (analytical manual).

Prinz M. (1967) Geochemistry of basaltic rocks trace elements in H. H. Hess and A. Poldervaart ed., *Basalts 1*, 271–323, John Wiley.

Reid A. M. Eidley W. I. Warner J. Harmon R. S. Brett R. and Jakeš P. (1971) Lunar basalts (abstract). *Trans. Amer. Geophys. Union* **52**, 857.

Reid A. M. Eidley W. I. Harmon R. S. Warner J. Brett R. Jakeš P. and Brown R. W. (in press) Feldspathic basalts in lunar soils and the nature of the lunar highlands.

Ringwood A. E. (1970) Petrogenesis of Apollo 11 basalts and implications for lunar origin. *J. Geophys. Res.* **75**, 6453–6479.

Shapiro L. and Brannock W. W. (1962) Rapid analysis of silicate, carbonate and phosphate rocks. U.S. Geol. Survey Bull. 1144 A, 56 p.

Short N. M. (1970) Evidence and implications of shock metamorphism in lunar samples. *Proc. Apollo 11 Lunar Sci. Conf., Geochim. Cosmochim. Acta* Vol. 1, 865–871. Pergamon.

Smith J. V. and Gay P. (1958) The powder patterns and lattice parameters of plagioclase feldspars II. *Mineral. Mag.* **31**, 744–762.

Smith J. V. Anderson A. T. Newton R. C. Olsen E. J. Wyllie P. J. Crewe A. V. Isaacson M. S. and Johnson D. (1970) Petrologic history of the moon inferred from petrography, mineralogy and petrogenesis of Apollo 11 rocks. *Proc. Apollo 11 Lunar Sci. Conf., Geochim. Cosmochim. Acta* Vol. 1, 897–925. Pergamon.

Tera F. Eugster O. Burnett D. S. and Wasserburg G. J. (1970) Comparative study of Li, Na, K, Rb, Cs, Ca, Sr and Ba abundances in achondrites and in Apollo 11 lunar samples. *Proc. Apollo 11 Lunar Sci. Conf., Geochim. Cosmochim. Acta* Vol. 2, 1637–1657. Pergamon.

Tilley C. E. (1950) Some aspects of magmatic evolution. *Quart. J. Geol. Soc. London* **106**, 37–61.

Tilley C. E. (1961) The occurrence of hypersthene in Hawaiian basalts. *Geol. J.* **96**, 257–260.

Wakita H. and Schmitt R. A. (1970) Elemental abundances in seven fragments from lunar rock 12013. *Earth Planet. Sci. Lett.* **9**, 169.

Wood J. A. (1970a) Petrology of the lunar soil and geophysical implication. *J. Geophys. Res.* **75**, 6497–6513.

Wood J. A. Dickey J. S. Jr. Marvin U. B. and Powell B. M. (1970b) Lunar anorthosites and geophysical model of the moon. *Proc. Apollo 11 Lunar Sci. Conf., Geochim. Cosmochim. Acta* Vol. 1, 965–988. Pergamon.

Yoder H. S. and Tilley C. E. (1962) Origin of basaltic magmas: An experimental study of natural and synthetic rock systems. *J. Petrol.* **3**, 342–532.

Proceedings of the Third Lunar Science Conference
(Supplement 3, *Geochimica et Cosmochimica Acta*)
Vol. 1, pp. 707–721
The M.I.T. Press, 1972

Chondrules of lunar origin

Gero Kurat*, Klaus Keil,
Martin Prinz, and C. E. Nehru†
Department of Geology and
Institute of Meteoritics
The University of New Mexico
Albuquerque, New Mexico 87106, U.S.A.

Abstract—Chondrules and glass spherules from Apollo 14 breccia 14318,4 were studied microscopically, and their bulk and mineral compositions were determined with the electron microprobe. Approximately 65% of all chondrules are of ANT (anorthositic-noritic-troctolitic) composition with all but one clustering around the compositional equivalent of an anorthositic norite. No ANT glass spherules were observed. KREEP and basaltic chondrules overlap in composition and abundance with KREEP and basaltic glass spherules. Conclusions: (i) Chondrules in 14318,4 are similar in texture to meteoritic chondrules but differ drastically in composition. Hence, the former were produced on the lunar surface by impact melting and splattering of lunar rocks of ANT, KREEP, and basaltic composition. (ii) Lunar chondrules formed by spontaneous crystallization of highly supercooled, freely floating, molten droplets. (iii) The cooling rates of impact produced molten droplets at Apollo 14 (large impact events) were generally lower than those of droplets produced at Apollo 11 (small impact events). Hence, the probability for homogeneous and heterogeneous nucleation from the supercooled state was higher at Apollo 14 than for most Apollo 11 events, resulting in the formation of many chondrules at Apollo 14 but overwhelming glass spherules at Apollo 11. (iv) ANT (anorthositic norite) molten droplets, for reasons of composition, appear to be more apt to nucleate from the supercooled state, and, hence, form chondrules than those of KREEP and basaltic composition. (v) The relatively restricted composition of the main cluster of ANT chondrules suggests that they were formed by impact melting and splattering of a relatively homogeneous, fine-grained anorthositic norite, or from a coarser-grained norite that was homogenized in the impact process. (vi) The similarity in texture of impact produced lunar and meteoritic chondrules suggests that if major impact events occurred on parent meteorite bodies, at least some of the meteoritic chondrules may have formed by impact melting and splattering.

INTRODUCTION

Although meteoritic chondrules have been studied extensively in the past, there is still not complete agreement as to their origin and history: one hypothesis suggests that chondrules are primary bodies that condensed from a nebula of solar composition and then agglomerated to form parent meteorite bodies (Suess, 1949; Wood, 1962; Blander and Katz, 1967; Blander and Abdel-Gawad, 1969). The other hypothesis suggests that chondrules are secondary objects that formed from pre-existing material by either volcanism (Tschermak, 1875; Merrill, 1920; Wahl, 1952; Ringwood, 1961; Fredriksson and Ringwood, 1963), lightning discharge (Whipple, 1966), shock melting of primitive dust (Wood, 1963), or impact splattering (Fredriksson, 1963; 1969; Fredriksson and Reid, 1965; Urey, 1967; Kurat *et al.*, 1969).

In the present study, chondrule-like spherules were examined microscopically,

*Permanent address: Naturhistorisches Museum, A–1014 Vienna, Austria.

†Permanent address: Department of Geology, Brooklyn College, City University of New York, Brooklyn, N.Y. 11210, U.S.A.

and their mineral constituents were analyzed with an electron microprobe X-ray analyzer, following procedures previously described (Keil, 1967). Furthermore, their bulk compositions were determined using the broad beam electron microprobe technique (Prinz *et al.*, 1971a).

Spherules and glass fragments with textures somewhat similar to chondrules were found in lunar microbreccias from Apollo 11 (Fredriksson *et al.*, 1970; Roedder, 1970) and Apollo 12 (Fredriksson *et al.*, 1971). Abundant chondrules and chondrule-like spherules were found in samples from Apollo 14 (Fredriksson *et al.*, 1971, Kurat *et al.*, 1971; King *et al.*, 1972), Luna 16 (Keil *et al.*, 1972), and, more recently, from Apollo 15 (unpublished data). The present paper, however, is devoted to a detailed study of these objects from microbreccia 14318, where they are more abundant than in any other Apollo 14 rock studied by us. Comparison of their textures to those of artificial chondrule-like spherules (Nelson *et al.*, 1972) and meteoritic chondrules (e.g., Tschermak, 1885; Kurat, 1967) as well as of their bulk compositions to those of meteoritic chondrules suggests that they originated on the lunar surface by impact splattering of lunar rocks and spontaneous crystallization from super-cooled, molten droplets. Hence, these objects are properly referred to as chondrules (of lunar origin).

Fig. 1. Chondrule No. 16 of ANT composition; breccia 14318,4. Transmitted light; largest diameter is approximately 280 μ. Plagioclase laths of various sizes occur in a fine-grained, fibrous matrix. The outline of the chondrule, for the most part, is clearly visible; in two places (upper left hand and lower left hand corners, respectively), however, the chondrule is somewhat deformed, apparently while still plastic; this suggests that crystallization took place during agglomeration.

RESULTS

Textures

Rock 14318 is a complex breccia consisting mainly of microbreccia igneous and metamorphic rock fragments of different sizes and degrees of metamorphism, mineral and glass fragments, glass spherules and chondrules that are embedded into a fine-grained matrix. Although different in bulk composition, this rock exhibits many of the textural characteristics of polymict-brecciated chondrites, i.e., chondrules, nonchondritic brecciated, metamorphosed and igneous rock fragments, fine-grained matrix, and a high degree of welding. A detailed description of this rock, including analyses of minerals, glasses, rock fragments, and chondrules is given elsewhere (Kurat *et al.*, 1972).

Chondrules occur exclusively in the fine-grained matrix of the rock in amounts that usually do not exceed a few volume percent. Their shapes vary from perfectly spherical to ellipsoidal and irregular, and their average apparent diameters in thin sections range from approximately 20–500 μ. Most chondrules have igneous (crypto-crystalline to intersertal) textures, i.e., large, elongated crystals, mostly of plagioclase (Figs. 1, 2), and sometimes of olivine (Fig. 3), are embedded into a fine-grained matrix consisting mainly of plagioclase and pyroxene. However, igneous textures are commonly modified due to various degrees of recrystallization (Fig. 4). Fibrous and excentroradial chondrules are rare.

Fig. 2. Chondrule No. 47 of ANT composition (Table 3); breccia 14318,4. Transmitted light; largest diameter approximately 240 μ. Plagioclase laths, in fine-grained matrix. Chondrule outlines are somewhat vague, indicating extensive post-agglomeration welding.

Fig. 3. Chondrule No. 51 of KREEP composition (Table 3); breccia 14318,4. Transmitted light; largest diameter approximately 140 μ. Four plagioclase and one olivine crystal (center) in a fine-grained opaque matrix.

Chondrule classification

Chondrules are classified on the basis of bulk and normative compositions into three main groups. These are the ANT (anorthositic-noritic-troctolitic) group, the KREEP group, and the basalt group. This classification of chondrules is analogous to that proposed by us for lithic fragments and glasses from Apollo 11, Apollo 12, Apollo 14, and Luna 16. The ANT group consists of an apparently continuous clan of rocks and their impact produced glass derivatives ranging from anorthosite, described by Wood *et al.* (1970), to noritic anorthosite, anorthositic norite, olivine norite, norite, anorthositic troctolite, and troctolite described by Prinz *et al.* (1971a). This group is represented in materials collected at the Apollo 11 (Prinz *et al.*, 1971a, 1971b), Apollo 12 (Keil *et al.*, 1971; Bunch *et al.*, 1972 and in preparation), Apollo 14 (Kurat *et al.*, 1972; Prinz *et al.*, in preparation), and Luna 16 (Keil *et al.*, 1972) landing sites. The term KREEP was introduced by Hubbard *et al.* (1971) to describe a group of K-, REE-, and P-rich glasses that are the compositional and normative equivalents of norites or low-Ca pyroxene basalts. This term has been extended to describe lithic fragments of similar composition that we interpret to be the percursors of the KREEP glasses. These KREEP lithic fragments mostly have clastic but some have igneous textures and occur in Apollo 11 (Albee and Chodos, 1970), Apollo 12 (Keil *et al.*, 1971; Meyer *et al.*, 1971 and Bunch *et al.*, 1972, in preparation); Apollo 14 (Kurat *et al.*, 1972; Prinz *et al.*, in preparation), and Apollo 15 (Meyer, 1972) samples. It

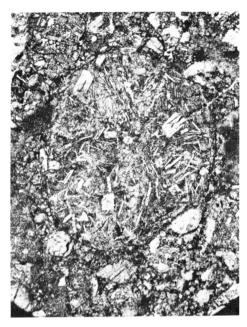

Fig. 4. Chondrule No. 20 of ANT composition; breccia 14318,4. Transmitted light; largest diameter approximately 400 μ. Plagioclase microphenocrysts and laths in fine-grained matrix. Matrix pyroxene is somewhat recrystallized and granular. Attached and tightly welded to lower right hand side of chondrule is a deformed (light) glass spherule of KREEP composition (note small crystals extending from chondrule into glass spherule). Chondrule and glass spherule apparently joined when both were still plastic.

should be noted that KREEP lithic fragments and glasses are mostly distinctly different from ANT norite in elements other than K, REE, and P: KREEP is mostly higher in FeO, MgO, and TiO_2 and lower in Al_2O_3 and CaO than ANT norite (Kurat et al., 1972). The basalt group is comprised of lithic fragments and glasses of basaltic composition that occur, in varying abundance, at all Apollo and the Luna 16 landing sites. Their compositions range somewhat from site to site and even at a specific site, particularly as far as TiO_2 and alkali contents are concerned [for example, high alkali, high-K, and low-K basalt glasses from Apollo 14 (Kurat et al., 1972; Prinz et al., in preparation) and high and low-K basaltic lithic fragments from Apollo 11 (Prinz et al., 1971a, 1971b)].

Chondrules of the ANT group. Eighteen chondrules of ANT composition, ranging in texture from intersertal to microporphyritic and granular (e.g., Figs. 1, 2, 4), make up approximately 65% of all chondrules studied here. They are characterized by their relatively narrow range in composition (Fig. 5): All but one chondrule have the bulk and normative compositions of anorthositic norite (Table 1). The chondrule that is outside this narrow compositional range is the compositional equivalent of anorthosite (Table 1; Fig. 5), having 95.5% normative plagioclase.

Within the main cluster of ANT group chondrules, two slight but apparently real compositional variations are noted: First, the proportion of normative plagioclase to ferromagnesian minerals ranges between 76:24 and 65:35, whereas the molecular ratio of $Mg/Mg + Fe + Mn$ is essentially constant (0.606 to 0.648). This relationship is indicated in Fig. 5, where the points for ANT group chondrules plot along a line of constant $Mg/Mg + Fe$ ratio but have a range of Ca content. Second, a slight variation in SiO_2 is observed at or near silica saturation, causing the presence of small amounts of normative quartz (up to 3%) or normative olivine (up to 4.5%).

The mineralogical composition of ANT chondrules is simple and monotonous: They consist of crystals of plagioclase (An_{98-90}, average An_{94}) and a fine-grained matrix of plagioclase and pyroxene. Individual plagioclase crystals within a given chondrule are usually homogeneous and only rarely was slight zoning detected very close to the margins of crystals (largest range observed was from An_{95-92} for plagioclase in chondrule 16; Fig. 1).

The composition of chondrule matrices varies strongly as a function of chondrule texture (Table 2). Chondrules exhibiting the most pronounced igneous textures (e.g., chondrule 16, Fig. 1; chondrule 47, Fig. 2) have the most differentiated matrices, i.e., matrices that are farthest removed in composition from the bulk composition of the chondrule. Because the crystallization sequence in ANT chondrules is simple, the relationship between bulk and matrix compositions, respectively, is also simple and governed exclusively by the crystallization of plagioclase; because ANT chondrules are very rich in feldspar component, plagioclase is the first phase to crystallize (Figs. 1, 2) and the residual liquid is progressively enriched in ferromagnesian minerals. In ANT chondrule 47, for example, SiO_2, TiO_2, FeO, MnO, MgO, Na_2O, and K_2O are higher and Al_2O_3 and CaO are lower in the chondrule matrix in comparison to

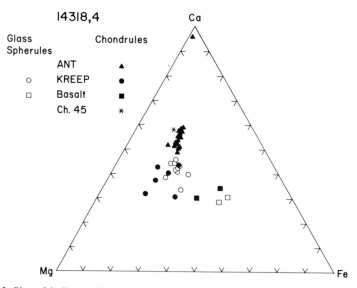

Fig. 5. Plot of bulk atomic proportions of Ca, Mg, and Fe for chondrules and glass spherules, obtained by broad electron beam microprobe techniques. Most abundant chondrules are of ANT composition, with no ANT glass spherule equivalents. KREEP and basaltic chondrules and glass spherules overlap in composition and abundance.

Table 1. Bulk compositions of chondrules in breccia 14318,4 (in weight percent) as obtained by broad beam electron microprobe techniques, and CIPW molecular norms*

	ANT group			KREEP group		Basalt group	
	Anorthosite Ch52	Anorthositic norite	Chondrule 45	Average	(Range)	High-K basalt Ch49	High-alkali basalt Ch43
SiO_2	45.4	47.2 (45.7 –48.5)	54.8	49.0	(46.2 –51.0)	45.2	49.5
TiO_2	0.05	0.30 (0.19– 0.79)	0.16	0.97	(0.30– 1.74)	3.4	2.93
Al_2O_3	34.1	24.8 (22.5 –26.6)	20.9	17.9	(14.9 –21.5)	12.1	7.9
Cr_2O_3	<0.02	0.15 (0.10– 0.21)	0.29	0.20	(0.14– 0.25)	0.37	0.16
FeO	0.25	6.0 (5.3 – 6.8)	4.1	7.9	(6.1 –11.1)	17.3	16.6
MnO	0.01	0.09 (0.06– 0.18)	0.08	0.15	(0.13– 0.20)	0.31	0.30
MgO	0.34	5.7 (4.6 – 6.8)	4.8	9.7	(7.1 –13.1)	9.3	5.2
CaO	18.4	14.9 (13.8 –15.8)	13.6	12.0	(9.7 –13.7)	11.4	10.6
Na_2O	1.19	0.51 (0.38– 0.93)	1.40	0.36	(0.10– 0.59)	0.34	0.80
K_2O	0.11	0.17 (0.09– 0.39)	0.78	0.73	(0.35– 1.15)	0.41	1.14
P_2O_5	0.04	0.11 (0.05– 0.25)	<0.02	0.22	(0.06– 0.49)	0.10	3.0
TOTAL	99.89	99.93	100.91	98.98		100.23	98.13
No. Specimens	1	17	1	6		1	1

CIPW Molecular Norms

q	—	0.42	7.57	1.36		—	10.60
or	0.65	1.01	4.58	4.36		2.51	7.28
ab	7.96	4.59	12.49	3.27		3.16	7.77
an	86.93	65.11	48.12	45.56		31.33	15.78
ne	1.61	—	—	—		—	—
di { wo	1.27	3.33	7.57	5.26		10.61	7.95
di { en	0.94	2.13	5.20	3.73		5.68	3.15
di { fs	0.34	1.21	2.37	1.53		4.93	4.81
wo	0.19	—	—	—		—	—
hy { en	—	13.66	7.97	23.33		15.04	12.38
hy { fs	—	7.73	3.64	9.56		13.05	18.95
ol { fo	—	—	—	—		4.39	—
ol { fa	—	—	—	—		3.81	—
cm	—	0.17	0.32	0.22		0.43	0.19
il	0.07	0.42	0.23	1.37		4.90	4.41
ap	0.09	0.23	—	0.47		0.22	6.78

*For individual analyses, see Prinz *et al.*, 1972.

the bulk chondrule composition. Differences in bulk and matrix compositions are less pronounced in fine-grained chondrules, where crystallization of large early plagioclase did not occur (note that the time interval between formation of early and late plagioclase by nucleation from the supercooled state was probably very short). Compositional trends between bulk chondrule and chondrule matrix compositions are somewhat different for the one chondrule containing plagioclase and olivine (Table 2, KREEP chondrule 51); in this chondrule, SiO_2, TiO_2, Cr_2O_3, MnO, CaO, K_2O, and P_2O_5 increase and Al_2O_3, FeO, MgO, and Na_2O decrease when comparing bulk to matrix compositions.

An unusual chondrule (No. 45) with apparent affinities to the ANT group should be mentioned here. This chondrule has a very fine-grained fibrous-radial texture and is similar in composition to the main cluster of ANT group chondrules (Fig. 5) with the exception of exceedingly high SiO_2, Na_2O, and K_2O contents (Table 1). Therefore, this chondrule is listed separately in Table 1 and Fig. 5.

Chondrules of the KREEP group. Six chondrules, ranging in texture from intersertal

Table 2. Comparison of chondrule matrix and bulk compositions (in weight percent), in microbreccia 14318,4 as obtained by broad beam electron microprobe techniques. CIPW molecular norms are also given.

		Chondrule 47 (ANT)		Chondrule 51 (KREEP)	
		Bulk	Matrix	Bulk	Matrix
	SiO_2	45.7	49.0	49.0	53.8
	TiO_2	0.19	0.44	1.74	5.0
	Al_2O_3	26.3	14.3	14.9	11.2
	Cr_2O_3	0.16	n.d.	0.14	0.28
	FeO	5.4	12.1	11.1	8.6
	MnO	0.07	0.21	0.16	0.19
	MgO	4.9	10.7	9.3	4.4
	CaO	15.3	11.7	9.7	11.5
	Na_2O	0.57	0.74	0.77	0.68
	K_2O	0.15	0.43	0.50	0.78
	P_2O_5	0.08	n.d.	0.49	1.36
	TOTAL	98.82	99.62	97.80	97.79

CIPW Molecular Norms

		Bulk	Matrix	Bulk	Matrix
	q	—	—	3.15	20.84
	or	0.90	2.57	3.07	4.95
	ab	5.19	6.71	7.17	6.55
	an	69.66	34.77	37.03	27.04
di {	wo	2.69	9.54	3.82	9.86
	en	1.68	5.88	2.42	6.44
	fs	1.01	3.67	1.41	3.42
hy {	en	9.70	17.43	24.20	6.60
	fs	5.80	10.89	14.06	3.51
ol {	fo	1.75	4.90	—	—
	fa	1.05	3.06	—	—
	cm	0.18	—	0.16	0.33
	il	0.27	0.62	2.52	7.47
	ap	0.17	—	1.07	3.05

n.d. = not determined

to cryptocrystalline and hornfelsic, are of KREEP composition. Their bulk compositions have a greater range than those of the main cluster of ANT chondrules (e.g., Mg/Mg + Fe + Mn ranges from 0.601–0.759) (Table 1; Fig. 5), but they can collectively be described as the compositional analogues of norite or low-Ca pyroxene basalt with intermediate ($\sim 50\%$) normative plagioclase content. Plagioclase from KREEP chondrules ranges widely in composition as a function of bulk alkali content from An_{98-78}. Some of the plagioclase crystals within a single intersertal chondrule are strongly zoned ranging, for example, from An_{96-80}.

One chondrule of KREEP composition (No. 51) is particularly interesting, because it is the only one observed that contains a ferromagnesian mineral (olivine) large enough for electron microprobe analysis. This chondrule contains four large plagioclase crystals, one large olivine crystal, and a fine-grained matrix (Fig. 3). The bulk (Table 2), matrix (Table 2), feldspar, and olivine compositions were determined in an attempt to decipher the origin of this chondrule. The olivine crystal is homogeneous ($Fo_{60.5}$) with the exception of very slight zoning (to Fo_{58}) at the very end of the "tail" of the crystal (Fig. 3). Plagioclase is also rather homogeneous, ranging in composition from An_{83-80}. The matrix (Table 2) is typically enriched in Fe in comparison to Mg and also high in SiO_2, resulting in a normative quartz content of 20.8%. Because of the simultaneous and early crystallization of plagioclase and olivine, the matrix composition trend in chondrule 51 is different from that in chondrules in which plagioclase is the only early phase.

Chondrules of the basalt group. Chondrules of the basalt group are represented by two specimens (Table 1, Fig. 5). Both chondrules are small (25 and 40 μ, respectively) and contain a few crystals too small for electron microprobe analysis in an extremely fine-grained dark red to opaque matrix. These chondrules differ considerably in composition (e.g., Mg/Mg + Fe + Mn is 0.596 for chondrule 49 and 0.355 for chondrule 43). Furthermore, both chondrules are rich in K_2O in comparison to many lunar lithic fragments, glasses, and chondrules of similar bulk composition. Chondrule 43, in addition, is rich in Na_2O and P_2O_5 (Table 1) (Prinz *et al.*, 1971a, 1971b; Keil *et al.*, 1971, 1972; Kurat *et al.*, 1972).

Glass spherule classification

Glass spherules are less common than chondrules in rock 14318,4: This is due to the absence of glass spherules of ANT composition (ANT chondrules are the most abundant chondrules). KREEP and basalt glass spherules are about as abundant as chondrules of that composition. Glass spherules are generally also somewhat smaller than chondrules (Fig. 6).

Glass spherules of the KREEP group are clear and commonly contain a few very small equidimensional feldspar crystals near their surfaces where they are tightly

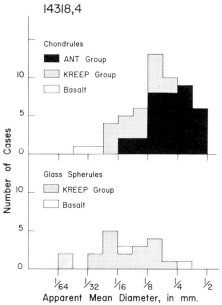

Fig. 6. Histogram of the average apparent diameters of chondrules and glass spherules of ANT, KREEP, and basaltic composition. Breccia 14318,4. On the average, chondrules are larger than glass spherules.

welded to the rock matrix. They overlap in composition with KREEP chondrules and are the compositional analogues of norites or low-Ca pyroxene basalts (Table 3; Fig. 5). KREEP glass spherules are homogeneous in composition with the exception of one spherule (GS1) that displays a regular decrease in K_2O content from its surface to the interior (Fig. 7). These gradients are similar to the ones reported for glass spherules from Apollo 11 rock 10019 (Kurat and Keil, 1972) but differ in detail: K_2O gradients in glass spherule GS1 are shallower than those reported for Apollo 11, corresponding to a $D_K \cdot t = 2.4 \times 10^{-7}$ (cm^2) (product of diffusion coefficient D_K for potassium and diffusion time t, calculated from the measured concentration C_s at the spherule surface; the measured concentration C_0 inside the spherule at the spherule center assumed to correspond to the concentration before diffusion started; and the concentration C_x at point x, using the diffusion equation for diffusion from a thin surface layer (Kingery, 1967)). The $D_K \cdot t$ for spherule GS1 is six times higher than for a glass spherule of similar composition from Apollo 11 (Kurat and Keil, 1972).

Only two glass spherules were found that have been classified as basaltic (Fig. 5). Their compositions, however, are somewhat unusual and different from those of

Table 3. Bulk compositions of glass spherules in breccia 14318,4 (in weight percent) as obtained by point electron microprobe techniques. CIPW molecular norms are also given.*

		KREEP group		Basalt group	
		Avg.	Range	GS10	GS12
SiO_2		49.7	(45.7 –52.9)	53.3	54.7
TiO_2		0.98	(0.25– 1.84)	3.0	3.4
Al_2O_3		19.7	(15.9 –20.9)	10.7	10.8
Cr_2O_3		0.19	(0.07– 0.32)	0.13	0.10
FeO		8.7	(6.9 –11.2)	16.2	15.6
MnO		0.12	(0.04– 0.17)	0.26	0.26
MgO		7.7	(6.3 – 9.5)	5.5	4.4
CaO		12.4	(10.7 –14.4)	8.1	8.1
Na_2O		0.67	(0.30– 1.31)	1.41	0.10
K_2O		0.31	(0.22– 0.43)	1.40	1.57
P_2O_5		0.19	(0.01– 0.35)	1.90	0.38
TOTAL		100.66		101.90	99.41
No. Specimens		10		1	1

CIPW Molecular Norms

q		2.59		9.56	17.29
or		1.83		8.51	9.91
ab		6.03		13.02	0.96
an		49.93		19.26	26.05
di { wo		4.18		3.72	5.70
di { en		2.66		1.56	2.17
di { fs		1.52		2.17	3.53
hy { en		18.63		14.06	10.81
hy { fs		10.66		19.66	17.69
ol { fo		—		—	—
ol { fa		—		—	—
cm		0.21		0.15	0.12
il		1.37		4.30	5.06
ap		0.40		4.09	0.85

*For individual analyses, see Prinz et al., 1972.

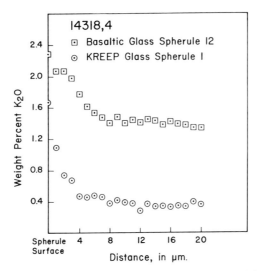

Fig. 7. Plot of K_2O content vs. distance from glass spherule surface towards interior. Breccia 14318,4, basaltic glass spherule 12 and KREEP glass spherule 1. Potassium content decreases from spherule surface towards interior and the slopes of the diffusion curves correspond to $D_K \cdot t$ (product of diffusion coefficient for potassium D_K in glass of that composition and diffusion time t) of 1.9×10^{-6} (cm^2) (for spherule 12) and 2.4×10^{-7} (cm^2) (for spherule 1).

basalt chondrules, particularly as far as their high SiO_2 and, hence, normative quartz contents are concerned (9.6 and 17.3%, respectively; Table 3). Although not typically basaltic, they are classed with this group for simplicity. Both spherules are inhomogeneous, displaying a regular decrease in K_2O content from their surfaces towards the interiors (Fig. 7). The $D_K \cdot t$ values for these basaltic spherules are 1.9×10^{-6} (cm^2).

DISCUSSION

The conclusions of this study may be summarized as follows:

(i) Chondrules described here have textures very similar to those observed in meteoritic chondrules (e.g., Tschermak, 1885; Kurat, 1967), but their bulk compositions are drastically different: Lunar chondrules are of ANT, KREEP, and basaltic compositions, i.e., they correspond in composition to lunar rocks represented in hand specimens and/or lithic fragments (Prinz et al., 1971a, 1971b; Keil et al., 1971, 1972; Kurat et al., 1972), whereas meteoritic chondrules normally have the composition of ultramafic rocks (Fredriksson, 1963; Walter, 1969). On the basis of composition, it is concluded that chondrules in lunar rock 14318 are of lunar origin and formed by impact melting and splattering of ANT, KREEP, and basaltic rocks; they are not trapped, stray meteoritic chondrules.

(ii) Many of the textures observed in lunar and meteoritic chondrules have been reproduced in the laboratory by spontaneous crystallization of highly supercooled, freely falling droplets of molten oxides, silicates, and silicate mixtures that were melted by a high power CO_2 laser beam (Nelson et al., 1972). Similarities in textures between synthetic, meteoritic, and lunar chondrules suggest that the latter also formed by spontaneous crystallization of freely floating supercooled molten droplets that were produced in a major impact event on the lunar surface.

Evidence for an origin of lunar chondrules from the highly supercooled state is contained in the lunar chondrules themselves. KREEP chondrule 51, for example, consists of four nearly homogeneous plagioclases and one olivine crystal ($Fo_{60.5}$) in contact with a fine-grained matrix rich in SiO_2 (20.8% normative quartz). The nearly homogeneous plagioclase-olivine crystals appear to suggest near-equilibrium crystallization, whereas the assemblage olivine-quartz normative matrix indicates a disequilibrium assemblage. This apparent contradiction is readily explained by assuming that the plagioclase-olivine assemblage crystallized in "psuedoequilibrium" from a supercooled molten droplet at temperatures far below the liquidus; there, nearly homogeneous crystals would grow spontaneously. The fine-grained matrix is interpreted to have solidified after the plagioclase-olivine crystals had grown. Because of the rapid cooling of the molten droplet and the high growth rate of crystals from the supercooled state, the early plagioclase-olivine crystals did not react with the later residual matrix material. It should be noted that similar assemblages commonly are observed in meteoritic chondrules and that they were interpreted to have formed in a similar process from highly supercooled molten droplets (Kurat, 1967).

(iii) Apollo 11 samples contain only very few chondrules but abundant glass spherules of compositions similar to ANT chondrules in rock 14318,4 (Prinz et al., 1971a, 1971b). The question arises why, in spite of the compositional similarities, only few chondrules formed in case of Apollo 11, and why no glass spherules of ANT composition formed in case of rock 14318,4. It is suggested that this apparent contradiction is readily explained by the difference in magnitudes of the impact events that formed the Apollo 11 and 14 materials, respectively. In case of Apollo 11, molten droplets were apparently formed in small-scale impact events that resulted in fast supercooling of individual droplets and, hence, low probability for homogeneous and heterogeneous nucleation (i.e., by collision with mineral and rock fragments, vibrations, etc.). Thus, in case of Apollo 11, the vast majority of molten droplets cooled rapidly, supercooled, and solidified to form glass spherules. In case of Apollo 14, on the other hand, molten droplets were formed in large-scale impact events where cooling rates were low due to shielding in the large impact plume and, hence, the probability for spontaneous crystallization of molten drops from the supercooled state is high due to a higher probability for homogeneous and heterogeneous nucleation. Thus, in case of Apollo 14, many molten droplets crystallized from the supercooled state and formed chondrules.

The proposition of slower cooling of molten droplets in the case of Apollo 14 in comparison to Apollo 11 is also supported by the different compositional gradients for potassium observed at and near the surfaces of a few glass spherules from the two sites. The product $D_K \cdot t$ of the diffusion coefficient D_K for potassium and diffusion time t in the Apollo 14 glass spherule GS1 was determined to be six times higher than that for a spherule of similar composition from Apollo 11 (Kurat and Keil, 1972). Because of their compositional similarities it may be assumed, as a first approximation, that the diffusion coefficients for K in these Apollo 14 and 11 spherules are approximately the same and, hence, that the higher value for $D_K \cdot t$ for the Apollo 14 spherule is largely the result of a considerably longer diffusion time t, i.e., the Apollo 14 spherule cooled slower than the Apollo 11 spherule. The proposition is further supported by the nature of the sodium gradients in the two spherules: In the Apollo 14 spherule, Na concentration is nearly homogeneous from the spherule surface to the interior, whereas in the Apollo 11 spherule, the decrease in the concentration of Na from the spherule surface to the interior is similar to the one observed for K. Because the diffusion coefficient for Na in these systems is considerably higher than that for K, it is concluded that in the relatively long cooling time of the Apollo 14 spherule Na was completely homogenized, whereas in the faster cooled Apollo 11 spherule, a Na gradient is retained.

(iv) In rock 14318, abundant ANT chondrules (anorthositic norite composition) but no ANT glass spherules are observed. KREEP and basalt glass spherules, on the other hand, are more abundant than chondrules of those compositions. Apparently, these ANT molten droplets, for reasons of composition, are more apt to nucleate from the supercooled state and, hence, form chondrules than those of KREEP or basaltic compositions.

(v) The relatively restricted composition of the main cluster ANT chondrules in breccia 14318,4 suggests that they formed by impact melting and splattering of a relatively homogeneous, fine-grained anorthositic norite, or from a coarser-grained norite that was homogenized in the impact process.

(vi) The similarity in texture of impact produced lunar and meteoritic chondrules suggests that if major impact events occurred on parent meteorite bodies, at least some meteoritic chondrules may have formed by impact melting and splattering.

Acknowledgments—We gratefully acknowledge the assistance of Messrs. G. H. Conrad and J. A. Green in the electron microprobe work and of Mrs. Julie Hultzen in the data reduction. This work is supported in part by NASA Grant NGL-32-004-063 (Klaus Keil, Principal Investigator).

REFERENCES

Albee A. L. and Chodos A. A. (1970) Microprobe investigations on Apollo 11 samples. *Proc. Apollo 11 Lunar Sci. Conf., Geochim. Cosmochim. Acta,* Suppl. 1, Vol. 1, 135–157.

Blander M. and Katz J. L. (1967) Condensation of primordial dust. *Geochim. Cosmochim. Acta* **31**, 1025–1034.

Blander M. and Abdel-Gawad M. (1969) The origin of meteorites and the constrained equilibrium theory. *Geochim. Cosmochim. Acta* **33**, 701–716.

Bunch T. E. Erlichman J. Quaide W. L. Busche F. D. Conrad G. H. Keil K. and Prinz M. (1972) Electron microprobe analyses of lithic fragments and glasses from Apollo 12 lunar samples. Spec. Publ. No. 4, Univ. of New Mexico-Institute of Meteoritics.

Bunch T. E. Prinz M. and Keil K. (in preparation) Lithic fragments and glasses from Apollo 12 samples. *Earth Planet. Sci. Lett.*

Fredriksson K. and Ringwood A. E. (1963) Origin of meteoritic chondrules. *Geochim. Cosmochim. Acta* **27**, 639–641.

Fredriksson K. (1963) Chondrules and the meteorite parent bodies. *Trans. N.Y. Acad. Sci.* **25**, 756–769.

Fredriksson K. and Reid A. M. (1965) A chondrule in the Chainpur meteorite. *Science* **149**, 856–860.

Fredriksson K. (1969) The Sharps chondrite: New evidence on the origin of chondrules and chondrites. in Symp. on Meteorite Research (ed. P. M. Millman), Reidel, Dordrecht, 155–165.

Fredriksson K. Nelen J. and Melson W. G. (1970) Petrology and origin of lunar breccias and glasses. *Proc. Apollo 11 Lunar Sci. Conf., Geochim. Cosmochim. Acta,* Suppl. 1, Vol. 1, 419–432.

Fredriksson K. Nelen J. and Noonan A. (1971) Lunar chondrules (abstract). *Meteoritics* **6**, 270.

Hubbard N. J. Meyer C. Gast P. W. and Wiesmann H. (1971) The composition and derivation of Apollo 12 soils. *Earth Planet. Sci. Lett.* **10**, 341–350.

Keil K. (1967) The electron microprobe X-ray analyzer and its application in mineralogy. *Fortschr. Mineral.* **44**, 4–66.

Keil K. Prinz M. and Bunch T. E. (1971) Mineralogy, petrology, and chemistry of some Apollo 12 samples. Proc. *Second Lunar Sci. Conf., Geochim. Cosmochim. Acta*, Suppl. 2, Vol. 1, 319–341.

Keil K. Kurat G. Prinz M., and Green J. A. (1972) Lithic fragments, glasses and chondrules from Luna 16 fines. *Earth Planet. Sci. Lett.* **13**, 243–256.

King E. A. Carman M. F. and Butler J. C. (1972) Chondrules in Apollo 14 samples: Implications for the origin of chondritic meteorites. *Science* **175**, 59–60.

Kingery W. D. (1967) *Introduction to Ceramics*, John Wiley (1960) (fourth printing, April 1967).

Kurat G. (1967) Zur Entstehung der Chondren. *Geochim. Cosmochim. Acta* **31**, 491–502.

Kurat G. Fredriksson K. and Nelen J. (1969) Der Meteorit von Siena. *Geochim. Cosmochim. Acta* **33**, 765–773.

Kurat G. Keil K. Prinz M. and Nehru C. E. (1971) Chondrules of lunar origin (abstract). *Meteoritics* **6**, 285.

Kurat G. Keil K. and Prinz M. (1972) Lunar breccia 14318,4: Composition of lithic fragments, glasses, chondrules, and textural similarities to chondrites. *Geochim. Cosmochim. Acta*, in preparation.

Kurat G. and Keil K. (1972) Effects of vaporization and condensation on Apollo 11 glass spherules: Implications for cooling rates. *Earth Planet. Sci. Lett.* **14**, 7–13.

Merrill G. P. (1920) On chondrules and chondritic structure in meteorites. *Proc. Nat. Acad. Sci.*, **6**, 449–472.

Meyer C. Brett R. Hubbard N. J. Morrison D. A. McKay D. S. Aitken F. K. Takeda H. and Schonfield E. (1971) Mineralogy, chemistry, and origin of the KREEP component in soil samples from the Ocean of Storms. *Proc. Second Lunar Sci. Conf., Geochim. Cosmochim. Acta*, Suppl. 2, Vol. 1, 393–411.

Meyer C. (1972) Mineral assemblages of lithic fragments of non-Mare lunar rock types (abstract). In *Lunar Science—III*, pp. 542–544, Lunar Science Institute Contr. No. 88.

Nelson L. S. Blander M. Skaggs S. R. and Keil K. (1972) Use of a CO_2 laser to prepare chondrule-like spherules from supercooled molten oxide and silicate droplets. *Earth Planet. Sci. Lett.* (in press).

Prinz M. Bunch T. E. and Keil K. (1971a) Composition and origin of lithic fragments and glasses in Apollo 11 samples. *Contrib. Mineral. Petrol.* **32**, 211–230.

Prinz M. Bunch T. E. and Keil K. (1971b) Electron microprobe analyses of lithic fragments and glasses from Apollo 11 lunar samples. Spec. Publ. No. 2, Univ. of New Mexico-Institute of Meteoritics.

Prinz M. Nehru C. E. Kurat G. Keil K. Conrad G. H. and Busche F. D. (1972) Electron microprobe analyses of lithic fragments, glasses, and chondrules from Apollo 14. Spec. Publ. No. 6, UNM-Institute of Meteoritics.

Prinz M. Keil K. and Kurat G. (in prep.) Lithic fragments and glasses from Apollo 14 samples. *Contrib. Mineral. Petrol.*

Ringwood A. E. (1961) Chemical and genetic relationships among meteorites. *Geochim. Cosmochim. Acta* **24**, 159–197.

Roedder E. (1970) Natural and laboratory crystallization of lunar glasses from Apollo 11. Proc. 7th General Meeting. Intern. Mineral. Ass., Spec. Paper No. 1, Mineral. Soc. Japan, 5–12.

Suess H. E. (1949) Zur Chemie der Planeten-und Meteoritenbildung. *Z. Elektrochem.* **53**, 237–241.

Tschermak G. (1875) Die Bildung der Meteoriten und der Vulcanismus. Sitzungsber. Akad. Wiss. Wien **71**, 661–673.

Tschermak G. (1885) Die mikroskopische Beschaffenheit der Meteoriten. E. Schweizerbart'sche Verlagsgesellschaft, Stuttgart. In *Smiths. Contrib. Astrophysics* Vol. 4, No. 6 (1964) 138–239 (J. A. Wood and E. M. Wood, translators).

Urey H. C. (1967) Parent bodies of the meteorites. *Icarus* **7**, 350–359.

Wahl W. (1952) The brecciated stony meteorites and meteorites containing foreign fragments. *Geochim. Cosmochim. Acta* **2**, 91–117.

Walter L. S. (1969) The major-element composition of individual chondrules of the Bjurbole meteorite. In Symp. on Meteorite Research (ed. P. M. Millman) 191, Reidel, Dordrecht.

Whipple F. L. (1966) Chondrules: Suggestion concerning the origin. *Science* **153**, 54.

Wood J. A. (1962) Metamorphism in chondrites. *Geochim. Cosmochim. Acta* **26**, 739–749.

Wood J. A. (1963) On the origin of chondrules and chondrites. *Icarus* **2**, 152–180.

Wood J. A. Dickey J. S. Marvin U. B. and Powell B. N. (1970) Lunar anorthosites and a geophysical model of the moon. *Proc. Apollo 11 Lunar Sci. Conf., Geochim. Cosmochim. Acta*, Suppl. 1, Vol. 1, 965–988.

Proceedings of the Third Lunar Science Conference
(Supplement 3, *Geochimica et Cosmochimica Acta*)
Vol. 1, pp. 723–737
The M.I.T. Press, 1972

Lunar glasses, breccias, and chondrules

Joseph Nelen, Albert Noonan, and Kurt Fredriksson

Department of Mineral Sciences,
Smithsonian Institution, Washington, D.C. 20560

Abstract—Glasses from Apollo 11, 12, 14, and 15 soil samples have been grouped on the basis of their chemistry and appear to provide an independent tool for identifying the presence of various rock types. The glasses also appear to constitute a record of complex differentiation processes that seem to have taken place within the moon. A green glass, abundant in some samples from Hadley Delta, could have been formed from a previously proposed differentiation model. Four breccias are described briefly and certain similarities between impact features, including chondrules, in lunar samples and meteorites are noted.

Soil Glasses

Apollo 14

A total of 130 glass fragments and spherules in soil samples from various locations (14148,19; 14149,36; 14156,20 respectively top, bottom, and middle trench—14163,86 bulk sample—14259,33; 14260,2 contingency sample) were analyzed with the electron microprobe. No major differences were found with respect to the sampling site, and therefore the Apollo 14 soil glasses are treated as one sample.

Plots of K_2O, MgO, Al_2O_3, and TiO_2 versus the FeO contents of these glasses are given in Figs. 1 to 4. The degree of silica saturation is also indicated. Various groups are evident in these plots, and average compositions are given in Table 1. Group 1 represents glasses chemically related to mare basalts. These glasses are relatively rare at the Fra Mauro site. Groups 2a and 2b represent mafic to ultramafic rock types. Type 2b has previously been described by Prinz *et al.* (1971), and 2a is similar but somewhat higher in plagioclase. Groups 3a, 3b, and 3c glasses relate to KREEP-type basalts. More than half of the Apollo 14 soil glasses analyzed by us belong to this Group. Groups 3a and 3b differ mainly in Si and Mg content, and 3b appears slightly higher in alkali. Group 3c is undersaturated in Si and seems to be analogous to the low-K basaltic lithic fragments previously described by various investigators. Group 4 glasses have a narrow compositional range and appear to be the equivalent of an anorthositic norite. Groups 5a and b are anorthositic in composition. Although group b represents only 2 glass fragments, it might be of some significance because these glasses evidently are derived from a high Na anorthosite. Group 6 represents two potash granite glasses.

Apollo 15

About 200 glass analyses have thus far been performed on Apollo 15 soil samples. They are glasses from the coarse fines and the 60–100 mesh size of the <1 mm fines from four locations. The samples are 15021,42 and 15022,9, the LM site; 15251,55

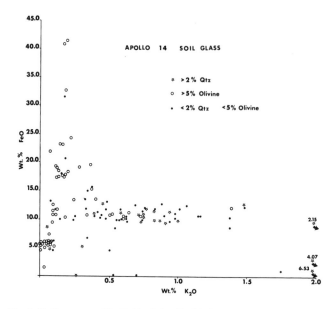

Fig. 1. K_2O content of individual Apollo 14 glasses versus FeO content.

and 15252,9, the east rim of a small fresh crater at station 6; 15301,118 and 15302,21, the northeast rim of Spur Crater station 7; and 15401,54 and 15402,6, fillet southside large boulder station 6a. Average compositions of various types from each sample site are given in Table 2. Group designations are similar to the ones used for the

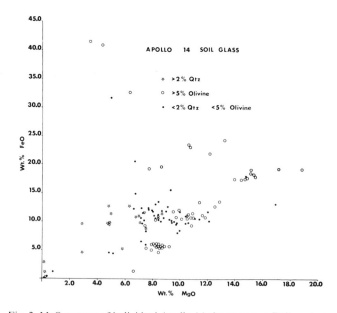

Fig. 2. MgO content of individual Apollo 14 glasses versus FeO content.

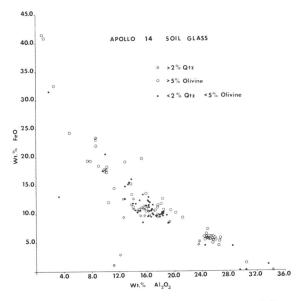

Fig. 3. Al₂O₃ content of individual Apollo 14 glasses versus FeO content.

Apollo 14 samples. Group 1 again represents what are believed to be mare basalt derived glasses. Group 1a was not observed by us among the Apollo 14 soil glasses. Iron and Ti contents are lower than in the so-called mare derived glasses, and Si, Al and Ca somewhat higher. The chemistry of these glasses closely resembles the

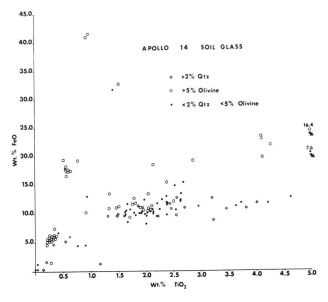

Fig. 4. TiO₂ content of individual Apollo 14 glasses versus FeO content.

Table 1. Apollo 14 Soil glasses. Range and average composition in weight percent.

Group	Number glasses	SiO_2	TiO_2	Al_2O_3	FeO	MgO	CaO	Na_2O	K_2O	Total
1	5	42.2(41.3–44.4)	4.4(2.1–7.6)	8.9(8.6–9.9)	21.6(18.6–23.5)	11.0(6.7–15.1)	9.2(8.8–10.4)	0.20(0.12–0.47)	0.1(0.07–0.18)	97.6
2a	8	46.0(45.0–47.0)	0.6(0.5–0.9)	10.0(9.8–10.4)	17.8(17.4–18.3)	14.7(13.9–15.4)	9.5(8.9–10.2)	0.20(0.12–0.30)	0.2(0.12–0.20)	99.0
2b	2	44.4(44.2–44.6)	0.6(0.5–0.7)	7.6(7.3–7.8)	19.3(19.3–19.3)	18.0(17.1–18.8)	8.2(8.0–8.5)	0.10(0.07–0.11)	0.1(0.11–0.11)	98.3
3a	58	47.0(44.9–49.7)	2.3(0.9–4.6)	16.0(12.8–20.1)	11.0(9.6–15.4)	10.4(6.6–12.8)	10.3(8.3–13.0)	0.7(0.34–1.48)	0.7(0.24–1.48)	98.4
3b	6	51.1(49.2–53.3)	1.7(1.3–2.6)	15.9(12.6–17.6)	10.5(9.6–12.7)	4.5(2.9–5.0)	10.1(9.2–10.9)	1.20(0.46–1.80)	1.2(0.73–2.15)	96.2
3c	9	43.2(41.1–46.2)	1.9(1.6–2.1)	18.7(17.9–21.2)	11.2(9.4–12.7)	10.9(9.5–12.5)	11.7(10.9–13.3)	<0.10(0.01–0.17)	0.1(0.09–0.18)	97.7
4	19	43.8(43.0–45.0)	0.3(0.2–0.3)	25.3(24.4–26.5)	6.3(4.9–7.5)	8.4(7.5–9.4)	14.8(14.4–15.4)	0.10(0.01–0.10)	0.1(0.01–0.10)	99.1
5a	4	45.9(42.3–49.2)	0.2(0.1–0.3)	32.4(30.4–34.8)	0.9(0.1–1.4)	2.5(0.1–6.6)	17.0(16.1–18.2)	0.8(0.1–1.5)	0.7(0.10–1.80)	100.4
5b	2	51.0(49.3–52.7)	0.1	30.0(29.8–30.4)	0.2	0.2(0.1–0.2)	13.2(12.7–13.6)	3.6(3.1–4.0)	0.6(0.53–0.70)	98.9
6	2	76.8(76.3–77.3)	0.7(0.2–1.2)	11.6(11.1–12.1)	2.0(1.2–2.8)	0.1	1.1(0.8–1.4)	0.5(0.23–0.77)	5.3(4.07–6.53)	98.1

Table 2. Apollo 15 soil glasses. Range and average composition in weight percent.

Group	Number* glasses	SiO_2	TiO_2	Al_2O_3	FeO	MgO	CaO	Na_2O	K_2O	Total
1	11	43.2(41.7–46.0)	3.3(2.8–3.9)	8.9(8.4–9.8)	21.4(20.6–22.0)	12.5(10.4–13.3)	8.0(7.4–8.8)	0.4(0.15–0.70)	0.1(0.01–0.19)	97.8
1a	9	45.8(43.1–48.7)	2.0(1.9–2.3)	14.3(13.0–15.8)	15.2(13.4–16.7)	11.3(8.8–13.0)	10.4(9.5–11.2)	0.2(0.00–0.55)	0.1(0.02–0.23)	99.3
2b	108	45.0(44.0–46.8)	0.5(0.42–0.52)	7.6(7.4–8.0)	20.2(19.4–20.5)	17.5(17.2–18.2)	7.4(7.1–7.7)	0.1(0.08–0.31)	<0.1(0.01–0.04)	98.3
3a	41	50.0(46.4–54.7)	1.5(1.48–1.54)	17.1(16.1–18.9)	9.9(8.0–12.8)	9.8(7.8–12.9)	10.0(8.6–11.9)	0.7(0.38–0.82)	0.4(0.10–0.69)	99.4
3c	2	40.1(39.8–40.3)	1.5(1.1–1.9)	19.6(19.5–19.7)	10.8(10.4–11.2)	12.8(11.6–14.0)	12.0(11.8–12.2)	0.1(0.01–0.04)	0.1	96.8
4	5	45.4(43.8–47.2)	0.4(0.2–0.6)	25.8(24.6–27.4)	5.7(4.8–6.6)	7.4(5.6–9.1)	14.6(13.6–16.1)	0.1(0.03–0.36)	<0.1(0.03–0.06)	99.4
5a	4	42.8(41.4–44.7)	0.1	35.2(34.6–35.5)	0.4(0.1–1.2)	0.6(0.1–1.7)	19.7(18.9–20.0)	0.6(0.25–0.70)	<0.1(0.01–0.05)	99.4
6	2	68.7(67.5–69.5)	0.8(0.1–1.7)	16.5(13.7–19.3)	4.4(2.2–6.7)	1.6(0.3–2.9)	3.8(2.2–5.4)	0.8(0.3–1.2)	2.5(1.9–3.1)	99.1

* Number of glasses of each group at the different sampling locations.

Sample 15021/022—Group	1–2		2b–2	3a–8	3c–0	4–2	5a–2	6–1
Sample 15251/252—Group	1–4	1a–1	2b–7	3a–15	3c–0	4–1	5a–2	6–0
Sample 15301/302—Group	1–4	1a–3	2b–41	3a–12	3c–2	4–2	5a–0	6–0
Sample 15401/402—Group	1–1	1a–0	2b–58	3a–6	3c–0	4–0	5a–0	6–1

bulk composition of Spur Crater comprehensive soil sample 15301 (LSPET 1972). Therefore they might be chemically analogous to the mare basalts, but diluted with anorthite. Group 2b glasses, observed but rarely in samples from previous missions, are very abundant in the Apennine Front samples. This type of glass is bright green, but nearly colorless in thin section. It is remarkably uniform in composition. The particles are mostly spherical. Although high in FeO content, the mare basalt seems an unlikely source for this glass, not only because of its uniformity, but also because of its low TiO_2 content and its high MgO/FeO ratio. Group 3a, KREEP-type glasses in the Apollo 15 soil samples closely resemble the ones in the Apollo 14 samples, but are slightly lower in titanium and potassium. Group 4 glasses (6% FeO, 25% Al_2O_3) were much less abundant here. A few glasses belonging in groups 5 and 6 were also analyzed and are reported (Table 2).

Comparison of lunar glasses

In addition to the Apollo 14 and 15 soil glasses we also have grouped the Apollo 11 and 12 soil glasses previously analyzed by us. The average can be found in Tables 3 and 4. In Fig. 5, the FeO/MgO ratios of most groups of glasses are plotted together with the FeO/MgO ratios in a number of analyzed hand specimens (bulk chemistry).

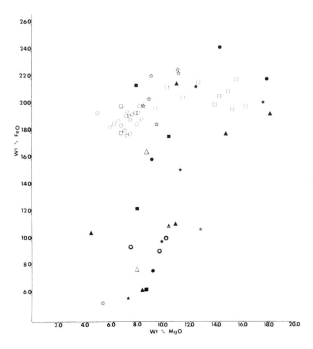

Fig. 5. Average FeO/MgO contents of groups of Apollo 11, 12, 14, and 15 glasses (Tables 1, 2, 3, and 4) with the FeO/MgO contents of igneous rocks from these missions (LSPET, *Science* 1972). Open symbols—rocks, closed symbols—glasses; circles—Apollo 11, squares —Apollo 12, triangles—Apollo 14, stars—Apollo 15; three open stars in solid circles— KREEP averages (LSPET diagram).

Table 3. Apollo 11 soil glasses. Range and average compositions in weight percent.

Group	Number glasses	SiO_2	TiO_2	Al_2O_3	FeO	MgO	CaO	Na_2O	K_2O	Total
1	7	38.7(36.4–40.6)	9.4(8.9–9.9)	5.7(4.7–6.8)	24.4(24.0–25.2)	14.2(12.0–15.6)	7.9(7.6–8.1)	0.3(0.22–0.35)	0.1(0.05–0.23)	100.7
1a	21	42.0(39.5–44.6)	7.2(5.2–8.9)	14.0(11.4–17.4)	16.0(13.4–18.3)	9.1(7.2–12.4)	12.3(10.2–14.0)	0.3(0.05–0.47)	0.2(0.11–0.30)	101.1
2b	1	46.4	0.5	8.2	21.9	17.7	8.2	0.2	<0.1	103.1
4	14	45.1(37.3–48.6)	1.5(0.1–3.9)	22.8(20.1–27.6)	7.8(5.2–9.7)	9.2(5.6–14.4)	14.0(10.9–15.3)	0.4(0.05–0.80)	0.1(0.05–0.13)	100.9
5a	6	45.7(43.2–49.7)	0.2(0.1–0.7)	33.3(30.3–36.7)	1.2(0.2–4.6)	1.2(0.1–5.1)	17.8(15.5–19.3)	0.1(0.05–1.47)	0.1(0.05–0.12)	99.6

Table 4. Apollo 12 soil glasses. Range and average compositions in weight percent.

Group	Number glasses	SiO_2	TiO_2	Al_2O_3	FeO	MgO	CaO	Na_2O	K_2O	Total
1	7	46.3(43.5–47.7)	3.8(3.0–5.4)	10.9(10.1–12.4)	21.5(20.1–24.0)	7.9(6.5–10.6)	11.2(10.0–12.0)	0.3(0.16–0.49)	0.3(0.20–0.53)	102.2
1a	11	41.0(38.2–44.0)	3.1(1.9–4.7)	13.6(9.7–16.2)	17.7(16.1–18.7)	10.4(8.8–12.1)	12.4(11.3–13.4)	0.1(0.02–0.21)	0.2(0.14–0.34)	98.5
3a	24	49.8(48.3–51.4)	2.1(1.8–3.1)	16.2(13.4–18.2)	12.4(11.0–14.2)	8.0(6.9–10.0)	11.0(10.5–11.6)	0.8(0.17–1.37)	0.7(0.23–1.46)	101.0
4	11	44.6(42.3–47.2)	0.3(0.2–0.3)	25.4(21.4–29.4)	6.3(4.7–7.2)	8.7(7.5–9.6)	15.7(14.5–17.3)	0.2(0.05–0.27)	0.1(0.05–0.23)	101.3

Group 1, the mare derived glasses have a FeO range of 18–24 wt% and a MgO range 8–14 wt%, reasonably well within the range of the analyzed basalts. The 1a group, however, definitely falls outside this range and seems to show dilution of mare basalts with other rock types. The dilution appears greater in the Apollo 11 and 15 samples, where these glasses chemically resemble bulk compositions of analyzed soil samples 10084 and 15301 (LSPET, 1972).

Two other major types of glasses, groups 3 and 4, did not have any igneous hand specimen counterparts until the Apollo 15 mission. One of the new samples, 15418, is a gabbroic anorthosite in composition. The group 4 glasses have a chemistry similar to this rock and their FeO/MgO ratios agree well (Fig. 5). Various investigators (Prinz *et al.*, 1971; Wood *et al.*, 1970), have reported lithic fragments in soil samples chemically similar to the group 4 glasses as well as the KREEP glasses. Meyer (1972) mentions many small igneous fragments related to both these groups of glasses in the Apollo 15 soils and breccias. In general, it can thus be said that selected groups of glasses reflect igneous rock types, also suggested by Reid *et al.* (1972) and other investigators. It also seems reasonable to assume that the abundance of certain groups of glasses at a particular site are a measure of the abundance of certain rock types or remnants thereof. Because KREEP glasses are the dominant group at the Fra Mauro sampling site and because the Fra Mauro formation is considered part of the Imbrian ejecta blanket, KREEP basalts appear to have been part of the lunar crust prior to formation of the Imbrium Basin.

Meyer (1972) recently proposed the existence of a plagioclase/pyroxene gradient in the pre-Imbrian lithosphere. In this model, large bodies of Al-rich rock cover more mafic equivalents to depths of 10 km or more. The glasses in groups 2, 4, and 5 would fit in well with such a model. The green glass (group 2b) abundant in some Apollo 15 samples could have been derived from the lower, more mafic portion of such a gradient by large impacts and deposited on the higher slopes of Hadley Delta, from which it is gradually moving downward (LSPET, 1972). Alternatively, the Imbrian impact might have tilted the pre-Imbrian bedrock (now the Apennines), exposing a mafic-rich portion to subsequent impact. The Apollo 14 group 2a glasses also could constitute remnants of such a gradient. The group 4 glasses, called "Highland Basalts" by Reid *et al.* (1972), and the plagioclase-rich glasses in group 5 could have been derived from the plagioclase-rich portions of Meyer's model.

BRECCIAS

Rock 14068 (Fig. 6) has matrix with typical basaltic texture, with Mg-rich olivine crystallites and mesostasis in which are embedded mineral, rock, and sialic glass fragments. The host olivine needles (Fa 20.1) have a narrow compositional range (Fa 19.1–21.6) and are distinctly different from the embedded large olivine fragments (Fa 10.9–13.0). Other foreign inclusions are broken plagioclase crystals (An 81.2–96.9) and anorthositic fragments with an average composition of An 80.2 (73.0–83.8), also metal (Co 0.5–0.7%; Ni 6–21%), probably of meteoritic origin, minor amounts of shocked high-K orthoclase and large (0.4–0.8 mm) patches of clear glass with granitic composition. The chemical composition of this rock, reported by Hubbard

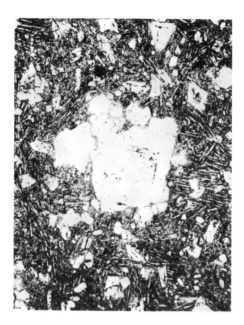

Fig. 6. Typical basaltic texture of rock 14068 with euhedral olivines and plagioclase needles
and with a large sialic glass inclusion. Length of field of view is 2 mm.

et al. (1972) indicates a source rich in magnesium and KREEP. The high magnesium
content accounts for the abundance of small olivine phenocrysts. Foreign mineral
and rock fragments, incorporated in the fine-grained basaltic host, suggest impact
origin followed by rapid crystallization of molten parts and no significant reheating.

Rock 14321 is a somewhat loosely compacted breccia. It contains a variety of
clasts (Grieve *et al.*, 1972; Duncan *et al.*, 1972) including fine- to coarse-grained
basaltic fragments, K-feldspar, quartz-feldspar intergrowth (9.9 % K_2O–4.8 % BaO),
zoned orthopyroxenes (Fs23–Fs32) and olivines (Fall–Fa24), but no glass. The
absence of glass as reported also by Jackson and Wilshire (1972) may indicate some
heating during or following impact and agglomeration.

Breccia 14315 is a dense, plagioclase-rich rock. It contains a fair number of
chondrules (5–10%) consisting of plagioclase needles in a glassy matrix (Fig. 7) and
glass spherules and fragments, most of which are devitrified. Fourteen glasses were
analyzed, all of them having a low (< 10%) iron content (Fig. 8). Several sharply
delineated, light-colored fragments consisting largely of fine-grained, partly recrystallized plagioclase were observed in thin section. We also observed one large
fragment (2 × 6 mm) that contained several clasts of quartz—K-feldspar intergrowth
and a large rounded grain of orthopyroxene. Generally, however, this breccia contains little pyroxene. Most of the many angular to rounded clasts are either olivine
(some zoned) or plagioclase.

Breccia 14318 is texturally similar to rock 14315 and also contains plagioclase
chondrules. The plagioclase needles have a uniform composition of An 87–90. In
addition to the chondrules, 36 matrix glass fragments were analyzed. A plot of the

Fig. 7. Chondrule cluster in rock 14315. Length of field of view is 1.5 mm.

FeO versus K_2O content is given in Fig. 9. Pyroxenes seem to fall in two groups, (En 40, Fs 25 and En 75, Fs 25—see also Fredriksson, *et al.*, 1972), and most of the olivine has a 30–35 % fayalite content. One small clast, however, has a Fa 11 composition and resembles the "dunite" reported by Kurat *et al.* (1972).

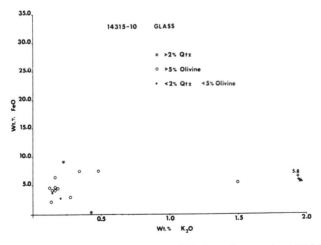

Fig. 8. FeO content versus K_2O content of 14 glasses from section 14315,10.

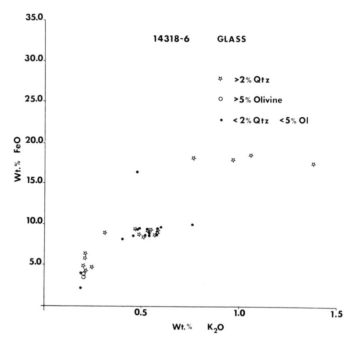

Fig. 9. FeO content versus K$_2$O content of 36 glasses from section 14318,6.

CHONDRULES

It has now become obvious that the impact products and features in lunar samples have many similarities with stony meteorites, especially chondrites (Fredriksson *et al.*, 1970; 1971; 1972, King *et al.*, 1972, Kurat *et al.* (in press), Christophe-Michel-Levi *et al.*, 1972). Chondrules, for example, which were previously thought to be uniquely meteoritic, are now known to occur on the moon and in some cases in fairly high concentrations (about 5–10% in rocks 14315 and 14318). Although chemically different, these lunar analogs have certain structural and chemical characteristics that are quite similar in many respects to those found in both equilibrated and unequilibrated chondrites. A rare, barred olivine (Fa3) chondrule in soil sample 14259 (Fig. 10) is nearly identical structurally to the common Mg-rich olivine chondrules in most stony meteorites such as Sharps (Fig. 11). Plagioclase (An 87–90) chondrules (Fig. 12) are quite similar to the rare plagioclase chondrules in the chondrite Hedjaz (Fig. 13). More complex porphyritic and rimmed lunar spherules (Fig. 14) appear similar to chondrules in, e.g., Clovis (H3) (Fig. 15) and Chainpur (LL3).

Glass spherules, with and without included mineral fragments and devitrified glass particles, are known to be common in the lunar regolith and, although rare, do exist in the carbonaceous and ordinary chondrites. An uncommon orange glass bead found in Murray (C2) is similar to those found on the moon (Fig. 16). Of particular interest are the irregular-shaped glass fragments attached to and including mineral

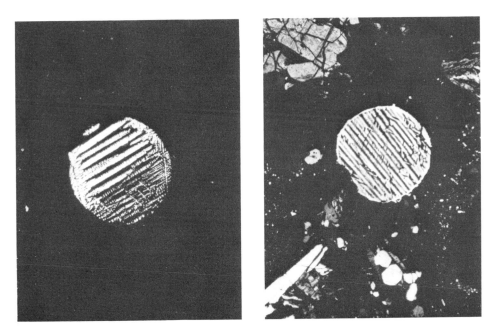

Fig. 10–11. The rare, barred olivine (Fa3) chondrule found in soil sample 14259 (Fig. 10, left) is similar to those commonly found in meteoritic chondrites (Sharps, Fig. 11, right). Lengths of fields of view are 0.2 mm and 1 mm, respectively.

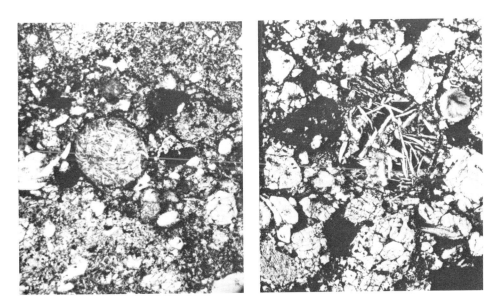

Fig. 12–13. Plagioclase (An_{87-90}) chondrule in breccia 14318 (Fig. 12, left) resembles those in the Hedjaz meteorite (Fig. 13, right—Courtesy of F. Kraut, Musée national d'histoire naturelle, Paris, France). Lengths of fields of view are 0.8 mm and 1.5 mm, respectively.

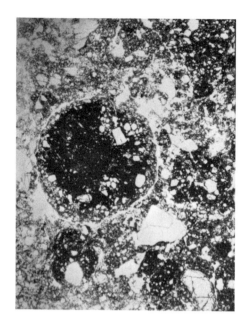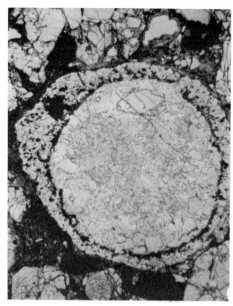

Fig. 14–15. A dust-covered, chondrule-like basaltic fragment (sample 14168) with peripheral mineral inclusions (Fig. 14, left) is similar to some of the more complex trimmed chondrules as in Clovis (Fig. 15, right). Length of fields of view are 1.2 mm and 1 mm, respectively.

and rock fragments, as in Murchison (C2) and Mezö-Madaras (L3) (Fig. 17). These are very similar in appearance to the relatively abundant lunar glass-coated lithic fragments.

Summary

Glasses from the Apollo 11, 12, 14, and 15 sampling sites, analyzed by us, have been grouped on the basis of their chemistry. Most groups appear to occur at all sites, but there are marked differences in abundances. At the Apollo 11 and 12 sites a large proportion of the glasses are made up of mare basalt components. This type of glass is much rarer at the Apollo 14 Fra Mauro sites. Glasses rich in KREEP, first observed in quantity in the Apollo 12 samples and extensively described (Hubbard *et al.*, 1971; Meyer *et al.*, 1971) represent the largest group here. Other important groups in the 14 samples are groups 2, 3c, and 4. Because the Fra Mauro formation is assumed to represent ejecta deposits from the Mare Imbrium Basin, it is speculated that these glasses represent pre-Imbrian crustal material. Apollo 15 glasses reflect the variety of regions found here. There is a good representation of mare basalt type glasses, especially in the samples from the LM site (sample 15021, 15022) and station 6 (sample 15251, 15252). The green glass from group 2 and assumed by us to have been derived from pre-Imbrian crustal material is abundant in the samples from the higher slopes of Hadley Delta. KREEP glass accounts for about 15% of all Apollo 15 glasses analyzed.

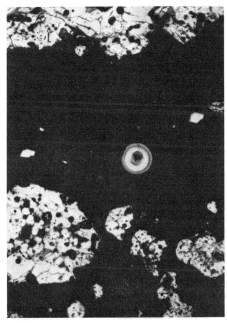

Fig. 16. An uncommon glass spherule in the matrix of the Murray meteorite, which resembles the common impact-produced lunar glass spheres. Length of field of view is 0.4 mm.

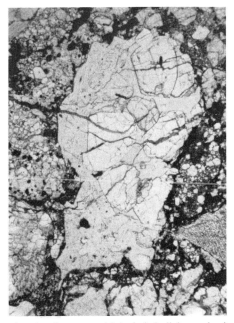

Fig. 17. A large, angular glass fragment with included olivine grains in the Mezö-Madaras chondrite. Similar type glass fragments are quite common in lunar samples. Length of field of view is 1.5 mm.

The four Apollo 14 rocks studied by us appear to have been formed as a result of impact processes. Rock 14068 has a fine-grained quench texture and includes mineral, rock and sialic glass fragments. Rock 14321 appears to be a somewhat recrystallized lithic breccia lacking glass. Breccias 14315 and 14318, which contain a number of well-developed chondrules, are texturally unique among the Apollo 14 rocks and similar in many respects to some chondrites. The occurrence of chondrules and glass spherules and fragments in lunar rocks and soil as well as in chondritic meteoritic material suggests that although the parent-body chemistry, cooling, and recrystallization histories were different, the impact process was an important mechanism operating on the surface of the moon and probably on other bodies early in the development of the solar system. The similarity between lunar chondrules and certain types of meteoritic chondrules supports theories by Fredriksson (1963) and others that meteoritic chondrules can be formed by impact.

Acknowledgments—We wish to thank W. G. Melson for a thorough review and helpful suggestions. G. Moreland and H. Moore prepared many excellent polished thin sections. Phyllis Brenner assisted with much of the technical work. We are also grateful to Ruth Rogers for typing the manuscript under considerable pressure. Financial support was provided under NASA Grant NGR 09–015–148.

References

Christophe-Michel-Levy M. Levy C. and Pierrot R. (1972) Mineralogical aspects of Apollo 14 samples: Lunar chondrules, pink spinel bearing rocks, ilmenites. In *Lunar Science—III* (editor C. Watkins), p. 136, Lunar Science Institute Contr. No. 88.

Duncan A. R. Lindstrom M. M. Lindstrom D. J. McKay S. M. Stoeser, J. W. Goles G. G. and Fruchter J. S. (1972) Comments on the genesis of breccia 14321. In *Lunar Science—III* (editor C. Watkins), p. 192, Lunar Science Institute Contr. No. 88.

Fredriksson K. (1963) Chondrules and the meteorite parent bodies. *Trans. N.Y. Acad. Sci. II* **25**, 756–769.

Fredriksson K. Nelen J. and Melson W. G. (1970) Petrography and origin of lunar breccias and glasses. *Proc. Apollo 11 Lunar Sci. Conf., Geochim. Cosmochim. Acta*, Suppl. 1, **1**, 419–432. Pergamon.

Fredriksson K. Nelen J. Noonan A. Anderson C. A. and Hinthorne J. R. (1971) Glasses and sialic components in Mare Procellarum soil. *Proc. Second Lunar Sci. Conf., Geochim. Cosmochim. Acta*, Suppl. 2, **1**, 727–735. M.I.T. Press.

Fredriksson K. Nelen J. Noonan A. and Kraut F. Apollo 14 (1972) Glasses, breccias, chondrules. In *Lunar Science—III* (editor C. Watkins), p. 280, Lunar Science Institute Contr. No. 88.

Grieve R. McKaye G. Smith H. Weill D. and McCallum S. (1972) Mineralogy and petrology of polymict breccia 14321. In *Lunar Science—III* (editor C. Watkins), p. 338, Lunar Science Institute Contr. No. 88.

Hubbard N. J. Meyer C. Jr. Gast P. W. and Wiesmann H. (1971) The composition and derivation of Apollo 12 soils. *Earth Planet Sci. Lett.* 10, 241–350.

Hubbard N. J. Gast P. W. Rhodes M. and Wiesmann H. (1972) Chemical composition of Apollo 14 materials and evidence for alkali volatilization. In *Lunar Science—III* (editor C. Watkins), p. 407, Lunar Science Institute Contr. No. 88.

Jackson E. D. and Wilshire H. G. (1972) Classification of the samples returned from the Apollo 14 landing site. In *Lunar Science—III* (editor C. Watkins), p. 418, Lunar Science Institute Contr. No. 88.

King E. A. Butler J. C. and Carman M. F. (1972) Chondrules in Apollo 14 breccias and estimation of lunar surface exposure ages from grain size analysis. In *Lunar Science—III* (editor C. Watkins), p. 449, Lunar Science Institute Contr. No. 88.

Kurat G. Keil K. Prinz M. and Bunch T. E. (1972) A "chondrite" of lunar origin: textures, lithic fragments, glasses, and chondrules. In *Lunar Science—III* (editor C. Watkins), p. 463, Lunar Science Institute Contr. No. 88.

Kurat G. Keil K. Prinz M. and Nehru C. E. Chondrules of lunar origin. *Proc. Third Lunar Sci. Conf.* (in press).

LSPET (1972) (Lunar Sample Preliminary Examination Team) Preliminary examination of lunar samples from Apollo 15. *Science* **175**, 363–375.

Meyer C. Jr. Brett R. Hubbard N. J. Morrison D. A. McKaye D. S. Aitken F. K. Takeda H. and Schonfeld E. (1971) Mineralogy, chemistry, and origin of the Kreep component in soil samples from the Ocean of Storms. *Proc. Second Lunar Sci. Conf.*, *Geochim. Cosmochim. Acta*, Suppl. 2, Vol. 1, pp. 393–411. M.I.T. Press.

Meyer C. Jr. (1972) Mineral assemblages and the origin of non-mare lunar rock types. In *Lunar Science—III* (editor C. Watkins), p. 542, Lunar Science Institute Contr. No. 88.

Prinz M. Bunch T. E. Keil K. (1971) Composition and origin of lithic fragments and glasses in Apollo 11 samples. *Mineral. and Petrol.* **32**, 211–230.

Reid A. M. Ridley W. I. Warner J. Harmon R. S. Brett R. Jakeš P. and Brown R. W. (1972) Chemistry of highland and mare basalts as inferred from glasses in the lunar soils. In *Lunar Science—III* (editor C. Watkins), p. 640, Lunar Science Institute Contr. No. 88.

Wood J. A. Dickey J. S. Marvin U. B. Powell B. N. (1970) Lunar anorthosites and a geophysical model model of the moon. *Proc. Apollo 11 Lunar Sci. Conf.*, *Geochim. Cosmochim. Acta*, Suppl. 1, **1**, 965–968. Pergamon.

Proceedings of the Third Lunar Science Conference
(Supplement 3, *Geochimica et Cosmochimica Acta*)
Vol. 1, pp. 739–752
The M.I.T. Press, 1972

Vapor phase crystallization in Apollo 14 breccia

DAVID S. McKAY, UEL S. CLANTON, AND DONALD A. MORRISON

NASA Manned Spacecraft Center
Houston, Texas 77058

GARTH H. LADLE

Lockheed Electronics Corp.
Houston, Texas 77058

Abstract—Many of the highly recrystallized breccias from Apollo 14 contain vugs with well-developed crystals of plagioclase, pyroxene, ilmenite, apatite, whitlockite, iron, nickel-iron, and troilite that extend from the vug walls and bridge open spaces. These crystals are interpreted as having formed by deposition from a hot vapor containing oxides, halides, sulfides, alkali metals, iron, and possibly other chemical species. The hot vapor was associated with the thermal metamorphism and subsequent cooling of the Fra Mauro formation after it had been deposited as an ejecta blanket by the Imbrian impact. The cooling of hot lunar ejecta blankets resemble the cooling of terrestrial ash flows that commonly have had vapor-phase crystallization in cavities. Redistribution of volatile species in a vapor phase appears to be a significant feature of large lunar ejecta blankets, and may have been more important in early lunar history when late stage accretion and mare basin excavation provided very large ejecta blankets.

INTRODUCTION

VUGS AND VESICLES are well known in basalts from each of the Apollo missions. Their mode of occurrence and mineral assemblages are summarized by Schmitt *et al.* (1970) for Apollo 11 basalts and by Warner (1971) for Apollo 12 basalts. Plagioclase, clinopyroxene, and ilmenite are the primary minerals lining the cavities in these basalts.

The Apollo 11 breccias are porous but no vugs have been reported. The porosity is a result of open spaces between fragments and by vesicles in glass clasts. Apollo 12 breccias are similar to the Apollo 11 breccias except for 12057 and 12034. These two breccias show matrix vesicularity that is interpreted as resulting from partial melting (Sclar, 1971; McKay *et al.*, 1971). Breccia 12013 also contains cavities in the dark lithology (Drake *et al.*, 1970).

Vugs and vesicles in some of the Apollo 14 breccias were reported in the Apollo 14 Lunar Sample Information Catalog (Warner and Duke, 1971; Samples 14006, 14068, 14264, and 14311). We have examined thin sections of Apollo 14 vuggy breccias and have investigated the minerals in the vugs with a scanning electron microscope (SEM) with attached energy dispersive x-ray system (EDX).

VUG ABUNDANCE AND TYPE

Vugs are present only in the more highly metamorphosed or recrystallized Apollo 14 breccias; the breccias are the high-grade facies of Warner (1971, 1972). Not all of the recrystallized breccias contain vugs nor are they present in the low or middle facies rocks except in rare clasts.

Vugs are most common in recrystallized breccia fragments in the coarse fines (1–4 mm) from the soil samples. The abundance of these recrystallized breccia fragments is discussed by McKay *et al.* (1972). 60 % of the 265 well-recrystallized breccia fragments examined have vugs that are visible with a binocular microscope. The Cone Crater sample has the highest abundance of recrystallized breccias (94 %), of which 78 % are vuggy.

The vugs can be classified, based on morphology, into two types. Type one is irregularly shaped; some vugs of this type occur as long narrow cavities. Figure 1 is a SEM photograph of a breccia fragment with small irregular vugs and a larger channel-like vug, shown at higher magnification in Fig. 2. Although some of these irregular vugs are self-contained, most appear to be interconnected to form a network, resulting in a permeable fragment. Irregular vugs commonly contain a profusion of euhedral crystals extending into the cavity. The sizes of irregular vugs range from less than 50 μ (Fig. 2) to greater than 1 mm (Fig. 3). Breccia 14311 contains vugs as large as 2 mm (Warner and Duke, 1971).

The second vug type typically has a spheroidal or ellipsoidal shape and could be properly termed a vesicle. Spheroidal cavities do not appear to be interconnected even though some porosity exists among the crystals lining the wall (Fig. 4). Plagioclase and pyroxene commonly form a smooth mat more or less tangential to the vug wall, but crystals may also extend into the cavity. All of the vuggy samples contain irregularly shaped vugs. In addition, approximately 25 % of these samples also contain spheroidal vugs.

VUG MINERALS

Minerals occur in the vugs as wall linings and as free-standing crystals that project into the cavity and bridge open spaces (Figs. 2 and 5). Crystals were identified by comparing an unknown EDX spectrum to standards. In a few cases, networks of plagioclase and pyroxene crystals nearly fill vugs (Figs. 6 and 7). A spectacular array of tabular crystals, needles, and equant crystals growing in profusion from all vug surfaces is shown in Fig. 8. Crystals may form on the faces of larger crystals; tabular and equant plagioclase crystals have grown on the growth steps of a larger pyroxene crystal (Fig. 7). In general, plagioclase crystals in irregular vugs are equant (Figs. 7, 8, and 9) with larger crystals having a more tabular habit (Fig. 10). We have not seen acicular needles of plagioclase in breccia vugs of the type reported by Skinner and Winchell (1972) in basalt vugs.

Most pyroxenes are low calcium varieties and are assumed to be orthopyroxene or pigeonite, but some calcium-rich pyroxenes also are present. Pyroxene crystals are common in both vug types with euhedral, free-standing crystals most common in irregular vugs (Figs. 8 and 11). The crystals are characterized by linear bunching of growth steps (Figs. 11 and 12). Pyroxene crystals smaller than about 2 μ in diameter occur as cylinders with hemispherical terminations (Fig. 11). Crystals larger than 2 μ have angular faces and show well-developed crystal form.

The most perfectly formed vug crystals are apatite. Apatite crystals preferentially are concentrated in the vugs relative to the matrix in many of the recrystallized

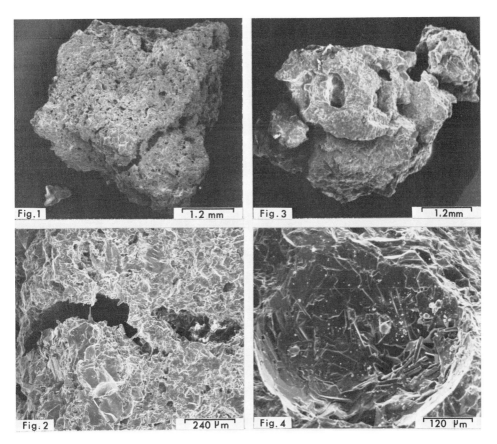

Fig. 1. SEM photograph of a chip of vuggy recrystallized breccia. Many small irregular vugs and several larger irregular vugs are apparent. (14261)

Fig. 2. Higher magnification SEM photograph of the large irregular vug in Fig. 1. Small crystals extend from the vug walls. Some vugs are less than 50 μ in size. (14261)

Fig. 3. SEM photograph of vuggy recrystallized breccia. Vugs in this breccia are considerably larger than most of the vugs in Fig. 1. Many of the vugs are interconnected. (14268)

Fig. 4. Spheroidal vug or vesicle. Spheroidal vugs are usually self-contained and not interconnected, but as shown in this SEM photograph, may have porosity beyond the vesicle wall. The small spheroidal objects are crystals of nickel-iron. (14001)

breccias. A high proportion of the larger vugs contain euhedral apatite crystals. Figure 13 shows a well-developed, doubly terminated, hexagonal apatite crystal containing about 2% chlorine. First $\{10\bar{1}0\}$ and second $\{11\bar{2}0\}$ order prisms; first, second, and third order hexagonal dipyramids $\{hh\bar{2}hl\}$; and basal pinacoids $\{0001\}$ are regularly observed on apatite crystals (Figs. 13, 14, and 15). The crystals appear to be late forming and are found only on a substrate of plagioclase and pyroxene.

Whitlockite also is found in vugs but occurrences are rare. It occurs as stubby, discoidal crystals with hexagonal outlines (Fig. 16). A tabular rhombohedral habit is

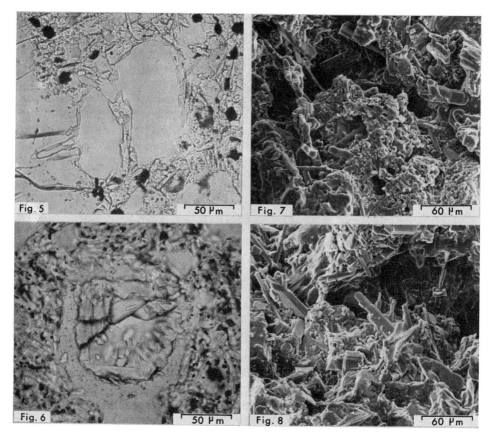

Fig. 5. Thin section photograph of a vug in the fragment shown in Fig. 1. Crystals of pyroxene and plagioclase bridge across the vug. (14261)

Fig. 6. Thin section photograph showing pyroxene and plagioclase crystals nearly filling a vug in the sample shown in Fig. 1. Note that some of the pyroxenes in the vug are considerably coarser than the matrix pyroxenes. (14261)

Fig. 7. SEM photograph of a bridge of plagioclase crystals partially filling a vug. Large crystals are apatite and pyroxene. (14261)

Fig. 8. SEM photograph of a profuse growth of vug crystals. Elongate crystals are mostly low calcium pyroxene; stubby or equant crystals are mostly plagioclase. (14261)

suggested, but crystals found to date have somewhat rounded faces that make indexing difficult.

Ilmenite typically occurs as small tabular or pseudo-octahedral crystals but is not particularly common (Figs. 11 and 17).

Many of the larger vugs contain metallic crystals of iron or nickel-iron. The iron crystal (Fig. 18) is a tetrahexahedron, which has an axis of four-fold symmetry projecting toward the upper right of the photograph. EDX analysis indicates that nickel and phosphorus, if present, are less than 1% in this crystal. Figure 19 shows

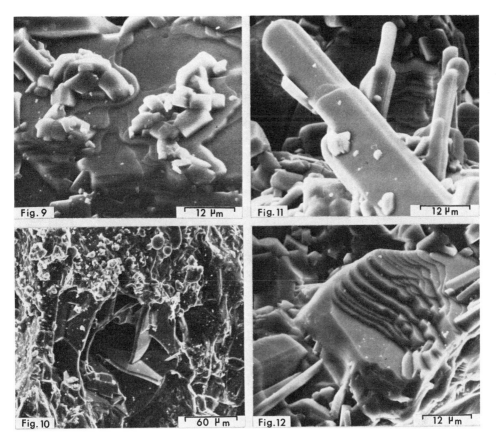

Fig. 9. SEM photograph of a host of small, mostly tabular plagioclase crystals that have grown on the growth steps of a pyroxene crystal. (14261)

Fig. 10. SEM photograph of tabular plagioclase crystals in a small vug in a recrystallized breccia. (14001)

Fig. 11. High magnification SEM photograph of low calcium, prismatic pyroxene crystals in Fig. 8. Pyroxene needles smaller than about 2μ in diameter are rounded and have hemispherical terminations; larger crystals have regular faces. A tabular crystal of ilmenite clings near the free end of the largest pyroxene crystal; an iron crystal is partially visible in the background behind the ilmenite crystal. (14261)

Fig. 12. SEM photograph of growth steps on a calcium pyroxene crystal; this linear bunching of growth lines is characteristic of many of the vug pyroxene crystals. (14161)

another crystal of relatively pure iron. The three large faces form part of an octahedron; the smaller rectangular form is one face of a dodecahedron. Figure 20 shows a nickel-iron crystal that contains about 12% nickel. This tetrahexahedron has been photographed along an axis of three-fold symmetry and this produces the pseudo-hexagonal shape. The crystal is partially covered with a coating of iron sulfide, presumably troilite. The rough texture of the nickel-iron crystal may have been

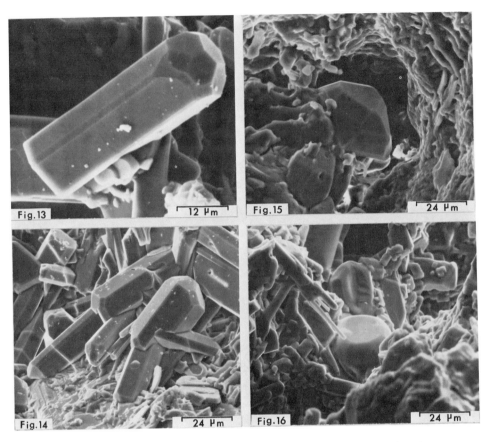

Fig. 13. SEM photograph of a doubly terminated hexagonal apatite crystal in a breccia vug. This apatite contains about 2 % chlorine based on EDX analysis. (14261)

Fig. 14. SEM photograph of a nest of apatite crystals. First- and second-order prisms; first-, second-, and third-order hexagonal dipyramids; and basal pinacoid faces are present on the apatite crystals in Figs. 13, 14, and 15. (14161)

Fig. 15. SEM photograph of a stubby apatite partially filling a small vug; this crystal is twinned. Apatite crystals are found only on substrates of plagioclase and pyroxene crystals. (14143)

Fig. 16. SEM photograph of a stubby discoidal crystal of whitlockite (center). A tabular rhombohedral habit is suggested but rounded faces make indexing uncertain. (14261)

caused by a former coating of troilite. Figure 21 shows another nickel-iron crystal which contains about 4% nickel. This near spheroidal shape is typical of most of the nickel-iron particles that were photographed. Only at magnifications above 1000X can crystal face development be observed. Faces that are observed at these magnifications are incompletely developed, repeated, and sometimes protrude from the spheroidal mass such that the crystal habit is not obvious.

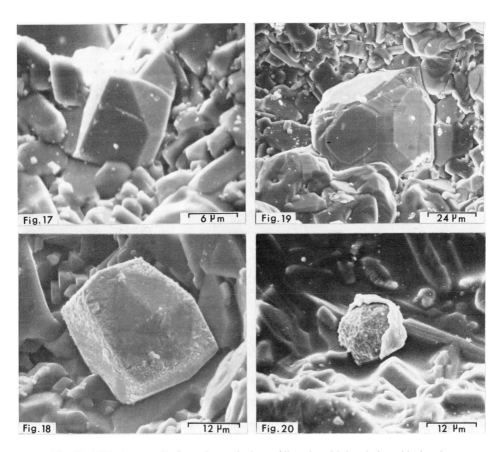

Fig. 17. SEM photograph of pseudo octahedron of ilmenite with basal pinacoid, rhombo-
hedron and prism faces. Ilmenite crystals are relatively uncommon in vugs; most tend to be
small and tabular as shown in Fig. 11. (14262)

Fig. 18. SEM photograph of a euhedral iron crystal. This tetrahexahedron has an axis of
four-fold symmetry projecting toward the upper right of the photograph. The crystal contains
no detectable nickel (less than 1°/₀). (14161)

Fig. 19. SEM photograph of an iron crystal with well-developed faces and bunched growth
lines. The three large faces form part of an octahedron; the smaller rectangular form is one
face of a dodecahedron. (14262)

Fig. 20. SEM photograph of a nickel-iron crystal with a partial coating of iron sulfide,
presumably troilite. The tetrahexahedron has been photographed along an axis of three-fold
symmetry resulting in the pseudo-hexagonal outline. The crystal contains about 12°/₀ nickel.
(14258)

Euhedral troilite crystals also are found in some of the vugs but are less abundant
than metallic crystals. The troilite crystal in Fig. 22 has first- and second-order
hexagonal prisms and first- and second-order hexagonal dipyramids. The troilite
crystal in Fig. 23 has pyramidal faces outlined by linear bunching of growth lines.

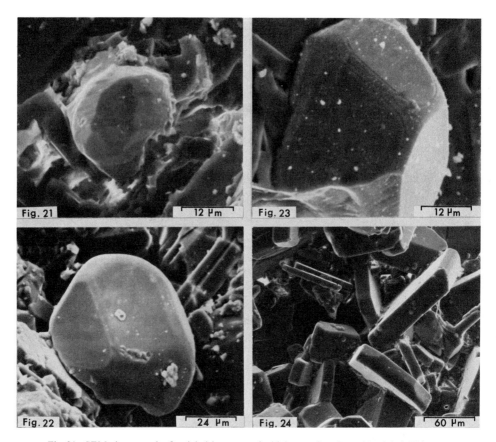

Fig. 21. SEM photograph of a nickel-iron crystal which contains about 4% nickel. This near spheroidal shape is typical of most of the nickel-iron particles that were photographed. The crystal habit commonly is not obvious. (14001)

Fig. 22. SEM photograph of a troilite crystal that shows development of first- and second-order hexagonal prisms; and first- and second-order hexagonal dipyramids. (14161)

Fig. 23. SEM photograph of a troilite crystal. Pyramidal faces are outlined by linear bunching of growth lines. (14001)

Fig. 24. SEM photograph of K-feldspar crystals in a cavity in the Bandolier ash-flow tuff of New Mexico. Vugs in lunar breccias are analogous to vugs in the vapor-phase zone of terrestrial ash-flow tuffs.

Both iron and troilite crystals commonly rest on a substrate of pyroxene and plagioclase. This relationship suggests that the former crystals may also have been among the last to form in the vugs.

DISCUSSION

Vugs in lunar basalts

All previously described euhedral crystals have been found in vugs within basalts. Evans (1970) described a euhedral troilite crystal in a vug of an Apollo 11 basalt and

Gay *et al.* (1970) reported an amphibole in a vug, which is discussed in detail by Charles *et al.* (1971) who suggest that chlorine may be an important constituent of the amphibole. Carter and MacGregor (1970) included a SEM photograph of pyroxene and plagioclase crystals lining a vug in an Apollo 11 basalt. Jedwab (1971) presents detailed observations on ilmenite crystals in vugs and concludes that they grew from a vapor phase. Skinner and Winchell (1972) describe vapor-phase growth of plagioclase needles in vugs and vesicles in Apollo 12 basalts.

Euhedral crystals

In addition to the euhedral crystals in vugs in basalts, other euhedral crystals have been reported. Euhedral iron has been found in troilite (Skinner, 1970; Brown *et al.*, 1970), in olivine (Cameron, 1971), and in glass (Engelhardt *et al.*, 1971). Euhedral apatite enclosed in pyroxferroite is illustrated by Fuchs (1970). The occurrence of euhedral crystals in lunar basalts and glasses is not uncommon; however, the occurrence of euhedral crystals in breccia vugs has not been reported previously although Agrell *et al.* (1972) also have observed them in the Apollo 14 breccias.

Vapor growth

Crystals in vugs generally are assumed to have been vapor-derived (Jedwab, 1971), and we so interpret the crystals described in this report. However, if the rocks were once molten it is possible that the crystals grew from a melt rather than from the vapor. Skinner and Winchell (1972) have considered this problem in discussing the growth of feldspar crystals in vesicular basalts. They conclude that if the vesicles are closed it was not possible for the melt to drain away and therefore projecting plagioclase crystals are vapor deposited. A similar argument can be applied to the Apollo 14 recrystallized breccias that contain crystals in closed cavities.

Some recrystallized breccias show evidence of a glassy or a crystallized granitic phase that commonly occurs as isolated patches or clasts. This partially remobilized granitic material texturally resembles the granitic material in rock 12013. A few of the Apollo 14 breccia vugs that are bounded by this granitic material have a "wetted" texture (Fig. 20) where an early generation of crystals has been glazed by a coating of glass. The possibility must be considered that some of the crystals grew from this partially molten phase. However, most of the vugs that we examined show no textural or chemical evidence of this granitic phase.

Additional features consistent with vapor growth are finely detailed crystal faces, abundant growth steps, and delicate networks of projecting crystals separated by open spaces. The layer-by-layer growth shown by many of the crystals is characteristic of slow deposition rates at low supersaturations (Jedwab, 1971).

Vapor composition

The composition of the vapor from which the crystals were deposited can only be inferred. Naughton *et al.* (1972) calculate, using standard thermodynamic relationships, the composition of a vapor in equilibrium associated with a magma of lunar composition at 1500°K. About 70% of this vapor consists of condensable components, including sulfur, titanium compounds, halides, sulfides, and elemental metals, including iron. Iron is generally considered to be a refractory element that, under appropriate equilibrium conditions, condenses at high temperatures (Larimer, 1967). However, in material of lunar composition it is relatively volatile compared to many other major components. De Maria *et al.* (1971) found, in volatilization studies of an Apollo 12 basalt, that iron vapor is detectable at 1300°K and is exceeded in volatility among the major elements only by sodium and potassium.

Carter (1971) shows evidence for the condensation of iron vapor on the vesicle walls of a lunar glass particle. Walter *et al.* (1971) propose iron vapor deposition as a mechanism for the filling of fractures by iron in a lunar pyroxene.

Sulfur and sulfide vapor are likely components of the vapor phase. The deposition of iron sulfide on iron surfaces, as observed in the vugs, is a likely result of vapor condensation because iron provides a preferred nucleation site for FeS (Larimer, 1967). This condensation of FeS occurs at 680°K in a cosmic gas (Larimer, 1967) but would in general occur at a different but still relatively low temperature in the breccia vugs. Goldstein *et al.* (1972) report sulfide husks around iron cores in particles separated from Apollo 14 soils, but they attribute this texture to the formation of two immiscible liquids.

Phosphates generally are considered to be volatile and apatite is commonly found in vugs in terrestrial rocks (Deer *et al.*, 1962). Well-developed apatite crystals projecting into vugs have been observed in an oceanic island basalt and are interpreted as vapor-deposited crystals by Bass (1972). The calculations of Naughton *et al.* (1972) indicate that P_4O_6 may be the major phosphorus species in a lunar vapor phase at 1500°K.

Chlorine found in the apatites suggests that chlorides may have been components of the vapor phase. An EDX spectrum of fine-grained material on the surface of a round iron mass in one of the breccia vugs contains a small chlorine peak. This raises the possibility that metal chlorides may be present in the vugs.

The major cations (Si, Al, Mg, Ca, Ti) of plagioclase, pyroxene, and ilmenite may be transported in the vapor as diverse chemical species such as oxides, oxyhalides, halides, and sulfides. The vapor phase also should contain alkali metals (Hubbard *et al.*, 1972). Thermodynamic calculations of the type performed by Naughton *et al.* (1972) for lower temperatures (700–1000°C) using Apollo 14 breccia compositions, would provide a better estimate of the composition of the vapor phase.

The more highly recrystallized breccias also are depleted in rare gases (Bogard and Nyquist, 1972) and these gases may have been a minor noncondensable component of the vapor. Water also is a possible minor component (Agrell *et al.*, 1972).

Vapor phase crystallization in the Fra Mauro formation

Considerable evidence exists that many of the Apollo 14 breccias underwent thermal metamorphism (Warner, 1972; Wilshire and Jackson, 1972). Anderson *et al.* (1972) estimate temperatures near 1000°C for the thermal event. Grieve *et al.* (1972), on the basis of Mg-Fe partitioning between coexisting pyroxenes, estimate that the most recrystallized breccia fragments in 14321 were metamorphosed at 1050°C and the other breccia types were metamorphosed at 700°C.

Most investigators conclude that at least some of this thermal metamorphism took place during and immediately after the deposition of the Fra Mauro ejecta blanket from the Imbrian impact. Nyquist *et al.* (1972) report that all four highly recrystallized breccias that they analyzed have identical ages of 3.9 ± 0.1 b.y.; they propose that this is the age of the Imbrian event and the time of major thermal and volatilization activity.

McKay *et al.* (1972) conclude that the highly recrystallized breccia is representative of material at depth within the Fra Mauro unit. This conclusion is based on the abundance of recrystallized breccias in the Cone Crater soil and coarse fines. These breccia fragments are from depths as great as 60–80 m beneath the surface at Cone Crater (Sutton *et al.*, 1972) and have been only slightly contaminated by other material (Burnett *et al.*, 1972). We suggest that many or most of the vug crystals discussed in this report were formed during the thermal metamorphism and cooling of the hot Fra Mauro ejecta blanket.

An analogy can be made to terrestrial ash flows, that commonly are partially thermally metamorphosed, and that devitrify during cooling. Ash flows may develop extensive zones of vapor phase crystallization (Smith, 1960) as vapors released during crystallization of hotter zones percolate up through cooler zones. Figure 24 is a SEM photograph of K-feldspar crystals formed by vapor deposition in the vapor-phase crystallization zone of the Bandolier ash-flow tuff of New Mexico. Terrestrial ash flows also typically develop fumaroles and sublimate minerals (Smith, 1960; Sheridan, 1970), and may develop alkali zonation from top to bottom resulting from hydrothermal vapor phase transport and exchange of alkalis (Scott, 1966).

Lacking H_2O as a major vapor component, cooling lunar ejecta blankets would not be expected to duplicate in detail features seen in terrestrial ash flows. However, as discussed above, a vapor phase would be expected and the cooling history would be analogous. Post-depositional temperatures in the thick (100 m or more) Fra Mauro ejecta blanket would be expected to range from low near the surface to highest (temperatures around 1000°C) in the deeper interior and to intermediate at the base. Vapor would be driven out of the hottest zones and would move upward through pores, cracks, and joints into and through cooler zones, eventually reaching the surface. Most of the noncondensable gases, including H_2, CO, the rare gases, and H_2O would be similarly driven out of the hotter zones and would eventually reach the surface and escape from the ejecta blanket. Condensable species in the vapor, including alkali metals, iron, volatile heavy elements, halides, sulfides, oxides, and oxyhalides, would be partially depleted in the hottest zones but would start condensing as the vapor passed through cooler zones and vapor phase crystallization would result. Some of this material may escape to the surface, but much of it would condense and be trapped in the cooler upper zones of the ejecta blanket. The escaping gases may have formed fumaroles in the ejecta blanket.

The more recrystallized breccias have lost most of their porosity and permeability during the recrystallization process, but they still retain some porosity. Rock 14311, for example, contains about 5% vesicle and vug porosity (Warner and Duke, 1971; Mizutani *et al.*, 1972). The observations that many of the vugs are interconnected demonstrates that the rocks are also permeable. When a 35 mg chip of a highly recrystallized breccia containing small irregular vugs was immersed in liquid Freon, tiny bubbles were emitted from visible vugs for about 15 minutes. A weight-loss measurement on the Freon-saturated chip, that was made as the Freon evaporated, indicated that the chip had a minimum of 4 ± 1 volume percent of interconnected pores open to the exterior vugs. This permeability would allow hot vapors to pass readily through the breccia.

SUMMARY AND CONCLUSIONS

1. Vugs are an important feature of the recrystallized Apollo 14 breccias. Many vugs contain crystals that are interpreted as vapor deposits. Euhedral crystals of plagioclase, clinopyroxene, orthopyroxene, apatite, whitlockite, iron, nickel-iron, troilite, and ilmenite have been observed.

2. The vug crystals were formed during the thermal metamorphism of a hot ejecta blanket. Most evidence indicates that this thermal event occurred when the Imbrian Basin was excavated by an impact and the Fra Mauro formation was deposited.

3. The cooling of a hot lunar ejecta blanket is somewhat analogous to the cooling of a hot terrestrial ash flow. Cooling and crystallization of a terrestrial ash flow includes vapor-phase crystallization in vugs and the transport and loss of volatiles from fumaroles. Large lunar ejecta blankets may also have lost noncondensable volatiles to the surface by vapor percolation through pores and cracks. Condensable volatiles may have been transported to upper zones of the blanket deposit and some may have escaped to the surface.

4. Vapor transport during the cooling of the thick, hot ejecta blankets associated with late-stage lunar accretion and mare excavation may have been a major process in the redistribution of many volatile components.

REFERENCES

Agrell S. O. Scoon J. H. Long J. V. P. and Coles J. N. (1972) The occurrence of goethite in a microbreccia from the Fra Mauro formation (abstract). In *Lunar Science—III* (editor C. Watkins), pp. 7–9, Lunar Science Institute Contr. No. 88.

Anderson A. T. Jr. Braziunas T. F. Jacoby J. and Smith J. V. (1972) Breccia populations and thermal history: Nature of pre-Imbrian crust and impacting body (abstract). In *Lunar Science—III* (editor C. Watkins), pp. 24–26, Lunar Science Institute Contr. No. 88.

Bass M. (1972) Personal communication.

Bogard D. D. and Nyquist L. E. (1972) Noble gas studies on regolith materials from Apollo 14 and 15 (abstract). In *Lunar Science—III* (editor C. Watkins), pp. 89–91, Lunar Science Institute Contr. No. 88.

Brown G. M. Emeleus C. H. Holland J. G. and Phillips R. (1970) Mineralogical, chemical and petrological features of Apollo 11 rocks and their relationship to igneous processes. *Proc. Apollo 11 Lunar Sci. Conf., Geochim. Cosmochim. Acta*, Suppl. 1, Vol. 1, pp. 195–219. Pergamon.

Burnett D. S. Haneke J. C. Podosek F. A. Russ G. P. III Turner G. and Wasserburg G. J. (1972) The irradiation history of lunar samples (abstract). In *Lunar Science—III* (editor C. Watkins), pp. 105–107, Lunar Science Institute Contr. No. 88.

Cameron E. N. (1971) Opaque minerals in certain lunar rocks from Apollo 12. *Proc. Second Lunar Sci. Conf., Geochim. Cosmochim. Acta*, Suppl. 2, Vol. 1, pp. 193–206. M.I.T. Press.

Carter J. L. (1971) Chemistry and surface morphology of fragments from Apollo 12 soil. *Proc. Second Lunar Sci. Conf., Geochim. Cosmochim. Acta*, Suppl. 2, Vol. 1, pp. 873–892. M.I.T. Press.

Carter J. L. and MacGregor I. D. (1970) Mineralogy, petrology, and surface features of some Apollo 11 samples. *Proc. Apollo 11 Lunar Sci. Conf., Geochim. Cosmochim. Acta*, Suppl. 1, Vol. 1, pp. 247–265. Pergamon.

Charles R. W. Hewitt D. A. and Wones D. R. (1971) H_2O in lunar processes: The stability of hydrous phases in lunar samples 10058 and 12013. *Proc. Second Lunar Sci. Conf., Geochim. Cosmochim. Acta*, Suppl. 2, Vol. 1, pp. 645–664. M.I.T. Press.

Deer W. A. Howie R. A. and Zussman J. (1962) *Rock Forming Minerals*, Vol. 5, Non-silicates. Longmans.

DeMaria G. Balducci G. Guido M. and Piacente V. (1971) Mass spectrometric investigation of the vaporization process of Apollo 12 lunar samples. *Proc. Second Lunar Sci. Conf., Geochim. Cosmochim. Acta*, Suppl. 2, Vol. 2, pp. 1367–1380. M.I.T. Press.

Drake M. J. McCallum I. S. McKay G. A. and Weill D. F. (1970) Mineralogy and petrology of Apollo 12 sample no. 12013: A progress report. *Earth Planet. Sci. Lett.* **9**, 103–123.

Engelhardt W. von Arndt J. Muller W. F. and Stoffler D. (1971) Shock metamorphism and origin of regolith and breccias at the Apollo 11 and Apollo 12 landing sites. *Proc. Second Lunar Sci. Conf., Geochim. Cosmochim. Acta*, Suppl. 2, Vol. 1, pp. 833–854. M.I.T. Press.

Evans H. T. Jr. (1970) The crystallography of lunar troilite. *Proc. Apollo 11 Lunar Sci. Conf., Geochim. Cosmochim. Acta*, Suppl. 1, Vol. 1, pp. 399–408. Pergamon.

Fuchs L. H. (1970) Fluorapatite and other accessory minerals in Apollo 11 rocks. *Proc. Apollo 11 Lunar Sci. Conf., Geochim. Cosmochim. Acta*, Suppl. 1, Vol. 1, pp. 475–479. Pergamon.

Gay P. Bancroft G. M. and Brown M. G. (1970) Diffraction and Mossbauer studies of minerals from lunar soils and rocks. *Proc. Apollo 11 Lunar Sci. Conf., Geochim. Cosmochim. Acta*, Suppl. 1, Vol. 1, pp. 481–497. Pergamon.

Goldstein J. I. Yen F. and Axon H. J. (1972) Metallic particles in the Apollo 14 lunar soil (abstract). In *Lunar Science—III* (editor C. Watkins), pp. 323–325, Lunar Science Institute Contr. No. 88.

Grieve R. McKay G. Smith H. Weill D. and McCallum S. (1972) Mineralogy and petrology of polymict breccia 14312 (abstract). In *Lunar Science—III* (editor C. Watkins), pp. 338–340, Lunar Science Institute Contr. No. 88.

Hubbard N. J. and Gast P. W. (1972) Chemical composition of Apollo 14 materials and evidence for alkali volatilization (abstract). In *Lunar Science—III* (editor C. Watkins), pp. 407–409, Lunar Science Institute Contr. No. 88.

Jedwab J. (1971) Surface morphology of free-growing ilmenites and chromites from vuggy rocks 10072,31 and 12036,2. *Proc. Second Lunar Sci. Conf., Geochim. Cosmochim. Acta*, Suppl. 2, Vol. 1, pp. 923–935. M.I.T. Press.

Larimer J. W. (1967) Chemical fractionations in meteorites—1. Condensation of the elements. *Geochim. Cosmochim. Acta* **31**, pp. 1215–1238.

McKay D. S. Morrison D. A. Clanton U. S. Ladle G. H. and Lindsay J. F. (1971) Apollo 12 soil and breccia *Proc. Second Lunar Sci. Conf., Geochim. Cosmochim. Acta*, Suppl. 2, Vol. 1, pp. 755–773. M.I.T. Press.

McKay D. S. Heiken G. H. Taylor R. M. Clanton U. S. Morrison D. A. and Ladle G. H. (1972) Apollo 14 soils: Size distribution and particle types. *Proc. Third Lunar Sci. Conf., Geochim. Cosmochim. Acta*, Suppl. 3, Vol. 1.

Mizutani H. Fujii N. Hamano Y. Osako M. and Kanamori H. (1972) Elastic wave velocities and thermal diffusivities of Apollo 14 rocks (abstract). In *Lunar Science—III* (editor C. Watkins), pp. 547–549, Lunar Science Institute Contr. No. 88.

Naughton J. J. Hammond D. A. Margolis S. V. and Muenow D. W. (1972) A study of the nature of the gas cloud produced by volcanic and impact events on the moon and its relation to alkali erosion (abstract). In *Lunar Science—III* (editor C. Watkins), pp. 578–580, Lunar Science Institute Contr. No. 88.

Nyquist L. E. Hubbard N. J. Gast P. W. Wiesmann H. and Church S. E. (1972) Rb-Sr relationships for some chemically defined lunar materials (abstract). In *Lunar Science—III* (editor C. Watkins), pp. 584–585, Lunar Science Institute Contr. No. 88.

Schmitt H. H. Lofgren G. Swann G. A. and Simmons G. (1970) The Apollo 11 samples: Introduction. *Proc. Apollo 11 Lunar Sci. Conf., Geochim. Cosmochim. Acta*, Suppl. 1, Vol. 1, pp. 1–54. Pergamon.

Sclar C. B. (1971) Shock-induced features of Apollo 12 microbreccias. *Proc. Second Lunar Sci. Conf., Geochim. Cosmochim. Acta*, Suppl. 2, Vol. 1, pp. 817–832. M.I.T. Press.

Scott R. (1966) Origin of chemical variations within ignimbrite cooling units. *Amer. J. Sci.* **264**, 273–288.

Sheridan M. F. (1970) Fumarolic mounds and ridges of the Bishop Tuff, California. *Bull. Geol. Soc. Amer.* **81**, 851–868.

Skinner B. J. (1970) High crystallization temperatures indicated for igneous rocks from Tranquillity Base. *Proc. Apollo 11 Lunar Sci. Conf., Geochim. Cosmochim. Acta*, Suppl. 1, Vol. 1, pp. 891–895. Pergamon.

Skinner B. J. and Winchell H. (1972) Vapor phase growth of feldspar crystals and fractionation of alkalis in feldspar crystals from 12038,22 (abstract). In *Lunar Science—III* (editor C. Watkins), pp. 710–711, Lunar Science Institute Contr. No. 88.

Smith R. L. (1960) Ash flows. *Bull. Geol. Soc. Amer.* **71**, 795–842.

Sutton R. L. Hait M. H. and Swann G. A. (1972) Geology of the Apollo 14 landing site (abstract). In *Lunar Science—III* (editor C. Watkins), pp. 732–734, Lunar Science Institute Contr. No. 88.

Walter L. S. French B. M. Heinrich K. F. J. Lowman P. D. Jr. Doan A. S. and Adler I. (1971) Mineralogical studies of Apollo 12 samples. *Proc. Second Lunar Sci. Conf., Geochim. Cosmochim. Acta*, Suppl. 2, Vol. 1, pp. 343–358. M.I.T. Press.

Warner J. L. (1971) Lunar crystalline rocks: Petrology and geology. *Proc. Second Lunar Sci. Conf., Geochim. Cosmochim. Acta*, Suppl. 2, Vol. 1, pp. 469–480. M.I.T. Press.

Warner J. L. (1971) Progressive metamorphism of Apollo 14 breccias (abstract). Geol. Soc. America, 1971 Ann. Mtg., Washington, D.C., Abs. with Programs **3**, p. 744.

Warner J. L. (1972) Apollo 14 breccias: Metamorphic origin and classification (abstract). In *Lunar Science—III* (editor C. Watkins), pp. 782–784, Lunar Science Institute Contr. No. 88.

Warner J. L. and Duke M. B. (1971) Apollo 14 lunar sample information catalog. NASA Tech. Memo X-58062, 114 pp.

Wilshire H. G. and Jackson E. D. (1972) Petrology of the Fra Mauro formation at the Apollo 14 landing site (abstract). In *Lunar Science—III* (editor C. Watkins), pp. 803–805, Lunar Science Institute Contr. No. 88.

Proceedings of the Third Lunar Science Conference
(Supplement 3, *Geochimica et Cosmochimica Acta*)
Vol. 1, pp. 753–770
The M.I.T. Press, 1972

Apollo 14 regolith and fragmental rocks, their compositions and origin by impacts

W. von Engelhardt, J. Arndt, D. Stöffler,* and H. Schneider

Mineralogisch-Petrographisches Institut der Universität Tübingen,
Tübingen, West Germany

Abstract—Fragmental rocks of Apollo 14 have been classified into three types according to their texture and modal composition: (1) glass-rich regolith breccias produced by impacts into the regolith; (2) glass-poor breccias with fragmental matrices, forming the lower member of the Fra Mauro formation, the ejecta blanket of the Imbrian event; (3) glass-poor fragmental rocks with crystalline matrices, forming the uppermost layer of the Fra Mauro ejecta blanket which is supposed to be the product of a hot base surge.

The Apollo 14 regolith and the regolith breccias contain pre-Imbrian material, derived from the Fra Mauro formation, and were mainly produced by local impacts. The admixture of mare basalts and highland material shows the influence of further distant impact events.

Shock effects in minerals and rock fragments from the regolith and fragmental rocks include: fragmentation and undulatory extinction of all transparent minerals; deformation lamellae in pyroxene, olivine, and ilmenite; isotropic lamellae, partial isotropization of plagioclase and diaplectic plagioclase glass; shock fusion of rocks.

Glasses from the regolith and the fragmental rocks are divided into six groups: (1) diaplectic plagioclase glasses; shock-fused glasses of (2) anorthositic, (3) basaltic low-alkali, (4) basaltic high-alkali, (5) mafic, and (6) granitic compositions.

Annealing behavior of fused glasses reveals their rapid cooling on the lunar surface. Two different kinds of annealing behavior have been observed with diaplectic glasses.

Rocks of the pre-Imbrian crust, as preserved in the fragmental Fra Mauro formation and in the regolith, are of noritic to anorthositic composition with mafic and rhyolitic products of magmatic differentiation and probably also some impact melt rocks from pre-Imbrian impacts.

Introduction

Photogeological mapping by Wilhelms and McCauley (1971) led to the conclusion that the area north of the old crater Fra Mauro and about 550 km south of the mountain ring bordering Mare Imbrium is underlain by material excavated by the large impact event that produced the Imbrium Basin. It was expected that samples of the Imbrian ejecta blanket (Fra Mauro formation) would be collected at the Apollo 14 landing site. The relatively young Cone Crater, 340 m in diameter, presumably has excavated materials from depths as great as 60 to 80 m, well below the regolith, which is estimated to be 10 to 20 m thick in this area. Fra Mauro material was therefore expected to occur abundantly around Cone Crater and within the ray of blocky ejecta that extends westward beyond the landing site (Sutton *et al.*, 1972).

Objectives of the present paper are (1) investigation of fragmental rocks which are supposed to represent the Fra Mauro formation and their interpretation as products of the Imbrian impact, (2) investigation of the Apollo 14 regolith and regolith breccias in regard to their formation by repeated post-Imbrian impacts, (3) detection and

---

* Present address: NASA Ames Research Center, Space Science Division, Moffett Field, Ca. 94035.

interpretation of shock effects in rocks and minerals, (4) composition and origin of shock-produced glasses, (5) investigation of cooling histories of shock-fused glasses by annealing experiments.

Fragmental Rocks

The term fragmental rocks was used by LSPET (1971) to describe rocks of various textures and compositions that are characterized by their content of rock, mineral and glass fragments (clasts). We adopt this general term and propose to define the following two subgroups of fragmental rocks: (1) breccias, consisting of rock and mineral fragments with or without glass fragments and regular glass bodies, set in matrices of fine-grained detritus, sometimes cemented by a small amount of glass; (2) fragmental rocks with crystalline matrices, consisting of rock and mineral fragments with or without glass fragments, set in matrices of fine- to medium-grained mosaics of interlocking crystals.

All Apollo 14 rock samples which we investigated in thin section are fragmental rocks. These also occur abundantly in the regolith. According to texture and composition we distinguish three major types of fragmental rocks which may be subdivided further into smaller subgroups. In the following paragraphs the three types are described in detail.

Glass-rich regolith breccias. 14307,3 (Station G); 14049,41 (Station Bg); 14055,9 (Station E); 14315,7 (Station H); 14318,44 (Station H).

This group corresponds to groups 1 and 2 of Warner (1972), groups F1 and F2 of Jackson and Wilshire (1972), groups 3 and 4 of Chao *et al.* (1972), and regolith breccias of Quaide (1972).

The breccias contain fragments of rocks, minerals, glasses, and regular glass bodies in a matrix of minute mineral and glass fragments. Matrices of 14307 and 14049 contain also a small amount of glass that welds the fragmental particles.

Rock fragments are predominantly fragmental rocks with crystalline matrices and medium- to fine-grained noritic rocks of hypautomorphic granular to fragmental textures as they are described below as inclusions of fragmental rocks with crystalline matrices. Intrasertal basalts of mare type, anorthosites (troctolites), and older breccias are minor constituents.

Mineral fragments are orthopyroxene, clinopyroxene (including pigeonite), olivine, plagioclase, ilmenite, troilite, and metallic iron.

Glass fragments are mostly brownish yellow. Greenish to colorless and opaque glasses are less abundant. Some of the colorless glasses are diaplectic plagioclase glasses. Many glasses are heterogeneous due to schlieren and mineral fragments. In 14049 and 14055, vesicular glass fragments are common, some of them being similar to the glassy agglomerates which form a major constituent of the regolith. 14307 contains no vesicular glass. Glass spherules, ellipsoids, and droplets occur in all of these samples. A few glass fragments are devitrified. The chemical composition of some analyzed glasses will be given in Table 5. They are interpreted as diaplectic plagioclase glasses, shock-fused anorthositic rocks, mare basalts, local regolith or pre-Imbrian material, and mafic rocks of unknown origin (see below).

Breccia 14307 is coated by brown vesicular glass of basaltic composition (belonging to group (4), which contains some rock and mineral fragments, identical to those of the breccia. Breccia 14055 is a mixture of a lighter and a darker component with fairly sharp boundaries. The darker material seems to be surrounded by the light one.

Modal compositions of mineral and glass fragments of three samples of this breccia type are presented in Table 1. Rock fragments amount to 30 to 50 % of all fragments.

The following shock effects have been observed in these breccias: fragmentation and undulatory extinction of plagioclase and pyroxene; deformation lamellae in pyroxene; fine twin lamellae in ilmenite, isotropic lamellae and partial isotropization in plagioclase; diaplectic plagioclase glass; deformation lamellae in olivine. A shocked troctolite fragment with partially isotropic plagioclase, diaplectic plagioclase glass, and olivine with lamellae was found in 14307 and a shocked basaltic rock fragment consisting of diaplectic plagioclase glass and pyroxene with lamellae in 14049.

The glass-rich regolith breccias are interpreted as being produced by small impacts into unconsolidated regolith. Although some differences exist between the three breccias themselves and between breccias and soils, the modal compositions of the three breccias are in general very similar to those of Apollo 14 soil samples, as it can be seen from Tables 1 and 4. Most breccias are richer in fragmental minerals than the soils, which contain more glassy agglomerates, breccias, and glasses. Because the amount of agglomerates, glasses, and breccia fragments in the regolith should increase with time, because of the effects of repeated impacts, the breccias may represent earlier stages of the regolith. The breccias contain mare type basalts and glasses of mare basalt composition. If it can be assumed that this basaltic material is of mare origin and not a component of the pre-Imbrian crust, the regolith breccias were formed after the filling of the maria with basaltic lavas.

The modal compositions of the samples listed in Table 1 are in general similar. Differences in the pyroxene: plagioclase ratios may indicate that 14307 was formed from a regolith richer in pyroxene than that from which 14049, 14055, and 14318 originated. 14307 contains more rock fragments than 14049. It may be that 14049 was formed from a more mature regolith. The glass crust draping 14307 is interpreted as melted regolith produced by a younger impact.

Table 1. Modal compositions of the mineral and glass content of glass-rich regolith breccias.

	(grain count)		(point count)		
	14307,3	14049,41	14055,9 dark	14055,9 light	14318,44
	(%)	(%)	(%)	(%)	(%)
Plagioclase	19	26	20	24	9
Pyroxene + Olivine	40	34	25	24	9
Opaque Minerals	3	3	4	1	3
Glass	38	38	51	51	78
(Pyroxene + Olivine): Plagioclase	2.1	1.3	1.3	1.0	1.0

Glass-poor breccias with fragmental matrices (White Rock type). 14082,10,11 and 13 (White Rock, station C1); 14063,56 (Station C1).

These breccias contain fragments of rocks, minerals, and subordinate glass in a fragmental matrix. This type corresponds to group 3 of Warner (1972), group F3 of Jackson and Wilshire (1972), and group 1 of Chao *et al.* (1972).

Fragments of a dark fragmental rock consist of a very fine-grained crystalline matrix with subrounded clasts of plagioclase, mostly fractured and with undulatory extinction, recrystallized plagioclase, and some ilmenite. Fine ilmenite grains are evenly distributed within the whole mass. Minor constituents are colorless glass with schlieren, clinopyroxene, and olivine.

Fragments of a light breccia are mostly subrounded and consist of large, small, and minute plagioclase fragments, largely fractured and with undulatory extinction, and some grains of recrystallized plagioclase. Some of the plagioclase fragments contain very thin multiple twin lamellae.

Other rock fragments are fine- to medium-grained noritic rocks of hypautomorphic granular to fragmental textures as they are described in the next section as constituents of fragmental rocks with crystalline matrices.

The larger mineral clasts are predominantly plagioclase, commonly fractured and with undulatory extinction. Other mineral fragments are clinopyroxene, ortho-pyroxene, and ilmenite. Red spinel is a rare constituent.

A few fragments of a brownish, mostly fluidal, slightly devitrified glass contain plagioclase detritus.

The modal composition of the breccia without the larger rock and breccia clasts has been determined by point counting and is represented in Table 2.

No strong shock effects, such as deformation lamellae, isotropization, or melting have been observed in mineral and crystalline rock fragments. Fracturing and undulatory extinction of many mineral particles probably are effects of weak shock.

On the other hand, the light and dark fragmental rock inclusions are interpreted as being produced by shock waves of high intensity. This is indicated by recrystallized plagioclase fragments, interpreted as annealed diaplectic plagioclase glass, which only occur in these fragmental rock clasts. The fine-grained crystalline matrix of the dark fragmental rock inclusions crystallized from a melt that was most probably produced by an impact. If this interpretation is correct, the dark fragmental rock inclusions are impact melt rocks, according to the nomenclature of Dence (1971). Such rocks, as known from many terrestrial impact craters, are characterized by abundant inclusions of shocked fragments and the lack of phenocrysts.

Table 2. Modal composition of breccia 14082,13 without clasts of rocks and breccias. (Point counting.)

Fine-grained matrix ($<5\ \mu$)	26%
Plagioclase	42%
Pyroxene	28%
Opaque minerals	3%
Glass	1%
Pyroxene:Plagioclase	0.6%

It is improbable that both the inclusions of highly shocked fragmental rocks and the enclosing breccia were formed by the same event. Otherwise the latter would also contain highly shocked material. We therefore assume that type 2 breccias represent weakly to moderately shocked breccias (Engelhardt, 1971) produced by the Imbrian event that ejected and mixed pre-Imbrian plagioclase-rich and noritic rocks together with highly shocked breccia fragments of a pre-Imbrian impact.

White Rock, from which sample 14082 was taken, is one of the large boulders on the rim of Cone Crater. 14063 also was collected at the rim of Cone Crater. 14064, the only other rock classified by Jackson and Wilshire (1972) in their group 3 was found at the same locality. It can therefore be inferred from their location that the weakly shocked breccias of type 2 were excavated by the Cone Crater impact from a relatively deep layer of the Fra Mauro formation, as was also supposed by Sutton *et al.* (1972). Another possibility is that these breccias represent the pre-Fra Mauro basement.

Glass-poor fragmental rocks with crystalline matrices (Turtle Rock type). 14312,12; 14312,15; 14314,12; 14319,16; 14320,8 (all from Turtle Rock, station H); 14006,8 (Contingency sample); 14066,50 (Station F); 14311,98; 14311,8 (Station D); 14321,215 (Station C1).

This group corresponds to group 7 of Warner (1972), group F4 of Jackson and Wilshire (1972), group 2 of Chao *et al.* (1972), and annealed breccias of Quaide (1972) and Dence *et al.* (1972).

The rocks consist of a fine-grained crystalline groundmass that contains fragments of various rocks of nonfragmental and fragmental textures, minerals, and a few particles of glass. The very intricate texture of these rocks is difficult to unravel. From the inspection of the thin sections available to us, it appears to us that these rocks are best represented by Turtle Rock samples. They may be described as a mixture of a dark and a light component, both of similar textures. The dark component forms fairly sharply bounded, irregular blebs or lapilli-like bodies within the light mass, or else it merges gradually into the light material.

The light groundmass consists of interlocking plagioclase and pyroxene crystals with grains of ilmenite. Other opaque components of the matrix are metallic iron, which in 14319,16 was observed to penetrate into fissures of a large plagioclase clast, and less abundant troilite, sometimes intergrown with iron. The dark groundmass is finer-grained and contains a larger number of evenly distributed small grains of euhedral platelets of ilmenite. In some areas, particularly in some of the lapilli-like bodies, the dark matrix contains abundant elongated and sometimes skeletal crystals of olivine that reflect fluidal textures, particularly around inclusions and parallel to the outlines of the lapilli. In general, the dark material contains fewer fragmental inclusions, particularly fewer rock fragments than the light matrix.

The rock fragments show a wide range of textural and modal variability. Predominant are noritic rocks of hypautomorphic granular to fragmental textures, as also occur within the breccias described in the last sections and in the regolith. They are composed of euhedral to subeuhedral crystals and crystal fragments of plagioclase, orthopyroxene, clinopyroxene (including pigeonite), and ilmenite in various proportions. Olivine, troilite, and iron are minor constituents. Some of these rocks may be interpreted as magmatic cumulates. Many contain angular fragments of plagioclase and also of pyroxene, and grade into fragmental rocks with crystalline matrices. Fragmentation may occur during the formation of magmatic cumulates. Because some of these rocks contain fragments of recrystallized plagioclase, which are probably recrystallized diaplectic glasses, it seems to us more likely that these rocks have to be interpreted as impact melt rocks (Dence, 1971). Other rock fragments consist exclusively of heavily brecciated plagioclase crystals. Less abundant are dunites, pyroxenites, and intrasertal basalts. Some plagioclase-pyroxene rocks grade continuously into the light matrix so that it is difficult to distinguish between matrix and inclusion.

The largest mineral fragments are plagioclase and recrystallized plagioclase. Clinopyroxene (including pigeonite) fragments are about three times more common in 14319,16 than those of orthopyroxene. Large crystal fragments of orthopyroxene commonly contain ilmenite and poikilitic plagioclase inclusions. Olivine is much rarer than pyroxene. In comparison with the light matrix, the dark material contains fewer, smaller, and, in general, subrounded fragments, in most areas nearly consisting exclusively of plagioclase. Olivine occurs preferentially within the dark matrix. The same holds for a transparent spinel of red or pinkish color. One spinel fragment from 14319,16 was analyzed with the microprobe. The composition corresponds to a nearly Ti-free Al-Cr-Mg-Fe-spinel $[(Al_{0.79}Cr_{0.21}Ti_{0.005})(Mg_{0.46}Fe_{0.54}Mn_{0.03})O_4$ not found in Apollo 11 and 12 rocks (Table 5 (7)]. Spinels of similar compositions have been found by Steele and Smith (1972) in breccia 14063 and by Grieve et al. (1972) in breccia 14321. Other minor components of the dark matrix are irregular, commonly slightly elongated patches of a colorless to light brown, more or less devitrified glass. Microprobe analysis of such a glass from 14319,16 demonstrated that it has a granitic composition, the brownish rim being enriched in SiO_2, K_2O, BaO, and ZrO_2 as compared with the colorless interior [Table 5, group 5) I, II]. Devitrification products are elongated crystals of feldspar and pyroxene, the random orientation of which indicates that crystallization took place when the mass was no longer in motion.

A few plagioclase and olivine inclusions within the dark matrix are surrounded by haloes consisting of pyroxene crystals larger than those of the matrix. Around the rhyolitic glass fragments, pyroxene crystals penetrate from the matrix into the glass.

Besides brecciation and fracturing of rock and mineral fragments and deformation lamellae in pyroxene and ilmenite, no shock effects have survived the high temperatures to which these fragmental rocks were exposed. Recrystallized plagioclase fragments are interpreted as annealed diaplectic plagioclase glasses, as was also inferred by Dence et al. (1972). Rhyolitic glass fragments are probably shock-melted rocks of granitic compositions.

Modal compositions of the fragments in glass-poor fragmental rocks with crystalline matrices, based on point counting of thin sections, are presented in Table 3.

Fragmental rocks with crystalline matrices cannot have been formed by thermal metamorphism of regolith or regolith breccias. This can be shown by a topological comparison of their textures. We measured the percentage of grains $> 20 \mu$ that are in direct contact with other grains in thin sections of regolith breccia 14055 and of frag-

Table 3. Modal compositions of glass-poor fragmental rocks with crystalline matrices, without matrix (Point counting).

	14006,8 (%)	14066,50 (%)	14311,98 (%)	14320,8 (%)	14321,215 (%)
Rocks	22	15	23	38	18
Plagioclase	44	26	45	36	49
Pyroxene + Olivine	28	45	26	18	26
Opaque Minerals	2	1	3	4	1
Glasses	3	12	3	4	1
(Pyroxene + Olivine): Plagioclase	0.64	1.7	0.58	0.50	0.53

mental rocks 14006, 14066, and 14320. Whereas in 14055, 20 to 50% of glass and mineral fragments are in direct contact, in 14006, 14066, and 14320 the percentages of mineral grains in contact are 6 to 8%, less than 1% and 6 to 10%, respectively. No annealing or recrystallization process could transform the regolith breccia texture into that of fragmental rocks with crystalline matrices where rock and mineral fragments are embedded in a fine-grained matrix.

Glass-poor fragmental rocks with crystalline matrices have been found within the whole Apollo 14 area visited by the astronauts (see the map in Wilshire and Jackson, 1972). They also are major components of the Apollo 14 regolith and of regolith breccias. It must therefore be assumed that these rocks represent the Fra Mauro formation underlying the regolith in the Apollo 14 area. Because they occur not only at the rim of Cone Crater but also in large blocks far from it, e.g., Turtle Rock at station H, it seems very probable that fragmental rocks with crystalline matrices form the uppermost layer of the Fra Mauro formation.

We interpret the dark-light texture of fragmental rocks with crystalline matrices as formed by a mechanical mixture of two hot masses which were both ejected from the Imbrium Basin. Source rocks of both materials are pre-Imbrian rocks that are characterized by predominant noritic rocks with anorthosites, minor amounts of basalts, mafic, and granitic products of magmatic differentiation. The noritic rocks may include impact melt rocks from large pre-Imbrian impacts. Dark and light materials differ in their contents of fragmental inclusions and also in the temperatures at which they became mixed. The dark material was embedded in the more mobile light mass as separate, irregularly shaped clods or lapilli-like bodies. The temperature of the dark masses was therefore presumably lower at the time of deposition than that of the light mass. Olivine crystals arranged along flow lines suggest that the dark material was melt, the crystallization of which started while the dark melt masses were still in motion. Inclusions of recrystallized plagioclase fragments, interpreted as annealed diaplectic plagioclase, indicate that the dark material represents presumably an impact melt rock. Both dark and light materials arrived in the Apollo 14 area at rather high temperatures and cooled so slowly that crystalline matrices could develop. Only the SiO_2-richest particles could anneal to glasses of granitic composition, due to their lower liquidus temperature and higher viscosity.

Several investigators (Duncan *et al.*, 1972; Floran *et al.*, 1972; Grieve *et al.*, 1972; Wilshire and Jackson, 1972, and others) have emphasized that more than two generations of fragmentation can be distinguished. From fragmental rock inclusions in fragmental rocks, visible in thin sections, they deduced time sequences of up to four brecciation events, which they attributed to the Imbrian impact and several pre-Imbrian events. It seems questionable that it is possible to decide with certainty from two-dimensional sections the true relationships in space, i.e., whether one particular fragmental rock is really included within another one in such a way that the first is necessarily older than the other.

It is possible that the light-dark texture may record more than one single impact event. The dark masses would then represent a fragmental rock from a pre-Imbrian impact that became incorporated into the Imbrian ejecta blanket. However, the alternative view that both, the dark and the light masses, are products of the Imbrian impact should not be excluded. It must be kept in mind that in highly shocked impact formations or suevites (Engelhardt, 1971)* of terrestrial craters which are much smaller than Mare Imbrium, mixing of materials coming from different parts of the target and transported on different ways is a well-known phenomenon. The suevite of the Ries Crater consists of a fragmental groundmass that contains aerodynamically shaped glass bombs. Whereas the main mass of suevite was moved along the ground as a hot base surge, the bombs represent impact melt splashes charged with rock fragments which were ejected in ballistic trajectories.

An analogous model may be adequate to explain the texture of the fragmental rocks with crystalline matrices of Fra Mauro formation: the light material may have been moved as a hot base surge composed of vapor and fractured and shocked rock debris intermixed with some melt along the surface. The dark masses may have been ejected from the crater in trajectories and later were incorporated into the moving base surge. Lapilli-like bodies may have been formed by accretion in an early stage or by mechanical attrition during the transport along the ground.

It was inferred from the distribution of fragmental rocks in the Apollo 14 area that the weakly shocked rocks with fragmental matrices form the lower member of the Fra Mauro formation and that they are overlain by the highly shocked rocks. This assumption is in agreement with the stratigraphy of impact formations around terrestrial craters where highly shocked suevites are underlain by weakly to moderately shocked breccias, such as the Bunte Breccie at the Ries Crater (Engelhardt, 1971).

Regolith

14003,39 (<1 mm, *Contingency sample*); *14230,71* (<1 mm, *Core 2043, Station G*); *14259,64* (<1 mm, *Comprehensive sample*); *14162,28* (1–2 mm, *Bulk sample*).

Grain-size fractions have been prepared by dry sieving (1 mm to 63 μ) and sedimentation in benzene (<63 μ). An average density of 3.09 was used for the calculation of diameters from settling velocities. The following median diameters have been found: 14259,64: 65 μ; 14003,39: 56 μ; 14230,71: 78 μ. All three Apollo 14 soils are coarser than Apollo 11 and 12 soils, for which we found median diameters of 48 μ and 54 μ, respectively (Engelhardt *et al.*, 1970, 1971). Differences in size distributions of

* Engelhardt assumed transportation in ballistic trajectories of all components of the Ries suevite. Recent investigations have led to a correction of this view.

14003, 14259, and 14230 show that the regolith is not quite uniform, both in lateral and vertical directions: 14003 has the highest content of material below 10 μ, probably due to a high amount of fine glass fragments. Core sample 14230,71, collected at station G, 14 to 15 cm below the surface, contains more coarse material than the two surface samples. What the differences in size distribution and modal composition mean in terms of the gardening model of regolith formation could probably be deduced from a comparison of all available surface and core samples.

Modal compositions of the regolith fractions >20 μ have been determined by optical microscopy. The results are presented in Table 4 and Fig. 1.

Less than 4% of all crystalline rock fragments are mare type basalts and anorthosites. Most abundant, in about equal proportion, are two rock types which were already described in the last section: (a) fine-grained fragmental rocks with crystalline matrices of the Fra Mauro formation; (b) rocks of hypautomorphic granular to fragmental textures of predominantly noritic composition. These two rock types have been distinguished by size and distribution of ilmenite which forms minute, evenly distribu-

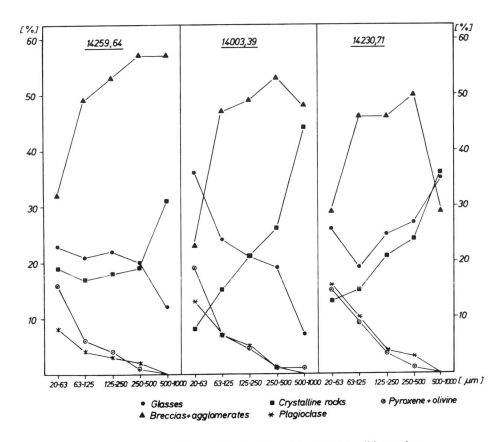

Fig. 1. Modal compositions of size fractions of Apollo 14 regolith samples.

Table 4. Modal compositions of Apollo 14 regolith samples. Averages of Apollo 11 and 12 regolith samples refer to 10084, 106 and 12001,84; 12070, 139, respectively (Engelhardt *et al.*, 1971).

	14259,64 20–1000 μ	14003,39 20–1000 μ	14230,71 20–1000 μ	14162,6 1–2 mm	Apollo 11 average 10–250 μ	Apollo 12 average 20–1000 μ	Apollo 14 average 20–1000 μ
Percentage of total soil	82	70	76	—	70	75	76
Crystalline rocks	20	19	19	46	4.3	10	19
Agglomerates + breccias	45	41	39	36	48	56	42
Glasses	21	25	25	17			24
Plagioclase	4.8	7.0	8.2	—	14	7	7
Pyroxene + Olivine	8.4	8.7	7.2	—	29	25	8
Opaque minerals	1.5	0.3	1.2	—	4	3	1
(Pyroxene + Olivine): Plagioclase	1.8	1.3	0.9	—	2.0	3.6	1.2

ted crystals in (a), but grains larger than 5 μ in (b). Due to their fine grain size, rock fragments are abundant also in fractions below 125 μ, in contrast to Apollo 11 and 12 soils (Engelhardt *et al.*, 1971).

As in Apollo 11 and 12 soils, glassy agglomerates are the main constituents of the Apollo 14 regolith. They consist of soil particles agglutinated by a small amount of vesicular glass. As produced by melt splashes from impacts into the local regolith, they record intensity and frequency of the bombardment of the lunar surface by small bodies. Agglomerates are difficult to distinguish from fragments of regolith breccias. Both constituents are therefore combined into one group in Fig. 1.

Glass fragments are predominantly yellow and brownish. Rarer are greenish, colorless, and opaque glasses. The abundance of regular glass bodies, which form 1 to 2% of the soil <1 mm, increases with decreasing grain size. Some glass fragments and regular glass bodies are very homogeneous. Most glasses contain schlieren, vesicles and mineral inclusions. Some glasses are devitrified. Devitrified spherules, as described by King *et al.* (1972) from Apollo 14 breccias, show textures that seem to be identical with those of meteoritic chondrules. The chemical composition and optical properties of the regolith glasses are described in the next section.

The main fragments are plagioclase, Ca-rich clinopyroxene, pigeonite, ortho-pyroxene, and ilmenite. The ratio (pyroxene + olivine):plagioclase is significantly lower than that of Apollo 11 and 12 soils. Mineral fragments in Apollo 11 and 12 soils are mostly from pyroxene-rich mare basalt sources, but plagioclase-rich pre-Imbrian rocks are the sources for mineral fragments in Apollo 14 soils. Recrystallized plagioclase fragments are interpreted as devitrified diaplectic plagioclase glasses, derived from the Fra Mauro formation.

The fraction <4 μ has been investigated by x rays. The following diffraction lines have been found: strong lines of plagioclase and pyroxene, weak lines of olivine, ilmenite, cristobalite, troilite, iron, and two lines not identified, corresponding to d = 3.64 Å and 3.04 Å, respectively.

The same shock effects have been observed in the regolith as have been described from the regolith breccias. Shock effects are rarer in Apollo 14 soil than in the regoliths of Apollo 11 and 12. The Apollo 14 regolith was produced by repeated impacts on the lunar surface which mixed together disintegrated, shocked, and shock-melted rocks of the local Fra Mauro formation with smaller amounts of material originating from larger impacts in mare and highland areas. Consequently, the Apollo 14 regolith is composed of magmatic rocks, fragmental rocks, minerals, and glasses that are products of four periods of lunar history (Wasserburg *et al.*, 1972):

(1) More than 4 billion years old: minerals and rocks of the pre-Imbrian crust, which include anorthosites, noritic, granitic, and basaltic rocks, basic differentiates, impact melt and impact fragmental rocks.

(2) About 3.9 billion years old: material from the Fra Mauro formation, the ejecta blanket of the Imbrian event.

(3) 3.7 to 3.2 billion years old: minerals and rocks from mare basalt areas.

(4) Glasses, shocked mineral and rock fragments, and regolith breccias produced by younger impacts into the regolith.

Glasses

Averages of microprobe analyses of 63 glasses from regolith and fragmental rock samples are presented in Table 5. According to their chemical compositions, six major groups of glasses can be distinguished. Groups (1), (2), (3 + 4), and (5) appear clearly separated on a MgO versus Al_2O_3 plot, Fig. 2.

(1) Diaplectic plagioclase glasses: colorless, characterized by monomineralic plagioclase composition.

Table 5. Electron microprobe analyses (wt. %).

	(1) Diaplectic plagioclase glasses[a] Average of 4		(2) Glasses of anorthositic composition[b] Average of 12		(3) Glasses of basaltic composition low alkali content[c] Average of 7	
	(%)	(σ)	(%)	(σ)	(%)	(σ)
SiO_2	46.3	1.00	44.6	1.19	43.6	2.66
TiO_2	0.10	0.01	0.45	0.48	4.06	3.20
Al_2O_3	34.1	0.21	25.1	2.10	14.4	3.76
FeO	0.09	0.04	5.7	1.25	14.1	3.81
MnO	0.00	—	0.06	0.02	0.12	0.03
MgO	0.06	0.09	7.5	2.07	9.7	1.89
CaO	17.6	0.90	15.2	1.26	12.3	1.34
Na_2O	1.04	0.09	0.21	0.36	0.11	0.08
K_2O	0.09	0.05	0.11	0.20	0.08	0.04
ZrO_2	0.00	—	0.08	0.07	0.12	0.07
Cr_2O_3	0.01	0.00	0.13	0.07	0.18	0.11
BaO	0.00	—	0.04	0.04	0.08	0.02
P_2O_5	0.07	0.00	0.23	0.34	—	—
TOTAL	99.46		99.41		98.85	

Table 5 (continued)

	(4) Glasses of basaltic composition high alkali content[d] Average of 37		(5) Opaque mafic glass* (14307,3)	(6) Brown rhyolitic glass (14319,16)[†] I	II	(7) Red spinel (14319,16)
	(%)	(σ)				
SiO_2	47.9	2.21	33.2	75.1	71.4	0.20
TiO_2	2.24	0.69	16.0	0.54	0.86	0.33
Al_2O_3	15.4	1.47	4.25	12.7	14.2	47.1
FeO	11.2	1.45	24.1	0.86	0.97	19.7
MnO	0.12	0.14	0.15	0.00	0.00	0.11
MgO	9.0	1.83	14.5	0.02	0.08	12.8
CaO	11.3	0.69	6.8	0.91	0.69	0.03
Na_2O	0.70	0.18	0.35	1.10	0.49	0.03
K_2O	0.61	0.19	0.14	8.2	13.2	0.00
ZrO_2	0.16	0.07	0.04	0.08	0.15	0.00
Cr_2O_3	0.18	0.13	0.89	0.00	0.00	18.6
BaO	0.09	0.03	0.12	0.03	0.24	0.00
P_2O_5	0.54	0.31	0.05	0.06	0.12	0.00
TOTAL	99.44		100.59	99.60	102.40	98.87

a. Averages from samples: 14049,41; 14259,64; 14290,8
b. Averages from samples: 14259,64; 14162,28; 14307,3; 14049,41
c. Averages from samples: 14049,41; 14162,28
d. Averages from samples: 14049,41; 14259,64; 14307,3; 14312,12; 14319,16

* Regular form of revolution
† I: Brown interior; II: Brown rim

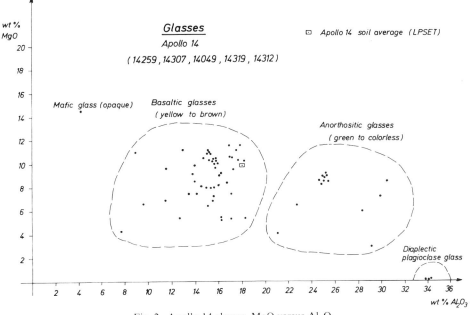

Fig. 2. Apollo 14 glasses. MgO versus Al₂O₃.

(2) Glasses of anorthositic composition in the broadest sense, including true anorthosites and anorthositic gabbros: most of these are colorless or greenish, some brownish; characterized by Al_2O_3 contents between 20 and 30%. Norms contain 80 to 64% plagioclase, wide ranges of hypersthene, diopside, and olivine, but no quartz. Glasses of this group are interpreted as shock-melted plagioclase-rich rocks from the highlands.

(3) Glasses of basaltic composition and low alkali content: yellow to brown, characterized by contents of Al_2O_3 between 9 and 16%, of SiO_2 between 39 and 45% and of K_2O below 0.1%. Norms contain 25 to 50% plagioclase, wide ranges of diopside, hypersthene, and olivine, 4 to 20% ilmenite, but no quartz. Glasses of this group are interpreted as shock-melted mare basalts.

(4) Glasses of basaltic composition and high alkali content: most are yellow to brown, a few colorless or greenish; characterized by contents of Al_2O_3 between 12 and 18%, of SiO_2 between 45 and 53%, and of K_2O above 0.3%. Norms contain 36 to 50% plagioclase, less diopside and more hypersthene than (3) glasses, olivine, or quartz and 0.3 to 6% ilmenite. Glasses of this group are most abundant in the regolith and in regolith breccias. They are interpreted as shock-melted local soil or Fra Mauro fragmental rocks, composed of pre-Imbrian crustal rocks.

(5) Mafic glasses: opaque, characterized by contents of Al_2O_3 below 5% and high concentrations of FeO and TiO_2. They may have been formed by shock melting of unknown mafic rocks.

(6) Glasses of granitic composition: colorless or light brown, characterized by contents of SiO_2 higher than 70% and of K_2O of 8% and more. They occur within

Fra Mauro fragmental rocks and are interpreted as shock melted granitic rocks of the pre-Imbrian crust.

The six glass groups are in general agreement with the classifications given by other investigators. (1) corresponds to group (1) of Glass (1972); (2) to group (2) of Glass and the high-Al, low-Ca group of Reid *et al.* (1972); (3) to group (3) of Glass and mare-derived glasses of Reid *et al.*; (4) to group (4) of Glass and high alumina tholeiite glasses of Reid *et al.*; (6) to group (6) of Glass.

Refractive indices of glasses increase with increasing FeO + TiO$_2$ and decreasing Al$_2$O$_3$ + SiO$_2$ contents (Figs. 3 and 4).

Annealing behavior of shock-fused and diaplectic glasses

We have investigated the annealing behavior of shocked fused glasses in order to reveal their cooling histories. The transition temperature (T$_g$) is defined as the temperature at which, during cooling, the high-temperature structure of a given glass melt is frozen in. T$_g$ increases with the rate of cooling. The room-temperature refractive index of a given glass is lower when T$_g$ was high because during rapid cooling a more open structure was frozen in. The results of annealing experiments with a reddish brown glass sphere and a green glass fragment are shown in Fig. 5. Both glasses were annealed at 630°C, i.e., within the glass transition region observed by Greene *et al.* (1971) on synthetic lunar glasses. Both glasses show an increase in refractive indices within the first 1.5 hours. This behavior indicates that the glass melts were very rapidly quenched on the lunar surface. After about 3.5 hours of annealing, unmixing occurred in both glasses, resulting in heterogeneity in refractive indices.

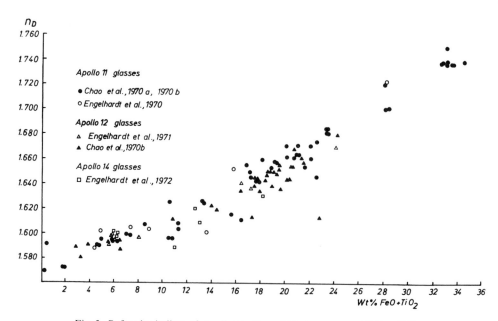

Fig. 3. Refractive indices of Apollo 11, 12, and 14 glasses versus FeO + TiO$_2$.

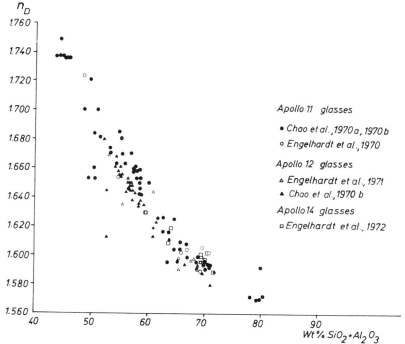

Fig. 4. Refractive indices of Apollo 11, 12, and 14 glasses versus $Al_2O_3 + SiO_2$.

Annealing of a diaplectic plagioclase glass from Apollo 14 at 800°C yielded a rapid decrease of its refractive index within the first 2 hours which then remained constant (Fig. 5). In contrast, we obtained on an Apollo 11 diaplectic plagioclase glass a constancy of refractive index during the first 10 hours of annealing at 800°C and a later increase within the following 30 hours (Engelhardt *et al.*, 1970). Both effects have already been reported in the literature: decrease of refractive indices after annealing was observed by Duke (1968) on maskelynite of the Shergotty meteorite at 450°C, by Bell and Chao (1968) on diaplectic plagioclase glass from the Ries, and on artificial diaplectic alkali feldspar glass, above 550°–660°C. Increase of refractive index on annealing at 800°C was observed by Bunch *et al.* (1968) on maskelynite of the Manicouagan Crater.

The T_g of fused albite and anorthite glass is about 750°C and 810°C, respectively (Arndt and Häberle, 1972). Annealing temperatures of 800°C are therefore well within the transition region of calcic plagioclase glasses. Apparently the observed increase of refractive index may correspond to the normal behavior of rapidly quenched fused glasses. The decrease of refractive index, observed with other samples of diaplectic glasses at temperatures as low as 450°C may be attributed to a relaxation process unknown from fused glasses.

CONCLUSIONS

(1) Fragmental rocks with crystalline matrices of the type represented by Turtle Rock form the uppermost member of the Fra Mauro formation underlying the

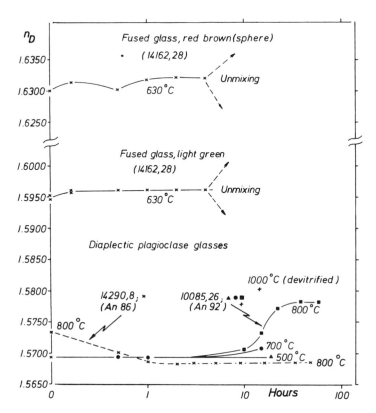

Fig. 5. Annealing behavior of fused and diaplectic glasses from Apollo 11 and 14.

regolith at the Apollo 14 landing site. They were formed by a hot base surge produced by the Imbrian event. Their light-dark texture is interpreted as a mixture of a light component moved more or less along the ground and a darker one ejected probably more in ballistic trajectories. Rock and mineral inclusions reveal that the pre-Imbrian crust was composed of plagioclase-rich rocks, K-rich noritic rocks, mafic and granitic products of magmatic differentiation and pre-Imbrian impact melt rocks.

(2) Breccias with fragmental matrices of the type represented by White Rock underlie the fragmental rocks with crystalline matrices. They belong either to a weakly shocked breccia member of the Fra Mauro formation, ejected from the Imbrium Basin, or they form the bedrock on which the Imbrian ejecta blanket was deposited.

(3) Regolith and regolith breccias were produced by repeated post-Imbrian impacts. They contain pre-Imbrian rocks and minerals, fragmental rocks of the Fra Mauro formation, minerals and rocks from mare basalt and highland areas, and shock-fused glasses.

(4) Shock-fused glasses in the regolith and in regolith breccias were formed from the regolith, Fra Mauro formation rocks, mare basalts, highland anorthosites, and

unknown mafic rocks. Annealing experiments show that shock-fused glasses were cooled in times not exceeding some hours.

(5) At the Apollo 12 site, especially around Head Crater, fragments of noritic breccias with glass matrices, noritic crystalline rocks with hypautomorphic to fragmental textures and glasses of the same chemical composition (KREEP) occur, which were supposed by several authors to belong to the Fra Mauro formation (Meyer *et al.*, 1971; Hubbard *et al.*, 1971; Anderson *et al.*, 1971; Marvin *et al.*, 1971; Engelhardt *et al.*, 1971; and others). This assumption can now be confirmed. The Apollo 12 fine-grained noritic, fragmental rocks (annealed breccias of Meyer *et al.*, 1971) are identical with fragmental rocks with crystalline matrices of Apollo 14. Apollo 12 medium-grained noritic rocks (gray-mottled basalt of Anderson *et al.*, 1971; A-norite-anorthosite of Marvin *et al.*, 1971) correspond to the noritic fragments with hypautomorphic to fragmental textures occurring in Apollo 14 regolith and fragmental rocks. Alkali-rich basaltic glasses of Apollo 14 have compositions similar to KREEP glasses of Apollo 12.

Acknowledgments—Samples 14006, 14055, 14063, 14066, 14082, 14311, 14312, 14315, 14318, and 14321 were studied through the courtesy of Dr. W. L. Quaide, Space Science Division, NASA Ames Research Center, to whom these samples were allocated.

We thank Mrs. E. Claviez for typing the manuscript, and Mrs. E. Lücke, Mr. D. Mangliers, and Mr. J. Mällich for technical assistance. We thank the staff of the Zentrum für Datenverarbeitung, University of Tübingen, especially Dr. P. Schmuck for the carrying out of computer work. Financial support from the Bundesministerium für Bildung und Wissenschaft, Federal Republic of Germany, is gratefully acknowledged. We thank the National Aeronautics and Space Administration for the generous supply of lunar samples.

References

Anderson A. T. and Smith J. V. (1971) Nature, occurrence, and exotic origin of "gray mottled" (Luny Rock) basalts in Apollo 12 soils and breccias. *Proc. Second Lunar Sci. Conf., Geochim. Cosmochim. Acta,* Suppl. 2, Vol. 3, pp. 431–438. M.I.T. Press.

Arndt J. and Häberle F. (1972) Thermal expansion and glass transition temperatures of synthetic plagioclase glasses. (To be published.)

Bell P. and Chao E. T. C. (1970) Annealing experiments with naturally and experimentally shocked feldspar glasses. *Carnegie Inst. Wash. Yearb.* **68**, 336–339.

Bunch T. E. Cohen A. I. and Dence M. R. (1968) Shock-induced structural disorder in plagioclase and quartz. In *Shock Metamorphism of Natural Materials* (editors B. M. French and N. M. Short) pp. 613–621. Mono.

Chao E. T. C. James O. B. Minkin J. A. Boreman J. A. Jackson E. D. and Raleigh C. B. (1970a) Petrology of unshocked crystalline rocks and evidence of impact metamorphism in Apollo 11 returned lunar sample. *Proc. Apollo 11 Lunar Sci. Conf., Geochim. Cosmochim. Acta,* Suppl. 1, Vol. 1, pp. 287–314.

Chao E. T. C. Boreman J. A. Minkin J. A. James O. B. and Desborough G. A. (1970b) Lunar glasses of impact origin: physical and chemical characteristics and geologic implications. *J. Geophys. Res.* **75**, 7445–7479.

Chao E. T. C. Minkin J. A. and Boreman J. A. (1972) The petrology of some Apollo 14 breccias. In *Lunar Science—III* (editor C. Watkins), p. 131, Lunar Science Institute Contr. No. 88.

Dence M. R. (1971) Impact melts. *J. Geophys. Res.* **76**, 552–565.

Dence M. R. Plant A. G. and Traill R. J. (1972) Impact-generated shock and thermal metamorphism in Fra Mauro lunar samples. In *Lunar Science—III* (editor C. Watkins), p. 174, Lunar Science Institute Contr. No. 88.

Duke M. B. (1968) The Shergotty Meteorite: Magmatic and shock metamorphic features. In *Shock Metamorphism of Natural Materials* (editors B. M. French and N. M. Short) pp. 613–621. Mono.

Duncan A. R. Lindstrom N. M. Lindstrom D. J. McKay S. M. Stoeser J. W. Goles G. G. and Fruchter J. S. (1972) Comments on the genesis of breccia 14 321. In *Lunar Science—III* (editor C. Watkins), p. 192, Lunar Science Institute Contr. No. 88.

Engelhardt W. von (1971) Detrital impact formations. *J. Geophys. Res.* **76**, 5566–5574.

Engelhardt W. von Arndt J. Müller W. F. and Stöffler D. (1970) Shock metamorphism of lunar rocks and origin of the regolith at the Apollo 11 landing site. *Proc. Apollo 11 Lunar Sci. Conf., Geochim. Cosmochim. Acta*, Suppl. 1, Vol. 1, pp. 363–384.

Engelhardt W. von Arndt J. Müller W. F. and Stöffler D. (1971) Shock metamorphism and origin of regolith and breccias at the Apollo 11 and Apollo 12 landing sites. *Proc. Second Lunar Sci. Conf., Geochim. Cosmochim. Acta*, Suppl. 2, Vol. 1, pp. 833–854. M.I.T. Press.

Floran R. J. Cameron K. Bence A. E. and Papike J. J. (1972) The 14313 Consortium: a mineralogic and petrologic report. In *Lunar Science—III* (editor C. Watkins), p. 268, Lunar Science Inst. Contr. No. 88.

Glass, B. P. (1972) Apollo 14 glasses. In *Lunar Science—III* (editor C. Watkins), p. 312, Lunar Science Institute Contr. No. 88.

Greene C. H. Pye L. D. Stevens H. J. Rase D. E. and Kay H. F. (1971) Compositions, homogeneity, densities, and thermal history of lunar glass particles. *Proc. Second Lunar Sci. Conf., Geochim. Cosmochim. Acta,* Suppl. 2, Vol. 3, pp. 2049–2055. M.I.T. Press.

Grieve R. McKay G. Smith H. and Weill D. (1972) Mineralogy and petrology of polymict breccia 14 321. In *Lunar Science—III* (editor C. Watkins), p. 338, Lunar Science Institute Contr. No. 88.

Hubbard N. J. Meyer C. and Gast P. W. (1971) The composition and derivation of Apollo 12 soils. *Earth Planet. Sci. Letters* **10**, 341–350.

Jackson E. D. and Wilshire H. G. (1972) Classification of the samples returned from the Apollo 14 landing site. In *Lunar Science—III* (editor C. Watkins), p. 418, Lunar Science Institute Contr. No. 88.

King E. A. Butler J. C. and Carman M. F. (1972) Chondrules in Apollo 14 breccias and estimation of lunar surface exposure ages from grain size analyses. In *Lunar Science—III* (editor C. Watkins), p. 449, Lunar Science Institute Contr. No. 88.

LSPET (1971) (Lunar Sample Preliminary Examination Team) Preliminary examination of lunar samples from Apollo 14, *Science* **173**, 681–693.

Marvin U. B. Wood J. A. Taylor G. J. Reid J. B. Powell B. N. Dickey J. S. and Bower J. F. (1971) Relative proportions and probable sources of rock fragments in the Apollo 12 soil samples. *Proc. Second Lunar Sci. Conf., Geochim. Cosmochim. Acta,* Suppl. 2, Vol. 1, pp. 675–699. M.I.T. Press.

Meyer C. Brett R. Hubbard N. J. Morrison D. A. McKay D. S. Aitken F. K. Tadeka H. and Schonfeld E. (1971) Mineralogy, chemistry, and origin of the KREEP-component from the Oceanus of Storms. *Proc. Second Lunar Sci. Conf., Geochim. Cosmochim. Acta*, Suppl. 2, Vol. 1, pp. 393–411. M.I.T. Press.

Quaide W. (1972) Mineralogy and origin of Fra Mauro fines and breccias. In *Lunar Science—III* (editor C. Watkins), p. 627, Lunar Science Institute Contr. No. 88.

Reid A. M. Ridley W. I. Warner J. Harmon R. S. Brett R. Jakes P. and Brown R. W. (1972) Chemistry of highland and mare basalts as inferred from glasses in the lunar soil. In *Lunar Science—III* (editor C. Watkins), p. 640, Lunar Science Institute Contr. No. 88.

Steele I. M. and Smith J. V. (1972) Mineralogy, petrology, bulk electron-microprobe analyses from Apollo 14, 15, and Luna 16. In *Lunar Science—III* (editor C. Watkins), p. 721, Lunar Science Institute Contr. No. 88.

Sutton R. L. Hait M. H. and Swann G. A. (1972) Geology of the Apollo 14 landing site. In *Lunar Science—III* (editor C. Watkins), p. 732, Lunar Science Institute Contr. No. 88.

Warner J. L. (1972) Apollo 14 breccias: metamorphic origin and classification. In *Lunar Science—III* (editor C. Watkins), p. 782, Lunar Science Institute Contr. No. 88.

Wasserburg G. J. Turner G. Tera F. Podosek F. A. Papanastassiou D. A. and Huneke J. C. (1972) Comparison of Rb-Sr, K-Ar, and U-Th-Pb ages; lunar chronology and evolution. In *Lunar Science—III* (editor C. Watkins), p. 788, Lunar Science Institute Contr. No. 88.

Wilhelms D. E. and McCauley J. F. (1971) Geologic map of the near side of the moon. U.S. Geol. Survey Misc. Geol. Invest. Map I-703.

Wilshire H. G. and Jackson E. D. (1972) Petrology of the Fra Mauro formation at the Apollo 14 landing site. In *Lunar Science—III* (editor C. Watkins), p. 803, Lunar Science Institute Contr. No. 88.

Proceedings of the Third Lunar Science Conference
(Supplement 3, *Geochimica et Cosmochimica Acta*)
Vol. 1, pp. 771–784
The M.I.T. Press, 1972

Mineralogy and origin of Fra Mauro fines and breccias

WILLIAM QUAIDE AND ROBERT WRIGLEY

NASA-Ames Research Center, Moffett Field, Calif. 94035

Abstract—Textural features, glassy spherule and glassy aggregate contents, and size-frequency distributions of lithic components of the regolith fines indicate that the Fra Mauro regolith of the smooth terrain is a highly reworked accumulation of debris derived primarily by impact comminution of annealed breccias. Analyses of the glassy spherules suggest that mare-derived debris must amount to less than 5% of the regolith, but exotic materials having the composition of the Fra Mauro breccias (highland-like) cannot be recognized. Breccia samples from the Fra Mauro site include (1) regolith breccias derived by lithification of local regolith material, (2) texturally and compositionally unique white rock breccias that may have been derived from ancient ejecta buried beneath the Fra Mauro formation, and (3) annealed breccias that are excavated samples of the Fra Mauro formation. The latter were deposited as heated impact ejecta, probably by avalanches produced by the Imbrian event and were annealed *in situ* in a thermal regime perhaps equivalent to that responsible for the pyroxene-hornfels facies of metamorphism in terrestrial rocks. Different degrees of annealing in different samples reflect their original positions in the cooling deposit. Lithic fragments in the annealed breccias reveal that the source area consisted of a plutonic complex of feldspathic and ultramafic rocks dominated by noritic types that had been highly cratered to yield a thick, complex regolith containing layers of reworked debris of thermally annealed ejecta and containing intercalated or capping flows of basalt of varied composition.

INTRODUCTION

THIS PAPER presents results of mineralogical studies and γ-ray spectrometric analyses of Apollo 14 regolith fines, breccias, and igneous rocks. The studies were undertaken to characterize the mineralogy and petrography of the fine (<1 mm) and breccia components of the regolith, to measure the concentrations of U and Th and the cosmogenic components ^{26}Al and ^{22}Na in certain samples and to use these properties and measurements to interpret the origin and history of the Fra Mauro formation and its blanketing regolith. Specific samples studied, their sizes, parent locations, and other pertinent data are listed in Table 1. Standard optical, x-ray, and electron microprobe techniques were used in the mineralogical studies. Techniques and equipment employed in the γ-ray spectrometric analyses were the same as those described by Wrigley (1971).

REGOLITH FINES

Regolith fines from bulk and comprehensive samples of Fra Mauro smooth terrain (14163,78; 14260,5) have identical size distributions (corrected for material removed by 1-mm sieving at the Lunar Receiving Laboratory) with a median diameter of 63 μ and log geometric quartile deviation of 0.570. They are thus coarser and more poorly sorted than most Apollo 11 and 12 surface samples (Quaide *et al.*, 1971). The differences may have no significance, however, for our measurements indicate that the content of ^{26}Al in sample 14163 (88 $\pm$ 6 dpm/kg) is a factor of two less than that

Table 1. Types, sizes, and parent locations of samples studied

Category	Sample	Type	Wt. and location of parent
< 1 mm fines	14163,0	299 g radiation counting	7129 g bulk sample
	14163,78	0.5 g aliquot	
	14260,5	0.5 g aliquot	282 g comprehensive sample
Regolith breccia	14055,9	Thin section	110 g, Station E
	14313,5	Probe section	144 g, Station G
	14318,48	Thin section	600 g, Station H
White rock breccia	14063,56	Thin section	135 g, Station C1
Annealed breccia	14006,8	Thin section	12 g, contingency
	14066,50	Probe section	509 g, Station F
	14066,8	Thin section	509 g, Station F
	14034,6	Probe section	2498 g, near comp. sample
	14306,3	Probe section	584 g, Station G
	14311,89	Thin section	3204 g, Station Dg
	14320,8	Probe section	65 g, Station H
	14321,215	Probe section	8998 g, Station C1
	14321,256	200 g sawdust	8998 g, Station C1
Basalt	14053,11	Probe mount	251 g, Station C2

in surface samples of Apollo 11 and 12 fines ($10084 = 147 \pm 16$ and $12070 = 171 \pm 18$ dpm/kg), and comparably less than that reported for sample 14259 (Keith et al., 1972). The concentration of ^{26}Al in sample 14163 is consistent with an average sampling depth of about 10 cm, and thus properties of this sample may be average properties of the upper 10 to 20 cm of the regolith. Furthermore, size distribution characteristics of sample 14260 suggest that it too is a mixed sample and therefore properties of neither of these samples should be compared strictly with those of samples taken from the immediate surface. Size distributions of component grains of these samples (Fig. 1) are generally similar to those of previously studied regolith fines with crystalline clasts, mainly monomineralic, predominating in small particle sizes and glassy aggregates and breccia clasts, mostly annealed types, being most abundant in coarser fractions. These data suggest that repetitive impacts have fragmented annealed breccia rocks giving rise to the observed size distributions of breccia clasts while releasing the finer crystalline grains and producing impact glass spatter that has bonded previously comminuted grains into larger cindery aggregates. Furthermore, impact generated glassy spheroids (Fig. 1) are much more abundant in the near surface Fra Mauro fines than in any Apollo 11 or 12 samples, indicating that repetitive impact reworking has been far more extensive at the Fra Mauro site than in any other area studied.

Although all regolith samples have certain textural similarities, they are compositionally distinct, one from another. The Fra Mauro fines are compositionally distinct from regolith samples of all other lunar areas yet sampled with significant relative enrichment of Al, Mg, and K, and depletion of Fe, Ti, and Cr. The differences are shown graphically in the normative composition diagram of Fig. 2. The normative compositions shown are consistent with modal mineral compositions of the fines. Plagioclase is more abundant among the monomineralic grains of the Fra Mauro fines (Fig. 1) than in any other lunar samples. Analyses of 64 grains reveal a range in composition from An_{72} to An_{97} with a mode of An_{93}. Bronzite is the most common

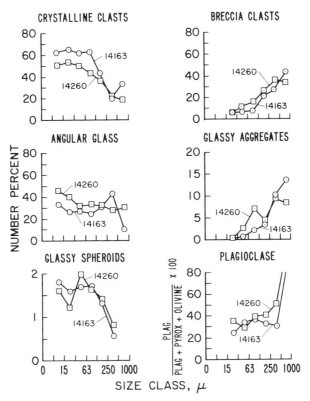

Fig. 1. Size-frequency distributions of major components of two samples of regolith fines. The percentage plagioclase/(plagioclase + pyroxene + olivine) was determined from mono-mineralic grains only.

variety of pyroxene, reflecting the enhanced Mg content although pigeonite, augite, subcalcic augite, subcalcic ferroaugite, and ferroaugite also are present. Olivine is present in amounts comparable to or slightly greater than that of other sites, but opaque minerals are depleted significantly. Crystalline rock fragments include microgranite in trace amounts and a few basalts. Anorthosite, olivinite, and pyroxenite clasts are common, but granulitic and hypidiomorphic granular rocks of noritic composition are most abundant. The latter are plagioclase-pyroxene rocks with or without olivine and opaque minerals. Plagioclase predominates and the pyroxenes commonly are bronzitic or pigeonitic. The hypidiomorphic norites, olivinites, pyroxenites, basalts, and anorthosites are igneous, but the granulitic clasts were derived from high-grade metamorphic rocks. Some clasts are transitional in texture to annealed Fra Mauro breccias, but most appear coarser and more thoroughly recrystallized.

The bulk compositions of the Fra Mauro fines are distinct from samples of probable local substrate origin, as is the case for samples from all areas studied. The relationship is well-illustrated in Fig. 2 in which the normative compositions of Apollo 11, 12, and 14 and Luna 16 fines lie close to the compositions of large rock

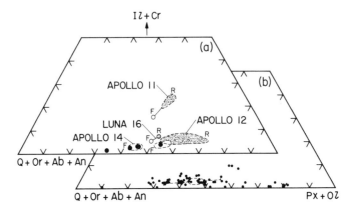

(a) ROCKS (R) AND FINES (F)

(b) GLASSY SPHEROIDS APOLLO 14

Fig. 2. Normative compositions of Apollo 11, 12, and 14 and Luna 16 rocks (R) and fines (F) and 62 glassy spheroids from samples 14163 and 14260. Compositions of Apollo 14 annealed breccias are indicated by filled circles. Sources: Apollo 11 fines (Agrell *et al.*, 1970); Apollo 11 rocks (Rose *et al.*, 1970; Maxwell *et al.*, 1970); Apollo 12 fines and rocks (Cuttitta *et al.*, 1971); Luna 16 rock and fines (Vinogradov, 1971); and Apollo 14 rocks and fines (LSPET, 1971).

fragments of probable local substrate origin. Major element compositions of the Fra Mauro fines are very similar to those of the annealed breccias (LSPET, 1971) as are REE abundances (LSPET, 1971) and U and Th contents (Table 2 and LSPET, 1971). The fines must have been derived from the breccias with little contribution of exotic material of dissimilar composition. The paucity of recognizable exotic components in the Fra Mauro fines contrasts strongly with samples of regolith fines of mare sites, as is shown in Fig. 2. It is noteworthy, however, that enrichment trends of mare fines defined by tie lines from rocks to fines converge at normative compositions typical of Fra Mauro breccia samples, suggesting that rocks with major element compositions similar to that of the Fra Mauro breccias have been the source of most of the exotic components in the mare fines. It is neither implied that Fra Mauro debris has been distributed to all the mare sites nor suggested that the fines at each mare site

Table 2. Uranium, Thorium, Potassium, ^{26}Al, and ^{22}Na in bulk fines sample 14163 and Rock 14321 sawdust

Sample	14321,256	14163,0
Mass	200.2 g	299.5 g
Uranium ppm	3.00 ± 0.06	3.75 ± 0.05
Thorium ppm	11.33 ± 0.19	13.20 ± 0.16
Potassium %	0.363 ± 0.009	0.456 ± 0.010
^{26}Al dpm/kg	79 ± 6	88 ± 6
^{22}Na dpm/kg	40 ± 6	$71 \pm 5^{*}$
K/U	1210 ± 40	1270 ± 30

* This value is high compared to the concentration measured by Keith *et al.* (1972), 46 ± 5, and may be in error because of small errors in the correction for thorium contribution to the ^{22}Na peak.

can be formed by simple mixing of Fra Mauro type debris and local mare material. It is implied, however, that the main exotic component at every site may be Fra Mauro-like in major element content and, therefore, if similar exotic debris has been delivered to the Fra Mauro site, it could not be recognized.

Exotic materials from mare sources should be recognizable. It was noted above that basaltic fragments are rare in Fra Mauro fines. Furthermore, fragments with basaltic textures are present in the annealed breccias and could represent the source of much of this debris. Exotic fragments with the composition of mare basalt appear to be rare. Nevertheless, attempts were made to identify such material by determining the compositions of a large number of glassy spheroids from the 63–125 μ size fraction (Fig. 2). The Fra Mauro spheroids, like the fines, are compositionally distinct from spherule populations of other Apollo samples. They have normative compositions that spread about the clustered norms of breccia and fine samples and an average composition that falls within the compositional field of Apollo 14 fines. They are thus provincial and were derived primarily by impact against local rocks. A few spheres have compositions that cluster at points distantly removed from the mode, but all these are not necessarily exotic. The group of spheroids enriched in pyroxene and olivine, for example, is identical to the most abundant type found in Apollo 15 Apennine Front samples and may have been inherited from Imbrium materials or produced locally from ultramafic clasts. A few spheroids, of the order of 5%, have compositions typical of Apollo 12 basalts and it is highly probable that these are mare derived contaminants. The same conclusions are reached by classifying the spherules on the basis of FeO and Al_2O_3 contents (Fig. 3). Approximately 15% of the spherules are high-Fe glasses, a smaller percentage are high-Al glasses and the

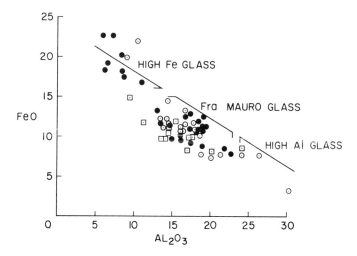

Fig. 3. Weight percent FeO as a function of Al_2O_3 for 63–125 μ glassy spheroids from samples 14163 and 14260. High-Fe glasses have FeO > 15%, high-Al glasses have Al_2O_3 > 23%. The remainder are termed Fra Mauro glasses. Filled circles indicate those spheroids with $K_2O + P_2O_5 < 0.5\%$; open circles have $1.0\% > K_2O + P_2O_5 > 0.5\%$ and squares, $K_2O + P_2O_5 > 1.0\%$.

remainder (Al < 23%, Fe < 15%) are here termed Fra Mauro types. The high-Fe glasses are divisible into two groups on the basis of FeO/MgO ratios, G and H of Table 3, and the Fra Mauro and high-Al glasses together are separable into six groups on the basis of FeO/MgO ratios and their $K_2O + P_2O_5$ contents (Fig. 4) with an average composition of each group as given in Table 3. The high-Fe glass of group G is comparable to that of groups 3A and 3B of Glass (1972) and has a composition similar to that of Apollo 12 basalts. Glasses of group H correspond to those of group 3C of Glass (1972). They are not considered to be mare derived as Glass (1971) indicates but rather are a group of mafic glasses of probable local origin as discussed above. Each of the groups A through F may represent specific source rock compositions, or they may be classification artifacts. The composition of group F or a combination of groups D and F, for example, is not unlike that of basalt 14310 (LSPET, 1971), and group C, or a combination of groups C and E, has a composition similar to that of many annealed breccias. Some, such as group A, appear to be unique, yet they are representative of no known rock types. Thus, some spherules in tightly clustering groups appear to reflect quite closely the compositions of apparent source rocks whereas others may not. The origins of glassy spheroids may be so complexly varied as to preclude simple interpretations of their compositions. All that can be said with certainty is that a few percent of the spherules were derived from mare terrain and that the remainder could all have been derived by impact against local annealed breccias and their contained igneous clasts. Scatter among their compositions may represent, in part, variations in compositions of the source rocks but it may also arise because of spherule forming processes; differences in volume of melt contributing to liquid jets, the effect of partial melting and the degree of mixing and fractionation prior to jetting, or even to formation of some groups by vapor condensation.

In summary, the Fra Mauro regolith is a highly reworked accumulation of debris derived primarily by impact comminution of annealed Fra Mauro breccias but with possible contributions of as much as 5% of the total debris from mare sources and an unknown amount from sources with mineralogy similar to that of the Fra Mauro terrain.

BRECCIAS

Studies of regolith fines have demonstrated that they were derived primarily by impact comminution of annealed breccias, the dominant rocks of the Fra Mauro

Table 3. Average compositions of glassy spheroid groups recognized in this study (wt. %)

	A	B	C	D	E	F	G	H
SiO_2	52.2	48.1	47.8	47.9	47.0	48.6	45.2	45.3
TiO_2	1.74	2.09	2.52	1.97	1.75	1.11	5.62	0.90
Al_2O_3	17.4	18.5	15.2	22.8	16.5	21.1	9.90	7.25
FeO	10.7	12.1	11.3	8.04	10.3	7.97	19.7	20.1
MgO	4.47	6.39	9.07	5.86	11.7	8.13	6.32	15.5
CaO	10.5	11.8	11.1	12.2	11.1	12.7	10.9	8.64
Na_2O	0.96	0.18	0.54	0.81	0.09	0.85	0.46	0.12
K_2O	0.88	0.06	0.62	0.40	0.05	0.19	0.24	0.06
P_2O_5	0.40	0.07	0.46	0.42	0.07	0.10	0.28	0.13
TOTAL	99.05	99.48	98.61	99.40	98.56	100.75	98.92	98.00
Number	5	8	20	5	7	4	4	5

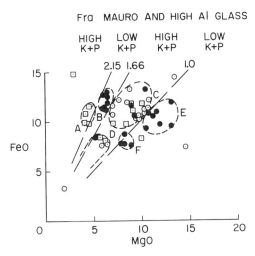

Fig. 4. Fra Mauro and high-Al glasses with weight percent FeO as a function of MgO. Filled circles indicate spheroids with $K_2O + P_2O_5 < 0.5\%$, open circles have $1.0\% > K_2O + P_2O_5 > 0.5\%$, and squares, $K_2O + P_2O_5 > 1.0\%$.

regolith. These breccias are so numerous on the ejecta apron of Cone Crater as to imply that they are the materials of the local substrate and therefore of the Fra Mauro formation. Mineralogical studies of the breccias were carried out, therefore, with the aim of characterizing the Fra Mauro formation and contributing information that can be used to interpret its origin. It must be remembered, however, that the samples studied have areas of but 1 to 3 sq cm. They may or may not be representative of the hand samples from which they were cut and they provide no information regarding the texture or structure of the rocks on a meter scale. Samples studied are those listed in Table 1.

The Apollo 14 breccias are a complex assemblage of clastic rocks. All contain monomineralic crystalline grains, light and dark microbreccia fragments of earlier generations, igneous and metamorphic rock clasts of various types, and devitrified glass. Undevitrified vesicular glass and glassy spheroids are present in some samples but are absent in most. Furthermore, glass-rich samples have no matrix and are characterized by a detrital texture with a continuum of grain sizes, whereas members of the glass-poor group have a true matrix that is even grained and finely crystalline. The matrix clearly has recrystallized with equilibration reactions between clasts and matrix apparent in many samples. Clearly, a two-fold subdivision into glass-rich and glass-poor types is indicated, or as they are termed here, regolith and annealed breccias.

Regolith breccias

Glass-rich breccias examined in this study include samples 14063, 14055, 14313, and 14318. Of these, all but sample 14063 have abundant glassy spherules ($>1.5\%$ of grains coarser than $20\,\mu$) and vesicular glass and have all the characteristics of regolith fines except that they have been lithified to varying degrees. None appear to have been annealed. They are characterized in section by high proportions of clastic glass of spherical and agglutinated forms, lack of matrix, and a continuum of grain sizes. Chemical and rare gas analyses (LSPET, 1971) reveal further, probably inherited, characteristics: high Ni, total C and solar-wind-implanted noble gas contents compared to annealed types. Furthermore, these samples contain far higher proportions of breccia clasts than do the annealed breccias, indicating that the various annealed rocks have not been produced from them by progressive thermal metamorphism as Warner (1972) has suggested. Rather, the glass-rich breccias have been formed by lithification of regolith deposits previously derived by comminution of annealed breccias. This group is here termed regolith breccias. One regolith breccia examined (14055) is extremely porous and corresponds to unshocked microbreccias of Chao *et al.* (1971). Samples 14313 and 14318, on the other hand, have numerous grain-to-grain contacts with all grains tightly packed, particularly those between large clasts, as if they have been mechanically compressed. Many glass grains containing delicate flow features are unbroken, however, and it is concluded that this group of regolith breccias are of the shock-compressed type of Chao *et al.* (1971), but from environments where peak pressure amplitudes have been low.

White rock breccia

Rock 14063 of the white rock group is a special type of breccia. Although it contains abundant angular glass, it has few characteristics of regolith breccias: no glassy or devitrified spheroids, no cindery aggregates, and no fragments of annealed breccias. Its texture is coarse grained and detrital with monomineralic grains, mainly plagioclase, that are exceedingly common. Other grains include pyroxene, opaque minerals, pink spinel, olivine, and alkali feldspar in order of decreasing abundance. Pink spinel is so common, 0.6% by volume, as to be characteristic. Another characterizing feature is purplish-brown glass, mostly homogeneous but commonly recrystallized containing mineral grain inclusions similar to those of the breccias as a whole. Igneous rock fragments are rare but include anorthositic and basaltic types. The rock does not appear to have been annealed. The unique textural and mineralogical features indicate that this rock is neither a regolith breccia nor an annealed breccia. Moreover, the U the Th contents of 14063 (LSPET, 1971) further attests to its uniqueness; they are less than 50% of the remarkably uniform U the Th contents of 14321 (Table 2) and other Fra Mauro breccias (LSPET, 1971). Its collection location at the rim of Cone Crater suggests an origin from deep within the substrate, perhaps from layers beneath the Fra Mauro formation. The white rock breccias are not ejecta from a buried, heavily reworked regolith layer, but must have come from a massive layer of ejecta. It is possible that such a layer could have been derived from pre-Fra Mauro rocks by impacts of high-speed projectiles produced by the Imbrian event, projectiles that were immediate precursors to the later arriving avalanche of Fra Mauro ejecta.

Annealed breccias

Most of the breccias examined (14006, 14066, 14304, 14306, 14311, 14320, and 14321) are characterized by a recrystallized matrix and are here termed annealed breccias. They are assumed to be samples of the Fra Mauro formation and studies of their textures and compositions were designed to contribute to an understanding of the history and origin of that body of rocks.

All annealed breccias are characterized by a recrystallized matrix, a rarity of glass other than fracture filings or rock coatings, an abundance of coarse monomineralic grains dominated by calcic plagioclase, the presence of breccia clasts of varied type of more than one generation and a paucity of rock clasts with primary igneous textures. The recrystallized matrix consists of a finely crystalline mosaic of interlocking low and moderate birefringent silicate grains, granular to platy and dendritic opaque minerals having reflected light characteristics of ilmenite and grains of Fe metal. The silicate grains are too small in all cases studied to permit positive identification using standard optical and electron microprobe techniques. However, the low birefringent grains probably are plagioclase and the others, pyroxene. The degree of recrystallization of matrix minerals, or of minerals of equivalent size in included clasts, and the extent to which clasts in the matrix have been modified vary among samples studied. The matrix of some breccias appears to have been recrystallized to an even grain size without observable reactions between grains and matrix and where opaque minerals, though widely disseminated, have no obvious growth habits, nor are they concentrated about clast margins. Many clasts of these breccias have granulitic and hornfelsic textures, but there is no evidence that they were acquired during matrix recrystallization.

Still other breccias (e.g., 14311) were observed wherein recrystallization has produced a matrix of the same grain size as that just described, but in which equilibration reactions of olivine and, to a lesser extent, orthopyroxene are common. All olivine clasts have undergone equilibration reactions. Large clasts characteristically have radiating plates and elongate blebs of ilmenite growing near the clast rim enclosed in a thin shell of granular pyroxene crystals with clustering fine-grained ilmenite at the immediate boundary and a wide enclosing zone of matrix in which opaque minerals have been depleted (Fig. 5). The same pattern is repeated for progressively smaller clasts except that the smaller the clast, the more complete the reaction with the smallest grains now consisting of ilmenite-eyed clinopyroxene aggregates set in a zone of matrix depleted in opaque minerals. Orthopyroxene crystals with the same but less thoroughly developed corona structures were observed: unmodified cores rimmed successively by ilmenite, pyroxene, fine-grained opaque minerals and opaque depleted matrix. Some large orthopyroxenes appear even less affected, however, having only an outer recrystallized rim of different optical orientation giving the clast a zoned appearance (Fig. 5). Rare pink spinel grains also have corona boundaries, one grain having an inner rim of low birefringent material (plagioclase?) surrounded by a shell of granular pyroxene all in a matrix area depleted in opaque minerals. Another pink spinel grain, color zoned with a more purplish rim, has a narrow border of low birefringent grains (plagioclase?) followed by a zone slightly enriched in fine-grained opaque minerals all in a matrix with no obvious opaque mineral depletion. Plagioclase grains appear little changed except in a few cases where they contain a thin outer rim of granular plagioclase, as if the clasts were merely sites of nucleation rather than grains undergoing equilibration reactions.

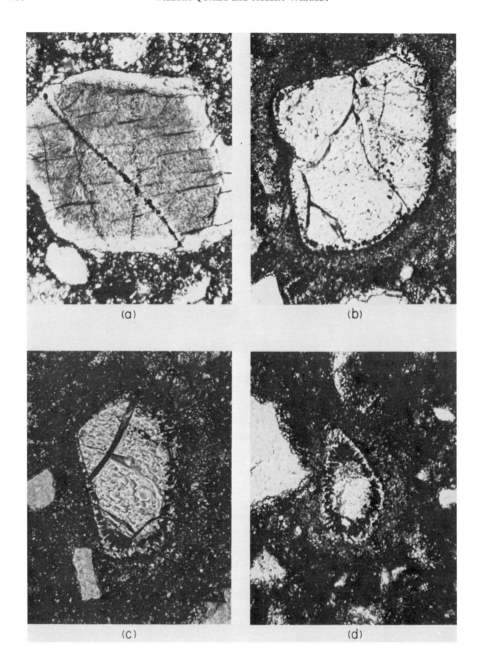

Fig. 5. Olivine and orthopyroxene clasts in breccia 14311 which have undergone equilibration reactions. (a) orthopyroxene; (b), (c), and (d) olivine. Note the narrow corona on the
large olivine clast and the typical structure of all the olivine grains: an olivine core, a zone of
radiating plates and blebs of ilmenite, a granular pyroxene rim and the enclosing matrix
depleted in opaque minerals. The large orthopyroxene pictured has only a recrystallized rim.

Small, fine-grained breccia clasts enclosed in dark matrix also have corona structures with zones identical to equilibrating olivine except that the cores consist of fine, matted plagioclase crystals. Additionally, a few breccia clasts lack the ilmenite enrichment and instead consist of felted plagioclase cores surrounded by successive zones of glass and granular pyroxene. In these cases there is no evidence of iron migration from surrounding matrix materials. Large light colored breccia clasts in these rocks show little evidence of matrix recrystallization and contain no minerals with corona structures, even in areas immediately adjacent to dark matrix where the features described above are common. The reasons for the absence of apparent recrystallization may be revealed by a study of their texture and mineralogy. The light clasts have no true matrix but rather consist of a continuum of particle sizes with an average size much coarser than that of the darker clasts or the common matrix. Minerals present are plagioclase with lesser amounts of clino- and ortho-pyroxene and rare large grains of ilmenite. Thus, the apparent lack of recrystallization could be due to either the much smaller surface areas or to the possibility that the mineral assemblage was stable in the annealing temperature regime. The latter explanation is preferred, and it is suggested that the assemblage is indicative of high-grade thermal metamorphism with temperatures probably comparable to those responsible for the pyroxene-hornfels facies in terrestrial metamorphism. The lack of attainment of equilibrium may be due in part to rapidly falling temperatures and in part to the rarity of water, a feature that must have retarded ionic diffusion. There is abundant evidence that diffusion did take place over distances approaching a millimeter, however, and the presence of goethite in at least one breccia (Agrell et al., 1972) indicates that some water was present. Furthermore, diffusion must have been aided by other vapor phases that undoubtedly were present, as indicated by the presence of drusy cavities in rock 14311 and in many annealed breccia fragments described by McKay et al. (1972).

The metamorphic features and the impactoclastic textures of the annealed breccias suggest that the rocks of the Fra Mauro formation have at one time been heated impact ejecta which annealed after deposition. It is impossible to prove conclusively that the metamorphism of the various rock fragments took place at the Fra Mauro site without studying large exposures of the rocks in outcrop. Dence et al. (1972), for example, believe that the metamorphism may have taken place elsewhere with subsequent ejection and deposition at the Fra Mauro site. There is indirect evidence, however, that suggests that the metamorphic reactions took place at the Fra Mauro site. Compressional wave velocities of the substrate rocks (Fra Mauro formation) are three times that of the regolith (Kovach and Watkins, 1972), and our studies of crater morphology suggest that this more cohesive substrate unit is widespread. Further-more, ejecta of craters large enough to penetrate the substrate unit consists primarily of annealed breccia, indicating that this is the rock of the substrate. We suggest, therefore, that the compressional wave velocities and the lateral continuity of the cohesive substrate breccia unit can be explained only if the impact debris of this unit were annealed in situ. Different degrees of annealing in different rock fragments could have been attained at different depths in the deposit or at different locations where the beds were of different thickness.

Source rocks

If the Fra Mauro breccias are a product of the Imbrian event, and evidence presented here is consistent with that hypothesis, the components of the breccias provide evidence of the composition and early history of that portion of the lunar crust. The presence of breccia clasts of more than one generation in the Fra Mauro rocks together with mineral clasts with hornfelsic textures, indicative of a higher grade of metamorphism than that attained in their latest annealing, demonstrates that surface rocks of this crustal region were already highly modified by impacting projectiles before the Imbrian event took place. The surface must have consisted of a thick, complex regolith containing layers of reworked debris of highly annealed ejecta. The coarse grain size of many of the monomineralic clasts in the Fra Mauro breccias, coarser than that of mare basalts, suggests further that many of them are ejecta from rock bodies that cooled more slowly than the basalts. They may have been derived from rocks at some depth. Furthermore, the high proportions of plutonic rock fragments among the crystalline clasts also indicates sources from within the crust. The types and relative percentages of igneous and high-grade metamorphic rocks observed in all sections studied are listed in Table 4. A study of their composition provides further insight into the nature of this early crustal terrain.

Basalt fragments, though rare, include a variety of textural and compositional types. Some are pigeonitic with textures and compositions similar to that of 14053 (Melson *et al.*, 1972). Others include augite basalts with anorthite phenocrysts (An_{90-93}) in a groundmass of intergranular augite ($En_{44.5}$, $Wo_{40.5}$). Still others have textures and composition similar to rock 14310 (Melson *et al.*, 1972; Ridley *et al.*, 1972). Most of the basaltic fragments appear remarkably fresh and unmetamorphosed, as if they have been involved in only the last annealing. It is suggested that most of these fragments were derived from flows from the immediate surface regions sampled by the Fra Mauro ejecta. The variety of types present suggests that several flows were sampled, however, and indicates further that "highland basalts" cannot be characterized by a single composition.

Anorthosite fragments are moderately abundant. They consist of coarse-grained anorthite ($An_{93.4-94.6}$) with gabbroic texture with many of the grains peripherally granulated and recrystallized to a fine mosaic. In two instances the fine-grained mosaic was found to contain both calcic plagioclase and alkali feldspar. Pale colored pyroxene or olivine generally is present, but only in small amounts. A similar percentage of coarse-grained ultramafic rocks was also observed. Monomineralic rocks

Table 4. Types and percentages of igneous and meta-igneous rock fragments in annealed breccias

Texture	Ophitic, subophitic, porphyritic	Gabbroic, peripherally granulated and recrystallized	Hypidiomorphic granular and granulitic	Allotriomorphic granular
Rock type	Augite basalt Pigeonite basalt Aluminous basalt	Anorthosite	Leucogabbro Norite Troctolite	Bronzitite Hyperthenite Dunite Clinopyroxenite
% of types	6	18	60	16
Average % in breccias	0.6	1.8	5.9	1.6

recognized include bronzitite (En_{82}, $Wo_{1.2}$), hypersthenite (En_{64}, Wo_4), and granulated dunite in fragments up to 2 mm long, containing numerous even-grained individuals. One very large (5 mm) fragment of shocked clinopyroxenite also was observed, containing 86% augite (En_{37}, $Wo_{41.5}$), 11.5% opaque minerals, mostly shocked ilmenite but also Fe metal and troilite and 2.5% anorthite in a pod-like inclusion. The relatively high percentage of the coarse-grained ultramafic rocks suggests that they are samples of rather extensive plutonic cumulates of the fractionating magmas that produced the anorthosites and norites of the highlands.

The most abundant crystalline rocks recognized include both hypidiomorphic granular and granulitic types with compositions of leucogabbro, norite, and troctolite. Few coarse-grained norites were observed. They consist of strongly sheared coarse-grained plagioclase enclosing large anhedral orthopyroxene grains with relatively coarse exsolution lamellae. Most of the fragments of noritic or troctolitic composition have granulitic textures and are meta-igneous rocks. They must have been intensively metamorphosed, but it is uncertain whether this took place within the crust or in the interiors of gigantic deposits of impact-heated ejecta. This suite of noritic rocks is so prominent among the crystalline rock clasts as to be considered characteristic of the pre-Imbrian terrain, regardless of their subsequent metamorphism.

Thus, the pre-Imbrian crust apparently consisted of a plutonic complex of feldspathic and ultramafic rocks, norites, anorthosites, and troctolites on one hand and pyroxenites and dunites on the other. The two rock families may have been derived by regional fractionation of a single, early parent liquid with flotational cumulates, the norites and anorthosites, contributing to the formation of the early lunar crust in this and perhaps other regions.

Acknowledgments — We appreciate the extensive technical assistance of George Polkowski, Jozef Erlichman, and Oneida Hammond. We are also grateful to Dr. Dieter Stöffler for supplying us with data that he accumulated in his studies of the breccias and for many stimulating discussions.

REFERENCES

Agrell S. O. Scoon J. H. Muir I. D. Long J. V. P. McConnell J. D. C. and Peckett A. (1970) Observations on the chemistry, mineralogy and petrology of some Apollo 11 lunar samples. *Proc. Apollo 11 Lunar Sci. Conf., Geochim. Cosmochim. Acta*, Suppl. 1, Vol. 1, pp. 93–128. Pergamon.

Agrell S. O. Scoon H. J. Long J. V. P. and Coles J. N. (1972) The occurrence of goethite in a microbreccia from the Fra Mauro formation (abstract). In *Lunar Science — III* (editor C. Watkins), pp. 7–9, Lunar Science Institute Contr. No. 88.

Chao E. C. T. Boreman J. A. and Desborough G. A. (1971) The petrology of unshocked and shocked Apollo 11 and Apollo 12 microbreccias. *Proc. Second Lunar Sci. Conf., Geochim. Cosmochim. Acta*, Suppl. 2, Vol. 1, pp. 797–816. M.I.T. Press.

Cuttitta F. Rose H. J. Jr. Annel C. S. Carron M. K. Christian R. P. Dwornik E. J. Greenland L. P. Helz A. W. and Legon D. T. Jr. (1971) Elemental composition of some Apollo 12 lunar rocks and soils. *Proc. Second Lunar Sci. Conf., Geochim. Cosmochim. Acta*, Suppl. 2, Vol. 2, pp. 1217–1229. M.I.T. Press.

Dence M. R. Plant A. G. and Traill R. J. (1972) Impact-generated shock and thermal metamorphism in Fra Mauro lunar samples (abstract). In *Lunar Science — III* (editor C. Watkins), pp. 174–176, Lunar Science Institute Contr. No. 88.

Glass B. P. (1972) Apollo 14 glasses (abstract). In *Lunar Science — III* (editor C. Watkins), pp. 312–314, Lunar Science Institute Contr. No. 88.

Keith J. E. Clark R. S. and Richardson K. A. (1972) Gamma ray measurements of Apollo 12, 14 and 15 lunar samples (abstract). In *Lunar Science—III* (editor C. Watkins), pp. 446–448, Lunar Science Institute Contr. No. 88.

Kovach R. L. and Watkins J. S. (1972) The near-surface velocity structure of the moon (abstract). In *Lunar Science—III* (editor C. Watkins), pp. 461–462, Lunar Science Institute Contr. No. 88.

LSPET (1971) (Lunar Sample Preliminary Examination Team) Preliminary examination of lunar samples from Apollo 14. *Science* **173**, 681–693.

Maxwell J. A. Peck L. C. and Wiik A. B. (1970) Chemical composition of Apollo 11 lunar samples 10017, 10020, 10072 and 10084. *Proc. Apollo 11 Lunar Sci. Conf., Geochim. Cosmochim. Acta*, Suppl. 1, Vol. 2, pp. 1369–1374. Pergamon.

Melson W. G. Mason B. Nelen J. and Jacobson S. (1972) Apollo 14 basaltic rocks (abstract). In *Lunar Science—III* (editor C. Watkins), pp. 535–536, Lunar Science Institute Contr. No. 88.

Quaide W. Oberbeck V. Bunch T. and Polkowski G. (1971) Investigations of the natural history of the regolith at the Apollo 12 site. *Proc. Second Lunar Sci. Conf., Geochim. Cosmochim. Acta*, Suppl. 2, Vol. 1, pp. 701–718. M.I.T. Press.

Ridley W. I. Williams R. J. Brett R. Takeda H. and Brown R. W. (1972) Petrology of lunar basalt 14310 (abstract). In *Lunar Science—III* (editor C. Watkins), pp. 648–650, Lunar Science Institute Contr. No. 88.

Rose H. J. Cuttitta F. Dwornik E. J. Carron M. K. Christian R. P. Lindsay J. R. Legon D. T. and Larson R. R. (1970) Semimicro x-ray fluorescence analysis of lunar samples. *Proc. Apollo 11 Lunar Sci. Conf., Geochim. Cosmochim. Acta*, Suppl. 1, Vol. 2, pp. 1493–1497. Pergamon.

Vinogradov A. D. (1971) Preliminary data on lunar ground brought to Earth by automatic probe "Luna-16." *Proc. Second Lunar Sci. Conf., Geochim. Cosmochim. Acta*, Suppl. 2, Vol. 1, pp. 1–16. M.I.T. Press.

Warner J. L. (1972) Apollo 14 breccias: Metamorphic origin and classification (abstract). In *Lunar Science—III* (editor C. Watkins), pp. 782–784, Lunar Science Institute Contr. No. 88.

Wrigley R. C. (1971) Some cosmogenic and primordial radionuclides in Apollo 12 lunar surface material. *Proc. Second Lunar Sci. Conf., Geochim. Cosmochim. Acta*, Suppl. 2, Vol. 2, pp. 1791–1796. M.I.T. Press.

Proceedings of the Third Lunar Science Conference
(Supplement 3, *Geochimica et Cosmochimica Acta*)
Vol. 1, pp. 785–796
The M.I.T. Press, 1972

Mineralogy, petrology, and chemical composition of lunar samples 15085, 15256, 15271, 15471, 15475, 15476, 15535, 15555, and 15556

Brian Mason, E. Jarosewich, W. G. Melson

Smithsonian Institution, Washington, D.C. 20560

G. Thompson

Woods Hole Oceanographic Institution, Woods Hole, Massachusetts 02543

Abstract—Chemical analyses have been made of seven Apollo 15 rocks and one sample of Apollo 15 fines. The rocks, all of mare basalt composition, range from olivine-normative to slightly quartz-normative compositions, and vary considerably in grain size and texture. The principal minerals in all of them are plagioclase (average An_{90}) and pyroxenes (pigeonite and augite). Most of the rocks contain some olivine; the quartz-normative ones contain no olivine, but have accessory tridymite and/or cristobalite. The basalts from the edge of Hadley Rille (15535, 15555, 15556) appear related to one another by olivine fractionation, whereas those from Dune Crater (15475, 15476) represent a compositionally distinct flow or flows. A rock from the Apennine Front (15256) compositionally is similar to the mare basalts, but is a heterogeneous welded breccia, presumably formed by impact on the mare surface and ejected to its Front site. The major-element composition of the fines at Dune Crater (15471) can be closely approximated by a mixture of 70% local basalt and 30% troctolite, suggesting that troctolite may be an important constituent of the pre-Imbrian crust.

Introduction

We have examined samples of the Apollo 15 collections as listed in the title. All but three of them are mare basalts of rather similar composition; 15271 and 15471 are < 1 mm fines, and 15256 is a thoroughly welded impact breccia. Chemical analyses of all of them except 15271 are given in Table 1, along with the norms calculated therefrom.

Approximately 0.5 g of sample was taken for each of the chemical analyses. Silica, MgO, CaO, total iron as FeO, TiO_2, and MnO were determined according to the general procedure for silicate analysis (Peck, 1964). Aluminum was determined gravimetrically by precipitation with 8-hydroxyquinoline on an aliquot from the R_2O_3 solution after removal of interfering ions by means of the mercury electrode. Sodium and potassium were determined by flame photometry, and phosphorus and chromium colorimetrically with molybdenum blue (Boltz, 1958) and diphenyl-carbazide (Sandell, 1954), respectively. Summations uniformly are somewhat greater than 100.00, except for the < 325 mesh fraction of 15471,27; this feature has been noted by other analysts of lunar rocks. Some of this excess can be ascribed to the presence of metallic iron, whereas all iron is reported as FeO. It is also possible that Ti and Cr may be present in lower valence states than reported. It was noted that ignition of the samples in air results in a weight gain, evidently the result of combination with oxygen. Sulfur was not determined, in order to economize on the consumption of material; analyses reported by LSPET (1972) show about 0.07% S in the Apollo 15 rocks, equivalent to approximately 0.2% troilite.

Table 1. Chemical analyses and norms of Apollo 15 samples (E. Jarosewich, analyst)

	15085,34	15256,10	15471,27	15471,27 (<325 mesh)	15475,33	15476,5	15535,32	15555,157	15556,26
SiO$_2$	46.39	45.32	46.43	46.22	48.32	48.46	44.46	44.75	46.18
TiO$_2$	3.07	2.54	1.61	1.65	1.57	1.75	2.19	2.07	2.64
Al$_2$O$_3$	5.79	9.20	13.43	15.35	9.23	9.54	8.68	8.67	9.85
Cr$_2$O$_3$	0.67	0.31	0.35	0.28	0.66	0.43	0.57	0.61	0.77
FeO	26.75	22.51	16.20	14.51	20.17	20.76	23.80	23.40	21.70
MnO	0.37	0.35	0.26	0.23	0.31	0.28	0.33	0.30	0.32
MgO	8.20	9.45	11.15	10.11	9.54	8.69	11.27	11.48	8.03
CaO	9.12	10.17	10.58	10.77	10.33	10.50	9.20	9.14	10.72
Na$_2$O	0.21	0.30	0.37	0.36	0.27	0.28	0.28	0.24	0.30
K$_2$O	0.07	0.12	0.17	0.18	0.05	0.07	0.04	0.05	0.09
P$_2$O$_5$	0.09	0.07	0.11	0.12	0.05	0.05	0.06	0.05	0.07
Total	100.73	100.34	100.66	99.78	100.50	100.81	100.88	100.76	100.67
Q	—	—	—	—	0.05	1.01	—	—	—
Or	0.41	0.71	1.00	1.06	0.30	0.41	0.24	0.30	0.53
Ab	1.78	2.54	3.13	3.05	2.28	2.37	2.37	2.03	2.54
An	14.65	23.40	34.48	39.74	23.83	24.57	22.31	22.43	25.27
Di	25.48	22.35	14.30	10.67	22.73	22.88	19.21	18.89	23.16
Hy	50.67	34.49	32.65	34.31	47.79	45.50	31.33	33.32	40.36
Ol	0.72	11.41	11.27	7.13	—	—	20.29	18.84	2.51
Cr	0.99	0.46	0.52	0.41	0.97	0.63	0.84	0.90	1.13
Il	5.83	4.82	3.06	3.13	2.98	3.32	4.16	3.93	5.01
Ap	0.21	0.16	0.25	0.28	0.12	0.12	0.14	0.12	0.16
Fem*	0.62	0.55	0.43	0.42	0.53	0.55	0.52	0.51	0.57

*Fem = FeO/(FeO + MgO) mole ratio in normative silicates.

Modal analyses were made of most of the rocks by the point-counting technique. Generally the modal analyses were in good agreement with the normative composition, which is to be expected in view of the close correspondence between normative and observed mineralogy. Where discrepancies were found, they could be ascribed to sampling problems, as discussed in the following section. The close correspondence between normative and observed mineralogy is documented by the occurrence of tridymite and cristobalite. Of the two rocks with free SiO$_2$ in the norm, 15475 contains about 0.5% tridymite and 15476 about 0.7% cristobalite. The only other rock containing these minerals is 15085, which contains about 0.7% tridymite and 0.4% cristobalite; the norm shows slight undersaturation in SiO$_2$, equivalent to 0.7% olivine, but the discrepancy is small and well within possible analytical and sampling errors.

Trace element analyses on aliquots of the previously analysed samples are reported in Table 2. The technique used is described by Thompson and Bankston (1969). These analyses show that the fines (15471), compared to the rocks, are enriched in Ba, Ni, Pb, Rb, Sr, Y, and Zr, and are depleted in Cr and V; these trends are most marked in the <325 mesh fraction. The enrichment of Ni in the fines is consistent with a meteoritic increment in the material; a 1% increment of carbonaceous chondrite composition will increase the Ni concentration by approximately 100 ppm, but the Co concentration will increase by only 5 ppm and thus not readily be detected. Trace element variations within the crystalline rocks are not great; the porphyritic olivine basalts show higher concentrations of Ni and V and lower concentrations of Ba than the porphyritic clinopyroxene basalts. Rock 15476 shows

Table 2. Trace element analyses by direct-reading emission spectrometry, in ppm; precision and accuracy ±5–10%. Analyses are mean of four separate burns except 15471, which are mean of two separate burns. Ag < 1, Cd, Bi < 2, Zn < 10 ppm in all samples. (G. Thompson, analyst.)

	15085,34	15256,10	15471,27	15471,27 (<325 mesh)	15475,33	15476,5	15535,32	15555,157	15556,26
Rb	<5	<5	5	5	<5	<5	<5	<5	<5
Ba	87	41	120	150	47	63	38	30	50
Sr	92	88	115	125	96	98	83	83	102
Pb	2	<2	6	4	<2	<2	<2	<2	<2
Cu	18	11	10	19	6	7	8	17	10
Y	54	48	55	65	37	42	42	47	50
Zr	150	100	205	265	65	105	85	60	100
Co	49	46	40	37	56	37	52	66	46
Ni	45	60	160	225	50	27	70	70	65
Li	8	8	9	9	8	8	7	7	9
V	110	135	100	75	130	90	140	145	165
Cr	4600	4200	3600	3300	4500	3600	4800	4500	5200
Ga	5	4	4	2	3	3	3	3	5
B	5	3	2	1	4	2	4	3	3

significantly lower concentrations of Co, Cr, Ni, and V. The metabasalt 15256 does not show any anomalous concentrations of these trace elements.

SAMPLING PROBLEMS

We attempted to assess how representative our analyses are, in view of the small size of our subsample (0.5 g) and large grain size of some of the samples, especially 15085. Even a cursory appraisal of sampling statistics shows that such small sub-samples cannot be even roughly representative of the whole for rocks like 15085. The size of subsamples required for representative analysis is proportional to the average volume of crystal (cube of the average grain size), and subsample size becomes prohibitively large (> 10 g) where average grain sizes exceed 2 mm. Average grain size for 15085 is greater than 3 mm, and our analysis thus has very low probability of being representative of the whole. The problem is less acute for the other samples, but our analyses of 15475, 15476, 15535, and 15555 must still be viewed with caution. On the other hand, our analyses of 15256 and 15556 are based on reasonably sized subsamples in view of their fine grain size (average <0.5 mm). For samples less than 0.5 mm average grain size, subsamples on the order of 0.10 grams could give reasonably meaningful results. An additional problem arises for 15256, a welded breccia, in that it contains mineral and rock clasts that commonly exceed 1 mm in maximum dimension, that is, it is not a petrologically homogeneous sample.

Some of the problems encountered in using nonrepresentative analyses can be well-demonstrated by comparing our analysis of 15085 with the others in Table 1 and with previous analyses of lunar basaltic rocks. Our analysis of 15085 is much lower in Al_2O_3 (5.79%) than practically all other lunar mare basalts. This is shown in Fig. 1, where we find a good direct correlation between the Fe/(Fe + Mg) ratio and total normative feldspar in most Apollo 12 and 15 mare basalts. Rock 15085 falls far off this trend for other Apollo 15 mare basalts. The analysis of 15076, also

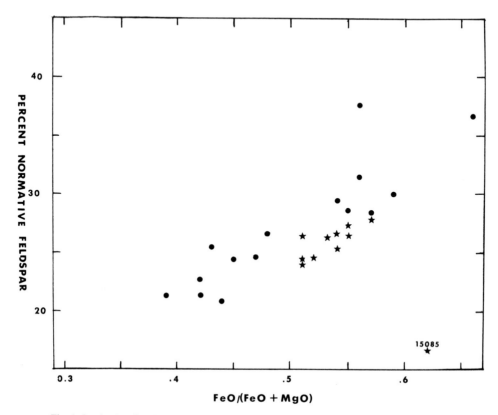

Fig. 1. In the Apollo 12 and 15 mare basalts, there is a rough direct correlation between Fe/(Fe + Mg) and total normative feldspar (An + Ab + Or). Some of the scatter in these data, like our result for 15085, is a result of sampling problems for coarse-grained rocks. Most samples that fall very far from this trend have average grain sizes greater than 1 mm. Stars: Apollo 15 analyses from LSPET (1972) and Jarosewich (this paper). Circles: Apollo 12 mare basalts.

a coarse-grained rock, departs significantly from this trend (LSPET, 1972), but not as much as our 15085 analysis.

There are significant variations in the major element analyses so far reported for rock 15555 (Table 3). The analysis by D. F. Nava (Schnetzler *et al.*, 1972) departs widely from the others for a number of elements, especially Al, Mg, and Ti. Sampling errors possibly are evident in the Fe/(Mg + Fe) ratio, reflecting variation in the olivine to pyroxene ratio, and in the up to 2 percent variation in alumina content, possibly reflecting differences in plagioclase contents of the subsamples, all of which are 0.5 g or less.

Much more meaningful analyses would be obtainable if subsamples were sized so as to be reasonably representative of the whole. This size would be prohibitively large for very coarse-grained samples, but would be reasonable for most mare basalts, normally less than 10 grams. Such a subsample ideally should be taken and powdered by the NASA curatorial facility and smaller aliquots of this could then be

Table 3. Different analyses* of 15555 (n.d. = not determined)

	1	2	3	4
SiO$_2$	44.75	43.82	44.24	45.0
TiO$_2$	2.07	2.63	2.26	1.60
Al$_2$O$_3$	8.67	7.45	8.48	9.37
Cr$_2$O$_3$	0.61	0.59	0.70	0.47
FeO	23.40	24.58	22.47	21.18
MnO	0.30	0.32	0.29	0.26
MgO	11.48	10.96	11.19	12.22
CaO	9.14	9.22	9.45	9.25
Na$_2$O	0.24	0.24	0.24	0.26
K$_2$O	0.05	0.04	0.03	0.03
P$_2$O$_5$	0.05	0.07	0.06	0.07
S	n.d.	0.06	0.05	n.d.
Fe/(Fe + Mg)	0.51	0.53	0.51	0.48

*Sources.

1. E. Jarosewich, this paper (0.5 g).
2. Chappell et al. (1972) (0.36 g).
3. LSPET (1972) (0.28 g).
4. Schnetzler et al. (1972) (0.5 g); D. F. Nava analyst.

distributed, rather than small rock chips, as has often been done. Analyses of small chips (< 1 g) of coarse-grained (> 1 mm) rocks give a large spread that does more to obscure than to reveal significant variations in lunar magmas.

PETROLOGY OF APOLLO 15 MARE BASALTS

In the following discussion, our analyses of 15256, 15475, 15476, 15535, 15555, and 15556 will be assumed to be representative of the whole rocks. The close agreement of our analyses of 15555, 15256, and 15471 (fines) with those of LSPET (1972) suggests that, in regard to sample size and analytical methods, both sets of analyses are meaningful. The samples considered include also 15016, 15058, 15076, and 15499, of which analyses from LSPET (1972) are available, bringing the total number of Apollo 15 mare basalt samples considered here to ten.

The samples from the Hadley Rille rim (Station 9A, samples 15535, 15555, 15556) appear related to one another by fractionation, with olivine as the dominant fractionating phase (Fig. 2). The samples from Dune Crater (15475, 15476, 15499) fall close to one another and off the Rille trend, presumably reflecting a different magmatic parent, and thus represent a flow or flows compositionally distinct from those sampled at the Rille. The basalt from the ALSEP site, 15058 (LSPET, 1972) appears to belong to the Dune Crater group, whereas 15016, from the mare surface, appears closer compositionally to the Rille basalts of Station 9A. The division into two major compositional groups correlates with topography. The lower FeO group (Fig. 2) corresponds to mare basalts nearer the Apennine Front and at slightly higher elevations than those from the ALSEP site and from the Rille.

The two analyses of basalts from Elbow Crater, 15085 and 15076, are of coarse-grained samples. One, 15085, as previously discussed, was of a chip too small to give meaningful results, and the other, 15076, also may have been too small to give meaningful results.

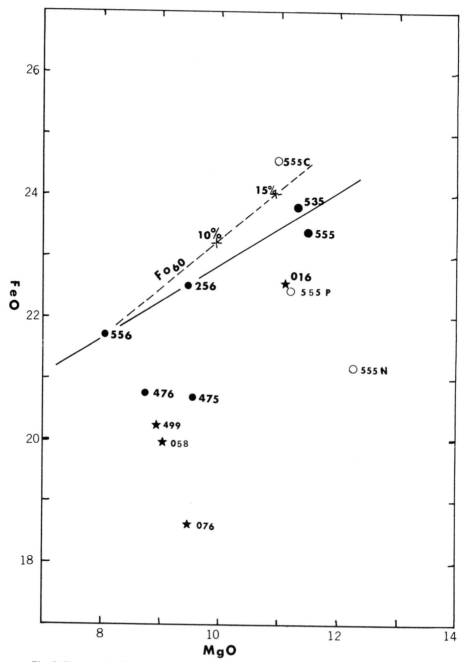

Fig. 2. The Apollo 15 mare basalts fall into at least two major groups. One group, those from near Hadley Rille (15535, 15555, and 15556) falls along a trend approximating an olivine fractionation trend, shown for addition of Fo_{60} to rock 15556. The other group, from Dune Crater and the ALSEP site, contain much less olivine, and cannot be simply related to those from Rille Station by olivine fractionation. Some of the divergence in 15555 analyses also is shown: 15555C from Chappell *et al.* (1972), 15555P from LSPET (1972), 15555N from Schnetzler *et al.* (1972).

The Apollo 15 mare basalts more closely resemble those of Apollo 12 than those of Apollo 11. They have low TiO_2 content and, like the Apollo 12 basalts, show evidences of olivine fractionation. So far, however, the analyses show much less compositional spread than those of Apollo 12 (e.g., Fig. 1). The age so far available, around 3.36 b.y. for 15555 (Podosek *et al.*, 1972), falls within the range of those for Apollo 12 basalts, suggesting that the volcanism at the Apollo 12 site was part of a very broad zone of volcanism, and one that led to the filling of Mare Imbrium as well as adjacent low areas.

Compositional trends are best shown by the results of analyses done by the same analyst, a not unexpected result and one noted for Apollo 12 samples by James and Wright (1972). Figure 3 shows this for our analyses and those of LSPET (1972). Both show an expected trend of increasing normative olivine with decreasing silica content, but our values are systematically displaced toward higher silica contents.

Sample 15555 has been studied extensively by us, whereas the others, some newly received, have not been examined in detail. Figure 4 shows the complex, and in places, oscillatory zoning in one of the large pigeonite-cored pyroxenes, presumably reflecting relative movement of crystal and magma during pyroxene growth. Hypothetically, movement of pyroxene into nearby magma already depleted by pyroxene growth accentuated zoning, giving a sharp increase in the Fe/Mg ratio. Movement in magma depleted in calcium by growth of plagioclase causes decrease in Ca content of the pyroxene at constant Mg. All these changes require some undercooling and slow diffusion rates of Ca, Mg, and Fe in the magma.

The olivine in 15555 commonly is enclosed in plagioclase, and is unusually iron-rich compared to that of other mare basalts, even for its most magnesian compositions (Fo_{58}). Coexisting pigeonite-cored pyroxene (approximately $Fs_{32}Wo_{11}En_{57}$) are too magnesian to have crystallized in equilibrium with such olivine. Chappell *et al.* (1972) note that a composition like 15555 would first crystallize olivine near Fo_{72}, and infer that the more iron-rich olivine reflects accumulation of about 20% olivine at a temperature around 1140°C. Figure 2 shows that addition of about 15% olivine near Fo_{60} to rock 15556 gives a reasonably good fit to the composition of rock 15555.

Rock 15256 is heterogeneous, and contains thoroughly welded mineral and rock clasts (Fig. 5). Large olivine clasts occur in some portions and typically have compositions around Fo_{65}. Some basaltic clasts are vitrophyric, with abundant olivine phenocrysts and microlites, typically ranging from Fo_{65} to Fo_{50}. Although texturally complex, this sample shows distinct mare basalt major element composition (Table 1 and Figs. 2 and 3) and falls on compositional trends established for the Rille basalts from Station 9A. We infer that it was ejected from the mare surface to its Front site by impact.

The trace elements contents (Table 2) can be compared with the proposed fractionation in the Rille series, in which we have included 15256 because of its close similarity in composition. If we assume that 15535 and 15555 are related to 15556 by the addition of 16% olivine, and 15256 by the addition of 7% olivine (Fig. 2), and further that olivine contains none of the incompatible elements, Ba, Sr, Zr, and K, we obtain the following model results, in ppm:

Ba	determined	50	38	30	41
	predicted	—	42	42	47
Sr	determined	102	83	83	88
	predicted	—	70	70	95
Zr	determined	100	85	60	100
	predicted	—	84	84	93
K	determined	900	400	500	1200
	predicted	—	760	760	840

There are some differences between predicted and determined values, although there is a general trend toward increasing incompatible element content with decreasing olivine content. The differences between the model and determined values may indicate (1) that the model is incorrect, that is, that each rock type represents an independent magma, or (2) the precision of the trace element determinations is inadequate for a comparison of this sort.

APOLLO 15 FINES

Two samples of < 1 mm fines have been studied, 15271 from Station 6 on the Apennine Front and 15471 from Station 4 near the rim of Dune Crater. Two analyses of 15471 were made, one of an "as received" aliquot, and one of a < 325 mesh sieve fraction (< 0.043 mm grain size). These analyses (Table 1) are quite similar, as might be expected. The principal difference is a somewhat higher Al_2O_3 content in the < 325 mesh fraction, giving a higher normative anorthite content. We have noted a similar enrichment in plagioclase content of the < 325 mesh fraction of Apollo 11, 12, and 14 fines.

Significant differences were observed between the 15271 material and the 15471 material. Although the bulk composition of each is essentially basaltic, 15471 appears to be produced largely from local mare basalt with some admixture of feldspar-rich non-mare material, whereas 15271 represents a mixture of mare and non-mare material in subequal amounts. This interpretation is supported by our preliminary examination of coarser fines from these same locations.

Both samples consist largely of black cindery glass particles with minor amounts of mineral grains, and a few rock fragments in the coarsest (0.5–1 mm) fraction. A hand magnet provides a quick, if crude, separation of the glassy material, because it contains minute metal particles. The mineral fraction consists largely of plagioclase, pyroxenes, and olivine; in 15271 the plagioclase approximately is equal in abundance to the ferromagnesian minerals, whereas in 15471 the latter are considerably more abundant than plagioclase. The compositional ranges (determined by microprobe analysis) of the ferromagnesian minerals also are distinctive. In 15471 the ferromagnesian minerals consist of approximately 23% olivine (Fa_{28-58}, av. Fa_{39}), 3% orthopyroxene, 30% pigeonite, 21% subcalcic augite, 20% augite, and 3% ferrohedenbergite plus pyroxferroite. In 15271 they consist of 24% olivine in two distinct compositional ranges (about 6% Fa_{10-20}, 18% Fa_{44-50}), 16% orthopyroxene (Fs_{12-25}), 30% pigeonite, 14% subcalcic augite, 15% augite, and 1% ferrohedenbergite plus pyroxferroite. The magnesium-rich olivine and orthopyroxene in 15271 distinguish it clearly from 15471, and probably are derived from non-mare rocks. We have found these minerals in feldspar-rich basalt fragments in coarse fines 15272

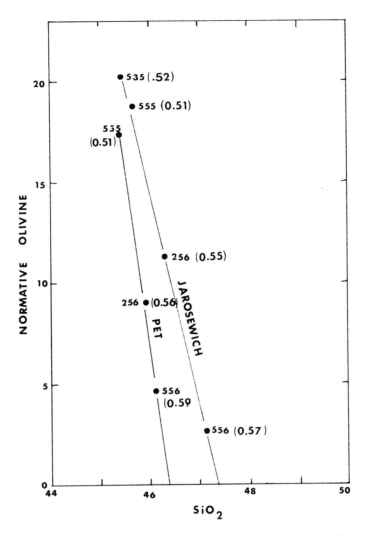

Fig. 3. Both LSPET (1972) and our major element analyses show a well-developed trend for normative olivine versus silica for the Rille basalts (Station 9A), but are systematically different in SiO₂ content. Numbers in parentheses are Fe/(Fe + Mg) in normative silicate, that is, with FeO in FeTiO₃ and FeCr₂O₄ subtracted. As expected in an olivine fractionation series, Fe/(Fe + Mg) decreases with increasing olivine content.

and 15273. Rare grains of red spinel were found in both 15271 and 15471; these probably have been derived from troctolitic rocks of non-mare origin (Mason *et al.*, 1972).

From our data it is possible to make a comparison between soil and rock compositions at Dune Crater (Station 4). The analyses of 15471 (< 1 mm fines) and rocks 15475 and 15476 are given in Table 1. These three samples were collected at the same place, described as approximately 28 m SSE of the rim crest of Dune Crater.

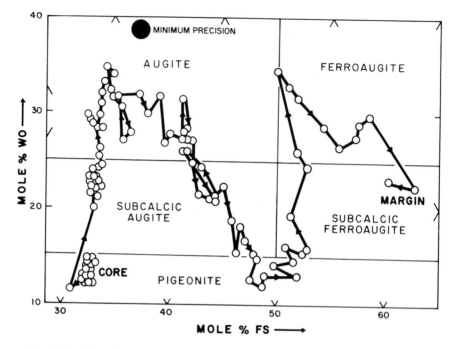

Fig. 4. Complex zoning, probably reflecting crystal growth during relative movement of crystal and magma either during turbulent flow or by crystal settling, is well-shown by large pyroxenes in 15555,29. Portion of the pyroxene ternary wollastonite (Wo), ferrosilite (Fs), enstatite (En).

The compositions of 15475 and 15476 essentially are identical, although the rocks differ somewhat in texture. The chemical composition of 15471 differs most notably from 15475 and 15476 in being higher in Al_2O_3 and MgO and lower in FeO and SiO_2. These facts can be accounted for by a higher content of plagioclase in the fines than in the local rocks, and the addition of a material with a relatively low FeO/(FeO + MgO) ratio and a low silica content—magnesian olivine is the obvious candidate. As far as the major components are concerned, a good match for 15471 is obtained by diluting the local rocks with 20% plagioclase (An_{90}) and 10% magnesium-rich olivine (Fa_{10}), as follows:

	SiO_2	Al_2O_3	FeO	MgO	CaO	Na_2O
15476 (70%)	33.9	6.68	14.5	6.08	7.34	0.20
An_{90} (20%)	8.8	6.95	—	—	3.64	0.20
Fa_{10} (10%)	4.1	—	1.0	4.90	—	—
Sum	46.8	13.63	15.5	10.98	10.98	0.40
15471	46.4	13.43	16.2	11.15	10.58	0.37

Thus, the composition of a mixture of 30% troctolite and 70% local rock approximates closely the composition of the <1 mm fines. Although we did not find any olivine as magnesium-rich as Fa_{10} in our microprobe analyses of the 15471 minerals,

Fig. 5. Rock 15256 is a texturally heterogeneous, shock-altered rock, probably ejected to the Apennine Front from a mare impact site. Some clasts, upper left, possess a well-developed vitrophyric texture containing phenocrysts or microlites of olivine and less commonly of pyroxene. Maximum width of section (15256,20) is 12 mm. Dark areas are devitrified glass, with the glass probably produced by the impact event.

these analyses were made on the coarse grains (60–100 mesh) and thus may not be representative of the total mineral content.

We have made a preliminary examination of the coarse fines at this locality but have not so far identified any troctolite fragments. However, the coarse fines contain numerous complex breccia fragments that may well incorporate such material. In addition, as mentioned above, 15471 contains sporadic grains of pink spinel, which we have suggested originate in a troctolitic rock. We therefore believe that troctolites may be significant components of the pre-Imbrian crust.

Acknowledgments—We wish to thank Mr. J. Nelen and Miss S. Jacobsen for microprobe analyses of pyroxenes and olivines in the Apollo 15 fines. Our work has been supported by a grant (NGR 09–015–146) from the National Aeronautics and Space Administration.

REFERENCES

Boltz D. L. (1958) *Colorimetric Determination of Nonmetals.* Interscience.
Chappell B. W. Compston W. Green D. H. and Ware N. G. (1972) Chemistry, geochronology, and petrogenesis of lunar sample 15555. *Science* **175**, 415–416.

James O. and Wright T. (1972) Chemical fractionation of Apollo 12 basalts. *Bull. Geol. Soc. Amer.* **83**, in press.

LSPET (1972) (Lunar Sample Preliminary Examination Team) The Apollo 15 lunar samples: a preliminary description. *Science* **175**, 363–375.

Mason B. Melson W. G. and Nelen J. (1972) Spinel and hornblende in Apollo 14 fines (abstract). In *Lunar Science—III* (editor C. Watkins), pp. 512–514, Lunar Science Institute Contr. No. 88.

Peck L. C. (1964) Systematic analysis of silicates. *U.S. Geol. Surv. Bull. 1170.*

Podosek F. A. Huneke J. C. and Wasserburg G. J. (1972) Gas-retention and cosmic-ray exposure ages of lunar rock 15555. *Science* **175**, 423–425.

Sandell E. B. (1954) *Colorimetric Determination of Trace Metals.* Interscience.

Schnetzler C. C. Philpotts J. A. Nava D. F. Schuhmann S. and Thomas H. H. (1972) Geochemistry of Apollo 15 basalt 15555 and soil 15531. *Science* **175**, 426–428.

Thompson G. and Bankston D. C. (1969) A technique for trace element analysis of powdered materials using the d.c. arc and photoelectric spectrometry. *Spectrochim. Acta* **24**, ser. B. 335–350.

Proceedings of the Third Lunar Science Conference
(Supplement 3, *Geochimica et Cosmochimica Acta*)
Vol. 1, pp. 797–817
The M.I.T. Press, 1972

Experimental petrology and origin of Fra Mauro rocks and soil

DAVID WALKER, JOHN LONGHI, and JAMES FRED HAYS

Department of Geological Sciences, Harvard University,
Cambridge, Massachusetts 02138

Abstract—Melting experiments over the pressure range 0 to 20 kbar on Apollo 14 igneous rocks 14310 and 14072, and on comprehensive fines 14259, demonstrate the following:

(1) Low pressure crystallization of rocks 14310 and 14072 proceeds as predicted from the textural realtionships displayed by thin sections of these rocks. The mineralogy and textures of these rocks are the result of near-surface crystallization.

(2) The chemical compositions of these lunar samples all show special relationships to multiply saturated liquids in the system anorthite-forsterite-fayalite-silica at low pressure.

(3) Partial melting of a lunar crust consisting largely of plagioclase, low calcium pyroxene, and olivine, followed by crystal fractionation at the lunar surface, is a satisfactory mechanism for the production of the igneous rocks and soil glasses sampled by Apollo 14. The KREEP component of other lunar soils may have a similar origin.

INTRODUCTION

THE APOLLO 14 MISSION to the Fra Mauro hills returned many breccias that have geological significance in establishing, by close-up observation, the clastic nature of the Fra Mauro formation that previously had been inferred from remote observation. Few of the rocks returned have textures that reasonably can be interpreted as resulting from direct crystallization of a silicate melt; however, many of the breccia clasts have igneous textures. Rock 14310 is a 3.4 kg rock that has been interpreted as igneous by other observers. Rock 14072 is a 45 g fragment that also appears to be igneous. Interpretation of the magmatic history of these rocks, apart from their subsequent mechanical adventures, may lead to a better understanding of lunar volcanic processes and the nature of the pre-Imbrian lunar crust, and may ultimately help to place constraints on the constitution and origin of the moon.

Our petrographic observations of polished thin sections 14310,30 and 14072,16 (Longhi *et al.*, 1972), coupled with phase equilibrium experiments on a homogenized powder sample 14310,138; a powdered rock chip 14072,3; and a sample of comprehensive fines 14259,85, have led to a reconstruction of the near-surface volcanism that produced the Apollo 14 igneous rocks. Consideration of the chemical compositions and phase relations of these and other lunar materials suggest a model for the nature of the lunar crust and for the production of silicate melts capable of yielding crystalline rocks such as those of the Fra Mauro formation. The origin of the KREEP component of lunar fines also is explained by our model.

EXPERIMENTAL PROCEDURES

About 1 g of each sample was ground finely under acetone in an agate mortar for 20 minutes. The resulting powder, of grainsize less than 40 μ, was stored in a stoppered glass vial in a desiccator. Approximately 2 mg of powder was used in each run. This powder consisted largely of the crystalline material of the original rock or soil plus the small percentage of glass also there. The homogeneity of the starting material was monitored by analysis of runs above the liquidus with an electron microprobe. Crystalline starting material was employed to avoid the changes in oxidation state inherent in preliminary fusions. This procedure has two additional benefits: (1) nucleation of phases present is no problem; (2) at temperatures well below the liquidus, crystals remain of size large enough for microprobe analysis. It is possible that experimental products might be biased in favor of the starting materials when using crystalline powders. To avoid such a possibility and to demonstrate the equilibrium nature of our products, some runs were equilibrated above the liquidus and then dropped to a lower temperature. In this manner, several important phase boundaries on the PT diagram were reversed. We had less success in reversing the liquidus at high pressures than at low pressure (where we were generally successful), no doubt a result of the longer run times.

Work done at low pressure was performed in Mo capsules. These were drilled from 2.54 mm Mo rod and had friction-fitted Mo stoppers to contain the charges. The capsules were sealed in evacuated silica-glass tubes. While the glass was being evacuated, the sample was gently heated with an oxygen-hydrogen torch to dry it and then the sample was sealed off under vacuum. The sealed-silica glass vessels were then suspended at the hot spot of a vertical tube, Pt-wound quenching furnace less than 5 mm from a Pt-Pt 10% Rh thermocouple junction. Temperature control was maintained by monitoring this thermocouple and adjusting the furnace power for temperature deviations from the desired temperature. The control is precise to $\pm 1/3^\circ$C over periods of weeks. Quenching was accomplished by dropping the silica-glass vessel out the bottom of the furnace into a dish of water. Quenching times were less than 2 seconds. Run times ranged from 4 hours to 5 days in length.

Our thermocouples were calibrated against the melting points of Au (1064.5°C) and Li metasilicate (1203°C). The assembly of silica tube and thermocouple was the same as during a quenching run. The calibration material was held at successively higher temperatures until melting was observed. Melting intervals were less than a degree. The corrections to normal thermocouple EMF/temperature tables were between 3 and 15 degrees, depending on the age of the thermocouple and the length of use above 1300°C.

Experiments conducted at high pressure were performed in a solid-medium, piston-cylinder apparatus (Boyd and England, 1960, 1963). Capsules drilled from Mo rod were used to contain the sample. Furnace assembly, capsule and sample were dried in dry nitrogen at 1100°C for $\frac{1}{2}$ h. Boron nitride was used as a sleeve and Pt-Pt 10% Rh as a thermocouple pair. All runs were carried out on piston-in strokes with a -8% friction correction being applied to the nominal pressure to determine the pressure of the run (Johannes et al., 1971). The pressure was raised before the internal graphite resistance furnace was heated, but the final piston-in pressure adjustment was made at run temperature. No correction was made for the effect of pressure on thermocouple EMF and the thermocouples were not individually calibrated. Quenching was carried out by turning off the furnace power. Quenching times were less than 10 sec. Before quenching, temperature control was maintained to $\pm 2^\circ$C by monitoring the thermocouple EMF and correcting the furnace power for temperature fluctuation. Run lengths could not profitably be made longer than a day due to thermocouple contamination. Increasing degrees of melting were observed in products equilibrated for longer periods of time at constant thermocouple EMF. Both vitreous and crystalline starting materials give this result, so the effect cannot be kinetic. Pt and Al_2O_3 discs were placed between the thermocouple junction and the sample capsule to protect against contamination from below. Either this precaution was ineffective or the contamination comes from elsewhere (cf. Mao and Bell, 1971).

Capsule material and oxidation states

Equilibrium studies on iron-bearing silicate systems suffer from difficulties introduced by the variable oxidation state of Fe. In the lunar samples very little, if any, Fe^{3+} is observed whereas Fe metal commonly is present with ferrous silicates. Experiments conducted under oxidizing conditions would not properly model the lunar environment of crystallization. It is difficult to find an ideal container on which to conduct experiments under reducing conditions. Pt has long been known to remove Fe metal from the sample, changing the total Fe content and oxidation state of the remaining Fe. Using a Pt capsule resulted in replacement of liquidus plagioclase by spinel in 14310 and changed the color of the glass from green to brown. Graphite, which might be satisfactory at high pressures were it not for its lack of mechanical integrity, is totally unsatisfactory at low pressure where smelting occurs. Any Fe^{2+} reacts with the graphite to produce metal, and CO, that bursts the silica tube. Other laboratories have used Fe capsules.

If the starting material is in a somewhat oxidized state, reaction with the metal container enriches the melt in FeO and may cause leaks in the container. If the starting material is reduced, we observed rather erratic concentrations of Fe in the melt in runs of less than 10 h. We have used Mo as a capsule material. Our investigations into the diffusion of Fe into Mo have shown that the alloy Fe_3Mo_2 might be a better capsule material. The commercial unavailability of this material prevents its use at this time. The high melting temperature of Mo and the low diffusion coefficient of Fe into Mo (Walker and Hays, 1972) in the temperature range of the experiments have prompted the use of Mo as a reasonable alternative.

Our experimental conditions (sealed tubes and piston-cylinder apparatus) allow the oxidation state of the Fe in the charge to vary only by interaction with the Mo. Rock 14310 and the homogeneous starting powder contain some metallic Fe as part of the total Fe reported as FeO in published analyses (Ringwood and Green, 1972; Kushiro, 1972). Low pressure runs in Mo, above the liquidus, do not show Fe droplets but electron probe analysis of the glass shows that about half the "FeO" reported in whole-rock analyses has gone. This Fe is found in the Mo capsules. Long runs in Fe capsules, however, show the same amount of Fe "loss" suggesting that immiscibility of the reduced fraction of the Fe is responsible for the low FeO content of the melts, rather than the extraction of Fe^{2+} by the Mo capsule. Long runs in Mo do not decrease FeO in the liquid significantly below the level reached in long runs in Fe capsules under these closed capsule conditions. It is only the metallic Fe droplets that have been removed by the Mo during the long runs. Figure 1 shows the results of analyses of glasses formed by equilibrating sample charges at

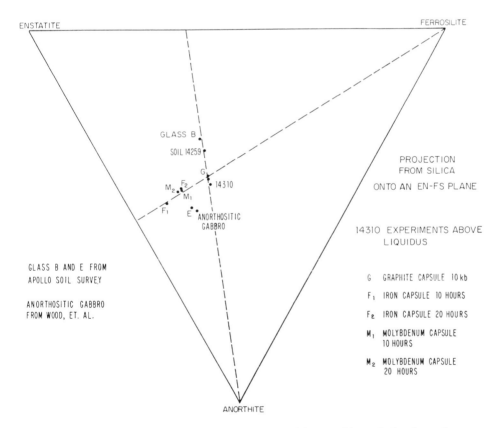

Fig. 1. Electron microprobe analyses of glasses formed by quenching melted rock powder from 14310,138 held above the liquidus temperature in various capsule materials. Other analyses are plotted for comparison. See captions to Fig. 7 and Fig. 9 for projection method.

temperatures just above liquidus. A pressure of 5 kb on the liquid in a graphite capsule was apparently sufficient to dissolve the metallic Fe, because the analysis of this run has a full complement of Fe in the silicate glass. The diagram shows that the low-pressure glasses in Mo and in Fe have less Fe than the whole rock value. (Fe plots at ferrosilite in this diagram.) It is also evident that short runs in Fe capsules give erratic results for unknown reasons but all differ from the whole rock in having too little Fe in the glass. We may conclude that for low pressures, the Mo capsule in a sealed-silica tube holds the oxidation state of the charge constant. The capsule is self-buffered. We shall see that in the case of 14310 this was perhaps not the most desirable circumstance, because the rock suffered substantial subsolidus reduction later than the events of the main crystallization.

The control of oxidation state at high pressures appears to be maintained either internally or by the capsule, depending on temperature. Two regions on our PT diagram could be distinguished on the basis of whether MoO_2 or Fe metal droplets (short runs) were present in the experimental product. The production of MoO_2 in a closed system requires a reaction of the sort:

$$2FeO \text{ (sample)} + 21Mo \text{ (capsule)} = Fe_2Mo_{20} \text{ (ss)} + MoO_2 \text{ (xtl)}.$$

Biggar's (1970) data suggests that the $Mo-MoO_2$ buffer is rather close to the Fe–FeO buffer in the temperature range of our experiments. The fact that Fe metal, which is present in the starting material, disappears in some cases, cannot be an oxidation effect because the Fe content of the glass actually decreases. The fact that at the same time as the Fe metal disappears, presumably by diffusion into the Mo, we observe MoO_2 to appear, implies that some reduction of the FeO is occurring by the above reaction. This in turn implies that the f_{O_2} (T, P) buffer curve for $Mo-MoO_2$ must lie below the Fe–FeO equilibrium in our charge, at least for temperatures above which diffusion becomes significant. The removal of the Fe metal phase is not a serious problem since its only practical effect, when present, was to buffer f_{O_2}. This function is assumed by the Mo. The fact that Mo is able to reduce FeO by dissolving Fe is analogous to the Pt problem. We are somewhat better off with Mo because the remaining iron is not oxidized but, instead, the Mo forms MoO_2. Then the only detrimental effect is the change in FeO content of the liquid caused by the slightly greater reducing power of Mo than of Fe. The amount of MoO_2 produced is quite small so this effect is thought to be unimportant. We may conclude that at temperatures sufficient for significant diffusion, oxidation state is controlled by $Mo-MoO_2$ reducing FeO. At lower temperatures the charge is self-buffered.

RESULTS OF CRYSTALLIZATION EXPERIMENTS

The products and conditions of our experimental runs are given in Table 1. The full PT diagrams for the compositions studied are given in Figs. 2, 3, and 4. The results of our crystallization experiments in sealed-silica tubes are summarized in Fig. 5.

The low pressure crystallization of sample 14310,138 begins with plagioclase (An 92) precipitating at 1310°C and precipitating alone until 1202°C where both orthopyroxene (En 88, Wo 3) and a trace of olivine (Fo 89) appear. Olivine disappears very quickly below 1200°C and at 1190°C the orthopyroxene has begun to react with the liquid to produce pigeonite (En 86, Wo 5). At 1100°C a silica phase, probably tridymite, is present in the residuum. With increasing pressure, plagioclase remains the liquidus phase up to 10 kb; however, olivine and orthopyroxene are replaced by pigeonite as the primary ferromagnesian mineral in this pressure range. In the neighborhood of 15 kb, spinel, and possibly aluminous orthopyroxene are on the liquidus. At about 12 kb, pigeonite must be very nearly on the liquidus and at about 18 kb, aluminous clinopyroxene must be near the liquidus. At 20 kb, the crystallization is dominated by pyrope-rich (py 57, alm 22, gross 21) garnet and very aluminous

Table 1. Results of quenching experiments.

P (kb)	Run	Starting material	T °C	Duration (hours)	Capsule	Results
0	71	Powder (1)	1320	21.5	Mo	glass
0	72	Powder (1)	1320	21.5	Fe	glass
0	73	Powder (1)	1320	10.7	Mo	glass
0	79	Powder (2)	1320	11.	Mo	glass
0	80	Powder (2)	1320	9.7	Fe	glass
0	48	Powder (1)	1319	10.8	Fe	glass
0	46†	Powder (1)	1319	5.0	Fe	glass + plag
0	21	Powder (1)	1317	10.5	Mo	glass
0	63	Powder (1)	1315	50.8	Pt	glass + sp + (?) plag
0	38	Glass (3)	1304	46.2	Mo	glass + plag
0	37	Powder (1)	1303	65.	Mo	glass + plag
0	20	Powder (1)	1302	3.8	Mo	glass + plag
0	16	Powder (1)	1277	4.5	Mo	glass + plag
0	28	Powder (1)	1252	21.2	Mo	glass + plag
0	15	Powder (1)	1227	10.7	Mo	glass + plag
0	40	Glass (4)	1204	12.	Mo	glass + plag
0	23	Powder (1)	1202	24.	Mo	glass + plag + opx + ol (trace)
0	68	Powder (1)	1200	21.	Fe	glass + plag + opx + ol
0	69	Powder (2)	1200	111.	Mo	glass + plag + opx + ol
0	67	(5)	1190	14.8	Mo	glass + plag + opx + pig
0	41	Glass (6)	1184	21.8	Mo	glass + plag + pig + opx
0	25	Glass (7)	1177	41.	Mo	glass + plag + pig + opx
0	42	Glass (8)	1164	72.	Mo	glass + plag + pig + (opx)
0	34	Powder (1)	1152	48.	Mo	glass + plag + opx + pig
0	49	Glass (4)	1104	82.5	Mo	glass + plag + (opx) + pig + trid

*(14310 composition modified)

P (kb)	Run	Starting material	T °C	Duration (hours)	Capsule	Results
0	65	(9)*	1200	47.5	Mo	glass + plag + opx + ol(tr) + sp(tr)
0	66	(9)*	1200	117.0	Mo	glass + plag + opx
0	62	(10)*	1200	48.5	Mo	glass + plag + ol + opx
0	81	(11)*	1200	28.5	Mo	glass + plag + ol
0	82	(12)*	1250	56.	Mo	glass + plag
0	83	(13)*	1250	48.	Mo	glass
5	71	Powder (1)	1325	1.8	Graphite	glass
5	17	Powder (1)	1300	2.	Mo	glass + plag
5	55	Powder (1)	1265	3.3	Mo	glass + plag
5	4	Powder (1)	1250	2.	Mo	glass + plag + pig + opx
5	57†	(14)	1200	4.5	Mo	glass + plag
5	7	Powder (1)	1150	12.	Mo	glass + plag + pig + (opx)
10	19	Powder (1)	1325	2.	Mo	glass
10	22	Powder (1)	1310	2.2	Mo	glass + plag
10	6	Powder (1)	1260	2.	Mo	glass + plag + pig + (opx)
15	9	Powder (1)	1350	2.	Mo	glass
15	12	Powder (1)	1340	2.	Mo	glass + sp
15	53	Powder (1)	1340	3.	Mo	glass + (?) opx
15	31	Powder (1)	1333	4.4	Mo	glass + cpx
15	45†	Glass (15)	1330	6.	Mo	glass
15	51†	Powder (1)	1326	3.5	Mo	glass
15	36†	Powder (1)	1325	6.	Mo	glass
15	54	Powder (1)	1325	5.	Mo	glass + sp + plag + (?) cpx
15	3	Powder (1)	1325	2.5	Mo	glass + sp + plag + cpx
15	39†	Glass (16)	1315	9.7	Mo	glass
15	35	Powder (1)	1300	5.2	Mo	glass + sp + (?) px
15	52	Powder (1)	1300	3.5	Mo	glass + sp + cpx + (?)opx + plag
15	13	Powder (1)	1225	13.5	Mo	glass + pig + (?)opx + plag
17.5	26	Powder (1)	1340	6.	Mo	glass + (?) opx
17.5	56	Powder (1)	1340	3.	Mo	glass + sp
17.5	32	Powder (1)	1250	24.	Mo	glass + garnet + cpx
20	5	Powder (1)	1405	2.	Mo	glass

Table 1 (continued)

P (kb)	Run	Starting material	T°C	Duration (hours)	Capsule	Results
20	11	Powder (1)	1375	2.	Mo	glass + (?) garnet
20	47†	Glass (15)	1365	6.2	Mo	glass
20	27	Powder (1)	1365	3.	Mo	glass + garnet + cpx
20	14	Powder (1)	1350	3.	Mo	glass + garnet + cpx
20	43†	Glass (18)	1325	6.	Mo	glass
20	8	Powder (1)	1250	11.5	Mo	glass + garnet + cpx
20	18†	Powder (1)	1200	12.5	Mo	glass + garnet + pig + cpx + plag
20	44	Powder (1)	1150	24.	Mo	glass + garnet + px + (?) sp
20	24	Glass (19)	1150	59.	Mo	(?)glass + garnet + cpx

Results of Experiments on 14259

P (kb)	Run	Starting material	T°C	Duration (hours)	Capsule	Results
0	56	Powder (20)	1252	1.5	Mo	glass
0	55	Powder (20)	1242	1.2	Mo	glass + plag
0	58	Powder (20)	1227	5.2	Mo	glass + plag
0	51	Powder (20)	1202	9.7	Mo	glass + plag + ol
0	57	Powder (20)	1166	8.	Mo	glass + plag + ol + opx + (?)pig
5	16	Powder (20)	1250	2.5	Mo	glass
5	22	Powder (20)	1240	3.	Mo	glass + plag
5	40†	Glass (21)	1240	3.	Mo	glass
5	45†	Glass (21)	1230	3.	Mo	glass
5	21	Powder (20)	1225	2.	Mo	glass + plag + opx
5	17	Powder (20)	1200	4.	Mo	glass + plag + pig + opx
5	19	Powder (20)	1175	4.	Mo	glass + plag + pig + opx
5	24	Powder (20)	1150	8.	Mo	glass + plag + pig + (?)opx
5	43	Glass (21)	1150	8.	Mo	glass + plag + pig
10	49	Powder (20)	1290	2.	Mo	glass
10	14	Powder (20)	1275	3.	Mo	glass
10	46	Powder (20)	1275	2.	Mo	glass + plag + opx
10	23	Powder (20)	1260	3.	Mo	glass + pig
10	41	Glass (22)	1260	3.	Mo	glass + plag + pig
10	10	Powder (20)	1250	6.5	Mo	glass + plag + pig + (?) opx
10	18	Powder (20)	1225	3.8	Mo	glass + plag + pig
10	25	Powder (20)	1175	6.5	Mo	glass + plag + pig
12.5	52	Powder (20)	1150	24.	Mo	glass + plag + garnet + cpx + opx
15	39	Powder (20)	1360	2.	Mo	glass
15	36	Powder (20)	1350	1.7	Mo	glass
15	42	Powder (20)	1340	3.2	Mo	glass
15	50	Powder (20)	1330	2.	Mo	glass + opx
15	47†	Glass (23)	1325	3.3	Mo	glass
15	37	Powder (20)	1325	3.2	Mo	glass + garnet + cpx + opx
15	28	Powder (20)	1250	10.	Mo	glass + garnet + cpx
20	35	Powder (20)	1425	2.	Mo	glass
20	38	Powder (20)	1410	2.7	Mo	glass
20	44	Powder (20)	1405	2.7	Mo	glass + garnet + cpx
20	48†	Glass (24)	1405	2.3	Mo	glass
20	34	Powder (20)	1400	3.	Mo	glass + garnet + cpx
20	33	Powder (20)	1375	1.5	Mo	glass + garnet + cpx
20	32	Powder (20)	1325	3.	Mo	glass + garnet + cpx
20	11	Powder (20)	1250	12.5	Mo	glass + garnet + cpx
20	27	Powder (20)	1200	12.7	Mo	glass + garnet + cpx

Results of Experiments on 14072

P (kb)	Run	Starting material	T°C	Duration (hours)	Capsule	Results
0	19	Powder (25)	1285	24.	Mo	glass
0	10	Powder (25)	1262	19.8	Mo	glass + ol
0	6	Powder (25)	1212	17.5	Mo	glass + ol
0	12	Powder (25)	1190	24.5	Mo	glass + ol + sp
0	20	Liquid (26)	1175	54.2	Mo	glass + sp + pig + (aug) + ol
0	18	Powder (25)	1175	48.	Mo	glass + sp + pig + plag + ol
0	26	Powder (25)	1170	148.	Mo	glass + sp + plag + pig + ol
0	29	Powder (27)	1150	44.	Mo	glass + (?)ol + plag + pig + sp
0	21	Glass (28)	1140	48.	Mo	glass + ol + sp + pig + (opx) + aug

Table 1 (continued)

P (kb)	Run	Starting material	T°C	Duration (hours)	Capsule	Results
0	11	(17)	1112	70.	Mo	glass + sp + pig + plag
5	30	Powder (25)	1325	2.	Graphite	glass
5	8	Powder (25)	1300	3.	Mo	glass
5	2	Powder (25)	1250	3.5	Mo	glass + ol + (?)sp
5	22	Powder (25)	1200	5.5	Mo	glass + ol + sp + pig
10	5	Powder (25)	1350	3.	Mo	glass
10	3	Powder (25)	1300	3.5	Mo	glass + ol + opx + pig
15	7	Powder (25)	1390	3.2	Mo	glass + (quench cpx)
15	4	Powder (25)	1375	3.	Mo	glass + (?) cpx
15	1	Powder (25)	1350	3.	Mo	glass + opx + (?cpx, trid)
15	23	Powder (25)	1150	15.	Mo	plag + ilm + cpx
20	17	Powder (25)	1440	3.	Mo	glass
20	16	Powder (25)	1420	3.2	Mo	glass + opx + (quench cpx)
20	9	Powder (25)	1400	3.2	Mo	glass + opx + (quench cpx)

Note: sp = spinel; plag = plagioclase feldspar; opx = orthopyroxene; ol = olivine; pig = pigeonitic clinopyroxene; px = pyroxene; trid = tridymite; aug = augitic clinopyroxene; ilm = ilmenite.

(1) Rock powder 14310,138
(2) Rock powder 14310,140 (Supplied by A. Muan)
(3) Glass product of Run 21
(4) Glass produced at 1315°C for 9.2 hours from (1)
(5) Liquid and plag produced at 1210°C for 4.5 hours from (1)
(6) Glass made in (4), treated like run 40
(7) Glass from fusing (1) under vacuum
(8) Glass made in (4), treated like runs 40 and 41
(9) Glass and ol and sp and plag and opx produced by hydrothermal crystallization of (1) for 3.5 hours at 5 kb in graphite
(10) Rock powder (1) plus 2% olivine (Fo 74)
(11) Rock powder (1) plus 0.34% Na_2O as $NaHCO_3$
(12) Rock powder (1) plus 1.5% Na_2O as $NaHCO_3$
(13) Rock powder (1) plus 2% Na_2O as $NaHCO_3$; held 3 hours at 1320°C
(14) Liquid prepared by fusion of (1) at 1325°C for 0.7 hours at 5 kb
(15) Glass prepared by fusion of (1) at 1315°C for 5 hours at 1 atm.
(16) Liquid prepared by fusion of (1) at 1400°C for 0.5 hours at 15 kb
(17) Liquid and olivine prepared by partial fusion of (25) at 1250°C for 1.5 hours
(18) Liquid prepared by fusion of (1) at 1400°C for 0.5 hours at 20 kb
(19) Liquid prepared by fusion of (1) at 1400°C for 0.2 hours at 20 kb
(20) Comprehensive fines 14259,85.
(21) Liquid prepared by fusion of (20) at 1250°C at 5 kb
(22) Liquid prepared by fusion of (20) at 1275°C at 10 kb
(23) Liquid prepared by fusion of (20) at 1340°C at 15 kb
(24) Liquid prepared by fusion of (20) at 1415°C at 20 kb
(25) Rock powder 14072,3
(26) Liquid prepared by fusion of (25) at 1285°C for 24 hours
(27) Sintered powder product of (25) at 1125°C for 205 hours
(28) Glass plus crystals prepared with run 20
† A run not used to construct P, T diagram because of reaction with capsule or failure to achieve equilibrium.

clinopyroxene (17.5% Al_2O_3). The density of this assemblage (3.5 g/cm^3) is too great to be consistent with the lunar mean density; hence 14310 cannot be the lunar bulk composition.

The orthopyroxene cores to pigeonite in rock 14310 contain a few percent Al_2O_3 and induced Ridley *et al.* (1971) to suggest they might be xenocrysts from depth. Reference to Fig. 6 demonstrates that our experimental orthopyroxenes, grown from melted 14310 in evacuated silica tubes, are quite as aluminous as the natural ones. We feel the xenocryst hypothesis of Ridley *et al.* is unnecessary.

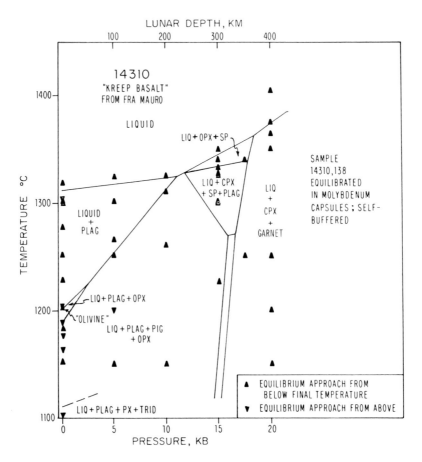

Fig. 2. Pressure-temperature diagram for 14310,138.

The low pressure crystallization of comprehensive fines sample 14259,85 is similar to that of 14310, because both compositions are colinear with that of anorthite. The smaller amount of anorthite in 14259 results in a lower liquidus temperature, 1250°C. Furthermore, the stability field of olivine is enlarged in 14259; a result, we believe, of a difference in oxidation state. With increasing pressure, increasingly aluminous orthopyroxene becomes the liquidus phase. At 20 kb, pyrope-rich garnet and aluminous clinopyroxene are again the principal crystalline products. The same argument used on 14310 excludes 14259 as the lunar bulk composition.

In rock 14072, the low pressure liquidus phase is olivine (Fo_{85}) appearing at 1275°C, followed by chrome spinel at 1200°C. Pigeonite and plagioclase appear simultaneously at 1180°C. Olivine is consumed with falling temperature, and at pressures above 10 kb it is replaced as the liquidus phase by aluminous orthopyroxene. The occurrence of early pigeonite rather than orthopyroxene in this rock is consistent with an Fe/Fe + Mg ratio greater than that of 14310 and 14259.

Low pressure crystallization of a synthetic glass prepared to study average Apollo

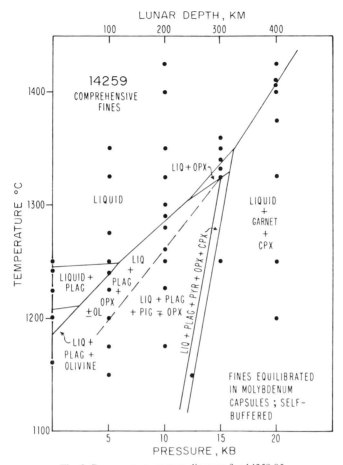

Fig. 3. Pressure-temperature diagram for 14259,85.

11 anorthositic gabbro (Wood *et al.*, 1970) showed an anorthite liquidus at 1485°C with a clear spinel joining at 1450°C. At about 1250°C, olivine appears at the expense of the spinel and final crystallization yields plagioclase, olivine and pyroxene with no spinel remaining. The solidus is near 1200°C.

INTERPRETATION OF RESULTS

Figure 5 displays these low pressure results as a function of the bulk Fe/Fe + Mg ratio of the compositions. These materials all show liquid saturated with olivine, anorthite and low-calcium pyroxene at temperatures near 1200°C. There is a tendency for the temperature of triple saturation to fall with increasing Fe/Fe + Mg. This behavior of these samples may be understood in terms of the synthetic system anorthite-silica-forsterite-fayalite (Andersen, 1915; Roeder and Osborn, 1966) (Fig. 7). Figure 8 is a projection of the boundary curves in this tetrahedron onto a plane perpendicular to the forsterite-fayalite join. It appears quite similar to Ander-

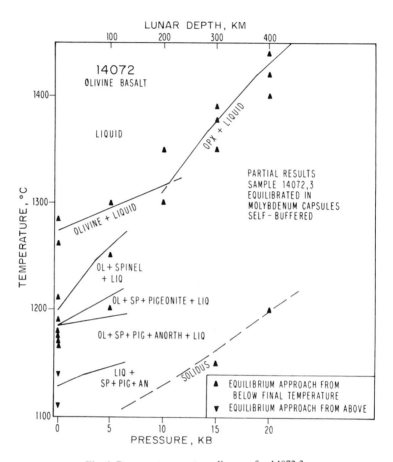

Fig. 4. Pressure-temperature diagram for 14072,3.

sen's iron-free, end-member system but the distortions are caused by plotting molar units and by choosing a section where Fe/Fe + Mg is about 0.3–0.4. The intersection of the silica, pyroxene, and anorthite primary phase fields lies inside the silica-pyroxene-anorthite compositional triangle and hence is a eutectic. The intersection of the olivine, pyroxene, and anorthite primary phase fields lies outside the olivine-pyroxene-anorthite compositional triangle and hence is a peritectic. Bulk compositions within the olivine-pyroxene-anorthite triangle complete equilibrium crystallization on this peritectic and produce their first liquids upon partial melting on this peritectic. Our crystallization experiments show that this point must be at about 1200°C in samples of Fe/Fe + Mg $\simeq$ 0.3–0.4 and at lower temperatures for higher values of Fe/Fe + Mg.

It can easily be seen that 14310 and 14259 fall into the plagioclase primary phase field while 14072 falls in the olivine field. It should be noted that the path away from anorthite through 14310 and 14259 projects very close to the peritectic, explaining the simultaneous appearance of olivine and pyroxene and subsequent disappearance

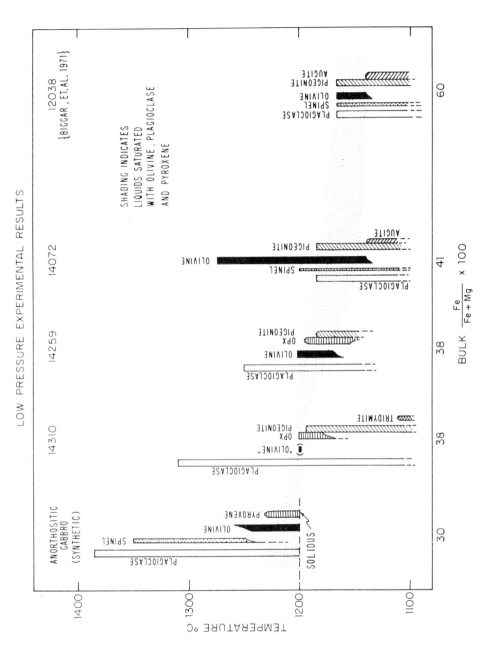

Fig. 5. Summary of results from low-pressure quenching experiments.

14310 ORTHOPYROXENES				
	NATURAL		EXPERIMENTAL	
SiO_2	51.6	54.5	54.8	54.6
TiO_2	.78	.47	.60	.51
Al_2O_3	3.70	1.26	3.04	2.20
Cr_2O_3	.43	–	.53	.47
FeO	13.8	14.1	5.90	6.98
MgO	27.2	27.2	33.4	33.0
CaO	2.26	1.88	1.54	1.77
K_2O	.09	–	–	–
Na_2O	.03	–	.02	–
SUM	99.9	99.4	99.8	99.5

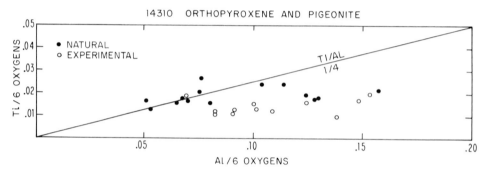

Fig. 6. Electron microprobe analyses of lunar pyroxene from 14310,30 and synthetic pyroxenes from experiments on 14310,138.

of olivine. On the other hand, the path through 14072 from olivine also projects near the peritectic, explaining the simultaneous appearance of pyroxene and plagioclase in the crystallization of 14072. The olivine is not consumed with falling temperature because the bulk composition of 14072 lies within the olivine-pyroxene-anorthite triangle. The slightly lower temperature of the peritectic in 14072 and the appearance of a pigeonitic, low-calcium pyroxene rather than orthopyroxene is a result of the higher Fe/Fe + Mg of 14072.

Oxidation state of 14310

El Goresy *et al.* (1971, 1972) have noted the assemblages ulvöspinel-ilmenite-iron metal, and fayalite-silica-iron metal in 14310 and other Apollo 14 igneous rocks, and have made an argument for extreme subsolidus reduction. We have also observed these features and concur with their interpretation as the most rational explanation of the following observations. Although our experimental crystallization sequence matches that deduced from textural study of 14310 and 14072, our experimental pyroxenes and olivines are considerably more magnesian than the natural ones (Fig. 6). This suggests that the reduction of the iron, which effectively lowers the FeO content of the silicate liquid from which the ferromagnesian phases crystallize, occurred after those phases had crystallized in the real rock.

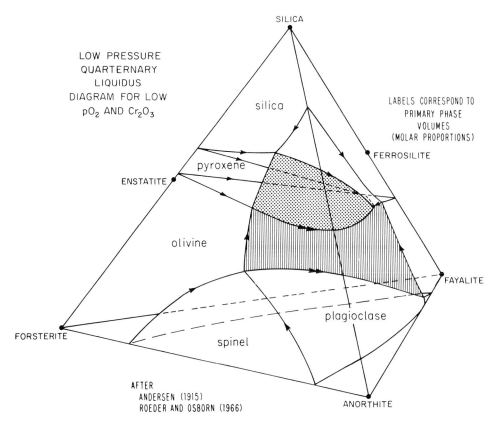

SILICA

LOW PRESSURE
QUARTERNARY
LIQUIDUS
DIAGRAM FOR LOW
pO₂ AND Cr₂O₃

silica

LABELS CORRESPOND TO
PRIMARY PHASE
VOLUMES
(MOLAR PROPORTIONS)

FERROSILITE

pyroxene

ENSTATITE

olivine

FAYALITE

FORSTERITE

plagioclase

spinel

ANORTHITE

AFTER
ANDERSEN (1915)
ROEDER AND OSBORN (1966)

Fig. 7. Liquidus tetrahedron at atmospheric pressure for the system anorthite-forsterite-fayalite-silica at very low oxygen fugacities (ca. 10^{-15} atm). Analyses may be plotted within this tetrahedron by recalculating the SiO_2, Al_2O_3, FeO, MgO and CaO in terms of SiO_2, $CaAl_2Si_2O_8$, Fe_2SiO_4, Mg_2SiO_4, and $CaSiO_3$. Because the Ca/Al ratio of the rocks plotted is close to that of anorthite, the small $CaSiO_3$ component is ignored. Analyses of rocks or glasses are plotted in Figs. 1, 8, 9, and 10 on projections within this tetrahedron.

As a corollary to these conclusions and the controlled f_{O_2} experiments performed at Edinburgh (Ford *et al.*, 1972) we may rather closely estimate the f_{O_2} at which 14310 actually crystallized. Ford *et al.* performed their experiments at $P_{O_2} = 10^{-12}$ atm and failed to produce either iron droplets or the correct sequence of phases crystallizing (orthopyroxene was missing and spinel appeared as an early phase). The true P_{O_2} of crystallization of 14310 is therefore below 10^{-12} atm. By 10^{-14} atm (M. J. O'Hara, personal communication) orthopyroxene has returned to the crystallization sequence but so have abundant iron droplets that would deplete the liquid in FeO and cause precipitation of pyroxenes more magnesian than the real ones. We may thus say that 14310 crystallized at a P_{O_2} between 10^{-14} and 10^{-12} atm but was later subjected to a significant reduction.

We point out that the longer interval of olivine crystallization in 14259 (which is colinear with 14310 and anorthite) is probably an oxidation effect. If the 14259 composition had not had its iron so strongly reduced or if it were reoxidized as suggested by Griscom and Marquardt (1972), olivine would have an enhanced crystallization interval according to the Ford *et al.* experiments.

DISCUSSION OF RESULTS

Near-surface crystallization

The two igneous rock samples that we studied, 14310 and 14072, have texturally determined crystallization sequences (Longhi *et al.*, 1972) which are duplicated in our crystallization experiments in evacuated silica tubes. Considering the vesicular nature of both rocks, it is evident that the mineralogy and texture of these two rocks were produced by low pressure crystallization. We shall now try to determine what processes were responsible for the compositions of these rocks.

Significance of major element chemistry

Let us consider the low pressure regime. Figure 8 is the appropriate low-pressure liquidus diagram as discussed above. Once again note the central role of the olivine-pyroxene-anorthite peritectic and the positions of 14310, 14259, and 14072 relative to this triple saturation point. As noted before, these compositions very nearly fall on the lines joining the peritectic to the liquidus phases of each rock. Can this be an accident? If 14310 were a primary magma as the Apollo Soil Survey (1971) first proposed, then this would be coincidental.

Plotted also on Fig. 8 are the average analyses of preferred glass compositions (B, C, D, E) found by the Apollo 14 Soil Survey. The abundance weighted average of these glasses is quite close to comprehensive fines sample 14259. The abundance pattern in which 14310 is so poorly represented is further evidence that 14310 is not a primary composition. By the same logic then some composition near 14259 or B would be expected to be a primary material at the Apollo 14 site. In fact, these compositions plot near the peritectic in Fig. 8. We also note that glasses C and D are easily interpreted as being controlled by the pyroxene-plagioclase saturation curve emanating from the peritectic. Ford *et al.* (1972) have shown experimentally that glass B is, indeed, doubly saturated with respect to olivine and plagioclase. The observation that 14259 lies slightly within the plagioclase primary phase field is a natural consequence of contamination of a near-peritectic composition with a small amount of glass Type E, the possible significance of which we will discuss below.

What, then, is the explanation of the compositions of 14310 and 14072 which are not close to this peritectic? Considering the positions of 14310 and 14072 relative to this peritectic noted above, it is not difficult to imagine that if plagioclase were added to a peritectic liquid, 14310 would result; and if olivine were added to the peritectic, then 14072 would result.

We have concluded that the peritectic composition is important at the Apollo 14 site, but can we argue that plagioclase enrichment of this liquid in feldspar or olivine took place? We have described elsewhere (Longhi *et al.*, 1972) the texturally anomalous, large feldspars in 14310 and their compositional pecularities with respect

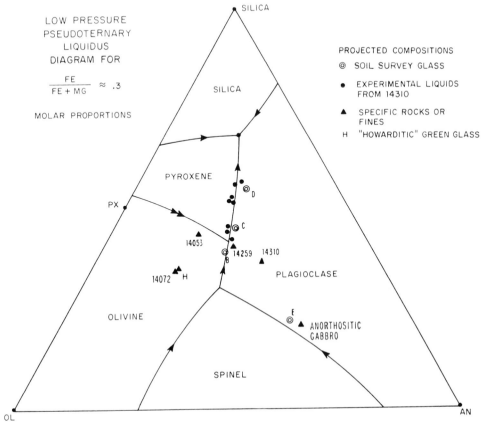

LOW PRESSURE
PSEUDOTERNARY
LIQUIDUS
DIAGRAM FOR

$$\frac{FE}{FE + MG} \approx .3$$

MOLAR PROPORTIONS

PROJECTED COMPOSITIONS
◎ SOIL SURVEY GLASS

•   EXPERIMENTAL LIQUIDS
    FROM 14310

▲   SPECIFIC ROCKS OR
    FINES

H   "HOWARDITIC" GREEN GLASS

SILICA

SILICA

PYROXENE

PX

D

C

14053

14259   14310

14072   H

PLAGIOCLASE

OLIVINE

E

ANORTHOSITIC
GABBRO

SPINEL

OL

AN

Fig. 8. Projection, onto the anorthite-forsterite-silica face, of equilibria within the tetrahedron of Fig. 7. Projection is parallel to the forsterite-fayalite join. Field boundaries are
located by analyses of multiply saturated liquids in our experimental runs and correspond
to an Fe/Fe + Mg ratio of 0.3 to 0.4. Boundary curves and their intersections in this figure
represent surfaces of two-fold saturation and curves of three-fold saturation, respectively,
within the tetrahedron of Fig. 7.

to Fe/Mg ratio. Brown and Peckett (1971) have also noted these large anorthite
grains and termed the rock "feldspar-phyric." The percentage of these phenocrysts
is difficult to estimate because of the lack of a clear cut-off separating them from the
largest lath-shaped crystals which are not so clearly phenocrystic. Kushiro (1972)
has suggested that there is not enough phenocrystic anorthite to displace 14310
sufficiently from the peritectic composition ($\sim 15\%$ phenocrysts necessary). However,
photographs of 30 thin sections of 14310 reveal that it is quite heterogeneous with
respect to phenocryst abundance and that 15% phenocrysts is well within the
observed range. We feel that the colinear nature of peritectic–14310–anorthite and
the textural and compositional peculiarities of the feldspar phenocrysts support the
feldspar addition hypothesis. In 14072 we have noted the conspicuous olivine

phenocrysts and the resorbed nature of the olivine in thin section (Longhi *et al.*, 1972).

We believe that the textural and chemical evidence of the rocks supports the hypothesis that olivine and plagioclase have been introduced into "proto" 14072 and 14310 liquids, respectively. In our experimental charges plagioclase remains suspended in 14310 liquid and olivine very quickly sinks in 14072 liquid. Flotation, or concentration by eddy currents, of plagioclase in a body of peritectic magma with complementary sinking of olivine is a convenient mechanism for producing the desired phenocrystic rocks. The Fe/Mg ratio of 14072 is slightly higher than that of 14310 and 14259 indicating that some fractional crystallization has occurred. Addition of olivine by settling from higher parts of the magma chamber would allow the 14072 parent liquid to remain triply saturated during fractional crystallization despite a reaction relationship with olivine along the three-fold saturation curve.

We do not necessarily believe that 14072 and 14310 were produced in the very same magma chamber, although the ages of the two are the same by Rb–Sr and Ar^{39}–Ar^{40} dating (Compston *et al.*, 1972; York *et al.*, 1972). Helmke and Haskin (1972) have determined that the REE of 14072, although enriched, do not show the spectacular concentrations found in sample 14310. This makes it unlikely that 14310 and 14072 are actually comagmatic.

Figure 8 shows the plotted composition of "howarditic" green glass (H) identified by Marvin *et al.* (1972) in various lunar soils. It is quite close to 14072 in composition and we feel that shock melting of material like 14072 is more likely than the survival of primitive nebular materials as an origin for this glass.

Origin of triply saturated liquids

It appears that the olivine-pyroxene-anorthite peritectic liquid can explain many features of Apollo 14 rock and soil chemistry. From the soil-glass abundance data and the indirect arguments from 14310 and 14072, we would expect a good deal of the liquid to have been produced at the same time. How was it produced? One property of a peritectic as opposed to a eutectic is that it is difficult to generate large amounts of peritectic liquid in any crystallization process. During equilibrium crystallization, a liquid composition, which reaches the peritectic from some higher temperature, either concludes its crystallization there or pauses for reaction of olivine and liquid and then continues to change composition. During fractional crystallization, a liquid that reaches the peritectic has no tendency to remain there, as it loses heat, but keeps changing in composition. In such a situation the only way to get a substantial quantity of peritectic liquid is to extract it from some larger body of liquid, which has just reached this delicate chemical and thermal balance. For this reason the norite layer postulated by Wood *et al.* (1971) and Wood (1972) (essentially this peritectic in composition) within the moon is highly improbable. First, to segregate so much peritectic liquid while cooling the moon would be difficult, and then to keep it from differentiating, given the tendency of olivine to sink and requiring plagioclase to float to form anorthosite, is hard to accomplish.

Partial melting of the lunar crust

Fortunately a peritectic is not a "no-way" street. Upon adding heat to an olivine-pyroxene-plagioclase assemblage, any amount of peritectic liquid can be generated until one of the phases, pyroxene or plagioclase, is exhausted. Peritectic liquid is the partial-melting product of any bulk composition in the olivine-pyroxene-plagioclase volume. Partial melting of an olivine, low calcium pyroxene-plagioclase assemblage seems the most reasonable mechanism for producing the primary material of the Apollo 14 landing site. Although the source material could be of any composition within the olivine-pyroxene-anorthite volume, there is one lunar rock type that is a particularly attractive candidate. In Fig. 8, glass E of the soil survey and the anorthositic gabbro of Wood *et al.* (1970) fall close together in the olivine-pyroxene-anorthite volume. Furthermore, this composition is reminiscent of the Surveyor 7 analysis of Tycho ejecta (Patterson *et al.*, 1969). Reid *et al.* (1972) have recently furnished us with a preprint noting the remarkable uniformity and distribution of this component around the moon. The collective conclusion is that this anorthositic gabbro is representative of the lunar highlands. Note that the solidus is in the neighborhood of 1200°C for our synthetic anorthositic gabbro. It is plausible that partial melting of this highland material, and subsequent differentiation of that liquid, would produce the materials sampled by Apollo 14. If such is the case, the highlands (mineralogically) should be olivine, low-calcium pyroxene and plagioclase. 14310 has orthopyroxene and, nearly simultaneously, pigeonite, but 14072 has only pigeonite. It is possible that both varieties of low-calcium pyroxene may be in the highlands as a consequence (or cause?) of the Fe/Fe + Mg of the peritectic liquid. (G. M. Brown has pointed out to us that if orthopyroxene is the dominant pyroxene in this proposed source rock, then the correct rock name is "feldspathic *norite*" or "anorthositic *norite*." Because this term "anorthositic *gabbro*" used by Wood *et al.* (1970) has been widely adopted in the literature of lunar geochemistry and geophysics, we have chosen to retain this term, at least until macroscopic samples of such rocks are available for study.)

When did this melting take place? Ganapathy *et al.* (1972) and others have suggested that 14310 crystallized from melted soil, on the basis of trace element data. Others have suggested this conclusion on the basis of the superficial major element similarity of fines sample 14259 and rock 14310. The Imbrian impact event is commonly supposed to be the excavation that sprinkled the Fra Mauro formation about the lunar surface and could provide the necessary energy. Two lines of evidence suggest that the Imbrian event is not the immediate cause of the 14310 melting. Ridley *et al.* (1972) noted many rock types like 14310 incorporated as clasts in the Fra Mauro breccias, which implies pre-Imbrian generation. Papanastassiou and Wasserburg (1971) find evidence for a pre-Imbrian crystallization date in the Rb/Sr systematics for 14310. This is not to say that a pre-Imbrian impact might not have produced the earlier heat input that caused the partial melting. This seems a very likely mechanism for incorporating the trace elements characteristic of a meteorite-contaminated soil. The alternate possibility is that some internal heating anomaly produced high-level partial melts prior to the Imbrian event.

Origin of KREEP

Hubbard and Gast (1972) have shown that 14310 has a trace element enrichment like the KREEP component first recognized at the Apollo 12 site. The peritectic composition and its differentiates are strongly reminiscent of KREEP. Meyer (1972) has shown that Apollo 15 KREEP is mineralogically quite like the Apollo 14 Fra Mauro basalts. Taylor *et al.* (1971) have shown that the Apollo 14 soil is heavily dominated by this KREEP material. The enormous enrichments in REE (200–300 × chondrites) but the compositionally primitive Fe/Mg ratio clearly favor a small degree of partial melting, rather than extensive fractional crystallization to produce the Fra Mauro type KREEP basalts of which 14310 is an example. The two-stage process of enrichment implicit in melting a lunar crust, which itself is a differentiate of some more primitive lunar material, is ideally suited to producing these enrichments.

Partial melting of the lunar interior?

One possible alternative explanation is that these two rocks are partial melts of the lunar interior, which have suffered only minor modification in transit to the lunar surface where they crystallized. If such is the case, rock 14310 is likely to be derived from a source region of orthopyroxene, clinopyroxene and spinel, as implied by the liquidus assemblages in the pressure interval 12–18 kb (corresponding to 240 to 360 km depth). Depths greater than this would imply a garnet-clinopyroxene residue of density too great (>3.5 g/cc) for the lunar interior, on the grounds of mean density and moment of inertia of the moon. If 14072 is a partial melt, it is most readily derived from the 10 kb (200 km) region, where an olivine-orthopyroxene source region is implied by the liquidus assemblage. Recent work on a synthetic glass of similar composition also shows pigeonite near the liquidus. These implied lunar interiors are not greatly different from the model proposed by Ringwood and Essene (1970) on the basis of their Apollo 11 work. We do not consider partial melting at depth to be a preferred mechanism for generating these lavas when the alternatives above are considered.

Effect of alkali loss

Following Brown and Peckett (1971), Kushiro (1972) has added 3.2% Na_2O and 1% K_2O to 14310 and finds cotectic behavior of 14310 at 3 kb with respect to olivine, plagioclase, spinel, and orthopyroxene. He suggests direct melting of this assemblage at depth as a possible source of 14310 with restored alkalis.

We have performed experiments by adding alkalis to 14310 and subjecting the sample to vacuum. Addition of 0.4% Na_2O (as $NaHCO_3$) causes orthopyroxene to disappear and greatly increases the amount of olivine at 1200°C. Addition of 2% Na_2O reduced the liquidus temperature below 1250°C, so it appears that the large plagioclase crystallization interval is reduced, perhaps approaching cotectic behavior. Addition of ~1.5% Na_2O allowed plagioclase of composition An86 to grow, presumably near the liquidus, at 1250°C. An86 is more sodic than the phenocrystic plagioclase cores in 14310. Addition of still more Na_2O would certainly make the plagioclase too sodic to be consistent with the natural cores. We feel that Brown and Peckett's estimate of alkali loss must be exaggerated because restoration of that much sodium produces plagioclase too sodic compared to the natural examples. Although 14310 may have lost much of the sodium associated with the very late interstitial liquids, its bulk sodium has not been so drastically reduced. We therefore believe that these experiments by Kushiro are not relevant.

SUMMARY AND CONCLUSIONS

Evidence is accumulating that the moon has a differentiated crust some 50 to 100 kilometers in thickness (Toksöz *et al.*, 1972; Langseth *et al.*, 1972; Gast and McConnell, 1972; Wood, 1972.) Density, compressional wave velocity, heat production, and major and trace element chemistry of the nonmare portion of this crust seem to be consistent with the hypothesis that a significant portion of the crust consists of anorthositic gabbro.

We have shown here that partial melting of anorthositic gabbro (or indeed of any mixture of the minerals plagioclase-olivine-low-calcium pyroxene) will tend to produce significant quantities of liquid having a major element chemistry resembling that of certain Apollo 14 materials. Furthermore, well-understood processes of crystal-liquid fractionation acting on such a liquid and its crystalline parents and/or products are capable of explaining in detail the major element chemistry of Apollo 14 crystalline rock samples 14310 and 14072, and of the preferred glass compositions in the Apollo 14 soil including so-called granitic or rhyolitic material. It seems likely to us that such partial melting events, whether generated by internal or external heat sources, were not uncommon in the early history of the moon, and that much of the material variously described as norites, feldspathic basalts, nonmare basalts, gray mottled fragments, and KREEP, had such an origin.

It also seems likely to us that the trace element and isotopic chemistry of these materials can be accounted for by such a process involving partial melting of pre-enriched crustal materials, but such models remain to be worked out in detail.

Acknowledgments—We thank D. Chipman, P. Lyttle, C. B. Ma, and M. Campot for assistance with various aspects of the experimental work; D. Weill, J. A. Wood, G. J. Taylor, J. Reid, U. Marvin, and J. L. Warner for their criticism and helpful discussions; and J. A. Wood, M. J. O'Hara, I. Kushiro, A. E. Ringwood, N. Toksöz, R. Williams, and the Apollo Soil Survey for making their results available to us in advance of publication. A. Muan arranged an exchange of starting materials, and J. L. Warner provided us with photographs of thirty thin-sections of rock 14310.

This work has been supported by the Committee on Experimental Geology and Geophysics, Harvard University, and by NASA grants NGR 22–007–175 and NGR 22–007–199. David Walker is an NSF predoctoral fellow.

REFERENCES

Andersen O. (1915) The system anorthite-forsterite-silica. *Amer. J. Sci.*, 4th ser. **39**, 407–454.
Apollo Soil Survey (1971) Apollo 14: Nature and origin of rock types in soil from the Fra Mauro formation. *Earth Planet. Sci. Lett.* **12**, 49–54.

Biggar G. M. (1970) Molybdenum as a container for melts containing iron oxide. *Ceramic Bull.* **49**, 286–288.

Biggar G. M. O'Hara M. J. Peckett A. and Humphries D. J. (1971) Lunar lavas and the achondrites: Petrogenesis of protohypersthene basalts and the maria lava lakes. *Proc. Second Lunar Sci. Conf., Geochim. Cosmochim. Acta* Suppl. 2, Vol. 1, pp. 617–643. MIT Press.

Boyd F. R. and England J. L. (1960) Apparatus for phase equilibrium measurements at pressures up to 50 kilobars and temperatures up to 1750°C. *J. Geophys. Res.* **65**, 741–748.

Boyd F. R. and England J. L. (1963) Effect of pressure on the melting point of diopside, $CaMgSi_2O_6$ and albite, $NaAlSi_3O_8$ in the range up to 50 kilobars. *J. Geophys. Res.* **68**, 311–323.

Brown G. M. and Peckett A. (1971) Selective volatilization on the lunar surface: Evidence from Apollo 14 feldspar-phyric basalts. *Nature* **234**, 262–266.

Compston W. Vernon M. J. Berry H. Rudowski R. Gray C. M. and Ware N. (1972) Age and petrogenesis of Apollo 14 basalts (abstract). In *Lunar Science—III* (editor C. Watkins), pp. 151–153, Lunar Science Institute Contr. No. 88.

El Goresy A. Ramdohr P. and Taylor L. A. (1971) The opaque mineralogy of Apollo 14 crystalline rock 14310 (abstract). In *Abstracts With Programs*, 1971 Annual Meetings, Geological Society of America, p. 555.

El Goresy A. Ramdohr P. and Taylor L. A. (1972) Fra Mauro crystalline rocks: Petrology, geochemistry, and subsolidus reduction of the opaque minerals (abstract). In *Lunar Science—III* (editor C. Watkins), pp. 224–226, Lunar Science Institute Contr. No. 88.

Ford C. E. Humphries D. J. Wilson G. Dixon D. Biggar G. M. and O'Hara M. J. (1972) Experimental petrology of high-alumina basalt, 14310, and related compositions (abstract). In *Lunar Science—III* (editor C. Watkins), pp. 274–276, Lunar Science Institute Contr. No. 88.

Ganapathy R. Laul J. C. Morgan J. W. and Anders E. (1972) Moon: Possible nature of the body that produced the Imbrian Basin, from the composition of Apollo 14 samples. *Science* **175**, 55–58.

Gast P. W. and McConnell R. K. Jr. (1972) Evidence for initial chemical layering of the moon (abstract). In *Lunar Science—III* (editor C. Watkins), pp. 289–290, Lunar Science Institute Contr. No. 88.

Griscom D. L. and Marquardt C. L. (1972) Electron spin resonance studies of iron phases in lunar glasses and simulated lunar glasses (abstract). In *Lunar Science—III* (editor C. Watkins), pp. 341–343, Lunar Science Institute Contr. No. 88.

Helmke P. A. and Haskin L. A. (1972) Rare earth and other trace elements in Apollo 14 lunar samples (abstract). In *Lunar Science—III* (editor C. Watkins), pp. 366–368, Lunar Science Institute Contr. No. 88.

Hubbard N. J. and Gast P. W. (1972) Chemical composition of Apollo 14 materials and evidence for alkali volatilization (abstract). In *Lunar Science—III* (editor C. Watkins), pp. 407–409, Lunar Science Institute Contr. No. 88.

Johannes W. Bell P. M. Mao H. K. Boettcher A. L. Chipman D. W. Hays J. F. Newton R. C. and Seifert F. (1971) An interlaboratory comparison of piston-cylinder pressure calibration using the albite-breakdown reaction. *Contrib. Mineral. Petrol.* **32**, 24–38.

Kushiro I. (1972) Petrology of high-alumina basalt (abstract). In *Lunar Science—III* (editor C. Watkins), pp. 466–468, Lunar Science Institute Contr. No. 88.

Langseth M. G. Jr. Clark S. P. Jr. Chute J. Jr. and Keihm S. (1972) The Apollo 15 lunar heat flow measurement (abstract). In *Lunar Science—III* (editor C. Watkins), pp. 475–477, Lunar Science Institute Contr. No. 88.

Longhi J. L. Walker D. and Hays J. F. (1972) Petrography and crystallization history of basalts 14310 and 14072. This volume.

Mao H. K. and Bell P. M. (1971) Behavior of thermocouples in the single-stage piston-cylinder apparatus. *Carnegie Inst. Wash. Yearbook* **69**, 207–216.

Marvin U. B. Reid J. B. Jr. Taylor G. J. and Wood J. A. (1972) Lunar mafic green glasses, howardites, and the composition of undifferentiated lunar material (abstract). In *Lunar Science—III* (editor C. Watkins), pp. 507–509, Luna: Science Institute Contr. No. 88.

Papanastassiou D. A. and Wasserburg G. J. (1971) Rb–Sr ages of igneous rocks from the Apollo 14 mission and the age of the Fra Mauro formation. *Earth Planet. Sci. Lett.* **12**, 36–48.

Patterson J. H. Franzgrote E. J. Turkevich A. L. Anderson W. A. Economov T. E. Griffin H. E. Grotch

S. L. and Sowinski K. P. (1969) Alpha-scattering experiment on Surveyor 7: Comparison with Surveyors 5 and 6. *J. Geophys. Res.* **74**, 6120–6148.

Reid A. M. Ridley W. I. Harmon R. S. Warner J. Brett R. Jakeš P. and Brown R. W. (1972) Feldspathic basalts in lunar soils and the nature of the lunar highlands: Preprint furnished by the authors.

Ridley W. I. Williams R. J. Takeda H. Brown R. W. and Brett R. (1971) Petrology of Fra Mauro basalt 14310 (abstract). In *Abstracts With Programs*, 1971 Annual Meetings, Geological Society of America, pp. 682–683.

Ridley W. I. Williams R. J. Brett R. Takeda H. and Brown R. W. (1972) Petrology of lunar basalt 14310 (abstract). In *Lunar Science—III* (editor C. Watkins), pp. 648–650, Lunar Science Institute Contr. No. 88.

Ringwood A. E. and Essene E. (1970) Petrogenesis of Apollo 11 basalts, internal constitution and origin of the moon. *Proc. Apollo 11 Lunar Sci. Conf., Geochim. Cosmochim. Acta* Suppl. 1, Vol. 1, pp. 769–799.

Ringwood A. E. Green D. H. and Ware N. G. (1972) Experimental petrology and petrogenesis of Apollo 14 basalts (abstract). In *Lunar Science—III* (editor C. Watkins), pp. 654–656, Lunar Science Institute Contr. No. 88.

Roeder P. L. and Osborn E. F. (1966) Experimental data for the system $MgO–FeO–Fe_2O_3–CaAl_2Si_2O_8–SiO_2$ and their petrologic implications. *Amer. J. Sci.* **264**, 428–480.

Taylor S. R. Muir P. and Kaye M. (1971) Trace element chemistry of Apollo 14 lunar soil from Fra Mauro. *Geochim. Cosmochim. Acta* **35**, 975–981.

Toksöz M. N. Press F. Anderson K. Dainty A. Latham G. Ewing M. Dorman J. Lammlein D. Sutton G. Duennebier F. and Nakamura Y. (1972) Velocity structure and properties of the lunar crust (abstract). In *Lunar Science—III* (editor C. Watkins), pp. 758–760, Lunar Science Institute Contr. No. 88.

Walker D. and Hays J. F. (1972) Diffusion of iron into molybdenum at 10 kilobars. *Met. Trans.* in press.

Wood J. A. Dickey J. S. Jr. Marvin U. B. and Powell B. N. (1970) Lunar anorthosites. *Science* **167**, 602–604.

Wood J. A. Marvin U. B. Reid J. B. Jr. Taylor G. J. Bower J. F. Powell B. N. and Dickey J. S. Jr. (1971) Mineralogy and petrology of the Apollo 12 lunar sample. Smithsonian Astrophysical Observatory Special Report 333.

Wood J. A. (1972) The nature of the lunar crust and composition of undifferentiated lunar material. Submitted to *The Moon* for publication.

York D. Kenyon W. J. and Doyle R. J. (1972) $^{40}Ar–^{39}Ar$ ages of Apollo 14 and 15 samples (abstract). In *Lunar Science—III* (editor C. Watkins), pp. 822–824, Lunar Science Institute Contr. No. 88.

Proceedings of the Third Lunar Science Conference
(Supplement 3, *Geochimica et Cosmochimica Acta*)
Vol. 1, pp. 819–835
The M.I.T. Press, 1972

Thermal and mechanical history of breccias 14306, 14063, 14270, and 14321

A. T. Anderson, T. F. Braziunas, J. Jacoby, and J. V. Smith

Dept. of the Geophysical Sciences, The University of Chicago,
Chicago, Illinois 60637

Abstract—Breccia 14306 has three generations: I, light-gray basaltic (noritic) metabreccia; II, dark-gray polymict metabreccia; and III, medium-gray polymict host matrix. Type I fragments are partly glassy and have Mg-rich ilmenite indicating recrystallization arrested at 850 to 1000°C. Type II fragments contain abundant subrounded, weak to strongly shocked plagioclase clasts and minor variously shocked granitic rocks, coarsely-exsolved orthopyroxenes and round, rimmed rock fragments suggestive of ash flow transport of ejecta derived from plutonic and volcanic sources. Preservation of concentration gradients in rhyolite and devitrification of diaplectic plagioclase glass suggest brief annealing near 800°C. Type III matrix is crystal-rich with many contorted grains but is otherwise similar to II. Tentatively we regard I as Pre-Imbrian, and II as the top of the Imbrian ejecta blanket (Fra Mauro formation). Type III possibly formed as II was violently mixed with underlying crystal-rich material, possibly by impact while hot.

Assuming ash flow transport without radiation loss, our data are consistent with initial impact heating amounting to 300°C or less, followed by heating during transport and deposition of less than 400°C, and a pre-impact temperature of the ejecta near or above 100°C.

INTRODUCTION

This paper describes those features of some Apollo 14 breccias that permit arguments about the thermal and mechanical history. The first section reviews observations bearing on the mechanical and thermal history, while the second section tries to assess the various contributions to the heat budget. In principle, the initial, pre-shock temperature of material ejected by the Imbrian impact can be obtained by estimating the temperature of material forming the ejecta blanket, subtracting contributions from shock heating and from kinetic energy converted to heat during transport, and adding losses by radiation and by incorporation of cold material during transport. Although there are many practical uncertainties, we felt it helpful to suggest one possible explanation of the features observed.

Briefly, 14306 is a multigeneration metamorphosed breccia. The oldest fragments are basaltic (noritic) metabreccias that are coarsely crystalline and partly glassy. The second-generation fragments are dark gray, micro- to crypto-crystalline polymict metabreccias. Fragments of plagioclase, pyroxene (including coarsely exsolved varieties), Fe metal, rhyolite (concentrically zoned), granophyre and granite (grain size 0.2 to 0.4 mm) are erratically distributed, variably rounded and variably shocked. The third-generation is similar to the second-generation breccia, and is the host matrix. It is medium- to light-gray, rich in irregular crystals of plagioclase and pyroxene, with locally subparallel elongation, and poor in metal.

The first generation of breccia may predate the impact. It was fragmented and combined with plutonic and volcanic material to produce the second generation. The third generation resulted from some process, possibly an impact, that embedded second-generation material in a crystal-rich matrix. Finally, the 3-generation breccia was thrown three crater diameters from Cone Crater.

Breccias 14063, 14270, and 14321 had histories differing from that of 14306, but documentation is still incomplete.

Petrographic Observations and Interpretations

Evidence on pre- and post-shock thermal histories of breccias 14306, 14063, 14321, and 14270

Fragments of rhyolite, granophyre, and granite. Fragments of glassy rhyolite, granophyre, and granite are relatively abundant in 14306, 14321, and 14270 (Table 1, Figs. 1, 2, and 3). Granitic fragments in 14306 are subrounded and one large fragment has a margin of glassy rhyolite (Fig. 4). Several rhyolite fragments contain cores of quartz and feldspar. The K-feldspar and quartz associated with rhyolitic glass contain fractures, and some grains have low birefringence and planar elements (Fig. 5). Glassy rhyolite and apparently unshocked granophyre occur together in the same matrix. The glassy rhyolite fragments are subrounded to subangular and commonly are zoned continuously with tan interiors and colorless rims up to 60 μ wide (Fig. 1). The rims are rich in crystals of metallic iron (Fig. 6), and are richer in Na_2O and poorer in K_2O than the interior (Table 2).

Fragments of glassy and vesicular rhyolite in 14270 differ from their counterparts in 14306; they are colorless throughout, bordered by rims of pyroxene (Fig. 7), and poor in Fe metal. One fragment contains a large subrounded crystal of quartz. Probable rhyolitic glasses in 14321 are similar to those in 14306 but appear unzoned and contain little metal.

Fragments of basaltic metabreccia. Light-gray, fine-grained, mottled, basaltic ("noritic") fragments with a range of textures occur in all our Apollo 14 breccias. Textures include microgranular, subophitic, ophitic, and poikilitic. A little interstitial glass is common. Almost without exception, fragments lack vesicles and vugs. Except for large anhedral crystals, analyzed pyroxenes and ilmenites are compositionally uniform within individual fragments, but differ from one fragment to another (Table 3). Some large included plagioclase grains are zoned concentrically. Although widely varied and subject to poor statistics, the large clastic grains apparently do not include exsolved orthopyroxene (Table 1).

The distribution of Fe, Mg between pyroxene and ilmenite in four fragments of 14306 yielded temperature estimates near 850 to 1000°C (Fig. 8).

Matrix crystallinity. Warner (1971, 1972a) established a useful petrographic classification of Apollo 14 breccias based on grain size, texture, and mineral-chemical equilibration of the matrix. The crystallinity of the matrix of the fragments of light-gray basaltic metabreccia (Fig. 9) corresponds approximately to Warner's two highest groups. The matrix of the dark basaltic polymict metabreccia fragments in 14306 (Fig. 10) corresponds to Warner's groups 5 and 6, although the general matrix (Fig. 11) of 14306 falls in group 4. The scale of in situ matrix recrystallization (5 μ— e.g., pyroxene rims on silica grains—Fig. 10) of the dark polymict breccia fragments corresponds to the groundmass grain size of the upper centimeters of a terrestrial

Table 1. Abundance of Rock and Mineral Fragments in Apollo 14 Breccias.[1]

Sample	Matrix[2] type	Plagioclase[3]				Opx[4]	Cpx[4]	Spinel	Olivine	Zircon	Quartz[5]	Rhyolite	Grano-phyre	Granite	Fe metal	Brown devit. glass
		0	1	2	3											
14306,53-F,4	A	6	2	—	—	?1	?1	—	3	—	—	—	—	—	—	—
14306,53-G,3-4	A	5	3	—	2	—	?2	—	?2	—	—	3	5	—	—	—
14306,53-F,3	A	—	2	3	—	—	—	—	—	tr	—	?	—	—	0.04	—
14306,53-E,1-2	A	4	2	2	—	—	8	—	—	tr	—	—	—	—	—	—
14306,53-D,4-5	A	2	5	1	—	—	1ex?	—	—	—	—	—	—	—	—	—
14306,53-C,1-2	A	4	1	—	—	—	2ex	—	—	—	—	—	—	—	—	—
14306,53-B,5	A	5	—	3	—	—	?1	tr	6	—	—	—	—	—	—	—
14306,53-D,5	A	2	10	—	—	—	?tr	—	2	—	—	—	—	—	tr	—
14306,53-B,1	B	4	1	1	tr	1	2ex	tr	2	tr	1	1	1	2	1	—
14306,53-B,3	B	6	1	1	1	3ex	1	tr	tr	tr	1	4	1	tr	0.1	—
14306,53-C,3-4	C	6	2	—	1	2ex	2	tr	tr	tr	1	2	tr	tr	0.02	0.4
14306	host	8	15	6	2	—	tr	—	tr	—	?tr	tr	1	tr	tr	tr
14321,231	?host	7	—	—	2	1ex	tr	—	—	—	—	1	—	—	1	—
14270,10	?host	9	2	1	3	1ex	—	—	3	—	—	1	—	—	1	—
14063	host	15	5	—	tr[6]	1ex	—	—	1	—	?tr	—	—	?tr	tr	1[6]

[1] Visual estimates of volume percent for grains larger than 0.08 mm.

[2] 14306 subdivided into:
A = light-gray basaltic metabreccia, e.g., Fig. 9.
B = medium-gray polymict metabreccia.
C = dark-gray polymict metabreccia, e.g., Fig. 10.
Other rocks not subdivided.

[3] Stages of shock:
0 = no shock—straight extinction, normal birefringence
1 = weak shock—undulatory extinction
2 = moderate shock—mosaic—patchy extinction, e.g., Fig. 21
3 = strong shock—sutured, e.g., Fig. 20.

[4] ex = exsolution lamellae present in some grains.

[5] Possibly includes some tridymite and cristobalite-identification based on relief and reflectivity, uniaxial positive.

[6] Mostly diaplectic plagioclase glass and undevitrified brown glass.

tr = trace (<1%).

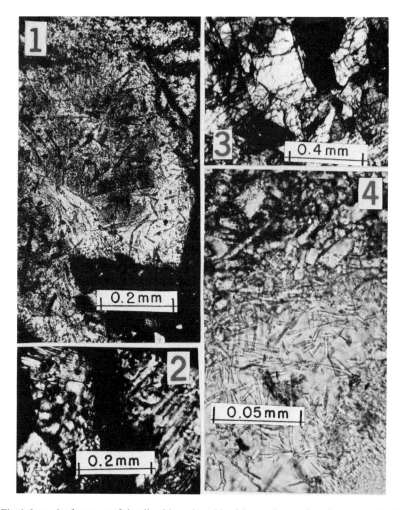

Fig. 1. Irregular fragment of rhyolite. Note clear, Na-rich margin, needles of pyroxene, black cubes of Fe metal, dark matrix. Section 14306,55. Plane polarized light.

Fig. 2. Granophyric aggregate of quartz and potassium feldspar in dark gray polymict breccia. Note regular intergrowth. Section 14306,53. Crossed polarizers.

Fig. 3. Granitic fragment in 14306,53. Note regular grain boundaries. Crossed polarizers.

Fig. 4. Rhyolitic margin of granitic fragment of Fig. 3. Granular dark-gray matrix in top 1/3. Strongly shocked granite in lower right corner with hint of lamellar structures. Glassy rhyolite in middle section has microphenocrysts of pyroxene needles. Plane polarized light.

basalt flow (~ 1200°C) or the upper ten meters or more of a dacite flow (~ 1000°C) or ash flow tuff (~ 900°C).

Porosity. The recrystallized breccias contain various kinds and amounts of pores. Dark metabreccia in 14306 lacks obvious pores. Breccia 14270 contains about 30 vol. % vesicles while 14321 has about 5 % of small vugs. Warner (1972b) estimates

Fig. 5. Planar elements probably caused by shock in granite, same fragment as in Fig. 3.

Fig. 6. Marginal rim on rhyolite fragment enriched in cubes of Fe metal (white squares). Same fragment as in Fig. 1. Reflected light.

Fig. 7. Rhyolite fragment in 14270,10. Note reaction zone of pyroxene (light-gray across top of picture) between rhyolite (lower half of picture) and matrix (upper quarter of picture). Note oval grain of quartz (center, lower third of picture) and edge of vesicle filled with mottled epoxy (across bottom tenth of picture). Reflected light.

Table 2. Microprobe Analyses (wt %) of Phases in Rhyolite and Granite.

	1	2	3	4
SiO_2	75	75	62.5	53.0
Al_2O_3	12.2	12.4	18.4	2.3
FeO	1.7	0.7	—	18.1
MgO	0.08	0.03	—	7.7
CaO	1.0	0.7	0.8	14.5
Na_2O	0.25	0.99	0.5	≤0.04
K_2O	9.1	8.1	13.7	—
P_2O_5	0.12	0.08	—	—
BaO	0.80	0.11	2.6	—
ZrO_2	0.13	0.12	—	—
TiO_2	0.49	0.46	—	0.09
Cl	0.03	—	—	—
S	0.004	—	—	—
Cu	0.005	—	—	—
TOTAL	100.91	98.69	98.5	96.5

Key: (1) Brown glass interior of rhyolite fragment 14306,55-B. (2) Clear glass exterior of rhyolite fragment 14306,55-B. (3) Potassium feldspar crystal in granite fragment 14306,53-20A. (4) Pyroxene microphenocryst in interior of rhyolite fragment 14306,55-B.

Table 3. Microprobe Analyses of Coexisting Pyroxenes and Ilmenites in Breccia 14306.

	1	2	3	4	5	6	7
FeO[1]	42.0	38.9	41.9	40.6	41.8	—	—
MgO[1]	3.6	5.5	3.7	4.6	4.1	—	—
FeSiO$_3$[2]	34.0	31.7	37.4	33.8	41.3	42.6	16.9
MgSiO$_3$[2]	58.3	47.5	52.1	57.0	43.7	47.0	33.2
CaSiO$_3$[2]	4.5	7.7	5.7	5.6	8.6	2.6	45.3
$(Mg/Fe)_{Px}/(Mg/Fe)_{Il}$	14.8	7.8	11.6	11.0	8.0	—	—

Key: (1) Averages of 3 ilmenite and 3 pyroxene grains; 14306,56-A. (2) Averages of 3 ilmenite and 2 pyroxene grains; 14306,56-C. (3) Adjacent ilmenite and pyroxene; 14306,56-D. (4) Averages of 3 ilmenite and 3 pyroxene grains; 14306,56-E. (5) Averages of 2 ilmenite and 5 pyroxene grains; 14306,56-M. (6 and 7) Exsolved host and lamellae in pyroxene grain in dark gray polymict breccia: 14306,53.

[1] Wt. % FeO and MgO in ilmenite.
[2] Wt. % of FeSiO$_3$, MgSiO$_3$, and CaSiO$_3$ in pyroxene based on Fe, Mg, Ca analyses.

20 to 33% pores for all Apollo 14 breccias. Watkins and Kovach (1972) estimate 33 to 60% porosity for the second layer (299 m/sec = Fra Mauro formation?). Porosity apparently is unevenly distributed and requires further study.

Reaction halos. Large (100 μ) angular to subrounded grains of quartz (or other high-SiO$_2$ phase with low reflectivity and moderately low refringence), pyroxene, plagioclase, and spinel commonly have recrystallized or zoned margins 0 to 50 μ wide (Figs. 12–16). Because complementary halos around quartz and spinel occur (Figs. 12 and 15) in the surrounding matrix, recrystallization of these minerals undoubtedly occurred in the present matrix. Reaction rims and halos are most pronounced in 14270 and the dark and host breccias in 14306. Warner (1972a,b) noted systematic

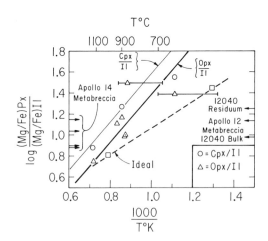

Fig. 8. Distribution of Fe and Mg between pyroxene and ilmenite. Plotted points are selected from Anderson (1966, 1968), Carmichael (1967), Anderson and Wright (in press). Ideal line is based on thermochemical data given by Robie and Waldbaum (1968) and Lindsley *et al.* (1964). Data for Apollo 12 rocks are from Smith *et al.* (1971). Apollo 14 data are given in Table 3. The Apollo 14 pyroxenes are low-Ca pigeonites and probably should be considered to be orthopyroxenes for purposes of this plot. The evident inverse correlation between Ca content of pyroxene and distribution factor (Table 3) probably is a thermal phenomenon.

Fig. 9. Matrix of light-gray basaltic metabreccia (generation I) in 14306,53 showing sub-hedral grains of pyroxene (light gray), euhedral plagioclase grains (medium-gray), and interstitial glass (dark gray). Few white specks of Fe metal in glass. Reflected light.

Fig. 10. Matrix of dark-gray polymict metabreccia (generation II) in 14306,53. Plagioclase-dark gray, pyroxene-light gray. White specks include carbon coat in cracks and ilmenite grains. Note quartz (darker than plagioclase in upper right quarter) rimmed by pyroxene. Reflected light.

Fig. 11. Host matrix (generation III) of breccia 14306,53. Plagioclase, pyroxene, and quartz as in Fig. 10. Large dark-gray grain at top center probably is rhyolitic glass. Note pyroxene rims on probable quartz grains (small grains in upper right quarter, medium sized grain in lower left corner). Note contorted shape of large plagioclase grain in center. Reflected light.

development of compositionally distinct rims on large pyroxene and plagioclase grains in progressively metamorphosed breccias.

Exsolution lamellae. Many pyroxenes in the breccias contain exsolution lamellae up to 20 μ wide (Table 1, Figs. 17 and 18). The compositions of the exsolved pyroxenes are consistent with equilibration below about 1000°C. By analogy with terrestrial occurrences, the dimensions of the lamellae imply slow cooling in a plutonic environment blanketed by a few kilometers of solid rock. Rock fragments that contain the exsolved pyroxenes (e.g., Fig. 17) are not particularly coarse-grained and lack cumulate textures, suggesting that crystallization occurred in a thin sill or dike followed by cooling in a plutonic environment.

Devitrification of diaplectic glass. Breccias 14306, 14321, and 14270 apparently lack diaplectic glass. However, feathery, cryptocrystalline fragments of plagioclase (Figs. 19 and 20) were interpreted as devitrified diaplectic plagioclase glass (Engelhardt

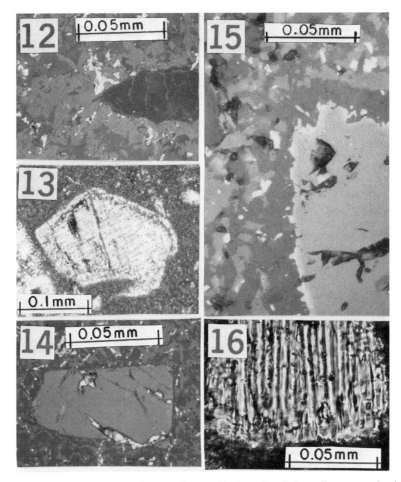

Fig. 12. Pyroxene (light gray) halo around quartz (dark gray) grain in medium-gray polymict metabreccia (unenumerated generation) in 14306,53. Reflected light.

Fig. 13. Recrystallized rim on pyroxene grain in dark-gray polymict metabreccia in 14306,53. Transmitted light superimposed on reflected light.

Fig. 14. Spinel grain (medium-gray) in dark-gray polymict metabreccia of 14306,53 (generation II). Note *lack* of reaction rim. Reflected light.

Fig. 15. Spinel grain (medium-gray, lower-right third of picture) with iron-rich rim (light-gray) and plagioclase halo in adjacent matrix of 14270,10. Reflected light.

Fig. 16. Recrystallized (?) feldspar rim on quartz-feldspar intergrowth in breccia 14306,53. Plane polarized transmitted light.

et al., 1970; Chao *et al.*, 1970; Short, 1970; Sclar, 1970). Radial patterns of such fragments in 14321 (Fig. 19) demonstrate that devitrification occurred after shaping of the particles and presumably in situ. Such patterns are absent in 14306, and prior devitrification may have occurred for these fragments. Annealing experiments revealed that heating at 800 to 900°C for many hours is required to devitrify diaplectic

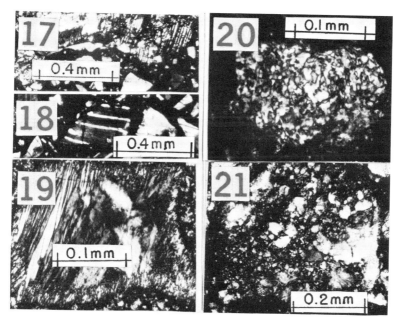

Fig. 17. Margin of large fragment of gabbro in dark-gray monomict breccia fragment in 14306,54. Note coarse exsolution lamellae in orthopyroxene (upper right quarter of picture). Crossed polarizers.

Fig. 18. Crystal of exsolved orthopyroxene in same fragment as Fig. 17. Crossed polarizers.

Fig. 19. Fragment of concentrically-devitrified diaplectic plagioclase glass in 14321,213. Crossed polarizers.

Fig. 20. Fragment of coarsely-devitrified diaplectic plagioclase glass in 14306,53. Crossed polarizers. Compare Fig. 1E of Short (1970), Fig. 9b of Sclar (1970).

Fig. 21. Fragment of shock-granulated plagioclase in 14306,53. Compare Fig. 11 of Quaide and Bunch (1970). Crossed polarizers.

plagioclase glass (Engelhardt *et al.*, 1970; Chao *et al.*, 1970; Engelhardt *et al.*, 1972). Thus absence of diaplectic glass and presence of its devitrified counterpart suggests annealing at temperatures near or above 800°C for several hours or days.

Evidence bearing on the amount of initial shock heating

Surface-tension rounded particles. Our recrystallized breccias contain less than 1% glass beads or their devitrified and crystalline counterparts. In 14306, 14063, and 14321 round fragments of devitrified or recrystallized glass are rare (Table 1). Round fragments of devitrified glass are more common in 14270 but are still far less abundant than in soil (Carr and Meyer, 1972).

Shock metamorphism of plagioclase. Several distinguishable generations of Apollo 14 breccias contain erratically distributed, variously shocked plagioclase (Table 1). Most plagioclase fragments in the dark-gray polymict basaltic metabreccias are unshocked, but many resemble the weakly shocked patchy or mosaic aggregate (Fig. 21) illustrated by Quaide and Bunch (1970). About 10 to 20% of all plagioclase clasts in 14306, 14321, and 14270 are microcrystalline aggregates (Fig. 20), such as those illustrated by Short (1970, Fig. 1e) and Sclar (1970, Fig. 9b). A dark-gray monomict basaltic metabreccia fragment in 14306, which is rich in coarsely-exsolved pyroxene, contains only plagioclase with normal birefringence and uniform extinction. Plagioclase in the light-gray recrystallized host matrix of breccia 14306 is heterogeneously shocked: some regions are rich in microcrystalline plagioclase, but others contain little. Rare grains of diaplectic plagioclase glass occur in 14063.

In breccias 14306, 14063, 14270, and 14321 there are rare polycrystalline fragments of subhedral plagioclase, minor glass and traces of other minerals which may be crystallized shock melts (cf. Sclar, 1970, Fig. 8).

Evidence bearing on the mode of transportation

Roundness and size of plagioclase grains. The roundest grains (Table 4) are found in the dark-gray polymict breccia fragments in 14306. They are largely unshocked and presumably had the longest history of gentle abrasion.

Occurrence of Fe metal. Metallic iron is heterogeneously distributed throughout breccia 14306. Large grains (≥ 0.022 mm) are most common in the dark-gray polymict metabreccia fragments, where the total metal content is about 0.1% by volume. The medium-gray host matrix is very poor in large metal grains and has only 0.02 vol.% of total metal. Rhyolite fragments have metal-rich margins and about 0.1 to 0.3% metal on the average per total fragment.

Evidence bearing on mode of deposition

Sorting of plagioclase. Sorting of plagioclase clasts coarser than $4\,\mu$ (Table 5) in the breccias is similar to that in terrestrial ash flow deposits (Sheridan, 1971), suggesting a transport mechanism similar to that proposed by Masursky (1968), Fisher and Waters (1969), McKay and Morrison (1971), and Pai *et al.* (1972). The size distribution over a limited range fits a power law predicted by Gault, Shoemaker, and Moore (1963) for crushed rock, as was found by others who studied the grain size distribution in lunar soil and breccia (Quaide and Bunch, 1970; Heywood, 1971; Görz *et al.*,

Table 4. Roundness of Plagioclase Clasts (>0.06 mm).

Sample	Matrix[1]	Percent of grains in roundness category[2]							Total No. of grains
		0.1	0.2	0.3	0.4	0.5	0.6	0.7	
14306,53-E,1–2[3]	A	35	35	22				8	14
14306,53-C,3[3]	C	12	35	47	6				17
14306,53-F,16[3]	D	37	19	35	9				32
14306,53-G,16[3]	E	11	36	22	14	14	3		36
14306,53-I,1–2[4]	C	0	13	41	26	11	4	2	46

[1](A) Light-gray basaltic metabreccia, e.g., Fig. 9. (C) Dark-gray polymict metabreccia, e.g., Fig. 10. (D) Dark-gray basaltic metabreccia, Figs. 17 and 18. (E) Medium-gray host matrix, Fig. 11.

[1] See Table 5, next page.
[2] Visual estimates based on comparison with Fig. 4–10, p. 111 of Krumbein and Sloss (1968).
[3] J. Jacoby, operator.
[4] T. Braziunas, operator.

Table 5. Size distribution of plagioclase clasts.

Sample[1]	Matrix[1] type	No. of grains per 2.75 mm² above size limit (μ) (maximum dimension as seen in section)								Sorting coefficient
		500[2]	250[2]	125[2]	62[2]	31[3]	16[3]	8[4]	4[4]	
14306,53-E,12	A	0	2	5	36	—	—	—	—	—
14306,53-C,3	C	0	1	8	22	—	—	—	—	—
14306,53-F,16	D	0	4	17	52	—	—	—	—	—
14306,53-G,16 host		0	5	19	55	—	—	—	—	—
14306,53-I,1–2	C	0	1	7	46	270	1326	5926	≥11726	3.6[8]
Base Surge Dune[5]		0	0	6	40	189	1500	5320	45720	2.8
Bishop Ash Flow[5]		0	0	4	38	330	1500	8360	32360	3.7
Apollo 11 Breccia[6]		0	2	7	10	—	—	—	—	—
Apollo 12 Soil 12057,72[7]		0	0	1	25	300	1500	4000	25000	2.4

[1] (A) Light-gray basaltic metabreccia, e.g., Fig. 9. (C) Dark-gray polymict metabreccia, e.g., Fig. 10. (D) Dark-gray basaltic metabreccia, Figs. 17 and 18.
[2] Measured in transmitted light photomicrographs at 40 × magnification.
[3] Measured in reflected light photomicrographs at 160 × magnification.
[4] Measured in reflected light photomicrographs at 400 × magnification.
[5] From Sheridan (1971, Fig. 1) normalized to 1500 at 16 μ.
[6] From Quaide and Bunch (1970, Fig. 7) normalized to 1500 at 16 μ.
[7] From Heywood (1971, Fig. 5) normalized to 1500 at 16 μ.
[8] Obtained by converting to area (vol.) % and log-linear extrapolation of cumulative frequency curve, assuming 0.55 = average width-to-length ratio as obtained for 79 grains.

1971). Evidence of the presence of gas needed for winnowing during transport was presented by Carter and MacGregor (1970), and McKay *et al.* (1970).

SOURCE OF THE BRECCIAS

It is difficult to establish which matrix and recrystallization is indigenous to the Fra Mauro formation, because multiple stages of breccia formation are evident (LSPET, 1971). More than one stage may be indigenous to the Fra Mauro formation because of a possible complex and prolonged depositional history (hours) including partially contemporaneous ash flow deposition and bombardment by ballistic projectiles. (Time of flight estimates for ballistic trajectories from the Imbrian Crater range from 15 minutes to at least several hours. A ground-hugging ash flow, with only a small fraction of momentum carried by gas, would have a similarly short travel time.)

Dence *et al.* (1972), Duncan *et al.* (1972), Engelhardt *et al.* (1972), Floran *et al.* (1972), Grieve *et al.* (1972) Hutcheon *et al.* (1972), Quaide (1972), and Wilshire and Jackson (1972) attributed early brecciations to pre-Imbrian events. Hait (1972) identified Fra Mauro matrix with contorted portions of the white rocks. On the other hand, the preponderance of dark-clast-dominant breccias at the Apollo 14 site suggests a near-surface origin; whereas the restriction of light-clast-dominant breccias to the rim of Cone Crater (Swann *et al.*, 1971b) suggests a deeper origin, possibly from beneath Fra Mauro formation. Tentatively we accept the interpretation of Swann *et al.* that the dark clast breccias (including 14306 and 14321) are pieces of Fra Mauro formation.

We attempted to reconstruct the history of dark-clast breccias by coordinating information on the populations, textures, and topologic relations bearing in mind the orthodox interpretation of the Fra Mauro formation as the product of the Imbrian impact. A preferred history was summarized in the Introduction and is now discussed in detail.

Tentatively, we identify the dark-gray polymict basaltic metabreccia clasts (generation II) in 14306 as broken pieces of the uppermost parts of the Imbrian ejecta blanket (Fra Mauro formation), because of (1) maximum roundness of plagioclase clasts, (2) presence of helter-skelter clasts with variable shock features and compositions that include plutonic members (granite and coarsely exsolved orthopyroxene), (3) agreement between expected near-surface derivation of 14306 from Cone Crater and an independently inferred history of rapid cooling, and (4) maximum content of metallic iron. The light gray (generation I) basaltic (noritic) metabreccias probably are pre-Fra Mauro (pre-Imbrian) breccias. The host matrix of 14306 (generation III) may be a late stage feature indigenous to the local Fra Mauro formation. The reasons for these assignments follow.

The relatively round shapes of plagioclase grains in the dark polymict metabreccia fragments in 14306 are qualitatively consistent with expectations for an ash flow transport mechanism such as has been proposed for the Fra Mauro formation (Eggleton, 1965).

Because turbulent ash flow probably would mix the particles rather well (Fisher and Waters, 1970), any candidate matrix for the original Imbrian ejecta blanket should contain a diverse assemblage of variously shocked rocks and minerals. Rocks derived from 5 to 10 and perhaps even 50 km, deep within the moon should be present in the Imbrian ejecta blanket at the Apollo 14 site (Eggleton and Offield, 1970). The granite and the coarsely exsolved orthopyroxene in the dark-gray polymict breccia fragments in 14306 require slow cooling and thermal blanketing by several km of rock. Many different rock types occur in the generation II and III breccias. However, the light-gray basaltic metabreccias (generation I) contain a less diverse assemblage of clasts with a paucity or total absence of slowly cooled rocks (Table 1), and may be fragments of pre-Imbrian breccias recrystallized prior to the Imbrian event. A cursory examination by A. T. Anderson of nineteen Apollo 15 breccias revealed minor amounts of light-gray basaltic (noritic) metabreccia fragments and apparent absence of both the plutonic rock suite (granite and exsolved orthopyroxene) and dark-gray polymict metabreccia similar to that in 14306. This agrees with the pre-mission predictions of dominant pre-Imbrian material derived from the highly eroded Apennine massifs (Carr *et al.*, 1971) and with a local (Fra Mauro) construction of generation II breccia from Imbrian ejecta.

The erratic occurrence and degree of shock in plagioclase from the generation II and III breccias imply mixing together of unshocked to strongly shocked materials (Chao *et al.*, 1970). About 10% of the plagioclase clasts in breccias 14306, 14321, and 14270 were strongly shocked, compared to 4% in breccias 10020 and 10061 (Quaide and Bunch, 1970) and to 1 to 20% in Apollo 12 regolith (Quaide *et al.*, 1971). Gault and Heitowit (1963) and Quaide and Oberbeck (1969) suggest that less than 1% of all ejecta from a single impact will be noticeably shocked. The Apollo 14 breccias apparently are either a strongly biased aliquot (consistent with high ejection velocity for long travel) or show combined effects from multiple impacts (consistent with orthodox ideas of multiple impacts on the early moon).

The Apollo 14 dark-clast breccias do not all have the same amounts and proportions of rock and mineral fragments (Jackson and Wilshire, 1972). The evidently poor mixing of the components and the heterogeneity of some large blocks (Swann *et al.*, 1971a,b) suggest multiple episodes of accumulation (Wilshire and Jackson, 1972). Some of these episodes possibly occurred in pre-Imbrian time. Late-stage disturbance of an initially heterogeneous ejecta blanket could partially mix separate portions and explain the formation of the third-generation matrix in 14306.

Breccia 14306 has a Cone Crater exposure age (Crozaz *et al.*, 1972) and was found 3 crater diameters away from the center of Cone Crater; hence it probably came from the upper few meters of the Fra Mauro formation (Gault *et al.*, 1963). Rhyolitic fragments in 14321 (from the rim of Cone Crater) are clear and apparently unzoned. Zoning of the rhyolite glass in 14306 would develop in about one minute at 1000°C but would require about an hour at 700°C (Harry Smith, University of Oregon, written communication, 1972). The inferred near-surface initial position of 14306 supports an especially rapid cooling that could quench in the concentration gradients not found in 14321. These facts constitute the only known evidence that relates the recrystallization of particular breccias to a probable local thermal history, but further study of annealed breccias containing rhyolitic glass is required to test this tentative conclusion.

Thermal History of the Breccias

The generation I breccias recrystallized near 850° to 1000°C as evidenced by glass and the pyroxene-ilmenite geothermometer (see above). This agrees with the independent estimate of Grieve *et al.* (1972).

The generation II dark, fine-grained polymict metabreccias of 14306 as well as the lighter host matrix of 14306 and 14321 recrystallized at about 800 to 900°C as evidenced by the devitrification of diaplectic plagioclase glass and by the development of composition gradients in rhyolitic fragments.

Mechanical Contribution to Heating

The mechanical contribution from impact consists of the prompt heating from passage of the shock wave(s) and conversion of kinetic energy into heat during transport and deposition.

Prompt heating

Such heating from shock waves appears to be minor, because devitrified glasses rounded by surface tension are rare, devitrified diaplectic glass is not abundant, and many granitic fragments survived with little shock damage.

The microcrystalline plagioclase aggregates are assumed to be recrystallized diaplectic glasses indicating shock pressures over 300 kbars (Chao *et al.*, 1970; Stöffler, 1971). Lamellar and planar features are anomalously rare by comparison with terrestrial ejecta blankets, perhaps explained by the highly calcic composition (Engelhardt *et al.*, 1970). Cold solid anorthite should melt under shock at 700 kbars, but warm anorthite in a high-impedance matrix would melt at lower pressure (Chao *et al.*, 1970). The average prompt heating for plagioclase would be $\leq 310°C$, assuming 80% low shock ($\leq 200°C$), 15% strong shock ($\leq 500°C$), and 5% very strong shock ($\sim 1500°C$) and the calibrations of Chao *et al.* (1970) and Stöffler (1971). The average prompt heating for the whole breccia may differ from that estimated for the larger plagioclase clasts: reshocked material may be present; more highly shocked material may be concentrated in the fine-grained matrix; shock heating of pyroxene and ilmenite associated with plagioclase probably would be less because of their higher impedance; large and small grains and different minerals may derive from differently shocked regions. Tentatively, we assume 300°C for the bulk prompt shock heating.

Heating during transport

This factor cannot be estimated accurately without determination of the mode of transport. Ballistic particles smaller than a few centimeters should lose considerable heat by radiation. Ash flows may suffer little radiation loss, but should undergo cooling from incorporation of cold regolith. Both gain heat from kinetic energy. Although the breccias contain only little evidence of transport in a turbulent ash flow (e.g., rounding of fragments), the topographic and geologic evidence strongly support such a mode for the Fra Mauro formation (Eggleton, 1965; Masursky, 1968; Swann *et al.*, 1971a,b). Tentatively we use the circular velocity (1.7 km/sec) and 50% conversion efficiency of kinetic energy to heat (Braslau, 1970) to yield 100 cal/gm. The resultant heating of about 400°C is regarded as a generous upper limit. Further experimental and theoretical work is needed to refine this estimate.

DISCUSSION

The textural evidence for temperatures near 800°C for brief thermal metamorphism in the upper part of the ejecta blanket, plus generous estimates of mechanical contributions totaling 700°C, imply that the upper part of the Imbrian crust averaged at least 100°C at the time of impact. The preferential shock melting of granitic (as compared to granophyric) fragments is consistent with pre-impact granite hotter than granophyre, with the former at greater depth and temperature. Although these suggestions are highly speculative, the present data are consistent with a pre-Imbrian (?warm) crust dominated by basalts and minor feldspar-rich rocks, at least some of which had cooled slowly, rather than a simple anorthosite layer some tens of kilometers thick.

Acknowledgments—We thank Edward Anders, Robert Clayton, and John Jamieson of the University of Chicago and Jeffrey Warner of NASA-MSC, Houston for helpful discussions. Harry Smith of the University of Oregon kindly evaluated our analytical data on zoned rhyolite fragments in the light of his unpublished experimental data on diffusion of alkalis in rhyolitic glass. Elbert King (University of Houston), S. I. Pai (University of Maryland), and David Stewart (U.S. Geol. Survey, Washington, D.C.) contributed helpful correspondence. We thank W. L. Quaide (NASA-Ames) for a thorough, although discouraging, review. We thank I. Baltuska, O. Draughn, and R. Zechman for technical help, and NASA for grant NGL-14-001-171.

REFERENCES

Anderson A. T. (1966) Mineralogy of the Labrieville anorthosite, Quebec. *Amer. Mineral.* **51**, 1671–1711.

Anderson A. T. (1968) Oxidation of the LaBlache Lake titaniferous magnetite deposit, Quebec. *J. Geol.* **76**, 528–547.

Braslau, D. (1970) Partitioning of energy in hypervelocity impact against loose sand targets. *J. Geophys. Res.* **75**, 3987–3999.

Carmichael I. S. E. (1967) The iron-titanium oxides of salic volcanic rocks and their associated ferromagnesian silicates. *Contr. Mineral. Petrol.* **14**, 36–64.

Carr M. H. Howard K. A. and El-Baz F. (1971) Geologic maps of the Apennine-Hadley region of the moon. U.S. Geol. Survey, Geologic Atlas of the Moon, Apennine-Hadley Region—Apollo 15, Map. I-723.

Carr M. H. and Meyer C. E. (1972) Petrologic and chemical characterization of soils from the Apollo 14 landing site (abstract). In *Lunar Science—III* (editor C. Watkins), pp. 116–118, Lunar Science Institute Contr. No. 88.

Carter J. L. and MacGregor I. D. (1970) Mineralogy, petrology and surface features of some Apollo 11 samples. *Proc. Apollo 11 Lunar Sci. Conf., Geochim. Cosmochim. Acta*, Suppl. 1, Vol. 1, pp. 247–265. Pergamon.

Chao E. C. T. James O. B. Minkin J. A. Boreman J. Jackson E. D. and Raleigh C. B. (1970) Petrology of unshocked crystalline rocks and evidence of impact metamorphism in Apollo 11 returned lunar sample. *Proc. Apollo 11 Lunar Sci. Conf., Geochim. Cosmochim. Acta*, Suppl. 1, Vol. 1, pp. 287–314. Pergamon.

Crozaz G. Drozd R. Hohenberg C. M. Hoyt H. P. Jr. Ragan D. Walker R. M. and Yuhas D. (1972) Solar flare and galactic cosmic ray studies of Apollo 14 samples (abstract). In *Lunar Science—III* (editor C. Watkins), pp. 167–169, Lunar Science Institute Contr. No. 88.

Dence M. R. Plant A. G. and Traill R. J. (1972) Impact-generated shock and thermal metamorphism in Fra Mauro lunar samples (abstract). In *Lunar Science—III* (editor C. Watkins), pp. 174–176, Lunar Science Institute Contr. No. 88.

Duncan A. R. Lindstrom M. M. Lindstrom D. J. McKay S. M. Stoeser J. W. Goles G. G. and Bruchter J. S. (1972) Comments on the genesis of breccia 14321 (abstract). In *Lunar Science—III* (editor C. Watkins), pp. 192–194, Lunar Science Institute Contr. No. 88.

Eggleton R. E. (1965) Geologic map of the Riphaeus Mountains region of the moon. U.S. Geol. Survey Misc. Geol. Inv. Map I-458.

Eggleton R. E. and Offield T. W. (1970) Geologic maps of the Fra Mauro region of the moon. U.S. Geol. Survey, Geologic Atlas of the Moon, Fra Mauro Region—Apollo 14, Map I-708.

Engelhardt W. von Arndt J. Müller W. F. and Stöffler D. (1970) Shock metamorphism of lunar rocks and origin of the regolith at the Apollo 11 landing site. *Proc. Apollo 11 Lunar Sci. Conf., Geochim. Cosmochim. Acta*, Suppl. 1, Vol. 1, pp. 363–384. Pergamon.

Engelhardt W. von Arndt J. and Stöffler D. (1972) Apollo 14 soils and breccias, their compositions and origin by impacts (abstract). In *Lunar Science—III* (editor C. Watkins), pp. 233–235, Lunar Science Institute Contr. No. 88.

Fisher R. V. and Waters A. C. (1970) Bed forms in base surge deposits. Lunar implications. *Science* **165**, 1349–1351.

Floran R. J. Cameron K. Bence A. E. and Papike J. J. (1972) The 14313 consortium: A mineralogic and petrologic report (abstract). In *Lunar Science—III* (editor C. Watkins), pp. 268–270, Lunar Science Institute Contr. No. 88.

Gault D. E. and Heitowit E. D. (1963) The partition of energy for hypervelocity impact craters formed in rock. In *Proc. 6th Hypervelocity Impact Symp.*, Cleveland, Ohio, Vol. 2, pp. 419–456.

Gault D. E. Shoemaker E. M. and Moore H. J. (1963) Spray ejected from the lunar surface by meteroid impact. *NASA Pub. TN D-1767*, 39 pp.

Görz H. White E. W. Roy R. and Johnson G. G. Jr. (1971) Particle size and shape distributions of lunar fines by CESEMI. *Proc. Second Lunar Sci. Conf., Geochim. Cosmochim. Acta*, Suppl. 2, Vol. 3, pp. 2021–2025. M.I.T. Press.

Grieve R. McKay G. Smith H. and Weill D. (1972) Mineralogy and petrology of polymict breccia 14321 (abstract). In *Lunar Science—III* (editor C. Watkins), pp. 338–340, Lunar Science Institute Contr. No. 88.

Hait M. H. (1972) The white rock group and other boulders of the Apollo 14 site: A partial record of Fra Mauro history (abstract). In *Lunar Science—III* (editor C. Watkins), p. 353, Lunar Science Institute Contr. No. 88.

Heywood H. (1971) Particle size and shape distribution for lunar fines sample 12057–72. *Proc. Second Lunar Sci. Conf., Geochim. Cosmochim. Acta*, Suppl. 2, Vol. 3, pp. 1989–2001. M.I.T. Press.

Hutcheon I. D. Phakey P. P. Price P. B. and Rajan R. S. (1972) History of lunar breccias (abstract). In *Lunar Science—III* (editor C. Watkins), pp. 415–417, Lunar Science Institute Contr. No. 88.

Jackson E. D. and Wilshire H. G. (1972) Classification of the samples returned from the Apollo 14 landing site (abstract). In *Lunar Science—III* (editor C. Watkins), pp. 418–420, Lunar Science Institute Contr. No. 88.

Lindsley D. H. Davis B. T. C. and MacGregor I. D. (1964) Ferrosilite ($FeSiO_3$): Synthesis at high pressures and temperatures. *Science* **144**, 73–74.

LSPET (1971) (Lunar Sample Preliminary Examination Team) Preliminary examination of lunar samples from Apollo 14. *Science* **173**, 681–693.

Masursky H. (1968) Preliminary geologic interpretation of Lunar Orbiter photography. Hearings before the Subcommittee on Space Science and Applications, U.S. House of Representatives, Ninetieth Congress, H. R. Doc. 15086, pp. 664–691.

McKay D. S. Greenwood W. R. and Morrison D. A. (1970) Origin of small lunar particles and breccia from the Apollo 11 site. *Proc. Apollo 11 Lunar Sci. Conf., Geochim. Cosmochim. Acta*, Suppl. 1, Vol. 1, pp. 673–694. Pergamon.

McKay D. S. and Morrison D. A. (1971) Lunar breccias. *J. Geophys. Res.* **76**, 5658–5669.

Pai S. I. Hsieh T. and O'Keefe J. A. (1972) Lunar ash flows, how they work (abstract). In *Lunar Science—III* (editor C. Watkins), pp. 593–595, Lunar Science Institute Contr. No. 88.

Quaide W. (1972) Mineralogy and origin of Fra Mauro fines and breccias (abstract). In *Lunar Science—III* (editor C. Watkins), pp. 627–629, Lunar Science Institute Contr. No. 88.

Quaide W. L. and Oberbeck V. R. (1969) Geology of the Apollo landing sites. *Earth Sci. Rev.* **5**, 255–278.

Quaide W. and Bunch T. (1970) Impact metamorphism of lunar surface materials. *Proc. Apollo 11 Lunar Sci. Conf., Geochim. Cosmochim. Acta*, Suppl. 1, Vol. 1, pp. 711–729. Pergamon.

Quaide W. Oberbeck V. Bunch T. and Polkowski G. (1971) Investigations of the natural history of the regolith at the Apollo 12 site. *Proc. Second Lunar Sci. Conf., Geochim. Cosmochim. Acta*, Suppl. 2, Vol. 1, pp. 701–718. M.I.T. Press.

Robie R. A. and Waldbaum D. R. (1968) Thermodynamic properties of minerals and related substances at 298.15°K (25.0°C) and one atmosphere (1.013 bars) pressure and at higher temperature. U.S. Geol. Survey, Bull. 1259, 256 pp.

Sclar C. B. (1970) Shock metamorphism of lunar rocks and fines from Tranquillity Base. *Proc. Apollo 11 Lunar Sci. Conf., Geochim. Cosmochim. Acta*, Suppl. 1, Vol. 1, pp. 849–864. Pergamon.

Sheridan M. F. (1971) Particle size characteristics of pyroclastic tuffs. *J. Geophys. Res.* **76**, 5627–5634.

Short N. M. (1970) Evidence and implications of shock metamorphism in lunar samples. *Proc. Apollo 11 Lunar Sci. Conf., Geochim. Cosmochim. Acta*, Suppl. 1, Vol. 1, pp. 865–871. Pergamon.

Smith J. V. Anderson A. T. Jr. and Newton R. C. (1971) Final report for contract NAS 9-8086, submitted to NASA.

Stöffler D. (1971) Progressive metamorphism and classification of shocked and brecciated crystalline rocks at impact craters. *J. Geophys. Res.* **76**, 5541–5551.

Swann G. A. Trask N. J. Hait M. H. and Sutton R. L. (1971a) Geologic setting of the Apollo 14 samples. *Science* **173**, 716–719.

Swann G. A. Bailey N. G. Batson R. M. Eggleton R. E. Hait M. H. Holt H. E. Larson K. B. McEwen M. C. Mitchell E. D. Schafer J. P. Shepard A. B. Sutton R. L. Trask N. J. Ulrich G. E. Wilshire H. G. and Wolfe E. W. (1971b) Preliminary geologic investigations of the Apollo 14 landing site. *NASA Interagency Report 29*, 124 pp.

Warner J. L. (1971) Progressive metamorphism of Apollo 14 breccias. *Geol. Soc. Amer. Abstracts with Programs* **3**, No. 7, p. 744.

Warner J. L. (1972a) Apollo 14 breccias: Metamorphic origin and classification (abstract). In *Lunar Science—III* (editor C. Watkins), pp. 782–784, Lunar Science Institute Contr. No. 88.

Warner J. L. (1972b) Metamorphism of Apollo 14 breccias. *Proc. Third Lunar Sci. Conf., Geochim. Cosmochim. Acta*, Suppl. 3, Vol. 1. M.I.T. Press.

Watkins J. S. and Kovach R. L. (1972) Apollo 14 active seismic experiment. *Science* **175**, 1244–1245.

Wilshire H. G. and Jackson E. D. (1972) Petrology of the Fra Mauro formation at the Apollo 14 landing site (abstract). In *Lunar Science—III* (editor C. Watkins), pp. 803–805, Lunar Science Institute Contr. No. 88.

Proceedings of the Third Lunar Science Conference
(Supplement 3, *Geochimica et Cosmochimica Acta*)
Vol. 1, pp. 837–852
The M.I.T. Press, 1972

Petrology and origin of lithic fragments in the Apollo 14 regolith

Benjamin N. Powell

Department of Geology, Rice University
Houston, Texas 77001

and

Paul W. Weiblen

Department of Geology and Geophysics
University of Minnesota
Minneapolis, Minnesota 55455

Abstract—A systematic particle-by-particle study of Apollo 14 soil samples established six broad groups of lithic particle types, including: A. primary igneous basalts; B. microbreccias and glass-rich particles of mare basalt affinity; C. feldspathic microbreccias, including related anorthosites, norites, troctolites (ANT); D. complex multigenerational ANT microbreccias; E. mixed (basaltic + ANT) microbreccias; F. ultramafic particles (olivine- and/or pyroxene-rich).

Population studies indicate the regolith at the Apollo 14 site is characterized by a higher ratio of fragmental (degraded) to primary igneous lithic particles and a much higher ANT/basalt ratio than Apollo 11, 12, 15, and Luna 16 soil samples.

Detailed studies of groups A, C, D, and F permit several petrogenetic conclusions: (1) A few Apollo 14 basalts resemble mare basalts and probably have similar origins as thin lava flows (from Oceanus Procellarum). (2) Most Apollo 14 basaltic particles show affinities to KREEP basalts and are presumed to have a nonmare origin. (3) Rare basalts with the most primitive Mg/Fe ratios yet reported for lunar materials may represent ancient crustal material and/or melt compositions from the lunar interior from which many lunar materials were derived (including ANT). (4) ANT materials were produced by cumulate igneous differentiation processes in early lunar history and were probably complemented by settled ultramafic cumulates. (5) Rare primary igneous material of late-stage differentiated character resembles upper layers of terrestrial layered gabbroic complexes; liquid immiscibility appears to be an important mechanism in late-stage lunar differentiation. (6) ANT and ultramafic materials were excavated from depth in the lunar crust by a major cratering event (Imbrium). (7) Residence in a thick hot ejecta blanket (Fra Mauro formation) caused annealing and recrystallization of fragmental textures.

Introduction

In this paper we describe the results of our systematic study of the mineralogy, petrology, and chemistry of Apollo 14 soil samples. We have characterized and classified lithic particles and determined the relative abundances of the various rock types present in the samples. Specific rock types were studied in detail in order to determine their petrogenesis and, insofar as possible, their source in the moon.

Samples Studied and Analytical Procedures

Samples received for study are listed in Table 1. The two sample collection sites represented are both situated in what has been interpreted as "smooth terrain material of the Fra Mauro Formation" (Eggleton and Offield, 1970; Sutton *et al.*, 1971). After sieving, size fractions $> 149\ \mu$ were made into polished thin sections for transmitted and reflected light microscopy and electron microprobe analysis. Microprobe data reduction procedures used are essentially those described by Wood *et al.* (1970 and 1971). Most

Table 1. Apollo 14 soil samples studied in this investigation.

Sample Number	Sample	Nominal Size (mm)	Weight (g)
14161,29	bulk	2–4	1.0955
14162,41	bulk	1–2	0.5189
14163,77	bulk	<1	0.2521
14258,28	comprehensive	1–2	0.4828

petrogenetic conclusions are based largely on size fractions ($>233\ \mu$) in which lithic particles with representative mineralogy and textures predominate over single mineral and glass fragments and incomplete lithic fragments. Population studies were made on the >0.5 mm size fractions.

CLASSIFICATION OF LITHIC PARTICLES

Most rock types present as lithic particles are fragmental in character (though commonly recrystallized) or show other evidence of shock metamorphism, including deformation twins and kink bands in certain minerals, mosaicism, and fusion. Of 343 particles >0.5 mm only 14 ($\sim 4\%$) are homogeneous crystalline rocks with primary igneous textures. Six broad groups, distinguished on the basis of mineralogy and texture, were established. These are described briefly below.

Group A—Basalts

These rocks are homogeneous fine- to medium-grained hypocrystalline to holocrystalline basalts with primary igneous textures, including intersertal, subophitic, variolitic, and microporphyritic varieties (Figs. 1a–c). Basaltic particles are described more fully in the section on petrology.

Group B—Soil breccias and agglutinates with basaltic affinities

This group includes several specific lithic varieties, all of which exhibit a common affinity to the basaltic rocks. Included in this group are dark soil breccias, agglutinates and glass-rich particles. Breccia matrices range from opaque to transparent, from glassy to crystalline (recrystallized). Clasts are dominated by basaltic lithic varieties and basaltic mineral fragments, including notable amounts of ilmenite (Fig. 2). Minor amounts of feldspathic clasts (similar to group C) are present. Glasses are typically dark colored, usually in shades of yellow, orange, and brown and are very similar in appearance to those glasses with basaltic compositions from Apollo 11 samples (*e.g.*, Wood *et al.*, 1970, and Keil *et al.*, 1970).

Group C—Feldspathic rocks

This group is comprised of plagioclase-rich crystalline rocks with variably recrystallized fragmental textures and an apparent mutual affinity. Specific varieties include feldspathic norites, anorthosites, and troctolites (Fig. 3). These are described more fully below.

Group D—Complex feldspathic microbreccias

These particles are similar to those of group C except that they show two or more cycles of brecciation and aggregation. Clasts themselves commonly are microbreccias, containing their own breccia clasts (Fig. 4). Type C materials predominate, and typically two or all three feldspathic lithic varieties occur as clasts within a single particle. Clasts of basaltic material are very rare, though present in very minor amounts. Matrices typically are partially recrystallized. Interstitial glass, if present, is typically pale tan to colorless, with a feldspathic composition.

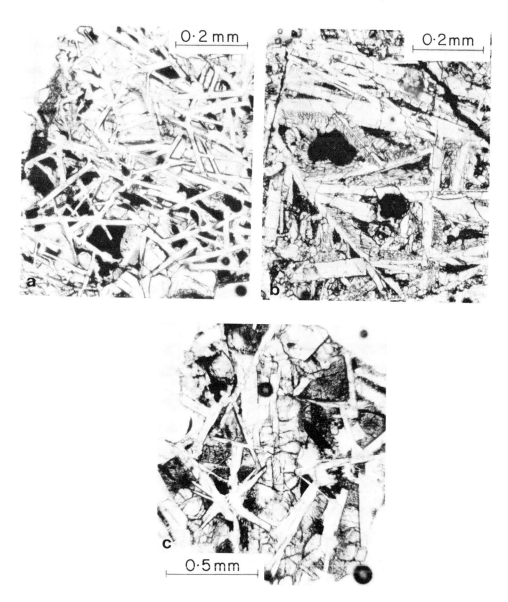

Fig. 1. Photomicrographs (transmitted light) of typical primary igneous textures in basalt particles (group A) from Apollo 14 soil samples 14162,41 and 14163,77. (a), upper left: subophitic texture in basalt like 14310; see also Fig. 6 (1407–4). (b), upper right: variolitic texture; note semi-radiating sheafs of pyroxene and plagioclase (1407–13). (c), lower: intersertal texture; hypocrystalline (pyroxene) mesostasis between plagioclase, olivine and large zoned clinopyroxene grains (1408–8).

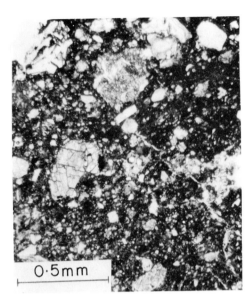

Fig. 2. Group B microbreccia (basalt affinity): clasts include basalt mineral fragments and orange-yellow glass; matrix is dark brown, largely glass (1425–12). (Transmitted light.)

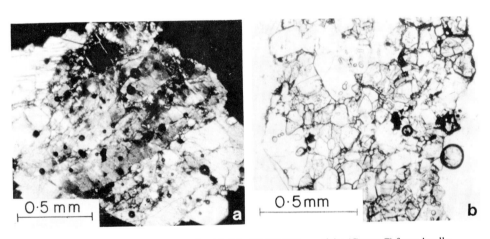

Fig. 3. Photomicrographs (transmitted light) of ANT lithic particles (Group C) from Apollo 14 soils. (a) Anorthosite (1419–7), 98% shocked plagioclase with accessory primitive diopside (see Table 6). (Dark rounded features are bubbles in epoxy.) (Crossed nicols.) (b) Strongly recrystallized norite with granulitic texture (1431–3). (See Table 6.)

Group E—Mixed microbreccias

These are complex (multigenerational) microbreccias consisting of approximately equal portions of group C components on the one hand and group A and B components on the other (Fig. 5). Matrices range from glassy to slightly recrystallized and appear to be similar to group B matrices. The distinctive characteristic of these particles is the admixing of two components that otherwise are generally rather well separated, *i.e.*, basaltic materials versus ANT (anorthosite-norite-troctolite) materials. The minor admixing noted in group B particles may indicate a continuous gradation between groups B and E, although we did not observe this directly. However, the very rare basaltic material and the "feldspathic" matrices of group D particles suggest that no such continuum exists between groups D and E.

Group F—Ultramafic lithic fragments

These particles consist largely of olivine and/or pyroxene. They appear to have been derived from relatively coarse-grained rocks not represented by the other groups. Group F materials occur as clasts occasionally in group D and E complex microbreccias. This group is described more fully in the petrology section.

POPULATION STUDY

The distribution of lithic particles >525 μ among the six groups is shown in Table 2. Two general features of the Fra Mauro formation indicated in Table 2 readily distinguish it from the regolith at other sampled sites. The soil samples we studied, compared to Apollo 11, 12, 15, and Luna 16 samples, contain a much smaller percentage of "undegraded" primary igneous particles (all basaltic), as shown in Table 3. They also show a much higher abundance of anorthosites, norites, and related troctolites (ANT), as shown in Table 3. Considering the abundances of such material as constituents of group E and B particles, ANT material may comprise as much as 55% by volume of the soil samples we studied.

PETROLOGY OF CRYSTALLINE ROCKS

In our detailed petrologic studies we have emphasized crystalline rock types, including their fragmental derivatives (groups A, C, and F). The petrology of specific rock types is discussed below.

Basalts

The major phases (>10 vol. %) present in Apollo 14 basalts include calcic plagioclase (typically An_{75-90}) and clinopyroxene, including augite, subcalcic augite, and pigeonite. Olivine (Fa_{25-40}) and orthopyroxene are major phases in some examples, minor (2–10%) in others. Other typical minor or accessory ($<2\%$) phases include ilmenite, troilite, Fe metal, Cr-spinel, apatite, K-feldspar and two immiscible glasses (Fig. 6). Textural varieties are mentioned above, some of which are illustrated (Fig. 1). The grain size range is very similar to the Apollo 11 basalts. We have seen little evidence in our samples of coarser-grained gabbroic materials typically present in Apollo 12 soils. Like Apollo 12 basaltic rocks (Wood *et al.*, 1971a; Keil *et al.*, 1971; LSPET, 1970), the Apollo 14 examples generally are poor in ilmenite and rich in olivine by comparison to basalts from Mare Tranquillitatis. A few basaltic particles have relatively low plagioclase (30%) and high ilmenite

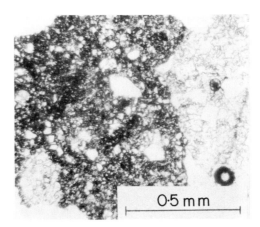

Fig. 4. Complex multigenerational ANT microbreccia (group D); granulitic norite and troctolite lithic clasts in slightly recrystallized noritic microbreccia with matrix of pale brown glass and partially crystalline plagioclase (1424). (Transmitted light.)

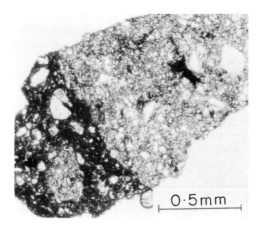

Fig. 5. Mixed microbreccia (basaltic + ANT) (group E) with noritic breccia and basalt mineral fragments in matrix of dark brown and yellow-orange glass (1431–1). (Transmitted light.)

Table 2. Distribution according to lithic type of Apollo 14 soil
particles in the size range 0.5 to 4 mm.
(See text for Group descriptions.)

Sample No.	GROUP						
	A	B	C	D	E	F	TOTAL
14161,–162,–163	14	107	71	46	24	4	266
14258	—	15	20	28	14	—	77
Total Particles	14	122	91	74	38	4	343
% of Total	4	36	26	22	11	1	100

Table 3. Relative abundances in lunar soil samples of primary ("undamaged") igneous (mostly basaltic) particles and fragmental or otherwise "damaged" particles (microbreccias, shocked materials, glasses). Also shown are relative abundances of basaltic, ANT and "other" particles.

	Apollo 14	Apollo 11	Apollo 12	Apollo 15	Luna 16
Primary Igneous	4%	37%	23%	14%	22%
Fragmental, "Damaged"	96	63	77	86	78
	100%	100%	100%	100%	100%
Basalts and Basaltic Microbreccias	40%	89%	50%	14%*	~74%
Anorthosites, Norites, and Troctolites (ANT)	48	4	11	4	~1.5
Other (glasses and/or mixed microbreccias)	12	7	39	82	~24.5
	100%	100%	100%	100%	100%
No. of Particles Surveyed	343	1676	497	914	316
Size Range (mm)	0.5–4	0.6–3	0.6–3	4–10	0.24–0.42

Data Sources:
   Apollo 14: This study.
   Apollo 11, 12: Marvin, et al. (1971).
   Apollo 15: Powell (1972).
   Luna 16: Wood, et al. (1971b).

*Figure probably too low and should include some particles listed as "other"; these particles were identified by stereomicroscope, not thin section.

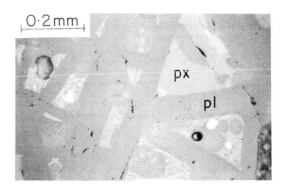

Fig. 6. Immiscible glasses in the mesostasis of basalt particle 1407-4 (see Fig. 1a). (Reflected light.) Note high-Fe blebs (medium gray) in high-Si glass (darker gray) and *vice versa*. Some devitrification to pyroxene and ilmenite is visible in some of the high-Fe glass. Labeled crystals are plagioclase (pl) and clinopyroxene (px).

(10%) contents and thus are similar to Apollo 11 and 12 mare basalts. However, most of the Apollo 14 basaltic particles have a greater plagioclase content (45–65%) than Apollo 11 or 12 basalts (15–45%). The presence of orthopyroxene in these Apollo 14 basalts is a further distinction. Such basalt particles generally are similar to the nonmare KREEP basalts described by Meyer (1972), although we do not have trace element data to verify this.

Microprobe analyses of some typical pyroxenes from Apollo 14 basaltic particles are shown in Table 4 and Fig. 7. Several varieties of pyroxene are represented, and typically two or more are present in a given particle, in some cases as discrete phases, in others in a zoning relationship. Pyroxene zoning typically is not as pronounced as in Apollo 11 and 12 samples, although some extreme zoning has been reported in rocks 14310 (Brown *et al.*, 1972) and 14053 (Gancarz *et al.*, 1971). Zoned pyroxene cores are typically Mg-rich and Ca-poor relative to grain rims. Small matrix grains in inequigranular rocks are typically enriched in Fe and Ca relative to rims of larger zoned grains (Fig. 7). With a few exceptions, Ti, Al, and Cr contents in pyroxenes are lower in Apollo 14 basalts than in Apollo 11 and 12, although some overlap in range occurs (see Wood *et al.*, 1970, 1971a; Keil *et al.*, 1970, 1971; Busche *et al.*, 1971).

Two immiscible glasses of high (K + Si) and of high Fe-low Si composition are typically present in the mesostasis of Apollo 14 basalts (Fig. 6). These glasses are similar in composition to those present in basalts from Apollo 11, 12, and Luna 16 samples (Roedder and Weiblen, 1971, 1972).

We found one basalt particle of unusual phase chemistry: a microporphyritic olivine-rich ($\sim$30%) basalt with highly magnesian (Fo_{88}) skeletal olivine phenocrysts and pyroxene ($\sim$10%) including the most Mg- and Ca-rich compositions (close to pure diopside) reported in lunar materials (1407–11, Table 4; solid circles, Fig. 7). The pyroxenes are unusually rich in Al_2O_3 (3 to 4.5%) Cr_2O_3 (1.4%) and TiO_2 (3%) for Apollo 14 basalts. Most TiO_2 in the rock appears to be contained in

Table 4. Selected pyroxene analyses (wt%) from basaltic particles in Apollo 14 soil sample 14162,41. (Numbers in parentheses are section and particle numbers used in our laboratory.) (n.d. = not determined.)

	(1407–4)		(1408–8)		(1407–11)	
	hy	pig	subcal aug	Fe–aug	di	aug
SiO_2	53.7	52.5	50.2	49.3	49.8	47.3
TiO_2	0.58	0.6	1.10	0.94	2.99	3.20
Al_2O_3	2.6	0.9	1.7	1.7	4.4	3.2
Cr_2O_3	0.66	0.7	0.60	0.29	1.42	1.31
FeO	10.9	16.5	22.5	23.9	2.9	5.6
MnO	0.22	0.3	0.55	0.54	0.21	0.18
MgO	29.5	24.0	12.7	7.4	16.5	24.7
CaO	1.81	4.3	9.9	15.7	20.7	14.8
Na_2O	0.0	0.2	n.d.	n.d.	0.4	0.4
K_2O	0.0	0.0	n.d.	n.d.	n.d.	n.d.
TOTAL	100.0	100.0	99.3	99.8	99.3	100.7
mol% EN	79.6	65.6	38.9	22.9	49.8	64.0
mol% FS	16.9	25.9	39.4	42.3	5.2	8.5
mol% WO	3.5	8.5	21.7	34.8	45.0	27.5

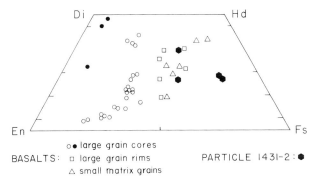

BASALTS: o• large grain cores
          □ large grain rims   PARTICLE 1431-2: ●
          △ small matrix grains

Fig. 7. Representative pyroxene compositions (microprobe) from basalts and late-stage differentiate material (particle 1431–2, see Table 7). Filled circles are from primitive basalt particle 1407–11 (see text and Table 4).

pyroxene, as the opaque content is low ($<1\%$). The remainder of the rock consists of plagioclase ($\sim 50\%$) and a glassy mesostasis ($\sim 10\%$).

The special importance of this particle lies in its primitive Mg/Fe ratio, which implies derivation from a lunar interior with a terrestrial-type $Mg/(Mg + Fe)$ of at least 0.90 (as suggested by Brown *et al.*, 1972, for isolated magnesian olivines and pyroxenes in Apollo 14 soil). It is intriguing that this material may represent a magma composition from which the mare basalts were derived by fractionation processes. It is also intriguing that the olivine composition (with 0.1% Cr_2O_3) is closely related to that in Apollo 12 norite-anorthosites (Marvin *et al.*, 1971) (unlike other basalts), suggesting the possibility that this basalt represents a melt composition from which the former were derived by cumulate processes.

Feldspathic rocks (ANT)

These particles include anorthositic (principally plagioclase), noritic (plagioclase + orthopyroxene), and troctolitic (plagioclase + olivine) assemblages, which generally include as minor or accessory phases clinopyroxene (commonly Ca-poor), ilmenite, Fe metal, troilite, apatite (and/or whitlockite) and K-feldspar. Some particles also contain accessory glass. These three subgroups appear to have a common affinity: types grade into one another and occur mixed together as clasts in complex (group D) breccias. All have fragmental textures, obscured in many cases by some degree of recrystallization, which often has produced a granulitic texture (Fig. 3). The relatively large size of many mineral grains in these particles indicate they were derived from rocks very coarse-grained by comparison to mare basalts. By further contrast to the basalts, the pyroxenes as a group are Mg-rich, Ca- and Ti-poor, much less strongly zoned, and paler in color (see Fig. 8, Table 5). Pyroxenes in the less recrystallized varieties occasionally show optically visible ex-solution lamellae, a feature lacking in the rapidly cooled mare basalts.

Olivine compositions are similar to those from Apollo 11 and 12 anorthosites and norites, as are the typically calcic plagioclase compositions (An_{85-95}). Phase compositions in a typical strongly recrystallized norite (Fig. 3b) and an anorthosite

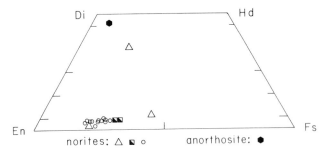

Fig. 8. Representative pyroxene compositions (microprobe) from ANT lithic particles (see also Table 5). The diopside composition is previously unreported from lunar anorthosites and reveals the primitive character of this particular particle (1419–7) (see text and Table 6). Norite analyses are from three different particles (different symbols).

Table 5. Typical orthopyroxene and accessory clinopyroxene compositions (wt%) in norite particles from Apollo 14 soil sample 14258,28. (Numbers in parentheses are section and particle numbers used in our laboratory.)

	hy (1403–5)	hy (1403–1)	pig (1403–1)	aug (1403–1)
SiO_2	55.3	54.1	51.4	52.5
TiO_2	0.61	0.40	0.39	0.62
Al_2O_3	1.70	1.03	0.36	0.80
Cr_2O_3	0.55	1.86	0.28	1.53
FeO	12.3	12.8	25.4	12.1
MnO	0.24	0.34	0.37	0.16
MgO	28.3	28.3	17.7	16.2
CaO	1.91	1.05	3.15	18.0
TOTAL	100.9	99.9	99.1	101.9
mol% EN	77.0	77.7	51.4	45.2
mol% FS	19.3	20.2	42.0	19.0
mol% WO	3.7	2.1	6.6	35.8

Table 6. Phase compositions (wt%) in a typical granulitic norite and in a primitive anorthosite from Apollo 14 soil samples 14162,41 and 14258,28, respectively (Numbers in parentheses are section and particle numbers used in our laboratory.)

	Recrystallized Norite (Particle 1431–3).				Anorthosite (Particle 1419–7).		
	Pyroxene	Plagioclase	Orthoclase	Ilmenite	Pyroxene	Plagioclase	Ilmenite
SiO_2	56.0	45.9	65.1	—	55.8	46.8	0.3
TiO_2	0.57	0.31	0.38	48.3	0.70	0.04	51.7
Al_2O_3	0.54	34.0	18.7	0.48	0.96	34.2	0.19
Cr_2O_3	0.29	—	—	0.52	0.37	—	0.58
FeO	18.1	0.15	0.14	42.8	3.8	0.05	37.7
MnO	0.23	—	—	0.32	0.14	—	0.20
MgO	23.7	—	—	3.2	16.4	—	8.2
CaO	2.0	18.1	0.6	—	22.0	17.1	0.31
Na_2O	—	1.0	1.0	—	—	1.4	—
K_2O	—	0.13	12.7	—	—	0.29	—
TOTAL	101.4	99.6	98.6	95.6	100.2	99.9	99.2
mol% EN	66.9	AB 9.4	0.9		EN 47.5	AB 12.3	
mol% FS	29.0	AN 89.8	3.5		FS 6.5	AN 86.0	
mol% WO	4.1	OR 0.8	86.6		WO 46.0	OR 1.7	

(Fig. 3a) are shown in Table 6. This anorthosite particle is of special interest for its primitive Mg/Fe ratio. The accessory ($\sim 1\%$) pyroxene, which occurs as small rounded grains interstitial to and included in much larger plagioclase crystals, is close to diopside in composition. The ilmenite ($< 1\%$) has an unusually high MgO content, which distinguishes it from ilmenite of most other lunar materials, including Apollo 12 and 14 norites (Fig. 9). The plagioclase has an FeO content (0.05%) lower than in plagioclase of most lunar feldspathic materials and much lower than in typical mare basalt plagioclase (see Wood *et al.*, 1970, 1971a, and Quaide, 1972). The primitive phase compositions suggest the intriguing possibility that this particle represents a tiny sample of a primitive lunar crust representing an earlier differentiate than the by now familiar norite-anorthosites, which have higher Fe/Mg ratios and more "evolved" phase compositions. The K_2O content (0.29 wt%) of the plagioclase places this particle in the "high K anorthosite" group of Hubbard *et al.* (1972), which they suggest crystallized from a melt of KREEP affinity.

All of the characteristics of the particles in this general group indicate a different original lunar crystallization environment and lack of any close genetic relationship to the mare basalts. Group C rocks have specialized compositions suggestive of cumulate differentiation processes. We consider these materials to represent fragments of the pre-Imbrian terra crust deposited at the Apollo 14 site within the Fra Mauro formation by the Imbrian event. Slow annealing within this hot ejecta blanket caused recrystallization of fragmental textures and obscured primary igneous textures previously modified by processes related to the impact event.

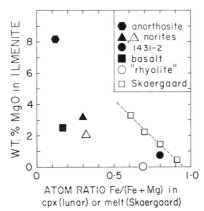

Fig. 9. Plot of MgO contents of lunar ilmenites versus Fe/(Fe + Mg) atom ratio in coexisting clinopyroxenes. Note primitive character of ilmenite from anorthosite 1419–7 compared to typical norites and mare basalt. Also note highly differentiated character of ilmenite from particle 1431–2 (see text), which is similar to ilmenite in "rhyolite" particle from Apollo 12 soil described by Wood (1970) and Wood *et al.* (1971a). Filled symbols, Apollo 14 (this study); open triangle, typical Apollo 12 norite (Wood *et al.*, 1971a); open squares, ilmenites from the layered series of the Skaergaard intrusion: MgO contents from Vincent and Phillips (1954), Table 4; Fe ratios derived for parent *melts* from Wager and Brown (1967), Table 5 and Fig. 119. The Skaergaard data show a characteristic anticorrelation between MgO in ilmenite and crystallization (differentiation) of the melt and reveal in a qualitative fashion the relative primitive and evolved character of certain lunar materials.

Ultramafic materials

Materials of ultramafic affinity, chiefly olivine- and/or pyroxene-rich lithic fragments, are rare in the Apollo 14 soils that we studied. Their potential petrogenetic importance as mafic cumulates within a differentiated lunar crust makes them worthy of special attention despite their statistical unimportance.

The relatively simple assemblages expected in such mafic cumulates complicates their unequivocal identification in the fines samples because of the ever-present possibility that a mafic assemblage is merely an incomplete, nonrepresentative sample of a medium or coarse-grained basalt. One must rely on association and minor element phase chemistry. We have analyzed several "dunite" (polycrystalline olivine) fragments, most of which occur as clasts within feldspathic (group D) breccias. The fayalite content (Fa_{16-36}) overlaps that of basaltic olivines and is thus not distinctive, but the uniformly low Cr_2O_3 content (0.0 to 0.13%) coupled with the major element compositions show these clasts to resemble olivines from Apollo 11 and 12 norite-anorthosites and differ from typical mare basalt olivines (Marvin *et al.*, 1971). Such clasts cannot with complete certainty be identified as ultramafic cumulates. However, their phase chemistry and occurrence in group D breccias reveal their affinity to ANT rather than basaltic materials.

One lithic fragment from group F was discovered which we feel has special significance. The assemblage includes 30% subhedral unzoned subcalcic ferroaugite (Fs_{61}), 50% subhedral to euhedral unzoned fayalitic olivine (Fa_{68}), 5% interstitial plagioclase (An_{69}), 5% ilmenite (0.7% MgO) and 10% mesostasis, a mixture of SiO_2 and K-feldspar (Fig. 10). All phases show evidence of shock deformation: lamellae in pyroxene, twinning in ilmenite, granulation of olivine, isotropism of plagioclase. However, primary igneous textures are preserved. The euhedral olivines poikilitically enclose highly rounded blebs of high (Si + K) material compositionally similar to the mesostasis (Fig. 10b). The pyroxene contains moderately coarse exsolution lamellae (verified by microprobe), indicative of deep-seated slow-cooling magmatic environment (Fig. 10c). All phase compositions (Table 7) reveal the late-stage differentiated character of this assemblage: the high Fe/Mg ratios and low Cr contents of pyroxene and olivine; the relatively high Na, K, and Ba contents of plagioclase and mesostasis and the high SiO_2 content of the latter; the low MgO content of the ilmenite (Fig. 9).

Phase compositions and textural relationships suggest liquid immiscibility was important in the petrogenesis of this material. The textures and specialized (ultramafic) bulk composition also suggest a cumulate process involving settling of Fe-rich ferromagnesians which trapped minor amounts of immiscible high (Si + K) liquid. The bulk composition of this particle is listed in Table 7, estimated from defocused beam microprobe analyses. Average compositions of high Fe and high Si glasses from Apollo 12 basalts (Roedder and Weiblen, 1971) are also shown for comparison, as is an average rock composition from the upper Zone C of the layered series of the Skaergaard intrusion, estimated by Wager and Brown (1967). The similarity between the lunar particle and the Skaergaard composition suggests (but doesn't prove) that the former could represent a sampling of an upper portion of a layered gabbroic plutonic complex.

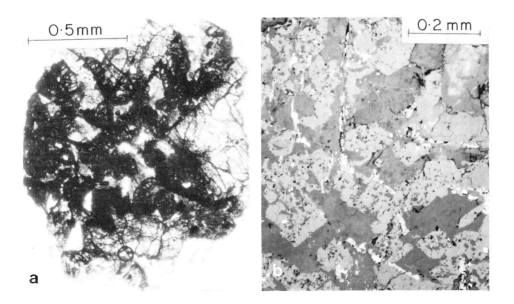

Fig. 10. Photomicrographs of late-stage differentiate particle 1431–2. (a), upper left: large light areas are clinopyroxene; dark granular areas, olivine + mesostasis; small bright grains, plagioclase (transmitted light). (b), upper right: reflected light: subhedral and euhedral olivine (medium gray) poikilitically enclosing blebs of high-(Si + K) material identical to mesostasis (dark gray); bright grains are ilmenite. (c), lower: exsolution lamellae in clinopyroxene (transmitted light, crossed nicols).

Table 7. Phase compositions (wt%) in late-stage differentiate, particle 1431–2 (our number). Also listed is an estimated bulk composition (DBA), along with Apollo 12 immiscible glasses and the Skaergaard Upper Zone C average rock.

	Phase compositions, particle 1431–2				Bulk compositions			
	Pyroxene	Olivine	Plag	Matrix	Apollo 12 Hi–Fe Glass[1]	Particle 1431–2[3]	Skaergaard Rock UZc[2]	Apollo 12 Hi–Si Glass[1]
SiO_2	50.1	31.0	51.3	81.3	42.2	47.7	44.2	76.3
TiO_2	0.79	0.10	0.07	0.31	4.2	2.83	2.28	0.68
Al_2O_3	1.0	0.17	28.6	9.2	5.4	6.5	8.1	11.5
Cr_2O_3	0.12	0.04	—	—	n.d.	n.d.		n.d.
FeO	32.6	53.0	1.7	1.8	32.9	32.3	29.2*	3.0
MnO	0.52	0.95	—	—	0.32	—	0.48	—
MgO	4.8	14.	—	—	0.79	0.6	0.3	0.07
CaO	10.0	0.46	13.2	1.1	10.7	6.5	8.8	1.6
BaO	n.d.	n.d.	1.31	1.1	0.0	n.d.		0.62
Na_2O	n.d.	n.d.	2.8	0.53	0.16	—	2.6	0.14
K_2O	n.d.	n.d.	0.75	6.0	0.41	1.1	0.4	6.7
TOTAL	99.9	99.8	99.7	100.2	97.1	97.5	97.9†	100.6
mol%	EN 15.6 FS 60.8 WO 23.6	FO 31.6 FA 68.4	AB 26.4 AN 68.9 OR 4.7					

[1] Roedder and Weiblen (1971).
[2] Wager and Brown (1967), Table 6.
[3] Average of several defocused beam analyses.
*All Fe calculated as FeO.
†Includes 1.5% P_2O_5.

The superficial compositional similarity between this late-stage assemblage and the high Fe interstitial glasses found in lunar basalts raises the question of whether the former may have been broken out of a typical basalt. Although this question may not be answered with total certainty, the mineralogical and textural features of the particle, which indicate slow cooling of an evolved melt and not quenching of a basalt mesostasis, argue in favor of the alternative that the particle is representative of an appreciable volume of melt (rock). Whether such rocks are volumetrically important in the lunar crust remains to be seen. In any case, the differentiated character and probable deep-seated origin (relative to mare basalts) of this rock fragment provide more information regarding lunar differentiation trends and processes and represent a small but significant piece in the puzzle that must be assembled to tell the complex story of lunar crustal evolution.

CONCLUSIONS

Several conclusions can be drawn about the origin and evolution of lithic fragments in the regolith at the Apollo 14 site on the basis of this study:

(1) The regolith materials are texturally mature by comparison to those of other sampled sites, as shown by the very low abundance of primary, undegraded lithic particles, and the high abundance of complex multigenerational breccias indicating multiple cycles of brecciation and agglomeration.

(2) Many materials in the regolith, principally the feldspathic varieties (Groups C, D), have undergone slow annealing, causing recrystallization of fragmental textures.

(3) The dominant rock types are of specialized (differentiated) character and show evidence suggestive of origins as igneous cumulates crystallized at depth in the lunar crust.

(4) Ultramafic lithologies may represent settled cumulates complementary to feldspathic flotational cumulates, developed during an early large-scale melting episode in the lunar crust. Late-stage differentiation may have produced materials similar to upper layers in terrestrial gabbroic complexes.

(5) A few basaltic particles in the Apollo 14 soils are broadly similar to Apollo 11 and 12 mare basalts and probably have a similar origin as thin surficial lava flows. These basalts may have originated from nearby Oceanus Procellarum, deposited on the Fra Mauro formation by post-Imbrian impact events. (Age determinations on these particles could not be made to support this contention.)

(6) Most Apollo 14 basaltic particles resemble KREEP basalts and are presumed to have a nonmare origin.

(7) Rare basaltic materials show the most primitive compositions yet observed in lunar materials and may represent melt compositions from which the norites and anorthosites crystallized at an early stage in lunar crustal evolution.

Acknowledgments—We wish to acknowledge the valuable assistance provided by Mr. James Kavanaugh in connection with computer reduction of our microprobe data. We are grateful to J. A. Wood for kindly permitting us to use a computer program written by him for microprobe data reduction. This work was largely supported by NASA grant NGR–44–006–142.

REFERENCES

Brown G. M. Emeleus C. H. Holland J. G. Peckett A. and Phillips R. (1972) Mineral fractionation patterns between Apollo 14 primitive feldspathic rocks and Apollo 15 and other basalts (abstract). In *Lunar Science—III* (editor C. Watkins), p. 95, Lunar Science Institute Contr. No. 88.

Busche F. D. Conrad G. H. Keil K. Prinz M. Bunch T. E. Erlichman J. and Quaide W. L. (1971) Electron microprobe analyses of minerals from Apollo 12 lunar samples. University of New Mexico Institute of Meteoritics Spec. Publ. No. 3, 61 pp.

Eggleton R. E. and Offield T. W. (1970) Geologic Maps of the Fra Mauro region of the moon: U.S. Geol. Survey Misc. Geol. Inv. Map I–708.

Gancarz A. J. Albee A. L. and Chodos A. A. (1971) Petrologic and mineralogic investigation of some crystalline rocks returned by the Apollo 14 mission. *Earth Planet. Sci. Lett.* **12**, 1–18.

Hubbard N. J. Gast P. W. and Meyer C. Jr. (1972) Chemical composition of lunar anorthosites and their parent liquids (abstract). In *Lunar Science—III* (editor C. Watkins), p. 404, Lunar Science Institute Contr. No. 88.

Keil K. Bunch T. E. and Prinz M. (1970) Mineralogy and composition of Apollo 11 lunar samples. *Proc. Apollo 11 Lunar Sci. Conf., Geochim. Cosmochim. Acta* Suppl. 1, Vol. 1, 561–598. Pergamon.

Keil K. Prinz M. and Bunch T. E. (1971) Mineralogy, petrology, and chemistry of some Apollo 12 samples. *Proc. Second Lunar Sci. Conf., Geochim. Cosmochim. Acta* Suppl. 2, Vol. 1, 319–341. MIT Press.

LSPET (1970) (Lunar Sample Preliminary Examination Team) Preliminary examination of the lunar samples from Apollo 12. *Science* **167**, 1325–1339.

Marvin U. B. Wood J. A. Taylor G. J. Reid J. B. Jr. Powell B. N. Dickey J. S. Jr. and Bower J. (1971) Relative proportions and probable sources of rock fragments in the Apollo 12 soil samples. *Proc. Second Lunar Sci. Conf., Geochim. Cosmochim. Acta* Suppl. 2, Vol. 1, 679–699. MIT Press.

Meyer C. Jr. (1972) Mineral assemblages and the origin of non-mare lunar rock types (abstract). In *Lunar Science—III* (editor C. Watkins), p. 542, Lunar Science Institute Contr. No. 88.

Powell B. N. (1972) Apollo 15 coarse fines (4–10 mm): Sample classification, description and inventory. NASA MSC Publ. 03228.

Quaide W. (1972) Fe and Mg in bytownite-anorthite plagioclase in lunar basaltic rocks (abstract). In *Lunar Science—III* (editor C. Watkins), p. 624, Lunar Science Institute Contr. No. 88.

Roedder E. and Weiblen P. W. (1971) Petrology of silicate melt inclusions, Apollo 11 and 12 and terrestrial equivalents. *Proc. Second Lunar Sci. Conf., Geochim. Cosmochim. Acta* Suppl. 2, Vol. 1, pp. 507–528. MIT Press.

Roedder E. and Weiblen P. W. (1972) Silicate melt inclusions and glasses in lunar science fragments from the Luna 16 core sample. *Earth Planet. Sci. Lett.* **13**, 272–285.

Sutton R. L. Batson R. M. Larson K. B. Schafer J. P. Eggleton R. E. and Swann G. A. (1971) Documentation of the Apollo 14 samples: U.S. Geol. Survey, Interagency Report 28, 37 pp.

Vincent E. A. and Phillips R. (1954) Iron-titanium oxide minerals in layered gabbros of the Skaergaard intrusion, East Greenland. Part I: Chemistry and ore microscopy. *Geochim. Cosmochim. Acta* **6**, 1–26.

Wager L. R. and Brown G. M. (1967) *Layered Igneous Rocks*, W. H. Freeman and Company, San Francisco, 588 pp.

Wood J. A. (1970) Petrology of the lunar soil and geophysical implications. *Jour. Geophys. Res.* **75**, 6497–6513.

Wood J. A. Marvin U. B. Powell B. N. and Dickey J. S. Jr. (1970) Mineralogy and petrology of the Apollo 11 lunar sample. Smithsonian Astrophysical Observatory Special Report 307, 99 pp.

Wood J. A. Marvin U. B. Reid J. B. Jr. Taylor G. J. Bower J. Powell B. N. and Dickey J. S. Jr. (1971a) Mineralogy and petrology of the Apollo 12 lunar sample. Smithsonian Astrophysical Observatory Special Report 333, 272 pp.

Wood J. A. Reid J. B. Jr. Taylor G. J. and Marvin U. B. (1971b) Petrological character of the Luna 16 sample from Mare Fecunditatis. *Meteoritics* **6**, 181–193.

Proceedings of the Third Lunar Science Conference
(Supplement 3, *Geochimica et Cosmochimica Acta*)
Vol. 1, pp. 853–864
The M.I.T. Press, 1972

Inclusions and interface relationships between glass and breccia in lunar sample 14306,50

J. F. Wosinski, J. P. Williams, E. J. Korda, W. T. Kane,
G. B. Carrier, and J. W. H. Schreurs

Corning Glass Works, Corning, New York 14830

Abstract—Preliminary observations of a 2 mm wide glass-filled fracture of breccia sample # 14306,50 indicate numerous spherical inclusions, ranging from about 30 Å to 100 μ in size, together with many voids. The inclusions in the 0.1 to 1μ range are abundant and evenly distributed throughout the glass phase, but the particles at the larger and smaller ends of the size range are more sparsely distributed. The spherical bodies are mainly iron-nickel with segregated zones of iron-nickel sulfide (troilite) and iron-nickel phosphide (schreibersite), although some of the inclusions appear to be nearly all sulfide and/or phosphide. As many as five magnetic phases may be present, and there is evidence of single crystal formation. Magnetic analysis indicates magnetic centers as small as 40 to 50Å.

The glass-breccia interface is sharply defined, with the transition band approximately 150 μ wide. The transition zone contains a few irregularly shaped metallic inclusions, ample evidence of flow lines and moderate numbers of breccia grains with rounded edges. Isometric concentration maps of glass-breccia grain interfaces indicate no diffusion of breccia constituents into the glass within the resolution of the electron microprobe. The average chemical composition of the glass phase is: SiO_2, 46.63 ± 0.24; FeO, 11.60 ± 0.47; Na_2O, 0.72 ± 0.04; TiO_2, 1.32 ± 0.03; CaO, 8.67 ± 0.35; MgO, 13.53 ± 0.75; Al_2O_3, 15.87 ± 0.36; K_2O, 0.61 ± 0.04; Total 98.95 wt%. No indication of crystallization from the glass was found at the interface or in the main body of the glass region. These observations suggest rapid flow and cooling of the glass melt.

Introduction

THE CHEMICAL, mineralogical, and physical relationships between the glass and breccia of sample 14306,50 have been determined to assist in understanding the formation and history of the glass portion of the specimen. The unique feature of the sample is a 2 mm wide glass-filled fracture, first reported by Swann *et al.* (1971). The breccia portion of the sample is in the F–4 classification of Jackson and Wilshire (1972). Glass coated rock surfaces, glass spheres, and shards previously have been described by a number of investigators, but lunar glass in a breccia fissure has not been thoroughly studied.

The glass portion of sample 14306,50 contains many spherical metallic inclusions. Metallic inclusions in lunar glass were first noted in the Apollo 11 mission samples and were reported in *Science* (1970) by several authors. Among these were Ramdohr and El Goresy (1970), who described iron spheres in a glass globule, and Simpson and Bowie (1970), who identified troilite and native iron in a glassy lunar fragment. Chao *et al.* (1970, 1971) described and classified impact glasses with nickel-iron inclusions. Duke *et al.* (1970) discussed metal spheres in glass that contained 4 to 15% nickel and noted flow structures in a lunar glass which had strings of metallic spheres concentrated along flow bands. McKay *et al.* (1970) suggest glass droplets in a molten state passed through an impact plume containing molten nickel-iron and troilite droplets, and some of the spheres may also be aerosol condensates from iron-rich vapor generated by an iron meteorite. Mason *et al.* (1970) report on nickel-iron pellets in the soil and give a good description of a single pellet containing taenite (nickel-iron) and troilite. Goldstein and Yakowitz (1971) characterized the reduced metal components of the Apollo 12 rocks, breccia, and soils. Carter (1971) described metallic fragments that were present on glass surfaces or totally buried. Most of these metallic fragments are spheroidal and contain iron, nickel, or phosphorus or some combination of the three elements. Glass (1971) listed electron microprobe analyses of metallic spheres found in Apollo 12 glasses. These contained various amounts of iron, nickel, phosphorus and some sulfur. Wood *et al.* (1971) discuss "yellowish" colored cindery glasses, which contain nickel-iron metal, and ropy glasses which contain metal grains relatively rich in cobalt (0.8–1.5 %) and nickel (0.6–5%). McKay *et al.* (1971) and Carter (1971) described welded dust observed in scanning electron and optical microscope studies. The welded dust was made up of small mineral and glass fragments and appeared to be the result of a pervasive lunar process.

Experimental Results

Optical, scanning, and transmission electron microscopy examination, together with electron microprobe and magnetic analyses, were employed to study the glass intrusion into the breccia. Optical micrographs of the glass-filled fracture of the breccia and of a thin section of the glass-filled fracture are presented in Fig. 1. The presence of many voids and a large population of spherical inclusions can be seen in the glass phase. The figure also displays two scanning electron micrographs of the surface topography of the glass zone of the polished thin section. Large voids, up to several hundred microns (μ) in diameter and numerous inclusions, commonly spherical and ranging from about 0.1 to 50 μ in size, are visible.

Scanning electron micrographs of the glass-breccia interface are presented in Fig. 2A and 2B. The interface zone, outlined by arrows, is 100 to 150 μ thick. The prominent spherical voids and inclusions in the glass phase do not extend into the breccia. Some breccia grains, completely surrounded by glass, are present near the interface region. A metallic inclusion, which appears to have been formed by spatter action, is delineated in the scanning electron micrographs of Fig. 2C. An x-ray energy spectrum of the metallic phase, also shown in the same figure, indicates that the inclusion consists of iron and nickel.

The wide variety of the metallic inclusions is demonstrated in the scanning electron micrographs displayed in Figs. 1, 2, and 3. The inclusions in the glass phase predominantly are spherical, although several oblong-shaped bodies, one with a distinct nodule, also are visible. A large population of inclusions less than 1μ in size is evident. The larger particles show the presence of more than one phase with segregation in the interior as well as at the rims of the bodies. X-ray energy spectra from several representative metallic inclusions, as well as adjacent glass, all obtained on polished thin sections, are presented in Fig. 3. Major amounts of iron, together with minor concentrations of nickel and variable levels of sulfur and phosphorus, were detected in the inclusions. The dark phase resolved in segregated zones in the

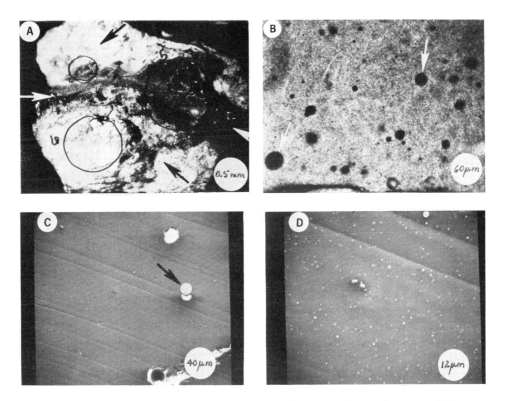

Fig. 1. Optical and scanning electron micrographs of glass. (A) Glass (white arrows) filled fracture in breccia (black arrows). (B) Thin section of glass region in transmitted light showing vesicles (white arrows) and inclusions randomly distributed. (C and D) Surface topography of polished thin section showing large spherical inclusion (arrow in C) and many small inclusions.

interior and at the outer surface of the larger spherules appears to be rich in sulfur. The light matrix is composed primarily of iron and nickel.

A view of the transition zone between the glass and breccia is delineated in the scanning electron micrograph in Fig. 4. Two kinds of breccia grains are visible in the glass matrix, and x-ray energy spectra from each of the three phases also are illustrated. The glass appears to contain a major amount of silicon, substantial concentrations of aluminum, calcium, magnesium, and iron, and low levels of titanium and potassium. One type of breccia grain is rich in silicon, aluminum and calcium (identified petrographically as bytownite) and the other in silicon, magnesium and iron (an olivine, suspected to be forsterite based on petrographic data).

Optical micrographs of a polished thin section of the glass-breccia interface obtained by transmitted, transmitted with crossed-polarizers, and reflected light are shown in Fig. 5. Each of the three composite micrographs picture the same region of the interface. Flow lines can be seen clearly in the transmitted light micrographs, round breccia grains are visible in crossed-polarized light, and the metallic inclusions

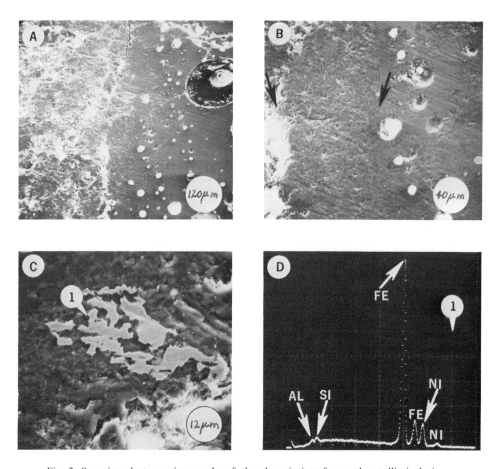

Fig. 2. Scanning electron micrographs of glass-breccia interface and metallic inclusion. (A and B) Surface topography at glass-breccia interface. Interface band about 100 μ wide (between arrows in B). (C) Spatter-type metallic inclusion. (D) X-ray energy spectrum of inclusion.

are evident in reflected light. These features are designated by white arrows in each photograph.

Thin edges of small chips of the glass phase were examined by direct electron transmission with an electron microscope. The major portions of the chips studied were reasonably homogeneous glass, where no heterogeneities greater than approximately 25 Å were observed. A very few scattered particles of greater density than the glass phase were found (Fig. 6A). In addition to the regions containing no or very few heterogeneities, several chip edges contained numerous small particles ranging from about 25 to several hundred Å in size. One such area is shown in Fig. 6B. Contrast effects (black arrows) as well as morphology of selected particles (white arrows) strongly suggest that at least some of the heterogeneities are crystalline. The detection of these density heterogeneities suggest that the metallic inclusions occur

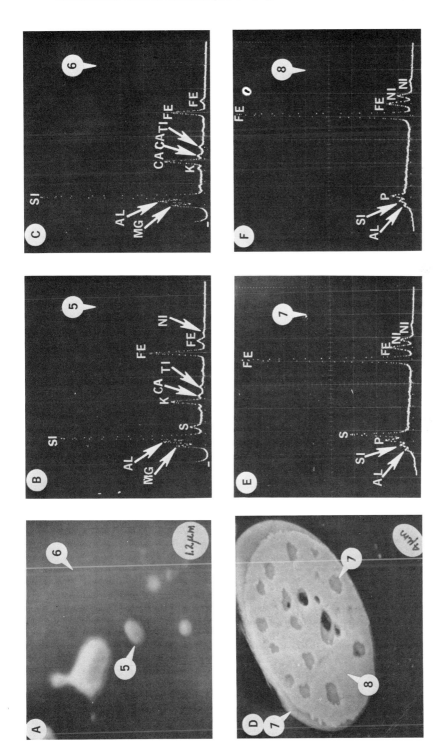

Fig. 3. Scanning electron micrographs and x-ray spectra of metallic inclusions. (A) Small inclusions (5) in glass phase (6) with corresponding x-ray spectra (B and C). (D) Large inclusion showing dark phase (7) in light matrix (8) with respective x-ray spectra (E and F).

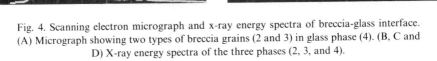

Fig. 4. Scanning electron micrograph and x-ray energy spectra of breccia-glass interface.
(A) Micrograph showing two types of breccia grains (2 and 3) in glass phase (4). (B, C and
D) X-ray energy spectra of the three phases (2, 3, and 4).

as particles as small as 25 Å and that at least some are crystalline.

Magnetic analysis of the glass phase containing the inclusions was carried out, and a hysteresis loop at room temperature was obtained with a vibrating sample magnetometer as demonstrated in Fig. 7A. At 11 kilogauss the moment was 15 gauss cm^3/g, the coercive force was 45 Oe and the remanence only a few percent. The shape of the curve suggests the presence of particles down to superparamagnetic particle size (~ 100 Å). Paramagnetic iron and/or nickel ions presumably are present in the glass phase. A magnetization versus temperature curve of the glass phase with an applied field of 500 Oe is shown in Fig. 7B. Four Curie temperatures, representing four magnetic phases, were observed at approximately $-125°C$ (phase 1), $-15°C$ (phase 2), 345°C (phase 3), and 670°C (phase 4). The shape of the curve indicates that a fifth phase also is present, with a Curie temperature $>670°C$. The Curie temperatures were obtained from an increasing temperature cycle. On cooling,

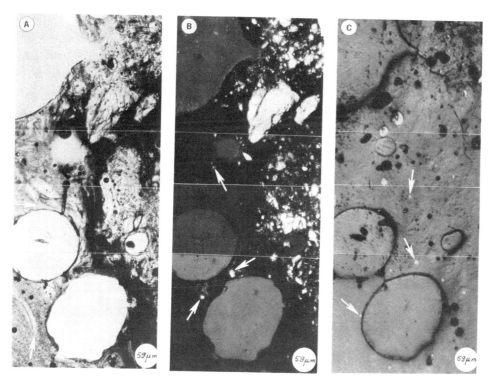

Fig. 5. Optical micrographs of glass-breccia interface. Micrographs of identical region taken
with transmitted light (A), transmitted light with crossed polarizers (B), and reflected light
(C). Flow lines, breccia grains and metallic inclusions are indicated by arrows.

phases 4 and 3 did not reacquire magnetism until much lower temperatures were
reached. After further recycling the original Curie temperatures were obtained. This
behavior suggests that phases 4 and 3 are iron-nickel, which are contaminated with
other elements. Phase 5 presumably is an iron phase. Phases 2 and 1 may possibly
be two different forms of iron phosphide with phase 1 containing less iron than
phase 2.

Electron microprobe backscattered electron images and element concentration
maps of the spherical particles in the glass matrix indicated the inclusions consist of
iron-nickel (Fe, Ni), schreibersite [$(Fe,Ni)_3P$] and troilite (FeS). Some particles
appear to have a core of schreibersite, which contains troilite inclusions and is rimmed with troilite. This is illustrated in Fig. 8A and 8B in which the backscatter image
shows a large spherical particle with a schreibersite core and troilite rim and interior
regions, and the element concentration maps for iron and phosphorus outline similar
features. A microprobe x-ray image for aluminum of a selected breccia grain is
recorded in Fig. 8C. Isometric concentration maps for aluminum, calcium, and
magnesium taken at the interface between the breccia grain, shown in the x-ray
image, and the glass matrix are displayed in Fig. 8D. The diffusion interfaces between
the two phases appear to be quite sharp, and no diffusion of the chemical constituents

Table 1. Chemical analysis of glass phase.

Element oxide	Weight percent as oxide		
	Center region	Edge region	Average
SiO_2	46.71 ± 0.30	46.54 ± 0.40	46.63 ± 0.24
FeO	11.04 ± 0.42	12.20 ± 0.46	11.60 ± 0.47
Na_2O	0.73 ± 0.05	0.72 ± 0.08	0.72 ± 0.04
TiO_2	1.29 ± 0.03	1.33 ± 0.05	1.32 ± 0.03
CaO	8.52 ± 0.57	8.84 ± 0.43	8.67 ± 0.35
MgO	12.90 ± 1.04	14.20 ± 0.80	13.53 ± 0.75
Al_2O_3	15.50 ± 0.40	16.20 ± 0.34	15.87 ± 0.36
K_2O	0.60 ± 0.06	0.63 ± 0.04	0.61 ± 0.04
TOTAL	97.29	100.66	98.95

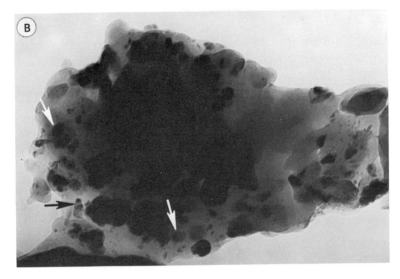

Fig. 6. Transmission electron micrographs of glass phase. (A) Note few heterogeneities designated by arrows. (B) Note numerous particles showing contrast effects (black arrows) and crystal like morphology (white arrows).

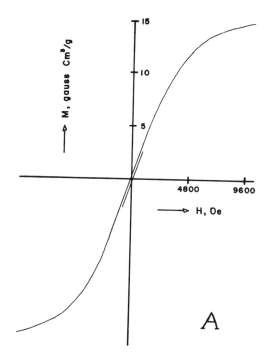

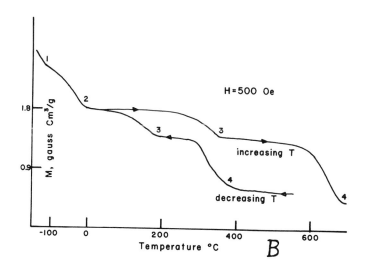

Fig. 7. Magnetic analysis of glass phase containing metallic inclusions. (A) Hysteresis loop from glass phase. (B) Magnetization versus temperature curve.

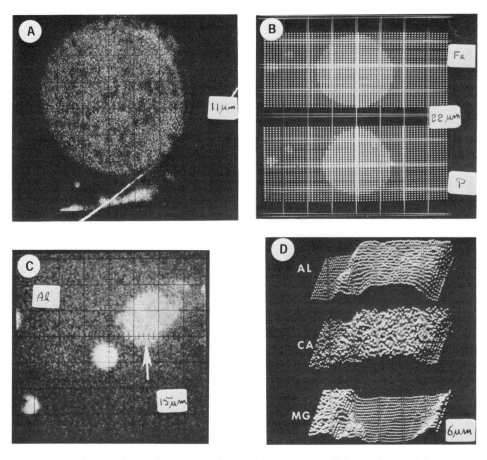

Fig. 8. Electron microprobe images and maps. (A) Backscattered electron image of large metallic inclusion. (B) Concentration maps of large and two small inclusions for iron and phosphorus. (C) X-ray image of breccia grain for aluminum. (D) Isometric concentration maps of breccia grain-glass interface for aluminum, calcium, and magnesium.

of the grain into the surrounding glass can be observed within the resolution of the microprobe (~ 1 to $2\ \mu$).

Quantitative chemical analyses of two general areas of glass in a polished thin section were carried out with the electron microprobe. One area was near the center of the 2 mm wide glassy vein and the other near the glass-breccia interface, but not in a flow line region. Microprobe data were processed by means of Colby's MAGIC–4 computer program with a reference glass (95 GPL) used for all elements determined. The composition of the reference glass was established by analysis using the CORSET program for x-ray emission analysis described by Stephenson (1971). Results of the microprobe chemical analyses are summarized in Table 1. Some variation in FeO and possibly CaO, MgO, and Al_2O_3 may exist between the center and edge of the glass vein.

SUMMARY

This preliminary survey of lunar sample 14306,50 has established the variety of the metallic inclusions in the glass-filled fracture, as well as the phases present in the breccia. The glass phase may not be homogeneous in chemical composition. The glass-breccia interface appears to be sharply defined, and diffusion between the breccia and glass was not detected. The large population of metallic inclusions limits the measure of the physical properties of the glass phase.

The study suggests a low viscosity, fluid glass was produced by meteoritic impact on the lunar regolith. The glass contained several condensed phases of a vaporized meteorite. Once formed, the glass had adequate time to flow into fissures present in the breccia. It cooled very rapidly for there are no signs of devitrification or liquid-liquid phase separation, which is characteristic of more moderate thermal treatments. The presence of glass flow lines, a sharp glass-breccia interface and no evidence of diffusion or crystallization of glass constituents into the breccia is in direct support of the meteoritic impact theory of glass formation.

Acknowledgments—The technical expertise of N. J. Binkowski, G. A. Geiger, K. M. Wilson, Jr., H. J. Holland, J. D. LaBarre, and L. H. Pruden in obtaining much of the data for this report is gratefully acknowledged. This work was jointly sponsored by NASA Contract NAS 9–11551 and the Corning Glass Works.

REFERENCES

Carter J. L. (1971) Chemistry and surface morphology of fragments from Apollo 12 soil. *Proc. Apollo 12 Lunar Science Conf., Geochim. Cosmochim. Acta*, Suppl. 2, Vol. 1, pp. 873–892.

Chao E. T. C. Boreman J. A. Minkin J. A. James O. B. and Desborough G. A. (1970) Lunar glasses of impact origin: physical and chemical characteristics and geologic implications. *J. Geophys. Res.* **75**, no. 35, pp. 7445–7479.

Chao E. T. C. Boreman J. A. and Desborough G. A. (1971) The petrology of unshocked and shocked Apollo 11 and Apollo 12 microbreccias. *Proc. Apollo 12 Lunar Science Conf., Geochim. Cosmochim. Acta*, Suppl. 2, Vol. 1, pp. 787–816.

Duke M. B. Woo C. C. Bird M. L. Sellers G. A. and Finkelman R. B. *Science* **167**, 648–650.

Glass B. P. (1971) Investigation of glass recovered from Apollo 12 sample no 12057. *J. Geophys. Res.* **76**, no. 23.

Goldstein J. E. and Yakowitz H. (1971) Metallic inclusions and metal particles in the Apollo 12 lunar soil. *Proc. Second Lunar Science Conf., Geochim. Cosmochim. Acta*, Vol. 1, pp. 177–191. M.I.T. Press.

Jackson E. D. and Wilshire H. G. (1972) Classification of the samples returned from the Apollo 14 landing site. In *Lunar Science—III* (editor C. Watkins), pp. 371–373, Lunar Science Institute Contr. No. 88.

Mason G. Fredrickson K. Henderson E. P. Jarosewich E. Melson W. G. Towe K. M. White J. S. Mineralogy and petrography of lunar samples. *Science* **167**, 664–666.

McKay D. S. Greenwood W. R. Morrison D. A. (1970) Morphology and related chemistry of small lunar particles from Tranquillity Base. *Science* **167**, 654–656.

McKay D. S. Morrison D. A. Clanton U. S. Ladle G. H. and Lindsay J. F. (1971) Apollo 12 soil and breccia. *Proc. Apollo 12 Lunar Sci. Conf., Geochim. Cosmochim. Acta*, Suppl. 2, Vol. 1, pp. 755–773. M.I.T. Press.

Ramdohr P. and El Goresy A. (1970) Opaque minerals of the lunar rocks and dust from Mare Tranquillitatis. *Science* **167**, 615–618.

Science (1970) **167**, Apollo 11 results from, No. 3918.

Simpson P. R. and Bowie S. H. U. (1970) Quantitative optical and electron probe studies of the opaque phases. *Science* **167**, 619–621.

Stephenson D. A. (1971) Theoretical analysis of quantitative x-ray emission data. *Anal. Chem.* **43**, 1761.

Swann G. A. *et al.* (1971) Preliminary geologic investigation of the Apollo 14 landing site. Apollo 14 Preliminary Science Report, NASA–SP–272, Section 3.

Wood J. A. Marvin U. B. Reid J. B. Jr. Taylor G. J. Bower J. F. Powell B. N. Dickey J. S. Jr. (1971) Mineralogy and petrology of Apollo 12 lunar sample. Smithsonian Astrophysical Observatory Special Report 333.

Proceedings of the Third Lunar Science Conference
(Supplement 3, *Geochimica et Cosmochimica Acta*)
Vol. 1, pp. 865–886
The M.I.T. Press, 1972

Rock 14068: An unusual lunar breccia*

ROSALIND TUTHILL HELZ

U. S. Geological Survey, Washington, D.C. 20242

Abstract—Rock 14068 consists chiefly of an olivine-rich groundmass (25% olivine in the norm) with the texture of a rapidly cooled igneous rock. This texture is tentatively ascribed to the quenching of a melt produced by meteorite impact. The groundmass includes a variety of mineral and lithic clasts. It is bordered on three sides by a fringe of black microbreccia, which contains a suite of mineral and lithic clasts quite different from those included in the melted groundmass. The presence of these two components suggests that rock 14068 is too heterogeneous to be included in Warner's (1972) classification.

The apparent crystallization sequence of the olivine-rich groundmass, with average observed compositions, is: olivine (Fo_{79}), plagioclase ($Ab_{20}An_{72}Or_8$), and low-calcium pyroxene ($Wo_6En_{75}Fs_{19}$). Absence of high-calcium pyroxene indicates noritic rather than basaltic affinities; however, olivine- and soda-rich members of a lunar noritic suite have not yet been described.

INTRODUCTION

ROCK 14068 IS A walnut-sized clast of dark breccia from station C1 near Cone Crater. Its dominant component is an olivine-rich groundmass (25% olivine in the norm), which has a bulk feldspar composition considerably more sodic than any common lunar rock type hitherto described. The texture of this component is that of a rapidly quenched igneous rock. The composition and texture are so unusual that it seemed best not to give it a terrestrial igneous rock name: It will be referred to in this paper as "the olivine-rich groundmass," or simply "the groundmass."

Detailed petrographic and chemical studies were made of polished thin section 14068,10, with additional work on sections 14068,7 and 14068,11.

All microprobe analyses were performed on an ARL-EMX microanalyzer at 15 kV, 0.01 μA sample current. Beam diameter was $<0.5\,\mu$. Mineral silicate and oxide standards were used throughout. Elements analyzed were Si, Ti, Al, Cr, Ni, Fe, Mg, Ca, Na, K, P, and Ba, where appropriate. The data were reduced using the program of Boyd *et al.* (1969).

Tables of analytical data for individual minerals are available from LSI on request. The contents of these tables are listed in Table 1.

THE OLIVINE-RICH GROUNDMASS OF 14068

Figure 1 is a detailed map of section 14068,10, traced from a photomosaic of the section. Figure 2 shows smaller sketches of sections 14068,7 and 14068,11. The most striking feature of these sections is that the olivine-rich groundmass (the white areas) dominates the core of the slides, but the border zones are charged with large mineral clasts and lithic inclusions. The "north" edge in particular is heavily contaminated in

---

*Publication authorized by the Director, U. S. Geological Survey.

Table 1. List of analytical tables available from L.S.I. The numbers of these tables are used to identify the corresponding objects and analyses in Figs. 1, 4, 14, and 15.

Table Ia. Groundmass plagioclases
    Ib. Groundmass olivines
Table IIa. Plagioclases in metanorite clast
    IIb. Pyroxenes in metanorite clast
Table IIIa. Plagioclases in allivalite clast
    IIIb. Olivine, pyroxenes in allivalite clast
Table IV. Pyroxenes in mafic breccia fragment
Table Va. Feldspars from basalt clast
    Vb. Pyroxenes, olivines from basalt clast
Table VIa. Zoned, remelted plagioclase clast
    VIb. Homogeneous subhedral plagioclase clasts
    VIc. Quench-textured anorthosite
    VId. Small plagioclase clasts in groundmass
Table VIIa. Orthopyroxene clasts, olivine in dunite
    VIIb. Spinel in dunite
    VIIc. Olivine clasts
Table VIII. Nickel-iron blebs
Table IX. Plagioclases from 14053,19

all three slides. The most abundant component of this border zone is a dark micro-breccia (hatched areas), which is second only to the olivine-rich groundmass* in area.

The bulk composition of the "olivine-rich groundmass" (Table 2, col. 1) was obtained by scanning over the groundmass extensively and averaging all data obtained as described above. (The prevalence of vesicles and small xenocrysts of calcic plagioclase made defocussed-beam analysis unreliable.) This analysis compares fairly well with the x-ray fluorescence analysis reported by Hubbard and Gast (1972).

Olivine is the most conspicuous mineral in the groundmass (Figs. 3, 10, 17b). It is more or less skeletal in habit, with hollow cores even where the crystal outline is relatively blocky (Fig. 3). There is no evidence of unmelted cores with new overgrowths. The grains analyzed range from $Fo_{80.3}$ to $Fo_{78.1}$ in composition (cf. the bulk normative olivine composition of Fo_{78}).

Plagioclase makes up 46% of the groundmass. It is variolithic in texture ("sheath-like", Warner, 1972; see especially Fig. 3), finer grained and has a greater range of composition than the olivine. The compositions obtained here range from $Ab_{23}An_{69}Or_8$ to $An_{19}An_{75}Or_5$ (Fig. 4). Warner (this volume) obtained groundmass plagioclase compositions ranging from $Ab_{21}An_{69}Or_{10}$ to $Ab_{15}An_{81}Or_4$. Although the Or content of the more sodic feldspar is very high, no discrete grains of K-feldspar large enough to analyze were found. The bulk normative feldspar, from Table 2, col. 1, is $An_{21}An_{71}O_8$ (G in Fig. 4), which compares well with the average feldspar composition obtained in this study. Because the traverses made in determining the bulk groundmass analysis were on a part of slide 14068,10 not used for any of the analyses of individual groundmass feldspars, the resemblance indicates that

---

*This use of the term "groundmass" is clearly relative. It is quite possible that on a larger scale, the olivine-rich material would appear as a clast in the dark microbreccia.

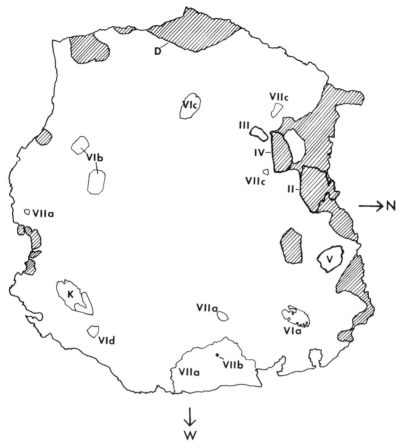

Fig. 1. Map of section 14068,10. E–W (vertical) dimension of fragment = 1.5 cm. Hatched areas = dark microbreccia; white area = olivine-rich groundmass, with analyzed inclusions indicated. K = analyzed potassic granitic glass (Table 2, Col. 4). D = analyzed microbreccia matrix (Table 2, Col. 3). Other features are labelled with the number of the appropriate analytical table(s) as listed in Table 1.

feldspar composition is quite homogeneous over slide 14068,10. Warner's data, with a slightly different composition range, is from another slide. The bulk normative feldspar for the x-ray fluorescence analysis is $Ab_{11.3}An_{75.3}Or_{8.4}$ (B in Fig. 4). It plots at somewhat higher An and Or values than the average groundmass plagioclase found in this study or by Warner (this volume). This may result from minor admixture of clasts, in the sample analyzed by Hubbard and Gast (1972), of (1) calcic plagioclase and (2) K-granitic glass.

No analyses of groundmass pyroxenes were made in this study. Warner (this volume) reports a cluster at about $Wo_6En_{75}Fs_{19}$, with no high-calcium pyroxene. This compares closely with the normative groundmass pyroxene of $Wo_2En_{77}Fs_{21}$ (Table 2, col. 1). The lower Wo content in the normative pyroxene is an artifact of

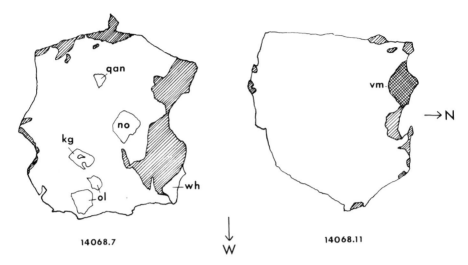

Fig. 2. Sketch maps of sections 14068,7 and 14068,11, in the same orientation as 14068,10 in Fig. 1. Vertical dimension of fragment = 1.5 cm. Hatched areas = dark microbreccia, white areas = olivine-rich groundmass, as in Fig. 1. Other labelled features are: In 14068,7: qan = quench-textured anorthosite; no = noritic anorthosite; kg = potassic-granitic glass; ol = olivine clasts; wh = "white rock" fragments. In 14068,11: vm = variolitic mafic rock.

Table 2. Microprobe analyses of various components of rock 14068,10, with CIPW norms, in weight percent.

	1 G Melted groundmass	2 B Bulk analysis*	3 D Matrix of dark breccia	4 K Granitic glass
SiO_2	47.3	47.2	47.0	75.1
TiO_2	1.36	1.39	1.68	0.48
Al_2O_3	14.7	13.3	15.8	12.2
FeO	10.06	10.0	10.40	0.42
MnO	—	0.13	—	—
MgO	17.56	17.6	12.16	0.08
CaO	7.61	8.28	9.37	0.73
Na_2O	1.10	0.75	0.73	1.34
K_2O	0.64	0.59	0.21	7.48
P_2O_5	0.37	0.55	0.44	0.07
TOTAL	100.70	99.79	97.79	97.90
Or	3.78	3.49	1.24	44.21
Ab	9.31	6.35	6.18	11.34
An	33.31	31.18	39.11	3.17
Qz	—	—	—	37.23
Ol	25.58	20.07	3.94	—
Hy	23.65	29.58	39.39	0.20
Di	1.65	5.09	3.68	—
Ap	0.86	1.27	1.02	0.16
Il	2.58	2.64	3.19	0.89
Cor	—	—	—	0.74

*X-ray fluorescence analysis reported by Hubbard and Gast (1972, Table 2, Col. 3)

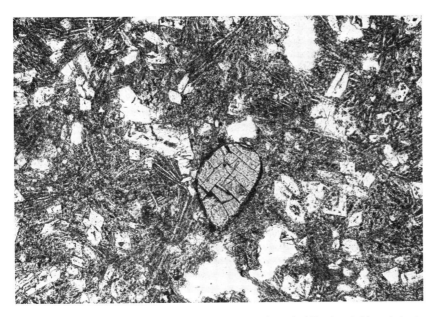

Fig. 3. Pink translucent spinel clast in 14068,11 (approximately 150 μ long). Note skeletal habit of olivine microphenocrysts, variolitic texture of plagioclase in the groundmass.

the assumption made in calculating the norm, that pyroxene contains no Al_2O_3, all Al_2O_3 being assigned to anorthite.

Other minerals in the groundmass include ilmenite, iron, a sulfide (probably troilite), and a phosphate, all too fine-grained to analyze.

INCLUSIONS IN THE OLIVINE-RICH GROUNDMASS

Tables 3a and b present frequency data on all included material, mineral, glassy or lithic, that appear to be within the olivine-rich groundmass of sections 14068,7, 14068,10, and 14068,11. In general, the material in Table 3a is monomineralic (except for some of the metal blebs) and monocrystalline (except for some of the plagioclase clasts*).

Mineral clasts

A noteworthy feature of Table 3a is that the relative proportions of the four main clast types (plagioclase, pyroxene, olivine, and nickel-iron) are quite constant, for the three sections investigated. This is curious, because the distribution of these clasts is different within each section. Plagioclase and nickel-iron are distributed

---

*The distinction between feldspar clasts and small anorthosite fragments is arbitrary. The object in Fig. 6, with two large grains of feldspar and a few small ones was classed as a feldspar clast, for instance. Only inclusions with twenty grains or so, or a distinctive polygranular microtexture, were classified as anorthosites.

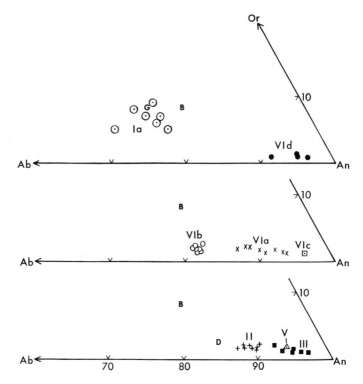

Fig. 4. Anorthite corner of the Ab-An-Or ternary, showing the distribution of plagioclase compositions in 14068,10, in mole percent. Top diagram includes groundmass feldspars and very small clasts in groundmass. Middle diagram includes analyses from large plagioclase clasts. Bottom diagram includes analyses of plagioclases in lithic fragments. B = normative plagioclase, bulk (Table 2, Col. 2). G = normative plagioclase, groundmass (Table 2, Col. 1). D = normative plagioclase, dark breccia (Table 2, Col. 3). Other labels correspond to analytical tables as listed in Table 1.

evenly throughout the groundmass, but pyroxene, and to a lesser extent olivine, tends to be concentrated near the margins of the thin sections.

The most abundant clast type is plagioclase. All analyzed plagioclase clasts are more calcic than the groundmass feldspar (Fig. 4). Many of them are surrounded by

Table 3a. Mineral clasts and/or inclusions found within the melted groundmass of 14068.

| | 14068,7 | | 14068,10 | | 14068,11 | |
	Number	Percent	Number	Percent	Number	Percent
Plagioclase	58	48	78	53	81	49
Olivine	19	16	26	18	21	13
Pyroxene	5	4	7	5	9	5
Iron	37	30	34	23	46	28
Spinel						
Opaque	2	2	1	1	7	4
Pink	0	0	1	1	2	1
TOTAL	121		147		166	

Table 3b. Lithic clasts that appear to be included within the melted groundmass of 14068, plus all glass clasts observed. Clasts marked with an asterisk are shown in Figs. 1 and 2.

	14068,7	14068,10	14068,11
Potassic-granitic			
glass ± plagioclase	7	8	2
Anorthosite			
"quench"-textured	1*	1*	0
plutonic/cumulate	0	1	1
"trachytic"	0	0	1
shocked	0	0	2
Noritic anorthosite	1*	0	1
Allivalite	0	1*	0
Basalt	0	1*	0
Dark breccia	2*	1*	1*

areas of oriented groundmass plagioclase, as shown for grain VIa in Fig. 5. In contrast to grain VIa with its lacy, moth eaten appearance, most feldspar clasts have smooth rounded outlines that may even transgress grain boundaries (Fig. 6). The complete lack of preferential reaction along grain boundaries of the clast shown in Fig. 6 suggests that the rounding was produced by mechanical abrasion rather than chemical reaction.

The second most abundant mineral inclusion is metallic nickel-iron. The nickel-iron blebs have subrounded, splattery outlines, with stringers of metal and smaller satellite blebs surrounding them. Fewer than 10% contain appreciable sulfide. Their nickel content ranges from 4.4% to 11.3%. This is the peak frequency range of

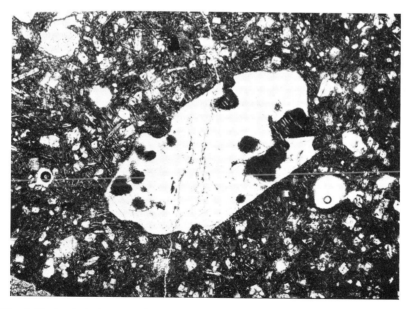

Fig. 5. Single plagioclase crystal, 1.2 mm long (VIa in 14068,10), which appears to have reacted with groundmass. It is surrounded by an area in which all the groundmass plagioclase laths are perpendicular to the c-axis of the large crystal. Plane polarized light.

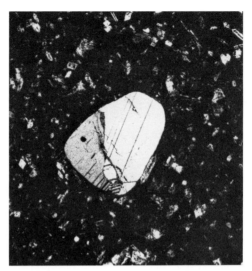

Fig. 6. Polygranular feldspar clast from 14068,7 (0.5 mm long) with smooth rounded outline. Note olivine microphenocryst against right side of clast. Crossed polarizers.

nickel content in meteorites (Mason, 1962, p. 132), which suggests that these blebs might be meteoritic material, though this is not conclusive evidence.

The third most frequent mineral clast type is olivine. In contrast to the plagioclase clasts with their rounded outlines, the olivine clasts are very angular (Fig. 7 shows one of the larger olivine clasts in 14068,10). Their most peculiar feature is that all olivine clasts analyzed are more iron-rich than the groundmass olivine ($Fo_{69.4}$ to $Fo_{74.4}$ for clasts versus Fo_{78-80} for groundmass).

Clasts of pyroxene are relatively scarce in 14068. All are large ($\sim$0.5 mm) single grains, euhedral to subhedral. The two analyzed, one of which is shown in Fig. 8, are highly magnesian orthopyroxene; the other pyroxene clasts appear to be the

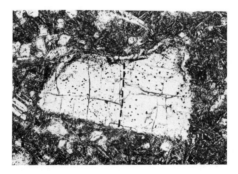

Fig. 7. Olivine clast (0.6 mm long) from 14068,10, broadly zoned from Fo_{69} (core) to Fo_{73} (rim). Electron microprobe traverse shown by dashed line. Note reaction with groundmass (melting?) around top half of clast. Plane polarized light.

Fig. 8. Orthopyroxene clast (0.5 mm long) from 14068,10. Note lack of reaction with groundmass. Plane polarized light.

same. There are no high-calcium pyroxene clasts in the melted groundmass of 14068, in the three sections studied.

Spinel clasts are rare in the melted groundmass of 14068. Three of the 13 grains seen are the translucent pink spinel reported to occur in the Apollo 14 "white rocks" and in 14321, which, like 14068, were picked up near Cone Crater (e.g. Christophe-Michel-Levy *et al.*, 1972). Figure 3 shows a grain of this pink spinel in the groundmass of 14068,11. There is a very thin rim of feldspar between the spinel and groundmass, which cannot be seen in the photograph.

The other grains are an opaque spinel, which is relatively abundant in section 14068,11. It is euhedral and, in strong illumination, is not strictly opaque, having a deep purplish color in thin edges.

Glass clasts

The most abundant inclusions listed in Table 3b are the inclusions of potassic-granitic glass. These are sometimes associated with plagioclase (as in Fig. 9). This is the inclusion from which the analysis in Table 2 was taken. The plagioclase in this inclusion is zoned, ranging from $Ab_{21}An_{75}Or_4$ at the core to $Ab_{35}An_{55}Or_{10}$ adjacent

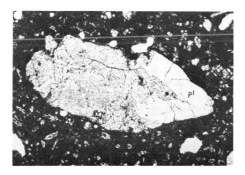

Fig. 9. Large glass clast, K from 14068,10 (1.9 mm long), surrounded on right end by plagioclase (pl). Glass contains needles of K-feldspar and nickel-poor iron spherules. Plane polarized light.

to the granitic melt. The glass contains tiny needles of K-feldspar and spherules of iron.

Most of the glass inclusions observed are vesicular; commonly a single large vesicle occupies most of the central area of the clast. The two inclusions shown in Figs. 9 and 16 are atypical in this respect.

Lithic clasts

Lithic material that is clearly within the olivine-rich groundmass is scarce and is always plagioclase-rich. The most abundant class of inclusions are pure feldspar rocks showing a variety of textures: a "quench" texture, with radiating feldspar crystals (Fig. 10), a plutonic or cumulate texture, a "trachytic" texture, and shock or fragmental textures. The first and third might be textures produced by devitrification of maskelynite, rather than true igneous textures.

There are several clasts that, though showing the effects of shock or thermal metamorphism, appear to be of igneous composition and origin:

(1) Orthopyroxene-plagioclase (2 clasts), referred to as noritic anorthosite. The orthopyroxene is euhedral and resembles the isolated orthopyroxene clasts in grain size and general appearance. These fragments are both quite coarse-grained. The larger ("no" in 14068,7, Fig. 2) is somewhat brecciated. Neither shows signs of thermal metamorphism.

(2) Olivine-anorthite (1 clast), referred to as allivalite. The texture of this fragment (Fig. 11) suggests it has been thermally metamorphosed.

One clast (V, Fig. 1) has basaltic texture and mineralogy. It is feldspar-rich (65% by volume plagioclase) has iron-rich mafic silicates (30% by volume, counting opaque minerals) and a substantial amount of crystalline K-feldspar and a silica mineral (5%). This K-feldspar is interstitial but occurs in fairly large laths (5–10 μ across). It contains 1.4–1.8% BaO. The chemistry of the pyroxene ($Mg/(Mg + Fe) = 0.45$ with high CaO content) and the presence of abundant K-feldspar suggest this fragment crystallized from a fairly differentiated basaltic melt. The absence of thermal metamorphic effects (e.g., exsolution) in the augite is remarkable, because some areas of the clast contain mosaics of high-calcium augite and olivine* that appear to replace the primary augite.

OTHER LITHIC MATERIAL INCLUDED IN ROCK 14068

The bulk of the lithic material, which rims the olivine-rich groundmass on three sides in sections 7 and 10, is a black microbreccia different from the olivine-rich groundmass in almost all respects. First, the composition of the matrix of this breccia, presented in Table 2, col. 3, differs from that of the groundmass. It has much lower MgO, Na_2O and K_2O contents than the analysis in col. 1, and the normative feldspar

---

*This mineral pair probably was not produced by isochemical breakdown, which should have given the assemblage augite-hypersthene. (Huebner and Ross, 1972). It may reflect some contamination by the olivine-rich groundmass, leading to the replacement of hypersthene by an olivine.

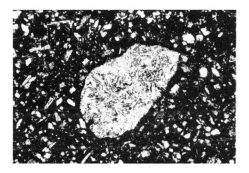

Fig. 10. "Quench"-textured anorthosite (VIc in 14068,10) 1.2 mm long. Crossed polarizers.

is much more calcic (D in Fig. 4). This dark breccia compositionally is similar to a
dark breccia in 14321 described by Grieve *et al.* (1972). (Rock 14321 is from station
C1 near Cone Crater, as is rock 14068.) The analysis they give differs from the matrix
analysis presented here chiefly in having more calcic feldspar in it.

Second, the dark breccia is irresolvably fine-grained and shows no signs of incipient melting even where surrounded by the olivine-rich groundmass. It is much
more friable than the melted groundmass and is badly plucked on the edges of the
sections. The dominant mineral clast type again is plagioclase, but here it is very
angular, in contrast to the rounded plagioclase clasts dominant in the groundmass.
Opaque spinel clasts are abundant.

A common type of lithic fragment in the black breccia is a thermally metamorphosed noritic anorthosite, with rounded, blebby pyroxenes. Lithic fragment II in
Fig. 1 (Fig. 12) is one example; there are three others in 14068,10 alone. Anorthosite
is less common. Lithic fragment IV in Fig. 1 (shown in Fig. 13) is a thermally metamorphosed breccia itself, making the dark breccia at least a two-generation breccia.

Less abundant lithic types found bordering the olivine-rich groundmass include:

(1) A nearly pure feldspar rock with a curious seriate, detrital texture ("wh" in
14068,7, Fig. 2). This rock contains clasts of pink spinel.

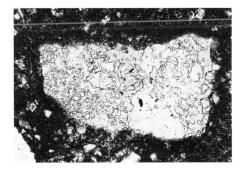

Fig. 11. Allivalite fragment (III in 14068,10), 1 mm long. Right half is plagioclase rich;
left half, olivine rich. Plane polarized light.

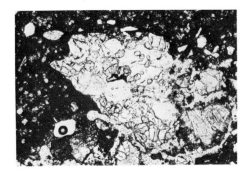

Fig. 12. Metamorphosed noritic anorthosite (II in 14068,10). Note blebby pyroxene crystals
in feldspar.

(2) A variolitic mafic rock ("vm" in 14068,11, Fig. 2) with phenocrysts of olivine
and spinel in a groundmass of plumose olivine and clinopyroxene. Feldspar is interstitial. Similar material has been described from 14321 (Grieve *et al.*, 1972).

(3) A "dunite" ("west" side of 14068,10, Fig. 1). This is a polygranular mosaic
of Fo_{89} with a single spinel grain ($Mg_{0.59}Fe_{0.49}Ti_{0.08}Cr_{1.14}Al_{0.70}O_4$). Its texture
appears to be metamorphic rather than cumulate.

RELATIONSHIP BETWEEN CLASTS AND GROUNDMASS

Warner (1972, this volume) developed a classification system for the Apollo 14
breccias, which is based on the degree of equilibration between clasts and matrix.
The extent of equilibration increases in going from his Class 1 to Class 8. This is of
concern here because (1) rock 14068 is the sole representative of Warner's Class 8,
and (2) the clast-matrix relations observed in this study do not entirely coincide with
those reported by Warner.

Mineral compositions of 14068,10, for both groundmass and clasts, are summarized in Figs. 14 and 4. Figure 14 shows the olivine and pyroxene compositions.
Each of the three lithic clasts of igneous origin (III, II, and V) has its own bulk Mg/Fe

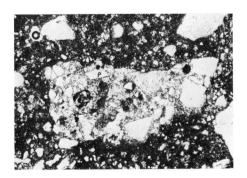

Fig. 13. Breccia fragment within dark microbreccia matrix (IV in 14068,10). Grain labelled
p is large, zoned (relict?) pigeonite.

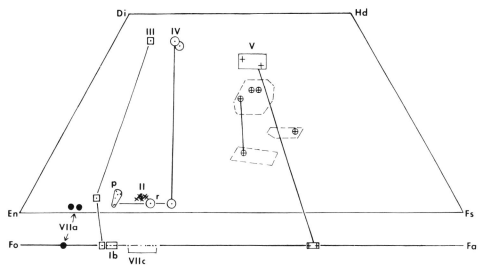

Fig. 14. Mineral compositions of mafic silicates in 14068,10, plotted on pyroxene quadri-lateral (mole percent). Labels correspond to analytical tables as listed in Table 1. Lines drawn completely from symbol to symbol are tielines from adjacent analyzed points. Lines broken near symbols indicate phases from within one lithic fragment that are not in direct contact with each other. For clast V (the basalt) the areas outlined show the observed ranges in pyroxene composition (Ca, Fe, Mg only); the symbols within the dashed areas show complete analyses. The crosses are analyses of the metamorphic (?) augite-olivine pair which locally replaces the igneous subcalcic augite-pigeonite pair.

ratio, which is reflected in the composition of all mafic silicates within the clast.* The analyzed grains conform to the clast bulk composition even if they are in direct contact with the olivine-rich groundmass. For clasts III and V, both of which lie completely within the groundmass, this implies the existence of substantial local gradients in Mg/Fe ratio.

Considering the monomineralic clasts, we see that some, the orthopyroxene and dunite clasts (VIIa), are far more magnesian than the groundmass. These are very homogeneous clasts, showing no zonation toward groundmass compositions; however, one of the pyroxenes has a corona of olivine around it. Since this olivine has the same composition (Fo_{78-80}) as that in the groundmass away from the orthopyroxene, there is again no evidence of chemical transport between clast and groundmass.

The isolated olivine clasts (VIIc) are all more iron-rich than the groundmass olivines. One of these clasts shows a reverse zoning that might suggest equilibration

---

*In contrast, fragment IV is a detrital rock with a range of pyroxene compositions. The magnesian pyroxene (p) is a zoned, presumably relict pigeonite with a very low-calcium core, probably of ortho-pyroxene. It has a rim composition (r) that approaches the composition of the pyroxenes in the matrix of fragment IV (points connected by the vertical tieline), but not that of the pyroxenes in the olivine-rich groundmass. This tieline and that shown for the coexisting pyroxenes in the allivalite (IV) have orientations predicted by Huebner and Ross (1972) to be stable at subsolidus (i.e., metamorphic) temperatures (about 900°C).

with the melt (core Fo_{69} to rim Fo_{73}). The probe traverse was taken as indicated in Fig. 7. The zoning is very broad, however, and this, together with the orientation of the traverse relative to the concave fracture of the edge of the olivine clast suggest that the zoning had developed before inclusion in the present groundmass.

Figure 4 shows all plagioclase compositions found in 14068,10. There is a marked tendency for symbols from the same lithic clast to cluster together. Thus the feldspar in the allivalite (III) varies in composition but is all more calcic than that in the metanoritic anorthosite (II). None approaches groundmass compositions (Ia).

These observations also hold for monomineralic clasts, such as VIa, VIb, VIc, and VId. Of these only clast VIa (Fig. 5) shows extensive zoning. This zoning, which might suggest equilibration with the groundmass, is broad and continuous. There is no narrow rim of a strongly contrasting (sodic) composition on this or any other analyzed feldspar clast in 14068,10. Clast VIc (Fig. 11) is homogeneous An_{96}, right to the edge. Clasts VIb (2 crystals) are An_{81}, again homogeneous to the edge of the grains. Clasts VId include extremely small polycrystalline slivers of An_{95} with texture similar to clast VIc. These also show no equilibration with the groundmass.

FeO and K_2O in feldspar

Figure 15 is a plot of K_2O versus FeO in all feldspars plotted in Fig. 4, plus four from 14053,19*. The K_2O content of plagioclase depends on several factors. In lunar rocks it tends to increase with decreasing An content (Fig. 4; see also Smith, 1971). The general level of K_2O in feldspar also depends, within limits, on the bulk K_2O content of the rock, and lastly, should tend to increase with increasing temperature of formation of the plagioclase. The FeO content of plagioclase depends on rock bulk composition (Quaide, 1972) and should tend to increase with increasing temperature.

The feldspars from 14053,19 (only the compositional extremes are shown, but zoning is continuous) show the pattern Quaide (1972) predicts for plagioclases crystallizing in a slightly differentiating basaltic melt. As the An content drops from 91% to 65%, both FeO and K_2O increase but K_2O more rapidly than FeO. Note that even at low potassium contents, the FeO content of these basaltic feldspars is quite high.

The feldspar from the basaltic fragment in 14068 (the small triangle just outside the area outlined for 14053) plots near those from 14053. This suggests that its cooling regime and the bulk composition from which it crystallized were similar to those of 14053.

The groundmass feldspars have both high iron contents and very high potassium contents. This suggests they crystallized from a more or less basaltic bulk composition with a fairly high K_2O content at temperatures in the melting range. They do not show the positive correlation between iron and potassium shown by the plagioclases from 14053. However, these analyses are not from large strongly zoned crystals like those in 14053 but from small laths with a narrow range of An content. The lack of correlation between K_2O and FeO may be due to much faster cooling in 14068.

---

*Sample 14053 is the only large basalt clast in the Apollo 14 collection.

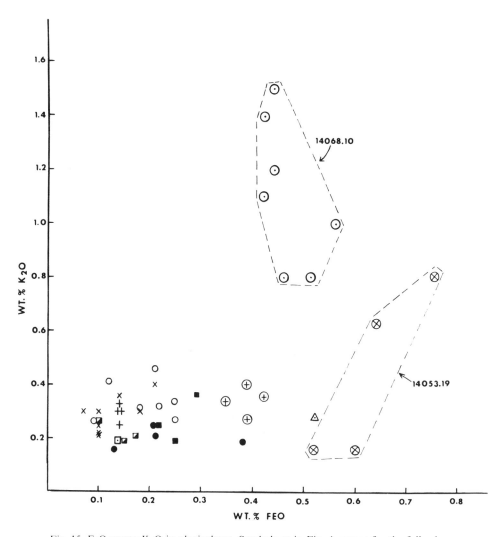

Fig. 15. FeO versus K_2O in plagioclases. Symbols as in Fig. 4, except for the following:
(1) Solid squares = small feldspars in allivalite (III), half-filled squares = large feldspars
in allivalite (III); (2) Crosses in circles = small feldspars in metanorite (II) crosses = large
feldspars in metanorite (II); (3) *xs* in circles = analyses of feldspars from lunar basalt
14053,19.

All other feldspars in 14068,10 contain very little K_2O, and the levels observed
for both FeO and K_2O show no correlation with An content over the range 97%
to 81% An. These feldspars must have formed (1) in low-potassium environments
and/or (2) at metamorphic, rather than igneous temperatures. In any case, they stand
in clear contrast to the (igneous?) feldspars from 14053,19, from the basaltic fragment
(V), and from the groundmass. They do not approach groundmass feldspar compositions in either major or minor components.

The place of 14068 in Warner's classification

Rock 14068 is the sole representative of Warner's Class 8, but several features were observed in this study that are not consistent with the characteristics of Class 8 as described by Warner (1972, this volume):

(1) There are no unequivocal examples of rimming of plagioclase or olivine monomineralic clasts in 14068,10.

(2) There are no rims on pyroxene clasts in 14068,10.

(3) Clasts of potassic granitic glass are present in all sections examined.

(4) There is a dark microbreccia component in 14068,7, 14068,10, and 14068,11 that differs in texture, bulk composition, etc. from the "sheathlike" olivine-rich groundmass on which Warner based his description of rock 14068.

Apparently rock 14068, though small, is quite inhomogeneous in respect to precisely those characteristics Warner used in setting up his classification. It therefore seems premature to include it in this classification.

Summary of Evidence that the Olivine-Rich Groundmass of 14068 Was Once Molten

There are many aspects of rock 14068 that suggest that the olivine-rich groundmass was once molten. These include (1) intrinsic features of the groundmass (2) the textural relations between the groundmass and its inclusions and (3) the nature of the contacts between the groundmass and the surrounding dark breccia.

Considering the olivine-rich groundmass alone, the following features suggest this component once existed as a melt:

(1) Skeletal olivine and variolitic plagioclase commonly are interpreted as evidence of crystallization from a rapidly quenched melt. For instance, rock 12009 shows textures similar to 14068, especially the olivines (Brett *et al.*, 1971).

(2) The narrow compositional ranges of the principal minerals, especially notable for the olivines (Fo_{78-80}), suggest that these minerals crystallized from a fairly homogeneous melt. The clast olivines and plagioclases show a much wider range of composition than the corresponding groundmass minerals.

(3) In the Fo-An-SiO_2 system (Fig. 16) the olivine-rich groundmass projects at G, with a liquidus temperature of just over 1500°C. Olivine is the first phase to crystallize from this composition, followed by plagioclase and low-calcium pyroxene in that order. This is the sequence suggested by the textures observed in 14068, which implies that rock 14068 crystallized from a melt at low pressures.

(4) The groundmass is highly vesicular. The vesicles are subrounded or oval (Fig. 17a, b) and commonly are rimmed by olivine. Where a bordering olivine crystal is elongate, its long dimension is parallel to the vesicle wall (Fig. 17b). The texture suggests the vesicles were still growing after some of the olivine had crystallized and imply that a considerable gas phase was present in the rock while it was molten.

Relationships between the groundmass and its inclusions that suggest that the former was molten include:

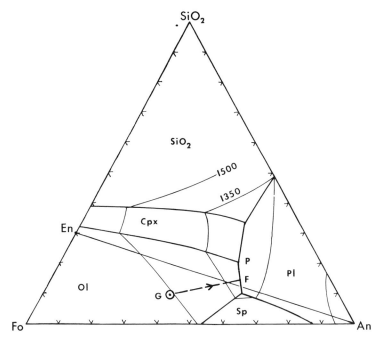

Fig. 16. The Fo-An-SiO$_2$ ternary system after Andersen (1915). Primary phase fields are olivine (Ol), clinoenstatite (Cpx), spinel (Sp), plagioclase (pl), and a silica mineral (SiO$_2$). Two isotherms (1500°, 1350°C) are shown. G-bulk composition of melted groundmass, ignoring Ab, Or in feldspar and treating all pyroxene as En. F-melt composition at which feldspar begins to precipitate from original composition G. P-the peritectic, where melt G finishes crystallizing.

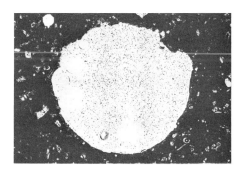

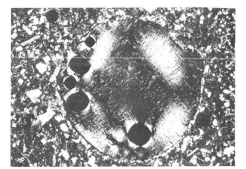

Fig. 17a. Large, nearly spherical vesicle (~ 1.5 mm across) in 14068,10; plane polarized light.

Fig. 17b. Same, with crossed polarizers. Note concentration of olivine, around vesicle rim. Round objects within vesicle are artifacts, visible because the epoxy is slightly birefringent.

(1) There are areas of oriented groundmass plagioclase around plagioclase clasts (Fig. 5). This texture suggests that the clasts were picked up before the melt began to crystallize plagioclase.

(2) The lacy texture of grain VIa (Fig. 5) suggests resorption by a melt not in equilibrium with calcic plagioclase rather than some kind of subsolidus reaction.

(3) The amoeboid contacts the potassic granitic glass clast and the groundmass, shown in Fig. 18, suggests that the groundmass and this potassic granite clast were both molten at the same time. Because these melt compositions probably are miscible (the mafic melt is very different in composition from Roedder's immiscible mafic melt described in Roedder and Weiblen, this volume), the lack of homogenization must be attributed to (a) a large temperature contrast or (b) a large viscosity contrast between the two melts, or both. In any case, the patches of potassic granitic glass are clasts, not pools of residual glass from the crystallization of the mafic melt.

(4) The single large vesicle found in most clasts of potassic granitic glass is in the present center of the clast. The walls of these vesicles are sometimes very thin. This fragile arrangement suggests that the vesicle formed after these clasts were included and implies that the clasts were heated sufficiently to soften and vesiculate after being included.

Lastly, contacts between the olivine-rich groundmass and the black microbreccia seen on the "north" side of sections 14068,7 and 14068,10 are extremely complex. There are long fingers of breccia extending into the groundmass, plus isolated bodies of groundmass in breccia and vice versa. These contacts virtually require one component or the other to have been fairly fluid. The weight of the evidence suggests that the olivine-rich groundmass intruded the breccia.

Another possible indication of the former plasticity of the olivine-rich groundmass is the nature of its contact with the smaller inclusions of dark microbreccia occurring on the "south" side of sections 7 and 10. This fringe of microbreccia has a cusp-like structure, locally outlined by olivine crystals. The impression created is that of a relatively friable material pushed against, or into, a more mobile groundmass, which tended to flow into cracks in the breccia, leaving a slight local accumulation of larger olivine crystals behind. The groundmass appears to have been much less fluid when the "south" contact formed than it was when the "north" contact formed.

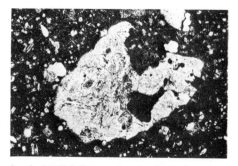

Fig. 18. Large glass clast in 14068,7 (2.1 mm long), with bulbous intrusion of groundmass melt (top), with another cut-off (?) finger of melt below it. Glass contains a variety of quench phases, chiefly pyroxene, K-feldspar, and iron spherules.

Petrogenesis of Rock 14068

Rock 14068 consists principally of the olivine-rich groundmass and a dark microbreccia. These two components differ in bulk composition, in texture, and in characteristic mineral and lithic clast population. The clast populations of the two components are so different that it seems unlikely the two formed close to each other in space. The difference in texture suggests very different temperatures were involved in their formation. The difference in bulk composition makes it clear that the first is not simply the melted equivalent of the second. Here we will be concerned with the origin of the olivine-rich groundmass only.

The first feature of its origin that must be considered is: What is the source of the heat necessary to melt this extremely refractory composition? The liquidus temperature for point G shown in Fig. 16 is about 1500°C. Because the feldspar is An_{72}, not An_{100}, the real temperature would be somewhat lower. Still, it appears that 14068 once contained a melt with 25% olivine dissolved in it. The most olivine-rich lunar melts (i.e., rocks containing no cumulate olivine) known so far are 12004 and 12009 (Green et al., 1971). Rock 12009 has 11% normative olivine, 12004 less than that. The liquidus for 12009 is given by Green et al., as 1230°C. The temperature required to produce a homogeneous noritic melt containing 25% normative olivine is probably well beyond the normal range observed for basaltic rocks.

Given the apparent temperature requirements for complete melting of such a mafic composition, the general lack of chemical homogenization between groundmass and clasts is remarkable. This can be resolved if the molten body was a small bleb, not much larger than the present size of 14068, and if all the associated debris was picked up after the melt had begun to cool and crystallize.

At this point, it should be noted that the phase relations shown in Fig. 16 require olivine to be the first phase to crystallize from composition G. This implies that (1) all feldspar clasts are accidental, (2) all olivines more iron-rich than the groundmass are accidental, (3) the silica-saturated granitic glasses are accidental. Here "accidental" means that they were picked up after this strongly olivine-normative groundmass had been formed.

A tentative outline of the history of rock 14068, in light of the above, is:

(1) The melting, involving temperatures beyond the igneous range, was the result of meteorite impact. The molten bleb produced was fairly small.

(2) The melt picked up plagioclase clasts and nickel-iron blebs. There was sufficient turbulence for the clasts to be mixed throughout the bleb.

(3) The melt picked up the olivine, orthopyroxene, lithic and glassy fragments. Most of the glassy clasts softened and vesiculated.

(4) The melt made contact with some of the dark breccia ("north" side) while still quite fluid.

(5) The melt made contact with the dark breccia on the "south" side while still plastic, but it was not as fluid as in step 4.

(6) Rock 14068 has not been involved in any major brecciation or thermal metamorphic events since solidification of the olivine-rich groundmass was completed.

The problem of the origin of the bulk composition of the olivine-rich groundmass, shown in Table 2, cols. 1 and 2, remains. This composition, with its very low-calcium bulk pyroxene has noritic, rather than basaltic affinities. The high TiO_2, K_2O, and P_2O_5 contents further suggest a relationship with KREEP-rich noritic rocks rather than ordinary orthopyroxene-plagioclase cumulates. However, no olivine-rich, high-Na_2O members of this widespread suite of lunar rocks have yet been reported. The association of very high olivine content with unusually sodic plagioclase (for a lunar rock) is anomalous. Possible origins include:

(1) The melt was produced by remelting a pre-existing lunar rock of the same composition. In this case, rock 14068 would presumably lie toward the high-temperature end of a hitherto unknown lunar magma series. Because its differentiates would be more sodic, theories advanced to explain the low soda contents of lunar rocks would need reevaluation. Also this would provide another reason to reconsider the present habit of applying terrestrial rock names such as basalt, norite, gabbro, troctolite, etc., to lunar rocks containing plagioclase in the range An_{80-100}. The chief problem with this explanation is, why was such an extremely rare and refractory lunar rock melted intact, without contamination?

(2) The composition is a mixture of at least two rock types. Because there are no known lunar rocks that could be mixed to produce this composition, meteorite compositions were examined. The only possibly useful meteorite composition discovered was that of the silicate fraction of chondritic meteorites. The composition selected is shown in Table 4. This composition was then used in a set of mixing calculations, following the technique of Wright and Doherty (1970). The basic equation used was: melt = lunar surface component + meteorite + K-granite. The results, for different lunar materials, are presented in Table 5. Note that substantial amounts of the meteorite component are involved in all cases. The calculation producing the best calculated melt composition was that using a KREEP-rich noritic anorthosite (Table 6). A surprising second best was that using the Apollo 14 bulk fines, analyzed by Compston et al. (1972). The most obvious problem with this model is the difficulty of segregating the silicate fraction of the chondritic meteorite from the iron-troilite fraction.

Given our limited knowledge of the composition of the moon, it is difficult to choose between these alternatives. In the event that no soda-rich rocks are ever

Table 4. Average composition of the silicate fraction of 26 chondritic meteorites from Urey and Craig (1953, Table 2), renormalized to 100%. Analyses taken were those in which (1) $Na_2O > 1.15\%$ and (2) Na_2O and K_2O were analyzed and reported separately.

SiO_2	46.93
TiO_2	0.15
Al_2O_3	3.39
FeO	15.12
MnO	0.32
MgO	28.98
CaO	2.73
Na_2O	1.75
K_2O	0.33
P_2O_5	0.29
TOTAL	99.99

Table 5. Results of mixing calculations, in weight percent.

	Lunar surface component	Meteorite	Potassic granite
D (Table 3, this paper)	77.4	22.6	0.0
Breccia (1)	66.8	33.0	0.2
Noritic anorthosite (1)	50.9	48.3	0.7
Apollo 14 fines (2)	66.4	33.4	0.2
14310(2)	57.9	41.8	0.4
KREEP norite (lithic) (3)	65.6	34.4	0.1
KREEP norite (glass) (3)	59.7	40.2	0.1
KREEP norite (chondrule) (3)	68.5	31.4	0.1
Anorthositic gabbro (4)	55.2	44.3	0.4
Norite (4)	53.4	46.6	0.0
Norite (4)	66.0	33.9	0.1

Analyses of lunar surface components are from:
(1) Grieve *et al.* (1972), Table 1, Cols. 2 and 3.
(2) Compston *et al.* (1972), Table 1, Cols. 3 and 5.
(3) Kurat *et al.* (1972), Table 1, Cols. 2, 6, 9.
(4) Marvin *et al.* (1971), Table 4, Cols. A, B, E.

Table 6. Best composition (wt%) obtained in mixing calculations, using anorthositic gabbro (Col. A), with melted groundmass for comparison.

	(1) Mix	(2) Melt (from Table 2)
SiO_2	47.1	47.3
TiO_2	1.40	1.36
Al_2O_3	14.0	14.7
FeO	10.55	10.06
MgO	16.82	17.56
CaO	7.93	7.61
Na_2O	1.12	1.10
K_2O	0.64	0.64
P_2O_5	0.36	0.37

found on the moon, or those found always contain very abundant olivine and generally resemble rock 14068, one might look harder at model (2). A most appropriate line of future investigation in that case would be to analyze 14068 for the siderophile trace elements.

Acknowledgments—The author profited by many discussions with Jeff Warner (NASA–MSC) who very kindly made his analytical data available. The suggestion that 14068 might contain a meteorite component was originally made by D. B. Stewart (U.S. Geological Survey, Washington). The manuscript has been improved by D. Appleman and O. James (U.S. Geological Survey, Washington), who acted as reviewers. This work was supported in part by NASA contract T–2354–A.

REFERENCES

Andersen O. (1915) The system anorthite-forsterite-silica. *Amer. J. Sci.* (4th series) **39**, 407–454.

Boyd F. R. Finger L. W. and Chayes F. (1969) Computer reduction of electron-probe data. Carnegie Inst. Yearbook **67**, 210–215.

Brett R. Butler P. Jr. Meyer C. Jr. Reid A. M. Takeda H. and Williams R. (1971) Apollo 12 igneous rocks 12004, 12008, 12009 and 12022: a mineralogical and petrological study. *Proc. Second Lunar Sci. Conf., Geochim. Cosmochim. Acta.* Vol. 1, pp. 301–317.

Christophe-Michel-Levy M. Levy C. and Pierrot R. (1972) Mineralogical aspects of Apollo XIV samples:

lunar chondrules, pink spinel bearing rocks, ilmenites (abstract). In *Lunar Science—III* (editor C. Watkins), pp. 136–138. Lunar Science Institute Contr. No. 88.

Compston W. Vernon M. J. Berry H. Rudowski R. Gray C. M. Ware N. Chappell B. W. and Kaye M. (1972) Age and petrogenesis of Apollo 14 basalts (abstract). In *Lunar Science—III* (editor C. Watkins), pp. 151–153, Lunar Science Institute Contr. No. 88.

Green D. H. Ringwood A. E. Ware N. G. Hibberson W. O. Major A. and Kiss E. (1971) Experimental petrology and petrogenesis of Apollo 12 basalts. *Proc. Second Lunar Sci. Conf., Geochim. Cosmochim. Acta*, Vol. 1, pp. 601–615.

Grieve R. McKay A. Smith H. and Weill D. (1972) Mineralogy and petrology of polymict breccia 14321 (abstract). In *Lunar Science—III* (editor C. Watkins), pp. 338–340, Lunar Science Institute Contr. No. 88.

Hubbard N. J. and Gast P. W. (1972) Chemical composition of Apollo 14 materials and evidence for alkali volatilization (abstract) in *Lunar Science—III* (editor C. Watkins), pp. 407–409, Lunar Science Institute Contr. No. 88.

Huebner J. S. and Ross M. (1972) Phase relations of lunar and terrestrial pyroxenes at one atm. (abstract). In *Lunar Science—III* (editor C. Watkins), pp. 410–412, Lunar Science Institute Contr. No. 88.

Kurat G. Keil K. and Prinz M. (1972) A "chondrite" of lunar origin: textures, lithic fragments, glasses and chondrules (abstract). In *Lunar Science—III* (editor C. Watkins), pp. 463–465, Lunar Science Institute Contr. No. 88.

Marvin U. B. Wood J. A. Taylor G. J. Reid J. B. Jr. Powell B. N. Dickey J. S. Jr. and Bower J. F. (1971) Relative proportions and probable sources of fragments in the Apollo 12 soil samples. *Proc. Second Lunar Sci. Conf., Geochim. Cosmochim. Acta*, Vol. 1, pp. 679–699.

Mason B. (1962) *Meteorites*. 274 pp. John Wiley.

Quaide W. (1972) Mineralogy and origin of Fra Mauro fines and breccias (abstract). In *Lunar Science—III* (editor C. Watkins), pp. 627–629, Lunar Science Institute Contr. No. 88.

Roedder E. and Weiblen P. W. (1972) Petrographic features and petrologic significance of melt inclusions in Apollo 14 and 15 rocks. Third Lunar Science Conference, unpublished proceedings.

Smith J. V. (1971) Minor elements in Apollo 11 and Apollo 12 olivine and plagioclase. *Proc. Second Lunar Sci. Conf., Geochim. Cosmochim. Acta*, Vol. 1, pp. 143–150.

Urey H. C. and Craig H. (1953) The composition of the stone meteorites and the origin of the meteorites. *Geochim. Cosmochim. Acta* **4**, 36–82.

Warner J. L. (1972) Apollo 14 breccias: metamorphic origin and classification (abstract). In *Lunar Science—III* (editor C. Watkins), pp. 782–784, Lunar Science Institute Contr. No. 88.

Warner J. L. (1972) Apollo 14 breccias: metamorphic origin and classification (abstract). In *Lunar Science* proceedings.

Wright T. L. and Doherty P. C. (1970) A linear programming and least squares computer method for solving petrologic mixing problems. *Bull. Geol. Soc. Am.* **81**, 1995–2008.

Proceedings of the Third Lunar Science Conference
(Supplement 3, *Geochimica et Cosmochimica Acta*)
Vol. 1, pp. 887–894
The M.I.T. Press, 1972

The magnesian spinel-bearing rocks from the Fra Mauro formation

M. Christophe-Michel-Levy and C. Levy

Laboratoire de Minéralogie-Cristallographie
associé au C.N.R.S., Université de Paris VI

R. Caye and R. Pierrot

Service géologique national
B.R.G.M., 45–Orléans

Abstract—The magnesium spinels found in many Fra Mauro breccias are slightly chromiferous pleonastes. They are particularly abundant in breccia 14063. This mineral, in isolated fragments, is one of the products of brecciation of various rocks, mostly anorthositic, which are present as clasts in the breccia. It also is found as euhedral crystallites in glassy debris. The large number of spinel-bearing rock types probably is indicative of a sequence of hyperaluminous rocks not observed in the mare regions. This sequence could be characteristic of the substratum, brought to the surface by the Cone Crater impact event.

INTRODUCTION

THE STUDY OF THE Apollo 14 rocks has shown that pleaonaste, a magnesium-iron-aluminium spinel, is an accessory mineral that occurs in a number of samples, and which is particularly abundant in breccia 14063. Several papers presented at the Third Lunar Science Conference give partial results that supplement our own observations. A review of this topic is presented here and some implications of the presence of this spinel are discussed.

COMPOSITION OF THE SPINEL

The spinel grains in breccias 14063, 14321, 14318 are light to bright pink in thin-sections, and in fines 14003, 14163 (one grain found in each of these soils) and red in breccia 14066. We analyzed (electron microprobe) four grains in sections 14063,23 and 14063,56. Their composition corresponds to the formula $(Mg, Fe)(Al, Cr)_2O_4$ with the following atomic ratios:

No.	Mg/(Mg + Fe)	Cr/(Cr + Al)		
1	0.65	0.05	14063,56	anorthosite clast
2	0.62	0.08	14063,23	isolated grain
3	0.62	0.14	14063,23	isolated grain
4	0.55	0.24	14063,56	isolated grain with Ti-rich lamellae

The TiO_2 content is $<1\%$. Grain 4 (Fig. 1), with Ti-rich lamellae, contains only $0.3\% TiO_2$. Several grains have been studied for their optical properties: we measured their reflectance in polished sections in order to deduce the index of refraction. This technique can apply to *in situ* grains and can be used for grains as small as a few

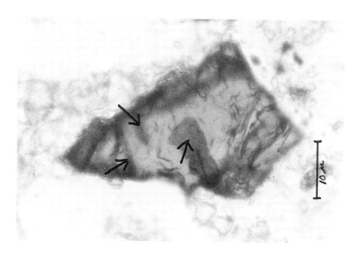

Fig. 1. A spinel grain containing titanium-rich lamellae (arrows). The boundaries are blurred because the microscope was focused on one of the lamellae. Scale bar is 10 μ.

microns. Most of the indices of refraction measured by us are within the 1.78–1.81 range but one measurement yielded 1.84 and another 1.75. The chemical compositions reported at the Third Lunar Science Conference (Brown *et al.*, 1972; Drever *et al.*, 1972; Engelhardt *et al.*, 1972; Haggerty, 1972; Mason *et al.*, 1972; Roedder and Weiblen, 1972; Steele and Smith, 1972; Walter *et al.*, 1972) indicate a range of Mg/(Mg + Fe) from 0.45 to 0.82, for Cr/(Cr + Al) from 0.02 to 0.21 with a grouping of the figures between 0.50 and 0.75 for the first ratio, below 0.1 for the second ratio. The titanium contents are all quite low. Only Walter *et al.* (1972) report 1.2% TiO_2 in a hercynitic spinel. Thus, the chemical data are consistent with our optical measurements.

As emphasized by Haggerty, (1972), there is no continuous solid-solution trend towards chromite or ulvöspinel. The coexistence of the pink spinel with other types of spinels appears to be due to the polymict character of the breccias and not to cogenetic crystallization. For instance, in section 14063,23, chromite is found in a troctolitic breccia coexisting with the main pleonaste-bearing breccia.

The compositions of the spinels differ slightly from grain to grain and commonly show a compositional range within a given crystal. Several authors have reported zoning of some grains. According to Brown *et al.* (1972), who observe a more reddish rim "due to a rise in the Fe/Mg ratio", this zoning is not the consequence of metamorphism. As far as Roedder and Weiblen (1972) are concerned, the grains they observed have darker edges, with a reaction rim of plagioclase in the breccia matrix; they attribute this effect to metamorphism. A few zoned grains have been observed by us in the unrecrystallized breccia 14063 and probably are the type described by Brown *et al.* (1972). Some grains with a feldspathic reaction corona in the recrystallized breccia 14321 possess a rim rich in tiny opaque crystals that darken the edges of the spinel if observed in transmitted light. In this case, the zoning probably is the type described by Roedder and Weiblen (1972).

THE SPINEL-BEARING BRECCIAS

The rock samples (all breccias) in which the existence of spinel has been reported are the following: 14049, 14055, 14063, 14066, 14171, 14303, 14305, 14312, 14314, 14318, 14319, 14320, 14321; breccia 14063 deserves a special study because of the abundance of spinel and because of the variety of petrographic types of spinel-bearing rocks. This sample has been collected on the slope of Cone Crater, at station C_1 together with the "football rock" 14321. It is likely that these two rocks have been excavated from Cone Crater during the same impact event. Because no dating has been performed on 14063, we shall rely on 14321 as this rock contains some spinel too.

The exposure age of rock 14321 is approximately 23×10^6y (Berdot *et al.*, 1972). Two clasts measured by Compston *et al.* (1972) have a Rb-Sr age of 4.02 and 4.15×10^9y, respectively. It is conceivable that the age of the 14321 breccia is 3.9–4.0×10^9y and that 14063 is at least as old as that. Furthermore, Dran *et al.* (1972) have shown that the crystals of the latter sample have low track density; breccia 14063 has not been recrystallized, and therefore major track annealing is improbable. It must have been weakly irradiated and not much exposed on the lunar surface prior to its relatively recent ejection.

Most of the other spinel-bearing samples (14312, 14, 18, 19, 20) come from station H, which is located on a bouldery ray ejecta of Cone Crater. The existence of spinel in the other six samples occurring on the extension of these ejecta towards the S–W raises the question of whether the ejecta are not of greater dimensions than indicated on the maps (Sutton *et al.*, 1972).

Breccia 14063 is complex. Three successive stages of brecciation can be identified in thin sections; spinel is present in each stage, either as isolated crystals or enclosed in rocks containing the following associations:

1. Plagioclase—spinel; these anorthosites are slightly to moderately shocked (Fig. 2); the spinel, present in grains of the order of 0.1 mm, has been broken up by shock into numerous fragments. This rock type is common.

2. Olivine-spinel-plagioclase; one very small clast of such crystalline rock has been found, the mineral proportions of which cannot be estimated adequately; the spinel shows signs of euhedral growth as indicated by rounded octahedral shapes.

3. Spinel-glass, and spinel-olivine-glass. Debris of a brownish glass with an index of refraction of 1.62 to 1.65, rarely containing fragments of feldspar and pyroxene, contain euhedral spinel crystallites as small octahedra (Fig. 3); euhedral or skeletal olivine crystals may also be present.

4. Various types of clasts of spinel-bearing breccias are found; one resembles a stratified sediment; another one, rich in brown interstitial glass, contains debris of spinel and glass with spinel octahedra; finally, another one is a recrystallized breccia.

The spinel-bearing rock clasts are accompanied by other rocks without spinel: clasts of basalts, KREEP norites, xenoliths of breccias with mainly plagioclase and olivine, or plagioclase and orthopyroxene, etc. Dunite clasts were not found in 14063 but are present in 14066. Some rather rare mineral species also are found: a yellow transparent baddeleyite grain, a euhedral hyalophane crystal (Fig. 4) with the formula $K_{0.765}Na_{0.082}Ca_{0.027}Ba_{0.126}Al_{1.153}Si_{2.84}O_8$.

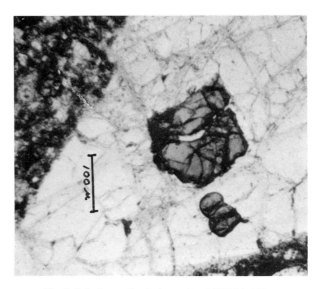

Fig. 2. Spinel-anorthosite in section 14063,56, 150 ×.

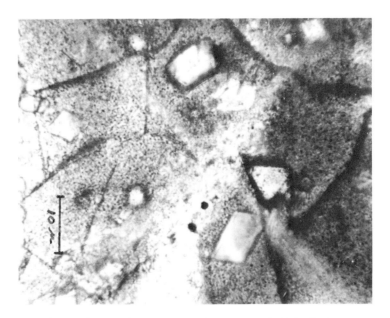

Fig. 3. Euhedral spinels in glass, section 14063,23. Scale bar is 10 μ.

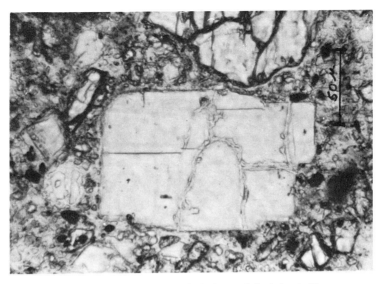

Fig. 4. A hyalophane large euhedral crystal. Scale bar is 50 μ.

DISCUSSION

If we look at the "selenographical" distribution of the spinel, we see that four spinel grains only have been found in Apollo 11 and 12 mare samples, and they are either more magnesium-rich (Keil *et al.*, 1970) or more iron-rich (Keil *et al.*, 1970; Reid, 1971; Wood *et al.*, 1971) than the Apollo 14 specimens. On the other hand, some spinel, analogous in composition, has been discovered in the soil of Luna 16 (Jakeš *et al.*, 1972) and Apollo 15 (Mason *et al.*, 1972; Christophe *et al.*, unpublished).

Luna 16 soil is thought to be rather rich in non-mare material. The Apollo 15 spinel grains are found in the soil 15231,66, which is rich in feldspathic crystals and comes from near St. George Crater, which may have (like Cone Crater) brought to the surface some material from the substratum.

The observations lead us to believe that the spinel-bearing rocks are old diversified rocks coming from the basement of the Fra Mauro formation. Several matters remain to be investigated: (1) whether the chemical composition of these rocks is peculiar, (2) whether the presence of spinel can be explained by a high alumina content, or (3) whether the spinel could appear in rocks analogous in composition to other Fra Mauro rocks (14310 basalt or spinel-free breccias), and finally (4) whether one can find indications of the physical conditions of crystallization as for plutonic rocks.

In the absence of chemical analyses of the anorthosite clasts and of the spinel-rich glasses, our discussion can hardly be more than conjectural. Because of the great heterogeneity of the constituents of the breccia, the major-element analysis of the bulk of the matrix (in fact, an incomplete analysis) indicating a rather high alumina content (21.5% to 23.5% Al_2O_3; Laul *et al.*, 1972) cannot be of great assistance. On the basis of trace element analysis of igneous fragments from 14063, Helmke and

Haskin (1972), suggest that they might be representatives of a new "strain" of igneous rocks. According to Hollister *et al.* (1972), the chemical compositions of the hypersthene and coexisting augite in the mineral fragments of breccia 14063 suggest that they are samples of plutonic rocks that formed part of the lunar crust.

It is well known that spinel appears on the liquidus in systems involving anorthite and forsterite. Basaltic rock 14310 and several fragments of high-alumina basalts from soils (Steele and Smith, 1972) correspond to compositions close to these systems but spinel after being formed would have been resorbed during the crystallization. If such resorption is prevented, either by crystal-settling or by rapid cooling, spinel would survive. Crystal-settling is a process that could have acted for spinel-olivine rocks, like a clast mentioned by Steele and Smith (1972), but not for spinel anorthosites, which are more abundant. Rapid cooling could explain the presence of spinel crystallites in the glasses, but no spinel appears in any of the high-alumina glasses of group E (Apollo Soil Survey, 1971), although their composition is suitable. The spinel-rich glasses strictly are connected to the spinel-bearing crystalline rocks and it is probable that, as a whole, these materials correspond to a calcic hyperaluminous formation. In such conditions, spinel would crystallize as a stable phase, and other characteristic minerals could be formed. For instance, *corundum*: this mineral, discovered by Kleinmann and Ramdohr (1971) as highly shocked crystals at the Apollo 11 and 12 sites is found at the Apollo 14 site as an undisturbed crystal, possibly in the vicinity of its parent rock. We found a grain approximately 200 μ in size in fines 14163. An electron microprobe investigation has revealed that this grain is composed of very pure alumina, without iron or any other noticeable impurities. Its crystallographic parameters have been determined by x-rays (precession technique): $a = 4.75_8$ Å, $c = 12.97$ Å, and the crystal is twinned according to the plane $\{0001\}$. This is an unusual twinning in the case of terrestrial corundum. Another characteristic mineral could be possible *melilite* reported by Masson *et al.* (1972).

Magnesian spinel, anorthite, crystalline alumina and possible melilite constitute a well-known association not only in synthetic systems, but also in meteorites (not to mention terrestrial rocks, which are too different from lunar materials to allow a constructive comparison with metamorphism of calcareous shales). Indeed, spinel is found in carbonaceous chondrites of type III and IV, always associated with a calcic silicate, gehlenite, anorthite, diopside, commonly with titaniferous concentrations (small grains of perovskite), sometimes with a phase analogous to β alumina (hibonite) and some sodalite. A spinel of the same type also is found, although not commonly, in chondrule-rich ordinary chondrites or in polymict breccias as isolated grains or grains associated with olivine.

Whatever the explanation proposed for the associations found in carbonaceous chondrites (Larimer and Anders, 1970; Kurat, 1970; Arrhenius and Alfven, 1971), would one of them be applicable to spinel-rocks of the Moon?

The presence of spinel in ordinary chondrites has not been an object of much discussion. It is interesting to note its presence in a meteorite such as *Hedjaz* and in a lunar chondritic breccia such as 14318, both of which Fredriksson *et al.* (1972) suspect to be impact ignimbrites.

CONCLUSION

Hubbard *et al.* (1972) consider that two distinct types of anorthosites and related high alumina basalts have been found among the lunar samples. We suggest (Christophe *et al.*, 1972) that a third series exists that is hyperaluminous and characterized particularly by the occurrence of magnesian spinel. It is likely that the corresponding formation is not exposed at the surface, because nowhere has a high enough Al/Si ratio been detected by the x-ray fluorescence experiment of Adler *et al.* (1972). In any case, it is almost certain that the presence of these high alumina rocks at the Fra Mauro site is related to the Cone Crater impact event. If one accepts that the Fra Mauro rocks belong to the Imbrian ejecta blanket, the spinel-rocks would come from the Imbrium Basin basement. It would be worth while to study other samples presumed to be of relatively deep origin to check whether this type of material represents more than the result of a local differentiation and, perhaps, belongs to an anorthosite shell in the crust of the moon.

Acknowledgments—We are indebted to A. Parfenoff for the discovery of the corundum grain, to C. Desnoyers for its x-ray study with advice of R. Chevalier, to M. Pasdeloup for optical measurements of spinels and to R. Giraud for microprobe analyses. Thanks are also due to Dr. D. Y. Jerome for help for the critical reading of the manuscript.

REFERENCES

Adler I. Trombka J. Gerard J. Lowman P. Yin L. Blodgett H. Gorenstein P. and Bjorkholm P. (1972) Preliminary results from the S–161 x-ray fluorescence experiment (abstract). In *Lunar Science—III* (editor C. Watkins), pp. 4–6, Lunar Science Institute Contr. No. 88.

Apollo Soil Survey (1971) Apollo 14: Nature and origin of rock types in soil from the Fra Mauro formation. *Earth Planet. Sci. Lett.* **12**, 49–54.

Arrhenius G. and Alfven H. (1971) Fractionation and condensation in space. *Earth Planet. Sci. Lett.* **10**, 253–267.

Berdot J. L. Chetrit G. C. Lorin J. C. Pellas P. Poupeau G. and Reeves H. (1972) Preliminary track data on some rocks from Apollo 14 (abstract). In *Lunar Science—III* (editor C. Watkins), pp. 62–64, Lunar Science Institute Contr. No. 88.

Brown G. M. Emeleus C. H. Holland J. G. Peckett A. and Phillips R. (1972) Mineral fractionation patterns between Apollo 14 primitive feldspathic rocks and Apollo 15 and other basalts (abstract). In *Lunar Science—III* (editor C. Watkins), pp. 95–97, Lunar Science Institute Contr. No. 88.

Christophe-Michel-Levy M. Levy C. Caye R. and Pierrot R. (1972) Existence de reches à spinelle magnésien dans la formation lunaire Fra Mauro. *C. R. Acad. Sci.* **274** D, 155–157.

Compston W. Vernon M. J. Berry H. Rudowski R. Gray C. M. Ware N. Chappel B. W. and Kaye M. (1972) Age and petrogenesis of Apollo 14 basalts (abstract). In *Lunar Science—III* (editor C. Watkins), pp. 151–153, Lunar Science Institute Contr. No. 88.

Dran J. C. Duraud J. P. Maurette M. Durrieu L. and Legressus C. (1972) The high resolution track and texture record of lunar breccias and gas-rich meteorites (abstract). In *Lunar Science—III* (editor C. Watkins), pp. 183–185, Lunar Science Institute Contr. No. 88.

Drever H. I. Johnston R. and Gibb F. G. F. (1972) A note on three Imbrium spinels, and a twinned pigeonite in high alumina basalt 14310 (abstract). In *Lunar Science—III* (editor C. Watkins), pp. 186–188, Lunar Science Institute Contr. No. 88.

Engelhardt W. von Arndt J. and Stoffler D. (1972) Apollo 14 soils and breccias, their compositions and origin by impacts (abstract). In *Lunar Science—III* (editor C. Watkins), pp. 233–235, Lunar Science Institute Contr. No. 88.

Fredriksson K. Nelen J. Noonan A and Kraut F. (1972) Apollo 14: Glasses, breccias, chondrules (abstract). In *Lunar Science—III* (editor C. Watkins), pp. 280–282, Lunar Science Institute Contr. No. 88.

Grieve R. McKay G. Smith H. Weill D. and McCallum S. (1972) Mineralogy and petrology of polymict breccia 14321 (abstract). In *Lunar Science—III* (editor C. Watkins), pp. 338–340, Lunar Science Institute Contr. No. 88.

Haggerty S. E. (1972) Subsolidus reduction and compositional variations of lunar spinels (abstract). In *Lunar Science—III* (editor C. Watkins), pp. 347–349, Lunar Science Institute Contr. No. 88.

Helmke P. A. and Haskin L. A. (1972) Rare earths and other trace elements in Apollo 14 lunar samples (abstract). In *Lunar Science—III* (editor C. Watkins), pp. 366–368, Lunar Science Institute Contr. No. 88.

Hollister L. Trzcienski W. Jr. Dymek R. Kulick C. Weigand P. and Hargraves R. (1972) Igneous fragment 14310, 21 and the origin of the mare basalts (abstract). In *Lunar Science—III* (editor C. Watkins), pp. 386–388, Lunar Science Institute Contr. No. 88.

Hubbard N. J. Gast P. W. and Meyer C. (1972) Chemical composition of lunar anorthosites and their parent liquids (abstract). In *Lunar Science—III* (editor C. Watkins), pp. 404–406, Lunar Science Institute Contr. No. 88.

Jakeš P. Warner J. Ridley W. I. Reid A. M. Harmon R. S. Brett R. and Brown R. W. (1972) Petrology of a portion of the Mare Fecunditatis regolith. *Earth Planet. Sci. Lett.* **13**, 257–271.

Keil K. Bunch I. E. and Prinz M. (1970) Mineralogy and composition of Apollo 11 lunar samples. *Proc. Apollo 11 Lunar Sci. Conf., Geochim. Cosmochim. Acta* Suppl. 1, Vol. 1, pp. 561–598. Pergamon.

Kleinmann B. and Ramdohr P. (1971) Corundum from the lunar dust. *Earth Planet. Sci. Lett.* **13**, 19–22.

Kurat G. (1970) Zur genese der Ca-Al-reichen Einschlüsse im chondriten von Lancé. *Earth Planet. Sci. Lett.* **9**, 225–231.

Larimer J. W. and Anders E. (1970) Chemical fractionation in meteorites III. Major element fractionation in chondrites. *Geochim. Cosmochim. Acta* **34**, 367–387.

Laul J. C. Boynton W. V. and Schmitt R. A. (1972) Bulk, REE and other elemental abundances in four Apollo 14 clastic rocks and three cove samples, two Luna 16 breccias and four Apollo 15 soils (abstract). In *Lunar Science—III* (editor C. Watkins), pp. 480–482, Lunar Science Institute Contr. No. 88.

Mason B. Melson W. G. and Nelen J. (1972) Spinel and hornblende in Apollo 14 fines (abstract). In *Lunar Science—III* (editor C. Watkins), pp. 512–514, Lunar Science Institute Contr. No. 88.

Masson C. R. Smith I. B. Jamieson W. D. and McLachlan J. L. (1972) Chromatographic and mineralogical study of Apollo 14 fines (abstract). In *Lunar Science—III* (editor C. Watkins), pp. 515–516, Lunar Science Institute Contr. No. 88.

Reid J. B. Jr. (1971) Apollo 12 spinels as petrogenetic indicators. *Earth Planet. Sci. Lett.* **10**, 351–356.

Roedder E. and Weiblen P. W. (1972) Petrographic and petrologic features of Apollo 14, 15 and Luna 16 samples (abstract). In *Lunar Science—III* (editor C. Watkins), pp. 657–659, Lunar Science Institute Contr. No. 88.

Steele I. M. and Smith J. V. (1972) Mineralogy, petrology, bulk electron-microprobe analyses from Apollo 14, 15 and Luna 16 (abstract). In *Lunar Science—III* (editor C. Watkins), pp. 721–723, Lunar Science Institute Contr. No. 88.

Sutton R. L. Hait M. H. and Swann G. A. (1972) Geology of the Apollo 14 landing site (abstract). In *Lunar Science—III* (editor C. Watkins), pp. 732–734, Lunar Science Institute Contr. No. 88.

Walter L. S. French B. M. and Doan A. S. Jr. (1972) Petrographic analysis of lunar samples 14171 and 14171 and 14305 (breccias) and 14310 (melt rock) (abstract). In *Lunar Science—III* (editor C. Watkins), pp. 773–775, Lunar Science Institute Contr. No. 88.

Wood J. A. Marvin U. B. Reid Y. B. Taylor G. J. Bower Y. F. Powell B. N. and Dickey J. S. Jr. (1971) Mineralogy and petrology of the Apollo 12 lunar sample. *Smiths. Astroph. Observatory S.R. 333.*

Proceedings of the Third Lunar Science Conference
(Supplement 3, *Geochimica et Cosmochimica Acta*)
Vol. 1, pp. 895–906
The M.I.T. Press, 1972

Deformation of silicates in some Fra Mauro breccias

Hans G. Avé Lallemant

Department of Geology,
Rice University, Houston, Texas 77001

Neville L. Carter

Department of Earth and Space Sciences,
State University of New York,
Stony Brook, New York 11790

Abstract—A total of eighteen thin sections of Fra Mauro breccias have been studied in detail, using optical techniques, for evidence of static and dynamic deformational processes in the silicates. Microstructures due to shock deformation are similar to those found in silicates from Mare Tranquillitatis and Oceanus Procellarum, but commonly they appear to have been modified by high temperature annealing. Notable among annealed features are healed microfractures, planar arrays of cavities along pre-existing planar features (shock lamellae), and partial to total recrystallization of highly deformed olivines and clinopyroxenes. Lack of evidence for the orthopyroxene-clinopyroxene inversion also may indicate prolonged high temperature annealing. The annealing may have taken place at depth by burial under a hot ejecta blanket thrown out during the excavation of Mare Imbrium.

There is no unequivocal evidence for appreciable static deformation in silicates from Fra Mauro, in accord with previous results for Mare Tranquillitatis and Oceanus Procellarum. Thus it appears as if at least the outer layer of the moon at these three sites has not undergone substantial tectonic activity since crystallization *ca.* 3 to 4 billion years ago.

Introduction

The primary purpose of our investigations is to determine the presence or absence of structures in lunar silicates that have been produced by static deformational processes. Correlations of deformational processes that have operated in recent static experiments with those in naturally deformed terrestrial and meteoritic silicates have provided new insights as to physical conditions during the natural deformations. Similar comparisons have provided and will continue to provide important information concerning lunar tectonic activity and hence on the thermal evolution of the moon. In these studies, it is essential to characterize, using meteorites, terrestrial impactites, and controlled shock experiments, structures induced by dynamic deformation and to distinguish these carefully from those due to static deformation. There is some ambiguity and overlap but, in general, sufficient criteria exist for distinguishing between these two extreme types of natural deformation (e.g. Carter, 1971) that differ in strain rate by about 20 orders of magnitude.

Because of the paucity of crystalline rocks returned by the Apollo 14 mission, we have concentrated on studies of static and dynamic deformation of silicates in the breccias. We shall compare and contrast deformational structures in silicates of the Fra Mauro formation, which presumably is composed mostly of ejecta from Mare Imbrium, with those observed in breccias from Mare Tranquillitatis and Oceanus Procellarum. There is no unequivocal evidence for static deformation in the silicates from all three areas other than processes associated with cooling or annealing. However, there are significant differences in the nature of structures induced by shock deformation of Fra Mauro material and those found for the other sites. These differences, we believe, are due to postshock annealing at high temperatures, and we shall discuss the basis for this interpretation below.

DEFORMATION OF SILICATES

We have examined in detail, using petrographic and U-stage techniques, a total of 18 thin sections of Fra Mauro breccias from our own allotment, those of our colleagues and a suite of sections at Manned Spacecraft Center. The specimens studied are: 14068,19; 14162,41; 14167,0; 14301,17,83; 14304,2; 14305,15,21; 14307,14; 14313,41,42,48; 14314,6; 14315,7; 14319,5; 14321,24,36,75. We do not intend to describe these sections in detail but we wish to emphasize representative structures in the silicates of these breccias, treating in turn, plagioclase, the pyroxenes, and olivine.

PLAGIOCLASE

Deformation twins according to albite, pericline, and, in some instances, Ala, laws are common in the plagioclases of these breccias. The twins are recognized as deformation twins by their association with, and progressively more intense development near, faults, mosaic zones, and zones of maskelynite. Ala twins in the fragment shown in Fig. 1a increase in intensity in the more highly deformed zones in the lower right and then disappear in the zone showing a mosaic structure which grades into finely recrystallized (or devitrified) plagioclase at the boundary. In another grain (Fig. 1b), plagioclase lamellae of one optical orientation occur in a matrix of finely recrystallized plagioclase of another optical orientation but whether the lamellae are twins or host can not be determined. Twins rarely are developed in crystals that show a well-developed mosaic structure throughout such as that shown in Fig. 1c.

Planar features (shock lamellae)—lamellar structures composed of material having short-range order (dense glass; Chao, 1967) (diaplectic plagioclase; Stöffler, 1967 and von Englehardt et al., 1968) that have originated by small shear displacements (Carter, 1968) during shock deformation—are rare in plagioclases of these breccias. These features are common in terrestrial impactites, and Short (1970) has suggested that differences in the composition of plagioclase in the terrestrial silicic rocks and that in these basic rocks could account for differences in their development. Planar arrays of spherical inclusions, such as shown in Fig. 1d are probably cavities formed during annealing of planar features (von Engelhardt and Bertsch, 1969), or they are healed microfractures, features common to terrestrially deformed quartz and plagioclase. In general, these arrays are not parallel to rational crystal planes, although in one crystal they are parallel to (010) and (0$\bar{2}$1). Drake et al. (1970) and Sclar (1971) described similar features as arrays of glass inclusions, formed by incipient melting due to shock deformation. This interpretation is based on a microprobe analysis (Drake et al., 1970), showing no appreciable compositional differences

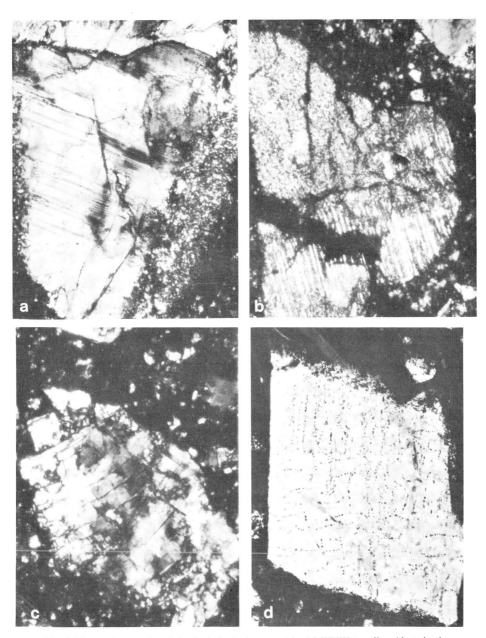

Fig. 1. Photomicrographs of shocked plagioclase crystals. (a) WNW-trending Ala-twins in plagioclase crystal in sample 14321,24. Strongly twinned areas are bounded by cracks. Shock deformation is most intense, causing mosaicism, at lower right grain boundary near finely recrystallized plagioclase coating. (b) Finely recrystallized plagioclase in sample 14305,2. NNW-trending features are twins or host lamellae. (c) Plagioclase crystal showing well-developed mosaic structure throughout in sample 14313,41. (d) Arrays of cavities decorating healed microfractures or planar features in specimen 14313,42. a–c width of field is approximately 0.4 mm. d width of field is approximately 0.24 mm.

between the host crystal and the inclusions. This result also might be expected
however, if the inclusions are cavities, certainly because their size (less than 2 μ) is
at the very limit of resolution of a microprobe.

With the exception of deformation twins, the fragment shown in Figs. 2a and 2b
shows all of the shock effects we have observed in plagioclase of Fra Mauro breccias.
This anorthositic fragment has an elliptical shape with a maximum dimension of
1 mm and contains a perfectly round hole near the center. Outlining the hole in an
asymmetric fashion are prismatic crystallites with interstitial glass; two trails of this
material extend from the hole toward the grain boundary (Fig. 2a). The intensity
of deformation diminishes away from the hole from intense mosaicism accompanied
by fine-grained recrystallization, mosaicism, faults and cracks (some partly healed)
sparse cracks, to no deformation at all at the two extremities. This enormous
change in intensity of deformation over a distance less than 0.5 mm clearly is due to
shock deformation accompanying impact of the projectile that formed the microcrater in this fragment. Shock pressures near the crater must have been of the order
of 300 kb (Chao *et al.*, 1970) to cause melting, dropping off to less than 10 kb at the
undeformed tips. The distance over which the shock intensity has decayed from
300 kb to <10 kb is about 0.5 mm and if the crater radius ($r = 0.05$ mm) is about
the size of the projectile, the shock stress level has attenuated approximately as
$1/r^2$. Rapid changes in intensity of shock damage in plagioclase were described by
von Engelhardt *et al.* (1970; 1971).

Much larger projectiles are required to produce the well-developed mosaic
texture throughout the fragment shown in Fig. 1c and the features shown in Figs.
2c and 2d. In the latter, a crystal twinned heavily according to albite and pericline
laws has been converted largely to maskelynite. An intermediate stage or possibly
devitrification (von Engelhardt *et al.*, 1972) is shown by the feathery intergrowth of
crystallites and glass in the extreme upper right near the grain boundary.

PYROXENES

Clinopyroxenes

Clinopyroxenes in the Fra Mauro breccias studied by us show a wide range of
structures apart from the extensively described domain structure (e.g. Ross *et al.*,
1970; Carter *et al.*, 1970; Boyd and Smith, 1971) which is not a deformation structure.
Twins according to the system {001}[100], resulting from high strain rate and low
temperature deformation in static tests (Raleigh and Talbot, 1967), are very common
and may occur as microtwins (Fig. 3a) or as somewhat broader twins. Twins also
occur frequently parallel to {100} but most of these are growth twins. In one crystal
(in 14162,41), however, undulatory extinction is caused by an abrupt increase in
number of {100} twins, indicating that the twins were induced mechanically. This
crystal also is unevenly fractured and contains mosaic structure (Carter *et al.*, 1968;
James, 1969), indicating a dynamic rather than static origin of the twins.

Planar features in clinopyroxenes are not nearly so abundant as the deformation
twins and they are much less common than in the pyroxenes in breccias from Apollo

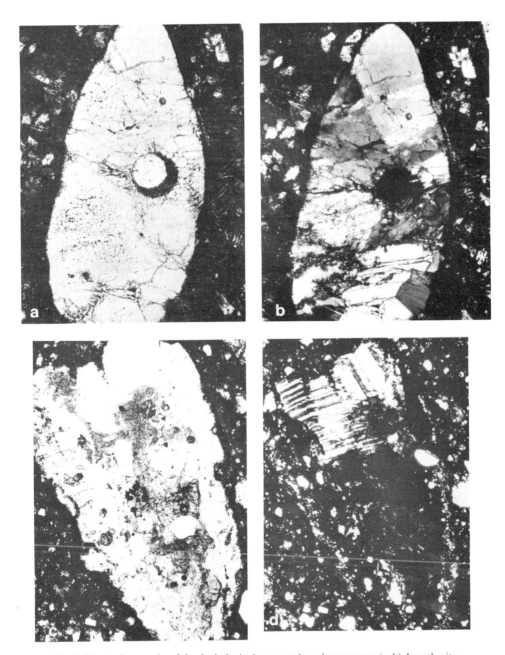

Fig. 2. Photomicrographs of shocked plagioclase crystals and aggregates. (a, b) Anorthosite particle in 14068,9. Except for cracks the two extremities of the particle are undeformed. Toward the center, near the microcrater, the intensity of shock damage increases from mosaicism, to mosaicism with recrystallization, to partial melt. (c, d) Plagioclase grain, partly transformed to maskelynite in sample 14313,42. At the upper right grain boundary the maskelynite has partly recrystallized. Widths of fields of view are approximately 0.64 mm.

Fig. 3. Photomicrographs of shocked clinopyroxene. (a) EW-trending microtwins parallel to (001) in clinopyroxene sample 14301,17. (b) ENE-trending planes of cavities probably decorating pre-existing (001) planar features in specimen 14305,15. (c) Recrystallization along zones parallel to annealed planar features in clinopyroxene specimen 14305,15. (d) Totally recrystallized, granular, polygonal clinopyroxene aggregate in sample 14301,83. The trails of inclusions are probably annealed fractures of a later shock event. According to Wilshire and Jackson (1972) this sample underwent at least two brecciation events. Widths of fields are approximately 0.5 mm.

11 and 12 sites. However, clinopyroxenes containing planar arrays of cavities, such as the crystal shown in Fig. 3b, are quite common. These arrays are most commonly parallel to {001}, but some are parallel to {100}, and we believe that they are annealed planar features.

Recrystallization has taken place along zones parallel to the planes of cavities in some of the crystals (Fig. 3c) and along zones of intense twinning in others. Some clinopyroxene crystals, such as that in Fig. 3d, have recrystallized entirely to a mosaic of small polygonal grains free of any strain (as deduced from optical studies). Still other crystals have recrystallized only at their borders and have preserved a small core of highly distorted host crystal near their center.

Orthopyroxene

Apart from fractures and faults, orthopyroxenes in these breccias, such as that in Fig. 4a, commonly show irregular undulatory extinction. In general, the undulatory extinction has been produced by translation gliding on {100} or on {hk0} in the [001] direction. In many of these grains, the intensity of the deformation ranges greatly and some grains showing undulatory extinction also show a mosaic structure and planes of inclusions parallel to {100}, which might be annealed shock lamellae. These associations and the irregular nature of the undulatory extinction lead us to believe that the deformation is of dynamic origin.

A puzzling feature of the orthopyroxene in the breccias studied is its failure to have inverted extensively to clinopyroxene. This inversion, which is nearly martensitic in nature, requires small shear displacements in the {100} plane parallel to [001] and commonly is observed in static experiments at high strain rates (Raleigh *et al.*, 1971). At a strain rate of 10^{-4}/sec., orthopyroxene inverts during deformation to clinopyroxene at temperatures below 1200°C; this temperature increases about 100°C for a ten-fold increase in strain rate. Thus, we would have expected a great deal of the orthopyroxene to have inverted to clinopyroxene (as is observed in chondritic meteorites) during shock deformation but we were unable to detect any clinopyroxene formed in this manner. Perhaps the inversion did take place and then the clinopyroxene re-inverted to orthopyroxene during high temperature annealing.

OLIVINE

Many olivine crystals in these breccias, in addition to containing fractures and faults, also show undulatory extinction and a few crystals contain well-defined kink bands (Fig. 4b). Most of the kink bands and zones of undulatory extinction are parallel to (001) and are consistent with slip on {110} and (100) in the [001] direction. These systems are those observed most commonly in olivines of chondritic meteorites and in static experiments at high strain rates and low temperatures (Raleigh, 1968; Carter and Avé Lallemant, 1970). Extrapolation of the static data to the higher strain rates of shock deformation indicates that slip parallel to [001] should take place to very high temperatures (1300°C to 1500°C). For terrestrial olivines the most important slip system is {0kl}[100], the high-temperature–low-strain rate system in static experiments (Raleigh, 1968; Carter and Avé Lallemant, 1970). A

Fig. 4. Photomicrographs of orthopyroxene and olivine. (a) NS-trending (100) exsolution lamellae in orthopyroxene in sample 14315,7. Kinks and zones of undulatory extinction are approximately parallel to (001). Width of field is approximately 0.64 mm. (b) EW-trending kinkbands subparallel to (001) in olivine in sample 14319,5. Alternating kink bands have approximately the same orientation. Width of field is approximately 1.0 mm. (c) ENE-trending shock lamellae parallel to {110} in olivine in sample 14321,24. Width of field is approximately 0.13 mm. (d) Recrystallized, granular, polygonal olivine aggregate in sample 14305,15. Relict near the center is strongly mosaicized. Width of field is approximately 0.42 mm.

few olivine crystals in the Fra Mauro breccias contain kink bands parallel to (100) and these may have originated by static plastic flow on {0kl}[100], but the evidence is not unequivocal.

Planar features (Fig. 4c) are observed rarely in olivine and the single set measured is parallel to {110}. Such features have not been observed in statically deformed olivine but have been observed parallel to (100), (010), (001), {130}, and {hkl} in experimentally shocked olivines (Müller and Hornemann, 1969; Sclar, 1969; James, 1969) and in chondritic meteorites (Müller and Hornemann, 1969; Sclar and Morzenti, 1971; Levi-Donati, 1971).

Olivines in these breccias commonly show the mosaic structure and these olivines commonly are partially recrystallized to mosaics of new strain-free crystals along their borders. Figure 4d shows an example of such crystals in which the highly distorted olivine grain is preserved near the center. Many of the crystals have recrystallized totally so that the host grain is no longer recognizable, as is also observed for the clinopyroxenes. The orientations of 21 recrystallized grains relative to the host crystal shown in Fig. 4d are given in Fig. 5. The orientations of the new grains are related strongly to that of the host, the various crystal axes being inclined, on the average, at 35° to respective crystal axes of the host. Similar orientations have

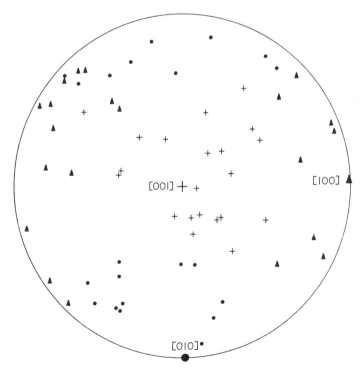

Fig. 5. Equal-area projection, showing the orientation of 21 recrystallized, granular, polygonal olivines with respect to the orientation of the central relict in the fragment from sample 14305,15. (Fig. 4d) (triangles are [100]; dots are [010]; crosses are [001]; the orientation of the relict grain is [100] EW, [010] NS, and [001] vertical).

been observed for intragranular recrystallization of experimentally deformed olivine (Avé Lallemant and Carter, 1970), quartz (Hobbs, 1968) and for a variety of other materials.

DISCUSSION AND CONCLUSIONS

Microstructures in silicates, similar to those induced by shock deformation at the Apollo 11 and 12 sites, also have been observed in the Fra Mauro breccias but, in general, they appear to have been modified by high temperature annealing. Notable among the annealed features are healed microfractures and dense planar arrays of cavities probably marking pre-existing planar features. Inasmuch as planar features are composed of glass less dense than the host crystal, conversion of the glass back to crystalline form during annealing would leave arrays of cavities in their planes, as observed. At higher temperatures, recrystallization has taken place along zones parallel to the planes of inclusions. Similar planes of inclusions out-lining pre-existing planar features have been observed for shocked quartz in the Sudbury (French, 1968), Vredefort (Carter, 1965, 1968) and Ries (von Engelhardt and Bertsch, 1969) structures. Recrystallization along zones parallel to these features also has been observed for quartz in the Vredefort structure. Temperatures required for annealing of planar features and other structures in terrestrial impactites are, however, probably appreciably lower than for annealing in the lunar materials because of the presence of H_2O that promotes recovery and recrystallization (Hobbs, 1968; Green et al., 1970; Carter, 1971).

Extensive recrystallization also has taken place along the borders of heavily shocked crystals of olivine and clinopyroxene. In many instances, entire crystals have been recrystallized so that the host crystal is no longer recognizable. Annealing recrystallization of olivine due to high residual temperatures following experimental shock deformation to stresses in excess of 1 mb has been observed (Carter et al., 1968), but the recrystallized grains are much smaller in size and uniform throughout. Shocked pyroxenes are much more resistant to recrystallization than olivine (Carter et al., 1968), as are statically deformed pyroxenes (Avé Lallemant and Carter, 1970; Carter, 1971; Carter et al., 1972). In the Fra Mauro breccias, however, clino-pyroxenes commonly have recrystallized totally to coarse-grained mosaics of polygonal, strain-free grains.

All of the features discussed above, we believe, have originated by prolonged high temperature annealing of silicates following intense shock deformation. This is in agreement with the classification of Jackson and Wilshire (1972) and Wilshire and Jackson (1972); all samples are of their F_2 and F_4 types. In addition, the lack of evidence for the orthopyroxene-clinopyroxene inversion also may indicate very high temperature annealing allowing reinversion. The annealing may have taken place at depth by burial under a hot ejecta blanket thrown out during the excavation of Mare Imbrium. The high strain energy in the crystals could have been imparted during the Imbrian event or during impact with pre-Imbrian crust. These con-clusions are in accord with those of several other authors, based generally on different types of evidence, in this proceedings.

We conclude, as before for Mare Tranquillitatis (Carter *et al.*, 1970) and Oceanus Procellarum (Carter *et al.*, 1971), that there is no unequivocal evidence for appreciable static deformation (other than annealing) in the Fra Mauro formation. These rocks probably have been derived from depths to *ca.* 10 km and our results suggest, therefore, that there has been no substantial tectonic activity from the surface to 10 km since these rocks and those of the other Apollo sites crystallized some 3 to 4 billion years ago (e.g. Wasserburg *et al.*, 1972). This negative conclusion implies that at least the outer layer of the moon has been static for this length of time.

Acknowledgments—This research was supported by NASA Grant NGR 07-004-135.

References

Avé Lallemant H. G. and Carter N. L. (1970) Syntectonic recrystallization of olivine and modes of flow in the upper mantle. *Bull. Geol. Soc. Amer.* **81**, 2203–2220.

Boyd F. R. and Smith D. (1971) Compositional zoning in pyroxenes from lunar rock 12021, Oceanus Procellarum. *J. Petrol.* **12**, 439–464.

Carter N. L. (1965) Basal quartz deformation lamellae, a criterion for recognition of impactites. *Amer. J. Sci.* **263**, 786–806.

Carter N. L. (1968) Dynamic deformation of quartz. In *Shock Metamorphism of Natural Materials* (editors B. M. French and N. M. Short), pp. 453–474. Mono Book Corp., Baltimore, Md.

Carter N. L. (1971) Static deformation of silica and silicates. *J. Geophys. Res.* **76**, 5514–5540.

Carter N. L. Raleigh C. B. and De Carli P. S. (1969) Deformation of olivine in stony meteorites. *J. Geophys. Res.* **73**, 5439–5461.

Carter N. L. and Avé Lallemant H. G. (1970) High temperature flow of dunite and peridotite. *Bull. Geol. Soc. Amer.* **81**, 2181–2202.

Carter N. L. Leung I. S. Avé Lallemant H. G. and Fernandez L. A. (1970) Growth and deformational structures in silicates from Mare Tranquillitatis. *Proc. Apollo 11 Lunar Sci. Conf.*, Geochim. Cosmochim. Acta Suppl. 1, Vol. 1, pp. 267–285. Pergamon.

Carter N. L. Fernandez L. A. Avé Lallemant H. G. and Leung I. S. (1971) Pyroxenes and olivines in crystalline rocks from the Ocean of Storms. *Proc. Second Lunar Sci. Conf.*, Geochim. Cosmochim. Acta Suppl. 2, Vol. 1, pp. 775–795. MIT Press.

Carter N. L. Baker D. W. and George R. P. Jr. (1972) Seismic anisotropy, flow and constitution of the upper mantle. Griggs' volume, A.G.U. Monograph, in press.

Chao E. C. T. (1967) Shock effects in certain rock-forming minerals. *Science* **156**, 192–202.

Chao E. C. T. James O. B. Minkin J. A. Boreman J. A. Jackson E. D. and Raleigh C. B. (1970) Petrology of unshocked crystalline rocks and shock effects in lunar rocks and minerals. *Science* **167**, 644–647.

Drake M. J. McCallum I. S. McKay G. A. and Weill D. F. (1970) Mineralogy and petrology of Apollo 12 sample no. 12013: A progress report. *Earth Planet. Sci. Lett.* **9**, 103–123.

Engelhardt W. von Hörz W. V. F. Stöffler D. and Bertsch W. (1968) Observations on quartz deformation in the breccias of West Clearwater Lake, Canada and the Ries Basin, Germany. In *Shock Metamorphism of Natural Materials* (editors B. M. French and N. M. Short), pp. 475–482. Mono Book Corp., Baltimore, Md.

Engelhardt W. von and Bertsch W. (1969) Shock induced planar deformation structures in quartz from the Ries Crater, Germany. *Contrib. Mineral. Petrol.* **20**, 203–234.

Engelhardt W. von Arndt J. Müller W. F. and Stöffler D. (1970) Shock metamorphism of lunar rocks and origin of the regolith at the Apollo 11 landing site. *Proc. Apollo 11 Lunar Sci. Conf.*, Geochim. Cosmochim. Acta Suppl. 1, Vol. 1, pp. 363–384. Pergamon.

Engelhardt W. von Ardnt J. Müller W. F. and Stöffler D. (1971) Shock metamorphism and origin of regolith and breccias at the Apollo 11 and 12 landing sites. *Proc. Second Lunar Sci. Conf.*, Geochim. Cosmochim. Acta Suppl. 2, Vol. 1, pp. 833–854. MIT Press.

Engelhardt W. von Arndt J. and Stöffler D. (1972) Apollo 14 soils and breccias, their composition and origin by impacts (abstract). In *Lunar Science—III* (editor C. Watkins), pp. 233–235, Lunar Science Institute Contr. No. 88.

French B. M. (1968) Sudbury structure, Ontario: Some petrographic evidence for an origin by meteorite impact. In *Shock Metamorphism of Natural Materials* (editors B. M. French and N. M. Short), pp. 373–412. Mono Book Corp., Baltimore, Md.

Green H. W. Griggs D. T. and Christie J. M. (1970) Syntectonic and annealing recrystallization of fine-grained quartz aggregates. In *Experimental and Natural Rock Deformation* (editor P. Paulitsch) pp. 272–335. Springer, New York.

Hobbs B. E. (1968) Recrystallization of single crystals of quartz. *Tectonophysics* **6**, 353–401.

Jackson E. D. and Wilshire H. G. (1972) Classification of the samples returned from the Apollo 14 landing site (abstract). In *Lunar Science—III* (editor C. Watkins), pp. 418–420, Lunar Science Institute Contr. No. 88.

James O. B. (1969) Shock and thermal metamorphism of basalt by nuclear explosion, Nevada test site. *Science* **166**, 1615–1620.

Levi-Donati G. R. (1971) Petrological features of shock metamorphism in chondrites: Alfianello *Meteoritics* **6**, 225–235.

Müller W. F. and Hornemann U. (1969) Shock-induced planar deformation structures in experimentally shock-loaded olivines and in olivines from chondritic meteorites. *Earth Planet. Sci. Lett.* **7**, 251–264.

Raleigh C. B. (1968) Mechanisms of plastic deformation of olivine. *J. Geophys. Res.* **73**, 5391–5406.

Raleigh C. B. and Talbot J. L. (1967) Mechanical twinning in naturally and experimentally deformed diopside. *Amer. J. Sci.* **265**, 151–165.

Raleigh C. B. Kirby S. H. Carter N. L. and Avé Lallemant H. G. (1971) Slip and the clinoenstatite transformation as competing rate processes in enstatite. *J. Geophys. Res.* **76**, 4011–4022.

Ross M. Bence A. E. Dwornik E. J. Clark J. R. and Papike J. J. (1970) Lunar clinopyroxenes: Chemical composition, structural state, and texture. *Science* **167**, 628–630.

Sclar C. B. (1969) Shock-wave damage in olivine (abstract). *Trans. Amer. Geophys. Union* **50**, 219.

Sclar C. B. (1971) Shock-induced features of Apollo 12 microbreccias. *Proc. Second Lunar Sci. Conf.*, *Geochim. Cosmochim. Acta* Suppl. 2, Vol. 1, pp. 817–832. MIT Press.

Sclar C. B. and Morzenti S. P. (1971) Shock-induced planar deformation structures in olivine from the Chassigny meteorite. *Meteoritics* **6**, 310–311.

Short N. M. (1970) Evidence and implications of shock metamorphism in lunar samples. *Science* **167**, 673–675.

Stöffler D. (1967) Deformation und Umwandlung von Plagioklas durch Stosswellen in den Gesteinen des Nördlinger Ries. *Contrib. Mineral. Petrol.* **16**, 51–83.

Wasserburg G. J. Turner G. Tera F. Podosek F. A. Papanastassiou D. A. and Huneke J. C. (1972) Comparison of Rb–Sr, K–Ar and U–Th–Pb ages; lunar chronology and evolution. In *Lunar Science—III* (editor C. Watkins), pp. 788–790, Lunar Science Institute Contr. No. 88.

Wilshire H. G. and Jackson E. D. (1972) Petrology of the Fra Mauro formation at the Apollo 14 landing site. In *Lunar Science—III* (editor C. Watkins), pp. 803–805, Lunar Science Institute Contr. No. 88.

Proceedings of the Third Lunar Science Conference
(Supplement 3, *Geochimica et Cosmochimica Acta*)
Vol. 1, pp. 907–925
The M.I.T. Press, 1972

Apollo 14 glasses of impact origin and their parent rock types*

E. C. T. Chao, Judith B. Best, and Jean A. Minkin

U.S. Geological Survey
Washington, D.C. 20242

Abstract—The color, refractive indices (R.I.), and chemical composition have been determined for more than 200 Apollo 14 glass particles of impact origin from the fines and breccias. As was found for Apollo 11 and 12, the Apollo 14 glasses fall into distinct chemical groups with characteristic colors and R.I. ranges. We interpret these chemical groups as representing specific parent rock types. They may be fused fines or breccias, fused uncontaminated or contaminated igneous rocks, or fused metamorphic rocks.

From the examination of plots of refractive indices, chemical variations, and CIPW norms, we find that there are at least eight distinct chemical groups of Apollo 14 rock glasses of impact origin. (Thetomorphic plagioclase glass produced by shock is present in Apollo 14 samples but is not considered in this paper.) The most abundant group, with a compositional range overlapping that of the Apollo 14 fines and regolith microbreccias, is similar to hornfelsed noritic microbreccias or micronorite hornfels in composition. Only one distinct group, with an anorthositic gabbro composition, also is present among Apollo 11 and 12 fines. The other parent rock types indicated by our study include rocks of anorthositic, feldspathic peridotite, troctolitic, mare basalt, and peridotite compositions. Glasses of salic composition also are present but are very rare.

A major conclusion from this study is that the dominant parent rock type of the Apollo 14 highland site is similar in composition to annealed noritic rocks rather than anorthosite.

Introduction

Our principal objective in the detailed studies of the compositions of lunar glasses of impact origin is to identify the major lunar rock types they represent. This in turn allows us to compare the assemblages of rock types or suites at the different Apollo landing sites. That this is a reasonably attainable objective has already been demonstrated in studies of impactite glasses from Apollo 11 and 12 fines and regolith microbreccias (Chao *et al.*, 1970a).

Evidence for the impact origin of both homogeneous and heterogeneous Apollo 14 glasses is identical to that presented for the Apollo 11 and 12 glasses (Chao *et al.*, 1970a). The probability that the bulk compositions of the glasses represent approximately the bulk compositions of their parent rock types also was summarized in the previous work. We therefore interpret the chemical groups as representing specific rock types that may be fused breccias, fines, uncontaminated or contaminated igneous rocks, or metamorphic rocks.

The Apollo 14 glasses of impact origin that we have studied are from the following samples: bulk fines 14163,36; comprehensive fines 14259,42 and 14260,1; trench samples 14148,18, 14149,35, and 14156,19; coarse fines 14161,3; and breccias 14301,62, 14063,46, and 14321,241. The refractive indices (R.I.) and chemical compositions of more than 200 glass particles have been determined by interference microscopy and by electron microprobe, respectively. We distinguish eight chemical

---

* Publication authorized by the Director, U.S. Geological Survey.

groups among the Apollo 14 glasses on the basis of plots of R.I., CIPW normative diagrams, and major element variation diagrams. We compare these chemical groups with analyzed Apollo 14 basalts, breccias and fines, and the results provide a survey of the various rock types and suites that can be considered representative of the Fra Mauro highland site.

Occurrences of Apollo 14 Glasses

The Apollo 14 glass particles are similar to the Apollo 11 and 12 glasses of impact origin (Chao *et al.*, 1970a). They include both homogeneous and heterogeneous glasses, and they have a wide range of color, transparency, and R.I. They are an important constituent of fines samples (Fig. 1) and porous regolith microbreccias (Fig. 2). Devitrified glasses also are found in both fines and regolith microbreccias, but they are most abundant in the thermally metamorphosed Apollo 14 breccias of the Fra Mauro formation. In fact the only glass particles these latter breccias contain are partly or completely devitrified and are of two compositional types, one with a high content of silica minerals and potassic feldspar (Fig. 3) and the other with a high plagioclase content (Fig. 4). (For the nomenclature and description of lithic fragment types in Fra Mauro breccias see Chao *et al.*, 1972, this volume.)

The most common types of glasses, which are found in both the fines and the porous regolith microbreccias, range in color from colorless to pale green and greenish-yellow to gold. More deeply colored green homogeneous glass particles are common in the fines, but they are rare in the porous regolith microbreccias. Reddish-brown and brownish-red to red, homogeneous glass particles are of about equal occurrence in the fines and in regolith microbreccias. Neither the green nor the reddish colored glasses have been found in the annealed Apollo 14 breccias of the Fra Mauro formation (Chao *et al.*, 1972, this volume).

Indices of Refraction

The refractive index of each glass particle selected for study was measured with an interference microscope (Chao *et al.*, 1970a), except for those particles that were devitrified, opaque, or highly heterogeneous or embedded in polished thin sections of breccias. The precision is ± 0.0002 in the range 1.460–1.700 and ± 0.0005 in the range greater than 1.700.

The frequency distribution of the refractive index of 231 glass particles picked at random from Apollo 14 fines is shown in Fig. 5. Similar plots for Apollo 11 and 12 glasses are included for comparison. The range of R.I. for the Apollo 14 glasses is 1.4961 to 1.741. The R.I. of the major population of Apollo 14 glasses lie to the low index side of those of Apollo 12, and those of Apollo 11 lie to the high index side of Apollo 12 glasses. As with the Apollo 11 and 12 glasses, the variation of R.I. of Apollo 14 glasses is discontinuous and has several distinct modes. One prominent mode that is common to all three Apollo site glasses lies at 1.5950, representing a group of colorless to pale green glasses of anorthositic gabbro composition (Chao *et al.*, 1970a). Two other prominent modes for Apollo 14 lie at 1.6350 and 1.6590. They are represented by yellowish-green to green glasses.

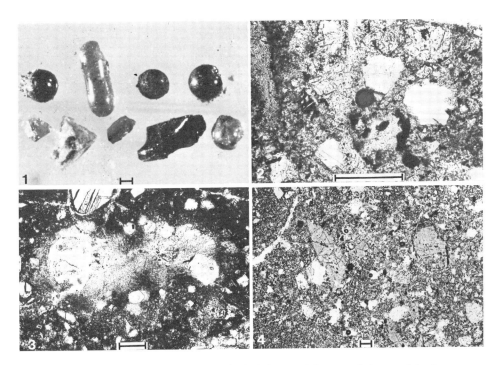

Fig. 1 (upper left). Photomicrograph of splash-form and fragmental glass particles from Apollo 14 fines 14259,42. Bar scale 100 μ. Fig. 2 (upper right). Photomicrograph of glass particles in a thin section of regolith microbreccia 14301,62. The glass sphere in the center of the photomicrograph contains 15.9 weight percent TiO_2 (sample 14301,62–27, Table 2). Plane polarized light. Bar scale 100 μ. Fig. 3 (lower left). Photomicrograph of a devitrified glass fragment with a high silica and potash content. It is enclosed in a dark microbreccia fragment bonded by devitrified glass in a thin section of breccia 14321,241. Plane polarized light. Bar scale 100 μ. Fig. 4 (lower right). Photomicrograph showing fragments of plagioclase-rich devitrified glass (mottled gray patches) in a micronorite hornfels fragment in a thin section of porous regolith microbreccia 14301,84. Plane polarized light. Bar scale 100 μ.

The R.I. of the broadest major glass group lie between 1.5960 and 1.6340. This group may include several subgroups, among which are the glasses referred to as basaltic and KREEP glasses by other investigators (Reid *et al.*, 1972). Most heterogeneous glasses fall within this group.

Two fragments of colorless glass have very low R.I., 1.4961 and 1.4969, respectively. Glasses with R.I. in excess of 1.680 are brownish-red to red in color.

Thetomorphic anorthite glass has R.I. near 1.580. It is rare and represents the only mineral glass we have found. It is not included in this study.

In general, R.I. reflects the major chemical characteristics of the glasses as previously pointed out (Chao *et al.*, 1970a). Clustering is evident when R.I. is plotted against TiO_2 + FeO and against Al_2O_3 (Fig. 6). The refractive index varies directly with TiO_2 + FeO and inversely with Al_2O_3. The clustering at R.I. 1.5950, 1.6350, and 1.6590 is identical to that shown in the histogram of Fig. 5.

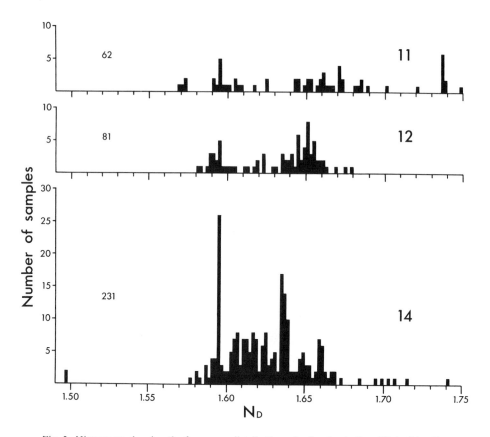

Fig. 5. Histograms showing the frequency distribution of refractive indices (N_D) of Apollo 11, 12, and 14 glasses. Note the three distinct modes for the Apollo 14 glasses at N_D 1.5950, 1.6350, and 1.6590. Total number of samples measured is shown at left of each histogram.

Chemical Groups and Possible Parent Rock Types

In order to distinguish compositional groups among glasses, we have plotted three types of ternary diagrams that use CIPW norm proportions of the five most common lunar minerals. Feldspar and pyroxene are two of the components in all three plots; the third component is, in the different cases, olivine, quartz, or ilmenite (plus chromite). Figures 7a and 7b show the plots of CIPW norms for the Apollo 14 glasses. For comparison, the CIPW norms of analyzed Apollo 14 basalts, breccias, and fines (see Table 1 for selected analyses) are plotted on the same diagrams. Intragroup variation, however, is not always clearly evident nor truly represented on normalized plots. To complement the norm plots, we have also examined major element or major element ratio variation diagrams to check the consistency of grouping.

Eight compositional groups of Apollo 14 glasses are apparent in the plots. The physical and compositional characteristics of each group are summarized in Tables 2 and 3 and are discussed briefly below, in the order of increasing average R.I.

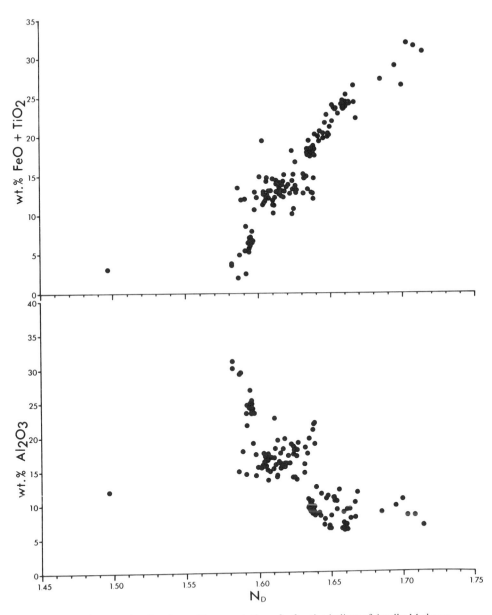

Fig. 6. Diagrams showing the positive correlation of refractive indices of Apollo 14 glasses with FeO + TiO$_2$ and the negative correlation of refractive indices with Al$_2$O$_3$. The clusters of points correspond to the principal Apollo 14 refractive index modes in Fig. 5, and to the chemical groups summarized in Table 2.

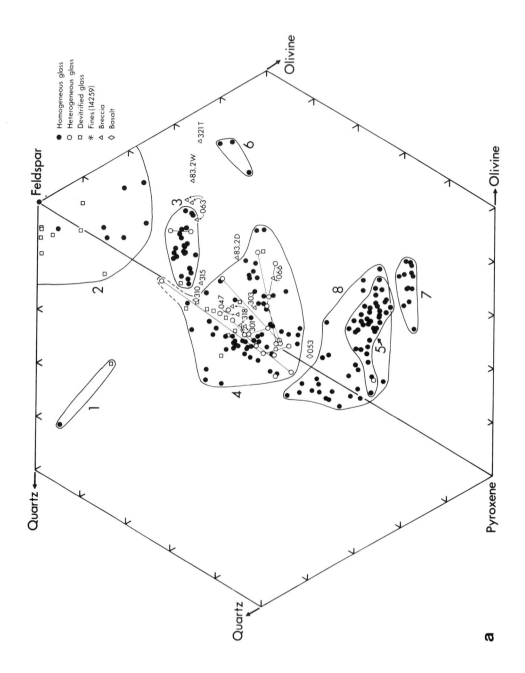

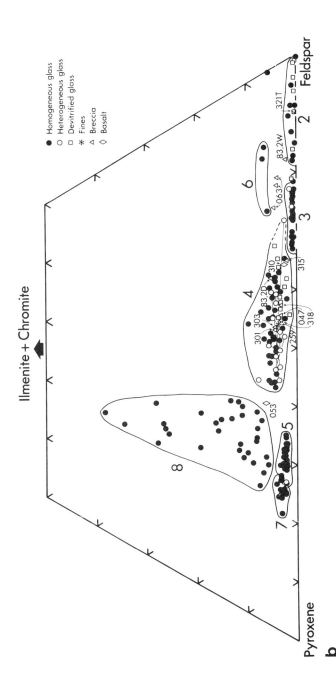

Fig. 7. (a) Feldspar-olivine-pyroxene-quartz normative diagram, and (b) feldspar-pyroxene-ilmenite + chromite normative diagram of Apollo 14 glasses. Analyses of various parts of the same particle of heterogeneous glass are joined by straight lines to indicate extent of composition range. Groups are outlined by heavy solid lines (broken where boundary is uncertain). (Many points in group 4 were omitted for clarity.) Group 1 is not distinct in 7b (it plots within group 2). The feldspar-pyroxene base line of the quadrilateral in 7b coincides with the principal diagonal of the hexagon in 7a and thus the two figures may be viewed simultaneously. Together they illustrate the bulk chemistry of the Apollo 14 glasses. Analyzed fines, breccias, and basalts are identified by three digits (e.g. 14053 = 053). Analyses of 14053 and 14321 (troccolite clast) are taken from Compston et al. (1972), and are given in Table 1. The remaining analyses are by H. J. Rose, Jr. et al., U.S. Geological Survey, and are given in either Table 1 (this paper) or in Table 5 (Chao et al., 1972, this volume).

Table 1. Chemical composition (in weight percent) of Apollo 14 basalts, breccias, and fines.

	Basalts		Troctolite	Breccias					Fines
	14310,114	14053*	14321*	14063,37A (white clasts)	14083,2W (dark clasts)	14083,2D	14303,34	14066,21	14163,54
SiO$_2$	47.81	46.18	43.50	45.02	44.20	46.19	47.49	47.59	47.97
TiO$_2$	1.11	2.94	0.19	1.58	0.80	2.21	1.98	1.68	1.77
Al$_2$O$_3$	21.54	12.84	23.29	21.53	22.06	17.16	16.05	14.61	17.57
FeO	7.62	17.09	4.56	7.00	7.16	9.66	10.96	10.82	10.41
MgO	7.48	8.59	15.82	10.79	11.48	11.55	10.99	13.67	9.18
CaO	12.92	11.18	12.27	12.40	12.66	10.76	10.03	9.07	11.15
Na$_2$O	0.68	0.44	0.28	0.93	0.65	1.01	0.87	1.01	0.68
K$_2$O	0.48	0.11	0.06	0.20	0.23	0.40	0.46	0.45	0.58
P$_2$O$_5$	0.43	0.13	0.03	0.29	0.45	0.63	0.56	0.55	0.52
MnO	0.10	0.26	0.06	0.09	0.06	0.11	0.15	0.13	0.14
Cr$_2$O$_3$	0.25	0.37	—	0.19	0.14	0.21	0.21	0.26	0.26
TOTAL	100.42	100.13	100.06	100.02	99.89	99.89	99.75	99.84	100.23

Analysts: H. J. Rose, Jr., et al., U.S. Geological Survey
* Taken from W. Compston et al. (1972)

Table 2. The chemical groups of Apollo 14 glasses.

Group	1		2			3	4			
Color*	colorless		colorless			colorless, Pgr	Y, Go, Gy, B, Gr	Y, Go-B, Y-Gr	Y-B, Go	Gy, Gr-Y, Y, Go
Range of refractive index	1.4961–1.4969		1.5817–1.5935			1.5916–1.6211	1.5860–1.6235	1.6028–1.6182	1.6110–1.6336	1.6124–1.6313
Number analyzed	2		16			29	100			
Representative sample	I3	14321-7-2	I4	14301-2-1	14063-5	O3	H6	N10	F15	H4
					Chemical composition (in weight percent)†					
SiO_2	77.8	66.5	45.6	49.6	44.9	45.6	48.6	46.9	45.7	46.4
TiO_2	0.09	0.39	0.13	0.19	0.04	0.28	3.04	2.04	1.76	2.25
Al_2O_3	11.7	14.5	29.8	32.7	36.5	24.2	15.1	15.6	18.6	13.4
FeO	2.7	4.1	3.2	0.29	1.69	6.2	10.9	10.0	10.8	14.3
MgO	—	3.4	3.6	0.18	2.4	9.1	8.3	12.0	10.8	10.4
CaO	1.30	3.0	16.7	16.3	15.9	14.7	10.8	10.4	12.9	10.3
Na_2O	2.4	0.50	0.50	1.13	0.52	0.10	0.81	0.83	0.03	0.44
K_2O	5.1	7.1	0.12	0.90	0.66	—	0.66	0.26	0.16	0.37
P_2O_5	—	0.15	0.08	0.07	0.03	0.03	0.18	0.14	—	—
MnO	—	0.05	0.06	—	0.03	0.19	0.08	0.20	0.13	—
Cr_2O_3	—	—	0.05	—	0.16	0.10	0.26	0.23	0.20	0.11
NiO	—	—	—	—	0.02	—	—	—	—	0.32
TOTAL	101	100	100	101	103	100	99	99	101	98

Table 2 (continued)

Group	5	6	7	7a	8		
Color*	Gr, Y-Gr	Go, Pgr-Go	Gr, Y, Gr	Y-Gr, Gr	Go-B, R-B, R-O		
Range of refractive index	1.6339–1.6453	1.6305–1.6404	1.6579–1.6610	1.6451–1.6493	1.6394–1.714		
Number analyzed	34	13	9	4	32		
	Chemical composition (in weight percent)†						
Representative sample	F2	I9	H8	K11	C2	A4	14301, 62-27
SiO_2	48.0	40.0	45.8	46.3	45.8	42.6	34.7
TiO_2	0.67	2.31	0.96	0.62	2.12	7.7	15.9
Al_2O_3	9.0	19.4	6.2	6.2	7.6	9.4	4.6
FeO	17.1	10.8	22.5	19.3	22.0	21.2	23.1
MgO	14.4	12.5	14.9	18.8	13.4	5.4	12.7
CaO	9.8	13.7	8.3	8.4	9.2	11.3	7.1
Na_2O	0.27	—	0.17	0.12	0.31	0.28	0.87
K_2O	0.28	0.03	0.28	0.10	0.14	0.07	0.31
P_2O_5	0.11	0.03	—	—	0.08	0.15	n.d.
MnO	0.30	0.15	0.14	0.12	0.22	0.22	n.d.
Cr_2O_3	0.58	0.12	0.54	0.51	0.48	0.20	0.94
NiO	—	—	—	—	0.04	0.04	n.d.
TOTAL	100	99	100	100	101	99	100

* Color code: P = pale, Y = yellow, B = brown, O = orange, R = red, Gr = green, Gy = gray, Go = gold
† Electron microprobe analyses by Jean A. Minkin and E. C. T. Chao
— Below limit of detection
n.d. Not determined

Table 3. CIPW norms of representative samples of each of the chemical groups of Apollo 14 glasses

Group	1	1	1	2	2	3	4	4	4	4	5	6	7	7a	8	8	8
Sample no.	13	14321 -7-2	14	14301 -2-1	14063-5	O3	H6	N10	F15	H4	F2	I9	H8	K11	C2	A4	14301, 62-27
Q	38.78	22.24	0	4.42	0.43	0	3.23	0	0	0	0	0	0	0	0	3.54	0
C	0	0.90	0	0.39	5.93	0	0	0	0	0	0	0	0	0	0	0	0
Or	29.80	42.09	0.71	5.25	3.79	0	3.95	1.56	0.94	2.22	1.65	0	1.66	0.59	0.82	0.42	1.83
Ab	20.41	4.24	4.24	9.43	4.28	0.84	6.94	7.12	0.25	3.79	2.27	0	1.44	1.01	2.59	2.40	7.35
An	5.84	13.95	78.84	79.33	76.52	65.26	36.08	38.61	49.61	34.08	22.40	53.36	15.36	16.01	18.68	24.55	7.71
Lc	0	0	0	0	0	0	0	0	0	0	0	0.14	0	0	0	0	0
Wo	0.22	0	1.51	0	0	2.97	7.10	5.34	5.72	7.48	10.54	4.85	10.82	10.64	10.79	13.09	11.45
En	0	8.49	7.79	0.44	5.81	14.06	20.94	23.26	18.63	23.12	25.30	3.00	21.15	23.41	20.13	13.65	15.14
Fs	4.76	7.00	5.04	0.22	2.87	6.94	15.11	11.80	11.77	20.05	21.41	1.56	22.52	17.10	22.25	26.85	7.35
Fo	0	0	0.80	0	0	5.95	0	4.94	5.59	2.27	7.28	19.92	11.24	16.26	8.97	0	11.51
Fa	0	0	0.57	0	0	3.24	0	2.76	3.90	2.17	6.79	11.42	13.19	13.08	10.92	0	6.15
Cs	0	0	0	0	0	0	0	0	0	0	0	1.07	0	0	0	0	0
Cm	0.02	0	0.07	0	0.23	0.15	0.39	0.34	0.29	0.48	0.85	0.18	0.80	0.75	0.70	0.30	1.38
Il	0.17	0.74	0.25	0.36	0.07	0.53	5.85	3.93	3.31	4.35	1.27	4.43	1.83	1.17	3.97	14.84	30.13
Ap	0	0.36	0.19	0.16	0.07	0.07	0.43	0.34	0	0	0.26	0.07	0	0	0.19	0.36	—
Feldspar Or	52.0	69.5	0.8	5.6	4.5	0	8.3	3.3	1.8	5.5	6.2	0	8.9	3.3	3.7	1.5	10.5
Feldspar Ab	37.8	7.4	5.4	10.6	5.4	1.4	15.5	15.8	0.5	10.0	9.1	0	8.2	6.1	12.4	9.3	45.0
Feldspar An	10.2	23.1	93.8	83.9	90.2	98.6	76.1	80.9	97.6	84.5	84.7	100.0	82.8	90.6	84.0	89.2	44.5
Pyroxene Ca	5.0	0	10.1	0	0	11.7	15.9	12.5	.15.2	14.4	18.0	50.0	19.6	20.2	20.1	24.9	32.3
Pyroxene Mg	0	61.5	60.2	72.9	72.7	64.2	54.3	63.1	57.3	51.6	49.9	35.8	44.4	51.3	43.4	30.1	49.4
Pyroxene Fe	95.0	38.5	29.6	27.1	27.3	24.1	29.8	24.4	27.5	34.0	32.1	14.2	36.0	28.5	36.5	45.0	18.2
Olivine Mg	0	0	67.0	0	0	72.7	0	72.1	67.5	60.2	60.8	71.6	55.2	64.3	54.3	0	73.0
Olivine Fe	0	0	33.0	0	0	27.3	0	27.9	32.5	39.8	39.2	28.4	44.8	35.7	45.7	0	27.0

Group 1. Glasses of granitic or salic composition. Glasses in this group contain the highest normative quartz of any group, and their feldspars contain the highest normative orthoclase (52–69.5%). These glasses occur as loose particles in the fines and the chemical analysis of one of these (I3) is given in Table 2. In addition, several yellowish-brown devitrified glass fragments of similar composition were found in black, extremely fine-grained annealed fragment-laden glass (Fig. 3); a representative analysis (14321-7-2) is presented in Table 2 (Grieve *et al.*, 1972, have analyzed similar glass particles in 14321). Other investigators who have reported high silica and potassic glasses are the Apollo Soil Survey Team (1971), Reid *et al.* (1972), Glass (1972), Fredriksson *et al.* (1972), and Engelhardt *et al.* (1972).

The composition of these glasses is analogous to the average of 33 high-silica glasses from the mesostasis of Apollo 11 basalts (Roedder and Weiblen, 1970) and to the salic component in 12013 (Drake *et al.*, 1970). Hence, this group represents a fused salic parent rock type, possibly crystallized from a residual igneous melt. This would be a highly differentiated lunar rock, characterized by a very high K_2O/Na_2O ratio.

Group 2. Glasses of anorthositic composition. These glasses plot near the feldspar corner in the CIPW normative diagram (Figs. 7a and 7b), with some containing normative olivine and others normative quartz (Table 3 and Fig. 7a). They are found as colorless undevitrified fragments in the fines (I4, Table 2) and as devitrified fragments from two sources: yellowish-brown angular xenolith fragments in micronorite hornfels (14301-2-1, Table 2, Fig. 4) and dark brownish-gray fragment in friable white feldspathic microbreccia 14063 (14063-5, Table 2, Fig. 3).

Because of the very high normative anorthite content, the parent rock type fused to produce these glasses could be a crystalline igneous or metamorphic anorthositic rock or a plagioclase-rich breccia. The glasses in this group that are high in normative olivine appear to indicate sources of troctolite composition and those high in normative quartz and K-feldspars may be related to sources of salic composition, like those of group 1.

Group 3. Glasses of anorthositic gabbro composition. These glasses are characterized by high alumina and lime, moderately low iron and magnesia, and low titania and alkalis (O3, Table 2). Compositions form a tightly clustered group on the CIPW plots (Figs. 7a and 7b) characterized by a nearly constant normative content of about 70% feldspar (almost entirely calcic plagioclase). This narrow range of variation of normative feldspar may indicate a plagioclase-rich crystalline rock of cumulative origin with a range of proportions of olivine and pyroxene or an olivine- and pyroxene-bearing feldspathic breccia.

Glasses identical in composition to these also are found at the Apollo 11 and 12 sites (verified also by Glass, 1972; Reid *et al.*, 1972), and this wide distribution points to a single parent rock type. Chemically it has an anorthositic gabbro composition (Chao *et al.*, 1970b; Wood *et al.*, 1970). It may be a member of a suite of rocks (possibly from the highlands) that ranges in composition from basaltic (including micronorites), to troctolitic, to anorthositic. Thus, the occurrence of such glasses in the Apollo 11 and 12 fines must be attributed to an exotic or deep-seated source or sources at the mare site.

Group 4. Glasses of basaltic composition. These samples form the largest and most abundant group analyzed, and particles show greater variability between samples than in most other groups. This group may include several subgroups. Four representative analyses are shown in Table 2 (H6, N10, F15, and H4), and many more covering the range of R.I. and chemical composition are shown in Table 4. CIPW norm calculations show that both quartz normative and olivine normative samples are represented (Table 3 and Fig. 7a). Most samples of heterogeneous glasses fall within this group, and some whole group compositional ranges are repeated by the ranges found within a single heterogeneous sample (Figs. 7a and 7b). Many of the group 4 glasses contain appreciable amounts of alkalis (*e.g.* H6, Table 2) and could be considered KREEP glasses (Apollo Soil Survey, 1971; and Reid *et al.*, 1972).

As shown by example N10, in which magnesia is greater than iron, F15, which has high alumina and lime, and H4 in which iron is much greater than magnesia (Table 2), there are glasses in group 4 that differ greatly from those represented by H6. Dividing the group into subgroups is difficult, however, because of the scatter on the normative pyroxene plot. It is conceivable that glasses of these different compositions may be derived from ray materials criss-crossing the Fra Mauro site at various times throughout the build-up of the regolith.

We interpret the majority of group 4 glasses, however, as fused products of regolith breccias and fines, because (1) these glasses are so abundant in the fines and in porous regolith microbreccias, (2) most of the heterogeneous glasses fall within this group, and (3) their bulk chemical compositions are similar to analyzed Apollo 14 regolith breccias and fines (Figs. 7a and 7b and Tables 4 and 1). Thus, they probably represent mixtures of crushed breccia fragments derived from the Fra Mauro formation. They do not necessarily represent a specific crystalline rock type or basalt type, as proposed by other investigators (Apollo Soil Survey, 1971; Reid *et al.*, 1972; and Engelhardt *et al.*, 1972). As seen from the available analyses of Apollo 14 basalts (Table 1), the 14053 type cannot contribute much to the high alkali and phosphorus content of the fines and regolith breccias. Although basalts of the 14310 type are higher in alkalis and phosphorus, these fragments are rare in the fines and breccias and thus their contribution also must be minor. Therefore, a specific component such as micronorite hornfels or hornfelsed noritic microbreccia from the Fra Mauro formation may be an important contributor to the KREEP component in the Apollo 14 fines (see Chao *et al.*, 1972, this volume).

Group 5. Glasses of feldspathic peridotite composition. A representative analysis (F2) is shown in Table 2, and others are given in Table 5 to show details of intragroup variation. They are characterized by high normative pyroxene and olivine and low normative feldspar, and tight clustering (Figs. 7a and 7b). The normative feldspar (mostly calcic plagioclase) covers a narrow range of 25 to 35%. The MgO/(MgO + FeO) ratios (0.42 to 0.48) plotted against weight percent SiO_2 (46.1–49.3) show no trend that is indicative of differentiation.

The compositions of these glasses are similar to howardites (Marvin *et al.*, 1972), but most howardites do not contain high normative olivine. A check of the USGS rock analyses file of terrestrial igneous rocks failed to reveal any picritic rocks of this composition. However, two analyses of eclogites (Williams, 1932) resemble these glasses closely in composition except that they are slightly higher in alkalis. One of

Table 4. Selected electron microprobe analyses (in weight percent) of Apollo 14 group 4 glasses*

Sample number, group 4

	O10	H5	F1	N4	N8	A1	K1	D1	B12	M10	B7
SiO_2	54.8	48.9	51.0	50.6	47.9	47.8	50.1	48.2	47.4	47.9	46.2
TiO_2	1.95	1.65	1.89	1.23	2.41	1.72	2.48	2.08	2.57	2.20	1.90
Al_2O_3	14.6	17.3	17.6	17.1	16.0	16.3	14.8	16.7	19.2	16.1	17.9
FeO	11.3	9.4	9.8	9.2	10.0	9.8	11.6	10.8	10.2	10.6	11.0
MgO	4.6	9.3	4.9	8.8	9.4	9.7	7.7	7.8	8.0	10.3	9.9
CaO	10.0	11.7	11.4	11.8	10.7	10.0	10.5	11.0	11.2	10.5	10.8
Na_2O	0.92	0.37	1.28	0.54	0.58	0.84	0.91	0.94	0.10	0.59	0.14
K_2O	0.97	0.40	1.00	0.45	0.60	0.56	0.73	0.62	0.10	0.30	0.12
P_2O_5	0.49	0.08	0.91	0.05	0.55	0.27	0.51	0.51	—	0.17	—
MnO	0.37	0.07	0.12	0.02	0.12	0.12	0.13	0.12	0.12	0.05	0.14
Cr_2O_3	0.16	0.12	0.11	0.31	0.22	0.20	0.17	0.20	0.14	0.27	0.14
NiO	—	—	—	—	—	0.04	—	0.06	0.06	—	0.05
TOTAL	100	99	100	100	98	97	100	99	99	99	98

Table 4 (continued)

Sample number, group 4

	A10	G6	N10	O5	O9	C10	F15	J8	O11	P7	H4
SiO_2	47.7	46.6	46.9	47.6	47.6	45.5	45.7	44.6	43.8	48.6	46.4
TiO_2	2.52	1.88	2.04	1.93	2.22	2.08	1.76	1.38	2.28	1.79	2.25
Al_2O_3	13.7	15.6	15.6	15.8	18.0	19.4	18.6	18.0	18.1	13.8	13.4
FeO	15.5	9.2	10.0	13.0	8.8	10.0	10.8	9.2	11.9	12.0	14.3
MgO	7.2	12.4	12.0	10.5	11.9	9.8	10.8	11.8	12.3	12.0	10.4
CaO	10.4	10.3	10.4	10.1	12.2	12.6	12.9	12.7	12.4	9.7	10.3
Na_2O	0.28	0.80	0.83	0.53	0.26	0.07	0.03	—	0.04	0.67	0.44
K_2O	0.18	0.65	0.26	0.34	0.14	0.08	0.16	0.09	—	0.60	0.37
P_2O_5	0.09	0.35	0.14	0.06	0.10	—	—	—	—	0.29	—
MnO	0.16	0.12	0.20	0.19	0.07	0.12	0.13	—	0.07	0.24	0.11
Cr_2O_3	0.26	0.18	0.23	0.18	0.32	0.12	0.20	0.13	0.36	0.31	0.32
NiO	0.04	—	—	—	—	0.04	—	0.13	0.17	—	—
TOTAL	98	98	99	100	102	100	101	98	101	100	98

* Analysts: Jean A. Minkin and E. C. T. Chao
— Below limit of detection

Table 5. Selected electron microprobe analyses (in weight percent) of Apollo 14 groups 5 and 6 glasses*

	Group 5											Group 6				
	D4	H1	F6	M7	B10	A7	F2	D3	K3	P5	G11	I9	I11	C1	J7	I1
SiO_2	47.2	47.5	48.8	48.2	47.3	46.5	48.0	47.2	46.9	47.3	48.1	40.0	41.0	39.0	37.6	41.7
TiO_2	0.63	0.60	0.68	0.74	0.62	0.56	0.67	0.60	0.75	0.56	0.53	2.31	2.27	2.10	2.18	2.14
Al_2O_3	10.5	8.9	9.0	9.0	10.1	9.6	9.0	10.2	8.9	8.6	8.4	19.4	20.7	21.5	21.7	18.6
FeO	17.1	16.9	17.2	17.1	16.9	17.3	17.1	17.1	17.5	17.3	17.7	10.8	10.4	10.5	9.7	12.4
MgO	12.5	13.1	13.6	13.7	13.4	13.8	14.4	13.3	14.7	15.1	15.2	12.5	11.9	12.0	12.8	12.5
CaO	10.1	10.2	10.6	9.2	9.6	9.2	9.8	10.0	9.0	9.3	10.0	13.7	14.5	14.3	14.4	12.8
Na_2O	0.32	0.20	0.27	0.23	0.24	0.24	0.27	0.29	0.18	0.20	0.17	0.03	0.21	—	—	—
K_2O	0.11	0.32	0.28	0.19	0.12	0.08	0.28	0.10	0.20	0.05	0.23	0.03	0.02	—	0.08	0.08
P_2O_5	0.08	0.17	—	0.15	0.04	0.04	0.11	0.08	0.03	0.12	0.27	0.15	0.07	—	—	0.08
MnO	0.18	0.16	0.16	0.13	0.20	0.18	0.30	0.19	0.16	0.18	0.21	0.12	0.09	0.12	0.16	0.14
Cr_2O_3	0.52	0.59	0.52	0.44	0.50	0.51	0.58	0.54	0.57	0.62	0.60	—	—	0.12	0.08	0.10
NiO	0.06	—	—	—	0.03	0.04	—	0.04	—	—	—	—	—	0.03	—	—
TOTAL	99	99	101	99	99	98	101	100	99	99	101	99	101	100	99	101

Table 5 (continued)

	Group 5					
	D2	F5	M8	M5	M1	M2
SiO_2	47.1	47.2	46.9	46.1	47.2	47.4
TiO_2	0.48	0.60	0.66	0.60	1.12	1.02
Al_2O_3	9.4	8.6	8.2	8.3	8.4	7.5
FeO	17.2	17.5	17.6	17.6	18.8	18.6
MgO	14.6	15.3	15.2	15.8	14.5	15.9
CaO	9.5	10.3	9.3	9.7	9.6	9.2
Na_2O	0.32	0.29	0.17	0.17	0.21	0.22
K_2O	0.10	0.19	0.24	0.24	0.14	0.20
P_2O_5	0.06	—	0.07	—	—	—
MnO	0.18	0.23	0.09	0.11	0.20	0.27
Cr_2O_3	0.52	0.63	0.57	0.57	0.46	0.45
NiO	0.06	—	—	—	—	—
TOTAL	100	101	99	99	101	101

* Analysts: Jean A. Minkin and E. C. T. Chao
— Below limit of detection

these with 1 % H_2O recalculated on a water-free basis is as follows (in weight percent): SiO_2, 48.53; TiO_2, 1.03; Al_2O_3, 7.50; total iron as FeO, 18.14; MgO, 14.00; CaO, 8.79; Na_2O, 1.16; K_2O, 0.32; P_2O_5, 0.08; MnO, 0.30; Cr_2O_3, 0.14.

The tight cluster suggests that the parent rock type of this group of green glasses may be a crystalline (either igneous or metamorphic) rock. The intragroup variation suggests a mixing trend, which is consistent with a cumulative origin, although a breccia of this composition cannot be ruled out. Glasses of this composition have not been reported from the Apollo 11 and 12 mare sites. No crystalline rock or breccia of this composition has been described to date among the returned lunar samples.

Group 6. Glasses of troctolitic composition. Representative analyses are given in Table 2 (19) and in Table 5. This group has very high normative feldspar and high normative olivine, and its normative ilmenite content also is unusually high relative to its feldspar content. Hence, these glasses are troctolitic in composition. A troctolite clast in 14321 has been analyzed by Compston *et al.* (1972). It is shown in Table 1 and included in the CIPW diagrams (Figs. 7a and 7b) for comparison; it is distinctly lower in iron and titania than the group 6 glasses.

Group 7 and subgroup 7a. Glasses of peridotite composition. These glasses are similar to those of group 5 except that FeO/MgO is notably higher, the magnesia content is slightly higher, and the alumina content is notably lower.

Subgroup 7a has a higher magnesia content and higher MgO/FeO ratio than does group 7 (Tables 2 and 6). Although 7 and 7a cannot be separated on the CIPW diagrams, they are distinct on plots of MgO against any major oxide and on plots showing the normative composition of the pyroxenes.

Group 8. Glasses of mare basalt composition. These samples have relatively high ratios of pyroxene to feldspar, contain appreciable normative ilmenite, and have high ratios of FeO/MgO. They vary from high normative olivine content to moderate amounts of normative quartz (see Fig. 7a). The highest TiO_2 content (about 16 weight %) occurs in glass spheroids in a porous regolith microbreccia (Fig. 2); a parent rock type with such a high titania content has not yet been reported.

Miscellaneous glasses of rare composition

Some glass particles we analyzed have odd compositions. One of these is a brown devitrified glass particle in regolith breccia 14301,62 which has a very high phosphorus content. The analysis in weight % is as follows: SiO_2, 40.0; TiO_2, 0.98; Al_2O_3, 26.1; FeO, 6.8; MgO, 7.7; CaO, 13.1; Na_2O, 0.94; K_2O, 0.45; P_2O_5, 4.3; MnO, 0.09; Cr_2O_3, 0.08.

Intergroup variations and parent rock suites

Groups 5 and 7 and subgroup 7a may be related and could represent a mafic rock suite, but we have not examined the relationship in detail. Because all are low in normative feldspar, one could speculate that these rocks are products of segregation or accumulation, and may be complementary to rock types such as those of anorthosite (group 2), anorthositic gabbro (group 3), or troctolitic composition (group 6).

Table 6. Electron microprobe analyses (in weight percent) of Apollo 14 glasses: Group 7 and 7a*

	Group 7										Group 7a		
	H10	C7	H8	P4	I13	J12	M3	C11	P2	P6	K11	M4	P1
SiO$_2$	46.7	45.4	45.8	46.8	47.0	45.5	45.2	45.2	46.9	47.6	46.3	45.8	46.8
TiO$_2$	0.90	0.88	0.96	0.92	0.85	0.86	1.13	0.92	1.12	0.76	0.62	0.77	0.52
Al$_2$O$_3$	6.1	7.0	6.2	5.9	6.3	5.9	6.1	6.7	6.0	6.8	6.2	6.5	6.2
FeO	23.1	22.7	22.5	23.2	23.1	22.6	23.0	22.8	23.2	19.6	19.3	19.5	19.6
MgO	14.5	14.5	14.9	14.6	15.3	15.6	16.1	15.0	15.7	17.0	18.8	18.4	18.8
CaO	8.7	8.8	8.3	8.0	8.6	8.5	8.0	8.6	8.4	8.8	8.4	8.2	8.3
Na$_2$O	0.16	0.20	0.17	0.14	0.14	0.17	0.17	0.14	0.14	0.18	0.12	0.14	0.14
K$_2$O	0.29	0.08	0.28	0.07	0.31	0.14	0.15	0.08	0.28	0.04	0.10	0.14	0.07
P$_2$O$_5$	0.09	0.06	—	0.20	0.07	0.05	—	0.06	0.14	0.16	—	0.13	0.16
MnO	0.21	0.22	0.14	0.32	0.26	0.24	0.21	0.23	0.37	0.32	0.12	0.12	0.22
Cr$_2$O$_3$	0.58	0.51	0.54	0.58	0.49	0.54	0.55	0.50	0.55	0.47	0.51	0.55	0.53
NiO	—	0.07	—	—	—	—	—	0.05	—	—	—	—	—
TOTAL	101	100	100	101	102	100	101	100	103	102	100	100	101

* Analysts: Jean A. Minkin and E. C. T. Chao
— Below limit of detection

We believe that the group 4 glasses are derived from fused regolith breccias or fines which themselves are mixtures of fragments derived from hornfelsed noritic micro-breccias or micronorite hornfels with Si–K-rich mesostasis, both of the Fra Mauro formation (see Chao *et al.*, 1972, this volume).

Summary

Eight chemical groups can be recognized on the basis of detailed R.I. and chemical studies of more than 200 Apollo 14 glass particles of impact origin. Because impactite glasses reflect essentially the bulk composition of their parent rock types, we have a survey of the various rock types they represent:

Group 1. Granitic or salic composition.
Group 2. Anorthositic composition.
Group 3. Anorthositic gabbro composition.
Group 4. Basaltic composition.
Group 5. Feldspathic peridotite composition.
Group 6. Troctolitic composition.
Group 7. Peridotite composition.
Group 8. Mare basalt composition.

This survey allows us to focus attention on several possible major rock suites. The principal rock suite, represented by the majority of glasses of group 4, probably is derived essentially from hornfelsed noritic microbreccias and micronorite hornfels described in a companion paper (Chao *et al.*, 1972, this volume).

Glasses of anorthositic (group 2), anorthositic gabbro (group 3), and troctolitic composition (group 6) are similar in bulk composition to lithic fragments that occur as inclusions in various types of Apollo 14 breccias. Glasses of groups 5 and 7 and subgroup 7a may be similar in bulk composition to rare fragments of olivine-rich microbreccias and metamorphic rocks and rare pyroxene-rich rocks which we have observed also in the Apollo 14 breccias. These latter glasses have not been found at the Apollo 11 and 12 Mare sites. Hence, it is possible that rocks represented by groups 2, 3, 5, 6, and 7 and subgroup 7a may be associated with the dominant rock types of the Fra Mauro formation (group 4). Source rocks characterized by high silica and high potash also are indicated (group 1) but they are rare, as are mare basalt glasses (group 8). We believe that this is a good representation of the range of rock types in the Fra Mauro formation collected from the Apollo 14 site.

An important consequence of this study is the conclusion that the major rock type of a highland site is dominated by annealed noritic rocks rather than by anorthosites as had previously been suggested by some investigators (Wood *et al.*, 1970), and that both mafic and salic rock types are associated with it.

Acknowledgments—We thank Marjorie Hooker, USGS Washington, D.C., for aid in searching the USGS rock analysis file of terrestrial igneous rocks, and George A. Desborough, USGS Denver and Benjamin A. Morgan, USGS Washington, D.C., for providing electron microprobe facilities. We thank our colleague Odette B. James for critical review of this paper. This work was carried out under NASA Contracts T-75412 and W13,130.

REFERENCES

Apollo Soil Survey (1971) Apollo 14: Nature and origin of rock types in soil from the Fra Mauro formation. *Earth Planet. Sci. Lett.* **12**, No. 1, 49–54.

Chao E. C. T. Boreman J. A. Minkin J. A. James O. B. and Desborough G. A. (1970a) Lunar glasses of impact origin: Physical and chemical characteristics and geologic implications. *J. Geophys. Res.* **75**, No. 35, 7445–7479.

Chao E. C. T. James O. B. Minkin J. A. Boreman J. A. Jackson E. D. and Raleigh C. B. (1970b) Petrology of unshocked crystalline rocks and evidence of impact metamorphism in Apollo 11 returned lunar sample. *Proc. Apollo 11 Lunar Sci. Conf., Geochim. Cosmochim. Acta*, Suppl. 1, Vol. 1, pp. 287–314. Pergamon.

Chao E. C. T. Minkin J. A. and Best J. B. (1972) Apollo 14 breccias: General characteristics and classification. *Proc. Third Lunar Sci. Conf., Geochim. Cosmochim. Acta,* Suppl. 3, Vol. 1, M.I.T. Press.

Compston W. Vernon M. J. Berry H. Rudowski R. Gray C. M. and Ware N. (1972) Age and petrogenesis of Apollo 14 basalts (abstract). In *Lunar Science—III* (editor C. Watkins), pp. 151–153, Lunar Science Institute Contr. No. 88.

Drake M. J. McCallum I. S. McKay G. A. and Weill D. F. (1970) Mineralogy and petrology of Apollo 12 sample no. 12013: A progress report. *Earth Planet. Sci. Lett.* **9**, No. 2, 103–123.

Engelhardt W. von Arndt J. and Stöffler D. (1972) Apollo 14 soils and breccias, their compositions and origin by impacts (abstract). In *Lunar Science—III* (editor C. Watkins), pp. 233–235, Lunar Science Institute Contr. No. 88.

Fredriksson K. Nelen J. and Noonan A. (1972) Apollo 14: Glasses, breccias, chondrules (abstract). In *Lunar Science—III* (editor C. Watkins), pp. 280–282, Lunar Science Institute Contr. No. 88.

Glass B. P. (1972) Apollo 14 glasses (abstract). In *Lunar Science—III* (editor C. Watkins), pp. 312–314, Lunar Science Institute Contr. No. 88.

Grieve R. McKay G. Smith H. Weill D. and McCallum S. (1972) Mineralogy and petrology of polymict breccia 14321 (abstract). In *Lunar Science—III* (editor C. Watkins), pp. 338–340, Lunar Science Institute Contr. No. 88.

Marvin U. B. Reid J. B. Jr. Taylor G. J. and Wood J. A. (1972) Lunar mafic green glasses, howardites, and the composition of undifferentiated lunar material (abstract). In *Lunar Science—III* (editor C. Watkins), pp. 507–509, Lunar Science Institute Contr. No. 88.

Reid A. M. Ridley W. I. Warner J. Harmon R. S. and Brett R. (1972) Chemistry of highland and mare basalts as inferred from glasses in the lunar soils (abstract). In *Lunar Science—III* (editor C. Watkins), pp. 640–642, Lunar Science Institute Contr. No. 88.

Roedder E. and Weiblen P. W. (1970) Lunar petrology of silicate melt inclusions, Apollo 11 rocks. *Proc. Apollo 11 Lunar Sci. Conf., Geochim. Cosmochim. Acta*, Suppl. 1, Vol. 1, pp. 801–837. Pergamon.

Williams A. F. (1932) *The Genesis of the Diamond*, Vol. 1, Ernest Benn Ltd., London.

Wood J. A. Dickey J. S. Jr. Marvin U. B. and Powell B. N. (1970) Lunar anorthosites and a geophysical model of the moon. *Proc. Apollo 11 Lunar Sci. Conf., Geochim. Cosmochim. Acta*, Suppl. 1, Vol. 1, pp. 965–988. Pergamon.

Proceedings of the Third Lunar Science Conference
(Supplement 3, *Geochimica et Cosmochimica Acta*)
Vol. 1, pp. 927–937
The M.I.T. Press, 1972

Chemistry and particle track studies of Apollo 14 glasses

B. P. Glass

Geology Department, University of Delaware, Newark, Delaware 19711

Dieter Storzer

Max-Planck-Institut für Kernphysik, 69 Heidelberg, West Germany

Günther A. Wagner*

Department of Geology, University of Pennsylvania, Philadelphia, Pennsylvania 19104

Abstract—Apollo 14 glasses from the <1 mm fines can be divided into five major groups based on their major element composition. Of the 282 glass particles analyzed, most (group 4) have compositions similar to the Apollo 14 soils and breccias. These glasses are similar in composition to the KREEP glasses found in Apollo 12 fines. Feldspathic glasses (group 2) with high Al_2O_3 (>21%) and CaO (>15%) contents make up about 20% of the glass particles analyzed. This is the only glass type that has been found at all of the lunar landing sites, and it was probably derived from the highlands. Glasses with high FeO (>14%) and low Al_2O_3 (<15%) contents make up 17 to 30% of the analyzed glasses. These glasses are probably derived from maria regions. Glasses with low Na_2O and K_2O (<0.4%) contents (group 5) may have been affected by vapor fractionation. A small number (<5%) of the glasses have anorthite compositions. Track analyses of twelve of the glass particles were carried out in order to determine their uranium contents and radiation histories. The uranium contents were found to range between 0.2 and 17 ppm with uranium correlating with K_2O content. Assuming a mean burial depth of 5 cm, radiation ages between 10^5 and 10^7 years were found.

Introduction

Major element analyses and track studies of glass particles from Apollo 14 fines have been made in order to gain a better insight into the chemistry and history of the lunar surface. Previous studies have shown that glass compositions provide a useful guide for determining the major rock types (and abundances) contributing to the soil. Most of the glass particles were probably formed by impact melting of the local and surrounding regolith. Many of the glass particles apparently represent homogenized portions of the lunar rock or soil from which they were formed. A 500 milligram sample of <1 mm fines will contain hundreds or thousands of such glass particles >50 μ in diameter. A significant proportion of these glass particles may have been derived from locations as far as several hundred kilometers from the collection site (Shoemaker *et al.*, 1970). Thus, the investigation of lunar glass particles from a single sample of <1 mm fines can yield a great deal of information about the history, composition, and variability of the lunar surface material within a radius of several hundred kilometers of the collection site.

This paper is divided into two parts. Part 1, by B. P. Glass, deals with glass content of the Apollo 14 fines (<1 mm) and major element analyses of glass particles recovered from the fines. Part 2, by G. A. Wagner and D. Storzer, concerns particle track analysis of glass particles from Apollo 14 sample 14163. Prior to track analyses

---

* Present address: Max-Planck-Institut für Kernphysik, 69 Heidelberg, West Germany.

the composition of each glass particle was determined by electron microprobe analysis, so that the track studies could be correlated to glass type.

Part 1. Abundance and Composition of Apollo 14 Glasses

Six Apollo 14 fines samples have been investigated (14148; 14149; 14156; 14163; 14230,75; and 14230,82). Sample 14163 is the <1 mm size fraction of the bulk fines. The bulk fines were collected as a scoop sample from a mean depth of several centimeters below the surface. Samples 14148, 14156, and 14149 are, respectively, fines from the surface, middle, and bottom of a trench dug on EVA–2. Samples 14230,75 and 14230,82 are fines from Apollo 14 core tube 14230. Sample 14230,82, which was studied in the greatest detail, is from the bottom of the core ($\sim$22.8 cm below top of drive-tube liner). The 0.149 to 1.000 mm size fraction of these fines samples contains from 13 to 27% glass particles. In general, the finer grained the soil sample the higher the percentage of glass that it contains (Table 1). Likewise, there appears to be an inverse correlation between glass content and percent lithic and mineral fragments. Glass spherules (spheres, dumbbells, etc.) make up approximately one percent of the samples (Table 1). The number of spherules increases with decreasing grain size. The average number of spherules per gram of soil recovered from the 0.505 to 1.000 mm, 0.295 to 0.505 mm, 0.149 to 0.295 mm, and 0.074 to 0.149 mm size fractions of three Apollo 14 soil samples (14163, 14148, and 14156) was 85, 125, 423, and 4450, respectively. The spherules have surface features similar to those observed on Apollo 11 and 12 glasses (i.e., splash silicate glass, metallic beads, and impact pits).

Chemical composition and glass types

Two hundred eighty-two glass particles (187 fragments and 95 spherules) from samples 14148, 14149, 14156, 14163, and 14230,82 were analyzed for Si, Ti, Al, Fe, Mn, Mg, Na, and K by electron microprobe analysis (for method of analysis see Glass, 1971a). Each glass particle was described and individually mounted for electron microprobe analysis so that the physical properties and appearance could be correlated with composition. The refractive indices of twenty-six particles were determined by the oil immersion method.

The glasses exhibit a wide range in composition. However, histograms of oxide abundances (Fig. 1) and various oxide plots show that the analyses tend to cluster into groups. Two hundred twenty of the glass particles appear to be homogeneous.

Table 1. Weight percent greater than 295 μ and modal composition of 149 to 295 μ size fraction of six Apollo 14 fines.

Sample no.	Wt. % >295 μ	Lithic	Mineral grains	Glassy agglutinates	Glass	Spherules	No. spherules per gram
14163	9.9	46	7	26	20	1.0	395
14148	7.0	36	3	34	24	1.7	496
14156	10.5	36	7	39	17	0.8	379
14149	13.5	52	9	22	16	0.5	365
14230,82	16.8	51	10	20	18	0.5	513
14230,75	10.5	48	8	19	24	1.4	645

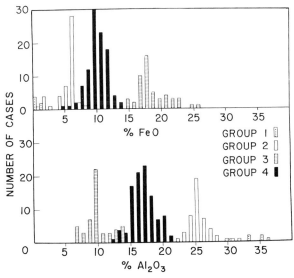

Fig. 1. Histogram of FeO and Al_2O_3 abundance in Apollo 14 glasses.

Table 2. Modal composition of glass types in Apollo 14 soils and other analyses.

	SiO_2	TiO_2	Al_2O_3	FeO	MnO	MgO	CaO	Na_2O	K_2O	N_d
Apollo 14										
Group 1 (6)	46.	0.1	34.4	0.4	0.02	0.1	18.4	1.4	0.13	
Group 2A (41)	45.	0.4	25.	6.0	0.07	8.6	15.	0.27	0.11	1.580–1.596 (5)
Group 2B (5)	46.	1.	24.	6.0	0.08	7.1	14.	0.33	0.19	1.595 (1)
Group 3A (13)	43.	4.6	9.4	22.	0.23	9.7	9.4	0.6	0.35	1.657–1.712 (2)
Group 3B (12)	46.	2.9	13.	17.	0.18	9.2	10.5	0.7	0.36	1.624 (1)
Group 3C (25)	47.	0.8	9.8	18.	0.21	15.	9.5	0.4	0.3	1.629–1.636 (5)
Group 4B (53)	48.	2.0	16.6	11.	0.12	9.9	10.4	0.95	0.7	1.598–1.604 (2)
Group 4C (32)	49.	2.2	17.	10.	0.11	7.6	10.6	1.3	0.9	1.580–1.603 (6)
Group 4D (16)	52.	1.8	17.	9.8	0.12	5.0	10.7	1.4	1.0	1.583–1.589 (3)
Group 5 (17)	45.	1.9	18.	10.6	0.13	10.7	11.7	0.17	0.17	
High silica glass fragment										
	73.9	0.64	12.5	3.2	0.06	1.1	2.0	1.3	4.9	
Transparent yellow-green glasses (from group 3C)										
	46.5	0.71	10.0	17.9	0.20	14.5	9.6	0.40	0.26	1.629–1.636 (5)
Apollo 14 fines (ave.) (LSPET, 1971)										
	48.	1.8	18.	10.	0.19	9.9	11.0	0.57	0.52	
Transparent pale green glass fragments										
Apollo 14 (12)	45.	0.34	25.6	5.8	0.07	8.56	15.0	0.18	0.11	
Apollo 12 (11)	45.	0.35	25.	5.7	0.07	8.89	15.8	0.15	0.05	1.594–1.596 (7)
*Apollo 11 (5)	44.	0.37	25.7	5.4	—	8.52	14.6	0.09	0.06	1.593–1.596 (5)

( ) Number of analyses or measurements.
*Average of five analyses from Chao et al. (1970).

Analyses of these particles (140 fragments and 80 spherules) were used to divide the glasses into five main groups and several subgroups (Table 2).

Group 1 The six glass particles in this group are mono mineralic with anorthite compositions. Four of them are colorless to pale green fragments and are probably

Table 4. Major element composition* refractive index, uranium content,

Sample no.	Group†	SiO$_2$	TiO$_2$	Al$_2$O$_3$	FeO	MnO	MgO
379F	4C	49.0	2.28	17.2	9.61	0.12	8.02
384F	4C	53.6	1.85	15.6	9.42	0.10	6.66
386S	2A	45.8	0.36	26.9	4.92	0.06	8.62
402F	4C	48.9	1.92	17.4	9.96	0.11	7.98
417F	4B	49.6	3.31	15.0	11.4	0.11	8.84
419F	4C	49.2	2.80	14.4	11.0	0.13	8.42
423F	4C	49.2	1.93	17.9	9.58	0.11	6.98
434F	4C	49.7	0.99	20.0	8.24	0.09	7.60
435F	4D	52.5	1.49	17.5	9.85	0.12	5.08
437F	4D	50.2	2.25	17.9	10.1	0.12	4.36
446F	3B	46.7	2.70	12.2	18.6	0.21	7.54
457F	4B	49.7	1.70	13.9	12.7	0.16	9.50

F = fragment   S = sphere.
*Values in weight percent unless otherwise specified.

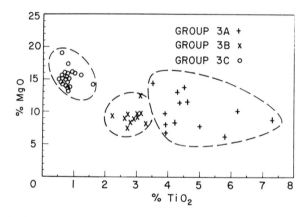

Fig. 2. Variation diagram of MgO versus TiO$_2$ content for group 3 glasses.

diaplectic plagioclase glasses. The remaining two are spherules (one colorless and one white opaque).

Group 2 The glasses in this group are characterized by their high Al$_2$O$_3$ (>22%) and CaO (>14%) contents. This group has been divided into two subgroups based on TiO$_2$ contents. The glasses in group 2A have TiO$_2$ contents less than 0.5% and the glasses in group 2B have TiO$_2$ contents greater than 0.8%. The glasses in group 2A are mostly transparent pale green fragments or translucent ropy fragments covered with rock flour. Group 2B glasses are colorless to yellow-green fragments. The uranium content was determined for only one group 2 glass particle (see Part 2). It had the lowest uranium content measured (Table 4).

Group 3 The glasses in this group are distinguished by their high FeO (>14%) and low Al$_2$O$_3$ (<15%) contents (Fig. 1). They have been divided into three subgroups based on their Ti and Mg contents (Fig. 2). The glasses in group 3A have TiO$_2$ contents greater than 3.5%. These glasses are generally transparent reddish brown or

track density, and VH radiation age of twelve Apollo 14 glass particles.

CaO	Na$_2$O	K$_2$O	R.I.	Uranium contents (ppm)	Fossil tracks	Residence time (10^{6}y) at depth of 5 cm	at top of soil
10.5	1.35	0.96	1.601	11.5	3.0	10	—
9.68	1.33	1.66	1.580	15.1	—	—	—
15.4	0.23	0.07	—	0.22	—	—	—
10.9	1.21	0.84	—	11.3	1.1	4	—
9.64	1.20	1.18	—	9.9	1.1	4	—
9.27	1.42	1.43	—	11.3	0.03	0.1	—
10.9	1.54	1.01	—	17.1	1.4	5	—
12.3	0.52	0.43	1.590	4.4	0.27	0.9	—
11.2	0.97	0.70	1.583	13.8	1.9–38	6	0.2(2)‡
11.2	1.74	1.11	1.589	16.6	2.5	8	—
10.6	0.81	0.41	—	4.0	5.0–40	0.4	2 (20)‡
9.03	1.44	1.08	—	11.3	—	—	—

†See Table 3 and text for definition of groups.
‡Corrected values in parentheses.

deep red fragments or spherules. The glasses in group 3B have TiO$_2$ contents between 2.7 and 3.5% and low MgO contents (<10%). These glasses are generally orange or reddish brown or red fragments. The group 3C glasses have low TiO$_2$ (<1.5%) and high MgO (>14%) contents. These glasses are generally transparent yellow-green fragments without relict crystalline inclusions or bubble cavities. The uranium content of one group 3B glass particle (see Part 2) was found to be 3.96 ppm. This is approximately one-third the amount found for most group 4 glasses (Table 4).

Group 4 These glasses are characterized by their low FeO (<14%) and high K$_2$O (>0.4%) or Na$_2$O (>0.4%) contents. The glasses in this group have been divided into three subgroups based on their MgO contents for the purpose of comparison with types B, C, and D as defined by the Apollo Soil Survey (1971). The glasses in this group are mostly light-colored fragments and some spherules. Many are covered with rock flour. Ten group 4 glasses were found to have uranium contents (see Part 2) ranging from 4.38 to 17.05 ppm. However, nearly all of the group 4 glasses have uranium contents greater than 10 ppm with the uranium content roughly proportional to the K$_2$O content.

Group 5 These glasses are distinguished from groups 1 and 2 by their lower Al$_2$O$_3$ contents (<21%), from group 3 by their lower FeO contents (<11%), and from group 4 by their lower K$_2$O and Na$_2$O contents (<0.4%). The glasses in this group are generally transparent yellow, yellow-brown, or brown spherules.

All but four of the homogeneous glasses and nearly all of the heterogeneous glasses in this study can be assigned to one of the above groups.

Discussion of chemical studies

For each soil sample investigated the percent abundance of the glass groups discussed above is given in Table 3 (for modal compositions of each group see Table 2). The groups defined are similar to those defined by the Apollo Soil Survey (1971), and the relative abundances of each group are similar to those found by the Apollo Soil Survey for sample 14259. The major difference is in the group 3 glasses. The

Table 3. Percent abundance of glass types in Apollo 14 and Apollo 12 samples.
For modal composition of each group, see Table 2.

| | Fines | Bulk trench samples | | | Core sample | Comprehensive sample | |
| | | Surface | Middle | Bottom | | | |
	14163	14148	14156	14149	14230,82	14259*	Apollo 12†
Group	(64)	(40)	(32)	(43)	(41)	(386)	(161)
1	2	0	6	2	5	0	1
2A	20	25	19	5	24	} 29	19
2B	2	0	3	5	2		3
3A	5	5	6	11	2		} 46
3B	6	10	6	0	5	} 11	
3C	6	15	9	19	10		0
4B	20	15	22	32	32	28	8
4C	25	12	16	5	10	28	15
4D	11	8	0	5	5	3	2
5	2	8	12	16	5	0	7
Misc.	2	2	1‡	0	0	2‡	0

*Apollo Soil Survey (1971).
†Based on analyses published by Glass (1971).
‡High silica and high potash glass.

percentage of these glasses analyzed in this report is about twice that found by the Apollo Soil Survey (Table 3). (These glasses are believed to be derived from a mare area and will be discussed later.) Another difference is that the Apollo Soil Survey did not report finding any glasses with compositions similar to the group 5 glasses. Otherwise, the relative abundances are in essential agreement.

Core sample There appears to be no consistent difference between the glass types and abundances found in the surface samples (14163, 14148, and 14259) and the glass types and abundances found in the trench samples (14156 and 14149) and the core sample (14230,82) (see Table 3). No glass types are unique to the trench or core samples. Thus the fines from the bottom of the trench and from the bottom of core 14230 were most likely derived from the same source as the surface fines.

Fra Mauro or KREEP Glasses The most abundant glasses in each of the Apollo 14 samples are the group 4 glasses (Table 3). They are similar in composition to the fines and breccias from the Apollo 14 site (Table 2). Thus these glasses were probably derived from the regolith at the Apollo 14 site. The Apollo Soil Survey (1971) refers to glasses with this composition as Fra Mauro basalts. Glasses with similar compositions are common in the Apollo 12 fines (Glass, 1971b; Meyer *et al.*, 1971). Material of this composition in Apollo 12 samples has been variously called KREEP, norite, gray mottled fragments, and nonmare basalts (Apollo Soil Survey, 1971).

Feldspathic glasses The group 2 glasses (with $Al_2O_3 > 21\%$) are probably best described as feldspathic. Feldspathic glasses are one of the major types of glasses at the Apollo 14 site (Table 3). Glasses with similar compositions are also common at the Apollo 11, Apollo 12, and Luna 16 sites (Reid *et al.*, 1971; Glass, 1971b; Chao *et al.*, 1971). This is the only glass type that is common to all of the lunar sites that have been sampled. Thus, material of this composition is probably very abundant on the lunar surface. A close similarity between the composition of these feldspathic glasses and the composition of ejecta from the highland crater Tycho (Patterson *et al.*, 1969), analyzed by Surveyor VII, suggests that the feldspathic glasses are derived from the highlands. This suggestion is supported by the preliminary Apollo 15 x-ray fluorescence data from lunar orbit (Adler *et al.*, 1972).

Many of the feldspathic glasses are transparent pale green fragments without relict crystalline inclusions, metallic spherules, or bubble cavities. Some, however, have partly devitrified to plagioclase; otherwise, they are homogeneous. Such glasses have been observed at the Apollo 11, 12, and 14 sites. The pale green feldspathic glasses from these three sites are remarkably similar in composition (Table 2). The uniformity in composition and the lack of relict minerals and metallic spherules suggests that these glasses may be of igneous rather than impact origin.

Mare-derived glasses The high FeO ($>14\%$) contents and low Al_2O_3 ($<15\%$) contents (Table 2) of the group 3 glasses suggest that they are derived from a mare area. Group 3 glasses make up from 17 to 30% of the analyzed glasses in the soil samples investigated (Table 3). This suggests a rather large contribution of mare-derived material at the Apollo 14 site; however, the selection of glass particles for analysis was not random, and there was probably a bias towards the transparent red fragments and distinctive green spherules of the group 3 glasses. Only about 11% of the glass particles analyzed by the Apollo Soil Survey (1971) from sample 14259 were believed to have been derived from mare areas.

The group 3C glasses are distinguished by their high MgO contents (Table 2). Glasses with this composition have not been observed in samples from the Apollo 11, Apollo 12, and Luna 16 sites. Most of the group 3C glasses are transparent yellow-green fragments of spherules that are remarkably uniform in composition from sample to sample. These green glasses may be similar to the green glass spheres that have been reported as being so common in Apollo 15 samples (LSPET, 1972). Like the Apollo 15 green glasses, they are remarkably homogeneous and free of relict crystalline inclusions (although several are partially devitrified) and vesicles. However, they have somewhat lower refractive indices (1.629 to 1.636) than those reported for the Apollo 14 green glass spheres (1.65). Again, the lack of any evidence indicating that these glasses were formed by impact suggests that they may be igneous glasses.

Possible vapor fractionation There is an almost continuous range in compositions between the group 4 glasses and the group 5 glasses. The division between these two groups is, in fact, rather arbitrary. However, on a plot of K_2O versus Na_2O there does appear to be a natural break at about 0.4% Na_2O and 0.4% K_2O (Fig. 3). Whereas most of the group 4 glasses (with Na_2O or K_2O contents greater than 0.4%) are translucent fragments or opaque spherules, all but one of the group 5 glasses (with Na_2O and K_2O contents of less than 0.4%) are transparent homogeneous spherules (Fig. 3). This suggests that the group 5 glasses were heated more intensely and that the difference in chemistry between the group 5 glasses and the group 4 glasses may be the result of vapor fractionation.

Potash granite glasses A single glass fragment with a high silica ($\sim 74\%$) content and K_2O ($\sim 5\%$) content (Table 2) was found among the 282 glass particles analyzed. Glass particles with similar compositions were reported by the Apollo Soil Survey (1971). According to the Apollo Soil Survey, glasses of this composition make up approximately 1.6% of the glasses in sample 14259,26. They point out that the composition is similar to potash granite.

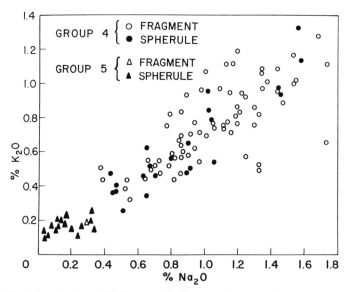

Fig. 3. Variation diagram of K_2O versus Na_2O content for group 4 and group 5 glasses.

Comparison with Apollo 12 glasses Based on analysis of 161 glass particles reported by Glass (1971), an attempt has been made to determine for the Apollo 12 fines the percent abundance of each glass group as defined above (Table 3). The percent abundance of feldspathic glasses and group 5 glasses with low alkali contents are similar to that found for the Apollo 14 samples; however, the Apollo 12 samples contain a much higher percentage of mare-derived (group 3) glasses and a much lower percentage of Fra Mauro or KREEP (group 4) glasses than the Apollo 14 samples. This is to be expected because the Apollo 12 site was on a mare surface and the Apollo 14 site was on the Fra Mauro formation.

Part 2. Particle Track Studies

Track analyses have been carried out in order to determine the uranium content (Price and Walker, 1963) and the radiation history of glass particles from the bulk fines sample 14163. The selected glass fragments and spherules are between 100 and 500 μ in size and come from the uppermost 10 cm of the lunar soil (scoop sample).

The following procedure was used in the track analyses. The polished sections of the glass particles were etched at 23°C in a mixture of 4 parts H_2O, 2 parts HF, and 1 part H_2SO_4 for 20 to 35 secs. The tracks were observed in reflected light. Thereafter the polished sections were covered with Lexan detectors and irradiated with thermal neutrons for uranium analysis, together with NBS-standard glasses 615 and 617. The irradiated glass particles were repolished and etched under the same conditions. The diameters of the induced fission tracks were compared with the fossil tracks. Finally, the glasses were annealed at 200°C for various periods of time and the fading of the tracks was studied.

Fossil tracks were found in most fragments, except fragments 384, 386, and 457. The track densities range from 3×10^4 to 4×10^7 tracks/cm^2 (Table 4). These tracks are thought to be caused nearly entirely by VH nuclei of the solar and the galactic component of cosmic radiation. Uranium fissions as a track source is negligible. The VH nuclei tracks in glass can be identified by their track-density distribution, track-etch-pit size, and annealing behavior. They should not be related to the uranium content and the uranium distribution. Our track studies have resulted in the following findings:

(a) In most glass particles the fossil tracks are homogeneously distributed. How-ever, fragment 435 has an increased density of small ($<2\,\mu$) tracks within 30 μ of the edge. The most distinct track-density gradient was found in fragment 446 (Fig. 4). Only VH solar flare nuclei can cause such steep gradients (Pellas *et al.*, 1968; Pellas *et al.*, 1969; Lal and Rajan, 1969).

(b) In all glass particles studied, the fossil tracks have smaller average diameters than the induced fission tracks. This is due to the fact the VH nuclei tracks have lower track etching rates, i.e., smaller etch-pit diameters than fission tracks (for example, in one silica glass, the diameters of the etched iron tracks are approximately 80% of the fission tracks; Krätschmer, 1971), and because they have undergone thermal fading. From our annealing experiments, track fading in the glass fragments must be expected in the uppermost 5 cm of the regolith based on knowledge of lunar surface temperatures (Creemers *et al.*, 1971).

(c) VH nuclei tracks in glasses have a characteristic annealing behavior in that they fade faster than fission tracks.

(d) The uranium content of the glass particles ranges between 0.2 and 17 ppm (Table 4). Uranium content is roughly proportional to the K_2O content of the glasses and therefore seems to correlate with glass type. Inhomogeneous uranium distribution was found in glass particles 384, 417 (in clusters), 423, and 457 (in schlieren). The distributions and densities of the fossil tracks are not parallel to the uranium dis-

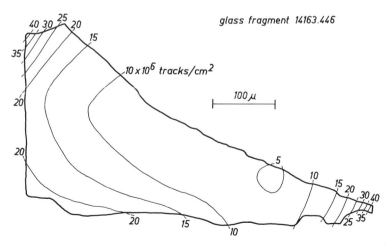

Fig. 4. Solar flare track density distribution in a glass fragment from bulk fines sample 14163.

tribution or contents in these fragments. This excludes a uranium-fission origin for the fossil tracks.

The proper identification of the galactic and solar flare VH nuclei tracks allows us to estimate residence times of the glass particles in the top few centimeters of the lunar soil. The homogeneously distributed tracks in the glass particles are attributed to the galactic VH nuclei. Assuming a mean burial depth of 5 cm and track-production rates from Barber *et al.* (1971), radiation ages between 10^5 and 10^7 years are found (Table 4). Because the fossil tracks have been more or less thermally affected, these ages are minimal ages. Additional uncertainty is introduced by the low and heterogeneous track-registration efficiency in glass. The estimated residence times of the glass particles within the top 5 centimeters of the lunar soil are similar to those found for minerals in the soil (Fleischer *et al.*, 1971; Arrhenius *et al.*, 1971).

All solar flare particle tracks are strongly annealed. Only this fact allows the optical counting of track densities as high as 4×10^7 tracks/cm^2. The fading of tracks was probably coincident with the formation of the tracks as the glasses were directly exposed to the sun. From the solar flare track densities, surface exposure ages for glass particles 435 and 446 are found to be 0.2×10^6 years and 2.0×10^6 years, respectively, using track-production rates by Barber *et al.* (1971). The track density versus depth relationship indicates erosion (and fragmentation for glass particle 446). Due to track-diameter reductions, these solar flare exposure ages can be corrected by a factor of about 10, assuming the diameter versus track density relationship of solar flare tracks is similar to that of fission tracks (Storzer and Wagner, 1969). In glass particle 446 there is an additional homogeneous background (1.2×10^5 tracks/cm^2) of thermally unaffected VH galactic tracks. From this it follows that glass fragment 446 was first directly exposed to the sun for about 20 m.y. and then 4×10^5 years ago was buried deep enough (>5 cm) to be protected from track fading. From the track studies it also follows that this glass particle has been broken after the solar flare irradiation. The lines of equal solar flare track densities are intersected by the grain boundary (Fig. 4). This fragmentation probably occurred simultaneously with the burying event.

Acknowledgments—Electron microprobe analyses were done at the Planetology Branch, Goddard Space Flight Center with the kind permission of L. S. Walter. F. Wood helped with the analyses and data reduction. J. A. Glass helped with manuscript preparation. This work was supported by NASA grant NGR 08–001–029.

REFERENCES

Adler I. Trombka J. Gerard J. Lowman P. Yin L. and Blodgett H. (1972) Preliminary results from the S–161 x-ray fluorescence experiment (abstract). In *Lunar Science—III* (editor C. Watkins), p. 4, Lunar Science Institute Contr. No. 88.

Apollo Soil Survey (1971) Apollo 14: Nature and origin of rock types in soil from the Fra Mauro formation. *Earth Planet. Sci. Lett.* **12**, 49–54.

Arrhenius G. Liang S. Macdougall D. Wilkening L. Bhandari N. Bhat S. Lal D. Rajagopalan G. Tamhane A. S. and Venkatavaradan (1971) The exposure history of the Apollo 12 regolith. *Proc. Second Lunar Sci. Conf., Geochim. Cosmochim. Acta* Suppl. 2, Vol. 3, pp. 2583–2598. MIT Press.

Barber D. J. Cowsik R. Hutcheon I. D. Price P. B. and Rajan R. S. (1971) Solar flares, the lunar surface, and gas-rich meteorites. *Proc. Second Lunar Sci. Conf., Geochim. Cosmochim. Acta* Suppl. 2, Vol. 3, pp. 2705–2714. MIT Press.

Chao E. C. T. Boreman J. A. Minkin J. A. James O. B. and Desborough G. A. (1970) Lunar glasses of impact origin: Physical and chemical characteristics and geologic implications. *J. Geophys. Res.* **75**, 7445–7479.

Creemers C. J. Birkebak R. C. and White E. J. (1971) Lunar surface temperatures from Apollo 12. *The Moon* **3**, 346–351.

Fleischer R. L. Hart H. R. Jr. Comstock G. M. and Evwaraye A. O. (1971) The particle track record of the Ocean of Storms. *Proc. Second Lunar Sci. Conf., Geochim. Cosmochim. Acta* Suppl. 2, Vol. 3, pp. 2559–2568. MIT Press.

Glass B. P. (1971a) Investigation of glass particles recovered from Apollo 11 and 12 fines: Implications concerning the composition of the lunar surface. *NASA Goddard Space Flight Center Doc. X–644–71–414.*

Glass B. P. (1971b) Investigation of glass recovered from Apollo 12 sample 12057. *J. Geophys. Res.* **76**, 5649–5657.

Krätschmer W. (1971) Die anatzbaren Spuren kunstlich beschleunigter schwerer Ionen in Quarzglas. Doctoral thesis, Heidelberg.

Lal D. and Rajan R. S. (1969) Observations relating to space irradiation of individual crystals of gas-rich meteorites. *Nature* **223**, 269–271.

LSPET (1971) (Lunar Sample Preliminary Examination Team) Preliminary examination of lunar samples from Apollo 14. *Science* **173**, 681–693.

LSPET (1972) (Lunar Sample Preliminary Examination Team) The Apollo 15 lunar samples: A preliminary description. *Science* **175**, 363–375.

Meyer C. Jr. Aitken F. K. Brett R. McKay D. S. and Morrison D. A. (1971) Mineralogy, chemistry, and origin of the KREEP component in soil samples from the Ocean of Storms. *Proc. Second Lunar Sci. Conf., Geochim. Cosmochim. Acta* Suppl. 2, Vol. 1, pp. 393–411. MIT Press.

Patterson J. H. Franzgrote E. K. Turkevich A. L. Anderson W. A. Economa T. E. Griffin H. E. Grotch S. L. and Sowinski K. P. (1969) Alpha-scattering experiment on Surveyor 7: Comparison with Surveyors 5 and 6. *J. Geophys. Res.* **74**, 6120–6143.

Pellas P. Popeau G. and Lorin J. C. (1968) Irradiation primitive par des ions lourds d'origine solaire des cristaux meteritiques. *Bull. Soc. fr. Mineral. Cristallog.* **91**, L.

Pellas P. Popeau G. Lorin J. Reeves H. and Adouze J. (1969) Primitive low-energy particle irradiation of meteoritic crystals. *Nature* **223**, 272–274.

Price P. B. and Walker R. M. (1963) A simple method of measuring low uranium concentrations in natural crystals. *Appl. Phys. Lett.* **2**, 23–25.

Reid A. M. Ridley W. I. Harmon R. S. Warner J. Brett R. Jakeš P. and Brown R. W. (1971) Feldspathic basalts in lunar soils and the nature of the lunar highlands. *Earth Planet. Sci. Lett.* (in press).

Shoemaker E. M. Hait M. H. Swann G. A. Schleicher D. L. Schaber G. G. Sutton R. L. Dahlem D. H. Goddard E. N. and Waters A. C. (1970) Origin of the lunar regolith at Tranquillity Base. *Proc. Apollo 11 Lunar Sci. Conf., Geochim. Cosmochim. Acta* Suppl. 1, Vol. 3, pp. 2399–2412. Pergamon.

Storzer D. and Wagner G. A. (1969) The correction of thermally fission track ages of tektites. *Earth Planet. Sci. Lett.* **5**, 463–468.

Proceedings of the Third Lunar Science Conference
(Supplement 3, *Geochimica et Cosmochimica Acta*)
Vol. 1, pp. 939–951
The M.I.T. Press, 1972

Structure of lunar glasses by Raman and soft x-ray spectroscopy

George W. Fabel, William B. White, Eugene W. White, and Rustum Roy

Materials Research Laboratory,
The Pennsylvania State University,
University Park, Pennsylvania 16802

Abstract—X-ray spectra and Raman spectra have been measured from individual particles of lunar glasses. Each particle was analyzed separately by electron microprobe. Silicon emission shifts vary between parent rock types and can be interpreted as a range of Si–O distances from 1.612 to 1.637Å. Aluminum emission shifts relate to the amount of 4- and 6-coordinated Al in the glass. Raman spectra show broadened bands. Certain bands recur in many specimens and relate to the main normative minerals for the glass bulk composition, olivine, pyroxene, and anorthite.

Introduction

The very large number of chemical analyses now available for glass particles and lithic fragments in the lunar soils cluster into distinct populations that permit the identification of the same number of lunar rock types. Various classifications have been proposed. We have adapted the work of Reid *et al.* (1972) to classify the glass particles examined in our investigation. It seems to be recognized that the composition of the glass particles also mirrors an average composition of parent rock types or individual lunar minerals with little modification during melting and transport. There are, therefore, a number of kinds of lunar glass, identifiable by their chemical composition. Our objectives in the present paper are to relate as far as possible the structure-sensitive parameters of these glasses to their chemical composition.

The approach is to measure Raman spectra and x-ray emission spectra on individual glass particles. Electron microprobe analyses of the same particles permit correlation of each particle with a known composition, and the interpretation of the structural measurements can be made on this basis.

Experimental Methods

Selection and mounting of specimens

A representative assortment of glass particles was selected under an optical microscope from Apollo 11–15 fines. These glassy specimens consisted primarily of opaque spherules, opaque glass coatings on rock fragments, and angular fragments of transparent and opaque glass. With the exception of three large specimens (15017,9, 15245,56, and 15015,26, all ~2 cm across), particle size ranged from 0.3 to 2 mm. All specimens were prepared as polished grain mounts using Koldmount self-curing resin. For the three large samples, small glass fragments (~3 mm) were broken off and mounted.

Chemical analysis

The chemical analyses shown in Table 1 were obtained with an Applied Research

Laboratories (ARL) electron microprobe. When two glass standards are used, results are good to approximately $\pm 10\%$ relatively. The specimens were first checked for chemical homogeneity by scanning the electron beam over a $360 \times 360\ \mu$ area and observing the absorbed electron image. The samples were found to fall into

Table 1. Microprobe analyses (wt%) of

Sample number	SiO_2	Al_2O_3	CaO	MgO	FeO
10084,110,1A	49	20	15	9.6	4.7
10084,110,1B	48	22	16	9.6	4.8
10084,110,2.8	50	23	14	7.8	4.5
10084,110,2.9	50–57	25–28	14–19	<0.1	<0.1
	53–58	0–2	3–18	15–40	7–12
10084,110,III	51–57	17–21	12–18	2–5	0–6
	29–56	9–18	1–12	8–17	6–23
10084,110,3.1	48–55	8.5–10	10–11	8–8.5	15–15.5
12070,42–1	52–61	11–19	1.5–8.5	3.5–11	10–14.5
12070,42–2	46–58	2–26	6–15	8–21	6–19
12070,42–3	55–65	0–4	5–17	8–21	12–27
12042,32–1	50	19	14	10.7	5.1
12042,32–2	42–63	2.5–13.5	5–14.5	2–10.5	4–16.5
12042,32–3	34–49	31–35	0–0.5	0–0.5	0.5–1.5
	40–54	0.5–3.0	3.5–7.0	16–21	14–21
12042,32–4	34–49	27–33	13–17	0–2.5	0.5–3
	14–44	2–13	3–9.5	6–25	18–29
14162,47–1	46	25	14	9.5	5.2
14162,47–2B	46–58	2–26	6–15	8–21	6–19
14162,47–2C	42–60	36–39	15–18	<0.1	0–2
	50–62	0–7	3–8	15–21	10–24
14162,47–3	45–53	32–38	10–15	0–1.5	0–1
	34–58	6–13	2–7	18–30	19–35
14162,47–5	45–49	32–38	13–17	0–1	0–2
	37–55	5–10	3–7	18–23	11–20
14162,61–1	50	18	10	9.8	6.3
14162,61–2	46–53	16–22	12–17	0–1	0–3
	44–51	9–12	5–11	9–12	4–8
14162,61–3	43–54	14–42	8–14	—	0–7
	44–50	5–12	2–5	20–23	9–15
14162,61–4	41–50	31–32	15–17	0.5–2	0–1
	32–44	0.1	0.2	28–34	27–28
14162,61–5	43–60	19–36	11–17	—	0–4
	46–54	7–12	4–7	16–22	7–11
15501,25	42–50	9–11	8–10	11–12	14–16
15531,40	39–43	<0.1	0–0.3	27–31	25–30
15231,49–1	44–50	15.5–19.5	12.5–14	7.5–11	8.5–11.5
15231,49–2	42–40	22–29	11–18	1.5–5.5	2.5–7.5
	47–59	4–13	9–13	8–13	9–19
15231,49–3	42–50	11–16	10–16	—	12–16
	43–68	0.5–1.0	5–12	13–16	15–25
15015,26	44–47	23–35	11–16	0–7.5	0.5–8.5
	43–61	0–8.5	0.5–12	16–31	13–24
15245,56	32–42	30–39	16–20	<0.1	0–3.5
	57–69	0–2	1–6.5	22–32	10–11
15017,9	25–58	7.5–12.5	9.5–10.8	9.7–11.8	9–13.2

three principal categories: (1) homogeneous, (2) heterogeneous with no clear phase separation, and (3) heterogeneous with separation into Al, Ca zones and Fe, Mg zones. With some care it was possible to focus the electron beam into each zone and obtain individual chemical analyses as shown in Table 1.

lunar glasses used for spectral measurements.

TiO$_2$	K$_2$O	Form	Color	Homogeneity*	Rock type†
—	0.1	Shard	Clear green	H	H
—	0.1	Shard	Clear green	H	H
—	0.1	Vesicular	Clear yellow	H	H
<0.1	<0.1	Shard	Clear brown	IH	H-M
0–0.5	<0.1				
0 2.5	<0.1	Vesicular	Black	IH	H-M
0–1.5	<0.1				
4.5–5.5	<0.1	Shard	Black	H	M
0.5–2	0.2–1.0	Spatter	Brown-black	IH	K
0.5–3.5	0–0.2	Spatter	Brown-black	IH	M
0.5–1.5	<0.1	Spatter	Brown-black	IH	M
<0.2	<0.1	Shard	Clear green	H	H
0–2.5	0–0.8	Sphere	Black	IH	M
<0.1	<0.1	Spatter	Clear yellow	IH	A-M
0–1	<0.1				
0–1	<0.1	Spatter	Brown-black	IH	A-M
0.5–11	0–0.2				
0.2	0.1	Shard	Brown-black	H	H
0.5–3.5	0–0.2	Vesicular	Brown-black	IH	K
<0.1	<0.1	Vesicular	Brown-black	IH	A-M
0–0.5	0–0.2				
—	—	Vesicular	Brown-black	IH	A-M
—	—				
—	—	Vesicular	Brown-black	IH	A-M
—	—				
1.7	0.3	Sphere	Black	H	K
0–1	0–0.3	Ropy strand	Brown-black	IH	K
3–5	0–0.5				
0–1	0–0.4	Shard	Brown-black	IH	K
4–7	<0.1				
0–0.2	0–0.3	Vesicular	Brown-black	IH	A-M
0–0.5	<0.1				
0–2	0–1	Vesicular	Brown-black	IH	A-K
0–2	0–0.2				
1.0–1.5	0–0.2	Ropy strand	Brown-black	H	M
<0.1	0.1	Shard	Brown	H	M
0.5–1	0.1	Sphere	Brown-black	H	K-M
0.1–0.6	0–0.3	Vesicular	Black	IH	H-K
0.5–1.5	0–0.4				
—	0–0.2	Vesicular	Brown-black	IH	M
0.5–1.0	<0.1				
0–0.5	<0.1	Glazed rock	Brown	IH	A-M
0–0.5	0–0.2				
<0.1	<0.1	Spatter	Brown-black	IH	A-M
0–0.3	<0.1				
0.9–1.2	0.2–0.6	Spatter	Brown-black	IH	K

*H = homogeneous; IH = inhomogeneous.
†Rock types are: M = Mare basalts undifferentiated; K = Fra Mauro (KREEP) basalts; H = Highland basalts; A = Anorthosite compositions.

Emission shifts

The AlK_β, SiK_β, and OK_α x-ray emission bands were recorded for all specimens using an ARL electron microprobe model EMX operated at 20 keV and a specimen current of 0.10 to 0.30 μA. An ADP crystal was used for resolving the AlK_β and SiK_β, and clinochlore was used for the OK_α. The peak shift was measured from 2/3 height position. The SiK_β and OK_α emission shifts (Δ) were measured with respect to an α-quartz standard and the AlK_β with respect to α-Al_2O_3. The microprobe was operated with a large spot diameter of $\sim 100 \mu$. The spectra were measured at several points on each sample. The shift (Δ) is defined here as

$$\Delta = \lambda \text{ (specimen)} - \lambda \text{ (standard)}.$$

The precision of the shift measurements is about $\pm 10\%$. Details of the experimental procedure employed here are reported elsewhere (White and Gibbs, 1967, 1969, and Gigl *et al.*, 1970).

Raman spectra

The Raman spectra of the mounted specimens were measured using a Spex Ramalog double grating spectrometer. Excitation was provided by the 488 or 514.5 nm line of an ionized-argon laser source with a power output of approximately 200 mW. The sample was positioned at the focal point of the laser beam and the Raman scattered light observed at 90° to the incident direction. Because the laser beam can be focused to a diameter less than 0.1 mm, it was possible to record spectra at several different points on each specimen. In a few instances, spectra from the specimen mounting material (Koldmount) were superimposed on the glass spectra. However, this was easily subtracted out.

Raman spectra of some particles, particularly those measured directly on unmounted and unpolished glass fragments, exhibited considerable sharp fine structure in the low-frequency region. Prolonged investigation has identified at least three sources for the fine structure:

1. Phase separation resulting in regions of crystalline material in the glass.
2. Instrument artifacts resulting from direct reflection of incident beam into spectrometer.
3. Unknown optical effects associated with use of small spherical specimens.

Because these effects have not yet been fully investigated, we limit the present paper to spectra without fine structure.

Composition of Glasses

Electron microprobe analyses of selected glass particles are listed in Table 1. Analyses for inhomogeneous glasses are shown as a range. The glasses, which are clearly separated into Ca-Al-rich regions and Fe-Mg-rich regions, are listed with an analysis of each region. Glasses with two or more phases are common, although only two phases are present in large quantity. Figure 1 shows the typical morphology of the biphasic glasses and the extent to which the two regions are interlocked. The

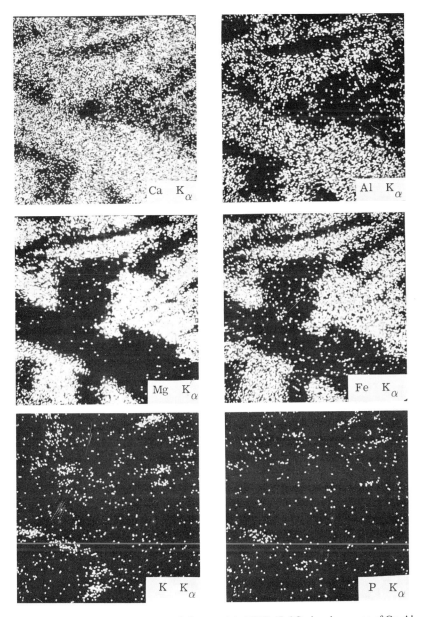

Fig. 1. Electron microprobe images of glass particle 14162,47–2C, showing zones of Ca–Al-
rich and Fe–Mg-rich glass. Field of view is 180 × 180 μ.

heterogeneity of these samples creates difficulties for both Raman and x-ray emission
spectroscopy because both utilize beams with about 100 μ spot sizes. The beam spans
the entire sample area to give an average spectrum, whereas two distinct compositions
and structures are present. The measured Al shifts for the biphasic glasses, however,

should be meaningful because the bulk of the Al contributing to the x-ray peak is contained in the Al-rich phase.

Many classifications of the parent rock type have now been proposed for the lunar fines. For the present work we have used the composition ranges of Reid *et al.* (1972). The separation into compositional populations by these workers is in general agreement with other analyses of lunar glasses (Glass, 1971; Prinz *et al.*, 1971; Apollo Soil Survey, 1971; Meyer *et al.*, 1971). The glasses selected for spectra measurement were assigned as derivatives of: (i) Mare basalts. (ii) Fra Mauro (KREEP) basalts, (iii) Highland (aluminum-rich basalts), and (iv) anorthosites. No particular attempt was made to separate out the several subtypes of Mare basalts because of the inhomogeneity of the specimens.

X-Ray Emission Spectra

All measurement on SiK_β, AlK_β, and OK_α are given in Table 2. All shifts are in units of 10^{-4} angstroms with respect to the listed standard. Table 3 lists several relevant minerals for comparison. Jadeite and spodumene are included only to provide an AlK_β shift for aluminum in 6-coordination in a pyroxene host.

Table 2. X-ray emission shift data for lunar glasses.

Sample	αSiO_2 Std. SiK_β $\Delta (A \times 10^{-4})$	αAl_2O_3 Std. AlK_β $\Delta (A \times 10^{-4})$	αSiO_2 Std. OK_α $\Delta (A \times 10^{-4})$
10084,110,1A (Green)	−15	+35	+262
10084,110,1B (Green)	−20	+16	+259
10084,110,2.8	−22	+37	+219
10084,110,2.9	−27	+36	+198
10084,110,III	−20	+23	+235
10084,110,3.1	−9	+35	+270
12070,42–1	−35	+35	+219
12070,42–2	−28	+36	+224
12070,42–3	−44	Insufficient Al	+235
12070,42–4	−25	+39	+256
12042,32–1	−26	+37	+222
12042,32–2	−28	+46	+187
12042,32–3	−34	+52	+200
12042,32–4	−29	+48	+184
14162,47–1	−39	+37	+227
14162,47–2B	−35	+36	+192
14162,47–2C	−25	+49	+182
14162,47–3	−44	+38	+206
14162,47–5	−40	+37	+171
14162,61–1	−42	+35	+208
14162,61–2	−38	+39	+214
14162,61–3	−49	+48	+206
14162,61–4	−37	+34	+214
14162,61–5	−31	+47	+216
15501,25	−39	+29	+240
15531,40	−52	No Al detected	+283
15231,49–1	−54	+21	+219
15231,49–2	−37	+35	+208
15231,49–3	−39	+30	+254
15015,26 (Rock)	−26	+49	+246
15017,9	−32	+34	+230
15245,56	−33	+37	+248

Table 3. X-ray emission shift data for reference minerals.

Sample	αSiO_2 Std. SiK_β Δ (A × 10^{-4})	αAl_2O_3 Std. AlK_β Δ (A × 10^{-4})	αSiO_2 Std. OK_α Δ (A × 10^{-4})
Anorthite (Cry.)	−41	+24	+203
Anorthite (Glass)	−48	+39	+192
Amelia Albite	−14	+63	+136
Orthoclase	−15	+62	+120
Olivine	−15	—	+219
Jadeite 1347	−26	+13	+112
Spodumene 1307	−4	+14	+136

The span of SiK_β shifts covers much of the range known for silicate minerals (White and Gibbs, 1967). The Highland and Fra Mauro basalt glasses have only a small spread in their shift values. The averages and corresponding Si–O distances (scaled from White and Gibbs's plot) are

	SiK_β	Range	Si–O(Å)
Highland basalt glass	–24.4	(–15:–39)	1.621
Fra Mauro basalt glass	–38.5	(–35:–49)	1.630

There is a good correlation between SlK_β and the Si–O distance, because all Si is in 4-coordination. There is some ambiguity between Al coordination and Al–O distances if only AlK_β shift data are available. In contrast, the SiK_β shifts of the Mare basalts vary widely between specimens. The range of shifts is from –9 to –52, which corresponds to a range of Si–O distances of 1.612 to 1.637. Because most of the Mare basalt glasses examined were very inhomogeneous, this result is not surprising. Two of the largest shift values were found in grains 12070,42–3 and 15531,40, which had extremely low aluminum concentrations. All glasses with anorthositic compositions occurred as second phases with lower aluminum glasses, and no interference-free SiK_β shift; were measured for these glasses.

The AlK_β shifts are plotted against aluminum concentration in Fig. 2. The Fra Mauro basalt glasses fall into a narrow range, whereas there is a large scatter for the other compositions. The anorthosite glasses have definitely larger shifts than the more Fe-Mg-rich glasses. The mean; are

Anorthosite basalt glass	43.4
Fra Mauro basalt glass	35.5
Mare basalt glass	35.2
Highland basalt glass	32.0

The ranges are shown by the scatter of the data in Fig. 2. According to the criteria proposed by White and Gibbs (1969), the aluminum in the anorthosite composition glasses should be mainly in fourfold coordination. Some of the Highland and Mare basalt composition glasses have low AlK_β shifts, implying aluminum mainly in 6-coordination, whereas the bulk of the compositions fall into a range that would imply mixed aluminum coordination.

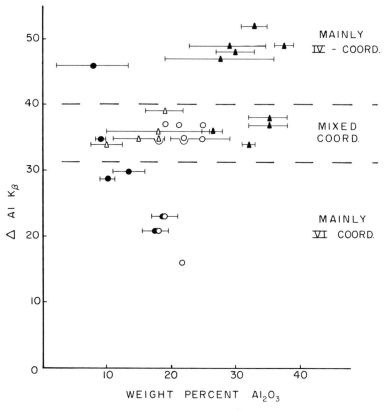

Fig. 2. X-ray emission shift for AlK_β (in units of $10^{-4}Å$) plotted against aluminum concentration. Solid circles: Mare basalt compositions; open circles: Highland basalt compositions; open triangles: Fra Mauro basalt compositions; solid triangles: anorthosite compositions.

Raman Spectra

Representative Raman spectra of the lunar glasses are shown in Figs. 3, 4, and 5. The spectra consist of a variable number of weak bands, mostly broader than expected from crystalline materials and narrower than most spectra of synthetic glasses. In general, the spectra vary considerably from grain to grain and also from different spots on the same grain, as might be expected from such heterogeneous samples. Although the overall spectra have a wide range, certain bands reappear consistently throughout the data. The spectra can be discussed in terms of four regions: 100 to 400 cm^{-1}, 400 to 800 cm^{-1} containing mainly bending motions of silicate structural units and stretching motions of octahedrally coordinated structural units, 800 to 1100 cm^{-1} containing stretching motions of SiO_4 tetrahedra, and the region above 1200 cm^{-1}.

The 800 to 1100 cm^{-1} region contains a commonly recurring band in the range of 1002 to 1006 cm^{-1}, a pair of bands at 819 to 823 and 850 to 853 cm^{-1}, and a few

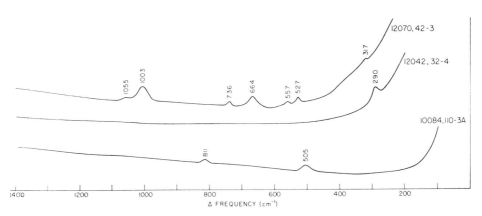

Fig. 3. Raman spectra of Apollo 11 and 12 glasses.

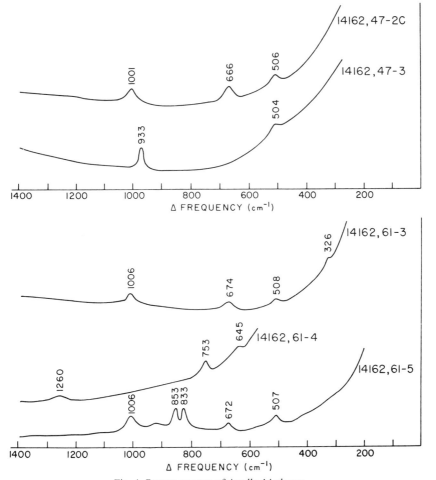

Fig. 4. Raman spectra of Apollo 14 glasses.

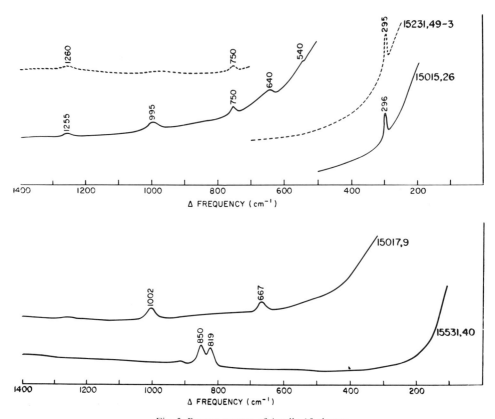

Fig. 5. Raman spectra of Apollo 15 glasses.

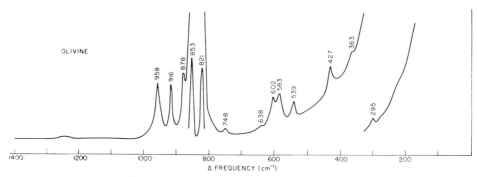

Fig. 6. Raman spectrum of polycrystalline olivine.

other less systematic bands. The 820 to 850 pair in particular seems unusually sharp for glass spectra. There is a good correlation between these bands and the stretching frequencies of olivine and pyroxene. Spectra of powder olivine and four pyroxenes are shown in Figs. 6 and 7. The strongest bands in the olivine spectrum are the two at 821 and 853. All pyroxene spectra contain a sharp band in the range of 1000 to

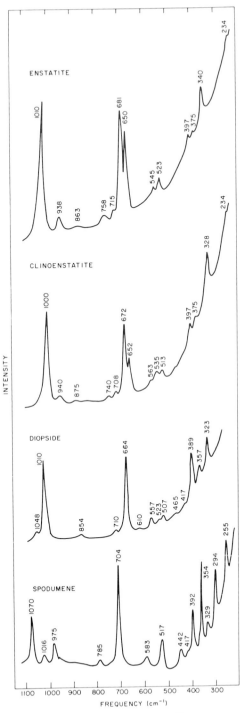

Fig. 7. Raman spectra of polycrystalline pyroxene.

1010 cm^{-1} in close agreement with the glass spectra. The second strongest band occurs as a doublet in the 650 to 670 cm^{-1} region for enstatite and clinoenstatite and as a single band at 664 in diopside. A band occurs in this frequency range for each specimen for which the 1002 to 1006 cm^{-1} band appears. However, these bands are broadened, and there is little evidence for doublet structure.

The other crystalline or quasi-crystalline phase that might contribute to the spectra is feldspar, particularly in those glasses with anorthosite composition zones. Because of the low alkali content of the lunar glasses, anorthite is the only feldspar likely to be present. Raman spectra for three end-members feldspars are shown in Fig. 8. The alkali feldspars have a strong band at 506 to 513 cm^{-1}, but the corresponding band in anorthite is relatively weak. The agreement between this frequency of the crystalline feldspars and the 506 to 508 band in the lunar glass spectra is rather good. The Si–O stretching frequencies of the feldspars are all weak and do not appear in the glass spectra.

The broad band at 1250 to 1260 cm^{-1} that appears in several of the spectra is difficult to interpret. It also appeared in the spectra of Apollo 11 glasses reported earlier (White *et al.*, 1971). This band is in the correct frequency range to be due to tetrahedral stretching vibrations but lies at a higher frequency than any Si–O vibra-

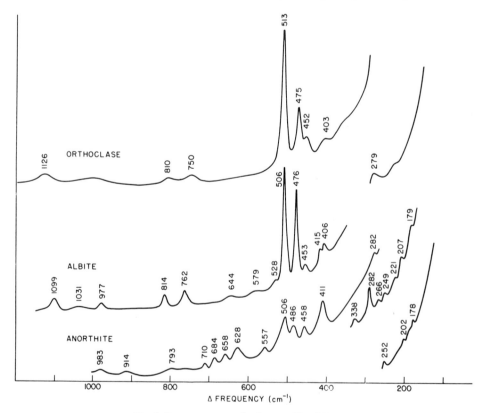

Fig. 8. Raman spectra of polycrystalline feldspars.

tions so far observed. When the Si–O stretching frequencies for olivine, pyroxene, and feldspar are compared, it is apparent that the Raman frequencies increase from 850 to 1000 to 1100 to cm^{-1} as the amount of corner linking of the tetrahedra increases. A similar effect in the infrared spectra of silicates has long been known. The highest-frequency mode in the feldspar structures with complete corner sharing is 1126 cm^{-1}, and the highest frequency in quartz is 1162 cm^{-1} (Scott and Porto, 1967).

The Raman spectra of the lunar glasses contain bands from the following sources:

1. Submicroscopic crystalline inclusions of olivine and pyroxene.
2. Regions of the glass in which there is incipient crystallization to one of the silicate mineral phases. These give rise to the strong bands characteristic of the mineral but with considerable band broadening.
3. Broad bands characteristic of the glass itself, most important of these in the 1260 cm^{-1} that may represent the Si–O stretching frequency of glass of higher density.

Acknowledgments—This work was supported under NASA Grant NGR–39–009(152) and through the Space Sciences and Engineering Laboratory of the Pennsylvania State University operated under NASA Grant NGL–39–009–015.

REFERENCES

Apollo Soil Survey (1971) Apollo 14: Nature and origin of rock types in soil from the Fra Mauro formation. *Earth Planet. Sci. Lett.* **12**, 49–54.

Gigl P. D. Savanick G. A. and White E. W. (1970) Characterization of corrosion layers on aluminum by shifts in the aluminum and oxygen x-ray emission bands. *J. Electrochem. Soc.* **117**, 15–17.

Glass B. P. (1971) Investigation of glass recovered from Apollo sample 12057. *J. Geophys. Res.* **76**, 5649–5657.

Meyer C. Jr. Brett R. Hubbard N. J. Morrison D. A. McKay D. S. Aiken F. K. Takeda A. and Schonfeld E. (1971). Mineralogy, chemistry and origin of the KREEP component in soil samples from the Ocean of Storms. *Proc. Second Lunar Sci. Conf.*, Geochim. Cosmochim. Acta Suppl. 2, Vol. 1, pp. 393–411. MIT Press.

Prinz M. Bunch E. E. and Keil K. (1971) Composition and origin of lithic fragments and glasses in Apollo 11 samples. *Contrib. Mineral. Petrol.* **32**, 211–230.

Reid A. M. Ridley W. I. Warner J. Harmon R. S. Brett R. Jakeš P. and Brown R. (1972) Chemistry of Highland and Mare basalts as inferred from glasses in the lunar soils (abstract). In *Lunar Science—III* (editor C. Watkins), p. 640, Lunar Science Institute Contr. No. 88.

Scott J. F. and Porto S. P. S. (1967) Longitudinal and transverse optical lattice vibrations in quartz. *Phys. Rev.* **161**, 903–910.

White E. W. and Gibbs G. V. (1967) Structural and chemical effects on the SiK_β x-ray line for silicates. *Amer. Mineral.* **52**, 985–993.

White E. W. and Gibbs G. V. (1969) Structural and chemical effects on the AlK_β x-ray emission band among aluminum-containing silicates and aluminum oxides. *Amer. Mineral.* **54**, 931–936.

White W. B. White E. W. Görz H. Henisch H. K. Fabel G. W. Roy R. and Weber J. N. (1971) Physical characterization of lunar glasses and fines. *Proc. Second Lunar Sci. Conf.*, Geochim. Cosmochim. Acta Suppl. 2, Vol. 3, pp. 2213–2221. MIT Press.

Proceedings of the Third Lunar Science Conference
(Supplement 3, *Geochimica et Cosmochimica Acta*)
Vol. 1, pp. 953–970
The M.I.T. Press, 1972

Metallic mounds produced by reduction of material of simulated lunar composition and implications on the origin of metallic mounds on lunar glasses*

James L. Carter

University of Texas at Dallas, Geosciences Division, Division of Earth and
Planetary Sciences, P.O. Box 30365, Dallas, Texas 75230

David S. McKay

National Aeronautics and Space Administration, Manned Spacecraft Center,
Houston, Texas 77058

Abstract—Silicate glass of composition similar to brown lunar glass was reduced with carbon and hydrogen. The typical complex iron sulfide and metallic iron mound formed by reduction with carbon is zoned. Its interior is metallic iron or a mixture of iron sulfide and metallic iron. The outer layer is pure metallic iron which is generally discontinuous, but the surface of the mound next to the silicate host is iron sulfide. This type of mound commonly has a waist of metallic iron and a void beneath it. The silicate surface of the void is covered with droplets or stringers of iron sulfide.

The complex iron sulfide and metallic iron mounds formed by reduction with hydrogen generally are zoned. The outer layer is iron sulfide or a mixture of iron sulfide and metallic iron, but the interior is metallic iron and iron sulfide, and the mound material next to the silicate host is iron sulfide. On the surface of some complex mounds, globules of silicate material are present. In one example, the globules consist of particles of aluminum oxide surrounded by silicate material. In turn, the margin of the globules is surrounded by iron sulfide. Dimples are present and the surface of the dimples is covered by dendritic sheaths of iron sulfide and isolated metallic iron globules.

These data suggest that: (1) the ratio of sulfur to iron has direct bearing on the morphology of metallic mounds; (2) the growth time also influences the nature of metallic mounds; and (3) mounds produced by reduction with carbon are different from mounds produced by reduction with hydrogen.

The laboratory data suggest that most metallic mounds on lunar glasses did not form by reducing processes *in situ*. However, during the melting of lunar soil, for example during a meteoroid impact event, hydrogen and to a lesser extent carbon, as a result of trapped solar winds, may play at least a secondary role in the formation of metallic iron, and may be responsible in a large part for the formation of mounds that are low in nickel which commonly occur in trains or patterns on the glass-bonded agglutinates. In addition, it is inferred from these data that the iron mounds rich in nickel, cobalt, sulfur, and phosphorus may be remobilized components of meteorites and probably formed in the impact-generated debris cloud.

Introduction

Several distinct morphological and chemical types of metallic mounds occur on lunar glass spheres and irregularly shaped glasses: (1) FeS (McKay *et al.*, 1970; Carter and McGregor, 1970); (2) metallic Fe (Ramdohr and El Goresy, 1970; Agrell *et al.*, 1970; Carter and MacGregor, 1970); (3) mixtures of metallic Fe and FeS (McKay *et al.*, 1970; Carter and McGregor, 1970); (4) metallic Fe and Ni (Agrell *et al.*, 1970; Carter and MacGregor, 1970; Frondel *et al.*, 1970; McKay *et al.*, 1970); (5) mixtures of Fe, Ni, and P (Carter and MacGregor, 1970; Goldstein *et al.*, 1970;

---

* Contribution No. 210, Geosciences Division, University of Texas at Dallas, P.O. Box 30365, Dallas, Texas 75230.

Carter, 1971); (6) mixtures of Fe, Ni, P, S, and C (Carter and MacGregor, 1970); (7) Fe and Ni with waists of FeS (Duke *et al.*, 1970; Agrell *et al.*, 1970; Frondel *et al.*, 1970; Goldstein *et al.*, 1970; Carter, 1971). In addition, as many as five magnetic phases may be present within glass spheres (Adler *et al.*, 1970; Agrell *et al.*, 1970; Duke *et al.*, 1970; McKay *et al.*, 1970; Ramdohr and El Goresy, 1970; Simpson and Bowie, 1970; Winchell and Skinner, 1970; Griscom and Marquardt, 1972; Wosinski *et al.*, 1972).

The iron sulfide mounds occur mainly on glass-bonded agglutinates and dust-welded glass spheres (McKay *et al.*, 1970; Carter and MacGregor, 1970). Trains of metallic iron mounds occur on glass-bonded agglutinates and irregularly shaped glassy objects (Ramdohr and El Goresy, 1970; Carter and MacGregor, 1970; Carter, 1971). The complex masses of iron sulfide and metallic iron, or iron sulfide and metallic iron-nickel occur as distinct mounds on glass spheres and irregularly shaped glassy objects (Agrell *et al.*, 1970; Carter and MacGregor, 1970; McKay *et al.*, 1970). Some metallic iron-nickel mounds are surrounded by waists of iron sulfide. The iron, nickel, and phosphorus mounds occur as individual mounds, groups of mounds or complex amoeboid-shaped mounds (Carter and MacGregor, 1970; Goldstein *et al.*, 1970; Carter, 1971). The iron, nickel, phosphorus, sulfur, and carbon mounds occur as complex masses on irregularly shaped glass fragments (Carter and MacGregor, 1970).

The origin of the various types of mounds on silicate surfaces has not been clearly established. There are at least three possible origins for the mounds: (1) reduction of silicate *in situ*; (2) splashes; and (3) vapor deposition and subsequent growth of mounds. The source of the material forming mounds of the latter two categories may be (1) reduction of iron-bearing material (both lunar and meteoritic), (2) remelted components of meteoritic material, or (3) remelted components of lunar material.

In order to better understand the possible origins of the various types of mounds, a survey project was initiated at NASA/MSC, Houston, to test the possibility of producing complex metallic iron and iron sulfide mounds similar to those observed on the surfaces of lunar glass particles by reduction processes *in situ*. A glass of composition similar to dark brown lunar glass was made from reagent grade chemicals (Table 1). The glass was ground in a tungsten-carbide mixer-mill for five minutes. Various amounts of elemental sulfur were added to aliquots of this homogenized ground glass.

The preliminary experiment involved placing approximately one gram of ground glass with various amounts of sulfur in carbon crucibles, heating them in a glo-bar furnace with an argon atmosphere at 1450°C for five minutes, and quenching the totally liquid silicate globule in air to form a glass spheroid approximately 9 mm by 7 mm. There was approximately 60 mm² of liquid silicate surface in contact with the carbon crucible. Other samples were placed in alumina boats in a glo-bar furnace for three minutes at 1450°C and flushed with argon; subsequently, a gas consisting of 5% hydrogen and 95% argon at one atmosphere was flowed over the samples for two minutes. One sample was reduced with the hydrogen mixture for fifteen minutes. The surface of the liquid silicate available for reduction by hydrogen was approximately four times greater than in the carbon experiment.

Table 1. Chemical composition in weight percent
of simulated brown lunar glass*

SiO$_2$	40.37
Al$_2$O$_3$	14.49
Cr$_2$O$_3$†	0.3
TiO$_2$	7.50
FeO	15.36
MgO	8.13
MnO†	0.2
CaO	13.25
Na$_2$O	0.59
K$_2$O	0.03
P$_2$O$_5$	0.03
S	0.05
TOTAL	100.30

* X-ray fluorescence analysis by J. M. Rhodes.
† Added to glass but not analyzed for by x-ray
fluorescence.

EXPERIMENTAL RESULTS

Results of scanning electron microscope and electron microprobe examinations are shown in Table 2. A JEOLCO JSM-1 scanning electron microscope and an ARL EMX-SM electron microprobe was used in the examinations. A wide variety of mounds and dimples or voids were observed on the glass surface (Table 2; Fig. 1). In addition, metallic iron globules up to 4 μ in diameter occur to a depth of approximately 300 μ into both the glass spheroid and the glass produced in the hydrogen reduction experiment. In this 300 μ zone of reduction, the iron content of the glass in the carbon-reduced samples is 9 % to 11 % less than in the starting material, and in the hydrogen-reduced samples it is 11 % to 12 % less than in the starting material.

Mounds

The sulfur content of the glass influences the composition and morphology of the mound. In general, mounds formed from samples rich in sulfur contain more iron sulfide and are larger than mounds formed from low sulfur samples (Table 2).

Reduction with carbon. When the glass with no sulfur added was reduced by contact with carbon, simple iron mounds were formed which display either a finely convoluted surface texture (Fig. 2) or, in some cases, a pattern of triangular-shaped ridges (Figs. 3 and 4). The finely convoluted structure may be iron oxide formed during quenching of the liquid silicate spheroid in air. A similar convoluted structure was seen on some lunar metallic mounds (Carter and MacGregor, 1970).

More complex zoned metallic iron and iron sulfide mounds are formed when sulfur was added to the glass (Table 2; Figs. 1, 5 to 12). The interior zone or core of the mound is predominantly metallic iron (Fig. 5) or a mixture of metallic iron and iron sulfide (Figs. 6 and 7). The outer zone consists of a discontinuous coating of metallic iron on the free surface (Figs. 5, 6, 8, 9, 11, and 12) and a waist of metallic iron surrounds the mound at the contact of the free surface of the mound with the silicate glass host in some specimens (Figs. 8 to 10). The underside of the mound in contact with glass lacks the metallic iron coating and is iron sulfide (Table 2; Figs. 5

Table 2. Description of mounds

Reducing agent	Wt. % S	Description of mounds
C	0.0	Numerous individual metallic iron mounds occur up to 50 μ in diameter, which are surrounded by smaller metallic iron mounds.
H$_2$	0.0	Mounds occur as irregular stringers or web-like metallic iron objects with crinkled surfaces. Others are flat, porous, fan-shaped metallic iron objects up to 20 μ in longest dimension. A fifteen-minute run resulted in a massive network of connected circular metallic iron mounds with crinkled surfaces.
C	0.24	Numerous individual mounds occur up to 100 μ in diameter. Some of the larger mounds are surrounded by masses of coalesced small metallic iron mounds. Some mounds are complex mixtures of iron sulfide and metallic iron; others are porous, dendritic, iron sulfide. Dimples with an inner dimple are common. The silicate surface of the inner dimple is covered with iron sulfide mounds.
H$_2$	0.24	Mounds occur as trains of connected metallic iron octahedra. Irregularly shaped, amoeboid-like, complex metallic iron, and iron sulfide mounds also occur.
C	0.49	Numerous mounds occur up to 150 μ in diameter. Some of the larger mounds are surrounded by masses of coalesced small metallic iron mounds. The mounds consist of metallic iron that is surrounded by a mixture of iron sulfide and metallic iron. Metallic iron margins are common. Dimples with an inner dimple are common. The silicate surface of the inner dimple is covered with mounds and stringers of iron sulfide.
H$_2$	0.49	Individual complex iron sulfide and metallic iron mounds occur up to 200 μ in diameter. Trains of connecting circular metallic iron mounds (2 to 15 μ in diameter) with rough surfaces are present also. The larger mounds in the trains have six-sided, flat-topped metallic iron objects on their surface. The metallic iron trains grade into areas of dendritic metallic iron mounds. Amoeboid-shape complex iron sulfide and metallic iron mounds up to 300 μ in longest dimension rarely are seen. Shrinkage cracks around their margins are poorly developed. Dimples are common. Some dimples have isolated circular metallic iron mounds and irregular dendritic areas of iron sulfide on their surface. One to 5 μ diameter spherules occur on the surface of some of the amoeboid-shape mounds. The spherules consist of a particle that appears to be aluminum oxide, which is surrounded by silicate material, and in turn the margin of the spherule is surrounded by iron sulfide.
C	1.0	Numerous complex individual mounds of metallic iron and iron sulfide occur up to 200 μ in diameter and are sometimes surrounded by masses of coalesced small metallic iron mounds. Individual mounds have metallic iron margins. Dimples with an inner dimple are common. The silicate surface of the inner dimple is covered with iron sulfide mounds.
H$_2$	1.0	Individual mounds occur up to 300 μ in diameter and are complex mixtures of dendritic iron sulfide and metallic iron. Some mounds 10 to 50 μ in diameter have a patchy layer of metallic iron over a core consisting of a mixture of iron sulfide and metallic iron. Some mounds have a thin waist of iron sulfide. The larger mounds have well-developed cooling cracks around their margins. Some larger mounds contain spherules of silicate material. Dimples are common. The upper margin of the dimple is textured and has dendritic areas of iron sulfide on its surface. Individual octahedra of metallic iron approximately 5 μ in diameter occur.

and 6). The presence of tungsten (approximately ≤ 56 wt. %) and cobalt (approximately ≤ 6 wt. %) in the mounds (Fig. 5) shows that the molten iron acts as a scavenger for these elements even though they occur in small amounts in the silicate host (W, 0.08 wt. %; Co, 0.017 wt. %, as determined by electron microprobe techniques). Lunar metallic iron contains relative high concentrations of tungsten (Wänke *et al.*, 1970).

Reduction with hydrogen. When the glass with no sulfur added is reduced by hydrogen, metallic iron is formed (Table 2; Figs. 13 to 16). The metallic iron may occur as a massive coating (Fig. 13), as isolated spongy masses (Fig. 14), or more typically as interconnected amoeboid-like stringers and mounds (Figs. 15 and 16). These

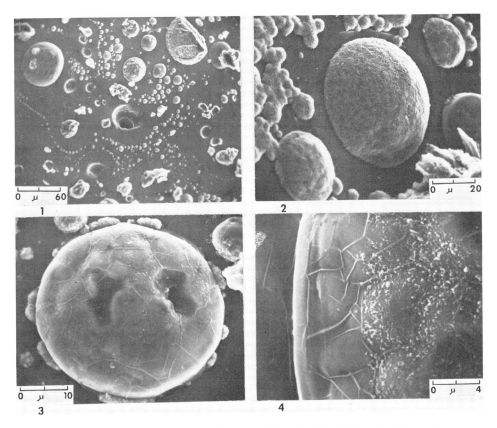

Fig. 1. View of part of a silicate surface (containing 0.23% sulfur) reduced by carbon showing dimples, individual mounds, and trains of mounds. The large dimple has an inner dimple (center of photograph) which was formed by trapped gas.

Fig. 2. Dominant type of metallic iron mound formed by carbon reduction on silicate host that contained 0.06% sulfur. The finely convoluted surface of the mounds is similar to certain mound surfaces observed in Apollo 11 material.

Fig. 3. Typical metallic iron mound on silicate surface (no sulfur present) that was reduced by carbon. Note pattern of interconnected triangular ridges on surface of mound.

Fig. 4. Enlarged view of interconnected triangular ridges on the surface of a metallic iron mound seen on a carbon-reduced silicate sphere that contained 0.06% sulfur.

mounds are usually zoned and contain a core of massive iron surrounded by a thin coating of convoluted iron (Fig. 16). The convoluted surface is similar to that observed with carbon reduction (Fig. 12) and may be iron oxide formed during quenching of the liquid silicate in air.

When sulfur is added to the glass, more complex zoned mounds are formed by hydrogen reduction (Table 2; Figs. 17 to 24). The interior zone may be either metallic iron or a mixture of metallic iron and iron sulfide (Table 2; Fig. 17). Mixed iron sulfide cores are more common in mounds formed from glasses with higher sulfur contents (Table 2). This core is covered by a mixture of iron sulfide and metallic iron on the free

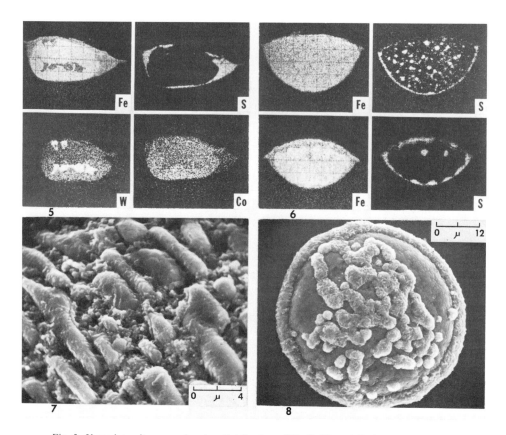

Fig. 5. X-ray intensity maps showing distribution of Fe, S, W, and Co in cross section of mound produced by carbon reduction of silicate containing 0.49% S. Scale is 15 μ per division.

Fig. 6. X-ray intensity maps showing distribution of Fe and S in cross section of two mounds produced by carbon reduction of silicate containing 0.49% S. Scale in upper section is 15 μ per division and 10 μ per division in the lower section.

Fig. 7. Upside-down enlarged view of largest mound in upper right portion of Fig. 1. Large metallic iron masses are seen in a fine-grained iron sulfide matrix.

Fig. 8. Complex metallic mound on silicate surface (containing 0.16% sulfur) that was reduced with carbon. The waist is metallic iron, while the central mass is iron sulfide that is blanketed by a mixture of iron sulfide and metallic iron.

surface (Figs. 17 to 19) and by pure iron sulfide in the area next to the glass host (Table 2; Fig. 17). In some cases, a thin waist of iron sulfide is present at the contact between the free surface of the mounds and the glass host (Figs. 20 to 22).

In addition to isolated well-developed mounds (Figs. 20 and 21), interconnecting and partially coalesced amoeboid-like masses may develop (Figs. 23 to 26). Some of the individual mounds and coalesced mounds display a poorly developed octahedral form (Fig. 26). On Fig. 26, tiny isolated mounds are reduced in size and abundance

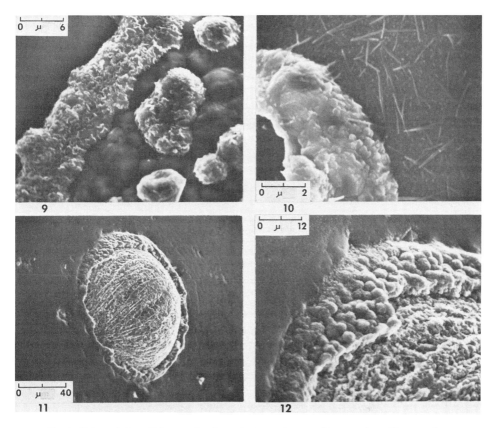

Fig. 9. Enlarged view of the margin of a typical complex metallic mound on silicate surface (containing 0.49% sulfur) that was reduced with carbon. Note the finely crinkled surface on the metallic iron margin.

Fig. 10. Enlarged view of a portion of the carbon reduced silicate surface (containing 1% sulfur) adjacent to a metallic iron mound. Needles of metallic iron and spherules of metallic iron (0.1 μ in diameter) formed on the silicate surface.

Fig. 11. Complex mound on silicate surface (containing 0.24% sulfur) that was reduced with carbon. The outer metallic iron and iron sulfide layer was partially stripped away during quenching, revealing an interior of dendritic iron.

Fig. 12. Enlarged view of upper left margin of mound in Fig. 11 showing details of the outer metallic iron and iron sulfide layer and the dendritic iron core. The outer layer consists of interlocking spherules in contrast to the dendritic structure of the iron core.

in the vicinity of larger partially coalesced mounds, suggesting that the larger mounds were being supplied with iron and sulfur from the surrounding area during growth.

Silicate spherules

Some complex mounds contain silicate spherules within the mounds and protruding from their surfaces (Table 2).

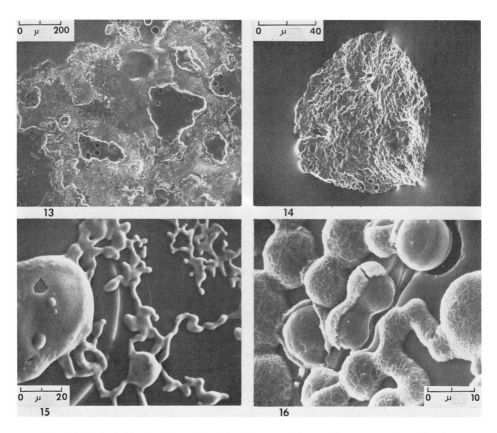

Fig. 13. Typical silicate surface reduced by hydrogen with no sulfur present; the surface is massive metallic iron.

Fig. 14. Isolated fan-shaped area of spongy metallic iron on silicate surface reduced by hydrogen (no sulfur present). Spherules of silicate material are present in the lower right portion of the fan-shaped object.

Fig. 15. Interconnected amoeboid-like masses of metallic iron on silicate surface (no sulfur present) that was reduced by hydrogen. Note globules of silicate material on surface of the largest iron mass in the left center of the photograph.

Fig. 16. Enlarged view of interconnected metallic iron spherules formed on silicate surface (no sulfur present) that was reduced by hydrogen for fifteen minutes. Note layered nature of the mounds and the finely convoluted surface on the outer layer.

Reduction with carbon. A very small fraction of the mounds contain silicate spherules protruding from their surfaces (Figs. 27 and 28). This glass is of similar composition to the host glass and it may be simply silicate material trapped in the mound during formation and growth. No mound on lunar samples has been observed with spherules of silicate material on its surface.

Reduction with hydrogen. Many of the larger amoeboid-like masses produced when larger amounts of sulfur are present contain silicate spherules on their surfaces (Table 2; Figs. 14, 15, 23, 24, 29, and 30). The globules on the surfaces of metallic

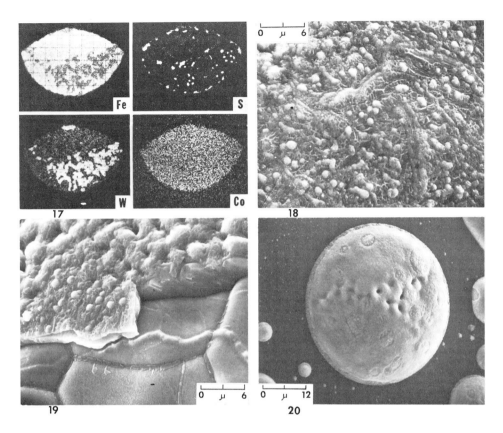

Fig. 17. X-ray intensity maps showing distribution of Fe, S, W, and Co in cross section of mound produced by hydrogen reduction of silicate containing 0.49% S. Grid is 20 μ per division.

Fig. 18. Enlarged view of the surface of a complex metallic mound formed on a silicate surface (containing 1% sulfur) that was reduced by hydrogen. Globules of iron sulfide are embedded in the finely convoluted surface of the same composition.

Fig. 19. Enlarged view of the surface of a complex metallic mound formed on a silicate surface (containing 1% sulfur) that was reduced by hydrogen. The iron sulfide skin was broken during quenching revealing an interior of interlocking metallic iron crystals.

Fig. 20. Isolated mound with waist of iron sulfide on silicate surface (containing 1% sulfur) that was reduced by hydrogen. The textured circular patches are iron sulfide suggesting that the mound has an incomplete coating of metallic iron.

mounds shown in Figs. 14 and 15 are silicate in composition and, like those seen on metallic mounds produced by reduction with carbon, appear to be simply silicate material trapped in the mounds during formation and growth. However, the silicate globules on the central massive metallic body in Fig. 23 apparently are different. An enlarged view of a silicate globule is shown in Fig. 29. The spherules less than 0.5 μ in diameter on the surface of the silicate globule are metallic iron. The composition of the silicate globules is similar to the host silicate, but there are areas that are

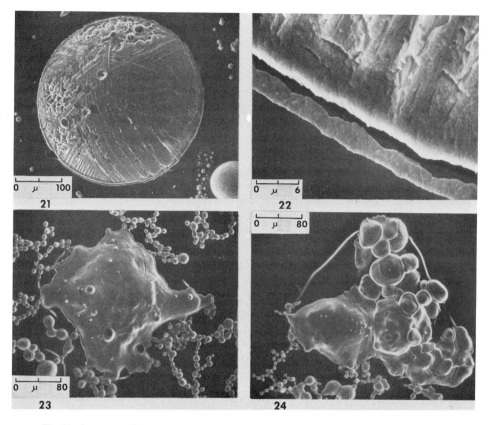

Fig. 21. Large complex mound on silicate surface (containing 1% sulfur) that was reduced by hydrogen. Iron sulfide skin covers dendritic sheaths of metallic iron. Dark globules of silicate material can be seen on surface of the mound.

Fig. 22. Enlarged view of the lower left portion of Fig. 21 showing a waist of iron sulfide. Note the well-developed cooling crack around the margin of the mound.

Fig. 23. Complex amoeboid-like mass of metallic iron and iron sulfide on silicate surface (containing 0.49% sulfur) that was reduced by hydrogen. The globules on the amoeboid-like mass are composed of silicate material. Note the trains of interconnected metallic iron mounds.

Fig. 24. Complex mass of metallic iron and iron sulfide mounds on silicate surface (containing 0.49% sulfur) that was reduced by hydrogen. Silicate material forms globules on the large center left mass.

enriched in alumina suggesting that the silicate spherules are not simply incorporated molten droplets of the silicate host, but may be condensates. This latter view is supported by the nature of the silicate droplets shown in Fig. 24 and enlarged in Fig. 30. The globules that are 1 to 5 μ in diameter are silicate in composition. The globules consist of a particle of what appears to be aluminum oxide surrounded by silicate material. In turn, the margin of the silicate globules is surrounded by iron sulfide.

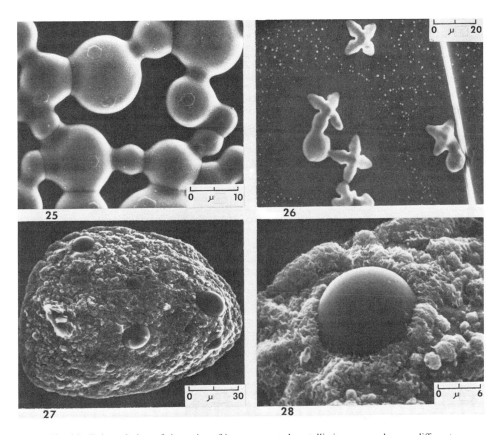

Fig. 25. Enlarged view of the trains of interconnected metallic iron mounds on a different area of the surface shown in Fig. 23. Note the pseudohexagonal platelets of metallic iron on the spherules; some platelets are twinned.

Fig. 26. Metallic iron octahedra on silicate surface (containing 1% sulfur) that was reduced by hydrogen. Note the small size and lower concentration of metallic iron mounds near the octahedra.

Fig. 27. Rare metallic iron mound containing spherules of silicate material formed during carbon reduction on silicate surface that contained 0.16% sulfur.

Fig. 28. Enlarged view of the upper left portion of Fig. 27 showing silicate spherule surrounded by metallic iron.

Dimples and voids

Some of the mounds were lost from the glass surface during quenching, leaving depressions or dimples (Table 2; Figs. 1, 31 to 34, and 37 to 40).

Reduction with carbon. The dimples of carbon-reduced samples sometimes contain an interior depression or void formed before the mound was removed (Figs. 31 and 32). This void probably was gas filled. The material formed on the surface of the void is metallic iron if no sulfur has been added to the glass (Fig. 31), whereas very small iron sulfide mounds commonly are present on the surface of these voids (Fig. 32) and

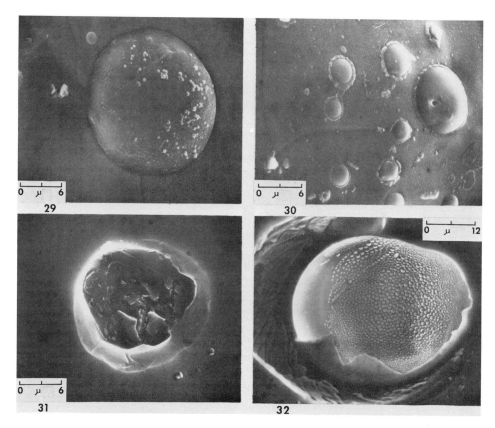

Fig. 29. Enlarged view of the silicate globule in center left of the amoeboid-like mass shown in Fig. 23; metallic iron spherules can be seen on its surface.

Fig. 30. Enlarged view of the left center portion of Fig. 24 showing silicate spherules that are surrounded by waists of iron sulfide. Note the triangular-shaped inclusion of apparently aluminum oxide in the largest silicate globule.

Fig. 31. Dimple in a silicate surface (no sulfur present) that was reduced by carbon. Note that the surface of the inner dimple contains blades of metallic iron.

Fig. 32. Dimple in a silicate surface (containing 0.24% sulfur) that was reduced by carbon. The surface of the dimple is covered with sheaths of iron sulfide, while the inner dimple is covered with globules of iron sulfide that are arranged in geometrical patterns.

on the surface of the silicate host which was in contact with the mound (Figs. 32 to 34) if sulfur has been added to the glass. Very similar dimples containing central voids are seen on some lunar glass surfaces (Fig. 35; also Carter and MacGregor, 1970, Figs. 28 and 33). However, they generally do not contain deposits on their surfaces (Fig. 36; also Carter and MacGregor, 1970).

Reduction with hydrogen. As with carbon reduction, dimples are formed when mounds are lost from the cooling silicate glass host. In both cases the loss of mounds is enhanced by differential contraction of the mound and the silicate host during

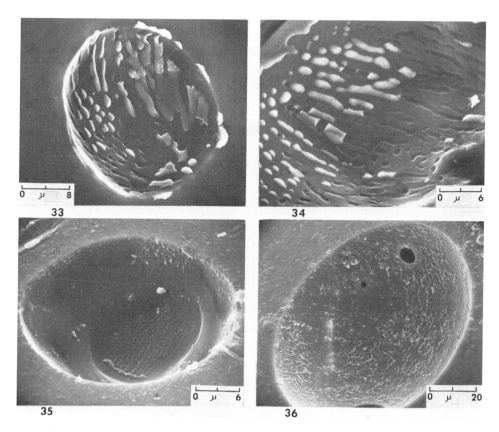

Fig. 33. Dimple in a silicate surface (containing 0.16% sulfur) that was reduced by carbon. Note that the surface of the dimple is covered with isolated globules and sheaths of iron sulfide.

Fig. 34. Enlarged view of the surface of a dimple seen in a silicate surface (containing 0.49% sulfur) that was reduced by carbon. The surface of the dimple is covered with isolated globules and sheaths of iron sulfide. Note the pattern on the dimple surface left by the loss of the iron sulfide.

Fig. 35. Dimple in an Apollo 11 brown silicate glass sphere. Note pattern of iron sulfide spherules on surface of inner dimple.

Fig. 36. Dimple in another Apollo 11 brown silicate glass sphere. Note pattern on surface of dimple.

cooling. This contraction may form a circumferential cooling crack (Figs. 21 and 22). Similar cooling cracks are present in lunar samples (Carter and MacGregor, 1970; McKay *et al.*, 1970; Carter, 1971).

Dimples formed by the loss of mounds on hydrogen-reduced glass display attached fragments of the dendritic iron sulfide mound and isolated globules of metallic iron (Figs. 37 to 40). Similar dimples are present on lunar glass spherules (Figs. 35 and 36; also Carter and MacGregor, 1970, Figs. 28, 31, 33 to 35, and 40).

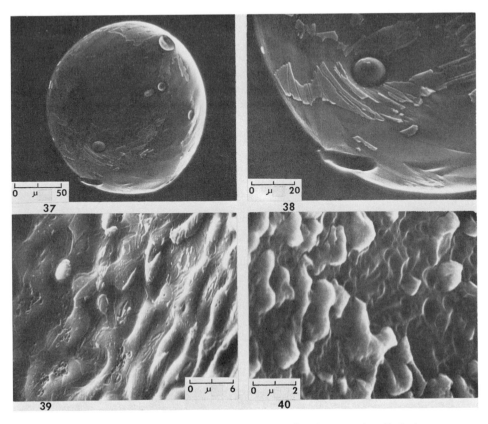

Fig. 37. Dimple in a silicate surface (containing 0.49 % sulfur) that was reduced by hydrogen. Globules on surface of dimple are metallic iron. Fragments of the dendritic iron sulfide mound are attached to the dimple surface.

Fig. 38. Enlarged view of the lower left portion of the dimple surface in Fig. 37. Note the metallic iron globule and the attached fragments of the dendritic iron sulfide mound.

Fig. 39. Enlarged view of a dimple surface in a silicate surface (containing 0.49 % sulfur) that was reduced by hydrogen showing a textured surface.

Fig. 40. Enlarged view of another dimple surface on silicate surface (containing 0.49 % sulfur) that was reduced by hydrogen. Note adhering masses of iron sulfide and impressions left by loss of iron sulfide.

Dɪsᴄᴜssɪᴏɴ ᴀɴᴅ Cᴏɴᴄʟᴜsɪᴏɴs

These data suggest that: (1) the ratio of sulfur to iron has a direct bearing on the morphology of metallic mounds; (2) the growth time also influences the nature of metallic mounds; and (3) mounds produced by reduction with carbon are different from mounds produced by reduction with hydrogen.

The typical mound formed by *in situ* reduction of sulfur-bearing silicate by contact with carbon is a mass of metallic iron or a mixture of metallic iron and iron sulfide with an outer layer of metallic iron that commonly is discontinuous (Table 2; Figs. 5 to 9, 11, and 12). The mound commonly has a waist of metallic iron (Figs. 8 to 10). No complex metallic mounds with waists of metallic iron have been recognized on lunar glasses.

The typical isolated mound formed by *in situ* reduction of silicate material containing sulfur with hydrogen is a complex mixture of iron sulfide and metallic iron (Table 2; Fig. 17). The outer layer is a complex mixture of iron sulfide and metallic iron (Figs. 17 to 21), and the mound itself commonly is encircled with a thin waist of iron sulfide (Figs. 20 to 22). Waists of iron sulfide around metallic iron mounds rich in nickel have been reported on lunar glasses (Duke *et al.*, 1970; Agrell *et al.*, 1970; Frondel *et al.*, 1970; Goldstein *et al.*, 1970; Carter, 1971). Reduction with hydrogen of glass with no sulfur added produces metallic iron. This metallic iron may occur as a massive coating (Fig. 13), as isolated spongy masses (Fig. 14) or more typically as interconnected amoeboid-like stringers and mounds (Figs. 15 and 16). Interconnected amoeboid-like stringers and mounds are present on lunar glasses (Carter and MacGregor, 1970; McKay *et al.*, 1970; Carter, 1971).

The nature of the interface between the metallic mound and the silicate host is important to the understanding of the origin of a mound. Globules of iron sulfide characteristically cover the surfaces of dimples formed by reduction with carbon *in situ*, if sulfur is present in the silicate (Figs. 31 to 34). In contrast, the surfaces of dimples formed by reduction with hydrogen have imprints and fragments of dendritic iron sulfide sheaths and isolated metallic iron globules that comprised the mound (Figs. 37 to 40). A comparison of these dimples with the dimples seen on lunar glasses (Figs. 35 and 36; also Carter and MacGregor, 1970, Figs. 33 to 35; Carter, 1971, Figs. 7, 14 to 16, and 29) suggests that carbon was not a significant contributor to the formation of lunar metallic mounds, with the possible exception of the mound that left the imprint shown by Fig. 35, but that hydrogen may have contributed to the formation of lunar metallic mounds. The surface of the inner dimple in Fig. 35 has droplets of iron sulfide on it. The surface of all other inner dimples observed on lunar materials is smooth (Carter and MacGregor, 1970) suggesting that the voids under the mounds did not contain sulfur vapor.

The wide variation in mound morphology produced by reduction processes *in situ* (which includes the amoeboid-like mounds formed by reduction with hydrogen) suggests that some of the morphological types of mounds seen on lunar glasses could have formed by reduction processes *in situ* (e.g., those seen on the irregularly shaped glassy objects, Carter, 1971, Figs. 27 to 31). The most likely origin for the mounds that are rich in iron, nickel, cobalt, sulfur, and phosphorus is that they are remobilized components of meteorites (Goldstein and Yakowitz, 1971). The mounds that are rich in iron, nickel, cobalt, sulfur, phosphorus, and carbon probably also are remobilized components of meteorites with some possible addition of solar wind carbon (Holland *et al.*, 1972; Pillinger *et al.*, 1972). However, of interest is the observation that the metallic iron formed in the reduction experiments acts as a scavenger incorporating the tungsten and cobalt from the tungsten-carbide mixer-mill (Figs. 5 and 17). This suggests that other siderophile elements may be incorporated in the iron metal during reduction of silicate materials containing those elements. Thus alloyed iron metal may form by *in situ* reduction processes on the lunar surface.

From Figs. 5, 6, and 17, it is seen that isolated mounds formed by *in situ* reduction on the surface of molten silicate have unequal radii of curvature. The surface in contact with the silicate melt has the shortest radius of curvature. Lunar metallic globules with unequal radii of curvatures are shown by Frondel *et al.* (1970, Fig. 12) and Mason *et al.* (1970, Figs. 2 and 3), and as suggested by Frondel *et al.* (1970) and McKay *et al.* (1971) they may have solidified as droplets on cooling silicate surfaces.

From these data it would appear that most metallic mounds on lunar glass spheres did not form by reduction processes *in situ* with the possible exception of the glass-bonded agglutinates, but formed as splashed, condensates or some origin not specified. Glass-bonded agglutinates are formed when impact-produced silicate liquid penetrates lunar soil. These agglutinates are typically vesicular and the release of implanted solar wind gases (primarily hydrogen and helium) from the heated soil particles may be the main vesicle-forming mechanisms (McKay and Ladle, 1971). This solar wind hydrogen, and to a lesser extent solar wind carbon, may play a major role in the formation of mounds low in nickel which often occur in trains or patterns on the glass-bonded agglutinates. Reduction processes, at least in part, may be responsible for the metallic iron present on the surface of some vesicles (Carter, 1971).

In addition to reduction by hydrogen and carbon, lunar samples heated by impact would be expected to show some reduction caused by loss of molecular oxygen at high temperatures. Gibson and Johnson (1971) and de Maria (1971) show evolution of molecular oxygen at temperatures of about 1400°C when lunar samples are heated in vacuum. However, Gibson and Johnson (1971) ascribed this oxygen evolution to the solution of iron into the platinum crucible. Gibson and Moore (1972) reported that oxygen does not evolve appreciably at temperatures up to 1400°C if an alumina crucible is used. At some higher temperature, however, metallic iron would be expected to form by the release of molecular oxygen from molten glass and iron globules within the glass and iron mounds on glass surfaces would be produced (Agrell *et al.*, 1970; Housley *et al.*, 1970). Metallic iron may be produced also by vacuum reduction of molten iron sulfide (Housley *et al.*, 1970).

A by-product of hydrogen reduction of iron-bearing silicate material is water [$FeO(glass) + H_2(gas) \rightarrow Fe(metal) + H_2O(gas)$]. This water, if produced in the hot ejecta blanket or base surge deposit of an impact (Pearce and Williams, 1972), may be retained in the deposit for a short time and thus may be the source of the water in the lunar "goethite" reported by Agrell *et al.* (1972). Pearce and Williams (1972) produced metallic iron by reduction of the simulated brown lunar glass powder at temperatures of 800°C to 1000°C, times between 5 hours and 74 hours, and oxygen fugacities between 1 and 2 orders of magnitude below the IW buffer curve.

The conclusions are reached that: (1) complex metallic and sulfide mounds are produced from silicate glass of lunar composition by carbon or hydrogen reduction, (2) such mounds and the globules on the surfaces of the related dimples are similar in chemistry and morphology to some of those found on lunar silicate glass particles, and (3) reduction may be an important mechanism for the formation of some lunar complex metallic and sulfide mounds.

Acknowledgments—Supported by NASA Contract NAS 9-10221 and NASA Grant NGR-44-004-116 and NGL-44-004-001. We thank J. B. Toney for technical assistance and J. M. Rhodes, who performed the x-ray fluorescence analysis of the simulated brown lunar glass. Critical review by H. Axon, J. Goldstein, A. Hales, E. Padovani, D. Presnall, and H. Taylor has contributed to the improvement of the manuscript.

REFERENCES

Adler I. Walter L. S. Lowman P. D. Glass B. P. French B. M. Philpotts J. A. Heinrich K. J. F. and Goldstein J. I. (1970) Electron microprobe analysis of Apollo 11 lunar samples. *Proc. Apollo 11 Lunar Sci. Conf., Geochim. Cosmochim. Acta*, Suppl. 1, Vol. 1, pp. 87–92. Pergamon.

Agrell S. O. Scoon J. H. Muir I. D. Long J. V. P. McConnell J. D. C. and Peckett A. (1970) Observations on the chemistry, mineralogy and petrology of some Apollo 11 lunar samples. *Proc. Apollo 11 Lunar Sci. Conf., Geochim. Cosmochim. Acta*, Suppl. 1, Vol. 1, pp. 93–128. Pergamon.

Agrell S. O. Scoon J. H. Long J. V. P. and Coles J. N. (1972) The occurrence of goethite in a microbreccia from the Fra Mauro formation (abstract). In *Lunar Science—III* (editor C. Watkins), pp. 7–9, Lunar Science Institute Contr. No. 88.

Carter J. L. (1971) Chemistry and surface morphology of fragments from Apollo 12 soil. *Proc. Second Lunar Sci. Conf., Geochim. Cosmochim. Acta*, Suppl. 2, Vol. 1, pp. 873–892. M.I.T. Press.

Carter J. L. and MacGregor I. D. (1970) Mineralogy, petrology, and surface features of some Apollo 11 samples. *Proc. Apollo 11 Lunar Sci. Conf., Geochim. Cosmochim. Acta*, Suppl. 1, Vol. 1, pp. 247–275. Pergamon.

De Maria G., Balducci G., Guido M., and Piacente V. (1971) Mass spectrometric investigation of the vaporization process of Apollo 12 lunar samples. *Proc. Second Lunar Sci. Conf., Geochim. Cosmochim. Acta*, Suppl. 2, Vol. 2, pp. 1367–1380. M.I.T. Press.

Duke M. B. Woo C. C. Bird M. L. Sellers G. A. and Finkelman R. B. (1970) Lunar soil: size distribution and mineralogical constituents. *Science* **167** 648–650.

Frondel C. Klein C. Jr. Ito J. and Drake J. C. (1970) Mineralogical and chemical studies of Apollo 11 lunar fines and selected rocks. *Proc. Apollo 11 Lunar Sci. Conf., Geochim. Cosmochim. Acta*, Suppl. 1, Vol. 1, pp. 445–474. Pergamon.

Gibson E. K. Jr. and Johnson S. M. (1971) Thermal analysis-inorganic gas release studies of lunar samples. *Proc. Second Lunar Sci. Conf., Geochim. Cosmochim. Acta*, Suppl. 2, Vol. 2, pp. 1351–1366. M.I.T. Press.

Gibson E. K. Jr. and Moore G. W. (1972) Inorganic gas release and thermal analysis of Apollo 14 and 15 soils. *Proc. Third Lunar Sci. Conf., Geochim. Cosmochim. Acta*, Suppl. 3, Vol. 2. M.I.T. Press.

Goldstein J. I. Henderson E. P. and Yakowitz H. (1970) Investigation of lunar metal particles. *Proc. Apollo 11 Lunar Sci. Conf., Geochim. Cosmochim. Acta*, Suppl. 1, Vol. 1, pp. 499–512. Pergamon.

Goldstein J. I. and Yakowitz H. (1971) Metallic inclusions and metal particles in the Apollo 12 lunar soil. *Proc. Second Lunar Sci. Conf., Geochim. Cosmochim. Acta*, Suppl. 2, Vol. 1, pp. 177–191. M.I.T. Press.

Griscom D. L. and Marquardt C. L. (1972) Electron spin resonance studies of iron phases in lunar glasses and simulated lunar glasses (abstract). In *Lunar Science—III* (editor C. Watkins), pp. 341–343, Lunar Science Institute Contr. No. 88.

Holland P. T. Simoneit B. R. Wszolek P. C. McFadden W. H. and Burlingame A. L. (1972) Carbon chemistry of the lunar surface. *Nature* **235**, 252–253.

Housley R. M. Blander M. Abdel-Gawad M. Grant R. W. and Muir A. H. Jr. (1970) Mössbauer spectroscopy of Apollo 11 samples. *Proc. Apollo 11 Lunar Sci. Conf., Geochim. Cosmochim. Acta*, Suppl. 2, Vol. 2, pp. 1351–1366. Pergamon.

Housley R. M. Grant R. W. and Abdel-Gawad M. (1972) Study of excess Fe metal in the lunar fines by magnetic separation (abstract). In *Lunar Science—III* (editor C. Watkins), pp. 392–394, Lunar Science Institute Contr. No. 88.

McKay D. Morrison D. Lindsay J. and Ladle G. (1970) Origin of small lunar particles and breccia from the Apollo 11 site. *Proc. Apollo 11 Lunar Sci. Conf., Geochim. Cosmochim. Acta*, Suppl. 1, Vol. 1, pp. 673–694. Pergamon.

McKay D. S. and Ladle G. (1971) Scanning electron microscope study of particles in the lunar soil. In *Proc. 4th Annual Scanning Electron Microscope Symposium*, pp. 177–184, I.I.T. Research Inst., Chicago.

McKay D. S. Carter J. L. and Greenwood W. R. (1971) Lunar metallic particle ("mini-moon"): An interpretation. *Science*. **171**, 479–480.

Mason B. Fredriksson K. Henderson E. P. Jarosewich E. Melson W. G. Towe K. M. and White J. S. Jr. (1970) Mineralogy and petrology of lunar samples. *Proc. Apollo 11 Lunar Sci. Conf., Geochim. Cosmochim. Acta*, Suppl. 1, Vol. 1, pp. 655–660. Pergamon.

Pearce G. W. and Williams R. J. (1972) Excess iron in lunar breccias and soils: possible origin (abstract). *Trans. Amer. Geophys. Union* **53**, No. 4, 360.

Pillinger C. T. Cadogan P. H. Eglinton G. Maxwell J. R. and Mays B. J. (1972) Simulation study of lunar carbon chemistry. *Nature Phys. Sci.* **235**, 108–109.

Ramdohr P. and El Goresy A. (1970) Opaque minerals of the lunar rocks and dust from Mare Tranquillitatis. *Science* **167**, 615–618.

Simpson P. R. and Bowie S. H. U. (1970) Quantitative optical and electron-probe studies of opaque phases in Apollo 11 samples. *Proc. Apollo 11 Lunar Sci. Conf., Geochim. Cosmochim. Acta*, Suppl. 1, Vol. 1, pp. 873–890. Pergamon.

Wänke H. Wlotzka F. Jagoutz E. and Begemann F. (1970) Composition and structure of metallic iron particles in lunar "fines." *Proc. Apollo 11 Lunar Sci. Conf., Geochim. Cosmochim. Acta*, Suppl. 1, Vol. 1, pp. 931–935. Pergamon.

Winchell H. and Skinner B. J. (1970) Glassy spherules from the lunar regolith returned by Apollo 11 expedition. *Proc. Apollo 11 Lunar Sci. Conf., Geochim. Cosmochim. Acta*, Suppl. 1, Vol. 1, pp. 957–964. Pergamon.

Wosinski J. F. Williams J. P. Korda E. J. Kane W. T. Carrier G. B. and Schreurs J. W. H. (1972) Inclusions and interface relationships between glass and breccia in lunar sample no. 14306.50 (abstract). In *Lunar Science—III* (editor C. Watkins), pp. 811–813, Lunar Science Institute Contr. No. 88.

Proceedings of the Third Lunar Science Conference
(Supplement 3, *Geochimica et Cosmochimica Acta*)
Vol. 1, pp. 971–981
The M.I.T. Press, 1972

Compositions and mineralogy of lithic fragments in 1–2 mm soil samples 14002,7 and 14258,33

I. M. Steele and J. V. Smith

Department of the Geophysical Sciences
The University of Chicago
Chicago, Illinois 60637

Abstract—Broad-beam electron microprobe analyses of igneous-textured fragments from 1–2 mm fines 14002,7 and 14258,33 as well as from larger rock sections 14310,175 and 14072,10 suggest three principal igneous rock types at the Fra Mauro site. The first (I) is a plagioclase-rich basalt characterized by high CaO and Al_2O_3; the second (II) is a picrobasalt characterized by high MgO relative to FeO resulting from abundant high-Mg, low-Ca pyroxene and minor olivine; the third rock Type (III) is an Fe-rich basalt with coarse texture similar to some Apollo 12 mare-type basalts.

The bulk composition of the plagioclase-rich rock type closely matches the analysis of Tycho material obtained by Surveyor VII. Averaged Apollo 14 soils analyses are well represented by 39% I + 35% II + 26% III. Glass compositions from 14259 soil (Apollo Soil Survey) do not match these three groups of analyses, suggesting a different origin.

Types I and II belong to the KREEP group, and differ mainly in proportions of plagioclase and Ca-poor pyroxene. They could be related by crystal-liquid fractionation involving these two minerals.

An olivine-chromite fragment has an olivine with a composition (88–89 mole% Fo; CaO 0.05 wt%) consistent with early differentiation under plutonic conditions.

INTRODUCTION

MINERAL AND BULK compositional data have been obtained from 1–2 mm fragments selected for apparent igneous textures from fines 14002,7 and 14258,33 as well as from igneous clasts in breccia section 14063,14. These compositional data were combined with existing data from larger rock samples such as 14053, 14072 and 14310 and compared with bulk analyses of soils to obtain compositional ranges and dominant igneous rock types present in the regolith and breccias. It was hoped that fragments with igneous-looking textures would be likely candidates for igneous rocks at the pre-Imbrian surface, though derivation from the impacting body or from melted regolith must be borne in mind. The bulk compositions are compared with those of glasses determined by the Apollo Soil Survey (1971) to test whether the latter represented the same rock types.

GENERAL PROCEDURE

The 1–2 mm fines were divided first into groups by stereomicroscope study; as found also by Powell and Weiblen (1972) and Carr and Meyer (1972) igneous fragments are much less abundant ($\sim 5\%$) than light and dark breccias. From polished sections, 27 fragments were selected because of an apparent igneous origin based on general texture, absence of clasts, and presence of only residual glass. Two igneous clasts from polished breccia section 14063,14 as well as large polished sections of 14072,10 and 14310,175 were also included. All samples except one olivine-rich fragment contained sufficient grains to yield fairly representative analyses.

Bulk microprobe analyses were obtained by translating the sample under a $\sim 50 \mu$ electron beam taking care to avoid cracks and fragment boundaries as far as possible. Data were collected for Na, Mg, Al, Si, P, K, Ca, Ti, Cr, Mn, and Fe for all samples as well as standards, and compositions were calculated using standard microprobe correction procedures. Repeated analyses of several fragments indicated a reproducibility of $<5\%$ of the amount present for most major elements; this value may be higher for elements contained in relatively uncommon minerals, e.g., Ti in ilmenite and spinel. Oxide sums totaled between 87% and 102% with an average of 94%. Because most of this discrepancy was considered to result from cracks and holes, all analyses were recalculated to 100%.

The accuracy was tested by analysing polished section 14310,175 using the above technique. The present analysis (Table 1, analysis 1) agrees only moderately with published analyses of 14310 given in Table 1, analyses 2 and 3 (LSPET, 1971; Compston et al., 1972). The apparent discrepancies in these two published analyses suggest that 14310 may not be a very homogeneous rock, and indeed there have been suggestions that 14310 was formed by melting of regolith. The higher values of Na_2O, CaO and Al_2O_3 and lower values of MgO and FeO are consistent with a higher plagioclase content of 14310,175 than the chemically analysed samples.

The bulk compositions as well as the calculated CIPW norm (Cr_2O_3 and MnO were not included in the norm calculation) are given in Table 1.

COMPOSITIONAL VARIATION AND GROUPINGS

Examination of computer-plotted variation diagrams indicated that the data of Table 1 could be placed in three groups plus several unique single analyses. The

Table 1. Bulk compositions and CIPW Norms (wt%) of Apollo 14 igneous fragments and rocks normalized to sum 100%. Specific designations of fragments are given at the end of this table.

	(1)	(2)	(3)	(4)	(5)	(6)	(7)	(8)	(9)	(10)	(11)	(12)	(13)	(14)	(15)	(16)	(17)
SiO_2	46.7	50.0	47.57	48.0	50.1	45.3	38.8	45.8	46.8	47.6	49.2	47.8	46.1	45.4	47.8	46.5	45.1
Al_2O_3	25.1	20.0	20.70	20.1	20.0	27.6	0.2	29.8	23.8	25.8	19.9	23.6	18.5	15.1	18.8	19.0	23.7
MgO	5.6	8.0	7.59	7.5	7.6	4.8	49.4	0.9	6.1	4.0	4.6	6.8	10.0	10.1	8.8	8.9	6.5
FeO	6.1	7.7	8.22	7.4	7.7	6.0	10.7	3.4	6.0	5.0	5.8	7.1	11.3	12.6	8.6	8.5	7.2
Cr_2O_3	0.15	0.16	0.15	0.20	—	0.13	0.47	0.03	—	—	0.22	0.14	0.22	0.42	0.25	0.17	0.19
MnO	0.09	0.14	0.12	0.08	—	0.07	0.11	0.05	—	—	0.09	0.10	0.13	0.18	0.12	0.10	0.10
TiO_2	1.0	1.3	1.24	1.5	1.3	0.9	0.0	1.1	0.8	0.7	2.9	0.5	1.6	2.2	1.8	2.1	0.8
CaO	13.3	11.0	12.54	11.5	10.7	13.6	0.1	16.3	14.6	15.2	13.5	11.4	9.6	11.0	10.5	10.5	13.7
K_2O	0.6	0.53	0.49	1.1	1.2	0.4	0.0	0.2	0.3	0.3	2.0	1.1	0.8	0.6	1.0	1.7	0.6
Na_2O	0.8	0.63	0.73	1.1	1.0	0.8	0.2	1.7	0.8	0.8	1.1	1.4	1.1	0.9	1.2	1.5	0.8
P_2O_5	0.5	—	0.41	1.5	0.4	0.4	0.0	0.7	0.8	0.6	0.7	0.1	0.7	1.0	1.2	1.0	1.3
Or	3.3	3.1	2.9	6.4	7.1	2.4	—	1.2	1.8	1.8	11.8	6.5	4.7	3.6	5.9	10.1	3.6
Ab	6.8	5.4	6.2	9.3	8.5	6.5	—	14.4	6.8	6.8	9.3	11.9	9.4	7.7	10.2	12.7	7.0
An	63.2	50.6	52.0	46.8	46.5	65.3	—	73.0	60.4	65.9	43.7	54.9	43.3	36.4	43.2	40.4	59.4
Qz	0.2	4.2	0.1	1.0	1.9	0.0	—	0.0	0.0	2.4	2.9	0.0	0.0	0.0	0.0	0.0	0.0
Ol	0.0	0.0	0.0	0.0	0.0	1.0	—	1.2	0.3	0.0	0.0	11.0	14.2	9.1	1.8	15.9	7.7
Hy	23.4	30.6	29.0	29.4	29.4	20.2	—	3.7	21.8	15.6	10.0	13.9	23.6	26.6	31.8	10.3	17.6
Di	0.0	3.6	6.4	0.7	3.2	0.0	—	2.8	5.5	4.8	15.2	0.6	0.2	10.1	0.9	4.3	0.0
Ap	1.1	—	1.0	3.5	0.9	0.9	—	1.6	1.9	1.4	1.6	0.2	1.6	2.3	2.8	2.3	3.0
Il	1.9	2.5	2.4	2.9	2.5	1.7	—	2.1	1.5	1.3	5.5	1.0	3.0	4.2	3.4	4.0	1.5
C	0.1	0.0	0.0	0.0	0.0	2.0	—	0.0	0.0	0.0	0.0	0.0	0.0	0.0	0.0	0.0	0.2

Table 1. (continued)

	(18)	(19)	(20)	(21)	(22)	(23)	(24)	(25)	(26)	(27)	(28)	(29)	(30)	(31)	(32)	(33)
SiO$_2$	50.7	60.8	45.5	47.4	47.2	44.1	46.6	50.2	48.8	47.4	47.4	48.8	46.7	47.2	52.3	53.6
Al$_2$O$_3$	18.4	12.1	23.7	24.1	15.5	16.3	19.9	23.4	16.0	18.2	15.5	17.5	19.6	14.6	22.0	20.0
MgO	8.6	1.6	6.7	0.7	14.3	16.9	12.4	6.6	13.4	12.8	17.3	12.0	8.8	10.3	4.8	6.2
FeO	7.8	11.4	6.8	5.6	9.9	9.1	7.7	5.4	9.1	7.6	9.2	8.1	8.2	14.2	5.1	5.4
Cr$_2$O$_3$	0.17	0.04	0.17	0.07	0.14	0.22	0.10	0.09	0.12	0.13	0.13	0.13	0.02	0.64	—	—
MnO	0.09	0.14	0.08	0.06	0.12	0.16	0.10	0.08	0.11	0.11	0.11	0.12	0.11	0.20	—	—
TiO$_2$	1.3	1.0	1.2	3.8	2.5	2.6	1.2	0.8	1.2	1.3	0.0	1.6	2.2	2.2	1.0	0.6
CaO	9.0	7.5	13.0	12.4	8.2	9.6	10.0	11.6	9.1	10.0	8.4	9.8	12.1	9.6	12.1	12.2
K$_2$O	1.3	2.3	0.6	2.9	0.6	0.1	0.6	0.4	0.7	0.5	0.5	0.8	0.8	0.7	1.4	0.7
Na$_2$O	1.4	1.4	0.9	1.6	0.8	0.5	0.9	0.9	0.8	1.0	0.8	0.4	1.0	0.4	0.9	1.0
P$_2$O$_5$	1.2	1.7	1.4	1.4	0.7	0.4	0.5	0.5	0.7	0.9	0.7	0.8	0.5	0.1	0.4	0.3
Or	7.7	13.6	3.6	17.1	3.3	0.6	3.6	2.4	4.1	3.6	3.0	4.7	4.7	4.1	8.3	4.1
Ab	11.9	12.3	7.6	13.5	6.8	4.2	7.6	7.6	6.8	8.6	6.8	3.4	8.5	3.0	7.6	8.5
An	37.6	19.6	55.3	50.3	36.7	42.0	46.2	54.2	37.7	43.4	37.3	43.6	46.7	36.5	52.0	47.9
Qz	3.3	22.8	0.0	2.0	0.0	0.0	0.0	6.1	0.0	0.0	0.0	1.1	0.0	0.0	7.3	8.9
Ol	0.0	0.0	1.7	0.0	7.1	22.2	12.7	0.0	9.4	10.2	22.7	0.0	7.8	2.5	0.0	0.0
Hy	33.1	20.0	24.8	4.8	39.6	22.6	25.1	25.2	36.0	29.2	28.6	42.1	18.7	41.0	17.4	20.2
Di	0.0	5.9	0.0	1.9	0.1	2.5	0.0	0.0	2.1	0.4	0.0	0.1	8.2	8.5	4.6	8.6
Ap	2.8	3.9	3.3	3.3	1.6	0.9	1.2	1.2	1.6	2.1	1.6	1.9	1.2	0.2	0.9	0.7
Il	2.5	1.9	2.3	7.2	5.2	5.0	2.3	1.5	2.3	2.5	0.0	3.0	4.2	4.2	1.9	1.1
C	1.1	0.0	1.4	0.0	0.0	0.0	0.9	1.5	0.0	0.0	0.0	0.0	0.0	0.0	0.0	0.0

(1) 14310,175; (2) 14310–LSPET (1971); (3) 14310–Compston *et al.* (1972); (4)–(18) are from 14002,7. Specific designations are: (4) D–6; (5) D–11; (6) D–23; (7) E–1–8; (8) E–1–11; (9) F–9; (10) F–12; (11) G–2; (12) G–3; (13) G–6; (14) G–13; (15) G–15; (16) G–16; (17) H–1–1; (18) H–1–4; (19)–(30) are from 14258,33. Specific designations are: (19) B–1–1; (20) B–2a; (21) B–3–1; (22) D–1–3; (23) D–1–4; (24) D–1–5; (25) D–1–7; (26) D–1–8; (27) D–2a–1; (28) D–2a–2; (29) D–2a–8; (30) D–2a–17; (31) 14072,10; (32) 14063,14; TS–IIa; (33) 14063,14 TS–IIb.

FeO–MgO variation diagram shown in Fig. 1 best shows these compositional groupings that are described as follows:

Group I. FeO and MgO are present in nearly equal amounts (marked by crosses) with values ranging from 5% to 9%; CaO as indicated in Fig. 2 (same notation) ranges between 9% and 15%; K$_2$O (range 0.3%–2.0%) and P$_2$O$_5$ (range 0.1%–1.5%) values are consistent with KREEP compositions (Meyer *et al.*, 1971); the Fe/(Fe + Mg) atom ratio for the bulk analysis is almost constant with an average value of 0.37 (Fig. 3).

Group II. MgO values are decidedly greater than FeO values (solid circles on Fig. 1) with an MgO range of 12% to 18% and an FeO range of 7% to 10%; CaO values range from 8% to 10% and are either equal to or lower than the lowest values from Group I (Fig. 2); as in Group I, the Group II values for K$_2$O (range 0.1%–0.7%) and P$_2$O$_5$ (range 0.4%–0.9%) are consistent with KREEP compositions; compositions are with one exception olivine-normative, whereas Group I had about an equal number of olivine- and quartz-normative compositions; the average Fe/(Fe + Mg) atom ratio for the bulk analyses is notably lower (0.26) than that for Group I, reflecting the higher MgO content.

Group III. This group is represented in this study only by 14072,10 although published data indicate that 14053 is essentially equivalent to 14072. The average grain size of these two samples precludes finding representative samples in the 1–2 mm fines. The FeO value (open circle, Fig. 1) is decidedly greater than the MgO value, and is greater than the FeO values for Groups I and II; the K$_2$O and P$_2$O$_5$

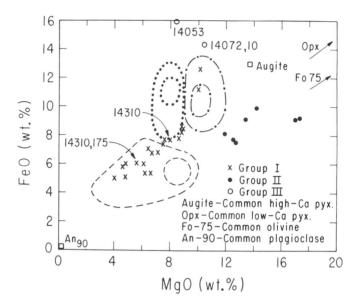

Fig. 1. FeO versus MgO for analyses in Table 1. The approximate range of preferred glass compositions recognized by the Apollo Soil Survey (1971) are indicated as follows: Type B ─·─·─; Type C ·····; Type E ─ ─ ─. Inner circles indicate greatest concentration for each glass type.

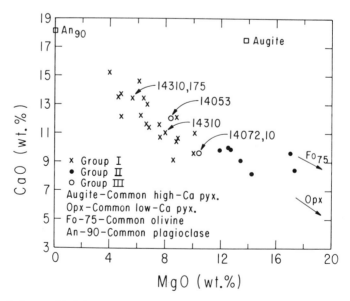

Fig. 2. CaO versus MgO for analyses in Table 1. Common minerals are plotted for reference. Symbols same as Fig. 1.

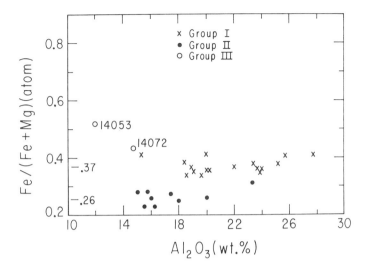

Fig. 3. Fe/(Fe + Mg) versus Al_2O_3 for analyses in Table 1. Symbols same as in Figs. 1 and 2.

values (both equal to 0.1%) are too low for KREEP and are similar to values for
Apollo 12 mare basalts; the bulk composition of both 14072 and 14053 are similar
to Apollo 12 compositions (high FeO, low K_2O and P_2O_5) and thus are considered
to be mare basalts.

   The compositional data for these three groups are summarized in columns 1 thru
5 of Table 2. The range and average compositions are given for Groups I and II and
the average composition of 14072 and 14053 is given for Group III.

   Three unique compositions were found among these fragments. Table 1, Analysis
7 corresponds to an olivine-chromite fragment; analysis 19 corresponds to a silica-
rich fragment with high values of K_2O and BaO (0.4 wt%); analysis 8 represents a
CaO and Al_2O_3-rich fragment with an anorthositic composition.

Table 2. Mean and ranges of compositions for Group I, II and III analyses, comparison of observed and
calculated average fines composition, and Surveyor VII analysis.

| | Group I | | Group II | | Group III | Calculated | Average | Surveyor |
	Mean	Range	Mean	Range	Mean[a]	Ave. fines[b]	Fines[c]	VII[d]
Na_2O	1.0	0.8– 1.5	0.7	0.5– 1.0	0.36	0.7	0.57	0.7 ± 0.3
MgO	6.6	4.0– 8.9	14.1	11.9–17.3	9.4	9.9	9.9	6.9 ± 3.0
Al_2O_3	22.0	18.5–27.7	16.7	15.0–20.0	13.3	17.9	18.0	22.2 ± 1.0
SiO_2	48.2	44.1–53.7	47.0	46.6–48.7	47.8	47.7	48.0	46.5 ± 3.3
P_2O_5	0.7	0.1– 1.5	0.6	0.4– 0.9	0.1	0.7	—	—
K_2O	0.9	0.3– 2.0	0.5	0.1– 0.7	0.1	0.8	0.52 ⎫	18.3 ± 1.5
CaO	12.2	9.1–15.2	9.1	8.1–10.0	10.8	10.7	11.0 ⎭	
TiO_2	1.2	0.5– 2.9	2.0	0.0– 3.0	1.8	1.6	1.8	0.0
FeO	6.7	5.0– 8.6	8.8	7.5–10.0	15.1	9.6	10.0	5.4 ± 1.9

[a]Average of 14053 (LSPET, 1971) and 14072,10 (Table 1) analyses.
[b]Best fit: 39% of Group 1 + 35% of Group 2 + 26% of Group 3.
[c]Analyses of average fines. LSPET (1971).
[d]Turkevich (1971): oxide interpretation.

Mineralogy and Texture

The previously defined groups each have characteristic mineralogy and textures that further support the original groupings.

Group I. Modal composition is 55–65% plagioclase, 35% pyroxene, 4% ilmenite with minor metal, phosphate, spinel, and troilite. Pyroxene compositions range widely with both high- and low-Ca varieties present in any one fragment. These compositions were checked by making detailed analyses of a few grains in most fragments (plotted on Fig. 4 as crosses) and by visually estimating compositions of a large number of grains by rate-meter deflections for major elements. It should be noted that there appears to be a compositional gap for those pyroxenes with high-Mg content. Plagioclase grains are characteristically lath-shaped with interstitial pyroxene, and have compositions ranging from An 95 (mole) to An 80 with Fe (wt%) ranging from 0.1 to 0.4. This range of An and Fe contents is consistent with the Fe content of plagioclase from exotic fragments in Apollo 12 samples (Smith, 1971). No olivine was found in most Group I fragments but others showed very minor olivine. In general, the textures and mineralogy of these fragments are similar to that of 14310 and because of the similarity in bulk compositions, they might be described as 14310-type.

Group II. Texturally, Group II fragments are finer grained than Group I with an average grain size less than 0.05 mm. Subophitic to poikilitic pyroxenes are common and in contrast to Group I, the pyroxene compositions have one distinct low-Ca and high-Mg composition (solid circles, Fig. 4). Also, in contrast to Group I, olivine is a minor phase common to all fragments. Modal analysis is about 40–50% plagioclase, 50% low-Ca pyroxene, 5% olivine and minor ilmenite and phosphate. Olivine compositions are rather constant with Fo (mole%) values of 60–70. Plagioclase compositions were indistinguishable from those of Group I. The lower modal plagioclase content and the appearance of only high-Mg pyroxene and olivine are consistent with the bulk composition of lower CaO and Al_2O_3 and a lower FeO/MgO ratio compared to Group I.

Group III. The two samples placed in this group, 14072 and 14053, show a much coarser texture than either Group I or II. Phenocrysts of both pyroxene and olivine

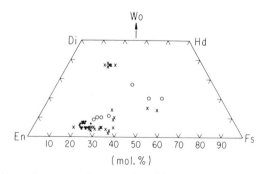

Fig. 4. Compositions of pyroxenes from samples of three groups. Symbols same as previous figures.

range above 5 mm, and the matrix pyroxene, plagioclase, and ilmenite average about
0.5 mm. Pyroxene compositions (e.g., Finger *et al.*, 1972) include both low- and
high-Ca varieties and show greater Fe enrichment compared to Group I and II
pyroxenes. Representative analyses are shown on Fig. 4 as open circles. Modal
composition is 20% pyroxene as phenocrysts, 30% plagioclase, 40% matrix pyroxene,
5% ilmenite, 5% olivine both as phenocrysts and matrix. The abundance of high-Fe
pyroxene in contrast to Groups I and II undoubtedly accounts for the high FeO
value in the bulk composition.

The mineralogy of the three unique fragments is as follows: Analysis 7 represents
a fragment composed of high-Mg olivine (Fo 88–89, mole%) and minor chromite
(Cr_2O_3 = 50.9, FeO = 22.3, Al_2O_3 = 16.6, MgO = 9.3, TiO_2 = .9, MnO = .2).
The low CaO content of this olivine (0.05 wt%) suggests crystallization at depth
(Simkin and Smith, 1970) and the high Mg content would be consistent with an
earlier stage of crystal-liquid differentiation than for olivines in mare-type basalts.
Analysis 19 represents a fine-grained fragment composed of a silica mineral, Ba-K-
feldspar, Fe-rich olivine (Fa 83) and relatively low-Ca plagioclase intergrown with
the K-feldspar. This mineralogy probably represents either a late-stage liquid
differentiate or an early fraction of partial melting. The relatively large size of this
fragment (1.5 mm) is notable, because it implies a concentration of residuum from
basaltic rocks. Analysis 8 is of a fragment composed of about 90% plagioclase, and
is unusual in that the plagioclase is relatively Ca-poor (An 83 mole %) compared to
other lunar anorthosite fragments. The remaining volume is pigeonitic pyroxene.

DISCUSSION

In the previous sections, criteria were given for the recognition of three types of
igneous fragments in a limited sample of Apollo 14 material. Rock types recognized
by other workers will now be briefly compared to our three groups.

Gancarz *et al.*, (1971) recognized many clasts and rock fragments that they
suggested were similar to rock 14310 and further suggested that ". . . rock fragments
that may be correlative with sample 14310 on the basis of mineralogic and petrologic
similarities are important constituents of the soil and breccia at the Apollo 14 site."
We agree. Gancarz *et al.*, also recognize a rock similar to 14310 except for the
abundant orthopyroxene (14073). Although this rock does contain clinopyroxene,
it may belong to the same suite as our Group II. LSPET (1971) recognized three
types of nonfragmental lithic clasts that were abundant in fragmental rocks. These
three types approximately correspond to Group I (feldspathic clasts), Group II
(plagioclase-orthopyroxene clasts), and Group III (clinopyroxene-plagioclase clasts).
Melson *et al.*, (1972) recognized two basalt types (14310-type, 14053-type) as components in soils, breccias and among walnut-size samples. Walter *et al.*, (1972)
presented a CaO–Al_2O_3 variation diagram for bulk analyses of breccia clasts, and
many analyses cluster about the 14310 analysis. Kurat *et al.*, (1972) recognized two
common "igneous" lithic fragment types in breccia 14318 that were both high in
Al_2O_3 relative to our Group II and were in general similar to Group I; however,
neither composition agrees well in detail with our Group I.

Because three igneous rock types appear to dominate our Apollo 14 material, we tested whether a linear combination of the average compositions of our three groups (Table 2, columns 1, 3, and 5) would reproduce the composition of the average fines (LSPET, 1971) and thus give an indication of the contribution of each group to the Fra Mauro regolith. Excellent agreement was found (compare Table 2, columns 6 and 7) for a composition consisting of 37% Group I, 35% Group II, and 26% Group III. Obviously such a calculation ignores minor components such as granitic, anorthositic and ultrabasic fragments that have been recognized repeatedly, but their contribution to the bulk composition will be minor. Absent also from this comparison would be fragments of any preexisting coarse-grained rocks now found only as individual mineral grains. That such coarse-grained rocks did exist is implied by coarse exsolution textures in some pyroxene grains and the large size (~ 0.5 mm) of some single mineral fragments. The latter, however, could be derived from Group III rocks.

An alternative method of deriving rock compositions has been to make electron microprobe analyses of a large number of glass fragments from the Apollo 14 comprehensive sample 14259 (Apollo Soil Survey, 1971). Three types of basaltic compositions (B, C, E) were recognized, as well as less common glass compositions corresponding to anorthosite, potash granite and mare basalt. None of these glass compositions correspond to 14310-like (Group I) or Group II compositions (Fig. 1), although these are common fragment types at Fra Mauro. A similar conclusion was reached by Chao et al., (1972) for 14310 compositions. In addition, only 11% of the glass analyses are similar to Group III compositions.

There are several ways to explain the difference between glass and rock compositions: (1) glasses originate from rock types other than those recognized at the Fra Mauro site; (2) the impact process may yield glasses of composition different from those of the original rocks; (3) glasses formed from well-mixed regolith or a large volume of regolith may have common compositions. Explanation (2) is suggested by Fig. 5 which shows the normative compositions of glasses B, C, D, and

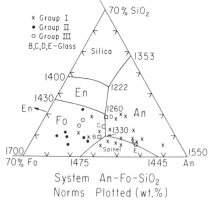

System An–Fo–SiO$_2$
Norms Plotted (wt.%)

Fig. 5. Normative compositions for analyses of Table 1 plotted in system SiO$_2$–Fo–An (after Anderson, 1915). Norms of average glass compositions (Apollo Soil Survey, 1971) for Types B, C, D, and E also are plotted.

E plotted in the system SiO_2–Fo–An (Anderson, 1915). It is noted that B, C, D lie nearly on a cotectic and thus may represent a partial melt of a bulk composition represented by this system (Walker *et al.*, 1972). Explanation (3) is suggested by Chao *et al.*, (1972) who recognized a frequent glass composition that correlates with analysis of bulk fines and breccias.

If the preferred glass compositions represent rock types common at Fra Mauro, then breccia and soil analyses should be expressible as a combination of these glasses. This is not possible for a number of samples that have a higher MgO content than the most MgO-rich glass (B–average MgO–10.2%). For examples of high MgO samples see Wänke *et al.*, (1972) analysis of fines 15471,29 (11.7% MgO) and breccia 14321, 184,25 (11.3% MgO); Rose *et al.*, (1972) analysis of average breccia (11.03% MgO); and LSPET (1971) analysis of fines 14305 (13% MgO). A similar problem arises for CaO values of breccias that are too low to be explained by glass compositions (less common case, but see LSPET analysis of 14321,9). These analyses indicate the presence of a high-MgO, low-CaO component in the fragmental material that would correspond to Group II compositions.

It is interesting to consider possible origins of these compositional groups recognized at Fra Mauro. The differences in bulk chemistry of compositions other than mare-type can be interpreted principally as a variation in the relative proportions of Ca-rich plagioclase and Mg-rich pyroxene or olivine. This suggests an origin by crystal-liquid separation of either or both minerals. The recognition of an olivine-rich fragment (analysis 7, Table 1) and of various specimens of anorthosite (e.g., Steele and Smith, 1971) supports this origin. On a ternary plot of the liquidus phases in the SiO_2–Fo–An system (Fig. 5) the group I and Group II normative compositions plot in the fields of primary anorthite and forsterite, respectively, with only a few exceptions. This indicates that the different compositions could originate by crystal cumulation or by partial melting of bulk compositions represented approximately by this system. We conclude that the rock compositions are consistent with an early major differentiation of the Moon that produced an Al-rich crust represented by feldspathic basalts and minor anorthosite and gabbroic anorthosite together with a more basic interior represented by olivine- and pyroxene-rich rocks. Later partial melting either by internal processes or impacts further modified these compositions. Even later, partial melting at depth of basic material might have produced the high-Fe mare basalts.

Any attempt to estimate quantitatively the proportions of the rock and magma compositions will best await completion of the Apollo missions. Nevertheless, the circum-lunar study of Adler *et al.*, (1972) showed that feldspar-rich basalts probably dominate the entire surface of the Moon, as proposed in the model of Smith *et al.*, (1970) based just on Apollo 11 material. A later paper will describe in detail rare basic and ultrabasic fragments that might derive from the lunar interior, together with speculation on their origin.

Rock 14310 (and probably Group I) is an enigma due to certain features such as its cognate inclusions, occurrence of schreibersite (El Goresy and Ramdohr, 1972) and trace elements (Morgan *et al.*, 1972) that indicate possible formation by melting of near-surface material, possibly by impact melting rather than occurrence as a true lunar liquid (Brown and Peckett, 1971). If this rock does represent an impact melt, then Group I compositions may represent merely "average" surface materials prior to impact. However, it is possible that rock 14310 represents crystallization of a melt formed from a nearly monomict regolith, and that Type I compositions represent actual magma compositions.

Lastly, a comparison of the average compositions of the three groups (Table 2) with the most recent interpretation of the Surveyor VII data (Turkevich, 1971) shows a striking correspondence between Group I and Surveyor VII analyses (Table 2, cols. 1 and 8). The only major discrepancy is the higher $(CaO + K_2O)$ value for Surveyor VII. Possibly the Tycho region, and perhaps other parts of the lunar highlands, may be dominated by Group I material modified by a more diopsidic pyroxene that supplies additional calcium.

Acknowledgments—We thank A. T. Anderson and A. M. Reid for helpful discussion and I. Baltuska, O. Draughn and R. Zechman for technical help. We thank NASA for grant NGL–14–001–171.

References

Adler I. Trombka J. Gerard J. Lowman P. Yin L. and Blodgett H. (1972) Preliminary results from the S-161 X-ray fluorescence experiment (abstract). In *Lunar Science—III* (editor C. Watkins), pp. 4–6, Lunar Science Institute Contr. No. 88.

Anderson O. (1915) The system anorthite-forsterite-silica. *Am. J. Sci.* **39,** 407–454.

Apollo Soil Survey (1971) Apollo 14: Nature and origin of rock types in soil from the Fra Mauro Formation. *Earth Planet. Sci. Lett.* **12,** 49–54.

Brown G. M. and Peckett A. (1971) Selective volatilization on the lunar surface: evidence from Apollo 14 feldspar-phyric basalts. *Nature* **234,** 262–266.

Carr M. H. and Meyer C. E. (1972) Petrologic and chemical characterization of soils from the Apollo 14 landing site (abstract). In *Lunar Science—III* (editor C. Watkins), pp. 116–118, Lunar Science Institute Contr. No. 88.

Chao E. C. T. Boreman J. A. and Minkin J. A. (1972) Apollo 14 glasses of impact origin (abstract). In *Lunar Science—III* (editor C. Watkins), pp. 133–134, Lunar Science Institute Contr. No. 88.

Compston W. Vernon M. J. Berry H. Rudowski R. Gray C. M. Ware N. Chappell B. W. and Kaye M. (1972) Age and petrogenesis of Apollo 14 basalts (abstract). In *Lunar Science—III* (editor C. Watkins), pp. 151–153, Lunar Science Institute Contr. No. 88.

El Goresy A. and Ramdohr P. (1972) Fra Mauro crystalline rocks: petrology, geochemistry, and subsolidus reduction of the opaque minerals (abstract). In *Lunar Science—III* (editor C. Watkins), pp. 224–226, Lunar Science Institute Contr. No. 88.

Finger L. W. Hafner S. S. Schürmann K. Virgo D. and Warburton D. (1972) Distinct cooling histories and reheating of Apollo 14 rocks (abstract). In *Lunar Science—III* (editor C. Watkins), pp. 259–261, Lunar Science Institute Contr. No. 88.

Gancarz A. J. Albee A. L. and Chodos A. A. (1971) Petrologic and mineralogic investigation of some crystalline rocks returned by the Apollo 14 mission. *Earth Planet. Sci. Lett.* **12,** pp. 1–18.

Kurat G. Keil K. and Prinz M. (1972) A "chondrite" of lunar origin: textures, lithic fragments, glasses and chondrules (abstract). In *Lunar Science—III* (editor C. Watkins), pp. 463–465, Lunar Science Institute Contr. No. 88.

LSPET (1971) (Lunar Sample Preliminary Examination Team) Preliminary Examination of Lunar Samples from Apollo 14. *Science* **173,** 681–693.

Melson W. G. Mason B. and Nelen J. (1972) Apollo 14 basaltic rocks (abstract). In *Lunar Science—III* (editor C. Watkins), pp. 535–536, Lunar Science Institute Contr. No. 88.

Meyer C. Jr. Brett R. Hubbard N. J. Morrison D. A. McKay D. S. Aitken F. K. Takeda H. and Schonfeld E. (1972) Mineralogy, chemistry, and origin of the KREEP component in soil samples from the Ocean of Storms. *Proc. Apollo 12 Lunar Sci. Conf., Geochim. Cosmochim. Acta* Suppl. 2, Vol, 1, pp. 393–411. MIT Press.

Morgan J. W. Krähenbühl U. Ganapathy R. and Anders E. (1972) Volatile and siderophile elements in Apollo 14 and 15 rocks (abstract). In *Lunar Science—III* (editor C. Watkins), pp. 555–557, Lunar Science Institute Contr. No. 88.

Powell B. N. and Weiblen P. W. (1972) Petrology and origin of rocks in the Fra Mauro Formation (abstract). In *Lunar Science—III* (editor C. Watkins), pp. 616–618, Lunar Science Institute Contr. No. 88.

Rose H. J. Cuttitta F. Annell C. S. Carron M. K. Christian R. P. Dwornik E. J. and Ligon D. T. (1972) Compositional data for fifteen Fra Mauro Lunar Samples (abstract). In *Lunar Science—III* (editor C. Watkins), pp. 661–662, Lunar Science Institute Contr. No. 88.

Simkin T. and Smith J. V. (1970) Minor-element distribution in olivine. *J. Geol.* **78**, 304–325.

Smith J. V. Anderson A. T. Newton R. C. Olsen E. J. Wyllie P. J. Crewe A. V. Isaacson M. S. and Johnson D. (1970) Petrologic history of the Moon inferred from petrography, mineralogy, and petrogenesis of Apollo 11 rocks. *Proc. Apollo 11 Lunar Sci. Conf., Geochim. Cosmochim. Acta* Suppl. 1, Vol. 1, pp. 897–925. Pergamon.

Smith J. V. (1971) Minor elements in Apollo 11 and Apollo 12 olivine and plagioclase. *Proc. Second Lunar Sci. Conf., Geochim. Cosmochim. Acta* Suppl. 2, Vol. 1, pp. 143–150. MIT Press.

Steele I. M. and Smith J. V. (1971) Mineralogy of Apollo 15415 "Genesis rock": Source of Anorthosite on Moon. *Nature* **234**, 138–140.

Turkevich A. L. (1971) Comparison of the analytical results from the Surveyor, Apollo and Luna Missions. *Proc. Second Lunar Sci. Conf., Geochim. Cosmochim. Acta* Suppl. 2, Vol. 2, pp. 1209–1215. MIT Press.

Walker D. Longhi J. and Hays J. F. (1972) Experimental petrology and origin of Fra Mauro rocks and soil (abstract). In *Lunar Science—III* (editor C. Watkins), pp. 770–772, Lunar Science Institute Contr. No. 88.

Walter L. S. French B. M. Doan A. S. and Heinrich K. F. J. (1972) Petrographic analysis of lunar samples 14171 and 14305 (breccias) and 14310 (melt rock) (abstract). In *Lunar Science—III* (editor C. Watkins), pp. 773–775, Lunar Science Institute Contr. No. 88.

Wänke H. Baddenhausen H. Balacescu A. Teschke F. Spettel B. Dreibus G. Quijano M. Kruse H. Wlotzka F. and Begemann F. (1972) Multielement analyses of lunar samples (abstract). In *Lunar Science—III* (editor C. Watkins), pp. 779–781, Lunar Science Institute Contr. No. 88.

Proceedings of the Third Lunar Science Conference
(Supplement 3, *Geochimica et Cosmochimica Acta*)
Vol. 1, pp. 983–994
The M.I.T. Press, 1972

Apollo 14 soils: Size distribution and particle types

David S. McKay, Grant H. Heiken, Ruth M. Taylor, Uel S. Clanton,
and Donald A. Morrison

Planetary and Earth Sciences Division,
NASA Manned Spacecraft Center,
Houston, Texas 77058

Garth H. Ladle

Lockheed Electronics Corp.,
Houston, Texas 77058

Abstract—Eleven soil samples from Apollo 14 have been studied by sieving and petrographic methods for sizes down to about 1 μ. Average mean grain size (M_z) of the less than 1 mm size fraction is 3.62 ϕ (81 μ) with a sorting value (σ_1) of 2.06 ϕ. Most of the Apollo 14 soils are slightly coarser and more poorly sorted than soil from the Apollo 11 and 12 landing sites. A graphic mean grain size of about 4.08 ϕ (60 μ) may represent an equilibrium grain size at which destructional processes are balanced by constructional processes. Agglutinates are the primary product of constructional processes in the lunar soil. They are produced for the most part by melting during micrometeorite impacts. Most of the glass produced on the lunar surface is in the form of agglutinates.

Soil from the "smooth terrain" at the Apollo 14 site contains abundant agglutinates (as much as 60% of some size fractions) and is low in breccia fragment content. Cone Crater soil is low in agglutinates and has the highest recrystallized breccia content (94%) of any of the Apollo 14 soil samples. It is probably the most representative sample of the deeper parts of the Fra Mauro formation, which suggests that these zones consist mainly of recrystallized breccia.

INTRODUCTION

WE HAVE STUDIED 0.25 g samples of each of the soils returned by the Apollo 14 mission. Samples from core 14230 were 0.10 g. Size distribution of the less than 1 mm soil fraction was determined by sieving down to 20 μ using electro-formed sieves with square holes of the following sizes: 500 μ, 250 μ, 150 μ, 90 μ, 75 μ, 60 μ, 45 μ, 30 μ, and 20 μ, and by using a millipore particle measurement computer system for sizes down to about 1 μ.

Each sample was gently dry sieved to prevent breakage of fragile particles. Samples were considered to be fractionated only after one hour of sieving produced no change in weight of material on each screen. The total sieving time for each sample ranged from 10 to 20 hours. Material examined under the binocular microscope at this stage appeared free of adhering finer dust. A portion of the soil in the less than 20 μ fraction was analyzed with the millipore particle measurement computer system and the data were converted from number percent to weight percent by graphical integration assuming spherical particles of uniform density. The combined data were plotted as cumulative weight percent curves and the Folk and Ward (1957) graphical size parameters were determined (Table 1). Table 1 includes both the parameters of the less than 1 mm size fraction and the parameters for the entire size range. The latter

Table 1. Folk and Ward (1957) graphical size parameters.

Sample		Median grain size		Graphic mean		Inclusive graphic standard deviation	Inclusive graphic skewness	Graphic kurtosis
14003,28-Contingency	(1)	3.00 φ	125 μ	2.85 φ	139 μ	2.33 φ	−0.19	0.99
	(2)	2.38 φ	104 μ	3.31 φ	100 μ	1.75 φ	−0.03	0.89
14141,30-Cone Crater	(1)	0.40 φ	758 μ	0.71 φ	612 μ	3.30 φ	+0.12	0.84
	(2)	3.04 φ	122 μ	3.00 φ	125 μ	2.17 φ	−0.04	0.96
14148,23-Trench Top	(1)	4.03 φ	62 μ	3.69 φ	78 μ	2.09 φ	−0.28	0.97
	(2)	4.30 φ	51 μ	4.02 φ	62 μ	2.78 φ	−0.24	0.96
14149,39-Trench Bottom	(1)	0.78 φ	583 μ	0.32 φ	802 μ	2.16 φ	−0.21	0.85
	(2)	3.56 φ	85 μ	3.41 φ	94 μ	2.86 φ	−0.13	0.92
14156,23-Trench Middle	(1)	3.73 φ	76 μ	3.40 φ	95 μ	2.39 φ	−0.28	1.05
	(2)	4.10 φ	58 μ	3.93 φ	66 μ	1.78 φ	−0.16	0.83
14163,76- Bulk	(1)	3.95 φ	65 μ	3.65 φ	80 μ	2.35 φ	−0.26	1.05
	(2)	4.36 φ	49 μ	4.14 φ	57 μ	2.85 φ	−0.20	0.90
14259,52-Comprehensive	(1)	3.98 φ	64 μ	3.74 φ	75 μ	1.92 φ	−0.20	1.01
	(2)	4.11 φ	58 μ	3.95 φ	65 μ	1.74 φ	−0.14	1.13
14260,4-Weigh Bag	(1)	3.40 φ	95 μ	3.08 φ	118 μ	2.13 φ	−0.35	1.19
	(2)	3.60 φ	83 μ	3.54 φ	86 μ	1.51 φ	−0.12	0.87
14230,130-Core	(1)	3.16 φ	112 μ	2.94 φ	131 μ	2.31 φ	−0.18	0.89
	(2)	3.35 φ	97 μ	3.23 φ	107 μ	2.04 φ	−0.12	0.87
14230,121-Core	(1)	3.89 φ	68 μ	3.75 φ	74 μ	1.64 φ	−0.18	1.03
	(2)	3.92 φ	66 μ	3.81 φ	72 μ	1.48 φ	−0.12	0.92
14230,113-Core	(1)	3.58 φ	84 μ	3.19 φ	110 μ	2.08 φ	+0.09	0.90
	(2)	3.71 φ	77 μ	3.52 φ	87 μ	1.67 φ	−0.17	0.89

Note: (1) refers to entire sample and includes Curator weight data on material greater than 1 mm combined with our data on material less than 1 mm, (2) refers to our data on material less than 1 mm only.

incorporates data from the Curator's office on weights of samples greater than 1 mm. The true size distribution must include the entire sample, but it is also important to have the size distribution of the less than 1 mm size fraction because this is the fraction that was analyzed chemically and studied in detail by most investigators.

Selected size fractions were thin-sectioned and particles were classified and counted with a petrographic microscope. The coarse soil fraction (1–4 mm) was classified and counted using a binocular microscope; representative fragments were then thin-sectioned for additional petrographic study. Selected particles also were examined with a scanning electron microscope (SEM).

Particle Size Characteristics

As shown in Table 1, most of the soil samples have similar size characteristics. The Cone Crater sample, 14141, shows the largest deviation; it is the coarsest and most poorly sorted. Sample 14003, the contingency sample collected in the immediate vicinity of the LM, is second in coarseness only to Cone Crater soil in the less than 1 mm size fraction. However, other parameters such as agglutinate content and carbon content (Moore *et al.*, 1972) show that this sample is more nearly like the typical "smooth terrain" samples 14259 and 14263. It is possible that 14003 was biased in collection, sample handling, or the soil in the vicinity of the LM was eroded by the LM descent engine exhaust and some fine-grained material was preferentially blown away (Scott, 1972).

The trench bottom sample, 14149, also is considerably coarser than the average soil sample if the entire size range is considered; however, this deviation disappears if only the material finer than 1 mm is considered. By comparison with Apollo 11 and 12 soils (Table 2), the Apollo 14 samples are the most poorly sorted of the three sets of samples as indicated by the higher graphic standard deviation. Apollo 14 samples are, on average, slightly coarser than samples returned from Apollo 11 and 12, but if the coarse Cone Crater, trench bottom and the disturbed contingency samples are excluded from the average, the Apollo 14 soils become intermediate in grain size between Apollo 11 and 12. It should be pointed out, however, that considerable overlap exists among the mean grain size of soils from the three missions. Apollo 14 soils include samples that are the finest (14163) and the coarsest (14141) soils collected during the first three missions.

If only the sample with the smallest mean grain size from each mission is considered, the graphic mean grain sizes are remarkably close: Apollo 11, 4.03 ϕ (62 μ) (King *et al.*, 1971, an average of three 10084 analyses); Apollo 12, 4.07 ϕ (59 μ) (King *et al.*, 1971, sample 12001), and Apollo 14, 4.14 ϕ (57 μ) (this paper, sample 14163).

Particle Types and Abundances

Particles in the Apollo 14 soils were divided into four general groups: (1) agglutinates (glass-bonded aggregates) and inhomogeneous glass droplets, (2) lithic fragments, (3) mineral fragments, and (4) miscellaneous glasses.

Most particles in group 1 are agglutinates (Fig. 1) that consist of mineral, lithic, and glassy debris bonded together by inhomogeneous glass. The group also includes inhomogeneous, schlieren-bearing glass droplets.

Group 2 consists primarily of vitric and recrystallized or thermally metamorphosed (Warner, 1972) breccias. The small particle size of our samples makes difficult a classification of breccia fragments based on clast content, as suggested by Jackson and Wilshire (1972). For the purpose of this study, a distinction was made only

Table 2. Comparison of Apollo 11, 12, and 14 size parameters (less than 1 mm size fractions).

Sample	Graphic mean		Inclusive Graphic standard deviation
Apollo 11 (King *et al.*, 1971, average of three)	4.03 ϕ	61 μ	1.95 ϕ
Apollo 12 (King *et al.*, 1971, average of twelve)	3.68 ϕ	78 μ	1.83 ϕ
Apollo 14 (this paper, average of eight surface samples and three core samples)	3.62 ϕ	81 μ	2.06 ϕ
Apollo 14 (this paper, average of five surface samples and three core samples—omits 14141, 14003 and 14149)	3.77 ϕ	73 μ	1.98 ϕ
Apollo 14 (this paper, average of three core samples)	3.52 ϕ	87 μ	1.73 ϕ

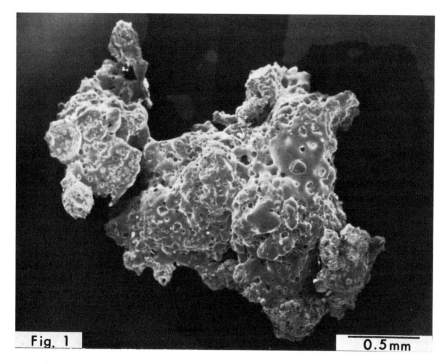

Fig. 1. SEM photograph of an agglutinate or glazed aggregate from sample 14002,5.
Agglutinates are formed when micrometeorites impact lunar soil (see text).

between vitric and recrystallized breccia fragments. The former consist of angular glass, mineral, and lithic fragments in a vitric matrix. The latter breccias have a glass-free matrix (classes 5, 6, 7, and 8 of Warner, 1972). Other less abundant types of lithic fragments include feldspathic rocks, pyroxenites, tachylites, and basalts.

In group 3, the most abundant material consists of mineral fragments of ortho-pyroxene, clinopyroxene (augite, pigeonite), plagioclase, ilmenite, and olivine. Many of the feldspars and pyroxenes are intensely fractured. Less common in the soil are grains of red spinel, tridymite, K-feldspar, and quartz.

Group 4 consists of all other glass particles. These include pale brown, pale green, colorless, gray-brown, and ropy glasses. The ropy glasses have schlieren of aligned inclusions and are often twisted around the long axis. Similar ropy glasses from Apollo 12 soils have been discussed in detail by Meyer *et al.* (1971) and McKay *et al.* (1971).

VARIATION IN THE SOILS

Soils collected at the Fra Mauro site are derived from two types of terrain: (1) the soils from the LM-ALSEP-Triplet Crater area or the "smooth terrain" of Swann *et al.* (1971); (2) the soil from the ejecta blanket surrounding Cone Crater. The pro-portion of different particle types from both areas is shown in Table 3 and Table 4;

Table 3. Comparison of a mature soil ("smooth terrain," 14003) to an immature soil
(Cone Crater ejecta, 14141) for four size fractions. All values except bottom row are in percent.

Components	14141,30				14003,28			
	150–250μ	90–150μ	60–75μ	20–30μ	150–250μ	90–150μ	60–75μ	20–30μ
Agglutinates	5.3	5.2	6.5	12.5	54.2	60.3	56.5	43.5
Microbreccias								
Recrystallized	57.5	47.2	36.5	15.5	19.2	20.5	16.5	7.0
Vitric	6.9	6.8	16.5	5.5	4.4	3.0	1.0	0.5
Angular Glass Fragments								
Brown	2.3	6.2	4.0	7.0	7.2	4.3	8.0	6.5
Colorless	0.3	2.6	1.5	5.0	2.0	3.0	3.0	3.5
Pale Green	0.3	—	—	—	0.2	—	—	—
Glass Droplets								
Brown	0.8	2.0	0.5	2.0	2.2	0.3	2.0	7.0
Colorless	—	0.2	—	—	0.2	0.3	—	—
Ropy Glasses	—	—	—	—	—	1.0	0.5	—
Clinopyroxene	4.1	8.0	15.5	18.5	3.0	2.3	5.5	18.5
Orthopyroxene	1.2	2.0	2.5	9.5	1.2	1.3	4.5	5.0
Plagioclase	5.6	6.6	11.5	21.5	3.6	2.3	1.5	7.0
Olivine	0.5	0.4	1.5	0.5	1.0	—	—	—
Opaque Minerals	—	—	—	—	—	—	—	1.5
Basalt	5.9	4.2	0.5	—	0.8	1.3	—	—
Anorthosite	0.3	1.2	—	—	0.2	—	—	—
Tachylite	8.9	7.0	3.0	2.5	—	—	1.0	—
No. of Grains								
Counted	400	500	200	200	500	300	200	200

Table 4. Comparison of the 90–150 μ fractions from Apollo 14 soil samples.
All values except bottom row are in percent.

Components	14259	14148	14156	14149	14230,113	14230,121	14230,130	14141
	Surface comprehensive samples	Trench Top (Sta. G)	Trench middle (Sta. G)	Trench bottom (Sta. G)	Core Sta. G (−8.0 cm)	Core Sta. G (−13.0 cm)	Core Sta. G (−17.0 cm)	Cone Crater surface
Agglutinates	51.7	50.2	47.7	26.4	53.2	57.0	51.5	5.2
Breccias								
Recrystallized	20.3	24.2	23.4	27.0	13.5	16.3	27.0	49.5
Vitric	5.0	4.6	—	8.2	—	4.3	0.4	7.8
Angular Glass Fragments								
Brown	6.7	6.0	7.8	6.0	8.4	7.3	5.2	7.6
Colorless	3.3	2.4	3.2	2.8	1.9	3.0	0.4	2.6
Pale Green	—	0.8	0.9	—	—	1.3	0.9	—
Glass Droplets								
Brown	1.3	1.8	3.2	1.6	4.8	0.3	2.6	2.0
Colorless	0.6	—	0.4	2.0	—	—	—	0.2
Pale Green	0.3	0.2	1.5	0.4	—	—	—	—
Clinopyroxene	3.7	1.8	4.3	7.4	6.8	3.0	3.0	8.0
Orthopyroxene	0.6	1.4	1.3	3.6	0.6	0.6	—	3.8
Plagioclase	4.7	3.0	5.4	7.8	6.5	3.3	3.4	7.6
Olivine	—	0.8	0.2	—	0.6	—	0.4	0.4
Opaque Minerals	—	0.4	0.4	0.4	—	—	—	0.4
Ropy Glasses	0.6	0.2	0.2	—	2.9	1.7	2.1	—
Basalt	1.0	2.0	—	6.4	0.3	1.6	2.1	4.2
No. of Grains								
Counted	300	500	500	500	300	300	234	500

the most notable contrast is that the former is rich in agglutinates and the latter rich in lithic and mineral fragments.

It is possible to study variations in the soil at Station G with depth. A 35 cm deep trench dug by the astronauts exhibited visible layering. A core sample (14230), that penetrated about 17 cm into the soil, was collected near the trench. The core is described by Fryxell and Heiken (1971). The core samples and the top and middle layers of the trench are similar to surface samples collected in the "smooth terrain" and contain abundant agglutinate fragments. The soil sample collected from the trench bottom (about 30 cm below the surface) has less agglutinates, more breccia and mineral fragments, and the highest content of basalt fragments of any Apollo 14 soil. This trench bottom soil is similar in many respects to the soil collected at Cone Crater (14141), and this layer may be part of a ray from Cone Crater, overlain by 20 to 30 cm of ejecta from other nearby cratering events. This correlation is supported by particle track studies (Crozaz *et al.*, 1972) which show that track densities in feldspars from Cone Crater and trench bottom soils are similar (less than $10^8 t/cm^2$). All other soil samples have track densities greater than $10^8 t/cm^2$. Crozaz *et al.* (1972) also suspect that the soil in the bottom of the trench is from Cone Crater and has been covered by older, more irradiated material.

Sᴏɪʟ Mᴀᴛᴜʀɪᴛʏ

The maturity of a lunar soil reflects the amount of meteorite reworking it has undergone and this in turn is a function of exposure age or residence time on the lunar surface. Total exposure age can be determined directly by particle track methods (Arrhenius *et al.*, 1971; Crozaz *et al.*, 1971; Crozaz *et al.*, 1972). Correlation of exposure ages with characteristics of particle content and grain size parameters provide criteria for indirect determinations of soil maturity.

Hartung *et al.* (1972) show that most of the mass impacting the moon from space is in the form of micrometeorites, which make millimeter-size craters. Most of the micrometeorites impact lunar soil rather than bare rock surfaces. Agglutinates (glazed aggregates of Duke *et al.*, 1970; McKay *et al.*, 1971) are the primary constructional product of these impacts; they are 10 to 100 times more abundant than glass spheres in most lunar soils. Some of the agglutinates are bowl shaped or doughnut shaped and may be the original glassy crater formed by a micrometeorite striking the soil. Others are more dendritic and branching and may result from molten splash glass penetrating the soil. Agglutinates are the major type of constructional particle and the major, glassy component now being formed on the lunar surface.

Inasmuch as agglutinates are a direct result of the reworking of soil by micrometeorites, they are also an index of maturity or exposure age. Studies of Apollo 12 particle track densities (Fig. 2) show that those soils that have low track densities also have low agglutinate contents, (Arrhenius *et al.*, 1971; McKay *et al.*, 1971). The data of Crozaz *et al.* (1972) indicate that this correlation also is true for Apollo 14 samples. Cone Crater surface soil (14141) and trench bottom sample (14149) are lowest in particle track densities and are also lowest in agglutinate content.

It should be pointed out that total glass content of a soil *cannot* be used as an index of exposure age or maturity as suggested by Lindsay (1971). Sample 12033 is

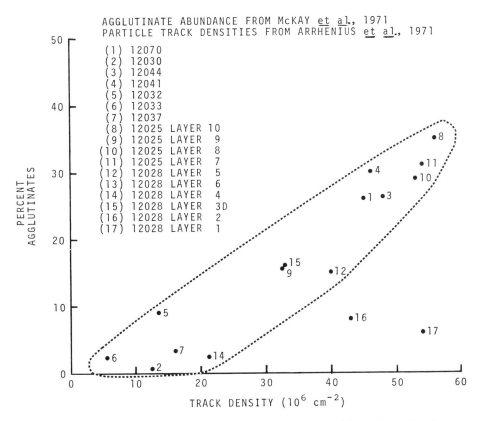

Fig. 2. Plot of agglutinate (glazed aggregate) content versus particle track densities for Apollo 12 samples. The core samples for the two measurements were not taken from exactly the same horizons and this may explain some of the scatter, particularly points 16 and 17. The dotted line emphasizes the trend of high agglutinate content correlating with high particle track density.

rich in ropy glass and sample 12030 is rich in breccia glass, yet both have low track densities, low agglutinate contents, and consequently are immature soils.

Size parameters also may be used as an indication of maturity. Figure 3 is a plot of graphic standard deviation (σ_1) against mean grain size (M_z) for Apollo 14 soils and two Apollo 12 core samples. The mature soils with high agglutinate content and high particle track densities (Crozaz *et al.*, 1971; Arrhenius *et al.*, 1971; Crozaz *et al.*, 1972) cluster around a mean grain size of 3 to 4 ϕ (125–63 μ) and a sorting of 1.5 to 2.5 ϕ. The immature soil samples (14141, 14149, and Layer VI of 12028) are coarser and, except for 12028 VI, more poorly sorted. With time, rock and mineral fragments are comminuted, agglutinates are produced, and the resulting mature soils become finer and better sorted and more negatively skewed (Fig. 4). The increased negative skewness indicates excess coarse material and may reflect the incorporation of fine particles into agglutinates (Lindsay, 1971).

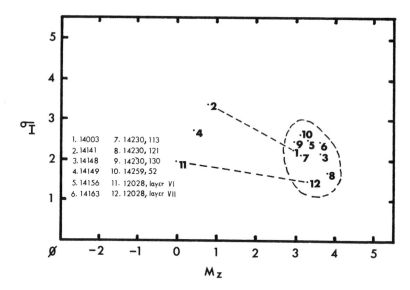

Fig. 3. Mean grain size (M_z) plotted against graphic standard deviation (σ_1) for Apollo 14 soils and two Apollo 12 core samples (Apollo 12 core data from J. Lindsay, unpublished data). Dashed circle encloses mature soil samples, i.e., high agglutinate contents. Lines connect examples of mature and immature soils (see text).

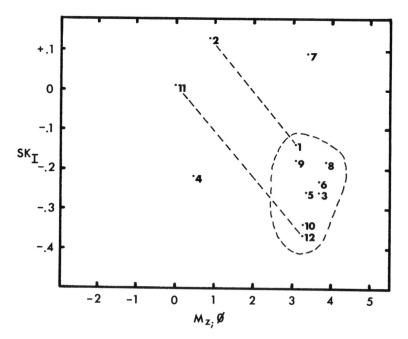

Fig. 4. Mean grain size (M_z) plotted against skewness (SK_1) for samples in Fig. 1. Dashed circle encloses mature soil samples (those with high agglutinate contents). Lines connect examples of mature and immature soil (see text).

The mean grain size of about 60 μ for mature soils from all Apollo missions may in fact represent an equilibrium grain size at which the destructional processes pulverizing the soil are balanced by constructional processes that form agglutinates and other glassy particles (Duke *et al.*, 1970; Lindsay, 1971; McKay and Ladle, 1971).

The production of agglutinates with time also must contribute to the darkening of bright halo craters, rays, and other fresh features. Adams and McCord (1971) showed that vitrification of ilmenite-bearing basalt significantly lowers its albedo and on the lunar surface this vitrification is accomplished primarily through the production of agglutinates.

COARSE FINES

Particles from the 1–4 mm fraction of each sample were classified and the results are presented in Table 5. Material in this size fraction is dominated by breccia fragments. Agglutinates are not as common as in finer fractions; discrete glass fragments and single mineral grains are rare. The abundance of basalt in the bottom of the trench is notable and may imply that a cratering event in the vicinity had penetrated the Fra Mauro formation to an underlying basalt. However, the basalt may be derived from clasts in the Fra Mauro.

Breccias were classified, as previously discussed, into vitric and recrystallized groups. The vitric breccias include medium to dark gray friable breccias, some of which have light clasts. They are similar to the F1 and F2 breccias of Jackson and Wilshire (1972) and the low to medium facies of Warner (1972). The recrystallized breccias are tough, light to dark gray, lack glassy clasts, and show an annealed matrix lacking appreciable glass. These correspond approximately to the F3 and F4 breccias of Jackson and Wilshire (1972) and the high facies of Warner (1972). This group includes some fragments that compositionally are anorthosite, but most fragments are noritic with a predominance of plagioclase and low-calcium pyroxene and subordinate ilmenite and clinopyroxene. Clasts are primarily plagioclase but also include pyroxene, basalt, and breccia. Many of the breccia fragments in this group contain vugs, which are discussed in more detail by McKay *et al.* (this volume).

As shown by the ratio of recrystallized to vitric breccia in Table 5, recrystallized breccias predominate in the comprehensive, bulk, and contingency samples but vitric breccias predominate in the top of the trench and upper part of the core, both from Station G. Material from Cone Crater surface samples is 94% recrystallized breccia. According to Sutton *et al.* (1972) this represents Fra Mauro material excavated from 60 to 80 m deep at Cone Crater. The slight enrichment of recrystallized breccia over vitric breccia at the comprehensive, bulk, and contingency sites as compared to Station G may indicate that the west ray from Cone Crater (Sutton *et al.*, 1972) has contributed recrystallized breccia to the former sites but not to Station G, which is farther off the trend of the ray.

ROPY GLASSES

Ropy glasses of the type present in Apollo 12 samples 12033 and 12032 (McKay *et al.*, 1971; Meyer *et al.*, 1971) are relatively rare in the Apollo 14 soil samples.

Table 5. Modal composition of the coarse fine (1–4 mm) fragments in Apollo 14 soils. All values except bottom two rows are in percent.

Components	14257,6 14258,25 Comprehensive	14001,3 14002,5 Contingency	14161,28 14162,40 Bulk	14146,3 14147,2 Trench top	14153,2 14154,3 Trench middle	14151,13 14152,3 Trench bottom	14230,113 Core* (−8.0 cm)	14230,121 Core* (−13.0 cm)	14230,130 Core* (−17.0 cm)	14142,2 14143,2 Cone Crater
Agglutinates	17	18	13	8	22	0	9	25	8	0
Breccias										
Vitric	29	29	30	56	22	28	45	50	23	4
Recrystallized	39	37	41	20	42	22	18	25	61	94
Glass Fragments	13	10	12	12	4	6	9	0	4	2
Plagioclase	0	3	1	0	0	0	0	0	0	0
Pyroxene	0	0	0	0	0	0	9	0	0	0
Basalt	1	0	0	0	0	44	9	0	4	0
No. of grains counted	69	115	143	25	45	18	11	4	26	123
Ratio recrystallized/ vitric breccias	1.3	1.3	1.4	0.4	1.9	0.8	0.4	0.5	2.7	24

* Core samples include 0.5–1 mm size fraction.

Table 6. Ropy glass and agglutinate content of 0.25–1 mm size fraction. Values except bottom row are in percent.

Components	14259,22 Comprehensive	14003,28 Contingency	14163,76 Bulk	14148,23 Trench Top	14156,23 Trench Middle	14149,39 Trench Bottom	14230,113 Core (−8.0 cm)	14230,121 Core (−13.0 cm)	14230,130 Core (−17.0 cm)	14141,30 Cone Crater
Ropy glasses	1.0	0.6	1.1	1.0	0.8	0.2	2.1	3.5	5.7	0.4
Agglutinates	30	24	20	32	25	16	—	—	—	1
No. of grains counted	406	500	363	496	508	485	141	115	175	553

Table 6 shows the ropy glass content in the 0.25–1 mm size fraction. The average ropy glass content for the surface samples and trench is 0.6% and for the three core samples is 3.8%. Apparently the event that provided the ropy glasses at the Apollo 12 site did not contribute significantly to the Apollo 14 site.

Conclusions

1. Agglutinates (glazed aggregates) are formed primarily by micrometeorite impact into lunar soil; in some cases the agglutinates may represent the glassy crater formed in the soil. Agglutinates appear to be the major particle type now being formed on the lunar surface and most of the glass now being produced on the moon forms as agglutinates.

2. Agglutinate content of a soil increases with particle track densities and with surface exposure time; consequently, it is a measure of soil maturity. Moreover, a mature lunar soil has finer particle size, better sorting and is more coarsely skewed than an immature lunar soil.

3. Prolonged micrometeorite bombardment produces an equilibrium soil with a mean grain size of about 60 μ. The equilibrium results when the destructional effects of the impacts are balanced by the constructional effects that produce agglutinates.

4. Surface material collected in the "smooth terrain" at the Apollo 14 site represent mature soils. These soils contain abundant agglutinates, high track densities, and are relatively low in lithic fragments.

5. The Apollo 14 soils all are characterized by a high breccia content and clearly are derived in large part from the underlying Fra Mauro formation.

6. The soil from Cone Crater is immature, contains only rare agglutinates, and consists mainly of highly recrystallized breccia fragments (94% of coarse fines) and mineral grains. This material may be the most representative sample of the deep interior of Fra Mauro formation.

7. Ropy glasses of the Apollo 12 type are present but are relatively rare in the Apollo 14 soils. If a Copernican ray provided the ropy glasses to the Apollo 12 site, a similar ray did not cover the Apollo 14 site.

8. Basalt fragments and large mineral grains are rare in the Apollo 14 soils in contrast to the mare soils of Apollo 11 and 12. Only the trench bottom sample contains appreciable basalt; this basalt may be clast material from Fra Mauro breccia or it may be derived from a cratering event that penetrated basalt bedrock.

Acknowledgments—We would like to thank John Lindsay, LSI, Houston, Texas, for permission to use his unpublished data on the Apollo 12 core. We also thank Odette B. James for critical review and helpful suggestions.

References

Adams J. B. and McCord T. B. (1971) Optical properties of mineral separates, glass, and anorthositic fragments from Apollo mare samples. *Proc. Second Lunar Sci. Conf., Geochim. Cosmochim. Acta,* Suppl. 2, Vol. 3, pp. 2183–2195. M.I.T. Press.

Arrhenius G. Liang S. Macdougall D. Wilkening L. Bhandari N. Bhat S. Lal D. Rajagopalan G. Tamhane A. S. and Venkalavaradan V. S. (1971) The exposure history of the Apollo 12 regolith. *Proc. Second Lunar Sci. Conf., Geochim. Cosmochim. Acta,* Suppl. 2, Vol. 3, pp. 2583–2598. M.I.T. Press.

Crozaz G. Walker R. and Woolum D. (1971) Nuclear track studies of dynamic surface processes on the moon and the constancy of solar activity. *Proc. Second Lunar Sci. Conf., Geochim. Cosmochim. Acta,* Suppl. 2, Vol. 3, pp. 2543–2558. M.I.T. Press.

Crozaz G. Drozd R. Hohenberg C. M. Hoyt H. P. Ragan D. Walker R. M. and Yuhas D. (1972) Solar flare and galactic cosmic ray studies of Apollo 14 samples (abstract). In *Lunar Science—III* (editor C. Watkins), pp. 167–169, Lunar Science Institute Contr. No. 88.

Duke M. B. Woo C. C. Sellers G. A. Bird M. L. and Finkelman R. B. (1970) Genesis of lunar soil at Tranquillity Base. *Proc. Apollo 11 Lunar Sci. Conf., Geochim. Cosmochim. Acta,* Suppl. 1, Vol. 1, pp. 347–361. Pergamon.

Folk R. L. and Ward W. C. (1957) Brazos River bar, a study in the significance of grain-size parameters. *J. Sediment. Petrol.* **27,** 3–26.

Fryxell R. and Heiken G. (1971) Description, dissection and subdivision of Apollo 14 core sample 14230. *NASA Tech. Memo X-58070,* 16 pp.

Hartung J. B. Horz F. and Gault D. E. (1972) The origin and significance of lunar microcraters (abstract). In *Lunar Science—III* (editor C. Watkins), pp. 363–365, Lunar Science Institute Contr. No. 88.

Jackson E. D. and Wilshire H. G. (1972) Classification of the samples returned from the Apollo 14 landing site (abstract). In *Lunar Science—III* (editor C. Watkins), pp. 418–420, Lunar Science Institute Contr. No. 88.

King E. A. Jr. Butler J. C. and Carmen M. F. Jr. (1971) The lunar regolith as sampled by Apollo 11 and Apollo 12: Grain size analyses, modal analyses, and origins of particles. *Proc. Second Lunar Sci. Conf., Geochim. Cosmochim. Acta,* Suppl. 2, Vol. 1, pp. 737–746. M.I.T. Press.

Lindsay J. F. (1971) Sedimentology of Apollo 11 and 12 lunar soils. *J. Sediment. Petrol.* **41,** 780–797.

McKay D. S. and Ladle G. H. (1971) Scanning electron microscope study of particles in the lunar soil. *Proc. Fourth Annual Scanning Electron Microscope Symposium,* IIT Res. Inst., 177–184.

McKay D. S. Morrison D. A. Clanton U. S. Ladle G. H. and Lindsay J. F. (1971) Apollo 12 soil and breccia. *Proc. Second Lunar Sci. Conf., Geochim. Cosmochim. Acta,* Suppl. 2, Vol. 1, pp. 755–773. M.I.T. Press.

Meyer C. Brett R. Hubbard N. J. Morrison D. A. McKay D. S. Aitken F. K. Takeda H. and Schonfeld E. (1971) Mineralogy, chemistry, and origin of the KREEP component in soil samples from the Ocean of Storms. *Proc. Second Lunar Sci. Conf., Geochim. Cosmochim. Acta,* Suppl. 2, Vol. 1, pp. 393–411. M.I.T. Press

Moore C. B. Lewis C. F. Cripe J. Kelly W. R. and Delles F. (1972) Total carbon, nitrogen, and sulfur abundances in Apollo 14 lunar samples (abstract). In *Lunar Science—III* (editor C. Watkins), pp. 550–551, Lunar Science Institute Contr. No. 88.

Scott R. F. (1972) Soil erosion during lunar module landing (abstract). In *Lunar Science—III* (editor C. Watkins), pp. 692–693, Lunar Science Institute Contr. No. 88.

Sutton R. L. Hait M. H. and Swann G. A. (1972) Geology of the Apollo 14 landing site (abstract). In *Lunar Science—III* (editor C. Watkins), pp. 732–734, Lunar Science Institute Contr. No. 88.

Swann G. A. Bailey N. G. Batson R. M. Eggleton R. E. Hait M. H. Holt H. E. Larson K. B. McEwen M. C. Mitchell E. D. Schaber G. G. Schafer J. P. Shephard A. B. Sutton R. L. Trask N. J. Ulrich G. E. Wilshire H. G. and Wolfe E. W. (1971) Preliminary geological investigations of the Apollo 14 landing site. *Apollo 14 Preliminary Science Report,* National Aeronautics and Space Administration. NASA SP-235, pp. 38–85.

Warner J. L. (1972) Apollo 14 breccias: Metamorphic origin and classification (abstract). In *Lunar Science—III* (editor C. Watkins), pp. 782–784, Lunar Science Institute Contr. No. 88.

Proceedings of the Third Lunar Science Conference
(Supplement 3, *Geochimica et Cosmochimica Acta*)
Vol. 1, pp. 995–1014
The M.I.T. Press, 1972

Noritic fragments in the Apollo 14 and 12 soils and the origin of Oceanus Procellarum

G. J. Taylor, Ursula B. Marvin, J. B. Reid, Jr., and J. A. Wood

Smithsonian Astrophysical Observatory, Cambridge, Massachusetts 02138

Abstract—Fragments of noritic breccias having a range of recrystallization textures and noritic basalts comprise about 80% of the 1–2 mm soil samples from Cone Crater (14142) and the bottom of the trench (14151). In five other samples, noritic fragments range in abundance from 42–56%. The remaining particles in each case are glasses, soil breccias, and glass-bonded aggregates; other crystalline rock types such as anorthosites constitute less than 1% of the soil. Noritic materials, i.e., Apollo 12 and 14 norites and ropy (KREEP) glass particles, can be divided into two broad groups on a chemical basis. The particles in one group have relatively high bulk TiO_2 contents and high $Fe/(Fe + Mg)$ ratios in their normative pyroxenes; the ropy glasses from Apollo 12 and 14 and a few Apollo 12 noritic rock fragments are in this category. Particles in the other class contain pyroxenes with a lower $Fe/(Fe + Mg)$ and have a range of TiO_2 contents; most noritic rock fragments from Apollo 12 and 14 are in this group. The ropy glasses may be from Copernicus or other rayed craters, but almost all the norites derive from the Fra Mauro formation or other noritic occurrences associated with Oceanus Procellarum. Non-KREEP components in the Apollo 14 soil include glasses having the compositions of anorthositic gabbro, mare basalts, and howarditic meteorites; but there are no crystalline equivalents of these glasses in our soil samples. Apparently, cratering projects glassy debris greater distances on the moon than it does lithic fragments.

Lunar norite appears to be associated uniquely with Oceanus Procellarum, the largest uninterrupted mare surface on the moon. We suggest that the Procellarum Basin was excavated by multiple impacts of planetesimals, focused on the earth-facing side of the moon by terrestrial gravity; and that heat generated by this concerted bombardment promoted partial melting in or under the local anorthositic crust, giving rise to noritic magmas.

Introduction

We have sectioned, examined, and classified 679 soil particles in the size range 1–2 mm from seven Apollo 14 soil samples, and 549 particles in the range 420 μ–1 mm from two other samples (Table 1). Each soil sample was weighed, then rinsed with methyl alcohol through nitex sieves with openings of 420 and 600 μ. All particles coarser than 600 μ were photographed, hand-picked into lithologic groups, and cut into polished thin sections for petrographic surveys and electron-microprobe analysis. In our text, a number such as (230–15) refers to particle 15 on our numbered photographic map of thin section no. 230.

In this paper, we summarize the results of this survey and briefly describe the petrology and chemistry of noritic rocks and glasses. Our observations and conclusions are presented in more detail in Taylor *et al.* (1972).

Table 1. Soil samples examined from Fra Mauro.

Sample no.	Particle size (mm)	Weight (mg)	SAO thin sections	Source
				EVA 1 West of LM. Smooth Terrain of Fra Mauro fm.
14002,8	1–2	204	205-210	Contingency sample, immediately outside LM. Scooped to depth of several cm.
14162,24	1–2	1000	242-248	Bulk fines, 5 m west of contingency sample. Depth several cm.
14163,145	<1	939	201, 203	Bulk fines, 5 m west of contingency sample. Depth several cm.
14258,36	1–2	520	212 & 219-224	Comprehensive fines, 140 m west of LM. Top 1 cm of soil.
14259,32	<1	950	202, 204	Comprehensive fines, 140 m west of LM. Top 1 cm of soil.
14421,20	<10	401		Comprehensive fines, 140 m west of LM. Top 1 cm of soil. (unsieved reserve)
14262,2	1–2	229	213-218	Residue of rocks in bag 1039 from site of comprehensive sample.
				EVA 2 East of LM.
14142,1	1–2	261	229-232	South rim of Cone Crater, Station C'.
				Trench, 75 cm long, 30 cm deep. Station G, Fra Mauro fm.
14146,2	1–2	255	233-237	Surface
14154,2	1–2	233	238-241	Middle
14151,12	1–2	261	225-228	Bottom

Petrology and Chemistry of Apollo 14 Soil Particles

Norites

The great majority of rock fragments in the Fra Mauro soil samples consist of calcic plagioclase and Ca-poor pyroxenes in approximately equal amounts. Lunar rocks having this mineralogy are known as norites. The Apollo 14 KREEP-rich noritic rock fragments fall into two broad categories. Most are recrystallized breccias very similar to those described in the Apollo 12 soil samples and termed Type A by Marvin et al., 1971; basaltic rocks chemically similar to the noritic breccias comprise a second, less abundant group. Three textural groups of recrystallized noritic breccias occur in our soil samples. (The loosely consolidated soil breccias are not among these breccia types.) Most abundant are a group of breccia fragments that appear to have recrystallized at relatively low, subsolidus temperatures (Figs. 1a, b, c). These consist of angular plagioclase and pyroxene clasts embedded in fine-grained matrices. Little intergrowth between clasts and matrix has occurred during recrystallization. Groundmass ilmenite characteristically has a finely disseminated texture. A continuous range of textures exists between these *low-temperature* norites and a second group of less porous breccias whose matrices contain the sort of euhedral plagioclase and pyroxene crystals characteristic of crystallization from a melt (Fig. 1d); the *high-temperature* noritic breccias apparently have undergone partial melting before recrystallization. A third textural type consists of clasts, commonly showing euhedral overgrowths, embedded in a matrix of small, uniform, equant pyroxene and plagioclase euhedra and commonly contain considerable K- and Si-rich glass (Fig. 1e, f). Judging from their coarser matrices, these *granulitic* noritic breccias have either cooled more slowly than the high-temperature breccias from temperatures above the solidus or perhaps have been subjected to protracted or repeated near-solidus recrystallizations. Modal analysis of 68 noritic particles has shown that low-temperature norites have the largest proportion of vugs (up to approximately 24 vol. %). High-temperature norites are somewhat less porous (up to 16% vugs), but the granulitic rocks are much more consolidated (11 of 14 examined have less than 4% vugs by volume).

Some chemical differences exist within and between these groups of rocks (see Chemical Distinctions Among Noritic Materials, below), but for the most part they seem to represent material from a fairly homogeneous source that has experienced a variety of recrystallization histories. Wilshire and Jackson (1972) have found evidence in the large Apollo 14 fragmental rocks for at least four impact recrystallization events of the sort implied from our petrographic study. We agree with Warner (1972) that these textures probably result from autometamorphism in thick, hot ejecta blankets of the scale of the Fra Mauro formation, rather than in local small-scale events of the sort responsible for soil-breccia formation. Warner's scheme, on the other hand, includes glass-rich soil breccias as part of the metamorphic sequence involving the recrystallized norites. We feel that soil breccias are products of local small-scale impact events, unrelated to the large-scale events responsible for the Fra Mauro formation.

Pyroxenes in all three textural types are dominantly of the low-Ca variety; ground-mass pyroxenes tend to fall in the pigeonite compositional range, but the larger pyroxene clasts tend to be orthopyroxene. Huebner and Ross (1972) have determined solidus and liquidus surfaces for the pyroxene quadrilateral and the stability fields for orthopyroxene, pigeonite, and augite at 900°C. It is pertinent that the stability field for pigeonite shrinks markedly with falling temperatures to Fe-rich compositions at 900°C, which are inaccessible to norite bulk compositions. It appears that the clastic material in the norites, which is dominantly orthopyroxene, has recorded a slow sub-solidus cooling history that has eliminated pigeonite compositions. The groundmass grains of these rocks, on the other hand, have recrystallized at or above solidus temperatures (approximately 1100°C) where, for these compositions, pigeonite is a stable phase.

Nearly all noritic plagioclase falls in the compositional range $Or_{0.4}Ab_{4.6}An_{95}$ to $Or_6Ab_{16}An_{78}$. The highest K_2O content measured ($Or_{12}Ab_{16}An_{76}$) was in an interstitial plagioclase grain in a granulitic norite. Interstitial glass ranges in composition from a high-SiO_2 material (82% SiO_2, 6.5% K_2O) to a lower SiO_2, high-K_2O glass close to orthoclase in composition (55% SiO_2, 15% K_2O). In these glasses, BaO ranges from 0.0–5% and tends to be inversely related to the SiO_2 content. Both whitlockite and apatite are found in the recrystallized norites. Whitlockites fall into two groups based on REE concentrations. REE-rich whitlockites have Nd concentrations between 1.5 and 2.5 wt%, while REE-poor whitlockites have Nd contents less than 0.4 wt%. Apatites contain on the order of 0.2% Nd.

The basaltic fragments show a wide range in groundmass grain size, ranging from plagioclase vitrophyres to ophitic orthopyroxene basalts, though in almost all cases the crystallization sequence is the same. Plagioclase is the liquidus phase, and together with pyroxenes ranging from pigeonite to subcalcic augite, it accounts for 90–95% of the volume of a given rock. A large number of interstitial phases are found

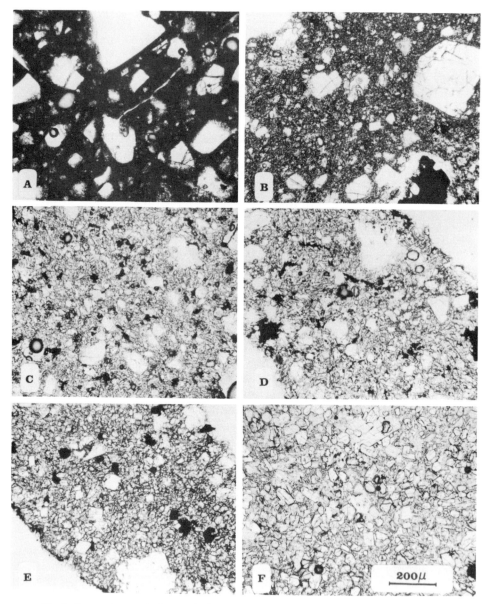

Fig. 1. Photomicrographs in unpolarized transmitted light of Apollo 14 noritic breccias; all to the same scale. A, B, C are examples of a textural spectrum of rock types termed "low-temperature noritic breccias" in the text. (a) Particle (210–8); representative of a class of breccias within the low-temperature norites that have the finest grained and least re-crystallized matrices, and hence seem to have been least affected by heat. Large angular clasts of pyroxene and plagioclase are embedded in a very fine-grained groundmass con-sisting of plagioclase, pyroxene, and finely disseminated ilmenite. (b) Particle (205–8); another low-temperature norite that has a coarser and somewhat more recrystallized matrix than (210–8) (Fig. 1a), and falls about in the middle of the textural spectrum we see in the

in the basalts; for example, at a single plagioclase-pyroxene boundary in a coarse noritic basalt particle (228–10), ilmenite, ulvöspinel, zirkelite, zircon, troilite, metal, phosphate minerals, and glass were observed to coexist. The interstitial glass shows the same range in composition as it does in the noritic breccias. Mineral compositions in the basalts are less uniform than in the recrystallized norites. Average values of Fe/(Fe + Mg), TiO_2, and Al_2O_3 in the pyroxenes of noritic basalts are higher than corresponding values in the pyroxenes of recrystallized rocks, suggesting that most of the noritic basalt fragments either represent primary igneous rocks or result from the remelting of soils with a different composition than that of the Fra Mauro soil. Whether these soil particles are primary igneous rocks or not, the presence of basalt clasts in large Apollo 14 breccias (LSPET, 1971) proves that at least some noritic basalts predate the Fra Mauro formation.

We measured the bulk chemical compositions of 56 norite particles by defocused-beam microprobe analysis (DBA). The results are displayed in Fig. 2, a triangular plot of normative mineral proportions that is designed to separate the main classes of lunar rocks: anorthosites, rich in plagioclase; norites, characterized by high contents of K and P and hence of orthoclase (K-feldspar) and phosphates; and mare basalts, notable for their high quantities of mafic silicates and oxides. We can make several observations from this diagram. The Apollo 14 low- and high-temperature noritic breccias have essentially the same range of compositions as do Apollo 12 norites. The granulitic norites, however, appear to fall into two groups: one is high in normative orthoclase plus apatite, and the other lower. Six of the eight basalts measured by us cluster in the lower left of the norite region of Fig. 2. (Rock 14310 would also plot in this group.) Finally, in contrast to our previous conclusions (Marvin et al., 1971; Wood, 1972), we note that no norite from either site has a composition that is gradational with the anorthositic rocks present in the Apollo 11 and, to a far lesser extent, the Apollo 12 soil samples. The REE distribution patterns (Hubbard et al., 1971) and the relatively high K_2O contents (>0.1%; Hubbard et al., 1971; Taylor et al., 1972) of most Apollo 14 anorthosites, however, indicate that they are derived from magmas of KREEP or noritic composition. The low K_2O (<0.05%) type of anorthosite, such as those in the Apollo 11 soil, apparently is unrelated to noritic, KREEP rocks.

low-temperature norites. (c) Particle (227–13); a low-temperature norite that has a still coarser and more recrystallized groundmass and probably was reheated to temperatures just below the solidus during recrystallization. (d) A representative particle (227–12) of a high-temperature breccia; euhedral groundmass phases suggest that the maximum temperature reached during recrystallization exceeded the solidus, allowing some of the rock to melt. (e) Granulitic noritic breccia (229–2) exhibiting a sugary groundmass that we feel either is the result of slower cooling than the high-temperature norites (Fig. 1d) experienced, or was developed during repeated impact recrystallization events. (f) A more advanced example of granulitic norite formation (229–16) in which there is little distinction between clasts and groundmass grains.

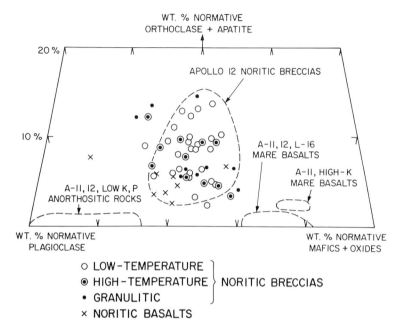

Fig. 2. Triangular plot, stretched vertically and truncated, of normative mineral compositions (in wt%) of Apollo 14 noritic breccia and basalt fragments analyzed by the defocused-
beam microprobe technique. Apollo 14 norites are compositionally similar to the Apollo 12
norites previously analyzed by us (unpublished data). No norites from either site, however,
are gradational to anorthosites.

Glass

In our Apollo 14 soil samples, four major types of glass particles can be distinguished on the basis of color and texture. We refer to these by the descriptive terms
ropy, cindery, green with metal droplets, and homogeneous.

Ropy glasses are particles whose microscopic appearance suggests they have
been pulled and twisted while hot. In thin sections, they are greenish-yellow and
commonly contain streaks of fine-grained opaque material. In many cases, crystallites
of plagioclase have precipitated and become aligned with the opaque streaks. Most
of the ropy glass particles have inclusions of orthopyroxene and/or plagioclase;
occasionally, fragments of noritic rock also are included. Many of the clasts are shock-
damaged, suggesting an impact rather than a volcanic origin for the glass.

Ropy glass compositions are displayed in Fig. 3. The glass is clearly noritic in
composition but occupies only the lower half of the norite field. Its average composition is given in Table 2. As many authors have noted (e.g., Meyer *et al.*, 1971;
Marvin *et al.*, 1971), ropy glasses also are present in the Apollo 12 soils. Some of these
glasses are greenish yellow, but most, particularly in samples 12032 and 12033, are
red brown to yellow brown. The average compositions of these two groups are very
similar (Table 2). The Apollo 14 greenish-yellow ropy glasses are slightly richer in

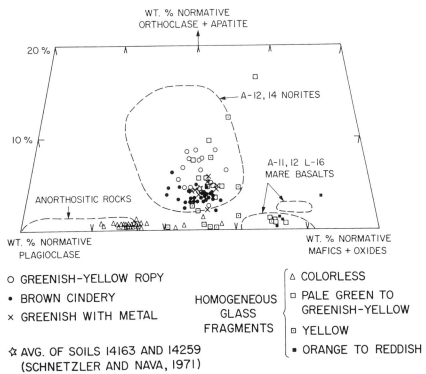

O GREENISH-YELLOW ROPY

• BROWN CINDERY

× GREENISH WITH METAL

HOMOGENEOUS
GLASS
FRAGMENTS

⎧
⎪
⎨
⎪
⎩

△ COLORLESS

□ PALE GREEN TO
   GREENISH-YELLOW

◨ YELLOW

■ ORANGE TO REDDISH

☆ AVG. OF SOILS 14163 AND 14259
   (SCHNETZLER AND NAVA, 1971)

Fig. 3. Five main types of glass can be distinguished on this plot of normative components:
homogeneous colorless glasses (anorthositic gabbros), greenish-yellow ropy glasses (noritic),
cindery glasses (melted Apollo 14 soil), green glass with metal droplets (also melted soil),
and other homogeneous glasses (mainly noritic or mare derived).

Al_2O_3 and marginally lower in FeO and MgO, but otherwise are not greatly different
from the Apollo 12 ropy glasses.

The apparent color difference between the greenish-yellow and red-brown
varieties could be due to differences in the oxidation states of the metals in the glasses
or to variable amounts of obscuration or diffraction of the light by submicron opaque
grains or vesicles. Whatever the reason, the dissimilarity in color strongly suggests
that the red-brown and greenish-yellow glasses have different source areas on the
moon.

Cindery glasses are irregular in shape and highly vesicular. In thin sections, they
are brownish and almost without exception contain inclusions of shocked mineral
and rock fragments. This type of glass occurs at all the lunar landing sites and approximates the composition of the local soil (see Marvin *et al.*, 1971; J. B. Reid *et al.*, 1972).
The same holds for Apollo 14 cindery glasses. In Fig. 3, they cluster near, though
slightly below, the average for two soils (Schnetzler and Nava, 1971). In Table 3, we
have listed the average composition of the cindery glasses and, for comparison, the
average of Schnetzler and Nava's (1971) soil analyses. The glass is lower in alkalis,
but otherwise has very nearly the same composition as the Apollo 14 soils.

Table 2. Average chemical compositions of some major types of lunar glasses (wt%).

	A	B	C	D	E	F	G
SiO$_2$	48.8	48.4	49.9	47.1	48.4	48.2	45.2
TiO$_2$	2.4	2.3	2.4	1.9	1.9	1.7	0.31
Al$_2$O$_3$	17.1	15.4	15.2	17.9	16.7	17.3	25.4
Cr$_2$O$_3$	0.20	0.23	0.23	0.22	0.20	0.19	0.15
FeO	10.6	11.2	11.6	10.2	9.9	10.2	5.9
MnO	0.15	0.17	0.18	0.16	0.17	0.12	0.10
MgO	8.0	8.2	8.3	9.8	9.6	9.3	8.4
CaO	10.4	10.3	10.1	11.0	11.1	10.6	14.5
Na$_2$O	0.73	0.84	0.48	0.46	0.52	0.62	0.16
K$_2$O	0.83	0.80	1.2	0.42	0.55	0.54	0.01
P$_2$O$_5$	0.65	0.67	0.98	0.47	0.39	0.48	0.11
NiO	0.01	0.01	0.01	0.04	0.07	—	0.01
SO$_3$	0.05	0.17	0.20	0.11	0.26	—	0.10
TOTAL	100.0	98.8	100.8	99.8	99.7	99.3	100.4

(A) A-14 greenish-yellow ropy glasses (15 analyses).
(B) A-12 greenish-yellow ropy glasses (21 analyses) (Marvin et al., 1971).
(C) A-12 red-brown ropy glasses (41 analyses) (Marvin et al., 1971).
(D) A-14 brown cindery glasses (29 analyses).
(E) A-14 green glass with metal droplets (8 analyses).
(F) Average of soils 14163 and 14259 (Schnetzler and Nava, 1971).
(G) A-14 colorless glass (31 analyses).

Table 3. Relative proportions of noritic breccias and basalts in Apollo 14 soils.
(Each entry is in percent of the total number of norites.)

	14142	14151	14154	14002	14162	14259
Breccias						
Low temperature	67.6	63.1	44.4	53.3	36.9	45.1
High temperature	12.3	19.5	11.0	23.3	20.7	10.0
Granulitic	11.9	4.9	25.9	16.6	34.4	35.1
Noritic Basalts	9.3	12.2	18.5	6.7	7.8	10.0
Total % norite in sample	81.3	80.4	41.6	47.6	52.5	27.1

Another type of glass also has the composition of the local soil. In thin sections, the glass is grayish with a slight greenish tint, and innumerable NiFe-metal and metal-troilite droplets are dispersed throughout each particle. The droplets range in size from ~0.5–50 μ. The composition of the glass is given in Fig. 3 and Table 2. The similarity to the Apollo 14 soils is quite obvious.

As above, we have the interesting problem of two glasses with different colors but the same composition. We measured absorption spectra for these glasses by using a Zeiss monochromator (range of 400–700 μ) and a Zeiss photometer with a cesium-antimony photocell. The spectra for the green-with-metal glasses are essentially flat, though there is a very slight peak in the green region of the spectrum (490–550 μ). Cindery glasses show a gradual increase in the amount of light transmitted from about 550–700 μ (yellow to red). It appears likely that the color of the green-with-metal glasses is governed largely by obscuration in an essentially colorless glass, but the brown cindery glass is genuinely colored by ions or diffraction effects.

Some glass particles are homogeneous throughout. They may be free forms (spheres, etc.) or angular fragments, with colors that range from colorless to orange and reds. In Fig. 3, we have plotted compositions of the homogeneous glasses, using different symbols to denote their colors. There appear to be five groups of glass particles. One consists of colorless glasses. These are low in K and P, and so plot along the base of the diagram; they have the normative composition of anorthositic gabbros (64–78% plagioclase). The average composition is given in Table 2. A. M. Reid et al. (1972) report that this is the dominant type of anorthositic glass in the soils from all the landing sites. They argue that it represents the composition of the lunar highlands. As we have shown above, however, at least two kinds of glass from Apollo 14 are samples of melted soils. It is also possible that the anorthositic gabbro glasses are melted samples of highland soils rather than a primary lunar rock type.

A second group of homogeneous glasses is greenish yellow to yellow and noritic in composition (Fig. 3). Glasses in this group probably represent melted noritic bedrock or nonropy equivalents of the ropy glasses.

A group of mafic glasses poor in K and P divides into two families on the basis of color. One group is deeply colored, high in Ti, and clearly derived from mare basalts. The others are weakly colored, pale-green glasses that are unusual because of their low TiO_2 content ($<1\%$) and low normative plagioclase ($<30\%$). Mafic glasses Ti- and KREEP-poor like these were also reported by Brown et al. (1971) in Apollo 14 soil 14259, and by J. B. Reid et al. (1972) and Wood et al. (1971) in the Luna 16 soil.

Finally, there is a group of yellowish glasses whose norms plot along the base of Fig. 3, below the norite field. They contain $\sim 2.5\%$ TiO_2 and $\sim 50\%$ normative plagioclase. Their low levels of K and P and their moderate TiO_2 content suggest that they are derived from a mare region.

Other types of particles

Like the soils returned by previous missions, the Apollo 14 fines contain abundant rock material that has been shocked and remelted. In the Fra Mauro region, the rocks so affected were mainly noritic breccias rather than mare basalt, but the textural types of the degraded rocks are the same as we have observed before: porous but cohesive soil breccias, glass-bonded aggregates of rock, mineral, and glass fragments, highly vesicular cindery glasses, and all gradations between these categories.

We did not observe any particles that are unequivocally mare basalt; particles (204–94) and (232–9) contain more mafic silicates and ilmenite than do most noritic basalts, but they are small in size and coarse-grained, so we cannot be certain we have seen representative assemblages. Particle (204–94) has minor amounts of a phosphate mineral, which suggests that the rock is noritic.

The Apollo 14 soil samples contain little other than norite and its degradation products. Only 14 of the 1228 particles we examined are fragments of other rock types: anorthosites (high-K type), granites, and ultrabasic rocks. These are described in Taylor et al. (1972).

CHEMICAL DISTINCTIONS AMONG KREEP-RICH MATERIALS

For the most part, noritic lithic fragments and glasses are remarkably uniform in composition. We have noted, however, that there are small but real compositional differences between certain of the classes of norite particles: differences in whole-rock TiO_2 content and in the ratio $Fe/(Fe + Mg)$ in norite pyroxenes.

Figure 4 is a histogram of bulk TiO_2 contents of noritic particles. All the categories of noritic material we have studied are included: lithic and glassy particles from Apollo 12 and 14 soils. The ropy noritic glasses from both sites contain about twice as much TiO_2 as do the crystalline norites; the two classes of material cannot have been simply derived from the same rock parent. The category of low-temperature Apollo 14 noritic breccias is an exception to this rule, however. Here, TiO_2 contents are evenly distributed between the ranges favored by crystalline and glassy norites. It is a curious fact that low-temperature noritic breccias in the samples collected from within a few centimeters of the lunar surface contain systematically less TiO_2 than do samples collected at greater depth. The near-surface breccias resemble other

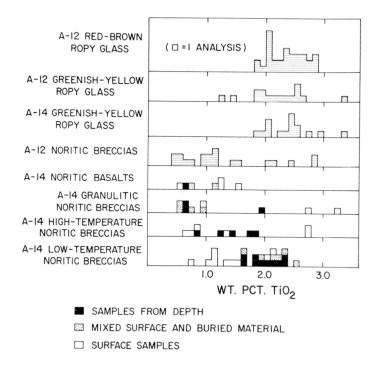

Fig. 4. Histogram of bulk TiO_2 contents, showing that the ropy glasses from Apollo 12 and 14 are richer in this oxide than are most noritic rocks. The low-temperature norites occurring in samples taken from the top few centimeters of the regolith contain less TiO_2 than do those from greater depth. The Apollo 12 norites span the entire range seen in the Apollo 14 norites, but most are lower in TiO_2 than are ropy glass. Apollo 12 analyses were made by us (unpublished data); Apollo 12 norite compositions are from Marvin et al. (1971) and Meyer et al. (1971).

crystalline norites in this regard, but breccias from depth correspond approximately to ropy noritic glasses in TiO_2 content.

Figure 5 is a histogram of the $Fe/(Fe + Mg)$ ratio in norite pyroxenes as determined directly by microprobe measurement and as calculated from the normative pyroxene compositions in defocused-beam whole-rock analyses. The essential similarity of the Apollo 14 noritic rocks is apparent. Low-temperature and high-temperature norites are quite restricted in $Fe/(Fe + Mg)$; some granulitic norites are considerably more Fe-rich and may come from a separate source. Again, the compositional range of Apollo 12 norites overlies that of Apollo 14 norites, suggesting that some of the Apollo 12 norites may originate from the same source material as do the Apollo 14 suite.

Meyer *et al.* (1971) noted a compositional similarity between the ropy glass fragments and the KREEP-rich noritic rocks in the Apollo 12 soil samples. There are striking similarities between the ropy glasses of the Apollo 12 and 14 samples (see Table 2, and Figs. 4 and 5), but it is clear from a consideration of Figs. 4 and 5 that the ropy glasses in these samples are not identical in composition to the noritic rocks

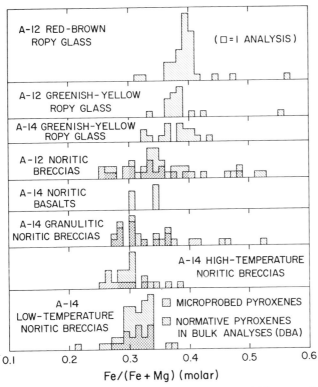

Fig. 5. Histogram of molar $Fe/(Fe + Mg)$ in pyroxenes, in which two main groups of noritic materials can be distinguished: those high in this ratio (ropy glass, a few Apollo 12 and 14 norites) and those low in $Fe/(Fe + Mg)$ (mostly noritic rocks). The high-Fe granulitic norites contain less TiO_2 than do high-Fe ropy glasses. Data from the same sources as in Fig. 4.

from Fra Mauro or the Apollo 12 soils. Although TiO_2 contents of the glasses are only slightly higher than those in the high-Ti low-temperature norites, the two groups of materials have quite different $Fe/(Fe + Mg)$ ratios in their respective pyroxenes. A few Apollo 14 granulitic noritic breccias contain pyroxenes with as high a $Fe/(Fe + Mg)$ ratio as do the normative pyroxenes of ropy glasses, but the former are impoverished in TiO_2. Some Apollo 12 norite particles are rich in TiO_2 and have high $Fe/(Fe + Mg)$ ratios; they may be related to the ropy glasses. On balance, however, it seems that the ropy glasses are a distinct compositional type among the lunar norites and may be derived either from a separate noritic parent rock type or from a melted regolith consisting of noritic rock and a small proportion of mare basalt relatively rich in Fe and Ti.

Relative Proportions and Probable Sources of Soil Fragments

The results of petrographic surveys of our thin sections are illustrated in Fig. 6. The most abundant particles in the regolith at Fra Mauro Base are recrystallized breccias and noritic basalts, making up 80% of the samples taken at the rim of Cone Crater and the bottom of the trench and 42–56% of the other soil samples. Anorthosites comprise less than 1% of the particles represented in Fig. 6. Colorless glasses of anorthositic composition are present at levels of 0–3%; ropy glasses, 0–4%; and various types of colored or devitrified glasses, 4–14%. The remaining particles are materials generated in the regolith: soil breccias, glass-welded aggregates, and cindery glasses rich in rock and mineral fragments. We found no particles that we could classify with confidence as mare basalt.

The geologic map of Wilhelms and McCauley (1971) shows that Fra Mauro base

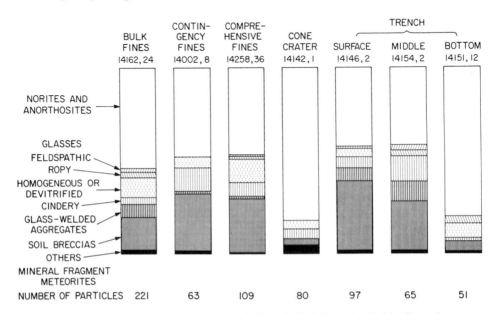

Fig. 6. Relative proportions of rock types and glasses in the 1–2 mm Apollo 14 soil samples.

lies near the eastern margin of a large north-south trending area, some 360 km long and 50–100 km broad, that is underlain by the Fra Mauro formation. A small exposure of the Cayley formation, a light-colored plains-forming unit, lies 12 km to the east. The nearest mare basalt flows lie about 50 km to the east and west. Small exposures of pre-Imbrian terra materials occur within 20–50 km to the east and south.

In order to compare our survey results with those predicted for soils randomly mixed by meteoroid impact, we adopted the method used by Marvin *et al.* (1971) and applied the range-frequency data of Shoemaker *et al.* (1970) to the Apollo 14 landing site. These authors calculated the proportions of ejecta particles that would be propelled various distances on the lunar surface by single- and by multiple-cratering events. Their results are equally useful for predicting the proportions of particles projected from various distances and mixed into the regolith at any given site.

Using Shoemaker's relationships we determined the percentages of particles that should have been contributed to the Fra Mauro soil samples by each of the local geological formations. Our results show that if each particle has been moved once, the regolith at the Apollo 14 landing site should contain fragments of Fra Mauro formation (90%), terra materials (3%). Cayley formation (3%), and mare basalts (2%). If each particle has moved three times, the predicted proportions are the following: Fra Mauro formation (82%), terra materials (2%), Cayley formation (5%), and mare basalts (10%). The small remainder in each case consists of undefined crater materials and 0.1–0.6% of dark mantling materials.

The range of rock types in our coarse fines is in fact more limited than that predicted by either of the above models, because we found no mare basalt particles. Nor were we able to distinguish a category of rock particles that could be specifically identified with the Cayley formation. It appears that the meteoroid bombardment has been substantially less effective in moving rock debris from one area to another than Shoemaker's relationships predict.

The Fra Mauro formation underlying the regolith is the main source of soil particles. Cone Crater, which is surrounded by blocks up to 15 m in width, has clearly penetrated the regolith; so we conclude that sample 14142, taken from the crater rim, yields the most direct evidence of the petrological character of the Fra Mauro formation. Norites of various types make up 83% of our 1–2 mm fraction of sample 14142. Mineral fragments and devitrified glasses constitute 8%, and soil breccias and cindery glasses make up the remaining 9% of this sample. We found only one particle of anorthosite (high-K type) and none at all of anorthositic or ropy glass.

The proportions of each type of noritic rock in the Apollo 14 soils are given in Table 3. The Cone Crater fines (14142) are particularly significant because this sample suggests the relative proportions of noritic rock types in the Fra Mauro formation. The most abundant are low-temperature noritic breccias, which constitute almost 70% of the norites in the sample. The other norite types are present in roughly equal amounts, at ~10%. The sample taken from the trench bottom (14151), like 14142, is relatively "fresh," containing 80% lithic fragments. It, too, is composed predominantly of low-temperature noritic breccias. The remaining samples contain far more degraded materials, which suggests that they have been subjected to meteoroid impacts for a much longer time than 14142 or 14151. Significantly, the percentage of low-temperature noritic breccias is much lower in these samples. This may mean that these porous, rather loosely consolidated rocks are most easily converted to soil breccias in the regolith. Alternatively it could mean that high-temperature and granulitic noritic breccias have been added to the local surface soil from distant noritic sources.

As we showed in the previous section, the ropy glasses have a composition distinct from the Apollo 14 norites and from most of the Apollo 12 norites. Although the red-brown and greenish-yellow ropy glasses have nearly the same composition, they have different colors and therefore probably come from different sources. The red-brown glass conspicuous in Apollo 12 samples 12032 and 12033 are considered by most authors (e.g., Meyer *et al.*, 1971; Marvin *et al.*, 1971; Eberhardt *et al.*, 1972) to be ray material from the Crater Copernicus. (Some authors, e.g., Quaide *et al.*, 1971, think the ropy glasses are locally derived, presumably from beneath the Procellarum basalts.) Possibly the Apollo 12 and 14 greenish-yellow ropy glasses are ray materials from other craters. The nearest rayed crater to Fra Mauro base is Lalande, which lies in a highland region 240 km to the east. A ray from Copernicus crosses the Apollo 14 site also, but only two brownish ropy glasses were found in our samples. Cone Crater cannot be a source for the Apollo 14 ropy glasses because the predominant material it ejected has too low an $Fe/(Fe + Mg)$ ratio (Fig. 5) for ropy glass.

The Apollo 12 norites show much greater compositional variability than do those from Apollo 14 (Figs. 4 and 5). In broad terms, these norites represent two main types: the minority are high in TiO_2 and $Fe/(Fe + Mg)$, like the ropy glasses, but most are like the Apollo 14 norites. Some of the latter may actually be particles ejected to the Apollo 12 landing site by impacts on the Fra Mauro formation. They may also represent rocks derived from beneath the local Oceanus Procellarum basalts.

Chemically, the most distinctive material in our Apollo 14 samples is the low-K colorless glass that is compositionally similar to anorthosites of the type we found in our Apollo 11 samples. This glass has no crystalline counterpart in the Apollo 14 soils that we have examined. We suggest that it derives from pre-Imbrian regoliths lying at such distances that the particles are projected to Fra Mauro base only by large, glass-forming impacts.

The mare basalt component, which is absent from our crystalline particles, does occur at Fra Mauro base in the form of small fragments of deeply colored glass (see Fig. 3). The Apollo Soil Survey (1971) reported that 11% of the glass particles in soil 14259 are mare derived. It appears that the mare component, like the anorthositic one, is represented by glass, not crystalline fragments.

From these observations we conclude that a given impact disperses glass fragments over much greater distances than it does crystalline fragments. Interestingly enough, shock effects short of melting, which many investigators felt intuitively would be ubiquitous among regolith particles, have proved far less common than expected. We are beginning to see that aside from aggregated soil breccias, the lunar regolith on both mare and highland sites consists mainly of crystalline rocks from the general vicinity and glasses from greater distances.

LUNAR NORITES AND OCEANUS PROCELLARUM

Source of the Fra Mauro norites

It is widely agreed that the Fra Mauro formation consists of debris ejected from the Imbrium Basin; the brecciated character of the rocks collected at Fra Mauro base is consistent with their having suffered a cataclysmic event of this scale. Cratering studies indicate that debris ejected as far from the impact point as Fra Mauro is from Mare Imbrium was sited at relatively shallow depths (0–10 km) in the lunar crust before the impact (Wilhelms, 1965). Presumably energy of the Imbrian event and of deposition of the Fra Mauro debris generated heat that produced most of or all the thermal recrystallization effects described above.

A substantial thickness of norite must have underlain the surface to be impacted, since the debris at Fra Mauro is virtually uncontaminated by nonnoritic rock types, but it is unlikely that the lunar crust in what is now Mare Imbrium ever consisted wholly or largely of norite, as was suggested previously by Wood (1972). The low-density crustal rock in this area was not completely stripped away by the Imbrian impact; a 15–30 km thickness of it still underlies the basalts of Mare Imbrium (Fig. 7). This low-density material cannot be norite, as the high noritic levels of K, U, and Th would, in decaying, maintain the local crust at too high a temperature to be strong enough to support the Imbrian mascon (see Fig. 21, Wood, 1972). Presumably the crust beneath Mare Imbrium is anorthositic rock, which appears to be the dominant class of lunar crustal material.

Formation of noritic magma

Wood (1972) proposed that extensive crystal fractionation in an early lunar-surface magma system produced an anorthositic layer by plagioclase flotation and a noritic, KREEP-rich residual liquid. It now seems unlikely that lunar norites were generated by such a process. Crystal fractionation extensive enough to produce the observed differences in concentration of rare-earth and other trace lithophile elements, between lunar anorthositic and noritic rocks, should also have produced much greater differences in Fe/(Fe + Mg) than are observed. Noritic liquids are much more easily accounted for as the product of partial melting in some suitable parent rock.

Walker *et al.* (1972) have proposed that the parent rock was anorthositic gabbro, such as appears to comprise the bulk of the lunar crust. The difficulty with models based on this proposition is that a surface layer of anorthositic gabbro, once formed by differentiation in a surface magma system, will thereafter cool monotonically; it is very hard to contrive circumstances, based on decay of radioactivity and internal warming of the moon, that would reheat and partially remelt the lunar crust very soon after it formed.

Gast and McConnell (1972) suggest that noritic liquids were produced by partial melting in a subcrustal zone of lunar rock not previously melted, by virtue of radio-active decay, soon after the moon was formed. Here (and also with the model of Wood, 1972) the difficulty is that noritic magmas should have been generated and

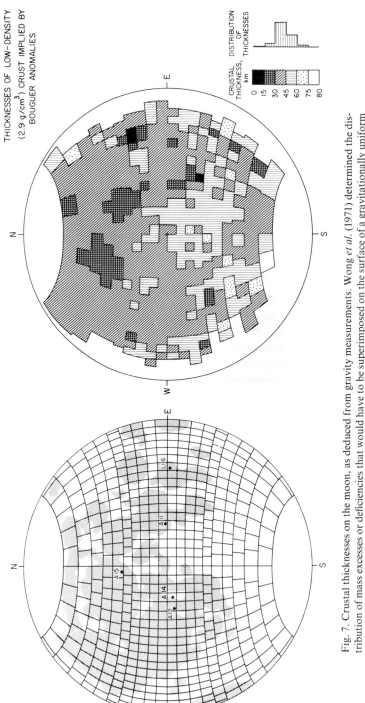

Fig. 7. Crustal thicknesses on the moon, as deduced from gravity measurements. Wong et al. (1971) determined the distribution of mass excesses or deficiencies that would have to be superimposed on the surface of a gravitationally uniform moon, in the cells defined in the left half of the drawing (equal-area projection), in order to reproduce the observed motions of spacecraft orbiting the moon. After a correction for topography is made, residual mass excesses/deficiencies can be interpreted in terms of variable thicknesses of underdense and/or superdense types of rock beneath the surface of the moon. Figure at right above interprets the data of Wong et al. entirely in terms of a crust made of one type of underdense rock; 2.9 g/cm³ is the density of typical anorthositic gabbro. Thicknesses are uncertain by the amount of an additive constant; the crust shown above is of minimal thickness possible without requiring <0 or >3.35 g/cm³ noncrustal rock in mascon maria. Derivation of model is discussed in detail in Taylor et al. (1972). Crustal thickness near Fra Mauro is consistent with that determined seismically by Latham et al. (1972).

erupted in areas distributed over the whole moon; whereas, in fact, the Apollo 15 orbital gamma-ray experiment has shown that soils radioactive enough to contain substantial proportions of noritic rock are highly localized on the moon, occurring only in Mare Imbrium and Oceanus Procellarum (Metzger *et al.*, 1972).

Localization of norite on the moon

It is important to recognize that the occurrence of lunar norite correlates primarily with Oceanus Procellarum, rather than with Mare Imbrium. The radioactivity high observed by Metzger *et al.* (1972) does not span Mare Imbrium, but falls off some 10° inside the eastern rim of the mare; on the other hand it continues for ∼25° west of the western mare rim, across much of northwestern Oceanus Procellarum. Soil samples at the Apollo 12 site, farther south in Oceanus Procellarum, contain abundant crystalline norite. Most of or all this norite is locally derived, not Copernican ray material; it is present in comparable amounts in all the Apollo 12 soil samples, whereas the ropy brown noritic glass that generally is believed to be Copernican ray material is abundant in only two of the soils (Marvin *et al.*, 1971). The Apollo 12 crystalline norite could, of course, derive from a stratum of Fra Mauro formation beneath the local mare basalts. The Imbrian impact did not generate the Fra Mauro norites; it merely delivered to Fra Mauro samples of norite that had occupied the upper levels of the pre-Imbrian crust in the impact area, which before the impact was an extension of Oceanus Procellarum.

Oceanus Procellarum is by far the largest uninterrupted mare surface on the moon. Now it appears to be unique not only in this respect, but also as the source of lunar norite. It is possible that the two exceptional properties of this area of the moon are related. Maria are topographically low regions, and lowness on an isostatically equilibrated early moon means the low-density crust is thinner beneath them than beneath the highlands (∼40 km versus ∼50 km; Fig. 7). The crust may have had nonuniform thickness from the outset, but we also know it can be thinned by major impacts that excavate light crustal material and distribute it elsewhere on the moon (Imbrium, Serenitatis, Crisium Basins).

Perhaps all mare basins originated in this way, and Oceanus Procellarum represents the focal point for an early planetesimal bombardment that favored the earth-facing side of the moon. Then the Procellarum Basin is actually composed of many overlapping impact basins, and if this focused bombardment occurred over a short period of time it would have deposited a non-negligible amount of energy in the lunar crust in this one area. Possibly this provided the pulse of heat needed to generate noritic magmas in the remaining anorthositic crust, or beneath it.

Nonrandom bombardment of the moon by planetesimals

The question of why planetesimal impacts would have focused on Oceanus Procellarum is part of the larger problem of the concentration of mare basins on the earth-facing side of the moon. The only apparent way of understanding the latter phenomenon is by assuming that spin-orbit coupling aligned the moon in its present orientation very early; thereafter the gravitational influence of the earth precluded

Table 4. Asymmetry of bombardment of moon by planetesimals in heliocentric orbit (ratio of impact rates on Earth-facing side of moon, to back of moon; and lunar longitude at which maximum impact rate obtains).

Earth-moon distance (Earth radii)	Intrinsic velocity of planetesimals (km/sec)		
	2	5	10
2.9[a]	1.17	1.71	1.87
	90°W	55°W	40°W
10	1.87	1.87	
	50°W	45°W	
60[b]	1.21	1.36	
	~85°W	~75°W	

[a]Immediately outside Roche limit.
[b]Present earth-moon distance.

an isotropic bombardment of the moon by planetesimals and worked to favor impacts on its near side.

It is unlikely that planetesimals in geocentric orbit impacted preferentially on the lunar nearside. Tidal friction has caused the moon's orbit to expand with time, so most geocentric planetesimals swept would be objects orbiting outside the lunar orbit, and these would tend to impact the far side of the moon.

The distribution of impacts by heliocentric planetesimals is controlled by two effects: the tendency of the leading edge of the moon to collect most impacts (like insects on the windshield of a car), and the gravitational focusing effect of the earth, which enhances the planetesimal flux on the earth-facing side of the moon. Interplay of these two effects leads to a maximum impact rate in the west-nearside quadrant of the moon, where Oceanus Procellarum also lies.

The longitude of maximum bombardment is a function of planetesimal velocity and earth-moon distance. We investigated the distribution of impacts for several velocities and distances, by use of a computer procedure that numerically determines the trajectories and impact points of large numbers of planetesimals fired into the earth-moon system (Taylor et al., 1972). Results appear in Table 4. For the maximum impact rate to occur as far east as the center of Oceanus Procellarum (~40°W), it appears that relatively high-velocity planetesimals must have bombarded the moon while it was still close to the earth.

Nearside impacts are not favored over those on the far side by a large factor (<2). Thus the Oceanus Procellarum Basin could not consist of a small number of overlapping mare-sized craters, or about half as many craters (now filled with basalt) should also be visible on the back of the moon. We think it more likely that the crust under Procellarum was thinned by the cumulative effect of a very large number of impacts by moderate-sized planetesimals. This would have promoted a net transfer of crustal debris from the front to the back of the moon.

Acknowledgments—We thank J. F. Bower for making most of our microprobe analyses. This work was supported in part by NASA grant NGL 09–015–150.

REFERENCES

Brown R. W. Reid A. M. Ridley W. I. Warner J. L. Jakes P. Butler P. Williams R. J. and Anderson D. H. (1971) Microprobe analysis of glasses and minerals from Apollo 14 sample 14259. *NASA Tech. Mem. TM X–58080.*

Eberhardt P. Eugster O. Geiss J. Grögler N. Schwarzmüller J. Stettler A. and Weber L. (1972) When was the Apollo 12 KREEP ejected? (abstract). In *Lunar Science—III* (editor C. Watkins), p. 206, Lunar Science Institute Contr. No. 88.

Gast P. W. and McConnell R. K. Jr. (1972) Evidence for initial chemical layering of the moon (abstract). In *Lunar Science—III* (editor C. Watkins), p. 289, Lunar Science Institute Contr. No. 88.

Hubbard N. J. Gast P. W. Meyer C. Nyquist L. E. and Shih C. (1971) Chemical composition of lunar anorthosites and their parent liquids. *Earth Planet. Sci. Lett.* **13**, 71–75.

Huebner J. S. and Ross M. (1972) Phase relations of lunar and terrestrial pyroxenes at one atm. (abstract). In *Lunar Science—III* (editor C. Watkins), p. 410, Lunar Science Institute Contr. No. 88.

Latham G. Ewing M. Press F. Sutton G. Dorman J. Nakamura Y. Toksoz N. Lammlein D. and Duennebier F. (1972) Moonquakes and lunar tectonism—results from the Apollo passive seismic experiment (abstract). In *Lunar Science—III* (editor C. Watkins), p. 478, Lunar Science Institute Contr. No. 88.

LSPET (1971) (Lunar Sample Preliminary Examination Team) Preliminary examination of Lunar samples from Apollo 14. *Science* **173**, 681–693.

Marvin U. B. Wood J. A. Taylor G. J. Reid J. B. Jr. Powell B. N. Dickey J. S. Jr. and Bower J. F. (1971) Relative proportions and probable sources of rock fragments in the Apollo 12 soil samples. *Proc. Second Lunar Sci. Conf., Geochim. Cosmochim. Acta* Suppl. 2, Vol. 1, pp. 679–699. MIT Press.

Metzger A. E. Trombka J. I. Peterson L. E. Reedy R. C. and Arnold J. R. (1972) A first look at the lunar orbital gamma ray data (abstract). In *Lunar Science—III* (editor C. Watkins), p. 540, Lunar Science Institute Contr. No. 88.

Meyer C. Jr. Brett R. Hubbard N. J. Morrison D. A. McKay D. S. Aiken F. K. Takeda H. and Schonfield E. (1971) Mineralogy, chemistry, and origin of KREEP component in soil samples from the Ocean of Storms. *Proc. Second Lunar Sci. Conf., Geochim. Cosmochim. Acta* Suppl. 2, Vol. 1, pp. 679–699. MIT Press.

Quaide W. Overbeck V. Bunch T. and Polkowski G. (1971) Investigations of the natural history of the regolith at the Apollo 12 site. *Proc. Second Lunar Sci. Conf., Geochim. Cosmochim. Acta* Suppl. 2, Vol. 1, pp. 701–718. MIT Press.

Reid A. M. Ridley W. I. Warner J. Harmon R. S. and Brett R. (1972) Chemistry of highland and mare basalts as inferred from glasses in the lunar soils (abstract). In *Lunar Science—III* (editor C. Watkins), p. 640, Lunar Science Institute Contr. No. 88.

Reid J. B. Jr. Taylor G. J. Marvin U. B. and Wood J. A. (1972) Luna 16: Relative proportions and petrologic significance of particles in the soil from Mare Fecunditatis. *Earth Planet. Sci. Lett.* **13**, 286–298.

Schnetzler C. C. and Nava D. F. (1971) Chemical composition of Apollo 14 soils 14163 and 14259. *Earth Planet. Sci. Lett.* **11**, 345–350.

Shoemaker E. M. Hait M. H. Swann G. A. Schleicher D. L. Schaber G. G. Sutton R. L. Dahlem D. H. Goddard E. N. and Waters A. C. (1970) Origin of the lunar regolith at Tranquillity Base. *Proc. Apollo 11 Lunar Sci. Conf., Geochim. Cosmochim. Acta* Suppl. 2, Vol. 3, pp. 2399–2412. Pergamon.

Taylor G. J. Marvin U. B. Wood J. A. Reid J. B. Jr. and Bower J. F. (1972) Mineralogy and petrology of the Apollo 14 lunar sample. *Smithsonian Astrophysical Observatory Special Report* No. 345.

Walker D. Longhi J. and Hays J. F. (1972) Experimental petrology and origin of Fra Mauro rocks and soils. *Proc. Third Lunar Science Conference, Geochim. Cosmochim. Acta* Suppl. 3, Vol. 1, pp. 797–817. MIT Press.

Warner J. L. (1972) Apollo 14 breccias: Metamorphic origin and classification (abstract). In *Lunar Science—III* (editor C. Watkins), p. 782, Lunar Science Institute Contr. No. 88.

Wilhelms D. D. (1965) Fra Mauro and Cayley formations in the Mare Vaporum and Julius Caesar quadrangles. *Astrogeol. Stud. Ann. Progr. Rept.,* July 1964–July 1965, Pt. A: U.S. Geol. Survey open-file report, pp. 13–28.

Wilhelms D. E. and McCauley J. F. (1971) Geologic map of the near side of the moon. *U.S. Geol. Surv. Misc. Geol. Inv.* Map I 703.

Wilshire H. G. and Jackson E. D. (1972) Petrology of the Fra Mauro formation at the Apollo 14 landing site (abstract). In *Lunar Science—III* (editor C. Watkins), p. 803, Lunar Science Institute Contr. No. 88.

Wong L. Buechler G. Downs W. Sjogren W. Muller P. and Gottleib P. (1971) A surface-layer representation of the lunar gravitational field. *J. Geophys. Res.* **76**, 6220–6236.

Wood J. A. (1972) Fragments of terra rock in the Apollo 12 soil samples and a structural model of the moon. *Icarus* **16**, in press.

Wood J. A. Reid J. B. Jr. Taylor G. J. and Marvin U. B. (1971) Petrological character of the Luna 16 sample from Mare Fecunditatis. *Meteoritics* **6**, 181–193.

Proceedings of the Third Lunar Science Conference
(Supplement 3, *Geochimica et Cosmochimica Acta*)
Vol. 1, pp. 1015–1027
The M.I.T. Press, 1972

Chemical and petrographic characterization of Fra Mauro soils

M. H. CARR and C. E. MEYER

U.S. Geological Survey,
345 Middlefield Road,
Menlo Park, California 94025

Abstract—The 70–1000 μ size fractions of six regolith samples from the Apollo 14 site were characterized in terms of their components. Glass fragments are most common (50–82%), being either a transparent homogeneous glass or dark cloudy glass laden with mineral debris. Meta breccia fragments constitute 17–18% of the samples and two types are recognized, dark metabreccias, most common in samples close to Cone Crater rays, and light metabreccia fragments, which greatly predominate elsewhere. 0.6–6% of the fragments are igneous rock, 4–14% are mineral grains. Most of the glasses and all the metabreccia fragments appear to be derived locally and compositionally are similar to previously described KREEP material. Their relative proportions are consistent with derivation from a rock sequence in which light metabreccias are more common close to the surface and dark metabreccias are more common at greater depths. Included also are exotic components from two main sources, the nearby mare and a unit with the composition of feldspathic basalt, possibly the Cayley formation.

INTRODUCTION

THE LUNAR REGOLITH IS composed of a wide variety of fragment types, the relative proportions of which vary according to the fragment size and the local geology (Marvin *et al.*, 1971; Quaide *et al.*, 1971). The purpose of this paper is to describe the nature of the regolith at several locations at the Apollo 14 landing site and to determine what this implies as regards the formation of the regolith and the depositional sequence of the local rocks. We suggest that most of the regolith materials were derived locally from a stratigraphic sequence consisting of light clast dominant rocks immediately beneath the regolith and dark clast dominant rocks at greater depths.

DESCRIPTION OF SAMPLES AND ANALYTICAL TECHNIQUES

Six regolith samples from the Apollo 14 landing site were examined; all are fines in the <1 mm size range. Three of the samples (14230,78; 14230,84; and 14230,103) are from different depths in the core from Station G, the other three are from the contingency (14003), bulk (14163) and comprehensive (14259) samples. The samples were first sieved, then the 70–1000 μ fraction made into polished thin sections. All the work in this report refers to material in this size fraction. The fragments making up the regolith were classified and counted according to their petrographic characteristics. The classification scheme followed was similar to that used by Jackson and Wilshire (1972) in classifying clasts in the Apollo 14 rocks, slightly modified to account for the peculiarities of the regolith, particularly as regards glassy fragments. Approximately 90% of the fragments could be unambiguously classified. Problems arose only in the case of composite particles, such as lithic fragments in a glassy matrix. These were assigned to the dominant type within the fragments. Two or three thin sections were examined from each sample and approximately 500 particles were classified and counted in each thin section. After counting, representative particles from each group were analyzed in the microprobe so that the chemical type and range of chemical variability could be established for each group. The coarser grained lithic fragments were analyzed by rapidly scanning a large area of the particle with the electron beam, to minimize sampling errors. Only analyses totalling between 98 and 102% were used.

Description of Particle Types

Four main groups of particles are recognized: glasses, metabreccias, igneous fragments, and mineral fragments.

Glasses

Glass fragments constitute 50–60% of each sample except 14259, which is composed of 82% glass fragments. Three types of glasses are recognized by their petrographic characteristics: homogeneous glass, vesicular glass, and dark cloudy glass. The homogeneous glass is transparent and ranges from colorless through light yellow to dark brown or green. Light yellow glass is the most common. Vesicles, mineral fragments, and other inclusions are rare but opaque spheres occur in some particles. As the name implies, most particles are homogeneous but some have a banding caused by different colored glasses, alined vesicles or microlites, or bands of fine mineral debris. The homogeneous glass fragments are normally angular, but nearly all spheres also are of this type of glass. Homogeneous glass may be enclosed in the other two glass types but vesicular or dark cloudy glass has not been observed within the homogeneous glass. Some of the homogeneous glass fragments have textures suggestive of quenching or partial devitrification. They commonly have sheaf-like arrays of finely fibrous crystals, such as described by Lofgren (1971), or dense opaque bands of randomly oriented microlites. Incipient devitrification is indicated in some fragments by numerous star shaped microlite clusters.

Vesicular glass has a blotchy appearance and is mostly dark brown although some fragments in thin section range from colorless to opaque. The fragments are highly vesicular and have smooth outlines suggestive of flow. Enclosed within the glass are numerous fragments of all other material present in the soil. The glass appears to correspond to the glazed aggregates of Duke *et al.* (1970) and the glass agglutinates of McKay *et al.* (1971). Many particles have a marked flow structure, generally caused by different colored bands in the glass that conform to the smooth outline of the particle. Typical examples of vesicular glass are shown in Figs. 1a and 1b.

The dark cloudy glass closely resembles the vesicular glass and its distinction as a separate fragment type may be artificial. Like the vesicular glass, it is full of lithic and mineral debris, is usually a cloudy dark brown, and has a considerable range of opacity. It is distinguished from vesicular glass by the more angular outlines of the particles and the much lower vesicularity.

The compositional variability of the Fra Mauro glasses has been described in detail by Apollo Soil Survey (1971) and the data presented here are consistent with their findings. Figure 2 shows CaO–Al_2O_3 ratios of the glasses we analyzed. The grouping suggested by Reid *et al.* (1972) is immediately obvious. Those glasses (group 1), with less than 14% Al_2O_3, all have compositions corresponding to mare basalts with FeO in excess of 15%. They represent approximately 10% of all glasses

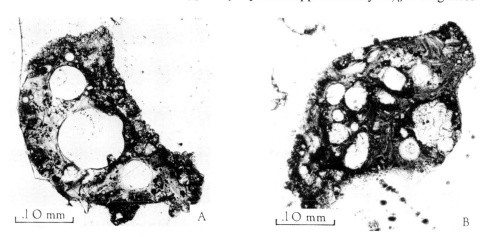

Fig. 1a. Dark vesicular glass fragment with smooth flow-shaped outline. Glass transparent in places, elsewhere almost opaque and laden with mineral debris.

Fig. 1b. Dark vesicular glass fragment with light metabreccia inclusion. Glass is finely banded and particle has smooth outline.

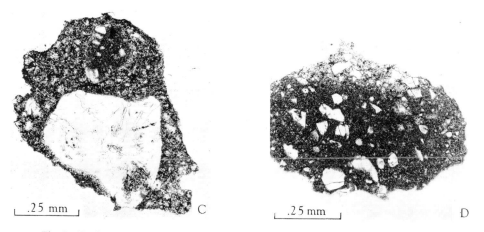

Fig. 1c. Dark metabreccia fragment containing an inclusion of recrystallized feldspar and a dark metabreccia fragment. The matrix appears to have been mobilized and injected into the feldspar.

Fig. 1d. Dark metabreccia fragment. Mineral debris, mainly plagioclase and clinopyroxene in a fine-grained matrix of plagioclase, orthopyroxene, and ilmenite. The fragment is more coarsely crystalline on one side and here resembles a light metabreccia.

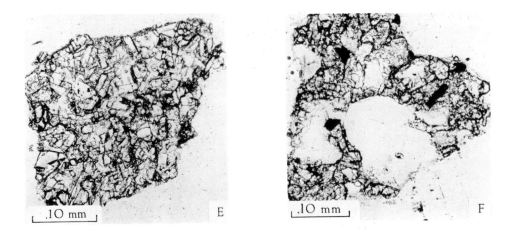

Fig. 1e. Typical light metabreccia fragment. Particle is a subequigranular intergrowth of short feldspar laths and equant orthopyroxene crystals. No mineral or lithic debris is present; the fragment is wholly of matrix material.

Fig. 1f. Light metabreccia fragment similar to the one shown in Fig. 1e but containing large feldspar fragments.

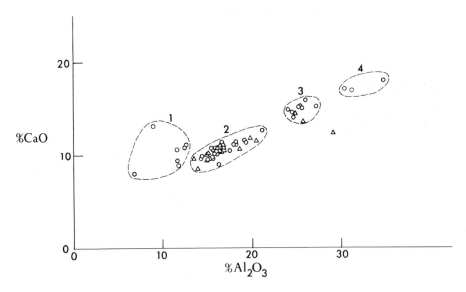

Fig. 2. Glass analyses on a CaO–Al$_2$O$_3$ plot. Triangles are dark inhomogeneous glasses, circles are homogeneous glasses. Group 1 glasses all have FeO greater than 15% and are compositionally similar to mare basalts. Group 2 glasses have KREEP composition (Apollo Soil Survey, 1971). Group 3 glasses have the composition of feldspathic basalts and Group 4 glasses are essentially feldspathic.

analyzed and all are the homogeneous type. Most homogeneous glasses and nearly all the vesicular and dark cloudy glasses have Al_2O_3 contents in the 13–21% range, forming a tight group on the Al_2O_3–CaO diagram. They have been termed "Fra Mauro basalts" by Reid *et al.* (1972) because they are basaltic in composition and typical of the Fra Mauro site. Their compositions are identical to the KREEP component in the soils from the Apollo 12 landing site (Hubbard *et al.*, 1971). The other two groups in Fig. 2 correspond approximately in composition to feldspathic basalts (22.5–35% mafic minerals) and feldspars and are almost all homogeneous glasses. In general the dark cloudy glasses, both vesicular and nonvesicular, have a restricted range in composition, mostly that of "Fra Mauro basalt." The homogeneous glasses are more variable in composition ranging from those similar to mare basalts to those similar to feldspars.

Metabreccias

In the samples examined, 12 to 18% of the fragments were metabreccias. The breccias are polymict consisting of glass, lithic and mineral clasts embedded in a fine-grained plagioclase-pyroxene matrix throughout which ilmenite is finely disseminated. Two main types are recognized: a dark colored breccia with a fine-grained ($< 1 \mu$) matrix and a light colored breccia with a more coarsely crystalline (5–20 μ) matrix of plagioclase and orthopyroxene. The dark breccias appear to correspond to the "annealed brecciated KREEP" of Meyer *et al.* (1971) the "gray mottled basalts" of Anderson and Smith (1971) and are similar to the dark clasts in the rocks at the Fra Mauro site (Jackson and Wilshire, 1972). Typical examples are shown in Figs. 1c and 1d. All fragment types except igneous occur as clasts in these breccias; some contain breccia clasts similar to the host rock itself.

The light colored metabreccias (Figs. 1e and 1f) consist of angular mineral fragments embedded in an interlocking mosaic of plagioclase and orthopyroxene. In contrast to the dark breccias, the light ones appear to contain no lithic or glass clasts. Both the plagioclase and orthopyroxene of the matrix tend to form equant grains, giving it a mosaic or hornfelsic texture. The ilmenite is localized in relatively large irregular grains. Shock effects are evidenced in some fragments by hazy crystal boundaries and undulatory extinction, but most fragments are free of obvious shock damage.

The two breccia types are chemically very similar to each other and to the KREEP glass or Fra Mauro basaltic glass of Reid *et al.* (1972) and the group 2 glasses above. The average compositions of all the breccias analyzed are compared with various glass types in Table 1. Figure 3 shows the relation of the metabreccia analyses to the glass groupings on the CaO–Al_2O_3 plot. The metabreccias show a wider range of Ca/Al ratios than the glasses, although most appear to cluster around the mean value of the Fra Mauro basaltic glasses. There is a suggestion that the Ca/Al ratio in the light glasses is lower than in the dark glasses but the difference, if real, is small. Figure 4 indicates that there is a strong covariance between FeO and MgO in the breccias, in contrast to the glasses (Apollo Soil Survey, 1972) which show poor covariance. The covariance between Fe and Mg probably results from variation in

Table 1. Averages of analyses (wt%) of glass and lithic fragments.

	Glasses*			Lithic	
	Mare-type basalt (8 analyses)	Fra Mauro basalt (35 analyses)	Feldspathic basalt (14 analyses)	Dark metabreccia (32 analyses)	Light metabreccia (19 analyses)
SiO$_2$	44.6 ± 1.5	49.1 ± 2.3	45.2 ± 0.8	49.2 ± 2.0	48.8 ± 2.9
Al$_2$O$_3$	11.1 ± 2.3	16.8 ± 1.8	26.1 ± 1.7	16.8 ± 5.0	17.9 ± 2.4
FeO	18.8 ± 2.7	11.0 ± 1.5	5.7 ± 0.8	10.4 ± 2.9	9.6 ± 2.8
MgO	11.7 ± 3.5	8.3 ± 2.2·	7.3 ± 1.6	9.6 ± 2.9	10.4 ± 2.6
CaO	9.8 ± 1.2	10.6 ± 0.9	15.0 ± 1.2	10.2 ± 2.2	9.9 ± 2.2
Na$_2$O	0.4 ± 0.2	0.7 ± 0.3	0.2 ± 0.2	0.9 ± 0.2	0.9 ± 0.2
K$_2$O	0.1 ± 0.1	0.7 ± 0.4	0.1 ± 0.1	0.6 ± 0.4	0.6 ± 0.3
TiO$_2$	3.6 ± 2.0	2.8 ± 0.9	0.5 ± 0.2	2.3 ± 1.0	1.67 ± 0.8

*The glasses are divided into three categories according to their natural grouping on the CaO–Al$_2$O$_3$ plot (Fig. 2). The names have no genetic significance; they are used only to signify compositional differences.

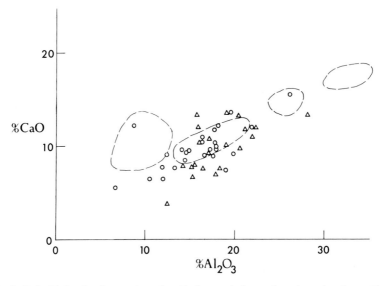

Fig. 3. CaO–Al$_2$O$_3$ plot for metabreccias. Circles are dark metabreccias; triangles are light metabreccias. Dashed lines indicate the glass groupings from Fig. 2.

the plagioclase-orthopyroxene ratio. Such variation could be caused, in part, by sampling errors. The area analyzed may not be representative of the fragment and the fragment may not be representative of the parent rock. Although this is a possible explanation of the variance in the coarser grained light breccias, it is unlikely that sampling errors are the cause for covariance in the dark breccias because they are so fine grained.

Igneous fragments

Each sample contained approximately 1% igneous fragments except sample 14230,103, which contained 5%. The fragments were identified from their textures.

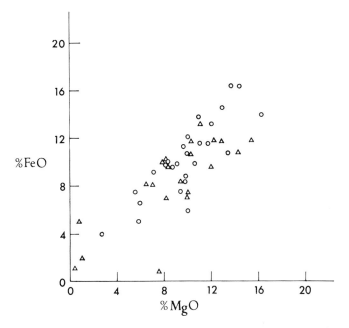

Fig. 4. FeO–MgO plot for metabreccias. Circles are dark metabreccias; triangles are light metabreccias. Plot shows strong covariance of Fe and Mg in contrast to the glasses (Apollo Soil Survey, 1971).

Most are intersertal and subophitic basalts, and troctolites were recognized. There are, however, too few in number for any statistically meaningful work to be done and none was attempted. Although only 1% of the fragments were identified as igneous, the igneous component may be somewhat larger. The igneous fragments generally have a grain size ($> 50\,\mu$) that is coarse in relation to the size of the fragments being examined. They are identifiable as igneous only in the case of the larger fragments. Further comminution would result only in mineral debris and much of that observed in the soil may be of igneous origin. The relatively large grain size also prevents accurate identification of the igneous rock type, because the relative proportion of the constituent minerals cannot be accurately determined.

Mineral fragments

Mineral grains constitute 4–15% of the fragments in the Fra Mauro soils, a much smaller percentage than in the same size range at the Apollo 11 and 12 sites (Quaide *et al.*, 1971; McKay *et al.*, 1971; Marvin *et al.*, 1971; Engelhardt *et al.*, 1971). Feldspar grains are the most common; of 22 analyzed, 18 had compositions in the An_{89} to An_{96} range. They differ from the feldspars in the metabreccias in that they are generally larger and subhedral rather than anhedral. They show a wide variety of strain and recrystallization textures. Pyroxene fragments are the next most common (3–8%). Clinopyroxenes are more common than orthopyroxenes, in contrast to the pyroxenes in the metabreccia matrices, which are preponderantly orthopyroxene

(Fig. 5). Less than 1% of the regolith material consists of olivine fragments; they range in composition from Fo_{58} to Fo_{70}.

High K particles

A small number of fragments with very high Si (63 to 75%) and K (2.3 to 6.9%) were encountered. They are of several types (glass fragments within normal dark metabreccias, K feldspar fragments, and K-feldspar-SiO_2 intergrowths) but are too few in number for any systematic work.

INTERPRETATION

As a first approximation, the regolith can be thought of as a mixture of two components, a local component controlled by the local geology and an exotic component that is more dependent on the regional geologic setting (Shoemaker *et al.*, 1970). Around a relatively young crater that penetrates the regolith the local component will be enhanced; at sites of more mature regolith development, away from young features, one can expect a larger exotic component. The regolith at the Apollo 14 landing site is affected by Cone Crater, a feature that is relatively young compared to the Fra Mauro formation upon which the regolith has formed (Eggleton and Offield, 1970). In regolith affected by the formation of Cone Crater, we therefore expect a larger local component than at sites less affected by the crater, and we would further expect that the local component be biased toward the composition of ejecta from Cone Crater. The nature and source of the exotic component are more difficult to evaluate but the homogeneous glasses, the igneous fragments, and the mineral debris provide clues that the component includes material derived from at least two igneous sources.

The geologic relations in Fig. 6 suggest that samples 14003, 14163, and 14259 should be more affected by the ejecta from Cone Crater than the samples from

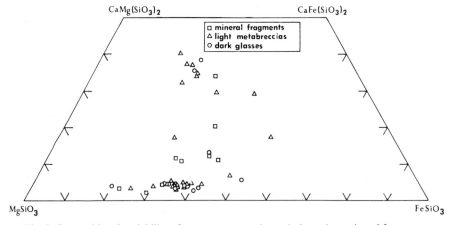

Fig. 5. Compositional variability of pyroxenes occurring as independant mineral fragments or as grains within glasses and light metabreccias.

Station G. The shaded pattern indicates the area over which blocks larger than 0.5 m across are abundant (Swann *et al.*, 1971). The blocks are clearly derived from Cone Crater and it is reasonable to assume that the finer debris from the crater had a similar but more extensive distribution pattern. Sample 14259 was collected along a ray from Cone Crater but somewhat beyond the zone of abundant blocks. Samples 14003 and 14163 were close to the side of the same ray. Each of these samples should have a larger Cone Crater component than the samples from Station G that are further removed from the ray. The main difference between the Station G samples and the others is the great predominance of light metabreccias over the dark, the latter constituting only 1.4 to 4.7% of the total sample (Fig. 7 and Table 2). The samples closer to the ray have approximately the same percentage of light meta-breccia fragments but the proportion of dark fragments is higher by a factor of 2 to 3. This suggests that the Cone Crater event has added to the local soil a component consisting predominantly of dark metabreccias. There may also be an added glass component. Sample 14259, which lies along the line of a ray, has a higher proportion of dark glass than the other samples and this enhancement may be due to Cone Crater ray material.

It is reasonable to assume that, in general, most of the locally derived regolith materials comes from relatively shallow depth as a result of repeated impact. Only around large impact craters are significant amounts of deeper material added to the soil in the form of debris excavated from the crater. This being so, then Station G samples indicate that rocks in which light metabreccias predominate are near the surface at the Fra Mauro site. Rocks in which dark metabreccias predominate are at greater depths and provide the dominant lithic component in the regolith only

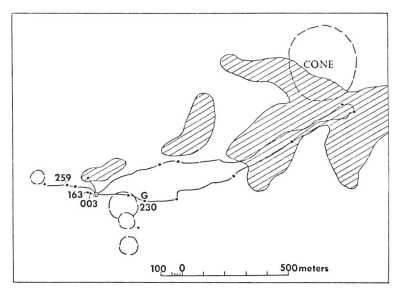

Fig. 6. The location of samples discussed in this report. In the shaded areas are abundant blocks 0.5 m across that are believed to have been ejected from Cone Crater.

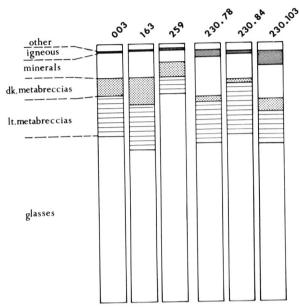

Fig. 7. Relative proportions of various regolith components in the six samples examined.

Table 2. Relative percentages of different components in six regolith samples.

	14003	14163	14259	14230,78	14230,84	14230,103
Homogeneous glass	15.2	20.1	14.3	16.6	9.7	12.2
Dark cloudy glass	51.3	41.0	68.0	47.1	56.3	50.9
Total glass	66.5	61.1	82.3	63.7	66.0	63.1
Light metabreccias	15.9	17.9	6.4	16.2	19.9	13.0
Dark metabreccias	6.8	10.0	6.1	2.6	1.4	4.7
Total metabreccias	22.7	27.9	12.5	18.8	21.3	17.7
Igneous	0.4	1.3	1.3	2.6	0.6	5.7
Plagioclase	4.6	5.1	2.6	8.3	5.5	7.6
Clinopyroxene	2.6	2.8	1.0	5.0	5.7	2.3
Orthopyroxene	1.8	1.3	0.3	—	0.4	2.8
Olivine	0.6	0.4	—	1.5	0.4	—
Total mineral fragments	9.6	9.6	3.9	14.8	12.0	12.7

where significant amounts of material excavated from Cone Crater are present. Wilshire and Jackson (1972), from a study of the distribution of rock types on the surface, similarly inferred that immediately below the regolith are light clast dominant breccias and at greater depths dark clast dominant types. Data on the regolith samples and rocks are, therefore, entirely consistent.

It should be emphasized that if the light and dark metabreccias in the regolith are derived from metabreccia clasts in the Fra Mauro formation then the meta-breccias are samples of the pre-Imbrian rocks that were in the Imbrium Basin region before the basin formed. During the formation of the basin they were incorporated as clasts into the Fra Mauro breccias and deposited around the Imbrium Basin. Subsequent comminution of the Fra Mauro breccias resulted in the clasts being introduced into the regolith, where they are now identified as dark and light metabreccia fragments. The Imbrian-age Fra Mauro matrix has not been recognized in the regolith.

Most of the dark vesicular glass fragments appear to be of local origin, because over 90% of those analyzed have a composition similar to the locally derived dark and light metabreccias. The remainder have compositions similar to feldspathic basalts; none have mare basalt compositions. They appear in large part to be agglutinates of regolith material welded together by splash glass, or vitrified regolith, or vitrified ejecta from local impact craters. The homogeneous glasses are of more diverse origin. Of 43 homogeneous glasses analyzed, only about half (25) have the composition of the Fra Mauro rocks. Of the 18 remaining, 3 are feldspathic and the rest are almost equally divided between those with mare basalt composition (7) and those with feldspathic basalt composition (8). If we include both the homogeneous and dark cloudy type of glass then the feldspathic basalts are about twice as abundant as the mare basalt composition (Table 1). The glasses of mare basalt composition are presumably derived from the extensive mare deposits that occur both to the east and west of the site. The source of the glasses of feldspathic basalt composition is more difficult to assess. Reid *et al.* (1972) call this component "highland basalt" and show that it occurs at all the lunar sites so far sampled so that the source rock must have wide distribution. A possible source is the Cayley formation which occurs widely throughout the lunar uplands filling large craters and other depressions. The unit is mostly Imbrian in age but predates the main period of mare development. Extensive areas of Cayley occur to the south-east of the Apollo 14 landing site in craters Bonpland and Perry and in the eastern half of the crater Fra Mauro (Wilhelms and McCauley, 1971). Cayley outcrop areas are situated at approximately the same distance from the landing site as the closest mare areas, and are roughly equivalent in extent to the maria at this distance (30–150 km range from the landing site) so the comparable abundances of the mare-type and highland-type basaltic glasses among the homogeneous glasses is consistent with the Cayley formation as a source for the highland basalts. We have suggested above that the dark cloudy glasses are of local origin. No dark cloudy glasses of mare basalt composition were found but several have "highland basalt" compositions. This implies that there are sources of "highland basalt" closer to the landing site than the mare basalt sources, which occur not closer than 30 km. Although no Cayley formation is mapped within 30 km of the site, a unit called "smooth terrain material" (Eggleton, 1970) occurs within several small low lying areas. This unit strongly resembles the Cayley formation and may indeed be merely a very thin and local deposit of Cayley. The "smooth terrain material" is thus suggested as the local source of the "highland basalt" glasses.

Of the 57 glasses analyzed here, 3 have highly feldspathic compositions. One has a composition almost identical with anorthite and is almost certainly a vitrified feldspar. The other 2 have compositions similar to gabbroic anorthosite. We prefer, however, not to speculate on the possible origin of gabbroic anorthosite rocks from these meager data.

The type and relative proportions of igneous rock and mineral debris are consistent with the model presented above. The Station G samples have a larger proportion of both igneous and mineral fragments which is expected since the regolith here is not significantly diluted by debris from Cone Crater. Unfortunately a distinction cannot be made in the igneous debris between mare basalt fragments and the more feldspathic highland basalt fragments. The grain size ($>50\ \mu$) prevents meaningful assessment of the relative proportions of the component minerals in the source rocks, so assignment of individual fragments to the two compositional types recognized in the glasses is not practical. The mineral debris must be derived largely from igneous fragments or from fragments within the local breccias because the breccia matrix itself is too fine grained to contribute significantly to the $>70\ \mu$ fraction. In contrast to the Fra Mauro breccias, the igneous fragments have clinopyroxene in excess of orthopyroxene. The same is true of the mineral debris in the regolith (Table 2), suggesting that much of the mineral debris is derived from the igneous fragments.

References

Anderson A. T. and Smith J. V. (1971) Nature, occurrence and exotic origin of "gray mottled" (Luny Rock) basalts in Apollo 12 soils and breccias. *Proc. Second Lunar Sci. Conf., Geochim. Cosmochim. Acta* Suppl. 2, Vol. 2, pp. 431–438. MIT Press.

Apollo Soil Survey (1971) Apollo 14: Nature and origin of rock types in soil from the Fra Mauro formation. *Earth Planet. Sci. Lett.* **12**, 49–54.

Duke M. B. Wood C. C. Sellers G. A. Bird M. L. and Finkelman R. B. (1970) Genesis of lunar soil at Tranquillity Base. *Proc. Apollo 11 Lunar Sci. Conf., Geochim. Cosmochim. Acta* Suppl. 1, Vol. 1, pp. 347–361. Pergamon.

Eggleton R. E. and Offield T. W. (1970) Geologic maps of the Fra Mauro region of the moon. U.S. Geol. Surv., Misc. Inv. Map I–708.

Engelhardt W. von Arndt J. Müller W. F. and Stöffler D. (1971) Shock metamorphism and origin of regolith and breccias at the Apollo 11 and Apollo 12 landing sites. *Proc. Second Lunar Sci. Conf., Geochim. Cosmochim. Acta* Suppl. 2, Vol. 1, pp. 833–854. MIT Press.

Hubbard N. J. Meyer C. Gast P. W. and Weismann H. (1971) The composition and derivation of Apollo 12 soils. *Earth Planet. Sci. Lett.* **10**, 341–350.

Jackson E. D. and Wilshire H. G. (1972) Classification of the samples returned from the Apollo 14 landing site (abstract). In *Lunar Science—III* (editor C. Watkins), pp. 419–421, Lunar Science Institute Contr. No. 88.

Lofgren G. (1971) Devitrified glass fragments from Apollo 11 and Apollo 12 lunar samples. *Proc. Second Lunar Sci. Conf., Geochim. Cosmochim. Acta* Suppl. 2, Vol. 1, pp. 949–955. MIT Press.

McKay D. S. Morrison D. A. Clanton U. S. Ladle G. H. and Lindsay J. F. (1971) Apollo 12 soil and breccia. *Proc. Second Lunar Sci. Conf., Geochim. Cosmochim. Acta* Suppl. 2, Vol. 1, pp. 755–773. MIT Press.

Marvin U. B. Wood J. A. Taylor G. J. Reid J. B. Powell B. M. Dickey J. S. and Bower J. F. (1971) Relative proportions and probable sources of rock fragments in the Apollo 12 soil samples. *Proc. Second Lunar Sci. Conf., Geochim. Cosmochim. Acta* Suppl. 2, Vol. 1, pp. 679–699. MIT Press.

Meyer C. Brett R. Hubbard N. J. Morrison D. A. McKay D. S. Aitken F. K. Takeda H. and Schonfeld E. (1971) Mineralogy, chemistry and origin of KREEP component in soil samples from the Ocean of Storms. *Proc. Second Lunar Sci. Conf., Geochim. Cosmochim. Acta* Suppl. 2, Vol. 1, pp. 393–411. MIT Press.

Quaide W. Oberbeck V. Bunch T. and Polkowski G. (1971) Investigation of the natural history of the regolith at the Apollo 12 site. *Proc. Second Lunar Sci. Conf., Geochim. Cosmochim. Acta* Suppl. 2, Vol. 1, pp. 701–718. MIT Press.

Reid A. M. Ridley W. I. Warner J. Harman R. S. Brett R. Jakeš P. and Brown R. W. (1972) Chemistry of highland and mare basalts as inferred from glasses in the lunar soils (abstract). In *Lunar Science—III* (editor C. Watkins), pp. 640–642, Lunar Science Institute Contr. No. 88.

Shoemaker E. M. Hait M. H. Swann G. A. Schleicher D. L. Schaber G. G. Sutton R. L. Dahlem D. H. Goddard E. N. and Waters A. C. (1970) Origin of the lunar regolith at Tranquillity Base. *Proc. Apollo 11 Lunar Sci. Conf., Geochim. Cosmochim. Acta* Suppl. 1, Vol. 3, pp. 2399–2412. Pergamon.

Swann G. A. Bailey N. G. Batson R. M. Eggleton R. E. Hait M. H. Holt H. E. Larson K. B. McEwen M. C. Mitchell E. D. Schaber G. G. Schafer J. P. Shepard A. B. Sutton R. L. Trask N. J. Ulrich G. E. Wilshire H. G. and Wolfe E. W. (1971) Preliminary geologic investigations of the Apollo 14 landing site. U.S. Geol. Survey Interagency Rept. No. 29.

Wilhelms D. E. and McCauley J. F. (1971) Geologic map of the near side of the moon. U.S. Geol. Survey, Misc. Geol. Inv. Map I–703.

Wilshire H. G. and Jackson E. D. (1972) Petrology of the Fra Mauro formation at the Apollo 14 landing site (abstract). In *Lunar Science—III* (editor C. Watkins), pp. 803–804, Lunar Science Institute Contr. No. 88.

Proceedings of the Third Lunar Science Conference
(Supplement 3, *Geochimica et Cosmochimica Acta*)
Vol. 1, pp. 1029–1036
The M.I.T. Press, 1972

Chromatographic and mineralogical study of Apollo 14 fines*

C. R. Masson, I. B. Smith, W. D. Jamieson, and J. L. McLachlan

Atlantic Regional Laboratory, National Research Council of Canada,
Halifax, Nova Scotia, Canada

and

A. Volborth

Department of Geology, Dalhousie University,
Halifax, Nova Scotia, Canada

Abstract—Trimethylsilylation of lunar fines from Apollo 14 and chromatographic separation of the products revealed the trimethylsilyl derivatives of the ions SiO_4^{4-}, $Si_2O_7^{6-}$, and $Si_3O_{10}^{8-}$. The derivative of the cyclic ion $Si_4O_{12}^{8-}$ barely was detectable, in contrast with results for Apollo 11 and 12 fines. Derivatives of higher anions were not detected in chromatograms at temperatures as high as 280°C. The yield and chromatographic pattern for Apollo 14 fines was consistent with the presence of approximately 7% olivine, as established by modal analysis. Volume percentages of the main mineral constituents in samples 14003,27 and 14163,70 were similar, as determined by point counting. These samples contain less pyroxene and fewer opaque minerals but more glass and many more metallic opaque beads than the Apollo 12 samples examined previously. The number of glass spheres, tear drops, and dumbbell-shaped objects per milligram is of the same order as in the Apollo 12 fines. As in previous work, the ratio of integrated peak areas Si_2O_7/SiO_4 due to the dimeric and monomeric derivatives was slightly higher for the lunar fines than could be accounted for by side reactions in the trimethylsilylation of olivine alone. The difference may be due partly to the other constituents in the fines and/or the presence of very small quantities of dimeric (e.g., melilite-group) minerals in the lunar material. Of interest in this connection is the presence in sample 14003,27 of some uniaxial-negative and -positive grains with low birefringence and refractive index and cleavage characteristic of such minerals.

INTRODUCTION

THIS PAPER DESCRIBES A continuation of previous work (Masson *et al.*, 1971) on the determination of anions in lunar fines and glass by the method of trimethylsilylation. The technique (Götz and Masson, 1970, 1971a, 1971b) is a modification of a procedure introduced by Lentz (1964) for the study of silicate structures and used mainly for the determination of lower silicate anions in extracts of minerals and glasses. Recent work (Jamieson *et al.*, 1972) has shown that ions as large as $Si_8O_{21}^{10-}$, as well as phosphate and silicophosphate ions, can be detected by this method. A technique of trimethylsilylation has been described recently (Butts and Rainey, 1971) for other common inorganic anions.

Because of its high sensitivity and the variety of anions that can be detected by this technique, application of the method to lunar material seems worthwhile. The absence of chemical weathering on the moon suggests that anions normally subject to leaching from terrestrial soils may be present in lunar materials. Evidence for this is found in the wide variety of glasses in lunar soil. Terrestrial glasses, in contrast, are comparatively rare and of comparatively recent origin. Although this may be attributed partly to the less favorable opportunities for impact metamorphic pro-

---

*Issued as NRCC No. 12596.

cesses, it is clear that chemical weathering must also be an important factor. Discrete anions are now known to be present in at least some silicate glasses (Götz and Masson, 1971b; Götz et al., 1972). Their presence in phosphate glasses has long been established (Westman, 1960).

EXPERIMENTAL

The technique of trimethylsilylation was similar to that described previously (Masson et al., 1971). Difficulty was experienced in obtaining reagents of high purity, particularly trimethylchlorosilane, which contained hexamethyldisiloxane and other unidentified impurities difficult to separate by fractional distillation. Prolonged and repeated distillation of this reagent led to reorganization reactions, products of which formed azeotropes with Me_3SiCl and appeared as spurious peaks in the blank at low attenuations. The presence of small amounts of hexamethyldisiloxane, the chief product of hydrolysis of trimethylchlorosilane, was unimportant as this compound is itself a constituent of the reaction mixture; larger amounts, however, unless allowed for, caused changes in the relative proportions of the reagents.

A synthetic lunar-like glass was prepared by melting reagent-grade chemicals in a platinum container in a molybdenum-wound resistance furnace held at 1200°C. A gas mixture of CO_2-H_2 was used to generate the required partial pressure of oxygen of $10^{-13.5}$ atm; the partial pressure of oxygen was measured using a lime-stabilized zirconia solid electrolyte cell. The gas flow rate through the furnace was 500 ml min^{-1}. The glass was held under these conditions for 9 hours, then pulled quickly into the water-cooled end of the furnace tube and quenched with a blast of helium.

MINERALOGICAL RESULTS

Modal analysis of Apollo 14 samples 14003,27 and 14163,70 was performed in polarized light using the Zeiss Integrating Micrometer-disk. About 1.5 mg of lunar material was used. The powder was mounted on gelatin-coated slides by dispersing it in a droplet of acetone diluted in half with distilled water (Masson et al., 1971). The analyses (Table 1) are based on 6300 counts for sample 14003,27 and 1600 counts for sample 14163,70.

Table 1. Modal analyses of lunar dust (volume %).

	14003,27 (6300/cts)	14163,70 (1600/cts)
Olivine	8 ± 2	6 ± 2
Plagioclase	12 ± 2	16 ± 2
Pyroxene	10 ± 2	13 ± 2
Glass	51 ± 2	51 ± 2
Aggregates	12 ± 2	9 ± 2
Opaques	2 ± 1	2 ± 1
Unknowns	5 ± 3	3 ± 2
TOTALS	100%	100%
Glass beads, dumbbells, etc.	1473/mg	700/mg
Opaque beads	88/mg	—

Uncertainties exist due to varying grain sizes and the difficulties in recognizing mineral grains coated with glass and adhering mineral and glass particles. The dust appears more heterogeneous than the Apollo 11 and 12 samples, and contains a considerable percentage of minerals that could not be identified positively by purely optical means during counting. In addition, the presence of complex-glass-cemented clusters of minerals and glass particles makes such counting uncertain. In comparison with Apollo 12 dust these samples seem to contain more glass (compare with Table 1, Masson *et al.*, 1971), less pyroxene, fewer opaque minerals, but many more metallic (Fe–Ni?) opaque beads (some 90 opaque beads per milligram in sample 14003,27), whereas the total number of glass spheres, tear drops, and "dumbbell"-shaped objects per milligram is of the same order as in Apollo 12 fines.

Mineralogically of interest is the discovery in sample 14003,27 of six optically uniaxial-negative grains with low birefringence of the order of 0.010 and $n_\omega \simeq 1.67$, three of which had two cleavages at 90° angle, as well as five optically uniaxial-positive grains, with lower refractive index, $n_\omega \simeq 1.64$, also with birefringence of about 0.010.

The small size of these grains (10 to 20 μ in diameter) prevented more exact optical characterization, but the suggestion is made that these may be melilite-group minerals, the possible presence of which was indicated in our previous studies (Masson *et al.*, 1971). Positive identification of these grains was not possible with the equipment available.

A few individual particles merit further attention. In sample 14003,27 a glass sphere some 20 μ in diameter with a burst spherical cavity was photographed (Fig. 1a), also a narrow-necked half of a dumbbell some 70 μ long (Fig. 1b) and a large nonhomogeneous glass fragment with strings of minute cavities—bubbles (Fig. 1c). In sample 14163,70 several nonhomogeneous glass fragments with elongated cavities and flow structures were detected (Fig. 1d). These structures may indicate volcanic origin rather than impact melted rock.

RESULTS OF TRIMETHYLSILYLATION

Apollo 14 fines

Trimethylsilylation of 0.5 g portions of samples 14003,27 and 14163,70 followed by chromatographic separation of the products revealed the ions SiO_4^{4-}, $Si_2O_7^{6-}$, and $Si_3O_{10}^{8-}$ as their trimethylsilyl (TMS) derivatives, as shown in Figs, 2a and 2b. The identity of these derivatives was checked by mass spectrometry. The TMS derivative of the cyclic anion $Si_4O_{12}^{8-}$ was barely detectable, in contrast with the results for Apollo 11 and 12 fines. No peaks due to higher anions were detected in chromatograms up to 280°C. Integrated peak areas are compared in Table 2.

Olivine

For comparison, Table 2 and Fig. 2c show the results, under the same conditions, for a 0.034 g sample of olivine. This is approximately 7% by weight of the lunar material taken for trimethylsilylation and corresponds to the mean value given by the modal analyses in Table 1.

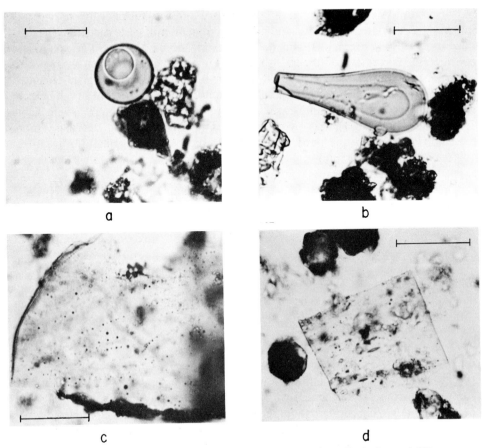

Fig. 1. (a) A glass spherule with a burst cavity on top with luminous Becke line and diffuse centered Becke line ring underneath on the boundary of the gelatin layer and the glass. 14003,27. Scale bar = 30 μ. (b) Broken, narrow-necked dumbbell shaped glass particle showing a diffuse Becke line underneath, between the gelatin and the glass. 14003,27. Scale bar = 30 μ. (c) Angular glass fragment showing string-like arrangement of small cavities and some randomly distributed opaque (?) spherical particles. 14003,27. Scale bar = 50 μ. (d) Rectangular glass fragment with colored flow lines and equally oriented elongated cavities indicating flow structure in lunar volcanic (?) glass. 14163,70. Scale bar = 50 μ.

Synthetic glass

The composition of the initial blend of reagents used to prepare the glass is given in Table 3. The glass is brown-green in reflected light. Microscopic examination of coarsely ground fragments demonstrated that the bulk of the glass, which is pale straw colored in transmitted light, has a refractive index of 1.640 $\pm$ 0.002. The glass contains some nearly opaque, rounded crystallite aggregates and some small cubic or octahedral crystals. These are similar in general appearance to the octahedral crystals of opaque iron–titanium oxide reported by O'Hara (1971) in brown glass obtained by rapid cooling of lunar rock held at 1190°C for several hours.

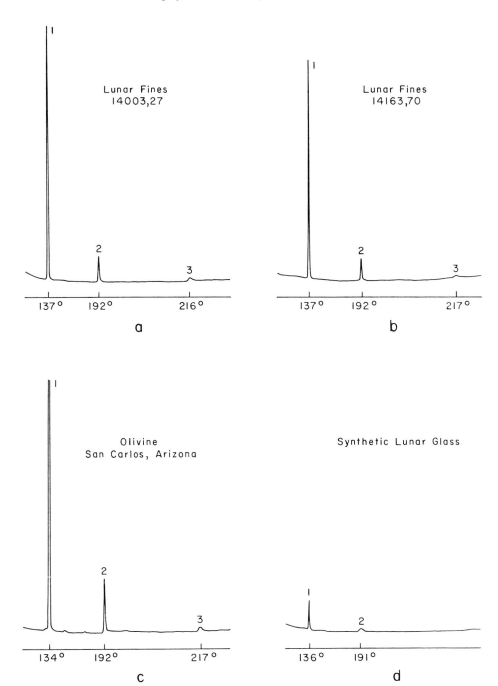

Fig. 2. Chromatograms obtained from extracts of (a) sample 14003,27, (b) sample 14163,70, (c) olivine, and (d) synthetic lunar glass. The peaks are due to the TMS derivatives of (1) SiO_4^{4-}, (2) $Si_2O_7^{6-}$, and (3) $Si_3O_{10}^{8-}$ anions.

Table 2. Integrated peak areas (× 100).

		Lunar fines	
Ratio	Olivine	14003,27	14163,70
Si_2O_7/SiO_4	15.8	16.6	16.7
Si_3O_{10}/SiO_4	2.1	2.4	2.7

Table 3. Composition of synthetic lunar glass.

Principal components	Wt % of principal components
SiO_2	42.098
Al_2O_3	14.799
FeO	14.766
CaO	12.580
MgO	7.614
TiO_2	5.425
Cr_2O_3	1,001
Na_2O	0.483
P_2O_5	0.304
NiO	0.297
FeS	0.264
K_2O	0.187
MnO	0.183

Trimethylsilylation of 0.255 g of this glass, corresponding to 51% by weight of the lunar samples (cf. Table 1) yielded the chromatogram shown in Fig. 2d.

Discussion

It is evident from Fig. 2 that olivine is the dominant source of SiO_4^{4-} ion in the lunar fines and that side reactions in the trimethylsilylation of olivine account largely for the remaining peaks in the chromatograms in Figs. 2a and 2b. This is shown in a comparison of the ratios of integrated peak areas (Table 2). The chromatographic patterns for samples 14003,27 and 14163,70 are almost identical, in line with their close similarity in modal composition (Table 1).

The yield of TMS derivatives from the lunar samples was somewhat lower than expected on the basis of 7% olivine, as evident from Fig. 2. Assessment of quantitative yields proved difficult, however, because of the heterogeneous nature of the lunar fines. To study this problem, experiments were performed with augite, calcic plagioclase, and synthetic mixtures. No derivatives of discrete silicate ions were detected with augite, as expected, and only trace quantities of silicate derivatives were observed with plagioclase. With mixtures of augite, plagioclase, olivine, and synthetic glass in approximately the same proportions as indicated by the modal analyses, trimethylsilylation of 0.5 g samples gave lower yields of SiO_4^{4-} derivative than predicted. Analysis of a mixture of olivine and finely ground silica, however, gave the expected yield, indicating that the lowering of the yield in the complex mixture probably was due to depletion of available reagents with formation of TMS derivatives of higher silicate structures that remain as solids in the reaction vessel or are of too high molecular weight to be detected chromatographically. The results with mixtures confirmed that the observed yield of SiO_4^{4-} derivative from the lunar material was consistent with the presence of approximately 7% olivine as established by the modal analyses.

The ratios of peak areas Si_2O_7/SiO_4 for the lunar fines (Table 2) are slightly higher than the corresponding ratio for olivine alone. The small differences may be due to the influence of the other constituents in the fines and/or the presence of very small quantities of dimeric (e.g., melilite-group) minerals, as suggested by the mineralogical observations.

The main difference between the present results and those for Apollo 11 and 12 fines is the virtual absence of the cyclic $Si_4O_{12}^{8-}$ derivative in the present extracts. The range of amounts of this derivative in different extracts support the view that the derivative is due to traces of another mineral, particularly in sample 12001, where the contribution due to this peak was highest.

The main conclusion that emerges from this and the previous study is that the orthosilicate ion is by far the most abundant discrete silicate ion in lunar fines from three separate locations. No evidence has yet been found for the presence of discrete phosphate or silicophosphate ions, which are clearly observable under the conditions of the present work, of cyclic $Si_3O_9^{6-}$ ions and of more complex silicate ions of charge 10^-, which also can be identified by this technique. These results are consistent with mineralogical studies. For all samples investigated there is evidence for the presence of discrete $Si_2O_7^{6-}$ ions in trace amounts and this is now supported by the detection of possible melilite-type minerals in sample 14003,27. There is also clear evidence for the presence of cyclic $Si_4O_{12}^{8-}$ anions in small amount in sample 12001,76, though this awaits confirmation by other techniques.

Acknowledgments—We are grateful to Brian Mason, Smithsonian Institution, Washington, D.C., for specimens of calcic plagioclase and augite. We thank G. W. Caines, W. R. Crosby, D. J. Embree, and D. R. Johnson for technical assistance.

References

Butts W. C. and Rainey W. T. Jr. (1971) Gas chromatography and mass spectrometry of the trimethylsilyl derivatives of inorganic anions. *Anal. Chem.* **43**, 538–542.

Götz J. and Masson C. R. (1970) Trimethylsilyl derivatives for the study of silicate structures. I. A direct method of trimethylsilylation. *J. Chem. Soc.* A, 2683–2686.

Götz J. and Masson C. R. (1971a) Trimethylsilyl derivatives for the study of silicate structures. II. Orthosilicate, pyrosilicate and ring structures. *J. Chem. Soc.* A, 686–688.

Götz J. and Masson C. R. (1971b) Trimethylsilylation of silicate anions: A method of studying the structure of crystalline silicates and glasses. *Proc. IXth Int. Congress on Glass, Versailles*, Sept. 27–Oct. 2, pp. 261–276.

Götz J. Masson C. R. and Castelliz L. M. (1972) Crystallization of lead silicate glass as studied by trimethylsilylation and chromatographic separation of the anionic constituents. In *Amorphous Materials* (editors R. W. Douglas and B. Ellis), pp. 317–325. Wiley.

Jamieson W. D. Mason F. G. and Masson C. R. (1972) Trimethylsilyl derivatives for the study of silicate structures. III. Identification of silicate, phosphate, and silicophosphate derivatives in extracts of metallurgical slags. In preparation.

Lentz C. W. (1964) Silicate minerals as sources of trimethylsilyl silicates and silicate structure analysis of sodium silicate solutions. *Inorg. Chem.* **3**, 574–579.

Masson C. R. Götz J. Jamieson W. D. McLachlan J. L. and Volborth A. (1971) Chromatographic and mineralogical study of lunar fines and glass. *Proc. Second Lunar Sci. Conf., Geochim. Cosmochim. Acta* Suppl. 1, Vol. 1, pp. 957–971. MIT Press.

Westman A. E. R. (1960) Constitution of phosphate glasses. In *Modern Aspects of the Vitreous State* (editor J. D. MacKenzie), Vol. 1, Chap. 4, pp. 63–91. Butterworths.

Proceedings of the Third Lunar Science Conference
(Supplement 3, *Geochimica et Cosmochimica Acta*)
Vol. 1, pp. 1037–1064
The M.I.T. Press, 1972

Metallic particles in the Apollo 14 lunar soil

J. I. Goldstein, H. J. Axon

Metallurgy and Materials Science Department, Lehigh University, Bethlehem,
Pennsylvania 18015

and

C. F. Yen

Metallurgy and Materials Science Department, Massachusetts Institute of
Technology, Cambridge, Massachusetts

Abstract—The metallographic structures, silicate associations and bulk compositions (Ni, Co, P, S, and Si) were determined by microprobe analysis for 205 metal particles selected by magnetic separation from the <1 mm, > 125 μ fraction of soil samples 14003,18 and 14163,165. A small number of exotic particles were encountered but almost all the metal could be accommodated within one of four sub-populations, the characteristics and genetic significance of which may be stated as follows:

I. The major proportion has Ni and Co contents corresponding to the range of meteoritic metal. This metal appears to have been liberated to the soil after undergoing thermal processing in the fragmental rocks that now constitute the Fra Mauro formation. As a consequence of its thermal history, it has lost its meteoritic microstructure and has undergone limited redistribution of trace elements. This metal had its origin in the predominantly chondritic projectiles that bombarded and accumulated on the pre-Imbrian lunar surface.

II. Small proportion (∼5%) that not only has meteoritic contents of Ni and Co but also shows still distinguishable indications of meteoritic microstructure. This metal forms part of the meteoritic material that fell on the Apollo 14 site subsequent to the emplacement of the Fra Mauro formation.

III. Another small population (∼10%) consists of metal excavated from lunar mare basalts by cratering events, and thrown onto the Apollo 14 site.

IV. Metallic spheroids (∼5%), analogous to those encountered in the vicinity of the Barringer Meteorite Crater, Arizona. We identify these particles as explosively melted portions of meteoritic projectiles.

We estimate that the maximum possible contribution to the metallic portion of the soil, that would arise if the projectile that excavated the Imbrium Basin were an iron meteorite or a chondritic meteorite, would be 4% and 6%, respectively. We have not been able to make any unambiguous identification of such material in our present studies.

Introduction

Metallic particles are a minor but ubiquitous component of the lunar rocks, breccia, and soil. Because of the relatively simple chemistry and phase equilibria relationships within the metallic phases, it is possible in many cases to determine unique-time-temperature-pressure histories for the metallic particles. The purpose of this investigation was to study the chemistry and structure of metallic particles in order to determine their origin (meteoritic vs. lunar) and to describe the effects of shock-impact-reheating events at the Imbrium and Fra Mauro locations.

Method

We have made a chemical and metallographic study of 205 of the most magnetic particles that were present in two 15 g samples (14003,18 and 14163,165) of the <1 mm Apollo 14 soil. Each soil sample was separated into 3 sieve fractions $> 125\,\mu$, $> 354\,\mu$ and $> 707\,\mu$ and each size fraction was then separated into more and less magnetic fractions by a Frantz separator. Finally, individual particles were selected from the magnetic fractions with the assistance of a hand magnet. Some of these particles were examined by scanning electron microscopy and all 205 were mounted in transparent epoxy and prepared for metallo-graphic and microprobe examination.

METALLIC PARTICLES

General description

In most of the magnetic particles the major phase was metallic, although only about 25% of the particles consisted of freestanding metal without visible silicate associations. In about 15% of the cases the metal existed as small metallic inclusions embedded in a silicate assemblage. Almost all of the pieces of metal in the magnetic fraction were in the size range of $100\,\mu$ to $150\,\mu$, and only 8 pieces were larger than $300\,\mu$. The weight fraction of clean magnetic particles was about 0.05 wt%, but this value greatly underestimates the total metallic content of the soil because the metal in the $< 125\,\mu$ sieve fraction was not measured and in the less magnetic fractions there is a great deal of metal incorporated as very small inclusions in predominantly silicate particles. The total metallic iron content of soil 14163 has been determined by Wlotzka, et al. (1972) as 0.5 wt%.

RESULTS

Metal particles in the Apollo 14 lunar soil—chemistry, metallography, and silicate associations

Bulk chemical analyses for Fe, Ni, Co, P, and S in 205 metal particles were measured with an electron microprobe. Silicon contents also were measured in a number of samples, but in all cases were less than 0.05 wt%. The bulk analyses varied widely from < 0.02 to 21.1 wt% Ni, < 0.02 to 3.0 wt% Co, < 0.02 to 7.7 wt% P, and < 0.02 to 4.2 wt% S. Some particles were chemically zoned, but the vast majority were homogeneous and had compositions in the range < 0.02–10.0 wt% Ni, < 0.02–0.9 wt% Co, < 0.20–0.5 wt% P, and < 0.02–0.1 wt% S. The Ni and Co analyses for all the particles are tabulated according to their metallographic structure and silicate mineral association in Table 1 for single phase metal particles and Table 2 for more

Table 1. The compositions and silicate-metal associations of magnetic particles from soil samples 14003 and 14163 in which the metal is essentially single phase.

Soil Sample 14003					Soil Sample 14163					
Sample number	Weight %			Type of silicate	Comments	Sample number	Weight %			Type of silicate	Comments
	Ni	Co	P				Ni	Co	P		
Structureless Metal in Idiomorphic Silicate											
27.7	1.3	0.4	—	II 1(I)		14.13	0.12	0.17	0.4	II 1(O)	
26.22	2.2	0.1	0.1	II 1(I)	0.12% S	M3.5	0.15	0.2	0.07	II 1(I)	
26.23	3.1	0.3	0.14	II 1(I)	sulphide	14.16	4.8	0.45	0.19	II 1	
26.21	5.1	0.4	0.1	II 1(O)		24.3	6.1	0.6	0.14	II 1	0.04% S
26.10	5.2	0.3	0.1	II 1(I)	sulphide						
26.33	5.5	0.6	—	II 1(I)	sulphide	13.2	6.7	0.5	0.05	II 1(I)	
26.28	5.7	0.3	0.26	II 1(I)	sulphide						
26.26	6.7	0.3	0.12	II 1	sulphide						

Table 1 (continued)

| Soil Sample 14003 | | | | | Soil Sample 14163 | | | | |
Sample number	Ni	Co	P	Type of silicate	Comments	Sample number	Ni	Co	P	Type of silicate	Comments

Structureless Metal in Nonidiomorphic Silicate

Sample number	Ni	Co	P	Type of silicate	Comments	Sample number	Ni	Co	P	Type of silicate	Comments
26.13	3.0	0.02	—	II 3(I)		14.8	4.8	0.4	0.16	II 2(O)	
26.30	4.5	0.2	—	II 2(O)	Ni zoned	16.1	5.2	0.4	—	II 2(I)	
28.29	5.0	0.4	0.1	II 2–3		14.10	5.4	0.5	0.15	II 2–3	P variable, high at edge
26.18	5.2	0.5	0.2	II 2					0.5		
26.31	5.4	0.1	0.32	II 3–2(I)		24.8	5.4	0.5	0.17	II 2	α
27.2	5.9	0.45	—	II 3–4			14.0	—	n.d		γ (trace)
26.35	6.1	0.4	0.17	II 2(I)		22.1	5.9	0.7	—	II 2–3	
26.20	6.2	0.4	—	II 3(I)		13.9	6.3	0.25	0.1	II 3	
26.3	6.3	0.3	0.07	II 3–2							
28.20	6.5	0.6	—	II 3(I)							
26.17	6.6	0.6	—	II 2(I)							
26.1	6.7	0.3	0.05	II 2							
27.16	7.2	0.3	0.3	II 3							
26.32	11.0	0.6	0.2	II 3–1(CZ)							

Structureless Metal with Idiomorphic Silicate

Sample number	Ni	Co	P	Type of silicate	Comments	Sample number	Ni	Co	P	Type of silicate	Comments
28.10	0.02	0.5	—	I 1–4		14.11	3.6	0.5	—	I 1	
M7.3	4.6	0.4	0.17	I 1		22.4	4.6	0.35	0.31	I 1	
26.9	4.7	0.3	0.25	I 1		22.8	4.7	0.4	—	I 1	
28.32	4.9	0.5	—	I 1		22.3	4.8	0.35	0.15	I 1	
26.11	5.1	0.4	0.25	I 1		23.10	4.9	0.4	—	I 1–2	
27.18	5.1	0.5	0.05	I 1		24.5	5.4	0.4	0.12	I 1	
27.30	5.1	0.65	—	I 1		14.22	5.4	0.5	0.08	I 1	
27.3	5.6	0.4	0.07	I 1		18.8	5.5	0.4	0.15	I 1–3	
27.22	5.8	0.4	0.1	I 1		24.1	5.7	0.45	—	I 1	
28.19	6.4	0.75	0.15	I 1–4		18.10	5.9	0.45	0.05	I 1–3	
27.32	6.5	0.5	—	I 1		14.7	6.1	0.53	0.1	I 1	
M7.6	6.5	0.6	0.17	I 1		14.9	6.2	0.52	0.11	I 1	
						14.24	6.3	0.5	0.17	?	identification of silicate uncertain
						14.14	6.6	0.75	—	?	
						14.1	6.7	0.32	0.19	I 1	

Structureless Metal with Breccia

Sample number	Ni	Co	P	Type of silicate	Comments	Sample number	Ni	Co	P	Type of silicate	Comments
26.15	0.3	≤0.02	—	I 2–4		23.6	4.5	0.4	0.15	I 2	
26.4	7.1	0.75	—	I 2		23.2	4.7	0.6	0.15	I 2	
28.22	7.5	0.75	—	I 2		23.18	6.2	0.5	0.25	I 2–4	
26.5	19.8	1.3	0.2	I 2–4							

Structureless Metal with Glass

Sample number	Ni	Co	P	Type of silicate	Comments	Sample number	Ni	Co	P	Type of silicate	Comments
26.24	4.8	0.3	0.17	I 3		14.29	5.0	0.5	0.1	I 3	
28.28	7.4	0.7	—	I 3							

Structureless Metal Freestanding

Sample number	Ni	Co	P	Type of silicate	Comments	Sample number	Ni	Co	P	Type of silicate	Comments
28.23	0.05	0.2	—	silicate inclusions in the metal		24.6	0.02	0.02	0.2	0	
						14.4	~5.0	0.3	—	0	
26.29	1.1	0.1	—	0		23.7	5.5	0.4	0.05	0	
26.7	3.5	0.9	—	0	small splinter	18.5	5.6	0.5	0.17	0	
28.5	5.5	0.35	—	0	some 4?	14.20	5.7	0.7	0.1	0	
28.31	6.7	0.5	—	0		14.12	6.1	0.5	0.15	0	
						18.11	6.6	0.75	—	0	
						14.3	6.8	0.65	0.04	0	

Single Crystal α with Neumann Lines

Sample number	Ni	Co	P	Type of silicate	Comments	Sample number	Ni	Co	P	Type of silicate	Comments
26.19	0.1	0.02	—	I 1–II 1		13.8	0.02	0.1	0.05	I 2	
27.4	0.3	0.4	—	0		14.6	0.04	0.05	0.16	I 1	
27.17	5.2	0.4	0.1	I 2		22.12	2.5	0.3	—	0	
M5.2	5.4	0.4	0.07	0		18.13	4.7	0.6	—	I 2	
27.19	5.6	0.5	—	I 1		M3.3	5.0	0.35	—	I 1	
27.11A	5.8	0.45	—	0	partly annealed	23.1	5.9	0.5	0.15	I 3	
27.15	5.9	0.45	—	I 1		18.7	6.4	0.5	0.1	I 3	
	6.6					14.4	6.5	0.61	0.05	0	
27.25	6.7	0.6	0.05	I 2–II 2 (4)							

Table 1 (continued)

	Soil Sample 14003						Soil Sample 14163				
Sample number	Weight %			Type of silicate	Comments	Sample number	Weight %			Type of silicate	Comments
	Ni	Co	P				Ni	Co	P		
Polycrystalline α											
M7.8	0.1	0.02	—	0	sulphide, ε	13.7	1.1	0.2	0.1	I 1	annealed Neumanns
27.12A	2.3	0.25	—	II 1		23.15	1.7	0.25	—	I 2–4	mixed g.s.
26.2	3.5	0.3	—	II 3(O)	mixed g.s.	16.12	2.3	0.2	—	I 1	mixed g.s.
28.2	3.6	0.25	0.17	I 3		16.16	2.4	0.4	0.2	0	sulphide
27.9	4.7	0.35	0.15	0	bicrystal with Neumanns	18.15	3.7	0.4	0.15	I 2	
26.16	5.8	0.5	0.04	II 3(I)	mixed g.s.	M3.2	4.6	1.3	0.12	I 1–3	0.05 wt% S ε
28.25	7.1	0.6	—	I 1–4	fractured metal	13.1	5.6	0.5	0.05	I 2	cracked metal
						13.4	5.6	0.35	0.2	I 1	sulphide cracked metal
						16.14	5.9	0.4	—	I 1	
Heavily deformed α											
27.14	4.0	0.4	—	I 2	ring of inclusions in metal	16.15	0.08	0.1	—	I 2–3	may be polycrystalline α
26.14	3.7	0.3	0.62	I 3	hot tears	18.3	0.08	0.3	—	0	splinter
α_2 (or Ragged $\gamma \to \alpha$ Transformation Product)											
27.13	4.2	0.3	0.05	I 4		18.2	0.02	0.02	—	I 3	
27.29	4.5	0.35	—	I 1		18.6	4.8	0.3	0.1	I 3	
27.21	5.4 } 6.0	0.6	—	I 4		16.11	5.7	0.4	—	I 2	
						14.2	5.9	0.55	—	0	
27.5	5.6	0.5	0.05	I 4		23.5	6.2 ± 0.3	0.4	0.4 } 0.75	0	phosphate inclusions in remelted metal
26.8	6.1	0.45	0.12	I 2–4							
						16.10	6.0	0.35	—	I 2	
						16.3	7.1	0.75	0.07	I 2	reheated, identification as α_2 uncertain.
Blocky Martensite											
27.23	6.3	0.4	0.05	I 4		16.2	6.2	0.5	—	I 3	
28.12	6.4	0.4	—	I 4		14.25	7.6	0.4	—	?	sulphide
28.16	6.5 } 7.5	0.75	—	I 2		14.23	8.3	0.7	—	I 2	
27.11	7.1	0.65	—	I 1	partly dissolved carbide						
M7.7	7.4	0.9	—	I 1							
27.28	8.1	0.65	—	I 3							

complex metal particles. A distinction is made between particles from soil 14003 and 14163. In Table 1 sulphur contents were nearly always at or below the detectability limits, < 0.02 wt%, and are recorded only in special cases. Similarly, in Table 1 only those phosphorus values > 0.02 wt% are tabulated. More detailed P and S data are presented in Table 2. The silicate-metal associations are recorded in Tables 1 and 2 according to the following scheme: freestanding metal, showing essentially no attached silicate in the plane of section examined in the present work, is designated 0. Cases in which the metal is the major component but is associated with relatively small amounts of peripheral silicate are designated I1, I2, I3, I4 to indicate that the silicate is (1) well crystallized, idiomorphic, (2) breccia, (3) glass or that the metal-silicate interface shows a (4) "corroded" interaction whereby the metallic and non-

Table 2. The compositions and silicate-metal associations of more complex metal particles
in soil samples 14003 and 14163. Bulk analyses only.

Sample number	Weight %				Type of silicate	Comments	
	Ni	Co	P	S			
					Two Phase α + γ Structures in Soil 14003		
27.12	6.1	0.45	0.05	—	I 1	isothermal taenite	M
28.9	6.2	0.45	0.2	—	0	isothermal taenite	M
28.1	6.5	0.6	—	—	I 2	clear taenite	
26.12	6.9	0.75	—	—	I 3–4	chondritic plessite	M
28.27	7.1	0.65	0.05	—	I 1		
27.26	7.2	0.65	—	—	I 3	chondritic plessite	M
28.13	7.2	0.5	—	—	I 1–4	clear taenite	M
26.27	7.7	0.4	0.16	—	I 3	clear taenite	M
27.31	8.0	0.65	0.13	—	I 2		
M9.0	8.5	0.75	—	—	I 3–2	tempered martensite	
					Two Phase α + γ Structures in Soil 14163		
16.4	5.9	0.6	0.05	—	I 1	clear γ?	
18.16	6.0	0.6	0.07	—	I 2	clear γ?	
M3.1	6.6	0.45	0.1	—	I 1	incomplete Widmanstatten structure	M
14.5	7.3	0.75	0.1	—	I 2	chondritic plessite	M
					Two Phase Metal Plus Compound Structures in Soil 14003		
28.6	0.1	0.05	0.05	—	I 2	carbide?	
28.21	0.25	0.4	0.005	—	0–I 2		
27.27	1.3	1.4	0.12	0.07	I 4	carbide containing 0.4 wt% Ni	
28.17	2.2	0.45	—	—	I 2	FeS + graphite nodule in α	
M7.4	2.3	0.75	—	0.05	FeS	α + massive FeS	
27.6	2.7	0.65	—	0.05	FeS	α + massive FeS	
28.24	2.8	0.65	—	—	I 3	annealed Neumanns	
28.26	2.9	0.3	0.1	—	I 4	phosphide: FeS eutectic in trace quantities.	
28.15	3.7	0.5	—	—	I 2		
28.14	5.3	0.5	0.17	—	I 1–4	? compound suspected	
M8.1	5.4	0.45	0.15	—	I 2	carbide, reheated.	M
27.33	6.5	0.25	0.3	—	II 2	minute sulphides	
27.1	6.5	0.5	0.12	—	I 1	α₂	
25.2	6.5	0.75	—	—	I 3	glass contains globules of Fe–FeS "eutectic."	
27.24	6.8	0.65	0.12	—	I 1	α₂-phosphide	
26.25	7.8	0.85	—	0.05	I 1–II 1	minute sulphides?	
25.1	8.1	0.4	0.1	—	I 2–4		
28.30	9.6	0.75	—	0.1	I 2	minute sulphides?	
M5.1	9.6	0.6	0.75	—	I 1	many small oriented phosphide precipitates	
					Two Phase Metal Plus Compound Structures in Soil 14163		
22.9	1.7	0.6	—	0.04	FeS	α + massive FeS	
23.9	2.8	0.45	0.2	—	Phosphide	α + massive phosphide	
14.18	4.8	0.9	—	0.07	FeS	α + massive FeS	
18.4	5.0	0.4	0.07	—	I 3		
24.4	5.2	0.3	0.37	—	0	P in solid solution, minute sulphides	
M3.4	5.3	0.45	0.12	—	I 1–2	carbide	M
13.6	6.3	0.6	—	—	0	small polycrystal FeS in α	M
23.16	6.7	0.5	—	—	0	α₂	
23.11	6.7	0.65	0.05	—	I 1	Neumanns in α	
					Special Case: Dendritic Metal Sulphide Spherules in Soil 14003		
28.3	6.1	0.25	0.9	2.3			M
M11.11	6.6	0.3	2.9	2.7			M
M6.1	~7.0	0.6	0.2	0.15			M
26.6	8.1	0.4	4.5	0.3			M
M6.3	8.9	0.3	6.5	4.2			M
27.10	13.0	0.45	6.4	1.0			M

Table 2 (continued)

Sample number	Weight %				Type of silicate	Comments
	Ni	Co	P	S		
Dendritic Metal Sulphide Spherules in Soil 14163						
M2.6	6.1	0.4	1.65	1.05		M
14.12	6.2	0.3	2.2	1.7		M
14.17	6.7	0.3	2.3	2.0		M
M2.7	7.1	0.35	3.5	1.0		M
16.5	12.1	0.45	6.5	2.4		M
M2.9	14.5	0.75	5.7	2.9		M
M10.1	15.6	0.45	7.7	2.0		M
Special Case: Phosphide Metal Particles in Soil 14003						
28.4	5.9	0.35	2.9	—	I 4	phosphide laths in α
M7.1	7.8	0.65	1.7	—	I 4	phosphide globules in polycrystalline α
Phosphide Metal Particles in Soil 14163						
23.20	4.3	0.4	0.3	—	0	α barred and rimmed with phosphide
22.7	12.8	0.9	0.17	0.05	0	γ rimmed and with precipitates of phosphide
22.5	13.9	0.8	0.29	—	0	γ rimmed with phosphide, some swathing α.
Special Case: Metal with Phosphide Eutectic in Soil 14003						
M7.5	6.1	0.45	0.2	—	I 2	annealed
M12.1	6.1	0.5	0.7	~1.5	?	cracked metal
27.20	6.2	0.35	0.6	—	I 2	cracked metal
27.8	6.6	0.3	0.8	—	0	smooth particle of α_2 not cracked. Thermally diffused eutectic rim.
Metal with Phosphide Eutectic in Soil 14163						
22.10	6.1	0.3	0.55	0.02	I 4	cracked metal
18.14	7.9	0.5	2.2	0.7	I 3	
Special Case: Exotic Particles						
28.11	~7.0	0.7	—	—	I 1	α, ilmenite, microcracks, probe Ni content variable down to ~6% in microcracked area.
28.8	21.1	3.0	—	0.15	I 4	small crystals of silicate and ilmenite in duplex α γ metal of high-cobalt content.
23.8	14.9 ±1.0	0.9	0.5	—	I 4	Polycrystalline martensite, perhaps remelted.
22.6	—	—	—	—	0	17% Cr 81% Fe

metallic material is finely intermixed. Occurrences with the metal present as an inclusion within a mass of silicate are designated II 1, II 2, II 3, according as the silicate is (1) crystalline, (2) breccia, (3) glass, and in these cases the location of the metal within the silicate mass is further specified as (I) inside the silicate mass, (O) at or near the outside of the silicate mass and, in one instance, (CZ) signifying that the metal resides at the contact zone between glass and crystalline rock. If gradations between metal or silicate types are observed, these variations are noted in the Tables with the predominant type given first. In a few instances the identification is uncertain.

Figure 1 shows a freestanding metal particle with little or no attached silicate, Fig. 2 shows Type I metal with well-crystallized, idiomorphic, silicate at the periphery of a softly rounded metal particle, and Fig. 3 shows an example of typical Type II metal particles of irregular ragged shape included in a silicate matrix. In the most

Fig. 1. Particle 14.2, field of view 167 × 214 μ. Etched 1% Nital. Freestanding metal particle [Type 0] with α_2 structure.

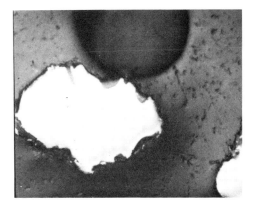

Fig. 2. Particle 27.11, field of view 207 × 266 μ. Etched 1% Nital. Softly rounded particle of structureless metal with well crystallized, idiomorphic, silicate at the periphery [Type I 1].

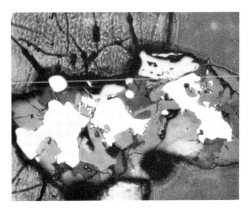

Fig. 3. Particle M3.5, field of view 267 × 343 μ. Etched 1% Nital. Irregular ragged particles of structureless metal embedded in a matrix of crystalline silicates [Type II 1].

highly altered silicate, breccia or glass, the Type II metal has a structure and dis-tribution that is quite similar to that found in the vein material of shocked chondrites. However the sulphide phase that is so abundant in chondrites is not found commonly in the lunar soil.

In many instances, the metal develops no etching structure when attacked by 1% nital for up to 40 seconds. Such metal is designated structureless in Table 1. In a number of cases, chemically homogeneous metal reveals a range of microstructures. These structures are classified in Table 1 as single crystal α with Neumann lines (Fig. 4), polycrystalline α (Fig. 5), heavily deformed α, α_2 (Fig. 1), or blocky martensite (Fig. 6). From Table 1 it may be seen that there is no essential difference between soil samples 14003 and 14163 in the range and distribution of metallic structures encountered in the sieved and magnetically separated fractions of our samples. Consequently, the Co and Ni data for metal particles with similar metallographic structures from both soils are combined and plotted together in Figs. 7–9. The Ni + Co range for meteoritic metal is also plotted on the figures following Goldstein and Yakowitz (1971).

The structureless metal in idiomorphic silicate (Fig. 7a) shows two ranges of composition, one of which is appropriate to meteoritic metal (Goldstein and Yakowitz, 1971) and a second range at low-Ni, low-Co compositions <1 wt% appropriate to metal in Apollo 11 lunar basalts or remelted lunar rock (Keil *et al.*, 1970). It is probable that many of the particles in Fig. 7a are of meteoritic origin but have lost their meteoritic structure by reheating during lunar residence. The structureless metal in non-idiomorphic silicate Fig. 7b, is essentially of meteoritic Ni + Co content and probably originated in meteoritic projectiles that became incorporated into the breccia or glass of the lunar fragmental rocks.

The structureless metal with minor association of idiomorphic silicate, Fig. 8a, commonly has size and shape similar to the metallic particles in the thin section of fragmental rock 14304 that we have examined. In addition the particles have a Ni + Co composition that, in general, is appropriate to meteoritic metal. It is therefore probable that the metal particles in Fig. 8a had their immediately previous residence in fragmental lunar rock, from which they were liberated by lunar "weathering" processes. The metal shows no meteoritic or shock structure and has a significant phosphorus content in solid solution. As discussed below, it is probable that the metal in this category entered the crust of the moon by way of chondritic meteorites that bombarded the surface of the moon in the region that was later to be excavated to form the Imbrium Basin. Residence within the lunar crust, both at the pre-Imbrium location and later at the Fra Mauro site was accompanied by thermal annealing effects (Anderson *et al.*, 1972). This reheating could allow minor changes of trace element chemistry and alterations of metallographic structure.

Most of the structureless metal with minor associations of breccia or glass (Fig. 8b) may have had a similar history to that already discussed for structureless metal incorporated into breccia or glass (Fig. 7b), except that the metal in Fig. 8b was more completely liberated from its associated silicate. The structureless free-standing metal falls into at least two subcategories (Fig. 8c). One category is based on the meteoritic Ni + Co chemistry and, in general, the outlines of these metal

Fig. 4. Particle M3.3, field of view 435 × 559 μ. Etched 1% Nital. Single crystal α with Neumann lines.

Fig. 5. Particle 16.12, field of view 312 × 401 μ. Etched 1% Nital. Polycrystalline α.

Fig. 6. Particle M9, field of view 171 × 219 μ. Etched 1% Nital. Blocky martensite, tempered.

particles suggest that they have been completely liberated from fragmental lunar rock. There is one peculiar particle with 3.5% Ni, 0.9% Co that has the form of a minute splinter of metal. The low-Ni, low-Co particles have block-like shapes that are not inconsistent with an origin in remelted or igneous lunar rocks, from which they were later liberated by fracture of the silicates.

The single phase α particles that show microstructural detail on etching, are collected in Table 1 and their Ni + Co contents are plotted in Fig. 9 according to metallographic type. The single crystal samples of α with Neumann lines are plotted in Fig. 9a with symbols to indicate their silicate associations. The solid symbols refer to large particles (>300 μ in size), the composition and structure of which are in fact not different from the general run of other particles. Metal with this

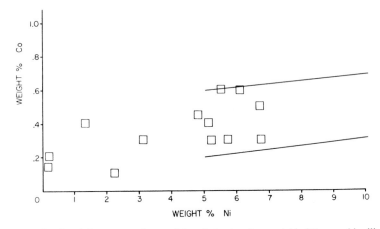

Fig. 7a. Bulk Ni and Co contents for particles of structureless metal in idiomorphic silicate (II 1).

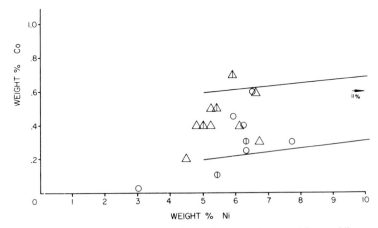

Fig. 7b. Bulk Ni and Co contents for particles of structureless metal in non-idiomorphic silicate. II 2 and II 3. Silicate associations breccia or predominantly breccia △ or ▲, glassy or predominantly glassy ○ or ⬤. Note one particle at 11% Ni, 0.6% Co. II 3–1 (CZ).

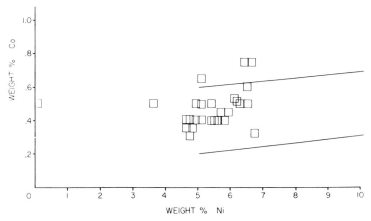

Fig. 8a. Bulk Ni and Co contents for particles of structureless metal with attachments of idiomorphic silicate, □, I 1.

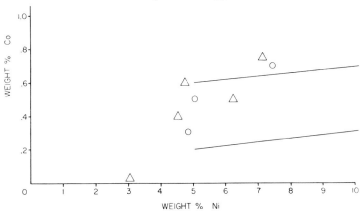

Fig. 8b. Bulk Ni and Co contents for particles of structureless metal with minor associations of breccia △ or glass ○. Note one particle at 19.8% Ni, 1.3% Co. I 2–4.

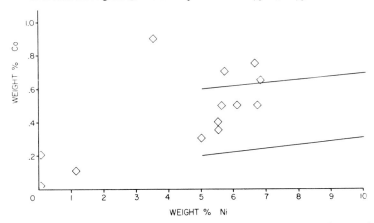

Fig. 8c. Bulk Ni and Co contents for freestanding particles of structureless metal ◇.

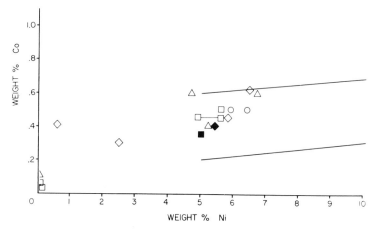

Fig. 9a. Bulk Ni and Co contents for single crystal particles of metal with Neumann bands. Silicate associations are noted by □ idiomorphic silicate, △ breccia, ○ glass as in Fig. 8. Freestanding particles are noted by ◇.

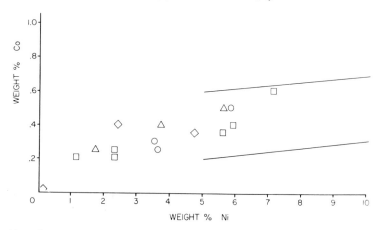

Fig. 9b. Bulk Ni and Co contents for polycrystalline particles of metal. Silicate associations as in Fig. 8. Note one particle at 4.6% Ni, 1.3% Co. I 1–3. ε.

microstructure has either the characteristic meteoritic or the Apollo 11-lunar Ni + Co content. One meteoritic particle shows a range of Ni content from 5.9 to 6.6 wt% Ni, but the remainder are chemically homogeneous.

In our description of the structureless metal we have drawn attention to the thermal annealing processes to which metal particles may have been subject within the moon's crust. By contrast, the metal particles in Fig. 9a contain indications of shock deformation that has not been annealed out. The deformation twins in the meteoritic metal may already have been present in the meteorite before it hit the moon or may have been produced by the meteorite-moon impact. However, in the metal of lunar composition the Neumann bands must have arisen from meteoritic impact.

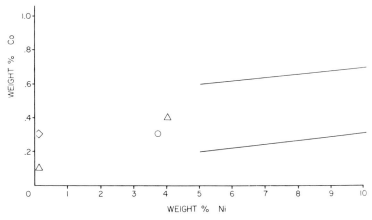

Fig. 9c. Bulk Ni and Co contents for very heavily deformed particles, silicate associations as in Fig. 8.

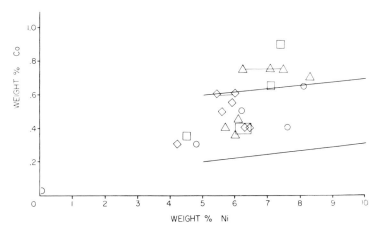

Fig. 9d. Bulk Ni and Co contents for metal particles with α_2 or blocky martensite structures. Silicate associations as in Fig. 8.

Very similar considerations of deformation effects and partly annealed deformation effects, including the shock generated ε polymorph of iron, cracking of the metal and recrystallization, apply to both the lunar and meteoritic metal that show a polycrystalline structure of either uniform or mixed grain size (Fig. 9b). The significance of polycrystalline α phase is that such metal has been subject to variable but small degrees of reheating after it experienced its most recent shock or deformation event. A small number of very heavily distorted particles were found. None is of unambiguously meteoritic Ni + Co content and they are plotted in Fig. 9c.

Figure 9d illustrates the silicate associations and Ni + Co contents of metal particles with an α_2 or blocky martensite structure. Reference to Table 1 indicates that a number of the particles in Fig. 9d are inhomogeneous in relation either to

Ni or P, and in one instance (#27.11) carbide, probably cohenite, is present and shows signs of incomplete dissolution in the metal. Reheating to 800°C for periods of a few seconds or longer would cause partial dissolution of cohenite (Brentnall and Axon, 1962) and allow for P or Ni redistributions in the solid state.

Table 2 contains information on the composition and structure of complex, mainly two-phase, metal particles, collected under the categories I α + γ structures, II metal plus compound structures, III special cases. The values in Table 2 are bulk compositions for the metallic portion of the particles. No attempt has been made to derive whole-body analyses for those fragments that consist of metal attached to massive sulphide or phosphide (Fig. 10). Our detailed microprobe investigation of the nickel distribution between phases will be presented in a separate paper.

I. α + γ structures

Figure 11a shows one of the three most obvious instances of essentially unaltered kamacite-plessite metal from chondritic meteorites. Figure 11b shows an unusual metal structure. We have not found an exactly similar particle in our studies of metal in stony meteorites. However we regard it as very probably a metallic particle from a stony meteorite. The center of the particle shows an incomplete Widmanstatten structure of kamacite bands with etched kamacite-kamacite boundaries between the bands. A minute quantity of residual taenite of plessitic morphology occurs in the smaller nodule of metal. Most of the metal-silicate interface shows swathing kamacite ranging in thickness, but there is a thin boundary of residual γ at portions of the interface furthest removed from the small nodule. Small particles of isothermal taenite appear to have exsolved at the centers of the kamacite bands.

Two other two-phase α + γ particles in Table 2 contain isothermal taenite such as could be obtained by mild reheating of meteoritic kamacite on the lunar surface. In addition, there are several examples of kamacite in contact with homogeneous γ or clear taenite of the type discussed by Taylor and Heymann (1971) for chondritic meteorites. We consider that the two-phase α + γ metal fragments that have been described (Table 2) originated in the meteoritic projectiles that produced craters on or near the Fra Mauro formation close to the Apollo 14 landing site. In Table 2 we identify such particles by the symbol *M*. Figure 12a shows the Ni + Co contents of all the two-phase α + γ particles. All of the particles have meteoritic compositions as well as appropriate silicate associations.

II. Two-phase metal + compound structures

The Ni + Co contents of two-phase metal and compounds (Fig. 12b) shows that in this category there are examples of material in both meteoritic and non-meteoritic ranges. There are several cases in which metal of low nickel content is found in contact with massive sulphide or phosphide fragments (Fig. 10), giving structures similar to some encountered in Apollo 14 rock samples. Of the particles in the meteoritic Ni + Co range of compositions, the two large particles plotted as adjacent solid symbols in Fig. 12b both contain cohenite, as shown in Fig. 13. We designate both of the carbide-containing particles as meteoritic (*M*) in Table 2.

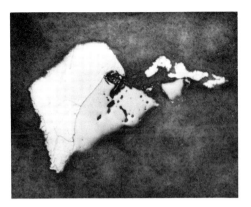

Fig. 10. Particle 22.9, field of view 285 × 366 μ. Etched 1% Nital. Low nickel α phase in association with massive sulphide.

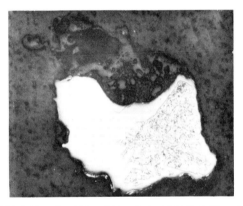

Fig. 11a. Particle 26.34, field of view 283 × 364 μ. Etched 1% Nital. Chondritic type structure of plessite in association with kamacite.

Fig. 11b. Particle M3.1, field of view 445 × 572 μ. Swathing kamacite and residual taenite around an incomplete Widmanstatten structure. An unusual metal particle, probably meteoritic.

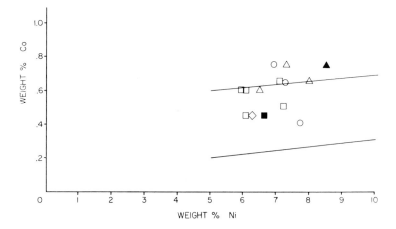

Fig. 12a. Bulk Ni and Co contents for particles with two-phase α + γ structures. Silicate associations are noted by □ idiomorphic silicate, △ breccia, and ○ glass as in Figs. 8 and 9. Freestanding particles are noted by ◇.

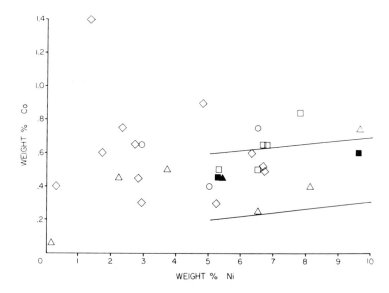

Fig. 12b. Bulk Ni and Co contents of particles with two-phase structures of metal with compounds. Silicate associations as in Fig. 12a.

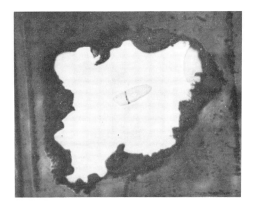

Fig. 13. Particle M3.4, field of view 286 × 368 μ. Cracked, boat-shaped particle of meteoritic carbide, cohenite, in kamacite. A portion of a larger cohenite particle is present at the edge of the metal.

The other particle so designated (#13.6) has polycrystalline troilite and kamacite and is structurally similar to metal found in some chondrites. In several instances, minute sulphides, or other phases, have been discovered or suspected in association with metal of meteoritic Ni + Co content, but in none of these cases is the structure unambiguously unaltered meteoritic. The large (solid symbol M5.1) particle with 9.6 wt% Ni contains an unusual distribution of very small phosphide precipitates oriented with respect to the metal matrix in a way that might have occurred through reheating. There is some uncertainty as to whether or not this particular fragment should be regarded as a special case of phosphide metal association.

Thus, for the metal plus compound particles a deliberately conservative estimate has been made of the number of unaltered meteoritic structures. It is possible that several of the particles with meteoritic Ni + Co content are of meteoritic origin with their original structures altered to an extent that makes identification somewhat uncertain.

III. Special cases

In this category there are 13 dendritic metal-sulphide spherules of meteoritic Ni + Co content. Their compositions are plotted in Fig. 14a and a particularly informative example of this group is illustrated in Fig. 15. The structure of the metallic globule is essentially dendritic with different amounts of interdendritic phosphide-sulphide eutectic phase and it is surrounded by a complete shell of sulphide-rich material. Other dendritic spherules have lost their sulphide shells to a greater or less extent and some have even suffered fragmentation of the interdendritic eutectic material. The metallic spherule particles have been found by many investigators in Apollo 11 and 12 soils and are very similar to the spheroids from the vicinity of the Barringer Meteorite Crater that are associated with the Cañon Diablo meteorites. According to Blau *et al.* (1972), the origin of these dendritic spheroids may be ascribed to the shock melting of sulphide-metal-phosphide areas within a

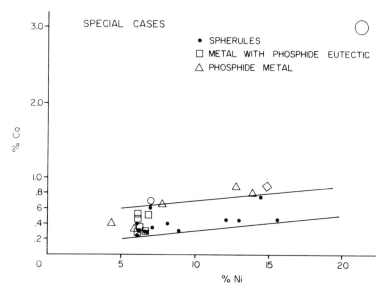

Fig. 14a. Bulk Ni and Co contents of particles that are regarded as "special cases," Table 2. Silicate associations are not represented in this figure. Exotic particles are noted by the symbols ◇ or ○. See Table 2 for other information. Note exotic particle at 21.1% Ni, 3.0% Co, 14.

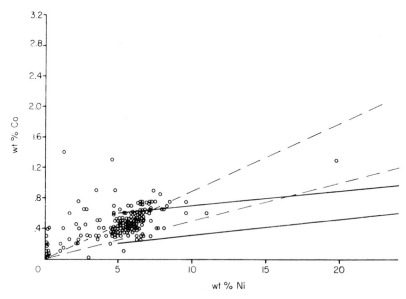

Fig. 14b. Bulk Ni and Co analyses for all metal particles of the present study except for the "special cases" given in Table 2 and Fig. 14a. Silicate associations and metallographic structures for the particles in Fig. 14b are presented in Figs. 7, 8, 9, and 12.

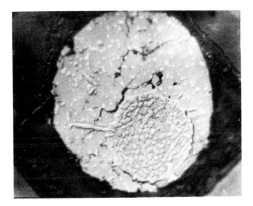

Fig. 15. Particle M2.9, field of view 170 × 218 μ. Dendritic globule of metal with phosphide eutectic embedded in a larger globule of troilite-rich eutectic. Note that the troilite is cracked. In other occurrences, only small amounts of troilite remain at the surface of the dendritic metal globule and we presume that for these globules the major quantity of troilite has been lost by mechanical processes.

meteoritic projectile when it hits its target with crater forming energy. In Table 2, we have recorded these spherules as meteoritic (M), in the sense that we regard them essentially as impact-melt products from meteoritic projectiles that produced cratering locally on the Fra Mauro formation. A detailed account of the dendritic metal-sulphide spherules will be presented in a separate paper.

The next group of special cases considered is designated phosphide-metal particles. Five examples have been found, with a possible sixth member represented by particle #M5.1 which has been discussed above. In general the phosphide metal particles have meteoritic compositions (Fig. 14a), a rounded appearance and are either freestanding metal or show minor surface corrosion effects. The phosphide may be encountered as laths or as more globular particles in an α matrix, or may form rims of phosphide around α or γ with more or less bar-like precipitates of phosphide within the particle. The origin of these particles is uncertain but they could be associated with a drastic reheating or remelting of meteoritic metal within the lunar rocks, such that phosphorus is reduced from lunar phosphate and enters the metal phase.

Another six particles have been found with a surface layer of phosphide eutectic. They are meteoritic in composition (Fig. 14a) and Fig. 16 illustrates such a structure in M12#1. This particle is about 1 mm in size, the largest particle obtained in our study. In most cases the metal shows signs of both fragmentation and heating, and it is common to find the eutectic located not only at the outside of the metal particle but also along penetrating cracks in the metal. In one instance (#27.8), the metal appears to have been dissolved away by contact with the hot phosphorous eutectic liquid. This particle is small and shows a smooth bean-shaped outline with no cracking. However, in most other cases the fragments of metal appear to have suffered only a small loss by dissolution in the melt. Some of these particles appear to have been annealed after the eutectic solidified. The structure of these particles is

Fig. 16. Particle M12.1, field of view 1242 × 1596 μ. Large particle of kamacite showing cracking. Phosphide eutectic is present at the surface of the particle and has penetrated some of the cracks.

very similar to that of an inclusion in breccia rock 10046–18A (Goldstein *et al.*, 1970) and to that of metal particle F2 from the Apollo 12 soil (Goldstein *et al.*, 1971). A meteoritic origin has been proposed for these particles, however, it also is possible that the structures could have originated by shock metamorphism in breccia rocks. In the present study we do not list these particles as meteoritic in the sense employed in Table 2.

Finally, there are a number of exotic particles including one (#22.6) that appears to be free of nickel and cobalt and consists of ~17% Cr ~81% Fe. This particle may be alien to the lunar population. The Ni + Co compositions of the rest of the exotic particles are given in Fig. 14a and are shown by the symbols ◇ or ○. A roughly globular particle (#23.8) consists of several austenite grains that have now transformed to martensite and which, because of its distribution of impurities, may have had a history of remelting and complicated heat treatment. One fragment (#28.11) has a Ni + Co content in the meteoritic range but has associated with it a crystal of ilmenite and it also has a range of fine microcracks in the metal that result in an apparent range of Ni content as analyzed by the microprobe. The remaining particle (#28.8) contains a high content of both nickel and cobalt similar to that found for Apollo 12 metal (Reid *et al.*, 1970) and its structure consists of small crystals of silicate and ilmenite included in a two-phase $\alpha + \gamma$ metal structure.

Figure 14b shows the bulk Ni + Co analyses for all the metal particles of the present study except for the "special cases" given in Table 2 and Fig. 14a. It may be seen that the vast majority of particles have Ni + Co contents in the meteoritic range but there is a secondary distribution of metal at low-Ni, low-Co contents, similar to the compositions found for some Apollo 11 metal particles. Many of the particles between 1 and 4 wt% Ni are from the metal plus compound group. The metal compositions of the nonmeteoritic particles are not, in general, similar to the metal from Apollo 12 rocks, in which the Co contents are mostly in excess of 1 wt%. In addition to the range of meteoritic Ni + Co contents (Fig. 14b), we have plotted the cosmic Ni/Co ratio of 20:1. It is apparent that our analyses do not follow this cosmic trend. The best straight line through our data points appears to represent a Ni:Co ratio of 9:1, but we are not certain whether or not this ratio has any particular significance. Very similar Ni + Co measurements have been obtained by Wlotzka et al. (1972) on metal particles less than 100 μ grain size. They propose an origin for the metal in soil 14163 by a process of reduction from meteoritic silicates. It should be noted that our own work has been concentrated on a study of the larger, $> 100 \mu$, metal particles. In addition, no attention has been paid to the shock melted metal-sulphide eutectics that are found as small globules in glass and which, in our view, provide the most likely candidates for metal produced by the Wlotzka mechanism.

DISCUSSION

We have made a detailed metallographic and microprobe study of the larger ($> 100 \mu$) metal particles in soil samples 14003 and 14163. In addition, we have made an optical study of the metallic particles in our polished thin section of rock 14304 which was collected from a location about 50 meters from the soil samples and has been classified by Jackson and Wilshire (1972) as a coherent breccia, corresponding to the thermally metamorphosed microbreccia of Chao et al. (1971). The metal particles in this rock are structurally very similar to those metal particles that are surrounded by idiomorphic silicate in our sample. From examination of 14304 and other polished thin sections, we conclude that the major portion of metal particles in the soils show every sign of having been liberated from the fragmental rocks by mechanical disintegration on the lunar surface. Also we find that the soils contain a number of other populations of metal. In addition to a few exotic particles, one population of metal appears to have come from lunar mare basalts of the Apollo 11 type and another large population has both meteoritic structure and composition and shows no sign of having been reprocessed as a constituent of a fragmental rock.

Quaide (1972), Wilshire and Jackson (1972), Dence et al. (1972), and Anderson et al. (1972) have paid particular attention to the silicate petrology of the thermally metamorphosed microbreccias and have concluded that the range of textures and the complexity of the clasts indicate perhaps half a billion years of impact and accretion on the pre-Imbrian crustal area before excavation by the Imbrian event and deposition as the Fra Mauro formation. These studies leave unresolved the nature of the projectiles that bombarded the pre-Imbrian lunar crust.

Anderson et al. (1972) suggest that the material at Fra Mauro comes from the upper 5 to 10 km of the pre-Imbrian "crust". It is emplaced by the high velocity jetting action that arises when the leading edge of the projectile first encounters the lunar surface. We would expect particles from the meteoritic projectiles that bombarded the original surface of the moon to be contained in these upper layers of the pre-Imbrian crust. The presence of meteoritic metal in the fragmental rocks of the Fra Mauro formation and in the soil at Fra Mauro is therefore more consistent with material from near the surface of the target body than with rock that has been excavated from a great depth. Thus, we conclude that a large proportion of the metallic particles in the lunar soils had their ultimate origin in the meteoritic projectiles that bombarded the pre-Imbrian crust and became incorporated in it.

The metamorphic processes to which this metal was subjected while incorporated in the fragmental lunar rocks have removed the meteoritic microstructure of the metal and have provided the opportunity for minor redistributions of some trace elements between the metallic and nonmetallic phases but have not been sufficiently severe to alter the major element (Ni, Co) chemistry of the meteoritic metal. We suggest that it is not necessary, on the basis of the available evidence, to assume that all the larger metallic particles in the fragmental rocks were reduced in situ from ferruginous oxides or silicates of meteoritic origin (Wlotzka *et al.*, 1972).

The P contents of most of the nonremelted metal particles listed in Tables 1 and 2 are high compared to the measured P contents in kamacite of unreheated ordinary chondrites (<0.01 wt%) Reed (1969). Also as reported by Wänke *et al.* (1970, 1971) and Wlotzka *et al.* (1972), the W content in many of the metal particles is in excess of the measured meteoritic values. On the other hand, we did not measure any increase of Si in the metallic particles all of which contained <0.02 wt% Si.

During the shock-reheating process, some or all of the phosphates present in a chondrite are reduced (Taylor and Heymann, 1971). In a series of 5 reheated chondrites studied by Taylor and Heymann that were reheated from 480°C to 1200°C, the P content of the metal phase was found to be substantially higher than in un-reheated ordinary chondrites (0.08 to 0.6 wt%). A similar transfer of P was observed in the Ramsdorf chondrite (Begemann and Wlotzka, 1969) in which melting occurred. These studies show first of all that P can enter the metal phase on reheating at an oxygen pressure typical of chondritic meteorites, and secondly that the amount of P in the metal varies from meteorite to meteorite and depends on the temperature of reheating. The range of P contents in the metallic soil particles also can be ascribed to phosphate reduction in the lunar rocks. The reheating temperatures are provided by the host silicates during metamorphism as discussed by Quaide (1972), Wilshire and Jackson (1972) and Anderson *et al.* (1972). The very few particles with high P contents (>0.6 wt%; Table 2), probably either were remelted by shock processes or heated above 800°C.

The fact that Si does not enter the metallic phase shows that the conditions at the pre-Imbrian site were not reducing enough to form metallic Si as found in the enstatite chondrites. The presence of W in the metal probably is due to the same reheating-reduction process as for phosphorus. The variability of W content from particle to particle and from one Apollo site to another (Wänke *et al.*, 1971; Wlotzka *et al.*, 1972) is consistent with this process. Because the standard free energy of reaction of WO_2 and WO_3 to metallic W is somewhat less than that of SiO_2 to metallic Si (Kubaschewski *et al.*, 1967) and occurs at higher oxygen pressures, it appears that W in the metal can be derived from the lunar silicates by reduction and incorporation into the metal. No correlation between P and W in the metal and the state of metamorphism of the surrounding silicate has yet been made.

As noted earlier, we can identify a proportion of metal particles that have not only retained their meteoritic compositions but also show recognizable meteoritic microstructures. These particles have escaped massive reprocessing in the lunar crust and are fragments of meteorites that impacted the Apollo 14 site after the Imbrian event. This additional infall of meteoritic metal onto the Fra Mauro formation accounts for the enrichment of nickel in the soils relative to the fragmental rocks (LSPET, 1971). Metal particles may also be added to the soil from subsequent cratering events in Mare material. More than 10% of the glass in Apollo 14 is Mare glass (Apollo soil survey, 1971), indicating that Mare material is present in the soil. Probably many of the low-Ni, low-Co particles, which are similar in composition to Apollo 11 nonmeteoritic lunar metal, are part of this Mare addition to the soil.

Other metallic particles in the Apollo 14 soils show remelted dendritic structures and complex sulphide-phosphide eutectic associations similar to those found in the metallic spheroids that Nininger (1956) reported in the vicinity of the Barringer Meteorite Crater, Arizona. Blau et al. (1972) have discussed these particles and concluded that they are from the explosively melted portion of a crater-forming iron meteorite projectile. It will be remembered that these dendritic lunar spherules, together with those particles of recognizable meteoritic structure, were designated (M) in Table 2 to indicate their status as meteoritic material that had not been reprocessed in the fragmental rocks.

It is of interest to inquire whether or not we can detect any of the metallic particles that are associated with the asteroidal sized impacting body that formed the Imbrium Basin. On the basis of trace element chemistry of bulk soils and separates from the 1–2 mm fraction of four soils, Ganapathy et al. (1971), Morgan et al. (1972) attribute the meteoritic component in the Fra Mauro soil to debris from the planetismal that produced the Imbrium Basin. The iron in the mostly silicate projectile was postulated to be material similar to Group IVa irons.

We would expect that any Imbrian projectile material that was to be found in the 100 μ to 1 mm metal fraction of the soils would not show any evidence of being incorporated in the lunar fragmental rocks, and would have a recognizable meteoritic or remelted meteoritic structure. Twenty-four particles (12% of our sample) are included in such a group. However, no structurally observable fragments of Group IVa meteorites have been observed. Except for a few possible particles (M8.1, M3.4) with carbides, no structurally identifiable fragments of iron meteorites have been observed. If the Imbrian projectile contained metal within chondritic silicate, then most of the 11 identifiable meteoritic particles could be considered as possibilities. It is also conceivable that some of the remelted spheroids come from the impacting planetismal (iron or chondritic). Metallic meteoritic spherules however are not restricted to the Fra Mauro location, they have also been observed by various authors in the mare regions, Apollo 11 and 12 soils. The origins of the Apollo 11 and 12 spheroids presumably are from impacting projectiles metallic, chondritic or other which form the recent craters in the surrounding highlands regions. The number of spheroids that we have found in the Apollo 14 soils is not out of proportion to the number observed in the Mare soils. Therefore, an origin similar to that of the Mare spheroids is consistent with most of the Apollo 14 spheroids, i.e. projectile material that formed recent craters surrounding the Fra Mauro formation. However the possibility remains that some of our spheroids could have come from the impacting planetismal. In our sample of particles, we estimate a maximum possible contribution of 4% if the projectile had a major amount of iron meteorites or 6% if the projectile had a major amount of chondritic meteorites.

If it is permissible to extraploate the conditions around terrestrial meteorite craters to the much more gross situation at and around Mare Imbrium, then, in the larger fractions of metallic particles, >1 mm, we would expect to see less re-processed metal and a higher percentage of meteoritic material from the Imbrian planetismal and nearby cratering projectiles. It would, therefore, be advantageous to examine magnetic particles in the larger soil sizes.

Another source of meteoritic metal that is not available from impacts on the earth is the vapor and the very fine sprays of molten metal that are produced by the collision. On the earth, this type of metal is quickly oxidized and is not recovered. On the moon, some of the metallic vapor as well as very fine sprays of molten meteoritic metal are retained. Evidence for this meteoritic component has been observed by McKay (1972) for vapor phase growth of nickel-iron crystals and by Agrell et al. (1970), McKay et al. (1970), and Goldstein et al. (1970) for metallic mounds on lunar glass spheres. We have examined only the larger >100 μ metal particles and these constitute but 10 wt% of the total metallic component of the soil. It is likely that much of the heavily shocked metallic remains of the Imbrian planetis-mal will be found in the 90% of the metal of the <100 μ grain size.

A histogram of the Ni contents of all the 205 metal particles analyzed from our soil samples is plotted in Figure 17a. These include particles from all the sources described previously (Imbrium, Fra Mauro, and Mare). Almost all particles of meteoritic Ni contents (>4.5 wt%) have meteoritic Co contents (Fig. 14b). Metal particles with Ni contents <4.5 wt%, approximately 20% of our sample, are from Mare sources as described above, from metal-sulphide compounds and from native metal of non-meteoritic composition in the Imbrian material, (e.g., metal from crystalline rock 14053 that has <4.5 wt% Ni, Gancarz, et al., 1971).

We can try to obtain the original distribution of nickel in the metal that has been subject to reprocessing in the fragmental rocks of the Imbrian lunar crust. This can be done in two ways. First we can subtract those cases in which a clearly un-altered meteoritic structure is still identifiable in the metal particles, and those cases where complete remelting of meteoritic material can be identified. Almost all of these are particles from relatively recent projectiles at Fra Mauro. There are 24 such cases and when they are subtracted, the shape of the histogram is altered slightly and the peak is shifted to a slightly lower Ni content, as shown in Fig. 17a. However this estimate of the meteoritic fraction of metal particles is too low, because between the extremes of unaltered and completed remelted structures there lies a range of incompletely altered meteoritic structures that is less easy to recognize. If we take account of such particles the peak of the distribution must be lowered further.

Secondly, we can plot only those metal particles that are found associated with crystalline silicate, i.e., they were clearly associated with the original rocks at the Imbrian site. This process also lowers the peaks of the histogram and shifts the peak to lower Ni contents. The true Ni distribution of metal particles that were incorporated in the pre-Imbrian crust lies between the two distributions obtained.

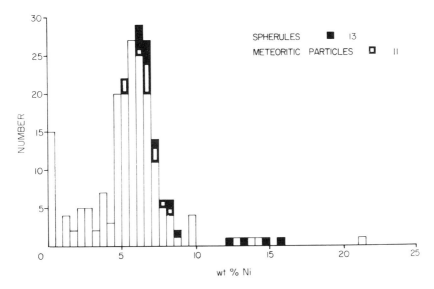

Fig. 17a. Histogram of bulk nickel analyses in 205 metal particles $> 100 \mu$ from soils 14003 and 14163. Special symbols are used for dendritic spherules and for particles that are considered on structural grounds (Table 2) to be fragments of meteorite that have not been processed by thermal metamorphism in the lunar fragmental rocks.

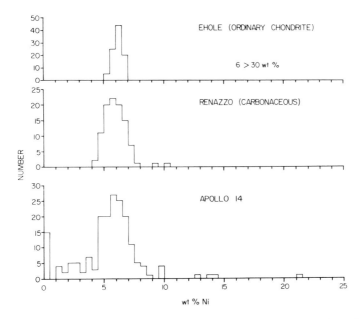

Fig. 17b. A comparison of nickel distributions in our samples of soils 14003 and 14163 with those separated in chondritic meteorites by Wood (1967).

The first distribution (subtracting the meteoritic component) is replotted on Fig. 17b. The histogram can be compared to Ni distributions in metal particles of chondrites as given by Wood (1967). The close resemblance between the Apollo 14-Imbrian metal histogram and the histograms for metal in chondrites indicates that the Imbrian site contained a major proportion of metal from this type of meteoritic source before the asteroidal body excavated the Imbrium Basin. Other sources of meteoritic metal may have been iron meteorites, pallasites or from sources that have not been observed in meteorites recovered in recent times.

CONCLUSIONS

1. The major proportion of the metallic particles in the lunar soils had their ultimate origin as metal in the meteorite projectiles that bombarded the pre-Imbrian crust and became incorporated in it. They are meteoritic in composition and show every sign of having been liberated from the fragmental rocks. It appears that the majority of metal added to the pre-Imbrian crust is from a chondritic source. Other sources of meteoritic metal may have been iron meteorites, pallasites or from sources that have not been observed in those meteorites that have been recovered in recent times.

2. Three other distinguishable populations of metal in the soil are found. One appears to have come from lunar mare basalts of the Apollo 11 type that have been subject to cratering. Another population has both meteoritic structure and composition and shows no sign of having been reprocessed as a constituent of a fragmental rock. They are fragments of meteorites that impacted the Apollo 14 site after the Imbrian event. The last population are metallic spheroids from the explosively melted portion of meteoritic projectiles that formed the relatively recent craters in the surrounding highlands regions.

3. Metamorphic processes to which the metal was subjected while incorporated in the fragmental lunar rocks have removed the meteoritic microstructure of the metal. In addition, these processes provided the reheating-reducing conditions that permitted trace elements such as P and W redistribute between nonmetallic and metallic phases.

4. We estimate a maximum possible contribution of 4% iron meteorite or 6% chondritic metal to the metallic portion of the soil due to the planetismal that formed the Imbrium Basin.

Acknowledgments—The authors acknowledge the assistance of Mr. D. Bush (Lehigh) for the electron probe microanalyses. Special thanks are due to Dr. B. Glass (U. of Delaware) for providing some lunar metallic particles for study and to Mr. H. W. Hitzrot (Bethlehem Steel Corporation) for providing the use of a magnetic separator. We also acknowledge helpful discussions with Dr. F. Wlotzka (Max Planck Institute, Mainz), Dr. L. S. Walter (NASA, Goddard) and Dr. J. A. Wood (Smithsonian Astrophysical Observatory). This project was supported by NASA Grant NGR 39-007-056.

REFERENCES

Agrell S. O. Scoon J. H. Muir I. D. Long J. V. P. McConnell J. D. C. and Peckett, A. (1970) Observations on the chemistry, mineralogy and petrology of some Apollo 11 lunar samples. *Proc. Apollo 11 Lunar Sci. Conf., Geochim. Cosmochim. Acta* Suppl. 1, Vol. 1, pp. 93–128, Pergamon.

Anderson A. T. Jr. Braziunas T. F. Jacoby J. and Smith J. V. (1972) Breccia populations and thermal history: Nature of pre-Imbrium crust and impacting body (abstract). In *Lunar Science—III* (editor C. Watkins), pp. 24–26, Lunar Science Institute Contr. No. 88.

Apollo Soil Survey (1971) Apollo 14: Nature and origin of rock types in soil from the Fra Mauro formation. *Earth Planet. Sci. Lett.* **12**, pp. 49–54.

Begemann F. and Wlotzka F. (1969) Shock induced thermal metamorphism and mechanical deformation in the Ramsdorf chondrite. *Geochim. Cosmochim. Acta* **33**, 1351–1370.

Blau P. Goldstein J. I. and Axon H. J. (1972) An investigation of the Canyon Diablo metallic spheroids and their relationship to the breakup of the Canyon Diablo meteorite, submitted to *J. Geophys. Res.*

Brentnall W. D. and Axon H. J. (1962) The response of Canyon Diablo meteorite to heat treatment. *J. Iron Steel Inst.* **200**, 947–955.

Chao E. C. T. Borman J. A. and Desborough G. A. (1971) *Proc. Second Lunar Sci. Conf., Geochim. Cosmochim. Acta* Suppl. 2, Vol. 1, pp. 797–816, MIT Press.

Dence M. R. Plant A. G. and Traill R. J. (1972) Impact generated shock and thermal metamorphism in Fra Mauro lunar samples (abstract). In *Lunar Science—III* (editor C. Watkins), pp. 174–176, Lunar Science Institute Contr. No. 88.

Ganapathy R. Laul J. C. Morgan J. W. and Anders E. (1971) Moon: Possible nature of body that produced Mare Imbrium from composition of Apollo 14 samples. *Science* **175**, 55–59.

Gancarz A. J. Albee A. L. Chodos A. A. (1971) Petrologic and mineralogic investigation of some crystalline rocks returned by the Apollo 14 mission. *Earth Planet. Sci. Lett.* **12**, 1–18.

Goldstein J. I. and Yakowitz H. (1971) Metallic inclusions and metal particles in the Apollo 12 lunar soil. *Proc. Second Lunar Sci. Conf., Geochim. Cosmochim. Acta* Suppl. 2, Vol. 1, pp. 177–191, MIT Press.

Goldstein J. I. Henderson E. P. and Yakowitz H. (1970) Investigation of lunar metal particles. *Proc. Apollo 11 Lunar Sci. Conf., Geochim. Cosmochim. Acta* Suppl. 1, Vol. 1, pp. 499–512, Pergamon.

Jackson E. D. and Wilshire H. G. (1972) Classification of the samples returned from the Apollo 14 landing site (abstract). In *Lunar Science—III* (editor C. Watkins), pp. 418–420, Lunar Science Institute Contr. No. 88.

Keil K. Bunch V. E. and Printz M. (1970) Mineralogy and composition of Apollo 11 lunar samples. *Proc. Apollo 11 Lunar Sci. Conf., Geochim. Cosmochim. Acta* Suppl. 1, Vol. 1, pp. 561–598, Pergamon.

Kubaschewski O. Evans E. and Alcock C. B. (1967) *Metallurgical Thermochemistry*, Pergamon.

LSPET (1971) (Lunar Sample Preliminary Examination Team) Preliminary examination of lunar samples from Apollo 14, *Science* **173**, 681–693.

McKay D. S. Greenwood W. R. Morrison D. A. (1970) Origin of small lunar particles and breccia from the Apollo 11 site. *Proc. Apollo 11 Lunar Sci. Conf., Geochim. Cosmochim. Acta* Suppl. 1, Vol. 1, pp. 673–694, Pergamon.

Morgan J. W. Laul J. C. Krähenbühl U. Ganapathy R. and Anders E. (1972) Major impacts on the moon: Chemical characterization of projectiles (abstract). In *Lunar Science—III* (editor C. Watkins), pp. 552–554, Lunar Science Institute Contr. No. 88.

Nininger H. H. (1956) *Arizona's Meteorite Crater*. American Meteorite Museum.

Quaide W. (1972) Mineralogy and origin of Fra Mauro fines and breccias (abstract). In *Lunar Science—III* (editor C. Watkins), pp. 627–629, Lunar Science Institute Contr. No. 88.

Reed S. J. B. (1969) Phosphorus in meteoritic nickel-iron. *Meteorite Research* (P. M. Millman, ed.), pp. 750–762, Reidel, Holland.

Reid A. M. Meyer C. Harmon R. S. and Brett R. (1970) Metal grains in Apollo 12 igneous rocks. *Earth Planet. Sci. Lett.* **9**, 1–5.

Taylor G. J. and Heymann D. (1971) Postshock thermal histories of reheated chondrites. *J. Geophys. Res.* **76**, 1879–1893.

Wänke H. Wlotzka F. Baddenhausen H. Balacescu A. Spettel B. Teschke F. Jagoutz E. Kruse H. Quijano-

Rico M. and Rieder R. (1971) Apollo 12 samples: Chemical composition and its relation to sample locations and exposure ages, the two component origin of the various soil samples and studies on lunar metallic particles. *Proc. Second Lunar Sci. Conf., Geochim. Cosmochim. Acta* Suppl. 2, Vol. 2, pp. 1187–1208, MIT Press.

Wänke H. Wlotzka F. Jagoutz E. and Begemann F. (1970) Composition and structure of metallic iron particles in lunar "fines". *Proc. Apollo 11 Lunar Sci. Conf., Geochim. Cosmochim. Acta* Suppl. 1, Vol. 1, pp. 931–935, Pergamon.

Wilshire H. G. and Jackson E. D. (1972) Petrology of the Fra Mauro formation at the Apollo 14 landing site (abstract). In *Lunar Science—III* (editor C. Watkins), pp. 803–805, Lunar Science Institute Contr. No. 88.

Wlotzka F. Jagoutz E. Spettel B. Baddenhausen H. Balacescu A. and Wänke H. (1972). On lunar metal particles and their contribution to the trace element content of the Apollo 14 and 15 soils (abstract). In *Lunar Science—III* (editor C. Watkins), pp. 806–808, Lunar Science Institute Contr. No. 88.

Wood J. A. (1967) Chondrites: Their metallic minerals, thermal histories and parent planets, *Icarus* **6**, 1–49.

Proceedings of the Third Lunar Science Conference
(Supplement 3, *Geochimica et Cosmochimica Acta*)
Vol. 1, pp. 1065–1076
The M.I.T. Press, 1972

Study of excess Fe metal in the lunar fines by magnetic separation, Mössbauer spectroscopy, and microscopic examination

R. M. HOUSLEY, R. W. GRANT, and M. ABDEL-GAWAD

North American Rockwell Science Center
Thousand Oaks, California 91360

Abstract—A simple and convenient method of making quantitative magnetic separations has been applied to the lunar fines. The fractions obtained form groups containing distinctively different particle types; thus, it appears that magnetic separation in itself may be a useful way of characterizing lunar fines. Mössbauer studies of fines 10084 show that the metal cannot contain more than about 1.5% Ni, implying that by far the bulk of the metal results from reduction rather than from direct meteoritic addition. Mössbauer data also place an upper limit on the magnetic content of the fines approximately an order of magnitude below that required to account for the characteristic ferromagnetic resonance observed. Microscopic examination of magnetic separates from 15101 fines suggests that reduction of Fe accompanies every major impact event on the moon, and also that the bulk of the material at the collection site has at one or more times been in an impact plume.

INTRODUCTION

WE PREVIOUSLY HAVE shown (Housley *et al.*, 1970; 1971) that typical dark gray lunar fines at the Apollo 11 and 12 sites contain considerably more metallic Fe than the local igneous rocks, that most of this excess Fe metal is associated intimately with glassy material, and that a significant fraction of the metal grains are less than approximately 100 Å in diameter.

Table 1 summarizes all of our absolute determinations of Fe metal content in

Table 1. Metallic Fe in lunar fines and rocks.*

Sample	Size range (μ)	Ferromagnetic Fe (wt%)	Excess area as wt% Fe
77°K			
10084,85 (S.G. < 3.3)	0–75	0.61 ± 0.04	0.29 ± 0.02
295°K			
10084,85 (S.G. < 3.3)	0–75	0.51 ± 0.04	0.39 ± 0.02
12042,38 fines	0–1000 ground	not determined	0.332 ± 0.004
12042,38 (S.G. < 3.3)	45–150	0.34 ± 0.05	0.17 ± 0.02
12042,38 (S.G. > 3.3)	45–150	0.04 ± 0.07	—
12025,15 core tube	0–150	0.38 ± 0.06	0.35 ± 0.02
12025,42 core tube	0–150	0.37 ± 0.06	0.35 ± 0.02
12028,88 core tube	0–150	0.39 ± 0.08	0.22 ± 0.02
12028,133 core tube	0–150	0.31 ± 0.05	0.43 ± 0.01
12038,47 igneous rock	ground	0.07 ± 0.03	0.04 ± 0.01
12052,16 igneous rock	ground	0.11 ± 0.06	0.03 ± 0.01
14003,22 fines	0–45	0.40 ± 0.04	0.36 ± 0.02
14163,52 fines	0–45	0.34 ± 0.03	0.28 ± 0.02
14311,41 breccia	ground	0.14 ± 0.04	0.06 ± 0.02
15101,92 fines	0–75	0.20 ± 0.03	0.37 ± 0.02

*Error limits are one standard deviation due to counting statistics only. All data were analyzed by the method described by Housley *et al.* (1971).

lunar samples, including new data on the Apollo 14 fines 14003,22 and 14163,52, Apollo 14 recrystallized breccia 14311,41 and Apollo 15 fines 15101,92. The second column in Table 1 indicates the sieved size fraction that was studied. The third column gives the ferromagnetic Fe metal content determined from the area in the normal Mössbauer hyperfine pattern of Fe metal, and the fourth column presents the Fe content corresponding to the excess area observed near zero relative velocity in the Mössbauer spectra. A sizeable fraction of this area has been shown (Housley *et al.*, 1971) to be due to superparamagnetic Fe grains less than about 100 Å in diameter.

The widespread occurrence of excess Fe metal in the fine size fractions of lunar fines suggested to us that further study of the composition, physical form, and association of the excess Fe metal might yield valuable insights into the dynamical events associated with meteorite impacts on the moon, and perhaps yield information concerning the nature of the impacting bodies and/or the impacted lunar material.

With the initial objective of facilitating this study, we have developed a simple and effective method of magnetically separating small amounts of fine grained material quantitatively without loss or contamination. It now seems likely that this technique will be useful in a much wider range of studies.

Here we will describe the technique and report on studies of magnetic separates from the fines 10084,85 and 15101,92 by Mössbauer spectroscopy, optical microscopy, and scanning electron microscopy.

MAGNETIC SEPARATION

The sample is spread over the bottom of a flat bottomed aluminum container, which in turn sits on a flat supporting plate, between the poles of a strong permanent horseshoe magnet. A thin flat soft Fe plate in contact with the upper magnet pole can be lowered to any desired height above the bottom of the sample container with a micrometer screw. In order to reduce the tendency of grains to stick to each other or the container, the latter is filled with pure ethanol to a level above the bottom of the soft Fe plate. A separation is made by gently sliding the sample dish around until all parts of the sample have been directly under the soft Fe plate. The Fe plate is then raised and removed from the magnet. During this operation the surface tension of the ethanol holds the collected material on the plate. After removal from the magnet, the plate and adhering magnetic separate are held over a suitable container for a short time while the ethanol evaporates. When evaporation is complete, most of the collected separate simply falls off the plate and any that remains can be removed by gently tapping or brushing. It is clearly important at this point that the Fe plate is sufficiently soft magnetically that it does not retain significant magnetism after removal from the magnet. Further fractions are obtained by repeating the procedure with the soft Fe plate at successively lower positions during the collections.

A significant amount of material ($\sim 5\%$) was collected from samples 10084,85 and 15101,92 at a separation of 6 mm, allowing them to be quantitatively separated into a number of distinct magnetic fractions. Sieved size fractions 45–75 μ and 420–1000 μ were successfully separated. Repeat collection at a given separation yielded little additional material, whereas a given fraction spread on a clean aluminum dish was almost entirely recollected at the separation at which it was initially collected, demonstrating that the fractions obtained are magnetically distinct.

Field and field gradient values at different separations h measured with a Hall effect Gaussmeter are given in the second and third columns of Table 2. The size of the probe prevented useful measurements for h values less than about 1 mm and perhaps introduces some uncertainty into the values at larger separations.

In a given magnetic field H and field gradient dH/dh, the magnetic force on a sample depends on its magnetization. For paramagnetic material, this is simply proportional to H. For ferromagnetic material the magnetization in a given field strongly depends on grain size and shape. The limiting force for extremely small grains corresponds to that expected for free paramagnetic atoms. The force per unit mass at first increases linearly with particle size as $mIH/3kT$ (Bean and Livingston, 1959) up to a diameter d of about 30 Å for Fe metal in about 1000 Oe where m is the mass, I the saturation moment per unit mass, and kT the temperature in energy units. Also, for spherical Fe particles in $H = 1000$ Oe the magnetization approaches the saturation value of I as $1-(kT/mIH)$ (Bean and Livingston, 1959) for particles with $d \gtrsim 60$ Å. For a range of diameters greater than 60 Å, spherical Fe particles are expected to be fully magnetized along the field direction and hence to experience the maximum magnetic force. For d values somewhat greater than 200 Å, the uniform magnetization state will become unstable for low H values (Frei et al., 1957; Brown, 1963) and at 1000 Oe the force per unit mass will decrease with d. For $d \gg 1000$ Å bulk behavior will be approached and the magnetization at low fields will depend on the demagnetization factor.

In column 4 of Table 2, we give the ratio of the magnetic force to the gravitational force for a hypothetical sample composed entirely of Fe^{++} ions. In column 5 we give the same ratio for fully aligned single domain Fe metal particles and in column 6 we give the result for spherical Fe metal particles large enough to approach bulk behavior. The magnetic forces expressed in columns 4, 5, and 6 of Table 2 can be used to place limits on the Fe metal content of any magnetic separate. For example, assuming a bulk density of 2.4 gm/cm^3 and an Fe^{++} content of 20% and correcting for the buoyancy of the alcohol, one finds that the fraction collected at $h = 6$ mm must contain at least 0.57% Fe metal whereas the fraction collected at $h = 2$ mm need contain only 0.17%. If large grains behaving like bulk Fe metal are assumed, the limits respectively become 3.6% and 0.8%.

Table 2. Characteristics of magnetic separator

h mm	H Oe	dH/dh Oe/cm	Fe^{++} (F/mg)	60–200Å (F/mg)	bulk (F/mg)
2	1557	1525	0.421	340	73.2
3	1417	1270	0.334	284	55.6
4	1303	1015	0.245	226	40.8
5	1214	762	0.171	170	28.5
6	1151	508	0.108	113	18.0

MÖSSBAUER SPECTROSCOPY

Bulk samples of ferromagnetic material in zero applied field H generally assume a domain configuration such that there are no free poles at the surfaces and the demagnetization field is hence essentially zero. In this case the field at the atomic nuclei H_{int} is equal to the field H_{hf} produced by hyperfine interactions alone. For metallic Fe, the width of a domain wall is about 1000 Å (Nagata, 1961), and particles must be much larger than this before bulk behavior can be assumed. There is good evidence (Runcorn et al., 1970; Nagata et al., 1970) that much of the Fe in the lunar fines actually is present as particles considerably smaller than 1000 Å in diameter. There is no clear justification in this case for interpreting H_{int} values obtained at zero field in terms of the absence or presence of alloying elements. This ambiguity can be removed however by taking data at low temperatures and in high applied fields H_o where the particles all are fully saturated and then extrapolating H_{int} versus H_o to zero H_o.

To obtain a sample as rich in Fe as possible, we made a magnetic separation of the light fraction (S.G. < 3.3) of the 0–75 μ fines 10084,85, which we previously had shown to contain most of the Fe metal (Housley et al., 1970). A total of 80.3 mg of material was collected at separations ≥ 3 mm and was combined in the Mössbauer absorber. The remaining less magnetic material weighed 67.5 mg. The magnetic fraction subsequently proved to contain about 1% Fe metal, which is consistent with the expected behavior of the separator for an average grain size between the limits corresponding to saturated single domain and bulk behavior.

The sample was mounted between Be windows in a superconducting magnet assembly and had a thickness of 41.4 mg/cm^2. The Mössbauer source was Co^{57} in Cu at 22°C and zero field. Spectra were obtained using an absolutely calibrated constant velocity spectrometer operating in an automated mode. At several fields and temperatures, complete spectra were recorded; additional data were recorded for velocity intervals that spanned the outer Fe metal lines. Typical data are shown in Fig. 1. In our complete series of runs, data were collected at 295°K and $H_o = 0$, 11°K and $H_o = 0$, 11°K and $H_o = 55$ kOe, and then a series of decreasing field values ending with 11°K and $H_o = 0$. Line positions and areas were determined by least squares fitting the data in the vicinity of the Fe lines to a Lorentz curve of unconstrained width, depth, and position plus a straight line of unconstrained height and slope to account for the overlapping area contributed by other phases. Areas were corrected for background, which was measured for each run.

The H_{int} values obtained from the line positions are plotted against H_o in Fig. 2. The solid line is a least squares fit to the data for $H_o \geq 8$ kOe, for which the grains are expected to be magnetically saturated. In calculating the demagnetization field we assume the Fe metal particles are spherical, which was the case for the vast majority of the hundreds of μ size metal grains that we have observed microscopically, and obtain a value of 7.4 kOe. Subtracting this from the intercept of the solid line with the $H_o = 0$ axis yields a value of H_{hf} of 340.6 $\pm$ 1.0 kOe. This corresponds with our value at 5°K 339.7 $\pm$ 0.2 for pure Fe metal. The use of the data of Johnson et al. (1961), on the change in average hyperfine field produced by alloying Fe with Ni,

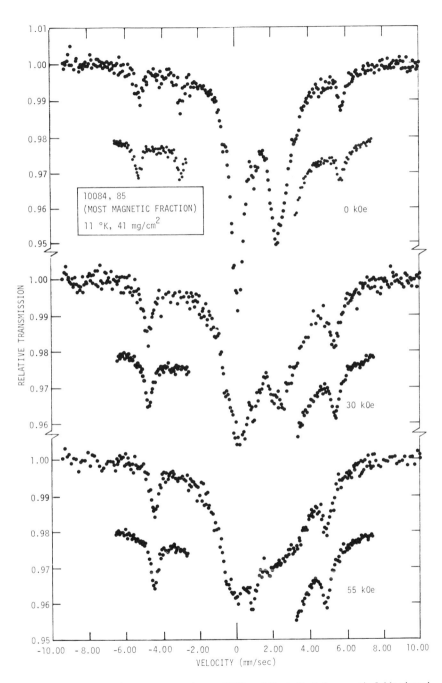

Fig. 1. Typical Mössbauer spectra taken at 11°K and the indicated magnetic fields plotted relative to Fe metal at 295°K as velocity zero. Displaced down 0.02 below each full spectrum, the corresponding higher statistical accuracy data taken in the velocity range spanning the outer Fe metal lines also are shown.

permits us to conclude that the metal grains cannot contain much more than 1.5% Ni on the average and need not contain any. This seems to be strong evidence that the origin of the metal is by a reduction process rather than by direct meteoritic addition.

The Mössbauer data also allow some rough conclusions to be drawn concerning the size distribution in the particles. If the particles are small enough that super-paramagnetic relaxation becomes rapid compared to the Larmor frequency of the nucleus in the hyperfine field then the hyperfine pattern is expected to collapse into

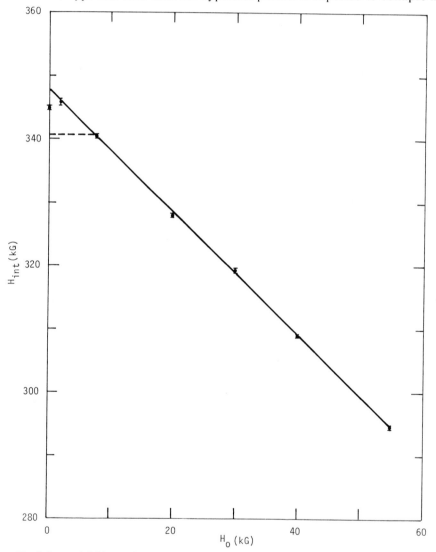

Fig. 2. Internal fields H_{int} for the metallic phase in 10084,85 at 11°K obtained by analysis of the Mössbauer spectra plotted versus the applied field H_0. The solid line is a least squares fit to the data for $H_0 \geq 8$ kOe. The dashed line would be followed by H_{int} for bulk Fe particles of spherical shape at low H_0.

a single line. The superparamagnetic relaxation time t for spherical Fe metal particles is expected to be given roughly by:

$$\frac{1}{t} = \frac{2\gamma_0 K_1 \, e^{-(K_1 V/kT)}}{J_s}$$

where γ_0 is the gyromagnetic ratio, K_1 is the first order anisotropy constant, and J_s is the saturation magnetization. Using values of the constants appropriate for pure Fe, the transition from ferromagnetic to superparamagnetic behavior in a Mössbauer experiment can be calculated to occur at a particle diameter of 134 Å at 295°K, 85 Å at 77°K, and 45 Å at 11°K. In addition, at 11°K and $H_0 = 55$ kOe grains as fine as 20 Å in diameter would be magnetically saturated and hence would appear in the ferromagnetic spectrum.

At 11°K the ratio of the area observed in the ferromagnetic component at $H_0 = 55$ kOe to that at $H_0 = 0$ is 1.53 ± 0.18, and the expected ratio due only to polarization of the originally unpolarized absorber is 1.5. The ratio between the areas obtained at 295°K and that obtained at 11°K in $H_0 = 0$, when corrected for the difference in Debye-Waller factors, is 1.23 ± 0.16, which is only slightly greater than the ratio obtained previously (Housley *et al.*, 1971) between 295°K and 77°K. These results imply that approximately 20% of the Fe metal is present as grains between ~134 and 85 Å in diameter and probably considerably fewer grains in the size range 85 to 20 Å.

We did not see evidence for magnetite or a similar spinel phase in any of our spectra, although weak Mössbauer evidence for it has been reported previously (Gay *et al.*, 1970). In view of the current interest this question has with regard to the interpretation of microwave resonance spectra of the lunar fines (Griscom and Marquardt, 1972; Weeks *et al.*, 1972), we collected a set of data at 11°K and $H_0 = 20$ kOe spanning the expected positions of the outer magnetite lines at low temperature (Kundig and Hargrove, 1969). Any magnetite grain $\gtrsim 30$ A in diameter would be

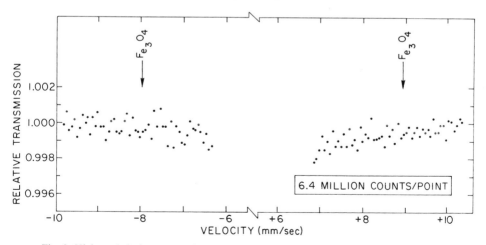

Fig. 3. High statistical accuracy data at 11°K and $H_0 = 20$ kOe in the velocity intervals where magnetite lines would be found. Arrows indicate expected absorption centroids. No evidence for magnetite can be seen.

sufficiently near magnetic saturation to be observed under these conditions, and any grain smaller would make only a weak contribution to the microwave resonance spectrum at room temperature and would have a temperature dependent intensity similar to paramagnetic materials. These data are shown in Fig. 3, and no magnetite lines are evident. The slight overall absorption in this region probably can be explained as the tails from resonance peaks at lower velocities. By summing all the area within ± 0.5 mm/sec of the expected magnetite line positions, we obtain an upper limit of 0.04% for the possible amount of Fe^{+++} in magnetite-like phases in this sample. Because this sample was composed largely of glass-welded aggregates, there seems to be strong evidence against the interpretation offered by Griscom and Marquardt (1972) of their microwave resonance results and strong support for the earlier interpretation of Tsay et al. (1971).

CHARACTERIZATION OF FINES 15101,92

Two size fractions 45–75 μ and 420–1000 μ from the Apollo 15 fines 15101,92, which were collected in an area that appears to consist largely of ejecta from St. George Crater (Swann et al., 1971), have been magnetically separated.

The coarse fraction material collected above 3 mm separation consisted entirely of glass welded aggregates. The glass beads present were collected between 1 and 3 mm. The nonmagnetic material, which could not be picked up even on contact, consisted largely of anorthite, but several dark, fine-grained rock fragments also were present.

The distribution of mass in the different magnetic separates of the 45–75 μ size fraction is shown in Table 3. One striking fact immediately evident from this table is that 63% of the mass is accounted for by particles containing at least 0.17% Fe metal.

Examination of the different magnetic fractions by optical microscopy, Fig. 4, shows that they form easily recognizable, distinct particle groups. The most magnetic particles are, except for very rare pure metal grains, all dark, semi-opaque, vesicular, inhomogeneous glass-welded aggregates of very fine mineral fragments. No particles of this type are found in the least magnetic fraction. No optically resolvable metallic phase can be found in the majority of particles in the most magnetic group; these particles, however, all appear to contain regions of welding glass, which has a milky appearance in reflected light as shown in Fig. 4d and which shows iridescence when

Table 3. Magnetic separation of 45–75 μ fraction
of 15101,92 fines.

Designation	Height (mm)	Weight (mg)	% of Total
1	6	7.4	5.2
2	5	16.2	11.4
3	4	14.8	10.4
4	3	14.6	10.3
5	2	36.5	25.7
1–5	>2	89.5	63.0
6	<2	52.5	37.0
TOTAL		142.0	100.0

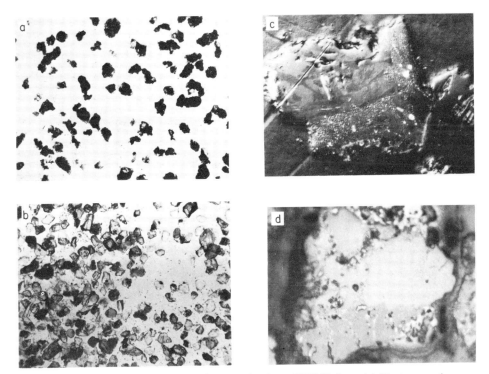

Fig. 4. Optical micrographs of polished grains from 15101,92 fines. (a) Most magnetic fraction in transmitted light. (b) Least magnetic fraction in transmitted light. (d) Grain from most magnetic fraction in reflected light. (c) Grain from most magnetic fraction in reflected light with surface relief emphasized by interference.

viewed with an oil immersion lens. Fig. 4c shows a similar region that contains some larger metal grains.

The least magnetic fraction consists largely of mineral grains containing about 50% plagioclase, about 35% mafic minerals, about 10% rock fragments and about 5% homogeneous glass spheres and shards.

Intermediate magnetic fractions seem to consist largely of mineral grains or aggregates with different amounts of attached welding glass. For material collected at 3 mm separation, welding glass is optically observable in all but about 10% of the grains in polished grain mounts.

Scanning electron micrographs, at medium and high magnification, of grains from the most and least magnetic fractions are shown in Figs. 5 and 6.

The numerous rounded surfaces, indicating solidification of an unconstrained liquid in the glass-welded aggregates constituting the most magnetic fraction, strongly suggests an origin in a hot impact plume (McKay *et al.*, 1970). Some welding glass can be seen on the surface of particles in the least magnetic fraction.

Microscopic examination of material from the most magnetic fraction of the Apollo 11 fines 10084,85 shows that it is similar in all respects to the description given above for the most magnetic fraction of 15101,92.

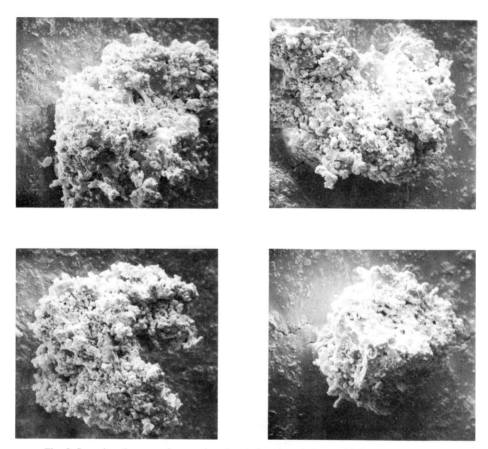

Fig. 5. Scanning electron micrographs of typical grains of glass-welded aggregates, which make up the magnetic fraction of fines 15101,92 at medium magnification. Smooth rounded glass surfaces indicate unconstrained solidification as would be expected in an impact plume.

Preliminary study of x-ray powder diffraction data on the most and least magnetic fractions from 15101,92 shows that these fractions contain the same crystalline phases in similar proportions.

Conclusions

The magnetic separator described allows clearly distinct groups of particles to be separated easily from the lunar fines. Because it is cheap and easily constructed and permits separation of small amounts of fine grain size material in a safe contamination free manner, it appears suitable for use in a number of other studies of the chemical mineralogical and physical properties of lunar fines.

The Mössbauer data on fines 10084,85 lead to the conclusion that by far the bulk of the Fe metal present resulted from reduction rather than being a direct meteoritic addition. The physical characteristics of the particles from the most magnetic fractions

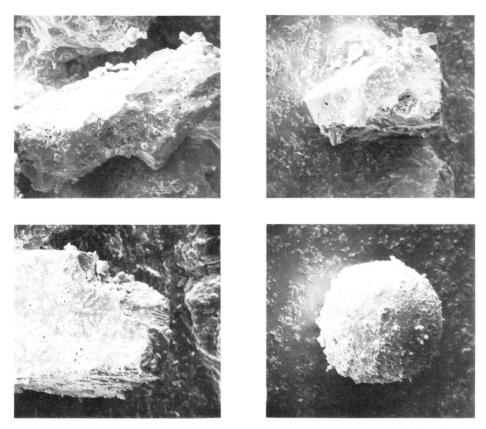

Fig. 6. Scanning electron micrographs of typical grains from the least magnetic fraction of fines 15101,92 at medium magnification. The fraction consists mostly of crystalline mineral grains but contains minor amounts of homogeneous transparent glass.

of fines 10084,85 and 15101,92 are completely consistent with this conclusion. The absence of glass-welded aggregates in the less magnetic fractions of 15101,92 further suggests that Fe is reduced in every impact plume of sufficient size to produce the glass-welded aggregates and hence does not require any special compositions or conditions, although this conclusion must be further confirmed by studying fines from other sites.

The facts that 63% of the material in 15101,92 was collected at separations that require at least 0.17% adhering Fe metal and that there seems to be a continuous gradation between glass-welded aggregates, which clearly formed in conditions similar to those expected in an impact plume, and minerals with small amounts of adhering welding glass suggest that the majority of fines at the collection site have been through an impact cloud rather than having gradually eroded out of rocks by some process such as thermal cycling.

The upper limit on the amount of magnetite or similar Fe^{+++} containing magnetic spinel present in the 10084,85 fines is about an order of magnitude lower than that required to account for the characteristic ferromagnetic resonance observed in the fines.

Acknowledgments—We wish to express our deep gratitude to L. H. Hackett who performed the scanning electron microscope studies for us. The technical assistance provided by P. B. Crandall and K. G. Rasmussen is also gratefully acknowledged. This work was largely supported by NASA Contract NAS 9-11539.

References

Bean C. P. and Livingston J. D. (1959) Superparamagnetism. *J. Appl. Phys.* **30**, 120S–129S.

Brown W. F. (1963) *Micromagnetics*. Interscience.

Frei E. H. Shtrikman S. and Treves D. (1957) Critical size and nucleation field of ideal ferromagnetic particles. *Phys. Rev.* **106**, 446–455.

Gay P. Bancroft G. M. and Brown M. G. (1970) Diffraction and Mössbauer studies of minerals from lunar soils and rocks. *Proc. Apollo 11 Lunar Sci. Conf., Geochim. Cosmochim. Acta* Suppl. 1, Vol. 1, pp. 481–497. Pergamon.

Griscom D. L. and Marquardt C. L. (1972) Electron spin resonance studies of iron phases in lunar glasses and simulated lunar glasses (abstract). In *Lunar Science—III* (editor C. Watkins), pp. 341–343, Lunar Science Institute Contr. No. 88.

Housley R. M. Blander M. Abdel-Gawad M. Grant R. W. and Muir A. H. Jr. (1970) Mössbauer spectroscopy of Apollo 11 samples. *Proc. Apollo 11 Lunar Sci. Conf., Geochim. Cosmochim. Acta*, Suppl. 1, Vol. 3, pp. 2251–2268. Pergamon.

Housley R. M. Grant R. W. Muir A. H. Jr. Blander M. and Abdel-Gawad M. (1971) *Proc. Second Lunar Sci. Conf., Geochim. Cosmochim. Acta* Suppl. 2, Vol. 3, pp. 2125–2136. MIT Press.

Johnson C. E. Ridout M. S. Cranshaw T. E. and Madsen P. E. (1961) Hyperfine field and atomic moment of iron in ferromagnetic alloys. *Phys. Rev. Lett.* **6**, 450–451.

Kundig W. and Hargrove R. S. (1969) Electron hopping in magnetite *Solid State Comm.* **7**, 223–227.

McKay D. S. Greenwood W. R. and Morrison D. A. (1970) Origin of small lunar particles and breccia from the Apollo 11 site. *Proc. Apollo 11 Lunar Sci. Conf., Geochim. Cosmochim. Acta* Suppl. 1, Vol. 1, pp. 673–694. Pergamon.

Nagata T. (1961) *Rock Magnetism*, revised edition. Maruzen Company Ltd.

Nagata T. Ishikawa Y. Kinoshita H. Kono M. Syono Y. and Fisher R. M. (1970) Magnetic properties and natural remanent magnetization of lunar materials. *Proc. Apollo 11 Lunar Sci. Conf., Geochim. Cosmochim. Acta* Suppl. 1, Vol. 3, pp. 2325–2340. Pergamon.

Runcorn S. K. Collinson D. W. O'Reilly W. Battey M. H. Stephenson A. Jones J. M. Manson A. J. and Readman P. W. (1970) Magnetic properties of Apollo 11 lunar samples. *Proc. Apollo 11 Lunar Sci. Conf., Geochim. Cosmochim. Acta* Suppl. 1, Vol. 3, pp. 2369–2387. Pergamon.

Swann G. A. Hait M. H. Schaber G. G. Freeman V. L. Ulrich G. E. Wolfe E. W. Reed V. S. and Sutton R. L. (1971) Preliminary description of Apollo 15 sample environments. United States Department of Interior Geological Survey, Interagency report: 36.

Tsay F. D. Chan S. I. and Manatt S. L. (1971) Ferromagnetic resonance of lunar samples. *Geochim. Cosmochim. Acta* **35**, 865–875.

Weeks R. A. Kolopus J. L. and Kline D. (1972) Magnetic phases in lunar material and their electron magnetic resonance spectra: Apollo 14 (abstract). In *Lunar Science—III* (editor C. Watkins), pp. 791–793, Lunar Science Institute Contr. No. 88.

Proceedings of the Third Lunar Science Conference
(Supplement 3, *Geochimica et Cosmochimica Acta*)
Vol. 1, pp. 1077–1084
The M.I.T. Press, 1972

On lunar metallic particles and their contribution to the trace element content of Apollo 14 and 15 soils

F. Wlotzka, E. Jagoutz, B. Spettel, H. Baddenhausen,
A. Balacescu, and H. Wänke

Max-Planck-Institut für Chemie (Otto-Hahn-Institut), Mainz, Germany

Abstract—Metal particles were separated from the fines 14163 and 15601 and an igneous fragment of rock 14321. The Co and Ni contents of approximately one hundred individual particles from each sample were measured. In the 14163 metal, a high proportion of particles with a meteoritic composition are found, which are nearly absent in the 15601 metal. By INAA, the trace elements Cu, Ga, Ge, As, Pd, W, Ir, and Au were measured in the bulk metal. The Co and Ni contents and the siderophile trace elements of the "meteoritic" 14163 metal can be matched by carbonaceous chondrite-metal or by hexahedrite metal of Ga–Ge-group II–A. Tungsten is enriched in the metal relative to Ni about 300 times over its cosmic abundance.

Introduction

THE LUNAR FINES contain metallic particles of various grain sizes, which are the most obvious candidates in the search for meteoritic matter on the lunar surface. We studied these particles in the fines from Apollo 11 (Wänke *et al.*, 1970) and Apollo 12 (Wänke *et al.*, 1971) and report here the continuation of this work on the Apollo 14 and 15 samples.

From 10 g of the fines 14163, we separated 4.5 mg of metallic particles with a hand magnet and purified them by successive grinding in an agate mortar and washing in preparation for determination of the trace element content by INAA methods (see Wänke *et al.*, 1970). No metallic tools were used during this procedure. After the INAA analysis, the particles were mounted in plastic, polished, and analyzed by electron microprobe. Contrary to the results from the Apollo 11 and 12 fines, we found only very small metal grains, most of them in the size range of 50 to 100 μ. With the same procedures we extracted 1 mg of metal out of 1 g of an igneous fragment from the clastic rock 14321 and 0.8 mg metal from 1 g of the fines 15601. Again, all of the particles were smaller than 100 μ. We found a few metal spherules and a few metal + sulfide eutectic intergrowths similar to those described by us from the Apollo 11 fines (Wänke *et al.*, 1970), and by Goldstein and Yakowitz (1971) from Apollo 12. We cannot say much about the original shape of most of the grains, as we used mechanical grinding during the separation.

Co and Ni Contents

In the lunar igneous rocks Ni is strongly depleted like other siderophile elements compared to terrestrial basalts (e.g. Wänke *et al.*, 1970, Anders *et al.*, 1971). On the other hand, Co is enriched relative to Ni so that the rocks have Ni/Co ratios of approximately 1 or less than 1, whereas the cosmic ratio is 20. In agreement with this, metal grains in the lunar igneous rocks commonly have low Ni contents and a low Ni/Co ratio.

Thus the Co-Ni contents of the metal from the lunar dust can be used to distinguish meteoritic metal from lunar metal (Fig. 1). The metal grains from Apollo 11 rocks have low Ni contents (0.0 to 1%) and Co contents in the same range, but mostly higher than Ni. This type of grain also is found in the fines from all four landing sites (Fig. 1). In the Apollo 12 rocks, the Ni and Co contents have a wider range; in some rocks they show a relation to the crystallization sequence of minerals in the rock, Ni-rich grains being associated with early formed minerals (Reid *et al.*, 1970). The grains of rock 12053 that we measured (Fig. 1) are rather typical for the Apollo 12 rocks and other authors report similar high Co and Ni contents with Ni/Co ratios commonly below one. (Cameron, 1971; Champness *et al.*, 1971; Keil *et al.*, 1971; Taylor *et al.*, 1971 and Walter *et al.*, 1971.)

In Apollo 14 rocks, a range of Ni contents was found again; we measured values in metal from 14321 between 0.1 and 1% Ni, and 0.0 and 0.1% Co, but El Goresy *et al.* (1972) reported up to 37% Ni in metal of rock 14310.

Despite these differences in Ni and Co values in the igneous rock metal, which are reflected in the three Co-Ni plots of 10084, 12001, and 14163, all three plots show clusters at approximately 5% Ni and 0.4% Co (Fig. 1), a concentration familiar from meteoritic metal. The composition of the metal from fines 14163 is especially restricted to this "meteoritic area," suggesting a higher proportion of meteoritic metal in this lunar soil.

In contrast, the Ni-Co plot of metals from 15601 shows a wide scatter of points, but nearly no entries in the "meteoritic" region. Apparently the meteoritic component is very small here, which may be due to the special situation at the edge of Hadley Rille. Similarly, Baedecker *et al.* (1972) found about a factor of 2 less meteoritic component for the similar Rille soil 15531 compared to soils near the Apennine front. This "anomaly" shows that regional mixing of soils can be very limited, and generalizations from data on only one soil should be made with caution.

Trace Elements

In Table 1 the concentrations of the trace elements Cu, Ga, Ge, As, Pd, W, Ir, and Au measured on bulk metal samples are given. In the same table, we calculated the amount of these trace elements in the bulk fines that is contained in the metal particles. For this calculation we measured the metal content of the bulk fines by a method described by Wänke *et al.* (1970): 14163 contains 0.5% metal and 15601 contains 0.35% metal. Thus, 14163 contains about four times as much metal as soil 12001, in accordance with its greater content of siderophile elements reported by many investigators (Baedecker *et al.*, 1972; Morgan *et al.*, 1972; Wänke *et al.*, 1972).

The data in Table 1 show that of the typical siderophilic elements, like Ni, Co, Ge, Pd, Ir, and Au, 50 to 100% of the amount found in the bulk fines is present in the metal. These figures are subject to a certain bias, because of sampling problems. We measured only metal grains above a certain grain size limit (~ 20 μ), whereas the very small metal grains observed in glass spheres and glass coatings were not

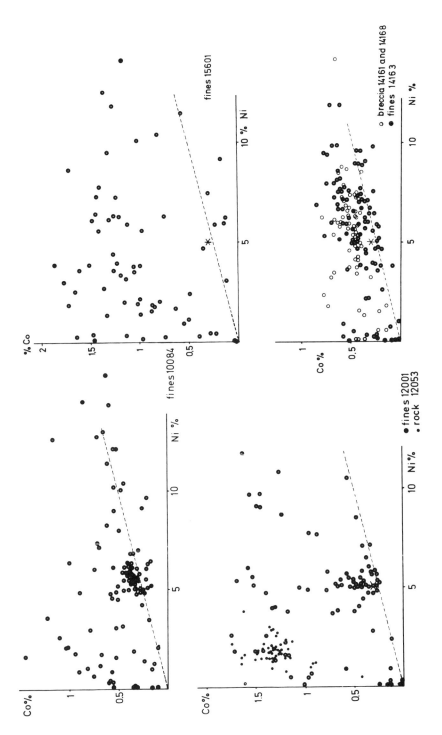

Fig. 1. Co versus Ni in metal from different lunar samples. The dashed line gives the cosmic Ni/Co ratio of 20. The asterisk * at 5% Ni and 0.3% Co stands for the Ni and Co content of metal made by reduction of the total iron of a normal or carbonaceous chondrite.

Table 1. Composition of metal from an igneous fragment of rock 14321 and of metal from fines 14163 and 15601. The trace element content in ppm is given and also the percentage of each element in the bulk fines that is contributed from the metal (14163:0.5% metal; 15601:0.35% metal). In the last columns, the atomic abundance of these 3 metal samples is compared to a carbonaceous chondrite or "cosmic" abundance (Cl values from the compilation by Cameron, 1968).

	Metal from rock 14321	Metal from fines 14163		Metal from fines 15601		Cl chond-rite	Metal from rock 14321	Metal from fines 14163	Metal from fines 15601
	(ppm)	(ppm)	(%) contribution to bulk fines	(ppm)	(%) contribution to bulk fines		(atomic abundance)		
Ni	4300	57400	73%	48500	78%	$=10^4$	$=10^4$	$=10^4$	$=10^4$
Co	1800	5500	65%	10350	71%	505	4100	950	2120
Cu	134	220	7%	863	37%	200	290	35	165
Ga	8.6	14	1%	9	0.9%	10.0	17	2.0	1.56
Ge	n.d.	138	117%	n.d.	—	27.6	—	19.5	—
As	n.d.	11	63%	n.d.	—	1.6	—	1.5	—
Pd	n.d.	2.8	50%	n.d.	—	0.32	—	0.27	—
W	46	223	56%	25	31%	0.035	34	12	1.65
Ir	1.3	2.0	53%	n.d.	—	0.094	0.9	0.10	—
Au	0.20	1.1	92%	0.35	55%	0.044	0.14	0.057	0.021

included and may have a different chemistry. However, we do not expect this effect to change our figures significantly.

In the last columns of Table 1 and in Fig. 2, the atomic abundance of the trace elements discussed above is compared to an average Type I carbonaceous chondrite abundance, normalized to Ni. It is obvious that the metal from 14163 matches the "cosmic" abundances for the siderophile elements quite well, whereas that of the igneous fragment of 14321 and that of fines 15601 does not. Thus, we come to the same conclusion reached before from the Co-Ni plot that the 14163 metal has the highest proportion of meteoritic particles (see also Goldstein et al., 1972).

THE TUNGSTEN ANOMALY

In contrast to the strongly siderophile elements Au, Ir, and Ge the behavior of W is different. It is not strongly depleted in the lunar igneous rocks, probably because it was not siderophile under the conditions of the first separation of siderophile elements from the moon. It is now found highly enriched in the lunar metal, but it is probably not derived from meteorites, as no meteorites or meteoritic metal is known to contain such high W abundances (Wänke et al., 1970). In the Apollo 12 and 14 soils, W shows a good correlation with La (see our companion paper Wänke et al., 1972), indicating that W is one of the elements enriched in KREEP, as was already indicated by the positive slope of the W correlation line in our Apollo 12 mixing diagram (Wänke et al. (1971), Fig. 6). The very high enrichment of W in the metal from all lunar fines (and rocks) compared to its cosmic abundance is shown in Fig. 2. Apparently W is now entering the metal phase as a strongly siderophile element, which may indicate more reducing conditions on the surface of the moon now than at the time of its formation or the first igneous differentiation.

Some metal grains from fines 12001 with high W contents were checked for other elements, which under reducing conditions might show siderophile tendencies.

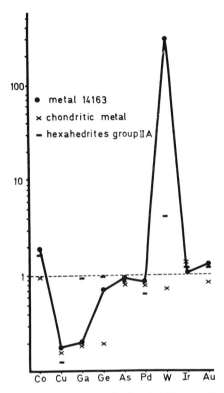

Fig. 2. Trace element content of metal from the fines 14163 normalized to a Type I carbonaceous chondrite and to Ni (values from Table 1). Also given is the composition of chondritic
metal (Co, Cu, Ga, W, Ir, and Au from Wänke *et al.*, 1970, Ge and Pd from Rieder and
Wänke, 1969; As from Mason and Graham, 1970) and of hexahedrites of Ga–Ge group
II–A (Co and Cu from Moore *et al.* (1969); Ga, Ge, and Ir from Wasson (1969); As from
Lipschutz (1971), Pd and Au from Goldberg *et al.* (1951), W from our own measurements
(average of 8 hexahedrites 2.5 ppm W).

The following values were obtained by microprobe analysis: $Si < 0.03\%$, $Ti <$
0.02%, $Mn < 0.02\%$, $Cr < 0.02\%$ (except one grain with 0.04% Cr).

DISCUSSION

The Co-Ni plots and the trace element content of the metallic particles from the
lunar fines have shown that they bear a certain relationship to meteorites, i.e., that
there is a meteoritic component in the soil consisting of metal particles.

If we want to trace this metal phase back to a certain class of meteorites, we find
that the Ni and Co contents do not fit very well to chondritic metal. Metal of chondrites consists of kamacite with about 6% Ni and 0.4% Co and taenite with 30 to
50% Ni and 0.3% Co (Reed, 1964; Wood, 1967), hence the bulk metal contains
about 10% Ni and 0.4% Co (see measurements by Wänke *et al.*, 1970) as we find
for instance in the remolten metal grains of a shocked chondrite (Begemann and
Wlotzka, 1969).

The Co-Ni cluster in our metal diagrams approximately fits chondritic kamacite, but no corresponding taenite is present. Among the several hundred grains measured only two were found with Ni contents above 30%, but with Co contents of about 2% in contrast to the 0.3% Co in chondritic taenite. A much better fit is obtained with a reduced chondrite. That is, if we calculate a metal composition made out of the total iron, nickel, and cobalt of a chondrite, we get a metal with 5% Ni and 0.3% Co. This composition is marked in the Co-Ni diagrams of Fig. 1.

A comparison of the trace element abundance of chondritic metal and metal from fines 14163 is made in Fig. 2. The agreement is satisfactory for Co, Cu, Ga, As, Pd, Ir, and Au, only Ge is off by a factor of 4. The same agreement would, of course, be found for metal from a reduced chondrite, because the values in Fig. 2 are normalized to Ni.

If, however, the metal is generated by reducing a *carbonaceous* chondrite, the Ge also would agree with our 14163 metal value (Fig. 2). The disagreement for Cu and Ga probably is caused by the less siderophilic character of these elements. The Co and Ni content of the "carbonaceous chondrite metal" would be the same as for the metal from the reduced normal chondrite and fit equally well to our Co-Ni clusters.

We thus reach the same conclusion as Baedecker *et al.* (1971) and Laul *et al.* (1971) from trace element abundances of bulk fines of Apollo 12. They found the meteoritic component to be of unfractioned cosmic composition, that probably is derived from Type I carbonaceous chondrite material. We must make the additional conclusion that this material consists in a large part of metal particles.

A comparison of the 14163 metal composition with iron meteorite values shows that for Co and Ni the hexahedrites come closest: about 0.4% Co and 5–6% Ni. The hexahedrites of Wasson's (1969) Ga–Ge group II–A agree fairly well also for Ge and Ir, if those with the lowest Ir content of 2 ppm are taken. Also Co, Cu, As, Pd, and Au do not deviate more than 50% from the 14163 metal values (Fig. 2).

The greatest deviation is found for Ga, which is a factor of 4 higher in the hexahedrites. The same discrepancy is found for Ga in relation to the "carbonaceous chondrite metal." Because Ga is not strongly siderophilic, it might not enter the metal phase during the assumed metal formation by reduction, or from hexahedrites it would be lost to the lunar silicates, in the same way as W was taken up.

Hence, a hexahedrite would fit equally well to our data for the 14163 metal as reduced carbonaceous chondrite metal. Metal from chondrites fits also in all elements determined except Ge. This conclusion that at least the majority of the "meteoritic" metal separated by us from the 14163 could be generated from hexahedrites, Ga–Ge group II–A, should not be confused with the deduction of Ganapathy *et al.* (1972), from the Apollo 14 fines that an iron meteorite of group IV–A could be the impacting body forming Mare Imbrium. Their conclusion was derived from the trace element abundance in glass and norite fragments, containing the record of the ancient meteoritic component from the time of their formation, whereas our results apply to the bulk soil and the meteoritic matter accumulating in it during its 3.9×10^9 years of history.

We do not know how much of our 14163 metal may predate the Fra Mauro formation. One clue, however, may be its conspicuously high W content (Table 1) of 223 ppm, which is ten times higher than in the metal from fines 15601. If we assume, as discussed above, that W was taken up from the lunar silicates, a high W content would be favored by a prolonged contact of metal and silicates at high temperatures, as would be expected in the Imbrian ejecta blanket. Therefore, most of the metal may have resided in breccias before entering the soil. Because the Imbrian event probably formed most of the Fra Mauro breccias, the metal in these breccias would predate this event or originate from the impacting body itself.

Acknowledgments—We are grateful to NASA for making available the lunar material for this investigation. The samples were irradiated in TRIGA-research reactor of the Institut für Anorganische Chemie und Kernchemie der Universität Mainz. We thank Dr. Kurt Fredriksson for lending us his thin-section of breccia 14161. We wish to thank the staff of the TRIGA-research reactor and the staff of our Institute, in particular Miss H. Prager, Mr. P. Deibele, Mr. H. Engler, and Mr. F. Rudolph. The financial support by the Bundesministerium für Bildung und Wissenschaft is gratefully acknowledged.

REFERENCES

Anders E. Ganapathy R. Keays R. R. Laul J. C. and Morgan J. W. (1971) Volatile and siderophile elements in lunar rocks: comparison with terrestrial and meteoritic basalt. *Proc. Second Lunar Sci. Conf. Geochim. Cosmochim. Acta*, Suppl. 2, Vol. 2, pp. 1021–1036. M.I.T. Press.

Baedecker P. A. Schaudy R. Elzie J. L. Kimberlin J. and Wasson J. T. (1971) Trace element studies of rocks and soils from Oceanus Procellarum and Mare Tranquillitatis. *Proc. Second Lunar Sci. Conf., Geochim. Cosmochim. Acta*, Suppl. 2, Vol. 2, pp. 1037–1061. M.I.T. Press.

Baedecker P. A. Chou C.-L. Kimberlin J. and Wasson J. T. (1972) Trace element studies on lunar rocks and soils (abstract). In *Lunar Science—III* (editor C. Watkins), pp. 35–37, Lunar Science Institute Contr. No. 88.

Begemann F. and Wlotzka F. (1969) Shock induced thermal metamorphism and mechanical deformations in the Ramsdorf chondrite. *Geochim. Cosmochim. Acta* 33, 1351–1370.

Cameron A. G. W. (1968) A new table of abundances of the elements in the solar system. In *Origin and Distribution of the Elements* (editor L. H. Ahrens), pp. 125–143, Pergamon.

Cameron E. N. (1971) Opaque minerals in certain lunar rocks from Apollo 12. *Proc. Second Lunar Sci. Conf., Geochim. Cosmochim. Acta*, Suppl. 2, Vol. 1, pp. 193–206. M.I.T. Press.

Champness P. E. Dunham A. C. Gibb F. G. F. Giles H. N. McKenzie W. S. Stumpfl E. F. and Zussman J. (1971) Mineralogy and petrology of some Apollo 12 lunar samples. *Proc. Second Lunar Sci. Conf., Geochim. Cosmochim. Acta*, Suppl. 2, Vol. 1, pp. 359–376. M.I.T. Press.

El Goresy A. Ramdohr P. and Taylor L. A. (1972) Fra Mauro crystalline rocks: petrology, geochemistry, and subsolidus reduction of the opaque minerals (abstract). In *Lunar Science—III* (editor C. Watkins), pp. 224–226, Lunar Science Institute Contr. No. 88.

Ganapathy R. Laul J. C. Morgan J. W. and Anders E. (1972) Moon: possible nature of the body that produced the Imbrian basin, from the composition of Apollo 14 samples. *Science* 175, 55–59.

Goldberg E. Uchiyama A. and Brown H. (1951) The distribution of Ni, Co, Ga, Pd, and Au in iron meteorites. *Geochim. Cosmochim. Acta* 2, 1–5.

Goldstein J. I. and Yakowitz H. (1971) Metallic inclusions and metal particles in the Apollo 12 lunar soil. *Proc. Second Lunar Sci. Conf., Geochim. Cosmochim. Acta*, Suppl. 2, Vol. 1, pp. 177–191. M.I.T. Press.

Goldstein J. I. Yen F. and Axon H. J. (1972) Metallic particles in the Apollo 14 lunar soil (abstract). In *Lunar Science—III* (editor C. Watkins), pp. 323–325, Lunar Science Institute Contr. No. 88.

Keil K. Prinz M. and Bunch T. E. (1971) Mineralogy, petrology, and chemistry of some Apollo 12 samples. *Proc. Second Lunar Sci. Conf., Geochim. Cosmochim. Acta*, Suppl. 2, Vol. 1, pp. 319–341. M.I.T. Press.

Laul J. C. Morgan J. W. Ganapathy R. Anders E. (1971) Meteoritic material in lunar samples: Characterization from trace elements. *Proc. Second Lunar Sci. Conf., Geochim. Cosmochim. Acta*, Suppl. 2, Vol. 2, pp. 1139–1158. M.I.T. Press.

Lipschutz M. E. (1971) Arsenic. In *Elemental Abundances in Meteoritic Matter* (editor B. Mason), pp. 261–269, Gordon and Breach, London.

Mason B. and Graham A. L. (1970) Minor and trace elements in meteoritic minerals. *Smithsonian Contr. Earth Sciences* **3**, 1–17.

Moore C. B. Lewis C. F. and Nava D. (1969) Superior analyses of iron meteorites. In *Meteorite Research* (editor P. Millman), pp. 738–748, D. Reidel.

Morgan J. W. Laul J. C. Krähenbühl U. Ganapathy R. and Anders E. (1972) Major impacts on the moon: chemical characterization of projectiles (abstract). In *Lunar Science—III* (editor C. Watkins), pp. 552–554, Lunar Science Institute Contr. No. 88.

Reed S. J. B. (1964) Composition of the metallic phases in some stone and stony-iron meteorites. *Nature* **204**, 374–375.

Reid A. M. Meyer C. Jr. Harmon R. S. and Brett R. (1970) Metal grains in Apollo 12 igneous rocks. *Earth Planet. Sci. Lett.* **9**, 1–5.

Rieder R. and Wänke H. (1969) Study of trace element abundance in meteorites by neutron activation. In *Meteorite Research* (editor P. Millman), pp. 75–86, D. Reidel.

Taylor L. A. Kullerud G. and Bryan W. B. (1971) Opaque mineralogy and textural features of Apollo 12 samples and a comparison with Apollo 11 rocks. *Proc. Second Lunar Sci. Conf., Geochim. Cosmochim. Acta*, Suppl. 2, Vol. 1, pp. 855–871. M.I.T. Press.

Walter L. S. French B. M. Heinrich K. F. J. Lowman P. D. Jr. Doan A. S. and Adler I. (1971) Mineralogical studies of Apollo 12 samples. *Proc. Second Lunar Sci. Conf., Geochim. Cosmochim. Acta*, Suppl. 2, Vol. 1, pp. 343–358. M.I.T. Press.

Wänke H. Wlotzka F. Jagoutz E. and Begemann F. (1970) Composition and structure of metallic iron particles in lunar "fines." *Proc. Apollo 11 Lunar Sci. Conf., Geochim. Cosmochim. Acta*, Suppl. 1, Vol. 1, pp. 931–935. Pergamon.

Wänke H. Wlotzka F. Baddenhausen H. Balacescu A. Spettel B. Teschke F. Jagoutz E. Kruse H. Quijano-Rico M. and Rieder R. (1971) Apollo 12 samples: chemical composition and its relation to sample locations and exposure ages and studies on lunar metallic particles. *Proc. Second Lunar Sci. Conf., Geochim. Cosmochim. Acta*, Suppl. 2, Vol. 2, pp. 1187–1208. M.I.T. Press.

Wänke H. Baddenhausen H. Balacescu A. Teschke F. Spettel B. Dreibus G. Quijano-Rico M. Kruse H. Wlotzka F. and Begemann F. (1972) Multielement analyses of lunar fines (abstract). In *Lunar Science—III* (editor C. Watkins), pp. 779–781, Lunar Science Institute Contr. No. 88.

Wasson J. T. (1969) The chemical classification of iron meteorites—III. Hexahedrites and other irons with germanium concentrations between 80 and 200 ppm. *Geochim. Cosmochim. Acta* **33**, 859–876.

Wood J. A. (1967) Chondrites: their metallic minerals, thermal histories, and parent planets. *Icarus* **6**, 1–49.

Proceedings of the Third Lunar Science Conference
(Supplement 3, *Geochimica et Cosmochimica Acta*)
Vol. 1, pp. 1085 1094
The M.I.T. Press, 1972

Glassy particles in Apollo 14 soil 14163,88:
Peculiarities and genetic considerations

G. Cavarretta, A. Coradini and R. Funiciello

Istituto di Geologia e Paleontologia dell'Università di Roma, Rome, Italy.

M. Fulchignoni

Laboratorio di Astrofisica Spaziale del CNR, Frascati, Italy.

A. Taddeucci

Istituto di Geochimica dell'Università di Roma, Rome, Italy.

R. Trigila

Istituto di Mineralogia e Petrografia dell'Università di Roma, Rome, Italy.

Abstract—Glassy particles with well-defined morphologies have been studied in the fraction greater than 62 μ from Apollo soil 14163,88. Besides spheres and sphere-derived forms, we have considered the "ropy strands glasses." They appear to comprise a well-defined chemical group clearly different from other lunar glasses. The mafic elements are constant, whereas potassium and phosphorus show a wide range of sympathetic variation, suggesting a variable content of the KREEP component.

The probable genetic process producing the ropy glasses and their coating has been investigated in terms of the energy distribution between projectile and target during an impact event. The coating on the ropy glasses is caused by the welding of small, still fluid particles on the main body within a long-lived plume that is generated during a catastrophic event involving energies greater than 10^{20} ergs.

Introduction

Sample 14163,88 represents lunar fines with particle diameters less than 1 mm. In the greater than 62 μ fraction, we have examined 300 glassy particles that could be selected by their distinctive shapes. We have found spheres and sphere-derived forms, namely elongated forms, dumbbells, tear drops, and another type of particle, already found in Apollo 12 soil: the "ropy strands glassy particles." (Carter, 1971) (Hubbard *et al.*, 1971) (McKay *et al.*, 1971) (Meyer *et al.*, 1971) (Wood *et al.*, 1971). The 300 selected particles can be divided into the following groups: opaque spheres and sphere derived forms (40%); coated, light yellow ropy strands glasses (35%); more or less coated, light yellow spheroids (15%); green, red, brown, and colorless spheres and sphere-derived forms (10%). A few chondrule-structured particles have been found. The one showing a complex-grated texture (Fig. 1a and 1b) is a bimineralic association of olivine and anorthite, as determined by x-ray diffraction analysis.

Data

Refractive index measurements show that the glasses can be classified into two groups. The frequency distribution versus refractive index for a few tens of Apollo 12 and 14 glassy spheroids are shown in Fig. 2. Apart from a few monomineralic

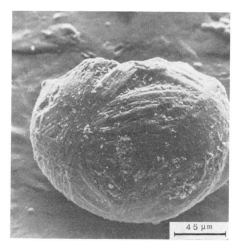

Fig. 1a. General view of an isolated chondrule-structured particle showing a complex-grated texture. The particle is a bimineralic association of anorthite and olivine.

glasses, two classes appear: The first has indices averaging 1.61–1.62, is most abundant in the Apollo 14 materials, and is rare in the Apollo 12 glasses; the second has higher indices, with a noticeable dispersion, occurs in the Apollo 12 samples, and is a distinctive family (n_D averaging 1.66–1.67) in the Apollo 14 spheroids.

These two groups seem to fit the classification proposed by Chao *et al.* (1972). The glasses in the class averaging $n_D = 1.61$–1.62 can be partly associated with those classified by Chao as "group 1," and partly with the "additional group of heterogeneous glasses." The other class with n_D averaging 1.66–1.67, corresponds to Chao's group 4; also the main chemical features seem to correspond in this case.

Electron microprobe (Cambridge Geoscan) analyses of 18 particles are plotted in Figs. 3 and 4. Two groups of glasses can be distinguished: The transparent

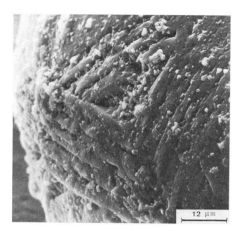

Fig. 1b. Enlarged detail of Fig. 1a.

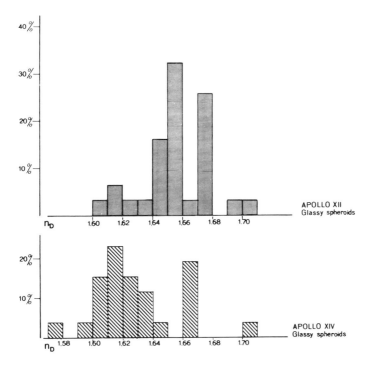

Fig. 2. Frequency distributions versus refractive index for Apollo 12 and Apollo 14 glassy spheroids (about 50 indices for each diagram have been plotted).

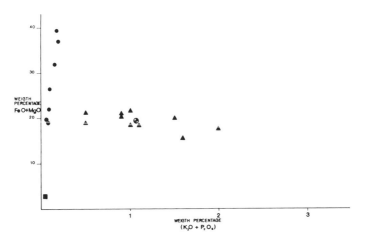

Fig. 3. $K_2O + P_2O_5$ versus $FeO + MgO$ distribution of some Apollo 14 glasses. Triangles represent ropy glasses and coated and opaque spheres. Solid circles represent transparent spheres and spheroids. The half-solid circle represents soil 14163. Solid square represents an anorthitic particle.

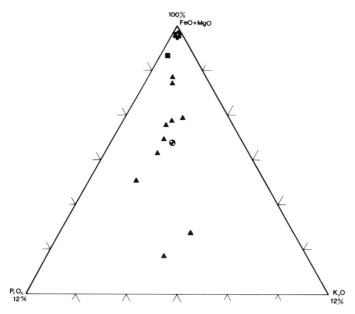

Fig. 4. FeO + MgO, P_2O_5, K_2O distribution of some Apollo 14 glasses. Triangles represent ropy glasses and coated and opaque spheres. Solid circles represent transparent spheres and spheroids. The half-solid circle represents soil 14163. Solid square represents an anorthitic particle.

spheres and sphere-derived forms belong to the first, the ropy glasses and the coated and opaque spheres belong to the second. The former is richer in iron and magnesium oxides, whereas the latter contain more potassium and phosphorus (Fig. 3).

In the second group the mafic components are quite constant, whereas the sum of potassium plus phosphorus shows a wide range. This is shown in Fig. 4, which illustrates the relationship between potassium and phosphorus.

This relationship is a good indication of the increasing KREEP component within the second group. The soil 14163 has been analyzed by several authors (Jackson et al., 1972) (Rose et al., 1972) (Schnetzler and Nava, 1971) (Wänke et al., 1972), and its bulk chemistry shows a similarity with the composition of the KREEP glasses studied by Meyer et al. (1971).

All of the ropy glasses and coated spheroids belong to the group of KREEP glasses. This correlation between the particle shape, surface morphology and chemistry of the glasses has been checked on 27 coated ropy glasses, by means of SEM with nondispersive detector.* We have found that 24 of the glasses clearly reflect the KREEP glass chemistry, and 3 have anorthositic compositions.

The general morphology of the coated ropy glasses with KREEP compositions is shown in Fig. 5; in most samples the particle appears elongated. The cross section

---

* The SEM used in this study was a Cambridge Stereoscan MARK II–a with an ORTEC non dispersive x-ray spectrometer attachment.

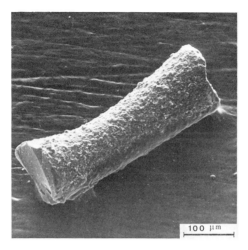

Fig. 5. General view of a coated ropy glass with subcircular section.

generally is subcircular but, obviously, in the fluted particles it is irregular (Fig. 6). In many samples, a range of the diameter of the cylindrical body is observable; consequently their morphology shows a sort of "boudinage" (Fig. 7).

The body of the ropy strands glasses generally appears truncated by transverse fractures, whereas no axial fractures have been found.

The surface is entirely coated by fines with a wide range of sizes, already noticed by McKay *et al.* (1971), and the fracture surfaces commonly appear quite clean and dust free (Fig. 8). On the fractures surface (Fig. 8) and in the body of the particles holes are commonly present, probably due to gas bubbles. Observation of the polished section of the ropy glasses under reflected light clearly shows typical flow

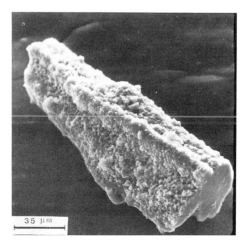

Fig. 6. Coated ropy glass showing an irregular section; this is the most common morphology for such glasses.

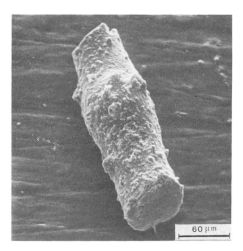

Fig. 7. The "boudinage" shown by this particle is probably due to some perturbations during its fluid stage.

structure (Fig. 9) and metallic inclusions also are visible; analyses by SEM with non dispersive detector show that the inclusions consist of Fe-Ti, Fe-S, and Fe-Ni-S.

Origin of Ropy Glass

It is expected that the morphological features of the ropy glasses can be explained in terms of their genetic process and we propose a possible model involving impact phenomena. From previous works on impact phenomena related to the genesis of melted material on the lunar surface (Gault, 1963) (Moore *et al.*, 1964) (Carusi *et al.*, 1971), it has been shown that a large fraction of the energy involved in such a process

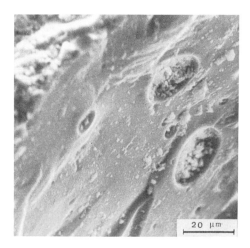

Fig. 8. Holes, probably due to gas bubbles, are shown on the fractures surfaces of the ropy glasses.

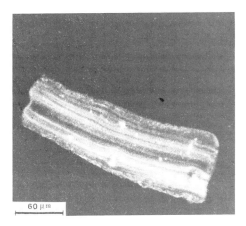

60 μm

Fig. 9. This typical flow-structure is shown by the ropy glasses. Polished section, reflected light.

is available as thermal and mechanical energy of the ejected material. Such consideration can be applied to the problem of the formation of ropy glasses. The presence of glassy filaments in the products of an impact may be related to any disturbance of the fluid material in a state of nonequilibrium. The momentum transferred to the fluid in the rarefaction phase by the system of shock waves is responsible for the formation as well as for the ejection of all possible kinds of glassy particles such as regular rotation forms, ropy glasses and irregular bodies, fragments, etc.

From a study of the equilibrium conditions for a cylindrical filament between the momentum transferred by the rarefaction wave and the surface tension for an inviscid fluid, we get

$$dQ = kS(t)\,dt$$

where k is the surface tension coefficient for the fluid, $S(t)$ is the surface area of the filament after t seconds from the beginning of the phenomenon, and dQ is the transferred momentum.

The formation energy is very low if compared with the energies involved in the process. Figure 10 plots the transferred energy Q versus the length h of the ropy strands; it appears that for masses of 5×10^{-4}g the energy is of some tens of ergs.

The cooling time of such a form on average can be assumed as

$$t = \frac{mc_v\alpha}{S\sigma}\left(\frac{1}{T_2^3} - \frac{1}{T_1^3}\right)$$

where m and S are, respectively, the mass and the surface area of the filament, c_v is the heat capacity for the considered material, α is its opacity (Isard, 1971), σ is the Stephan constant, T_1 and T_2 are, respectively, the initial and the final temperature.

Some experimental works (Gault *et al.*, 1963) (Moore *et al.*, 1964) give the relationships between the ejected mass from a hypervelocity impact crater and the energy of the projectile. Using these results together with the energy figure obtained from the

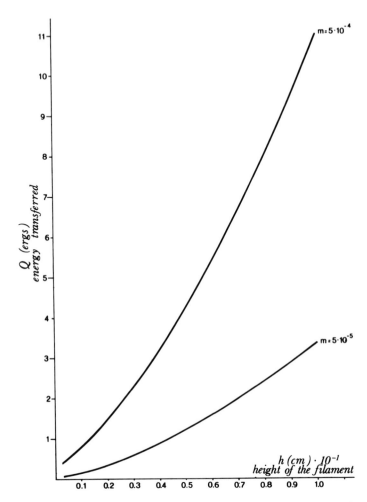

Fig. 10. Formation energy of cylindrical glassy filaments versus their length.

model of impact of an iron projectile on a basalt target (Carusi *et al.*, 1971), we find:

(a) for impact energies ranging from 10^8 ergs to 10^{20} ergs, the plume produced in the impact, which is quite instantaneous ($\sim 10^{-12}$ sec), has a very short lifetime. The cooling time obtained for the filaments in our model are of some tenths of a second, and therefore are greater than the lifetime of the plume.

(b) for energies greater than 10^{20} ergs, it is not realistic to consider the impact phenomenon as an instantaneous process: the projectile dimensions are large enough to allow the survival of the plume for many seconds (Barricelli and Metcalfe, 1969), a time longer than the cooling time of the filaments.

The capture process of the observed coating material may occur if the filament surface is at a temperature lower than the coating material temperature, consequently the coating occurs on the solidified body of the filament; in fact, it may be noted that the adhering dust appears isotropic as if it were deposited onto a body of fixed geometry and is not "streaky" as might be expected if the body were elongated after or during deposition. These observations are consistent with the hypothesis that the ropy glasses originate mainly during a large impact phenomena (Carter and McGregor, 1970) (Carter, 1971) (McKay *et al.*, 1970) (McKay *et al.*, 1971).

The fractures on solidified filaments only rarely show an extensive coating (Figs. 11a, 11b) and generally appear almost dust free (Figs. 5, 6). In the first case, the fragmentation of the filament is supposed to occur during the flight of the particle within the impact plume. In the second case, the fragmentation of the filament occurs after the flight of the particle within the plume.

Fig. 11a. This ropy glass was broken after solidifying. The sharp fracture surface was coated entirely by still fluid particles, during the flight of the body within the impact generated plume.

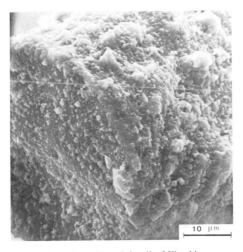

Fig. 11b. Enlarged detail of Fig. 11a.

Acknowledgments—We thank B. Accordi, M. Fornaseri, L. Gratton, and C. Lauro, who made this work possible, and all of the people who helped us during the researches and the compilation of data and manuscript: G. Bigi, F. Cinotti, G. Civitelli, J. Guidi, L. Spinozzi, P. Taddeucci. This work was supported by a C.N.R. grant.

References

Barricelli N. A. and Metcalfe R. (1969) The lunar surface and early history of Earth's satellite system. *Icarus* **10**, 144–163.

Carter J. L. (1971) Chemistry and surface morphology of fragments from Apollo 12 soil. *Proc. Second Lunar Sci. Conf., Geochim. Cosmochim. Acta*, Suppl. 2, Vol. 1, pp. 873–892. M.I.T. Press.

Carter J. L. and MacGregor I. D. (1970) Mineralogy, petrology and surface features of some Apollo 11 samples. *Proc. Apollo 11 Lunar Sci. Conf., Geochim. Cosmochim. Acta*, Suppl. 1, Vol. 1, pp. 247–265. Pergamon.

Carusi A. Coradini A. Fulchignoni M. and Magni G. (1971) Formation of lunar glassy spherules: a dynamical model. In *Proc. I.A.U. Symp. No. 47, The Moon*, pp. 100–104. Dordrecht.

Chao E. C. T. Boreman J. A. and Minkin J. A. (1972) Apollo 14 glasses of impact origin (abstract). In *Lunar Science—III* (editor C. Watkins), p. 133, Lunar Science Institute Contr. No. 88.

Gault D. E. Shoemaker E. M. and Moore H. J. (1963) Spray ejected from the lunar surface. *NASA Technical Note TN D-1767.*

Hubbard N. J. Meyer C. Jr. and Gast P. W. (1971) The composition and derivation of Apollo 12 soils. *Earth Planet. Sci. Lett.* **10**, 341–350.

Isard J. O. (1971) Formation of spherical glass particles on the lunar surface. *Proc. Second Lunar Sci. Conf., Geochim. Cosmochim. Acta*, Suppl. 2, Vol. 3, pp. 2003–2008. M.I.T. Press.

Jackson P. F. S. Coetzee J. H. J. Strasheim A. Strelow F. W. E. Gricius A. J. Wybonga F. and Kokot M. L. (1972) The analysis of lunar material returned by Apollo 14 (abstract). In *Lunar Science—III* (editor C. Watkins), p. 424, Lunar Science Institute Contr. No. 88.

McKay D. S. Greenwood W. R. and Morrison D. A. (1970) Origin of small lunar particles and breccia from the Apollo 11 site. *Proc. Apollo 11 Lunar Sci. Conf., Geochim. Cosmochim. Acta*, Suppl. 1, Vol. 1, pp. 673–694. Pergamon.

McKay D. S. Morrison D. A. Clanton U. S. Ladle G. H. and Lindsay J. F. (1971) Apollo 12 soil and breccia. *Proc. Second Lunar Sci. Conf., Geochim. Cosmochim. Acta*, Suppl. 2, Vol. 1, pp. 755–773. M.I.T. Press.

Meyer C. Jr. Brett R. Hubbard N. J. Morrison D. A. McKay D. S. Aitken F. K. Takeda H. and Schonfeld E. (1971) Mineralogy, chemistry, and origin of the KREEP component in soil samples from the Ocean of Storms. *Proc. Second Lunar Sci. Conf., Geochim. Cosmochim. Acta*, Suppl. 2, Vol. 1, pp. 393–411. M.I.T. Press.

Moore H. J. Gault D. E. and Heitowit E. D. (1964) Change of effective target strength with increasing size of hypervelocity impact craters. *Proc. 7th Hypervelocity Impact Symp.*, Vol. 4, pp. 35–46.

Rose H. J. Jr. Cuttitta F. Annell C. S. Carron M. K. Christian R. P. Dwornik E. J. and Liyon D. T. Jr. (1972) Compositional data for fifteen Fra Mauro lunar samples (abstract). In *Lunar Science—III* (editor C. Watkins), p. 660, Lunar Science Institute Contr. No. 88.

Schnetzler C. C. and Nava D. F. (1971) Chemical composition of Apollo 14 soils 14163 and 14259. *Earth Planet. Sci. Lett.* **11**, 345–350.

Wänke H. Baddenhausen H. Balacescu A. Teschke F. Spettel B. Dreibus G. Quijano M. Kruse H. Wlotzka F. and Begemann F. (1972) Multielement analyses of lunar samples (abstract). In *Lunar Science—III* (editor C. Watkins), p. 779, Lunar Science Institute Contr. No. 88.

Wood J. A. Marvin U. B. Reid J. B. Jr. Taylor G. J. Bower J. F. Powell B. N. and Dickey J. S. Jr. (1971) Mineralogy and petrology of the Apollo 12 lunar sample. *Smithsonian Astrophysical Observatory Special Report 333*, Smithsonian Institution, Washington, D.C.

Proceedings of the Third Lunar Science Conference
(Supplement 3, *Geochimica et Cosmochimica Acta*)
Vol. 1, pp. 1095–1113
The M.I.T. Press, 1972

Mineralogy, petrology, and surface features of some fragmental material from the Fra Mauro site*

CORNELIS KLEIN, JR.,† AND JOHN C. DRAKE‡

Department of Geological Sciences,
Harvard University, Cambridge, Mass. 02138

Abstract—Lithic fragments from a one gram sample of coarse fines (14257,2) and microbreccia 14305,4 were studied by optical, electron microprobe, and scanning electron microscope techniques. Five mineralogically and texturally distinct igneous rock fragments from the coarse fines were selected for study. These are microgabbro, olivine-ilmenite basalt, ophitic basalt, anorthosite, and norite. The basalt fragments show considerably less Fe-enrichment and zoning in single-crystal pyroxene grains than was observed in the pyroxenes of the Apollo 12 basalts. The basaltic pyroxene compositions range from pigeonite to subcalcic augite, augite, and ferroaugite. The plagioclases range from An_{76} to An_{90}. The anorthosite and norite fragments show strong granulation and shearing of the plagioclase (An_{86}–An_{96}), reflecting the impact metamorphism to which materials in the Fra Mauro formation were subjected.

Microbreccia 14305,4 and several fragments of breccia in the coarse fines show completely recrystallized matrices and their clasts exhibit the following shock and heat effects: (a) brecciation of single-crystal and rock fragments; (b) partial melting and reaction zones along outer edges of and fractures in single-crystal fragments; (c) microfaulting in mineral grains as shown by displaced twin lamellae; (d) devitrification and recrystallization of spheroidal clasts, probably originally glass spherules. Almost totally gradational contacts between some rock clasts and the matrix also were noted. Other microbreccia fragments show only partially devitrified matrices and some are completely unrecrystallized with glass clasts and glass matrix. It appears that a complete series exists from unrecrystallized to high-grade metamorphic breccias.

The outer surfaces of several of the rock and many of the glass fragments and spheroidal bodies in the coarse fines show fine-scale spallation features and microcraters due to the impact of hypervelocity particles. Similar craters occur on some of the iron-nickel fragments, which make up less than 0.1 wt % of the coarse fines.

INTRODUCTION

A COMBINED OPTICAL, electron microprobe and scanning electron microscope study of particles in a 2–4 mm sieve fraction of lunar soil (14257,2) reveals a variety of rock types and metal fragments and glasses of many colors and shapes. Table 1 gives the approximate proportions of the types of material in the one gram sample. The outer surfaces of the rock fragments, glass and metal show numerous small impact craters as were observed on material from the Apollo 11 and 12 missions. The rock fragments show considerable rounding and some show numerous closely spaced and irregular microfractures. Twenty-eight breccia, igneous rock, and glass fragments were selected for thin sectioning and subsequent optical study and electron microprobe analysis. A polished thin section of breccia 14305,4 also was studied optically and by electron microprobe techniques.

---

* Mineralogical Contribution, Department of Geological Sciences, Harvard University, no. 485.

† Present address: Department of Geology, Indiana University, Bloomington, Indiana 47401.

‡ Present address: Department of Geology, University of Vermont, Burlington, Vermont 05401.

Table 1. Proportions of types of material in a sieved, one gram sample of Apollo 14
lunar soil (14257,2; sieve fraction 2–4 mm).

	Fragments counted	Percentage
Breccias	36	62
Glass (angular fragments, irregular and spheroidal shapes)	14	24
Basaltic rocks	5	9
Anorthosite and norite fragments	3	5
Iron-nickel fragments	—	<0.1*
TOTAL	58	100

* Percentage obtained by weight because the size of the metal fragments is much
below 1 mm, ranging generally from 130 to 500 μ.

The electron microprobe procedures are outlined in Frondel *et al.* (1970). Quantitative electron probe analyses of opaque oxides and glasses were obtained through correction of the original x-ray counts for absorption, atomic number, and fluorescence effects according to the computer program by Goldstein and Comella (1969).

IGNEOUS ROCK FRAGMENTS IN COARSE FINES

Of the 58 fragments between 2 and 4 mm in our sample of coarse fines, only eight fragments were classified megascopically as igneous. Each of these apparently igneous fragments was thin sectioned and five fragments turned out to be basaltic (microgabbro, ilmenite basalt, and ophitic basalt), whereas two proved to be anorthositic and one noritic in composition (Table 1). Three of the basaltic rock types, one anorthosite, and one norite fragment were studied in detail optically and by electron microprobe. Figure 1 gives photomicrographs of each of these samples and Table 2 gives their modes.

Microgabbro fragment A consists of fine- to medium-grained subhedral clinopyroxenes, subhedral plagioclase laths and anhedral ilmenite. The texture of this microgabbro is very similar to that of the microgabbros from the Apollo 12 mission (e.g., Klein *et al.*, 1971; Dence *et al.*, 1971; and Brown *et al.*, 1971). In the literature these rocks are referred as both basalts and microgabbros (Warner, 1971). The chemical zoning of the clinopyroxenes in this microgabbro is much less distinct than in the pyroxenes of the microgabbros from Apollo 12, which range from magnesian pigeonite (or even orthopyroxene) in cores to hedenbergite on the edges of single

Table 2. Modes for igneous rock fragments in sample of coarse fines.

	A	B	C	D	E
Pyroxene	62	41	48*	6	27
Olivine	0	4	0	6	0
Plagioclase	31	19	48	86	73
Opaque (mostly ilmenite; minor ulvöspinel and troilite)	7	35	2	2	0
Residual phases (mostly glass)	<1	<1	2	0	0

* 40 % clear pigeonitic pyroxene and 8 % dark brown Ti-rich augite.
A: microgabbro; B: olivine-ilmenite basalt; C: ophitic basalt; D: anorthosite; E: norite.

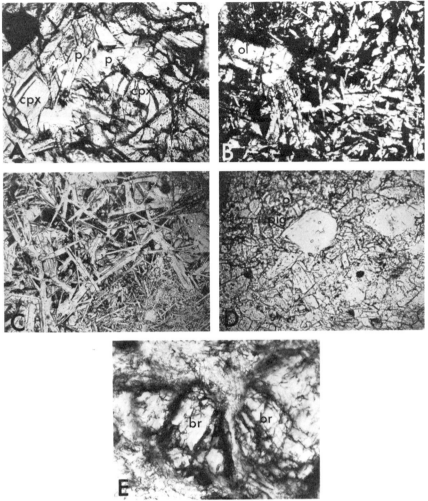

Fig. 1. Photomicrographs of igneous rock fragments in the coarse fines. (A) Microgabbro fragment A showing relatively small, subhedral plagioclase laths (p) and coarse, anhedral clinopyroxene (cpx) grains with pigeonitic cores and subcalcic augite rims (see Fig. 2 for pyroxene compositions). The opaque grain is ilmenite. Width of field is 0.47 mm. (B) Fine-grained olivine-ilmenite basalt fragment B. The large, colorless anhedral grain at the upper left hand is olivine (ol; Fo = 60), the opaque phase is ilmenite, and the remainder is a fine-grained intergrowth of clinopyroxene and plagioclase. The clinopyroxene is mainly subcalcic augite (see Fig. 2). Width of field is 1.2 mm. (C) Fine-grained ophitic basalt fragment C. The lower right-hand corner shows a finer grained cognate inclusion. The phase interstitial to the plagioclase is mainly pigeonite but some of the dark brown material is a second clinopyroxene, Ti-rich subcalcic augite. Width of field is 0.47 mm. (D) Anortho-site, fragment D showing granulation of the plagioclase grains. Minor olivine (ol; Fo = 60) and pigeonite (pig) are present. Width of field is 0.47 mm. (E) Highly sheared and granulated norite, fragment E. The two anhedral fragments with high relief in the central part are bronzite (br). Note the fracturing of these bronzites and the shearing and granulation of the plagioclase around and between them. Width of field is 0.47 mm.

crystals (e.g., Hollister *et al.*, 1971; Bence *et al.*, 1971; and Klein *et al.*, 1971). The optically poorly defined cores in the pyroxenes of this gabbro consist of pigeonite or subcalcic augite and the edges tend toward subcalcic ferroaugite (Fig. 2). The TiO_2 and Al_2O_3 contents in the clinopyroxenes range from 1.20 to 2.17 and 1.4 to 2.2 wt %, respectively. The plagioclase is almost constant in composition, and ranges from $An_{87.6}$ to $An_{88.2}$. Representative analyses of the clinopyroxenes and plagioclase are given in Tables 4 and 5, respectively. The predominant opaque phase is ilmenite, with traces of troilite and iron.

Olivine-ilmenite basalt fragment B consists of fine- to medium-grained, finely intergrown ilmenite, clinopyroxene, and plagioclase and relatively coarse, subrounded olivine grains, which make up 4 vol. % of the rock. The olivine (Fig. 2 and

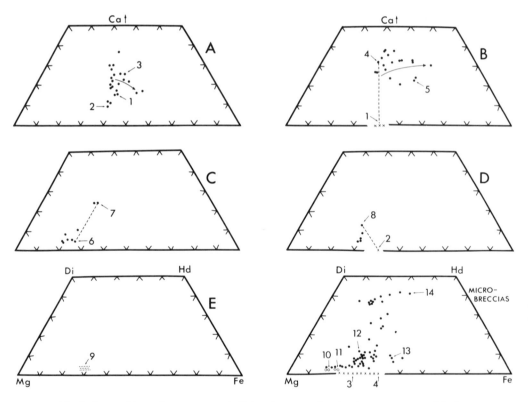

Fig. 2. Compositions of pyroxenes and olivines in terms of molecular percentages of CaO, MgO, and FeO. Clinopyroxenes are shown by large dots (●), orthopyroxenes by open circles (○) and small dots (•, in fragment E), and olivines by crosses (×). Dashed lines connect coexisting phases and curved arrows indicate compositional zonation from core to rim in a single grain. Numbers refer to the analyses in Tables 3 and 4, respectively. The igneous rock fragments are: A: microgabbro; B: olivine-ilmenite basalt; C: ophitic basalt; D: granulated anorthosite; E: sheared norite. The diagram in the lower right gives pyroxene and olivine compositions for microbreccia 14305,4 and for five other microbreccia fragments from the coarse fines. As the ranges in pyroxene compositions in these various breccias are very similar, no distinctions are made in the plot of their compositions.

Table 3) ranges from Fo_{57} to Fo_{60}. The clinopyroxenes are considerably more calcic and some tend toward more iron-rich compositions than those found in micro-gabbro fragment A (Fig. 2). The TiO_2 and Al_2O_3 contents of the clinopyroxenes range from 1.70 to 4.55 and 2.5 to 6.1 wt %, respectively. As noted in many studies of lunar clinopyroxenes (e.g., Hollister *et al.*, 1971; Frondel *et al.*, 1970; Klein *et al.*, 1971) the TiO_2 and Al_2O_3 zonation in the clinopyroxenes generally is sympathetic. The plagioclase feldspars are relatively fine-grained and show considerable in-homogeneity. Their compositions range from $An_{76.1}$ to $An_{83.8}$. Representative electron probe analyses of olivine, clinopyroxene and plagioclase are given in Tables 3, 4, and 5. The opaque phase is mostly ilmenite with very minor amounts of members of the chromite-ulvöspinel series and troilite.

Ophitic basalt fragment C consists of euhedral plagioclase laths and clear, trans-parent, anhedral clinopyroxene with lesser amounts of a darker clinopyroxene and ilmenite. This fragment contains several small cognate inclusions that are of finer grain size than the body of the rock (Fig. 1C). The texture and mineral composition of this basalt appear similar to those of rock 14310 as described by LSPET (1971) and Ridley *et al.* (1972). The clear, transparent and essentially colorless intragranular clinopyroxenes are magnesian pigeonite of relatively constant composition (Fig. 2 and Table 4, analysis no. 6); pyroxenes in rock 14310, however, have bronzite cores. A less abundant, coexisting, medium to dark brown clinopyroxene is augitic (Table 4, analysis no. 7). The plagioclase laths are relatively constant in composition and range from $An_{85.6}$ to $An_{89.9}$; this is in contrast with the plagioclase in rock 14310, which is highly zoned (Ridley *et al.*, 1972). Very small amounts of K-rich feldspar glass are present interstitially to the plagioclase and clinopyroxene grains.

Anorthosite fragment D consists mainly of granular, medium- to fine-grained plagioclase, lesser amounts of olivine and clinopyroxene, and traces of ilmenite

Table 3. Representative electron microprobe analyses (wt %) of olivines in some igneous rock fragments in the coarse fines and in microbreccia 14305,4.

	(1)	(2)	(3)	(4)
SiO_2	39.0	39.3	36.7	38.3
TiO_2	0.21	0.17	0.10	0.10
Al_2O_3	0.3	0.3	0.2	0.2
FeO*	32.6	32.6	33.6	25.5
MgO	27.2	27.0	29.7	36.6
CaO	0.3	0.0	0.3	0.0
TOTAL	99.61	99.37	100.6	100.7

Number of ions on the basis of 4 oxygens

Si	1.061		1.070		1.001		1.000	
Al	0.010 ⎤		0.010 ⎤		0.006 ⎤		0.006 ⎤	
Ti	0.004 ⎥		0.003 ⎥		0.002 ⎥		0.002 ⎥	
Fe	0.742 ⎬ 1.87		0.743 ⎬ 1.83		0.767 ⎬ 1.99		0.557 ⎬ 1.99	
Mg	1.103 ⎥		1.096 ⎥		1.208 ⎥		1.424 ⎥	
Ca	0.009 ⎦		— ⎦		0.009 ⎦		— ⎦	
mole % forsterite	59.5		59.6		60.9		71.9	

* Total Fe calculated as FeO only.

(1): Coarse-grained olivine crystal in olivine-ilmenite basalt fragment B. (2): Medium-grained olivine crystal in finer-grained, granulated plagioclase matrix of anorthosite fragment D. (3) and (4): In microbreccia 14305,4.

(Table 2). The plagioclase, although generally fine-grained, shows a very large range in crystal size (Fig. 1D). Large anhedral grains are set in a matrix of fine to very fine, granular plagioclase. This texture is suggestive of considerable granulation of the feldspar after crystallization. The plagioclase composition ranges from $An_{90.4}$ to $An_{95.9}$. The anhedral clinopyroxene and olivine grains that are sporadically present throughout this rock, are both Mg-rich. The clinopyroxene is magnesian pigeonite (analysis no. 8 in Table 4, and Fig. 2) and the olivine is Fo_{60} in composition. The

Table 4. Representative electron microprobe analyses (wt %) of pyroxenes in igneous rock fragments (A–E) in the coarse fines and in microbreccia 14305,4.

	(1)	(2)	(3)	(4)	(5)
SiO_2	51.7	51.0	51.0	49.6	46.9
TiO_2	1.20	1.76	1.45	2.84	4.55
Al_2O_3	1.5	2.2	1.8	4.5	6.1
FeO*	22.7	22.6	21.6	14.7	13.7
MgO	15.4	18.0	12.5	14.1	14.5
CaO	7.3	4.2	12.0	14.9	14.4
TOTAL	99.80	99.76	100.35	100.64	100.15

Number of ions on the basis of 6 oxygens

	(1)	(2)	(3)	(4)	(5)
Si	1.968 } 2.00	1.927 } 2.00	1.950 } 2.00	1.852 } 2.00	1.760 } 2.00
Al	0.032	0.073	0.050	0.148	0.240
Al	0.035	0.025	0.031	0.050	0.030
Ti	0.034	0.050	0.042	0.080	0.128
Fe	0.723 } 1.96	0.714 } 1.97	0.691 } 1.97	0.459 } 1.97	0.430 } 1.98
Mg	0.874	1.014	0.712	0.785	0.811
Ca	0.298	0.170	0.492	0.596	0.579
En	46.1	53.4	37.5	42.7	44.6
Fs	38.1	37.6	36.5	24.9	23.6
Wo	15.7	9.0	26.0	32.4	31.8

Table 4 (contd)

	(6)	(7)	(8)	(9)	(10)
SiO_2	55.1	47.3	53.9	55.0	56.1
TiO_2	0.82	4.38	1.16	0.53	0.30
Al_2O_3	2.0	6.3	1.5	1.3	1.0
FeO*	14.5	25.2	16.0	16.6	12.1
MgO	24.3	7.3	20.9	24.7	29.7
CaO	2.6	9.7	6.4	1.8	0.8
TOTAL	99.32	100.18	99.86	99.93	100.00

Number of ions on the basis of 6 oxygens

	(6)	(7)	(8)	(9)	(10)
Si	1.990 } 2.00	1.835 } 2.00	1.977 } 2.00	1.992 } 2.00	1.984 } 2.00
Al	0.010	0.165	0.023	0.008	0.016
Al	0.075	0.123	0.042	0.047	0.026
Ti	0.022	0.128	0.032	0.014	0.008
Fe	0.438 } 1.94	0.817 } 1.89	0.491 } 1.96	0.503 } 1.97	0.358 } 1.99
Mg	1.308	0.422	1.143	1.333	1.565
Ca	0.101	0.403	0.251	0.070	0.030
En	70.8	25.7	60.6	69.9	80.1
Fs	23.7	49.7	26.0	26.4	18.3
Wo	5.5	24.5	13.3	3.7	1.5

Table 4 (contd)

	(11)	(12)	(13)	(14)
SiO₂	55.7	52.9	53.0	50.4
TiO₂	0.30	1.31	0.60	1.10
Al₂O₃	0.8	2.5	0.7	1.2
FeO*	14.5	16.8	26.0	18.2
MgO	27.9	22.0	16.8	9.8
CaO	0.7	5.3	3.5	20.0
TOTAL	99.90	100.81	100.60	100.70

Number of ions on the basis of 6 oxygens

	(11)	(12)	(13)	(14)
Si	1.993 ⎱ 2.00	1.927 ⎱ 2.00	2.004 ⎱ 2.00	1.940 ⎱ 1.99
Al	0.007 ⎰	0.073 ⎰	0.000 ⎰	0.055 ⎰
Al	0.026	0.034	0.031	0.000
Ti	0.008	0.036	0.017	0.032
Fe	0.434 ⎱ 1.98	0.512 ⎱ 1.98	0.822 ⎱ 1.96	0.586 ⎱ 2.00
Mg	1.488	1.194	0.947	0.562
Ca	0.027 ⎰	0.207 ⎰	0.142 ⎰	0.825 ⎰
En	76.3	62.4	49.5	28.5
Fs	22.3	26.8	43.0	29.7
Wo	1.3	10.8	7.4	41.8

* Fe calculated as FeO only.

(1), (2) and (3): Clinopyroxenes in microgabbro fragment A. (4): Clinopyroxene edge in contact with olivine (Fo = 59.5) in olivine-ilmenite basalt fragment B. (5): Clinopyroxene in olivine-ilmenite basalt fragment B. (6): Transparent intergranular clinopyroxene in ophitic basalt fragment C. (7): Dark brown clinopyroxene in ophitic basalt fragment C. (8): Clinopyroxene coexisting with olivine (Fo = 59.6) in granulated anorthosite fragment D. (9): Orthopyroxene in highly sheared norite fragment E. (10) and (11): Orthopyroxenes in microbreccia 14305,4. (12), (13), and (14): Clinopyroxenes in microbreccia 14305,4.

Table 5. Representative electron microprobe analyses (wt%) of plagioclase feldspars in igneous rock fragments in coarse fines and in microbreccia 14305,4.

	(1)	(2)	(3)	(4)	(5)	(6)	(7)	(8)	(9)
SiO₂	47.2	47.8	46.7	46.4	46.2	46.0	46.6	49.0	54.2
Al₂O₃	34.1	34.0	34.2	35.3	34.9	35.4	34.7	33.0	29.2
CaO	17.9	16.9	17.5	17.6	17.5	18.4	18.0	16.3	12.8
Na₂O	1.30	1.80	1.51	0.90	1.50	0.60	0.90	1.90	3.54
K₂O	0.10	0.0	0.20	0.20	0.19	0.07	0.10	0.14	0.98
TOTAL	100.60	100.50	100.11	100.40	100.20	100.47	100.30	100.34	100.72

Number of ions on the basis of 8 oxygens

	(1)	(2)	(3)	(4)	(5)	(6)	(7)	(8)	(9)
Si	2.155	2.178	2.145	2.120	2.199	2.104	2.133	2.230	2.439
Al	1.835	1.826	1.851	1.901	1.887	1.908	1.872	1.771	1.549
Σ	3.99	4.00	3.99	4.02	4.01	4.01	4.00	4.00	3.99
Ca	0.876	0.825	0.861	0.862	0.862	0.902	0.883	0.795	0.617
Na	0.115	0.159	0.132	0.080	0.133	0.051	0.080	0.167	0.309
K	0.006	0.0	0.012	0.011	0.006	0.004	0.006	0.008	0.056
Σ	1.00	0.98	1.00	0.95	0.99	0.96	0.97	0.97	0.98
mole %									
An	87.8	83.8	85.6	90.4	86.1	94.3	91.1	81.9	62.8
Ab	11.5	16.2	13.3	8.3	13.3	5.3	8.2	17.2	31.4
Or	0.6	0.0	1.1	1.2	0.6	0.4	0.6	0.8	5.7

(1): Microgabbro fragment A. (2): Olivine-ilmenite basalt fragment B. (3): Ophitic basalt fragment C. (4): Granulated anorthosite fragment D. (5): Highly sheared norite fragment E. (6), (7), (8), and (9): From microbreccia 14305,4.

TiO$_2$ content is relatively low in the clinopyroxene and ranges from 1.16 to 1.25 wt%, and the Al$_2$O$_3$ content is a constant 1.5 wt%.

Norite fragment E consists of 27 vol. % relatively coarse-grained, subrounded, highly fractured bronzite with extensively developed undulatory extinction that is set in a matrix of highly sheared and granulated, fine- to medium-grained plagioclase. Figure 1E shows two of the bronzite grains separated by a thin shear zone of very fine-grained plagioclase. The compositions of the bronzites (2V$_\alpha \sim 75°$) are very similar and cluster about En$_{70}$Fs$_{26}$Wo$_4$ (Fig. 2 and Table 4, analysis no. 9). Their TiO$_2$ and Al$_2$O$_3$ contents are relatively low and constant, ranging from 0.48 to 0.58 and 1.3 to 1.4 wt%, respectively. The plagioclase is highly calcic and has a fairly wide range of composition, from An$_{86.1}$ to An$_{93.1}$.

Microbreccia 14305,4 and Glass-Rich Microbreccia from the Coarse Fines

A polished thin section of microbreccia 14305,4 and of six other smaller microbreccia fragments from the coarse fines were studied by optical and electron microprobe techniques.

Microbreccia 14305,4 consists of a fine-grained, granular, almost wholly crystalline matrix (grain size <0.01 mm) that is made up of a mixture of feldspar, pyroxene, and ilmenite and possibly minute amounts of glass. Set within this matrix are clasts of mineral fragments (plagioclase, pyroxene, olivine, and ilmenite), lithic fragments (igneous rocks and fragmental breccias) and recrystallized glass fragments and spherules. The lithic fragments include medium-grained microgabbro, fine-grained ilmenite basalt, fine-grained, granular anorthositic gabbro, norite, and breccia.

All of the breccia clasts in microbreccia 14305,4 consist of crystalline fragments enclosed in crystalline matrices. No undevitrified glass fragments were found. The boundaries between the lithic fragments and surrounding matrix range from extremely sharp, with easily discernible mineralogical and textural differences, to completely gradational zones that can, at best, be only approximated (compare Figs. 3B and G).

This breccia and several of the other smaller microbreccia fragments from the coarse fines show features that are clearly due to a combination of shock and heat effects. These features are (a) recrystallization and devitrification of spheroidal clasts, which probably represent glass spherules (Fig. 3A); (b) thin reaction zones between single-crystal fragments and the microbreccia matrix (Fig. 3C); (c) deformation of single-crystal pyroxene clasts as shown by abruptly bent cleavage traces (Fig. 3D); (d) complex and highly irregular, glass-lined fractures in plagioclase grains (Fig. 3E); (e) very thin, glass-lined outer edges of plagioclase clasts, along the contact zone of clasts and the breccia matrix (Fig. 3E); (f) microfaulting within mineral grains as shown by the displacement of twin lamellae (Fig. 3F); (g) undulatory extinction and shock twinning in some feldspar and pyroxene grains; and (h) the almost completely gradational contact between microbreccia matrix and some of the igneous rock clasts (Fig. 3G). The above features are the result of a metamorphic episode in the history of this microbreccia. Warner (1972) classified breccia 14305 in the high-grade metamorphic facies (his group 6) with an annealed to euhedral matrix texture.

In several parts of microbreccia 14305,4 clasts consist of an earlier microbreccia of gabbro and single-crystal mineral clasts, and as such, the present breccia represents a fragmental rock of a second generation. Analyses for several olivine, pyroxene, and plagioclase clasts in this breccia are given in Tables 3, 4, and 5, respectively. The olivine compositions range from Fo_{61} to Fo_{72}.

The pyroxene compositions are all relatively magnesian and none show the pronounced chemical zoning that is seen in most of the Apollo 12 and some of the Apollo 11 igneous clinopyroxenes. As pyroxene grains in a breccia represent the broken and reworked parts of primary igneous pyroxenes, one would expect most of the outer edges of such crystals, which show the strongest zoning toward Fe-enrichment, to be missing. Several of the pyroxenes are orthopyroxenes (bronzite). One such grain shows thin, acicular crystals of rutile aligned parallel to the c-axis of the bronzite (Fig. 3D). Some of the pigeonite grains show well-developed (001) lamellae of augite, up to several μ wide.

The total range in plagioclase compositions is from An_{63} to An_{96}, but the majority of compositions cluster between An_{82} and An_{94}. The most abundant opaque oxide mineral is ilmenite. Members of the ulvöspinel-chromite series are much less abundant. Representative analyses of a titanian chromite and several ilmenites are given in Table 6. The MgO content of the ilmenites varies sympathetically with their FeO content (compare analyses 4 and 5 in Table 6). This correlation was noted among others by Agrell et al. (1970) and Keil et al. (1970). As suggested by these authors, there appears to be an excess of titanium in some ilmenites (analysis 5 in Table 5) if it is recalculated as Ti^{4+} only. This may indicate the presence of some Ti^{3+}.

Table 6. Electron microprobe analyses (wt %) of titaniferous chromite and several ilmenites in microbreccia 14305,4 and two breccia fragments of the coarse fines.

	(1)	(2)	(3)	(4)	(5)
TiO_2	14.0	53.4	53.2	53.7	59.35
Al_2O_3	8.8	0.60	0.62	0.62	0.60
Cr_2O_3	35.4	0.73	0.64	0.58	0.85
FeO*	35.3	39.2	39.1	39.2	34.30
MgO	4.7	4.4	4.7	4.6	3.85
NiO	0.41	0.41	0.41	0.39	0.41
TOTAL	98.61	98.74	98.64	99.09	99.36
		Number of ions on basis of			
	4 oxygens	– – – – – – – – – 3 oxygens – – – – –			
Ti	0.365	0.989	0.985	0.989	1.062
Al	0.359	0.017	0.018	0.018	0.017
Cr	0.970	0.014	0.013	0.011	0.016
Σ	—	1.020	1.015	1.018	1.095
Fe	1.023	0.807	0.805	0.803	0.682
Mg	0.243	0.161	0.172	0.168	0.136
Ni	0.011	0.008	0.007	0.008	0.008
Σ	—	0.976	0.984	0.979	0.826
TOTAL CATIONS	2.971	1.996	1.999	1.997	1.921

* Total Fe recalculated as FeO only.
(1) and (2): In microbreccia 14305,4. (3), (4), and (5): In microbreccia fragments from coarse fines.

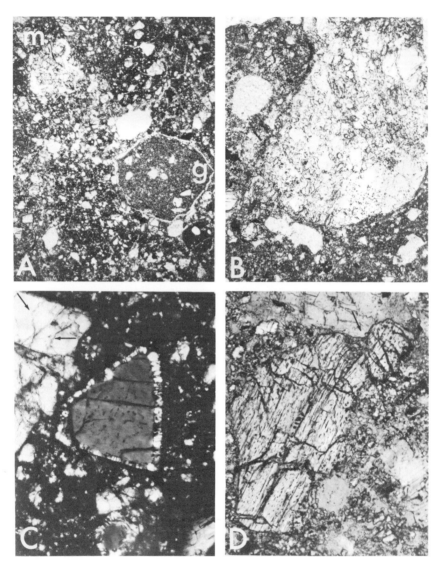

Fig. 3. Photomicrographs of various textures and fragments in microbreccia 14305,4 and of a glass-rich microbreccia fragment (F) in the coarse fines. (A) Large, subrounded, recrystallized glass fragment (g) at lower right, surrounded by subangular fragments of clinopyroxene and plagioclase and chips of microgabbro (m). Width of field is 2.2 mm. (B) Rounded anorthosite clast, consisting of granulated and sheared, anhedral plagioclase and small amounts of fine-grained, subhedral olivine dispersed throughout the plagioclase matrix. Width of field is 0.95 mm. (C) Angular, single-crystal olivine fragment (Fo = 61; see analysis 3 in Table 3) showing reaction rim of clinopyroxene. To the upper left of this olivine occurs a plagioclase grain with numerous glass-lined fractures (see arrows). Partially crossed nicols. Width of field is 0.24 mm. (D) Fractured and deformed single-crystal of bronzite (analysis no. 11 in Table 4) with thin exsolution lamellae of rutile aligned parallel to the c-axis of the orthopyroxene. Note deformation of crystal at upper right (see arrow).

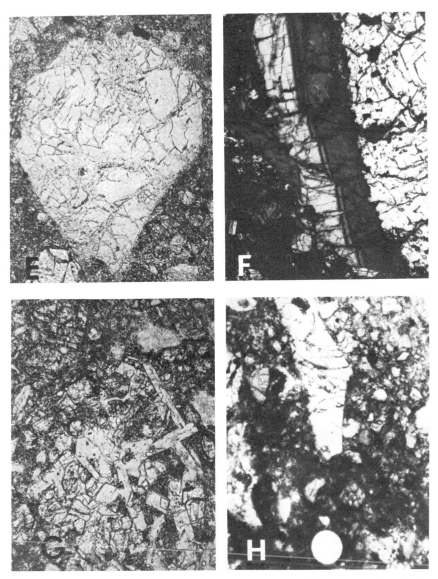

Width of field is 0.24 mm. (E) Subangular plagioclase clast with numerous glass-lined fractures and a very thin glassy outer edge. Width of field is 0.47 mm. (F) Deformed plagioclase grain in a coarse grained basalt clast in micrograbbro 14305,4. The twin lamellae in the plagioclase grain are offset by numerous small transverse faults. Light area on the right is clinopyroxene. Crossed nicols. Width of field is 0.47 mm. (G) Gradational contact between a medium-grained basaltic clast in the lower part of the photograph and the much finger grained, finely crystalline matrix of microgabbro 14305,4. Width of field is 0.47 mm. (H) Microbreccia fragment F with angular and spherical glass fragments which show no devitrification or recrystallization. This fragment is similar to many of the breccias of Apollo 11. Analyses of some of the glasses in this breccia are given in Table 7, and are shown graphically in Fig. 4. Width of field is 0.47 mm.

Of the six small microbreccia fragments from the coarse fines which were studied in detail, three fragments turned out to be very similar to parts of microbreccia 14305,4, whereas the other three fragments were found to be unrecrystallized fragmental rocks with large numbers of glass clasts set in an essentially unrecrystallized matrix. These microbreccias, of which fragment F (Fig. 3H) is an example, contain irregular fragments and spheroids of glass and show only negligible evidence of the combined shock and heat effects that are clearly displayed in microbreccia 14305,4, as described above. As suggested by Warner (1972) these unmetamorphosed breccias may be the product of local, relatively small-scale impacts, whereas the medium- to high-grade metamorphic breccias, such as 14305, probably are part of the Fra Mauro ejecta blanket, which may have been subjected to a process of autometamorphism. Electron probe analyses of several of the glass types in fragment F are given in Table 7, and are represented graphically in Fig. 4. As shown by the analyses of glass fragments in the fines and microbreccias of Apollo 11 and 12 (e.g., Frondel *et al.*, 1970 and Winchell and Skinner, 1970) the glass compositions represent mixtures of the anorthite, diopside, hypersthene and lesser olivine, K-feldspar, ilmenite, and silica components. The light colored and transparent glasses are richer in anorthite than the darker colored (light to dark brown) glasses which contain more of the pyroxene and ilmenite components (CIPW norms in Table 7). The matrix glass of microbreccia fragment F (analysis no. 6 in Table 7), which is medium grayish brown and only partially transparent, shows high contents of Na_2O (1.1 wt%), K_2O (0.2 wt%),

Table 7. Representative electron microprobe analyses (wt%) of angular glass fragments and spheroids in an unrecrystallized microbreccia fragment (F) from the coarse fines.

	(1)	(2)	(3)	(4)	(5)	(6) (average)	(7)
SiO_2	47.1	41.5	51.2	47.0	40.8	52.0	41.8
TiO_2	0.41	2.37	3.20	4.84	10.8	1.30	7.09
Al_2O_3	25.2	19.1	14.3	10.6	10.4	24.3	14.00
FeO^*	6.1	11.9	12.9	19.7	18.0	4.9	15.78
MgO	8.9	12.3	7.5	7.9	9.2	3.5	8.18
CaO	11.9	12.5	9.8	9.9	10.6	12.1	12.03
Na_2O	0.0	0.0	0.0	0.0	0.0	1.1	0.40
K_2O	0.0	0.0	0.7	0.0	0.1	0.2	0.14
TOTAL	99.6	99.7	99.6	99.9	99.9	99.4	99.42
				CIPW norm			
Q	3.54	—	10.48	5.54	2.35	10.57	—
Or	—	—	4.14	—	0.59	1.18	0.83
Ab	—	—	—	—	—	9.31	3.38
An	59.04	52.12	36.95	28.92	25.08	60.03	35.99
Di	—	8.05	9.65	16.91	20.01	—	19.50
Hy	32.69	9.28	32.31	39.38	28.35	15.57	21.22
Ol	—	25.73	—	—	—	—	5.12
Ilm	0.78	4.50	6.08	9.19	20.51	2.47	13.47

* Fe calculated as FeO only.

(1): Colorless, clear, transparent angular glass fragment in microbreccia fragment F (see Fig. 3H for a photomicrograph). (2): Light beige, clear, transparent spheroid of glass in microbreccia fragment F. (3): Light brown, transparent glass fragment in microbreccia fragment F. (4): Medium brown, transparent glass fragment in microbreccia fragment F. (5): Reddish-brown, transparent glass fragment in microbreccia fragment F. (6): Gray-brown, only partially transparent matrix of microbreccia fragment F; chemically somewhat inhomogeneous. (7): Wet chemical analysis of dark brown, scoriaceous glass from Frondel *et al.* (1970), sample no. 3. Elements also reported: $Cr_2O_3 = 0.30$, $MnO = 0.23$, $P_2O_5 = 0.1$, $ZrO_2 = 0.05$, $S = 0.1$, and $NiO \sim 0.01$, all wt%.

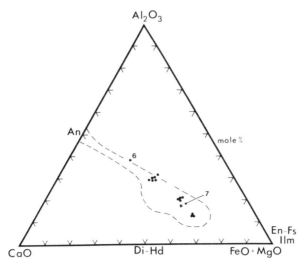

Fig. 4. Compositions of glasses in microbreccia fragment F from the coarse fines in terms of molecular percentages of Al_2O_3, CaO, and (MgO + FeO). The composition of the matrix of microbreccia fragment F is represented by 6. The area inside the dashed line represents the total compositional range shown by a large collection of glasses in the lunar fines of Apollo 11 (compiled from Frondel *et al.*, 1970 and Winchell and Skinner, 1970). The star (with no. 7) represents the average analysis of the most common type of glass in the Apollo 11 fines (from Frondel *et al.*, 1970). Numbers refer to analyses in Table 7. An (anorthite), Fs (ferrosilite), En (enstatite), Hd (hedenbergite), Di (diopside), Ilm (ilmenite).

and CaO (12.1 wt %), as compared with the enclosed glass fragments. This matrix glass is of a very anorthite-rich composition (CIPW norm, Table 7) with 10.5% normative SiO_2 and very considerable normative Or (1.2%) and Ab (9.3%) contents.

MICROCRATERS AND SPALLATION FEATURES

Microcraters caused by the impact of hypervelocity particles have been described on rock fragments, glass, and metal particles from the material returned by the Apollo 11 and 12 missions (Carter, 1971; Carter and McGregor, 1970; Frondel *et al.*, 1970; Fulchignoni *et al.*, 1971; McKay *et al.*, 1970 and 1971). In our present study of Apollo 14 fines similar impact features were found frequently on the surface of breccia and gabbro fragments, on glass fragments, coatings, and spheroidal forms and on metal particles.

Impact craters on rock fragments show smooth glass surfaces on the inside of the crater (Fig. 5A) and flowage features (Fig. 5B) around the central, originally molten area.

Spallation areas that lack the characteristic crater morphology are most easily recognized on relatively smooth glass surfaces, especially spheroids (Figs. 5C and D). Such spallation areas may represent intersecting conchoidal fractures, which generally form below a microcrater (Fig. 5H). In Figs. 5C and D the microcrater with central neck may have spalled off, leaving only the irregular pit-like impressions.

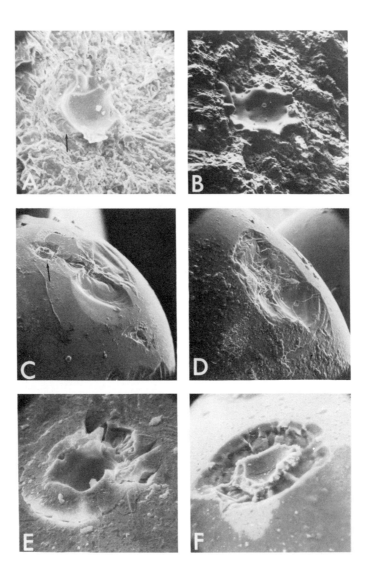

Fig. 5. SEM photographs of microcraters and spallation zones on rock fragments, glass, and metal particles. (A) Microcrater of dark brown glass on surface of fragment of micro-gabbro. Most of the outer rim of the crater is broken off as shown by the conchoidal fractures on the lower left corner of the crater (see arrow). Width of field is 160 μ. (B) Microcrater of dark brown glass on surface of a fragment of fine-grained breccia. In this crater the outer rim is preserved and shows flowage features. Width of field is 480 μ. (C) Spallation area and separate, later microcrater (shown by arrow) on dark brown glass spherule. Width of field is 0.8 mm. (D) Spallation area on glass sphere which is covered by fused-on fragments of lunar fines. Width of field is 0.8 mm. (E) Microcrater on dark brown glass sphere. The crater rim around the central depression shows flowage features similar to those found in metal fragments (see Figs. 5K and L). A fracture zone concentric with the crater outlines the region of incipient spallation. Width of field is 44 μ. (F) Microcrater on dark brown glass

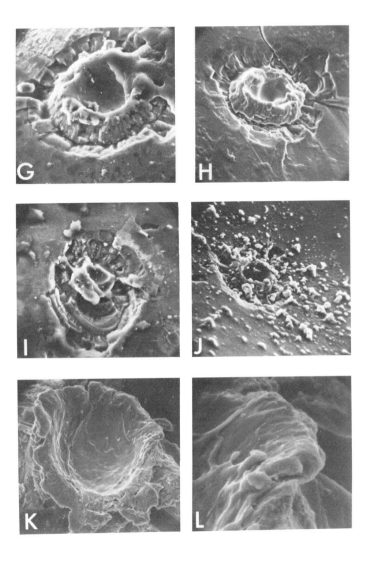

sphere. The central crater diameter is twice as large as that of the crater in Fig. 5E. The spallation zone is well-developed. Width of field is 88 μ. (G) Microcrater on dark brown glass sphere. This crater shows the effects of considerable flowage of glass from the center of the crater over the edge into the spallation zone. At the upper right a later splatter of glass covers part of the crater rim. Width of field is 88 μ. (H) Microcrater on brown glass sphere shown in Fig. 5C. Here a well-developed fracture zone is visible beneath and concentric with the inner fused part of the crater. Width of field is 160 μ. (I) Microcrater on brown glass spherule with broad and irregular spallation zone. Width of field is 100 μ. (J) Microcrater on dark brown, irregularly shaped mass of scoriaceous glass. Subsequent to the cratering event particles fused or condensed on the outer surface of the glass. Width of field is 100 μ. (K) Microcrater on irregularly shaped iron-nickel fragment. Note flowage of metal from the center over the outer, raised rim of the crater. Width of field is 100 μ. (L) Close-up of rim of crater shown in Fig. 5K. Width of field is 20 μ.

Very similar spallation areas were produced by Carter and McKay (1971) in their experimental study of crater morphology (especially *their* Fig. 15).

Well-developed microcraters show various stages and complexities in their morphological development on glass surfaces, as can be seen in Figs. 5E to J. Figure 5E shows a very well-developed set of conical fractures with only a small part spalled off. The central crater rim shows flowage features very similar to those found on metal particles (Figs. 5K and L). In Fig. 5F the central crater diameter, which is twice as large as the one shown in Fig. 5E, is surrounded by a completely spalled-off area. Figures 5G to I show well-developed conical fractures below and concentric with the crater as well as glass flowage features outward from the center of the crater. Figure 5J shows numerous silicate fragments and globules fused onto the outer surface of the glass around the central crater. These fragments may well represent the broken up parts and condensed products of the original impacting projectile. Carter and McKay (1971) in their experimental study of crater formation on glass surfaces show many crater phenomena that are very similar to the features shown here.

The microcraters on metallic particles show no spallation zone around the central crater area, but show pronounced rim development due to flowage of metal from the center to the outer rim of the crater. In many instances the outer edge of the rim is overturned (Figs. 5K and L). Thus, the morphology of these craters is directly related to the ductility of the metal phase.

It should be noted that microcraters can be found on all sides of most glass, rock, or metal fragments. Frondel *et al.* (1970) describe a 4 mm long, platy iron-nickel fragment that shows 42 craters over 0.1 mm in diameter and which are randomly distributed over the flat sides as well as the edges. Similarly, many of the angular glass fragments and spheroids in the present sample show numerous craters distributed over many parts of their surfaces. Such a crater distribution could result from secondary lunar particles ejected during large meteoritic impacts or from a combination of extralunar primary impacts and extensive gardening of the soil fragments. The experimental data of Carter and McKay (1971) support the hypothesis that most of the small craters on the glass fragments result from impacts of secondary particles in the debris cloud generated by a larger primary impact on the lunar surface. Hartung *et al.* (1972), however, provide experimental and observational data that would favor a primary impact origin by micrometeorites for all the craters displaying glass-lined pits in the 1 μ to cm range.

The shape and general morphology of a microcrater is a function of the velocity, angle of approach and size of the impacting projectile, as well as the physical characteristics of the impacted material. Carter and McKay (1971) have clearly shown that crater morphology on glasses is strongly influenced by the temperature of the target glass.

DISCUSSION AND CONCLUSIONS

Our small and statistically nonrepresentative sample of coarse fines contains a much higher percentage of igneous rock fragments and a larger variety of igneous rock types than was observed in the hand specimens (LSPET, 1971). A similar contrast between the types of larger size rocks and rock types in the coarse fines was noted also for the Apollo 11 and 12 materials (e.g., Wood *et al.*, 1970; Keil *et al.*, 1971; Marvin *et al.*, 1971 and Brown *et al.*, 1971). The igneous rock fragments, except for olivine-ilmenite basalt fragment B, show a considerably lower ilmenite content than was found generally in the Apollo 11 and 12 igneous rocks. Ophitic gabbro fragment C and anorthosite fragment D contain only 2% ilmenite and the norite fragment completely lacks ilmenite. The pyroxenes in the basaltic rocks show considerably less Fe zoning toward the edges of the grains and their average compositions are more magnesian than those of the Apollo 11 and 12 basalts. The textures of the three basaltic fragments studied range from fine-grained granular to ophitic without any indication of shock metamorphic effects. The anorthosite and norite fragments clearly show the effects of severe shearing and granulation. Similar rock types in the coarse fines of the Apollo 12 sample also showed considerable evidence of shock brecciation (Marvin *et al.*, 1971). Wood *et al.* (1970) conclude that the Apollo 11 anorthosite fragments are representative of the lunar highlands and Marvin *et al.* (1971) suggest that their Type A norites in the Apollo 12 sample which show effects of shock brecciation and recrystallization have moved over very considerable horizontal distances as a result of a large impact. The anorthosite and norite fragments from Apollo 14 have mineral compositions and textural features similar to those of the Apollo 12 brecciated norites and anorthosites. These plagioclase-rich rocks have undergone a much more complex postcrystallization history than the equigranular or ophitic basaltic rocks. It is probable that these plagioclase-rich fragments represent samples that have originated from areas at very considerable distance from the landing site and which have undergone shock brecciation during the Imbrian event. It is possible that the basaltic rocks (A, B, and C) represent fragments from lava flows that crystallized before the Imbrian impact.

The microbreccias in this study show textures that range from unrecrystallized through partially devitrified to completely recrystallized. Warner (1972) has shown, on the basis of his study of a large number of Apollo 14 breccias, that a complete range exists from unrecrystallized to high grade metamorphic breccias. The unrecrystallized breccias with numerous glass clasts and glassy matrix are very similar to those returned by the Apollo 11 and 12 missions. The higher grade metamorphic breccias probably formed during the large scale Imbrian event that produced the Fra Mauro ejecta blanket.

Acknowledgments—This research was supported by NASA grant NGR–22–007–191. We are grateful to Professor C. Frondel for his continued interest in this work and to Maurice Campot for keeping our analytical and optical equipment in excellent working order. We are indebted to Professor Robert E. Ogilvie at M.I.T. for making the scanning electron microscope available to us and to J. A. Adario and J. I. Herman for their technical assistance. We thank Drs. R. Brett and J. L. Warner for their constructive criticism.

REFERENCES

Agrell S. O. Scoon J. H. Muir I. D. Long J. V. P. McConnell J. D. and Peckett A. (1970) Observations on the chemistry, mineralogy, and petrology of Apollo 11 lunar samples. *Proc. Apollo 11 Lunar Sci. Conf., Geochim. Cosmochim. Acta*, Suppl. 1, Vol. 1, pp. 93–129. Pergamon.

Bence A. E. Papike J. J. and Lindsley D. H. (1971) Crystallization histories of clinopyroxenes in two porphyritic rocks from Oceanus Procellarum. *Proc. Second Lunar Sci. Conf., Geochim. Cosmochim. Acta*, Suppl. 2, Vol. 1, pp. 559–575. M.I.T. Press.

Brown G. M. Emeleus C. H. Holland J. G. Peckett A. and Phillips R. (1971) Picrite basalts, ferro-basalts, feldspathic norites, and rhyolites in a strongly fractionated lunar crust. *Proc. Second Lunar Sci. Conf., Geochim. Cosmochim. Acta*, Suppl. 2, Vol. 1, pp. 583–600. M.I.T. Press.

Carter J. L. (1971) Chemistry and surface morphology of fragments from Apollo 12 soil. *Proc. Second Lunar Sci. Conf., Geochim. Cosmochim. Acta*, Suppl. 2, Vol. 1, pp. 873–892. M.I.T. Press.

Carter J. L. and MacGregor I. D. (1970) Mineralogy, petrology, and surface features of some Apollo 11 samples. *Proc. Apollo 11 Lunar Sci. Conf., Geochim. Cosmochim. Acta*, Suppl. 1, Vol. 1, pp. 247–265. Pergamon.

Carter J. L. and McKay D. S. (1971) Influence of target temperature on crater morphology and implications on the origin of craters on lunar glass spheres. *Proc. Second Lunar Sci. Conf., Geochim. Cosmochim. Acta*, Suppl. 2, Vol. 3, pp. 2653–2670. M.I.T. Press.

Dence M. R. Douglas J. A. V. Plant A. G. and Traill R. J. (1971) Mineralogy and petrology of some Apollo 12 samples. *Proc. Second Lunar Sci. Conf., Geochim. Cosmochim. Acta*, Suppl. 2, Vol. 1, pp. 285–330. M.I.T. Press.

Frondel C. Klein C. Jr. Ito J. and Drake J. C. (1970) Mineralogical and chemical studies of Apollo 11 lunar fines and selected rocks. *Proc. Apollo 11 Lunar Sci. Conf., Geochim. Cosmochim. Acta*, Suppl. 1, Vol. 1, pp. 445–474. Pergamon.

Frondel C. Klein C. Jr. and Ito J. (1971) Mineralogy and chemical data on Apollo 12 lunar fines. *Proc. Second Lunar Sci. Conf., Geochim. Cosmochim. Acta*, Suppl. 2, Vol. 1, pp. 719–726. M.I.T. Press.

Fulchignoni M. Funiciello R. Taddeucci A. and Trigila R. (1971) Glassy spheroids in lunar fines from Apollo 12 samples 12070,37; 12001,73; and 12057,60. *Proc. Second Lunar Sci. Conf., Geochim. Cosmochim. Acta*, Suppl. 2, Vol. 1, pp. 937–948. M.I.T. Press.

Goldstein J. I. and Comella P. A. (1969) A computer program for electron probe microanalysis in the fields of metallurgy and geology. *NASA Special Publ.* X–642–69–115.

Hartung J. B. Horz F. and Gault D. E. (1972) The origin and significance of lunar microcraters (abstract). In *Lunar Science—III* (editor C. Watkins), pp. 363–365, Lunar Science Institute Contr. No. 88.

Hollister L. S. Trzciensky W. E. Hargraves R. B. and Kulick C. G. (1971) Petrogenetic significance of pyroxenes in two Apollo 12 samples. *Proc. Second Lunar Sci. Conf., Geochim. Cosmochim. Acta*, Suppl. 2, Vol. 1, pp. 529–559. M.I.T. Press.

Keil K. Bunch T. E. and Prinz M. (1970) Mineralogy and composition of Apollo 11 lunar samples. *Proc. Apollo 11 Lunar Sci. Conf., Geochim. Cosmochim. Acta*, Suppl. 1, Vol. 1, pp. 561–598. Pergamon.

Keil K. Prinz M. and Bunch T. E. (1971) Mineralogy, petrology, and chemistry of some Apollo 12 samples. *Proc. Second Lunar Sci. Conf., Geochim. Cosmochim. Acta*, Suppl. 2, Vol. 1, pp. 319–342. M.I.T. Press.

Klein C. Jr. Drake J. C. and Frondel C. (1971) Mineralogical, petrological, and chemical features of four Apollo 12 microgabbros. *Proc. Second Lunar Sci. Conf., Geochim. Cosmochim. Acta*, Suppl. 2, Vol. 1, pp. 265–285. M.I.T. Press.

Klein C. Jr. Drake J. C. Frondel C. and Ito J. (1972) Mineralogy and petrology of several Apollo 14 rock types and chemistry of the soil (abstract). In *Lunar Science—III* (editor C. Watkins), pp. 455–457, Lunar Science Institute Contr. No. 88.

LSPET (1971) (Lunar Sample Preliminary Examination Team) Preliminary examination of lunar samples from Apollo 14. *Science* **173**, 681–693.

Marvin U. B. Wood J. A. Taylor G. J. Reid J. B. Powell B. N. Dickey J. A. and Bower J. F. (1971) Relative proportions and probable sources of rock fragments in the Apollo 12 soil sample. *Proc. Second Lunar Sci. Conf., Geochim. Cosmochim. Acta*, Suppl. 2, Vol. 1, pp. 679–699. M.I.T. Press.

McKay D. S. Greenwood W. R. and Morrison D. A. (1970) Origin of small lunar particles and breccia from the Apollo 11 site. *Proc. Apollo 11 Lunar Sci. Conf., Geochim. Cosmochim. Acta*, Suppl. 1, Vol. 1, pp. 673–694. Pergamon.

McKay D. S. Morrison D. A. Clanton U. S. Ladle G. H. and Lindsay J. F. (1971) Apollo 12 soil and breccia. *Proc. Second Lunar Sci. Conf., Geochim. Cosmochim. Acta*, Suppl. 2, Vol. 1, pp. 755–773. M.I.T. Press.

Ridley W. I. Williams R. J. Brett R. and Takeda H. (1972) Petrology of lunar basalt 14310 (abstract). In *Lunar Science—III* (editor C. Watkins), pp. 648–650, Lunar Science Institute Contr. No. 88.

Warner J. L. (1971) Lunar crystalline rocks: petrology and geology. *Proc. Second Lunar Sci., Conf., Geochim. Cosmochim. Acta*, Suppl. 2, Vol. 1, pp. 469–480. M.I.T. Press.

Warner J. L. (1972) Apollo 14 breccias: metamorphic origin and classification (abstract). In *Lunar Science —III* (editor C. Watkins), pp. 782–784, Lunar Science Institute Contr. No. 88.

Winchell H. and Skinner B. J. (1970) Glassy spherules from the lunar regolith returned by Apollo 11 expedition. *Proc. Apollo 11 Lunar Sci. Conf., Geochim. Cosmochim. Acta*, Suppl. 1, Vol. 1, pp. 957–964. Pergamon.

Wood J. A. Dickey J. S. Marvin U. B. and Powell B. N. (1970) Lunar anorthosites and a geophysical model of the moon. *Proc. Apollo 11 Lunar Sci. Conf., Geochim. Cosmochim. Acta*, Suppl. 1, Vol. 1, pp. 965–988. Pergamon.

Proceedings of the Third Lunar Science Conference
(Supplement 3, *Geochimica et Cosmochimica Acta*)
Vol. 1, pp. 1115–1120
The M.I.T. Press, 1972

A new titanium and zirconium oxide from the Apollo 14 samples

C. Levy and M. Christophe-Michel-Levy

Laboratoire de minéralogie-cristallographie associé au C.N.R.S.,
University of Paris VI

P. Picot and R. Caye

Service Géologique National—B.R.G.M. Orleans

Abstract—This new phase was found in fines 14003,47. Its chemical composition is essentially that of an armalcolite, although the physical properties are different. Its formula is given below:

$$Ti_{1.91}Zr_{0.08}Fe_{0.27}Mg_{0.10}Cr_{0.25}Al_{0.07}Ca_{0.12}Si_{0.02}O_5$$

It has been observed in a polished section as an optically isotropic surface 30 μ across surrounded by ilmenite. Therefore no x-ray crystallographical study has been possible. Nevertheless, the dispersion curve of its reflectivities (from 400 to 700 nm) differs sufficiently in its shape and by its absolute reflectivity values from that of armalcolite to indicate that we are dealing with a different phase.

In a polished section made from a grain of soil sample 14003,47 (section 21645), an isotropic mineral has been observed with optical properties similar to that of the ilmenite, with which it is associated, (reflectivity, estimated visually in white light, between the maximum and minimum of the ilmenite) but with a different color gray, but less brownish than ilmenite) (Fig. 1).

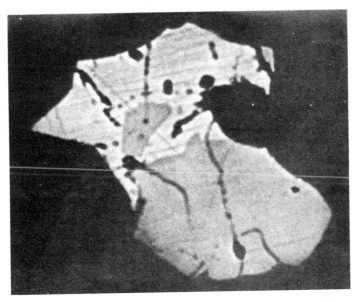

Fig. 1. Mineral described here (dark gray) and ilmenite (light gray). $\times 1200$—oil immersion.
(The striations are polishing scratches.)

Because the association of these properties does not correspond to the mineral ilmenite, which is anisotropic, nor a spinel, which has a lower reflectivity, qualitative microprobe analysis was performed. The results showed that the mineral contained, in addition to elements characteristic of lunar "oxides" (Ti, Fe, Cr), a significant quantity, more than 2%, of zirconium.

CHEMICAL COMPOSITION

Faced with the difficulty of distinguishing the two minerals (ilmenite and the new phase) under the microscope of the electron microprobe, the x-ray scanning images of zirconium, titanium, and iron were obtained in order to define homogeneous zones suitable for quantitative analysis (Fig. 2).

The zirconium image shows that this element is present only in part of the area, and is absent in the other part (ilmenite). The iron image is the inverse replica of that of zirconium; that is, the iron has a much lower concentration in the zirconium-rich mineral than in the associated ilmenite. The outline of the zirconium-rich mineral appears less distinctly on the titanium image, although on careful examination it seems to be richer in this element. The differences in contrast between the three x-ray

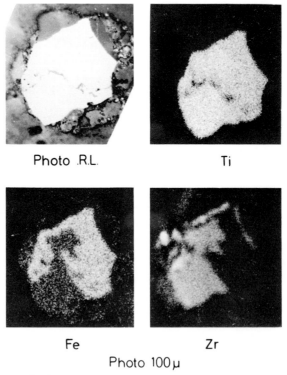

Photo .R.L.   Ti

Fe   Zr

Photo 100 μ

Fig. 2. X-ray scanning images obtained on the electron microprobe. Top left, photomicrograph of the specimen.

images can be explained readily by the ratios of the concentrations in the two minerals (100 for Zr, 4 for Fe, and 1.2 for Ti).

The results of quantitative analysis of the zirconium-rich mineral and of the associated ilmenite are presented in Table 1. Also given, for camparison, are those published by G. Brown *et al.* (1972) for a phase "X" discovered in rock 14310 and by Anderson *et al.* (1970) and Akimoto *et al.* (1970) for armalcolite.

It can be seen that the composition of the zirconium-rich mineral is very close to that of mineral "X", suggesting that it is the same mineral, and that the unique grain discovered in soil 14003 probably originates from rock similar to 14310, in which it should be more common. The similarity of this composition to that of armalcolite also is evident (very rich in TiO_2), although it differs slightly, notably by its richness in ZrO_2.

Calculation of the total number of atoms of metal on the basis of 5 oxygens gives a result close to 3, as for armalcolite, and this is also seen in the following formula:

$$Ti_{1.91}Zr_{0.08}Fe_{0.27}Mg_{0.10}Cr_{0.25}Al_{0.07}Ca_{0.12}Si_{0.02}O_5$$

The fact that the total number of metal atoms (2.82) calculated in this way is less than 3 may be explained by the compound being nonstoichiometric, but may also originate from experimental error that was due to the very small size of the mineral (30 μ) and which might have been diminished by a statistically more significant number of points suitable for analysis.

QUANTITATIVE OPTICAL PROPERTIES

The reflectivity dispersion curve of this isotropic mineral has been obtained using a microreflectometer fitted with a monochromator (Levy, 1967). Three measurements were made, every 20 nm, along the spectrum from 420 to 700 nm. The results are presented in Table 2 and in Fig. 3. Also included, for comparison, are the reflectivity curves, R_0 and R'_e, of the associated ilmenite, R_0 of another lunar ilmenite

Table 1. Composition in weight percent of the new phase.

	1	2	3	4	5
SiO_2	—	0.6	0.15	—	—
TiO_2	71.63	71.2	68.78	74.7	56.5
ZrO_2	—	4.4	6.06	—	0.05
Al_2O_3	2.18	1.7	0.89	—	—
Cr_2O_3	1.38	8.8	4.28	4.5	0.4
FeO	18.01	9.1	13.43	5.7	36.9
MnO	0.05	—	0.24	—	—
MgO	5.52	1.9	1.69	13.9	4.4
CaO	—	3.1	3.13	—	—
$La_2O_3 + RE_2O_3$	—	—	1.31	—	—
	98.77	100.8	99.96	98.8	98.2

1. Armalcolite in Anderson *et al.* (1970).
2. Mineral described here.
3. Phase "X" in Brown *et al.* (1972).
4. Armalcolite (1200°C) in Akimoto *et al.* (1970).
5. Associated ilmenite with mineral described here.

Table 2. Reflectivities of the new titanium and zirconium oxide.

λ, nm	420	440	460	480	500	520	540	
R, %	21.4	21.8	18.7	18.25	17.5	17.1	16.7	
λ, nm	560	580	600	620	640	660	680	700
R, %	16.5	16.35	16.2	16.1	16.1	16.3	16.2	16.2

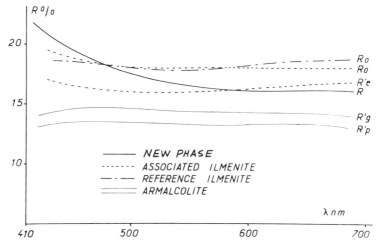

Fig. 3. Dispersion curves of the reflectivities of mineral described here, of armalcolite, and of ilmenite.

(but not R'_e in order to avoid overburdening the figure) (section 21384), and R'_g and R'_p of an armalcolite (section 21428) all found in the same soil sample 14003,47.

It can be seen that the curve of the zirconium-rich mineral expresses well the relationship observed in white light between its optical properties and those of the associated ilmenite. The height of the curve in the short to medium wavelengths accounts for the slightly less brownish but more bluish color than that of the associated ilmenite, but the reflectivity values are intermediate or close to those of R_o and R'_e of the ilmenite.

This curve, in its form and reflectivity values, is even more different from that of armalcolite, which is flatter and with lower reflectivity values (the values obtained for armalcolite agree almost exactly with those obtained by Cameron, 1970).

The curves of the associated ilmenite are quite normal. The curve of the ordinary ray reflectivities is almost identical to that of the reference lunar ilmenite.

DISCUSSION

It would have been possible to make an x-ray crystallographic study of this mineral, despite its very small size, provided that it was separated from the associated ilmenite (no collimator exists which could provide an x-ray beam of this dimension). This hazardous manipulation was not attempted for fear of loosing this unique grain.

Nevertheless, an attempt was made to obtain a Laue diagram by placing appropriate diaphrams over the polished section but, unfortunately, the results were negative. Without x-ray data the symmetry of this mineral cannot be established with certainty.

Its optical isotropy suggests that it probably is a cubic mineral, although the possibility that the crystal section is perpendicular to an optic axis of an anisotropic mineral, cannot be ruled out. Brown *et al.* (1972) did not give any indication of the optical properties of phase "X" which they analyzed. By means of a systematic examination of all the occurrences of this mineral in the polished thin sections of rock 14310 it should be possible to determine whether the optical isotropy of our grain is fortuitous or whether, alternatively, it is characteristic of this mineral.

The chemical composition of this mineral does not approach that of any described previously, apart from armalcolite. It is very different from that of tranquillityite (Lovering *et al.*, 1971) and other zirconium-rich silicates described in the same work, from phase "A" of Lovering and Wark (1971), and from terrestrial zirkelite, notably because of its richness in TiO_2.

It cannot be concluded from the analogy in composition between this mineral described and armalcolite that it is a zirconium-rich variety of this latter mineral. The difference between its quantitative optical properties and those of armalcolite is too important to be explained by the slight difference in chemical composition. From the singularity of these optical properties it is concluded that the mineral described here is a *new mineral* (Christophe *et al.*, 1972) left *unnamed* in the absence of crystallographic data.

If this mineral is considered to be a dimorph of armalcolite, its high content of chromium could be explained by the results of experimental work, notably by Akimoto *et al.* (1970), who showed that the high chromium content of synthetic armalcolite is a result of its high temperature of formation. The chromium content of the new mineral is greater than that of synthetic armalcolite (see Table 1) in which the content diminishes with the temperature of formation, and in which the content is greater than that of natural armalcolites.

If this were the case, it would be difficult to explain why the associated ilmenite contains so little chromium. Indeed, the partition coefficients for chromium in armalcolite and ilmenite (in the presence of 0.1% vanadium and at atmospheric pressure, however) are nearly identical (7 and 6) (N. Ware, in Ringwood and Essene, 1970).

An extremely rapid cooling (quenching) would explain why the unstable new mineral retains some elements normally present in differentiated minerals. It would explain especially the high content of zirconium of this mineral which, supposedly, in the conditions of "normal" cooling, would evolve towards an armalcolite poorer in Zr (0.1%, after Arrhenius *et al.*, 1970) associated with baddeleyite or some other zirconium-rich oxides. A close examination of the x-ray scanning image of Zr (Fig. 2) shows, in fact, some point segregations richer in this element.

Note Added in Proof. We recently received the section 14310–20 and attempted optical measurements on phase "X" discovered by Brown *et al.* (1972) (using the descriptions kindly supplied by these authors). The reflectivities obtained were identical to those of the mineral described here at wavelengths greater than 540 nm, but lower in the short wavelengths (curve of reflectivities flatter). An explanation of this slight discrepancy cannot be given with certainty, because the results on phase "X" are uncertain due to its very small dimensions. However, the conclusion reached on this new phase is unchanged, the differences between the optical properties (values of reflectivities and isotropy) of both specimens and those of armalcolite being too important to be explained by the difference in chemical composition.

Acknowledgments—We are indebted to R. Chevalier for the x-ray study, R. Giraud and M. Josephe for the microprobe analysis, M. Pasdeloup for the optical measurements, and A. Hall for the translation.

References

Akimoto S. Nakamura Y. Nishikawa M. Katsura T. and Kushiro I. (1970) Melting experiments of lunar crystalline rocks. *Proc. Apollo 11 Lunar Sci. Conf., Geochim. Cosmochim. Acta* Suppl. 1, Vol. 1, pp. 129–133. Pergamon.

Anderson A. T. Boyd F. R. Bunch T. E. Cameron E. N. El Goresy A. Finger L. W. Haggerty S. E. James O. B. Keil K. Prinz M. and Ramdohr P. (1970) Armalcolite: A new mineral from the Apollo 11 samples. *Proc. Apollo 11 Lunar Sci. Conf., Geochim. Cosmochim. Acta* Suppl. 1, Vol. 1, pp. 55–63. Pergamon.

Arrhenius G. Everson J. E. Fitzgerald R. W. and Fusita H. (1971) Zirconium fractionation in Apollo 11 and Apollo 12 rocks. *Proc. Second Lunar Sci. Conf., Geochim. Cosmochim. Acta* Suppl. 2, Vol. 1, pp. 169–176. MIT Press.

Brown G. M. Emeleus C. H. Holland J. G. Peckett A. and Phillips R. (1972) Mineralogical fractionation patterns between Apollo 14 primitive feldspathic rocks and Apollo 15 and other basalts (abstract). In *Lunar Science—III* (editor C. Watkins), pp. 95–97, Lunar Science Institute Contr. No. 88.

Cameron E. N. (1970) Opaque minerals in certain lunar rocks from Apollo 11. *Proc. Apollo 11 Lunar Sci. Conf., Geochim. Cosmochim. Acta* Suppl. 1, Vol. 1, pp. 221–245. Pergamon.

Christophe-Michel-Levy M. Levy C. and Pierrot R. (1972) Mineralogical aspects of Apollo 14 samples: Lunar chondrules, pink spinel bearing rocks, ilmenites (abstract). In *Lunar Science—III* (editor C. Watkins), pp. 136–138, Lunar Science Institute Contr. No. 88.

Levy C. (1967) Contribution à la minéralogie des sulfures complexes de cuivre du type $Cu_3 X S_4$. *Mem. Bur. Rech. Geol. Min.*, No. 54, 178 pp.

Lovering J. F. and Wark D. A. (1971) Uranium-enriched phases in Apollo 11 and Apollo 12 basaltic rocks. *Proc. Second Lunar Sci. Conf., Geochim. Cosmochim. Acta* Suppl. 2, Vol. 1, pp. 151–158. MIT Press.

Lovering J. F. Wark D. A. Reid A. F. Ware N. G. Keil K. Prinz M. Bunch T. E. El Goresy A. Ramdohr P. Brown G. M. Peckett A. Phillips R. Cameron E. N. Douglas J. A. V. and Plant A. G. (1971) Tranquillityite: A new silicate mineral from Apollo 11 and Apollo 12 basaltic rocks. *Proc. Second Lunar Sci. Conf., Geochim. Cosmochim. Acta* Suppl. 2, Vol. 1, pp. 39–45. MIT Press.

Ringwood A. E. and Essene E. (1970) Petrogenesis of Apollo 11 basalts, internal constitution and origin of the moon. *Proc. Apollo 11 Lunar Sci. Conf., Geochim . Cosmochim. Acta* Suppl. 1, Vol. 1, pp. 769–799. Pergamon.

Proceedings of the Third Lunar Science Conference
(Supplement 3, *Geochimica et Cosmochimica Acta*)
Vol. 1, pp. 1121–1132
The M.I.T. Press, 1972

Electron microscopy of some experimentally shocked counterparts of lunar minerals

CHARLES B. SCLAR AND STEPHEN P. MORZENTI

Department of Geological Sciences, Lehigh University,
Bethlehem, Pennsylvania 18015

Abstract—Particulate samples of experimentally shocked olivine, enstatite, and apatite and their un-shocked equivalents were thinned by ion bombardment and examined by transmission methods in a conventional (100 kV) electron microscope. Enstatite shock-loaded at a peak pressure of 250 kb was partly transformed to relatively dense inverse defect spinel. Magnesian olivine shocked at a peak pressure of 200 kb shows intense lattice deformation. Magnesian olivine shock-loaded at a peak pressure of 400 kb was recrystallized to very fine polycrystalline aggregates. Apatite shock-loaded at a peak pressure of 250 kb shows planar deformation elements and shock-induced imperfections.

These results indicate that this experimental approach has considerable potential as a means of developing shock-recognition criteria for minerals of lunar rocks.

INTRODUCTION

THE LUNAR BRECCIAS and fines returned from successive Apollo missions have shown abundant evidence of shock metamorphism (Apollo 11 samples: Chao *et al.*, 1970a; Chao *et al.*, 1970b; Dence *et al.*, 1970; Engelhardt *et al.*, 1970; McKay *et al.*, 1970; Quaide and Bunch, 1970; Sclar, 1970; Short, 1970; Apollo 12 samples: Carter *et al.*, 1971; Chao *et al.*, 1971; Engelhardt *et al.*, 1971; McKay *et al.*, 1971; Quaide *et al.*, 1971; Sclar, 1971; Apollo 14 and 15 samples: LSPET, 1971; LSPET, 1972; Ave'Lallemant and Carter, 1972; Chao *et al.*, 1972a; Chao *et al.*, 1972b; Dence *et al.*, 1972; Engelhardt *et al.*, 1972; Quaide, 1972; Wilshire and Jackson, 1972). Much of this evidence consists of microtextural relationships and microstructural features obtained by thin-section petrography supplemented by chemical data obtained by electron microprobe analysis. In addition, a single-crystal x-ray diffraction study of shocked lunar ilmenite was carried out by Minkin and Chao (1971) and a study of the fine structure of the matrix of two lunar breccias using the electron microscope was reported by Waters *et al.* (1971).

The advantages of transmission electron microscopy of thin foils of minerals and rocks as a means of revealing shock-induced structures and imperfections are self-evident, but the application of this method has been difficult because of the possibility of introducing damage into specimens prepared mechanically. In two pioneering papers, Radcliffe *et al.* (1970) and Christie *et al.* (1971) showed that satisfactory electron-transparent specimens for high-voltage (800 kV) microscopy could be prepared from lunar and terrestrial minerals and rocks by the ion-bombardment thinning technique. Since then, transmission electron microscopy studies of minerals using ion-thinned samples have been reported by Lally *et al.* (1972), Sclar and Morzenti (1972), Smith *et al.* (1972), and Wenk *et al.* (1972).

As the first step toward the development of microscopical criteria for recognition of shock effects in lunar minerals, we have prepared ion-thinned samples of terrestrial and meteoritic minerals and their experimentally shocked equivalents for transmission electron microscopy. The thinned samples were examined in a conventional (100 kV) electron microscope in bright and dark field and by selected-area electron diffraction for shock-induced features such as dislocation arrays, transformation textures, and high-pressure phases.

EXPERIMENTAL METHODS

Shock experiments

Many impacts on the lunar surface have occurred in incoherent or weakly coherent porous material. In view of this, a particulate sample ($-48 +150$ mesh) of each mineral was tamped into a cylindrical well ($\frac{1}{4}$-inch in diameter; $\frac{1}{4}$-inch high) in a cylindrical metal sample holder which, in turn, was enclosed in a massive metal block supported by metal spall plates. The initial sample porosity was high— probably of the order of 25%. The samples were shock loaded using chemical explosives in conjunction with a plane-wave generator arranged so that the shock waves were propagated axially through the cylindrical samples (see Sclar, 1968). High temperatures in excess of 1200°C may occur in porous particulate samples shock loaded under these experimental conditions (Sclar, 1969).

Preparation of electron-transparent foils

Samples were prepared by ion-bombardment thinning using two argon plasma guns. The apparatus employed is a commercial unit known as the Ion Micro Milling Instrument manufactured by Commonwealth Scientific Corporation, Alexandria, Virginia. This apparatus and its operating characteristics have been described in detail by Heuer *et al.* (1971); additional experimental detail is given by Radcliffe *et al.* (1970).

RESULTS OF TRANSMISSION ELECTRON MICROSCOPY

Enstatite

The enstatite (En_{98+} based on $\gamma = 1.658 \pm 0.002$) from the Mount Egerton enstatite achondrite is an orthoenstatite that is partly disordered in the sense of Brown and Smith (1963) as shown by x-ray powder diffraction data using the criteria of Pollack and Ruble (1964). This disordering consists of parallel lamellae of clinoenstatite arranged in a nonperiodic sequence along the *a* axis of the orthorhombic host. After explosive shock loading at a peak pressure of about 250 kb the degree of disordering is much greater as shown by an increase in the ratio of the intensities of the (610) reflection and the (420), (221) doublet from 1.0 to 1.4 and the appearance of a reflection of medium intensity at 2.97 Å. In effect, the x-ray powder diffraction data from the shocked enstatite may be interpreted as a mixture that is dominantly orthoenstatite and subordinately low clinoenstatite. Petrographically, the shocked enstatite shows diffuse inclined extinction positions; this may be due to optically irresolvable clinoenstatite lamellae, in accord with the x-ray data.

In the electron microscope the unshocked enstatite shows a well-developed lamellar structure (Figs. 1a and 1b); the individual lamellae are less than 150 Å thick.

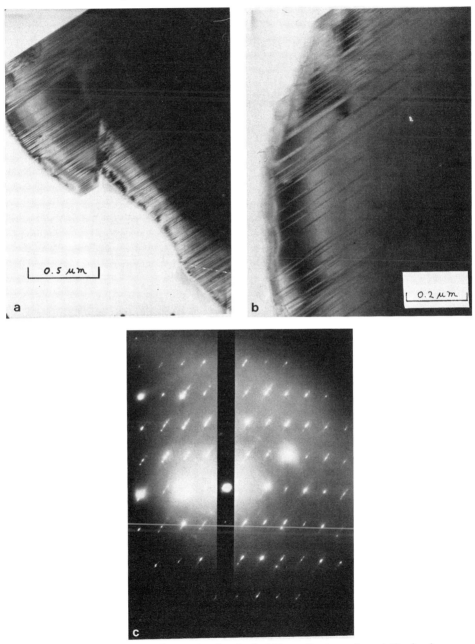

Fig. 1. (a and b) Electron micrographs of unshocked Mount Egerton orthoenstatite showing lamellar structure interpreted as disordering. (c) Selected-area electron diffraction pattern of unshocked Mount Egerton enstatite showing splitting of spots due to disordering.

Selected-area electron diffraction patterns show distinct splitting of diffraction spots (Fig. 1c) indicating the presence of both orthoenstatite and clinoenstatite. The lamellar structure is interpreted as the manifestation of the disordering indicated by x-ray diffraction.

Dispersed through the shocked enstatite, which no longer shows the simple lamellar structure, are minute crystallites that range in size from 10 Å to 5000 Å (Figs. 2a, b, and c). They are relatively opaque to the electron beam and may be thicker than the matrix due to differential sputtering and/or have a greater density than the matrix. Many of the dispersed crystallites are euhedral and show rectangular and six-sided outlines suggesting cubo-octahedral forms. In addition, the general morphology of the crystallites indicates that they constitute a single phase. Selected-area electron diffraction patterns show the single-crystal pattern of the enstatite matrix upon which is superimposed a polycrystalline ring pattern of the dispersed phase (Fig. 2d). The spots of the matrix pattern are highly streaked indicating intense damage to the lattice. The d values obtained from the ring pattern (Table 1) may be indexed on a cubic unit cell with $a_0 = 8.03 \pm 0.02$ Å and the indexed sequence of reflections and their relative intensities are compatible with the spinel structure. There is, however, one indexed reflection (332) that is not permitted in the FCC structure of spinel. This indicates that the lattice is primitive cubic and that the spinel has ordered cations and/or vacancies.

We have interpreted the dispersed phase in shocked enstatite as a shock-induced high-pressure polymorph of $MgSiO_3$ with the spinel structure. Starting with this concept, stoichiometric and structural considerations lead to the following formula:

$$[Si_1^{IV}(Mg_{4/3}Si_{1/3} \square_{1/3})^{VI}O_4].$$

This is an inverse cation-deficient defect spinel analogous to $\gamma - Fe_2O_3$ and $\gamma - Al_2O_3$. The calculated density is 3.43 gm/cc, which is 6.9% denser than low clino-enstatite.

Various ideas regarding the high-pressure stability of enstatite have been proposed on the basis of thermodynamic considerations, static high-pressure experimentation, and shock-compression studies. These include (1) enstatite disproportionates to forsterite plus stishovite (Ringwood and Seabrook, 1963; Sclar et al., 1964), (2)

Table 1. Powder electron diffraction data for the dispersed phase in shocked enstatite.

dÅ	I/I_0	hkl	
2.87	5	220	Indexed on a cubic
2.44	25	311	unit cell with
2.00	100	400	$a_0 = 8.03 \pm 0.02$ Å
1.72	70	332	
1.40	20	522,441	
1.23	50	533	
1.04	50	731,553	
1.00	20	800	
0.80	30	10,0,0;860	
0.77	30	10,2,0;862	
0.71	25	880	
0.67	20	12,0,0;884	

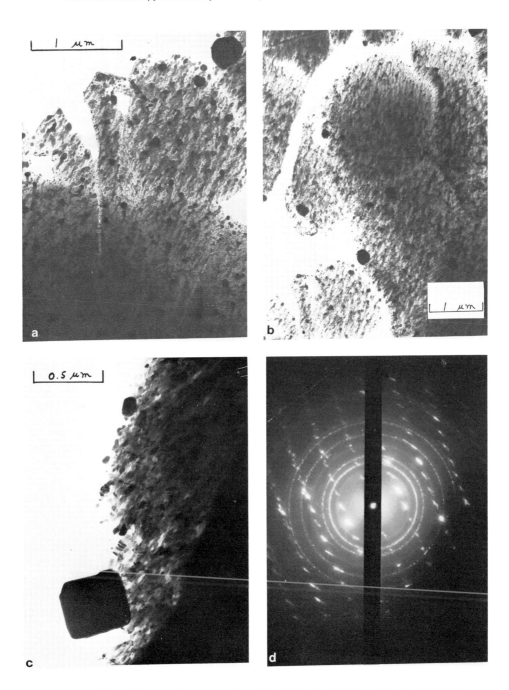

Fig. 2. (a, b, c) Electron micrographs of experimentally shocked Mount Egerton enstatite showing dispersed phase of relatively low electron transparency. (d) Selected-area electron diffraction pattern of field in 2b. Note streaked spots attributed to enstatite matrix and ring pattern attributed to dispersed phase.

enstatite disproportionation to $\beta - Mg_2SiO_4$ plus stishovite; bronzite disproportionates to $\gamma - (Mg, Fe)_2SiO_4$ (ringwoodite) plus stishovite (Ringwood and Major,
1970), (3) enstatite transforms to $MgSiO_3$ with corundum (ilmenite) structure
(Ringwood and Seabrook, 1962; Davies and Anderson, 1971), (4) enstatite transforms to $[Mg_3(Mg, Si)(SiO_4)_3]$ with the garnet structure in which $\frac{1}{4}$ of the silicon
atoms are in 6-fold coordination (Ringwood, 1972), and (5) enstatite transforms to
the perovskite structure (Ahrens et al., 1969; Gaffney and Ahrens, 1970). The results
reported in this paper indicate that, at least under dynamic conditions, enstatite
may transform to a defect spinel.

Ahrens and Gaffney (1971) recently have reported that Bamle bronzite (En_{86})
under dynamic compression starts to transform to a dense high-pressure phase at
about 135 kb and that the transformation is essentially complete above about
350 kb. They concluded that the high-pressure polymorph of bronzite is garnet with
a calculated zero-pressure density of 3.67 gm/cc based on the unit-cell edge of a
garnet (majorite) with the stoichiometry of pyroxene, which was discovered in a
shocked meteorite (Smith and Mason, 1970). They also report that the data from
their shock-compression experiments yield a zero-pressure density for the high-
pressure phase of 3.57 gm/cc or, if other constants and data sets are used, 3.80 gm/cc.
These calculated densities are too low for bronzite with either the ilmenite structure
or the perovskite structure, and too low for mixtures of either $\beta - Mg_2SiO_4$ and
stishovite or ringwoodite (spinel) and stishovite. The corresponding inverse defect
spinel with bronzite (En_{86}) composition, however, would have a density of 3.57
gm/cc in excellent agreement with a calculated zero-pressure density of the shock-
induced form of Bamle bronzite. We conclude that the defect spinel structure must
be considered as a possible high-pressure polymorph of enstatite and bronzite at
least under shock-metamorphic conditions. Conversely, the occurrence of such
a phase in shocked rocks would indicate shock pressures in excess of 135 kb.

Olivine

Thin foils of transparent yellow-brown, single-crystal fragments of olivine
(Fa_{11}) from the Admire pallasite essentially are featureless in the electron microscope
except for Bragg extinction contours which outline the edge (Fig. 3a). The high
perfection of this structure also is shown by Kikuchi lines (Fig. 3b). A thin foil of a
particulate sample of the same olivine shock-loaded at a peak pressure of about
200 kilobars, however, showed extensive lattice damage as revealed by regions of
very high dislocation density. The electron diffraction spots of such areas appear as
clusters of spots that are streaked and distorted (Fig. 4).

A thin foil of a transparent green olivine (Fa_{17}) from Hawaii (USNM # 44770)
was featureless in the electron microscope. A thin foil of a particulate sample of the
same olivine shocked at a peak pressure of about 400 kb, however, showed that much
of the olivine had been recrystallized to fine polycrystalline aggregates with an
average grain size of about 1000 Å (Fig. 5). This recrystallization probably is the
result of the intense thermal pulse which accompanies a shock event in a particulate
medium (Sclar, 1969). Similar recrystallization of olivine in dunite experimentally

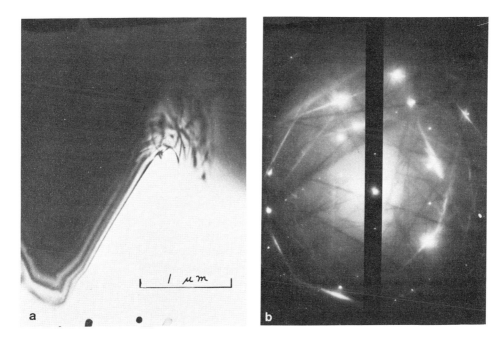

Fig. 3. (a) Electron micrograph of unshocked olivine from the Admire pallasite showing Bragg extinction contours. (b) Kikuchi lines in unshocked olivine from the Admire pallasite.

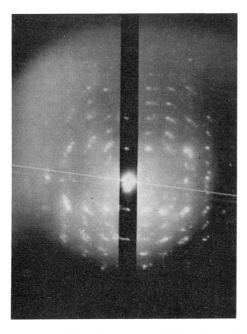

Fig. 4. Selected-area electron diffraction pattern of olivine from the Admire pallasite experimentally shocked at 200 kb. Note clustering and streaking of spots.

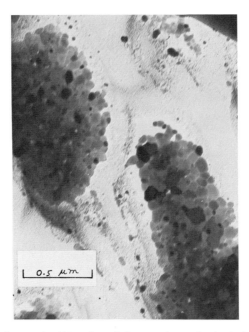

Fig. 5. Electron micrograph of Hawaiian olivine experimentally shocked at 400 kb showing
recrystallization texture.

shocked in the 1 Mb range was observed by Carter *et al.* (1968) with the optical microscope, and incipient recrystallization of a grain of shocked lunar olivine in breccia 10061,39 has been observed petrographically (Sclar, 1970). Recrystallization textures in olivine resolved by either light or electron microscopy may prove to be a very useful indicator of shock pressures in excess of several hundred kilobars.

Apatite

A gem-like transparent yellow-green crystal of fluorapatite from Durango, Mexico was thinned by ion bombardment for electron microscopy. The thin foil was featureless and the spots of the electron diffraction pattern were sharp and undistorted. A particulate sample of this apatite shock-loaded at a peak pressure of about 250 kb was also thinned and examined in the electron microscope. The shocked sample shows several kinds of shock-induced damage. An earlier petrographic study of this shocked apatite sample had revealed the presence of shock-induced sets of planar elements that subdivide each apatite crystal into a mosaic aggregate of blocks a few microns on an edge (Sclar and Usselman, 1970; Sclar, 1971). In the electron microscope, interplanar distances as small as 1000 Å can be observed (Fig. 6a). The structure of the apatite has been further damaged by the introduction of large helical dislocations (Fig. 6b) that were not observed in the unshocked crystals. Finally, as a result of shock-induced strain, the electron diffraction spots of the shocked apatite are streaked and distorted and the spots have a tendency to be arranged in crude ring patterns because of the mosaic block structure (Fig. 6c).

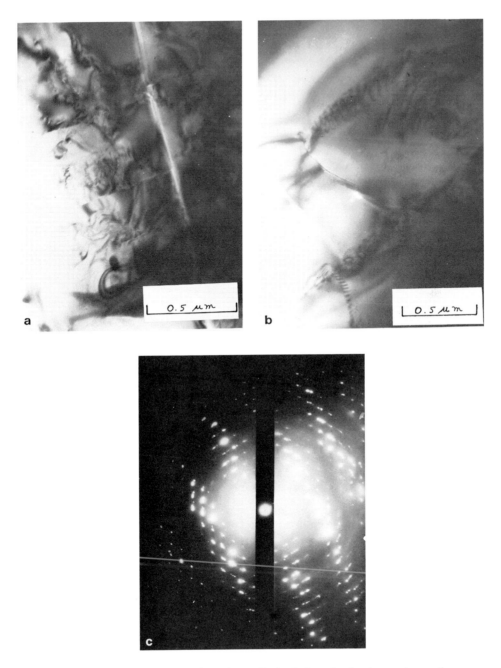

Fig. 6. (a) Electron micrograph of experimentally shocked apatite showing two intersecting sets of planar elements. Note offset of one set (NE–SW) by the other. (b) Electron micrograph of experimentally shocked apatite showing several large helical dislocations. (c) Selected-area electron diffraction pattern of shocked apatite showing ring-like arrangement of distorted spots attributed to shock-induced block structure.

CONCLUSIONS

Electron-transparent foils of minerals suitable for conventional (100 kV) transmission electron microscopy may be prepared readily by ion-bombardment thinning. Comparative electron microscopy of experimentally shocked samples and their unshocked equivalents has revealed shock-induced phenomena such as nucleation of high-pressure phases, lattice deformation, and dislocation arrays that may become useful criteria for recognition of shock effects in lunar minerals. The task now is to apply these results directly to the lunar samples.

This study has indicated that enstatite under peak shock pressures of about 250 kb may partly transform to an inverse defect spinel which involves a density increase of about 7%. Shocked magnesian olivine shows evidence of intense structural damage at about 200 kb peak pressure and a distinctive recrystallization texture at about 400 kb peak pressure. Apatite shows shock-induced lattice imperfections including planar features and large dislocations.

Acknowledgments—This study was supported by NASA Grant NGR 39-007-064. Mr. R. Korastinsky of the Materials Research Center, Lehigh University assisted the authors with the operation of the electron microscope. Mr. D. Laber of the Columbus Laboratories, Battelle Memorial Institute, carried out the shock experiments. Dr. Brian Mason, Smithsonian Institution, kindly provided the meteoritic minerals.

REFERENCES

Ahrens T. J. Anderson D. L. and Ringwood A. E. (1969) Equation of state and crystal structure of high pressure phases of shocked silicates and oxides. *Rev. Geophys.* **7**, 667–707.

Ahrens T. J. and Gaffney E. S. (1971) Dynamic compression of enstatite. *J. Geophys. Res.* **76**, 5504–5513.

Ave'Lallemant H. G. and Carter N. L. (1972) Deformation of silicates in some Fra Mauro breccias (abstract). In *Lunar Science—III* (editor C. Watkins), pp. 33–34, Lunar Science Institute Contr. No. 88.

Brown W. L. and Smith J. V. (1963) High-temperature x-ray studies on the polymorphism of $MgSiO_3$. *Z. Kristallogr.* **118**, 186–212.

Carter N. L. Raleigh C. B. and De Carli P. S. (1968) Deformation of olivine in stony meteorites. *J. Geophys. Res.* **73**, 5439–5461.

Carter N. L. Fernandez L. A. Ave'Lallemant H. G. and Leung I. S. (1971) Pyroxenes and olivines in crystalline rocks from the Ocean of Storms. *Proc. Second Lunar Sci. Conf., Geochim. Cosmochim. Acta*, Suppl. 2, Vol. 1, pp. 775–795. M.I.T. Press.

Chao E. C. T. James O. B. Minkin J. A. and Boreman J. A. (1970a) Petrology of unshocked crystalline rocks and evidence of impact metamorphism in Apollo 11 returned lunar sample. *Proc. Apollo 11 Lunar Sci. Conf., Geochim. Cosmochim. Acta*, Suppl. 1, Vol. 1, pp. 287–314. Pergamon.

Chao E. C. T. Boreman J. A. Minkin J. A. James O. B. and Desborough G. A. (1970b) Lunar glasses of impact origin: Physical and chemical characteristics and geologic implications. *J. Geophys. Res.* **75**, 7445–7479.

Chao E. C. T. Boreman J. A. and Desborough G. A. (1971) The petrology of unshocked and shocked Apollo 11 and Apollo 12 microbreccias. *Proc. Second Lunar Sci. Conf., Geochim. Cosmochim. Acta*, Suppl. 2, Vol. 1, pp. 797–816. M.I.T. Press.

Chao E. C. T. Minkin J. A. and Boreman J. A. (1972a) The petrology of some Apollo 14 breccias (abstract). In *Lunar Science—III* (editor C. Watkins), pp. 131–132, Lunar Science Institute Contr. No. 88.

Chao E. C. T. Boreman J. A. and Minkin J. A. (1972b). Apollo 14 glasses of impact origin (abstract). In *Lunar Science—III* (editor C. Watkins), pp. 133–134, Lunar Science Institute Contr. No. 88.

Christie J. M. Lally J. S. Heuer A. H. Fisher R. M. Griggs D. T. and Radcliffe S. V. (1971) Comparative electron petrography of Apollo 11, Apollo 12, and terrestrial rocks. *Proc. Second Lunar Sci. Conf., Geochim. Cosmochim. Acta*, Suppl. 2, Vol. 1, pp. 69–89. M.I.T. Press.

Davies G. F. and Anderson D. L. (1971) Revised shock-wave equations of state for high-pressure phases of rocks and minerals. *J. Geophys. Res.* **76**, 2617–2627.

Dence M. R. Douglas J. A. V. Plant A. G. and Traill R. J. (1970) Petrology, mineralogy, and deformation of Apollo 11 samples. *Proc. Apollo 11 Lunar Sci. Conf., Geochim. Cosmochim. Acta*, Suppl. 1, Vol. 1, pp. 315–340. Pergamon.

Dence M. R. Plant A. G. and Traill R. J. (1972) Impact-generated shock and thermal metamorphism in Fra Mauro lunar samples (abstract). In *Lunar Science—III* (editor C. Watkins), pp. 174–176, Lunar Science Institute Contr. No. 88.

Engelhardt W. von Arndt J. Müller W. F. and Stöffler D. (1970) Shock metamorphism of lunar rocks and origin of the regolith at the Apollo 11 landing site. *Proc. Apollo 11 Lunar Sci. Conf., Geochim. Cosmochim. Acta*, Suppl. 1, Vol. 1, pp. 363–384. Pergamon.

Engelhardt W. von Arndt J. Müller W. F. and Stöffler D. (1971) Shock metamorphism and origin of regolith and breccias at the Apollo 11 and Apollo 12 landing sites. *Proc. Second Lunar Sci. Conf., Geochim. Cosmochim. Acta*, Suppl. 2, Vol. 1, pp. 833–854. M.I.T. Press.

Engelhardt W. von Arndt J. and Stöffler D. (1972) Apollo 14 soils and breccias, their compositions and origin by impacts (abstract). In *Lunar Science—III* (editor C. Watkins), pp. 233–235, Lunar Science Institute Contr. No. 88.

Gaffney E. S. and Ahrens T. J. (1970) Stability of mantle minerals from lattice calculations and shock-wave data. *Phys. Earth Planet. Interiors* **3**, 205–212.

Heuer A. H. Firestone R. F. Snow J. D. Green H. W. Howe R. G. and Christie J. M. (1971) An improved ion thinning apparatus. *Rev. Sci. Instr.* **42**, 1177–1184.

Lally J. S. Fisher R. M. Christie J. M. Griggs D. T. Heuer A. H. Nord G. L. and Radcliffe S. V. (1972) Electron petrography of Apollo 14 and 15 samples (abstract). In *Lunar Science—III* (editor C. Watkins), pp. 469–471, Lunar Science Institute Contr. No. 88.

LSPET (1971) (Lunar Sample Preliminary Examination Team) Preliminary examination of lunar samples from Apollo 14. *Science* **173**, 681–693.

LSPET (1972) (Lunar Sample Preliminary Examination Team) The Apollo 15 lunar samples: A preliminary description. *Science* **175**, 363–375.

McKay D. S. Greenwood W. R. and Morrison D. A. (1970) Origin of small lunar particles and breccia from the Apollo 11 site. *Proc. Apollo 11 Lunar Sci. Conf., Geochim. Cosmochim. Acta*, Suppl. 1, Vol. 1, pp. 673–694. Pergamon.

McKay D. S. Morrison D. A. Clanton U. S. Ladle G. H. and Lindsay J. F. (1971) Apollo 12 soil and breccia. *Proc. Second Lunar Sci. Conf., Geochim. Cosmochim. Acta*, Suppl. 2, Vol. 1, pp. 755–773. M.I.T. Press.

Minkin J. A. and Chao E. C. T. (1971) Single crystal x-ray investigation of deformation in terrestrial and lunar ilmenite. *Proc. Second Lunar Sci. Conf., Geochim. Cosmochim. Acta*, Suppl. 2, Vol. 1, pp. 237–246. M.I.T. Press.

Pollack S. S. and Ruble W. D. (1964) X-ray identification of ordered and disordered ortho-enstatite. *Amer. Mineral.* **49**, 983–992.

Quaide W. (1972) Mineralogy and origin of Fra Mauro fines and breccias (abstract). In *Lunar Science—III* (editor C. Watkins), pp. 627–629, Lunar Science Institute Contr. No. 88.

Quaide W. and Bunch T. (1970) Impact metamorphism of lunar surface materials. *Proc. Apollo 11 Lunar Sci. Conf., Geochim. Cosmochim. Acta*, Suppl. 1, Vol. 1, pp. 711–729. Pergamon.

Quaide W. Overbeck V. Bunch T. and Polkowski G. (1971) Investigations of the natural history of the regolith at the Apollo 12 site. *Proc. Second Lunar Sci. Conf., Geochim. Cosmochim. Acta*, Suppl. 2, Vol. 1, pp. 701–718. M.I.T. Press.

Radcliffe S. V. Heuer A. H. Fisher R. M. Christie J. M. and Griggs D. T. (1970) High voltage (800 kV) electron petrography of type B rock from Apollo 11. *Proc. Apollo 11 Lunar Sci. Conf., Geochim. Cosmochim. Acta*, Suppl. 1, Vol. 1, pp. 731–748. Pergamon.

Ringwood A. E. (1972) Mineralogy of the deep mantle: current status and future developments. In *The Nature of the Solid Earth* (editor E. C. Robertson), pp. 67–92, McGraw-Hill.

Ringwood A. E. and Seabrook M. (1962) High-pressure transition of $MgGeO_3$ from pyroxene to corundum structure. *J. Geophys. Res.* **67**, 1690–1691.

Ringwood A. E. and Seabrook M. (1963) High-pressure phase transformations in germanate pyroxenes and related compounds. *J. Geophys. Res.* **68**, 4601–4609.

Sclar C. B. (1968) Shock-wave damage in olivine as revealed by light and electron microscopy. Final Report, NASA Contract NSR 36-002-062.

Sclar C. B. (1969) Shock-wave damage in olivine (abstract). *Trans. Amer. Geophys. Union* **50**, 219.

Sclar C. B. (1970) Shock metamorphism of lunar rocks and fines from Tranquillity Base. *Proc. Apollo 11 Lunar Sci. Conf.*, *Geochim. Cosmochim. Acta*, Suppl. 1, Vol. 1, pp. 849–864. Pergamon.

Sclar C. B. (1971) Shock-induced features of Apollo 12 microbreccias. *Proc. Second Lunar Sci. Conf.*, *Geochim. Cosmochim. Acta*, Suppl. 2, Vol. 1, pp. 817–832. M.I.T. Press.

Sclar C. B. Carrison L. C. and Schwartz C. M. (1964) High-pressure reaction of clinoenstatite to forsterite plus stishovite. *J. Geophys. Res.* **69**, 325–330.

Sclar C. B. and Usselman T. M. (1970) Experimentally induced shock effects in some rock-forming minerals (abstract). *Meteoritics* **5**, No. 4, 222–223.

Sclar C. B. and Morzenti S. P. (1972) Electron microscopy of experimentally shocked rock-forming minerals with application to lunar samples (abstract). In *Lunar Science—III* (editor C. Watkins), pp. 688–689, Lunar Science Institute Contr. No. 88.

Short N. M. (1970) Evidence and implications of shock metamorphism in lunar samples. *Science* **167**, 673–675.

Smith D. K. Thrower P. A. and Hoffman W. P. (1972) Electron microscope study of Apollo sample 14310 (abstract). In *Lunar Science—III* (editor C. Watkins), pp. 712–713, Lunar Science Institute Contr. No. 88.

Waters A. C. Fisher R. V. Garrison R. E. and Wax D. (1971) Matrix characteristics and origin of lunar breccia samples 12034 and 12073. *Proc. Second Lunar Sci. Conf.*, *Geochim. Cosmochim. Acta*, Suppl. 2, Vol. 1, pp. 893–907. M.I.T. Press.

Wenk H. R. Ulbrich M. and Müller W. (1972) Lunar plagioclase (a mineralogical study) (abstract). In *Lunar Science—III* (editor C. Watkins), pp. 797–799, Lunar Science Institute Contr. No. 88.

Wilshire H. G. and Jackson E. D. (1972) Petrology of the Fra Mauro formation at the Apollo 14 landing site (abstract). In *Lunar Science—III* (editor C. Watkins), pp. 803–805, Lunar Science Institute Contr. No. 88.

Lunar Sample Inventory for
Apollo 11, 12, 14, and 15

Following is a complete inventory of lunar samples as returned from the first four manned lunar landings. The weight of each sample is the best known value; previous lists, as in sample catalogues, are not as accurate. The indicated "Sample Type" serves only as a *general* guide. Location and comments are *generally* stated. More details on the specific location of each sample may be obtained from various publications, especially the Sample Catalogues. A glossary of terms for this inventory follows:

GLOSSARY

ALSEP: Apollo Lunar Surface Experiment Package: a location on the lunar surface specific for each mission.

ALSRC: Aluminum, vacuum sealable (indium seal) container used to transport rocks from the Lunar surface to the LRL.

Basalt: Used here to signify a crystalline rock with an igneous texture.

Bio Prep: A laboratory in the LRL used for processing rocks in a nitrogen atmosphere.

Bio Prime: Approx. 100 gm of lunar material used for biological protocol in the LRL.

Bio Pool: Approx. 500 gm of lunar material used for biological protocol in the LRL.

Bulk sample: Lunar material in the first ALSRC collected during Apollo 11.

Chip: A lunar sample less than 50 gm and greater than 1 cm.

Contingency sample: Sample collected during Apollo 11, 12, and 14 near LM.

Comprehensive sample: Approx. 50 chips and fines collected from a small area during EVA–1 on Apollo 14.

Core tube: Old name for a drive tube.

DB: Documented Bag: A small numbered teflon bag that lunar rocks are put into on the lunar surface. DBs are then put into SCBs or ALSRCs for transport to the LRC.

Documented sample: Lunar material in the second ALSRC collected during Apollo 11.

Drill stem: Tube of returned lunar regolith collected using a rotary drill; *see* Drive stem.

Drive stem: Tube of returned lunar regolith collected by hammering a hollow tube into the ground; *see* Drill stem.

EVA: Extravehicular activity: Lunar traverse.

F–201: Vacuum chamber in the LRL used for processing samples during Apollo 11 and 12.

Fines: Less than 1 cm lunar material.

GASC: Gas Analysis Sample Container: one of several special sample containers that were filled and sealed on the lunar surface using an indium seal.

LESC: Lunar Environment Sample Container, *see* GASC.

LM: Lunar Module: a location on the Lunar surface specific for each mission.

LRL: Lunar Receiving Laboratory.

Residue: Chips and fines of lunar material collected from the bottom of ALSRCs, SCBs, DBs, and LRL processing chambers.

Rock: A lunar sample greater than 50 gm. In the Apollo 14 list, all rocks are breccias unless otherwise stated.

SCB: Sample Collection Bag: cloth bag used to transport rocks from the lunar surface to the LRL.

SESC: Surface Environment Sample Container, *see* GASC.

Sta: Station.

Sweepings: Lunar material recovered from LRL sample processing chambers.

Weigh bag: Old name for a SCB.

York mesh: Stainless steel meshing used as packing material in ALSRCs.

APOLLO 11 INVENTORY

Sample number	Weight grams	Sample type	Location and Comments
10001	—		Documented sample container (ALSRC) studied in F–201 (vacuum laboratory).
10002	~ 350.		Bulk (selected) ALSRC; includes chips and fines; fines distributed as 10084–10086.
10003	213.	Basalt	Documented sample
10004	52.5	Fines	Core tube—second taken on Lunar surface
10005	64.0	Fines	Core tube—first taken on Lunar surface
10006	No sample		Solar Wind Experiment foil
10007	—	Fines	Gas Reaction Cell; from 10001; renumbered 10015
10008	—	Fines	Bio Prime sample; from 10001
10009	112.	Breccia	Documented sample
10010	504.	Fines	Contingency sample
10011	~ 200.	Fines	Documented sample; vacuum fines; distributed as 10087
10012	—	—	Organic monitor of York mesh; not a lunar sample
10013	—	—	Part of Bio Prime sample; from 10002
10014	50.	Chips	Documented sample
10015	—	Fines	Gas Reaction Cell; from 10001
10016	Number not used		
10017	973.	Basalt	Documented sample
10018	213.	Breccia	Documented sample
10019	297.	Breccia	Documented sample
10020	425.	Basalt	Documented sample
10021	255.0	Breccia	Contingency sample
10022	95.6	Basalt	Contingency sample; located
10023	66.0	Breccia	Contingency sample; located
10024	68.1	Basalt	Contingency sample
10025	8.6	Breccia	Contingency sample
10026	9.3	Breccia	Contingency sample
10027	9.9	Breccia	Contingency sample
10028	3.5	Breccia	Contingency sample
10029	5.5	Basalt	Contingency sample
10030	1.8	Breccia	Contingency sample
10031	2.7	Breccia	Contingency sample
10032	3.1	Basalt	Contingency sample
10033	1.1		Contingency sample
10034	< 1.	Residue	Contingency sample
10035	< 1.	Residue	Contingency sample
10082	50.5	Breccia	Documented sample
10083	Number not used		
10084	10391.	{ < 1 mm fines	Bulk sample—formally part of 10002
10085		{ 1 mm–1 cm fines	
10086			Part of 10084 distributed fines for organic analysis
10087	—	Fines	Distributed part of 10011
10088	Number not used		
10089	~ 50.	Breccia	Chip from 10002
10090	6.	Breccia	Chip from 10002
10091	~ 50.	Breccia	Chip from 10002
10092	Number not used		
10093	Number not used		
10094	Number not used		
10095	Number not used		
10096	Number not used		
10097	Number not used		

Apollo 11 Inventory (continued)

Sample number	Weight grams	Sample type	Location and Comments
10098	Number not used		
10099	Number not used		
10100	—	Residue	Fines, chips, sweepings from curator processing
10101	—	Residue	Fines, chips, sweepings from curator processing
10102	—	Residue	Fines, chips, sweepings from curator processing
10103	—	Residue	Fines, chips, sweepings from curator processing

APOLLO 12 INVENTORY

Sample number	Weight grams	Sample type	Location and Comments
12001	2216.0	Fines	NW of LM
12002	1529.5	Basalt	N of Head Crater
12003	300.0	Fines and chips	From 12001
12004	585.0	Basalt	80 m NNE of Head Crater rim
12005	482.0	Basalt	N of Head Crater
12006	206.4	Basalt	N of Head Crater
12007	65.2	Basalt	30 m NE of Head Crater rim
12008	58.4	Vitrophyre	30 m NE of Head Crater rim
12009	468.2	Vitrophyre	N of Head Crater
12010	360.0	Breccia	100 m N of Head Crater
12011	193.0	Basalt	N of Head Crater
12012	176.2	Basalt	N of Head Crater
12013	82.3	Breccia	100 m N of Head Crater rim
12014	159.4	Basalt	90 m NNE of Head Crater rim
12015	191.2	Basalt	100 m N of Head Crater rim
12016	2028.3	Basalt	100 m N of Head Crater rim
12017	53.0	Basalt	100 m NNE of Head Crater rim
12018	787.0	Basalt	N of Head Crater
12019	462.4	Basalt	100 m N of Head Crater rim
12020	312.0	Basalt	N of Head Crater
12021	1876.6	Basalt	40 m NE of Head Crater rim
12022	1864.3	Basalt	20 m N of Head Crater rim
12023	269.3	Fines	LESC; Sharp Crater
12024	56.5	Fines	GASC; Sharp Crater
12025	56.1	Fines	Double core tube; top; S of Surveyor Crater
12026	101.4	Fines	Core tube; N of LM
12027	80.0	Fines	Core tube; Sharp Crater
12028	189.6	Fines	Double core tube; bottom; S of Surveyor Crater
12029	6.5	Fines	Material in Surveyor scoop
12030	75.0	Fines	10 m NE of rim of Head Crater
12031	185.0	Basalt	NNW rim of Head Crater
12032	310.5	Fines	N rim of Bench Crater
12033	450.0	Fines	NNW rim of Head Crater
12034	155.0	Breccia	NNW rim of Head Crater
12035	71.0	Basalt	NW rim of Bench Crater
12036	75.0	Basalt	NW rim of Bench Crater
12037	145.0	Fines	NW rim of Bench Crater
12038	746.0	Basalt	WNW rim of Bench Crater
12039	255.0	Basalt	WNW rim of Bench Crater
12040	319.0	Basalt	WNW rim of Bench Crater
12041	24.8	Fines	75 m E of Bench Crater rim
12042	255.0	Fines	50 m SW of Surveyor Crater rim
12043	60.0	Basalt	S rim of Surveyor Crater
12044	92.0	Fines	S rim of Surveyor Crater
12045	63.0	Basalt	Rim of Block Crater
12046	166.0	Basalt	Rim of Block Crater
12047	193.0	Basalt	Rim of Block Crater
12048	136.0	Fines	NW rim of Bench Crater
12049	Number not used		
12050	1.0	Chip	EVA–2
12051	1660.0	Basalt	S rim of Surveyor Crater
12052	1866.0	Basalt	WSW rim of Head Crater
12053	879.0	Basalt	N rim of Bench Crater
12054	687.0	Basalt	S rim of Surveyor Crater

APOLLO 12 INVENTORY (continued)

Sample number	Weight grams	Sample type	Location and Comments
12055	912.0	Basalt	NW rim of Head Crater
12056	121.0	Basalt	WNW rim of Bench Crater
12057	650.0	Fines and chips	EVA–2
12058	Number not used		
12059	Number not used		
12060	20.7	Fines	EVA–2
12061	9.5	Chips	EVA–2
12062	738.7	Basalt	Near Surveyor III craft
12063	2426.0	Basalt	Rim of Block Crater
12064	1214.3	Basalt	Near Surveyor III craft
12065	2109.0	Basalt	Near Surveyor III craft
12066	Number not used		
12067	Number not used		
12068	Number not used		
12069	Number not used		
12070	1102.0	Fines	10 m NW of LM
12071	9.2	Chips	10 m NW of LM
12072	103.6	Basalt	10 m NW of LM
12073	407.7	Breccia	10 m NW of LM
12074	—	—	Sample part of 12073
12075	232.5	Basalt	10 m NW of LM
12076	54.6	Basalt	10 m NW of LM
12077	22.6	Basalt	10 m NW of LM
12078		Residue	Fines, chips, sweepings from curator processing
12079		Residue	Fines, chips, sweepings from curator processing
12080		Residue	

Apollo 14 Inventory

Sample number	Weight grams	Sample type	Location and Comments
14001	31.8	Fines 2–4 mm	Contingency sample
14002	42.1	Fines 1–2 mm	Contingency sample
14003	947.9	Fines < 1 mm	Contingency sample
14004	33.0	Fines 4–10 mm	Contingency sample
14006	12.13	Chip	Contingency sample
14007	3.67	Chip	Contingency sample
14008	4.35	Chip	Contingency sample
14009	1.09	Chip	Contingency sample
14010	1.00	Chip	Contingency sample
14011	.68	Chip	Contingency sample
14012	.103	Residue	Contingency sample
14041	166.27	Rock	Station A
14042	103.19	Rock	Station A
14043	5.94	Chip	Station A
14044	4.68	Residue	Station A
14045	65.24	Rock	Station A
14046	1.21	Residue	Station A
14047	242.01	Rock	Station B
14048	10.17	Residue	Station B
14049	200.13	Rock	80 m E of Station B
15050	6.91	Residue	80 m E of Station B
14051	191.31	Rock	Station C
14052	2.89	Residue	Station C
14053	251.32	Rock igneous	Station C2
14054	.52	Residue	Station C2
14055	110.99	Rock	Station E
14056	6.38	Chip	Station E
14057	5.51	Chip	Station E
14058	4.53	Chip	Station E
14059	8.68	Chip	Station E
14060	2.50	Chip	Station E
14061	3.11	Chip	Station E
14062	27.50	Residue	Station E
14063	135.45	Rock	Station C1
14064	107.53	Rock	Station C1
14065	7.72	Residue	Station C1
14066	509.80	Rock	Station F
14067	7.66	Residue	Station F
14068	35.47	Chip	Station C
14069	24.87	Chip	Station C
14070	36.56	Chip	Station C
14071	2.16	Chip	Station C
14072	45.06	Chip igneous	Station C
14073	10.35	Chip igneous	Station G bottom of trench
14074	5.16	Chip	Station G bottom of trench
14075	4.66	Chip	Station G bottom of trench
14076	2.00	Chip	Station G bottom of trench
14077	2.77	Chip	Station G bottom of trench
14078	8.30	Chip	Station G bottom of trench
14079	3.17	Chip	Station G bottom of trench
14080	1.94	Chip	Station G middle of trench
14081	0.84	Chip	Station G middle of trench
14082	62.63	Rock	Station C1
14083			Part of sample 14082

Apollo 14 Inventory (continued)

Sample number	Weight grams	Sample type	Location and Comments
14084	.83	Residue	Station C1
14140	12.57	Fines 4–10 mm	Station C
14141	28.50	Fines <1 mm	Station C
14142	5.35	Fines 1–2 mm	Station C
14143	6.73	Fines 2–4 mm	Station C
14144	4.77	Fines	Part of early bio sample
14145	.92	Fines 4–10 mm	Station G surface of trench
14146	2.82	Fines 1–2 mm	Station G surface of trench
14147	1.67	Fines 2–4 mm	Station G surface of trench
14148	71.65	Fines <1 mm	Station G surface of trench
14149	88.15	Fines <1 mm	Station G bottom of trench
14150	11.08	Fines 4–10 mm	Station G bottom of trench
14151	11.70	Fines 1–2 mm	Station G bottom of trench
14152	11.39	Fines 2–4 mm	Station G bottom of trench
14153	3.91	Fines 2–4 mm	Station G middle of trench
14154	5.49	Fines 1–2 mm	Station G middle of trench
14155	3.69	Fines 4–10 mm	Station G middle of trench
14156	137.98	Fines <1 mm	Station G middle of trench
14160	196.50	Fines 4–10 mm	Bulk sample (weigh bag no. 2 from EVA–1, near LM)
14161	197.10	Fines 2–4 mm	Bulk sample　　　　　　.DO.
14162	288.70	Fines 1–2 mm	Bulk sample　　　　　　.DO.
14163	7129.80	Fines <1 mm	Bulk sample　　　　　　.DO.
14165	9.10	Fines <1 mm	EVA–2 (Residue in bottom of weigh bag 1027)
14166	20.50	Fines 1–2 mm	EVA–2　　　　　　.DO.
14167	26.50	Fines 2–4 mm	EVA–2　　　　　　.DO.
14168	43.90	Fines 4–10 mm	EVA–2　　　　　　.DO.
14169	78.66	Rock	Comprehensive sample
14170	26.34	Chip	Comprehensive sample
14171	37.79	Chip	Comprehensive sample
14172	32.10	Chip	Comprehensive sample
14173	19.59	Chip	Comprehensive sample
14174	11.62	Chip	Comprehensive sample
14175	7.48	Chip	Comprehensive sample
14176	4.12	Chip	Comprehensive sample
14177	2.32	Chip	Comprehensive sample
14178	2.88	Chip	Comprehensive sample
14179	3.03	Chip	Comprehensive sample
14180	4.75	Chip	Comprehensive sample
14181	2.48	Chip	Comprehensive sample
14182	2.29	Chip	Comprehensive sample
14183	1.40	Chip	Comprehensive sample
14184	1.48	Chip	Comprehensive sample
14185	1.52	Chip	Comprehensive sample
14186	1.26	Chip	Comprehensive sample
14187	1.09	Chip	Comprehensive sample
14188	1.60	Chip	Comprehensive sample
14189	0.36	Residue	EVA–2 (Residue in bottom of weigh bag 1031)
14190	34.85	Fines <1 mm	EVA–2　　　　　　.DO.
14191	5.92	Fines 1–2 mm	EVA–2　　　　　　.DO.
14192	8.06	Fines 2–4 mm	EVA–2　　　　　　.DO.
14193	11.15	Fines 4–10 mm	EVA–2　　　　　　.DO.
14194	4.28	Chip	EVA–2　　　　　　.DO.
14195	2.77	Chip	EVA–2　　　　　　.DO.
14196	3.93	Chip	EVA–2　　　　　　.DO.

APOLLO 14 INVENTORY (continued)

Sample number	Weight grams	Sample type	Location and Comments
14197	1.63	Chip	EVA–2	.DO.
14198	1.63	Chip	EVA–2	.DO.
14199	1.88	Chip	EVA–2	.DO.
14200	1.24	Chip	EVA–2	.DO.
14201	1.56	Chip	EVA–2	.DO.
14202	.05	Residue	EVA–2	.DO.
14203	—		York mesh
14204	21.60	Residue	EVA–2 (Residue in bottom of weigh bag 1031)
14210	169.7	Fines	Station A Double core tube-bottom
14211	39.5	Fines	Station A Double core tube-top
14220	80.7	Fines	Station G First single core tube
14230	76.7	Fines	Station G Second single core tube
14240	168.0	Fines	Station G SESC
14241	—	Residue	—
14250	4.06	Chip	Comprehensive sample
14251	1.51	Chip	Comprehensive sample
14252	.86	Chip	Comprehensive sample
14254	1.01	Chip	Comprehensive sample
14253	1.23	Chip	Comprehensive sample
14255	22.15	Chip	Comprehensive sample
14256	13.71	Fines 4–10 mm	Comprehensive sample
14257	30.48	Fines 2–4 mm	Comprehensive sample
14258	64.33	Fines 1–2 mm	Comprehensive sample
14259	2694.10	Fines <1 mm	Comprehensive sample
14260	282.50	Fines <1 mm	Comprehensive sample
14261	8.20	Fines 2–4 mm	Comprehensive sample
14262	9.10	Fines 1–2 mm	Comprehensive sample
14263	16.20	Fines 4–10 mm	Comprehensive sample
14264	117.89	Rock	Comprehensive sample
14265	65.79	Rock	Comprehensive sample
14266	6.95	Chip	Comprehensive sample
14267	54.77	Rock	Comprehensive sample
14268	23.12	Chip	Comprehensive sample
14269	17.19	Chip	Comprehensive sample
14270	25.59	Chip	Comprehensive sample
14271	97.41	Rock	Comprehensive sample
14272	46.63	Chip	Comprehensive sample
14273	22.40	Chip	Comprehensive sample
14274	15.18	Chip	Comprehensive sample
14275	12.46	Chip	Comprehensive sample
14276	12.75	Chip	Comprehensive sample
14277	7.59	Chip	Comprehensive sample
14278	7.60	Chip	Comprehensive sample
14279	5.67	Chip	Comprehensive sample
14280	6.20	Chip	Comprehensive sample
14281	12.03	Chip	Comprehensive sample
14282	1.89	Chip	Comprehensive sample
14283	1.25	Chip	Comprehensive sample
14284	1.47	Chip	Comprehensive sample
14285	2.23	Chip	Comprehensive sample
14286	4.42	Chip	Comprehensive sample
14287	1.07	Chip	Comprehensive sample
14288	3.44	Chip	Comprehensive sample
14289	.20	Residue	Comprehensive sample

APOLLO 14 INVENTORY (continued)

Sample number	Weight grams	Sample type	Location and Comments
14290	23.63	Fines <1 mm	EVA–2 (Residue in bottom of weigh bag 1038)
14291	2.12	Fines 1–2 mm	EVA–2 .DO.
14292	3.53	Fines 2–4 mm	EVA–2 .DO.
14293	5.20	Fines 4–10 mm	EVA–2 .DO.
14294	3.43	Chip	EVA–2 .DO.
14295	1.24	Chip	EVA–2 .DO.
14296	2.26	Chip	EVA–2 .DO.
14297	1.73	Chip	EVA–2 .DO.
14298	200.00	Fines <1 mm	Reserve from 14259
14299	225.00	Fines <1 mm	Reserve from 14259
14300	4.06	Chip	Comprehensive sample
14301	1360.60	Rock	Station G1
14302			Part of sample 14305
14303	898.40	Rock	Comprehensive sample
14304	2498.90	Rock	40 m W of Comprehensive sample
14305	2497.50	Rock	40 m W of Comprehensive sample
14306	584.50	Rock	Station G
14307	155.00	Rock	Station G
14308			Part of sample 14311
14309	42.40	Chip	Unknown
14310	3439.00	Rock igneous	Station G
14311	3204.40	Rock	Station D
14312	299.00	Rock	Station H
14313	144.00	Chip	Station G1
14314	115.70	Rock	Station H
14315	115.00	Rock	Station H
14316	38.20	Chip	Station H
14317	16.10	Chip	Station H
14318	600.20	Rock	Station H
14319	211.60	Rock	Station H
14320	64.90	Rock	Station H
14321	8998.0	Rock	Station C1
14399	—	Residue	—
14401	0.0		Residue EVA–2 ALSRC
14402	0.20		Residue EVA–1 ALSRC
14408	—	Fines	Bio Prime sample
14411	5.5		Core bit double core
14414	5.5		Core bit first single core
14421	260.9	Fines <10 mm	Reserve from unsieved comprehensive sample
14422	251.00	Fines <1 mm	Reserve from 14163 (part to Bio Pool)
14423	—	Fines	Bio Pool sample
14425	.79	Chip	Bulk fines sample
14426	1.59	Chip	Bulk fines sample
14427	4.47	Chip	Bulk fines sample
14428	1.47	Chip	Bulk fines sample
14429	3.03	Chip	Bulk fines sample
14430	4.81	Chip	Bulk fines sample
14431	1.70	Chip	Bulk fines sample
14432	1.81	Chip	Bulk fines sample
14433	1.23	Chip	Bulk fines sample
14434	1.68	Chip	Bulk fines sample
14435	.92	Chip	Bulk fines sample
14436	3.76	Chip	Bulk fines sample
14437	2.65	Chip	Bulk fines sample

APOLLO 14 INVENTORY (continued)

Sample number	Weight grams	Sample type	Location and Comments
14438	3.35	Chip	Bulk fines sample
14439	1.00	Chip	Bulk fines sample
14440	1.50	Chip	Bulk fines sample
14441	.23	Chip	Bulk fines sample
14442	3.52	Chip	Bulk fines sample
14443	2.54	Chip	Bulk fines sample
14444	1.56	Chip	Bulk fines sample
14445	9.22	Chip	Bulk fines sample
14446	.82	Chip	Bulk fines sample
14447	.91	Chip	Bulk fines sample
14448	1.06	Chip	Bulk fines sample
14449	1.70	Chip	Bulk fines sample
14450	1.27	Chip	Bulk fines sample
14451	2.10	Chip	Bulk fines sample
14452	1.77	Chip	Bulk fines sample
14453	6.03	Chip	Bulk fines sample
14454	—	Residue	Chips, fines, sweepings from curator processing
14455	—	Residue	Chips, fines, sweepings from curator processing
14500	—	Residue	Chips, fines, sweepings from curator processing

Apollo 15 Inventory

Sample number	Weight grams	Description	Location and Comments
15001	232.8	Drill stem bottom	Sta 8–ALSEP
15002	210.1	Drill stem section	Sta 8–ALSEP
15003	223.0	Drill stem section	Sta 8–ALSEP
15004	210.6	Drill stem section	Sta 8–ALSEP
15005	239.1	Drill stem section	Sta 8–ALSEP
15006	227.9	Drill stem top	Sta 8–ALSEP
15007	768.2	Drive tube bottom	Sta 2–St. George
15008	510.2	Drive tube top	Sta 2–St. George
15009	622.0	Drive tube single	Sta 6–Front
15010	740.4	Drive tube bottom	Sta 9a–Rille
15011	653.6	Drive tube top	Sta 9a–Rille
15012	312.2	SESC 1	Sta 6–Front
15013	296.2	SESC (blank)	Lunar Module
15014	333.2	SESC 2	Sta 8–ALSEP
15015	4735.2	Glass coated breccia	Lunar Module
15016	923.7	Porphyritic vesicular basalt	Sta 3–Rhysling
15017	9.8	Glass spherical shell	Lunar Module
15018	5.7	Glass object	Lunar Module
15019	1.2	Glassy microbreccia	Lunar Module
15020	88.7	Unsieved fines	Lunar Module
15021	500.2	<1 mm fines	Lunar Module
15022	10.0	1–2 mm fines	Lunar Module
15023	5.0	2–4 mm fines	Lunar Module
15024	3.6	4–10 mm fines	Lunar Module
15025	77.3	Coherent breccia	Lunar Module
15026	1.1	Glass coated microbreccia	Lunar Module
15027	51.0	Breccia	Lunar Module
15028	59.4	Glassy breccia	Lunar Module
15030	75.3	Unsieved fines	Sta 8–ALSEP
15031	207.8	<1 mm fines	Sta 8–ALSEP
15032	7.0	1–2 mm fines	Sta 8–ALSEP
15033	6.6	2–4 mm fines	Sta 8–ALSEP
15034	7.0	4–10 mm fines	Sta 8–ALSEP
15040	113.4	Unsieved fines	Sta 8–ALSEP
15041	269.6	<1 mm fines	Sta 8–ALSEP
15042	5.1	1–2 mm fines	Sta 8–ALSEP
15043	2.8	2–4 mm fines	Sta 8–ALSEP
15044	1.5	4–10 mm fines	Sta 8–ALSEP
15058	2672.5	Porphyritic basalt	Sta 8–ALSEP
15059	1149.2	Glass coated breccia	Lunar Module
15065	1470.7	Gabbro	Sta 1–Elbow
15070	51.3	Unsieved fines	Sta 1–Elbow
15071	100.7	<1 mm fines	Sta 1–Elbow
15072	3.0	1–2 mm fines	Sta 1–Elbow
15073	1.4	2–4 mm fines	Sta 1–Elbow
15074	1.3	4–10 mm fines	Sta 1–Elbow
15075	792.1	Gabbro	Sta 1–Elbow
15076	400.5	Gabbro	Sta 1–Elbow
15080	73.5	Unsieved fines	Sta 1–Elbow
15081	106.9	<1 mm fines	Sta 1–Elbow
15082	2.0	1–2 mm fines	Sta 1–Elbow
15083	1.8	2–4 mm fines	Sta 1–Elbow
15084	1.1	4–10 mm fines	Sta 1–Elbow
15085	458.9	Basalt	Sta 1–Elbow

APOLLO 15 INVENTORY (continued)

Sample number	Weight grams	Description	Location and Comments
15086	216.5	Breccia	Sta 1–Elbow
15087	5.7	Gabbro	Sta 1–Elbow
15088	1.8	Breccia	Sta 1–Elbow
15090	39.3	Unsieved fines	Sta 2–St. George
15091	162.9	<1 mm fines	Sta 2–St. George
15092	2.7	1–2 mm fines	Sta 2–St. George
15093	0.6	2–4 mm fines	Sta 2–St. George
15095	25.5	Glass coated microbreccia	Sta 2–St. George
15100	281.0	Unsieved fines	Sta 2–St. George
15101	637.6	<1 mm fines	Sta 2–St. George
15102	12.2	1–2 mm fines	Sta 2–St. George
15103	4.1	2–4 mm fines	Sta 2–St. George
15104	1.5	4–10 mm fines	Sta 2–St. George
15105	5.6	Basalt	Sta 2–St. George
15115	4.0	Porphyritic basalt	Sta 2–St. George
15116	7.2	Gabbro	Sta 2–St. George
15117	23.3	Porphyritic basalt	Sta 2–St. George
15118	27.6	Porphyritic basalt	Sta 2–St. George
15119	14.1	Basalt with adhering breccia	Sta 2–St. George
15125	6.5	Basalt	Sta 2–St. George
15135	1.6	Glassy microbreccia	Sta 2–St. George
15145	15.1	Breccia	Sta 2–St. George
15146	1.0	Breccia	Sta 2–St. George
15147	3.7	Breccia	Sta 2–St. George
15148	3.0	Breccia	Sta 2–St. George
15200	7.7	Unsieved fines	Sta 2–St. George
15201	18.3	<1 mm fines	Sta 2–St. George
15202	0.4	1–2 mm fines	Sta 2–St. George
15203	0.2	2–4 mm fines	Sta 2–St. George
15204	0.1	4–10 mm fines	Sta 2–St. George
15205	337.3	Coarse breccia	Sta 2–St. George
15206	92.0	Glassy breccia	Sta 2–St. George
15210	221.2	Unsieved fines	Sta 2–St. George
15211	163.5	<1 mm fines	Sta 2–St. George
15212	3.6	1–2 mm fines	Sta 2–St. George
15213	2.4	2–4 mm fines	Sta 2–St. George
15214	0.2	4–10 mm fines	Sta 2–St. George
15220	160.5	Unsieved fines	Sta 2–St. George
15221	290.0	<1 mm fines	Sta 2–St. George
15222	2.4	1–2 mm fines	Sta 2–St. George
15223	5.8	2–4 mm fines	Sta 2–St. George
15224	7.0	4–10 mm fines	Sta 2–St. George
15230	99.1	Unsieved fines	Sta 2–St. George
15231	233.9	<1 mm fines	Sta 2–St. George
15232	5.2	1–2 mm fines	Sta 2–St. George
15233	3.8	2–4 mm fines	Sta 2–St. George
15234	1.8	4–10 mm fines	Sta 2–St. George
15240	67.1	Unsieved fines	Sta 6–Front
15241	197.4	<1 mm fines	Sta 6–Front
15242	18.9	1–2 mm fines	Sta 6–Front
15243	31.8	2–4 mm fines	Sta 6–Front
15244	32.6	4–10 mm fines	Sta 6–Front
15245	115.5	89 pieces—Glass coated breccias to agglutinates	Sta 6–Front
15250	207.0	Unsieved fines	Sta 6–Front

APOLLO 15 INVENTORY (continued)

Sample number	Weight grams	Description	Location and Comments
15251	380.9	<1 mm fines	Sta 6–Front
15252	8.3	1–2 mm fines	Sta 6–Front
15253	4.0	2–4 mm fines	Sta 6–Front
15254	1.2	4–10 mm fines	Sta 6–Front
15255	240.4	Glass coated breccia	Sta 6–Front
15256	201.0	Basalt	Sta 6–Front
15257	22.5	Microbreccia	Sta 6–Front
15259	0.7	Microbreccia	Sta 6–Front
15260	172.2	Unsieved fines	Sta 6–Front
15261	416.6	<1 mm fines	Sta 6–Front
15262	9.1	1–2 mm fines	Sta 6–Front
15263	6.7	2–4 mm fines	Sta 6–Front
15264	5.9	4–10 mm fines	Sta 6–Front
15265	314.1	Fine breccia	Sta 6–Front
15266	271.4	Fine breccia	Sta 6–Front
15267	1.8	Microbreccia	Sta 6–Front
15268	11.0	Microbreccia	Sta 6–Front
15269	6.0	Glassy microbreccia	Sta 6–Front
15270	319.0	Unsieved fines	Sta 6–Front
15271	798.3	<1 mm fines	Sta 6–Front
15272	20.7	1–2 mm fines	Sta 6–Front
15273	13.7	2–4 mm fines	Sta 6–Front
15274	4.4	4–10 mm fines	Sta 6–Front
15281	107.0	<1 mm (SCB 3 residue)	Sta 6–Front
15282	9.7	1–2 mm (SCB 3 residue)	Sta 6–Front
15283	13.3	2–4 mm (SCB 3 residue)	Sta 6–Front
15284	38.2	4–10 mm (SCB 3 residue)	Sta 6–Front
15285	259.2	Breccia	Sta 6–Front
15286	34.6	Vesicular basalt glass and microbreccia	Sta 6–Front
15287	44.9	Breccia	Sta 6–Front
15288	70.5	Glassy breccia	Sta 6–Front
15289	24.1	Breccia	Sta 6–Front
15290	55.0	Unsieved fines	Sta 6–Front
15291	169.0	<1 .nm fines	Sta 6–Front
15292	5.4	1–2 mm fines	Sta 6–Front
15293	6.7	2–4 mm fines	Sta 6–Front
15294	10.2	4–10 mm fines	Sta 6–Front
15295	947.3	Breccia	Sta 6–Front
15297	39.4	Breccia chips (SCB 3 residue)	Sta 6–Front
15298	1731.4	Microbreccia	Sta 6–Front
15299	1691.7	Breccia	Sta 6–Front
15300	390.7	Unsieved fines	Sta 7–Spur
15301	810.2	<1 mm fines	Sta 7–Spur
15302	23.2	1–2 mm fines	Sta 7–Spur
15303	12.7	2–4 mm fines	Sta 7–Spur
15304	7.3	4–10 mm fines	Sta 7–Spur
15305	2.9	<10 mm green soil conc.	Sta 7–Spur
15306	134.2	Breccia	Sta 7–Spur
15307	1.3	Hollow glass sphere	Sta 7–Spur
15308	1.7	Breccia	Sta 7–Spur
15310	140.6	Unsieved fines	Sta 7–Spur
15311	295.0	<1 mm fines	Sta 7–Spur
15312	10.1	1–2 mm fines	Sta 7–Spur
15313	9.8	2–4 mm fines	Sta 7–Spur

APOLLO 15 INVENTORY (continued)

Sample number	Weight grams	Description	Location and Comments
15314	8.4	4–10 mm fines	Sta 7–Spur
15315	35.6	Breccia containing clasts of brown pyroxene basalt	Sta 7–Spur
15316	6.1	Breccia containing clasts of brown pyroxene basalt	Sta 7–Spur
15317	0.6	Breccia containing clasts of brown pyroxene basalt	Sta 7–Spur
15318	5.4	Breccia containing clasts of brown pyroxene basalt	Sta 7–Spur
15319	8.0	Breccia containing clasts of brown pyroxene basalt	Sta 7–Spur
15320	4.7	Breccia containing clasts of brown pyroxene basalt	Sta 7–Spur
15321	0.3	Breccia containing clasts of brown pyroxene basalt	Sta 7–Spur
15322	8.4	Breccia lacking clasts brown pyroxene basalt	Sta 7–Spur
15323	4.4	Breccia lacking clasts brown pyroxene basalt	Sta 7–Spur
15324	32.3	Breccia lacking clasts brown pyroxene basalt	Sta 7–Spur
15325	57.8	Breccia lacking clasts brown pyroxene basalt	Sta 7–Spur
15326	2.5	Breccia lacking clasts brown pyroxene basalt	Sta 7–Spur
15327	12.4	Breccia lacking clasts brown pyroxene basalt	Sta 7–Spur
15328	0.3	Breccia lacking clasts brown pyroxene basalt	Sta 7–Spur
15329	2.2	Breccia lacking clasts brown pyroxene basalt	Sta 7–Spur
15330	57.8	Breccia lacking clasts brown pyroxene basalt	Sta 7–Spur
15331	2.6	Breccia lacking clasts brown pyroxene basalt	Sta 7–Spur
15332	2.3	Breccia lacking clasts brown pyroxene basalt	Sta 7–Spur
15333	0.3	Breccia lacking clasts brown pyroxene basalt	Sta 7–Spur
15334	7.5	Breccia lacking clasts brown pyroxene basalt	Sta 7–Spur
15335	6.0	Breccia lacking clasts brown pyroxene basalt	Sta 7–Spur
15336	0.2	Breccia lacking clasts brown pyroxene basalt	Sta 7–Spur
15337	4.3	Breccia lacking clasts brown pyroxene basalt	Sta 7–Spur
15338	11.1	Breccia lacking clasts brown pyroxene basalt	Sta 7–Spur
15339	0.4	Breccia lacking clasts brown pyroxene basalt	Sta 7–Spur
15340	0.9	Breccia lacking clasts brown pyroxene basalt	Sta 7–Spur
15341	1.6	Breccia lacking clasts brown pyroxene basalt	Sta 7–Spur
15342	7.5	Breccia lacking clasts brown pyroxene basalt	Sta 7–Spur
15343	6.9	Breccia lacking clasts of brown pyroxene basalt	Sta 7–Spur
15344	7.9	Breccia lacking clasts of brown pyroxene basalt	Sta 7–Spur
15345	12.3	Breccia lacking clasts of brown pyroxene basalt	Sta 7–Spur
15346	3.1	Breccia lacking clasts of brown pyroxene basalt	Sta 7–Spur
15347	3.2	Breccia lacking clasts of brown pyroxene basalt	Sta 7–Spur
15348	0.3	Breccia lacking clasts of brown pyroxene basalt	Sta 7–Spur
15349	2.3	Breccia lacking clasts of brown pyroxene basalt	Sta 7–Spur
15350	2.9	Breccia lacking clasts of brown pyroxene basalt	Sta 7–Spur
15351	4.2	Breccia lacking clasts of brown pyroxene basalt	Sta 7–Spur
15352	2.9	Breccia lacking clasts of brown pyroxene basalt	Sta 7–Spur
15353	10.6	Breccia lacking clasts of brown pyroxene basalt	Sta 7–Spur
15354	0.3	Breccia lacking clasts of brown pyroxene basalt	Sta 7–Spur
15355	5.2	Breccia lacking clasts of brown pyroxene basalt	Sta 7–Spur
15356	2.0	Tough microbreccia	Sta 7–Spur
15357	11.8	Tough microbreccia	Sta 7–Spur
15358	14.6	Tough microbreccia	Sta 7–Spur
15359	4.2	Tough microbreccia	Sta 7–Spur
15360	9.3	Tough microbreccia	Sta 7–Spur
15361	0.9	Pale green microcrystalline rock	Sta 7–Spur
15362	4.2	Anorthosite	Sta 7–Spur
15363	0.5	Anorthosite	Sta 7–Spur
15364	1.5	Anorthosite	Sta 7–Spur
15365	2.9	Green glass microbreccia	Sta 7–Spur
15366	3.3	Green glass microbreccia	Sta 7–Spur
15367	1.1	Green glas microbreccia	Sta 7–Spur

APOLLO 15 INVENTORY (continued)

Sample number	Weight grams	Description	Location and Comments
15368	0.4	Green glass microbreccia	Sta 7–Spur
15369	2.5	Green glass microbreccia	Sta 7–Spur
15370	2.9	Green glass microbreccia	Sta 7–Spur
15371	0.5	Green glass microbreccia	Sta 7–Spur
15372	0.8	Green glass microbreccia	Sta 7–Spur
15373	0.6	Green glass microbreccia	Sta 7–Spur
15374	1.0	Green glass microbreccia	Sta 7–Spur
15375	0.4	Green glass microbreccia	Sta 7–Spur
15376	1.0	Green glass microbreccia	Sta 7–Spur
15377	0.5	Green glass microbreccia	Sta 7–Spur
15378	3.3	Fine non-Mare basalt	Sta 7–Spur
15379	64.3	Fine non-Mare basalt	Sta 7–Spur
15380	5.2	Fine non-Mare basalt	Sta 7–Spur
15381	0.3	Fine non-Mare basalt	Sta 7–Spur
15382	3.2	Fine non-Mare basalt	Sta 7–Spur
15383	1.4	Fine non-Mare basalt	Sta 7–Spur
15384	1.4	Fine non-Mare basalt	Sta 7–Spur
15385	8.7	Coarse Mare basalt	Sta 7–Spur
15386	7.5	Coarse Mare basalt	Sta 7–Spur
15387	2.0	Coarse Mare basalt	Sta 7–Spur
15388	9.0	Coarse Mare basalt	Sta 7–Spur
15389	2.8	Glass	Sta 7–Spur
15390	3.5	Glass	Sta 7–Spur
15391	0.3	Glass	Sta 7–Spur
15392	0.4	Glass	Sta 7–Spur
15400	47.5	Unsieved fines	Sta 6a
15401	86.4	<1 mm fines	Sta 6a
15402	4.8	1–2 mm fines	Sta 6a
15403	6.1	2–4 mm fines	Sta 6a
15404	7.9	4–10 mm fines	Sta 6a
15405	513.1	Breccia, recrystallized	Sta 6a
15410	56.2	Unsieved fines	Sta 7–Spur
15411	103.3	<1 mm fines	Sta 7–Spur
15412	7.1	1–2 mm fines	Sta 7–Spur
15413	6.7	2–4 mm fines	Sta 7–Spur
15414	4.0	4–10 mm fines	Sta 7–Spur
15415	269.4	Anorthosite	Sta 7–Spur
15417	1.3	Breccia	Sta 7–Spur
15418	1140.7	Breccia, vitreous matrix	Sta 7–Spur
15419	17.7	Breccia, with glass	Sta 7–Spur
15421	254.7	<1 mm fines	Sta 7–Spur
15422	15.9	1–2 mm fines	Sta 7–Spur
15423	18.3	1–4 mm fines	Sta 7–Spur
15424	19.5	4–10 mm fines	Sta 7–Spur
15425	136.3	Green and gray clods	Sta 7–Spur
15426	223.6	Green and gray clods	Sta 7–Spur
15427	115.9	Green and gray clods	Sta 7–Spur
15431	475.7	<1 mm fines	Sta 7–Spur
15432	39.7	1–2 mm fines	Sta 7–Spur
15433	31.2	2–4 mm fines	Sta 7–Spur
15434	51.6	4–10 mm fines	Sta 7–Spur
15435	206.8	Gray clods, 32 splits	Sta 7–Spur
15445	287.2	Breccia with white clasts	Sta 7–Spur
15455	937.2	Black and white breccia	Sta 7–Spur

APOLLO 15 INVENTORY (continued)

Sample number	Weight grams	Description	Location and Comments
15459	5854.0	Breccia with large clasts	Sta 7–Spur
15465	376.0	Glass coated breccia	Sta 7–Spur
15466	119.2	Dark glass with clasts	Sta 7–Spur
15467	1.1	Microbreccia	Sta 7–Spur
15468	1.3	Glass and breccia	Sta 7–Spur
15470	8.2	Unsieved fines	Sta 4–Dune
15471	153.0	<1 mm fines	Sta 4–Dune
15472	6.1	1–2 mm fines	Sta 4–Dune
15473	4.5	2–4 mm fines	Sta 4–Dune
15474	4.7	4–10 mm fines	Sta 4–Dune
15475	406.8	Basalt	Sta 4–Dune
15476	257.5	Basalt	Sta 4–Dune
15485	104.9	Basalt	Sta 4–Dune
15486	46.8	Basalt with gray coatings	Sta 4–Dune
15495	908.9	Gabbro	Sta 4–Dune
15498	2339.8	Recrystallized breccia	Sta 4–Dune
15499	2024.0	Vesicular basalt	Sta 4–Dune
15500	24.8	Unsieved fines	Sta 9–Scarp
15501	103.0	<1 mm fines	Sta 9–Scarp
15502	4.4	1–2 mm fines	Sta 9–Scarp
15503	3.8	2–4 mm fines	Sta 9–Scarp
15504	4.1	4–10 mm fines	Sta 9–Scarp
15505	1147.4	Glass coated breccia	Sta 9–Scarp
15506	22.9	Glass coated microbreccia	Sta 9–Scarp
15507	3.9	Vesicular glass ellipsoid	Sta 9–Scarp
15508	1.4	Glass coated microbreccia	Sta 9–Scarp
15510	72.3	Unsieved fines	Sta 9–Scarp
15511	193.1	<1 mm fines	Sta 9–Scarp
15512	4.9	1–2 mm fines	Sta 9–Scarp
15513	4.4	2–4 mm fines	Sta 9–Scarp
15514	1.1	4–10 mm fines	Sta 9–Scarp
15515	144.7	Brownish gray clods, 48 pieces	Sta 9–Scarp
15528	4.7	Breccia	Sta 9a–Rille
15529	1531.0	Vesicular basalt	Sta 9a–Rille
15530	138.0	Unsieved fines	Sta 9a–Rille
15531	136.0	<1 mm fines	Sta 9a–Rille
15532	6.3	1–2 mm fines	Sta 9a–Rille
15533	5.4	2–4 mm fines	Sta 9a–Rille
15534	6.0	4–10 mm fines	Sta 9a–Rille
15535	404.4	Porphyritic olivine basalt	Sta 9a–Rille
15536	317.2	Basalt	Sta 9a–Rille
15537	1.9	Coarse grained basalt	Sta 9a–Rille
15538	2.6	Olivine microgabbro	Sta 9a–Rille
15545	746.6	Basalt	Sta 9a–Rille
15546	27.8	Basalt	Sta 9a–Rille
15547	20.1	Vuggy basalt	Sta 9a–Rille
15548	3.3	Basalt	Sta 9a–Rille
15555	9613.7	Vuggy basalt	Sta 9a–Rille
15556	1542.3	Vesicular basalt	Sta 9a–Rille
15557	2518.0	Basalt	Sta 9a–Rille
15558	1333.3	Breccia	Sta 9a–Rille
15561	112.5	<1 mm SCB 2 residue	Sta 9a–Rille
15562	20.6	1–2 mm SCB 2 residue	Sta 9a–Rille
15563	30.4	2–4 mm SCB 2 residue	Sta 9a–Rille

APOLLO 15 INVENTORY (continued)

Sample number	Weight grams	Description	Location and Comments
15564	50.0	4–10 mm SCB 2 residue	Sta 9a–Rille
15565	822.6	Breccia fragments, SCB 2 residue	Sta 9a–Rille
15595	237.6	Porphyritic basalt	Sta 9a–Rille
15596	224.8	Basalt	Sta 9a–Rille
15597	145.7	Basalt	Sta 9a–Rille
15598	135.7	Basalt	Sta 9a–Rille
15600	449.1	Reserve fines—comprehensive	Sta 9a–Rille
15601	802.0	<1 mm fines—comprehensive	Sta 9a–Rille
15602	32.9	1–2 mm fines	Sta 9a–Rille
15603	25.5	2–4 mm fines	Sta 9a–Rille
15604	21.5	4–10 mm fines	Sta 9a–Rille
15605	6.1	Basalt	Sta 9a–Rille
15606	10.1	Vesicular basalt	Sta 9a–Rille
15607	14.8	Coarse grained basalt	Sta 9a–Rille
15608	1.2	Microporphyritic basalt	Sta 9a–Rille
15609	1.1	Basalt	Sta 9a–Rille
15610	1.5	Basalt with clinopyroxene	Sta 9a–Rille
15612	5.9	Vesicular olivine basalt	Sta 9a–Rille
15613	1.0	Vesicular olivine basalt	Sta 9a–Rille
15614	9.7	Vesicular olivine basalt	Sta 9a–Rille
15615	1.7	Vesicular olivine basalt	Sta 9a–Rille
15616	8.0	Vesicular olivine basalt	Sta 9a–Rille
15617	3.1	Vesicular olivine basalt	Sta 9a–Rille
15618	0.8	Vesicular olivine basalt	Sta 9a–Rille
15619	0.6	Vesicular olivine basalt	Sta 9a–Rille
15620	6.6	Vesicular olivine basalt	Sta 9a–Rille
15621	1.6	Vesicular olivine basalt	Sta 9a–Rille
15622	29.5	Vesicular olivine basalt	Sta 9a–Rille
15623	3.0	Vesicular olivine basalt	Sta 9a–Rille
15624	0.2	Vesicular olivine basalt	Sta 9a–Rille
15625	0.5	Vesicular olivine basalt	Sta 9a–Rille
15626	0.6	Vesicular olivine basalt	Sta 9a–Rille
15627	0.4	Vesicular olivine basalt	Sta 9a–Rille
15628	0.4	Vesicular olivine basalt	Sta 9a–Rille
15629	0.4	Vesicular olivine basalt	Sta 9a–Rille
15630	23.2	Vesicular olivine basalt	Sta 9a–Rille
15632	2.3	Porphyritic basalt with clinopyroxene	Sta 9a–Rille
15633	7.4	Porphyritic basalt with clinopyroxene	Sta 9a–Rille
15634	5.2	Porphyritic basalt with clinopyroxene	Sta 9a–Rille
15635	0.5	Porphyritic basalt with clinopyroxene	Sta 9a–Rille
15636	336.7	Porphyritic basalt with clinopyroxene	Sta 9a–Rille
15637	0.9	Porphyritic basalt with clinopyroxene	Sta 9a–Rille
15638	3.6	Porphyritic basalt with clinopyroxene	Sta 9a–Rille
15639	7.0	Porphyritic basalt with clinopyroxene	Sta 9a–Rille
15640	0.5	Porphyritic basalt with clinopyroxene	Sta 9a–Rille
15641	6.9	Porphyritic basalt with clinopyroxene	Sta 9a–Rille
15642	1.9	Porphyritic basalt with clinopyroxene	Sta 9a–Rille
15643	17.9	Porphyritic basalt with clinopyroxene	Sta 9a–Rille
15644	0.4	Porphyritic basalt with clinopyroxene	Sta 9a–Rille
15645	0.5	Porphyritic basalt with clinopyroxene	Sta 9a–Rille
15647	58.1	Fine subophitic basalt	Sta 9a–Rille
15648	9.1	Fine subophitic basalt	Sta 9a–Rille
15649	6.2	Fine subophitic basalt	Sta 9a–Rille
15650	3.4	Fine subophitic basalt	Sta 9a–Rille

APOLLO 15 INVENTORY (continued)

Sample number	Weight grams	Description	Location and Comments
15651	1.6	Fine subophitic basalt	Sta 9a–Rille
15652	0.7	Fine subophitic basalt	Sta 9a–Rille
15653	0.4	Fine subophitic basalt	Sta 9a–Rille
15654	0.2	Fine subophitic basalt	Sta 9a–Rille
15655	0.4	Fine subophitic basalt	Sta 9a–Rille
15656	0.2	Fine subophitic basalt	Sta 9a–Rille
15658	11.6	Pyroxene phyric basalt	Sta 9a–Rille
15659	12.6	Pyroxene phyric basalt	Sta 9a–Rille
15660	8.9	Pyroxene phyric basalt	Sta 9a–Rille
15661	5.9	Pyroxene phyric basalt	Sta 9a–Rille
15662	4.9	Pyroxene phyric basalt	Sta 9a–Rille
15663	10.3	Pyroxene phyric basalt	Sta 9a–Rille
15664	7.4	Pyroxene phyric basalt	Sta 9a–Rille
15665	10.2	Olivine phyric basalt	Sta 9a–Rille
15666	3.9	Pigeonite phyric basalt	Sta 9a–Rille
15667	1.1	Pigeonite phyric basalt	Sta 9a–Rille
15668	15.1	Olivine phyric basalt	Sta 9a–Rille
15669	4.4	Olivine phyric basalt	Sta 9a–Rille
15670	2.0	Pyroxene phyric basalt	Sta 9a–Rille
15671	6.1	Vesicular subophitic basalt	Sta 9a–Rille
15672	21.4	Vesicular subophitic basalt	Sta 9a–Rille
15673	5.9	Vesicular subophitic basalt	Sta 9a–Rille
15674	35.7	Fine porphyritic basalt	Sta 9a–Rille
15675	34.5	Fine porphyritic basalt	Sta 9a–Rille
15676	25.3	Fine porphyritic basalt	Sta 9a–Rille
15677	6.4	Fine porphyritic basalt	Sta 9a–Rille
15678	7.5	Fine porphyritic basalt	Sta 9a–Rille
15679	0.7	Fine porphyritic basalt	Sta 9a–Rille
15680	0.3	Fine porphyritic basalt	Sta 9a–Rille
15681	0.3	Fine porphyritic basalt	Sta 9a–Rille
15682	50.6	Plumose porphyritic basalt	Sta 9a–Rille
15683	22.0	Fine porphyritic basalt	Sta 9a–Rille
15684	1.4	Glass cemented breccia fragments	Sta 9a–Rille
15685	0.8	Glass cemented breccia fragments	Sta 9a–Rille
15686	0.9	Glass cemented breccia fragments	Sta 9a–Rille
15687	1.4	Glass cemented breccia fragments	Sta 9a–Rille
15688	5.3	Glass cemented breccia fragments	Sta 9a–Rille
15689	2.8	Breccia with sugary clasts	Sta 9a–Rille
15901	13.6	DB residue	Sta 1–Elbow
15902	4.5	DB residue	Sta 1–Elbow
15903	2.7	DB residue	Sta 1–Elbow
15904	1.2	DB residue	Sta 2–St. George
15906	2.8	DB residue	Sta 2–St. George
15907	5.3	DB residue	Sta 2–St. George
15908	1.2	DB residue	Sta 2–St. George
15909	1.7	DB residue	Sta 2–St. George
15910	4.9	DB residue	Sta 2–St. George
15911	4.8	DB residue	Sta 2–St. George
15912	8.1	DB residue	Lunar module
15916	3.2	DB residue	Sta 6–Front
15917	26.6	DB residue	Sta 6–Front
15918	14.5	DB residue	Sta 6–Front
15924	2.7	DB residue	Sta 7–Spur
15925	0.8	DB residue	Sta 7–Spur

APOLLO 15 INVENTORY (continued)

Sample number	Weight grams	Description	Location and Comments
15926	12.4	DB residue	Sta 7–Spur
15927	20.1	DB residue	Sta 7–Spur
15931	1.7	DB residue	Sta 4–Dune
15932	20.5	DB residue	Sta 4–Dune
15933	15.9	DB residue	Sta 4–Dune
15936	6.7	DB residue	Sta 9–Scarp
15937	1.7	DB residue	Sta 9–Scarp
15938	16.6	DB residue	Sta 9a–Rille
15939	4.2	DB residue	Sta 9a–Rille
15940	3.5	DB residue	Sta 9a–Rille
15941	10.0	DB residue	Sta 9a–Rille
15942	58.8	DB residue	Sta 9a–Rille
15943	1.7	DB residue	Sta 9a–Rille
15951	30.0	SCB residue	EVA 1
15954	89.3	SCB residue	EVA 1
15955	8.6	SCB residue	EVA 2
15956	98.5	SCB residue	EVA 2
15957	75.5	SCB residue	EVA 3
15973	—	Residue	Fines, chips, sweepings from curator processing.

Author Index

Subject Index

(Where an index entry refers to the opening page of an article, the entry includes the entire article.)